STATISTICS
Informed Decisions Using Data

Practice
"Work the Problems"

Feature	Description	Benefit	Page
Concepts & Vocabulary	These Fill-in-the-Blank, True/False, and open-ended questions assess your understanding of vocabulary and statistical concepts.	Helps you to review the language of statistics. Plus, you are encouraged to explain statistics concepts in your own words.	130
Skill Building	"Drill and Practice" exercises that develop computational skills.	Working these exercises will increase your understanding of formulas and concepts.	130-131
Applying the Concepts	Exercises based in real data that ask a variety of questions, many of which require you to write a few sentences of explanation.	These exercises require a deeper level of thinking and analysis which, in turn, promotes statistical thinking.	131-137
Consumer Reports	A project that requires you to go through the same type of analysis that was used by researchers at Consumer Reports to help forms the basis for Product Ratings.	Gives you the chance to practice and apply the skills and concepts learned within the chapter and, also, to see how these skills are applied in every day life.	156
Activities	Short in-class projects that get students involved in collecting and analyzing their own data.	Promotes active learning in the classroom. Helps to generate enthusiasm and understanding of statistical concepts.	123

Review
"Study for Quizzes and Tests"

Feature	Description	Benefit	Page
Chapter Reviews at the end of each chapter contain...			
Summary, Formulas, and Vocabulary.	Provides an overview of important concepts and terms, and a listing of the formulas and vocabulary introduced in the chapter.	Review these and you'll know the most important material in the chapter!	184-186
Objectives	List of chapter objectives along with the corresponding review exercises that test your understanding of the objective.	Helps to get a "big picture" overview of the material presented in the chapter. Plus, makes review for the exam easier by organizing review problems by objective.	186
Review Exercises	These provide comprehensive review and practice of key skills, matched to the Learning Objectives for each section.	Practice makes perfect. These problems combine exercises from all sections, giving you a comprehensive review in one place.	186-191
Case Study	Each chapter concludes with a case study that requires you to use skills learned within the section to perform statistical analysis.	Brings together the chapter concepts within one case study.	192

Data Disk t/a Statistics: Informed Decisions Using Data 2e
Michael Sullivan, III
ISBN 0-13-173152-1
CD Licensing Agreement
© 2007 Pearson Education, Inc.
Pearson Prentice Hall
Pearson Education, Inc.
Upper Saddle River, NJ 07458
All rights reserved.
Pearson Prentice Hall™ is a trademark of Pearson Education, Inc.

STATISTICS
Informed Decisions Using Data

SECOND EDITION

Michael Sullivan, III
Joliet Junior College

PEARSON

Prentice Hall

Upper Saddle River, New Jersey 07458

Cataloging-in-Publication data is on file at the Library of Congress.

Sullivan, Michael III
   Statistics: informed decisions using data, 2nd edition
    p. cm
   Includes index.
   ISBN 0-13-187149-8

Editor-in-Chief: Sally Yagan
Executive Acquisitions Editor: Petra Recter
Project Manager: Michael Bell
Executive Managing Editor: Kathleen Schiaparelli
Senior Managing Editor: Linda Mihatov Behrens
Production Editor: Bob Walters, Prepress Management, Inc.
Managing Editor, Digital Supplements: Nicole M. Jackson
Media Production Editor: John Cassar
Editorial Assistant / Print Supplements: Joanne Wendelken
Manufacturing Buyer: Maura Zaldivar
Manufacturing Manager: Alexis Heydt-Long
Director of Marketing: Patrice Jones
Marketing Assistant: Jennifer de Leeuwerk
Development Editor: Don Gecewicz
Editor-in-Chief, Development: Carol Trueheart
Art Director: Jonathan Boylan
Interior Designers: Dina Curro, Jonathan Boylan
Cover Designers: Kristine Carney, Jonathan Boylan
Director of Creative Services: Paul Belfanti
Composition: ESM Inhouse Formatting
  Manager, Composition: Allyson Graesser

Assistant Manager, Formatting: Clara Bartunek
Electronic Production Specialist: Joanne Del Ben
Electronic Page Makeup: Jacqueline Ambrosius, Joanne Del Ben
Director, Image Resource Center: Melinda Reo
Manager, Rights and Permissions: Zina Arabia
Manager, Visual Research: Beth Brentzel
Cover Image Specialist: Karen Sanatar
Image Permission Coordinator: Joanne Dippel
Photo Researcher: Melinda Alexander
Cover Photo: David Mager; ©2007 Pearson Education
Art Studio: Artworks
  Senior Managing Editor, Art Production and Management: Patricia Burns
  Manager, Production Technologies: Matthew Haas
  Managing Editor, Art Management: Abigail Bass
  Manager, Illustration Production : Sean Hogan
  Assistant Manager, Illustration Production: Ronda C. Whitson
  Art Editor: Tom Benfatti
  Illustrations: Stacy Smith, Scott Wieber, Nathan Storck, Ryan Currier,
    Mark Landis, Audrey Simonetti, David Lynch
  Quality Assurance: Pamela Taylor, Timothy Nguyen, Cathy Shelley

© 2007, 2004 Pearson Education, Inc.
Pearson Prentice Hall
Pearson Education, Inc.
Upper Saddle River, New Jersey 07458

Printed in the United States of America
10   9   8   7   6   5   4   3   2   1

ISBN: 0-13-187149-8   (Student Edition)

Pearson Education LTD., *London*
Pearson Education Australia PTY, Limited, *Sydney*
Pearson Education Singapore, Pte. Ltd
Pearson Education North Asia Ltd. *Hong Kong*
Pearson Education Canada, Ltd., *Toronto*
Pearson Educación de Mexico, S.A. de C.V.
Pearson Education – Japan, *Tokyo*
Pearson Education Malaysia, Pte. Ltd

To My Wife Yolanda

and

My Children
Michael, Kevin, and Marissa

CONTENTS

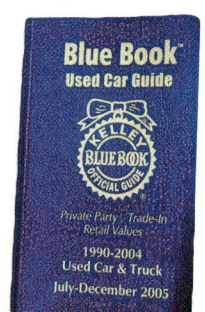

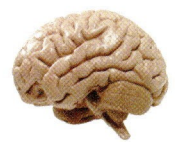

Part IV: Inference: From Samples to Population   415

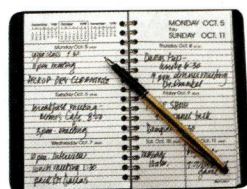

PREFACE TO THE INSTRUCTOR

Capturing a Powerful and Exciting Discipline in a Textbook

Statistics is a powerful subject. It is one of my passions. Bringing my passion for the subject together with my desire to create a text that would work for me, my students, and my school led me to write the first edition of this textbook.

I have taught introductory statistics for the past thirteen years at both 2-year and 4-year institutions, and during that time I have learned about the basic needs and issues confronting the introductory statistics course. The diversity of both the students taking the course and the instructors teaching the course heightens many of the challenges in presenting statistics at this introductory level. I brought these experiences and goals to the writing of the first edition.

The second edition of *Statistics: Informed Decisions Using Data* builds on the success of the first edition as well as on my continuing experience in teaching and shaping the course. Furthermore, as I was writing the second edition, I made a special effort to abide by the *Guidelines for Assessment and Instruction in Statistics Education* (GAISE) for the college introductory course endorsed by the American Statistical Association (ASA). The GAISE Report gives six recommendations for the college introductory statistics course:

1. Emphasize statistical literacy and develop statistical thinking
2. Use real data in teaching statistics
3. Stress conceptual understanding
4. Foster active learning
5. Use technology for developing conceptual understanding
6. Use assessments to improve and evaluate student learning

Emphasize Statistical Literacy and Develop Statistical Thinking

The definition of *statistical thinking* presented in the GAISE Report is "understanding the basic language of statistics and understanding some fundamental ideas of statistics." Among these fundamental ideas are the need to work with data, the importance of data production and collection, the ability to read statistical graphs, and the capacity to select an appropriate technique to answer a research question. *Statistics: Informed Decisions Using Data* addresses this recommendation in the following ways:

- **Extensive Discussion of Data Production**—Sections 1.2 through 1.4 discuss sampling techniques and pitfalls of sampling. Section 1.5 has a thorough discussion of the design of experiments.

- **Reading and Interpretation of Statistical Graphs**—The emphasis in Chapter 2 is on constructing and interpreting statistical graphs. In addition, we present Section 2.4, which illustrates the idea of misrepresenting data through graphs. The remainder of the text requires students to use graphs as a preliminary step in any statistical analysis. This continuing reliance on graphs is built into the examples and exercises.

- **Decision Making about Appropriate Techniques**—When a research question is posed, the researcher must decide on the type of analysis that is needed to answer the question. Different models may be used to answer the question. For example, in Section 1.5, students are often asked to design an experiment to answer a research question. Certainly, different designs may all be legitimate choices. The student is asked to justify his or her choice of design. Also, to help students in determining which technique to use, I have written new **Putting It All Together** sections in Chapters 9 and 10. The problems in these sections require that students decide which technique to use when performing inference, a key part of statistical thinking.

- **Decisions Projects**—Each chapter opens with a Decisions Project that requires students to collect, summarize, and interpret data to help make decisions we all must make, and would like to make intelligently, in our everyday lives. This edition highlights decision making by placing the Decisions Project on the chapter opener and weaving its resolution into the chapter itself.

- **Concepts and Vocabulary**—Each section exercise set has a list of problems that require students to learn the language of statistics and explain statistical concepts in their own words.

Use Real Data in Teaching Statistics

The examples and exercises both use real, sourced data to illustrate statistical concepts and procedures. The data come from a wide variety of disciplines such as psychology, sociology, finance, economics, biology, environmental studies, business, and chemistry so that all students can see the importance of data analysis in their field of study. A complete list of applications can be found in the List of Applications, pages xxiii-xxviii.

Among the many interesting real data sets that I have chosen are from my own students. Data sets from students serve several purposes: First, I like to show students that thinking statistically applies to ordinary situations. (Instructors may recall the famous story of Fisher and the lady tasting tea.) I also like to show students that they can act like a statistician and gather real and meaningful information about the environment, consumer products, and even golf. Demonstrating to students at my campus, and in your class, that real data are everywhere helps to validate the course and statistics as a discipline.

I have also learned that many instructors rely on classroom activities to teach statistical concepts and make data more vivid. Each chapter has at least one *optional* classroom activity that fosters active learning by allowing students to collect and analyze their own data.

Stress Conceptual Understanding

It is important that students understand statistical concepts because facility with concepts will contribute to their understanding of statistical procedures. Conceptual understanding will provide them with a strong problem-solving mindset that will help them to succeed in future statistics courses. In addition, conceptual understanding develops a student's ability to make decisions and think critically. *Statistics: Informed Decisions Using Data* helps students develop a conceptual understanding in the following ways:

- **Applying the Concepts Problems**—The Applying the Concepts problems in the exercise set not only ask students to analyze real data by drawing graphs or going through procedures, they also help to develop a student's ability to think statistically. Some examples of exercises that develop statistical thinking are: Problem 14 on page 109, Problem 25 on page 132, Problems 21 and 22 on page 149, Problem 31 on page 207, and the Problems in Sections 9.5 and 10.6 (the problems in these sections require students first to determine the appropriate procedure to use for a given scenario).

- **Simulations and Applets**—Throughout the text, you will find exercises that require students to perform simulations or use applets that will contribute to their conceptual understanding. For example, in Problem 33 on page 433, the students use simulation to enhance their understanding of the Central Limit Theorem. In Problems 50-52 on page 465, the students use an applet to enhance their understanding of confidence intervals. Simulations and applets are powerful because they let students work actively with data. In addition, many students who are visual learners find that applets give them the reinforcement that they need to master a concept.

- **The Presentation of Formulas**—When formulas are presented, we offer both a "computational formula" and a "definition formula." For example, see the definition for variance on page 139. The student develops an understanding that the variance is a mean of squared deviations from the definition formula. Or consider the definition for the correlation coefficient on page 198. The students develop a conceptual understanding of what correlation measures. It is important for students to know a formula, but students should also be able to explain a concept and the workings of a formula in their own words.

Foster Active Learning in the Classroom

To promote active learning, we have included numerous in-class activities that encourage students to collect, summarize, and analyze data. Doing these activities promotes conceptual understanding of statistical concepts. In addition to the classroom activities, we have incorporated applets that can be used in the classroom or as a group project to illustrate and explore statistical ideas. Plus, each chapter concludes with a case study that can be assigned as a small-group project. All these projects will help to foster active learning by encouraging students to discuss statistics among themselves and challenge each other's ideas.

Use Technology for Developing Concepts and Analyzing Data

We live in a digital world, and technology allows us to explore statistical ideas in new and exciting ways. Easily accessible technology now includes graphing calculators, statistical software packages, educational software, applets, and Web-based materials. For example, we can use graphing calculators or statistical software packages to explore "what if" scenarios. Doing so helps to develop the student's understanding of statistical concepts. The exercise sets in the text are written with "what if" scenarios built in. For example, Problem 24 in Section 4.1 asks the student to determine the impact of removing an observation from a data set. This problem foreshadows the idea of influential observations and is done easily with the help of technology.

Recognizing that various learning styles exist and that conceptual understanding can arise by working with formulas, each example is presented using computation "by hand." When appropriate, the results of the same example are obtained using a graphing calculator or software (MINITAB or Excel). The steps for obtaining the output using the TI-83/84 Plus graphing calculators, MINITAB, and Excel are included after the end-of-section exercise set.

To supplement the textbook, technology-specific manuals that contain tutorial instructions and worked examples and exercises are available (TI-83/84 Plus, MINITAB, Excel, and SPSS).

Use Assessments to Improve and Evaluate Student Learning

The exercises at the end of each section are divided into three parts: (1) Concepts and Vocabulary, (2) Skill Building, and (3) Applying the Concepts.

- The **Concepts and Vocabulary** exercises are fill-in-the-blank, true/false, and open-ended questions that assess students' understanding of vocabulary and statistical concepts. Formats such as true/false may be simple, but the Concepts and Vocabulary exercises have been written to review important concepts and to be thought provoking.

- The **Skill Building** exercises can be thought of as "drill and practice" problems that allow students to develop a level of comfort with the formulas and procedures in the section. These problems also serve as confidence-building problems, so that students experience early success and become more proficient with the quantitative aspects of a statistical analysis.

- The **Applying the Concepts** exercises are problems based in real data that ask a variety of questions to help develop statistical thinking. While these problems do ask the standard questions such as "find the mean" or "compute a 95% confidence interval," they also ask for students to explain the results or to respond to a variety of "what if" questions that broaden the scope of the concept and the analysis. For instance, the exercise may ask, "What is the impact of an outlier in the construction of a confidence interval?" Other exercises may highlight decision making about concepts by asking, "What is the best measure of central tendency for the given data set?" These "higher level of thinking" problems develop a student's statistical aptitude and teach the students to begin to frame their own questions and investigate problems on their own.

New to This Edition
Content Changes

- **Over 45% new and updated exercises,** with an emphasis on adding even greater variety, to give instructors an even larger range of exercises to choose from in selecting homework. The array of exercises includes many intriguing and meaningful topics, from M&Ms (always popular in statistics books) to the meaning of cholesterol counts, from identity theft to gas mileage, and from election results to cosmetic surgery. An interesting new exercise in Chapter 3 based on an experiment by Trina McNamara of Joliet Junior College compares a name-brand chocolate-chip cookie with a store-brand chocolate-chip cookie—one of the many interesting data sets contributed by students to the textbook.

- **Over 60% new examples**, most of which contain real world data, acting to further illustrate and reinforce concepts and techniques. Like the exercises, the choice of topics is practical, ranges widely, and will hold students' attention.

- **All sourced data have been updated** and carefully selected to engage the students.

- **Each chapter opens with a scenario,** *Making an Informed Decision,* that poses a question to the student and then, within the chapter, presents the content that is necessary to use statistics to make the decision. This feature engages the student in the statistical-thinking process and highlights the practicality of statistics.

- **Chapter 5 on probability has been revamped** to ease the introduction of topics in probability. The early sections take greater care in delineating disjoint events and the complement rule. The concept of independence gets fuller coverage in this new edition, too. At the request of reviewers, a new and optional topic has been added to the CD, Bayes's rule.

- **Updated and rewritten Chapter 6** on discrete probability distributions reflects reviewers' concerns. I focused especially on enhancing coverage of the binomial probability distribution. At the request of reviewers, binomial tables have been added to the appendices. Also at the request of reviewers, the Hypergeometric Probability Distribution has been added to the CD.

- **New Chapter 8 on sampling distributions** was added to strengthen the coverage of this important topic that tends to cause students to stumble. The decision to give more play to sampling distributions by placing them in their own chapter was heartily endorsed by the reviewers of this new edition.

- **Chapters 9 and 10** continue to discuss confidence intervals and hypothesis testing about a mean by presenting the "sigma known use z, sigma unknown use t" approach. I find that students benefit from some discussion of confidence intervals and hypothesis testing when the population standard deviation is known, even if such situations seldom arise in practice. In both of these chapters, I present the sigma-known case as a model or template, followed by sections on tests more likely to be used in practice.

- **Putting It All Together** sections have been added to Chapters 9 and 10. These sections require that students first determine the appropriate confidence interval or hypothesis test to conduct. This forces students to think about when each technique should be used, which is typically a stumbling block on exams.

- **New Chapter 13 on comparing three or more means reflects adopters' and reviewers' need** for a separate chapter on analysis of variance. This chapter now incorporates the reliable Tukey test. The last section gives the instructor the option of delving into two-way ANOVA, too.

- **New Chapter 14 is a more complete coverage of regression analysis**, and it also is a response to the concerns of adopters and reviewers. Section 14.3 is a new introduction to multiple regression for courses that include this topic.

Pedagogical Changes

- **New side-by-side presentation of hypothesis testing**, which allows instructors the flexibility for presenting the classical approach, P-value approach, or both approaches. Recognizing that there are advocates of each of these two approaches, this format allows an instructor to choose the method that suits the philosophy of the course.

- **Activities have been added to each chapter,** giving instructors the flexibility to engage the class in an activity or assign them for homework. Use of activities promotes participation in the course by the students, makes concepts more vivid, and underscores the value of learning and using statistical techniques.

- **More emphasis on statistical thinking and interpretation of results** appears throughout the chapters. I added more follow-up questions to exercises that require consideration and interpretation of what the students have calculated. Furthermore, examples and exercises encourage the use of boxplots and normal probability plots so that students can use descriptive statistics throughout the text to interpret data and understand the workings of techniques. I also redoubled my effort to include screen captures and computer printouts so that students will learn to "eyeball" and interpret results when using technology.

- **Using Technology examples show the way that software and calculators assist in analysis**, once students have mastered "by hand" calculations. The *Using Technology* examples also give the students a feeling for statistics in practice, demonstrate when to rely on different devices and software, and show the output or screens to expect.

- **A division into parts makes the structure of the book more transparent** to instructors and students. The parts show the student that the order of topics is logical and practical by following the statistical process.

- **MyMathLab**, offering course management, online homework and tutorial and assessment, is fully integrated with the text. Since 2001, over one million students have done better in math and statistics with MyMathLab!

Flexibility

Just as there are many different learning styles, there are many different teaching styles. One challenge in writing a text is to create a book that appeals to a broad audience. As I survey the halls of the mathematics department at Joliet Junior College, I see many high-quality instructors who present the same material in many different ways. Some like to incorporate as much technology as possible into their classroom, yet some prefer to minimize the technology. Some instructors prefer to use collaborative learning to present material, and others rely on lecture. For the introductory statistics course, some instructors are trained in the discipline of statistics, whereas others are trained in mathematics but teach statistics. Taking into consideration all these varied backgrounds, it is clear to me that a text should meet the needs of these diverse backgrounds and teaching philosophies.

"By Hand" Computation versus Technology Solutions

Let's consider how graphing calculators and statistical software are used in the course. Every concept in the text is first presented using the "by hand" approach. The reason for this is twofold. First, and probably most important- ly, when a student learns to arrive at a solution by working a problem by hand, the student's ability to understand the concept is enhanced. How else can a student understand the concept of linear correlation, except by seeing it as the product of z-scores? Second, it allows for flexibility in philosophies. If your particular philosophy is to present statistics using "by hand" solu- tions, it is there for you.

If you are more apt to use technology, then you can use the *Examples Using Technology* feature. Following most "by hand" examples is a *Using Tech- nology* example, which provides the output from a TI-83/84 Plus graphing calcu- lator, MINITAB, or Excel.

In addition, any problem that has 15 or more observations in the data set has a CD icon, which indicates to the instructor and student that the data set is available on the data CD.

Chapter-by-Chapter Content

Chapter 1   Data Collection

This chapter deals with the methods of obtaining data. There is a careful pres- entation of the various sampling techniques along with circumstances under which each is used. In addition, an entire section is dedicated to nonsampling errors and how to control them. The chapter ends with a detailed discussion of experimental design.

Chapter 2   Organizing and Summarizing Data

This chapter addresses methods for summarizing qualitative data (Section 2.1) and quantitative data (Sections 2.2 and 2.3). Instructors who do not wish to cover topics such as frequency polygons and ogives can skip Section 2.3 without loss of continuity. The chapter ends with a discussion of graphical misrepre- sentations of data. This section can be covered as a reading assignment.

Chapter 3   Numerically Describing Data

Sections 3.1 and 3.2 present numerical measures of central tendency and dis- persion. Section 3.3 is optional and can be skipped without loss of continuity. However, if it is skipped and Section 6.1 is covered, the instructor should pro- ceed slowly through the mean and standard deviation of a discrete random vari- able. Section 3.4 discusses measures of position, including the z-score, per- centiles, and outliers. Section 3.5 presents exploratory data analysis, especially the boxplot, which will get much use in later chapters.

Chapter 4   Describing the Relation between Two Variables

Section 4.1 introduces scatter diagrams and correlation. Section 4.2 presents the least-squares regression line. Section 4.3 presents the coefficient of determina- tion, residual analysis, and influential observations. This section is particularly important because it allows students to graphically assess whether or not the data they are analyzing are, in fact, linearly related. Section 4.4 has moved to a CD that accompanies the text and presents methods for using least squares to fit data to exponential and power models. It is optional and can be skipped without loss of continuity. In addition, the material in this chapter can be postponed until after Chapter 13.

Chapter 5 Probability

Section 5.1 introduces the basic concepts of probability and unusual events. Section 5.2 presents the Addition Rule and complements. Section 5.3 presents the Multiplication Rule for Independent Events. Section 5.4 presents conditional probability and the General Multiplication Rule and can be skipped without loss of continuity. Section 5.5 presents counting techniques and can be skipped without loss of continuity with the exception of Combinations. Finally, we have added Section 5.6, Bayes's Rule, which is located on the CD.

Chapter 6 Discrete Probability Distributions

The entire chapter is optional. Section 6.1 introduces the concept of a random variable and discrete probability distributions, along with expected value. Section 6.2 presents the Binomial Probability Distribution. If you intend to cover Section 6.2, it is a good idea to cover Section 5.3 and Combinations in Section 5.5 also. Section 6.3 presents the Poisson Probability Distribution. There is a new Section 6.4, which introduces the Hypergeometric Probability Distribution. The material in Section 6.4 is located on the CD.

Chapter 7 The Normal Probability Distribution

Sections 7.1 through 7.3 introduce the normal probability distribution. Section 7.4 presents normal probability plots as a means for assessing normality and is required to cover topics presented in Chapters 8 through 15. Section 7.4 does not require the use of technology because the output generated by MINITAB is included. This section is necessary to help students see that verifying normality is necessary before proceeding with inference for small samples. Section 7.5 discusses the normal approximation to the binomial and is optional.

Chapter 8 Sampling Distributions (New)

A new stand-alone chapter on sampling distributions has been added to place added emphasis on the importance of this concept. Section 8.1 introduces the sampling distribution of the sample mean. Section 8.2 discusses the sampling distribution of the sample proportion.

Chapter 9 Estimating the Value of a Parameter Using Confidence Intervals

Section 9.1 introduces the construction of a confidence interval when the population standard deviation is known, and Section 9.2 constructs confidence intervals when the population standard deviation is unknown. This approach is different from that in some other texts, but it is logical. It's as simple as "sigma known, use z; sigma unknown, use t." In both cases, small samples require that the population from which the sample was drawn be normal, so we check this requirement with a normal probability plot. Section 9.3 covers confidence intervals about a population proportion, and Section 9.4 covers confidence intervals about a population standard deviation. Section 9.4 is optional and can be skipped without loss of continuity. A new Section 9.5 has been added, entitled, "Putting It All Together: Which Method Do I Use?" When students are asked to construct a confidence interval in Section 9.2, they are fairly certain that they have to construct a confidence interval about a mean using the t-distribution. The material in this section does not provide hints, which play down the role of decision making; instead, it requires the student to choose the correct model first before employing any methods.

Chapter 10   Testing Claims Regarding a Parameter

Section 10.1 provides an introduction to the language of hypothesis testing. Sections 10.2 and 10.3 test claims regarding a population mean, again segmented by "sigma known, use z; sigma unknown, use t." New to this edition is the side-by-side "Classical" versus "P-value" approach to hypothesis testing. This allows instructors to choose the method that suits their philosophy. The side-by-side presentation continues for the rest of the text. Section 10.4 presents hypothesis testing about a population proportion. An interesting feature in this section is that it includes how to use the Binomial Probability Distribution to compute exact P-values. This is especially important if the requirement for using the normal approximation to the binomial is not satisfied. Section 10.5 discusses hypothesis testing about a population standard deviation. A new Section 10.6 has been added, entitled, "Putting It All Together: Which Method Do I Use?" Similar to the Putting It All Together from Chapter 9, students must first decide on the appropriate model to use before testing a claim. Section 10.7 discusses the power of the test and the probability of Type II errors. Sections 10.5 and 10.7 are optional.

Chapter 11   Inferences on Two Samples

Section 11.1 presents the analysis required for matched-pairs design. Section 11.2 presents the analysis for comparing two means from independent samples. Notice that the discussion regarding pooled estimates is absent. This discussion is not included because the pooled estimate approach requires that the two populations have a common variance, which is an extremely difficult requirement to test in practice. Because "pooling" versus "not pooling" often provides the same results, the pooled estimate approach is not necessary at this level to cloud the students' decision-making process. Section 11.3 discusses comparing two population proportions. Section 11.4 presents the comparison of two variances and is optional.

Chapter 12   Inference on Categorical Data

Section 12.1 presents the chi-square test for goodness of fit. The section presents a discussion of the Chi-square Distribution for those who skipped Sections 9.4 and 10.5. If the instructor did skip these sections, be sure to introduce students to the Chi-square table. Section 12.2 presents contingency tables, marginal distributions, and conditional distributions. If the instructor skipped Section 5.4, proceed at a slower pace. Section 12.3 discusses chi-square tests for independence and homogeneity. Again, if Sections 5.3 and/or 5.4 were skipped, proceed slowly.

Chapter 13   Comparing Three or More Means (New)

This new chapter was written to complete the discussion of the comparison of three or more means. Section 13.1 discusses one-way analysis of variance and was taken from Section 12.3 of the first edition. Section 13.2 discusses the Tukey test, which is used to determine which means differ significantly when the null hypothesis of equal means is rejected. Section 13.3 discusses the analysis of the randomized block design, and Section 13.4 discusses two-way analysis of variance.

Chapter 14   Inference on the Least-Squares Regression Model and Multiple Regression (New)

Sections 14.1 and 14.2 are based on Sections 12.1 and 12.2 from the first edition. Both sections require that Sections 4.1 and 4.2 be covered. Section 14.3 gives an introduction to multiple linear regression.

Chapter 15   Nonparametric Statistics (On CD)

This is the former Chapter 13 and has been updated and moved to the CD. Some instructors try to integrate nonparametrics tests at appropriate places in their course, as contrasts to the parametric tests. In this case, Section 15.2 on the runs test for randomness can be covered any time after Section 10.1. Section 15.3 can be presented right after Section 10.3. Section 15.4 can be covered after Section 11.1. Section 15.5 can be covered after Section 11.2. Section 15.6 can be covered after Section 14.1. Section 15.7 can be covered after Section 13.1.

Using the Text Efficiently with Your Syllabus

To meet the varied needs of diverse syllabi, this book has been organized with flexibility of use in mind. When structuring your syllabus, you will notice the topics listed in the "Preparing for This Section" material at the beginning of each section, which will tip you off to dependencies within the course.

The two most common variations within an introductory statistics course are the treatment of regression analysis and the treatment of probability.

- **Coverage of Correlation and Regression** The text was written with the descriptive portion of bivariate data (Chapter 4) presented after the descriptive portion of univariate data (Chapter 3). Instructors who prefer to postpone the discussion of bivariate data until later in the course can simply skip Chapter 4 and return to it before covering Chapter 14. Within Chapter 4, an instructor may skip Sections 4.3 and/or 4.4 without loss of continuity. (Because Section 4.4 on nonlinear regression is covered by a select few instructors, it is located on the CD that accompanies the text in Adobe pdf form, so that it can be easily printed.)
- **Coverage of Probability** The text allows for a course to present either an extensive introduction to probability or a light coverage of probability. Instructors wishing to present light coverage of probability may cover Section 5.1 and skip the remaining sections. A mid-level treatment of probability can be accomplished by covering Sections 5.1 through 5.3. Instructors who will be covering the chi-square test for independence will want to cover sections 5.1 through 5.3. In addition, any instructor who will be covering binomial probabilities will want to cover independence in Section 5.3 and combinations in Section 5.5.

Acknowledgments

Textbooks evolve into their final form through the efforts and contributions of many people. First and foremost, I would like to thank my family, whose dedication to this project was just as much as mine: my wife, Yolanda, whose words of encouragement and support were unabashed, and my children, Michael, Kevin, and Marissa, who would come and greet me every morning with smiles that only children can deliver. I owe each of them my sincerest gratitude.

I would also like to thank the entire Mathematics Department at Joliet Junior College, my colleagues who provided support, ideas, and encouragement to help me complete this project. From Prentice Hall: Petra Recter, who has worked so hard to help bring this text into its second edition; Joanne Wendelken for her talent and skill in coordinating the manuscript and reviews; Sally Yagan for her support, encouragement, and words of wisdom; Patrice Jones, for his marketing savvy and dedication to getting the word out; Michael Bell, for his hard work in coordinating all the supplements; and the Prentice Hall sales team, for their confidence and support of this book. I thank Bob Walters, my Production Editor, for his dedication, enthusiasm, pride, and uncanny knack for attention to detail. I also extend many thanks to Don Gecewicz for his ability to see the forest for the trees and his talent getting me thinking about each sentence; Kathleen Miranda for her insight, suggestions, and in creation of answers (I owe you one). Craig Johnson and Sudhir Goel for their dedication to reading pages and verifying the accuracy in the answers. I also would like to extend my thanks to Kevin Bodden and Randy Gallaher, whose ability to write solutions, check answers, and provide insightful suggestions is amazing.

Finally, we offer many thanks to all the reviewers, whose insights and ideas form the backbone of this text. We apologize for any omissions.

CALIFORNIA Charles Biles, *Humboldt State University* • Carol Curtis, *Fresno City College* • Freida Ganter, *California State University-Fresno* • Craig Nance, *Santiago Canyon College* **COLORADO** Roxanne Byrne, *University of Colorado-Denver* **CONNECTICUT** Kathleen McLaughlin, *Manchester Community College* • Dorothy Wakefield, *University of Connecticut* • Cathleen M. Zucco Teveloff, *Trinity College* **DISTRICT OF COLUMBIA** Jill McGowan, *Howard University* **FLORIDA** Randall Allbritton, *Daytona Beach Community College* • Franco Fedele, *University of West Florida* • Laura Heath, *Palm Beach Community College* Perrian Herring, *Okaloosa Walton College* • Marilyn Hixson, *Brevard Community College* • Philip Pina, *Florida Atlantic University* • Mike Rosenthal, *Florida International University* • James Smart, *Tallahassee Community College* **GEORGIA** Virginia Parks, *Georgia Perimeter College* • Chandler Pike, *University of Georgia* • Jill Smith, *University of Georgia* **IDAHO** K. Shane Goodwin, *Brigham Young University* • Craig Johnson, *Brigham Young University* • Brent Timothy, *Brigham Young University* • Kirk Trigsted, *University of Idaho* **ILLINOIS** Grant Alexander, *Joliet Junior College* • Kevin Bodden, *Lewis & Clark Community College* • Joanne Brunner, *Joliet Junior College* • James Butterbach, *Joliet Junior College* • Elena Catoiu, *Joliet Junior College* • Faye Dane, *Joliet Junior College* • Jason Eltrevoog, *Joliet Junior College* • Erica Egizio, *Joliet Junior College* • Randy Gallaher, *Lewis & Clark Community College* • Iraj Kalantari, *Western Illinois University* • Donna Katula, *Joliet Junior College* • Diane Long, *College of DuPage* • Jean McArthur, *Joliet Junior College* • Angela McNulty, *Joliet Junior College* • Judy Meckley, *Joliet Junior College* • Linda Padilla, *Joliet Junior College* • David Ruffato, *Joliet Junior College* • Patrick Stevens, *Joliet Junior College* • Robert Tuskey, *Joliet Junior College* • Stephen Zuro, *Joliet Junior College* **INDIANA** Jason Parcon, *Indiana University–Purdue University Ft. Wayne* **KANSAS** Ingrid Peterson, *University of Kansas* **MARYLAND** John Climent, *Cecil Community College* • Rita Kolb, *The Community College of Baltimore County* **MASSACHUSETTS** Daniel Weiner, *Boston University* **MICHIGAN** Margaret M. Balachowski, *Michigan Technological University* • Susan Lenker, *Central Michigan University* • Timothy D. Stebbins, *Kalamazoo Valley Community College* • Sharon Stokero, *Michigan Technological University* • Alana Tuckey, *Jackson Community College* **MINNESOTA** Mezbhur Rahman, *Minnesota State University* **MISSOURI** Farroll Tim Wright, *University of Missouri-Columbia* **NEBRASKA** JaneKeller, *Metropolitan Community College* **NEW YORK** Jacob Amidon, *Finger Lakes Community College* • Stella Aminova, *Hunter College* • Pinyuen Chen, *Syracuse University* • Bryan Ingham, *Finger Lakes Community College* • Anne M. Jowsey, *Niagara County Community College* • Maryann E. Justinger, *Erie Community College-South Campus* • Kathleen Miranda, *SUNY at OldWestbury* • Robert Sackett, *Erie Community College-North Campus* **NORTH CAROLINA** Fusan Akman, *Coastal Carolina Community College* • Mohammad Kazemi, *University of North Carolina-Charlotte* • Janet Mays, *Elon University* • Marilyn McCollum, *North Carolina State University* • Claudia McKenzie, *Central Piedmont Community College* • Said E. Said, *East Carolina University* • Karen Spike, *University of North Carolina-Wilmington* • Jeanette Szwec, *Cape Fear Community College* • Richard Einsporn, *The University of Akron* **OHIO** Michael McCraith, *Cuyaghoga Community College* **OREGON** Daniel Kim, *Southern Oregon University* **SOUTH CAROLINA** Diana Asmus, *Greenville Technical College* • Dr. William P. Fox, *Francis Marion University* • Cheryl Hawkins, *Greenville Technical College* • Rose Jenkins, *Midlands Technical College* • Lindsay Packer, *College of Charleston* **TENNESSEE** Nancy Pevey, *Pellissippi State Technical Community College* • David Ray, *University of Tennessee-Martin* **TEXAS** Jada Hill, *Richland College* • David Lane, *Rice University UTA* • Joe Gallegos, *Salt Lake City Community College* **VIRGINIA** Kim Jones, *Virginia Commonwealth University* • Vasanth Solomon, *Old Dominion University* **WEST VIRGINIA** Mike Mays, *West Virginia University* **WISCONSIN** William Applebaugh, *University of Wisconsin–Eau Claire* • Carolyn Chapel, *Western Wisconsin Technical College* • Beverly Dretzke, *University of Wisconsin-Eau Claire* • Jolene Hartwick, *Western Wisconsin Technical College* • Thomas Pomykalski, *Madison Area Technical*

Michael Sullivan, III
Joliet Junior College

Data CD

Text comes with CD containing data sets (ASCII, SPSS, Excel, MINITAB, TI, and JMP), statistical applets, which are useful for illustrating statistical concepts, and PDFs of additional topics, formulas, and tables. Included topics are Nonlinear Regression, Hypergeometric Distribution, Bayes' Rule and Nonparametric Statistics.

PRINTED RESOURCES
Student Study-Pack

Text and Student Study Pack bundle: 0-13-221659-0
Stand-alone ISBN: 0-13-221436-9

Everything a student needs to succeed in one place. Provided at no charge when packaged with a new book or available for purchase stand-alone. The Student Study-Pack contains:

Student Solutions Manual

Fully worked solutions to odd-numbered exercises

CD Lecture Videos

A comprehensive set of videos, tied to the textbook, in which examples from each chapter are worked out by a statistics instructor. The videos provide excellent support for students who require additional assistance, for distance learning and self-paced programs, or for students who missed class.

Technology Manual

Contains detailed tutorial instructions and worked out examples and exercises for:

- TI-83/84 Plus Graphing Calculator
- Excel (including PHStat, an Excel plug-in)
- SPSS
- MINITAB

Pearson Tutor Center

Tutors provide one-on-one tutoring for any problem with an answer at the back of the book. Students access the Tutor Center via toll-free phone, fax, or e-mail.

INTERNET RESOURCES
MyMathLab

Provided at no charge to college students when packaged with a new book or available for purchase stand-alone.

- Text and MyMathLab Bundle: 0-13-221660-4
- Stand-alone ISBN: 0-13-147894-X

MyMathLab is a text-specific, online course that provides unlimited tutorial exercises closely correlated to the exercises in the text, a multimedia textbook with links to learning aids, such as animations and videos, student supplements in electronic format—the student solutions manual, PowerPoint Presentations and much more! Since 2001, over one million students have done better in math and statistics with MyMathLab.

MathXL®

MathXL offers unlimited (algorithmically generated) practice exercises that are correlated directly to the exercises in the textbook.

- 12-month MathXL and Text Bundle: 0-13-148692-6
- 24-month MathXL and Text Bundle: 0-13-148695-0
- 12-month Stand-alone ISBN: 0-13-147892-3
- 24-month Stand-alone ISBN: 0-13-147897-4

LIST OF APPLICATIONS

PHOTO CREDITS

Getting the Information You Need

Statistics is a process—a series of steps that lead to a goal. This text is divided into parts to help the reader see the process of statistics.

The first step in the process is to determine the research objective or question to be answered. Then, information is obtained to answer the questions stated in the research objective.

Data Collection

Outline

DECISIONS

It is Monday morning and already you are thinking about Friday night—movie night. You don't trust the movie reviews published by professional critics, so you decide to survey "regular" people yourself. You need to design a questionnaire that can be used to help you make an informed decision about whether to attend a particular movie. See the Decision Activity on page 37.

●●● Putting It All Together

For most of you, this is your first Statistics course. Taking a Statistics course is different from taking a Mathematics course. While there are formulas and mathematical symbols in the course, Statistics is not Mathematics. So, whether you have struggled or had success in prior Mathematics courses, you can succeed in Statistics.

Before you begin the course, read "How to Use this Book" on the inside front cover of the text.

1.1 Introduction to the Practice of Statistics

Objectives

1. Define statistics and statistical thinking
2. Understand the process of statistics
3. Distinguish between qualitative and quantitative variables
4. Distinguish between discrete and continuous variables

1 Define Statistics and Statistical Thinking

What is statistics? When asked this question, many people respond that statistics is numbers. This response is only partially correct.

Definition

> **Statistics** is the science of collecting, organizing, summarizing, and analyzing information to draw conclusions or answer questions.

It is helpful to consider this definition in three parts. The first part of the definition states that statistics involves the collection of information. The second refers to the organization and summarization of information. Finally, the third states that the information is analyzed to draw conclusions or answer specific questions.

What is the information referred to in the definition? The information is *data*. According to the *American Heritage Dictionary*, **data** are "a fact or proposition used to draw a conclusion or make a decision." Data can be numerical, as in height, or they can be nonnumerical, as in gender. In either case, data describe characteristics of an individual. The reason that data are important in statistics can be seen in this definition: data are used to draw a conclusion or make a decision.

Analysis of data can lead to powerful results. Data can be used to offset anecdotal claims, such as the suggestion that cellular telephones cause brain cancer. After carefully collecting, summarizing, and analyzing data regarding this phenomenon, it was determined that there is no link between cell phone usage and brain cancer.

In Other Words

Anecdotal means that the information being conveyed is based on casual observation, not scientific research.

Because data are powerful, they can be dangerous when misused. The misuse of data usually occurs when data are incorrectly obtained or analyzed. For example, radio or television talk shows regularly ask poll questions in which respondents must call in or use the Internet to supply their vote. The only individuals who are going to call in are those that have a strong opinion about the topic. This group is not likely to be representative of people in general, so the results of the poll are not meaningful. Whenever we look at data, we should be mindful of where the data come from.

Even when data tell us that a relation exists, we need to investigate. For example, a study showed that breast-fed children have higher IQs than those who were not breast-fed. Does this study mean that mothers should breast-feed their children? Not necessarily. It may be that some other factor contributes to the IQ of the children. For example, it turns out that mothers who breast-feed generally have higher IQs than those who do not. Therefore, it may be genetics that leads to the higher IQ, not breast-feeding. This illustrates an idea in statistics known as the *lurking variable*. In statistics, we must consider the lurking variables because two variables most often are influenced by a third variable. A good statistical study will have a way of dealing with the lurking variable.

Another key aspect of data is that they vary. To help understand this variability, consider the students in your classroom. Is everyone the same height? No. Does everyone have the same color hair? No. So, among a group of individuals there is variation. Now consider yourself. Do you eat the same amount of food each day? No. Do you sleep the same number of hours each day? No. So, even looking at an individual there is variation. Data vary. The goal of statistics is to describe and understand the sources of variation.

Because of this variability in data, the results that we obtain using data can vary. This is a very different idea than what you may be used to from your mathematics classes. In mathematics, if Bob and Jane are asked to solve $3x + 5 = 11$, they will both obtain $x = 2$ as the solution, if they use the correct procedures. In statistics, if Bob and Jane are asked to estimate the average commute time for workers in Dallas, Texas, they will likely get different answers, even though they both use the correct procedure. The different answers occur because they likely surveyed different individuals, and these individuals have different commute times. Note: The only way Bob and Jane would get the same result is if they both asked *all* commuters or the same commuters how long it takes to get to work, but how likely is this?

So, in Mathematics when a problem is solved correctly, the results can be reported with 100% certainty. In Statistics when a problem is solved, the results do not have 100% certainty. In Statistics, we might say that we are 95% confident that the average commute time in Dallas, Texas is 21.5 minutes. While uncertain results may sound disturbing now, it will become more apparent what this means as we proceed through the course.

Without certainty, how can statistics be useful? Statistics can provide an understanding of the world around us because recognizing where variability in data comes from can help us to control it. Understanding the techniques presented in this text will provide you with powerful tools that will give you the ability to analyze and critique media reports, make investment decisions (such as what mutual fund to invest in), or conduct research on major purchases (such as what type of car you should buy). This will help to make you an informed consumer of information and guide you in becoming a critical and statistical thinker.

② Understand the Process of Statistics

The definition of statistics implies that the methods of statistics follow a process.

The Process of Statistics

1. *Identify the research objective.* A researcher must determine the question(s) he or she wants answered. The question(s) must be detailed so that it identifies a group that is to be studied and the questions that are to be answered. The group to be studied is called the **population**. An **individual** is a person or object that is a member of the population being studied. For example, a researcher may want to study the population of all 2005 model-year automobiles. The individuals in this study would be the cars.

2. *Collect the information needed to answer the questions posed in (1).* Gaining access to an entire population is often difficult and expensive. In conducting research, we typically look at a subset of the population, called a **sample**. For example, the U.S. population of people 18 years or older is about 218 million. Many national studies consist of samples of size 1100. The collection-of-information step is vital to the statistical process, because if the information is not collected correctly, the conclusions drawn are meaningless. Do not overlook the importance of appropriate data-collection processes.

3. *Organize and summarize the information.* This step in the process is referred to as *descriptive statistics*.

! CAUTION
Many nonscientific studies are based on *convenience samples*, such as internet surveys or phone-in polls. The results of any study performed using this type of sampling method are not reliable.

Definition
Descriptive statistics consists of organizing and summarizing the information collected.

Descriptive statistics describe the information collected through numerical measurements, charts, graphs, and tables. The main purpose of descriptive statistics is to provide an overview of the information collected.

4. *Draw conclusions from the information.* In this step the information collected from the sample is generalized to the population.

Definition

Inferential statistics uses methods that takes results obtained from a sample, extends them to the population, and measures the reliability of the result.

For example, if a researcher is conducting a study based on the population of Americans aged 18 years or older, she might obtain a sample of 1100 Americans aged 18 years or older. The results obtained from the sample would be generalized to the population. There is always uncertainty when using samples to draw conclusions regarding a population, because we can't learn everything about a population by looking at a sample. Therefore, statisticians will report a level of confidence in their conclusions. This level of confidence is a way of representing the reliability of results. If the entire population is studied, then inferential statistics is not necessary, because descriptive statistics will provide all the information that we need regarding the population.

The following example will illustrate the process of a statistical study.

EXAMPLE 1   Effectiveness of Antihypertensive Drugs

According to researchers, little information exists on the effects that antihypertensive drugs have on patients who have heart disease and normal blood pressure.* Blood pressure is the force of blood against the walls of arteries and is presented as two numbers: the systolic pressure (as the heart beats) over the diastolic pressure (as the heart relaxes between beats). A blood pressure measurement of 120/80 mm Hg (millimeters of mercury) is normal. Hypertension or high blood pressure exists in individuals with a systolic blood pressure above 160 mm Hg or a diastolic blood pressure above 100 mm Hg. Researchers Steven E. Nissan and his associates wanted to determine the effectiveness of an antihypertensive drug** on preventing cardiovascular events such as congestive heart failure, stroke, or other heart-related problems. The following statistical process allowed the researchers to measure the effectiveness of the drug:

Group 1   Group 2

1. *Identify the research objective.* Researchers wished to determine the effectiveness of the drug on preventing cardiovascular events in patients who have heart disease and normal blood pressure.

2. *Collect the information needed to answer the questions.* The researchers divided 1,317 patients with heart disease and diastolic blood pressure less than 100 mm Hg into two groups. Group 1 had 663 patients and group 2 had 654 patients. The patients in group 1 received 10 mg daily of the antihypertensive drug. The patients in group 2 received a *placebo*. A **placebo** is an innocuous drug such as a sugar tablet. Group 1 is called the **experimental group**. Group 2 is called the **control group**. Neither the doctor administering the drug nor the patient knew whether he or she was in the experimental or control group. This is referred to as a **double-blind** experiment. After 24 months of treatment, each patient's blood pressure was recorded. In addition, the number of patients in each group who experienced a cardiovascular event was counted.

3. *Organize and summarize the information.* Before administering any drugs, it was determined that both groups had similar blood pressure. After the

*The discussion is based on a study done by Steven E. Nissan, E. Murat Tuzcu, Peter Libby, Paul D. Thompson, Magdi Ghali, Dahlia Garza, Lance Berman, Harry Shi, Ethel Buebendorf, and Eric Topol published in the *Journal of the American Medical Association*, Vol. 292, No. 18.

**The drug used in the study was 10 mg of amlodipine.

24-month period ended, the experimental group's blood pressure decreased by 4.8/2.5 mm Hg, whereas the placebo group's blood pressure increased 0.7/0.6 mm Hg. In addition, 16.6% of patients in the experimental group experienced a cardiovascular event, while 23.1% of patients in the control (placebo) group experienced a cardiovascular event.

4. *Draw conclusions from the data.* We extend the results from the sample of 1,317 patients to all individuals who have heart disease and normal blood pressure. That is, the antihypertensive drug appears to decrease blood pressure and seems effective in reducing the likelihood of experiencing a cardiovascular event such as a stroke.

In the study presented in Example 1, notice that the population is clearly identified in the research objective as patients who have heart disease and normal blood pressure. However, the researchers collected the information that they needed by looking at a subset of this population, the 1,317 patients.

Now Work Problem 45.

③ Distinguish between Qualitative and Quantitative Variables

Once a research objective is stated, a list of the information the researcher desires about the individual must be created. **Variables** are the characteristics of the individuals within the population. For example, this past spring my son and I planted a tomato plant in our backyard. We decided to collect some information about the tomatoes harvested from the plant. The individuals we studied were the tomatoes. The variable that interested us was the weight of the tomatoes. My son noted that the tomatoes had different weights even though they all came from the same plant. He discovered that variables such as weight vary.

If variables did not vary, they would be constants, and statistical inference would not be necessary. Think about it this way: If all the tomatoes had the same weight, then knowing the weight of one tomato would be sufficient to determine the weights of all tomatoes. However, the weights of tomatoes vary from one tomato to the next. One goal of research is to learn the causes of the variability so that we can learn to grow plants that yield the best tomatoes.

Variables can be classified into two groups: *qualitative* or *quantitative*.

Definition

Qualitative or categorical variables allow for classification of individuals based on some attribute or characteristic.

Quantitative variables provide numerical measures of individuals. Arithmetic operations such as addition and subtraction can be performed on the values of a quantitative variable and will provide meaningful results.

Many examples in this text will include a suggested **approach**, or a way to look at and organize a problem so that it can be solved. The approach will be a suggested method of *attack* toward solving the problem. This does not mean that the approach given is the only way to solve the problem, because many problems have more than one approach leading to a correct solution. For example, if you turn the key on your car's ignition and it doesn't start, one approach would be to look under the hood and try to determine what is wrong. (Of course, this approach would work only if you know how to fix cars.) A second, equally valid approach would be to call an automobile mechanic to service the car.

In Other Words
Typically, there is more than one correct approach to solving a problem.

EXAMPLE 2   ### Distinguishing between Qualitative and Quantitative Variables

Problem: Determine whether the following variables are qualitative or quantitative.

(a) Gender

(b) Temperature

(c) Number of days during the past week a college student aged 21 years or older has had at least one drink

(d) Zip code

Approach: Quantitative variables are numerical measures such that arithmetic operations can be performed on the values of the variable. Qualitative variables describe an attribute or characteristic of the individual that allows researchers to categorize the individual.

Solution

(a) Gender is a qualitative variable because it allows a researcher to categorize the individual as male or female. Notice that arithmetic operations cannot be performed on these attributes.

(b) Temperature is a quantitative variable because it is numeric, and operations such as addition and subtraction provide meaningful results. For example, 70°F is 10°F warmer than 60°F.

(c) Number of days during the past week that a college student aged 21 years or older had at least one drink is a quantitative variable because it is numeric, and operations such as addition and subtraction provide meaningful results.

(d) Zip code is a qualitative variable because it categorizes a location. Notice that the addition or subtraction of zip codes does not provide meaningful results.

Now Work Problem 15.

On the basis of the result of Example 2(d), we conclude that a variable may be qualitative while having values that are numeric. Just because a variable is numeric does not mean that the variable is quantitative.

④ ## Distinguish between Discrete and Continuous Variables

We can further classify quantitative variables into two types.

Definition

In Other Words
If you count to get the value of a variable, it is discrete. If you measure to get the value of the variable, it is continuous. When deciding whether a variable is discrete or continuous, ask yourself if it is counted or measured.

A **discrete variable** is a quantitative variable that has either a finite number of possible values or a countable number of possible values. The term *countable* means that the values result from counting, such as 0, 1, 2, 3, and so on.

A **continuous variable** is a quantitative variable that has an infinite number of possible values that are not countable.

Figure 1 illustrates the relationship among qualitative, quantitative, discrete, and continuous variables.

Figure 1

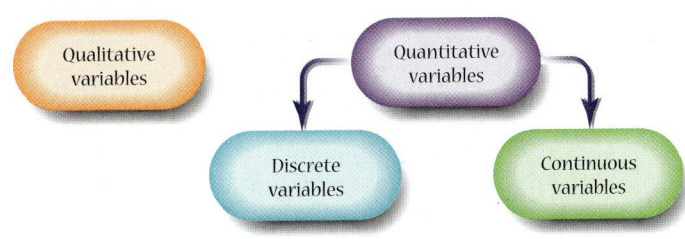

Recognizing the type of variable being studied is important because it dictates the type of analysis that can be performed. An example should help to clarify the definitions.

Distinguishing between Discrete and Continuous Variables

Problem: Determine whether the following quantitative variables are discrete or continuous.

(a) The number of heads obtained after flipping a coin five times.
(b) The number of cars that arrive at a McDonald's drive-through between 12:00 P.M and 1:00 P.M.
(c) The distance a 2005 Toyota Prius can travel in city driving conditions with a full tank of gas.

Approach: A variable is discrete if its value results from counting. A variable is continuous if its value is measured.

Solution

(a) The number of heads obtained by flipping a coin five times would be a discrete variable because we would count the number of heads obtained. The possible values of the discrete variable are {0, 1, 2, 3, 4, 5}.
(b) The number of cars that arrive at McDonalds drive-through between 12:00 P.M. and 1:00 P.M. is a discrete variable because its value would result from counting the cars. The possible values of the discrete variable are 0, 1, 2, 3, 4, and so on. Notice that there is no predetermined upper limit to the number of cars that may arrive.
(c) The distance traveled is a continuous variable because we measure the distance.

Continuous variables are often rounded. For example, when the miles per gallon (mpg) of gasoline for a certain make of car is given as 24 mpg, it means that the miles per gallon is greater than or equal to 23.5 and less than 24.5, or $23.5 \le \text{mpg} < 24.5$.

Now Work Problem 27.

The list of observed values for a variable is **data**. Gender is a variable, the observations male or female are data. **Qualitative data** are observations corresponding to a qualitative variable. **Quantitative data** are observations corresponding to a quantitative variable. **Discrete data** are observations corresponding to a discrete variable, and **continuous data** are observations corresponding to a continuous variable.

In Other Words
The singular of data is datum.

Distinguishing between Variables and Data

Problem: Table 1 presents a group of selected countries and information regarding these countries as of July, 2004. Identify the individuals, variables, and data in Table 1.

Approach: An individual is an object or person for whom we wish to obtain data. The variables are the characteristics of the individuals, and the data are the specific values of the variables.

Solution: The individuals in the study are the countries: Australia, Canada, and so on (in red ink). The variables measured for each country are *government type*, *life expectancy*, and *population* (in blue ink). The variable government type is qualitative because it categorizes the individual. The quantitative variables are life expectancy and population.

Country	Government Type	Life Expectancy (years)	Population (in millions)
Australia	Democratic	80.26	19.9
Canada	Confederation	79.96	32.5
France	Republic	79.44	60.4
Morocco	Constitutional Monarchy	70.35	32.2
Poland	Republic	74.16	38.6
Sri Lanka	Republic	72.89	19.9
United States	Federal Republic	77.43	293.0

Table 1

Source: CIA World Factbook

The quantitative variable life expectancy is continuous because it is measured. The quantitative variable population is discrete because we count people. The observations are the data (in green ink). For example, the data corresponding to the variable life expectancy are 80.26, 79.96, 79.44, 70.35, 74.16, 72.89, and 77.43. The following data correspond to the individual Poland: a *republic* government with residents whose *life expectancy* is 74.16 years and where *population* is 38.6 million people. Republic is an instance of qualitative data that results from observing the value of the qualitative variable government type. The life expectancy of 74.16 years is an instance of quantitative data that results from observing the value of the quantitative variable life expectancy.

Now Work Problem 51.

1.1 ASSESS YOUR UNDERSTANDING

Concepts and Vocabulary

1. Define statistics.
2. Explain the difference between a population and a sample.
3. Statistics is the science of collecting, organizing, summarizing, and analyzing _____ to answer questions or draw conclusions.
4. A(n) _____ is a person or object that is a member of the population being studied.
5. _____ statistics consists of organizing and summarizing information collected, while _____ statistics uses methods that generalize results obtained from a sample to the population and measure the reliability of the results.
6. What does it mean when an experiment is double-blind?
7. Discuss the differences between an *experimental* group and a *control* group.

8. What are placebos? Why do you think they are needed?
9. _____ are the characteristics of the individuals of the population being studied.
10. Contrast the differences between qualitative and quantitative variables.
11. Discuss the difference between discrete and continuous variables.
12. *True or False*: Both discrete and continuous variables are quantitative variables.
13. Explain the process of statistics.
14. The age of a person is commonly considered to be a continuous random variable. Could it be considered a discrete random variable instead? Explain.

Skill Building

In Problems 15–26, classify the variable as qualitative or quantitative.

15. Nation of origin
    NW
16. Number of siblings
17. Eye color
18. Number on a football player's jersey
19. Grams of carbohydrates in a doughnut
20. Number of unpopped kernels in a bag of ACT microwave popcorn

21. Assessed value of a house
22. Phone number
23. Population of a state
24. Cost (in dollars) to fill up a 2005 Chevrolet Corvette
25. Student ID number
26. Marital status

In Problems 27–38, determine whether the quantitative variable is discrete or continuous.

27. Runs scored in a season by Albert Pujols
NW

28. Volume of water lost each day through a leaky faucet

29. Length of a country song

30. At rest pulse rate of a 20-year-old college student

31. Number of sequoia trees in a randomly selected acre of Yosemite National Park

32. Weight of a randomly selected hog

33. Temperature on a randomly selected day in Memphis, Tennessee

34. Internet connection speed in kilobytes per second

35. Number of tornadoes in a year in the United States

36. Points scored in an NCAA basketball game

37. Number of donors at a blood drive

38. Air pressure in pounds per square inch in an automobile tire

In Problems 39–44, a research objective is presented. For each research objective, identify the population and sample in the study.

39. The Gallup Organization contacts 1,028 teenagers who are 13 to 17 years of age and live in the United States and asked whether or not they had been prescribed medications for any mental disorders, such as depression or anxiety.

40. A quality-control manager randomly selects 50 bottles of Coca-Cola that were filled on October 15 to assess the calibration of the filling machine.

41. A farmer wanted to learn about the weight of his soybean crop. He randomly sampled 100 plants and weighed the soybeans on each plant.

42. Every year the U.S. Census Bureau releases the Current Population Report based on a survey of 50,000 households. The goal of this report is to learn the demographic characteristics of all households within the United States, such as income.

43. **Folate and Hypertension** Researcher John P. Forman and co-workers wanted to determine whether or not higher fo-late intake is associated with a lower risk of hypertension (high blood pressure) in younger women (27 to 44 years of age). To make this determination, they looked at 7,373 cases of hypertension in younger women and found that younger women who consumed at least 1,000 micrograms per day (μg/d) of total folate (dietary plus supplemental) had a decreased risk of hypertension compared with those who consumed less than 200 μg/d.

(*Source:* "Folate intake and the risk of incident hypertension among US Women," John P. Forman, MD; Eric B. Rimm, ScD; Meir J. Stampfer, MD; Gary C. Curhan, MD, ScD; *Journal of the American Medical Association* 293(2005):320–329)

44. A large community college has noticed that an increasing number of full-time students are working while attending the school. The administration randomly selects 128 students and asks this question: How many hours per week do you work?

Applying the Concepts

For the studies in Problems 45–50, (a) identify the research objective, (b) identify the sample, (c) list the descriptive statistics, and (d) state the conclusions made in the study.

45. **A Cure for the Common Wart** A study conducted by re-
NW searchers was designed "to determine if application of duct tape is as effective as cryotherapy in the treatment of common warts." The researchers randomly divided 51 patients into two groups. The 26 patients in group 1 had their warts treated by applying duct tape to the wart for 6.5 days and then removing the tape for 12 hours, at which point the cycle was repeated for a maximum of 2 months. The 25 patients in group 2 had their warts treated by cryotherapy (liquid nitrogen applied to the wart for 10 seconds every 2 to 3 weeks) for a maximum of 6 treatments. Once the treatments were complete, it was determined that 85% of the patients in group 1 and 60% of the patients in group 2 had complete resolution of their warts. The researchers concluded that duct tape is significantly more effective in treating warts than cryotherapy.

(*Source:* "The Efficacy of Duct Tape vs. Cryotherapy in the Treatment of Verruca Vulgaris (The Common Wart)," Dean R. Focht III, Carole Spicer, Mary P. Fairchok; *Archives of Pediatrics and Adolescent Medicine*, Vol. 156, No. 10, October 2002)

46. **Early Epidurals** A study was conducted at Northwestern University in Chicago to determine if pregnant women in first-time labor could receive low-dose epidurals, an anesthesis to control pain during childbirth early without raising their chances of a Caesarean section. In the study, reported in the *New England Journal of Medicine*, "728 women in first-time labor were divided into two groups. One group received the spinal shot and then got epidurals when the cervix dilated to about 2 centimeters. The other group initially received pain-relieving medicine directly into their bloodstreams, and put off epidurals until 4 centimeters if they could tolerate the pain." In the end, the C-section rate was 18% in the early epidural group and 21% in the delayed group. The researchers concluded that pregnant women in first-time labor can be given a low-dose epidural early without raising their chances of a C-section.

(*Source: Associated Press*, Feb. 22, 2005)

47. **The Mozart Effect** Researchers at the University of California, Irvine, wished to determine whether "music cognition and cognitions pertaining to abstract operations such as mathematical or spatial reasoning" were related. To test the research question, 36 college students listened to Mozart's Sonata for Two Pianos in D Major, K488, for 10 minutes and then took a spatial reasoning test using the Stanford–Binet intelligence scale. The same students also took the test after sitting in a room for 10 minutes in complete silence. Whether the student experienced Mozart first or silence first was randomly determined. The average score on the test following the Mozart piece was 119, while the average test score following the silence was 110. The researchers concluded that subjects performed better on abstract and spatial reasoning tests after listening to Mozart.

(*Source:* "Music and Spatial Performance," Frances H. Rauscher et al.; *Nature* 365, 14 October 1993:611)

48. **Favorite Presidents** A poll commissioned by Washington College and conducted by Schulman, Ronca, and Bucuvalas, February 7–10, 2005, surveyed 800 U.S. adults. Asked who was the greatest president, 20% of respondents chose Abraham Lincoln, 15% Ronald Reagan, 12% Franklin Roosevelt, 11% John F. Kennedy, 10% Bill Clinton, 8% George W. Bush, and 6% George Washington. The remaining 18% selected other presidents. Washington College concluded that Abraham Lincoln ranks first among U.S. adults as the greatest president.

(*Source: Associated Press*, Feb. 22, 2005)

49. **Go to the Movies?** Gallup News Service conducted a survey of 1,003 American adults aged 18 years or older, December 5–8, 2004. The respondents were asked, "How many movies, if any, have you attended in a movie theater in the past 12 months?" Of the 1,003 adults surveyed, 65% said they attended at least one movie in a movie theater in the past 12 months. Gallup News Service concluded that 65% of all Americans aged 18 years or older attended at least one movie in a movie theater in the past 12 months.

50. **Morally Acceptable?** Gallup News Service conducted a survey of 1,000 U.S. adults aged 18 years or older, May 2–4, 2004. The respondents were asked, "Do you believe that having a baby outside of marriage is morally acceptable or morally wrong?" Of the 1,000 adults surveyed, 49% said that having a baby outside of marriage is morally acceptable. Gallup News Service concluded that 49% of all U.S. adults 18 years or older believe that having a baby outside of marriage is morally acceptable.

In Problems 51–54, identify the individuals, variables, and data corresponding to the variables. Determine whether each variable is qualitative, continuous, or discrete.

51. **Widescreen TVs** The following data relate to widescreen
NW high-definition televisions.

Model	Size (in.)	Screen Type	Price ($)
Sanyo #PDP42H2W	42	Plasma	2994
Panasonic #PT-47WX54	47	Projection	1072
Tatung #P50BSAT	50	Plasma	4248
RCA #HD50LPW42	50	Projection	2696
RCA #D52W19	52	Projection	1194
JVC #HD52Z575	52	Projection	2850
Sony #KDF-60XS955	60	Projection	4100

Sources: walmart.com and sears.com

52. **BMW Cars** The following information relates to the entire product line of BMW automobiles.

Model	Body Style	Weight (pounds)	Number of Seats
M/Z3 Coupe	Coupe	2945	2
M/Z3 Roadster	Convertible	2690	2
3 Series	Coupe	2780	5
5 Series	Sedan	3450	5
7 Series	Sedan	4255	5
Z8	Convertible	3600	2

Source: Car and Driver magazine

53. **Driver's License Laws** The following data represent driver's license laws for various states.

State	Minimum Age for Driver's License (Unrestricted)	Blood Alcohol Concentration Limit	Mandatory Belt-Use Law Seating Positions	Maximum Allowable Speed Limit, 2003
Colorado	17	0.10	Front	75
Missouri	18	0.08	Front	70
Montana	15	0.08	All	75
New York	17	0.08	All	65
Texas	16.5	0.08	Front	75

Source: Time Almanac, 2005

54. MP3 Players The following information concerns various MP3 players that can be purchased online at circuitcity.com:

Product	Memory Size	Weight (oz)	Price ($)
RCA Lyra™ 256 MB MP3 Player	256 MB	2 oz	$149.99
iRiver 256 MB MP3 Player	256 MB	2.1 oz	$99.99
Samsung 256 MB MP3 Player	256 MB	0.85 oz	$129.99
Creative® 256 MuVo® Micro N200	256 MB	0.8 oz	$99.99
iRiver 512 MB MP3 Player	512 MB	2.1 oz	$149.99
SanDisk 512 MB MP3 Player	512 MB	1.4 oz	$139.99
Creative® 512 MuVo® Micro N200	512 MB	0.8 oz	$139.99
SanDisk 1 GB MP3 Player	1 GB	1.4 oz	$169.99

Source: circuitcity.com

For Problems 55 and 56, read the newspaper article and identify (a) the research question the study addresses, (b) the population, (c) the sample, (d) the descriptive statistics, and (e) the inferences of the study.

55. By Jeff Donn
Associated Press
February 20, 2005

BOSTON–Levels of a stress-related protein in the blood could give doctors a powerful new tool for deciding which patients with clogged heart arteries are most in danger and need aggressive treatment, a study found.

This protein is yet another predictor of heart trouble, in addition to such substances as cholesterol and C-reactive protein.

The Danish study, published in Thursday's *New England Journal of Medicine*, focused on a substance called B-natriuretic peptide, or BNP, released into the blood when the heart is stressed. Some doctors already test for this protein to help evaluate patients with shortness of breath who may be suffering from congestive heart disease.

The researchers measured the protein levels in 1,034 patients and followed their health for nine years. People with the highest protein levels were $2\frac{1}{2}$ times more likely to die from any cause than those with the lowest.

This protein could help doctors decide which patients need angioplasty or heart-bypass surgery.

"It sort of tells you that no matter what condition you have, higher levels of these peptides are associated with worse outcomes," said cardiologist Dr. James de Lemos, at the University of Texas Southwestern Medical Center.

56. **Study: Exercise May Prevent Parkinson's** By Kathleen Fackelmann, *USA Today*

Men who engaged in regular, vigorous exercise as teens and young adults drastically cut their risk of developing Parkinson's disease later in life, a study reports Tuesday.

As many as 1 million people in the USA, including actor Michael J. Fox, have this progressive neurological disease, which commonly strikes after age 50.

In addition to suggesting that exercise could ward off the disease, the findings also raise the hope that physical activity might help hold the line on brain cell destruction in people who already have it.

Robin Elliott, executive director of the Parkinson's Disease Foundation in New York, said the study is promising because there is no cure for Parkinson's. Doctors today have no way to stop or delay the progression of the disease, which affects the brain region that controls movement. Common symptoms of Parkinson's are tremors and a shuffling gait.

Researcher Alberto Ascherio of the Harvard School of Public Health and his colleagues studied 48,000 men and 77,000 women who were relatively healthy and middle-aged or older at the study's start. Over the course of the study, 387 people developed the disease.

The team did a statistical analysis to look for a link between physical activity and the risk of Parkinson's.

Men who said they jogged, played basketball or participated in some other vigorous activity at least twice a week in high school, college and up to age 40 had a 60% reduced risk of getting Parkinson's, says the study, which was published today in *Neurology*.

The team found no such protection for women. But the women in this study came of age in the '50s and '60s, an era when young women did not have as many opportunities to participate in sports. Ascherio says. The study had so few female athletes that any protection may have been missed, he says.

This study's findings are supported by animal research: Michael Zigmond, a researcher at the University of Pittsburgh and his colleagues have reported that exercise protects rats from developing a Parkinson's-like disease.

No one knows for certain whether exercise can prevent or delay the symptoms of Parkinson's in humans. But there is plenty of evidence suggesting that a regular fitness regimen might ward off a host of ailments as diverse as heart disease and Alzheimer's.

That leads experts such as Zigmond to recommend a vigorous workout on most days of the week.

"We know that exercise is good for you in ways that go beyond any one condition."

Find this article at:

http://www.usatoday.com/news/health/
2005-02-21-parkinsons-exercise_x.htm

57. Nominal versus Ordinal Just as a quantitative variable can be further classified as discrete or continuous, we can classify a qualitative (or categorical) variable as *nominal* or *ordinal*. A **nominal variable** is a qualitative variable than describes an attribute of an individual. An **ordinal vari-**able is a qualitative variable that has all the properties of a nominal variable, but also has observations that can be ranked or put in order. For example, gender is a nominal variable, whereas a response to a customer satisfaction survey of poor, fair, good, or outstanding is considered ordinal. For each of the following qualitative variables, determine whether the variable is nominal or ordinal.

(a) Hair color
(b) Letter grade in a statistics class
(c) Make of a television set
(d) Rank of a military officer
(e) Method of payment (cash, check, debit card, credit card)

Observational Studies, Experiments, and Simple Random Sampling

Objectives

1 **Distinguish between an observational study and an experiment**

2 **Obtain a simple random sample**

We are now familiar with some of the terminology used in describing data. Now, we need to determine how to obtain data. When we defined the word *statistics*, we said it is a science that involves the collection of data. Data can be obtained from four sources:

1. A census
2. Existing sources
3. Survey sampling
4. Designed experiments

We start by defining a census.

Definition

> A **census** is a list of all individuals in a population along with certain characteristics of each individual.

If a census is available and the census data can be used to answer the questions posed in the research objective, then the census data should be used. This is because the results will answer the questions with 100% certainty.

The United States conducts a census every 10 years to learn the demographic makeup of the United States. Everyone whose usual residence is within the borders of the United States must fill out a questionnaire packet. The cost of obtaining the census in 2000 was approximately $6 billion. The census data provide information such as the number of members in a household, number of years at present address, household income, and more. Because of the cost of obtaining census data, most researchers obtain data through existing sources, survey samples, or designed experiments.

Have you ever heard this saying? *There is no point in reinventing the wheel.* Well, there is no point in spending energy obtaining data that already exist either. If a researcher wishes to conduct a study and a data set exists that can be used to answer the researcher's questions, then it would be silly to collect the data from scratch. For example, in the August 22, 2001, issue of the *Journal of the American Medical Association* ["Physical Activity, Obesity, Height, and the Risk of Pancreatic Cancer," Dominique S. Michaud, ScD, et al. Vol. 286, No. 8], researchers did a study in which they attempted to identify factors that increase the likelihood of an individual getting pancreatic cancer. Rather than conducting their own survey, they used data from two existing surveys: the Health Professionals Follow-up Study and the Nurses' Health Study. By doing this, they saved time and money. The moral of the story: **Don't collect data that have already been collected!**

 ### Distinguish between an Observational Study and an Experiment

Survey sampling is used in research when there is no attempt to influence the value of the variable of interest. For example, we may want to identify the "normal" systolic blood pressure of U.S. men aged 40 to 44. The researcher would obtain a sample of men aged 40 to 44 and determine their systolic blood pressure. No attempt is made to influence the systolic blood pressure of the men surveyed. Polling data are another example of data obtained from a survey sample because the respondent is asked his or her opinion. No attempt is made to influence this opinion. Data obtained from a survey sample lead to an *observational study*.

Definition   An **observational study** measures the characteristics of a population by studying individuals in a sample, but does not attempt to manipulate or influence the variable(s) of interest.

Observational studies are sometimes referred to as *ex post facto* (after the fact) studies because the value of the variable of interest has already been established. We distinguish an observational study with a *designed experiment*.

Definition   A **designed experiment** applies a treatment to individuals (referred to as **experimental units** or **subjects**) and attempts to isolate the effects of the treatment on a **response variable**.

Data obtained through experimentation will be thoroughly discussed in Section 1.5, but the main idea is that the researcher is able to control factors that influence the experimental units. For example, suppose my son has two types of fertilizer and wants to determine which results in better tomatoes. He might conduct an experiment in which he divides 20 tomato plants into two groups. Group 1 receives the recommended amount of the first fertilizer, and group 2 receives the recommended amount of the second fertilizer. All other factors that affect plant growth (amount of sunlight, water, soil condition, and so on) are kept the same for the two groups. The two fertilizers are the two treatments, and the tomatoes are the experimental units. The weight of the tomatoes is the response variable. The observed weights are the data.

An example should clarify the difference between an observational study and a designed experiment.

EXAMPLE 1   ### Observational Study versus Designed Experiment

In most types of research, the goal is to determine the relation, if any, that may exist between two or more variables. For example, a researcher may want to determine whether there is a connection between smoking and lung cancer.* This type of study is performed using an observational study because it is *ex post facto* (after the fact) research. The individuals in the study are examined after they have been smoking for some period of time. The individuals are not controlled in terms of the number of cigarettes smoked per day, eating habits, and

---

*The interested reader may wish to read Chapter 18, "Does Smoking Cause Cancer?" in David Salsburg's book *The Lady Tasting Tea.* W. H. Freeman and Co., 2001.

so on. A researcher simply interviews a sample of smokers and monitors their rate of cancer and compares it with a sample of nonsmokers. The nonsmokers serve as a *control group;* that is, the nonsmokers serve as the benchmark upon which the smokers' rate of cancer is judged. If a significant difference between the two groups' rates of cancer exists, the researcher might want to claim smoking *causes* cancer.

In actuality, the researcher determined that smoking is *associated* with cancer. It may be that the population of smokers has a higher rate of cancer, but this higher incidence rate is not necessarily a direct result of smoking. It is possible that smokers have some characteristic that differs from the nonsmoking group, other than smoking, that is the *cause* of cancer. The characteristics that may be related to cancer, but that have not been identified in the study, are referred to as **lurking variables**. For example, a lurking variable might be the amount of exercise. Maybe smokers generally exercise less than nonsmokers, and the lack of exercise is the cause of cancer. In this observational study, we might be able to state that smokers have a higher rate of cancer than nonsmokers and that smoking is *associated* with cancer, but we would not be able to definitively state that smoking causes cancer.

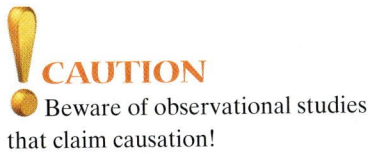

CAUTION

Beware of observational studies that claim causation!

Obtaining this type of data through experimentation requires randomly dividing a sample of people into two groups. We then require one group to smoke a pack of cigarettes each day for the next 20 years (the treatment) while the other group does not smoke (the control). We then compare the incidence rate of lung cancer (the response variable) in the smoking group to the nonsmoking (control) group. If the two cancer rates differ significantly, we could say that smoking causes cancer. By approaching the study in this way, we are able to control many of the factors that were beyond our control in the observational study. For example, we could make sure each group had the same diet and exercise regiment. This would allow us to determine whether smoking is a cause of cancer. Of course, moral issues preclude conducting this type of experiment. ▪

It is vital that we understand that observational studies do not allow a researcher to claim causation, only association.

> Observational studies are very useful tools for determining whether there is a relation between two variables, but it requires a designed experiment to isolate the cause of the relation.

Many observational studies, such as polls, are set up to learn the characteristics of a population. For example, the Gallup Organization routinely surveys U.S. residents in an attempt to identify opinion.

Observational studies are performed for two reasons:

1. To learn the characteristics of a population
2. To determine whether there is an association between two or more variables where the values of the variables have already been determined (again, this is often referred to as *ex post facto* research)

Ex post facto research is common when control of certain variables is impossible or unethical. For example, economists often perform research using observational studies because they do not have the ability to control many of the factors that affect our purchasing decisions. Some medical research requires that observational studies be conducted because of the risks thought to be associated with experimental study, such as the link between smoking and lung cancer.

Experiments, on the other hand, are used whenever control of certain variables is desired. This type of research allows the researcher to identify certain cause and effect relationships among the variables in the study.

The bottom line to consider is control. If control is possible, an experiment should be performed. However, if control is not possible or necessary, then observational studies are appropriate.

Now Work Problem 13.

Sampling

The goal in sampling is to obtain individuals for a study in such a way that accurate information about the population can be obtained. For example, the Gallup Organization typically polls a sample of about 1,000 adults from the population of adults aged 18 years or older. We want the sample to provide as much information as possible, but each additional piece of information has a price. So the question is this "How can the researcher obtain accurate information about the population through the sample while minimizing the costs in terms of money, time, personnel, and so on?" There is a balance between information and cost. An appropriate sample design can maximize the amount of information obtained about the population for a given cost.

We will discuss four basic sampling techniques: *simple random sampling*, *stratified sampling*, *systematic sampling*, and *cluster sampling*. These sampling methods are designed so that any biases introduced (knowingly and unknowingly) by the surveyor during the selection process are eliminated. In other words, the surveyor does not have a choice as to who is in the study. We discuss simple random sampling now and discuss the remaining three types of sampling in the next section.

② Obtain a Simple Random Sample

The most basic sample survey design is **simple random sampling**, which is often abbreviated as **random sampling**.

Definition

> A sample of size n from a population of size N is obtained through **simple random sampling** if every possible sample of size n has an equally likely chance of occurring. The sample is then called a **simple random sample**.

In Other Words
Simple random sampling is like selecting names from a hat.

The sample is always a subset of the population, meaning that the number of individuals in the sample is less than the number of individuals in the population.

In-Class Activity: Illustrating Simple Random Sampling

This activity illustrates the idea of simple random sampling.

(a) Choose 5 students in the class to represent a population. Number the students 1 through 5.

(b) Form all possible samples of size $n = 2$ from the population of size $N = 5$. How many different simple random samples are possible?

(c) Write the numbers 1 through 5 on five pieces of paper and place the paper in a hat. Select two of the numbers. The two individuals corresponding to these numbers are in the sample.

(d) Put the two numbers back in the hat. Select two of the numbers. The two individuals corresponding to these numbers are in the sample. Are the individuals in the second sample the same as the individuals in the first sample?

Obtaining a Simple Random Sample

How do we select the individuals in a simple random sample? To obtain a simple random sample from a population, we could write the names of the individuals in the population on different sheets of paper and then select names from the hat.

Often, however, the size of the population is so large that performing simple random sampling in this fashion is not practical. Typically, random numbers are used by assigning each individual in the population a unique number between 1 and N, where N is the size of the population. Then n random numbers from this list are selected. Because we must number the individuals in the population, we must have a list of all the individuals within the population, called a **frame**.

In Other Words
A frame lists all the individuals in a population. For example, a list of all registered voters in a particular precinct might be a frame.

EXAMPLE 2   Obtaining a Simple Random Sample

Problem: Senese and Associates has increased their accounting business. To make sure their clients are still satisfied with the services they are receiving, Senese and Associates decides to send a survey out to a simple random sample of 5 of its 30 clients.

Approach

Step 1: A list of the 30 clients must be obtained (the frame). Each client is then assigned a unique number from 01 to 30.

Step 2: Five unique numbers will be randomly selected. The client corresponding to the number is given a survey. This is called **sampling without replacement**. We sample without replacement so that we don't select the same client twice.

Solution

Step 1: Table 2 shows the list of clients. We arrange the clients in alphabetic order (although this is not necessary). Because there are 30 clients, we number the clients from 01 to 30.

Table 2		
01. ABC Electric	11. Fox Studios	21. R&Q Realty
02. Brassil Construction	12. Haynes Hauling	22. Ritter Engineering
03. Bridal Zone	13. House of Hair	23. Simplex Forms
04. Casey's Glass House	14. John's Bakery	24. Spruce Landscaping
05. Chicago Locksmith	15. Logistics Management, Inc.	25. Thors, Robert DDS
06. DeSoto Painting	16. Lucky Larry's Bistro	26. Travel Zone
07. Dino Jump	17. Moe's Exterminating	27. Ultimate Electric
08. Euro Car Care	18. Nick's Tavern	28. Venetian Gardens Restaraunt
09. Farrell's Antiques	19. Orion Bowling	29. Walker Insurance
10. First Fifth Bank	20. Precise Plumbing	30. Worldwide Wireless

Step 2: A table of random numbers can be used to select the individuals to be in the sample. See Table 3.* We select a starting place in the table of random numbers. This can be done by closing our eyes and placing a finger on the table. This may sound haphazard, but it accomplishes the goal of being random. Suppose we start in column 4, row 13. Because our data have two digits, we select two-digit numbers from the table using columns 4 and 5. We only select numbers greater than or equal to 01 or less than or equal to 30. Anytime we encounter 00, a number greater than 30, or a number already selected, we skip it and continue to the next number.

The first number in the list is 01, so the client corresponding to 01 will receive a survey. Moving down the list, the next number is 52. Because 52 is greater than 30, we skip it. Continuing down the list, the following numbers are selected from the list:

$$01, 07, 26, 11, 23$$

*Each digit is in its own column. The digits are displayed in groups of five for ease of reading. The digits in row 1 are 893922321274483 and so on. The first digit, 8, is in column 1; the second digit, 9, is in column 2; the ninth digit, 1, is in column 9.

Table 3

Row Number	Column Number									
	01–05	06–10	11–15	16–20	21–25	26–30	31–35	36–40	41–45	46–50
01	89392	**23**212	74483	36590	25956	36544	68518	40805	09980	00467
02	61458	17639	96252	95649	73727	33912	72896	66218	52341	97141
03	11452	74197	81962	48433	90360	26480	73231	37740	26628	44690
04	27575	04429	31308	02241	01698	19191	18948	78871	36030	23980
05	36829	59109	88976	46845	28329	47460	88944	08264	00843	84592
06	81902	93458	42161	26099	09419	89073	82849	09160	61845	40906
07	59761	55212	33360	68751	86737	79743	85262	31887	37879	17525
08	46827	25906	64708	20307	78423	15910	86548	08763	47050	18513
09	24040	66449	32353	83668	13874	86741	81312	54185	78824	00718
10	98144	96372	50277	15571	82261	66628	31457	00377	63423	55141
11	14228	17930	30118	00438	49666	65189	62869	31304	17117	71489
12	55366	51057	90065	14791	62426	02957	85518	28822	30588	32798
13	96**101**	30646	35526	90389	73634	79304	96635	6626	94683	16696
14	38**152**	55474	30153	26525	83647	31988	82182	98377	33802	80471
15	85**007**	18416	24661	95581	45868	15662	28906	36392	07617	50248
16	85**544**	15890	80011	18160	33468	84106	40603	01315	74664	20553
17	10**446**	20699	98370	17684	16932	80449	92654	02084	19985	59321
18	67**237**	45509	17638	65115	29757	80705	82686	48565	72612	61760
19	23**026**	89817	05403	82209	30573	47501	00135	33955	50250	72592
20	67**411**	58542	18678	46491	13219	84084	27783	34508	55158	78742

The clients corresponding to these numbers are

ABC Electric, Dino Jump, Travel Zone, Fox Studios, Simplex Forms

Each individual selected in the sample is set in boldface type in Table 3 to help you to understand where the numbers come from.

EXAMPLE 3   Obtaining a Simple Random Sample Using Technology

Problem: Find a simple random sample of five clients for the problem presented in Example 2.

Approach: The approach is similar to that given in Example 2.

Step 1: A list of the 30 clients must be obtained (the frame). The clients are then assigned a number from 01 to 30.

Step 2: Five numbers are randomly selected using a random number generator. The client corresponding to the number is given a survey. We sample without replacement so that we don't select the same client twice. To use a random-number generator using technology, we must first set the *seed*. The **seed** in a random-number generator provides an initial point for the generator to start creating random numbers. It is just like selecting the initial point in the table of random numbers. The seed can be any nonzero number. Statistical spreadsheets such as MINITAB or Excel can be used to generate random numbers, but we will use a TI-84 Plus graphing calculator. The steps for obtaining random numbers using MINITAB, Excel, or the TI-83/84 graphing calculator can be found in the Technology Step by Step on page 22.

Solution

Step 1: Table 2 on page 17 shows the list of clients and numbers corresponding to the clients.

Step 2: See Figure 2(a) for an illustration of setting the seed using a TI-84 Plus graphing calculator where the seed is set at 34. We are now ready to obtain the list of random numbers. Figure 2(b) shows the results obtained from a TI-84 Plus graphing calculator. If you are using a TI-83, your results will differ.

USING TECHNOLOGY

If you are using a different statistical package or type of calculator, the random numbers generated will likely be different. This does not mean you are wrong. There is no such thing as a wrong random sample as long as the correct procedures are followed.

Figure 2

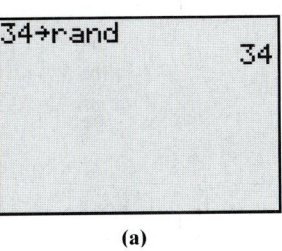

(a)

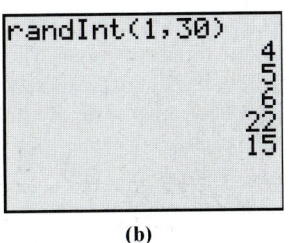

(b)

The following numbers are generated by the calculator:

$$4, 5, 6, 22, 15$$

The clients corresponding to these numbers are the clients to be surveyed: Casey's Glass House, Chicago Locksmith, DeSoto Painting, Ritter Engineering, and Logistics Management, Inc.

Now Work Problem 21.

CAUTION

Random-number generators are not truly random, because they are programs and programs do not act "randomly." The seed dictates the "random numbers" that are generated.

There is a very important consequence when comparing the by hand and technology solutions from Examples 2 and 3. Because both samples were obtained randomly, they resulted in different individuals in the sample! For this reason, each sample will likely result in different descriptive statistics. Any inference based on each sample *may* result in different conclusions regarding the population. This is the nature of statistics. Inferences based on samples will vary because the individuals in different samples vary.

1.2 ASSESS YOUR UNDERSTANDING

Concepts and Vocabulary

1. Explain the difference between an observational study and an experiment. In your explanation, be sure to discuss the circumstances in which each is appropriate.
2. Explain why a frame is necessary to obtain a simple random sample.
3. Discuss why sampling is used in statistics.
4. What does it mean when sampling is done without replacement?
5. A(n) _____ is a list of all individuals in a population along with certain characteristics of each individual.

6. *True or False*: An observational study measures the characteristics of a population by attempting to manipulate or influence the variables of interest.
7. *True or False*: Observational studies do not allow a researcher to claim causation.
8. When using a random-number generator to generate a simple random sample, why is it necessary to set a randomly selected seed?

Skill Building

In Problems 9–20, determine whether the study depicts an observational study or an experiment.

9. A study to determine whether there is a relation between the rate of cancer and an individual's proximity to high-tension wires.

10. Rats with cancer are divided into two groups. One group receives 5 mg of a medication that is thought to fight can-

cer, and the other receives 10 mg. After 2 years, the spread of the cancer is measured.

11. Seventh-grade students are randomly divided into two groups. One group is taught math using traditional techniques; the other is taught math using a reform method.

After 1 year, each group is given an achievement test to compare its proficiency with that of the other group.

12. A poll is conducted in which 500 people are asked whom they plan to vote for in the upcoming election.

13. A survey is conducted asking 400 people, "Do you prefer Coke or Pepsi?"
NW

14. While shopping, 200 people are asked to perform a taste test in which they drink from two unmarked cups. They are then asked which drink they prefer.

15. Sixty patients with carpal tunnel syndrome are randomly divided into two groups. One group is treated weekly with both acupuncture and an exercise regimen. The other is treated weekly with the exact same exercise regimen, but no acupuncture. After 1 year, both groups are questioned about the pain of carpal tunnel syndrome.

16. Conservation agents netted 250 large-mouth bass in a lake and determined how many were carrying parasites.

17. **Got Kidney Stones?** Researchers Eric N. Taylor and others wanted to determine if weight, weight gain, body mass index (BMI), and waist circumference are associated with kidney stone formation. To make this determination, they looked at 4,827 cases of kidney stones covering 46 years. Based on an analysis of the data, they concluded that obesity and weight gain increase the risk of kidney stone formation.
(*Source:* "Obesity, Weight Gain, and the Risk of Kidney Stones," Eric N. Taylor, MD; Meir J. Stampfer, MD, DrPH; and Gary C. Curhan, MD, ScD; *Journal of the American Medical Association* 293(2005):455–462)

18. **Cell Phones and Brain Tumors** Researchers Helle Collatz Christensen and co-workers examined the possible association between the use of cellular telephones and the development of acoustic neuroma (a noncancerous growth in the inner ear) by comparing 106 cases of individuals with acoustic neuroma to 212 cases of individuals who did not have acoustic neuroma. The data obtained included information on the use of cellular telephones from personal interviews, data from medical records, and the results of radiologic examinations.
(*Source:* "Cellular Telephone Use and Risk of Acoustic Neuroma," Helle Collatz Christensen, Joachim Schüz, Michael Kosteljanetz, Hans Skovgaard Poulsen, Jens Thomsen, and Christoffer Johansen; *American Journal of Epidemiology* 2004; 159:277–283)

Applying the Concepts

19. **Future Government Club** The Future Government Club wants to sponsor a panel discussion on the upcoming national election. The club wants four of its members to lead the panel discussion. Obtain a simple random sample of size 4 from the names to the right. Write a short description of the process you used to generate your sample.

Blouin	Fallenbuchel	Niemeyer	Rice
Bolden	Grajewski	Nolan	Salihar
Bolt	Haydra	Ochs	Tate
Carter	Keating	Opacian	Thompson
Cooper	Khouri	Pawlak	Trudeau
Debold	Lukens	Pechtold	Washington
De Young	May	Ramirez	Wright
Engler	Motola	Redmond	Zenkel

20. **Worker Morale** The owner of a private food store is concerned about employee morale at the store level. She decides to survey the employees to see if she can learn about work environment and job satisfaction. Obtain a simple random sample of size 5 from the names to the right. Write a short description of the process you used to generate your sample.

Archer	Foushi	Kemp	Oliver
Bolcerek	Gow	Lathus	Orsini
Bryant	Grove	Lindsey	Salazar
Carlisle	Hall	Massie	Ullrich
Cole	Hills	McGuffin	Vaneck
Dimas	Houston	Musa	Weber
Ellison	Kats	Nickas	Zavodny
Everhart			

21. **Obtaining a Simple Random Sample** The following table lists the 50 states.

NW (a) Obtain a simple random sample of size 10 using Table I in Appendix A, a graphing calculator, or computer software.

(b) Obtain a second simple random sample of size 10 using Table I in Appendix A, a graphing calculator, or computer software.

1.	Alabama	11.	Hawaii	21.	Massachusetts	31.	New Mexico	41.	South Dakota
2.	Alaska	12.	Idaho	22.	Michigan	32.	New York	42.	Tennessee
3.	Arizona	13.	Illinois	23.	Minnesota	33.	North Carolina	43.	Texas
4.	Arkansas	14.	Indiana	24.	Mississippi	34.	North Dakota	44.	Utah
5.	California	15.	Iowa	25.	Missouri	35.	Ohio	45.	Vermont
6.	Colorado	16.	Kansas	26.	Montana	36.	Oklahoma	46.	Virginia
7.	Connecticut	17.	Kentucky	27.	Nebraska	37.	Oregon	47.	Washington
8.	Delaware	18.	Louisiana	28.	Nevada	38.	Pennsylvania	48.	West Virginia
9.	Florida	19.	Maine	29.	New Hampshire	39.	Rhode Island	49.	Wisconsin
10.	Georgia	20.	Maryland	30.	New Jersey	40.	South Carolina	50.	Wyoming

22. **Obtaining a Simple Random Sample** The following table lists the 43 presidents of the United States.

(a) Obtain a simple random sample of size 8 using Table I in Appendix A, a graphing calculator, or computer software.

(b) Obtain a second simple random sample of size 8 using Table I in Appendix A, a graphing calculator, or computer software.

1.	Washington	10.	Tyler	19.	Hayes	28.	Wilson	37.	Nixon
2.	J. Adams	11.	Polk	20.	Garfield	29.	Harding	38.	Ford
3.	Jefferson	12.	Taylor	21.	Arthur	30.	Coolidge	39.	Carter
4.	Madison	13.	Fillmore	22.	Cleveland	31.	Hoover	40.	Reagan
5.	Monroe	14.	Pierce	23.	B. Harrison	32.	F.D. Roosevelt	41.	George H. Bush
6.	J.Q. Adams	15.	Buchanan	24.	Cleveland	33.	Truman	42.	Clinton
7.	Jackson	16.	Lincoln	25.	McKinley	34.	Eisenhower	43.	George W. Bush
8.	Van Buren	17.	A. Johnson	26.	T. Roosevelt	35.	Kennedy		
9.	W.H. Harrison	18.	Grant	27.	Taft	36.	L.B. Johnson		

23. **Sampling the Faculty** A small community college employs 87 full-time faculty members. To gain the faculty's opinions about an upcoming building project, the college president wishes to obtain a simple random sample that will consist of 9 faculty members. He numbers the faculty from 1 to 87.

(a) Using Table I from Appendix A, the president closes his eyes and drops his ink pen on the table. It points to the digit in row 5, column 22. Using this position as the starting point and proceeding downward, determine the numbers for the 9 faculty members who will be included in the sample.

(b) If the president uses the randInt(feature of a graphing calculator with a seed value of 47, determine the numbers for the 9 faculty members who will be included in the sample.

24. **Sampling the Students** The same community college from Problem 23 has 7,656 students currently enrolled in classes. To gain the student's opinions about an upcoming building project, the college president wishes to obtain a simple random sample of 20 students. He numbers the students from 1 to 7,656.

(a) Using Table I from Appendix A, the president closes his eyes and drops his ink pen on the table. It points to the digit in row 11, column 32. Using this position as the starting point and proceeding downward, determine the numbers for the 20 students who will be included in the sample.

(b) If the president uses the randInt(feature of a graphing calculator with a seed value of 142, determine the numbers for the 20 students who will be included in the sample.

25. **Obtaining a Simple Random Sample** Suppose you are the president of the student government. You wish to conduct a survey to determine the student body's opinion regarding student services. The administration provides you with a list of the names and phone numbers of the 19,935 registered students.

(a) Discuss the procedure you would follow to obtain a simple random sample of 25 students.

(b) Obtain this sample.

26. **Obtaining a Simple Random Sample** Suppose the mayor of Justice, Illinois, asks you to poll the residents of the village. The mayor provides you with a list of the names and phone numbers of the 5,832 residents of the village.

(a) Discuss the procedure you would follow to obtain a simple random sample of 20 residents.

(b) Obtain this sample.

Technology Step-by-Step	**Obtaining a Simple Random Sample**

TI-83/84 Plus

Step 1: Enter any nonzero number (the seed) on the HOME screen.

Step 2: Press the STO ⇒ button.

Step 3: Press the MATH button.

Step 4: Highlight the PRB menu and select 1: rand.

Step 5: From the HOME screen press ENTER.

Step 6: Press the MATH button. Highlight PRB menu and select
5: randInt(.

Step 7: With randInt(on the HOME screen enter 1, N, where N is the population size. For example, if $N = 500$, enter the following:

$$\text{randInt(1,500)}$$

Press ENTER to obtain the first individual in the sample. Continue pressing ENTER until the desired sample size is obtained.

MINITAB

Step 1: Select the **Calc** menu and highlight **Set Base** ....

Step 2: Enter any seed number you desire. Note that it is not necessary to set the seed, because MINITAB uses the time of day in seconds to set the seed.

Step 3: Select the **Calc** menu and highlight **Random Data** and select **Integer** ....

Step 4: Fill in the following window with the appropriate values. To obtain a simple random sample for the situation in Example 2, we would enter the following:

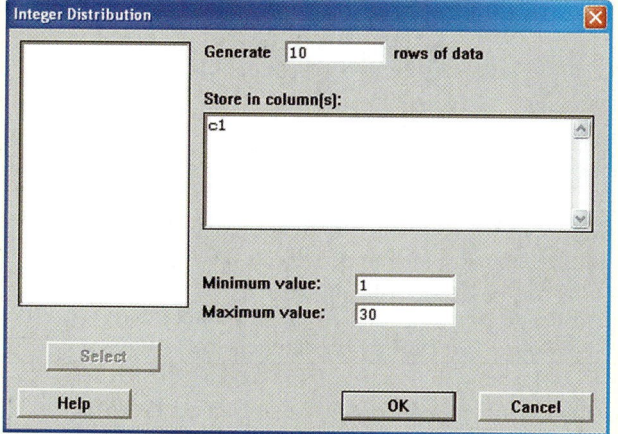

The reason we generate 10 rows of data (instead of 5) is in case any of the random numbers repeat. Select OK, and the random numbers will appear in column 1 (C1) in the spreadsheet.

Excel

Step 1: Be sure the Data Analysis Tool Pak is activated. This is done by selecting the **Tools** menu and highlighting **Add – Ins** .... Check the box for the Analysis ToolPak and select OK.

Step 2: Select **Tools** and highlight **Data Analysis** .... Highlight **Random Number Generation** and select OK.

Step 3: Fill in the window with the appropriate values. To obtain a simple random sample for the situation in Example 2, we would fill in the following:

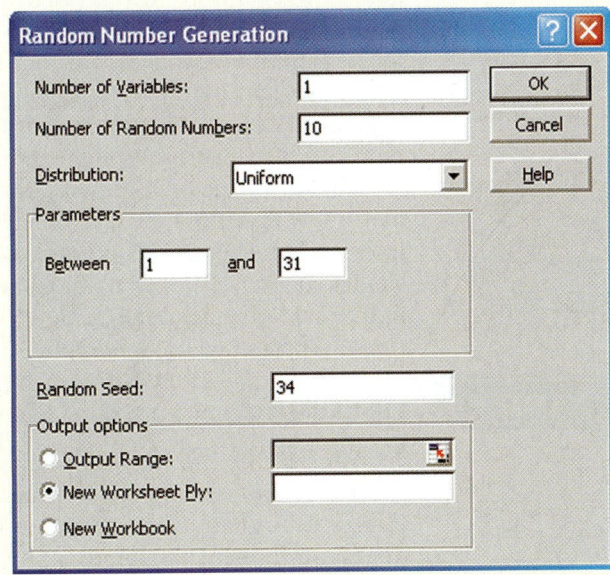

The reason we generate 10 rows of data (instead of 5) is in case any of the random numbers repeat. Notice also that the parameter is between 1 and 31, so any value less than or equal to 31 is possible. In the unlikely event that 31 appears, simply ignore it. Select OK and the random numbers will appear in column 1 (A1) in the spreadsheet. Ignore any values to the right of the decimal place.

1.3 Other Effective Sampling Methods

Objectives

1 Obtain a stratified sample
2 Obtain a systematic sample
3 Obtain a cluster sample

One goal of sampling is to obtain as much information as possible about the population at the least cost. Remember, we are using the word *cost* in a general sense. Cost includes monetary outlays, time, and other resources. With this goal in mind, we may find it advantageous to use sampling techniques other than simple random sampling.

1 **Obtain a Stratified Sample**

Under certain circumstances, *stratified sampling* provides more information about the population for less cost than simple random sampling.

Definition

A **stratified sample** is obtained by separating the population into nonoverlapping groups called *strata* and then obtaining a simple random sample from each stratum. The individuals within each stratum should be homogeneous (or similar) in some way.

For example, suppose Congress was considering a bill that abolishes estate taxes. In an effort to determine the opinion of her constituency, a senator asks a

In Other Words

Stratum is singular, while strata is plural. The word strata means divisions. So a stratified sample is a simple random sample of different divisions of the population.

pollster to conduct a survey within her district. The pollster may divide the population of registered voters within the district into three strata: Republican, Democrat, and Independent. This is because the members within the three party affiliations may have the same opinion regarding estate taxes. The main criterion in performing a stratified sample is that each group (stratum) must have a common attribute that results in the individuals being similar within the stratum.

An advantage of stratified sampling over simple random sampling is that it may allow fewer individuals to be surveyed while obtaining the same or more information. This result occurs because individuals within each subgroup have similar characteristics, so opinions within the group do not vary much from one individual to the next. In addition, a stratified sample guarantees that each stratum is represented in the sample.

EXAMPLE 1   Obtaining a Stratified Sample

Problem: The president of DePaul University wants to conduct a survey to determine the community's opinion regarding campus safety. The president thinks the DePaul community can be divided into three groups: resident students, nonresident (commuting) students, and staff (including faculty). So he will obtain a stratified sample. Suppose there are 6,204 resident students, 13,304 nonresident students, and 2,401 staff, for a total of 21,909 individuals in the population. The president wants to obtain a sample of size 100, with the number of individuals selected from each stratum weighted by the population size. So resident students make up 6,204/21,909 = 28% of the sample, nonresident students account for 61% of the sample, and staff constitute 11% of the sample. To obtain a sample of size 100, the president will obtain a stratified sample of $0.28(100) = 28$ resident students, $0.61(100) = 61$ nonresident students, and $0.11(100) = 11$ staff.

Approach: To obtain the stratified sample, conduct a simple random sample within each group. That is, obtain a simple random sample of 28 resident students (from the 6,204 resident students), a simple random sample of 61 nonresident students, and a simple random sample of 11 staff.

Solution: Using MINITAB, with the seed set to 4032 and the values shown in Figure 3, we obtain the following sample of staff:

$$240, 630, 847, 190, 2096, 705, 2320, 323, 701, 471, 744$$

Figure 3

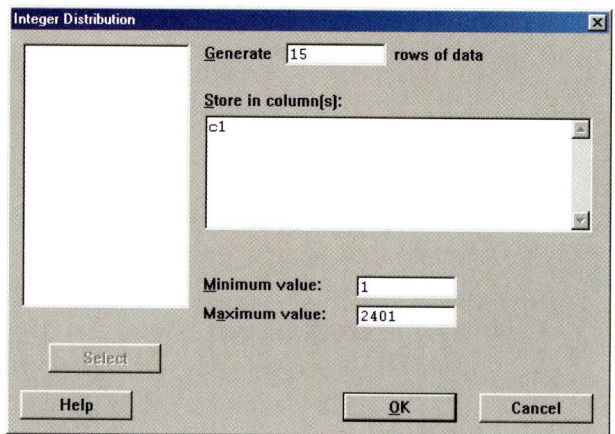

CAUTION

Do not use the same seed for all the groups in a stratified sample, because we want the simple random samples within each stratum to be independent of each other.

Repeat this procedure for the resident and nonresident students.

An advantage of stratified sampling over simple random sampling is that the researcher is able to determine characteristics within each stratum. This allows an analysis to be performed on each subgroup to see if any significant differences between the groups exist. For example, we could analyze the data obtained in Example 1 to see if there is a difference in the opinions of students versus faculty.

Now Work Problem 23.

② Obtain a Systematic Sample

In both simple random sampling and stratified sampling, it is necessary that a list of the individuals in the population being studied (the frame) exist. Therefore, these sampling techniques require some preliminary work before the sample is obtained. A sampling technique that does not require a frame is *systematic sampling*.

Definition

> A **systematic sample** is obtained by selecting every kth individual from the population. The first individual selected corresponds to a random number between 1 and k.

In Other Words

Systematic sampling is like selecting every fifth person out of a line.

Because systematic sampling does not require a frame, it is a useful technique when you can't obtain a list of the individuals in the population that you wish to study.

The idea behind obtaining a systematic sample is relatively simple: Select a number k, randomly select a number between 1 and k and survey that individual, then survey every kth individual thereafter. For example, we might decide to survey every $k = 8$th individual. We randomly select a number between 1 and 8 such as 5. This means we survey the 5th, $5 + 8 = 13$th, $13 + 8 = 21$st, $21 + 8 = 29$th, and so on, individuals until we reach the desired sample size.

EXAMPLE 2

Obtaining a Systematic Sample Without a Frame

Problem: The manager of Kroger Food Stores wants to measure the satisfaction of the store's customers. Design a sampling technique that can be used to obtain a sample of 40 customers.

Approach: A frame of Kroger customers would be difficult, if not impossible, to obtain. Therefore, it is reasonable to use systematic sampling by surveying every kth customer who leaves the store.

Solution: The manager decides to obtain a systematic sample by surveying every 7th customer. He randomly determines a number between 1 and 7, say 5. He then surveys the 5th customer exiting the store and every 7th customer thereafter, until a sample of 40 customers is reached. The survey will include customers 5, 12, 19, ..., 278. *

But how do we select the value of k? If the size of the population is unknown, there is no mathematical way to determine k. It must be chosen by determining a value of k that is not so large that we are unable to achieve our desired sample size, but not so small that we do not obtain a sample size that is representative of the population.

To clarify this point, let's revisit Example 2. Suppose we chose a value of k that was too large, say 30. This means that we will survey every 30th shopper, starting with the 5th. To obtain a sample of size 40 would require that 1,175 shoppers visit Kroger on that day. If Kroger does not have 1,175 shoppers, the desired sample size will not be achieved. On the other hand, if k is too small, say 4, we would survey the 5th, 9th, ..., 161st shopper. It may be that the 161st shopper exits the store at 3 P.M., which means our survey did not include any of the evening shoppers. Certainly, this sample is not representative of *all* Kroger patrons! An estimate of the size of the population would certainly help determine an appropriate value for k.

---

*Because we are surveying 40 customers, the first individual surveyed is the 5th, the second is the $5 + 7 = 12$th, the third is the $5 + (2) 7 = 19$th, and so on, until we reach the 40th, which is the $5 + (39) 7 = 278$th shopper.

To determine the value of k when the size of the population, N, is known is relatively straightforward. Suppose we wish to survey a population whose size is known to be $N = 20{,}325$. We desire a sample of size $n = 100$. To guarantee that individuals are selected evenly from the beginning as well as the end of the population (such as early and late shoppers), we compute N/n and round down to the nearest integer. For example, $20{,}325/100 = 203.25$, so $k = 203$. Then we randomly select a number between 1 and 203 and select every 203rd individual thereafter. So, if we randomly selected 90 as our starting point, we would survey the 90th, 293rd, 496th, ..., 20,187th individuals.

We summarize the procedure as follows:

> ### Steps in Systematic Sampling
>
> **Step 1:** If possible, approximate the population size, N.
> **Step 2:** Determine the sample size desired, n.
> **Step 3:** Compute $\frac{N}{n}$ and round down to the nearest integer. This value is k.
> **Step 4:** Randomly select a number between 1 and k. Call this number p.
> **Step 5:** The sample will consist of the following individuals:
>
> $$p, p + k, p + 2k, \ldots, p + (n - 1)k$$

Because systematic sampling does not require that the size of the population be known, it typically provides more information for a given cost than does simple random sampling. In addition, systematic sampling is easier to employ, so there is less possibility of interviewer error occurring, such as selecting the wrong individual to be surveyed.

Now Work Problem 27.

③ Obtain a Cluster Sample

A fourth sampling method is called *cluster sampling*. The previous three sampling methods discussed have benefits under certain circumstances. So does cluster sampling.

Definition

A **cluster sample** is obtained by selecting all individuals within a randomly selected collection or group of individuals.

In Other Words

Imagine a mall parking lot. Each subsection of the lot could be a cluster (Section F-4 for example).

Consider a quality-control engineer who wants to verify that a certain machine is filling bottles with 16 ounces of liquid detergent. To obtain a sample of bottles from the machine, the engineer could use systematic sampling by sampling every kth bottle from the machine. However, it would be time consuming waiting next to the filling machine for the bottles to come off the line. Suppose that, as the bottles come off the line, they are placed into cartons of 12 bottles each. An alternative sampling method would be to randomly select a few cartons and measure the contents of all 12 bottles from each carton selected. This would be an example of cluster sampling. It is a good sampling method in this situation because it would speed up the data-collection process.

EXAMPLE 3 **Obtaining a Cluster Sample**

Problem: A sociologist wants to gather data regarding the household income within the city of Boston. Obtain a sample using cluster sampling.

Approach: The city of Boston can be set up so that each city block is a cluster. Once the city blocks have been identified, we obtain a simple random sample of the city blocks and survey all households on the blocks selected.

Solution: Suppose there are 10,493 city blocks in Boston. First, we must number the blocks from 1 to 10,493. Suppose the sociologist has enough time and money to survey 20 clusters (city blocks). Therefore, the sociologist should obtain a simple random sample of 20 numbers between 1 and 10,493 and survey all households from the clusters selected. Cluster sampling is a good choice in this example because it reduces the travel time to households that is likely to occur with both simple random sampling and stratified sampling. In addition, there is no need to obtain a detailed frame with cluster sampling. The only frame needed is one that provides information regarding city blocks.

Recall that in systematic sampling we had to determine an appropriate value for k, the number of individuals to skip between individuals selected to be in the sample. We have a similar problem in cluster sampling. The following are a few of the questions that arise:

- How do I cluster the population?
- How many clusters do I sample?
- How many individuals should be in each cluster?

First, it must be determined whether the individuals within the proposed cluster are homogeneous (similar individuals) or heterogeneous (dissimilar individuals). Consider the results of Example 3. City blocks tend to have similar households. Surveying one house on a city block is likely to result in similar responses from another house on the same block. This results in duplicate information. We conclude the following: If the clusters have homogeneous individuals, it is better to have more clusters with fewer individuals in each cluster.

What if the cluster is heterogeneous? Under this circumstance, the heterogeneity of the cluster likely resembles the heterogeneity of the population. In other words, each cluster is a scaled-down representation of the overall population. For example, a quality-control manager might use shipping boxes that contain 100 light bulbs as a cluster, since the rate of defects within the cluster would closely mimic the rate of defects in the population assuming the bulbs are randomly placed in the box. Thus, when each of the clusters is heterogeneous, fewer clusters with more individuals in each cluster are appropriate.

> **Now Work Problem 11.**

The four sampling techniques just presented are sampling techniques in which the individuals are selected randomly. Often, however, sampling methods are used in which the individuals are not randomly selected, such as *convenience sampling*.

Convenience Sampling

Convenience sampling is probably the easiest sampling method.

Definition

> A **convenience sample** is a sample in which the individuals are easily obtained.

There are many types of convenience samples, but probably the most popular are those in which the individuals in the sample are **self-selected** (the individuals themselves decide to participate in a survey). These are also called **voluntary response** samples. Examples of self-selected sampling include phone-in polling; a radio personality will ask his or her listeners to phone the station to submit their opinions. Another example is the use of the Internet to conduct surveys. For example, *Dateline* will present a story regarding a certain topic and ask its viewers to "tell us what you think" by completing a questionnaire online or phoning in an opinion. Both of these samples are poor designs because the

! CAUTION

Stratified and cluster samples are different. In a stratified sample, we divide the population into two or more homogeneous groups. Then we obtain a simple random sample from each group. In a cluster sample, we divide the population into groups, obtain a simple random sample of some of the groups, and survey *all* individuals in the selected groups.

! CAUTION

Studies that use convenience sampling generally have results that are suspect. The results should be looked on with extreme skepticism.

individuals who decide to be in the sample generally have strong opinions about the topic. A more typical individual in the population will not bother phoning or logging on to a computer to complete a survey. Any inference made regarding the population from this type of sample should be made with extreme caution.

Multistage Sampling

In practice, most large-scale surveys obtain samples using a combination of the techniques just presented.

As an example of multistage sampling, consider Nielsen Media Research. Nielsen randomly selects households and monitors the television programs these households are watching through a People Meter. The meter is an electronic box placed on each TV within the household. The People Meter measures what program is being watched and who is watching it. Nielsen selects the households with the use of a two-stage sampling process.

Stage 1: Using U.S. Census data, Nielsen divides the country into geographic areas (strata). The strata are typically city blocks in urban areas and geographic regions in rural areas. About 6,000 strata are randomly selected.

Stage 2: Nielsen sends representatives to the selected strata and lists the households within the strata. The households are then randomly selected through a simple random sample.

Nielsen sells this information to television stations and companies. Their results are used to help determine prices for commercials.

As another example of multistage sampling, consider the sample used by the Census Bureau for the Current Population Survey. This survey requires five stages of sampling:

Stage 1: Stratified sample

Stage 2: Cluster sample

Stage 3: Stratified sample

Stage 4: Cluster sample

Stage 5: Systematic sample

This survey is very important because it is used to obtain demographic estimates of the United States in noncensus years. A detailed presentation of the sampling method used by the Census Bureau can be found in *The Current Population Survey: Design and Methodology*, Technical Paper No. 40.

Sample Size Considerations

Throughout the discussion of sampling, we did not mention how to determine the sample size. Determining the sample size is key in the overall statistical process. In other words, the researcher must ask this question: "How many individuals must I survey to draw conclusions about the population within some predetermined margin of error?" The researcher must find the correct balance between the reliability of the results and the cost of obtaining these results. The bottom line is that time and money determine the level of confidence a researcher will place on the conclusions drawn from the sample data. The more time and money the researcher has available, the more accurate will be the results of the statistical inference.

Nonetheless, techniques do exist for determining the sample size required to estimate characteristics regarding the population within some margin of error. We will consider some of these techniques in Sections 9.1 and 9.3. (For a detailed discussion of sample size considerations, consult a text on sampling techniques such as *Elements of Sampling Theory and Methods* by Z. Govindarajulu, Prentice Hall, 1999.)

Summary

Figure 4 provides a summary of the four sampling techniques presented.

Figure 4

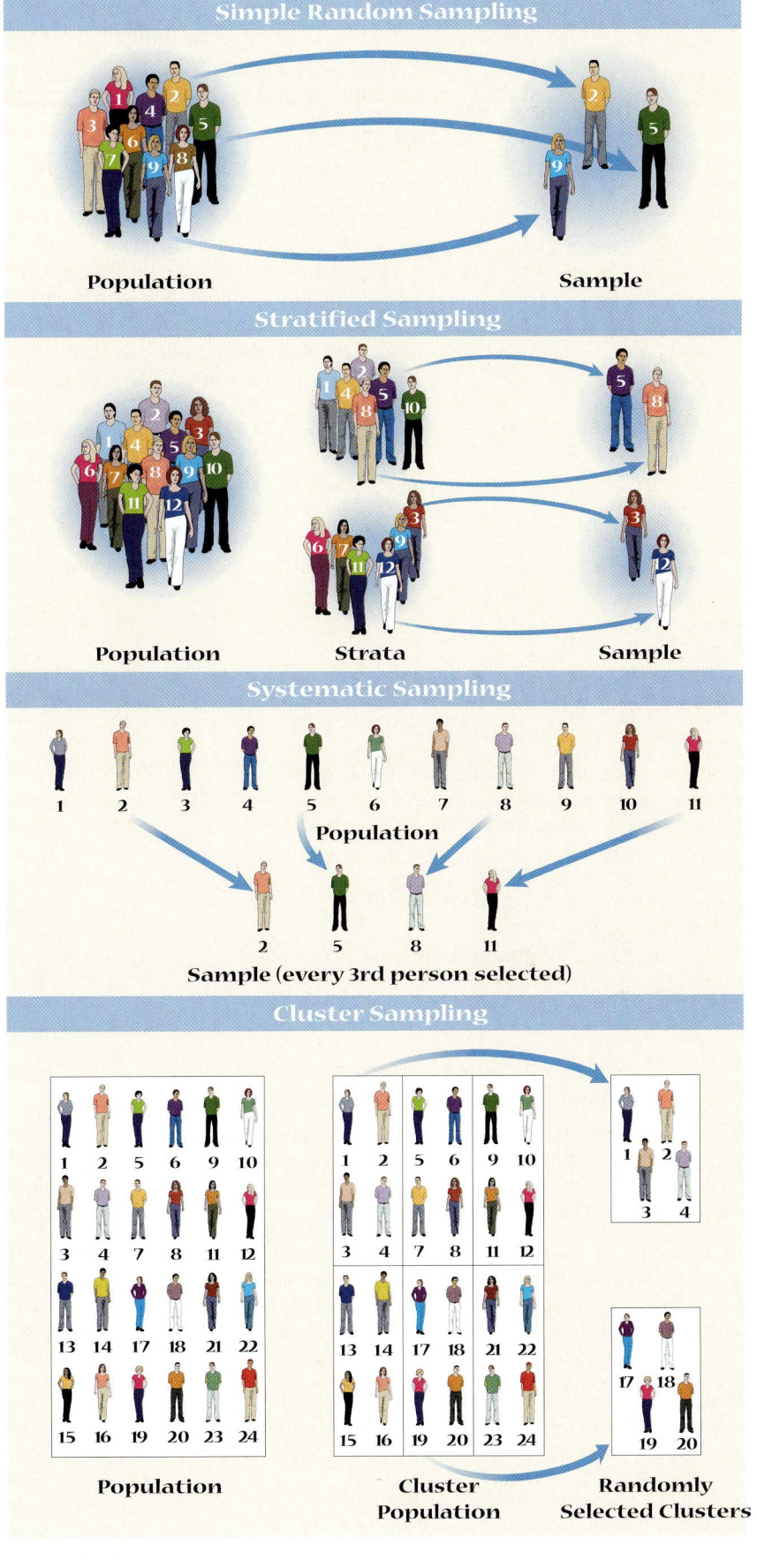

In-Class Activity: Different Sampling Methods

The following question was recently asked by the Gallup Organization.
"In general, are you satisfied or dissatisfied with the way things are going in the country?"

(a) Number the students in the class from 1 to N, where N is the number of students. Obtain a simple random sample and have them answer this question. Record the number of satisfied responses and the number of dissatisfied responses.

(b) Divide the students in the class by gender. Treat each gender as a stratum. Obtain a simple random sample from each stratum and have them answer this question. Record the number of satisfied responses and the number of dissatisfied responses.

(c) Treat each row of desks as a cluster. Obtain a simple random sample of clusters and have each student in the selected clusters answer this question. Record the number of satisfied responses and the number of dissatisfied responses.

(d) Number the students in the class from 1 to N, where N is the number of students. Obtain a systematic sample and have the selected students answer this question. Record the number of satisfied responses and the number of dissatisfied responses.

(e) Were there any differences in the results of the survey? State some reasons for any differences.

1.3 ASSESS YOUR UNDERSTANDING

Concepts and Vocabulary

1. Describe a circumstance in which stratified sampling would be an appropriate sampling method.
2. Which sampling method does not require a frame?
3. Why are convenience samples ill advised?
4. A _____ is obtained by dividing the population into groups and selecting all individuals from within a random sample of the groups.
5. A _____ is obtained by dividing the population into homogeneous groups and randomly selecting individuals from each group.

6. *True or False*: When taking a systematic random sample of size n, every group of size n from the population has the same chance of being selected.
7. *True or False*: A simple random sample is always preferred because it obtains the same information as other sampling plans but requires a smaller sample size.
8. *True or False*: When conducting a cluster sample, it is better to have fewer clusters with more individuals when the clusters are heterogeneous.

Skill Building

In Problems 9–22, identify the type of sampling used.

9. To estimate the percentage of defects in a recent manufacturing batch, a quality-control manager at Intel selects every 8th chip that comes off the assembly line starting with the 3rd until she obtains a sample of 140 chips.

10. To determine the average IQ of ninth-grade students, a school psychologist obtains a list of all high schools in the local public school system. She randomly selects five of these schools and administers an IQ test to all ninth-grade students at the selected schools.

11. To determine customer opinion of their boarding policy,
NW Southwest Airlines randomly selects 60 flights during a certain week and surveys all passengers on the flights.

12. A member of Congress wishes to determine her constituency's opinion regarding estate taxes. She divides her constituency into three income classes: low-income households, middle-income households, and upper-income households. She then takes a simple random sample of households from each income class.

13. In an effort to identify if an advertising campaign has been effective, a marketing firm conducts a nationwide poll by randomly selecting individuals from a list of known users of the product.

14. A radio station asks its listeners to call in their opinion regarding the use of U.S. forces in peacekeeping missions.

15. A farmer divides his orchard into 50 subsections, randomly selects 4, and samples all the trees within the 4 subsections to approximate the yield of his orchard.

16. A school official divides the student population into five classes: freshman, sophomore, junior, senior, and graduate student. The official takes a simple random sample from each class and asks the members' opinions regarding student services.

17. A survey regarding download time on a certain Web site is administered on the Internet by a market research firm to anyone who would like to take it.

18. A group of lobbyists has a list of the 100 senators of the United States. To determine the Senate's position regarding farm subsidies, they decide to talk with every seventh senator on the list, starting with the third.

19. A small-town newspaper reporter wants to get local reaction to a controversial new film. She waits outside the theater during an afternoon show and, starting with the second, asks every fifth patron leaving how much they liked the movie.

20. To determine his DSL Internet connection speed, Shawn divides up the day into four parts: morning, midday, evening, and late night. He then measures his Internet connection speed at 5 randomly selected times during each part of the day.

21. A statistics instructor with a large number of students attempts to reduce time spent grading by only grading a portion of assigned homework problems. He randomly selects one of the first four problems in the assignment and then grades that problem and every fourth problem thereafter.

22. 24 Hour Fitness wants to administer a satisfaction survey to its current members. Using their membership roster, the club randomly selects 40 club members and asks them about their level of satisfaction with the club.

Applying the Concepts

23. Stratified Sampling The Future Government Club wants to sponsor a panel discussion on the upcoming national election. The club wants to have four of its members lead the panel discussion. To be fair, however, the panel should consist of two Democrats and two Republicans. Below is a list of the Democrats and Republicans who are currently members of the club. Obtain a stratified sample of two Democrats and two Republicans to serve on the panel.

Democrats		Republicans	
Bolden	Motola	Blouin	Ochs
Bolt	Nolan	Cooper	Pechtold
Carter	Opacian	De Young	Redmond
Debold	Pawlak	Engler	Rice
Fallenbuchel	Ramirez	Grajewski	Salihar
Haydra	Tate	Keating	Thompson
Khouri	Washington	May	Trudeau
Lukens	Wright	Niemeyer	Zenkel

24. Stratified Sampling The owner of a private food store is concerned about employee morale. She decides to survey the managers and hourly employees to see if she can learn about work environment and job satisfaction. Below is a list of the managers and hourly workers at the store. Obtain a stratified sample of two managers and four hourly employees to survey.

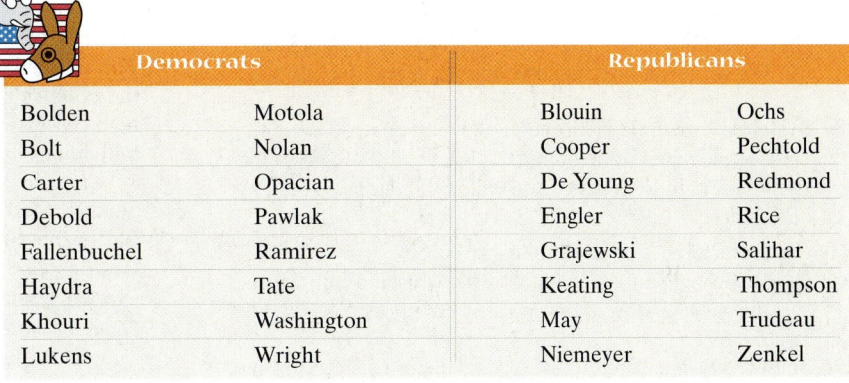

Managers		Hourly Employees		
Carlisle	Oliver	Archer	Foushi	Massie
Hills	Orsini	Bolcerek	Gow	Musa
Kats	Ullrich	Bryant	Grove	Nickas
Lindsey		Cole	Hall	Salazar
McGuffin		Dimas	Houston	Vaneck
		Ellison	Kemp	Weber
		Everhart	Lathus	Zavodny

25. **Stratified Sample** A local youth advisory committee is comprised of 18 youths and 6 adults. The committee needs to select 2 adults and 4 youths to meet with the mayor to discuss renovations of the community center. The members of the committee are:

    Youth: Patrick, Elise, Shawn, Lauren, Sandra, Josh, Erin, Payton, Michelle, Chad, Katy, Amanda, Jim, Janice, Jeff, Amy, Logan, Steve

    Adults: Harold, Kerry, Keith, Greg, Marcie, Debbie

    Obtain a stratified random sample of 4 youths and 2 adults.

26. **Systematic Sample** In 2005 the University of Illinois men's basketball team set a school record for wins in a row (29). Their points per game (listed in chronological order horizontally) for regular and postseason games are given next. Obtain a systematic random sample of size 6 from the scores. Be sure to describe your sampling method.

87	91	85	89	91	72	78	74	83	93
70	105	69	67	84	68	90	78	73	75
89	81	60	57	70	83	75	84	84	64

27. **Systematic Sample** The human resource department at a [NW] certain company wants to conduct a survey regarding worker morale. The department has an alphabetical list of all 4,502 employees at the company and wants to conduct a systematic sample.

    (a) Determine k if the sample size is 50.

    (b) Determine the individuals who will be administered the survey. More than one answer is possible.

28. **Systematic Sample** To predict the outcome of a county election, a newspaper obtains a list of all 945,035 registered voters in the county and wants to conduct a systematic sample.

    (a) Determine k if the sample size is 130.

    (b) Determine the individuals who will be administered the survey. More than one answer is possible.

29. **Sample Design** The city of Naperville is considering the construction of a new commuter rail station. The city wishes to survey the residents of the city to obtain their opinion regarding the use of tax dollars for this purpose. Design a sampling method to obtain the individuals in the sample. Be sure to support your choice.

30. **Sample Design** A school board at a local community college is considering raising the student services fees. The board wants to obtain the opinion of the student body before proceeding. Design a sampling method to obtain the individuals in the sample. Be sure to support your choice.

31. **Sample Design** Target wants to open a new store in the village of Lockport. Before construction Target's marketers want to obtain some demographic information regarding the area under consideration. Design a sampling method to obtain the individuals in the sample. Be sure to support your choice.

32. **Sample Design** The county sheriff wishes to determine if a certain highway has a high proportion of speeders traveling on it. Design a sampling method to obtain the individuals in the sample. Be sure to support your choice.

33. **Sample Design** A pharmaceutical company wants to conduct a survey of 30 individuals who have high cholesterol. The company has obtained a list from doctors throughout the country of 6,600 individuals who are known to have high cholesterol. Design a sampling method to obtain the individuals in the sample. Be sure to support your choice.

34. **Sample Design** A marketing executive for Coca Cola. Inc., wants to identify television shows that people in the Boston area who typically drink Coke are watching. The executive has a list of all households in the Boston area. Design a sampling method to obtain the individuals in the sample. Be sure to support your choice.

35. Suppose an individual who is selected to be in a survey is not at home when a researcher calls, but the individual's spouse offers to answer the questions. What should the researcher do and why?

36. Research the origins of the Gallup Poll and the current sampling method the organization uses. Report your findings to the class.

37. Research the sampling methods used by a market research firm in your neighborhood. Report your findings to the class. The report should include the types of sampling methods used, number of stages, and sample size.

1.4 Sources of Errors in Sampling

Objective   **1**   **Understand how error can be introduced during sampling**

1   ## Understand How Error Can Be Introduced During Sampling

Thus far, we have discussed *how* to obtain samples, but have neglected to look at any of the pitfalls that inevitably arise in sampling. In this section we look at problems that can occur in sampling. Some of these problems can be remedied. Some, however, have no solution. Collectively, these errors are called *nonsampling errors*.

Definition

In Other Words
We use the word bias in nonsampling error to mean that preference is given to selecting some individuals over others.

Nonsampling errors are errors that result from the survey process. They are due to the nonresponse of individuals selected to be in the survey, to inaccurate responses, to poorly worded questions, to bias in the selection of individuals to be given the survey, and so on.

We contrast nonsampling errors with *sampling errors*.

Definition

In Other Words
We can think of sampling error as error that results from using a subset of the population to describe characteristics of the population. Nonsampling error is error that results from obtaining and recording information collected.

Sampling error is the error that results from using sampling to estimate information regarding a population. This type of error occurs because a sample gives incomplete information about the population.

By incomplete information, we mean that the individuals in the sample cannot reveal all the information about the population. Consider the following: Suppose that we wanted to determine the average age of the students enrolled in an introductory statistics course. To do this, we obtain a simple random sample of 4 students and ask them to write their age on a sheet of paper and turn it in. The average age of these 4 students is found to be 23.25 years. Assume that no students lied about their age, nobody misunderstood the question, and the sampling was done appropriately. If the actual average age of all 30 students in the class (the population) is 22.91 years, then the sampling error is $23.25 - 22.91 = 0.34$ year. Now suppose that the same survey is conducted, but this time one individual lies about his age. Then the results of the survey will have nonsampling error.

A well-designed survey and sampling technique can minimize nonsampling error. However, sampling error is more difficult to control.

When a sampling design is done poorly, the descriptive statistics computed from the data obtained in the sample may not be close to the values that would be obtained if the entire population were surveyed. For example, the *Literary Digest* predicted that Alfred M. Landon would defeat Franklin D. Roosevelt in the 1936 presidential election. The *Literary Digest* conducted a poll by mailing questionnaires based on a list of its subscribers, telephone directories, and automobile owners. On the basis of the results, the *Literary Digest* predicted that Landon would win with 57% of the popular vote. However, Roosevelt won the election with about 62% of the popular vote. The incorrect prediction by the *Literary Digest* was the result of a poor sample design. In 1936, most subscribers to the magazine, households with telephones, and automobile owners were Republican, the party of Landon. Therefore, the choice of the frame used to conduct the survey led to an incorrect prediction. This is an example of nonsampling error.

We now list some sources of nonsampling error. These sources include the frame, nonresponse, data-entry error, and poorly worded questions.

The Frame

Recall that the frame is the list of all individuals in the population under study. For example, in a study regarding voter preference in an upcoming election, the frame would be a list of all registered voters. Sometimes, obtaining the frame would seem to be a relatively easy task, such as obtaining the list of all registered voters. Even under this circumstance, however, the frame may be incomplete. People who recently registered to vote may not be on the published list of registered voters.

Often, it is difficult to gain access to a *complete* list of individuals in a population. For example, in public opinion polls, random telephone surveys are frequently conducted, which implies that the frame is all households with telephones. This method of sampling will exclude any household that does not have a telephone as well as all homeless people. In such a situation, certain segments of the population are *underrepresented*. A part of the population is **underrepresented** when the sampling method used tends to exclude this segment.

In designing any sample, the hope is that the frame used is as complete as possible so that any results inferred regarding the population have as little error as possible.

Nonresponse

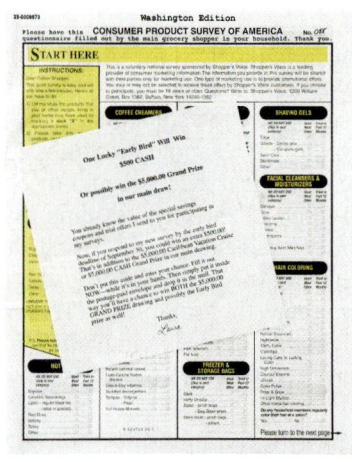

Nonresponse means that an individual selected for the sample does not respond to the survey. Nonresponse can occur because individuals selected for the sample do not wish to respond or because the interviewer was unable to contact them.

This type of error can be controlled using callbacks. The type of callback employed typically depends on the type of survey initially used. For example, if nonresponse occurs because a mailed questionnaire was not returned, a callback might mean phoning the individual to conduct the survey. If nonresponse occurs because an individual was not at home, a callback might mean returning to the home at other times in the day or other days of the week.

Another method to improve nonresponse is using rewards and incentives. Rewards may include cash payments for completing a questionnaire, made only upon receipt of the completed questionnaire. Incentives might also include a cover letter that states that the responses to the questionnaire will dictate future policy. For example, a village may send out questionnaires to households and state in a cover letter that the responses to the questionnaire will be used to decide pending issues within the village.

Interviewer Error

A trained interviewer is essential to obtain accurate information from a survey. A good interviewer will have the skill necessary to elicit responses from individuals within a sample and be able to make the interviewee feel comfortable enough to give truthful responses. For example, a good interviewer should be able to obtain truthful answers to questions as sensitive as "Have you ever cheated on your taxes?" Do not be quick to trust surveys that are conducted by poorly trained interviewers. Do not trust survey results if the sponsor has a vested interest in the results of the survey. Would you trust a survey conducted by a car dealer that reports 90% of customers say they would buy another car from the dealer?

Misrepresented Answers

Watch out for answers to survey questions that misrepresent facts or are flat-out lies. In fact, respondent may lie on surveys even when there is no problem with the interviewer. For example, a survey of recent college graduates may find that self-reported salaries are somewhat inflated.

Data Checks

Once data are collected, the results typically must be entered into a computer. Data entry inevitably results in input errors. It is imperative that data be checked for accuracy at every stage of the statistical analysis. In this text, we present some methodology that can be used to check for data-entry errors.

Questionnaire Design

Appropriate questionnaire design is critical in minimizing the amount of non-sampling error. We will concentrate on the main aspects in the design of a good questionnaire.

One of the first considerations in designing a question is determining whether the question should be open or closed.

An **open question** is one for which the respondent is free to choose his or her response. For example:	A **closed question** is one for which the respondent must choose from a list of predetermined responses.
What is the most important problem facing America's youth today?	What is the most important problem facing America's youth today? (a) Drugs (b) Violence (c) Single-parent homes (d) Promiscuity (e) Peer pressure

When designing an open question, be sure to phrase the question so that the responses are similar. (You don't want a wide variety of responses.) This allows for easy analysis of the responses. The benefit of closed questions is that they limit the number of respondent choices and, therefore, the results are much easier to analyze. However, this limits the choices and does not always allow the respondent to respond the way he or she might want to respond. If the desired answer is not provided as a choice, the respondent will be forced to choose a secondary answer or skip the question.

Survey designers recommend conducting pretest surveys with open questions and then using the most popular answers as the choices on closed question surveys. Another issue to consider in the closed question design is the number of responses the respondent may choose from. It is recommended that the option "no opinion" be omitted, because this option does not allow for meaningful analysis. The bottom line is to try to limit the number of choices in a closed-question format without forcing respondents to choose an option they otherwise would not.

! CAUTION
The wording of questions can significantly affect the responses and, therefore, the validity of a study.

Wording of Questions

The wording of a survey question is vital in obtaining data that are not misrepresentative. Questions must always be asked in balanced form. For example, the "yes/no" question

Do you oppose the reduction of estate taxes?

should be written

Do you favor or oppose the reduction of estate taxes?

The second question is balanced. Do you see the difference? Consider the following report based on studies from Schuman and Presser (*Questions and Answers in Attitude Surveys*, 1981, p. 277), who asked the following two questions:

(A) Do you think the United States should forbid public speeches against democracy?

(B) Do you think the United States should allow public speeches against democracy?

For those respondents presented with question A, 21.4% gave yes responses, while for those given question B, 47.8% gave no responses. The conclusion you may arrive at is that most people are not necessarily willing to forbid something, but more people are willing not to allow something. These results imply that the wording of the question can alter the outcome of a survey.

Another consideration in wording a question is not to be vague. For example, the question "How much do you study?" is too vague. Does the researcher mean how much do I study for all my classes or just for statistics? Does the researcher mean per day or per week? The question should be written "How many hours do you study statistics each week?"

The Order of the Questions, Words, and Responses

Many surveys will rearrange the order of the questions within a questionnaire so that responses are not affected by prior questions. Consider the following example from Schuman and Presser in which the following two questions were asked:

(A) Do you think the United States should let Communist newspaper reporters from other countries come in here and send back to their papers the news as they see it?

(B) Do you think a Communist country such as Russia should let American newspaper reporters come in and send back to America the news as they see it?

For surveys conducted in 1980 in which the questions appeared in the order (A, B), 54.7% of respondents answered yes to A and 63.7% answered yes to B. If the questions were ordered (B, A), then 74.6% answered yes to A and 81.9% answered yes to B. When Americans are first asked if U.S. reporters should be allowed to report Communist news, they are more likely to agree that Communists should be allowed to report American news. Questions should be rearranged as much as possible to help reduce the effects of this type.

Pollsters will also rearrange words within a question. For example, the Gallup Organization asked the following question of 1,017 adults aged 18 years or older:

> **Do you consider the first six months of the Bush administration to be a** [rotated: **success** (or a) **failure**]?

Notice how the words *success* and *failure* were rotated. The purpose of this is to remove the effect that may occur by writing the word *success* first in the question.

Not only should the order of the questions or certain words within the question be rearranged, but, in closed questions, the possible responses should also be rearranged. The reason is that respondents are likely to choose early choices in a list rather than later choices.

MAKING AN INFORMED DECISION

What Movie Should I Go To?

One of the most difficult tasks of surveying is phrasing questions so that they are not misunderstood. In addition, they must be phrased so that the researcher obtains answers that allow for meaningful analysis. We wish to create a questionnaire that can be used to make an informed decision about whether to attend a certain movie. Select a movie that you wish to see. If the movie is still in theaters, make sure that it has been released for at least a couple of weeks so it is likely that a number of people have seen it. Design a questionnaire to be filled out by individuals who have seen the movie. You may wish to include questions regarding the demographics of the respondents first (such as age, gender, level of education, and so on). Ask as many questions as you feel are necessary to obtain an opinion regarding the movie. The questions can be open or closed. Administer the survey to at least 20 randomly selected people who have seen the movie. While administering the survey, keep track of those individuals who have not seen the movie. In particular, keep track of their demographic information. After administering the survey, summarize your findings. On the basis of the survey results, do you think that you will enjoy the movie? Why? Now see the movie. Did you like it? Did the survey accurately predict whether you would enjoy the movie? Now answer the following questions:

(a) What sampling method did you use? Why? Did you have a frame for the population?

(b) Did you have any problems with respondents misinterpreting your questions? How could this issue have been resolved?

(c) What role did the demographics of the respondent have in forming your opinion? Why?

(d) Did the demographics of individuals who did not see the movie play a role while you were forming your opinion regarding the movie?

(e) Look up a review of the movie by a professional movie critic. Did the movie critic's opinion agree with yours? What might account for the similarities or differences in your opinions?

(f) Describe the problems that you had in administering the survey. If you had to do this survey over again, would you change anything? Why?

1.4 ASSESS YOUR UNDERSTANDING

Concepts and Vocabulary

1. Why is it rare for frames to be completely accurate?
2. What are some solutions to nonresponse?
3. What is a closed question? What is an open question? Discuss the advantages and disadvantages of each type of question.
4. What does it mean when a part of the population is under-represented?
5. Discuss the benefits of having trained interviewers.
6. What are the advantages of having a pretest when constructing a questionnaire that has closed questions?
7. Discuss the pros and cons of telephone interviews that take place during dinner time in the early evening.
8. Why is a high response rate desired? How would a low response rate affect survey results?
9. Discuss why the order of questions or choices within a questionnaire are important in sample surveys.
10. Suppose a survey asks, "Do you own any CDs?" Explain how this could be interpreted in more than one way. Suggest a way in which the question could be improved.

Skill Building

In Problems 11–22, the survey design is flawed. (a) Determine whether the flaw is due to the sampling method or the survey itself. For flawed surveys, identify the cause of the error (wording of question, nonresponse, and so forth). (b) Suggest a remedy to the problem.

11. A college vice-president wants to conduct a study regarding the achievement of undergraduate students. He selects the first 50 students who enter the building on a given day and administers his survey.

12. The village of Oak Lawn wishes to conduct a study regarding the income level of households within the village. The village manager selects 10 homes in the southwest corner of the village and sends an interviewer to the homes to determine household income.

13. An antigun advocate wants to estimate the percentage of people who favor stricter gun laws. He conducts a nationwide survey of 1,203 randomly selected adults 18 years old and older. The interviewer asks the respondents, "Do you favor harsher penalties for individuals who sell guns illegally?"

14. Suppose you are conducting a survey regarding students' study habits. From a list of full-time registered students, you obtain a simple random sample of 90 students. One survey question is "How many hours do you study?"

15. A polling organization conducts a study to estimate the percentage of households that speak a foreign language as the primary language. It mails a questionnaire to 1,023 randomly selected households throughout the United States and asks the head of household if a foreign language is the primary language spoken in the home. Of the 1,023 households selected, 12 responded.

16. Cold Stone Creamery is considering opening a new store in O'Fallon. Before opening the store, the company would like to know the percentage of households in O'Fallon that regularly visit an ice cream shop. The market researcher obtains a list of households in O'Fallon and randomly selects 150 of them. He mails a questionnaire to the 150 households that asks about ice cream eating habits and flavor preferences. Of the 150 questionnaires mailed, 4 are returned.

17. The owner of a shopping mall wishes to expand the number of shops available in the food court. She has a market researcher survey mall customers during weekday mornings to determine what types of food the shoppers would like to see added to the food court.

18. A produce buyer for a supermarket is inspecting crates of peaches. He examines 20 peaches from the top of each crate and purchases the crate if all 20 peaches are satisfactory.

19. A magazine is conducting a study on the effects of infidelity in a marriage. The editors randomly select 400 women whose husbands were unfaithful and ask, "Do you believe a marriage can survive when the husband destroys the trust that must exist between husband and wife?"

20. A textbook publisher wants to determine what percentage of college professors either require or recommend that their students purchase textbook packages with supplemental materials, such as study guides, digital media, and online tools. The publisher sends out surveys by email to a random sample of 320 faculty members who have registered with their web site and have agreed to receive solicitations. The publisher reports that 80% of college professors require or recommend that their students purchase some type of textbook package.

21. Suppose you are conducting a survey regarding illicit drug use among teenagers in the Baltimore School District. You obtain a cluster sample of 12 schools within the district and sample all sophomore students in the randomly selected schools. The survey is administered by the teachers.

22. To determine public opinion of the police department, the police chief obtains a cluster sample of 15 census tracts within his jurisdiction and samples all households in the randomly selected tracts. Uniformed police officers go door to door to conduct the survey.

Applying the Concepts

23. Order of the Questions Consider the following two questions.

A. *Suppose that a rape is committed in which the woman becomes pregnant. Do you think the criminal should or should not face additional charges if the woman becomes pregnant?*

B. *Do you think abortions should be legal under any circumstances, legal under certain circumstances, or illegal in all circumstances?*

Do you think the order in which the questions are asked will affect the survey results? If so, what can the pollster do to alleviate this response bias?

24. Order of the Questions Consider the following two questions:

A. *Do you believe that the government should or should not be allowed to prohibit individuals from expressing their religious beliefs at their place of employment?*

B. *Do you believe that the government should or should not be allowed to prohibit teachers from expressing their religious beliefs in public school classrooms?*

Do you think the order in which the questions are asked will affect the survey results? If so, what can the pollster do to alleviate this response bias? Discuss the choice of the word *prohibit* in the survey questions.

25. Rotating Choices Consider this question from a recent Gallup poll:

Thinking about how the abortion issue might affect your vote for major offices, would you vote only for a candidate who shares your views on abortion or consider a candidate's position on abortion as just one of many important factors? [rotated]

Why is it important to rotate the two choices presented in the question?

26. Overcoverage An article in the *Los Angeles Times* (Sept. 21, 1996) reported that results of Bosnian elections the previous week were suspect because an independent watchdog agency, International Crisis Group, estimated that voter turnout was 106.7%. Why do you think the agency was concerned?

27. Exit Polls Read the following article from the January 20, 2005 *USA Today*. What types of nonsampling errors led to incorrect exit polls?

FIRMS REPORT FLAWS THAT THREW OFF EXIT POLLS

Kerry backers' willingness, pollsters' inexperience cited

By Mark Memmott, *USA Today*

The exit polls of voters on Election Day so overstated Sen. John Kerry's support that, going back to 1988, they rank as the most inaccurate in a presidential election, the firms that did the work concede.

One reason the surveys were skewed, they say, was because Kerry's supporters were more willing to participate than Bush's. Also, the people they hired to quiz voters were on average too young and too inexperienced and needed more training.

The exit polls, which are supposed to help the TV networks shape their coverage on election night, were sharply criticized. Leaks of preliminary data showed up on the Internet in the early afternoon of Election Day, fueling talk that Kerry was beating President Bush. After the election, some political scientists, pollsters and journalists questioned their value.

In a report to the six media companies that paid them to conduct the voter surveys, pollsters Warren Mitofsky and Joseph Lenski, said Wednesday that "on average, the results from each precinct overstated the Kerry-Bush difference by 6.5 (percentage) points. This is the largest (overstatement) we have observed ... in the last five presidential elections."

Lenski said Wednesday that issuing the report was like "hanging out your dirty underwear. You hope it's cleaner than people expected."

Among the findings:

- They hired too many relatively young adults to conduct the interviews. Half of the 1,400 interviewers were younger than 35. That may explain in part why Kerry voters were more inclined to participate, since he drew more of the youth vote than did Bush. But Mitofsky and Lenski also found younger interviewers were more likely to make mistakes.

- Early results were skewed by a "programming error" that led to including too many female voters. Kerry outpolled Bush among women.

- Some local officials prevented interviewers from getting close to voters.

For future exit polls, Lenski and Mitofsky recommended hiring more experienced polltakers and giving them better training, and working with election officials to ensure access to polling places.

Lenski and Mitofsky noted that none of the media outlets they worked for—ABC, CBS, CNN, Fox News, NBC and the Associated Press—made any wrong "calls" on election night. Representatives of those six are reviewing the report.

Many other news media, including *USA Today*, also paid to get some of the data.

28. Increasing Response Rates Offering rewards or incentives is one way of attempting to increase response rates. Discuss a possible disadvantage of such a practice.

29. Wording Survey Questions Write a survey question that contains strong wording and a survey question that contains tempered wording. Present the strongly worded question to 10 randomly selected people and the tempered question to 10 different randomly selected people. How does the wording affect the response?

30. Order in Survey Questions Write two questions that could have different responses, depending on the order in which the questions are presented. Randomly select 20 people and present the questions in one order to 10 of the people and in the opposite order to the other 10 people. Did the results differ?

31. Research a survey method used by a company or government branch. Determine the sampling method used, the sample size, the method of collection, and the frame used.

32. Out-of-Class Activity People often respond to survey questions without any knowledge of the subject matter. A common example of this is the discussion on banning dihydrogen monoxide. The Centers for Disease Control (CDC) reports that there were 1,493 deaths due to asbestos in 2002, but over 3,200 deaths were attributed to dihydrogen monoxide in 2000. Articles and Web sites, such as www.dhmo.org tell how this substance is widely used despite the dangers associated with it. Many people have joined the cause to ban this substance without realizing that dihydrogen monoxide is simply water (H_2O). Their eagerness to protect the environment or their fear of seeming uninformed may be part of the problem. Put together a survey that asks individuals whether dihydrogen monoxide should or should not be banned. Give the survey to 20 randomly selected students around campus and report your results to the class. An example survey might look like the following:

> Dihydrogen monoxide is colorless, odorless, and kills uncounted thousands of people every year. Most of these deaths are caused by accidental inhalation, but the dangers of dihydrogen monoxide do not stop there. Prolonged exposure to its solid form can severely damage skin tissue. Symptoms of ingestion can include excessive sweating and urination and possibly a bloated feeling, nausea, vomiting, and body electrolyte imbalance. Dihydrogen monoxide is a major component of acid rain and can cause corrosion after coming in contact with certain metals.

> Do you believe that the government should or should not ban the use of dihydrogen monoxide?

1.5 The Design of Experiments

Objectives

1. Define designed experiment
2. Understand the steps in designing an experiment
3. Understand the completely randomized design
4. Understand the matched-pairs design
5. Understand the randomized block design

A major theme of this chapter has been data collection. Sections 1.2 through 1.4 discussed techniques for obtaining data through surveys. Data obtained from surveys lead to observational studies. Obtaining data through an experiment is discussed in this section.

1 Define Designed Experiment

When people hear the word *experiment*, they typically think of a laboratory with a controlled environment. Researchers also have control when they perform experiments.

Definition

> A **designed experiment** is a controlled study conducted to determine the effect that varying one or more **explanatory variables** has on a response variable. The explanatory variables are often called **factors**. The **response variable** represents the variable of interest. Control, manipulation, randomization, and replication are the key ingredients of a well-designed experiment.

Historical Note

Sir Ronald Fisher, often called the Father of Modern Statistics, was born in England on February 17, 1890. He received a BA in astronomy from Cambridge University in 1912. In 1914, he took a position teaching mathematics and physics at a high school. He did this to help serve his country during World War I. (He was rejected by the army because of his poor eyesight.) In 1919, Fisher took a job as a statistician at Rothamsted Experimental Station, where he was involved in agricultural research. In 1933, Fisher became Galton Professor of Eugenics at Cambridge University, where he studied Rh blood groups. In 1943 he was appointed to the Balfour Chair of Genetics at Cambridge. He was knighted by Queen Elizabeth in 1952. Fisher retired in 1957 and died in Adelaide, Australia, on July 29, 1962. One of his famous quotations is "To call in the statistician after the experiment is done may be no more than asking him to perform a postmortem examination: he may be able to say what the experiment died of."

A **treatment** is any combination of the values of each factor. The **experimental unit** is a person, object, or some other well-defined item to which a treatment is applied. We often refer to the experimental unit as a **subject** when he or she is a person. The subject (experimental unit) is analogous to the individual in a survey.

The goal in an experiment is to determine the effect various treatments have on the response variable. For example, suppose a researcher wants to measure the effect of sleep deprivation on a person's fine-motor skills. The researcher might take a group of 100 individuals and randomly divide them into four groups. The first group might sleep 8 hours a night for four nights, the second will sleep 6 hours per night for four nights, the third group will sleep 4 hours per night for four nights, while the fourth group will sleep 2 hours per night for four nights. The experimental unit or subject is the person, the explanatory variable or factor is the amount of sleep, the treatments are 8, 6, 4, and 2 hours, and the response variable might be the reaction time of the person to some stimulus.

Many designed experiments are **double-blind**. This means that neither the experimental unit nor the experimenter knows what treatment is being administered to the experimental unit. For example, in clinical studies of the cholesterol-lowering drug Lipitor, researchers administered either 10, 20, 40, or 80 mg of Lipitor or a **placebo**, an innocuous medication such as a sugar tablet, to patients (the subjects) with high cholesterol. Because the experiment was double-blind, neither the patients nor the researchers knew which medication was being administered. It is important that double-blind methods be used in this case so that the patients and researchers do not behave in such a way as to affect the results. For example, the researcher might not give as much time to a patient receiving the placebo.

② Understand the Steps in Designing an Experiment

The process of conducting an experiment requires a series of steps.

Step 1: Identify the Problem to Be Solved. The statement of the problem should be as explicit as possible. The statement should provide the experimenter with direction. In addition, the statement must identify the response variable and the population to be studied. Often, the statement is referred to as the **claim**.

Step 2: Determine the Factors That Affect the Response Variable. The factors are usually identified by an expert in the field of study. In identifying the factors, we must ask, "What things affect the value of the response variable?" Once the factors are identified, it must be determined which factors will be fixed at some predetermined level, which will be manipulated, and which will be uncontrolled.

Step 3: Determine the Number of Experimental Units. As a general rule, choose as many experimental units as time and money will allow. Techniques do exist for determining sample size, provided certain information is available. Some of these techniques are discussed later in the text.

Step 4: Determine the Level of Each Factor. There are three ways to deal with the factors:

1. **Control:** Fix their level at one predetermined value throughout the experiment. These are variables whose effect on the response variable is not of interest.

2. **Maniuplate:** Set them at predetermined levels. These are the variables whose effect on the response variable interests us. The combinations of the levels of these variables constitute the treatments in the experiment.

3. **Randomize:** Randomize the experimental units to various treatment groups so that the effect of factors whose levels cannot be controlled is minimized. The idea is that randomization averages out the effects of uncontrolled factors (explanatory variables).

It is difficult, if not impossible, to identify all factors in an experiment. This is why randomization is so important. It mutes the effect of variation attributable to factors not controlled or manipulated.

Step 5: Conduct the Experiment.

(a) The experimental units are randomly assigned to the treatments. **Replication** occurs when each treatment is applied to more than one experimental unit. By using more than one experimental unit for each treatment, we can be assured that the effect of a treatment is not due to some characteristic of a single experimental unit. It is a good idea to assign an equal number of experimental units to each treatment.

(b) Collect and process the data. Measure the value of the response variable for each replication. Then organize the results. The idea is that the value of the response variable for each treatment group is the same before the experiment because of randomization. Then any difference in the value of the response variable among the different treatment groups can be attributed to differences in the level of the treatment.

Step 6: Test the claim. This is the subject of inferential statistics. **Inferential statistics** is a process in which generalizations about a population are made on the basis of results obtained from a sample. In addition, a statement regarding our level of confidence in our generalization is provided. We study methods of inferential statistics in Chapters 9 through 14.

3 Understand the Completely Randomized Design

The steps just given apply to any type of designed experiment. We now concentrate on the simplest type of experiment.

Definition

> A **completely randomized design** is one in which each experimental unit is randomly assigned to a treatment.

An example will help clarify the process of experimental design.

EXAMPLE 1   **A Completely Randomized Design**

Problem: A farmer wishes to determine the optimal level of a new fertilizer on his soybean crop. Design an experiment that will assist him.

Approach: We follow the steps for designing an experiment.

Solution

Step 1: The farmer wants to identify the optimal level of fertilizer for growing soybeans. We define *optimal* as the level that maximizes yield. So the response variable will be crop yield.

Step 2: Some factors that affect crop yield are fertilizer, precipitation, sunlight, method of tilling the soil, type of soil, plant, and temperature.

Step 3: In this experiment, we will plant 60 soybean plants (experimental units).

Step 4: We list the factors and their levels.

- **Fertilizer.** We manipulate the level of this factor. We wish to measure the effect of varying the level of this variable on the response variable, yield. We will set the level of fertilizer (the treatment) as follows:

    Treatment A: 20 soybean plants receive no fertilizer.

    Treatment B: 20 soybean plants receive 2 teaspoons of fertilizer per gallon of water every 2 weeks.

    Treatment C: 20 soybean plants receive 4 teaspoons of fertilizer per gallon of water every 2 weeks.

    See Figure 5.

Figure 5

- **Precipitation.** Although we cannot control the amount of rainfall, we can control the amount of watering we do. This factor will be controlled so that each plant receives the same amount of precipitation.
- **Sunlight.** This is an uncontrollable factor, but it will be roughly the same for each plant.
- **Method of tilling.** We can control this factor. We agree to use the round-up ready method of tilling for each plant.
- **Type of soil.** We can control certain aspects of the soil such as level of acidity. In addition, each plant will be planted within a 1 acre area, so it is reasonable to assume that the soil conditions for each plant are equivalent.

- **Plant.** There may be variation from plant to plant. To account for this, we randomly assign the plants to a treatment.
- **Temperature.** This factor is not within our control, but will be the same for each plant.

Step 5:

(a) We need to assign each plant to a treatment group. To do this, we will number the plants from 1 to 60. To determine which plants get treatment A, we randomly generate 20 numbers. The plants corresponding to these numbers get treatment A. Now number the remaining plants 1 to 40 and randomly generate 20 numbers. The plants corresponding to these numbers get treatment B. The remaining plants get treatment C. Now till the soil, plant the soybean plants, and fertilize according to the schedule prescribed.

(b) At the end of the growing season, determine the crop yield for each plant.

Step 6: Determine whether any differences in yield exist among the three treatment groups.

Figure 6 illustrates the experimental design.

Figure 6

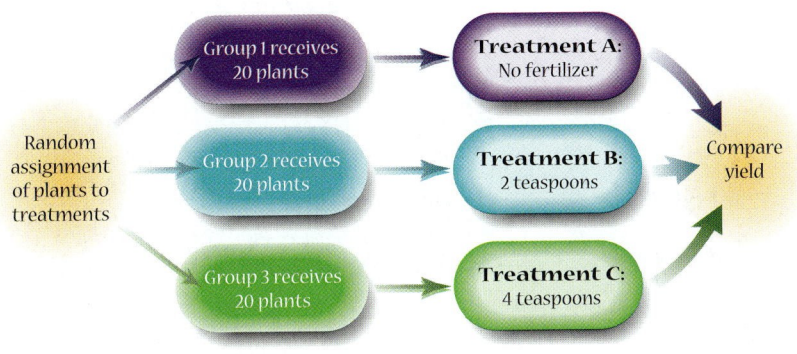

Example 1 is a completely randomized design because the experimental units (the plants) were randomly assigned the treatment. It is the most popular experimental design because of its simplicity, but it is not always the best. We discuss inferential procedures for the completely randomized design in which there are two treatments in Section 11.2 and in which there are three or more treatments in Section 13.1.

Now Work Problem 5.

4 **Understand the Matched-Pairs Design**

Another type of experimental design is called a *matched-pairs design*.

Definition

> A **matched-pairs design** is an experimental design in which the experimental units are paired up. The pairs are matched up so that they are somehow related (that is, the same person before and after a treatment, twins, husband and wife, same geographical location, and so on). There are only two levels of treatment in a matched-pairs design.

In matched-pairs design, one matched individual will receive one treatment and the other matched individual receives a different treatment. The assignment of the matched pair to the treatment is done randomly using a coin flip or a random-number generator. We then look at the difference in the results of each matched pair. One common type of matched-pairs design is to measure a response variable on an experimental unit before a treatment is applied, and then measure the response variable on the same experimental unit after the treatment is applied. In this way, the individual is matched against itself. These experiments are sometimes called before–after or pretest–posttest experiments.

EXAMPLE 2　A Matched-Pairs Design

Problem: An educational psychologist wanted to determine whether listening to music has an impact on a student's ability to learn. Design an experiment to help the psychologist answer the question.

Approach: We will use a matched-pairs design by matching students according to IQ and gender (just in case gender plays a role in learning with music).

Solution: We match students according to IQ and gender. For example, a female with an IQ in the 110 to 115 range will be matched with a second female with an IQ in the 110 to 115 range.

For each pair of students, we will flip a coin to determine whether the first student in the pair is assigned the treatment of a quiet room or a room with music playing in the background.

Each student will be given a statistics textbook and asked to study Section 1.1. After 2 hours, the students will enter a testing center and take a short quiz on the material in the section. We compute the difference in the scores of each matched pair. Any differences in scores will be attributed to the treatment. Figure 7 illustrates the design.

Figure 7

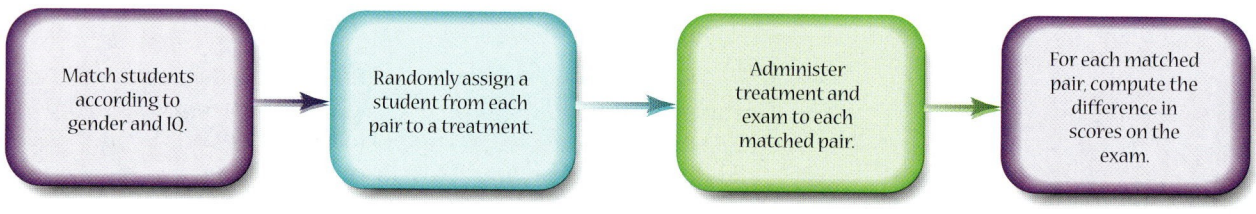

We discuss the statistical inference of matched-pairs design in Section 11.1.

Now Work Problem 7.

⑤ Understand the Randomized Block Design

The completely randomized design is the simplest experimental design. However, its simplicity can lead to flaws. Before we introduce a slightly more complicated experimental design, consider the following story.

I coach my son's soccer team, which is comprised of four 10-year-olds, six 9-year-olds, and four 8-year-olds. After each practice, I like to have a 15-minute scrimmage where I randomly assign seven players to each team. I quickly learned that randomly assigning players, without taking age into consideration, would sometimes result in very unequal teams (once I randomly created a team of four 10-year-olds and three 9-year-olds). So I learned that I should randomly assign two 10-year-olds to one team with the other two 10-year-olds going on the other team. I then assigned three 9-year-olds to each team and two 8-year-olds to each team. Using the language of statistics, I was *blocking* by age.

Definition　Grouping similar (homogeneous) experimental units together and then randomizing the experimental units within each group to a treatment is called **blocking**. Each group of homogeneous individuals is called a **block**.

With the soccer story in mind, let's revisit our crop-yield experiment from Example 1, where we assumed that the soybean plants were the same variety. Suppose we found out that we in fact had two varieties of soybeans: Chemgro and Pioneer. Using the completely randomized design, it is impossible to know whether any differences in yield were due to the differences in fertilizer or to the differences in variety.

It may be that the Chemgro variety naturally has higher yield potential than Pioneer. If a majority of the soybean plants that were randomly assigned to treatment C was Chemgro, then we would not know whether to attribute the different crop yields to the level of fertilizer or to the plant variety. This effect is known as *confounding*. **Confounding** occurs when the effect of two factors (explanatory variables) on the response variable cannot be distinguished.

To resolve issues of this type, we introduce a second experimental design.

Definition   A **randomized block design** is used when the experimental units are divided into homogeneous groups called blocks. Within each block, the experimental units are randomly assigned to treatments.

In a randomized block design, we do not wish to determine whether the differences between blocks result in any difference in the value of the response variable. Our goal is to isolate any variability in the response variable that may be attributable to the block. We discuss this design in the next example.

EXAMPLE 3   **Illustrating the Randomized Block Design**

Problem: Suppose that the 60 soybean plants were actually two different varieties: 30 Chemgro soybean plants and 30 Pioneer soybean plants. In addition, the Chemgro variety is known to have higher yields, on average, than Pioneer. Design an experiment that could be used to measure the effect of level of fertilizer while taking into account the variety of soybean.

Approach: Because the yield of Chemgro is higher than that of Pioneer, we will use a randomized block design to determine which level of fertilizer results in the highest yield. We will block by variety.

Solution: In a randomized block design, we divide 60 plants into two blocks. Block I will contain the 30 Chemgro plants and block 2 will contain the 30 Pioneer plants. Within each block, we randomly assign each plant to a treatment (A, B, or C). See Figure 8.

Figure 8

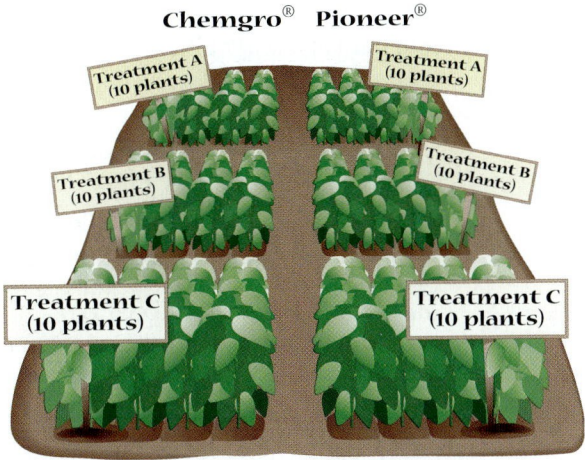

The idea is that in each block the plants are all the same variety of plant, so plant variety does not affect the value of the response variable. Figure 9 illustrates the design. Notice that we do not compare yields across plant variety because this was not our goal—we already know which variety yields more. The reason for blocking is to reduce variability due to plant variety by comparing yields within each variety.

Figure 9

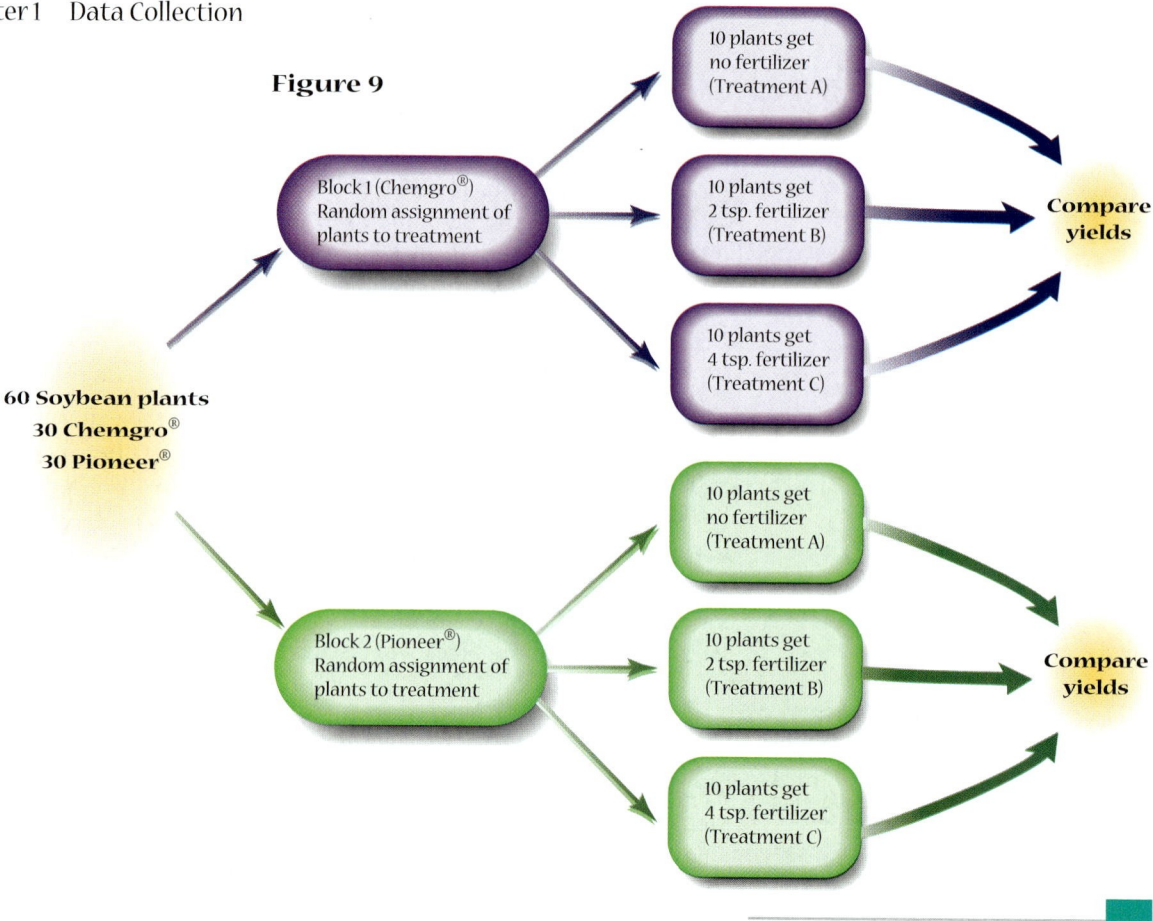

60 Soybean plants
30 Chemgro®
30 Pioneer®

Block 1 (Chemgro®)
Random assignment of
plants to treatment

10 plants get
no fertilizer
(Treatment A)

10 plants get
2 tsp. fertilizer
(Treatment B)

10 plants get
4 tsp. fertilizer
(Treatment C)

Compare
yields

Block 2 (Pioneer®)
Random assignment of
plants to treatment

10 plants get
no fertilizer
(Treatment A)

10 plants get
2 tsp. fertilizer
(Treatment B)

10 plants get
4 tsp. fertilizer
(Treatment C)

Compare
yields

Now Work Problem 15.

One note about the relation between a designed experiment and simple random sampling: It is often the case that the experimental units selected to participate in a study are not randomly selected. This is because we often need the experimental units to have some common trait, such as high blood pressure. For this reason, participants in experiments are recruited or volunteer to be in a study. However, once we have the experimental units, we use simple random sampling to assign them to treatment groups. With random assignment we assume that the participants are similar at the start of the experiment. Because the treatment is the only difference between the groups, we can say the treatment *caused* the difference observed in the response variable.

In-Class Activity: Experimental Design (Hippity-Hop)

You are commissioned by the board of directors of Paper Toys, Inc. to design a new paper frog for their Christmas catalog. The design for the construction of the frog has already been completed and will be provided to you. However, the material with which to make the frogs has not yet been determined. The Materials Department has narrowed the choices down to either newspaper or brown paper (such as that used in grocery bags). You have decided to test both types of paper. Management decided to build the frogs from sheets of paper 9 inches square.

The goal of the experiment is to determine the material that results in frogs that jump farther.

(a) As a class, design an experiment that will answer the research question.
(b) Make the frogs.
(c) Conduct the experiment.
(d) As a class discuss the strengths and weaknesses of the design. Would you change anything?

1.5 ASSESS YOUR UNDERSTANDING

Concepts and Vocabulary

1. Define the following:
   (a) Experimental unit
   (b) Treatment
   (c) Response variable
   (d) Factor
   (e) Double-blind
   (f) Placebo
   (g) Confounding

2. What is replication in an experiment?

3. A(n) _____ _____ design is one in which each experimental unit is randomly assigned to a treatment. A(n) _____ _____ design is one in which the experimental units are paired up.

4. Discuss how a randomized block design is similar to a stratified random sample.

Applying the Concepts

5. **School Psychology** A school psychologist wants to test
   NW the effectiveness of a new method for teaching reading. She recruits 500 first-grade students in District 203 and randomly divides them into two groups. Group 1 is taught by means of the new method, while Group 2 is taught via traditional methods. The same teacher is assigned to teach both groups. At the end of the year, an achievement test is administered and the results of the two groups compared.
   (a) What is the response variable in this experiment?
   (b) Think of some of the factors in the study. Which are controlled? Which factor is manipulated?
   (c) What are the treatments? How many treatments are there?
   (d) How are the factors that are not controlled or manipulated dealt with?
   (e) What type of experimental design is this?
   (f) Identify the subjects.
   (g) Draw a diagram similar to Figure 6, 7, or 9 to illustrate the design.

6. **Pharmacy** A pharmaceutical company has developed an experimental drug meant to relieve symptoms associated with the common cold. The company identifies 300 adult males 25 to 29 years old who have a common cold and randomly divides them into two groups. Group 1 is given the experimental drug, while group 2 is given a placebo. After 1 week of treatment, the proportions of each group that still have cold symptoms are compared.
   (a) What is the response variable in this experiment?
   (b) Think of some of the factors in the study. Which are controlled? Which factor is manipulated?
   (c) What are the treatments? How many treatments are there?
   (d) How are the factors that are not controlled or manipulated dealt with?
   (e) What type of experimental design is this?
   (f) Identify the subjects.
   (g) Draw a diagram similar to Figure 6, 7, or 9 to illustrate the design.

7. **Assessment** To help assess student learning in her devel-
   NW opmental math courses, a mathematics professor at a community college implemented pre- and posttests for her developmental math students. A knowledge-gained score was obtained by taking the difference of the two test scores.

   (a) What type of experimental design is this?
   (b) What is the response variable in this experiment?
   (c) What is the treatment?

8. **Whiter Teeth** An ad for Crest Whitestrips Premium claims that the strips will whiten teeth in 7 days and the results will last for 12 months. A researcher who wishes to test this claim studies 20 sets of identical twins. Within each set of twins, one is randomly selected to use Crest Whitestrips Premium in addition to regular brushing and flossing, while the other just brushes and flosses. Whiteness of teeth is measured at the beginning of the study, after 7 days, and every month thereafter for 12 months.
   (a) What type of experimental design is this?
   (b) What is the response variable in this experiment?
   (c) What factor is manipulated? What are the treatments?
   (d) What is another factor (controlled or uncontrolled) that could affect the response variable?
   (e) What might be an advantage of using identical twins as subjects in this experiment?

9. **Insomnia** Researchers Jack D. Edinger and associates wanted to test the effectiveness of a new cognitive behavioral therapy (CBT) compared with both an older behavioral treatment and a placebo therapy for treating insomnia. They identified 75 adults with chronic insomnia. Patients were randomly assigned to one of three treatment groups. Twenty-five patients were randomly assigned to receive CBT (sleep education, stimulus control, and time-in-bed restrictions), another 25 received muscle relaxation training (RT), and the final 25 received a placebo treatment. Treatment lasted 6 weeks, with follow-up conducted at 6 months. To measure the effectiveness of the treatment, researchers used wake time after sleep onset (WASO). Cognitive behavioral therapy produced larger improvements than did RT or placebo treatment. For example, the CBT-treated patients achieved an average 54% reduction in their WASO whereas RT-treated and placebo-treated patients, respectively, achieved only 16% and 12% reductions in this measure. Results suggest that CBT treatment leads to significant sleep improvements within 6 weeks, and these improvements appear to endure through 6 months of follow-up.

   (*Source:* "Cognitive Behavioral Therapy for Treatment of Chronic Primary Insomnia," Jack D. Edinger, PhD; William K. Wohlgemuth, PhD; Rodney A. Radtke, MD; Gail R. Marsh, PhD; Ruth E. Quillian, PhD; *Journal of the American Medical Association* 285(2001): 1856–1864)

(a) What type of experimental design is this?
(b) What is the population being studied?
(c) What is the response variable in this study?
(d) What is the factor? What are the treatments?
(e) Identify the experimental units.
(f) Draw a diagram similar to Figure 6, 7, or 9 to illustrate the design.

10. **Depression** Researchers wanted to compare the effectiveness and safety of an extract of St. John's wort with placebo in outpatients with major depression. To do this, they recruited 200 adult outpatients diagnosed as having major depression and having a baseline Hamilton Rating Scale for Depression (HAM-D) score of at least 20. Participants were randomly assigned to receive either St. John's wort extract (900 mg/d for 4 weeks, increased to 1200 mg/d in the absence of an adequate response thereafter) or a placebo for 8 weeks. The response variable was the change on the HAM-D over the treatment period. After analysis of the data, it was concluded that St. John's wort was not effective for treatment of major depression.

    (*Source:* "Effectiveness of St. John's Wort in Major Depression," Richard C. Shelton, MD; et. al.; *Journal of the American Medical Association* 285 (2001):1978–1986)

    (a) What type of experimental design is this?
    (b) What is the population that is being studied?
    (c) What is the response variable in this study?
    (d) What is the factor? What are the treatments?
    (e) Identify the experimental units.
    (f) Draw a diagram similar to Figure 6, 7, or 9 to illustrate the design.

11. **The Memory Drug?** Researchers wanted to evaluate whether ginkgo, an over-the-counter herb marketed as enhancing memory, improves memory in elderly adults as measured by objective tests. To do this, they recruited 98 men and 132 women older than 60 years and in good health. Participants were randomly assigned to receive ginkgo, 40 mg 3 times per day, or a matching placebo. The measure of memory improvement was determined by a standardized test of learning and memory. After 6 weeks of treatment, the data indicated that ginkgo did not increase performance on standard tests of learning, memory, attention, and concentration. These data suggest that, when taken following the manufacturer's instructions, ginkgo provides no measurable increase in memory or related cognitive function to adults with healthy cognitive function.

    (*Source:* "Ginkgo for Memory Enhancement," Paul R. Solomon, et. al.; *Journal of the American Medical Association* 288(2002): 835–840)

    (a) What type of experimental design is this?
    (b) What is the population being studied?
    (c) What is the response variable in this study?
    (d) What is the factor? What are the treatments?

(e) Identify the experimental units.
(f) Draw a diagram similar to Figure 6, 7, or 9 to illustrate the design.

12. **Treating Depression** Researchers wanted to test whether a new drug therapy results in a more rapid response in patients with major depression. To do this, they recruited 63 inpatients with a diagnosis of major depression. Patients were randomly assigned to two treatment groups receiving either placebo (31 patients) or the new drug therapy (32 patients). The response variable was the Hamilton Rating Scale for Depression score. After collecting and analyzing the data, it was concluded that the new drug-therapy is effective in the treatment of major depression.

    (*Source:* "Metyrapone as Additive Treatment in Major Depression," Holger Jahn, MD; et.al.; *Archives of General Psychiatry*, 61(2004): 1235–1244)

    (a) What type of experimental design is this?
    (b) What is the population that is being studied?
    (c) What is the response variable in this study?
    (d) What is the factor? What are the treatments?
    (e) Identify the experimental units.
    (f) Draw a diagram similar to Figure 6, 7, or 9 to illustrate the design.

13. **Dominant Hand** Professor Andy Neill wanted to determine if the reaction time in people differs in their dominant hand versus their nondominant hand. To do this, he recruited 15 students. Each student was asked to hold a yardstick between the index finger and thumb. The student was asked to open the hand, release the yardstick, and then asked to catch the yardstick between the index finger and thumb. The distance that the yardstick fell served as a measure of reaction time. A coin flip was used to determine whether the student would use their dominant hand first or the nondominant hand. Results indicated that the reaction time in the dominant hand exceeded that of the nondominant hand.

    (a) What type of experimental design is this?
    (b) What is the response variable in this study?
    (c) What is the factor? What is the treatment?
    (d) Identify the experimental units.
    (e) Why did Professor Neill use a coin flip to determine whether the student should begin with the dominant hand or the nondominant hand?
    (f) Draw a diagram similar to Figure 6, 7, or 9 to illustrate the design.

14. **Golf Anyone?** A local golf pro wanted to compare two styles of golf club. One golf club had a graphite shaft and the other had the latest style of steel shaft. It is a common belief that graphite shafts allow a player to hit the ball farther, but the manufacturer of the new steel shaft said the ball travels just as far with its new technology. To test this claim, the pro recruited 10 golfers from the driving range. Each player was asked to hit one ball with the graphite-shafted club and one ball with the new steel-shafted club.

The distance that the ball traveled was determined using a range finder. A coin flip was used to determine whether the player hit with the graphite club or the steel club first. Results indicated that the distance that hit the ball was hit with the graphite club was no different than the distance when using the steel club.

(a) What type of experimental design is this?
(b) What is the response variable in this study?
(c) What is the factor? What is the treatment?
(d) Identify the experimental units.
(e) Why did the golf pro use a coin flip to determine whether the golfer should hit with the graphite first or the steel first?
(f) Draw a diagram similar to Figure 6, 7, or 9 to illustrate the design.

15. Marketing A marketing research firm wishes to determine the most effective method of advertising: print, radio, or television. They recruit 300 volunteers to participate in the study. The chief researcher believes that level of education plays a role in the effectiveness of advertising, so she segments the volunteers by level of education. Of the 300 volunteers, 120 have a high school education, 120 have a college diploma, and 60 have advanced degrees. The 120 volunteers with a high school diploma are randomly assigned to either the print advertising group, the radio group, or the television group. The same procedure is followed for the college graduate and advanced-degree volunteers. Each group is exposed to the advertising. After 1 hour, a recall exam is given with the proportion of correct answers recorded.

(a) What type of experimental design is this experiment?
(b) What is the response variable in this experiment?
(c) What factor is manipulated? How many treatments are there?
(d) What variable serves as the block?
(e) Draw a diagram similar to Figure 6, 7, or 9 to illustrate the design.

16. Social Work A social worker wants to examine methods that can be used to deter truancy. Three hundred chronically truant students from District 103 volunteer for the study. Because the social worker believes that socioeconomic class plays a role in truancy, she divides the 300 volunteers according to household income. Of the 300 students, 120 fall in the low-income category, 132 fall in the middle-income category, and the remaining 48 fall in the upper-income category. The students within each income category are randomly divided into three groups. The students in group 1 receive no intervention. The students in group 2 are treated with positive reinforcement in which for each day the student is not truant he or she receives a star that can be traded in for rewards. The students in group 3 are treated with negative reinforcement such that each truancy results in a 1-hour detention. However, the hours of detention are cumulative, meaning that the first truancy results in 1 hour of detention, the second truancy results in 2 hours, and so on. After a full school year, the total number of truancies are compared.

(a) What type of experimental design is used in this experiment?
(b) What is the response variable in this experiment?
(c) What factor is manipulated? How many treatments are there?
(d) What variable serves as the block?
(e) Draw a diagram similar to Figure 6, 7, or 9 to illustrate the design.

17. Drug Effectiveness A pharmaceutical company wants to test the effectiveness of an experimental drug meant to reduce high cholesterol. The researcher at the pharmaceutical company has decided to test the effectiveness of the drug through a completely randomized design. She has obtained 20 volunteers with high cholesterol: Ann, John, Michael, Kevin, Marissa, Christina, Eddie, Shannon, Julia, Randy, Sue, Tom, Wanda, Roger, Laurie, Rick, Kim, Joe, Colleen, and Bill. Number the volunteers from 1 to 20. Use a random-number generator to randomly assign 10 of the volunteers to the experimental group. The remaining volunteers will go into the control group. List the individuals in each group.

18. Effects of Alcohol A researcher has recruited 20 volunteers to participate in a study. The researcher wishes to measure the effect of alcohol on an individual's reaction time. The 20 volunteers are randomly divided into two groups. Group 1 will serve as a control group in which participants drink four 1-ounce cups of a liquid that looks, smells, and tastes like alcohol in 15-minute increments. Group 2 will serve as an experimental group in which participants drink four 1-ounce cups of 80-proof alcohol in 15-minute increments. After drinking the last 1-ounce cup, the participants sit for 20 minutes. After the 20-minute resting period, the reaction time to a stimulus is measured.

(a) What type of experimental design is this?
(b) Use Table I in Appendix A or a random-number generator to divide the 20 volunteers into groups 1 and 2 by assigning the volunteers a number between 1 and 20. Then randomly select 10 numbers between 1 and 20. The individuals corresponding to these numbers will go into group 1.

19. Tomatoes An oncologist wants to perform a long-term study on the benefits of eating tomatoes. In particular, she wishes to determine whether there is a significant difference in the rate of prostate cancer among adult males after eating one serving of tomatoes per week for 5 years, eating three servings of tomatoes per week for 5 years, and after eating five servings of tomatoes per week for 5 years. Help the oncologist design the experiment. Include a diagram to illustrate your design.

20. Batteries An engineer wants to determine the effect of temperature on battery voltage. In particular, he is interested in determining if there is a significant difference in the voltage of the batteries when exposed to temperatures of 90°F, 70°F, and 50°F. Help the engineer design the experiment. Include a diagram to illustrate your design.

21. **The Better Paint** Suppose you are interested in comparing Benjamin Moore's MoorLife Latex house paint with Sherwin Williams' LowTemp 35 Exterior Latex paint. Design an experiment that will answer this question: Which paint is better for painting the exterior of a house? Include a diagram to illustrate your design.

22. **Tire Design** An engineer has just developed a new tire design. However, before going into production, the tire company wants to determine if the new tire reduces braking distance on a car traveling 60 miles per hour compared with radial tires. Design an experiment to help the engineer determine if the new tire reduces braking distance.

23. **Octane** Does the level of octane in gasoline affect gas mileage? To answer this question, an automotive engineer obtains 60 cars. Twenty of the cars are considered to be compact, 20 are full size, and 20 are sport utility vehicles (SUVs). Design an experiment for the engineer. Include a diagram to illustrate your design.

24. **School Psychology** The school psychologist first presented in Problem 5 worries that girls and boys may react differently to the two methods of instruction. Design an experiment that will eliminate the variability due to gender on the response variable, the score on the achievement test. Include a diagram to illustrate your design.

25. **Designing an Experiment** Researchers wish to know if there is a link between hypertension (high blood pressure) and consumption of salt. Past studies have indicated that the consumption of fruits and vegetables offsets the negative impact of salt consumption. It is also known that there is quite a bit of person-to-person variability as far as the ability of the body to process and eliminate salt. However, no method exists for identifying individuals who have a higher ability to process salt. The U.S. Department of Agriculture recommends that daily intake of salt should not exceed 2400 mg. The researchers want to keep the design simple, so they choose to conduct their study using a completely randomized design.

   (a) What is the response variable in the study?
   (b) Name three factors that have been identified.
   (c) For each factor identified, determine whether the variable can be controlled, cannot be controlled, or should be manipulated. If a factor cannot be controlled, what should be done to reduce variability in the response variable?
   (d) How many treatments would you recommend? Why?

26. Search a newspaper, magazine, or other periodical that describes an experiment. Identify the population, experimental unit, response variable, treatment, factors, and their levels.

27. Research the *placebo effect* and the *Hawthorne effect*. Write a paragraph that describes how each affects the outcome of an experiment.

Consumer Reports® | Emotional "Aspirin"

Americans have a long history of altering their moods with chemicals ranging from alcohol and illicit drugs to prescription medications, such as diazepam (Valium) for anxiety and fluoxetine (Prozac) for depression. Today, there's a new trend: the over-the-counter availability of apparently effective mood modifiers in the form of herbs and other dietary supplements.

One problem is that many people who are treating themselves with these remedies may be sufficiently anxious or depressed to require professional care and monitoring. Self-treatment can be dangerous, particularly with depression, which causes some 20,000 reported suicides a year in the United States. Another major pitfall is that dietary supplements are largely unregulated by the government, so consumers have almost no protection against substandard preparations.

To help consumers and doctors, *Consumer Reports* tested the amounts of key ingredients in representative brands of several major mood-changing pills. To avoid potential bias, we tested samples from different lots of the pills using a randomized statistical design. The following table contains a subset of the data from this study.

Each of these pills has a label claim of 200 mg of SAM-E. The column labeled Random Code contains a set of 3-digit random codes that were used so that the laboratory did not know which manufacturer was being tested. The column labeled Mg SAM-E contains the amount of SAM-E measured by the laboratory.

(a) Why is it important to label the pills with random codes?

(b) Why is it important to randomize the order in which the pills are tested instead of testing all of brand A first, followed by all of brand B, and so on?

(c) Sort the data by brand. Does it appear that each brand is meeting its label claims?

(d) Design an experiment that follows the steps presented to answer the following research question: "Is there a difference in the amount of SAM-E contained in brands A, B, C, and D?"

Note to Readers: *In many cases, our test protocol and analytical methods are more complicated than described in this example. The data and discussion have been modified to make the material more appropriate for the audience.*

Run Order	Brand	Random Code	Mg SAM-E
1	B	461	238.9
2	D	992	219.2
3	C	962	227.1
4	A	305	231.2
5	B	835	263.7
6	D	717	251.1
7	A	206	232.9
8	D	649	192.8
9	C	132	213.4
10	B	923	224.6
11	A	823	261.1
12	C	515	207.8

Review

Summary

We defined statistics as a science in which data are collected, organized, summarized, and analyzed to infer characteristics regarding a population. Descriptive statistics consists of organizing and summarizing information, while inferential statistics consists of drawing conclusions about a population based on results obtained from a sample. The population is a collection of individuals on which the study is made, and the sample is a subset of the population.

Data are the observations of a variable. Data can be either qualitative or quantitative. Quantitative data are either discrete or continuous.

Data can be obtained from four sources: a census, existing sources, survey sampling, or a designed experiment. A census will list all the individuals in the population, along with certain characteristics. Due to the cost of obtaining a census, most researchers opt for obtaining a sample. In observational studies, the variable of interest has already been established. For this reason, they are often referred to as *ex post facto* studies. Designed experiments are used when control of the individuals in the study is desired to isolate the effect of a certain treatment on a response variable.

We introduced five sampling methods: simple random sampling, stratified sampling, systematic sampling, cluster sampling, and convenience sampling. All the sampling methods, except for convenience sampling, allow for unbiased statistical inference to be made. Convenience sampling typically leads to an unrepresentative sample and biased results.

Vocabulary

Be sure you can define the following:

Statistics (p. 3)
Data (pp. 3, 8)
Population (p. 4)
Individual (p. 4)
Sample (p. 4)
Descriptive statistics (p. 4)
Inferential statistics (p. 5)
Placebo (pp. 5, 40)
Double-blind (pp. 5, 40)
Variable (p. 6)
Qualitative variable (p. 6)
Quantitative variable (p. 6)
Discrete variable (p. 7)
Continuous variable (p. 7)
Qualitative data (p. 8)
Quantitative data (p. 8)
Discrete data (p. 8)

Continuous data (p. 8)
Census (p. 13)
Observational study (p. 14)
Designed experiment (pp. 14, 40)
Lurking variable (p. 15)
Simple random sampling (p. 16)
Frame (p. 17)
Sampling without replacement (p. 17)
Seed (p. 18)
Stratified sample (p. 23)
Systematic sample (p. 25)
Cluster sample (p. 26)
Convenience sample (p. 27)
Nonsampling error (p. 33)
Sampling error (p. 33)
Underrepresented (p. 34)
Nonresponse (p. 34)

Open question (p. 35)
Closed question (p. 35)
Designed experiment (p. 40)
Response variable (p. 40)
Experimental unit (p. 40)
Explanatory variable (p. 40)
Treatment (p. 40)
Claim (p. 41)
Factors (p. 41)
Replication (p. 41)
Completely randomized design (p. 42)
Matched-pairs design (p. 43)
Block (p. 44)
Confounding (p. 45)
Randomized block design (p. 45)

Objectives

Section	You should be able to . . .	Example	Review Exercises
1.1	**1** Define statistics and statistical thinking (p. 3)		1
	2 Understand the process of statistics (p. 4)	1	6
	3 Distinguish between qualitative and quantitative variables (p. 6)	2	9–14
	4 Distinguish between discrete and continuous variables (p. 7)	3, 4	9–14
1.2	**1** Distinguish between an observational study and an experiment (p. 14)	1	15–18
	2 Obtain a simple random sample (p. 16)	2, 3	25, 29
1.3	**1** Obtain a stratified sample (p. 23)	1	26
	2 Obtain a systematic sample (p. 25)	2	27
	3 Obtain a cluster sample (p. 26)	3	28
1.4	**1** Understand how error can be introduced during sampling (p. 33)		7
1.5	**1** Define designed experiment (p. 40)		5
	2 Understand the steps in designing an experiment (p. 41)		8, 37
	3 Understand the completely randomized design (p. 42)	1	31, 32, 37(a), 38
	4 Understand the matched-pairs design (p. 43)	2	30, 33, 34, 37(b), 39
	5 Understand the randomized block design (p. 44)	3	33, 38

Review Exercises

In Problems 1–5, provide a definition using your own words.

1. Statistics

2. Population

3. Sample

4. Observational study

5. Designed experiment

6. What is meant by the *process of statistics*?

7. State some sources of error in sampling. Provide some methods for correcting these errors. Distinguish sampling and nonsampling error.

8. Describe the components in an appropriately designed experiment.

In Problems 9–14, classify the variable as qualitative or quantitative. If the variable is quantitative, state whether it is discrete or continuous.

9. Number of new automobiles sold at a dealership on a given day

10. Species of tree sold at a nursery

11. Weight in carats of an uncut diamond

12. Wait time for an amusement park ride

13. Brand name of a pair of running shoes

14. Frequency with which the variable x is used in an algebra book

In Problems 15–18, determine whether the study depicts an observational study or a designed experiment.

15. A parent group examines 25 recently released PG-13 movies and records the number of sexual innuendos and curse words that occur in each.

16. A sample of 10 new Dodge Neons is randomly divided into two groups. In the first group of cars, SAE 5W-30 motor oil is used. In the second group, SAE 10W-30 motor oil is used. All other variables are controlled to be the same in both groups. Each car is driven 3,000 highway miles. The gasoline mileages of the two groups are then compared.

17. A sample of 504 patients in early stages of Alzheimer's disease is divided into two groups. One group receives an experimental drug regimen while the other receives a placebo regimen. The advance of the disease in the patients from the two groups is tracked at 1-month intervals over the next year.

18. A sample of 1,000 registered voters is asked, "Do you approve or disapprove of the president's overall job performance?"

In Problems 19–22, determine the type of sampling used.

19. On election day, a pollster for Fox News positions herself outside a polling place near her home. She then asks the first 50 voters leaving the facility to complete a survey.

20. The manager of an automobile repair shop compiles a list of all its customers from the past 6 months. He assigns a unique number to each customer and uses a random-number generator to create a list of 40 unique numbers. The customers corresponding to the number are surveyed about satisfaction with services.

21. An Internet service provider randomly selects 15 residential blocks from a large city. It then surveys every household in those 15 blocks to determine the number that would use a high-speed Internet service if it were made available.

22. A hotel manager divides the population of customers into four groups depending on length of stay: 1 night, 2 to 3 nights, 4 to 5 nights, and 6 or more nights. She obtains a simple random sample from each group and conducts a survey on satisfaction.

23. For each error given, identify the nonsampling error.
    (a) A politician sends a survey about tax issues to a random sample of subscribers to a literary magazine.
    (b) An interviewer with little foreign language knowledge is sent to an area where her language is not commonly spoken.
    (c) A data-entry clerk mistypes survey results into his computer.

24. For each error given, identify the nonsampling error.
    (a) A questionnaire given to students asks, "Are you doing well in school?"
    (b) Of 1,240 surveys sent out in a study, 17 are returned.
    (c) A warehouse manager doing a study on job commitment surveys the first 20 workers who come in on Monday morning.

25. Obtaining a Simple Random Sample The mayor of a small town wants to conduct personal interviews with small business owners to determine if there is anything the mayor could to do to help improve business conditions. The following list gives the names of the companies in the town. Obtain a simple random sample of size 5 from the companies in the town.

Allied Tube and Conduit	Lighthouse Financial	Senese's Winery
Bechstien Construction Co.	Mill Creek Animal Clinic	Skyline Laboratory
Cizer Trucking Co.	Nancy's Flowers	Solus, Maria, DDS
D & M Welding	Norm's Jewelry	Trust Lock and Key
Grace Cleaning Service	Papoose Children's Center	Ultimate Carpet
Jiffy Lube	Plaza Inn Motel	Waterfront Tavern
Levin, Thomas, M.D.	Risky Business Security	WPA Pharmacy

26. Obtaining a Stratified Sample A congresswoman wants to survey her constituency regarding public policy. She asks one of her staff members to obtain a sample of residents of the district. The frame she has available lists 9,012 Democrats, 8,302 Republicans, and 3,012 Independents. Obtain a stratified random sample of 8 Democrats, 7 Republicans, and 3 Independents.

27. Obtaining a Systematic Sample A quality-control engineer wants to be sure that bolts coming off an assembly line are within prescribed tolerances. He wants to conduct a systematic sample by selecting every 9th bolt to come off the assembly line. The machine produces 30,000 bolts per day, and the engineer wants a sample of 32 bolts. Which bolts will be sampled?

28. Obtaining a Cluster Sample A farmer has a 500-acre orchard in Florida. Each acre is subdivided into blocks of 5. Altogether, there are 2500 blocks of trees on the farm. After a frost, he wants to get an idea of the extent of the damage. Obtain a sample of 10 blocks of trees using a cluster sample.

29. Obtaining a Simple Random Sample Based on the Military Standard 105E (ANS1/ASQC Z1.4, ISO 2859) Tables, a lot of 91 to 150 items with an acceptable quality level (AQL) of 1% and normal inspection plan would require a sample of size 13 to be inspected for defects. If the sample contains no defects, the entire lot is accepted. Otherwise, the entire lot is rejected. A shipment of 100 night vision goggles is received and must be inspected. Discuss the procedure you would follow to obtain a simple random sample of 13 goggles to inspect.

30. Effects of Music To study the effects of music on sleep quality in adults 18 years of age or older, 100 adults were given the Pittsburgh Sleep Quality Index (PSQI) and then asked to listen to 30 minutes of music at bedtime for 3 weeks. At the end of the 3-week study, subjects were given the PSQI again and the change in score was reported.

(a) What type of experimental design in this?
(b) What is the population that is being studied?
(c) What is the factor?
(d) What is the response variable?
(e) Identify the experimental units.
(f) Draw a diagram similar to Figure 6, 7, or 9 to illustrate the design.

31. Integrated Circuits An integrated circuit manufacturer has developed a new circuit for solid-state ballasts that will be more energy efficient than the circuit they currently manufacture. The company manufactures 100 of each type of circuit and randomly assigns them to be used in the ballast of fluorescent bulbs. The amount of energy required to light each bulb is measured.

(a) What type of experimental design is this?
(b) What is the response variable in this experiment?
(c) What is the factor? What are the treatments?
(d) Identify the experimental units.
(e) Draw a diagram similar to Figure 6, 7, or 9 to illustrate the design.

32. Atopic Dermatitis In September 2004, Nucryst Pharmaceuticals, Inc., announced the results of its first human trial of NPI 32101, a topical form of its skin ointment. A total of 224 patients diagnosed with skin irritations were randomly divided into three groups as part of a double-blind, placebo-controlled study to test the effectiveness of the new topical cream. The first group received a 0.5% cream, the second group received a 1.0% cream, and the third group received a placebo. Groups were treated twice daily for a 6-week period.

(*Source:* www.nucryst.com)

(a) What type of experimental design is this?
(b) What is the response variable in this experiment?
(c) What is the factor? What are the treatments?
(d) Why do you think it is important for this study to be double-blind?
(e) Identify the experimental units.

(f) Draw a diagram similar to Figure 6, 7, or 9 to illustrate the design.

33. **Exam Grades** A statistics instructor wants to see if allowing students to use a notecard on exams will affect their overall exam scores. The instructor thinks that the results could be affected by the type of class: traditional, web-enhanced, or online. He randomly selects half of his students from each type of class to use notecards. At the end of the semester, the average exam grades for those using notecards is compared to the average for those without notecards.
    (a) What type of experimental design is this?
    (b) What is the response variable in this experiment?
    (c) What is the factor? What are the treatments?
    (d) Identify the experimental units.
    (e) Draw a diagram similar to Figure 6, 7, or 9 to illustrate the design.

34. **Skylab** The four members of Skylab had their lymphocyte count per cubic millimeter measured 1 day before liftoff and measured again on their return to Earth.
    (a) What is the response variable in this experiment?
    (b) What is the treatment?
    (c) What type of experimental design is this?
    (d) Identify the experimental units.
    (e) Draw a diagram similar to Figure 6, 7, or 9 to illustrate the design.

35. **Multiple Choice** A common tip for taking multiple-choice tests is to always pick (b) or (c) if you are unsure. The idea is that instructors tend to feel the answer is more hidden if it is surrounded by distractor answers. An astute statistics instructor is aware of this and decides to use a table of random digits to select which choice will be the correct answer. If each question has five choices, use Table I in Appendix A or a random-number generator to determine the correct answers for a 20-question multiple-choice exam.

36. **Do Not Call** What percent of students on campus have their phone number (either home or cell) on the National Do Not Call Registry? Polling every student on campus is inefficient and would require a fair amount of resources. Design and implement a sampling method to obtain a sample of 50 students at your college or university. Ask each student whether he or she has registered a phone with the National Do Not Call Registry, and use the results to estimate the percentage of all students on campus who have registered with the Registry. Compare your sampling method and results with others in the class. Discuss similarities and differences, as well as the advantages or disadvantages, of your approach.

37. **Humor in Advertising** A marketing research firm wants to know whether information presented in a commercial is better recalled when presented using humor or serious commentary in adults between 18 and 35 years of age. They will use an exam that asks questions about information presented in the ad. The response variable will be percentage of information recalled.
    (a) Create a completely randomized design to answer the question. Be sure to include a diagram to illustrate your design.
    (b) The chief of advertising criticizes your design because she knows that women react differently from men to advertising. Redesign the experiment as a randomized block design assuming that 30 of the subjects in your study are female and the remaining 20 are male. Be sure to include a diagram to illustrate your design.

38. Describe what is meant by an experiment that is a completely randomized design. Contrast this experimental design with a randomized block design.

39. Describe what is meant by a matched-pairs design. Contrast this experimental design with a completely randomized design.

40. **The Better Design?** Suppose a psychologist wants to determine whether Monster Energy Drinks improve performance in sports. One measure of performance might be reaction time.
    (a) Design a completely randomized design to help the psychologist answer her question.
    (b) Design a matched-pairs design to help the psychologist answer her question.
    (c) In your opinion, which design is better? Justify your opinion.

Chrysalises for Cash

The colorful butterfly symbolizes the notion of personal change. Increasingly, people are turning to butterflies to consecrate meaningful events such as birthdays, weddings, and funerals. To fill this need, a new industry has hatched.

Suppliers of butterflies are closely monitored by governmental agencies to ensure that local environments are not subjected to the introduction of invasive species. In addition to following these regulations, butterfly suppliers need to ensure the quality and quantity of their product, while maintaining a profit. To this end, an individual supplier may hire a number of smaller independent contractors to hatch the varieties needed. These entrepreneurs are paid a small fee for each chrysalis delivered, with a 50% bonus added for each hatched healthy butterfly. This fee structure provides little room for profit. Therefore, it is important that these small contractors deliver a high proportion of healthy butterflies that emerge at a fairly predictable rate.

In Florida, one such entrepreneur specializes in harvesting the black swallowtail butterfly. In the southern United States, the black swallowtail has at least three broods a year. The female flutters through open fields seeking plants of the Apiaceae family, such as carrot and parsley, upon which to lay her eggs. The resulting catepillars are dark brown with a small white saddle mark. As the catepillars consume the leaves of their host plant, they increase in size, changing color to a vibrant green with intermittent stripes of yellow and black. Once a catepillar has eaten enough, it secures itself and sheds its skin, revealing an emerald chrysalis. During this resting phase, environmental factors such as temperature and humidity may affect the transformation process. Typically, the black swallowtail takes about 1 week to complete its metamorphosis and emerge from its chrysali. The transformation occasionally results in a deformed butterfly. Deformities range from wings that will not fully open to missing limbs.

The Florida contractor believes that there are differences in quality and emergence time among his broods. Not having taken a scientific approach to the problem, he relies on his memory of seasons past. It seems to him that late-season butterflies emerge sooner and with a greater number of deformities than their early-season counterparts. He also speculates that the type and nutritional value of the food consumed by the catepillar might contribute to any observed differences. This year he is committed to a more formal approach to his butterfly harvest.

Since it takes 2 days to deliver the chrysalises from the contractor to the supplier, it is important that the butterflies do not emerge prematurely. It is equally important that the number of defective butterflies be minimized. With these two goals in mind, the contractor seeks the best combination of food source, fertilizer, and brood season to maximize his profits. To examine the effects of these variables on emergence time and number of deformed butterflies, the entrepreneur designed the following experiment.

Eight identical pots were filled with equal amounts of a particular soil mixture. The watering of all the pots and the plants contained within them was carefully monitored for consistency. Four pots were set outside during the early part of the brood season. Of these, two contained carrot plants, while the remaining pair grew parsley. For the carrot pair, one pot was fed a fixed amount of liquid fertilizer, while the other pot was fed a nutritionally similar amount of solid fertilizer. The two pots containing parsley were similarly fertilized. All four pots were placed next to each other to ensure similar exposures to environmental conditions such as temperature and solar radiation. Five black swallowtail caterpillars of similar age were placed into each container. The caterpillars were allowed to mature and form a chrysalis. The time from chrysalis formation until emergence was reported to the nearest day. The occurrence of any defects was also noted. The same procedure was followed with the four pots that were placed outdoors during the late brood season.

Write a report describing the experimental goals and design for the entrepreneur's experiment. Follow the procedure outlined in the box on the steps in designing and conducting an experiment (p. 41). Step 5(b), *collect and process the data*, of this procedure is provided in the following table and should be included in your report.

Florida Black Swallowtail Chrysalis Experiment Data

Season	Food	Fertilizer	Number Deformed	Emergence Time (Days)
Early	Parsley	Solid	0	6,6,7,7,7
Early	Parsley	Liquid	0	6,7,7,8,8
Early	Carrot	Solid	1	3,6,6,7,8
Early	Carrot	Liquid	0	6,6,7,8,8
Late	Parsley	Solid	2	2,3,4,4,5
Late	Parsley	Liquid	1	2,3,4,5,5
Late	Carrot	Solid	2	3,3,3,4,5
Late	Carrot	Liquid	0	2,4,4,4,5

In your report, provide a general descriptive analysis of these data. Be sure to include recommendations for the combination of season, food source, and type of fertilizer that results in the fewest deformed butterflies while achieving a long emergence time.

Conclude your report with recommendations for further experiments. For each proposed experiment, be sure to do the following:

1. State the problem to be solved and define the response variables.
2. Define the factors that affect the response variables.
3. State the number of experimental units.
4. State the level of each factor.

PART 2

Descriptive Statistics

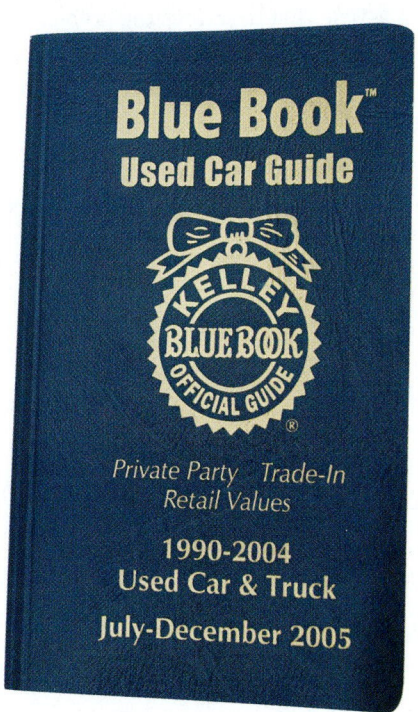

Remember, statistics is a process. The first chapter (Part I) dealt with the first two steps in the statistical process: (1) identify the research objective and (2) collect the information needed to answer the questions in the research objective. The next three chapters deal with organizing, summarizing, and presenting the data collected. This step in the process is called descriptive statistics.

Organizing and Summarizing Data

Outline

DECISIONS

Suppose that you work for the school newspaper. Your editor approaches you with a special reporting assignment. Your task is to write an article that describes the "typical" student at your school, complete with supporting information. How are you going to do this assignment? See the Decisions project on page 107.

●●● Putting It All Together

Chapter 1 discussed how to collect data. We learned that data can be obtained from four sources: (1) a census, (2) existing data sets, (3) surveys, or (4) designed experiments. When data are obtained, they are referred to as **raw data**. Raw data must be organized into a meaningful form.

Methods for organizing raw data include the creation of tables or graphs, which allow for a quick overview of the information collected. Describing data is the third step in the statistical process. The procedures used in this step depend on whether the data are qualitative, discrete, or continuous.

2.1 Organizing Qualitative Data

Preparing for This Section   Before getting started, review the following:

• Qualitative data (Section 1.1, p. 6)

Objectives   **1** **Organize qualitative data in tables**
2 **Construct bar graphs**
3 **Construct pie charts**

In this section we will concentrate on tabular and graphical summaries of qualitative data. Sections 2.2 and 2.3 discuss methods for summarizing quantitative data.

1 **Organize Qualitative Data in Tables**

In Other Words
Qualitative data is also known as catagorical data.

Recall that qualitative data provide measures that categorize or classify an individual. When qualitative data are collected, we are often interested in determining the number of individuals observed within each category.

Definition   A **frequency distribution** lists each category of data and the number of occurrences for each category of data.

EXAMPLE 1 | **Organizing Qualitative Data into a Frequency Distribution**

Problem: A physical therapist wants to get a sense of the types of rehabilitation required by her patients. To do so, she obtains a simple random sample of 30 of her patients and records the body part requiring rehabilitation. See Table 1. Construct a frequency distribution of location of injury.

Table 1					
Back	Back	Hand	Neck	Knee	Knee
Wrist	Back	Groin	Shoulder	Shoulder	Back
Elbow	Back	Back	Back	Back	Back
Back	Shoulder	Shoulder	Knee	Knee	Back
Hip	Knee	Hip	Hand	Back	Wrist

Source: Krystal Catton, student at Joliet Junior College

Approach: To construct a frequency distribution, we create a list of the body parts (categories) and tally each occurrence. Finally, we add up the number of tallies to determine the frequency.

Solution: See Table 2. From the table, we can see that the back is the most common body part requiring rehabilitation, with a total of 12.

Table 2		
Body Part	**Tally**	**Frequency**
Back	ⲐⲐⲐ ⲐⲐⲐ ||	12
Wrist	||	2
Elbow			1
Hip	||	2
Shoulder	||||	4
Knee	ⲐⲐⲐ	5
Hand	||	2
Groin			1
Neck			1

CAUTION
The data in Table 2 are still qualitative. The frequency simply represents the count of each category.

With frequency distributions, it is a good idea to add up the frequency column to make sure that it sums to the number of observations. In the case of the data in Example 1, the frequency column adds up to 30, as it should.

Often, rather than being concerned with the frequency with which categories of data occur, we want to know the *relative frequency* of the categories.

Definition

In Other Words

A frequency distribution shows how many observations are in each category. A relative frequency distribution shows the proportion of observations that belong in each category.

The **relative frequency** is the proportion (or percent) of observations within a category and is found using the formula

$$\text{Relative frequency} = \frac{\text{frequency}}{\text{sum of all frequencies}} \quad \textbf{(1)}$$

A **relative frequency distribution** lists each category of data together with the relative frequency.

EXAMPLE 2   **Constructing a Relative Frequency Distribution of Qualitative Data**

Problem: Using the data in Table 2, construct a relative frequency distribution.

Approach: Add all the frequencies, and then use Formula (1) to compute the relative frequency of each category of data.

Solution: We add the values in the frequency column in Table 2:

    Sum of all frequencies = 12 + 2 + 1 + 2 + 4 + 5 + 2 + 1 + 1 = 30

We now compute the relative frequency of each category. For example, the relative frequency of the category Back is

$$\frac{12}{30} = 0.4$$

After computing the relative frequency for the remaining categories, we obtain the relative frequency distribution shown in Table 3.

Table 3		
Body Part	**Frequency**	**Relative Frequency**
Back	12	$\frac{12}{30} = 0.4$
Wrist	2	$\frac{2}{30} \approx 0.0667$
Elbow	1	0.0333
Hip	2	0.0667
Shoulder	4	0.1333
Knee	5	0.1667
Hand	2	0.0667
Groin	1	0.0333
Neck	1	0.0333

From the table, we can see that the most common body part for rehabilitation is the back.

USING TECHNOLOGY

Some statistical spreadsheets such as MINITAB have a `Tally` command. This command will construct a frequency and relative frequency distribution of raw qualitative data.

It is a good idea to add up the entries in the relative frequency column to be sure they sum to 1. Sometimes the sum of the relative frequencies will differ slightly from 1 due to rounding. Nonetheless, the check is a good idea.

Now Work Problems 27(a) and (b).

2 Construct Bar Graphs

Once raw data are organized in a table, we can create graphs. Creating graphs allows us to see the data and get a sense of what the data are saying about the individuals in the study. In general, pictures of data result in a more powerful message than tables. Try the following exercise for yourself: Open a newspaper and look at a table and graph. Study each. Now put the paper away and close your eyes. What do you see in your mind's eye? Can you recall information obtained from the table or the graph? In general, people are more likely to recall information obtained from a graph than they are from a table.

One of the most common devices for graphically representing qualitative data is a bar graph.

Definition

> A **bar graph** is constructed by labeling each category of data on a horizontal axis and the frequency or relative frequency of the category on the vertical axis. Rectangles of equal width are drawn for each category. The height of each rectangle is equal to the category's frequency or relative frequency.

EXAMPLE 3

Constructing a Frequency and Relative Frequency Bar Graph

Problem: Use the data summarized in Table 3 to construct the following:

(a) Frequency bar graph
(b) Relative frequency bar graph

Approach: A horizontal axis is used to indicate the categories of the data (body parts, in this case), and a vertical axis is used to represent the frequency or relative frequency. Draw rectangles of equal width to the height that is the frequency or relative frequency for each category. The bars do not touch each other.

Solution

(a) Figure 1(a) shows the frequency bar graph.
(b) Figure 1(b) shows the relative frequency bar graph.

Figure 1

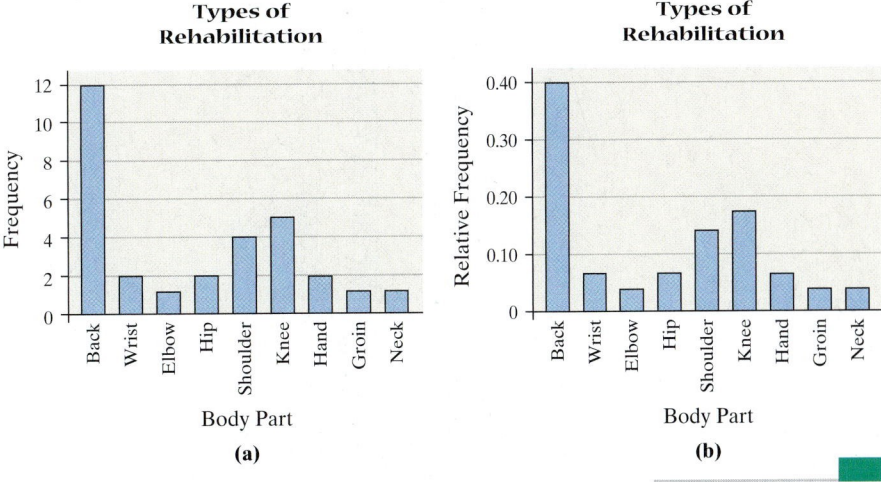

(a)

(b)

⚠ CAUTION
Watch out for graphs that start the scale at some value other than 0, have bars with unequal widths, or have bars with different colors, because they can misrepresent the data.

EXAMPLE 4

Constructing a Frequency or Relative Frequency Bar Graph Using Technology

Problem: Use a statistical spreadsheet to construct a frequency or relative frequency bar graph.

Approach: We will use Excel to construct the frequency and relative frequency bar graph. The steps for constructing the graphs using MINITAB or Excel are given in the Technology Step by Step on page 76. **Note:** The TI-83 and TI-84 Plus graphing calculators cannot draw frequency or relative frequency bar graphs.

Result: Figure 2(a) shows the frequency bar graph and Figure 2(b) shows the relative frequency bar graph obtained from Excel.

Figure 2

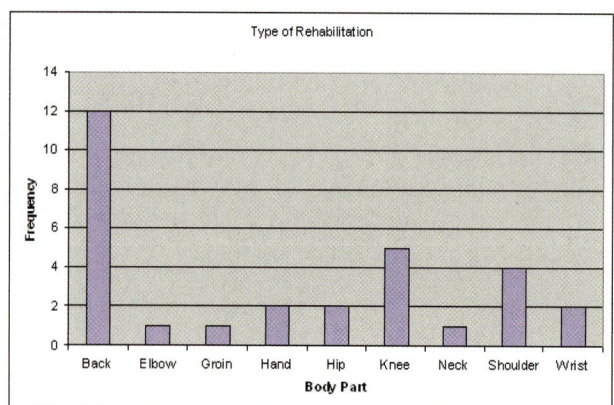
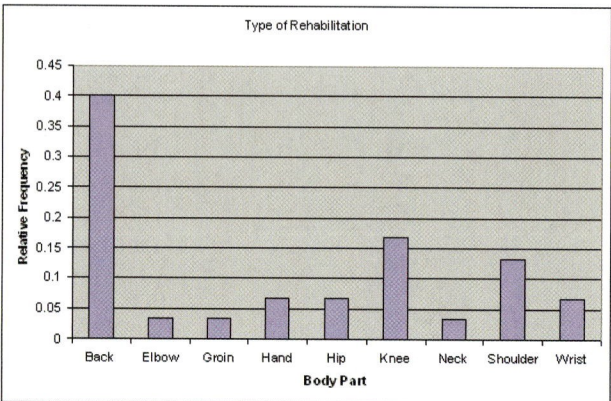

Now Work Problems 27(c) and (d).

Some statisticians prefer to create bar graphs with the categories arranged in decreasing order of frequency.

Definition   A **Pareto chart** is a bar graph whose bars are drawn in decreasing order of frequency or relative frequency.

Figure 3 illustrates a relative frequency Pareto chart for the data in Table 3.

Figure 3

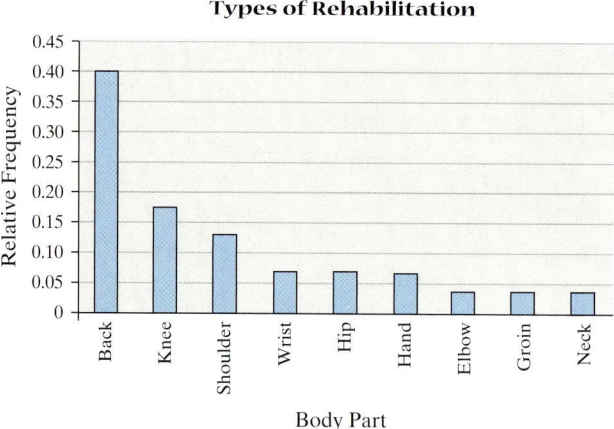

USING TECHNOLOGY

The graphs obtained from a different statistical package should not differ from those in Figure 2. Some packages use the word *count* in place of *frequency* or *percent* in place of *relative frequency*, however.

Side-by-Side Bar Graphs

Graphics provide insight when you are comparing two sets of data. For example, suppose we wanted to know if more people are finishing college today than in 1990. We could draw a **side-by-side bar graph** to compare the two data sets. Data sets should be compared by using relative frequencies, because different sample or population sizes make comparisons using frequencies difficult or misleading.

| EXAMPLE 5 | **Comparing Two Data Sets** |

Problem: The data in Table 4 represent the educational attainment in 1990 and 2003 of adults 25 years and older who are residents of the United States. The data are in thousands. So 16,502 represents 16,502,000.

(a) Draw a side-by-side relative frequency bar graph of the data.
(b) Are a greater proportion of Americans earning bachelor's degrees?

Table 4

Educational Attainment	1990	2003
Less than 9th grade	16,502	12,276
9th to 12th grade, no diploma	22,842	16,323
High school diploma	47,643	59,292
Some college, no degree	29,780	31,762
Associate's degree	9,792	15,147
Bachelor's degree	20,833	33,213
Graduate or professional degree	11,478	17,169
Totals	158,870	185,182

Source: U.S. Census Bureau

Approach: First, we determine the relative frequencies of each category for each year. To construct the side-by-side bar graphs, we draw two bars for each category of data. One of the bars will represent 1990 and the other will represent 2003.

Solution: Table 5 shows the relative frequency for each category.

(a) The side-by-side bar graph is shown in Figure 4.

Table 5

Educational Attainment	1990	2003
Less than 9th grade	0.1039	0.0663
9th to 12th grade, no diploma	0.1438	0.0881
High school diploma	0.2999	0.3202
Some college, no degree	0.1874	0.1715
Associate's degree	0.0616	0.0818
Bachelor's degree	0.1311	0.1794
Graduate or professional degree	0.0722	0.0927

Figure 4   **Educational Attainment in 1990 versus 2003**

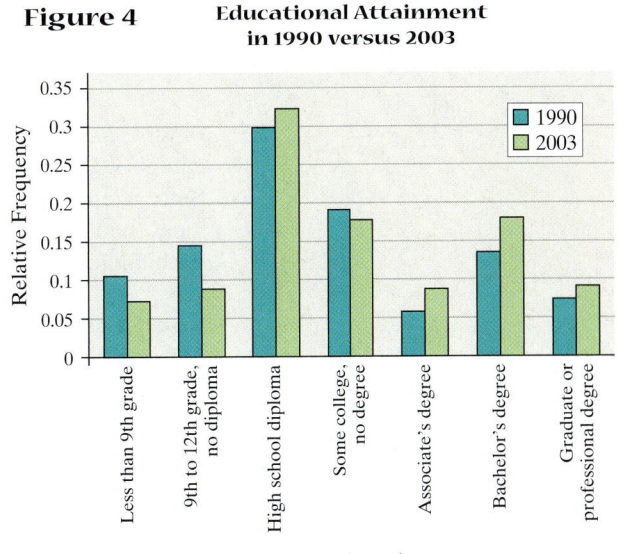

(b) From the graph, we can see that the proportion of Americans 25 years and older who earned a bachelor's degree is higher in 2003. This information is not clear from the frequency table, because the sizes of the populations are different. Increases in the number of Americans who earned a bachelor's degree are due partly to the increases in the sizes of the populations.

Now Work Problem 23.

3 Construct Pie Charts

Pie charts are typically used to present the relative frequency of qualitative data.

Definition

> A **pie chart** is a circle divided into sectors. Each sector represents a category of data. The area of each sector is proportional to the frequency of the category.

EXAMPLE 6 Constructing a Pie Chart

Problem: The data presented in Table 6 represent the educational attainment of residents of the United States 25 years or older in 2003, based on data obtained from the U.S. Census Bureau. The data are in thousands. Construct a pie chart of the data.

Table 6

Educational Attainment	2003
Less than 9th grade	12,276
9th to 12th grade, no diploma	16,323
High school diploma	59,292
Some college, no degree	31,762
Associate's degree	15,147
Bachelor's degree	33,213
Graduate or professional degree	17,169
Totals	**185,182**

Approach: The pie chart will have seven parts, or sectors, corresponding to the seven categories of data. The area of each sector is proportional to the frequency of each category. For example,

$$\frac{12{,}276}{185{,}182} = 0.0663$$

of all U.S. residents 25 years or older have less than a 9th-grade education. The category "less than 9th grade" will make up 6.63% of the area of the pie chart. Since a circle has 360 degrees, the degree measure of the sector for the category "less than 9th-grade" will be $(0.0663)360° \approx 24°$. Use a protractor to measure each angle.

Solution: We follow the approach presented for the remaining categories of data to obtain Table 7.

Table 7

Education	Frequency	Relative Frequency	Degree Measure of Each Sector
Less than 9th grade	12,276	0.0663	24
9th to 12th grade, no diploma	16,323	0.0881	32
High school diploma	59,292	0.3202	115
Some college, no degree	31,762	0.1715	62
Associate's degree	15,147	0.0818	29
Bachelor's degree	33,213	0.1794	65
Graduate or professional degree	17,169	0.0927	33

USING TECHNOLOGY

Most statistical spreadsheets are capable of drawing pie charts. See the Technology Step by Step on page 76 for instructions on how to obtain a pie chart using MINITAB or Excel. The TI-83 and TI-84 Plus graphing calculators do not draw pie charts.

Figure 5

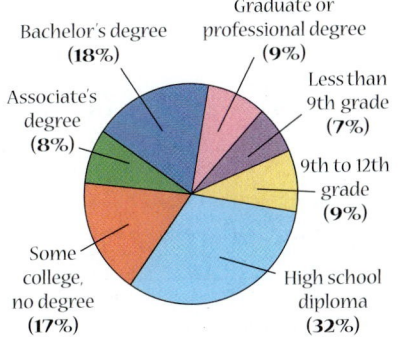

Educational Attainment, 2003

Bachelor's degree (18%)
Graduate or professional degree (9%)
Associate's degree (8%)
Less than 9th grade (7%)
9th to 12th grade (9%)
Some college, no degree (17%)
High school diploma (32%)

To construct a pie chart by hand, we use a protractor to approximate the angles for each sector. See Figure 5.

Pie charts can be created only if all the categories of the variable under consideration are represented. For example, from the data given in Example 1, we could create a bar graph that lists the proportion of patients requiring rehabilitation on their back, shoulder or knee only, but it would not make sense to construct a pie chart for this situation. Do you see why? Only 70% of the data would be represented.

When should a bar graph be used to display information? When should a pie chart be used? Pie charts are useful for showing the division all possible values of a qualitative variable into its parts. However, because angles are often hard to judge in pie charts, they are not as useful in comparing two specific values of the qualitative variable. Instead the emphasis is on comparing the part to the whole. Bar graphs are useful when we want to compare the different parts, not the parts to the whole. For example, if we wanted to get the "big picture" regarding educational attainment in 2003, then a pie chart is a good visual summary. However, if we want to compare bachelor's degrees to high school diplomas, then a bar graph is a good visual summary.

Now Work Problem 27(e).

2.1 ASSESS YOUR UNDERSTANDING

Concepts and Vocabulary

1. Define raw data in your own words.
2. A frequency distribution lists the _____ of occurrences of each category of data, while a relative frequency distribution lists the _____ of occurrences of each category of data.
3. When constructing a frequency distribution, why is it a good idea to add up the frequencies?
4. In a relative frequency distribution, what should the relative frequencies add up to?
5. What is a Pareto chart?
6. Why should relative frequencies be used when comparing two data sets?
7. Explain why Pareto charts might be preferred over bar graphs.

8. Consider the information in the "Why we can't lose weight" chart shown below which is in the *USA Today* style of graph. Could the information provided be organized into a pie chart? Why or why not.

Why we can't lose weight

63% Metabolism too slow
59% Don't exercise
50% Don't have self-discipline
49% Splurge on favorite foods

Skill Building

9. **Golf Driving Range** The pie chart at the right, the type we see in *USA Today*, depicts the bucket size golfers choose while at the driving range.

   (a) What is the most popular size? What percentage of golfers choose this size?
   (b) What is the least popular size? What percentage of golfers choose this size?
   (c) What percent of golfers choose a medium-sized bucket?

Most popular bucket sizes at the range

Large (52%)
X-Large (9%)
Small (16%)
Medium (23%)

10. Cosmetic Surgery The *USA Today*-type chart shows the most frequent cosmetic surgeries for women in 2003.

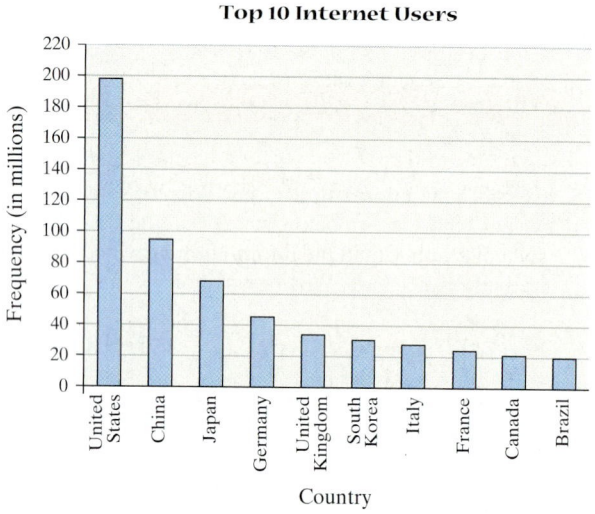

Top Cosmetic Surgeries for Women

Liposuction	21%
Breast augmentation	18%
Eyelid surgery	14%
Breast reduction	9%

(a) If women had 1,548,000 cosmetic surgeries in 2003, how many of them were for liposuction?
(b) How many were for eyelid surgery?
(c) What percentage of surgeries are not accounted for in the graph?

11. Internet Users The following Pareto chart represents the top 10 countries in Internet users as of February 2005. (*Source:* www.internetworldstats.com)

Top 10 Internet Users

(a) Which country has the most Internet users?
(b) Approximately what is the Internet usage in Canada?
(c) Approximately how many more users are in China than in Japan?

12. Poverty Every year the U.S. Census Bureau counts the number of people living in poverty. The bureau uses money income thresholds as its definition of poverty, so noncash benefits such as Medicaid and food stamps do not count toward poverty thresholds. For example, in 2002 the poverty threshold for a family of three was $14,348. The

bar chart represents the number of people living in poverty in the United States in 2003, by ethnicity, based on March 2004 estimates.

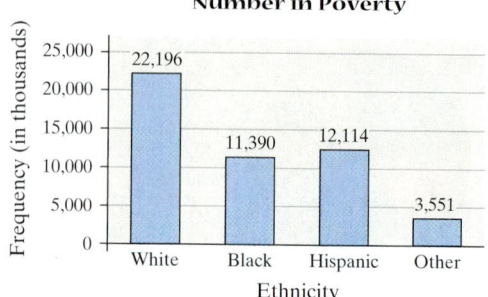

Number in Poverty

Source: The Henry Kaiser Family Foundation (www.statehealthfacts.kff.org)

(a) How many whites were living in poverty in 2003?
(b) Of the impoverished, what percent were Hispanic?
(c) How might this graph be misleading?

13. 2004 Presidential Election The following bar chart represents the number of voters who cast votes for the top five vote-getters and all others in the 2004 presidential election. The number of votes received for each candidate appears above the bar.

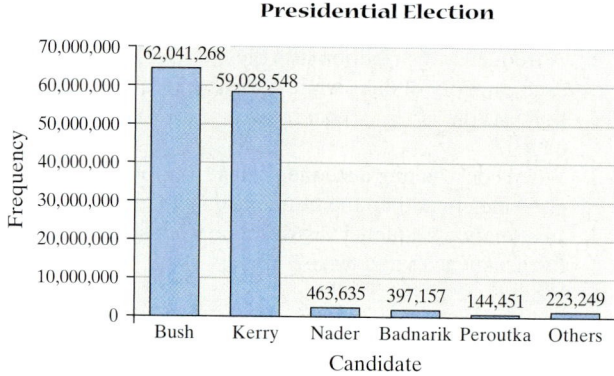

Votes in 2004 Presidential Election

Source: David Leip's Atlas of U.S. Presidential Elections (www.uselectionatlas.org)

(a) How many votes were cast for Michael Badnarik of the Libertarian Party?
(b) What percent of the votes was cast for George W. Bush? For John F. Kerry?
(c) To receive presidential election funds, a candidate must garner at least 5% of the vote. Will Ralph Nader receive presidential election funds?

14. **Identity Fraud** In a study conducted by the Better Business Bureau and Javelin Strategy and Research, victims of identity fraud were asked, "Who was the person who misused your personal information?" The following Pareto chart represents the results for cases in the year 2004 for which the perpetrator's identity was known.

Person Who Misused Personal Information

Source: Javelin Strategy & Research, 2004.

(a) Approximately what percentage of identity-fraud victims were victimized by a family member or relative?

(b) If there were 9.3 million identity-fraud victims in 2004, how many were victimized by someone at a company with access to personal information?

15. **Home Heating Fuel** The following side-by-side bar graph represents the proportion of households using four common sources of home heating fuel for the years 1978, 1987, 1997, and 2001.

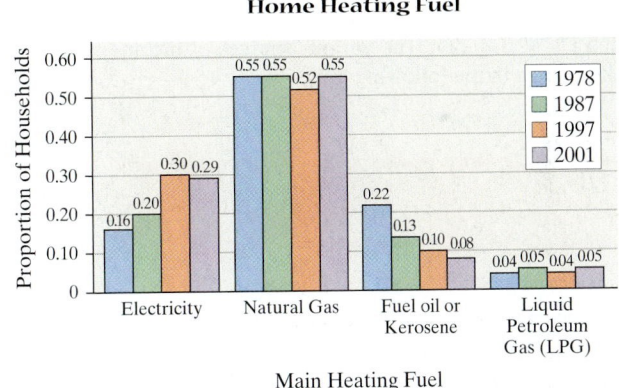

Home Heating Fuel

Source: Energy Information Administration, Residential Energy Consumption Survey, 1978, 1987, 1997, 2001

(a) What proportion of households used electricity as their main source of home heating fuel in 1978? in 1997?

(b) What was the most popular source of home heating fuel in 1987?

(c) If there were 107 million U.S. households in 2001, how many used LPG as their main source of home heating fuel?

(d) What might account for the rise in homes that use electricity as the main source of home heating fuel?

(e) Which sources of home heating fuel has been decreasing rather steadily? What might account for this?

(f) Which sources of home heating fuel have remained rather steady?

16. **Doctorate Recipients** The following side-by-side bar graph represents the number of doctorate recipients from U.S. universities within broad fields of study for the years 1983, 1993, and 2003.

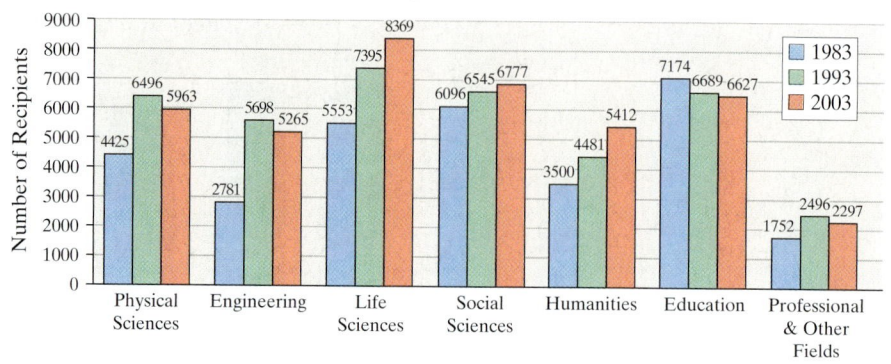

Doctorate Recipients from U.S. Universities

Source: NSF/NIH/USED/NEH/USDA/NASA, 2003 Survey of Earned Doctorates

(a) How many more engineering doctorates were awarded in 1993 than in 2003?
(b) In 2003, what percentage of doctoral recipients received degrees in physical science? in education?
(c) What field of study consistently decreased in the number of doctoral degree recipients in the 3 years of the survey?
(d) Which field of study had the largest increase in the number of doctoral degree recipients in the 3 years of the survey?

Applying the Concepts

17. **Government Income** For fiscal year 2003 (October 1, 2002 to September 30, 2003), the federal government's income was $1,782.3 billion. The various sources of income are broken down in the following table.

Source of Income	Amount (in billions of dollars)
Individual income taxes	793.7
Corporate income taxes	131.8
Social insurance taxes	713.0
Excise, estate and gift taxes, customs, and miscellaneous receipts	143.8

Source: Congressional Budget Office

(a) Construct a relative frequency distribution of the data shown.
(b) What percentage of total income is attributable to individual income taxes?
(c) Construct a frequency bar graph of the data.
(d) Construct a relative frequency bar graph of the data.
(e) Construct a pie chart of the data.
(f) In your opinion, which graph appears to place more emphasis on social insurance taxes as a source of income, the relative frequency bar graph or the pie chart? Why?

18. **Government Expenditures** For fiscal year 2003 (October 1, 2002 to September 30, 2003), the federal government spent $2,158 billion. The breakdown of expenditures is given in the following table.

Category	Expenditure (in billions of dollars)
National defense and foreign affairs	474.8
Social programs	1251.6
Physical, human, and community development	215.8
Net interest on the debt	151.1
Law enforcement and general government	64.7

Source: Budget of the United States and Internal Revenue Service

(a) Construct a relative frequency distribution of the data shown.
(b) What percentage of total expenditures is attributable to net interest on debt?
(c) Construct a frequency bar graph of government expenditures.
(d) Construct a relative frequency bar graph of government expenditures.
(e) Construct a pie chart of government expenditures.
(f) In your opinion, which graph appears to place more emphasis on net interest on the debt, the relative frequency bar graph or the pie chart? Why?

19. College Survey In a national survey conducted by the Centers for Disease Control to determine health-risk behaviors among college students, college students were asked, "How often do you wear a seat belt when riding in a car driven by someone else?" The frequencies were as follows:

Response	Frequency
Never	125
Rarely	324
Sometimes	552
Most of the time	1257
Always	2518

(a) Construct a relative frequency distribution.
(b) What percentage of respondents answered "Always"?
(c) What percentage of respondents answered "Never" or "Rarely"?
(d) Construct a frequency bar graph.
(e) Construct a relative frequency bar graph.
(f) Construct a pie chart.
(g) Suppose that a representative from the Centers for Disease Control says, "52.7% of all college students never wear a seatbelt." Is this a descriptive or inferential statement?

20. College Survey In a national survey conducted by the Centers for Disease Control to determine health-risk behaviors among college students, college students were asked, "How often do you wear a seat belt when driving a car?" The frequencies were as follows:

Response	Frequency
I do not drive a car	249
Never	118
Rarely	249
Sometimes	345
Most of the time	716
Always	3093

(a) Construct a relative frequency distribution.
(b) What percentage of respondents answered "Always"?
(c) What percentage of respondents answered "Never" or "Rarely"?
(d) Construct a frequency bar graph.
(e) Construct a relative frequency bar graph.
(f) Construct a pie chart.
(g) Compute the relative frequencies of "Never," "Rarely," "Sometimes," "Most of the time," and "Always," excluding those that do not drive. Compare with those in Problem 19. What might you conclude?
(h) Suppose that a representative from the Centers for Disease Control says, "2.5% of the college students in this survey responded that they never wear a seatbelt." Is this a descriptive or inferential statement?

21. Foreign-Born Population The following data represent the region of birth of foreign-born residents of the United States in 2003.

Region	Number (thousands)
Caribbean	3,384
Central America	12,362
South America	2,111
Asia	8,375
Europe	4,590
Other Regions	2,680

Source: U.S. Census Bureau

(a) Construct a relative frequency distribution.
(b) What percentage of foreign-born residents was born in Asia?
(c) Construct a frequency bar graph.
(d) Construct a relative frequency bar graph.
(e) Construct a pie chart.

22. Robbery The following data represent the number of offenses for various robberies in 2003.

Type of Robbery	Number (thousands)
Street or Highway	131
Commercial	61
Gas Station	10
Convenience Store	26
Residence	41
Bank	7

Source: U.S. Federal Bureau of Investigation

(a) Construct a relative frequency distribution.
(b) What percentage of robberies was of gas stations?
(c) Construct a frequency bar graph.
(d) Construct a relative frequency bar graph.
(e) Construct a pie chart.

23. Educational Attainment On the basis of the 2003 Current **NW** Population Survey, there were 88.7 million males and 96.6 million females 25 years old or older in the United States. The educational attainment of the males and females was as follows:

Educational Attainment	Males (in millions)	Females (in millions)
Not a high school graduate	14.1	14.5
High school graduate	27.4	31.9
Some college, but no degree	15.2	16.6
Associate's degree	6.4	8.8
Bachelor's degree	16.4	16.9
Advanced degree	9.2	7.9

Source: U.S. Census Bureau

(a) Construct a relative frequency distribution for males.
(b) Construct a relative frequency distribution for females
(c) Construct a side-by-side relative frequency bar graph.
(d) Compare each gender's educational attainment. Make a conjecture about the reasons for the differences.

24. Internet Access The following data represent the number of people who had Internet access in the years 2000 and 2003 by level of education. Data are in thousands of U.S. residents.

Educational Attainment	2000	2003
No college	24,662	65,862
Some college	31,462	50,931
Graduated college	34,379	49,106

Source: U.S. Statistical Abstract, 2004

(a) Construct a relative frequency distribution for 2000.
(b) Construct a relative frequency distribution for 2003.
(c) Construct a side-by-side relative frequency bar graph.
(d) Compare each year's Internet access. Make some conjectures about the reasons for any differences or similarities.

25. Murder Victims A criminologist wanted to know if there was any relation between age and gender of murder victims. The following data represent the number of male and female murder victims by age in 2002.

Age	Number of Males	Number of Females
Less than 17	650	444
17–24	3435	704
25–34	2990	704
35–54	2859	1085
55 or older	787	442

Source: U.S. Federal Bureau of Investigation

(a) Construct a relative frequency distribution for males.
(b) Construct a relative frequency distribution for females.
(c) Construct a side-by-side relative frequency bar graph.
(d) Compare each gender's age percentages. Make a conjecture about the reasons for the differences or similarities.

26. Car Color DuPont Automotive is a major supplier of paint to the automotive industry. It conducted a survey of 100 randomly selected autos in the luxury car segment and 100 randomly selected autos in the sports car segment that were recently purchased and obtained the following colors.

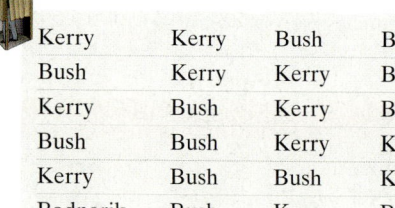

Color	Number of Luxury Cars	Number of Sports Cars
Silver	32	27
White	30	11
Blue	9	13
Black	9	16
Gray	7	7
Red	6	18
Gold	3	4
Green	2	0
Brown	2	4

(a) Construct a relative frequency distribution for each car type.
(b) Draw a side-by-side relative frequency bar graph.
(c) Compare the colors for the two types of car. Make a conjecture about the reasons for the differences.

27. 2004 Presidential Election An exit poll was conducted in Los Alamos County, New Mexico, in which a random sample of 40 voters revealed whom they voted for in the presidential election. The results of the survey are as follows:

Kerry	Kerry	Bush	Bush
Bush	Kerry	Kerry	Bush
Kerry	Bush	Kerry	Bush
Bush	Bush	Kerry	Kerry
Kerry	Bush	Bush	Kerry
Badnarik	Bush	Kerry	Bush
Kerry	Bush	Kerry	Bush
Bush	Bush	Kerry	Kerry
Bush	Bush	Bush	Nader
Bush	Kerry	Bush	Kerry

(a) Construct a frequency distribution.
(b) Construct a relative frequency distribution.
(c) Construct a frequency bar graph.
(d) Construct a relative frequency bar graph.

(e) Construct a pie chart.
(f) On the basis of the data, make a conjecture about which candidate will win Los Alamos County. Would your conjecture be descriptive statistics or inferential statistics? If George W. Bush wins Los Alamos County, what conclusions might be drawn, assuming that the sample was conducted appropriately? Would you be confident in making this prediction with a sample of 40? If the sample consisted of 100 voters, would your confidence increase? Why?

28. Hospital Admissions The following data represent the diagnoses of a random sample of 20 patients admitted to a hospital.

Cancer	Motor vehicle accident	Congestive heart failure
Gunshot wound	Fall	Gunshot wound
Gunshot wound	Motor vehicle accident	Gunshot wound
Assault	Motor vehicle accident	Gunshot wound
Motor vehicle accident	Motor vehicle accident	Gunshot wound
Motor vehicle accident	Gunshot wound	Motor vehicle accident
Fall	Gunshot wound	

Source: Tamela Ohm, student at Joliet Junior College

(a) Construct a frequency distribution.
(b) Construct a relative frequency distribution.
(c) Which diagnosis had the most admissions?
(d) What percentage of diagnoses was motor vehicle accidents?
(e) Construct a frequency bar graph.
(f) Construct a relative frequency bar graph.
(g) Construct a pie chart.
(h) Suppose that an admission specialist at the hospital stated that 40% of all admissions were gunshot wounds. Would this statement be descriptive or inferential? Why?

29. Which Position in Baseball Pays the Most? You are a prospective baseball agent and are in search of clients. You would like to recruit the highest-paid players as clients, so you perform a study in which you identify the 24 top-paid players for the 2004 season and their positions. The table shows the results of your study.

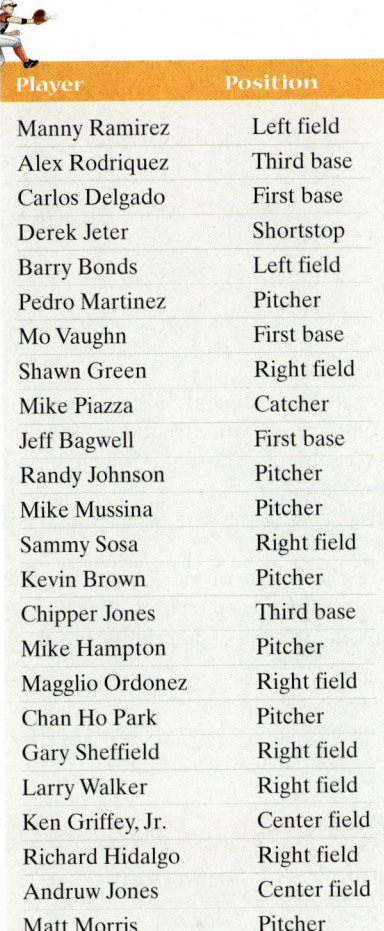

Player	Position
Manny Ramirez	Left field
Alex Rodriquez	Third base
Carlos Delgado	First base
Derek Jeter	Shortstop
Barry Bonds	Left field
Pedro Martinez	Pitcher
Mo Vaughn	First base
Shawn Green	Right field
Mike Piazza	Catcher
Jeff Bagwell	First base
Randy Johnson	Pitcher
Mike Mussina	Pitcher
Sammy Sosa	Right field
Kevin Brown	Pitcher
Chipper Jones	Third base
Mike Hampton	Pitcher
Magglio Ordonez	Right field
Chan Ho Park	Pitcher
Gary Sheffield	Right field
Larry Walker	Right field
Ken Griffey, Jr.	Center field
Richard Hidalgo	Right field
Andruw Jones	Center field
Matt Morris	Pitcher

Source: usatoday.com

(a) Construct a frequency distribution of position.
(b) Construct a relative frequency distribution of position.
(c) Which position appears to be the most lucrative? For which position would you recruit?
(d) Are there any positions that you would avoid recruiting? Why?
(e) Draw a frequency bar graph.
(f) Draw a relative frequency bar graph.
(g) Draw a pie chart.

30. Blood Type A phlebotomist draws the blood of a random sample of 50 patients and determines their blood types as shown:

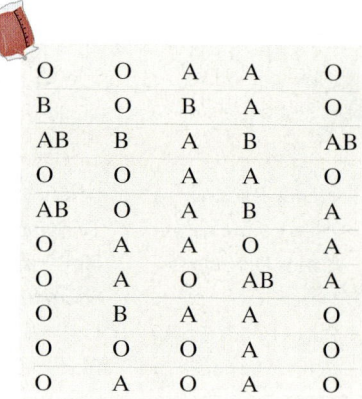

O	O	A	A	O
B	O	B	A	O
AB	B	A	B	AB
O	O	A	A	O
AB	O	A	B	A
O	A	A	O	A
O	A	O	AB	A
O	B	A	A	O
O	O	O	A	O
O	A	O	A	O

(a) Construct a frequency distribution.
(b) Construct a relative frequency distribution.
(c) According to the data, which blood type is most common?
(d) According to the data, which blood type is least common?
(e) Use the results of the sample to conjecture the percentage of the population that has type O blood. Is this an example of descriptive or inferential statistics?

(f) Contact a local hospital and ask them the percentage of the population that is blood type O. Why might the results differ?
(g) Draw a frequency bar graph.
(h) Draw a relative frequency bar graph.
(i) Draw a pie chart.

31. Foreign Language According to the Modern Language Association, the number of college students studying foreign language is increasing. The following data represent the foreign language being studied based on a simple random sample of 30 students learning a foreign language.

Spanish	Chinese	Spanish	Spanish	Spanish
Chinese	German	Spanish	Spanish	French
Spanish	Spanish	Japanese	Latin	Spanish
German	German	Spanish	Italian	Spanish
Italian	Japanese	Chinese	Spanish	French
Spanish	Spanish	Russian	Latin	French

Source: Based on data obtained from the Modern Language Association

(a) Construct a frequency distribution.
(b) Construct a relative frequency distribution.
(c) Construct a frequency bar graph.
(d) Construct a relative frequency bar graph.
(e) Construct a pie chart.

32. The following table lists the presidents of the United States and their state of birth.

Birthplace of U.S. President					
President	**State of Birth**	**President**	**State of Birth**	**President**	**State of Birth**
Washington	Virginia	Lincoln	Kentucky	Coolidge	Vermont
J. Adams	Massachusetts	A. Johnson	North Carolina	Hoover	Iowa
Jefferson	Virginia	Grant	Ohio	F. D. Roosevelt	New York
Madison	Virginia	Hayes	Ohio	Truman	Missouri
Monroe	Virginia	Garfield	Ohio	Eisenhower	Texas
J. Q. Adams	Massachusetts	Arthur	Vermont	Kennedy	Massachusetts
Jackson	South Carolina	Cleveland	New Jersey	L. B. Johnson	Texas
Van Buren	New York	B. Harrison	Ohio	Nixon	California
W.H. Harrison	Virginia	Cleveland	New Jersey	Ford	Nebraska
Tyler	Virginia	McKinley	Ohio	Carter	Georgia
Polk	North Carolina	T. Roosevelt	New York	Reagan	Illinois
Taylor	Virginia	Taft	Ohio	George H. Bush	Massachusetts
Fillmore	New York	Wilson	Virginia	Clinton	Arkansas
Pierce	New Hampshire	Harding	Ohio	George W. Bush	Connecticut
Buchanan	Pennsylvania				

(a) Construct a frequency bar graph for state of birth.
(b) Which state has yielded the most presidents?
(c) Explain why the answer obtained in part (b) may be considered to be misleading.

Consumer Reports® | Consumer Reports Rates Treadmills

A study that compared exercisers who worked out equally hard for the same time on several different types of machines found that they generally burned the most calories on treadmills. Our own research has shown that treadmills are less likely than other machines to sit unused. So it should come as no surprise that treadmills are the best-selling home exercise machine in the United States.

In a study by *Consumer Reports*, we tested 11 best-selling brands of treadmills ranging in price from $500 to $3000. The treadmills were rated on ease of use, ergonomics, exercise factors, construction, and durability. Ease of use is based on how straightforward the treadmill is to use. Ergonomics, including safety factors, belt size, and handrail placement, indicates how well the treadmill fits people of different sizes. Exercise includes evaluations of the minimum incline level, speed control, and heart-rate monitoring. Construction covers factors like the motor's continuous-duty horsepower rating and weld quality.

To help compare the treadmills, the individual attribute scores were combined into an overall score. The

figure below is a ratings chart for the 11 treadmills based on our test results. In addition to the performance ratings, other useful information, such as the models' price and belt size, is included.

(a) What type of graph is illustrated to display overall score in the figure?

(b) Which model has the highest construction score? Which models have the lowest ease of use score?

(c) For ease of use, how many treadmills rated excellent? very good? good? fair? poor?

(d) Draw a frequency bar graph for each rating category. In other words, draw a bar graph for ease of use, ergonomics, and so on.

(e) Does there appear to be a relationship between price and overall score? Explain your opinion.

Note to Readers: *In many cases, our test protocol and analytical methods are more complicated than described in these examples. The data and discussions have been modified to make the material more appropriate for the audience.*

Ratings Chart for Treadmills

KEY NO.	BRAND & MODEL; SIMILAR MODELS IN SMALL TYPE	PRICE	OVERALL SCORE	BELT (IN.)	EASE	ERGONOMICS	EXERCISE	CONSTRUCTION	FOLDS
1	Life Fitness T3 T3i	$2,200		19x52	Excellent	Excellent	Excellent	Very good	
2	Reebok ACD4	1,850		19x61	Very good	Excellent	Excellent	Very good	✔
3	Precor M9.33 M9.35	3,000		20x59	Good	Excellent	Excellent	Excellent	
4	Star Trac TR901	2,500		19x50	Very good	Very good	Excellent	Very good	
5	Image 10.6QL	1,500		20x61	Very good	Very good	Excellent	Good	✔
6	HealthRider S500xi	1,100		19x54	Very good	Good	Excellent	Good	✔
7	Tunturi J6F J6	2,100		18x55	Fair	Very good	Very good	Very good	✔
8	Trimline 2610	1,200		18x48	Very good	Good	Very good	Very good	
9	ProForm 785SS	750		20x55	Very good	Very good	Fair	Fair	✔
10	NordicTrack EXP 1000i	800		20x55	Very good	Very good	Fair	Fair	✔
11	ProForm 525E	500		18x52	Fair	Good	Fair	Fair	✔

Drawing Bar Graphs and Pie Charts

TI-83/84 Plus The TI-83 or TI-84 Plus does not have the ability to draw bar graphs or pie charts.

MINITAB **Frequency or Relative Frequency Distributions from Raw Data**

Step 1: Enter the raw data in C1.

Step 2: Select **Stat** and highlight **Tables** and select Tally Individual Variables...

Step 3: Fill in the window with appropriate values. In the "Variables" box, enter C1. Check "counts" for a frequency distribution and/or "percents" for a relative frequency distribution. Click OK.

Bar Graphs from Summarized Data

Step 1: Enter the categories in C1 and the frequency or relative frequency in C2.

Step 2: Select **Graph** and highlight **Bar Chart**.

Step 3: In the "Bars represent" pull-down menu, select "Values from a table" and highlight "Simple." Press OK.

Step 4: Fill in the window with the appropriate values. In the "Graph variables" box, enter C2. In the "Categorical variable" box, enter C1. By pressing Labels, you can add a title to the graph. Click OK to obtain the bar graph.

Bar Graphs from Raw Data

Step 1: Enter the raw data in C1.

Step 2: Select **Graph** and highlight **Bar Chart.**

Step 3: In the "Bars represent" pull-down menu, select "Counts of unique values" and highlight "Simple." Press OK.

Step 4: Fill in the window with the appropriate values. In the "Categorical variable" box, enter C1. By pressing Labels, you can add a title to the graph. Click OK to obtain the bar graph.

Pie Chart from Raw or Summarized Data

Step 1: If the data are in a summarized table, enter the categories in C1 and the frequency or relative frequency in C2. If the data are raw, enter the data in C1.

Step 2: Select **Graph** and highlight **Pie Chart.**

Step 3: Fill in the window with the appropriate values. If the data are summarized, click the "Chart values from a table" radio button; If the data are raw, click the "Chart raw data" radio button. For summarized data, enter C1 in the "Categorical variable" box and C2 in the "Summary variable" box. If the data are raw, enter C1 in the "Categorical variable" box. By pressing Labels, you can add a title to the graph. Click OK to obtain the pie chart.

Excel **Bar Graphs from Summarized Data**

Step 1: Enter the categories in column A and the frequency or relative frequency in column B.

Step 2: Select the chart wizard icon. Click the "column" chart type. Select the chart type in the upper-left-hand corner and hit "Next."

Step 3: Click inside the data range cell. Use the mouse to highlight the data to be graphed. Click "Next."

Step 4: Click the "Titles" tab to enter x-axis, y-axis, and chart titles. Click "Finish."

Pie Charts from Summarized Data

Step 1: Enter the categories in column A and the frequencies in column B. Select the chart wizard icon and click the "pie" chart type. Select the pie chart in the upper-left-hand corner.

Step 2: Click inside the data range cell. Use the mouse to highlight the data to be graphed. Click "Next."

Step 3: Click the "Titles" tab to the chart title. Click the "Data Labels" tab and select "Show label and percent." Click "Finish."

2.2 Organizing Quantitative Data: The Popular Displays

Preparing for This Section Before getting started, review the following:

- Quantitative data (Section 1.1, p. 6)
- Continuous data (Section 1.1, p. 7)
- Discrete data (Section 1.1, p. 7)

Objectives

1. Organize discrete data in tables
2. Construct histograms of discrete data
3. Organize continuous data in tables
4. Construct histograms of continuous data
5. Draw stem-and-leaf plots
6. Draw dot plots
7. Identify the shape of a distribution

The first step in summarizing quantitative data is to determine whether the data are discrete or continuous. If the data are discrete and there are relatively few different values of the variable, the categories of data will be the observations (as in qualitative data). If the data are discrete, but there are many different values of the variables, categories of data (called *classes*) must be created using intervals of numbers. If the data are continuous, the categories of data must be created using intervals of numbers. We will first present the techniques required to organize discrete quantitative data when there are relatively few different values and then proceed to organizing continuous quantitative data.

1 Organize Discrete Data in Tables

The values of a discrete variable are used to create the categories of data.

EXAMPLE 1 **Constructing Frequency and Relative Frequency Distributions from Discrete Data**

Problem: The manager of a Wendy's fast-food restaurant is interested in studying the typical number of customers who arrive during the lunch hour. The data in Table 8 represent the number of customers who arrive at Wendy's for 40 randomly selected 15-minute intervals of time during lunch. For example, during one 15-minute interval, seven customers arrived. Construct a frequency and relative frequency distribution.

Table 8				
Number of Arrivals at Wendy's				
7	6	6	6	4
5	6	6	11	4
2	7	1	2	4
6	5	5	3	7
2	2	9	7	5
6	2	6	5	7
6	8	2	6	5
4	6	9	8	5

Approach: The number of people arriving could be 0, 1, 2, 3, ... From Table 8, we see that there are 11 categories of data from this study: 1, 2, 3, ..., 11. We tally the number of observations for each category, add up each tally, and create the frequency and relative frequency distributions.

Solution: The frequency and relative frequency distributions are shown in Table 9.

Table 9				
Number of Customers	**Tally**	**Frequency**		**Relative Frequency**
1	|	1		$\dfrac{1}{40} = 0.025$
2	⊞ |	6		0.15
3	|	1		0.025
4	||||	4		0.1
5	⊞ ||	7		0.175
6	⊞ ⊞ |	11		0.275
7	⊞	5		0.125
8	||	2		0.05
9	||	2		0.05
10		0		0.0
11	|	1		0.025

On the basis of the relative frequencies, 27.5% of the 15-minute intervals had six customers arrive at Wendy's during the lunch hour.

Now Work Problems 27(a)–(d).

② Construct Histograms of Discrete Data

As with qualitative data, quantitative data may also be represented graphically. We begin our discussion with a graph called the *histogram*, which is similar to the bar graph drawn for qualitative data.

Definition

A **histogram** is constructed by drawing rectangles for each class of data. The height of each rectangle is the frequency or relative frequency of the class. The width of each rectangle is the same and the rectangles touch each other.

EXAMPLE 2 — Drawing a Histogram for Discrete Data

Problem: Construct a frequency histogram and a relative frequency histogram using the data summarized in Table 9.

Approach: On the horizontal axis, we place the value of each category of data (number of customers). The vertical axis will be the frequency or relative frequency of each category. Rectangles of equal width are drawn, with the center of each rectangle located at the value of each category. For example, the first rectangle is centered at 1. For the frequency histogram, the height of the rectangle will be the frequency of the category. For the relative frequency histogram, the height of the rectangle will be the relative frequency of the category. Remember, the rectangles touch for histograms.

Solution: Figure 6(a) shows the frequency histogram. Figure 6(b) shows the relative frequency histogram.

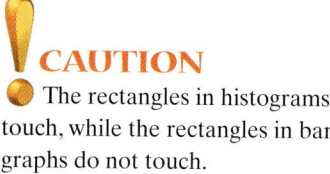

CAUTION

The rectangles in histograms touch, while the rectangles in bar graphs do not touch.

Figure 6

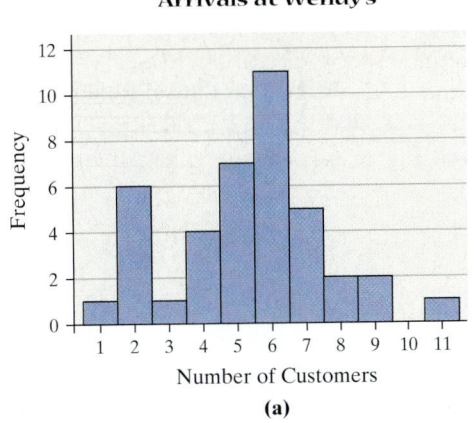

Arrivals at Wendy's

(a)

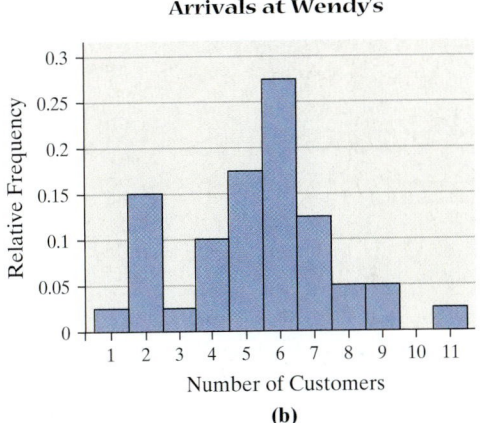

Arrivals at Wendy's

(b)

Now Work Problems 27(e) and (f).

③ Organize Continuous Data in Tables

Raw continuous data do not have any predetermined categories that can be used to construct a frequency distribution. Therefore, the categories must be created. Categories of data are created by using intervals of numbers called **classes**.

Table 10 is a typical frequency distribution created from continuous data. The data represent the number of U.S. residents between the ages of 25 and 74 who have earned a bachelor's degree. The data are based on the Current Population Survey conducted in 2003.

In the table, we notice that the data are categorized, or grouped, by intervals of numbers. Each interval represents a class. For example, the first class is 25- to 34-year-old residents of the United States who have a bachelor's degree. We read this interval as follows: "The number of residents of the United States in 2003 who were between 25 and 34 years of age and have a bachelor's degree was 8,849,000." There are five classes in the table, each with a *lower bound* and an *upper bound*. The **lower class limit** of a class is the smallest value within the class, while the **upper class limit** of a class is the largest value within the class. The lower class limit for the first class in Table 10 is 25; the upper class limit is 34. The **class width** is the difference between consecutive lower class limits. The class width for the data in Table 10 is $35 - 25 = 10$.

The classes in Table 10 do not overlap, so there is no confusion as to which class a data value belongs. Notice that the class widths are equal for all classes. One exception to this requirement is in open-ended tables. A table is **open ended** if the first class has no lower class limit or the last class does not have an upper class limit. The data in Table 11 represent the number of persons under sentence of death as of December 31, 2003, in the United States. The last class in the table, "60 and older," is open ended.

Table 10

Age	Number (in thousands)
25–34	8849
35–44	8915
45–54	7576
55–64	4254
65–74	2047

Table 11

Age	Number
20–29	533
30–39	1147
40–49	1090
50–59	493
60 and older	110

Source: U.S. Justice Department

EXAMPLE 3

Organizing Continuous Data into a Frequency and Relative Frequency Distribution

Problem: Suppose you are considering investing in a Roth IRA. You collect the data in Table 12, which represent the 3-year rate of return (in percent) for a simple random sample of 40 small-capitalization growth mutual funds. Construct a frequency and relative frequency distribution of the data.

Approach: To construct a frequency distribution, we first create classes of equal width. There are 40 observations in Table 12, and they range from 0.05 to 14.48, so we decide to create the classes such that the lower class limit of the first

In Other Words

For qualitative and many discrete data, the categories of data are formed by using the data. For continuous data, the categories are formed by using an interval of numbers, such as 30–39.

Table 12

Three-Year Rate of Return of Mutual Funds

5.37	4.31	4.13	8.58	5.99	7.90	9.11	6.11
3.06	14.48	12.50	8.33	10.10	8.21	6.83	10.94
2.34	0.97	8.33	8.89	6.07	6.50	5.99	9.38
0.05	13.88	3.71	10.07	9.88	4.93	6.38	10.34
2.27	11.91	11.69	12.06	9.84	7.75	2.86	6.68

Source: Morningstar.com

CAUTION

Watch out for tables with class widths that overlap, such as a first class of 20–30 and a second class of 30–40.

class is 0 (a little smaller than the smallest data value) and the class width is 2. There is nothing magical about the choice of 2 as a class width. We could have selected a class width of 8 (or any other class width, as well). We choose a class width that we think will nicely summarize the data. If our choice doesn't accomplish this, we can always try another one. The lower class limit of the second class will be 0 + 2 = 2. Because the classes must not overlap, the upper class limit of the first class is 1.99. Continuing in this fashion, we obtain the following classes:

$$0-1.99$$
$$2-3.99$$
$$\vdots$$
$$14-15.99$$

This gives us eight classes. We tally the number of observations in each class, add up the tallies, and create the frequency distribution. The relative frequency distribution would be created by dividing each class's frequency by 40, the number of observations.

Solution: We tally the data as shown in the second column of Table 13. The third column in the table shows the frequency of each class. From the frequency distribution, we conclude that a 3-year rate of return between 8.0% and 9.99% occurs with the most frequency. The fourth column in the table shows the relative frequency of each class. So 22.5% of the small-capitalization growth mutual funds had a 3-year rate of return between 8% and 9.99%.

Historical Note

Florence Nightingale was born in Italy on May 12, 1820. She was named after the city of her birth. Nightingale was educated by her father, who attended Cambridge University. Between 1849 and 1851, she studied nursing throughout Europe. In 1854, she was asked to oversee the introduction of female nurses into the military hospitals in Turkey. While there, she greatly improved the mortality rate of wounded soldiers. She collected data and invented graphs (the polar area diagram), tables, and charts to show that improving sanitary conditions would lead to decreased mortality rates. In 1869, Nightingale founded the Nightingale School Home for Nurses. After a long and eventful life as a reformer of heath care and contributor to graphics in statistics, Florence Nightingale died on August 13, 1910.

Table 13

Class (3-Year Rate of Return)	Tally	Frequency	Relative Frequency
0–1.99	||	2	2/40 = 0.05
2–3.99	⊞	5	5/40 = 0.125
4–5.99	⊞ |	6	6/40 = 0.15
6–7.99	⊞ |||	8	0.2
8–9.99	⊞ ||||	9	0.225
10–11.99	⊞ |	6	0.15
12–13.99	|||	3	0.075
14–15.99	|	1	0.025

Only one mutual fund had a 3-year rate of return between 14% and 15.99%. We might consider this mutual fund worthy of our investment. This type of information would be more difficult to obtain from the raw data. ■

Notice that the choice of the lower class limit of the first class and the class width was rather arbitrary. While formulas and procedures do exist for creating frequency distributions from raw data, they do not necessarily provide better summaries. It is incorrect to say that a particular frequency distribution is the correct one. Constructing frequency distributions is somewhat of an art form in which the distribution that seems to provide the best overall summary of the data should be used.

Consider the frequency distribution in Table 14, which also summarizes the 3-year rate of return data discussed in Example 3. Here, the lower class limit of the first class is 0 and the class width is 4. Do you think Table 13 or Table 14 provides a better summary of the distribution of 3-year rates of return? In forming your opinion, consider the following: Too few classes will cause a bunching effect. Too many classes will spread the data out, thereby not revealing any pattern.

In Other Words

Creating the classes for summarizing continuous data is an art form. There is no such thing as the correct frequency distribution. However, there can be less desirable frequency distributions. The larger the class width, the fewer classes a frequency distribution will have.

Table 14		
Class	**Tally**	**Frequency**
0–3.99	𝍷𝍷𝍷𝍷 ||	7
4–7.99	𝍷𝍷𝍷𝍷 𝍷𝍷𝍷𝍷 ||||	14
8–11.99	𝍷𝍷𝍷𝍷 𝍷𝍷𝍷𝍷 𝍷𝍷𝍷𝍷	15
12–15.99	||||	4

The goal in constructing a frequency distribution is to reveal interesting features of the data. With that said, when constructing frequency distributions, we typically want the number of classes to be between 5 and 20. When the data set is small, we want fewer classes. When the data set is large, we want more classes. Why do you think this is reasonable?

Now Work Problems 29(a) and (b).

4   Construct Histograms of Continuous Data

We are now ready to draw histograms of continuous data.

EXAMPLE 4   **Drawing a Histogram of Continuous Data**

Problem: Construct a frequency and relative frequency histogram of the 3-year rate of return data discussed in Example 3.

Approach: To draw the frequency histogram, we will use the frequency distribution in Table 13. We label the lower class limits of each class on the horizontal axis. Then, for each class, we draw a rectangle whose width is the class width and whose height is the frequency. To construct the relative frequency histogram, we let the height of the rectangle be the relative frequency, instead of the frequency.

Solution: Figure 7(a) represents the frequency histogram, and Figure 7(b) represents the relative frequency histogram.

Figure 7
Frequency and Relative
Frequency Histograms

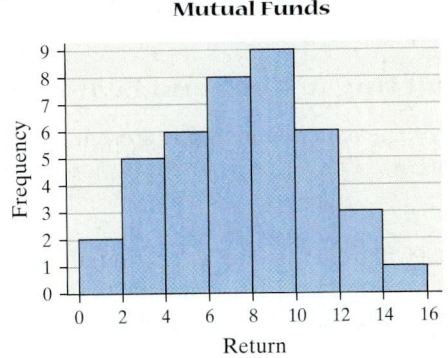

Three-Year Rate of Return
for Small Capitalization
Mutual Funds

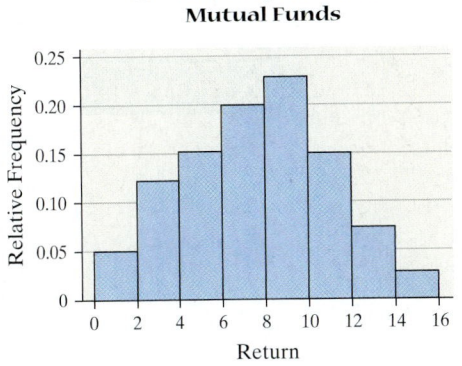

Three-Year Rate of Return
for Small Capitalization
Mutual Funds

EXAMPLE 5 **Drawing a Histogram for Continuous Data Using Technology**

Problem: Construct a frequency and relative frequency histogram of the 3-year rate of return data discussed in Example 3.

Approach: We will use MINITAB to construct the frequency and relative frequency histograms. The steps for constructing the graphs using the TI-83/84 Plus graphing calculators, MINITAB, and Excel are given in the Technology Step by Step on page 96.

Result: Figure 8(a) shows the frequency histogram and Figure 8(b) shows the relative frequency histogram obtained from MINITAB. Note that MINITAB expresses relative frequencies using percent.

Figure 8

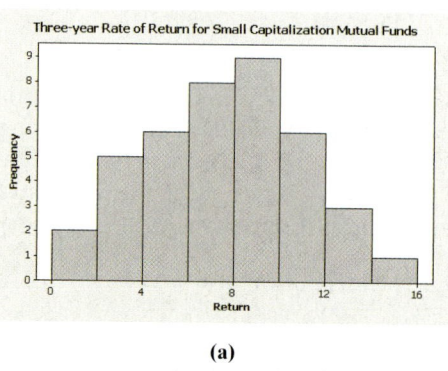

(a)

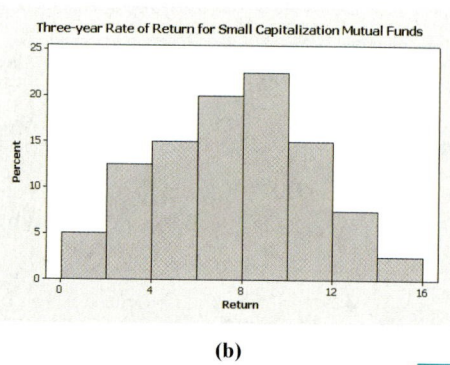

(b)

Now Work Problems 29(c)–(d).

5 Draw Stem-and-Leaf Plots

A **stem-and-leaf plot** is another way to represent quantitative data graphically. Stem-and-leaf plots have some advantages over histograms. Use the following steps to construct a stem-and-leaf plot.

Construction of a Stem-and-Leaf Plot

Step 1: The **stem** of the graph will consist of the digits to the left of the right-most digit. The **leaf** of the graph will be the rightmost digit. Sometimes it is necessary to modify the method of choosing the stem if a different class width is desired.

Step 2: Write the stems in a vertical column in increasing order. Draw a vertical line to the right of the stems.

Step 3: Write each leaf corresponding to the stems to the right of the vertical line.

Step 4: Write the leaves in ascending order.

EXAMPLE 6 **Constructing a Stem-and-Leaf Plot**

Problem: The data in Table 15 represent the percentage of persons in poverty, by state, in 2002. Draw a stem-and-leaf plot of the data.

Approach:

Step 1: We will treat the integer portion of the number as the stem and the decimal portion as the leaf. For example, the stem of Alabama will be 14 and the leaf will be 6. The stem of 14 will include all data from 14.0 to 14.9.

Step 2: Write the stems vertically in ascending order, and then draw a vertical line to the right of the stems.

Table 15

Percentage of Persons Living in Poverty

State	Percent	State	Percent	State	Percent
Alabama	14.6	Kentucky	13.1	North Dakota	11.9
Alaska	8.3	Louisiana	17.0	Ohio	10.1
Arizona	13.3	Maine	11.3	Oklahoma	14.7
Arkansas	18.0	Maryland	7.3	Oregon	11.2
California	12.8	Massachusetts	9.6	Pennsylvania	9.2
Colorado	9.4	Michigan	10.3	Rhode Island	10.3
Connecticut	7.8	Minnesota	6.5	South Carolina	13.5
Delaware	8.1	Mississippi	17.6	South Dakota	10.2
D.C.	16.8	Missouri	9.6	Tennessee	14.2
Florida	12.1	Montana	13.7	Texas	15.3
Georgia	12.1	Nebraska	9.5	Utah	9.3
Hawaii	10.6	Nevada	8.3	Vermont	9.9
Idaho	11.8	New Hampshire	5.6	Virginia	8.7
Illinois	11.2	New Jersey	7.8	Washington	10.8
Indiana	8.7	New Mexico	17.8	West Virginia	16.0
Iowa	8.3	New York	14.0	Wisconsin	8.6
Kansas	9.4	North Carolina	13.1	Wyoming	9.5

Source: Poverty in the United States, 2002, Current Population Reports

Step 3: Write the leaves corresponding to the stem.

Step 4: Write the leaves in ascending order.

Solution:

Step 1: The stem from Alabama is 14 and the corresponding leaf is 6. The stem from Alaska is 8 and its leaf is 3, and so on.

Step 2: We write the stems vertically in Figure 9(a), along with a vertical line to the right of the stem.

Step 3: We write the leaves corresponding to each stem. See Figure 9(b).

Step 4: Write the leaves in ascending order. See Figure 9(c).

Figure 9

```
 5 |                    5 | 6                     5 | 6
 6 |                    6 | 5                     6 | 5
 7 |                    7 | 8 3 8                 7 | 3 8 8
 8 |                    8 | 3 1 7 3 3 7 6         8 | 1 3 3 3 6 7 7
 9 |                    9 | 4 4 6 6 5 2 3 9 5     9 | 2 3 4 4 5 5 6 6 9
10 |                   10 | 6 3 1 3 2 8          10 | 1 2 3 3 6 8
11 |                   11 | 8 2 3 9 2            11 | 2 2 3 8 9
12 |                   12 | 8 1 1                12 | 1 1 8
13 |                   13 | 3 1 7 1 5            13 | 1 1 3 5 7
14 |                   14 | 6 0 7 2              14 | 0 2 6 7
15 |                   15 | 3                    15 | 3
16 |                   16 | 8 0                  16 | 0 8
17 |                   17 | 0 6 8                17 | 0 6 8
18 |                   18 | 0                    18 | 0
     (a)                    (b)                      (c)
```

EXAMPLE 7   Constructing a Stem-and-Leaf Plot Using Technology

Problem: Construct a stem-and-leaf plot of the poverty data discussed in Example 6.

Using Technology

In MINITAB, there is a column of numbers left of the stem. The (6) indicates that there are 6 observations in the class containing the middle value (called the *median*). The values above the (6) represent the number of observations less than or equal to the upper class limit of the class. For example, 12 states have percentage in poverty less than 8.9. The values in the left column below the (6) indicate the number of observations greater than or equal to the lower class limit of the class. For example, 7 states have percentages in poverty greater than 15.0.

Approach: We will use MINITAB to construct the stem-and-leaf plot. The steps for constructing the graphs using MINITAB are given in the Technology Step by Step on page 96. **Note:** The TI graphing calculators and Excel are not capable of drawing stem-and-leaf plots.

Result: Figure 10 shows the stem-and-leaf plot obtained from MINITAB.

Figure 10

```
 1     5    6
 2     6    5
 5     7    388
12     8    1333677
21     9    234455669
(6)   10    123368
24    11    22389
19    12    118
16    13    11357
11    14    0267
 7    15    3
 6    16    08
 4    17    068
 1    18    0
```

Now Work Problem 35.

If you look at the stem-and-leaf plot carefully, you'll notice that it looks much like a histogram turned on its side. The stem serves as the class. For example, the stem 10 contains all data from 10.0 to 10.9. The leaf serves as the frequency (height of the rectangle). Therefore, it is important to space the leaves equally when drawing a stem-and-leaf plot.

One advantage of the stem-and-leaf plot over frequency distributions and histograms is that the raw data can be retrieved from the stem-and-leaf plot.

In Other Words

The choice of the stem in the construction of a stem-and-leaf diagram is also an art form. It acts just like the class width. For example, the stem of 7 in Figure 10 represents the class 7.0–7.9. The stem of 8 represents the class 8.0–8.9. Notice that the class width is 1.0. The number of leaves is the frequency of each category.

Once a frequency distribution or histogram of continuous data is created, the raw data are lost. However, the raw data can be retrieved from the stem-and-leaf plot.

The steps listed for creating stem-and-leaf plots sometimes must be modified to meet the needs of the data. Consider the next example.

EXAMPLE 8

Constructing a Stem-and-Leaf Plot after Modifying the Data

Problem: Construct a stem-and-leaf plot of the 3-year rate of return data listed in Table 12 on page 80.

Approach

Step 1: If we use the approach from Example 6 and use the integer portion as the stem and the decimals as the leaves, the stems will be 0, 1, 2, ..., 14. This is fine. However, the leaves will be two digits (such as 37, 06, and so on). This is not acceptable. To address this problem, we will round the data to the nearest tenth. Then the stem can be the whole numbers 0, 1, 2, ..., 14, and the leaves will be the decimal portion.

Step 2: Create a vertical column of the whole-number stems in increasing order.

Step 3: Write the leaves corresponding to each stem.

Step 4: Write the leaves in ascending order.

Solution:

Step 1: We round the data to the nearest tenth as shown in Table 16.

Table 16							
5.4	4.3	4.1	8.6	6.0	7.9	9.1	6.1
3.1	14.5	12.5	8.3	10.1	8.2	6.8	10.9
2.3	1.0	8.3	8.9	6.1	6.5	6.0	9.4
0.1	13.9	3.7	10.1	9.9	4.9	6.4	10.3
2.3	11.9	11.7	12.1	9.8	7.8	2.9	6.7

Step 2: Write the stems vertically in ascending order as shown in Figure 11(a).
Step 3: Write the leaves corresponding to each stem as shown in Figure 11(b).
Step 4: Write the leaves in ascending order as shown in Figure 11(c).

Figure 11

```
 0 |                    0 | 1                   0 | 1
 1 |                    1 |                     1 |
 2 |                    2 | 3 3 9               2 | 3 3 9
 3 |                    3 | 1 7                 3 | 1 7
 4 |                    4 | 3 1 9               4 | 1 3 9
 5 |                    5 | 4                   5 | 4
 6 |                    6 | 0 1 5 8 0 4 1 7     6 | 0 0 1 1 4 5 7 8
 7 |                    7 | 9 8                 7 | 8 9
 8 |                    8 | 3 6 3 9 2           8 | 2 3369
 9 |                    9 | 9 8 1 4             9 | 1 4 8 9
10 |                   10 | 1 1 9 3            10 | 1 1 3 9
11 |                   11 | 9 7                11 | 7 9
12 |                   12 | 5 1                12 | 1 5
13 |                   13 | 9                  13 | 9
14 |                   14 | 5                  14 | 5
      (a)                     (b)                     (c)
```

Split Stems

Consider the data shown in Table 17. The data range from 11 to 48. If we drew a stem-and-leaf plot using the tens digit as the stem and the ones digit as the leaf, we would obtain the results shown in Figure 12. The data appear rather "bunched." To resolve this problem, we can use **split stems.** For example, rather than using one stem for the class of data 10–19, we could use two stems, one for the 10–14 interval and the second for the 15–19 interval. We do this in Figure 13.

In Other Words
Using split stems is like adding more classes to a frequency distribution.

Table 17				
27	17	11	24	36
13	29	22	18	17
23	30	12	46	17
32	48	11	18	23
18	32	26	24	38
24	15	13	31	22
18	21	27	20	16
15	37	19	19	29

Figure 12

```
1 | 1 1 2 3 3 5 5 6 7 7 7 8 8 8 8 9 9
2 | 0 1 2 2 3 3 4 4 4 6 7 7 9 9
3 | 0 1 2 2 6 7 8
4 | 6 8
```

Figure 13

```
1 | 1 1 2 3 3
1 | 5 5 6 7 7 7 8 8 8 8 9 9
2 | 0 1 2 2 3 3 4 4 4
2 | 6 7 7 9 9
3 | 0 1 2 2
3 | 6 7 8
4 |
4 | 6 8
```

The stem-and-leaf plot shown in Figure 13 reveals the distribution of the data better. As with the construction of class intervals in the creation of frequency histograms, judgment plays a major role. There is no such thing as a correct stem-and-leaf plot. However, a quick comparison of Figures 12 and 13 shows that some are better than others.

One final note: **Stem-and-leaf plots are best used when the data set is small.**

Now Work Problem 41.

Draw Dot Plots

One more graph! A **dot plot** is drawn by placing each observation horizontally in increasing order and placing a dot above the observation each time it is observed.

EXAMPLE 9   **Drawing a Dot Plot**

Problem: Draw a dot plot for the number of arrivals at Wendy's data from Example 1 on page 77.

Approach: The smallest observation in the data set is 1 and the largest is 11. We write the numbers 1 through 11 horizontally. For each observation, we place a dot above the value of the observation.

Solution: Figure 14 shows the dot plot.

Figure 14

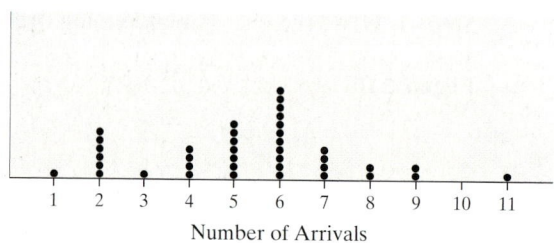

Number of Arrivals

Now Work Problem 45.

 Identify the Shape of a Distribution

One way that a variable is described is through the shape of its distribution. Distribution shapes are typically classified as symmetric, skewed left, or skewed right. Figure 15 displays various histograms and the shape of the distribution.

Figures 15(a) and (b) display symmetric distributions. These distributions are symmetric because, if we split the histogram down the middle, the right and left sides of the histograms are mirror images. Figure 15(a) is a **uniform distribution**, because the frequency of each value of the variable is evenly spread out across the values of the variable. Figure 15(b) displays a **bell-shaped distribution**, because the highest frequency occurs in the middle and frequencies tail off to the left and right of the middle. Figure 15(c) illustrates a distribution that is **skewed right**. Notice that the tail to the right of the peak is longer than the tail to the left of the peak. Finally, Figure 15(d) illustrates a distribution that is **skewed left**, because the tail to the left of the peak is longer than the tail to the right of the peak.

! CAUTION

We do not describe qualitative data as skewed left, skewed right, or uniform.

Figure 15

! CAUTION

It is important to recognize that data will not always exhibit behavior that perfectly matches any of the shapes given in Figure 15. To identify the shape of a distribution, some flexibility is required. In addition, people may disagree on the shape, since identifying shape is subjective.

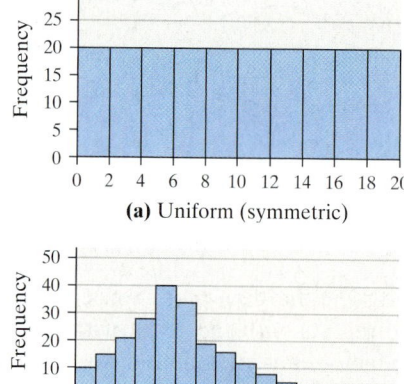

(a) Uniform (symmetric)

(c) Skewed Right

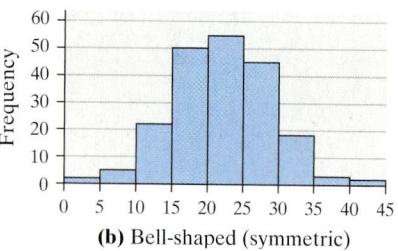

(b) Bell-shaped (symmetric)

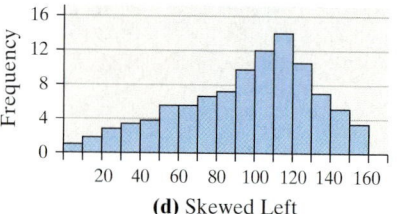

(d) Skewed Left

| EXAMPLE 10 | **Identifying the Shape of a Distribution** |

Problem: Figure 16 displays the histogram obtained for the 3-year rates of return for small-capitalization stocks. Describe the shape of the distribution.

Figure 16

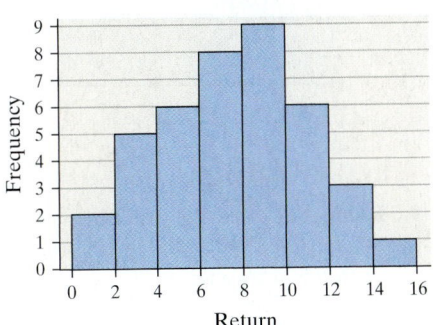

Three-Year Rate of Return for Small Capitalization Mutual Funds

Approach: We compare the shape of the distribution displayed in Figure 16 with those in Figure 15.

Solution: Since the histogram looks most like Figure 15(b), the distribution is bell-shaped.

Now Work Problem 29(e).

In-Class Activity: Random-Number Generators, Pulse Rate, and Household Size

1. We saw in Chapter 1 how to use a graphing calculator or statistical spreadsheet to generate random numbers. Using either of these, randomly generate 200 integers between 0 and 20, inclusive. That is, the integer can be any value greater than or equal to 0 or less than or equal to 20. Store these data in your calculator or spreadsheet.

2. Everyone in the class should determine their resting pulse rate. Collect these data for the class and store them in your calculator or spreadsheet.

3. Everyone in class should share how many people live in their household. Collect the data for the class, and input them into your calculator or spreadsheet.

   (a) What shape do you expect the distribution of random integers to have? Why?

   (b) What shape do you expect the distribution of pulse rates to have? Why?

   (c) What shape do you expect the distribution of household size to have? Why?

   (d) Draw a histogram of each data set. For the random integer data, use a class width of 2.

   (e) What shape did each have? Are you surprised?

2.2 ASSESS YOUR UNDERSTANDING

Concepts and Vocabulary

1. Discuss circumstances under which it is preferable to use relative frequency distributions, instead of frequency distributions.

2. Why shouldn't classes overlap when one summarizes continuous data?

3. The histogram to the right represents the total rainfall for each time it rained in Chicago during the month of August since 1871. The histogram was taken from the *Chicago Tribune* on August 14, 2001. What is wrong with the histogram?

4. State the advantages and disadvantages of histograms versus stem-and-leaf plots.

5. Contrast the differences between histograms and bar graphs.

6. *True or False*: There is not one particular frequency distribution that is correct, but there are frequency distributions that are less desirable than others.

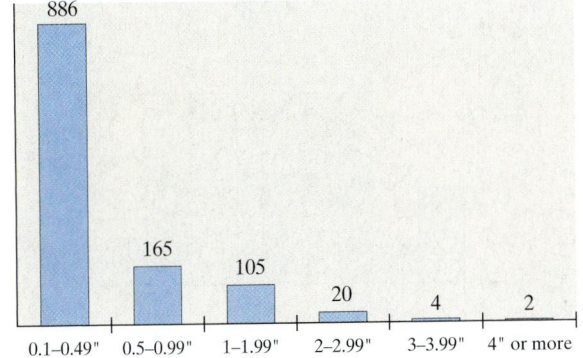

Total August Rain Events Since 1871 in Chicago

7. *True or False*: The shape of the distribution shown is best classified as skewed left.

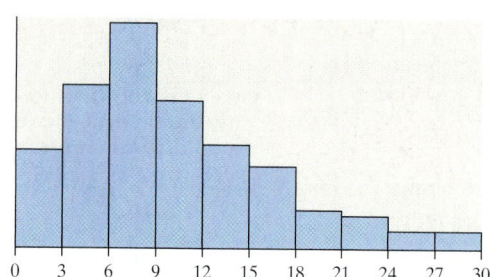

8. *True or False*: The shape of the distribution shown is best classified as uniform.

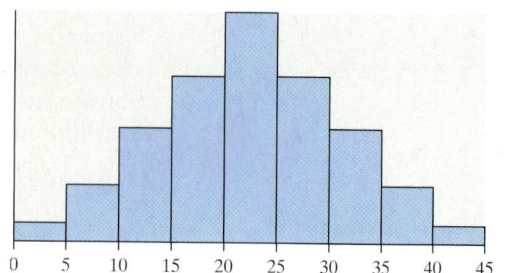

Skill Building

9. Rolling the Dice An experiment was conducted in which two fair dice were thrown 100 times. The sum of the pips showing on the dice was then recorded. The following frequency histogram gives the results.

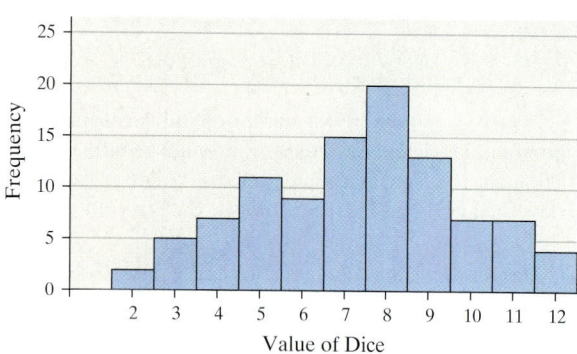

Sum of Two Dice

Value of Dice

(a) What was the most frequent outcome of the experiment?
(b) What was the least frequent?
(c) How many times did we observe a 7?
(d) Determine the percentage of time a 7 was observed.
(e) Describe the shape of the distribution.

10. Car Sales A car salesman records the number of cars he sold each week for the past year. The following frequency histogram shows the results.

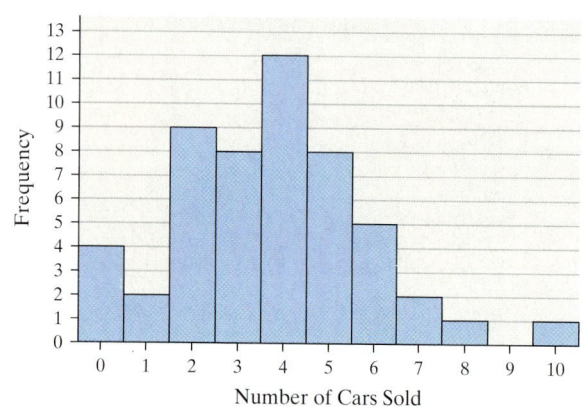

Cars Sold per Week

Number of Cars Sold

(a) What is the most frequent number of cars sold in a week?
(b) For how many weeks were two cars sold?
(c) Determine the percentage of time two cars were sold.
(d) Describe the shape of the distribution.

11. IQ Scores The following frequency histogram represents the IQ scores of a random sample of seventh-grade students. IQs are measured to the nearest whole number. The frequency of each class is labeled above each rectangle.

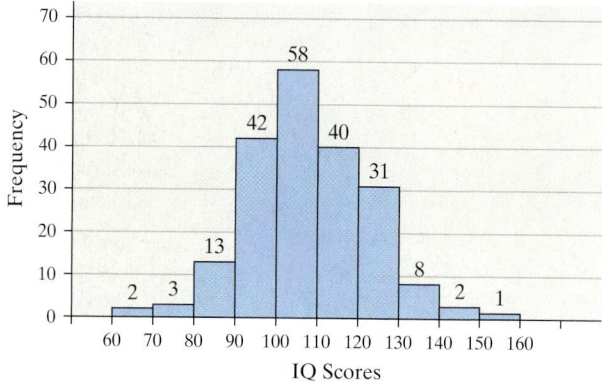

IQs of 7th Grade Students

IQ Scores

(a) How many students were sampled?
(b) Determine the class width.
(c) Identify the classes and their frequencies.
(d) Which class has the highest frequency?
(e) Which class has the lowest frequency?

12. Alcohol-Related Traffic Fatalities The following frequency histogram represents the number of alcohol-related traffic fatalities by state in 2003 according to data obtained from the National Highway Traffic Safety Administration.

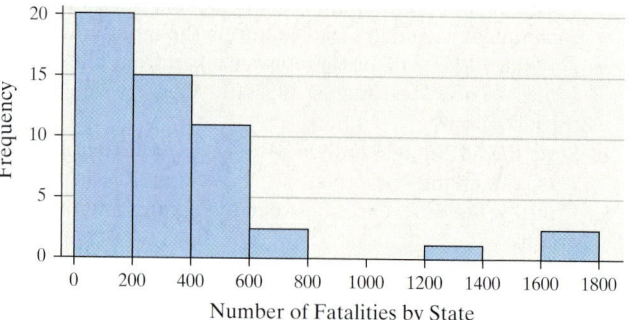

Alcohol-Related Traffic Fatalities

Number of Fatalities by State

(a) Determine the class width.
(b) Identify the classes.
(c) Which class has the highest frequency?
(d) Describe the shape of the distribution.

(e) A reporter writes the following statement: "According to the data, Texas had 1709 alcohol-related deaths, while Vermont had only 29. So the roads in Vermont are much safer." Explain what is wrong with this statement and how a fair comparison can be made between alcohol-related traffic fatalities in Texas versus Vermont.

Applying the Concepts

13. **Predicting School Enrollment** To predict future enrollment, a local school district wants to know the number of children under the age of 5. Fifty households within the district were sampled, and the head of household was asked to disclose the number of children under the age of 5 living in the household. The results of the survey are presented in the following table.

Number of Children under 5	Number of Households
0	16
1	18
2	12
3	3
4	1

(a) Construct a relative frequency distribution of the data.
(b) What percentage of households has two children under the age of 5?
(c) What percentage of households has one or two children under the age of 5?

14. **Free Throws** A basketball player habitually makes 70% of her free throws. In an experiment, a researcher asks this basketball player to record the number of free throws she shoots until she misses. The experiment is repeated 50 times. The following table lists the distribution of the number of free throws attempted until a miss is recorded.

Number of Free Throws Until a Miss	Frequency
1	16
2	11
3	9
4	7
5	2
6	3
7	0
8	1
9	0
10	1

(a) Construct a relative frequency distribution of the data.
(b) What percentage of the time did she miss on her fourth free throw?
(c) What percentage of the time did she make nine in a row and then miss the tenth free throw?
(d) What percentage of the time did she make at least five in a row?

In Problems 15 and 16, determine the original set of data. The stem represents the tens digit and the leaf represents the ones digit.

15.
```
1 | 0 1 4
2 | 1 4 4 7 9
3 | 3 5 5 5 7 7 8
4 | 0 0 1 2 6 6 8 9 9
5 | 3 3 5 8
6 | 1 2
```

16.
```
4 | 0 4 7
5 | 2 2 3 9 9
6 | 3 4 5 8 8 9
7 | 0 1 1 3 6 6
8 | 2 3 8
```

In Problems 17 and 18, determine the original set of data. The stem represents the ones digits and the leaf represents the tenths digit.

17.
```
1 | 2 4 6
2 | 1 4 7 7 9
3 | 3 3 3 5 7 7 8
4 | 0 1 1 3 6 6 8 8 9
5 | 3 4 5 8
6 | 2 4
```

18.
```
12 | 3 7 9 9
13 | 0 4 5 7 8 9 9
14 | 2 4 4 7 7 8 9
15 | 1 2 2 5 6
16 | 0 3
```

In Problems 19–22, find (a) the number of classes, (b) the class limits, and (c) the class width.

19. Health Insurance The following data represent the number of people aged 25 to 64 covered by health insurance in 2003.

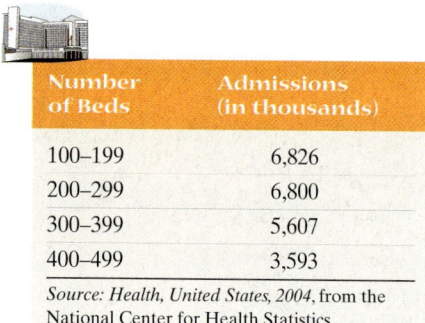

Age	Number (millions)
25–34	28.9
35–44	35.7
45–54	35.1
55–64	24.7

Source: U.S. Census Bureau, Current Population Survey, 2004 Annual Social and Economic Supplement

20. Earthquakes The following data represent the number of earthquakes worldwide whose magnitude was less than 8.0 in 2004.

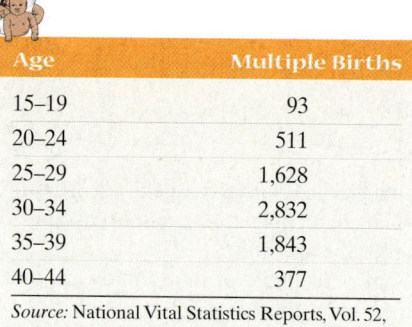

Magnitude	Number
0–0.9	3,179
1.0–1.9	1,340
2.0–2.9	6,297
3.0–3.9	7,826
4.0–4.9	10,975
5.0–5.9	1,430
6.0–6.9	139
7.0–7.9	13

Source: U.S. Geological Survey, Earthquake Hazards Program

21. Hospital Admissions The following data represent community hospital admissions for hospitals with 100 to 499 beds during the year 2002.

Number of Beds	Admissions (in thousands)
100–199	6,826
200–299	6,800
300–399	5,607
400–499	3,593

Source: Health, United States, 2004, from the National Center for Health Statistics

22. Multiple Births The following data represent the number of live multiple births (three or more babies) in 2002 for women 15 to 44 years old.

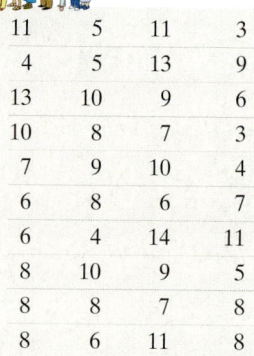

Age	Multiple Births
15–19	93
20–24	511
25–29	1,628
30–34	2,832
35–39	1,843
40–44	377

Source: National Vital Statistics Reports, Vol. 52, No. 10, December 17, 2003

In Problems 23–26, construct (a) a relative frequency distribution, (b) a frequency histogram, and (c) a relative frequency histogram for the given data. Then answer the questions that follow.

23. Using the data in Problem 19, of the people covered by health insurance, what percentage is 25 to 34 years old? Of the people covered by health insurance, what percentage is 44 years or younger?

24. Using the data in Problem 20, what percentage of earthquakes registered 4.0 to 4.9? What percentage of earthquakes registered 4.9 or less?

25. Using the data in Problem 21, what percentage of the admissions was in hospitals with 300 to 399 beds? What percentage of the admissions was in hospitals with 300 or more beds?

26. Using the data in Problem 22, what percentage of multiple births was to women 40 to 44 years old? What percentage of multiple births was to women 24 years or younger?

27. Waiting The following data represent the number of customers waiting for a table at 6:00 P.M. for 40 consecutive Saturdays at Bobak's Restaurant:

11	5	11	3
4	5	13	9
13	10	9	6
10	8	7	3
7	9	10	4
6	8	6	7
6	4	14	11
8	10	9	5
8	8	7	8
8	6	11	8

(a) Construct a frequency distribution of the data.
(b) Construct a relative frequency distribution of the data.
(c) What percentage of the Saturdays had 10 or more customers waiting for a table at 6:00 P.M.?
(d) What percentage of the Saturdays had five or fewer customers waiting for a table at 6:00 P.M.?
(e) Construct a frequency histogram of the data.
(f) Construct a relative frequency histogram of the data.
(g) Describe the shape of the distribution.

28. Highway Repair The following data represent the number of potholes on 50 randomly selected 1-mile stretches of highway in the city of Chicago.

2	7	4	7	2	7
2	2	2	3	4	3
1	2	3	2	1	4
2	2	5	2	3	4
4	1	7	10	3	5
4	3	3	2	2	
1	6	5	7	9	
2	2	2	1	5	
3	5	1	3	5	

(a) Construct a frequency distribution of the data.
(b) Construct a relative frequency distribution of the data.
(c) What percentage of the 1-mile stretches of highway had seven or more potholes?
(d) What percentage of the 1-mile stretches of highway had two or fewer potholes?
(e) Construct a frequency histogram of the data.
(f) Construct a relative frequency histogram of the data.
(g) Describe the shape of the distribution.

29. Average Income The following data represent the per capita (average) disposable income (income after taxes) for the 50 states and the District of Columbia in 2003.

24,028	30,641	24,293	22,123	29,798	30,507
36,726	28,960	42,345	27,610	26,356	27,837
23,584	30,063	25,929	26,409	27,033	23,567
23,889	25,900	32,637	34,570	27,275	30,397
21,677	26,317	23,528	27,865	28,188	31,251
35,411	23,301	31,527	25,307	26,902	26,684
24,169	26,102	28,557	28,365	23,753	27,149
26,314	26,922	22,581	27,750	29,683	30,288
22,252	27,508	29,600			

Source: U.S. Bureau of Economic Analysis, May 2004

With the first class having a lower class limit of 20,000 and a class width of 2500,

(a) Construct a frequency distribution.
(b) Construct a relative frequency distribution.
(c) Construct a frequency histogram of the data.
(d) Construct a relative frequency histogram of the data.
(e) Describe the shape of the distribution.
(f) Repeat parts (a)–(e) using a class width of 4000. Which frequency distribution seems to provide a better summary of the data?
(g) The highest per capita disposable income exists in the District of Columbia, yet the District of Columbia has one of the highest unemployment rates (7% unemployed). Is this surprising to you? Why?

30. Poverty Every year the federal government adjusts the income level that must be earned to be above the poverty level. The dollar amount depends on the number of people living in the household. In 2002, a family of four needed to earn $18,392 to be above the poverty line. The following data represent the percentage of each state's population (plus the District of Columbia) that is living in poverty.

14.6	8.3	13.3	18.0	12.8	9.4
7.8	8.1	16.8	12.1	12.1	10.6
11.8	11.2	8.7	8.3	9.4	13.1
17.0	11.3	7.3	9.6	10.3	6.5
17.6	9.6	13.7	9.5	8.3	5.6
7.8	17.8	14.0	13.1	11.9	10.1
14.7	11.2	9.2	10.3	13.5	10.2
14.2	15.3	9.3	9.9	8.7	10.8
16.0	8.6	9.5			

Source: U.S. Census Bureau

With the first class having a lower class limit of 5 and a class width of 1,

(a) Construct a frequency distribution.
(b) Construct a relative frequency distribution.
(c) Construct a frequency histogram of the data.
(d) Construct a relative frequency histogram of the data.
(e) Describe the shape of the distribution.
(f) Repeat parts (a)–(e) using a class width of 2. Which frequency distribution seems to provide a better summary of the data?
(g) From Problem 29, we learned that the highest per capita disposable income exists in the District of Columbia. The poverty rate of the District of Columbia is 16.8. Is this surprising to you? Why?

31. Serum HDL Dr. Paul Oswiecmiski randomly selects 40 of his 20- to 29-year-old patients and obtains the following data regarding their serum HDL cholesterol:

70	56	48	48	53	52	66	48
36	49	28	35	58	62	45	60
38	73	45	51	56	51	46	39
56	32	44	60	51	44	63	50
46	69	53	70	33	54	55	52

With the first class having a lower class limit of 20 and a class width of 10,

(a) Construct a frequency distribution.
(b) Construct a relative frequency distribution.
(c) Construct a frequency histogram of the data.
(d) Construct a relative frequency histogram of the data.
(e) Describe the shape of the distribution.
(f) Repeat parts (a)–(e) using a class width of 5.
(g) Which frequency distribution seems to provide a better summary of the data?

32. Volume of Altria Group Stock The volume of a stock is the number of shares traded on a given day. The following data, in millions, so that 3.78 represents 3,780,000 shares traded, represent the volume of Altria Group stock traded for a random sample of 35 trading days in 2004.

3.78	8.74	4.35	5.02	8.40
6.06	5.75	5.34	6.92	6.23
5.32	3.25	6.57	7.57	6.07
3.04	5.64	5.00	7.16	4.88
10.32	3.38	7.25	6.52	4.43
3.38	5.53	4.74	9.70	3.56
10.96	4.50	7.97	3.01	5.58

Source: Yahoo.finance.com

With the first class having a lower class limit of 3 and a class width of 2,

(a) Construct a frequency distribution.
(b) Construct a relative frequency distribution.
(c) Construct a frequency histogram of the data.
(d) Construct a relative frequency histogram of the data.
(e) Describe the shape of the distribution.
(f) Repeat parts (a)–(e) using a class width of 1.
(g) Which frequency distribution seems to provide a better summary of the data?

33. Dividend Yield A dividend is a payment from a publicly traded company to its shareholders. The dividend yield of a stock is determined by dividing the annual dividend of a stock by its price. The following data represent the dividend yields (in percent) of a random sample of 28 publicly traded stocks of companies with a value of at least $5 billion.

1.7	0	1.15	0.62	1.06	2.45	2.38
2.83	2.16	1.05	1.22	1.68	0.89	0
2.59	0	1.7	0.64	0.67	2.07	0.94
2.04	0	0	1.35	0	0	0.41

Source: Yahoo! Finance

With the first class having a lower class limit of 0 and a class width of 0.40,

(a) Construct a frequency distribution.
(b) Construct a relative frequency distribution.
(c) Construct a frequency histogram of the data.
(d) Construct a relative frequency histogram of the data.
(e) Describe the shape of the distribution.
(f) Repeat parts (a)–(e) using a class width of 0.8.
(g) Which frequency distribution seems to provide a better summary of the data?

34. Violent Crimes Violent crimes include murder, forcible rape, robbery, and aggravated assault. The following data represent the violent crime rate (crimes per 100,000 population) by state plus the District of Columbia in 2002.

444	563	553	424	593	352
311	599	1,633	770	459	262
255	621	357	286	377	279
662	108	770	484	540	268
343	539	352	314	638	161
375	740	496	470	78	351
503	292	402	285	822	177
717	579	237	107	291	345
234	225	274			

Source: U.S. Federal Bureau of Investigation

With the first class having a lower class limit of 0 and a class width of 150,

(a) Construct a frequency distribution.
(b) Construct a relative frequency distribution.
(c) Construct a frequency histogram of the data.
(d) Construct a relative frequency histogram of the data.
(e) Describe the shape of the distribution.
(f) Repeat parts (a)–(e) using a class width of 300. Which frequency distribution seems to provide a better summary of the data?
(g) Do you believe that the violent crime rate is a good measure of how safe a state is? Why or why not?

In Problems 35–38, construct stem-and-leaf plots.

35. Age at Inauguration The following data represent the
NW ages of the presidents of the United States on inauguration day.

57	61	57	57	58
54	68	51	49	64
65	52	56	46	54
47	55	55	54	42
55	51	54	51	60
55	56	61	52	69
57	50	49	51	62
61	48	50	56	43
64	46	54		

36. Divorce Rate The following data represent the divorce
rates (per 1000 population) for most states in the United
States in the year 2001. (**Note:** The list includes the District
of Columbia but excludes California, Colorado, Indiana,
and Louisiana because of failure to report.)

5.3	4.1	4.2	6.6	2.9	4.0	2.3
5.4	3.8	3.8	5.6	3.2	3.2	3.2
5.5	3.9	3.0	2.4	3.9	3.3	5.4
4.3	2.6	3.7	6.8	5.0	3.5	5.1
3.0	4.5	2.7	4.0	3.4	4.9	3.2
3.3	3.5	3.4	5.2	4.1	4.4	4.0
4.3	4.5	5.2	3.2	6.1		

Source: U.S. Census, Statistical Abstract of the United States,
2003

37. Grams of Fat in a McDonald's Breakfast The following
data represent the number of grams of fat in breakfast
meals offered at McDonald's.

12	23	28	2	28	33
31	11	23	40	35	1
23	33	23	16	11	8
8	17	16	15		

*Source: McDonald's Corporation, A Full
Serving of Nutrition Facts, April 2003*

38. Miles per Gallon The following data represent the number of miles per gallon achieved on the highway for compact cars for the model year 2005.

30	29	30	21	18	29	27	30	29
34	34	30	28	30	20	32	28	32
34	35	26	26	31	25	35	32	25
19	26	19	24	22	24	19	31	26
34	32	34	25	34	34	32	29	25
31	29	30	30	34	32	29	38	39
46	31	31	30	27	29	26	29	24

Source: U.S. Department of Energy

39. Energy from Coal The following data represent the percentage of energy derived
from coal for the 50 states plus the District of Columbia.

54.2	0	75.0	64.3	15.6	8.0	94.2
8.5	32.4	90.4	34.6	87.7	55.7	0
40.6	62.3	23.2	83.1	16.6	0	50.8
48.5	13.3	2.7	60.2	60.4	38.3	8.4
1.3	0.9	59.5	63.1	94.6	42.4	98.1
77.6	46.1	27.4	51.1	90.4	62.1	68.1
10.3	93.7	56.6	23.3	60.8	36.8	95.8
57.7	83.2					

Source: U.S. Energy Information Administration

(a) Round each observation to the nearest percent and draw a stem-and-leaf
diagram.
(b) Describe the shape of the distribution.
(c) West Virginia is one of the largest coal mining states. Which number in the data
set do you think is West Virginia?

40. Housing Prices The data at the right represent the percentage change in the price of housing from 1998 to 2003 for a random sample of 40 cities.

(a) Round each observation to the nearest percent and draw a stem-and-leaf diagram.

(b) Describe the shape of the distribution.

23.3	20.8	32.5	15.8	47.1	18.9	66.0	22.6
23.4	24.1	16.2	9.1	17.1	22.7	17.0	21.6
29.9	15.6	24.8	52.4	28.3	53.5	17.8	20.6
20.6	37.6	49.4	62.4	11.8	19.2	19.8	59.1
48.1	19.1	35.9	14.7	24.9	25.0	26.1	47.7

Source: Global Insight

41. Dependability Survey J.D. Power and Associates regularly surveys car owners and NW asks them about the reliability of their cars. The following data represent the number of problems per 100 vehicles for the 2000 model year for all makes.

Make	Problems per 100 Vehicles	Make	Problems per 100 Vehicles	Make	Problems per 100 Vehicles
Lexus	163	Subaru	266	Audi	318
Infiniti	174	Nissan	267	Mercedes Benz	318
Buick	179	GMC	269	Jeep	321
Porsche	193	Chevrolet	272	Volvo	330
Acura	196	Saturn	273	Mitsubishi	339
Toyota	201	Oldsmobile	283	Hyundai	342
Cadillac	209	Mazda	288	Isuzu	368
Lincoln	212	Pontiac	293	Volkswagen	391
Honda	218	Chrysler	295	Suzuki	403
Mercury	240	Ford	295	Daewoo	421
Jaguar	247	Plymouth	302	Land Rover	441
Saab	255	Dodge	312	Kia	509
BMW	262				

Source: Associated Press

(a) Round the data to the nearest tens (for example, round 163 as 160).

(b) Draw a stem-and-leaf diagram, treating the hundreds position as the stem and the tens position as the leaf. For example, for the observation 160, 1 is the stem and 6 is the leaf. Do you think this is a good summary of the data? Why?

(c) Redraw the stem-and-leaf diagram using split stems. For example, data between 200 and 240 is one stem and data between 250 and 290 is a second stem. Does this stem-and-leaf diagram better summarize the data? Why?

42. Crime Rates The data on the following page represent the violent crime rates per 100,000 population for the 50 states and the District of Columbia in 2002.

(a) Round the data to the nearest tens (for example, round 563 to 560).

(b) Draw a stem-and-leaf diagram, treating the hundreds position as the stem and the tens position as the leaf. For the observation 560, 5 is the stem and 6 is the leaf. Do you think this is a good summary of the data? Why?

(c) Redraw the stem-and-leaf diagram using split stems. For example, data between 200 and 240 is one stem and data between 250 and 290 is a second stem. Does this stem-and-leaf diagram better summarize the data? Why?

State	Violent Crime Rate	State	Violent Crime Rate	State	Violent Crime Rate
Alabama	444	Kentucky	279	North Dakota	78
Alaska	563	Louisiana	662	Ohio	351
Arizona	553	Maine	108	Oklahoma	503
Arkansas	424	Maryland	770	Oregon	292
California	593	Massachusetts	484	Pennsylvania	402
Colorado	352	Michigan	540	Rhode Island	285
Connecticut	311	Minnesota	268	South Carolina	822
Delaware	599	Mississippi	343	South Dakota	177
District of Columbia	1633	Missouri	539	Tennessee	717
Florida	770	Montana	352	Texas	579
Georgia	459	Nebraska	314	Utah	237
Hawaii	262	Nevada	638	Vermont	107
Idaho	255	New Hampshire	161	Virginia	291
Illinois	621	New Jersey	375	Washington	345
Indiana	357	New Mexico	740	West Virginia	234
Iowa	286	New York	496	Wisconsin	225
Kansas	377	North Carolina	470	Wyoming	274

Source: Federal Bureau of Investigation

In Problems 43 and 44, we compare data sets. A great way to compare two data sets is through back to back stem-and-leaf diagrams. The figure represents the number of grams of fat in 14 sandwiches served at McDonald's and 14 sandwiches served at Burger King. In the figure, 0|1|5 represents 10 for McDonald's and 15 for Burger King.
(*Source:* McDonald's Corporation, *A Full Serving of Nutrition Facts*, April 2003; Burger King Corporation, *Great Taste Table*, January 1999.

```
 McDonald's   Burger King
        740 | 1 | 5 9
     766631 | 2 | 4 6 7 8
       7420 | 3 | 6
          8 | 4 | 0 2 3 3 8
            | 5 | 9
            | 6 | 7
```

43. **Academy Award Winners** The following data represent the ages of the Academy Award winners for Best Actor and Best Actress in a leading role for the 30 years from 1975 to 2004.

Best Actor Ages					
38	60	30	40	42	37
76	39	52	45	35	61
43	51	32	42	54	52
37	38	31	45	60	45
40	36	47	29	43	37

Best Actress Ages					
41	35	31	41	33	31
74	33	49	38	61	21
41	25	80	42	28	33
35	45	49	39	34	24
25	33	35	35	28	30

(a) Construct a back to back stem-and-leaf display.
(b) Compare the two populations. What can you conclude from the back to back stem-and-leaf display?

44. Home Run Distances In 1998, Mark McGwire of the St. Louis Cardinals set the record for the most home runs hit in a season by hitting 70 home runs. Three years later in 2001, Barry Bonds of the San Francisco Giants broke McGwire's record by hitting 73 home runs. The following data represent the distances of each player's home runs in his record-setting season.

Mark McGwire						
360	370	370	430	420	340	460
410	440	410	380	360	350	527
380	550	478	420	390	420	425
370	480	390	430	388	423	410
360	410	450	350	450	430	461
430	470	440	400	390	510	430
450	452	420	380	470	398	409
385	369	460	390	510	500	450
470	430	458	380	430	341	385
410	420	380	400	440	377	370

Barry Bonds						
420	417	440	410	390	417	420
410	380	430	370	420	400	360
410	420	391	416	440	410	415
436	430	410	400	390	420	410
420	410	410	450	320	430	380
375	375	347	380	429	320	360
375	370	440	400	405	430	350
396	410	380	430	415	380	375
400	435	420	420	488	361	394
410	411	365	360	440	435	454
442	404	385				

(a) Construct a back to back stem-and-leaf display.
(b) Compare the two populations. What can you conclude from the back to back stem-and-leaf display?

NW 45. Waiting Draw a dot plot of the waiting data from Problem 27.

46. Highway Repair Draw a dot plot of the highway repair data from Problem 28.

Technology Step-by-Step   **Drawing Histograms and Stem-and-Leaf Plots**

TI-83/84 Plus   **Histograms**

Step 1: Enter the raw data in L1 by pressing STAT and selecting 1: Edit.

Step 2: Press 2nd Y = to access the StatPlot menu. Select 1: Plot1.

Step 3: Place the cursor on "ON" and press ENTER.

Step 4: Place the cursor on the histogram icon (see the figure) and press ENTER. Press 2nd QUIT to exit Plot 1 menu.

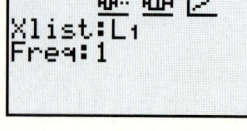

Step 5: Press WINDOW. Set Xmin to the lower class limit of the first class. Set Xmax to the lower class limit of the class following the class containing the largest value. For example, if the first class is 0–9, set Xmin to 0. If the class width is 10 and the last class is 90–99, set Xmax to 100. Set Xscl to the class width. Set Ymin to 0. Set Ymax to a value larger than the frequency of the class with the highest frequency.

Step 6: Press GRAPH.

Helpful Hints: To determine each class frequency, press TRACE and use the arrow keys to scroll through each class. If you decrease the value of Ymin to a value such as −5, you can see the values displayed on the screen easier. The TI graphing calculators do not draw stem-and-leaf plots or dotplots.

MINITAB   **Histograms**

Step 1: Enter the raw data in C1.

Step 2: Select the **Graph** menu and highlight **Histogram** . . .

Step 3: Highlight the "simple" icon and press OK.

Step 4: Put the cursor in the "Graph variables" box. Highlight C1, and press Select. Click SCALE and select the Y-Scale Type tab. For a frequency histogram, click the frequency radio button. For a relative frequency histogram, click the percent radio button. Click OK twice.

Note: To adjust the class width and to change the labels on the horizontal axis to the lower class limit, double-click inside one of the bars in the histogram. Select the "binning" tab in the window that opens. Click the cutpoint button and the midpoint/cutpoint position radio button. In the midpoint/cutpoint box, enter the lower class limits of each class. Click OK.

Stem-and-Leaf Plots

Step 1: With the raw data entered in C1, select the **Graph** menu and highlight **Stem-and-Leaf**.

Step 2: Select the data in C1 and press OK.

Dot Plots

Step 1: Enter the raw data in C1.

Step 2: Select the **Graph** menu and highlight **Dotplot**.

Step 3: Highlight the "simple" icon and press OK.

Step 4: Put the cursor in the "Graph variables" box. Highlight C1 and press Select. Click OK.

Excel ### Histograms

Step 1: Enter the raw data in column A.

Step 2: Select **Tools** and **Data Analysis** . . .

Step 3: Select Histogram from the list.

Step 4: With the cursor in the Input Range cell, use the mouse to highlight the raw data. Select the Chart Output box and press OK.

Step 5: Double-click on one of the bars in the histogram. Select the Options tab from the menu that appears. Reduce the gap width to zero.

Excel does not draw stem-and-leaf or dot plots.

2.3 Additional Displays of Quantitative Data

Objectives
1. Construct frequency polygons
2. Create cumulative frequency and relative frequency tables
3. Construct frequency and relative frequency ogives
4. Draw time-series graphs

In this section, we continue to organize and summarize quantitative data.

1 Construct Frequency Polygons

Another way of graphically representing quantitative data sets is through *frequency polygons*. They provide the same information as histograms. Before we can provide a method for constructing frequency polygons, we must learn how to obtain the class midpoint of a class.

Definition The **class midpoint** is found by adding consecutive lower class limits and dividing the result by 2.

The class midpoint is used to draw frequency polygons.

Definition

A **frequency polygon** is drawn by plotting a point above each class midpoint on a horizontal axis at a height equal to the frequency of the class. After the points for each class are plotted, straight lines are drawn between consecutive points.

USING TECHNOLOGY

Statistical spreadsheets and certain graphing calculators have the ability to create frequency polygons.

Suppose we wish to construct a frequency polygon of the data summarized in Table 13 on page 80 from Section 2.2. First, we need to determine the class midpoints of each class, as shown in Table 18.

We then plot points whose x-coordinate is the class midpoint and y-coordinate is the frequency. Finally, we connect the points with straight lines and obtain Figure 17.

Table 18			
Class (3-year rate of return)	**Class Midpoint**	**Frequency**	**Relative Frequency**
0–1.99	$\dfrac{0 + 2}{2} = 1$	2	0.05
2–3.99	3	5	0.125
4–5.99	5	6	0.15
6–7.99	7	8	0.2
8–9.99	9	9	0.225
10–11.99	11	6	0.15
12–13.99	13	3	0.075
14–15.99	15	1	0.025

Figure 17

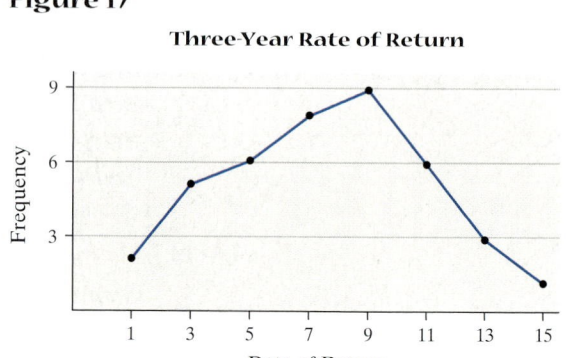

Now Work Problem 15(c).

2 **Create Cumulative Frequency and Relative Frequency Tables**

Since quantitative data can be ordered (that is, written in ascending or descending order), they can be summarized in a cumulative frequency distribution and a cumulative relative frequency distribution.

Definitions

A **cumulative frequency distribution** displays the aggregate frequency of the category. In other words, for discrete data, it displays the total number of observations less than or equal to the category. For continuous data, it displays the total number of observations less than or equal to the upper class limit of a class.

A **cumulative relative frequency distribution** displays the proportion (or percentage) of observations less than or equal to the category for discrete data and the proportion (or percentage) of observations less than or equal to the upper class limit for continuous data.

The cumulative frequency for a class is obtained by adding the frequencies of the classes less than or equal to the upper class limit of the class. For example, the cumulative frequency for the second class is the sum of the frequencies of classes 1 and 2, the cumulative frequency for the third class is the sum of the frequencies of classes 1, 2, and 3, and so on.

The cumulative relative frequency for continuous data is obtained using the same procedures used in creating the cumulative frequency distribution, except that we use the relative frequencies instead of the frequencies.

Table 19 displays the cumulative frequency and cumulative relative frequency of the data summarized in Table 13 in Section 2.2. From the table, we see

Table 19

Class (three-year rate of return)	Frequency	Relative Frequency	Cumulative Frequency	Cumulative Relative Frequency
0–1.99	2	0.05	2	0.05
2–3.99	5	0.125	2 + 5 = 7	0.05 + 0.125 = 0.175
4–5.99	6	0.15	7 + 6 = 13	0.175 + 0.15 = 0.325
6–7.99	8	0.2	21	0.525
8–9.99	9	0.225	30	0.75
10–11.99	6	0.15	36	0.9
12–13.99	3	0.075	39	0.975
14–15.99	1	0.025	40	1

that 36 of the 40 mutual funds had 3-year rates of return of 11.99% or less. The cumulative relative frequency distribution is shown in the fifth column of the table. We see that 90% of the mutual funds had a 3-year rate of return of 11.99% or less. We can also see that a mutual fund with a 3-year rate of return of 14% or higher is outperforming 97.5% of its peers.

Now Work Problems 15(a) and (b).

3   Construct Frequency and Relative Frequency Ogives

Recall that the cumulative frequency of a class is the aggregate frequency less than or equal to the upper class limit.

Definition

An **ogive** (read as "oh jīve") is a graph that represents the cumulative frequency or cumulative relative frequency for the class. It is constructed by plotting points whose x-coordinates are the upper class limits and whose y-coordinates are the cumulative frequencies or cumulative relative frequencies. After the points for each class are plotted, straight lines are drawn between consecutive points.

We can construct a relative frequency ogive using the data in Table 19 by plotting points whose x-coordinate is the class upper class limit and whose y-coordinate is the cumulative relative frequency of the class. We then connect the points with straight lines. See Figure 18.

Figure 18

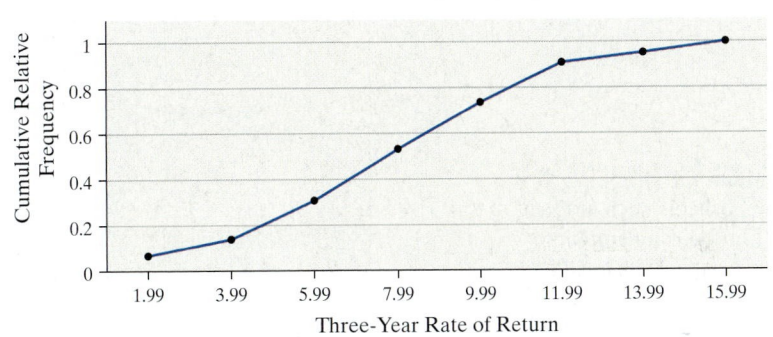

Relative Frequency Ogive

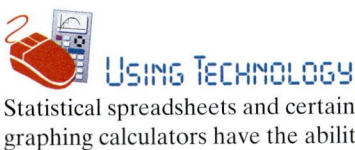

Using Technology:
Statistical spreadsheets and certain graphing calculators have the ability to draw ogives.

From Figure 18 we can see that 30% of the mutual funds had a 3-year rate of return less than or equal to 5.99%.

Now Work Problems 15(d) and (e).

4 Draw Time-Series Graphs

If the value of a variable is measured at different points in time, the data are referred to as **time-series data**. The closing price of Cisco Systems stock each month for the past 12 years is an example of time-series data.

Definition

> A **time-series plot** is obtained by plotting the time in which a variable is measured on the horizontal axis and the corresponding value of the variable on the vertical axis. Lines are then drawn connecting the points.

Time-series plots are very useful in identifying trends in the data.

EXAMPLE 1 Drawing a Time-Series Plot

Problem: The data in Table 20 represent the closing price of Cisco Systems stock at the end of each month from March 2003 through February 2005. Construct a time-series plot of the data.

Approach

Step 1: Plot points for each month, with the date on the horizontal axis and the closing price on the vertical axis.

Step 2: Connect the points with straight lines.

Solution: Figure 19 shows the graph of the time-series plot. The trend since June, 2004 does not bode well for investors of Cisco Systems stock.

Table 20			
Data	**Closing Price**	**Date**	**Closing Price**
3/03	12.98	3/04	23.57
4/03	15.00	4/04	20.91
5/03	16.41	5/04	22.37
6/03	16.79	6/04	23.70
7/03	19.49	7/04	20.92
8/03	19.14	8/04	18.76
9/03	19.59	9/04	18.10
10/03	20.93	10/04	19.21
11/03	22.70	11/04	18.75
12/03	24.23	12/04	19.32
1/04	25.71	1/05	18.04
2/04	23.16	2/05	17.42

Source: NASDAQ

Using Technology:
Statistical spreadsheets, such as Excel or MINITAB, and certain graphing calculators, such as the TI-83 or TI-84 Plus, have the ability to create time-series graphs.

Figure 19

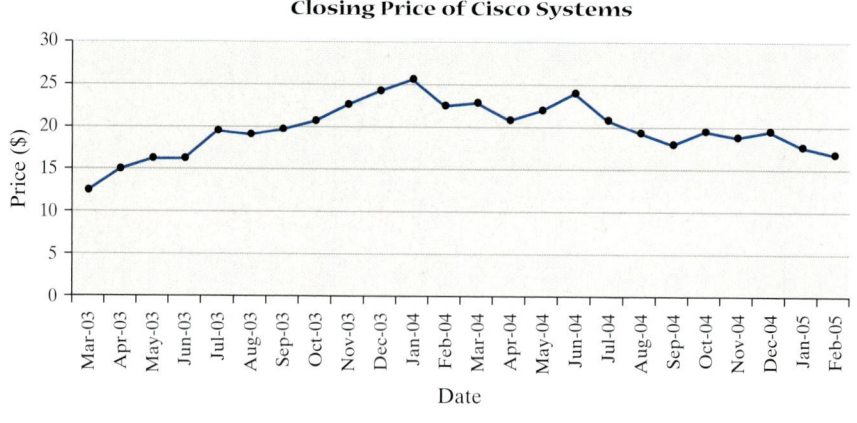

Now Work Problem 17.

2.3 ASSESS YOUR UNDERSTANDING

Concepts and Vocabulary

1. Which do you prefer, histograms, stem-and-leaf plots, or frequency polygons? Be sure to support your opinion. Are there circumstances in which one might be preferred over another?
2. The cumulative relative frequency for the last class must always be 1. Why?
3. What is an ogive?
4. What are time-series data?
5. *True or False*: When plotting an ogive, the plotted points have *x*-coordinates that are equal to the upper limits of each class.
6. *True or False*: When plotting a frequency polygon, we plot the percentages for each class above the midpoint and connect the points with straight line segments.

Skill Building

7. **Age of Population** The following frequency polygon shows the number of U.S. residents in 2003 from 20 to 79 years old (in millions).
   (*Source*: U.S. Census Bureau)

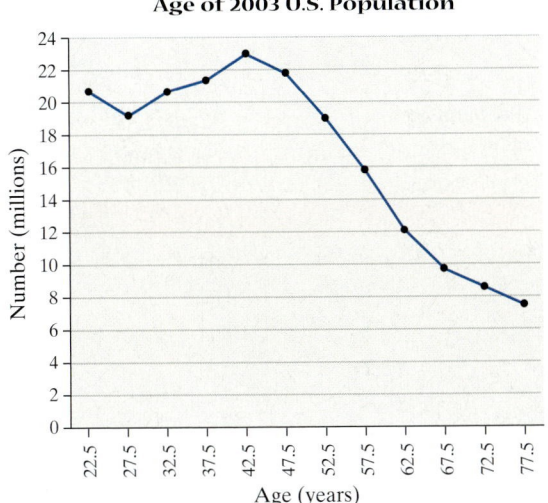

Age of 2003 U.S. Population

(a) What is the class width?
(b) What is the lower class limit of the fourth class?
(c) What is the upper class limit of the fourth class?
(d) Which class (age group) has the highest frequency?
(e) Which class (age group) has the lowest frequency?

8. **Cause of Death** The following frequency polygon represents the number of deaths due to accidents in 2001 for people 5 to 84 years old.
   (*Source:* U.S. National Center for Health Statistics)

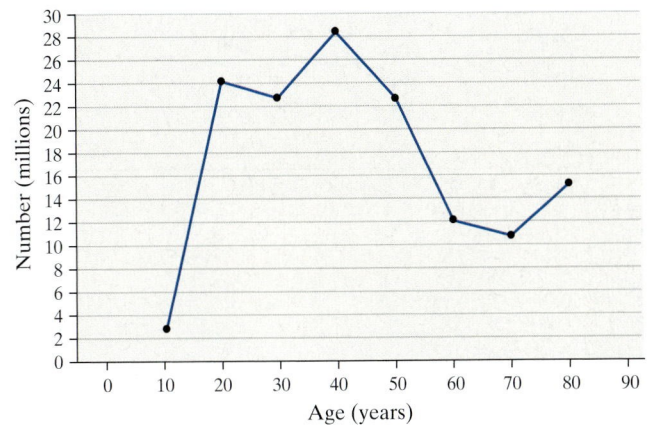

Number of Deaths Due to Accidents

(a) What is the class width?
(b) What is the lower class limit of the second class?
(c) What is the upper class limit of the second class?
(d) Which class had the highest number of deaths due to accidents?
(e) Which class had the lowest number of deaths due to accidents?

9. **ACT Scores** The following relative frequency ogive represents the ACT composite score for the high school graduating class of 2004.
   (*Source:* www.act.org)

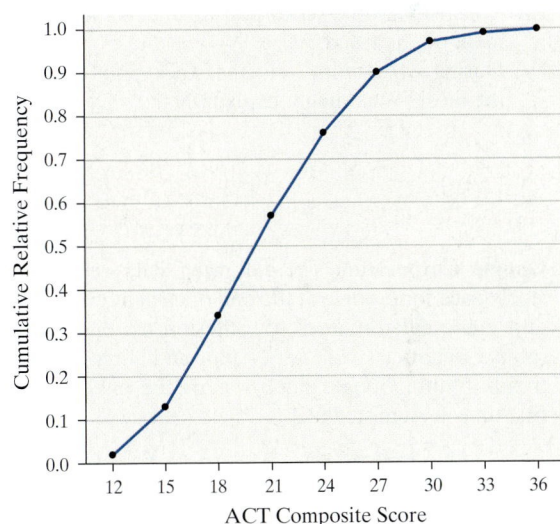

ACT Composite Score

(a) What is the class width?
(b) What is the lower class limit of the third class?
(c) Approximately what percent of students had an ACT composite score of 27 or below?
(d) Using your answer from part (c), approximately *how many* students had an ACT composite score of 27 or below if 1,171,460 students took the exam?

10. **SAT Scores** The frequency ogive to the right represents the SAT verbal scores of 120 randomly selected college-bound students.

   (a) What is the class width?

   (b) What is the upper class limit of the second class?

   (c) Approximately how many students scored 459 or less? 579 or less?

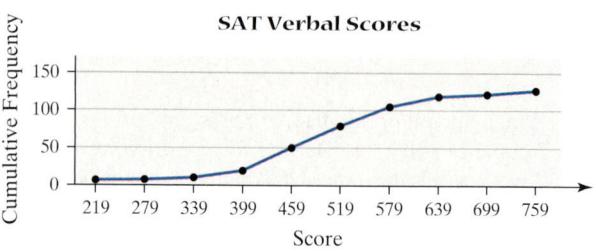

SAT Verbal Scores

Applying the Concepts

In Problems 11–16, use the frequency distributions in the problem indicated from Section 2.2 to do the following:

(a) *Construct a cumulative frequency distribution.*

(b) *Construct a cumulative relative frequency distribution.*

(c) *Draw a frequency polygon.*

(d) *Draw a frequency ogive.*

(e) *Draw a relative frequency ogive.*

11. Problem 19

12. Problem 20

13. Problem 21

14. Problem 22

NW 15. Problem 29

16. Problem 30

17. **Pixar Corporation** The data to the right represent the NW stock price for Pixar at the end of each month in 2004, adjusted for dividends and splits.

   (a) Construct a time-series plot of the data and comment on the trend.

   (b) During what month did the stock price increase the most? What might explain this?

Date	Closing Price	Date	Closing Price
1/04	66.39	7/04	68.24
2/04	65.76	8/04	77.72
3/04	64.46	9/04	78.90
4/04	68.27	10/04	80.42
5/04	67.85	11/04	90.67
6/04	69.51	12/04	85.61

18. **Google Corporation** The following data represent the stock price for Google at the end of each month from August 2004 to February 2005, adjusted for dividends and splits. Construct a time-series plot and comment on any trends. (**Note:** Google stock was traded publicly for the first time in August 2004.)

Google

Date	Closing Price
8/04	102.37
9/04	129.60
10/04	190.64
11/04	181.98
12/04	192.79
1/05	187.99
2/05	185.18

19. **College Enrollment** The following data represent the percentage of recent high school graduates (graduated within 12 months before the given year-end) who enrolled in college in the fall. Construct a time-series plot of the data.

Year	Percent Enrolled	Year	Percent Enrolled
1988	58.9	1996	65.0
1989	59.6	1997	67.0
1990	60.1	1998	65.6
1991	62.5	1999	62.9
1992	61.9	2000	63.3
1993	62.6	2001	61.7
1994	61.9	2002	65.2
1995	61.9	2003	63.9

Source: U.S. Center for Education Statistics

20. IRS Audits The following data represent the percentage of tax returns audited by the Internal Revenue Service. Construct a time-series plot of the data.

Year	Percent Audited	Year	Percent Audited
1988	1.57	1996	1.67
1989	1.29	1997	1.28
1990	1.04	1998	0.99
1991	1.17	1999	0.90
1992	1.06	2000	0.49
1993	0.92	2001	0.58
1994	1.07	2002	0.57
1995	1.67	2003	0.65

Source: U.S. General Accounting Office

21. Rates of Return of Stocks Stocks may be categorized by industry. The following data represent the 5-year rates of return for a simple random sample of financial stocks and energy stocks ending March 4, 2004.

Financial Stocks							
17.10	16.26	22.10	9.96	7.94	10.95	16.34	20.43
7.54	26.84	28.02	15.92	10.80	11.27	20.68	11.09
9.84	11.82	6.28	3.27	21.97	13.74	33.63	25.53
11.01	18.15	17.36	19.14	17.80	26.33	5.35	8.44

Energy Stocks							
11.43	14.52	22.14	7.03	42.31	15.15	9.43	7.39
30.88	19.50	21.17	16.03	53.61	15.38	42.74	26.34
7.51	45.62	19.67	15.17	8.39	43.50	29.97	6.11
23.84	26.18	38.79	15.35	18.42	16.67	20.93	28.23

Source: Morningstar.com

(a) Construct a frequency distribution for each industry. To make an easy comparison, create each frequency distribution so that the lower class limit of the first class is 0.00 and the class width is 5.00.
(b) Construct a relative frequency distribution for financial stocks, and a relative frequency distribution for energy stocks.
(c) On the same graph, construct a relative frequency polygon for the two industries.
(d) On the same graph, construct a relative frequency ogive for the two industries.
(e) Which industry appears to have the better performance for the 5-year period? Support your opinion.

22. American League versus National League The following data represent the earned-run average (ERA) of the top 40 pitchers in both the American League and National League in 2003. **Note**: ERA is the average number of runs given up per 9 innings.

(a) Why would it be appropriate to use frequencies to compare the two leagues?

(b) Construct a frequency distribution for each league. To make an easy comparison, create each frequency distribution so that the lower class limit of the first class is 2.0 and the class width is 0.50.

(c) On the same graph, construct a frequency polygon for the American and National Leagues.

(d) On the same graph, construct a frequency ogive for the American and National Leagues.

(e) Which league appears to have better pitchers? Support your opinion (**Note**: Be sure to take into account the fact that the National League does not have a designated hitter.)

American League				
2.22	2.70	2.90	3.13	3.25
3.27	3.30	3.40	3.57	3.60
3.77	3.78	3.78	3.87	3.91
4.02	4.09	4.14	4.14	4.18
4.21	4.29	4.34	4.43	4.47
4.49	4.51	4.51	4.57	4.59
4.61	4.63	4.67	4.68	4.85
5.15	5.20	5.56	5.73	5.75

National League				
2.34	2.39	2.43	2.84	2.95
3.09	3.11	3.20	3.20	3.24
3.28	3.54	3.59	3.62	3.68
3.76	3.78	3.81	3.82	3.84
3.87	3.96	3.99	4.00	4.01
4.08	4.11	4.11	4.13	4.13
4.16	4.19	4.23	4.30	4.43
4.45	4.52	4.52	4.59	4.64

Technology	Drawing Frequency Polygons, Ogives, and Time-Series Plots
TI-83/84 Plus	The TI-83 and TI-84 Plus are capable of drawing all three of these graphs.
MINITAB	MINITAB is capable of drawing all three of these graphs.
Excel	Excel is capable of drawing all three of these graphs.

2.4 Graphical Misrepresentations of Data

Objective Describe what can make a graph misleading or deceptive

1 Describe What Can Make a Graph Misleading or Deceptive

Often, statistics gets a bad rap for having the ability to manipulate data to support any position desired. One method of distorting the truth is through graphics. Sometimes graphics *mislead*; other times they *deceive*. We will call graphs misleading if they unintentionally create an incorrect impression. We consider graphs deceptive if they purposely create an incorrect impression. We have already discussed the power that graphical representations of data can have, so it is important to be able to recognize misleading and deceptive graphs. The most common graphical misrepresentation of data is accomplished through the manipulation of the scale of the graph.

EXAMPLE 1 **Misrepresentation of Data**

Problem: The bar graph illustrated in Figure 20 is a *USA Today* type graph. A survey was conducted by Impulse Research for Quilted Northern Confidential in which individuals were asked how they would flush a toilet when the facilities are not sanitary. What's wrong with the graphic?

Figure 20

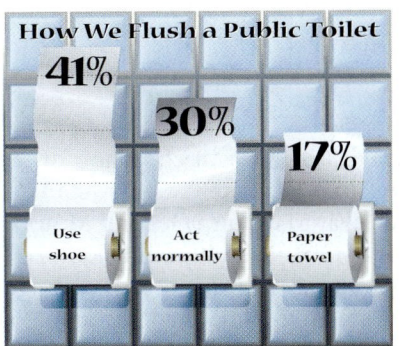

Approach: We need to compare the vertical scales of each bar to see if they accurately depict the percentages given.

Solution: First, it is unclear whether the bars include the roll of toilet paper or not. In either case, the roll corresponding to "use shoe" should be 2.4 (= 41/17) times longer than the roll corresponding to "paper towel." If we include the roll of toilet paper, then the bar corresponding to "use shoe" is less than double the length of "paper towel." If we do not include the roll of toilet paper, then the bar corresponding to "use shoe" is almost exactly double the length of the bar corresponding to "paper towel." The vertical scaling is incorrect.

EXAMPLE 2   **Misrepresentation of Data by Manipulating the Vertical Scale**

Problem: The bar graph shown in Figure 21 depicts the average SAT Math scores of college-bound seniors for the years 1991–2004, based on data from the College Board. Determine why this graph might be considered misrepresentative.

Approach: We need to look at the graph for any characteristics that may mislead a reader, such as manipulation of the vertical scale.

Figure 21

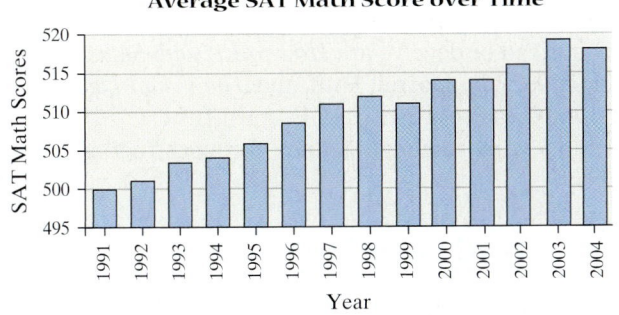

Solution: The graph in the figure may lead a reader to believe that SAT math scores have increased substantially since 1991. While SAT math scores have been increasing, they have not doubled or tripled, as may be inferred from the graph (since the bar for 1997 is three times as high as the bar for 1991). We notice in the figure that the vertical axis begins its labeling at 495 instead of 0. This type of scaling is common when the smallest observed data value is a rather large number. It is not necessarily done purposely to confuse or mislead the reader. Often, the main purpose in graphs is to discover a trend, rather than the actual differences in the data. The trend is clearer in Figure 21 than in Figure 22, where the vertical axis begins at 0.

Often, instead of beginning the axis of a graph at 0 as in Figure 22, the graph is begun at a value slightly less than the smallest value in the data set. However, special care must be taken to make the reader aware of the vertical-axis scaling. Figure 23 shows the proper construction of the graph of the SAT scores, with the graph beginning at 495. The symbol ϟ is used to signify that the graph has a gap in it.

Figure 22

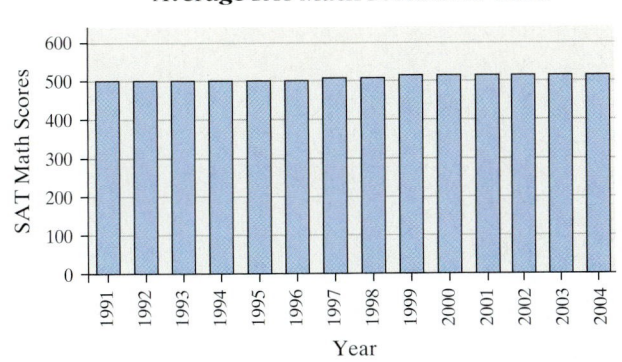

Figure 23

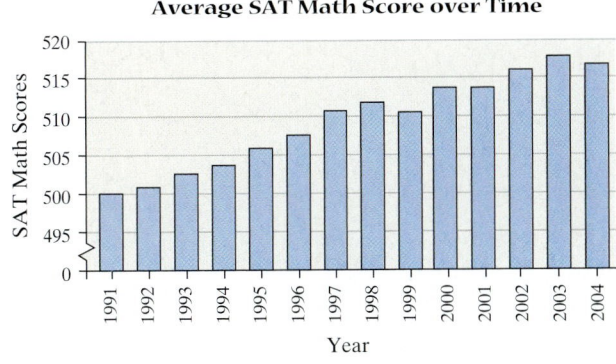

Now Work Problem 5.

In addition to vertical-axis scaling, figures can be misleading through pictures. Consider the next example.

EXAMPLE 3 | **Misleading Graphs**

Figure 24

February 20, 1996

February 28, 2005

Problem: The Dow Jones Industrial Average (DJIA) is a collection of 30 stocks from the stock market that are thought to be representative of the U.S. economy. It includes companies such as Intel, Wal-Mart, and General Motors. Had you invested $10,000 in the DJIA on February 20, 1996, it would have been worth $20,000 on February 28, 2005. To illustrate this investment, a brokerage firm might create the graphic shown in Figure 24. Describe how this graph is misleading.

Approach: Again, we look for characteristics of the graph that seem to manipulate the facts, such as an incorrect depiction of the size of the graphics.

Solution: The graphic on the right of the figure has been doubled in length, width, and height, causing an eightfold increase in the size, thereby misleading the reader into thinking that the size of the investment increased by eight times instead of by two times.

Now Work Problem 15.

There are many ways to create graphs that mislead. Two popular texts written about ways that graphs mislead or deceive are *How to Lie with Statistics* (W. W. Norton & Company, Inc., 1982), by Darrell Huff, and *The Visual Display of Quantitative Information* (Graphics Press, 2001), by Edward Tufte.

We conclude this section with some guidelines for constructing good graphics.

Characteristics of Good Graphics

- Label the graphic clearly and provide explanations if needed.
- Avoid distortion. Don't lie about the data.
- Avoid three dimensions. Three-dimensional charts may look nice, but they distract the reader and often result in misinterpretation of the graphic.
- Do not use more than one design in the same graphic. Sometimes graphs use a different design in a portion of the graphic to draw attention to this area. Don't use this technique. Let the numbers speak for themselves.

MAKING AN INFORMED DECISION

Tables or Graphs?

You work for the school newspaper. Your editor approaches you with a special reporting assignment. Your task is to write an article that describes the "typical" student at your school, complete with supporting information. To write this article, you have to survey at least 40 students and ask them to respond to a questionnaire. The editor would like to have at least two qualitative and two quantitative variables that describe the typical student. The results of the survey will be presented in your article, but you are unsure whether you should present tabular or graphical summaries, so you decide to perform the following experiment.

1. Develop a questionnaire that results in obtaining the values of two qualitative and two quantitative variables. Administer the questionnaire to at least 40 students on your campus.

2. Summarize the data in both tabular and graphical form.

3. Select 20 individuals. (They don't have to be students at your school.) Give the tabular summaries to 10 individuals and the graphical summaries to the other 10. Ask each individual to study the table or graph for 5 seconds. After 1 minute, give a questionnaire that asks various questions regarding the information contained in the table or graph. For example, if you summarized age data, ask the individual which age group had the highest frequency. Record the number of correct answers for each individual. Which summary results in a higher percentage of correct answers, the tables or the graphs? Write a report that discusses your findings.

4. Now use the data collected from the questionnaire to create a couple of misleading graphs. Again, select 20 individuals. Give 10 individuals the misleading graphs and 10 individuals the correct graphs. Ask each individual to study each graph for 5 seconds. After 1 minute has elapsed, give a questionnaire that asks various questions regarding the information contained in the graphs. Record the number of correct answers for each individual. Did the misleading graphs mislead? Write a report that discusses your findings.

2.4 ASSESS YOUR UNDERSTANDING

Applying the Concepts

1. **Inauguration Cost** The following is a *USA Today* type graph. Explain how it is misleading.

2. **Burning Calories** The following is a *USA Today* type graph.

(a) Explain how it is misleading.

(b) What could be done to improve the graphic?

3. **Median Earnings** The following graph shows the median earnings for females from 1998 to 2003.
(*Source:* U.S. Census Bureau, Income, Poverty, and Health Insurance Coverage in the United States, 2003.)

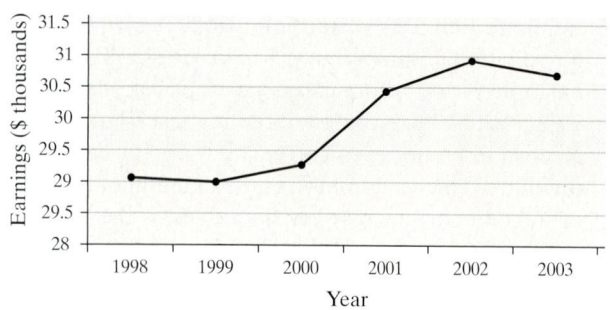

Median Earnings for Females

(a) How is the graph misleading? What does the graph seem to convey?
(b) Redraw the graph so that it is not misleading. What does the new graph seem to convey?

4. **Union Membership** The following relative frequency histogram represents the proportion of employed people aged 25 to 64 years old who were members of a union.
(*Source:* U.S. Bureau of Labor Statistics)

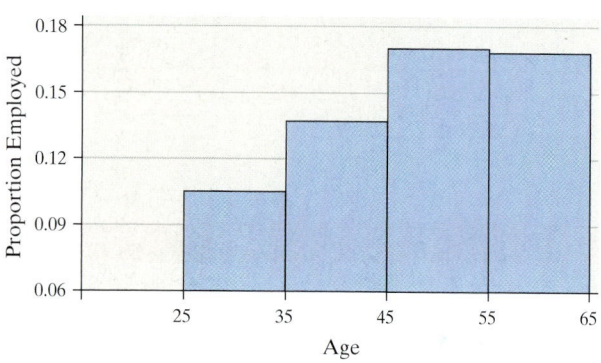

Union Membership

(a) Describe how this graph is misleading. What might a reader conclude from the graph?
(b) Redraw the histogram so that it is not misleading.

5. **Health Insurance** The following relative frequency histogram represents the proportion of people aged 25 to 64 years old not covered by any health insurance in 2002.
(*Source:* U.S. Census Bureau)

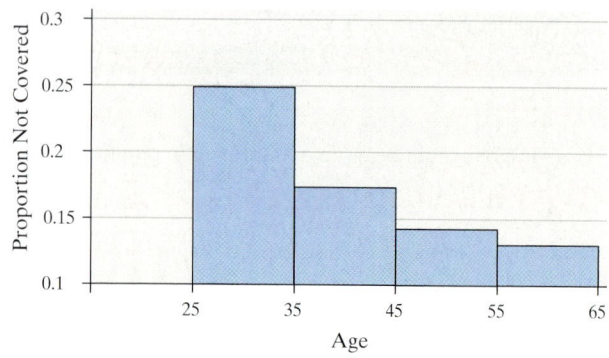

Proportion Not Covered by Health Insurance

(a) Describe how this graph is misleading. What might a reader conclude from the graph?
(b) Redraw the histogram so that it is not misleading.

6. **New Homes** The following time-series plot shows the number of new homes built in the Midwest from 1998 to 2004.
(*Source:* U.S. Census Bureau)

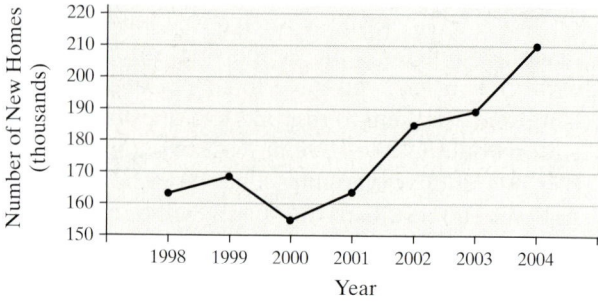

New Homes in Midwest

(a) Describe how this graph is misleading.
(b) What is the graph trying to convey?

7. **Median Income** The following time-series plot shows the median household income for the years 1998 to 2003.
(*Source:* U.S. Census Bureau)

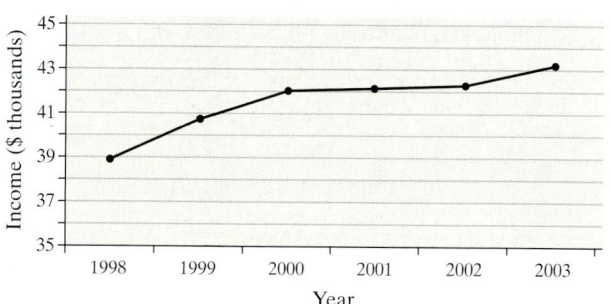

U.S. Median Household Income

(a) Describe how this graph is misleading.
(b) What is the graph trying to convey?

8. **National Debt** The following graphic is a *USA Today* type graph.

How many times larger should the graphic for 1999 be than the 1900 graphic?

9. **Cost of Kids** The following is a *USA Today* type graph based on data from the Department of Agriculture. It represents the percentage of income a middle-income family will spend on their children.

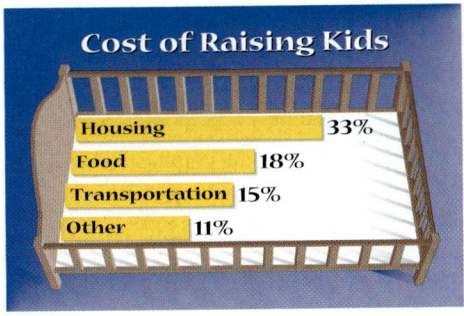

Cost of Raising Kids

Housing	33%
Food	18%
Transportation	15%
Other	11%

(a) How is the graphic misleading?
(b) What could be done to improve the graphic?

10. SAT Combined Scores The following table gives the average SAT combined scores for the years 1999–2003.

Year	Avg. SAT Combined Score
1999	1016
2000	1019
2001	1020
2002	1020
2003	1026

Note: Beginning with the tests administered in March 2005, the SAT test consists of three parts for a total possible score of 1800.

(a) Construct a misleading time-series plot that indicates the average SAT combined score has risen sharply over the given time period.
(b) Construct a time-series plot that is not misleading.

11. Engineering Degrees The following table gives the number of bachelor's degrees in engineering that were awarded from 1999 to 2003.

Year	Degrees Awarded
1999	62,372
2000	63,731
2001	65,113
2002	67,301
2003	70,949

(a) Construct a misleading graph indicating that the number of degrees awarded has more than doubled since 1999.
(b) Construct a graph that is not misleading.

12. Worker Injury The safety manager at Klutz Enterprises provides the following graph to the plant manager and claims that the rate of worker injuries has been reduced by 67% over a 12-year period. Does the graph support his claim? Explain why or why not.

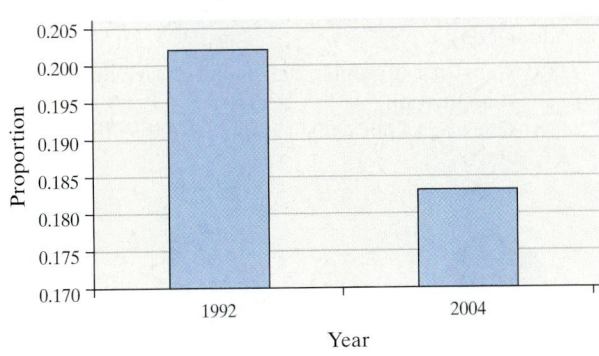

Proportion of Workers Injured

13. Health Care Expenditures The following data represent health care expenditures as a percentage of the U.S. gross domestic product (GDP) from 1997 to 2002. Gross domestic product is the total value of all goods and services created during the course of the year.

Year	Health Care as a Percent of GDP
1997	13.1
1998	13.1
1999	13.2
2000	13.3
2001	14.1
2002	14.9

Source: U.S. Health Care Financing Administration

(a) Construct a time-series plot that a politician would create to support the position that health care expenditures, as a percentage of GDP, are increasing and must be slowed.
(b) Construct a time-series plot that the health care industry would create to refute the opinion of the politician.
(c) Construct a time-series plot that is not misleading.

14. Motor Vehicle Accidents The following data represent the number of motor vehicle accidents and the traffic death rates (number of deaths per 100,000 licensed drivers) from 1998 to 2002.

Year	Motor Vehicle Deaths (in thousands)	Traffic Death Rate (per 100,000 licensed drivers)
1998	43.5	22.4
1999	43.0	22.3
2000	43.0	22.0
2001	43.7	22.1
2002	44.0	22.0

Source: National Safety Council

(a) Construct a time-series graph to support the belief that the roads are becoming less safe.
(b) Construct a time-series graph to support the belief that the roads are becoming safer.

15. Distance Learning Between 1995 and 2001, the enroll-
    NW ment in college-level distance education courses nearly
    doubled.
    (a) Construct a graphic that is not misleading to depict
        this situation.
    (b) Construct a misleading graphic to depict this situation.

16. Overweight Between 1980 and 2002, the number of ado-
    lescents in the United States who were overweight nearly
    tripled.
    (a) Construct a graphic that is not misleading to depict
        this situation.
    (b) Construct a misleading graphic to depict this situation.

Chapter 2 **Review**

Summary

Raw data are first organized into tables. Data are organized by
creating classes into which they fall. Qualitative data and discrete data have values that provide clear-cut categories of
data. However, with continuous data the categories, called
classes, must be created. Typically, the first table created is a
frequency distribution, which lists the frequency with which
each class of data occurs. Other types of distributions include
the relative frequency distribution and the cumulative frequency distribution.

Once data are organized into a table, graphs are created.
For data that are qualitative, we can create bar charts and pie
charts. For data that are quantitative, we can create histograms,
stem-and-leaf plots, frequency polygons, and ogives.
In creating graphs, care must be taken not to draw a graph
that misleads or deceives the reader. If a graph's vertical axis
does not begin at zero, the symbol ⌇ should be used to indicate
the gap that exists in the graph.

Vocabulary

Raw data (p. 60)
Frequency distribution (p. 61)
Relative frequency (p. 62)
Relative frequency distribution (p. 62)
Bar graph (p. 63)
Pareto chart (p. 64)
Side-by-side bar graph (p. 64)
Pie chart (p. 66)
Histogram (p. 78)
Class (p. 79)

Lower and upper class limit (p. 79)
Class width (p. 79)
Open ended (p. 79)
Stem-and-leaf plot (p. 82)
Stem (p. 82)
Leaf (p. 82)
Split stems (p. 85)
Dot plot (p. 86)
Uniform distribution (p. 86)
Bell-shaped distribution (p. 86)

Skewed right (p. 86)
Skewed left (p. 86)
Class midpoint (p. 97)
Frequency polygon (p. 98)
Cumulative frequency distribution (p. 98)
Cumulative relative frequency
    distribution (p. 98)
Ogive (p. 99)
Time-series data (p. 100)
Time series plot (p. 100)

Objectives

Section	You should be able to . . .	Example	Review Exercises
2.1	**1** Organize qualitative data in tables (p. 61)	1, 2	3(a), 4(a), 7(a) and (b), 8(a) and (b)
	2 Construct bar graphs (p. 63)	3 through 5	3(c) and (d), 4(c) and (d), 7(c), 8(c)
	3 Construct pie charts (p. 66)	6	3(e), 4(e), 7(d), 8(d)
2.2	**1** Organize discrete data in tables (p. 77)	1	9(a) and (b), 10(a) and (b)
	2 Construct histograms of discrete data (p. 78)	2	9(e) and (f), 10(e) and (f)
	3 Organize continuous data in tables (p. 79)	3	11(a) and (b), 12(a) and (b), 13 (a) and (b), 14(a) and (b)
	4 Construct histograms of continuous data (p. 81)	4, 5	11(e) and (f), 12(e) and (f), 13(c) and (d), 14(c) and (d)
	5 Draw stem-and-leaf plots (p. 82)	6 through 8	15, 16

Section	You should be able to . . .	Example	Review Exercises
	6 Draw dot plots (p. 86)	9	9(i), 10(i)
	7 Identify the shape of a distribution (p. 86)	10	11(e), 12(c), 13(e), 14(c), 15, 16
2.3	**1** Construct frequency polygons (p. 97)	pp. 97–98	5(f), 6(f)
	2 Create cumulative frequency and relative frequency tables (p. 98)	pp. 98–99	5(b) and (c), 6(b) and (c)
	3 Construct frequency and relative frequency ogives (p. 99)	p. 99	5(h) and (i), 6(h) and (i)
	4 Draw time-series graphs (p. 100)	1	17, 18
2.4	**1** Describe what can make a graph misleading or deceptive (p. 104)	1 through 3	19, 20, 21

Review Exercises

1. Energy Consumption The following bar chart represents the energy consumption of the United States (in quadrillion Btu) in 2003.

(*Source:* Energy Information Administration)

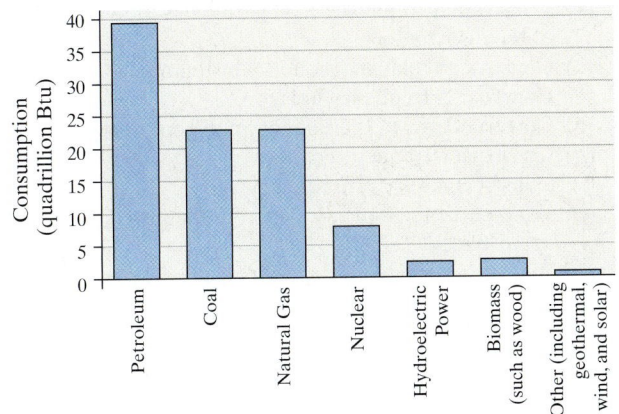

Energy Consumption

(a) Approximately how much energy did the United States consume from natural gas?
(b) Approximately how much energy did the United States consume from biomass?
(c) Approximate the total energy consumption of the United States in 2003.
(d) Which category has the lowest frequency?
(e) Is it appropriate to describe the shape of the distribution as skewed right?

2. Highway Funds The following frequency histogram represents the dollar amount each state and the District of Columbia gets back in federal highway and mass transit grants for each dollar of gasoline tax that their motorists pay into the federal highway trust fund, based on data obtained from the Federal Highway Administration.
(a) Determine the class width.
(b) Identify the classes.
(c) Which class has the highest frequency?

Federal Highway and Mass Transit Grants

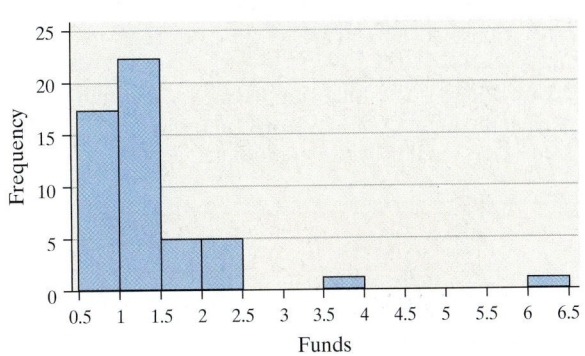

3. Weapons Used in Homicide The following frequency distribution represents the cause of death in homicides for the year 2002.

Type of Weapon	Frequency
Firearms	9369
Knives or cutting instruments	1767
Blunt objects (clubs, hammers, etc.)	666
Personal weapons (hands, fists, etc.)	933
Strangulation	143
Fire	104
Other weapon or not stated	1176

Source: Crime in the United States, 2002, FBI, Uniform Crime Reports

(a) Construct a relative frequency distribution.
(b) What percentage of homicides was committed using a blunt object?
(c) Construct a frequency bar graph.
(d) Construct a relative frequency bar graph.
(e) Construct a pie chart.

4. U.S. Greenhouse Emissions The following frequency distribution represents the total greenhouse emissions in millions of metric tons in 2003 in the United States.

Gas	Emissions
Carbon dioxide	1600.8
Methane	164.1
Nitrous oxide	87.3
Hydrofluorocarbons, perfluorocarbons, and sulfur hexafluoride	39.1

Source: Energy Information Administration

(a) Construct a relative frequency distribution.
(b) What percent of emissions was due to carbon dioxide?
(c) Construct a frequency bar graph.
(d) Construct a relative frequency bar graph.
(e) Construct a pie chart.

5. Live Births The following frequency distribution represents the number of live births (in thousands) in the United States in 2003 by age of mother.

Age of Mother (years)	Births (thousands)
10–14	7
15–19	415
20–24	1032
25–29	1087
30–34	976
35–39	468
40–44	101

Source: National Center for Health Statistics

(a) Construct a relative frequency distribution.
(b) Construct a cumulative frequency distribution.
(c) Construct a cumulative relative frequency distribution.
(d) Construct a frequency histogram. Describe the shape of the distribution.
(e) Construct a relative frequency histogram.
(f) Construct a frequency polygon.
(g) Construct a relative frequency polygon.
(h) Construct a frequency ogive.
(i) Construct a relative frequency ogive.
(j) What percentage of live births was to mothers aged 20 to 24?
(k) What percentage of live births was to mothers of age 30 or older?

6. Suicides The following frequency distribution represents the number of suicides in the United States in 2001, by age.

Age	Suicides	Age	Suicides
5–9	7	55–59	1985
10–14	272	60–64	1332
15–19	1611	65–69	1212
20–24	2360	70–74	1220
25–29	2389	75–79	1219
30–34	2681	80–84	973
35–39	3176	85–89	538
40–44	3459	90–94	200
45–49	3260	95–99	28
50–54	2682		

Source: National Center for Health Statistics

(a) Construct a relative frequency distribution.
(b) Construct a cumulative frequency distribution.
(c) Construct a cumulative relative frequency distribution.
(d) Construct a frequency histogram. Describe the shape of the distribution.
(e) Construct a relative frequency histogram.
(f) Construct a frequency polygon.
(g) Construct a relative frequency polygon.
(h) Construct a frequency ogive.
(i) Construct a relative frequency ogive.
(j) What percentage of suicides from the distribution was aged 40 to 44?
(k) What percentage of suicides from the distribution provided was 24 years or younger?

7. **Political Affiliation**   A sample of 100 randomly selected registered voters in the city of Naperville was asked their political affiliation: Democrat (D), Republican (R), or Independent (I). The results of the survey are as follows:

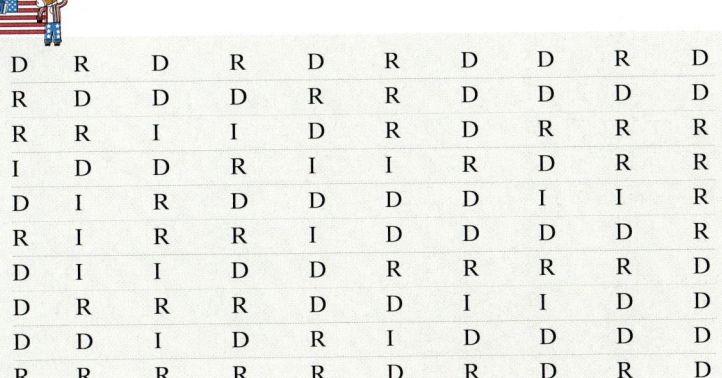

D	R	D	R	D	R	D	D	R	D
R	D	D	D	R	R	D	D	D	D
R	R	I	I	D	R	D	R	R	R
I	D	D	R	I	I	R	D	R	R
D	I	R	D	D	D	D	I	I	R
R	I	R	R	I	D	D	D	D	R
D	I	I	D	D	R	R	R	R	D
D	R	R	R	D	D	I	I	D	D
D	D	I	D	R	I	D	D	D	D
R	R	R	R	R	D	R	D	R	D

(a) Construct a frequency distribution of the data.
(b) Construct a relative frequency distribution of the data.
(c) Construct a relative frequency bar graph of the data.
(d) Construct a pie chart of the data.
(e) What appears to be the most common political affiliation in Naperville?

8. **Educational Attainment**   The Metra Train Company was interested in knowing the educational background of its customers. The company contracted a marketing firm to conduct a survey with a random sample of 50 commuters at the train station. In the survey, commuters were asked to disclose their educational attainment. The following results were obtained:

No high school diploma	Some college	Advanced degree	High school graduate	Advanced degree
High school graduate	High school graduate	High school graduate	High school graduate	No high school diploma
Some college	High school graduate	Bachelor's degree	Associate's degree	High school graduate
No high school diploma	Bachelor's degree	Some college	High school graduate	No high school diploma
Associate's degree	High school graduate	High school graduate	No high school diploma	Some college
Bachelor's degree	Bachelor's degree	Some college	High school graduate	Some college
Bachelor's degree	Advanced degree	No high school diploma	Advanced degree	No high school diploma
High school graduate	Bachelor's degree	No high school diploma	High school graduate	No high school diploma
Associate's degree	Bachelor's degree	High school graduate	Bachelor's degree	Some college
Some college	Associate's degree	High school graduate	Some college	High school graduate

(a) Construct a frequency distribution of the data.
(b) Construct a relative frequency distribution of the data.
(c) Construct a relative frequency bar graph of the data.
(d) Construct a pie chart of the data.
(e) What is the most common educational level of a commuter?

9. Family Size The members of a random sample of 60 couples married for 7 years were asked to give the number of children they have. The results of the survey are as follows:

0	0	3	1	2	3
3	4	3	3	0	3
1	2	1	3	0	3
4	2	3	2	2	4
2	1	3	4	1	3
0	3	3	3	2	1
2	0	3	1	2	3
4	3	3	5	2	0
4	2	2	2	3	3
2	4	2	2	2	2

(a) Construct a frequency distribution of the data.
(b) Construct a relative frequency distribution of the data.
(c) Construct a cumulative frequency distribution of the data.
(d) Construct a cumulative relative frequency distribution of the data.
(e) Construct a frequency histogram of the data. Describe the shape of the distribution.
(f) Construct a relative frequency histogram of the data.
(g) What percentage of couples married 7 years has two children?
(h) What percentage of couples married 7 years has at least two children?
(i) Draw a dot plot of the data.

10. Waiting in Line The following data represent the number of cars that arrived at a McDonald's drive-through between 11:50 A.M. and 12:00 noon each Wednesday for the past 50 weeks:

1	7	3	8	2	3	8	2	6	3
6	5	6	4	3	4	3	8	1	2
5	3	6	3	3	4	3	2	1	2
4	4	9	3	5	2	3	5	5	5
2	5	6	1	7	1	5	3	8	4

(a) Construct a frequency distribution of the data.
(b) Construct a relative frequency distribution of the data.
(c) Construct a cumulative frequency distribution of the data.
(d) Construct a cumulative relative frequency distribution of the data.
(e) Construct a frequency histogram of the data. Describe the shape of the distribution.
(f) Construct a relative frequency histogram of the data.
(g) What percentage of the time did exactly three cars arrive between 11:50 A.M. and 12:00 noon?
(h) What percentage of the time did three or more cars arrive between 11:50 A.M. and 12:00 noon?
(i) Draw a dot plot of the data.

11. Crime Rate by State The following data represent the crime rate (per 100,000 population) for each state in 2002.

State	Crime Rate	State	Crime Rate	State	Crime Rate
Alabama	4465.2	Kentucky	2902.6	North Dakota	2406.2
Alaska	4309.7	Louisiana	5098.1	Ohio	4107.3
Arizona	6386.3	Maine	2656.0	Oklahoma	4743.2
Arkansas	4157.5	Maryland	4747.4	Oregon	4868.4
California	3943.7	Massachusetts	3094.2	Pennsylvania	2841.0
Colorado	4347.8	Michigan	3874.1	Rhode Island	3589.1
Connecticut	2997.2	Minnesota	3535.1	South Carolina	5297.3
Delaware	3939.0	Mississippi	4159.2	South Dakota	2278.7
District of Columbia	8022.3	Missouri	4602.4	Tennessee	5018.0
Florida	5420.6	Montana	3512.9	Texas	5189.6
Georgia	4507.2	Nebraska	4256.7	Utah	4452.4
Hawaii	6043.7	Nevada	4497.5	Vermont	2530.0
Idaho	3172.5	New Hampshire	2220.0	Virginia	3140.3
Illinois	4016.4	New Jersey	3024.2	Washington	5106.8
Indiana	3750.0	New Mexico	5077.8	West Virginia	2515.2
Iowa	3448.2	New York	2803.7	Wisconsin	3252.7
Kansas	4087.0	North Carolina	4721.4	Wyoming	3580.9

Source: Crime in the United States, 2002. FBI, Uniform Crime Reports.

In (a)–(f), start the first class at a lower class limit of 2000 and maintain a class width of 400.
(a) Construct a frequency distribution.
(b) Construct a relative frequency distribution.
(c) Construct a cumulative frequency distribution.
(d) Construct a cumulative relative frequency distribution.
(e) Construct a frequency histogram. Describe the shape of the distribution.
(f) Construct a relative frequency histogram.
(g) Repeat (a)–(f), using a class width of 1000. In your opinion, which class width provides the better summary of the data? Why?

12. Towing Capacity   The following data represent the towing capacity (in pounds) for selected sport utility vehicles (SUVs).

SUV	Towing Capacity	SUV	Towing Capacity	SUV	Towing Capacity
Acura MDX	4,500	GMC Yukon	8,700	Land Rover Range Rover	7,700
BMW X5	6,000	GMC Yukon XL	12,000	Lincoln Navigator	8,800
Buick Rendezvous	3,500	Honda Passport	4,500	Mitsubishi Montero	5,000
Chevrolet Blazer	5,600	Hummer	8,300	Nissan Pathfinder	5,000
Chevrolet Suburban	12,000	Infiniti QX4	5,000	Pontiac Aztek	3,500
Chevrolet Tahoe	8,700	Isuzu Axiom	4,500	Suzuki XL-7	3,000
Dodge Durango	7,650	Isuzu Rodeo	4,500	Toyota 4Runner	5,000
Ford Escape	3,500	Jeep Cherokee	5,000	Toyota Highlander	3,500
Ford Excursion	10,000	Jeep Grand Cherokee	6,500	Toyota Land Cruiser	6,500
Ford Expedition	8,100	Jeep Liberty	5,000		
GMC Jimmy	5,900	Land Rover Discovery	7,700		

Source: Manufacturers

Start the first class at a lower class limit of 3000 and maintain a class width of 1000:
(a) Construct a frequency distribution.
(b) Construct a relative frequency distribution.
(c) Construct a frequency histogram. Describe the shape of the distribution.
(d) Construct a relative frequency histogram.

13. Diameter of a Cookie   The following data represent the diameter (in inches) of a random sample of 34 Keebler Chips Deluxe™ Chocolate Chip Cookies.

2.3414	2.3010	2.2850	2.3015	2.2850	2.3019	2.2400
2.3005	2.2630	2.2853	2.3360	2.3696	2.3300	2.3290
2.2303	2.2600	2.2409	2.2020	2.3223	2.2851	2.2382
2.2438	2.3255	2.2597	2.3020	2.2658	2.2752	2.2256
2.2611	2.3006	2.2011	2.2790	2.2425	2.3003	

Source: Trina S. McNamara, student at Joliet Junior College

In (a)–(f), start the first class at a lower class limit of 2.2000 and maintain a class width of 0.0200:
(a) Construct a frequency distribution.
(b) Construct a relative frequency distribution.
(c) Construct a cumulative frequency distribution.
(d) Construct a cumulative relative frequency distribution.
(e) Construct a frequency histogram. Describe the shape of the distribution.
(f) Construct a relative frequency histogram.
(g) Repeat (a)–(f) using a class width of 0.0400. In your opinion, which class width provides the better summary of the data? Why?

14. Home Sales The data to the right represent the closing price (in U.S. dollars) of homes sold in a midwest city.

Start the first class at a lower class limit of 85,000 and maintain a class width of 10,000.
(a) Construct a frequency distribution.
(b) Construct a relative frequency distribution.
(c) Construct a frequency histogram. Describe the shape of the distribution.
(d) Construct a relative frequency histogram.

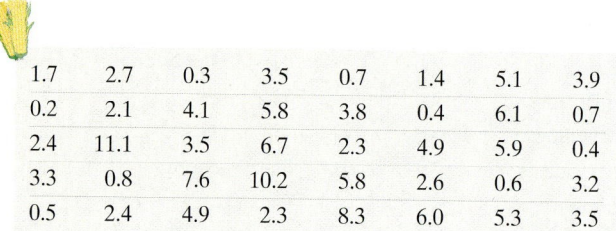

138,820	149,143	99,000	115,000	157,216
169,541	140,794	136,924	124,757	149,380
135,512	153,146	136,833	128,429	136,529
147,500	120,936	95,491	115,744	119,900
89,900	102,696	149,634	123,103	126,630
140,269	183,000	133,646	121,225	121,524
146,439	182,000	110,128	109,520	104,640
124,760	134,305	111,220	121,795	170,072
136,550	115,595	155,507	152,600	130,000
152,537	163,165			

Source: Transamerica Intellitech

15. Eat Your Vegetables! The following data represent the number of servings of vegetables per day that a random sample of forty 20- to 39-year-old females consumes.

1.7	2.7	0.3	3.5	0.7	1.4	5.1	3.9
0.2	2.1	4.1	5.8	3.8	0.4	6.1	0.7
2.4	11.1	3.5	6.7	2.3	4.9	5.9	0.4
3.3	0.8	7.6	10.2	5.8	2.6	0.6	3.2
0.5	2.4	4.9	2.3	8.3	6.0	5.3	3.5

The data are based on a survey conducted by the U.S. Department of Agriculture. Construct a stem-and-leaf diagram of the data, and comment on the shape of the distribution.

16. Fertility Rate The following data represent the fertility rate (births per 1000 women aged 15 to 44) for each state and the District of Columbia.

State	Fertility Rate	State	Fertility Rate	State	Fertility Rate
Alabama	61.2	Kentucky	60.5	North Dakota	58.7
Alaska	73.5	Louisiana	65.4	Ohio	61.7
Arizona	77.8	Maine	49.8	Oklahoma	68.8
Arkansas	66.6	Maryland	60.6	Oregon	61.9
California	68.3	Massachusetts	56.7	Pennsylvania	56.4
Colorado	69.3	Michigan	60.7	Rhode Island	54.6
Connecticut	58.8	Minnesota	62.0	South Carolina	60.7
Delaware	62.2	Mississippi	65.7	South Dakota	68.3
District of Columbia	52.9	Missouri	62.1	Tennessee	62.2
Florida	62.5	Montana	60.3	Texas	77.1
Georgia	68.4	Nebraska	69.5	Utah	90.6
Hawaii	68.6	Nevada	72.5	Vermont	48.9
Idaho	73.8	New Hampshire	52.4	Virginia	61.9
Illinois	66.1	New Jersey	63.5	Washington	60.2
Indiana	64.8	New Mexico	70.7	West Virginia	57.0
Iowa	61.7	New York	59.8	Wisconsin	59.0
Kansas	68.7	North Carolina	65.4	Wyoming	63.6

Source: U.S. National Center for Health Statistics

(a) Round each observation to the nearest whole number and draw a stem-and-leaf diagram.
(b) Describe the shape of the distribution.
(c) Redraw the stem-and-leaf diagram using split stems. For example, data between 60 and 64 is one stem and data between 65 and 69 is a second stem. Does this stem-and-leaf diagram better summarize the data? Why?

17. Federal Minimum Wage Rates The following data represent the value of the minimum wage for the years 1980 to 2003.

Year	Minimum Wage	Year	Minimum Wage
1980	3.10	1992	4.25
1981	3.35	1993	4.25
1982	3.35	1994	4.25
1983	3.35	1995	4.25
1984	3.35	1996	4.75
1985	3.35	1997	5.15
1986	3.35	1998	5.15
1987	3.35	1999	5.15
1988	3.35	2000	5.15
1989	3.35	2001	5.15
1990	3.80	2002	5.15
1991	4.25	2003	5.15

Source: Economic Policy Institute

(a) Construct a time-series plot of the data.
(b) Comment on the apparent trend.

18. Federal Minimum Wage Rates The following data represent the value of the minimum wage for the years 1980 to 2003 in constant 2003 dollars. Constant dollars are dollars adjusted for inflation.

Year	Minimum Wage	Year	Minimum Wage
1980	6.55	1992	5.46
1981	6.48	1993	5.33
1982	6.11	1994	5.22
1983	5.87	1995	5.09
1984	5.64	1996	5.54
1985	5.46	1997	5.89
1986	5.36	1998	5.80
1987	5.19	1999	5.68
1988	5.01	2000	5.50
1989	4.80	2001	5.35
1990	5.19	2002	5.27
1991	5.60	2003	5.15

Source: Economic Policy Institute

(a) Construct a time-series plot of the data.
(b) Comment on the apparent trend.
(c) Compare this time-series plot with the one in Problem 17. Which graph is misleading? Why?

19. Misleading Graphs The following graph was found in a magazine advertisement for skin cream. How is this graph misleading?

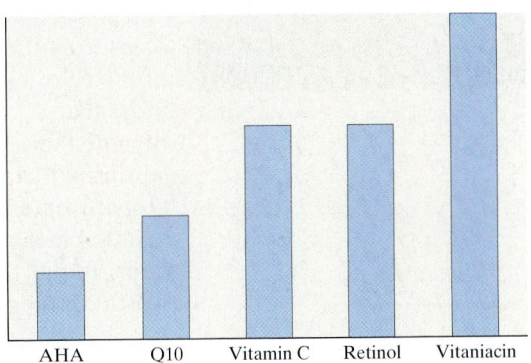

Skin Health
(Moisture Retention)

20. Misleading Graphs The following is a *USA Today* type graph.

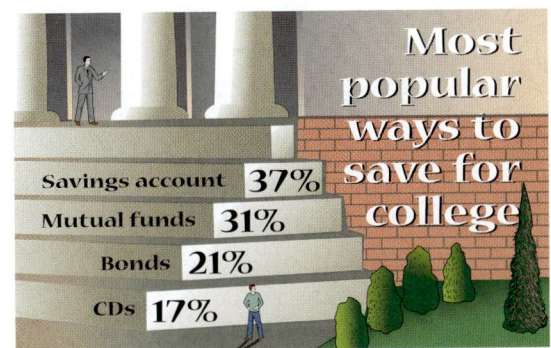

Most popular ways to save for college

Savings account	37%
Mutual funds	31%
Bonds	21%
CDs	17%

Do you think the graph is misleading? Why? If you think it is misleading, what might be done to improve the graph?

21. Misleading Graphs In 2002 the average earnings of a high school graduate were $27,280. At $51,194, the average earnings of a recipient of a bachelor's degree were about 88% higher.
(*Source:* U.S. Census Bureau, Current Population Survey, 2003)
(a) Construct a graph that a college recruiter might create to convince high school students that they should attend college.
(b) Construct a graph that does not mislead.

CASE STUDY

The Day the Sky Roared

Shortly after daybreak on April 3, 1974, thunder began to rumble through the dark skies that covered much of the midwestern United States. Lightning struck areas from the Gulf Coast states to the Canadian border. By the predawn hours of the next day, the affected region of around 490,000 acres was devastated by more than 100 tornadoes. This "super outbreak" was responsible for the deaths of more than 300 people in 11 states. More than 6100 people were injured by the storms, with approximately 27,500 families suffering some kind of loss. The total cost attributed to the disaster was more than $600 million. Amazingly, the storm resulted in six category 5 tornadoes with wind speeds exceeding 261 miles per hour. To put this figure in perspective, the region endured about one decade's worth of category 5 tornadoes in a single 24-hour period!

Fujita Wind Damage Scale		
F-scale	**Wind Speed (mph)**	**Damage**
F-0	Up to 72	Light
F-1	73 to 112	Moderate
F-2	113 to 157	Considerable
F-3	158 to 206	Severe
F-4	207 to 260	Devastating
F-5	Above 260	Incredible

Structural engineers and meteorologists are interested in understanding catastrophic events such as this tornado outbreak. Variables such as tornado intensity (as described by the F-scale), tornado duration (time spent by the tornado in contact with the ground), and death demographics can provide insights into these events and their impact on the human population. The following table lists the duration time and F-scale for each tornado in the April 1974 super outbreak.

Tornado Duration Times for Outbreak of April 3–4, 1974	
F-scale	**Tornado Duration (minutes)**
F-0	1,1,5,1,1,6,4,10,5,4,1,1,1,1,1,1,1,1,1,30,1,9
F-1	16,13,9,8,13,10,15,1,17,23,10,8,12,5,20,31,12,5,30,13,7,1,5,13,1,2,5,10,1,20,5
F-2	7,15,2,10,23,10,7,12,8,1,8,19,5,10,15,20,10,13,20,15,13,14,1,4,2,15,30,91,11,5
F-3	9,20,8,16,26,36,10,20,50,17,26,31,21,30,23,28,23,18,35,35,15,25,30,15,22,18,58, 19,23,31,13,26,40,14,11
F-4	120,23,23,42,47,25,22,22,34,50,38,28,39,29,28,25,34,16,40,55,124,30,30,31
F-5	37,69,23,52,61,122

The following tables present the number of deaths as a function of F-scale and community size:

Deaths as a Function of F-scale for April 3–4, 1974, Tornadoes	
F-Scale	**Deaths**
F-0	0
F-1	0
F-2	14
F-3	32
F-4	129
F-5	130

Deaths as a Function of Community Size for Tornado Super Outbreak of April 3–4, 1974	
Community Size	**Deaths**
Rural areas	99
Small communities	77
Small cities	63
Medium cities	56
Large cities	10

Create a report that graphically displays and discusses the tornado-related data. Your report should include the following:

1. A bar graph or pie chart (or both) that depicts the number of tornadoes by F-scale. Generally, only a little more than 1% of all tornadoes exceed F-3 on the Fujita Wind Damage Scale. How does the frequency of the most severe tornadoes of the April 3–4, 1974, outbreak compare with normal tornado formation?

2. A single histogram that displays the distribution of tornado duration for all the tornadoes.

3. Six histograms displaying tornado duration for each of the F-scale categories. Does there appear to be a relationship between duration and intensity? If so, describe this relationship.

4. A bar chart that shows the relationship between the number of deaths and tornado intensity. Ordinarily, the most severe tornadoes (F-4 and F-5) account for more than 70% of deaths. Is the death distribution of this outbreak consistent with this observation?

5. A bar chart that shows the relationship between the number of deaths and community size. Are tornadoes more likely to strike rural areas? Include a discussion describing the number of deaths as a function of community size.

6. A general summary of your findings and conclusions.

Data Source: Abbey, Robert F., and T. Theodore Fujita. "Tornadoes: The Tornado Outbreak of 3–4 April 1974." In *The Thunderstorm in Human Affairs*, 2nd ed, edited by Edwin Kessler, 37–66. Norman, OK: University of Oklahoma Press, 1983. The death figures presented in this case study are based on approximations made from charts by Abbey and Fujita. Additional descriptions of events and normal tornado statistics are derived from Jack Williams's *The Weather Book* (New York: Vintage Books, 1992).

Numerically Summarizing Data

Outline

DECISIONS

Suppose that you are in the market for a used car. To make an informed decision regarding your purchase, you decide to collect as much information as possible. What information is important in helping you make this decision? See the Decisions project on page 180.

●●● Putting It All Together

When we look at a distribution of data, we should consider three characteristics of the distribution: its shape, its center, and its spread. In the last chapter, we discussed methods for organizing raw data into tables and graphs. These graphs (such as the histogram) allow us to identify the shape of the distribution. Recall that we describe the shape of a distribution as symmetric (in particular, bell-shaped or uniform), skewed right, or skewed left.

The center and spread are numerical summaries of the data. The center of a data set is commonly called the

average. There are many ways to describe the average value of a distribution. In addition, there are many ways to measure the spread of a distribution. The most appropriate measure of center and spread depends on the shape of the distribution.

Once these three characteristics of the distribution are known, we can analyze the data for interesting features, including unusual data values, called *outliers*.

3.1 Measures of Central Tendency

Preparing for This Section   Before getting started, review the following:

- Quantitative data (Section 1.1, p. 6)
- Qualitative data (Section 1.1, p. 6)
- Population versus sample (Section 1.1, p. 4)
- Simple random sampling (Section 1.2, pp. 16–19)

Objectives

1 **Determine the arithmetic mean of a variable from raw data**

2 **Determine the median of a variable from raw data**

3 **Determine the mode of a variable from raw data**

4 **Use the mean and the median to help identify the shape of a distribution**

A measure of central tendency numerically describes the average or typical data value. We hear the word *average* in the news all the time:

- The average miles per gallon of gasoline of the 2005 Chevrolet Camaro in city driving is 19 miles.
- According to the U.S. Census Bureau, the national average commute time to work in 2005 was 24.3 minutes.
- According to the U.S. Census Bureau, the average household income in 2003 was $43,527.
- The average American woman is 5′4″ tall and weighs 142 pounds.

!**CAUTION**
Whenever you hear the word *average*, be aware that the word may not always be referring to the mean. One average could be used to support one position, while another average could be used to support a different position.

In this chapter, we discuss three measures of central tendency: the *mean*, the *median*, and the *mode*. While other measures of central tendency exist, these three are the most widely used. When the word *average* is used in the media (newspapers, reporters, and so on) it usually refers to the mean. But beware! Some reporters use the term average to refer to the median or mode. As we shall see, these three measures of central tendency can give very different results!

Before we discuss measures of central tendency, we must consider whether or not we are computing a measure of central tendency that describes a population or one that describes a sample.

Definitions

A **parameter** is a descriptive measure of a population.

A **statistic** is a descriptive measure of a sample.

For example, if we determine the average test score for *all* the students in a statistics class, our population, the average is a parameter. If we compute the average based on a simple random sample of five students, the average is a statistic.

1 **Determine the Arithmetic Mean of a Variable from Raw Data**

When used in everyday language, the word average often stands for the arithmetic mean. To compute the arithmetic mean of a set of data, the data must be quantitative.

Definitions

The **arithmetic mean** of a variable is computed by determining the sum of all the values of the variable in the data set, divided by the number of observations. The **population arithmetic mean**, μ (pronounced "mew"), is computed using all the individuals in a population. The population mean is a parameter. The **sample arithmetic mean**, $\bar{x}$ (pronounced "x-bar"), is computed using sample data. The sample mean is a statistic.

While other types of means exist (see Problem 51), the arithmetic mean is generally referred to as the **mean**. We will follow this practice for the remainder of the text.

In statistics, Greek letters are used to represent parameters, and Roman letters are used to represent statistics. Statisticians use mathematical expressions to describe the method for computing means.

Definitions

In Other Words

To help you remember the difference between a parameter and a statistic, think of the following:

p = parameter = population

s = statistic = sample

In Other Words

To find the mean of a set of data, add up all the observations and divide by the number of observations.

If $x_1, x_2, \ldots, x_N$ are the N observations of a variable from a population, then the population mean, μ, is

$$\mu = \frac{x_1 + x_2 + \cdots + x_N}{N} = \frac{\sum x_i}{N} \tag{1}$$

If $x_1, x_2, \ldots, x_n$ are n observations of a variable from a sample, then the sample mean, $\bar{x}$, is

$$\bar{x} = \frac{x_1 + x_2 + \cdots + x_n}{n} = \frac{\sum x_i}{n} \tag{2}$$

Note that N represents the size of the population, while n represents the size of the sample. The symbol Σ (the Greek letter capital sigma) tells us the terms are to be added. The subscript i is used to make the various values distinct and does not serve as a mathematical operation. For example, x_1 is the first data value, x_2 is the second, and so on.

Let's look at an example to help distinguish the population mean and sample mean.

EXAMPLE 1

Computing a Population Mean and a Sample Mean

Problem: The data in Table 1 represent the first exam score of 10 students enrolled in a section of Introductory Statistics.

Table 1

Student	Score
1. Michelle	82
2. Ryanne	77
3. Bilal	90
4. Pam	71
5. Jennifer	62
6. Dave	68
7. Joel	74
8. Sam	84
9. Justine	94
10. Juan	88

(a) Compute the population mean.
(b) Find a simple random sample of size $n = 4$ students.
(c) Compute the sample mean of the sample obtained in part (b).

Approach

(a) To compute the population mean, we add up all the data values (test scores) and then divide by the number of individuals in the population.
(b) Recall from Section 1.2 that we can use either Table I in Appendix A, a calculator with a random-number generator, or computer software to obtain simple random samples. We will use a TI-84 Plus graphing calculator.
(c) The sample mean is found by adding the data values that correspond to the individuals selected in the sample and then dividing by $n = 4$, the sample size.

Solution

(a) We compute the population mean by adding the scores of all 10 students:

$$\sum x_i = x_1 + x_2 + x_3 + \cdots + x_{10}$$

$$= 82 + 77 + 90 + 71 + 62 + 68 + 74 + 84 + 94 + 88$$

$$= 790$$

Divide this result by 10, the number of students in the class.

$$\mu = \frac{\sum x_i}{N} = \frac{790}{10} = 79$$

Although it was not necessary in this problem, we will agree to round the mean to one more decimal place than that in the raw data.

(b) To find a simple random sample of size $n = 4$ from a population whose size is $N = 10$, we will use the TI-84 Plus random-number generator with a seed of 54. (Recall that this gives the starting point that the calculator uses to generate the list of random numbers.) Figure 1 shows the students in the sample. Bilal (90), Ryanne (77), Pam (71), and Michelle (82) are in the sample.

(c) We compute the sample mean by first adding the scores of the individuals in the sample.

Figure 1

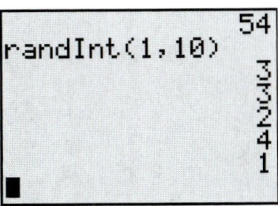

$$\sum x_i = x_1 + x_2 + x_3 + x_4$$
$$= 90 + 77 + 71 + 82$$
$$= 320$$

Divide this result by 4, the number of individuals in the sample.

$$\bar{x} = \frac{\sum x_i}{n} = \frac{320}{4} = 80$$

Now Work Problem 25.

In-Class Activity: Population Mean versus Sample Mean

Treat the students in the class as a population. All the students in the class should determine their pulse rates.

(a) Compute the population mean pulse rate.
(b) Obtain a simple random sample of $n = 4$ students and compute the sample mean. Does the sample mean equal the population mean?
(c) Obtain a second simple random sample of $n = 4$ students and compute the sample mean. Does the sample mean equal the population mean?
(d) Are the sample means the same? Why?

It is helpful to think of the mean of a data set as the center of gravity. In other words, the mean is the value such that a histogram of the data is perfectly balanced, with equal weight on each side of the mean. Figure 2 shows a histogram of the data in Table 1 with the mean labeled. The histogram balances at $\mu = 79$.

Figure 2

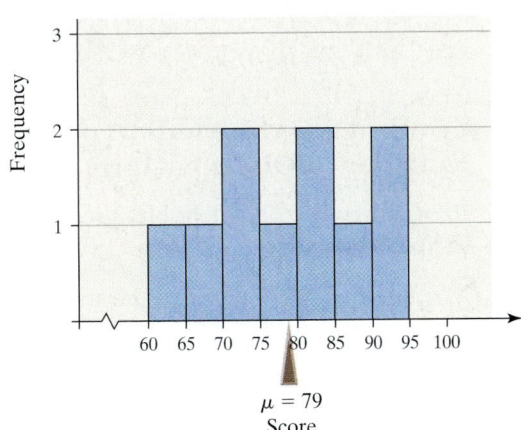

Scores on First Exam

$\mu = 79$
Score

In-Class Activity: The Mean as the Center of Gravity

Find a yardstick, a fulcrum, and three objects of equal weight (maybe 1-kilogram weights from the physics department). Place the fulcrum at 18 inches so that the yardstick balances like a teeter-totter. Now place one weight on the yardstick at 12 inches, another at 15 inches, and the third at 27 inches. See Figure 3.

Figure 3

Does the yardstick balance? Now compute the mean of the location of the three weights. Compare this result with the location of the fulcrum. Conclude that the mean is the center of gravity of the data set.

② Determine the Median of a Variable from Raw Data

A second measure of central tendency is the median. To compute the median of a set of data, the data must be quantitative.

Definition

The **median** of a variable is the value that lies in the middle of the data when arranged in ascending order. That is, half the data are below the median and half the data are above the median. We use M to represent the median.

In Other Words
To help remember the idea behind the median, think of the median of a highway; it divides the highway in half.

To compute the median of a set of data, we use the following steps:

Steps in Computing the Median of a Data Set

Step 1: Arrange the data in ascending order.
Step 2: Determine the number of observations, n.
Step 3: Determine the observation in the middle of the data set.

- If the number of observations is odd, then the median is the data value that is exactly in the middle of the data set. That is, the median is the observation that lies in the $\left(\dfrac{n+1}{2}\right)$ position.
- If the number of observations is even, then the median is the mean of the two middle observations in the data set. That is, the median is the mean of the data values on either side of the observation that lies in the $\left(\dfrac{n+1}{2}\right)$ position.

EXAMPLE 2

Computing the Median of a Data Set with an Odd Number of Observations

Problem: The data in Table 2 represent the length (in seconds) of a random sample of songs released in the 1970s. Find the median length of the songs.

Approach: We will follow the steps listed above.

Solution

Step 1: Arrange the data in ascending order:

179, 201, 206, 208, 217, 222, 240, 257, 284

Table 2	
Song Name	**Length**
"Sister Golden Hair"	201
"Black Water"	257
"Free Bird"	284
"The Hustle"	208
"Southern Nights"	179
"Stayin' Alive"	222
"We Are Family"	217
"Heart of Glass"	206
"My Sharona"	240

Step 2: There are $n = 9$ observations.

Step 3: Since there are an odd number of observations, the median will be the observation exactly in the middle of the data set. The median, M, is 217 seconds (the $\dfrac{n+1}{2} = \dfrac{9+1}{2} = $ 5th data value). We list the data in ascending order, with the median in blue.

$$179, 201, 206, 208, 217, 222, 240, 257, 284$$

Notice there are four observations to the left and four observations to the right of the median.

EXAMPLE 3 **Computing the Median of a Data Set with an Even Number of Observations**

Problem: Find the median score of the data in Table 1 on page 122.

Approach: We will follow the steps given on page 124.

Solution

Step 1: Arrange the data in ascending order:

$$62, 68, 71, 74, 77, 82, 84, 88, 90, 94$$

Step 2: There are $n = 10$ observations.

Step 3: Because there are $n = 10$ observations, the median will be the mean of the two middle observations. Because $\dfrac{n+1}{2} = \dfrac{10+1}{2} = 5.5$, the median is halfway between the 5th and 6th observations. We compute the median, M, by determining the mean of the 5th and 6th observations with the data written in ascending order. So the median is the mean of 77 and 82:

$$M = \dfrac{77 + 82}{2} = 79.5$$

Notice that there are five observations to the left and five observations to the right of the median, as follows:

$$62, 68, 71, 74, 77, 82, 84, 88, 90, 94$$

$$M = 79.5$$

We conclude that 50% of the students scored less than 79.5 and 50% of the students scored above 79.5.

Now compute the median of the data in Problem 19 by hand.

3 Determine the Mode of a Variable from Raw Data

A third measure of central tendency is the mode. The mode can be computed for either quantitative or qualitative data.

Definition The **mode** of a variable is the most frequent observation of the variable that occurs in the data set.

To compute the mode, tally the number of observations that occur for each data value. The data value that occurs most often is the mode. A set of data can have no mode, one mode, or more than one mode. If no observation occurs more than once, we say the data have **no mode**.

EXAMPLE 4 | **Finding the Mode of Quantitative Data**

Problem: The following data represent the number of O-ring failures on the shuttle Columbia prior to its fatal flight for its seventeen flights:

$$0, 0, 0, 0, 0, 0, 0, 0, 0, 0, 0, 1, 1, 1, 1, 2, 3$$

Find the mode number of O-ring failures.

Approach: We tally the number of times we observe each data value. The data value with the highest frequency is the mode.

Solution: The mode is 0 because it occurs most frequently (eleven times).

EXAMPLE 5 | **Finding the Mode of Quantitative Data**

Problem: Find the mode of the data listed in Table 1 on page 122.

Approach: Tally the number of times we observe each data value. The data value with the highest frequency is the mode. Although not necessary, it is helpful to find the mode of quantitative data by arranging the data in ascending order.

Solution: We arrange the data in ascending order:

$$62, 68, 71, 74, 77, 82, 84, 88, 90, 94$$

Since each data value occurs only once, there is no mode.

> **Now compute the mode of the data in Problem 19.**

A data set can have more than one mode. For example, suppose the instructor added the scores of Pam and Sam incorrectly and they actually scored 77 and 88, respectively. The data set in Table 1 would now have two modes: 77 and 88. In this case, we say the data are **bimodal**. If a data set has three or more data values that occur with the highest frequency, the data set is **multimodal**. Typically, the mode is not reported for multimodal data because it is not representative of a central tendency or typical value.

We cannot determine the value of the mean or median of data that are qualitative. The only measure of central tendency that can be determined for qualitative data is the mode.

EXAMPLE 6 | **Determining the Mode of Qualitative Data**

Problem: The data in Table 3 represent the location of injuries that required rehabilitation by a physical therapist. Determine the mode area of injury.

Table 3					
Back	Back	Hand	Neck	Knee	Knee
Wrist	Back	Groin	Shoulder	Shoulder	Back
Elbow	Back	Back	Back	Back	Back
Back	Shoulder	Shoulder	Knee	Knee	Back
Hip	Knee	Hip	Hand	Back	Wrist

Source: Krystal Catton, student at Joliet Junior College

Approach: Determine the location of injury that occurs with the highest frequency.

Solution: The mode location of injury is the back, with 12 instances.

Now Work Problem 39.

EXAMPLE 7   Finding the Mean, Median, and Mode Using Technology

Problem: Use a statistical spreadsheet or calculator to determine the population mean, median, and mode of the student test score data in Table 1 on page 122.

Approach: We will use Excel to obtain the mean, median, and mode. The steps for calculating measures of central tendency using the TI-83/84 Plus graphing calculator, MINITAB, or Excel are given in the Technology Step-by-Step on page 137.

Result: Figure 4 shows the output obtained from Excel. The #N/A in the output indicates that the data set has no mode.

Figure 4

	Student Scores
Mean	79
Standard Error	3.272783389
Median	79.5
Mode	#N/A

4   Use the Mean and the Median to Help Identify the Shape of a Distribution

Often, the mean and the median provide different values. Table 4 shows the mean and median scores on the exam for the data in Table 1 on page 122.

Notice that the median and the mean are close in value. Refer back to Table 1. Suppose Jennifer did not study for the exam and scored 28. The median would not change, but the mean would decrease from 79 to 75.6. We say that the median is **resistant** to extreme values (very large or small), but the mean is not resistant. Therefore, when data sets have unusually large or small values relative to the entire set of data or when the distribution of the data is skewed, the median is the preferred measure of central tendency over the mean because it is more representative of the typical observation.

In fact, the mean and median can be useful in determining the shape of a distribution. It can be shown that, if a distribution is perfectly symmetric and has one mode, then the median will equal the mean (and the mode). So symmetric distributions will have a median and a mean that are close in value. If the mean is substantially larger than the median, the distribution will be skewed right. Do you know why? In distributions that are skewed right, a few data values are substantially larger than the others. These larger data values cause the mean to be inflated while having little, if any, effect on the median. Similarly, distributions that are skewed left will have a mean that is substantially smaller than the median. We summarize these ideas in Table 5 and Figure 5.

Table 4

Mean	79
Median	79.5

! **CAUTION**
Because the mean is not resistant, it should not be reported as a measure of central tendency when the distribution of data is highly skewed.

Table 5

Relation Between the Mean, Median, and Distribution Shape	
Distribution Shape	**Mean versus Median**
Skewed left	Mean substantially smaller than median
Symmetric	Mean roughly equal to median
Skewed right	Mean substantially larger than median

Figure 5
Mean/median versus skewness

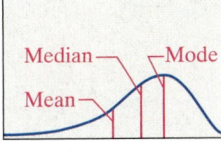

(a) Skewed Left
Mean < Median

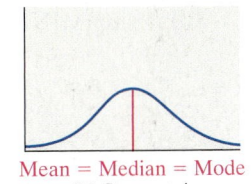

(b) Symmetric
Mean = Median

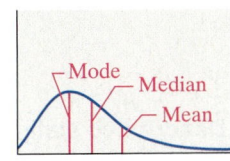

(c) Skewed Right
Mean > Median

EXAMPLE 8 **Describing the Shape of a Distribution**

Problem: In 2004, the New York Yankees had a record $184 million payroll. The data in Table 6 represent the salaries of the players on the opening-day roster in 2004 in thousands of dollars.

Table 6			
Player	**Salary**	**Player**	**Salary**
Brown, Kevin	15,714	Lofton, Kenny	3,100
Cairo, Miguel	900	Matsui, Hideki	7,000
Clark, Tony	750	Mussina, Mike	16,000
Contreras, Jose	8,500	Osborne, Donovan	450
Crosby, Bubba	301	Posada, Jorge	9,000
De Paula, Jorge	303	Quantrill, Paul	3,000
Flaherty, John	775	Rivera, Mariano	10,890
Giambi, Jason	12,429	Rodriguez, Alex	22,000
Gordon, Tom	3,500	Sheffield, Gary	13,000
Heredia, Felix	1,800	Sierra, Ruben	1,000
Hernandez, Orlando	500	Vazquez, Javier	9,000
Jeter, Derek	18,600	White, Gabe	1,925
Karsay, Steve	6,000	Williams, Bernie	12,357
Lee, Travis	2,000	Wilson, Enrique	700
Lieber, Jon	2,700		

Source: usatoday.com

(a) Draw a frequency histogram of the Yankee salaries.
(b) Find the mean and median Yankee salary.
(c) Describe the shape of the distribution of Yankee salaries.
(d) Which measure of central tendency better describes the average salary of a player on the Yankees, the mean or the median?

Approach

(a) We will use MINITAB to draw a histogram of the salaries.
(b) We will use MINITAB to determine the mean and median salary.
(c) We can identify the shape of the distribution by looking at the frequency histogram and comparing the mean to the median. Refer to Table 5 and Figure 5.
(d) If the data are skewed left or skewed right, the median is the better measure of central tendency. If the data are symmetric, the mean is the better measure of central tendency.

Solution

(a) Figure 6 shows a histogram of the data drawn using MINITAB.

Figure 6

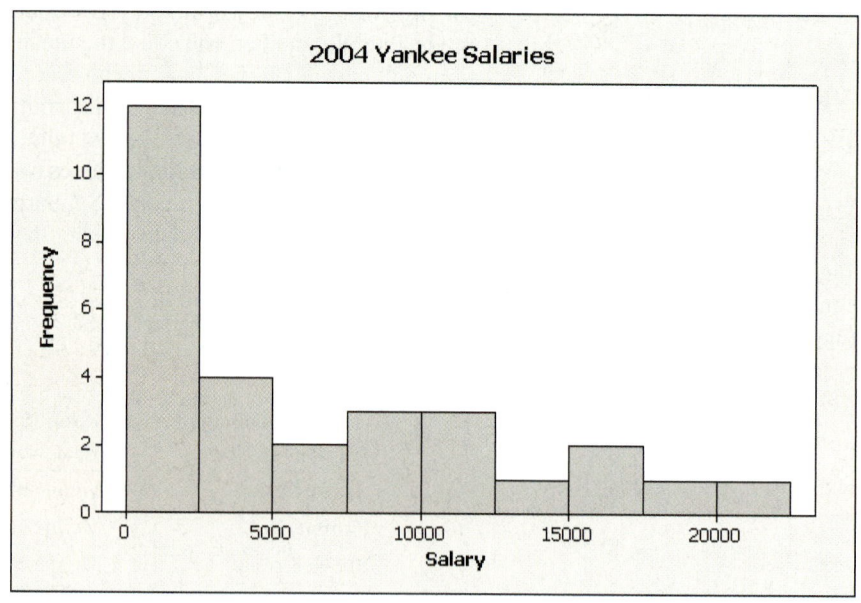

(b) Using MINITAB, we find $\mu = 6352$ and $M = 3100$. See Figure 7.

Figure 7

Descriptive statistics

Variable	N	N*	Mean	SE Mean	StDev	Minimum	Q1	Median	Q3	Maximum
Salaries	29	0	6352	1180	6353	301	838	3100	11624	22000

(c) The shape of the histogram drawn in Figure 6 is skewed right. Notice that the mean is substantially larger than the median because the high salaries (especially Alex Rodriguez and Derek Jeter) push up the value of the mean.

(d) Because the shape of the distribution is skewed right, the median is the better measure of central tendency.

EXAMPLE 9   **Describing the Shape of a Distribution**

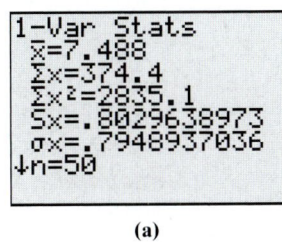

Problem: The data in Table 7 represent the birth weights (in pounds) of 50 randomly sampled babies.

(a) Find the mean and the median.

(b) Describe the shape of the distribution.

(c) Which measure of central tendency better describes the average birth weight?

Table 7

5.8	7.4	9.2	7.0	8.5	7.6
7.9	7.8	7.9	7.7	9.0	7.1
8.7	7.2	6.1	7.2	7.1	7.2
7.9	5.9	7.0	7.8	7.2	7.5
7.3	6.4	7.4	8.2	9.1	7.3
9.4	6.8	7.0	8.1	8.0	7.5
7.3	6.9	6.9	6.4	7.8	8.7
7.1	7.0	7.0	7.4	8.2	7.2
7.6	6.7				

Approach

(a) Use a TI-84 Plus to compute the mean and the median.

(b) The histogram, along with the mean and the median, is used to identify the shape of the distribution.

(c) If the data are roughly symmetric, the mean is a good measure of central tendency. If the data are skewed, the median is a good measure of central tendency.

Solution

(a) Using a TI-84 Plus, we find $\bar{x} = 7.49$ and $M = 7.35$. See Figure 8.

Figure 8

```
1-Var Stats
x̄=7.488
Σx=374.4
Σx²=2835.1
Sx=.8029638973
σx=.7948937036
↓n=50
```
(a)

```
1-Var Stats
↑n=50
minX=5.8
Q₁=7
Med=7.35
Q₃=7.9
maxX=9.4
```
(b)

(b) See Figure 9 for the frequency histogram with the mean and median labeled. The distribution is bell shaped. We have further evidence of the shape because the mean and median are close to each other.

Figure 9
Birth weights of 50 randomly selected babies

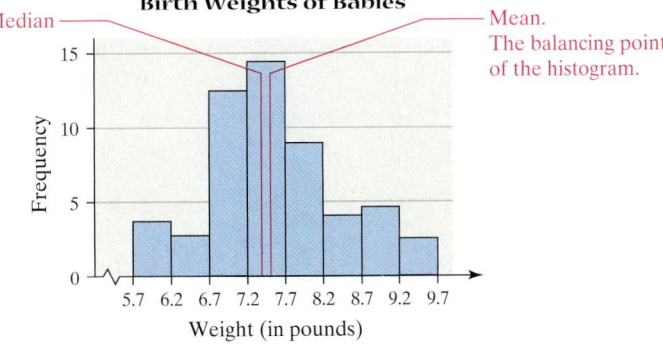

(c) Because the mean and median are close in value, we use the mean as the measure of central tendency.

Now Work Problem 29.

A question you may be asking yourself is "Why would I ever compute the mean?" After all, the mean and median are close in value for symmetric data, and the median is the better measure of central tendency for skewed data. The reason we compute the mean is that much of the statistical inference that we perform is based on the mean. We will have more to say about this in Chapter 8.

We conclude this section with the following chart, which addresses the circumstances under which each measure of central tendency should be used.

Measure of Central Tendency	Computation	Interpretation	When to Use
Mean	Population mean: $\mu = \dfrac{\Sigma x_i}{N}$ Sample mean: $\bar{x} = \dfrac{\Sigma x_i}{n}$	Center of gravity	When data are quantitative and the frequency distribution is roughly symmetric
Median	Arrange data in ascending order and divide the data set in half	Divides the bottom 50% of the data from the top 50%	When the data are quantitative and the frequency distribution is skewed left or skewed right
Mode	Tally data to determine most frequent observation	Most frequent observation	When the most frequent observation is the desired measure of central tendency or the data are qualitative

3.1 ASSESS YOUR UNDERSTANDING

Concepts and Vocabulary

1. What does it mean if a statistic is resistant? Why is the median resistant, but the mean is not? Is the mode a resistant measure of center?

2. Describe how the mean and the median can be used to determine the shape of a distribution.

3. In the 2000 census conducted by the U.S. Census Bureau, two average household incomes were reported: $41,349 and $55,263. One of these averages is the mean and the other is the median. Which is the mean? Support your answer.

4. The U.S. Department of Housing and Urban Development (HUD) uses the median to report the average price of a home in the United States. Why do you think HUD uses the median?

5. A histogram of a set of data indicates that the distribution of the data is skewed right. Which measure of central tendency will be larger, the mean or the median? Why?

6. If a data set contains 10,000 values arranged in increasing order, where is the median located?

7. Explain why the mode is used as the measure of central tendency for qualitative data.

8. A(n) _____ is a descriptive measure of a population, and a(n) _____ is a descriptive measure of a sample.

9. *True or False*: A data set will always have exactly one mode.

10. *True or False*: If the number of observations is odd, the median is $M = \dfrac{n+1}{2}$.

Skill Building

In Problems 11–14, find the population mean or sample mean as indicated.

11. Sample: 20, 13, 4, 8, 10

12. Sample: 83, 65, 91, 87, 84

13. Population: 3, 6, 10, 12, 14

14. Population: 1, 19, 25, 15, 12, 16, 28, 13, 6

15. For Super Bowl XXXIX, Fox television sold 59 ad slots for a total revenue of roughly $142 million. What was the mean price per ad slot?

16. The median for the given set of seven ordered data values is 26.5. What is the missing value? 7 12 21 _____ 41 50

17. **Crash Test Results** The Insurance Institute for Highway Safety crashed the 2001 Honda Civic four times at 5 miles per hour. The costs of repair for each of the four crashes were

$$\$420, \$462, \$409, \$236$$

Compute the mean, median, and mode cost of repair.

18. **Cell Phone Use** The following data represent the monthly cell phone bill for my wife's phone for six randomly selected months.

$$\$35.34, \$42.09, \$39.43, \$38.93 \ \$43.39, \$49.26$$

Compute the mean, median, and mode phone bill.

19. **Concrete Mix** A certain type of concrete mix is designed to **NW** withstand 3000 pounds per square inch (psi) of pressure. The strength of concrete is measured by pouring the mix into casting cylinders 6 inches in diameter and 12 inches tall. The cylin-

der is allowed to "set up" for 28 days. The cylinders are then stacked on one another until the cylinders are crushed. The following data represent the strength of nine randomly selected casts (in psi).

3960, 4090, 3200, 3100, 2940, 3830, 4090, 4040, 3780

Compute the mean, median, and mode strength of the concrete (in psi).

20. **Flight Time** The following data represent the flight time (in minutes) of a random sample of seven flights from Las Vegas, Nevada, to Newark, New Jersey, on Continental Airlines.

282, 270, 260, 266, 257, 260, 267

Compute the mean, median, and mode flight time.

21. For each of the three histograms shown, determine whether the mean is greater than, less than, or approximately equal to the median. Justify your answer.

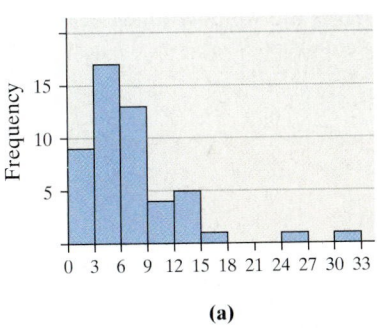

(a)

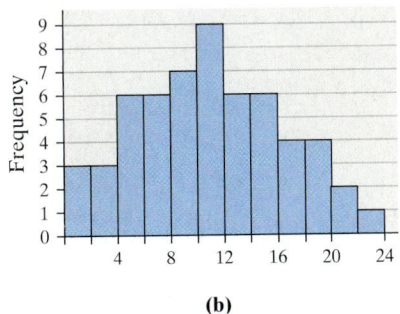

(b)

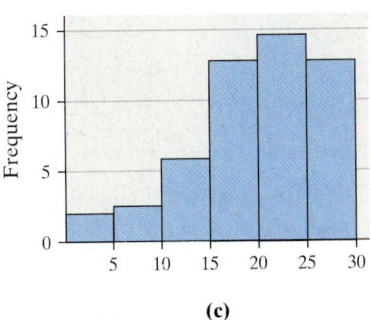

(c)

22. Match the histograms shown to the summary statistics:

	Mean	Median
I	42	42
II	31	36
III	31	26
IV	31	32

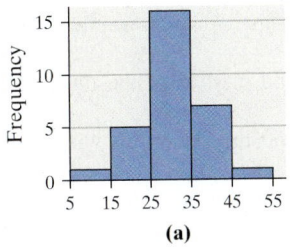

(a)

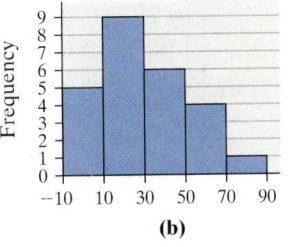

(b)

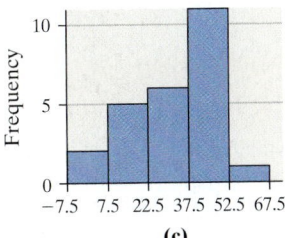

(c)

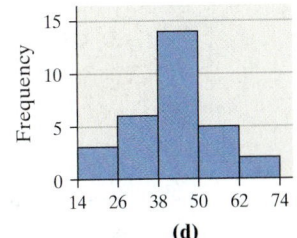

(d)

Applying the Concepts

23. **ATM Fees** The following data for a random sample of banks in Los Angeles and New York City represent the ATM fee for using another bank's ATM.

Los Angeles	2.00	1.50	1.50	1.00	1.50	2.00	0.00	2.00
New York City	1.50	1.00	1.00	1.25	1.25	1.50	1.00	0.00

Source: www.bankrate.com

Compute the mean, median, and mode ATM fee for each city. Does there appear to be a difference in the ATM fee charged in Los Angeles versus New York City? Why might this be the case?

24. **Reaction Time** In an experiment conducted online at the University of Mississippi, study participants are asked to react to a stimulus. In one experiment, the participant must press a key upon seeing a blue screen. The time (in seconds) to press the key is measured. The same person is

then asked to press a key upon seeing a red screen, again with the time to react measured. The table shows the results for six study participants. Compute the mean, median, and mode reaction time for both blue and red. Does there appear to be a difference in the reaction time? What might account for any difference? How might this information be used?

Participant Number	Reaction Time to Blue	Reaction Time to Red
1	0.582	0.408
2	0.481	0.407
3	0.841	0.542
4	0.267	0.402
5	0.685	0.456
6	0.45	0.533

Source: PsychExperiments at the University of Mississippi (www.olemiss.edu/psychexps/)

25. Pulse Rates The following data represent the pulse rates NW (beats per minute) of nine students enrolled in a section of Sullivan's Introductory Statistics course. Treat the nine students as a population.

Student	Pulse
Perpetual Bempah	76
Megan Brooks	60
Jeff Honeycutt	60
Clarice Jefferson	81
Crystal Kurtenbach	72
Janette Lantka	80
Kevin McCarthy	80
Tammy Ohm	68
Kathy Wojdyla	73

(a) Compute the population mean pulse.
(b) Determine two simple random samples of size 3 and compute the sample mean pulse of each sample.
(c) Which samples result in a sample mean that overestimates the population mean? Which samples result in a sample mean that underestimates the population mean? Do any samples lead to a sample mean that equals the population mean?

26. Travel Time The following data represent the travel time (in minutes) to school for nine students enrolled in Sullivan's College Algebra course. Treat the nine students as a population.

Student	Travel Time	Student	Travel Time
Amanda	39	Scot	45
Amber	21	Erica	11
Tim	9	Tiffany	12
Mike	32	Glenn	39
Nicole	30		

(a) Compute the population mean for travel time.
(b) Determine three simple random samples of size 4 and compute the sample mean for travel time of each sample.
(c) Which samples result in a sample mean that overestimates the population mean? Which samples result in a sample mean that underestimates the population mean? Do any samples lead to a sample mean that equals the population mean?

27. Soccer Goals Mia Hamm, who retired after the 2004 Olympics, is considered by some to be the most prolific player in international soccer. The following data represent the number of goals scored over her 18-year career.

0	0	0	4	10	1	10	10	19
9	18	20	13	13	2	7	8	13

Source: www.soccerhall.com

(a) Compute the population mean of the number of goals she scored.
(b) Determine two simple random samples of size 3 and compute the sample mean of the number of goals she scored.
(c) Which samples result in a sample mean that overestimates the population mean? Which samples result in a sample mean that underestimates the population mean? Do any samples lead to a sample mean that equals the population mean?

28. Tour de Lance Lance Armstrong won the Tour de France seven consecutive times (1999–2005). The following table gives the winning times, distances, speeds, and margin of victory.

Year	Winning Time (h)	Distance (km)	Winning Speed (km/h)	Winning Margin (min)
1999	91.538	3687	40.28	7.617
2000	92.552	3662	39.56	6.033
2001	86.291	3453	40.02	6.733
2002	82.087	3278	39.93	7.283
2003	83.687	3427	40.94	1.017
2004	83.601	3391	40.56	6.317
2005	86.251	3593	41.65	44.667

Source: cyclingnews.com

(a) Compute the mean and median of his winning times for the six races.
(b) Compute the mean and median of the distances for the six races.
(c) Compute the mean and median of his winning time margins.
(d) Compute the mean winning speed by finding the mean of the data values in the table. Next, compute the mean winning speed by finding the total of the six distances and dividing by the total of the six winning times. Finally, compute the mean winning speed by dividing the mean distance by the mean winning time. Do the three values agree or are there differences?

29. Connection Time The following data represent the con-NW nection time in seconds to an Internet service provider for 30 randomly selected connections.

39.76	36.13	36.61	38.80	39.04	39.09
37.24	35.62	40.07	38.76	39.23	38.38
38.24	36.34	35.89	42.86	36.03	37.03
38.64	41.86	41.22	37.19	40.50	39.81
39.84	39.45	40.91	43.12	40.54	42.02

Source: Nicole Spreitzer, student at Joliet Junior College

A histogram of the data is shown. The mean connection time is 39.007 seconds and the median connection time is 39.065 seconds. Use this information to identify the shape of the distribution. Which measure of central tendency better describes the "center" of the distribution?

Histogram of Time (seconds)

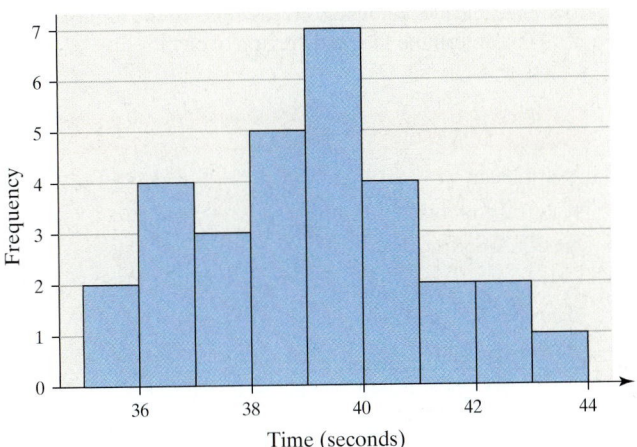

30. **Journal Costs** The following data represent the annual subscription cost (in dollars) for a random sample of 26 biology journals.

1188	778	1970	661	1294	1175
2033	3911	198	8415	796	1840
1141	1050	3643	1407	1092	585
1049	1092	1589	4115	1150	2799
707	2330				

Source: Carol Wesolowski, student at Joliet Junior College

A histogram of the data is shown. The mean subscription cost is $1846 and the median subscription cost is $1182. Use this information to identify the shape of the distribution. Which measure of central tendency better describes the "center" of the distribution?

Histogram of Cost

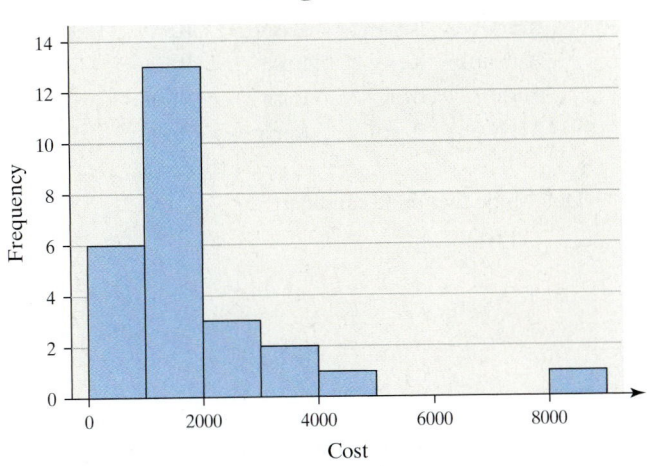

31. **Serum HDL** Dr. Paul Oswiecmiski randomly selects 40 of his 20- to 29-year-old patients and obtains the following data regarding their serum HDL cholesterol.

70	56	48	48	53	52	66	48
36	49	28	35	58	62	45	60
38	73	45	51	56	51	46	39
56	32	44	60	51	44	63	50
46	69	53	70	33	54	55	52

(a) Compute the mean and the median serum HDL.
(b) Identify the shape of the distribution based on the histogram drawn in Problem 31 in Section 2.2 and the relationship between the mean and the median.

32. **Volume of Altria Group Stock** The volume of a stock is the number of shares traded on a given day. The following data represents the volume of Altria Group stock traded for a random sample of 35 trading days in 2004. The data are in millions, so 3.78 represents 3,780,000 shares traded.

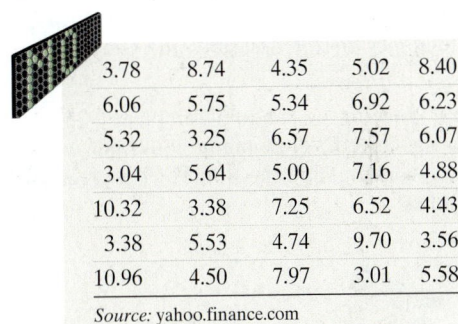

3.78	8.74	4.35	5.02	8.40
6.06	5.75	5.34	6.92	6.23
5.32	3.25	6.57	7.57	6.07
3.04	5.64	5.00	7.16	4.88
10.32	3.38	7.25	6.52	4.43
3.38	5.53	4.74	9.70	3.56
10.96	4.50	7.97	3.01	5.58

Source: yahoo.finance.com

(a) Compute the mean and the median number of shares traded.
(b) Identify the shape of the distribution based on the histogram drawn in Problem 32 in Section 2.2 and the relationship between the mean and the median.

33. **M&Ms** The following data represent the weights (in grams) of a simple random sample of 50 M&M plain candies.

0.87	0.88	0.82	0.90	0.90	0.84	0.84
0.91	0.94	0.86	0.86	0.86	0.88	0.87
0.89	0.91	0.86	0.87	0.93	0.88	
0.83	0.94	0.87	0.93	0.91	0.85	
0.91	0.91	0.86	0.89	0.87	0.93	
0.88	0.88	0.89	0.79	0.82	0.83	
0.90	0.88	0.84	0.93	0.76	0.90	
0.88	0.92	0.85	0.79	0.84	0.86	

Source: Michael Sullivan

Determine the shape of the distribution of weights of M&Ms by drawing a frequency histogram and computing the mean and median. Which measure of central tendency better describes the weight of a plain M&M?

34. Old Faithful We have all heard of the Old Faithful geyser in Yellowstone National Park. However, there is another, less famous, Old Faithful geyser in Calistoga, California. The following data represent the length of eruption (in seconds) for a random sample of eruptions of the California Old Faithful.

108	108	99	105	103	103	94
102	99	106	90	104	110	110
103	109	109	111	101	101	
110	102	105	110	106	104	
104	100	103	102	120	90	
113	116	95	105	103	101	
100	101	107	110	92	108	

Source: Ladonna Hansen, Park Curator

Determine the shape of the distribution of time between eruptions by drawing a frequency histogram and computing the mean and median. Which measure of central tendency better describes the time between eruptions?

35. Hours Working A random sample of 25 college students was asked, "How many hours per week typically do you work outside the home?" Their responses were as follows:

0	0	15	20	30
40	30	20	35	35
28	15	20	25	25
30	5	0	30	24
28	30	35	15	15

Determine the shape of the distribution of hours worked by drawing a frequency histogram and computing the mean and median. Which measure of central tendency better describes hours worked?

36. A Dealers Profits The following data represent the profits (in dollars) of a new car dealer for a random sample of 40 sales.

781	1,038	453	1,446	3,082
501	451	1,826	1,348	3,001
1,342	1,889	580	0	2,909
2,883	480	1,664	1,064	2,978
149	1,291	507	261	540
543	87	798	673	2,862
1,692	1,783	2,186	398	526
730	2,324	2,823	1,676	4,148

Source: Ashley Hudson, student at Joliet Junior College

Determine the shape of the distribution of new car-profits by drawing a frequency histogram and computing the mean and median. Which measure of central tendency better describes the profit?

37. Foreign-Born Population The following data represent the region of birth of foreign-born residents of the United States in 2003. Determine the mode region of birth.

Region	Number (thousands)
Caribbean	3,384
Central America	12,362
South America	2,111
Asia	8,375
Europe	4,590
Other Regions	2,680

Source: U.S. Census Bureau

38. Robbery The following data represent the number of offenses for various robberies in 2003. Determine the mode offense.

Type of Robbery	Number (thousands)
Street or highway	131
Commercial	61
Gas station	10
Convenience store	26
Residence	41
Bank	7

Source: U.S. Federal Bureau of Investigation

39. 2004 Presidential Election An exit poll was conducted in NW Los Alamos County, New Mexico, in which a random sample of 40 voters revealed whom they voted for in the presidential election. The results of the survey are as follows:

Kerry	Kerry	Bush	Bush	Bush
Bush	Kerry	Kerry	Bush	Bush
Kerry	Bush	Kerry	Bush	Kerry
Bush	Bush	Kerry	Kerry	Nader
Kerry	Bush	Bush	Kerry	Kerry
Badnarik	Kerry	Bush	Bush	Bush
Bush	Bush	Bush	Bush	Kerry
Kerry	Kerry	Kerry	Bush	Bush

Determine the mode candidate.

40. Hospital Admissions The following data represent the diagnosis of a random sample of 20 patients admitted to a hospital.

Cancer	Motor Vehicle Accident	Congestive Heart Failure
Gunshot wound	Fall	Gunshot wound
Gunshot wound	Motor vehicle accident	Gunshot wound
Assault	Motor vehicle accident	Gunshot wound
Motor vehicle accident	Motor vehicle accident	Gunshot wound
Motor vehicle accident	Gunshot wound	Motor vehicle accident
Fall	Gunshot wound	

Source: Tamela Ohm, student at Joliet Junior College

Determine the mode diagnosis.

41. Resistance and Sample Size Each of the following three data sets represents the IQ scores of a random sample of adults. IQ scores are known to have a mean and median of 100. For each data set, compute the mean and median. For each data set recalculate the mean and median, assuming that the individual whose IQ is 106 is accidentally recorded as 160. For each sample size, state what happens to the mean and the median. Comment on the role the number of observations plays in resistance.

Sample of Size 5				
106	92	98	103	100

Sample of Size 12					
106	92	98	103	100	102
98	124	83	70	108	121

Sample of Size 30					
106	92	98	103	100	102
98	124	83	70	108	121
102	87	121	107	97	114
140	93	130	72	81	90
103	97	89	98	88	103

42. Super Bowl XXXIX Champion New England Patriots The following table gives roster information for the offense of the Super Bowl XXXIX Champion New England Patriots

No.	Name	Position	Age (yr)	Weight (lb)	Years of Experience	College
12	Tom Brady	QB	27	225	5	Michigan
6	Rohan Davey	QB	26	245	3	LSU
13	Jim Miller	QB	33	225	10	Michigan State
27	Ribih Abdullah	RB	29	220	7	Lehigh
34	Cedric Cobbs	RB	24	225	Rookie	Arkansas
28	Corey Dillon	RB	30	225	8	Washington
33	Kevin Faulk	RB	28	202	6	LSU
35	Patrick Pass	RB	27	217	5	Georgia
83	Deion Branch	WR	25	193	3	Louisville
87	David Givens	WR	24	215	3	Notre Dame
81	Bethel Johnson	WR	25	200	2	Texas A&M
10	Kevin Kasper	WR	27	197	4	Iowa
86	David Patten	WR	30	190	8	Western Carolina
88	Christian Fauria	TE	33	250	10	Colorado
82	Daniel Graham	TE	26	257	3	Colorado
85	Jed Weaver	TE	28	258	6	Oregon
67	Daniel Koppen	C	25	296	2	Boston College
66	Lonie Paxton	C	26	260	5	Sacramento State
76	Brandon Gorin	OT	26	308	3	Purdue
72	Matt Light	OT	26	305	4	Purdue
63	Joe Andruzzi	OG	29	312	8	Southern Connecticut State
71	Russ Hochstein	OG	27	305	4	Nebraska
64	Gene Mruczkowski	OG	24	305	2	Purdue
61	Stephen Neal	OG	28	305	3	California State Bakersfield
74	Billy Yates	OG	24	305	2	Texas A&M

Source: ESPN.com

(a) Find the mean, median, and mode age.

(b) Find the mean, median, and mode weight

(c) Find the mean, median, and mode years of experience (**Note:** Rookie = 0 years).

(d) Find the mode college attended.

(e) Obtain a simple random sample of six members of New England's offense. Compute sample mean age, weight, and years of experience. How do the sample means compare to the population means?

(f) Compute the mean, median, and mode weights of the five offensive guards (OG). Compute the mean, median, and mode weights of the five running backs (RB). Does there appear to be a difference in the weights? What might account for any differences?

(g) Does it make sense to compute the mean player number? Why?

43. **Super Bowl XXXIX Champion New England Patriots Revisited** Using the data presented in Problem 42, answer the following.

(a) The skilled positions in football are quarterback (QB), running back (RB), wide receiver (WR), and tight end (TE). Obtain a stratified sample by using skilled positions as one stratum and the remaining positions as a second stratum. Randomly select four skilled players and two "nonskilled" players. Compute the mean weight of the sample data. Compare to the result in part (e) in Problem 42.

(b) Cluster the players by position. Obtain a cluster sample by randomly selecting two clusters. Compute the mean weight of the sample data. Compare the result in to part (e) in Problem 42. Can you think of any problems with obtaining a cluster sample?

44. You are negotiating a contract for the Players' Association of the NBA. Which measure of central tendency will you use to support your claim that the average player's salary needs to be increased? Why? As the chief negotiator for the owners, which measure would you use to refute the claim made by the Player's Association?

45. In January 2005, the mean amount of money lost per visitor to a local riverboat casino was $135. Do you think the median was more than, less than, or equal to this amount? Why?

46. **Missing Exam Grade** A professor has recorded exam grades for 20 students in his class, but one of the grades is no longer readable. If the mean score on the exam was 82 and the mean of the 19 readable scores is 84, what is the value of the unreadable score?

47. Suppose that the mean of a set of six data values is 34. What is the sum of the six data values?

48. For each of the following situations, determine which measure of central tendency is most appropriate and justify your reasoning.

(a) Average price of a home sold in Pittsburgh, Pennsylvania, in 2002

(b) Most popular major for students enrolled in a statistics course

(c) Average test score when the scores are distributed symmetrically

(d) Average test score when the scores are skewed right

(e) Average income of a player in the National Football League

(f) Most requested song at a radio station

49. **Linear Transformations** Benjamin owns a small Internet business. Besides himself, he employs nine other people. The salaries earned by the employees are given below in thousands of dollars (Benjamin's salary is the largest, of course):

$$30, 30, 45, 50, 50, 50, 55, 55, 60, 75$$

(a) Determine the mean, median, and mode for salary.

(b) Business has been good! As a result, Benjamin has a total of $25,000 in bonus pay to distribute to his employees. One option for distributing bonuses is to give each employee (including himself) $2500. Add the bonuses under this plan to the original salaries to create a new data set. Recalculate the mean, median, and mode. How do they compare to the originals?

(c) As a second option, Benjamin can give each employee a bonus of 5% of his or her original salary. Add the bonuses under this second plan to the original salaries to create a new data set. Recalculate the mean, median, and mode. How do they compare to the originals?

(d) As a third option, Benjamin decides not to give his employees a bonus at all. Instead, he keeps the $25,000 for himself. Use this plan to create a new data set. Recalculate the mean, median, and mode. How do they compare to the originals?

50. **Linear Transformations** Use the five test scores of 65, 70, 71, 75, and 95 to answer the following questions:

(a) Find the sample mean.

(b) Find the median.

(c) Which measure of central tendency best describes the typical test score?

(d) Suppose the professor decides to curve the exam by adding 4 points to each test score. Compute the sample mean based on the adjusted scores.

(e) Compare the unadjusted test score mean with the curved test score mean. What effect did adding 4 to each score have on the mean?

51. **Trimmed Mean** Another measure of central tendency is the trimmed mean. It is computed by determining the mean of a data set after deleting the smallest and largest observed values. Compute the trimmed mean for the data in Problem 33. Is the trimmed mean resistant? Explain.

52. Midrange The midrange is also a measure of central tendency. It is computed by adding the smallest and largest observed values of a data set and dividing the result by 2; that is,

$$\text{Midrange} = \frac{\text{largest data value } + \text{ smallest data value}}{2}$$

Compute the midrange for the data in Problem 33. Is the midrange resistant? Explain.

Technology Step-By-Step	**Determining the Mean and Median**

TI-83/84 Plus   *Step 1:* Enter the raw data in L1 by pressing STAT and selecting 1:Edit.
Step 2: Press STAT, highlight the CALC menu, and select 1:1-Var Stats.
Step 3: With 1-Var Stats appearing on the HOME screen, press 2nd 1 to insert L1 on the HOME screen. Press ENTER.

MINITAB   *Step 1:* Enter the data in C1.
Step 2: Select the **Stat** menu, highlight **Basic Statistics**, and then highlight **Display Descriptive Statistics**.
Step 3: In the **Variables** window, enter C1. Click OK.

Excel   *Step 1:* Enter the data in column A.
Step 2: Select the **Tools** menu and highlight **Data Analysis . . .**
Step 3: In the Data Analysis window, highlight **Descriptive Statistics** and click OK.
Step 4: With the cursor in the **Input Range** window, use the mouse to highlight the data in column A.
Step 5: Select the **Summary statistics** option and click OK.

3.2 Measures of Dispersion

Objectives
1. Compute the range of a variable from raw data
2. Compute the variance of a variable from raw data
3. Compute the standard deviation of a variable from raw data
4. Use the Empirical Rule to describe data that are bell shaped
5. Use Chebyshev's inequality to describe any set of data

In Section 3.1, we discussed measures of central tendency. The purpose of these measures is to describe the typical value of a variable. In addition to measuring the central tendency of a variable, we would also like to know the amount of dispersion in the variable. By dispersion, we mean the degree to which the data are "spread out." An example should help to explain why measures of central tendency are not sufficient in describing a distribution.

EXAMPLE 1   Comparing Two Sets of Data

Problem: The data in Table 8 represent the IQ scores of a random sample of 100 students from two different universities. For each university, compute the mean IQ score and draw a histogram, using a lower class limit of 55 for the first class and a class width of 15. Comment on the results.

Table 8

University A										University B									
73	103	91	93	136	108	92	104	90	78	86	91	107	94	105	107	89	96	102	96
108	93	91	78	81	130	82	86	111	93	92	109	103	106	98	95	97	95	109	109
102	111	125	107	80	90	122	101	82	115	93	91	92	91	117	108	89	95	103	109
103	110	84	115	85	83	131	90	103	106	110	88	97	119	90	99	96	104	98	95
71	69	97	130	91	62	85	94	110	85	87	105	111	87	103	92	103	107	106	97
102	109	105	97	104	94	92	83	94	114	107	108	89	96	107	107	96	95	117	97
107	94	112	113	115	106	97	106	85	99	98	89	104	99	99	87	91	105	109	108
102	109	76	94	103	112	107	101	91	107	116	107	90	98	98	92	119	96	118	98
107	110	106	103	93	110	125	101	91	119	97	106	114	87	107	96	93	99	89	94
118	85	127	141	129	60	115	80	111	79	104	88	99	97	106	107	112	97	94	107

Approach: We will use MINITAB to compute the mean and draw a histogram for each university.

Solution: We enter the data into MINITAB and determine that the mean IQ score of both universities is 100.0. Figure 10 shows the histograms.

Figure 10

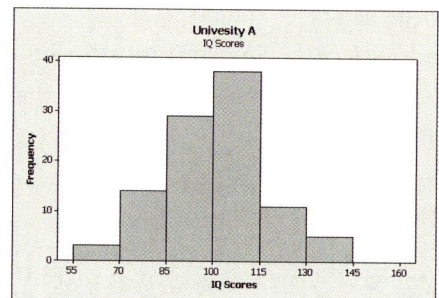 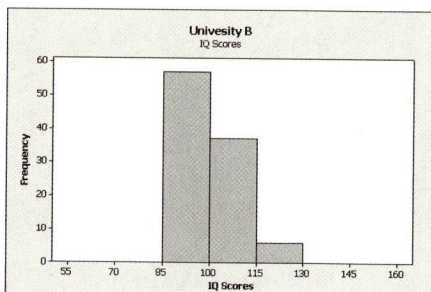

We notice that both universities have the same mean IQ, but the histograms indicate the IQs from University A are more spread out, that is, more dispersed. While an IQ of 100.0 is typical for both universities, it appears to be a more reliable description of the typical student from University B than from University A. That is, a higher proportion of students have IQ scores within, say, 15 points of the mean of 100.0 from University B than from University A.

Our goal in this section is to discuss numerical measures of dispersion so that we can quantify the spread of data. In this section, we discuss three numerical measures for describing the dispersion or spread of data: the range, variance, and standard deviation. In Section 3.4, we will discuss another measure of dispersion, the *interquartile range* (IQR).

① Compute the Range of a Variable from Raw Data

The simplest measure of dispersion is the range. To compute the range, the data must be quantitative.

Definition

The **range, R,** of a variable is the difference between the largest data value and the smallest data value. That is,

$$\text{Range} = R = \text{largest data value} - \text{smallest data value}$$

EXAMPLE 2 Computing the Range of a Set of Data

Problem: The data in Table 9 represent the scores on the first exam of 10 students enrolled in a section of Introductory Statistics. Compute the range.

Table 9	
Student	**Score**
1 Michelle	82
2. Ryanne	77
3. Bilal	90
4. Pam	71
5. Jennifer	62
6. Dave	68
7. Joel	74
8. Sam	84
9. Justine	94
10. Juan	88

Approach: The range is found by computing the difference between the largest and smallest data values.

Solution: The highest test score is 94 and the lowest test score is 62. The range, R, is

$$94 - 62 = 32$$

All the students in the class scored between 62 and 94 on the exam. The difference between the best score and the worst score is 32 points.

Now compute the range of the data in Problem 19.

In Other Words
The range is not resistant.

Notice that the range is affected by extreme values in the data set, so the range is not resistant. If Jennifer did not study and scored 28, the range becomes $R = 94 - 28 = 66$. In addition, the range is computed using only two values in the data set (the largest and smallest). The *variance* and the *standard deviation*, on the other hand, use all the data values in the computations.

② Compute the Variance of a Variable from Raw Data

Just as there is a population mean and sample mean, we also have a population variance and a sample variance. Measures of dispersion are meant to describe how spread out data are. Another way to think about this is to describe how far, on average, each observation is from the mean. Variance is based on the **deviation about the mean**. For a population, the deviation about the mean for the ith observation is $x_i - \mu$. For a sample, the deviation about the mean for the ith observation is $x_i - \bar{x}$. The further an observation is from the mean, the larger the absolute deviation.

The sum of all deviations about the mean must equal zero. That is,

$$\sum(x_i - \mu) = 0 \quad \text{and} \quad \sum(x_i - \bar{x}) = 0$$

In other words, observations larger than the mean are offset by observations smaller than the mean. Because the sum of deviations about the mean is zero, we cannot use the average deviation about the mean as a measure of spread. However, squaring a nonzero number always results in a positive number, so we could find the average squared deviation.

Definition

The **population variance** of a variable is the sum of the squared deviations about the population mean divided by the number of observations in the population, N. That is, it is the mean of the squared deviations about the population mean. The population variance is symbolically represented by σ^2 (lower case Greek sigma squared).

$$\sigma^2 = \frac{(x_1 - \mu)^2 + (x_2 - \mu)^2 + \cdots + (x_N - \mu)^2}{N} = \frac{\sum(x_i - \mu)^2}{N} \quad (1)$$

where $x_1, x_2, \ldots, x_N$ are the N observations in the population and μ is the population mean.

Note: In using Formula (1), do not round until the last computation. Use as many decimal places as allowed by your calculator to avoid round-off errors. ◄

A formula that is equivalent to Formula (1), called the **computational formula** for determining the population variance is

$$\sigma^2 = \frac{\sum x_i^2 - \dfrac{(\sum x_i)^2}{N}}{N}$$

where $\sum x_i^2$ means to square each observation and then sum these squared values, and $(\sum x_i)^2$ means to add up all the observations and then square the sum.

We illustrate how to use both formulas for computing the variance in the next example.

EXAMPLE 3   Computing a Population Variance

Problem: Compute the population variance of the test scores presented in Table 9.

Approach Using Formula (1)

Step 1: Create a table with four columns. Enter the population data in the first column. In the second column, enter the population mean.

Step 2: Compute the deviation about the mean for each data value. That is, compute $x_i - \mu$ for each data value. Enter these values in column 3.

Step 3: Square the values in column 3, and enter the results in column 4.

Step 4: Sum the squared deviations in column 4, and divide this result by the size of the population, N.

Solution

Step 1: See Table 10. Column 1 lists the observations in the data set, and column 2 contains the population mean.

Table 10

Score, x_i	Population Mean, μ	Deviation about the Mean, $x_i - \mu$	Squared Deviations about the Mean, $(x_i - \mu)^2$
82	79	$82 - 79 = 3$	$3^2 = 9$
77	79	$77 - 79 = -2$	$(-2)^2 = 4$
90	79	11	121
71	79	-8	64
62	79	-17	289
68	79	-11	121
74	79	-5	25
84	79	5	25
94	79	15	225
88	79	9	81
		$\sum(x_i - \mu) = 0$	$\sum(x_i - \mu)^2 = 964$

Step 2: Compute the deviations about the mean for each observation, as shown in column 3. For example, the deviation

Approach Using the Computational Formula

Step 1: Create a table with two columns. Enter the population data in the first column. Square each value in the first column and enter the result in the second column.

Step 2: Sum the entries in the first column. This is, find $\sum x_i$. Sum the entries in the second column. That is, find $\sum x_i^2$.

Step 3: Substitute the values found in Step 2 into the computational formula and simplify.

Solution

Step 1: See Table 11. Column 1 lists the observations in the data set, and column 2 contains the values in column 1 squared.

Table 11

Score, x_i	Score squared, x_i^2
82	$82^2 = 6724$
77	$77^2 = 5929$
90	8100
71	5041
62	3844
68	4624
74	5476
84	7056
94	8836
88	7744
$\sum x_i = 790$	$\sum x_i^2 = 63{,}374$

Step 2: The last row of columns 1 and 2 show that $\sum x_i = 790$ and $\sum x_i^2 = 63{,}374$.

about the mean for Michelle is $82 - 79 = 3$. It is a good idea to add up the entries in this column to make sure they sum to 0.

Step 3: Column 4 shows the squared deviations about the mean.

Step 4: We sum the entries in column 4 to obtain the numerator of Formula (1). We compute the population variance by dividing the sum of the entries in column 4 by the number of students, 10:

$$\sigma^2 = \frac{\sum(x_i - \mu)^2}{N} = \frac{964}{10} = 96.4$$

Step 3: We substitute 790 for Σx_i, 63,374 for Σx_i^2, and 10 for N into the computational formula.

$$\sigma^2 = \frac{\sum x_i^2 - \dfrac{\left(\sum x_i\right)^2}{N}}{N} = \frac{63,374 - \dfrac{(790)^2}{10}}{10}$$

$$= \frac{964}{10}$$

$$= 96.4$$

The unit of measure of the variance in Example 3 is points squared. This unit of measure results from squaring the deviations about the mean. Because points squared does not have any obvious meaning, the interpretation of variance is limited.

The sample variance is computed using sample data.

Definition

The **sample variance**, s^2, is computed by determining the sum of the squared deviations about the sample mean and dividing this result by $n - 1$. The formula for the sample variance from a sample of size n is

$$s^2 = \frac{\sum(x_i - \overline{x})^2}{n - 1} = \frac{(x_1 - \overline{x})^2 + (x_2 - \overline{x})^2 + \cdots + (x_n - \overline{x})^2}{n - 1} \qquad \textbf{(2)}$$

where $x_1, x_2, \ldots, x_n$ are the n observations in the sample and $\overline{x}$ is the sample mean.

CAUTION

When using Formula (2), be sure to use $\overline{x}$ with as many decimal places as possible to avoid round-off error.

A computational formula that is equivalent to Formula (2) for computing the sample variance is

$$s^2 = \frac{\sum x_i^2 - \dfrac{\left(\sum x_i\right)^2}{n}}{n - 1}$$

CAUTION

When computing the sample variance, be sure to divide by $n - 1$, not n.

where Σx_i^2 means to square each observation and then sum these squared values, whereas $(\Sigma x_i)^2$ means to add up all the observations and then square the sum.

Notice that the sample variance is obtained by dividing by $n - 1$. If we divided by n, as we might expect, the sample variance would consistently underestimate the population variance. Whenever a statistic consistently overestimates or underestimates a parameter, it is called **biased**. To obtain an unbiased estimate of the population variance, we divide the sum of the squared deviations about the mean by $n - 1$.

To help understand the idea of a biased estimator, consider the following situation: Suppose you work for a carnival in which you must guess a person's age. After 20 people come to your booth, you notice that you have a tendency to underestimate people's age. (You guess too low.) What would you do about this? In all likelihood, you would adjust your guesses higher so that you don't underestimate anymore. In other words, before the adjustment, your guesses were biased. To remove the bias, you increase your guess. That is what dividing by $n - 1$ in the sample variance formula accomplishes. Dividing by n results in an underestimate, so we divide by a smaller number to increase our "guess."

Although a proof that establishes why we divide by $n - 1$ is beyond the scope of the text, we can provide an explanation that has intuitive appeal. We already know that the sum of the deviations about the mean, $\Sigma(x_i - \overline{x})$, must equal zero. Therefore, if the sample mean is known and the first $n - 1$

observations are known, then the nth observation must be the value that causes the sum of the deviations to equal zero. For example, suppose $\bar{x} = 4$ based on a sample of size 3. In addition, if $x_1 = 2$ and $x_2 = 3$, then we can determine x_3.

$$\frac{x_1 + x_2 + x_3}{3} = \bar{x}$$

$$\frac{2 + 3 + x_3}{3} = 4 \qquad x_1 = 2, x_2 = 3, \bar{x} = 4$$

$$5 + x_3 = 12$$

$$x_3 = 7$$

In Other Words

We have $n - 1$ degrees of freedom in the computation of s^2 because an unknown parameter, μ, is estimated with $\bar{x}$. For each parameter estimated, we lose 1 degree of freedom.

We call $n - 1$ **degrees of freedom** for the following reasons. The first $n - 1$ observations have freedom to be whatever value they wish, but the nth value has no freedom. It must be whatever value forces the sum of the deviations about the mean to equal zero.

Again, you should notice that Greek letters are used for parameters, while Roman letters are used for statistics. Do not use rounded values of the sample mean in Formula (2).

EXAMPLE 4   Computing a Sample Variance

Problem: Compute the sample variance of the sample obtained in Example 1(b) on page 122 from Section 3.1.

Approach: We follow the same approach that we used to compute the population variance, but this time using the sample data. In looking back at Example 1(b) from Section 3.1, we see that Bilal (90), Ryanne (77), Pam (71), and Michelle (82) are in the sample.

Solution Using Formula (2)

Step 1: Create a table with four columns. Enter the sample data in the first column. In the second column, enter the sample mean. See Table 12.

Table 12

Score, x_i	Sample Mean, $\bar{x}$	Deviation about the Mean, $x_i - \bar{x}$	Squared Deviations about the Mean, $(x_i - \bar{x})^2$
90	80	$90 - 80 = 10$	$10^2 = 100$
77	80	-3	9
71	80	-9	81
82	80	2	4
		$\sum (x_i - \bar{x}) = 0$	$\sum (x_i - \bar{x})^2 = 194$

Step 2: Compute the deviations about the mean for each observation, as shown in column 3. For example, the deviation about the mean for Bilal is $90 - 80 = 10$. It is a good idea to add up the entries in this column to make sure they sum to 0.

Step 3: Column 4 shows the squared deviations about the mean.

Step 4: We sum the entries in column 4 to obtain the numerator of Formula (1). We compute the population variance by dividing the sum of the entries in column 4 by one fewer than the number of students, $4 - 1$:

$$s^2 = \frac{\sum (x_i - \bar{x})^2}{n - 1} = \frac{194}{4 - 1} = 64.7$$

Solution Using the Computational Formula

Step 1: See Table 13. Column 1 lists the observations in the data set, and column 2 contains the values in column 1 squared.

Table 13

Score, x_i	Score squared, x_i^2
90	$90^2 = 8{,}100$
77	$77^2 = 5{,}929$
71	5,041
82	6,724
$\sum x_i = 320$	$\sum x_i^2 = 25{,}794$

Step 2: The last rows of columns 1 and 2 show that $\sum x_i = 320$ and $\sum x_i^2 = 25{,}794$.

Step 3: We substitute 320 for $\sum x_i$, 25,794 for $\sum x_i^2$, and 4 for n into the computational formula.

$$s^2 = \frac{\sum x_i^2 - \dfrac{\left(\sum x_i\right)^2}{n}}{n - 1} = \frac{25{,}794 - \dfrac{(320)^2}{4}}{4 - 1}$$

$$= \frac{194}{3}$$

$$= 64.7$$

Notice that the sample variance obtained for this sample is an underestimate of the population variance we found in Example 3. This discrepancy does not violate our definition of an unbiased estimator, however. A biased estimator is one that *consistently* under- or overestimates.

3 Compute the Standard Deviation of a Variable from Raw Data

The standard deviation and the mean are the most popular methods for numerically describing the distribution of a variable. This is because these two measures are used for most types of statistical inference.

Definitions

The **population standard deviation**, σ, is obtained by taking the square root of the population variance. That is,

$$\sigma = \sqrt{\sigma^2}$$

The **sample standard deviation**, s, is obtained by taking the square root of the sample variance. That is,

$$s = \sqrt{s^2}$$

EXAMPLE 5   **Obtaining the Standard Deviation for a Population and a Sample**

Problem: Use the results obtained in Examples 3 and 4 to compute the population and sample standard deviation score on the statistics exam.

Approach: The population standard deviation is the square root of the population variance. The sample standard deviation is the square root of the sample variance.

Solution: The population standard deviation is

$$\sigma = \sqrt{\sigma^2} = \sqrt{\frac{\sum (x_i - \mu)^2}{N}} = \sqrt{\frac{964}{10}} = 9.8 \text{ points}$$

The sample standard deviation for the sample obtained in Example 1 from Section 3.1 is

$$s = \sqrt{s^2} = \sqrt{\frac{\sum (x_i - \bar{x})^2}{n-1}} = \sqrt{\frac{194}{4-1}} = 8.0 \text{ points}$$

! CAUTION
Never use the rounded variance to compute the standard deviation.

To avoid round-off error, never use the rounded value of the variance to compute the standard deviation.

Now Work Problem 25.

In-Class Activity: The Sample Standard Deviation

Using the pulse data from Section 3.1, page 123, do the following:

(a) Obtain a simple random sample of $n = 4$ students and compute the sample standard deviation.
(b) Obtain a second simple random sample of $n = 4$ students and compute the sample standard deviation.
(c) Are the sample standard deviations the same? Why?

EXAMPLE 6

Determining the Variance and Standard Deviation Using Technology

Problem: Use a statistical spreadsheet or calculator to determine the population standard deviation of the data listed in Table 9. Also determine the sample standard deviation of the sample data from Example 4.

Approach: We will use a TI-84 Plus graphing calculator to obtain the population standard deviation and sample standard deviation score on the statistics exam. The steps for determining the standard deviation using the TI-83 or TI-84 Plus graphing calculator, MINITAB, or Excel are given in the Technology Step-by-Step on page 157.

Solution: Figure 11(a) shows the population standard deviation, and Figure 11(b) shows the sample standard deviation. Notice the TI graphing calculators provide both a population and sample standard deviation as output. This is because the calculator does not know whether the data entered are population data or sample data. It is up to the user of the calculator to choose the correct standard deviation. The results agree with those obtained in Example 5. To get the variance, we need to square the standard deviation. For example, the population variance is $9.818350167^2 = 96.4$ points^2.

Figure 11

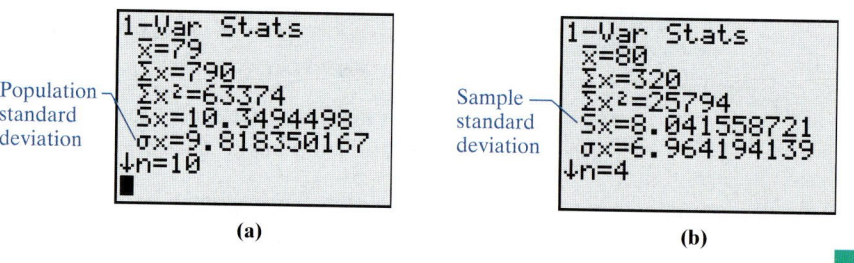

(a)

(b)

Interpretations of the Standard Deviation

The standard deviation is used in conjunction with the mean to numerically describe distributions that are bell-shaped and symmetric. The mean measures the center of the distribution, while the standard deviation measures the spread of the distribution. So how does the value of the standard deviation relate to the dispersion of the distribution? If we are comparing two populations, then **the larger the standard deviation, the more dispersion the distribution has**. This rule is true provided that the variable of interest from the two populations has the same unit of measure. The units of measure must be the same so that we are comparing apples with apples. For example, a standard deviation of $100 is not the same as 100 Japanese yen, because $1 is equivalent to about 109 yen. This means a standard deviation of $100 is substantially higher than a standard deviation of 100 yen.

EXAMPLE 7

Comparing the Standard Deviation of Two Data Sets

Problem: Refer to the data in Example 1. Use the standard deviation to determine whether University A or University B has more dispersion in the IQ scores of its students.

Approach: We will use MINITAB to compute the standard deviation of IQ for each university. The university with the higher standard deviation will be the university with more dispersion in IQ scores. Recall that, on the basis of the histograms, it was apparent that University A had more dispersion. Therefore, we would expect University A to have a higher sample standard deviation.

Solution: We enter the data into MINITAB and compute the descriptive statistics. See Figure 12.

Figure 12
Descriptive Statistics

Descriptive statistics

Variable	N	N*	Mean	SE Mean	StDev	Minimum
Univ A	100	0	100.00	1.61	16.08	60
Univ B	100	0	100.00	0.83	8.35	86

Variable	Q1	Median	Q3	Maximum
Univ A	90	102	110	141
Univ B	94	98	107	119

The sample standard deviation is larger for University A (16.1) than for University B (8.4). Don't forget that we agreed to round the mean and standard deviation to one more decimal place than the original data. Therefore, University A has IQ scores that are more dispersed.

4 Use the Empirical Rule to Describe Data That Are Bell Shaped

If data have a distribution that is bell shaped, the following rule can be used to determine the percentage of data that will lie within k standard deviations of the mean.

The Empirical Rule

If a distribution is roughly bell shaped, then

- Approximately 68% of the data will lie within 1 standard deviation of the mean. That is, approximately 68% of the data lie between $\mu - 1\sigma$ and $\mu + 1\sigma$.
- Approximately 95% of the data will lie within 2 standard deviations of the mean. That is, approximately 95% of the data lie between $\mu - 2\sigma$ and $\mu + 2\sigma$.
- Approximately 99.7% of the data will lie within 3 standard deviations of the mean. That is, approximately 99.7% of the data lie between $\mu - 3\sigma$ and $\mu + 3\sigma$.

Note: We can also use the empirical rule based on sample data with $\bar{x}$ used in place of μ and s used in place of σ. ◄

Figure 13 illustrates the Empirical Rule.

Figure 13

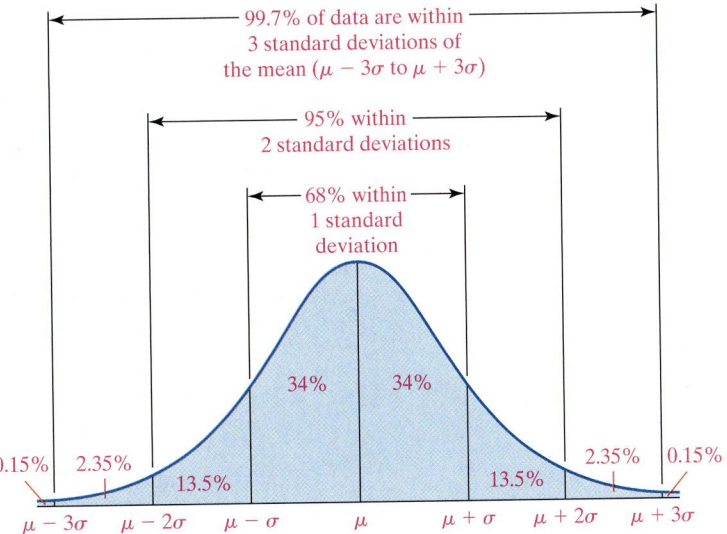

Let's revisit the data from University A in Table 8.

EXAMPLE 8   Using the Empirical Rule

Problem: Using the data from University A in Table 8.

(a) Determine the percentage of students who have IQ scores within 3 standard deviations of the mean according to the Empirical Rule.

(b) Determine the percentage of students who have IQ scores between 67.8 and 132.2 according to the Empirical Rule.

(c) Determine the actual percentage of students who have IQ scores between 67.8 and 132.2.

(d) According to the Empirical Rule, what percentage of students will have IQ scores above 132.2?

Approach: To use the Empirical Rule, a histogram of the data must be roughly bell shaped. Figure 14 shows the histogram of the data from University A.

Solution: The histogram of the data drawn in Figure 14 is roughly bell shaped. From Example 7 we know that the mean IQ score of the students enrolled in University A is 100 and the standard deviation is 16.1. To help organize our thoughts and make the analysis easier, we draw a bell-shaped curve like the one in Figure 13, with the $\bar{x} = 100$ and $s = 16.1$. See Figure 15.

Figure 14

University A IQ Scores

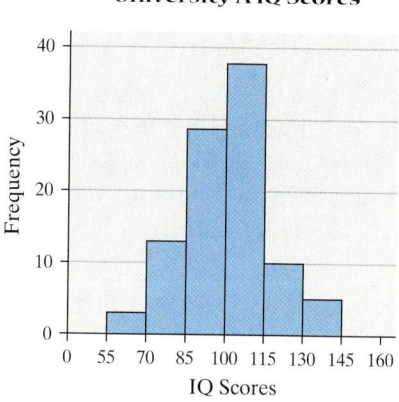

Figure 15

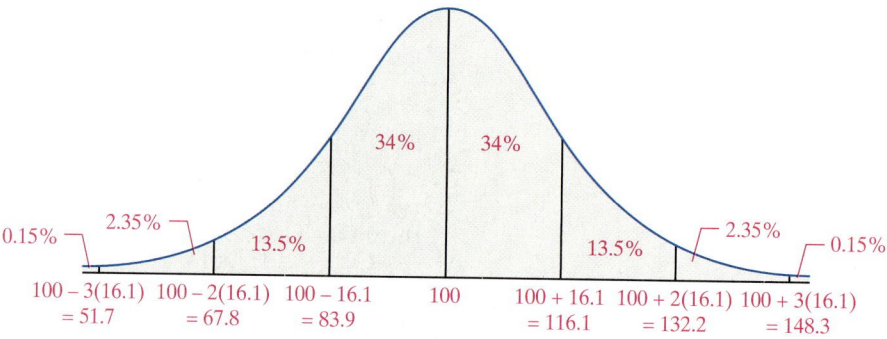

(a) According to the Empirical Rule, approximately 99.7% of the IQ scores will be within 3 standard deviations of the mean. That is, approximately 99.7% of the data will be greater than or equal to $100 - 3(16.1) = 51.7$ and less than or equal to $100 + 3(16.1) = 148.3$.

(b) Since 67.8 is exactly 2 standard deviations below the mean $[100 - 2(16.1) = 67.8]$ and 132.2 is exactly 2 standard deviations above the mean $[100 + 2(16.1) = 132.2]$, we use the Empirical Rule to determine that approximately 95% of all IQ scores lies between 67.8 and 132.2.

(c) Of the 100 IQ scores listed in Table 8, 96, or 96%, are between 67.8 and 132.2. This is very close to the approximation given by the Empirical Rule.

(d) Based on Figure 15, approximately $2.35\% + 0.15\% = 2.5\%$ of students at University A will have IQ scores above 132.2.

Now Work Problem 39.

 Use Chebyshev's Inequality to Describe Any Set of Data

Chebyshev's Inequality was developed by the Russian mathematician Pafnuty Chebyshev (1821–1894). The inequality is used to determine a lower bound on the percentage of observations that lie within k standard deviations of the mean, where $k > 1$. What's amazing about this result is that these bounds are arrived at regardless of the basic shape of the distribution (skewed left, skewed right, or symmetric).

Chebyshev's Inequality

For any data set, regardless of the shape of the distribution, at least $\left(1 - \dfrac{1}{k^2}\right)100\%$ of the observations will lie within k standard deviations of the mean, where k is any number greater than 1. That is, at least $\left(1 - \dfrac{1}{k^2}\right)100\%$ of the data will lie between $\mu - k\sigma$ and $\mu + k\sigma$ for $k > 1$.

Note: We can also use Chebyshev's Inequality based on sample data.

Caution

The Empirical Rule holds only if the distribution is bell shaped. Chebyshev's Inequality holds regardless of the shape of the distribution.

For example, at least $\left(1 - \dfrac{1}{2^2}\right)100\% = 75\%$ of all observations will lie within $k = 2$ standard deviations of the mean and at least $\left(1 - \dfrac{1}{3^2}\right)100\% = 88.9\%$ of all observations will lie within $k = 3$ standard deviations of the mean.

Notice the result does not state that exactly 75% of all observations lie within 2 standard deviations of the mean, but instead states that 75% or more of the observations will lie within 2 standard deviations of the mean.

EXAMPLE 9   Using Chebyshev's Inequality

Problem: Using the data from University A in Table 8.

(a) Determine the minimum percentage of students who have IQ scores within 3 standard deviations of the mean according to Chebyshev's Inequality.

(b) Determine the minimum percentage of students who have IQ scores between 67.8 and 132.2, according to Chebyshev's Inequality.

(c) Determine the actual percentage of students who have IQ scores between 67.8 and 132.2.

Approach

(a) We use Chebyshev's Inequality with $k = 3$.

(b) We have to determine the number of standard deviations 67.8 and 132.2 are from the mean of 100.0. We then substitute this value of k into Chebyshev's Inequality.

Historical Notes

Pafnuty Chebyshev was born on May 16, 1821, in Okatovo, Russia. In 1847, he began teaching mathematics at the University of St. Petersburg. Some of his more famous work was done on prime numbers. In particular, he discovered a way to determine the number of prime numbers less than or equal to a given number. Chebyshev also studied mechanics, including rotary motion. Chebyshev was elected a Fellow of the Royal Society in 1877. He died November 26, 1894, in St. Petersburg.

(c) We refer to Table 8 and count the number of observations between 67.8 and 132.2. We divide this result by 100, the number of observations in the data set.

Solution

(a) We use Chebyshev's Inequality with $k = 3$ and determine that at least $\left(1 - \dfrac{1}{3^2}\right)100\% = 88.9\%$ of all students have IQ scores within 3 standard deviations of the mean. Since the mean of the data set is 100.0 and the standard deviation is 16.1, at least 88.9% of the students have IQ scores between $\bar{x} - ks = 100.0 - 3(16.1) = 51.7$ and $\bar{x} + ks = 100 + 3(16.1) = 148.3$.

(b) Since 67.8 is exactly 2 standard deviations below the mean $[100 - 2(16.1) = 67.8]$ and 132.2 is exactly 2 standard deviations above the mean $[100 + 2(16.1) = 132.2]$, we use Chebyshev's Inequality with $k = 2$

to determine that at least $\left(1 - \dfrac{1}{2^2}\right)100\% = 75\%$ of all IQ scores lie between 67.8 and 132.2.

(c) Of the 100 IQ scores listed, 96 or 96% is between 67.8 and 132.2. Notice that Chebyshev's inequality provides a rather conservative result.

Now Work Problem 43.

Because the Empirical Rule requires that the distribution be bell shaped, while Chebyshev's inequality applies to all distributions, the Empirical Rule provides results that are more precise.

3.2 ASSESS YOUR UNDERSTANDING

Concepts and Vocabulary

1. Would it be appropriate to say that a distribution with a standard deviation of 10 centimeters is more dispersed than a distribution with a standard deviation of 5 inches? Support your position.
2. What is meant by the phrase *degrees of freedom* as it pertains to the computation of the sample variance?
3. Is the standard deviation resistant?
4. The sum of the deviations about the mean always equals _____.
5. What does it mean when a statistic is biased?
6. The simplest measure of dispersion is the _____.
7. Discuss the relationship between variance and standard deviation.

8. The standard deviation is used in conjunction with the _____ to numerically describe distributions that are bell shaped. The _____ measures the center of the distribution, while the standard deviation measures the _____ of the distribution.
9. *True or False*: When comparing two populations, the larger the standard deviation, the more dispersion the distribution has, provided that the variable of interest from the two populations has the same unit of measure.
10. *True or False*: Chebyshev's inequality applies to all distributions regardless of shape, but the Empirical Rule holds only for distributions that are bell shaped.

Skill Building

In Problems 11–16, find the population variance and standard deviation or the sample variance and standard deviation as indicated.

11. Sample: 20, 13, 4, 8, 10
12. Sample: 83, 65, 91, 87, 84
13. Population: 3, 6, 10, 12, 14
14. Population: 1, 19, 25, 15, 12, 16, 28, 13, 6
15. Sample: 6, 52, 13, 49, 35, 25, 31, 29, 31, 29
16. Population: 4, 10, 12, 12, 13, 21
17. **Crash Test Results** The Insurance Institute for Highway Safety crashed the 2001 Honda Civic four times at 5 miles per hour. The cost of repair for each of the four crashes is as follows:

    $420, $462, $409, $236

    Compute the range, sample variance, and sample standard deviation cost of repair.
18. **Cell Phone Use** The following data represent the monthly cell phone bill for my wife's phone for six randomly selected months:

    $35.34, $42.09, $39.43, $38.93, $43.39, $49.26

    Compute the range, sample variance, and sample standard deviation phone bill.

19. **Concrete Mix** A certain type of concrete mix is designed to withstand 3000 pounds per square inch (psi) of pressure. The strength of concrete is measured by pouring the mix into casting cylinders 6 inches in diameter and 12 inches tall. The cylinder is allowed to set up for 28 days. The cylinders are then stacked on one another until the cylinders are crushed. The following data represent the strength of nine randomly selected casts:

    3960, 4090, 3200, 3100, 2940, 3830, 4090, 4040, 3780

    Compute the range, sample variance, and sample standard deviation for strength of the concrete (in psi).

20. **Flight Time** The following data represent the flight time (in minutes) of a random sample of seven flights from Las Vegas, Nevada, to Newark, New Jersey, on Continental Airlines.

    282, 270, 260, 266, 257, 260, 267

    Compute the range, sample variance, and sample standard deviation of flight time.

21. Which histogram depicts a higher standard deviation? Justify your answer.

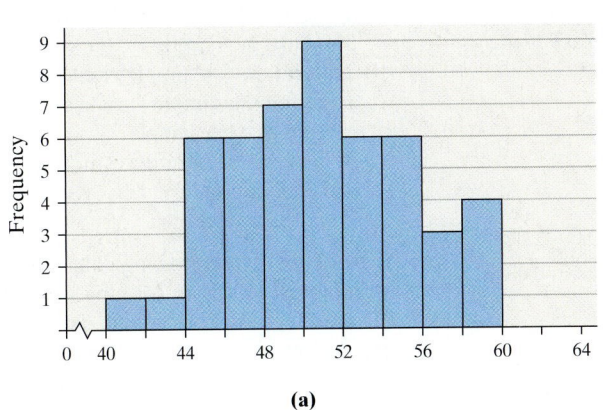

(a)

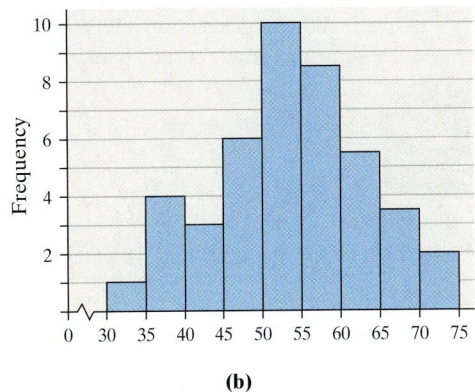

(b)

22. Match the histograms to the summary statistics given.

	Mean	Median	Standard Deviation
I	53	53	1.3
II	60	60	11
III	53	53	9
IV	53	53	0.12

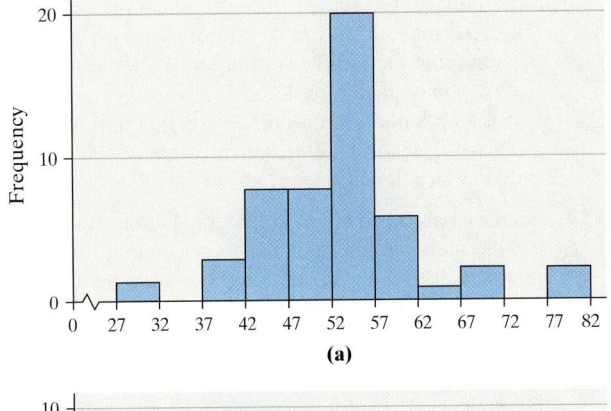

(a)

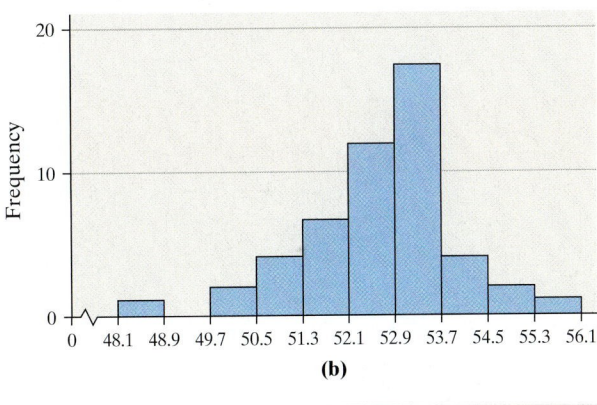

(b)

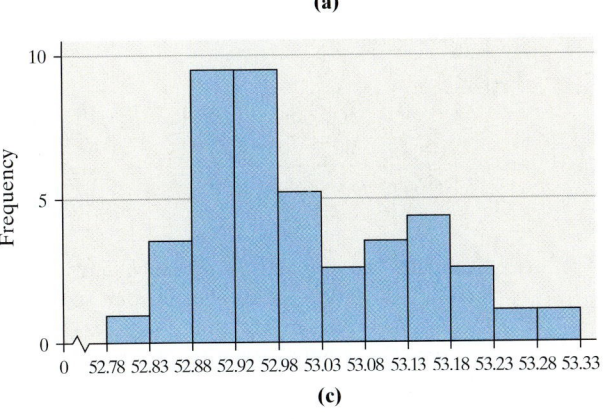

(c)

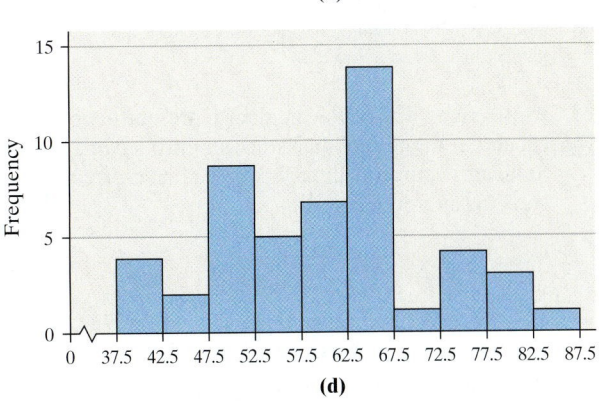

(d)

Applying the Concepts

23. ATM Fees The following data for a random sample of banks in Los Angeles and New York City represent the ATM fees for using another bank's ATM.

| Los Angeles | 2.00 | 1.50 | 1.50 | 1.00 | 1.50 | 2.00 | 0.00 | 2.00 |
| New York City | 1.50 | 1.00 | 1.00 | 1.25 | 1.25 | 1.50 | 1.00 | 0.00 |

Source: www.bankrate.com

Compute the range and sample standard deviation for ATM fees for each city. Which city has the most dispersion based on range? Which city has more dispersion based on the standard deviation?

24. Reaction Time In an experiment conducted online at the University of Mississippi, study participants are asked to react to a stimulus. In one experiment, the participant must press a key upon seeing a blue screen. The time (in seconds) to press the key is measured. The same person is then asked to press a key upon seeing a red screen, again with the time to react measured. The results for six study participants are listed in the table. Compute the range and sample standard deviation for reaction time for both blue and red. Does there appear to be a difference in the variability of reaction time? What might account for any difference?

Participant Number	Reaction Time to Blue	Reaction Time to Red
1	0.582	0.408
2	0.481	0.407
3	0.841	0.542
4	0.267	0.402
5	0.685	0.456
6	0.450	0.533

Source: PsychExperiments at the University of Mississippi (www.olemiss.edu/psychexps/)

25. Pulse Rates The following data represent the pulse rates (beats per minute) of nine students enrolled in a section of Sullivan's course in Introductory Statistics. Treat the nine students as a population.

Student	Pulse
Perpectual Bempah	76
Megan Brooks	60
Jeff Honeycutt	60
Clarice Jefferson	81
Crystal Kurtenbach	72
Janette Lantka	80
Kevin McCarthy	80
Tammy Ohm	68
Kathy Wojdyla	73

(a) Compute the population variance and population standard deviation.
(b) Determine two simple random samples of size 3, and compute the sample variance and sample standard deviation of each sample.
(c) Which samples underestimate the population standard deviation? Which overestimate the population standard deviation?

26. Travel Time The following data represent the travel time (in minutes) to school for nine students enrolled in Sullivan's College Algebra course. Treat the nine students as a population.

Student	Travel Time	Student	Travel Time
Amanda	39	Scot	45
Amber	21	Erica	11
Tim	9	Tiffany	12
Mike	32	Glenn	39
Nicole	30		

(a) Compute the population variance and population standard deviation.
(b) Determine three simple random samples of size 4, and compute the sample variance and sample standard deviation of each sample.
(c) Which samples underestimate the population standard deviation? Which overestimate the population standard deviation?

27. Soccer Goals Mia Hamm, considered by some to be the most prolific player in international soccer, retired after the 2004 Olympics. The following data represent the number of goals scored over her 18-year career.

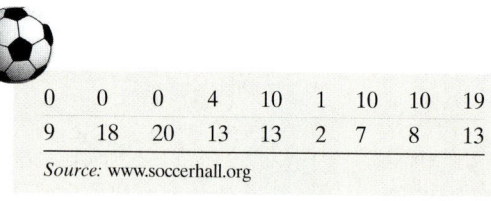

| 0 | 0 | 0 | 4 | 10 | 1 | 10 | 10 | 19 |
| 9 | 18 | 20 | 13 | 13 | 2 | 7 | 8 | 13 |

Source: www.soccerhall.org

(a) Compute the population variance and population standard deviation for number of goals scored.
(b) Determine three simple random samples of size 3, and compute the sample variance and sample standard deviation of each sample.
(c) Which samples underestimate the population standard deviation? Which overestimate the population standard deviation?

28. Tour de Lance Lance Armstrong won the Tour de France seven consecutive times (1999–2005). The table gives the winning times, distances, speeds, and margin of victory.

Year	Winning Time (h)	Distance (km)	Winning Speed (km/h)	Winning Margin (min)
1999	91.538	3687	40.28	7.617
2000	92.552	3662	39.56	6.033
2001	86.291	3453	40.02	6.733
2002	82.087	3278	39.93	7.283
2003	83.687	3427	40.94	1.017
2004	83.601	3391	40.56	6.317
2005	86.251	3593	41.65	4.667

Source: www.cyclinynews.com

(a) Compute the range, population variance, and population standard deviation for winning times for the six races.
(b) Compute the range, population variance, and population standard deviation for distances for the six races.
(c) Compute the range, population variance, and population standard deviation for winning time margins.
(d) Compute the range, population variance, and population standard deviation for winning speeds.

29. A Fish Story Ethan and Drew went on a 10-day fishing trip. The number of smallmouth bass caught and released by the two boys each day was as follows:

Ethan:	9	24	8	9	5	8	9	10	8	10
Drew:	15	2	3	18	20	1	17	2	19	3

(a) Find the population mean and the range for the number of smallmouth bass caught per day by each fisherman. Do these values indicate any differences between the two fishermen's catches per day? Explain.
(b) Find the population standard deviation for the number of smallmouth bass caught per day by each fisherman. Do these values present a different story about the two fishermen's catches per day? Which fisherman has the more consistent record? Explain.
(c) Discuss limitations of the range as a measure of dispersion.

30. 2004 NFC Champion Philadelphia Eagles The following data represent the weights (in pounds) of the 33 offensive players and the 24 defensive players for the 2004 NFC Champion Philadelphia Eagles.

Offense			Defense		
195	218	215	281	272	265
240	222	212	265	264	298
210	250	200	303	293	306
243	205	180	294	240	241
210	195	226	254	245	262
180	245	258	200	196	194
255	244	312	210	177	210
300	305	310	211	206	202
340	330	327			
349	330	310			
320	330	325			

Source: ESPN.com

(a) Compute the population mean, the range, and the population standard deviation for the Philadelphia offense.
(b) Compute the population mean, the range, and the population standard deviation for the Philadelphia defense.
(c) Which player type has more dispersion? Explain how you know.

In Problems 31 and 32, compute the range, sample variance, and sample standard deviation.

31. Serum HDL Dr. Paul Oswiecmiski randomly selects 40 of his 20- to 29-year-old patients and obtains the following data regarding their serum HDL cholesterol:

70	56	48	48	53	52	66	48
36	49	28	35	58	62	45	60
38	73	45	51	56	51	46	39
56	32	44	60	51	44	63	50
46	69	53	70	33	54	55	52

32. Volume of Altria Group Stock The volume of a stock is the number of shares traded on a given day. The following data, given in millions so that 3.78 represents 3,780,000 shares traded, represent the volume of Altria Group stock traded for a random sample 35 trading days in 2004.

3.78	8.74	4.35	5.02	8.40
6.06	5.75	5.34	6.92	6.23
5.32	3.25	6.57	7.57	6.07
3.04	5.64	5.00	7.16	4.88
10.32	3.38	7.25	6.52	4.43
3.38	5.53	4.74	9.70	3.56
10.96	4.50	7.97	3.01	5.58

Source: Yahoo.finance.com

33. The Empirical Rule The following data represent the weights (in grams) of a random sample of 50 M&M plain candies.

0.87	0.88	0.82	0.90	0.90	0.84	0.84
0.91	0.94	0.86	0.86	0.86	0.88	0.87
0.89	0.91	0.86	0.87	0.93	0.88	
0.83	0.94	0.87	0.93	0.91	0.85	
0.91	0.91	0.86	0.89	0.87	0.93	
0.88	0.88	0.89	0.79	0.82	0.83	
0.90	0.88	0.84	0.93	0.76	0.90	
0.88	0.92	0.85	0.79	0.84	0.86	

Source: Michael Sullivan

(a) Determine the sample standard deviation weight. Express your answer rounded to two decimal places.
(b) On the basis of the histogram drawn in Section 3.1, Problem 33, comment on the appropriateness of using the Empirical Rule to make any general statements about the weights of M&Ms.
(c) Use the Empirical Rule to determine the percentage of M&Ms with weights between 0.79 and 0.95 gram. *Hint:* $\bar{x} = 0.87$.
(d) Determine the actual percentage of M&Ms that weigh between 0.79 and 0.95 gram, inclusive.
(e) Use the Empirical Rule to determine the percentage of M&Ms with weights more than 0.91 gram.
(f) Determine the actual percentage of M&Ms that weigh more than 0.91 gram.

34. The Empirical Rule The following data represent the length of eruption for a random sample of eruptions at the Old Faithful geyser in Calistoga, California.

108	108	99	105	103	103	94
102	99	106	90	104	110	110
103	109	109	111	101	101	
110	102	105	110	106	104	
104	100	103	102	120	90	
113	116	95	105	103	101	
100	101	107	110	92	108	

Source: Ladonna Hansen, Park Curator

(a) Determine the sample standard deviation length of eruption. Express your answer rounded to the nearest whole number.
(b) On the basis of the histogram drawn in Section 3.1, Problem 34, comment on the appropriateness of using the Empirical Rule to make any general statements about the length of eruptions.
(c) Use the Empirical Rule to determine the percentage of eruptions that last between 92 and 116 seconds. *Hint:* $\bar{x} = 104$.

(d) Determine the actual percentage of eruptions that last between 92 and 116 seconds, inclusive.
(e) Use the Empirical Rule to determine the percentage of eruptions that last less than 98 seconds.
(f) Determine the actual percentage of eruptions that last less than 98 seconds.

35. Which Car Would You Buy? Suppose that you are in the market to purchase a car. With gas prices on the rise, you have narrowed it down to two choices and will let gas mileage be the deciding factor. You decide to conduct a little experiment in which you put 10 gallons of gas in the car and drive it on a closed track until it runs out gas. You conduct this experiment 15 times on each car and record the number of miles driven.

Car 1				
228	223	178	220	220
233	233	271	219	223
217	214	189	236	248

Car 2				
277	164	326	215	259
217	321	263	160	257
239	230	183	217	230

Describe each data set. That is, determine the shape, center, and spread. Which car would you buy and why?

36. Which Investment Is Better? You have received a year-end bonus of $5000. You decide to invest the money in the stock market and have narrowed your investment options down to two mutual funds. The following data represent the historical quarterly rates of return of each mutual fund for the past 20 quarters (5 years).

Mutual Fund A				
1.3	−0.3	0.6	6.8	5.0
5.2	4.8	2.4	3.0	1.8
7.3	8.6	3.4	3.8	−1.3
6.4	1.9	−0.5	−2.3	3.1

Mutual Fund B				
−5.4	6.7	11.9	4.3	4.3
3.5	10.5	2.9	3.8	5.9
−6.7	1.4	8.9	0.3	−2.4
−4.7	−1.1	3.4	7.7	12.9

Describe each data set. That is, determine the shape, center, and spread. Which mutual fund would you invest in and why?

37. Rates of Return of Stocks Stocks may be categorized by industry. The following data represent the 5-year rates of return for a simple random sample of financial stocks and energy stocks ending March 4, 2004.

(a) Compute the mean and the median rate of return for each industry. Which sector has the higher mean rate of return? Which sector has the higher median rate of return?

Financial Stocks							
17.10	16.26	22.10	9.96	7.94	10.95	16.34	20.43
7.54	26.84	28.02	15.92	10.80	11.27	20.68	11.09
9.84	11.82	6.28	3.27	21.97	13.74	33.63	25.53
11.01	18.15	17.36	19.14	17.80	26.33	5.35	8.44

Energy Stocks							
11.43	14.52	22.14	7.03	42.31	15.15	9.43	7.39
30.88	19.50	21.17	16.03	53.61	15.38	42.74	26.34
7.51	45.62	19.67	15.17	8.39	43.50	29.97	6.11
23.84	26.18	38.79	15.35	18.42	16.67	20.93	28.23

Source: Morningstar.com

(b) Compute the standard deviation for each industry. In finance, the standard deviation rate of return is called **risk**. Which sector is riskier?

38. American League versus National League The following data represent the earned-run average of a random sample of pitchers in both the American League and the National League during the 2004 season. **Note:** Earned-run average (ERA) is the mean number of runs given up per nine innings pitched. A higher ERA is indicative of a worse pitcher.

American League				
2.22	2.70	2.90	3.13	3.25
3.27	3.30	3.40	3.57	3.60
3.77	3.78	3.78	3.87	3.91
4.02	4.09	4.14	4.14	4.18
4.21	4.29	4.34	4.43	4.47
4.49	4.51	4.51	4.57	4.59
4.61	4.63	4.67	4.68	4.85
5.15	5.20	5.56	5.73	5.75

National League				
2.34	2.39	2.43	2.84	2.95
3.09	3.11	3.20	3.20	3.24
3.28	3.54	3.59	3.62	3.68
3.76	3.78	3.81	3.82	3.84
3.87	3.96	3.99	4.00	4.01
4.08	4.11	4.11	4.13	4.13
4.16	4.19	4.23	4.30	4.43
4.45	4.52	4.52	4.59	4.64

(a) Compute the mean and the median earned-run average for each league. Which league has the higher mean earned-run average? Which league has the higher median earned-run average?

(b) Compute the standard deviation for each league. Which league has more dispersion?

39. The Empirical Rule One measure of intelligence is the
NW Stanford–Binet Intelligence Quotient (IQ). IQ scores
have a bell-shaped distribution with a mean of 100 and a
standard deviation of 15.
   (a) What percentage of people has an IQ score between
       70 and 130?
   (b) What percentage of people has an IQ score less than
       70 or greater than 130?
   (c) What percentage of people has an IQ score greater
       than 130?

40. The Empirical Rule SAT Math scores have a bell-shaped
distribution with a mean of 518 and a standard deviation
of 114.

Source: College Board, 2004

   (a) What percentage of SAT scores is between 404 and
       632?
   (b) What percentage of SAT scores is less than 404 or
       greater than 632?
   (c) What percentage of SAT scores is greater than 746?

41. The Empirical Rule The weight, in grams, of the pair of
kidneys in adult males between the ages of 40 and 49 have
a bell-shaped distribution with a mean of 325 grams and a
standard deviation of 30 grams.
   (a) About 95% of kidneys will be between what weights?
   (b) What percentage of kidneys weighs between 235
       grams and 415 grams?
   (c) What percentage of kidneys weighs less than 235
       grams or more than 415 grams?
   (d) What percentage of kidneys weighs between 295
       grams and 385 grams?

42. The Empirical Rule The distribution of the length of
bolts has a bell shape with a mean of 4 inches and a standard deviation of 0.007 inch.
   (a) About 68% of bolts manufactured will be between
       what lengths?
   (b) What percentage of bolts will be between 3.986 inches
       and 4.014 inches?
   (c) If the company discards any bolts less than 3.986 inch-
       es or greater than 4.014 inches, what percentage of
       bolts manufactured will be discarded?
   (d) What percentage of bolts manufactured will be be-
       tween 4.007 inches and 4.021 inches?

43. Chebyshev's Inequality In December 2004, the average
NW price of regular unleaded gasoline excluding taxes in the
United States was $1.37 per gallon according to the Ener-
gy Information Administration. Assume that the standard
deviation price per gallon is $0.05 per gallon to answer the
following.
   (a) What percentage of gasoline stations had prices with-
       in 3 standard deviations of the mean?
   (b) What percentage of gasoline stations had prices with-
       in 2.5 standard deviations of the mean? What are the
       gasoline prices that are within 2.5 standard deviations
       of the mean?
   (c) What is the minimum percentage of gasoline stations
       that had prices between $1.27 and $1.47?

44. Chebyshev's Inequality According to the U.S. Census Bureau, the mean of the commute time to work for a resident
of Boston, Massachusetts, is 27.3 minutes. Assume that the

standard deviation of the commute time is 8.1 minutes to
answer the following:
   (a) What percentage of commuters in Boston has a
       commute time within 2 standard deviations of the
       mean?
   (b) What percentage of commuters in Boston has a com-
       mute time within 1.5 standard deviations of the mean?
       What are the commute times within 1.5 standard devia-
       tions of the mean?
   (c) What is the minimum percentage of commuters who
       have commute times between 3 minutes and 51.6
       minutes?

45. Comparing Standard Deviations The standard deviation
of batting averages of all teams in the American League is
0.008. The standard deviation of all players in the American League is 0.02154. Why is there less variability in team
batting averages?

46. Linear Transformations Benjamin owns a small Internet
business. Besides himself, he employs nine other people.
The salaries earned by the employees are given next in thousands of dollars (Benjamin's salary is the largest, of course):

$$30, 30, 45, 50, 50, 50, 55, 55, 60, 75$$

   (a) Determine the range, population variance, and popu-
       lation standard deviation for the data.
   (b) Business has been good! As a result, Benjamin has a
       total of $25,000 in bonus pay to distribute to his employ-
       ees. One option for distributing bonuses is to give each
       employee (including himself) $2500. Add the bonuses
       under this plan to the original salaries to create a new
       data set. Recalculate the range, population variance, and
       population standard deviation. How do they compare
       to the originals?
   (c) As a second option, Benjamin can give each employee
       a bonus of 5% of his or her original salary. Add the
       bonuses under this second plan to the original salaries
       to create a new data set. Recalculate the range, popu-
       lation variance, and population standard deviation.
       How do they compare to the originals?
   (d) As a third option, Benjamin decides not to give his
       employees a bonus at all. Instead, he keeps the
       $25,000 for himself. Use this plan to create a new data
       set. Recalculate the range, population variance, and
       population standard deviation. How do they compare
       to the originals?

47. Resistance and Sample Size Each of the following three
data sets represents the IQ scores of a random sample of
adults. IQ scores are known to have a mean and median of
100. For each data set, determine the sample standard deviation. Then recompute the sample standard deviation
assuming that the individual whose IQ is 106 is accidentally recorded as 160. For each sample size, state what happens to the standard deviation. Comment on the role that
the number of observations plays in resistance.

Sample of Size 5				
106	92	98	103	100

Sample of Size 12					
106	92	98	103	100	102
98	124	83	70	108	121

Sample of Size 30					
106	92	98	103	100	102
98	124	83	70	108	121
102	87	121	107	97	114
140	93	130	72	81	90
103	97	89	98	88	103

48. Compute the sample standard deviation of the following test scores: 78, 78, 78, 78. What can be said about a data set in which all the values are identical?

49. Coefficient of Variation The coefficient of variation is a measure that allows for the comparison of two or more variables measured on a different scale. It measures the relative variability in terms of the mean and does not have a unit of measure. The lower the coefficient of variation, the less the data vary. The coefficient of variation is defined as

$$CV = \frac{\text{standard deviation}}{\text{mean}} \cdot 100\%$$

For example, suppose the senior class of a school district has a mean ACT score of 23 with a standard deviation of 4 and a mean SAT score of 1100 with a standard deviation of 150. Which test has more variability? The coefficient of variation for the ACT exam is $\frac{4}{23} \cdot 100\% = 17.4\%$; the coefficient of variation for the SAT exam is $\frac{150}{1100} \cdot 100\% = 13.6\%$. The ACT has more variability.

(a) Suppose the systolic blood pressure of a random sample of 100 students before exercising has a sample mean of 121, with a standard deviation of 14.1. The systolic blood pressure after exercising is 135.9, with a standard deviation of 18.1. Is there more variability in systolic blood pressure before exercise or after exercise?

(b) An investigation of intracellular calcium and blood pressure was conducted by Erne, Bolli, Buergisser, and Buehler in 1984 and published in the *New England Journal of Medicine*, 310:1084–1088. The researchers measured the free calcium concentration in the blood platelets of 38 people with normal blood pressure and 45 people with high blood pressure. The mean and standard deviation of the normal blood pressure group were 107.9 and 16.1, respectively. The mean and standard deviation of the high blood pressure group were 168.2 and 31.7, respectively. Is there more variability in free calcium concentration in the normal blood pressure or the high blood pressure group?

50. Mean Absolute Deviation Another measure of variation is the mean absolute deviation. It is computed using the formula

$$MAD = \frac{\sum |x_i - \bar{x}|}{n}$$

Compute the mean absolute deviation of the data in Problem 17 and compare the results with the sample standard deviation.

51. Coefficient of Skewness Karl Pearson developed a measure that describes the skewness of a distribution, called the **coefficient of skewness**. The formula is

$$\text{Skewness} = \frac{3(\text{mean} - \text{median})}{\text{standard deviation}}$$

The value of this measure generally lies between −3 and +3. The closer the value lies to −3, the more the distribution is skewed left. The closer the value lies to +3, the more the distribution is skewed right. A value close to 0 indicates a symmetric distribution. Find the coefficient of skewness of the following distributions and comment on the skewness.

(a) Mean = 50, median = 40, standard deviation = 10
(b) Mean = 100, median = 100, standard deviation = 15
(c) Mean = 400, median = 500, standard deviation = 120
(d) Compute the coefficient of skewness for the data in Problem 33.
(e) Compute the coefficient of skewness for the data in Problem 34.

52. Diversification A popular theory in investment states that you should invest a certain amount of money in foreign investments to reduce your risk. The risk of a portfolio is defined as the standard deviation of the rate of return. Refer to the following graph, which depicts the relation between risk (standard deviation of rate of return) and reward (mean rate of return).

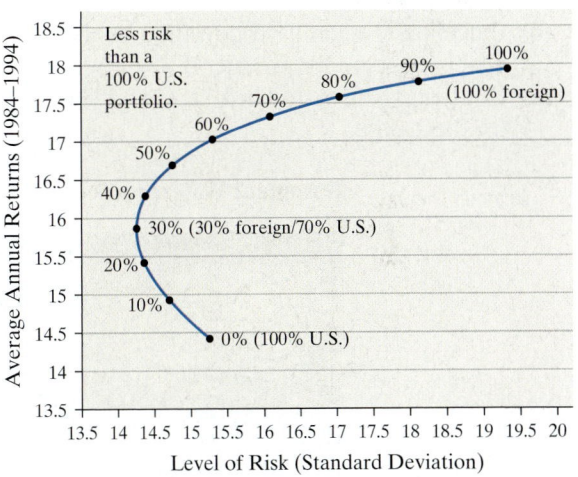

How Foreign Stocks Benefit a Domestic Portfolio

Source: T. Rowe Price

(a) Determine the average annual return and level of risk in a portfolio that is 10% foreign.
(b) Determine the percentage that should be invested in foreign stocks to best minimize risk.
(c) Why do you think risk initially decreases as the percent of foreign investments increases?
(d) A portfolio that is 30% foreign and 70% American has a mean rate of return of about 15.8%, with a standard deviation of 14.3%. According to Chebyshev's inequality, at least 75% of returns will be between what values? According to Chebyshev's inequality, at least 88.9% of returns will be between what two values? Should an investor be surprised if she has a negative rate of return? Why?

Consumer Reports® — BASEMENT WATERPROOFING COATINGS

A waterproofing coating can be an inexpensive and easy way to deal with leaking basements. But how effective are they? In a study, *Consumer Reports* tested nine waterproofers to rate their effectiveness in controlling water seepage though concrete foundations.

To compare the products' ability to control water seepage, we applied two coats of each product to slabs cut from concrete block. For statistical validity, this process was repeated at least six times. In each test run, four blocks (each coated with a different product) were simultaneously placed in a rectangular aluminum chamber. See the picture.

The chamber was sealed and filled with water and the blocks were subjected to progressively increasing hydrostatic pressures. Water that leaked out during each period was channeled to the bottom of the chamber opening, collected, and weighed.

The table contains a subset of the data collected for two of the products tested. Using these data,

(a) Calculate the mean, median, and mode weight of water collected for product A.

(b) Calculate the standard deviation of the weight of water collected for product A.

(c) Calculate the mean, median, and mode weight of water collected for product B.

(d) Calculate the standard deviation of the weight of water collected for product B.

(e) Construct a back-to-back stem-and-leaf diagrams for these data.

Product	Replicate	Weight of Collected Water (in grams)
A	1	91.2
A	2	91.2
A	3	90.9
A	4	91.3
A	5	90.8
A	6	90.8
B	1	87.1
B	2	87.2
B	3	86.8
B	4	87.0
B	5	87.2
B	6	87.0

Does there appear to be a difference in these two products' ability to mitigate water seepage? Why?

Note to Readers: *In many cases, our test protocol and analytical methods are more complicated than described in these examples. The data and discussions have been modified to make the material more appropriate for the audience.*

Basement Waterproofer Test Chamber

Determining the Range, Variance, and Standard Deviation
The same steps followed to obtain the measures of central tendency from raw data can be used to obtain the measures of dispersion.

3.3 Measures of Central Tendency and Dispersion from Grouped Data

Preparing for This Section   Before getting started, review the following:

- Organizing discrete data in tables (Section 2.2, pp. 77–78)
- Organizing continuous data in tables (Section 2.2, pp. 79–81)
- Class midpoint (Section 2.3, pp. 97–98)

Objectives

1 **Approximate the mean of a variable from grouped data**

2 **Compute the weighted mean**

3 **Approximate the variance and standard deviation of a variable from grouped data**

We have discussed how to compute descriptive statistics from raw data, but many times the data that we have access to have already been summarized in frequency distributions (grouped data). While we cannot obtain exact values of the mean or standard deviation without raw data, these measures can be approximated using the techniques discussed in this section.

1 **Approximate the Mean of a Variable from Grouped Data**

Since raw data cannot be retrieved from a frequency table, we assume that, within each class, the mean of the data values is equal to the class midpoint. We then multiply the class midpoint by the frequency. This product is expected to be close to the sum of the data that lie within the class. We repeat the process for each class and sum the results. This sum approximates the sum of all the data.

Definition

Approximate Mean of a Variable from a Frequency Distribution

Population Mean

$$\mu = \frac{\sum x_i f_i}{\sum f_i} = \frac{x_1 f_1 + x_2 f_2 + \cdots + x_N f_N}{f_1 + f_2 + \cdots + f_N} \tag{1a}$$

Sample Mean

$$\bar{x} = \frac{\sum x_i f_i}{\sum f_i} = \frac{x_1 f_1 + x_2 f_2 + \cdots + x_n f_n}{f_1 + f_2 + \cdots + f_n} \tag{1b}$$

where x_i is the midpoint or value of the ith class

f_i is the frequency of the ith class

n is the number of classes

In Formula (1), $x_1 f_1$ approximates the sum of all the data values in the first class, $x_2 f_2$ approximates the sum of all the data values in the second class, and so on. Notice that the formulas for the population mean and sample mean are essentially identical, just as they were for computing the mean from raw data.

EXAMPLE 1

Approximating the Mean for Continuous Quantitative Data from the Frequency Distribution

Problem: The frequency distribution in Table 14 represents the 3-year rate of return of a random sample of 40 small-capitalization growth mutual funds. Approximate the mean 3-year rate of return.

Approach: We perform the following steps to approximate the mean.

Step 1: Determine the class midpoint of each class. Recall that the class midpoint is found by adding consecutive lower class limits and dividing the result by 2.

Step 2: Compute the sum of the frequencies, Σf_i.

Step 3: Multiply the class midpoint by the frequency to obtain $x_i f_i$ for each class.

Step 4: Compute $\Sigma x_i f_i$.

Step 5: Substitute into Formula (1b) to obtain the mean from grouped data.

Table 14

Class (3-year rate of return)	Frequency
0–1.99	2
2–3.99	5
4–5.99	6
6–7.99	8
8–9.99	9
10–11.99	6
12–13.99	3
14–15.99	1

Solution

Step 1: The class midpoint of the first class is $\dfrac{0 + 2}{2} = 1$, so $x_1 = 1$. The remaining class midpoints are listed in column 2 of Table 15.

Step 2: We add the frequencies in column 3 to obtain $\Sigma f_i = 2 + 5 + \cdots + 1 = 40$.

Step 3: Compute the values of $x_i f_i$ by multiplying each class midpoint by the corresponding frequency and obtain the results shown in column 4 of Table 15.

Step 4: We add the values in column 4 of Table 15 to obtain $\Sigma x_i f_i = 304$.

Table 15

Class (3-year rate of return)	Class Midpoint, x_i	Frequency, f_i	$x_i f_i$
0–1.99	$\dfrac{0 + 2}{2} = 1$	2	$(1)(2) = 2$
2–3.99	3	5	$(3)(5) = 15$
4–5.99	5	6	30
6–7.99	7	8	56
8–9.99	9	9	81
10–11.99	11	6	66
12–13.99	13	3	39
14–15.99	15	1	15
		$\Sigma f_i = 40$	$\Sigma x_i f_i = 304$

Step 5: Substituting into Formula (1b), we obtain

$$\bar{x} = \frac{\Sigma x_i f_i}{\Sigma f_i} = \frac{304}{40} = 7.6$$

The approximate mean 3-year rate of return is 7.6%.

CAUTION

We computed the mean from grouped data in Example 1 even though the raw data is available. The reason for doing this was to illustrate how close the two values can be. In practice, use raw data whenever possible.

The mean 3-year rate of return from the raw data listed in Example 3 on page 79 from Section 2.2 is 7.5%. The approximate mean from grouped data is pretty close to the actual mean.

Note To compute the mean from a frequency distribution where the data are discrete, treat each category of data as the class midpoint. For discrete data, the mean from grouped data will equal the mean from raw data. ◀

Now compute the mean of the frequency distribution in Problem 3.

2   ## Compute the Weighted Mean

Sometimes, certain data values have a higher importance or weight associated with them. In this case, we compute the *weighted mean*. For example, your grade-point average is a weighted mean, with the weights equal to the number of credit hours in each course. The value of the variable is equal to the grade converted to a point value.

Definition   The **weighted mean**, $\bar{x}_w$, of a variable is found by multiplying each value of the variable by its corresponding weight, summing these products, and dividing the result by the sum of the weights. It can be expressed using the formula

$$\bar{x}_w = \frac{\sum w_i x_i}{\sum w_i} = \frac{w_1 x_1 + w_2 x_2 + \cdots + w_n x_n}{w_1 + w_2 + \cdots + w_n} \tag{2}$$

where w_i is the weight of the ith observation
        x_i is the value of the ith observation

EXAMPLE 2   ## Computing the Weighted Mean

Problem: Marissa just completed her first semester in college. She earned an A in her 4-hour statistics course, a B in her 3-hour sociology course, an A in her 3-hour psychology course, a C in her 5-hour computer programming course, and an A in her 1-hour drama course. Determine Marissa's grade-point average.

Approach: We must assign point values to each grade. Let an A equal 4 points, a B equal 3 points, and a C equal 2 points. The number of credit hours for each course determines its weight. So a 5-hour course gets a weight of 5, a 4-hour course gets a weight of 4, and so on. We multiply the weight of each course by the points earned in the course, sum these products, and divide the sum by the number of credit hours.

Solution

$$\text{GPA} = \bar{x}_w = \frac{\sum w_i x_i}{\sum w_i} = \frac{4(4) + 3(3) + 3(4) + 5(2) + 1(4)}{4 + 3 + 3 + 5 + 1} = \frac{51}{16} = 3.19$$

Marissa's grade-point average for her first semester is 3.19.

   Now Work Problem 15.

3   ## Approximate the Variance and Standard Deviation of a Variable from Grouped Data

The procedure for approximating the variance and standard deviation from grouped data is similar to that of finding the mean from grouped data. Again, because we do not have access to the original data, the variance is approximate.

Definition   **Approximate Variance of a Variable from a Frequency Distribution**

Population Variance	**Sample Variance**	
$\sigma^2 = \dfrac{\sum (x_i - \mu)^2 f_i}{\sum f_i}$	$s^2 = \dfrac{\sum (x_i - \bar{x})^2 f_i}{\left(\sum f_i\right) - 1}$	(3)

where x_i is the midpoint or value of the ith class
        f_i is the frequency of the ith class

An algebraically equivalent formula for the population variance is $\dfrac{\sum x_i^2 f_i - \dfrac{(\sum x_i f_i)^2}{\sum f_i}}{\sum f_i}$.

We approximate the standard deviation by taking the square root of the variance.

EXAMPLE 3

Approximating the Variance and Standard Deviation from a Frequency Distribution

Problem: The data in Table 14 on page 158 represent the 3-year rate of return of a random sample of 40 small-capitalization growth mutual funds. Approximate the variance and standard deviation of the 3-year rate of return.

Approach: We will use the sample variance Formula (3).

Step 1: Create a table with the class in the first column, the class midpoint in the second column, the frequency in the third column, and the unrounded mean in the fourth column.

Step 2: Compute the deviation about the mean, $x_i - \bar{x}$, for each class, where x_i is the class midpoint of the ith class and $\bar{x}$ is the sample mean. Enter the results in column 5.

Step 3: Square the deviation about the mean and multiply this result by the frequency to obtain $(x_i - \bar{x})^2 f_i$. Enter the results in column 6.

Step 4: Add the entries in columns 3 and 6 to obtain Σf_i and $\Sigma (x_i - \bar{x})^2 f_i$.

Step 5: Substitute the values obtained in Step 4 into Formula (3) to obtain an approximate value for the sample variance.

Solution

Step 1: We create Table 16. Column 1 contains the classes. Column 2 contains the class midpoint of each class. Column 3 contains the frequency of each class. Column 4 contains the unrounded sample mean obtained in Example 1.

Step 2: Column 5 of Table 16 contains the deviation about the mean, $x_i - \bar{x}$, for each class.

Step 3: Column 6 contains the values of the squared deviation about the mean multiplied by the frequency, $(x_i - \bar{x})^2 f_i$.

Step 4: We add the entries in columns 3 and 6 and obtain $\Sigma f_i = 40$ and $\Sigma (x_i - \bar{x})^2 f_i = 465.6$.

Table 16

Class (3-year rate of return)	Class Midpoint, x_i	Frequency, f_i	$\bar{x}$	$x_i - \bar{x}$	$(x_i - \bar{x})^2 f_i$
0–1.99	1	2	7.6	−6.6	87.12
2–3.99	3	5	7.6	−4.6	105.8
4–5.99	5	6	7.6	−2.6	40.56
6–7.99	7	8	7.6	−0.6	2.88
8–9.99	9	9	7.6	1.4	17.64
10–11.99	11	6	7.6	3.4	69.36
12–13.99	13	3	7.6	5.4	87.48
14–15.99	15	1	7.6	7.4	54.76
		$\Sigma f_i = 40$			$\Sigma (x_i - \bar{x})^2 f_i = 465.6$

Step 5: Substitute these values into Formula (3) to obtain an approximate value for the sample variance.

$$s^2 = \frac{\Sigma (x_i - \bar{x})^2 f_i}{\left(\Sigma f_i \right) - 1} = \frac{465.6}{39} \approx 11.94$$

Take the square root of the unrounded estimate of the sample variance to obtain an approximation of the sample standard deviation.

$$s = \sqrt{s^2} = \sqrt{\frac{465.6}{39}} \approx 3.46\%$$

We approximate the sample standard deviation 3-year rate of return to be 3.46%.

EXAMPLE 4   **Approximating the Mean and Standard Deviation Using Technology**

Problem: Approximate the mean and standard deviation of the 3-year rate of return data in Table 14 using a TI-83/84 Plus graphing calculator.

Approach: The steps for approximating the mean and standard deviation of grouped data using the TI-83 or TI-84 Plus graphing calculator are given in the Technology Step-by-Step on page 165.

Result: Figure 16 shows the result from the TI-84 Plus.

Figure 16

Approximate mean

Approximate sample standard deviation

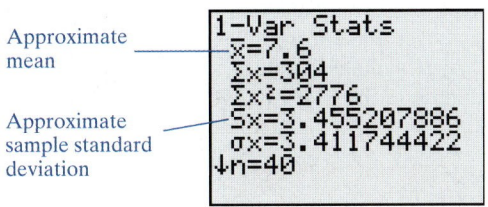

```
1-Var Stats
x̄=7.6
Σx=304
Σx²=2776
Sx=3.455207886
σx=3.411744422
↓n=40
```

From the output, we can see that the approximate mean is 7.6% and the approximate standard deviation is 3.46%. The results agree with our by-hand solutions.

From the raw data listed in Example 3, we find that the sample standard deviation is 3.46%. The approximate sample standard deviation from grouped data equals the sample standard deviation from the raw data!

> **Now compute the standard deviation from the frequency distribution in Problem 3.**

3.3 ASSESS YOUR UNDERSTANDING

Concepts and Vocabulary

1. Explain the role of the class midpoint in the formulas to approximate the mean and the standard deviation.

2. In Section 3.1, the mean is given by $\bar{x} = \dfrac{\Sigma x_i}{n}$. Explain how this is a special case of the weighted mean, $\bar{x}_w$.

Applying the Concepts

3. **Cell Phones** A sample of college students was asked how much they spent monthly on a cell phone plan (to the nearest dollar). Approximate the mean and standard deviation for the cost.

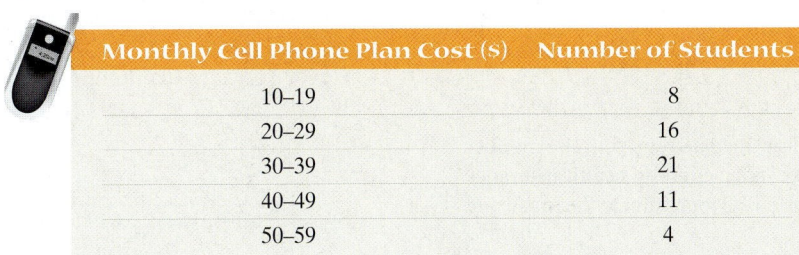

Monthly Cell Phone Plan Cost ($)	Number of Students
10–19	8
20–29	16
30–39	21
40–49	11
50–59	4

4. **Bowl Games** The following data represent the difference in scores between the winning and losing teams in the 2004–2005 college football bowl games. Approximate the mean and standard deviation for the point difference.

Point Difference	Number of Bowl Games
1–5	11
6–10	0
11–15	5
16–20	6
21–25	1
26–30	2
31–35	1
36–40	2

Source: espn.com

5. **100-Degree Days** The following data represent the annual number of days over 100°F for Dallas–Fort Worth from 1905 to 2004. Approximate the mean and standard deviation annual number of days over 100°F.

Number of 100° + Days	Number of Years
0–9	31
10–19	39
20–29	17
30–39	6
40–49	4
50–59	2
60–69	1

Source: National Weather Service

6. **Working Students** The following data represent the number of hours (on average) worked each week for a sample of community college students. Approximate the mean and standard deviation of the number of hours.

Hours Worked (per week)	Number of Students
0–9	24
10–19	14
20–29	39
30–39	18
40–49	5

7. **Health Insurance** The following data represent the number of people aged 25 to 64 years covered by health insurance (private or government) in 2003. Approximate the mean and standard deviation for age.

Age	Number (millions)
25–34	28.9
35–44	35.7
45–54	35.1
55–64	24.7

Source: U.S. Census Bureau

8. **Earthquakes** The following data represent the magnitude of earthquakes in the United States in 2004. Approximate the mean and standard deviation of the magnitude.

Magnitude	Number
0–0.9	539
1.0–1.9	1
2.0–2.9	1336
3.0–3.9	1363
4.0–4.9	289
5.0–5.9	21
6.0–6.9	2

Source: U.S. Geological Survey

9. **Meteorology** The following data represent the high-temperature distribution for the month of August in Chicago since 1872.

Temperature (°)	Days
50–59	1
60–69	308
70–79	1519
80–89	1626
90–99	503
100–109	11

Source: National Oceanic and Atmospheric Administration

(a) Approximate the mean and standard deviation for temperature.
(b) Draw a frequency histogram of the data to verify that the distribution is bell shaped.
(c) According to the Empirical Rule, 95% of days in the month of August will be between what two temperatures?

10. **Rainfall** The following data represent the annual rainfall distribution for St. Louis, Missouri, from 1870 to 2004.

Rainfall (inches)	Number of years
20–24	4
25–29	15
30–34	27
35–39	40
40–44	28
45–49	15
50–54	4
55–59	2

Source: National Oceanic and Atmospheric Administration

(a) Approximate the mean and standard deviation for rainfall.
(b) Draw a frequency histogram of the data to verify that the distribution is bell shaped.
(c) According to the Empirical Rule, 95% of annual rainfalls in St. Louis will be between what two amounts?

11. **Multiple Births** The following data represent the number of live multiple-delivery births (three or more babies) in 2002 for women 15 to 44 years old.

Age	Number of Multiple Births
15–19	93
20–24	511
25–29	1628
30–34	2832
35–39	1843
40–44	377

Source: National Vital Statistics Reports, Vol. 52, No. 10, December 17, 2003

(a) Approximate the mean and standard deviation for age.
(b) Draw a frequency histogram of the data to verify that the distribution is bell shaped.
(c) According to the Empirical Rule, 95% of mothers of multiple births will be between what two ages?

12. **SAT Scores** The following data represent SAT Verbal scores for ISACS college-bound seniors in 2003.

SAT Verbal Score	Number
400–449	281
450–499	577
500–549	840
550–599	1120
600–649	1166
650–699	900
700–749	518
750–800	394

Source: www.isacs.org

(a) Approximate the mean and standard deviation of the score.
(b) Draw a frequency histogram of the data to verify that the distribution is bell shaped.
(c) According to the Empirical Rule, 95% of these ISACS college-bound seniors will have SAT Verbal scores between what two values?

13. **Serum HDL** Use the frequency distribution whose class width is 10 obtained in Problem 31 in Section 2.2 to approximate the mean and standard deviation for serum HDL. Compare these results to the values obtained in Problem 31 in Sections 3.1 and 3.2.

14. **Volume of Altria Group Stock** Use the frequency distribution whose class width is 2 obtained in Problem 32 in Section 2.2 to approximate the mean and standard deviation of the number of shares traded. Compare these results to the values obtained in Problem 32 in Sections 3.1 and 3.2.

15. **Grade-Point Average** Marissa has just completed her second semester in college. She earned a B in her 5-hour calculus course, an A in her 3-hour social work course, an A in her 4-hour biology course, and a C in her 3-hour American literature course. Assuming that an A equals 4 points, a B equals 3 points, and a C equals 2 points, determine Marissa's grade-point average for the semester.

16. **Computing Class Average** In Marissa's calculus course, attendance counts for 5% of the grade, quizzes count for 10% of the grade, exams count for 60% of the grade, and the final exam counts for 25% of the grade. Marissa had a 100% average for attendance, 93% for quizzes, 86% for exams, and 85% on the final. Determine Marissa's course average.

17. **Mixed Chocolates** Michael and Kevin want to buy chocolates. They can't agree on whether they want chocolate-covered almonds, chocolate-covered peanuts, or chocolate-covered raisins. They agree to create a mix. They bought 4 pounds of chocolate-covered almonds at $3.50 per pound, 3 pounds of chocolate-covered peanuts for $2.75 per pound, and 2 pounds of chocolate-covered raisins for $2.25 per pound. Determine the cost per pound of the mix.

18. **Nut Mix** Michael and Kevin return to the candy store, but this time they want to purchase nuts. They can't decide among peanuts, cashews, or almonds. They again agree to create a mix. They bought 2.5 pounds of peanuts for $1.30 per pound, 4 pounds of cashews for $4.50 per pound, and 2 pounds of almonds for $3.75 per pound. Determine the price per pound of the mix.

19. **Population** The following data represent the male and female population by age of the United States for residents under 100 years old in July 2003.

Age	Male Resident Pop (in thousands)	Female Resident Pop (in thousands)
0–9	20,225	19,319
10–19	21,375	20,295
20–29	20,437	19,459
30–39	21,176	20,936
40–49	22,138	22,586
50–59	16,974	17,864
60–69	10,289	11,563
70–79	6,923	9,121
80–89	3,053	5,367
90–99	436	1,215

Source: U.S. Census Bureau

(a) Approximate the population mean and standard deviation of age for males.
(b) Approximate the population mean and standard deviation of age for females.
(c) Which gender has the higher mean age?
(d) Which gender has more dispersion in age?

20. **Age of Mother** The following data represent the age of the mother at childbirth for 1980 and 2002.

Age of Mother	Number of Births 1980 (thousands)	Number of Births 2002 (thousands)
10–14	1.1	0.7
15–19	53.0	43.0
20–24	115.1	103.6
25–29	112.9	113.6
30–34	61.9	91.5
35–39	19.8	41.4
40–44	3.9	8.3
45–49	0.2	0.5

Source: National Vital Statistics Reports, Vol. 52, No. 10

(a) Approximate the population mean and standard deviation of age for mothers in 1980.
(b) Approximate the population mean and standard deviation of age for mothers in 2002.
(c) Which year has the higher mean age?
(d) Which year has more dispersion in age?

Problems 21–24 use the following steps to approximate the median from grouped data.

Approximating the Median from Grouped Data

Step 1: Construct a cumulative frequency distribution.

Step 2: Identify the class in which the median lies. Remember, the median can be obtained by determining the observation that lies in the middle.

Step 3: Interpolate the median using the formula

$$\text{Median} = M = L + \frac{\frac{n}{2} - CF}{f}(i)$$

where L is the lower class limit of the class containing the median

n is the number of data values in the frequency distribution

CF is the cumulative frequency of the class immediately preceding the class containing the median

f is the frequency of the median class

i is the class width of the class containing the median

21. Approximate the median of the frequency distribution in Problem 5.

22. Approximate the median of the frequency distribution in Problem 6.

23. Approximate the median of the frequency distribution in Problem 7.

24. Approximate the median of the frequency distribution in Problem 8.

Problems 25–28 use the following definition of the modal class.

The **modal class** of a variable can be obtained from data in a frequency distribution by determining the class that has the largest frequency.

25. Determine the modal class of the frequency distribution in Problem 5.

26. Determine the modal class of the frequency distribution in Problem 6.

27. Determine the modal class of the frequency distribution in Problem 7.

28. Determine the modal class of the frequency distribution in Problem 8.

29. Unequal Class Widths Thus far, we have only looked at grouped data for which the class width is the same. However, many data reports have groups of different widths, often because the frequencies of several groups may be small relative to the rest. For example, consider the undergraduate student age profile information provided by the University of Northern Colorado for Fall 2004.

Age	Number of Students
Less than 18	139
18–19	4089
20–21	3357
22–24	1661
25–29	470
30–34	145
35–39	95
40–49	117
50 and above	21

Source: www.unco.edu

(a) What is a reasonable midpoint for the first class? Explain.
(b) What is a reasonable midpoint for the last class? Explain.
(c) Approximate the mean age of undergraduates at UNC and compare this to the reported age of 20.9.

Technology Step by Step	**Determining the Mean and Standard Deviation from Grouped Data**
TI-83/84 Plus	**Step 1:** Enter the class midpoint in L1 and the frequency or relative frequency in L2 by pressing STAT and selecting 1:Edit.
	Step 2: Press STAT, highlight the CALC menu and select 1:1-Var Stats
	Step 3: With 1-Var Stats appearing on the HOME screen, press 2nd 1 to insert L1 on the HOME screen. Then press the comma and press 2nd 2 to insert L2 on the HOME screen. So, the HOME screen should have the following:
	1-Var Stats L1, L2
	Press ENTER to obtain the mean and standard deviation.

3.4   Measures of Position

Objectives

1 Determine and interpret *z*-scores

2 Determine and interpret percentiles

3 Determine and interpret quartiles

4 Check a set of data for outliers

In Section 3.1, we were able to find measures of central tendency. Measures of central tendency are meant to describe the "typical" data value. Section 3.2 discussed measures of dispersion, which describe the amount of spread in a set of data. In this section, we discuss measures of position; that is, we wish to describe the relative position of a certain data value within the entire set of data.

1   Determine and Interpret *z*-Scores

At the end of the 2004 season, the Boston Red Sox led the American League with 949 runs scored, while the St. Louis Cardinals led the National League with 855 runs scored. A quick comparison might lead one to believe that the Red Sox are the better run-producing team. However, this comparison is unfair because the two teams play in different leagues. The Red Sox play in the American League, where the designated hitter bats for the pitcher, whereas the Cardinals play in the National League, where the pitcher must bat (pitchers are typically poor hitters). To compare the two teams' scoring of runs we need to determine their relative standings in their respective leagues. This can be accomplished using a *z-score*.

Definition

The **z-score** represents the distance that a data value is from the mean in terms of the number of standard deviations. It is obtained by subtracting the mean from the data value and dividing this result by the standard deviation. There is both a population *z*-score and a sample *z*-score; their formulas follow:

$$\text{Population } z\text{-Score} \qquad \text{Sample } z\text{-Score}$$

$$z = \frac{x - \mu}{\sigma} \qquad\qquad z = \frac{x - \overline{x}}{s} \qquad \textbf{(1)}$$

The *z*-score is unitless. It has mean 0 and standard deviation 1.

In Other Words

z-Scores provide a way to compare apples to oranges by converting variables with different centers and/or spreads to variables with the same center (0) and spread (1).

If a data value is larger than the mean, the *z*-score will be positive. If a data value is smaller than the mean, the *z*-score will be negative. If the data value equals the mean, the *z*-score will be zero. Z-scores measure the number of standard deviations an observation is above or below the mean. For example, a *z*-score of 1.24 is interpreted as "the data value is 1.24 standard deviations above the mean." A *z*-score of −2.31 is interpreted as "the data value is 2.31 standard deviations below the mean."

We are now prepared to determine whether the Red Sox or Cardinals had a better year in run production.

EXAMPLE 1   **Comparing z-Scores**

Problem: Determine whether the Boston Red Sox or the St. Louis Cardinals had a relatively better run producing season. The Red Sox scored 949 runs and play in the American League, where the mean number of runs scored was $\mu = 811.3$. The standard deviation was $\sigma = 73.7$. The Cardinals scored 855 runs and play in the National League, where the mean number of runs scored was $\mu = 751.1$. The standard deviation was $\sigma = 78.6$.

Approach: To determine which team had the relatively better run-producing season, we compute each team's *z*-score. The team with the higher *z*-score had the better season. Because we know the values of the population parameters, we will compute the population *z*-score.

Solution: First, we compute the *z*-score for the Red Sox. Z-scores are typically rounded to two decimal places.

$$z\text{-score} = \frac{x - \mu}{\sigma} = \frac{949 - 811.3}{73.7} = 1.87$$

Next, we compute the *z*-score for the Cardinals.

$$z\text{-score} = \frac{x - \mu}{\sigma} = \frac{855 - 751.1}{78.6} = 1.32$$

So, the Red Sox had run production 1.87 standard deviations above the mean, while the Cardinals had run production 1.32 standard deviations above the mean. Therefore, the Red Sox had a relatively better year at scoring runs.

Now Work Problem 7.

2 ## Determine and Interpret Percentiles

Recall that the median divides the lower 50% of a set of data from the upper 50%. In general, the ***k*th percentile**, denoted P_k, of a set of data divides the lower $k\%$ of a data set from the upper $(100 - k)\%$. Percentiles divide a data set that is written in ascending order into 100 parts, so 99 possible percentiles can be computed. For example, P_1 divides the bottom 1% of the data from the top 99% while P_{99} divides the lower 99% of the data from the top 1%. Figure 17 displays the 99 possible percentiles.

Figure 17

If a data value lies at the 40th percentile, then approximately 40% of the data are less than this value and approximately 60% are higher than this value.

Percentiles are often used to give the relative standing of a data value. Many standardized exams, such as the SAT college entrance exam, use percentiles to provide students with an understanding of how they scored on the exam in relation to all other students who took the exam. For example, in 2004, an SAT verbal score of 580 was at the 73rd percentile. This means approximately 73% of the scores are below 580 and 27% are above 580. Pediatricians use percentiles to describe the progress of a newborn baby's weight gain relative to other newborn babies. A 3- to 5-month-old male child who weighs 14.3 pounds would be at the 15th percentile.

The following steps can be used to compute the kth percentile:

Determining the *k*th Percentile, P_k

Step 1: Arrange the data in ascending order.

Step 2: Compute an index i using the formula

$$i = \left(\frac{k}{100}\right)(n + 1) \tag{2}$$

where k is the percentile of the data value and n is the number of individuals in the data set.

Step 3:

(a) If i is an integer, the kth percentile, P_k, is the ith data value.

(b) If i is not an integer, find the mean of the observations on either side of i. This number represents the kth percentile, P_k.

! **CAUTION**

Don't forget to write the data in ascending order before finding the percentile.

An example should clarify the procedure.

EXAMPLE 2   ### Determining the Percentile of a Data Value, Index an Integer

Problem: The data in Table 17 represent the violent crime rate (violent crimes per 100,000 population) for the 50 states and the District of Columbia in 2003. Find the state that corresponds to the 75th percentile.

	Table 17						
State	**Crime Rate**	**State**	**Crime Rate**	**State**	**Crime Rate**
1. North Dakota	77.8	18. Oregon	295.5	35. Missouri	472.8
2. Maine	108.9	19. Connecticut	308.2	36. Oklahoma	505.7
3. Vermont	110.2	20. Mississippi	325.5	37. Michigan	511.2
4. New Hampshire	148.8	21. Ohio	333.2	38. Arizona	513.2
5. South Dakota	173.4	22. Colorado	345.1	39. Texas	552.5
6. Wisconsin	221.0	23. Washington	347.0	40. Illinois	556.8
7. Idaho	242.7	24. Indiana	352.8	41. California	579.3
8. Utah	248.6	25. Montana	365.2	42. Alaska	593.4
9. West Virginia	257.5	26. New Jersey	365.8	43. Nevada	614.2
10. Kentucky	261.7	27. Kansas	395.5	44. Louisiana	646.3
11. Wyoming	262.1	28. Pennsylvania	398.0	45. Delaware	658.0
12. Minnesota	262.6	29. Alabama	429.5	46. New Mexico	665.2
13. Hawaii	270.4	30. Georgia	453.9	47. Tennessee	687.8
14. Iowa	272.4	31. North Carolina	454.9	48. Maryland	703.9
15. Virginia	275.8	32. Arkansas	456.1	49. Florida	730.2
16. Rhode Island	285.6	33. New York	465.2	50. South Carolina	793.5
17. Nebraska	289.0	34. Massachusetts	469.4	51. District of Columbia	1608.1

Source: Federal Bureau of Investigation, Uniform Crime Reports, 2003

Approach: We will follow the steps given on page 167.

Solution

Step 1: The data provided in Table 17 are already listed in ascending order.

Step 2: To find the 75th percentile, P_{75}, we compute the index i with $k = 75$ and $n = 51$.

$$i = \left(\frac{75}{100}\right)(51 + 1) = 39$$

Step 3: The 75th percentile is the 39th observation of the data set written in ascending order. The 39th observation, which corresponds to the state of Texas, is 552.5. Approximately 75% of the states have a violent crime rate less than 552.5 crimes per 100,000 population, and approximately 25% of the states have a violent crime rate above 552.5 crimes per 100,000 population.

EXAMPLE 3 **Determining the Percentile of a Data Value, Index Not an Integer**

Problem: Find the crime rate that corresponds to the 90th percentile for the data in Table 17.

Approach: We will follow the steps given on page 167.

Solution

Step 1: The data provided in Table 17 are listed in ascending order.

Step 2: To find the 90th percentile, P_{90}, we compute the index i with $k = 90$ and $n = 51$.

$$i = \left(\frac{90}{100}\right)(51 + 1) = 46.8$$

Step 3: Because the index, $i = 46.8$, is not an integer, the 90th percentile is the mean of the 46th and 47th data value.

$$P_{90} = \frac{665.2 + 687.8}{2} = 676.5$$

Approximately 90% of the states have violent crime rates below 676.5 crimes per 100,000 population. Approximately 10% of the states have violent crime rates above 676.5 crimes per 100,000 population.

> **Now Work Problem 13(a).**

Often we are interested in knowing the percentile to which a specific data value corresponds. The kth percentile of a data value, x, from a data set that contains n values is computed by using the following steps:

Finding the Percentile That Corresponds to a Data Value

Step 1: Arrange the data in ascending order.
Step 2: Use the following formula to determine the percentile of the score, x.

$$\text{Percentile of } x = \frac{\text{number of data values less than } x}{n} \times 100 \qquad (3)$$

Round this number to the nearest integer.

EXAMPLE 4 **Finding the Percentile of a Specific Data Value**

Problem: Find the percentile rank for the state of Kentucky using the data provided in Table 17.

Approach: We will follow the steps given above.

Solution

Step 1: The data provided in Table 17 are in ascending order.
Step 2: Nine states have a violent crime rate that is less than Kentucky's violent crime rate. So

$$\text{Percentile rank of Kentucky} = \frac{9}{51} \cdot 100 \approx 17.6$$

We round 17.6 to 18. Kentucky's violent crime rate is at the 18th percentile. Approximately 18% of the states have violent crime rates that are less than that of Kentucky, and approximately 82% of the states have violent crime rates that are larger than that of Kentucky.

> **Now Work Problem 13(d).**

 Determine and Interpret Quartiles

The most common percentiles are quartiles. **Quartiles** divide data sets into fourths, or four equal parts. The first quartile, denoted Q_1, divides the bottom 25% of the data from the top 75%. Therefore, the first quartile is equivalent to the 25th percentile. The second quartile divides the bottom 50% of the data from the top 50%, so the second quartile is equivalent to the 50th percentile, which is equivalent to the median. Finally, the third quartile divides the bottom

75% of the data from the top 25%, so that the third quartile is equivalent to the 75th percentile. Figure 18 illustrates the concept of quartiles.

In Other Words
The first quartile, Q_1, is equivalent to the 25th percentile, P_{25}. The 2nd quartile, Q_2, is equivalent to the 50th percentile, P_{50}, which is equivalent to the median, M. Finally, the third quartile, Q_3, is equivalent to the 75th percentile, P_{75}.

Figure 18

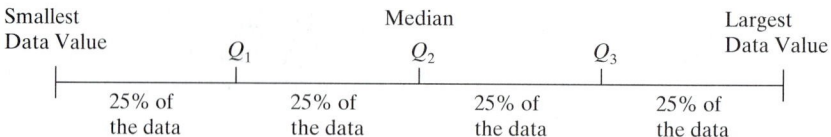

Smallest Data Value | Q_1 | Median Q_2 | Q_3 | Largest Data Value

25% of the data | 25% of the data | 25% of the data | 25% of the data

EXAMPLE 5 Finding the Quartiles of a Data Set

Problem: Find the first, second, and third quartiles for the violent crime rates listed in Table 17.

Approach

Step 1: The first quartile, Q_1, is the 25th percentile, P_{25}. We let $k = 25$ in Formula (2) to obtain the index, i.

Step 2: The second quartile, Q_2, is the 50th percentile, P_{50}. We let $k = 50$ in Formula (2) to obtain the index, i.

Step 3: The third quartile, Q_3, is the 75th percentile, P_{75}. We let $k = 75$ in Formula (2) to obtain the index, i.

In Other Words
To find Q_2, determine the median of the data set. To find Q_1, determine the median of the "lower half" of the data set. To find Q_3, determine the median of the "upper half" of the data set.

Solution

Step 1: The index for the first quartile, Q_1, is

$$i = \left(\frac{25}{100}\right)(51 + 1) = 13$$

The 13th observation will be the first quartile. So $Q_1 = P_{25} = 270.4$.

Step 2: The index for the second quartile, Q_2, is

$$i = \left(\frac{50}{100}\right)(51 + 1) = 26$$

The 26th observation will be the second quartile. So $Q_2 = P_{50} = M = 365.8$

Step 3: The third quartile, Q_3, is the 75th percentile, which we found in Example 2. The third quartile is $Q_3 = P_{75} = 552.5$.

EXAMPLE 6 Finding Quartiles Using Technology

Problem: Find the quartiles of the violent crimes data in Table 17.

Approach: We will use a TI-84 Plus graphing calculator to obtain the quartiles. The steps for obtaining quartiles using a TI-83/84 Plus graphing calculator, MINITAB, or Excel are given in the Technology Step-by-Step on page 175.

Figure 19

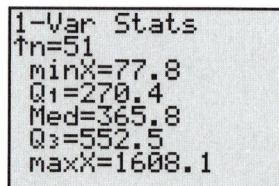

```
1-Var Stats
↑n=51
minX=77.8
Q₁=270.4
Med=365.8
Q₃=552.5
maxX=1608.1
```

Result: Figure 19 shows the results obtained from a TI-84 Plus graphing calculator. Notice that the calculator uses Med for the 2nd quartile.

Using Technology
Statistical packages may use different formulas for obtaining the quartiles, so results may differ slightly if the index is not an integer.

Now Work Problem 15(b).

4 Check a Set of Data for Outliers

Whenever performing any type of data analysis, we should always check for extreme observations in the data set. Extreme observations are referred to as **outliers**. Whenever outliers are encountered, their origin must be investigated. They can occur by chance, because of error in the measurement of a variable, during data entry, or from errors in sampling. For example, in the 2000 presidential election, a precinct in New Mexico accidentally recorded 610 absentee ballots for Al Gore as 110. Workers in the Gore camp discovered the data-entry error through an analysis of vote totals.

Sometimes extreme observations are common within a population. For example, suppose we wanted to estimate the mean price of a European car. We might take a random sample of size 5 from the population of all European automobiles. If our sample included a Ferrari 360 Spider (approximately $170,000), it probably would be an outlier, because this car costs much more than the typical European automobile. The value of this car would be considered *unusual* because it is not a typical value from the data set.

We can use the following steps to check for outliers using quartiles.

Checking for Outliers by Using Quartiles

Step 1: Determine the first and third quartiles of the data.

Step 2: Compute the interquartile range. The **interquartile range** or **IQR** is the difference between the third and first quartile. That is,

$$IQR = Q_3 - Q_1$$

Step 3: Determine the fences. **Fences** serve as cutoff points for determining outliers.

$$\text{Lower fence} = Q_1 - 1.5(IQR)$$
$$\text{Upper fence} = Q_3 + 1.5(IQR)$$

Step 4: If a data value is less than the lower fence or greater than the upper fence, it is considered an outlier.

EXAMPLE 7   **Checking for Outliers**

Problem: Check the data that represent the violent crime rates of the 50 states and the District of Columbia for outliers.

Approach: We follow the preceding steps. Any data value that is less than the lower fence or greater than the upper fence will be considered an outlier.

Solution

Step 1: The quartiles were found in Examples 5 and 6. So $Q_1 = 270.4$ and $Q_3 = 552.5$.

Step 2: The interquartile range, IQR, is

$$\begin{aligned} IQR &= Q_3 - Q_1 \\ &= 552.5 - 270.4 \\ &= 282.1 \end{aligned}$$

Step 3: The lower fence, LF, is

$$\begin{aligned} LF &= Q_1 - 1.5(IQR) \\ &= 270.4 - 1.5(282.1) \\ &= -152.75 \end{aligned}$$

The upper fence, UF, is

$$\begin{aligned} UF &= Q_3 + 1.5(IQR) \\ &= 552.5 + 1.5(282.1) \\ &= 975.65 \end{aligned}$$

Step 4: There are no outliers below the lower fence. However, we do have an outlier above the upper fence corresponding to the District of Columbia (1608.1 violent crimes per 100,000 population).

Figure 20 shows a histogram of the data. We can easily identify the outlier corresponding to the District of Columbia.

Figure 20

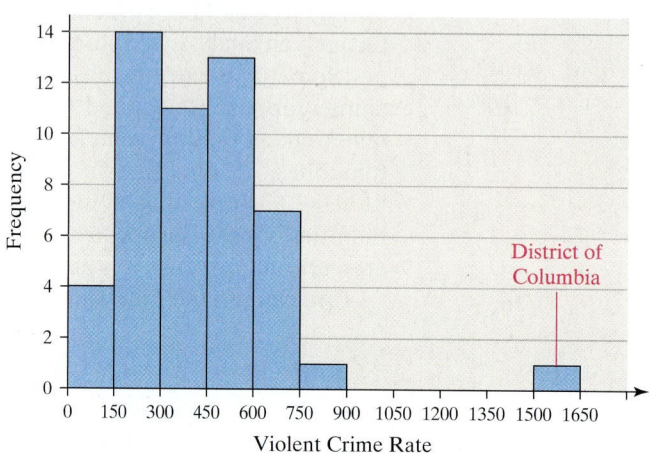

Violent Crime Rate per
100,000 Population, 2003

Now Work Problems 15(c) and (d).

3.4 ASSESS YOUR UNDERSTANDING

Concepts and Vocabulary

1. Write a paragraph that explains the meaning of percentiles.
2. Suppose you received the highest score on an exam. Your friend scored the second-highest score, yet you both were in the 99th percentile. How can this be?
3. Morningstar is a mutual fund rating agency. They rank a fund's performance by using one to five stars. A one-star mutual fund is in the bottom 20% of its investment class; a five-star mutual fund is in the top 20% of its investment class. Interpret the meaning of a four-star mutual fund.

4. When outliers are discovered, should they always be removed from the data set before further analysis?
5. Mensa is an organization designed for people of high intelligence. One qualifies for Mensa if one's intelligence is measured at or above the 98th percentile. Explain what this means.
6. Explain the advantage of using z-scores to compare observations from two different data sets.

Applying the Concepts

7. **Birth Weights** In 2003, babies born after a gestation period of 32 to 35 weeks had a mean weight of 2600 grams and a standard deviation of 670 grams. In the same year, babies born after a gestation period of 40 weeks had a mean weight of 3500 grams and a standard deviation of 475 grams. Suppose a 34-week gestation period baby weighs 2400 grams and a 40-week gestation period baby weighs 3300 grams. Which baby weighs less relative to the gestation period?

8. **Birth Weights** In 2003, babies born after a gestation period of 32 to 35 weeks had a mean weight of 2600 grams and a standard deviation of 670 grams. In the same year, babies born after a gestation period of 40 weeks had a mean weight of 3500 grams and a standard deviation of 475 grams. Suppose a 34-week gestation period baby weighs 3000 grams and a 40-week gestation period baby weighs 3900 grams. Which baby weighs less relative to the gestation period?

9. **Men versus Women** The average 20- to 29-year old man is 69.6 inches tall, with a standard deviation of 2.7 inches, while the average 20- to 29-year old woman is 64.1 inches tall, with a standard deviation of 2.6 inches. Who is relatively taller, a 75-inch man or a 70-inch woman?

Source: CDC Vital and Health Statistics, Advance Data, Number 347, October 27, 2004

10. **Men versus Women** The average 20- to 29-year-old man is 69.6 inches tall, with a standard deviation of 2.7 inches, while the average 20- to 29-year-old woman is 64.1 inches tall, with a standard deviation of 2.6 inches. Who is relatively taller, a 68-inch man or a 62-inch woman?

Source: CDC Vital and Health Statistics, Advance Data, Oct. 2004

11. **ERA Champions** In 2004, Jake Peavy of the San Diego Padres had the lowest ERA (earned-run average, mean number of runs yielded per nine innings pitched) of any

pitcher in the National League, with an ERA of 2.27. Also in 2004, Johann Santana of the Minnesota Twins had the lowest ERA of any pitcher in the American League with an ERA of 2.61. In the National League, the mean ERA in 2004 was 4.198 and the standard deviation was 0.772. In the American League, the mean ERA in 2004 was 4.338 and the standard deviation was 0.785. Which player had the better year relative to his peers, Peavy or Santana? Why?

12. **Batting Champions** The highest batting average ever recorded was by Ted Williams in 1941 when he hit 0.406. That year, the mean and standard deviation for batting average were 0.28062 and 0.03281. In 2004, Ichiro Suzuki was the American League batting champion, with a batting average of 0.372. In 2004, the mean and standard deviation for batting average were 0.26992 and 0.02154. Who had the better year relative to their peers, Williams or Suzuki? Why?

13. **Violent Crime Rates** Use the data in Table 17 regarding NW the violent crime rates in 2003 to answer the following:
    (a) Find and interpret the 40th percentile.
    (b) Find and interpret the 95th percentile.
    (c) Find and interpret the 10th percentile.
    (d) What is the percentile rank of the state of Flordia?
    (e) What is the percentile rank of the state of California?

14. **Violent Crime Rates** Use the data in Table 17 regarding the violent crime rates in 2003 to answer the following.
    (a) Find and interpret the 30th percentile.
    (b) Find and interpret the 85th percentile.
    (c) Find and interpret the 5th percentile.
    (d) What is the percentile rank of the state of New Mexico?
    (e) What is the percentile rank of the state of Rhode Island?

15. **April Showers** The following data represent the number NW of inches of rain in Chicago, Illinois, during the month of April for 20 randomly selected years.

0.97	2.78	4.00	5.50
1.14	3.41	4.02	5.79
1.85	3.48	4.11	6.14
2.34	3.94	4.77	6.28
2.47	3.97	5.22	7.69

Source: NOAA, Climate Diagnostics Center

(a) Compute the z-score corresponding to the rainfall in 1971 of 0.97 inch. Interpret this result.
(b) Determine the quartiles.
(c) Compute the interquartile range, IQR.
(d) Determine the lower and upper fences. Are there any outliers, according to this criterion?

16. **Hemoglobin in Cats** The following data represent the hemoglobin (in g/dL) for 20 randomly selected cats.

5.7	8.9	9.6	10.6	11.7
7.7	9.4	9.9	10.7	12.9
7.8	9.5	10.0	11.0	13.0
8.7	9.6	10.3	11.2	13.4

Source: Joliet Junior College Veterinarian Technology Program

(a) Compute the z-score corresponding to the hemoglobin of Blackie, 7.8 g/dL. Interpret this result.
(b) Determine the quartiles.
(c) Compute the interquartile range, IQR.
(d) Determine the lower and upper fences. Are there any outliers, according to this criterion?

17. **Concentration of Dissolved Organic Carbon** The following data represent the concentration of organic carbon (mg/L) collected from organic soil.

22.74	29.8	27.1	16.51	6.51
8.81	5.29	20.46	14.9	33.67
30.91	14.86	15.91	15.35	9.72
19.8	14.86	8.09	17.9	18.3
5.2	11.9	14	7.4	17.5
10.3	11.4	5.3	15.72	20.46
16.87	15.42	22.49		

Source: Lisa Emili, Ph.D. candidate, University of Waterloo, Ontario

(a) Compute the z-score corresponding to 20.46. Interpret this result.
(b) Determine the quartiles.
(c) Compute the interquartile range, IQR.
(d) Determine the lower and upper fences. Are there any outliers, according to this criterion?

18. **Concentration of Dissolved Organic Carbon** The following data represent the concentration of organic carbon (mg/L) collected from mineral soil.

8.5	3.91	9.29	21	10.89
10.3	11.56	7	3.99	3.79
5.5	4.71	7.66	11.72	11.8
8.05	10.72	21.82	22.62	10.74
3.02	7.45	11.33	7.11	9.6
12.57	12.89	9.81	17.99	21.4
8.37	7.92	17.9	7.31	16.92
4.6	8.5	4.8	4.9	9.1
7.9	11.72	4.85	11.97	7.85
9.11	8.79			

Source: Lisa Emili, Ph.D. candidate, University of Waterloo, Ontario

(a) Compute the z-score corresponding to 17.99. Interpret this result.
(b) Determine the quartiles.

(c) Compute the interquartile range, IQR.

(d) Determine the lower and upper fences. Are there any outliers, according to this criterion?

19. Fraud Detection As part of its "Customers First" program, a cellular phone company monitors monthly phone usage. The goal of the program is to identify unusual use and alert the customer that their phone may have been used by an unscrupulous individual. The following data represent the monthly phone use in minutes of a customer enrolled in this program for the past 20 months.

346	345	489	358	471
442	466	505	466	372
442	461	515	549	437
480	490	429	470	516

The phone company decides to use the upper fence as the cutoff point for the number of minutes at which the customer should be contacted. What is the cutoff point?

20. Stolen Credit Card A credit card company decides to enact a fraud-detection service. The goal of the credit card company is to determine if there is any unusual activity on the credit card. The company maintains a database of daily charges on a customer's credit card. Any day when the card was inactive is excluded from the database. If a day's worth of charges appears unusual, the customer is contacted to make sure that the credit card has not been compromised. Use the following daily charges (rounded to the nearest dollar) to determine the amount the daily charges must exceed before the customer is contacted.

143	166	113	188	133
90	89	98	95	112
111	79	46	20	112
70	174	68	101	212

21. Student Survey of Income A survey of 50 randomly selected full-time Joliet Junior College students was conducted during the Fall 2005 semester. In the survey, the students were asked to disclose their weekly income from employment. If the student did not work, $0 was entered.

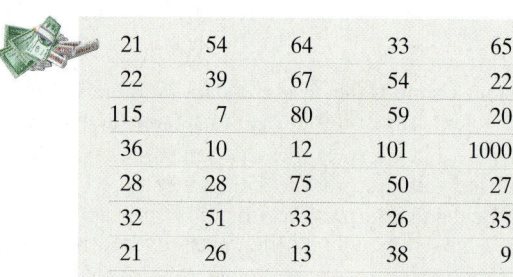

0	262	0	635	0
244	521	476	100	650
12,777	567	310	527	0
83	159	0	547	188
719	0	367	316	0
479	0	82	579	289
375	347	331	281	628
0	203	149	0	403
0	454	67	389	0
671	95	736	300	181

(a) Check the data set for outliers.

(b) Draw a histogram of the data and label the outliers on the histogram.

(c) Provide an explanation for the outliers.

22. Student Survey of Entertainment Spending A survey of 40 randomly selected full-time Joliet Junior College students was conducted in the Fall 2005 semester. In the survey, the students were asked to disclose their weekly spending on entertainment. The results of the survey are as follows:

21	54	64	33	65
22	39	67	54	22
115	7	80	59	20
36	10	12	101	1000
28	28	75	50	27
32	51	33	26	35
21	26	13	38	9
16	14	36	8	48

(a) Check the data set for outliers.

(b) Draw a histogram of the data and label the outliers on the histogram.

(c) Provide an explanation for the outliers.

23. Pulse Rate Use the results of Problem 25 in Sections 3.1 and 3.2 to compute the z-scores for all the students. Compute the mean and standard deviation of these z-scores.

24. Travel Time Use the results of Problem 26 in Sections 3.1 and 3.2 to compute the z-scores for all the students. Compute the mean and standard deviation of these z-scores.

Technology Step-by-Step	**Determining Percentiles**
TI-83/84 Plus	To compute the quartiles, follow the same steps given to compute the mean and median from raw data.
MINITAB	MINITAB computes only the quartiles. Follow the same steps given to compute the mean and median from raw data.
Excel	***Step 1:*** Enter the raw data into column A.
	Step 2: With the data analysis Tool Pak enabled, select the **Tools** menu and highlight **Data Analysis. . . .**
	Step 3: Select **Rank and Percentile** from the Data Analysis window.
	Step 4: With the cursor in the **Input Range** cell, highlight the data. Press OK.

3.5 The Five-Number Summary and Boxplots

Objectives   Compute the five-number summary

 Draw and interpret boxplots

Some aspects of statistical analysis attempt to verify a conjecture by means of observational studies or designed experiments. In other words, a theory is conjectured, and then data are collected to test the theory. For example, a dietitian might conjecture that exercise will lower an individual's cholesterol. The dietitian would carefully design an experiment that randomly divides study participants into two groups: the control group and the experimental group. She would impose a treatment (exercise or no exercise) on the two groups and then measure the effect on the response variable, cholesterol levels.

Another aspect of statistics looks at data to spot any interesting results that might be concluded from the data. In other words, rather than develop a theory and use data to support or disprove the theory, a researcher starts with data and looks for a theory. This area of statistics is referred to as **exploratory data analysis (EDA)**. The idea behind exploratory data analysis is to draw graphs of data and obtain measures of central tendency and spread to form some conjectures regarding the data.

Many of the methods of exploratory data analysis were developed by John Tukey (1915–2000). A complete presentation of the materials found in this chapter can be found in his text *Exploratory Data Analysis* (Addison-Wesley, 1977).

Compute the Five-Number Summary

Remember that the median is a measure of central tendency that divides the lower 50% from the upper 50% of the data. This particular measure of central tendency is resistant to extreme values and is the preferred measure of central tendency when data are skewed right or left.

The three measures of dispersion presented in Section 3.2 (range, variance, and standard deviation) are not resistant to extreme values. However, the interquartile range, $Q_3 - Q_1$, is resistant. It measures the spread of the data by determining the difference between the 25th and 75th percentiles. It is interpreted as the range of the middle 50% of the data. However, the median, Q_1, and Q_3 do not provide information about the tails of the distribution of the data. To get this information, we need to know the smallest and largest values in the data set.

The **five-number summary** of a set of data consists of the smallest data value, Q_1, the median, Q_3, and the largest data value. Symbolically, the five-number summary is presented as follows:

Five-Number Summary

MINIMUM Q_1 M Q_3 MAXIMUM

EXAMPLE 1   **Obtaining the Five-Number Summary**

Problem: The data shown in Table 18 show the finishing times (in minutes) of the men in the 60- to 64-year-old age group in a 5-kilometer race.

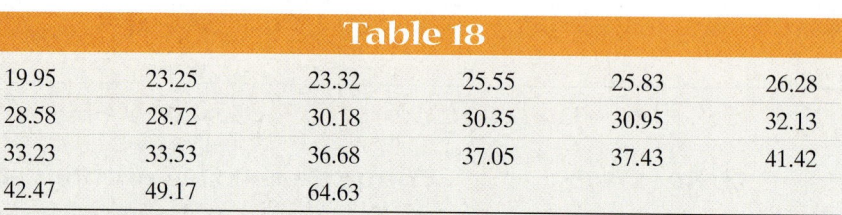

Table 18					
19.95	23.25	23.32	25.55	25.83	26.28
28.58	28.72	30.18	30.35	30.95	32.13
33.23	33.53	36.68	37.05	37.43	41.42
42.47	49.17	64.63			

Source: Laura Gillogly, student at Joliet Junior College

Approach: The five-number summary requires that we determine the minimum data value, Q_1 (the 25th percentile), M (the median), Q_3 (the 75th percentile), and the maximum data value. We need to arrange the data in ascending order, and then use the procedures introduced in Section 3.4 to obtain Q_1, M, and Q_3.

Solution: The data in ascending order are as follows:

19.95, 23.25, 23.32, 25.55, 25.83, 26.28, 28.58, 28.72, 30.18, 30.35, 30.95, 32.13, 33.23, 33.53, 36.68, 37.05, 37.43, 41.42, 42.47, 49.17, 64.63

The smallest number (the fastest time) in the data set is 19.95. The largest number in the data set is 64.63. The first quartile, Q_1, is 26.06. The median, M, is 30.95. The third quartile, Q_3, is 37.24. The five-number summary is

19.95 26.06 30.95 37.24 64.63

EXAMPLE 2   **Obtaining the Five-Number Summary Using Technology**

Problem: Using a statistical spreadsheet or graphing calculator, determine the five-number summary of the data presented in Table 18.

Approach: We will use MINITAB to obtain the five-number summary. The steps for obtaining the five-number summary using a TI-83 or TI-84 Plus graphing calculator, MINITAB, or Excel are given in the Technology Step-by-Step on page 184.

Result: Figure 21 shows the output supplied by MINITAB. The five-number summary is highlighted.

Figure 21

Descriptive statistics: Times

Variable	N	N*	Mean	SE Mean	StDev	Minimum	Q1	Median	Q3	Maximum
Times	21	0	33.37	2.20	10.10	19.95	26.06	30.95	37.24	64.63

2   **Draw and Intepret Boxplots**

The five-number summary can be used to create another graph, called the **boxplot**.

Drawing a Boxplot

Step 1: Determine the lower and upper fences:

$$\text{Lower fence} = Q_1 - 1.5(\text{IQR})$$
$$\text{Upper fence} = Q_3 + 1.5(\text{IQR})$$

Remember, $\text{IQR} = Q_3 - Q_1$.

Step 2: Draw vertical lines at Q_1, M, and Q_3. Enclose these vertical lines in a box.

Step 3: Label the lower and upper fences.

Step 4: Draw a line from Q_1 to the smallest data value that is larger than the lower fence. Draw a line from Q_3 to the largest data value that is smaller than the upper fence.

Step 5: Any data values less than the lower fence or greater than the upper fence are outliers and are marked with an asterisk (*).

EXAMPLE 3   **Constructing a Boxplot**

Problem: Use the results from Example 1 to a construct a boxplot of the finishing times of the men in the 60- to 64-year-old age group.

Approach: Follow the steps presented above.

Solution: From the results of Example 1, we know that $Q_1 = 26.06$, $M = 30.95$, and $Q_3 = 37.24$. Therefore, the interquartile range = IQR = $Q_3 - Q_1 = 37.24 - 26.06 = 11.18$. The difference between the 75th percentile and 25th percentile is a time of 11.18 minutes.

Step 1: We compute the lower and upper fences:

$$\text{Lower fence} = Q_1 - 1.5(\text{IQR}) = 26.06 - 1.5(11.18) = 9.29$$
$$\text{Upper fence} = Q_3 + 1.5(\text{IQR}) = 37.24 + 1.5(11.18) = 54.01$$

Step 2: Draw a horizontal number line with a scale that will accommodate our graph. Draw vertical lines at $Q_1 = 26.06$, $M = 30.95$, and $Q_3 = 37.24$. Enclose these lines in a box. See Figure 22(a).

Figure 22(a)

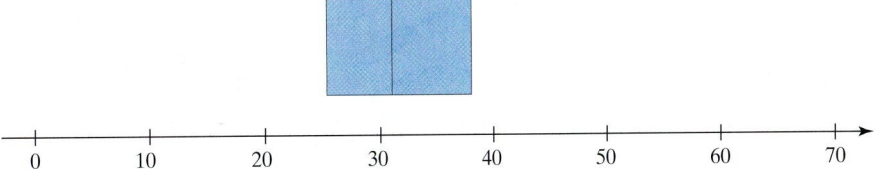

Step 3: Temporarily mark the location of the lower and upper fence with brackets ([and]). See Figure 22(b).

Figure 22(b)

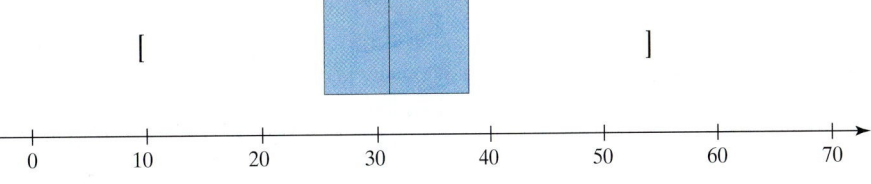

Step 4: The smallest data value that is larger than 9.29 (the lower fence) is 19.95. The largest data value that is smaller than 54.01 (the upper fence) is 49.17. We draw horizontal lines from Q_1 to 19.95 and from Q_3 to 49.17. See Figure 22(c).

Figure 22(c)

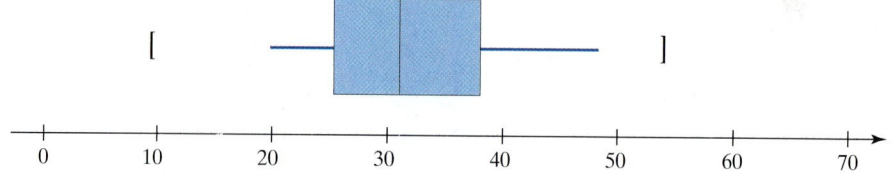

Step 5: Plot any values less than 9.29 (the lower fence) or greater than 54.01 (the upper fence) using an asterisk (*). These values are outliers. So, 64.63 is an outlier. Remove the brackets from the graph. See Figure 22(d).

Figure 22(d)

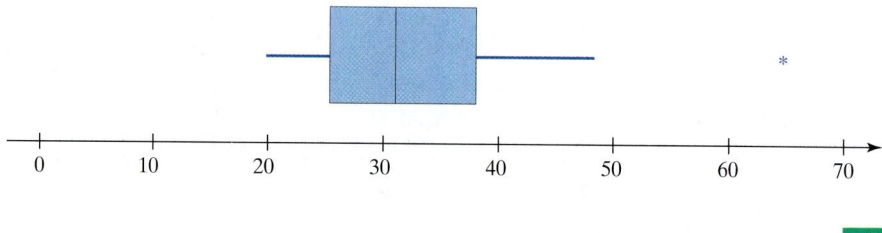

We can describe the shape of the distribution using the boxplot.

Distribution Shape Based on the Boxplot

1. If the median is near the center of the box and each horizontal line is of approximately equal length, the distribution is roughly symmetric.
2. If the median is to the left of the center of the box or the right line is substantially longer than the left line, the distribution is skewed right.
3. If the median is to the right of the center of the box or the left line is substantially longer than the right line, the distribution is skewed left.

!**CAUTION**

● Identifying the shape of a distribution from a boxplot (or from a histogram, for that matter) is subjective. When identifying the shape of a distribution from a graph, be sure to support your opinion.

Figure 23 on page 179 provides examples of boxplots that are (a) symmetric, (b) skewed right, and (c) skewed left, along with the corresponding histograms.

The boxplot in Figure 22(d) suggests that the distribution is skewed right, since the right line is longer than the left and the median is to the left of the center of the box.

We stated that the interquartile range (IQR) is a measure of dispersion (just like the standard deviation is a measure of dispersion). If the median is used as the measure of central tendency, then the IQR should be used as the measure of spread. Remember, we use the median to describe the "center" of a set of data when the shape of the distribution is skewed because the mean is distorted by skewness. Likewise, the standard deviation is distorted by skewness. Because the IQR is resistant to extreme values, it is a better measure of spread for skewed distributions.

Summary: Which Measures to Report

Shape of Distribution	Measure of Central Tendency	Measure of Dispersion
Symmetric	Mean	Standard deviation
Skewed left or skewed right	Median	Interquartile range

Figure 23

Historical Note

John Tukey was born on July 16, 1915, in New Bedford, Massachusetts. His parents graduated numbers 1 and 2 from Bates College and were elected "the couple most likely to give birth to a genius." In 1936, Tukey graduated from Brown University with an undergraduate degree in chemistry. He went on to earn a master's degree in chemistry at Brown. In 1939, Tukey earned his doctorate in mathematics from Princeton. He remained at Princeton and in 1965 became the founding chair of the Department of Statistics. Among his many accomplishments, Tukey is credited with coining the terms *software* and *bit*. In the early 1970s, he discussed the negative effects of aerosol cans on the ozone layer. Tukey recommended that the 1990 Census be adjusted by means of statistical formulas. John Tukey died in New Brunswick, New Jersey, July 26, 2000.

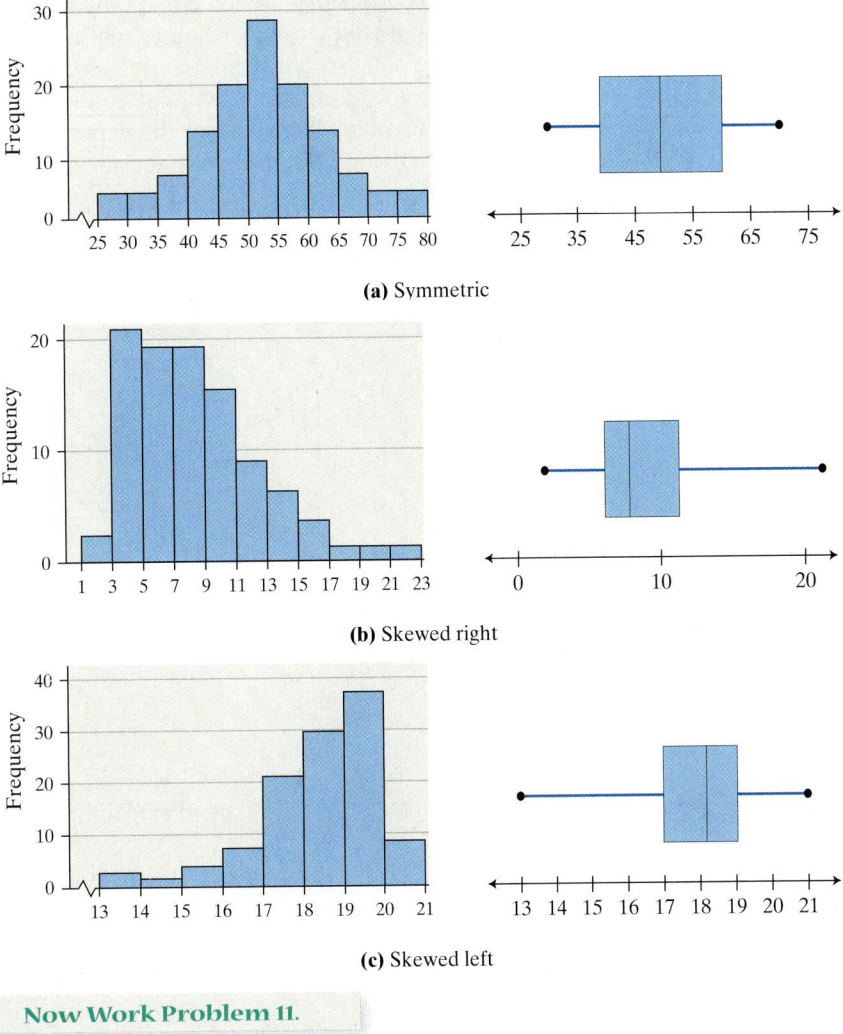

(a) Symmetric

(b) Skewed right

(c) Skewed left

Now Work Problem 11.

EXAMPLE 4   **Comparing Two Distributions by Using Boxplots**

Problem: In the Spacelab Life Sciences 2, 14 male rats were sent to space. Upon their return, the red blood cell mass (in milliliters) of the rats was determined. A control group of 14 male rats was held under the same conditions (except for spaceflight) as the space rats, and their red blood cell mass was also determined when the space rats returned. The project was led by Paul X. Callahan. The data in Table 19 were obtained. Construct boxplots for red blood cell mass for the flight group and control group. Does it appear that the flight to space affected the red blood cell mass of the rats?

Table 19			
Flight		**Control**	
8.59	8.64	8.65	6.99
6.87	7.89	7.62	7.44
7.00	8.80	7.33	8.58
6.39	7.54	7.14	9.14
7.43	7.21	8.40	9.66
9.79	6.85	8.55	8.70
9.30	8.03	9.88	9.94

Source: NASA Life Sciences Data Archive

Approach: When comparing two data sets, we draw the boxplots on the same horizontal number line to make the comparison easy. Graphing calculators with advanced statistical features, as well as statistical spreadsheets such as MINITAB and Excel, have the ability to draw boxplots. We will use MINITAB to draw the boxplots. The steps for drawing boxplots using a TI-83 or TI-84 Plus graphing calculator, MINITAB, or Excel are given in the Technology Step-by-Step on page 184.

Solution: Figure 24 shows the boxplots drawn in MINITAB. From the boxplots, it appears that the space flight has reduced the red blood cell mass of the rats.

Figure 24

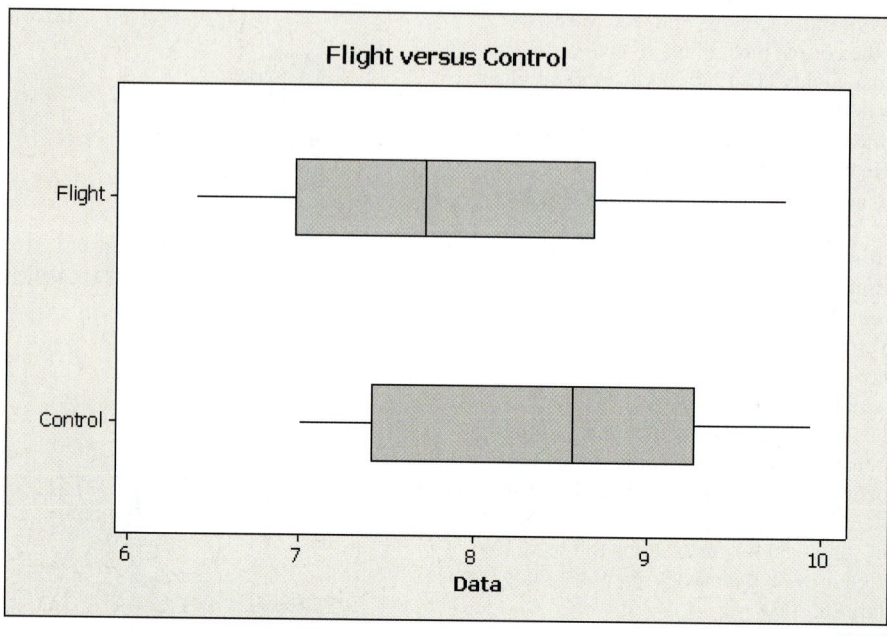

Now Work Problem 15.

MAKING AN INFORMED DECISION

What Car Should I Buy?

Suppose you are in the market to purchase a used car. To make an informed decision regarding your purchase, you would like to collect as much information as possible. Among the information you might consider are the typical price of the car, the typical number of miles the car should have and its crash test results, insurance costs, and expected repair costs.

1. Make a list of at least three cars that you would consider purchasing. To be fair, the cars should be in the same class (such as compact, midsize, and so on). They should also be of the same age.

2. Collect information regarding the three cars in your list by finding at least eight cars of each type that are for sale. Obtain such information as the asking price and the number of miles the car has. Sources of data include your local newspaper, classified ads, and car Web sites (such as www.cars.com/ and www.vehix.com). Compute summary statistics for asking price, number of

miles, and other variables of interest. Using the same scale, draw boxplots of each variable considered.

3. Go to the Insurance Institute for Highway Safety Web site (www.hwysafety.org). Select the Vehicle Ratings link. Choose the make and model for each car you are considering. Obtain information regarding crash testing for each car under consideration. Compare cars in the same class. How does each car compare? Is one car you are considering substantially safer than the others? What about repair costs? Compute summary statistics for crash tests and repair costs.

4. Obtain information about insurance costs. Contact various insurance companies to determine the cost of insuring the cars you are considering. Compute summary statistics for insurance costs and draw boxplots.

5. Write a report supporting your conclusion regarding which car you would purchase.

3.5 ASSESS YOUR UNDERSTANDING

Concepts and Vocabulary

1. Explain the circumstances under which the median and interquartile range would be better measures of central tendency and dispersion than the mean and standard deviation.

2. In a boxplot, if the median is to the left of the center of the box or the right line is substantially longer than the left line, the distribution is skewed _____.

Skill Building

In Problems 1 and 2, (a) identify the shape of the distribution, and (b) determine the five-number summary. Assume that each number in the five-number summary is an integer.

3.

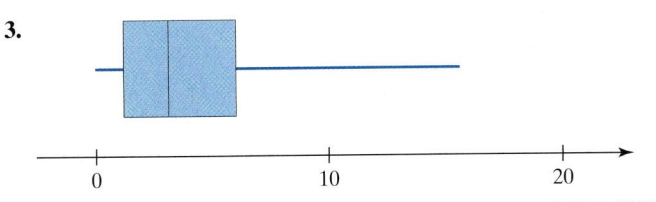

4.

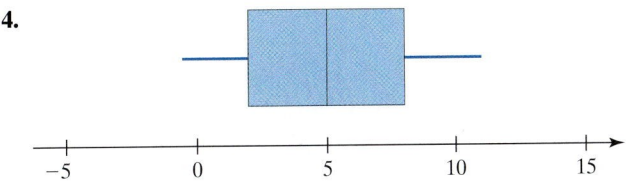

Applying the Concepts

In Problems 5–10, find the five-number summary, and construct a boxplot for the data in the indicated problem. Comment on the shape of the distribution.

5. Age at Inauguration The following data represent the age of U.S. presidents on their respective inauguration days.

57	61	57	57	58	57	61
54	68	51	49	64	50	48
65	52	56	46	54	49	50
47	55	55	54	42	51	56
55	51	54	51	60	62	43
55	56	61	52	69	64	46
54						

6. Grams of Fat in a McDonald's Breakfast The following data represent the number of grams of fat in breakfast meals offered at McDonald's.

12	23	28	2	28	33
31	11	23	40	35	1
23	33	23	16	11	8
8	17	16	15		

Source: McDonald's Corporation, A Full Serving of Nutrition Facts, April 2003

7. Super Bowl Point Spreads The following data represent the number of points by which the winning team won Super Bowls I to XXXIX.

25	19	9	16	3	21	7	17
10	4	18	17	4	12	17	5
10	29	22	36	19	32	4	45
1	13	35	17	23	10	14	7
15	7	27	3	27	3	3	

Source: superbowl.com

8. Miles per Gallon The following data represent the number of miles per gallon achieved on the highway for compact cars for model year 2005.

30	29	30	21	18	29	27	30	29
34	34	30	28	30	20	32	28	32
34	35	26	26	31	25	35	32	25
19	26	19	24	22	24	19	31	26
34	32	34	25	34	34	32	29	25
31	29	30	30	34	32	29	38	39
46	31	31	30	27	29	26	29	24

Source: U.S. Department of Energy, Office of Energy Efficiency and Renewable Energy, U.S. Environmental Protection Agency, Fuel Economy Guide, Model Year 2005 (www.fueleconomy.gov)

9. Got a Headache? Drugs are made of active ingredients and inactive ingredients. The Food and Drug Administration states that a drug should have the same amount of active ingredient in over-the-counter drugs. The following data represent the weight (in grams) of a random sample of 25 Tylenol tablets. What do you think is the source of the variability in weight?

0.608	0.601	0.606	0.602	0.611
0.608	0.610	0.610	0.607	0.600
0.608	0.608	0.605	0.609	0.605
0.610	0.607	0.611	0.608	0.610
0.612	0.598	0.600	0.605	0.603

Source: Kelly Roe, student at Joliet Junior College

10. Gasoline Expenditures The following data represent the mean gasoline expenditures per person for each state and the District of Columbia. Wyoming has the highest mean expenditure. What might explain this? New York has the lowest mean expenditure. What might explain this?

971	787	713	704	688	675	660	643	581
830	741	711	698	684	672	654	618	480
821	740	711	698	683	669	653	616	421
816	737	707	692	682	667	649	611	
802	726	707	692	679	666	646	598	
791	715	706	688	678	664	645	583	

Source: Energy Information Administration

11. Serum HDL Dr. Paul Oswiecmiski randomly selects forty **NW** of his 20 to 29-year-old patients and obtains the following data regarding their serum HDL cholesterol.

70	56	48	48	53	52	66	48
36	49	28	35	58	62	45	60
38	73	45	51	56	51	46	39
56	32	44	60	51	44	63	50
46	69	53	70	33	54	55	52

(a) Compute the five-number summary.
(b) Draw a boxplot of the data.
(c) Determine the shape of the distribution from the boxplot. Refer to the histogram drawn in Problem 31 in Section 2.2 to test your answer.
(d) Which measures of central tendency and dispersion should be reported for this data?

12. Volume of Altria Group Stock The volume of a stock is the number of shares traded on a given day. The following data, given in millions so that 3.78 represents 3,780,000 shares traded, represent the volume of Altria Group stock traded for a random sample 35 trading days in 2004.

3.78	8.74	4.35	5.02	8.40
6.06	5.75	5.34	6.92	6.23
5.32	3.25	6.57	7.57	6.07
3.04	5.64	5.00	7.16	4.88
10.32	3.38	7.25	6.52	4.43
3.38	5.53	4.74	9.70	3.56
10.96	4.50	7.97	3.01	5.58

Source: yahoo.finance.com

(a) Compute the five-number summary.
(b) Draw a boxplot of the data.

(c) Determine the shape of the distribution from the boxplot. Refer to the histogram drawn in Problem 32 in Section 2.2 to test your answer.
(d) Which measures of central tendency and dispersion should be reported for this data?

13. Dividend Yield A dividend is a payment from a publicly traded company to its shareholders. The dividend yield of a stock is determined by dividing the annual dividend of a stock by its price. The following data represent the dividend yields (in percent) of a random sample of 28 publicly traded stocks with a value of at least $5 billion.

1.7	0	1.15	0.62	1.06	2.45	2.38
2.83	2.16	1.05	1.22	1.68	0.89	0
2.59	0	1.7	0.64	0.67	2.07	0.94
2.04	0	0	1.35	0	0	0.41

Source: Yahoo! Finance

(a) Compute the five-number summary.
(b) Draw a boxplot of the data.
(c) Determine the shape of the distribution from the boxplot. Refer to the histogram drawn in Problem 33 in Section 2.2 to test your answer.
(d) Which measures of central tendency and dispersion should be reported for this data?

14. Violent Crimes Violent crimes include murder, forcible rape, robbery, and aggravated assault. The following data represent the violent crime rate (per 100,000 population) by state and the District of Columbia in 2002.

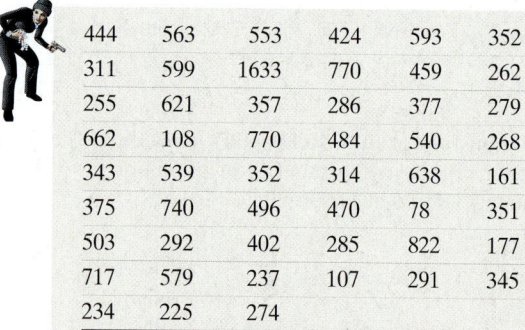

444	563	553	424	593	352
311	599	1633	770	459	262
255	621	357	286	377	279
662	108	770	484	540	268
343	539	352	314	638	161
375	740	496	470	78	351
503	292	402	285	822	177
717	579	237	107	291	345
234	225	274			

Source: U.S. Federal Bureau of Investigation

(a) Compute the five-number summary.
(b) Draw a boxplot of the data.
(c) Determine the shape of the distribution from the boxplot. Refer to the histogram drawn in Problem 34 in Section 2.2 to test your answer.
(d) Which measures of central tendency and dispersion should be reported for this data?

In Problems 15–17, compare the data sets by determining the five-number summary and constructing boxplots on the same scale.

15. Chips Per Cookie ber of chips per cookie in a random sample of Keebler Chips Deluxe Chocolate Chip Cookies and the number of chips per cookie in a store brand's chocolate chip cookies. Does there appear to be a difference in the number of chips per cookie? Does one brand have a more consistent number of chips per cookie?

Keebler			Store Brand		
32	23	28	21	23	24
28	28	29	24	25	27
25	20	25	26	26	21
22	21	24	18	16	24
21	24	21	21	30	17
26	28	24	23	28	31
33	20	31	27	33	29

Source: Trina McNamara, student at Joliet Junior College

16. Tornadoes The following data give the number of tornadoes in Oklahoma, Kansas, and Nebraska for the years 1990 to 2004. Which state appears to have the higher number of tornadoes per year?

Year:	1990	1991	1992	1993	1994	1995	1996	1997	1998	1999	2000	2001	2002	2003	2004
Oklahoma:	30	73	64	64	40	79	47	55	83	145	44	61	18	78	62
Kansas:	88	116	92	113	42	73	68	62	71	64	59	101	95	91	122
Nebraska:	88	63	74	69	55	26	60	30	65	102	61	62	28	81	110

Source: National Oceanic and Atmospheric Administration

17. Home-run Distances During the 1998 major league baseball season, Mark McGwire of the St. Louis Cardinals and Sammy Sosa of the Chicago Cubs thrilled fans across the country in a race to set the record for the most home runs hit in a season. Sosa ended the season with 66 home runs, and McGwire set the record with 70 home runs. Only 3 years later in 2001, Barry Bonds of the San Francisco Giants broke McGwire's record by hitting 73 home runs. The following data represent the distances of each player's home runs in his record-setting season. Which player appears to have the longest distances? Which player appears to have the most consistent distances?

Mark McGwire						
360	370	370	430	420	340	460
410	440	410	380	360	350	527
380	550	478	420	390	420	425
370	480	390	430	388	423	410
360	410	450	350	450	430	461
430	470	440	400	390	510	430
450	452	420	380	470	398	409
385	369	460	390	510	500	450
470	430	458	380	430	341	385
410	420	380	400	440	377	370

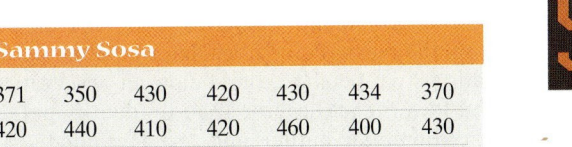

Sammy Sosa						
371	350	430	420	430	434	370
420	440	410	420	460	400	430
410	370	370	410	380	340	350
420	410	415	430	380	380	366
500	380	390	400	364	430	450
440	365	420	350	420	400	380
380	400	370	420	360	368	430
433	388	440	414	482	364	370
400	405	433	390	480	480	434
344	410	420				

Barry Bonds						
420	417	440	410	390	417	420
410	380	430	370	420	400	360
410	420	391	416	440	410	415
436	430	410	400	390	420	410
420	410	410	450	320	430	380
375	375	347	380	429	320	360
375	370	440	400	405	430	350
396	410	380	430	415	380	375
400	435	420	420	488	361	394
410	411	365	360	440	435	454
442	404	385				

Technology Step by Step	**Drawing Boxplots Using Technology**

TI-83/84 Plus

Step 1: Enter the raw data into L1.

Step 2: Press 2nd Y= and select 1:Plot 1.

Step 3: Turn the plots ON. Use the cursor to highlight the modified boxplot icon. Your screen should look as follows:

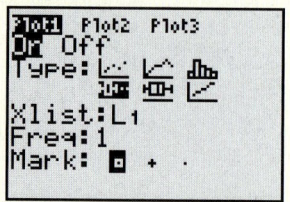

Step 4: Press ZOOM and select 9: ZoomStat.

MINITAB

Step 1: Enter the raw data into column C1.

Step 2: Select the **Graph** menu and highlight **Boxplot . . .**

Step 3: For a single boxplot, select One Y, simple. For two or more boxplots, select Multiple Y's, simple.

Step 4: Select the data to be graphed. If you want the boxplot to be horizontal rather than vertical, select the Scale button, then transpose value and category scales. Click OK.

Excel

Step 1: Start the PHStat Add-in.

Step 2: Enter the raw data into column A.

Step 3: Select the **PHStat** menu and highlight **Box-and-Whisker Plot** . . . With the cursor in the Data Variable Cell Range cell, highlight the data in column A.

Step 4: Click OK.

Chapter  **Review**

Summary

This chapter concentrated on describing distributions numerically. Measures of central tendency are used to indicate the typical value in a distribution. Three measures of central tendency were discussed. The mean measures the center of gravity of the distribution. The median separates the bottom 50% of the data from the top 50%. Both measures require that the data be quantitative. The mode measures the most frequent observation. The data can be either quantitative or qualitative to compute the mode. The median is resistant to extreme values, while the mean is not. A comparison between the median and mean can help determine the shape of the distribution.

Measures of dispersion describe the spread in the data. The range is the difference between the highest and lowest data value. The variance measures the average squared deviation about the mean. The standard deviation is the square root of the variance. The mean and standard deviation are used in many types of statistical inference.

The mean, median, and mode can be approximated from grouped data. The variance and standard deviation can also be approximated from grouped data.

We can determine the relative position of an observation in a data set using Z-scores and percentiles. Z-scores denote how many standard deviations on observation is from the mean. Percentiles determine the percent of observations that lie above and below an observation. The upper and lower fences can be used to identify potential outliers. Any potential outlier must be investigated to determine whether it was the result of a data entry error, of some other error in the data collection process, or of an unusual value in the data set.

The interquartile range is also a measure of dispersion. The five-number summary provides an idea about the center and spread of a data set, through the median and the interquartile range. The length of the tails in the distribution can be determined from the smallest and largest data values. The five-number summary is used to construct boxplots. Boxplots can be used to describe the shape of the distribution.

Chapter 3 Review **185**

Formulas

Population Mean

$$\mu = \frac{\sum x_i}{N}$$

Sample Mean

$$\bar{x} = \frac{\sum x_i}{n}$$

Population Variance

$$\sigma^2 = \frac{\sum (x_i - \mu)^2}{N} = \frac{\sum x_i^2 - \frac{\left(\sum x_i\right)^2}{N}}{N}$$

Sample Variance

$$s^2 = \frac{\sum (x_i - \bar{x})^2}{n - 1} = \frac{\sum x_i^2 - \frac{\left(\sum x_i\right)^2}{n}}{n - 1}$$

Population Standard Deviation

$$\sigma = \sqrt{\sigma^2}$$

Sample Standard Deviation

$$s = \sqrt{s^2}$$

Range = Largest Data Value − Smallest Data Value

Weighted Mean

$$\bar{x}_w = \frac{\sum w_i x_i}{\sum w_i}$$

Population Mean from Grouped Data

$$\mu = \frac{\sum x_i f_i}{\sum f_i}$$

Sample Mean from Grouped Data

$$\bar{x} = \frac{\sum x_i f_i}{\sum f_i}$$

Population Variance from Grouped Data

$$\sigma^2 = \frac{\sum (x_i - \mu)^2 f_i}{\sum f_i}$$

Sample Variance from Grouped Data

$$s^2 = \frac{\sum (x_i - \bar{x})^2 f_i}{\left(\sum f_i\right) - 1}$$

Population z-Score

$$z = \frac{x - \mu}{\sigma}$$

Sample z-Score

$$z = \frac{x - \bar{x}}{s}$$

Percentile of $x = \dfrac{\text{number of data values less than } x}{n} \cdot 100$

Interquartile Range

$$\text{IQR} = Q_3 - Q_1$$

Lower and Upper Fences

$$\text{Lower Fence} = Q_1 - 1.5(\text{IQR})$$
$$\text{Upper Fence} = Q_3 + 1.5(\text{IQR})$$

Vocabulary

Parameter (p. 121)	Deviation about the mean (p. 139)	kth percentile (p. 167)
Statistic (p. 121)	Population variance (p. 139)	Quartiles (p. 169)
Arithmetic Mean (p. 121)	Sample variance (p. 141)	Outlier (p. 170)
Median (p. 124)	Biased (p. 141)	Interquartile range (p. 171)
Mode (p. 125)	Degrees of freedom (p. 142)	Fences (p. 171)
bimodal (p. 126)	Population standard deviation (p. 143)	Exploratory data analysis (p. 175)
multimodal (p. 126)	Sample standard deviation (p. 143)	Five-number summary (p. 176)
Resistant (p. 127)	Weighted mean (p. 159)	Boxplot (p. 177)
Range (p. 138)	Z-score (p. 166)	

Objectives

Section	You should be able to . . .	Example	Review Exercises
3.1	1 Determine the arithmetic mean of a variable from raw data (p. 121)	1, 7	1(a)–8(a)
			1(a)–8(a)
	2 Determine the median of a variable from raw data (p. 124)	2, 3, 7	5(a) and 6(a)
	3 Determine the mode of a variable from raw data (p. 125)	4, 5, 6, 7	15(a)–(c), 16(a)–(c)
	4 Use the mean and median to help identify the shape of a distribution (p. 127)	8, 9	
3.2	1 Compute the range of a variable from raw data (p. 138)	2	1(b)–8(b)
	2 Compute the variance of a variable from raw data (p. 139)	3, 4, 6	1(b)–8(b), 15(d), 16(d)
	3 Compute the standard deviation of a variable from raw data (p. 143)	5, 6, 7	1(b)–8(b), 15(d), 16(d)
	4 Use the Empirical Rule to describe data that are bell shaped (p. 145)	8	9(a)–(d), 10 (a)–(c)
	5 Use Chebyshev's inequality to describe any set of data (p. 146)	9	9(e) and (f), 10(d) and (e)
3.3	1 Approximate the mean of a variable from grouped data (p. 157)	1, 4	11(a), 12(a)
	2 Compute the weighted mean (p. 159)	2	13, 14
	3 Approximate the variance and standard deviation of a variable from grouped data (p. 159)	3, 4	11(b), 12(b)
3.4	1 Determine and interpret z-scores (p. 165)	1	19
	2 Determine and interpret percentiles (p. 167)	2–4	17, 18
	3 Determine and interpret quartiles (p. 169)	5, 6	15(e), 16(e)
	4 Check a set of data for outliers (p. 170)	7	15(h), 16(h)
3.5	1 Compute the five-number summary (p. 175)	1, 2	15(e), 16(e)
	2 Draw and interpret boxplots (p. 177)	3, 4	15(f), 16(f), 20

3 Review Exercises

1. Muzzle Velocity The following data represent the muzzle velocity (in meters per second) of rounds fired from a 155-mm gun.

793.8	793.1	792.4	794.0	791.4
792.4	791.7	792.3	789.6	794.4

Source: Christenson, Ronald, and Blackwood, Larry; "Tests for Precision and Accuracy of Multiple Measuring Devices." Technometrics, Nov. 93, Vol. 35, Issue 4, pp. 411–421.

(a) Compute the sample mean and median muzzle velocity.
(b) Compute the range, sample variance, and sample standard deviation.

2. Pulse Rates The following data represent the pulse rate of eight randomly selected females after stepping up and down on a 6-inch platform for 3 minutes. Pulse is measured in beats per minute.

136	169	120	128	129
143	115	146	96	86

Source: Michael McCraith, Joliet Junior College

(a) Compute the sample mean and median pulse.
(b) Compute the range, sample variance, and sample standard deviation.

3. Price of Chevy Cavaliers   The following data represent the sales price in dollars for nine 2-year-old Chevrolet Cavaliers in the Los Angeles area.

14,050	13,999	12,999	10,995	9,980
8,998	7,889	7,200	5,500	

Source: cars.com

(a) Compute the sample mean and median price.
(b) Compute the range and sample standard deviation.
(c) Redo (a) and (b) if the data value 14,050 was incorrectly entered as 41,050. How does this change affect the mean? The median? The range? The standard deviation? Which of these values is resistant?

4. Home Sales   The following data represent the closing prices (in U.S. dollars) of 15 randomly selected homes sold in Joliet, Illinois, in December 2004.

138,820	140,794	136,833	157,216
169,541	153,146	115,000	149,380
135,512	99,000	124,757	136,529
149,143	136,924	128,429	

Source: Transamerica Intellitech

(a) Compute the sample mean and median sale price.
(b) Compute the range and sample standard deviation.

5. Chief Justices   The following data represent the ages of chief justices of the U.S. Supreme Court when they were appointed.

Justice	Age
John Jay	44
John Rutledge	56
Oliver Ellsworth	51
John Marshall	46
Roger B. Taney	59
Salmon P. Chase	56
Morrison R. Waite	58
Melville W. Fuller	55
Edward D. White	65
William H. Taft	64
Charles E. Hughes	68
Harlan F. Stone	69
Frederick M. Vinson	56
Earl Warren	62
Warren E. Burger	62
William H. Rehnquist	62

Source: Information Please Almanac

(a) Compute the population mean, median, and mode ages.
(b) Compute the range and population standard deviation ages.
(c) Obtain two simple random samples of size 4, and compute the sample mean and sample standard deviation ages.

6. National League Home Runs   The following data represent the number of home runs hit by all teams in the National League in 2004.

Team	Home Runs	Team	Home Runs
1. St. Louis Cardinals	214	9. Los Angeles Dodgers	203
2. San Francisco Giants	183	10. Cincinnati Reds	194
3. Philadelphia Phillies	215	11. Florida Marlins	148
4. Colorado Rockies	202	12. New York Mets	185
5. Atlanta Braves	178	13. Pittsburgh Pirates	142
6. Houston Astros	187	14. Montreal Expos	151
7. Chicago Cubs	235	15. Milwaukee Brewers	135
8. San Diego Padres	139	16. Arizona Diamondbacks	135

Source: Major League Baseball

(a) Compute the population mean, median, and mode for number of home runs.
(b) Compute the range and population standard deviation for number of home runs.
(c) Obtain two simple random samples of size 3, and compute the sample mean and sample standard deviation for number of home runs.
(d) If a sports reporter stated that the average number of home runs hit by teams in the National League in 2004 was 135, is he lying? Is he being deceptive?

7. Family Size A random sample of 36 married couples who had been married 7 years were asked the number of children they had. The results of the survey follow:

0	0	3	1	2	3
3	4	3	3	0	3
1	2	1	3	0	3
4	2	3	2	2	4
2	1	3	4	1	3
0	3	3	3	2	1

(a) Compute the sample mean and the median number of children.
(b) Compute the range and the sample standard deviation number of children.

8. Waiting in Line The following data represent the number of cars that arrived at a McDonald's drive-through between 11:50 A.M. and 12:00 noon each Wednesday for the past 30 weeks:

1	3	2	8	6
6	6	3	3	1
5	6	3	3	1
4	9	5	3	5
2	6	7	5	8
7	8	3	2	3

(a) Compute the sample mean and the median number of cars.
(b) Compute the range and the sample standard deviation number of cars.

9. Chebyshev's Inequality and the Empirical Rule Suppose that a random sample of 200 light bulbs has a mean life of 600 hours and a standard deviation of 53 hours.
(a) A histogram of the data indicates the sample data follow a bell-shaped distribution. According to the Empirical Rule, 99.7% of light bulbs have lifetimes between _____ and _____ hours.
(b) Assuming the data are bell shaped, determine the percentage of light bulbs that will have a life between 494 and 706 hours.
(c) Assuming the data are bell shaped, what percentage of light bulbs will last between 547 and 706 hours?
(d) If the company that manufactures the light bulb guarantees to replace any bulb that does not last at least 441 hours, what percentage of light bulbs can the firm expect to have to replace, according to the Empirical Rule?
(e) Use Chebyshev's inequality to determine the minimum percentage of light bulbs with a life within 2.5 standard deviations of the mean.
(f) Use Chebyshev's inequality to determine the minimum percentage of light bulbs with a life between 494 and 706 hours.

10. Chebyshev's Inequality and the Empirical Rule In a random sample of 250 toner cartridges, the mean number of pages a toner cartridge can print is 4302 and the standard deviation is 340.

(a) Suppose a histogram of the data indicates that the sample data follow a bell-shaped distribution. According to the Empirical Rule, 99.7% of toner cartridges will print between _____ and _____ pages.
(b) Assuming that the distribution of the data is bell shaped, determine the percentage of toner cartridges whose print total is between 3622 and 4982 pages.
(c) If the company that manufactures the toner cartridges guarantees to replace any cartridge that does not print at least 3622 pages, what percent of cartridges can the firm expect to be responsible for replacing, according to the Empirical Rule?
(d) Use Chebyshev's inequality to determine the minimum percentage of toner cartridges with a page count within 1.5 standard deviations of the mean.
(e) Use Chebyshev's inequality to determine the minimum percentage of toner cartridges that print between 3282 and 5322 pages.

11. Vehicle Fatalities The frequency distribution listed in the table represents the number of drivers in fatal crashes in 2003, by age, for males 20 to 84 years old.

Age	Number of Drivers	Age	Number of Drivers
20–24	6035	55–59	2355
25–29	4352	60–64	1664
30–34	4083	65–69	1173
35–39	3933	70–74	1025
40–44	4194	75–79	895
45–49	3716	80–84	744
50–54	3005		

Source: NHTSA

(a) Approximate the mean age of a male involved in a traffic fatality.
(b) Approximate the standard deviation age of a male involved in a traffic fatality.

12. Vehicle Fatalities The frequency distribution listed in the table represents the number of drivers in fatal crashes in 2003, by age, for females 20 to 84 years old.

Age	Number of drivers	Age	Number of drivers
20–24	1903	55–59	784
25–29	1415	60–64	599
30–34	1364	65–69	415
35–39	1430	70–74	482
40–44	1409	75–79	456
45–49	1242	80–84	372
50–54	1008		

Source: NHTSA

(a) Approximate the mean age of a female involved in a traffic fatality.
(b) Approximate the standard deviation age of a female involved in a traffic fatality.
(c) Compare the results to those obtained in Problem 11. How do you think an insurance company might use this information?

13. Weighted Mean   Michael has just completed his first semester in college. He earned an A in his 5-hour calculus course, a B in his 4-hour chemistry course, an A in his 3-hour speech course, and a C in his 3-hour psychology course. Assuming an A equals 4 points, a B equals 3 points, and a C equals 2 points, determine Michael's grade-point average if grades are weighted by class hours.

14. Weighted Mean   Yolanda wishes to develop a new type of meat loaf to sell at her restaurant. She decides to combine 2 pounds of ground sirloin (cost $2.70 per pound), 1 pound of ground turkey (cost $1.30 per pound), and $\frac{1}{2}$ pound of ground pork (cost $1.80 per pound). What is the cost per pound of the meat loaf?

15. Mets versus Yankees   The following data represent the 2004 salaries (in dollars) of the players on the rosters of the New York Mets and the New York Yankees.

Yankees		Mets	
Player	**Salary**	**Player**	**Salary**
Bubba Crosby	301,400	Tyler Yates	300,000
Jorge De Paula	302,550	Eric Valent	302,500
Donovan Osborne	450,000	Jose Reyes	307,500
Orlando Hernandez	500,000	Dan Wheeler	311,500
Enrique Wilson	700,000	Ty Wigginton	316,000
Tony Clark	750,000	Orber Moreno	317,500
John Flaherty	775,000	Jason Phillips	318,000
Miguel Cairo	900,000	Grant Roberts	319,500
Ruben Sierra	1,000,000	Joe McEwing	500,000
Felix Heredia	1,800,000	Shane Spencer	537,500
Gabe White	1,925,000	Scott Strickland	650,000
Travis Lee	2,000,000	Scott Erickson	700,000
Jon Lieber	2,700,000	Vance Wilson	715,000
Paul Quintrill	3,000,000	Karim Garcia	800,000
Kenny Lofton	3,100,000	John Franco	1,000,000
Tom Gordon	3,500,000	Todd Zeile	1,000,000
Steve Karsay	6,000,000	Braden Looper	2,000,000
Hideki Matsui	7,000,000	Mike Stanton	3,000,000
Jose Contreras	8,500,000	David Weathers	3,933,333
Jorge Posada	9,000,000	Ricky Gutierrez	4,166,667
Javier Vazquez	9,000,000	Mike Cameron	4,333,333
Mariano Rivera	10,890,000	Steve Trachsel	5,000,000
Bernie Williams	12,357,143	Kazuo Matsui	5,033,333
Jason Giambi	12,428,571	Cliff Floyd	6,500,000
Gary Sheffield	13,000,000	Al Leiter	10,295,600
Kevin Brown	15,714,286	Tom Glavine	10,765,608
Mike Mussina	16,000,000	Mike Piazza	16,071,429
Derek Jeter	18,600,000	Mo Vaughn	17,166,667
Alex Rodriguez	22,000,000		

Source: USATODAY.com

(a) Compute the population mean salary for each team.

(b) Compute the median salary for each team.

(c) Given the results of (a) and (b), decide whether the distributions are symmetric, skewed right, or skewed left.

(d) Compute the population standard deviation salary for each team. Which team has more dispersion in its salaries?

(e) Compute the five-number summary for each team.

(f) On the same graph, draw boxplots for the two teams. Annotate the graph with some general remarks comparing the team salaries.

(g) Describe the shape of the distribution of each team, as illustrated by the boxplots. Does this confirm the result obtained in (c)?

(h) Which measure of central tendency is the better measure of central tendency? Why?

16. Bearing Failures An engineer is studying bearing failures for two different materials in air-craft gas turbine engines. The following data are failure times (in millions of cycles) for samples of the two material types.

Material A	3.17	4.31	4.52	4.66	5.69	5.88	6.91	8.01	8.97	11.92
Material B	5.78	6.71	6.84	7.23	8.20	9.65	13.44	14.71	16.39	24.37

(a) Compute the sample mean of the failure time for each material.
(b) Compute the median failure time for each material.
(c) Given the results of (a) and (b), decide whether the distributions are symmetric, skewed right, or skewed left.
(d) Compute the sample standard deviation of the failure time for each material. Which material has its failure times more dispersed?
(e) Compute the five-number summary for each material.
(f) On the same graph, draw boxplots for the two materials. Annotate the graph with some general remarks comparing the failure times.
(g) Describe the shape of the distribution of each material, as illustrated by the boxplots. Does this confirm the result obtained in (c)?

17. NASCAR Earnings The following data represent the total earnings (in dollars) of drivers in the 2004 Nextel Cup Series.

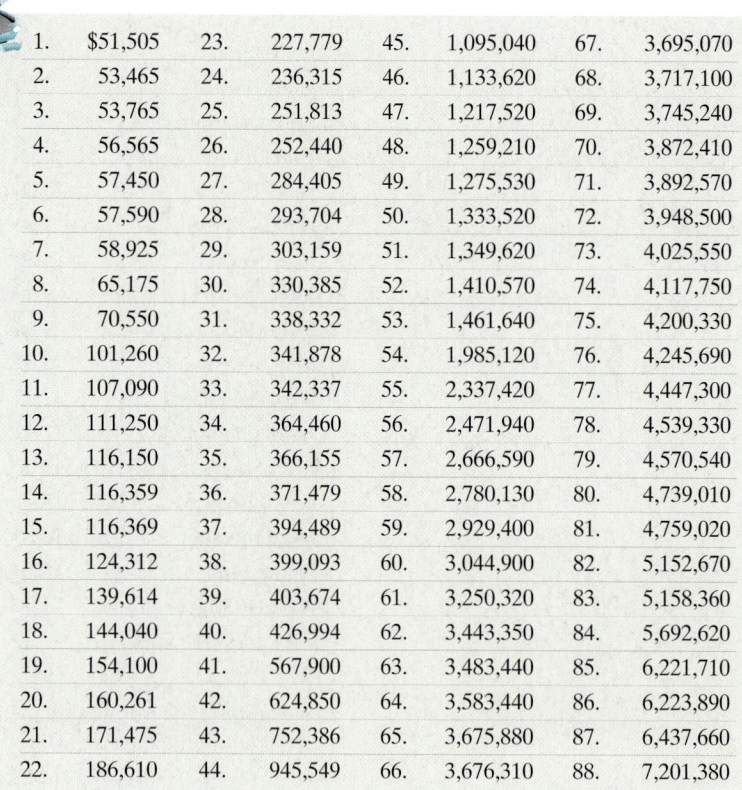

1.	$51,505	23.	227,779	45.	1,095,040	67.	3,695,070
2.	53,465	24.	236,315	46.	1,133,620	68.	3,717,100
3.	53,765	25.	251,813	47.	1,217,520	69.	3,745,240
4.	56,565	26.	252,440	48.	1,259,210	70.	3,872,410
5.	57,450	27.	284,405	49.	1,275,530	71.	3,892,570
6.	57,590	28.	293,704	50.	1,333,520	72.	3,948,500
7.	58,925	29.	303,159	51.	1,349,620	73.	4,025,550
8.	65,175	30.	330,385	52.	1,410,570	74.	4,117,750
9.	70,550	31.	338,332	53.	1,461,640	75.	4,200,330
10.	101,260	32.	341,878	54.	1,985,120	76.	4,245,690
11.	107,090	33.	342,337	55.	2,337,420	77.	4,447,300
12.	111,250	34.	364,460	56.	2,471,940	78.	4,539,330
13.	116,150	35.	366,155	57.	2,666,590	79.	4,570,540
14.	116,359	36.	371,479	58.	2,780,130	80.	4,739,010
15.	116,369	37.	394,489	59.	2,929,400	81.	4,759,020
16.	124,312	38.	399,093	60.	3,044,900	82.	5,152,670
17.	139,614	39.	403,674	61.	3,250,320	83.	5,158,360
18.	144,040	40.	426,994	62.	3,443,350	84.	5,692,620
19.	154,100	41.	567,900	63.	3,483,440	85.	6,221,710
20.	160,261	42.	624,850	64.	3,583,440	86.	6,223,890
21.	171,475	43.	752,386	65.	3,675,880	87.	6,437,660
22.	186,610	44.	945,549	66.	3,676,310	88.	7,201,380

(a) Find and interpret the 40th percentile.
(b) Find and interpret the 95th percentile.
(c) Find and interpret the 10th percentile.
(d) What is the percentile rank of $4,117,750?
(e) What is the percentile rank of $116,359?

18. **NASCAR Earnings** Use the data in problem 17 to answer the following:
    (a) Find and interpret the 30th percentile.
    (b) Find and interpret the 90th percentile.
    (c) Find and interpret the 5th percentile.
    (d) What is the percentile rank of $1,333,520?
    (e) What is the percentile rank of $139,614?

19. **Weights of Males versus Females** According to the National Center for Health Statistics, the mean weight of a 20- to 29-year-old female is 156.5 pounds, with a standard deviation of 51.2 pounds. The mean weight of a 20- to 29-year-old male is 183.4 pounds, with a standard deviation of 40.0 pounds. Who is relatively heavier: a 20- to 29-year-old female who weights 160 pounds or a 20- to 29-year-old male who weighs 185 pounds?

20. **Crime Rate** Answer the accompanying questions regarding the boxplot, which illustrates crime-rate data per 100,000 population for the 50 states and the District of Columbia in 2002. (Source:www.infoplease.com)

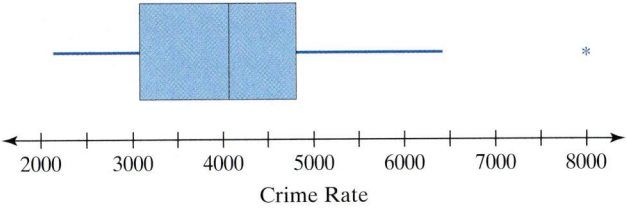

(a) Approximately, what is the median crime rate in the United States?
(b) Approximately, what is the 25th percentile crime rate in the United States?
(c) Are there any outliers? If so, identify their value(s).
(d) What is the lowest crime rate?

FEDERALIST:

A COLLECTION

OF

E S S A Y S,

WRITTEN IN FAVOUR OF THE

NEW CONSTITUTION,

AS AGREED UPON BY THE FEDERAL CONVENTION, SEPTEMBER 17, 1787.

CASE STUDY

Who Was "A MOURNER"?

Which colonial patriot penned the VINDEX essay in the January 8, 1770, issue of *The Boston Gazette and Country Journal*? Who wrote the 12 contested *Federalist* papers? Such questions about authorship of unattributed documents are one of the many problems confronting historians. Statistical analyses, when combined with other historical facts, often help resolve these mysteries.

One such historical conundrum is the identity of the writer known simply as A MOURNER. A letter appeared in the February 26, 1770, issue of *The Boston Gazette and Country Journal*. The text of the letter is as follows:

The general Sympathy and Concern for the Murder of the Lad by the base and infamous Richardson on the 22d Instant, will be sufficient Reason for your Notifying the Publick that he will be buried from his Father's House in Frogg Lane, opposite Liberty-Tree, on Monday next, when all the Friends of Liberty may have an Opportunity of paying their last Respects to the Remains of this little Hero and first Martyr to the noble Cause—Whose manly Spirit (after this Accident happened) appear'd in his discreet Answers to his Doctor, his Thanks to the Clergymen who prayed with him, and Parents, and while he underwent the greatest Distress of bodily Pain; and with which he met the King of Terrors. These Things, together with the several heroic Pieces found in his Pocket, particularly *Wolfe's Summit of human Glory*, gives Reason to think he had a *martial Genius*, and would have made a clever Man.

A MOURNER.

CASE STUDY

The Lad the writer refers to is Christopher Sider, an 11-year-old son of a poor German immigrant. Young Sider was shot and killed on February 22, 1770, in a civil disturbance involving schoolboys, patriot supporters of the nonimportation agreement, and champions of the English crown. This event preceded the bloody Boston Massacre by just a couple of weeks. Of Sider's funeral, John Adams wrote, "My eyes never beheld such a funeral. The Procession extended further than can be well imagined." (Diary of John Adams entry for 1770. MONDAY FEB. 26 OR THEREABOUTS.)

From a historical perspective, the identity of A MOURNER remains a mystery. However, it seems clear from the letter's text that the author supported the patriot position. This assumption somewhat narrows the field of possible writers.

Ordinarily, a statistical analysis of the frequencies of various words contained in the contested document, when compared with frequency analyses for known authors, would permit an identity inference to be drawn. Unfortunately, in this case, the letter is too short for this strategy to be useful. Another possibility is based on the frequencies of word lengths. In this instance, a simple count of letters is generated for each word in the document. Proper names, numbers, abbreviations, and titles are removed from consideration, because these do not represent normal vocabulary use. Care must be taken in choosing comparison texts, since colonial publishers habitually incorporated their own spellings into the essays they printed. Therefore, it is desirable to get comparison texts from the same printer the contested material came from.

The following table contains the summary analysis for six passages printed in *The Boston Gazette, and Country Journal* in early 1770. The Tom Sturdy (probably a pseudonym) text was included because of the use of the phrase "Friends of Liberty," which appears in A MOURNER's letter, as opposed to the more familiar "Sons of Liberty." John Hancock, James Otis, and Samuel Adams are included because they are well-known patriots who frequently wrote articles that were carried by the Boston papers. In the case of Samuel Adams, the essay used in this analysis was signed VINDEX, one of his many pseudonyms. The table presents three summaries of work penned by Adams. The first two originate from two separate sections of the VINDEX essay. The last summary is a compilation of the first two.

Summary Statistics of Word Length from Sample Passages by Various Potential Authors of the Letter Signed A MOURNER

	Tom Sturdy	John Hancock	James Otis	Samuel Adams-1	Samuel Adams-2	Samuel Adams-1&2
Mean	4.08	4.69	4.58	4.60	4.52	4.56
Median	4	4	4	3	4	4
Mode	2	3	2	2	2	2
Standard deviation	2.17	2.60	2.75	2.89	2.70	2.80
Sample variance	4.70	6.76	7.54	8.34	7.30	7.84
Range	13	9	14	12	16	16
Minimum	1	1	1	1	1	1
Maximum	14	10	15	13	17	17
Sum	795	568	842	810	682	1492
Count	195	121	184	176	151	27

1. Acting as a historical detective, generate a data set consisting of the length of each word used in the letter signed by A MOURNER. Be sure to disregard any text that uses proper names, numbers, abbreviations, or titles.

2. Calculate the summary statistics for the letter's word lengths. Compare your findings with those of the known authors, and speculate about the identity of A MOURNER.

3. Compare the two Adams summaries. Discuss the viability of word-length analysis as a tool for resolving disputed documents.

4. What other information would be useful to identify A MOURNER?

Describing the Relation between Two Variables

CHAPTER

Outline

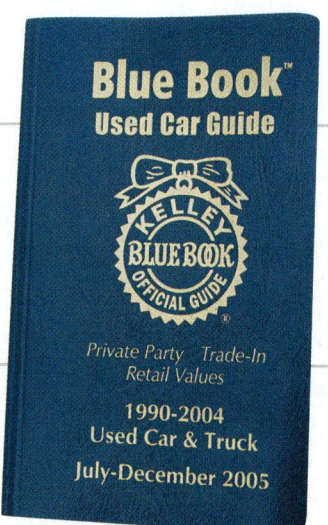

DECISIONS

You are still in the market to buy a car. Because cars lose value over time at different rates, you want to look at the depreciation rates of cars you are considering. After all, the higher the depreciation rate, the more value the car loses each year. See the Decision Project on page 221.

●●● Putting It All Together

In Chapters 2 and 3 we examined data in which a single variable was measured for each individual in the study (**univariate data**), such as the 3-year rate of return (the variable) for various mutual funds (the individuals). We obtained descriptive measures for the variable that were both graphical and numerical.

However, much research is designed to describe the relation that may exist between two variables. For example, a researcher may be interested in the relationship between the club-head speed of a golf club and the distance the golf ball travels. Here, each swing represents an individual, and the two variables are club-head speed and distance. This type of data is referred to as *bivariate data*. **Bivariate data** are data in which two variables are measured on an individual. To describe the relation between the two quantitative variables, we first graphically represent the data and then obtain some numerical descriptions of the data, just as we did when analyzing univariate data.

4.1   Scatter Diagrams and Correlation

Preparing for This Section   Before getting started, review the following:

- Mean (Section 3.1, pp. 121–124)
- Standard deviation (Section 3.2, pp. 143–144)
- *z*-Scores (Section 3.4, pp. 165–166)

Objectives

1. **Draw and interpret scatter diagrams**
2. **Understand the properties of the linear correlation coefficient**
3. **Compute and interpret the linear correlation coefficient**

Before we can graphically represent bivariate data, a fundamental question must be asked. Am I interested in using the value of one variable to predict the value of the other variable? For example, it seems reasonable to think that as the speed at which a golf club is swung increases, the distance the golf ball travels also increases. Therefore, we might use club-head speed to predict distance. We call distance the *response* (or *dependent*) *variable* and club-head speed the *explanatory* (or *predictor* or *independent*) *variable*.

Definition   The **response variable** is the variable whose value can be explained by the value of the **explanatory** or **predictor variable**.

In Other Words
We use the term explanatory variable because it helps to explain variability in the response variable.

! **CAUTION**
If bivariate data are observational, then we cannot conclude that any relation between the explanatory and response variables are due to cause and effect.

It is important to recognize that, if the data used in the study are observational, we cannot conclude there is a causal relationship between the two variables. We cannot say that changes in the level of the explanatory variable *cause* changes in the level of the response variable. In fact, it may be that the two are related through some *lurking variable*.

Recall that a **lurking variable** is a variable that may affect the response variable, but is excluded from the analysis. For example, air-conditioning bills can be used to predict lemonade sales. As air-conditioning bills rise, the sales of lemonade rise. This relation does not mean that high air-conditioning bills cause high lemonade sales, because both high air-conditioning bills and high lemonade sales are associated with high summer temperatures. Therefore, air temperature is a lurking variable.

1 **Draw and Interpret Scatter Diagrams**

The first step in identifying the type of relation that might exist between two variables is to draw a picture. Bivariate data can be represented graphically through a *scatter diagram*.

Definition   A **scatter diagram** is a graph that shows the relationship between two quantitative variables measured on the same individual. Each individual in the data set is represented by a point in the scatter diagram. The explanatory variable is plotted on the horizontal axis and the response variable is plotted on the vertical axis. Do not connect the points when drawing a scatter diagram.

EXAMPLE 1   **Drawing a Scatter Diagram**

Problem: A golf pro wanted to learn the relation between the club-head speed of a golf club (measured in miles per hour) and the distance (in yards) that the ball will travel. He realized that there are other variables besides club-

Club-Head Speed (mph)	Distance (yards)
100	257
102	264
103	274
101	266
105	277
100	263
99	258
105	275

Table 1

Source: Paul Stephenson, student at Joliet Junior College

head speed that determine the distance a ball will travel (such as club type, ball type, golfer, and weather conditions). To eliminate the variability due to these variables, the pro used a single model of club and ball. One golfer was chosen to swing the club on a clear, 70-degree day with no wind. The pro recorded the club-head speed and measured the distance that the ball traveled and collected the data in Table 1. Draw a scatter diagram of the data.

Approach: Because the pro wants to use club-head speed to predict the distance the ball travels, club-head speed is the explanatory variable (horizontal axis) and distance is the response variable (vertical axis). We plot the ordered pairs (100, 257), (102, 264), and so on, in a rectangular coordinate system.

Solution: The scatter diagram is shown in Figure 1.

Figure 1

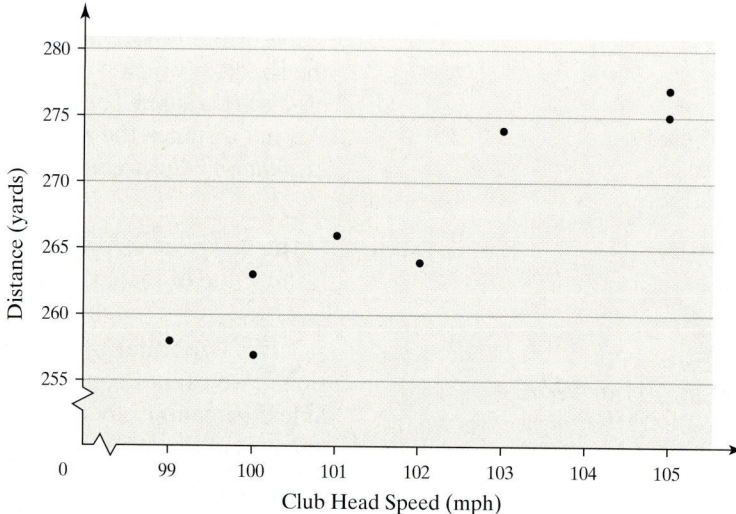

It would appear from the graph that as club-head speed increases the distance that the ball travels increases as well.

It is not always clear which variable should be considered the response variable and which should be considered the explanatory variable. For example, does high school GPA predict a student's SAT score or can the SAT score be used to predict GPA? The researcher must determine which variable plays the role of explanatory variable based on the questions he or she wants answered. For example, if the researcher is interested in predicting SAT scores on the basis of high school GPA, then high school GPA will play the role of explanatory variable.

Now Work Problems 23(a) and 23(b).

Scatter diagrams show the type of relation that exists between two variables. Our goal in interpreting scatter diagrams will be to distinguish scatter diagrams that imply a linear relation from those that imply a nonlinear relation and those that imply no relation. Figure 2 displays various scatter diagrams and the type of relation implied.

As we compare Figure 2(a) with Figure 2(b), we notice a distinct difference. In Figure 2(a), the data follow a linear pattern that slants upward to the right; the data in Figure 2(b) follow a linear pattern that slants downward to the right. Figures 2(c) and 2(d) show scatter diagrams of nonlinear relations. Figure 2(e) shows a scatter diagram in which there is no relation between the explanatory and response variables.

Figure 2

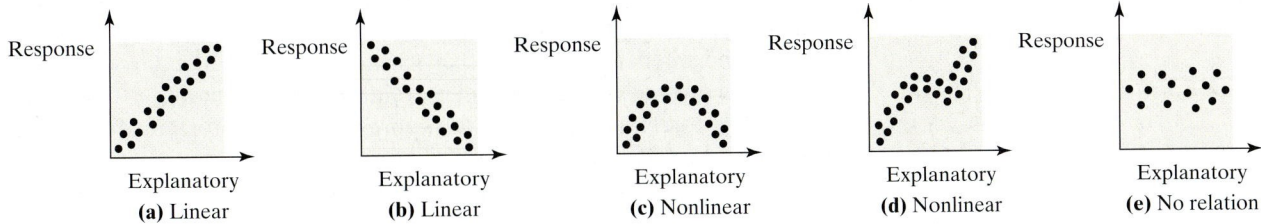

(a) Linear   (b) Linear   (c) Nonlinear   (d) Nonlinear   (e) No relation

Definitions

In Other Words

If two variables that are linearly related are positively associated, then as one goes up the other also tends to go up. If two variables that are linearly related are negatively associated, then as one goes up the other tends to go down.

Two variables that are linearly related are said to be **positively associated** when above-average values of one variable are associated with above-average values of the other variable. That is, two variables are positively associated if, whenever the value of one variable increases, the value of the other variable also increases.

Two variables that are linearly related are said to be **negatively associated** when above-average values of one variable are associated with below-average values of the other variable. That is, two variables are negatively associated if, whenever the value of one variable increases, the value of the other variable decreases.

So the scatter diagram from Figure 1 implies that club-head speed is positively associated with the distance a golf ball travels.

Now Work Problem 11.

② Understand the Properties of the Linear Correlation Coefficient

It is dangerous to use only a scatter diagram to decide whether two variables follow a linear relation. Suppose we redraw the scatter diagram in Figure 1 using a different scale as shown in Figure 3.

Figure 3

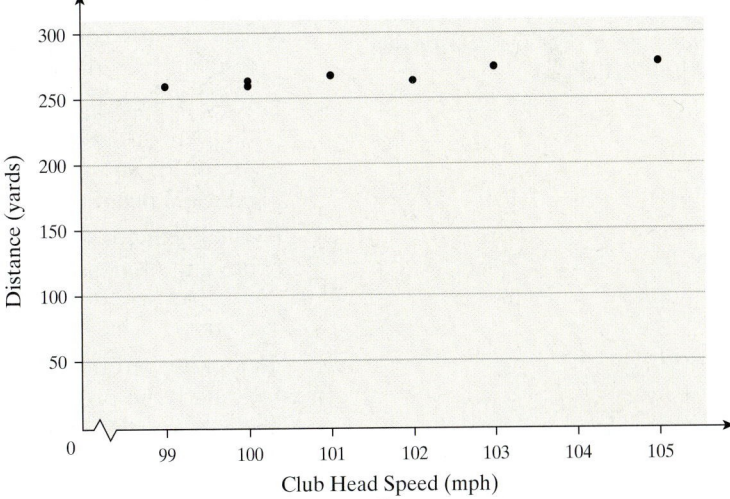

CAUTION

The horizontal or vertical scale of a scatter diagram should be set so that the scatter diagram does not mislead a reader.

From Figure 3, we might conclude that club-head speed and distance are not related. The moral of the story is this: Just as we can manipulate the scale of graphs of univariate data, we can also manipulate the scale of the graphs of bivariate data, thereby encouraging incorrect conclusions. Therefore, numerical summaries of bivariate data should be used in conjunction with graphs to determine the type of relation, if any, that exists between two variables.

Definition

The **linear correlation coefficient** or **Pearson product moment correlation coefficient** is a measure of the strength of linear relation between two quantitative variables. We use the Greek letter ρ (rho) to represent the population correlation coefficient and r to represent the sample correlation coefficient. We present only the formula for the sample correlation coefficient.

Sample Correlation Coefficient*

$$r = \frac{\sum \left(\dfrac{x_i - \overline{x}}{s_x} \right) \left(\dfrac{y_i - \overline{y}}{s_y} \right)}{n - 1} \tag{1}$$

where $\overline{x}$ is the sample mean of the explanatory variable

s_x is the sample standard deviation of the explanatory variable

$\overline{y}$ is the sample mean of the response variable

s_y is the sample standard deviation of the response variable

n is the number of individuals in the sample

The Pearson linear correlation coefficient is named in honor of Karl Pearson (1857–1936).

Historical Note

Karl Pearson was born March 27, 1857. Pearson's proficiency as a statistician was recognized early in his life. It is said that his mother told him not to suck his thumb, because otherwise his thumb would wither away. Pearson analyzed the size of each thumb and said to himself, "They look alike to me. I can't see that the thumb I suck is any smaller than the other. I wonder if she could be lying to me."

Karl Pearson graduated from Cambridge University in 1879. From 1893 to 1911, he wrote 18 papers on genetics and heredity. Through this work, he developed ideas regarding correlation and the chi-square test. (See Chapter 12.) In addition, Pearson came up with the term *standard deviation*.

Pearson and Ronald Fisher didn't get along. The dispute between the two was bad enough to have Fisher turn down the post of chief statistician at the Galton Laboratory in 1919 on the grounds that it would have meant working under Pearson. Pearson died on April 27, 1936.

Properties of the Linear Correlation Coefficient

1. The linear correlation coefficient is always between -1 and 1, inclusive. That is, $-1 \le r \le 1$.

2. If $r = +1$, there is a perfect positive linear relation between the two variables. See Figure 4(a).

3. If $r = -1$, there is a perfect negative linear relation between the two variables. See Figure 4(d).

4. The closer r is to $+1$, the stronger is the evidence of positive association between the two variables. See Figures 4(b) and 4(c).

5. The closer r is to -1, the stronger is the evidence of negative association between the two variables. See Figures 4(e) and 4(f).

6. If r is close to 0, there is little or no evidence of a *linear* relation between the two variables. Because the linear correlation coefficient is a measure of the strength of the linear relation, **r close to 0 does not imply no relation, just no linear relation.** See Figures 4(g) and 4(h).

7. The linear correlation coefficient is a unitless measure of association. So the unit of measure for x and y plays no role in the interpretation of r.

CAUTION

A linear correlation coefficient close to 0 does not imply that there is no relation, just no linear relation. For example, although the scatter diagram drawn in Figure 4(h) indicates that the two variables are related, the linear correlation coefficient of these data is close to 0.

In looking carefully at Formula (1), we should notice that the numerator of the formula is the product of z-scores for the explanatory (x) and response (y) variables. A positive linear correlation coefficient means that the sum of the product of the z-scores for x and y must be positive. Under what circumstances

*An equivalent formula for the linear correlation coefficient is

$$r = \frac{\sum x_i y_i - \dfrac{\sum x_i \sum y_i}{n}}{\sqrt{\left(\sum x_i^2 - \dfrac{(\sum x_i)^2}{n} \right)} \sqrt{\left(\sum y_i^2 - \dfrac{(\sum y_i)^2}{n} \right)}} = \frac{S_{xy}}{\sqrt{S_{xx}} \sqrt{S_{yy}}}$$

Figure 4

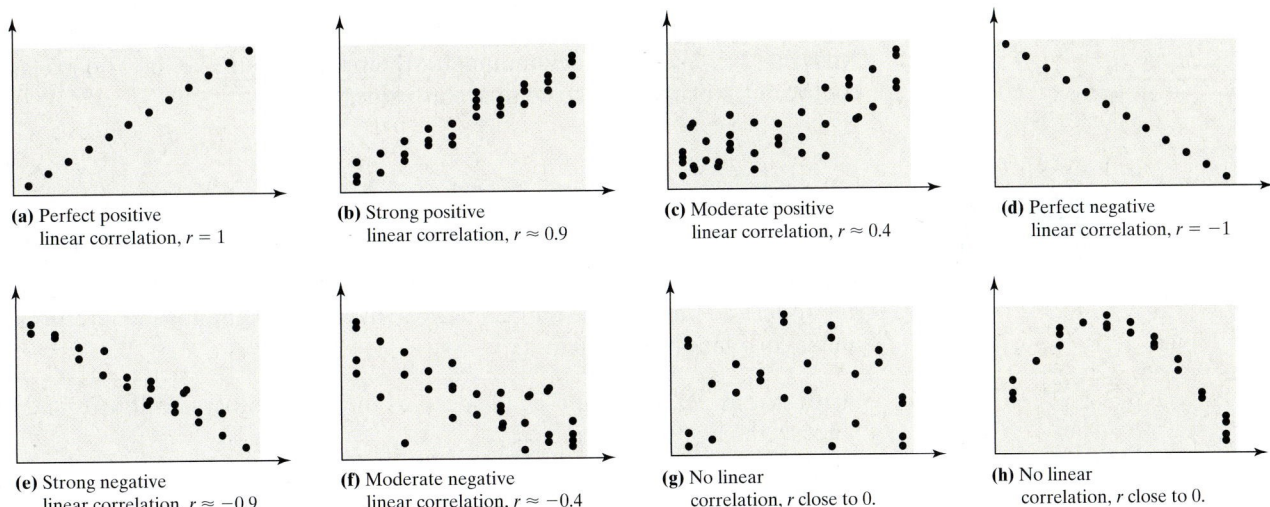

(a) Perfect positive
linear correlation, $r = 1$

(b) Strong positive
linear correlation, $r \approx 0.9$

(c) Moderate positive
linear correlation, $r \approx 0.4$

(d) Perfect negative
linear correlation, $r = -1$

(e) Strong negative
linear correlation, $r \approx -0.9$

(f) Moderate negative
linear correlation, $r \approx -0.4$

(g) No linear
correlation, r close to 0.

(h) No linear
correlation, r close to 0.

does this occur? Figure 5 shows a scatter diagram that implies a positive association between x and y. The vertical dashed line represents the value of $\overline{x}$, and the horizontal dashed line represents the value of $\overline{y}$. These two dashed lines divide our scatter diagram into four quadrants, labeled I, II, III, and IV.

Figure 5

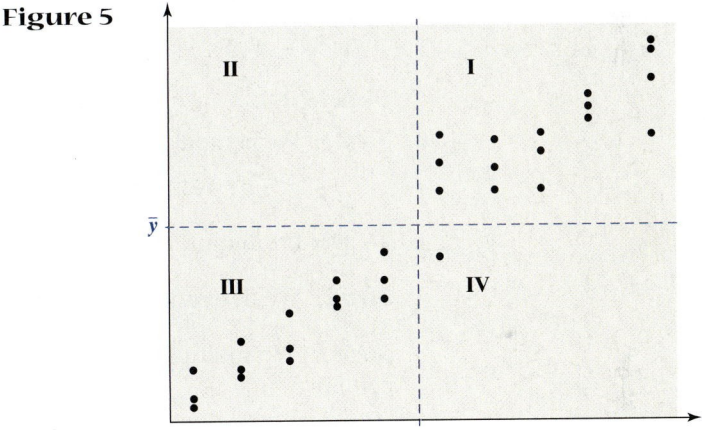

Consider the data in quadrants I and III. If a certain x-value is above its mean, $\overline{x}$, then the corresponding y-value will be above its mean, $\overline{y}$. If a certain x-value is below its mean, $\overline{x}$, then the corresponding y-value will be below its mean, $\overline{y}$. Therefore, for data in quadrant I, we have $\dfrac{x_i - \overline{x}}{s_x}$ positive and $\dfrac{y_i - \overline{y}}{s_y}$ positive, so their product is positive. For data in quadrant III, we have $\dfrac{x_i - \overline{x}}{s_x}$ negative and $\dfrac{y_i - \overline{y}}{s_y}$ negative, so their product is positive. The sum of these products is positive, and therefore we have a positive linear correlation coefficient. A similar argument can be made for negative correlation.

Now suppose the data are equally dispersed in the four quadrants. Then the negative products (resulting from data in quadrants II and IV) will offset the positive products (resulting from data in quadrants I and III). The result is a linear correlation coefficient close to 0.

In Other Words
The correlation coefficient describes the strength and the direction of the linear relationship between two variables.

Now Work Problem 15.

3 ## Compute and Interpret the Linear Correlation Coefficient

Now that we have an understanding of the properties of the linear correlation coefficient, we are ready to compute its value.

EXAMPLE 2 ## Computing and Interpreting the Correlation Coefficient

Problem: In Table 2, columns 1 and 2 represent the club-head speed (in miles per hour) and the distance the ball travels (in yards). Compute and interpret the linear correlation coefficient.

Approach: We treat club-head speed as the explanatory variable, x, and distance as the response variable, y.

Step 1: Compute $\bar{x}$, s_x, $\bar{y}$, and s_y.

Step 2: Determine $\dfrac{x_i - \bar{x}}{s_x}$ and $\dfrac{y_i - \bar{y}}{s_y}$ for each observation.

Step 3: Compute $\left(\dfrac{x_i - \bar{x}}{s_x}\right)\left(\dfrac{y_i - \bar{y}}{s_y}\right)$ for each observation.

Step 4: Determine $\sum\left(\dfrac{x_i - \bar{x}}{s_x}\right)\left(\dfrac{y_i - \bar{y}}{s_y}\right)$ and substitute this value into Formula (1).

Solution

Step 1: We compute $\bar{x}$, s_x, $\bar{y}$, and s_y:

$$\bar{x} = 101.875, \quad s_x = 2.29518, \quad \bar{y} = 266.75, \quad s_y = 7.74135$$

To avoid round-off error when using Formula (1), do not round the statistics.

Step 2: We determine $\dfrac{x_i - \bar{x}}{s_x}$ and $\dfrac{y_i - \bar{y}}{s_y}$ in columns 3 and 4 in Table 2.

Step 3: We multiply the entries in columns 3 and 4 to obtain the entries in column 5.

		Table 2		
Club-Head Speed, x_i	**Distance, y_i**	$\dfrac{x_i - \bar{x}}{s_x}$	$\dfrac{y_i - \bar{y}}{s_y}$	$\left(\dfrac{x_i - \bar{x}}{s_x}\right)\left(\dfrac{y_i - \bar{y}}{s_y}\right)$
100	257	−0.816929	−1.259470	1.028898
102	264	0.054462	−0.355235	−0.0193347
103	274	0.490158	0.936529	0.459047
101	266	−0.381234	−0.096882	0.036935
105	277	1.361549	1.324058	1.802770
100	263	−0.816929	−0.484412	0.395726
99	258	−1.252625	−1.130294	1.415835
105	275	1.361549	1.065706	1.451011

$$\sum\left(\dfrac{x_i - \bar{x}}{s_x}\right)\left(\dfrac{y_i - \bar{y}}{s_y}\right) = 6.570887$$

Step 4: We add the entries in column 5 to obtain

$$\sum\left(\frac{x_i - \overline{x}}{s_x}\right)\left(\frac{y_i - \overline{y}}{s_y}\right) = 6.570887.$$

Substitute this value into Formula (1) to obtain the correlation coefficient.

$$r = \frac{\sum\left(\dfrac{x_i - \overline{x}}{s_x}\right)\left(\dfrac{y_i - \overline{y}}{s_y}\right)}{n - 1} = \frac{6.570887}{8 - 1} = 0.9387$$

The linear correlation between club-head speed and distance is 0.9387, indicating a strong positive association between the two variables. The higher the club-head speed, the farther the golf ball tends to travel.

Notice in Example 2 that we carry many decimal places in the computation of the correlation coefficient to avoid rounding error. Also, compare the signs of the entries in columns 3 and 4. Notice that negative values in column 3 correspond to negative values in column 4 and that positive values in column 3 correspond to positive values in column 4 (except for the second trial of the experiment). This means that above-average values of x are associated with above-average values of y, and below-average values of x are associated with below-average values of y. This is why the linear correlation coefficient is positive.

EXAMPLE 3 **Drawing a Scatter Diagram and Determining the Linear Correlation Coefficient Using Technology**

Problem: Use a statistical spreadsheet or a graphing calculator with advanced statistical features to draw a scatter diagram of the data in Table 1. Then determine the linear correlation between club-head speed and distance.

Approach: We will use Excel to draw the scatter diagram and obtain the linear correlation coefficient. The steps for drawing scatter diagrams and obtaining the linear correlation coefficient using MINITAB, Excel, or the TI-83 and TI-84 Plus graphing calculators are given in the Technology Step by Step on page 212.

Result: Figure 6(a) shows the scatter diagram and Figure 6(b) shows the linear correlation coefficient obtained from Excel. Notice that Excel provides a **correlation matrix**, which means that for every pair of columns in the spreadsheet it will compute and display the correlation in the bottom triangle of the matrix.

Figure 6

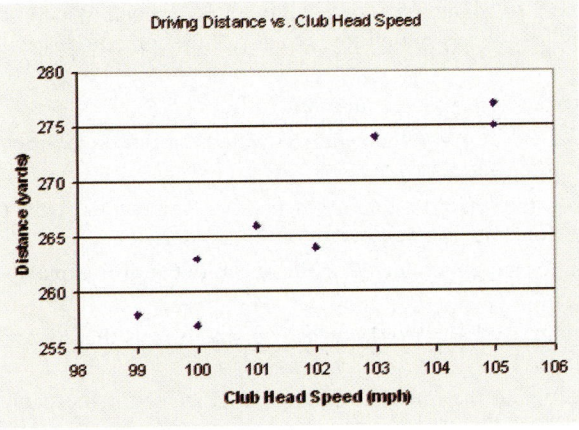

	Driving Distance	Club Head Speed
Driving Distance	1	
Club Head Speed	0.938695838	1

(a)

(b)

Correlation versus Causation

In Chapter 1 we stated that there are two types of studies: observational and experimental. The data given in Examples 1 through 3 are the result of an experiment. Therefore, we can claim that a higher club-head speed causes the golf ball to travel a longer distance. However, if data result from an observational study, we cannot claim causation. Consider the scatter diagram shown in Figure 7, which shows the relation between the birthrate (births per 1000 women) of teenagers and the homicide rate (homicides per 100,000 inhabitants) for the years 1993 to 2000.

Figure 7

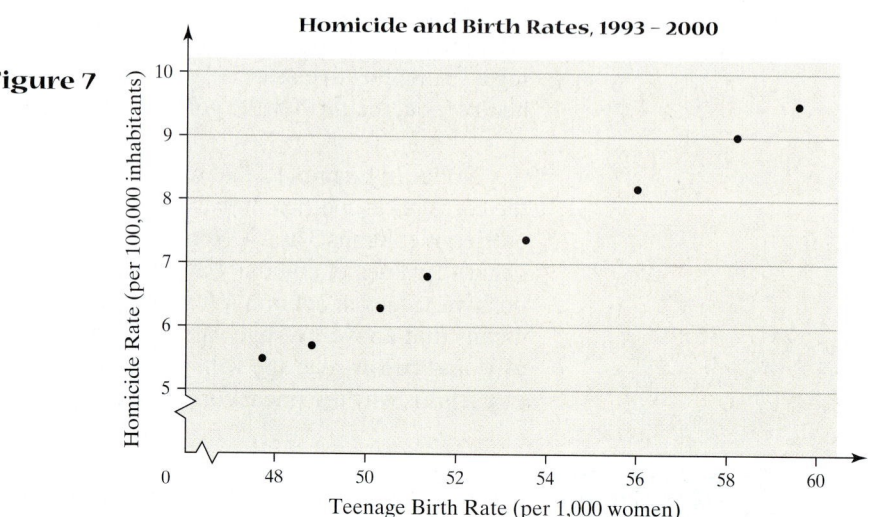

The linear correlation coefficient between these two variables is 0.9987. Does this mean that higher birthrates among teenagers cause a higher homicide rate? Certainly not!

In Chapter 1, we introduced lurking variables, a lurking variable is one that has not been considered in your analysis, but is related to both variables in the study. Can you think of any variables that might be related to teen birthrates and homicide rates? Perhaps there is an economic variable, such as poverty rate, proportion of homes with a single parent, or high school dropout rate, that is related to both teenage birthrate and homicide rate.

In-Class Activity: Correlation

Randomly select six students from the class and have them determine their at-rest pulse and then discuss the following:

1. When determining the at-rest pulse rate, would it be better to count beats for 30 seconds and multiply by 2 or count beats for 1 full minute? Explain. What are some other ways to find the at-rest pulse rate? Do any of these methods have an advantage?

2. What effect will physical activity have on pulse rate?

3. Do you think the at-rest pulse rate will have any effect on the pulse rate after physical activity? If so, how? If not, why not?

Have the same six students jog in place for 3 minutes and then immediately determine their pulse rate using the same technique as for the at-rest pulse rate.

4. Draw a scatter diagram for the pulse data using the at-rest data as the explanatory variable.

5. Comment on the relationship, if any, between the two variables. Is this consistent with your expectations?

6. Based on the graph, estimate the linear correlation coefficient for the data. Then compute the correlation coefficient using a graphing utility and compare to your estimate.

4.1 ASSESS YOUR UNDERSTANDING

Concepts and Vocabulary

1. Describe the difference between univariate and bivariate data.
2. Explain what is meant by a lurking variable. Provide an example.
3. What does it mean to say that two variables are positively associated?
4. What does it mean to say that the linear correlation coefficient between two variables equals 1? What would the scatter diagram look like?
5. What does it mean if $r = 0$?
6. Is the linear correlation coefficient a resistant measure? Support your answer.
7. Explain what is wrong with the following statement: "We have concluded that there is a high correlation between the gender of drivers and rates of automobile accidents."
8. Write a statement that explains the concept of correlation. Include a discussion of the role that $x_i - \bar{x}$ and $y_i - \bar{y}$ play in the computation.
9. Explain what is wrong with the following statement: "A recent study showed that the correlation between the number of acres on a farm and the amount of corn produced was 0.93 bushel."
10. Explain the difference between correlation and causation. When does a linear correlation coefficient that implies a strong positive correlation also imply causation?

Skill Building

In Problems 11–14, determine whether the scatter diagram indicates that a linear relation may exist between the two variables. If the relation is linear, determine whether it indicates a positive or negative association between the variables.

11.
NW
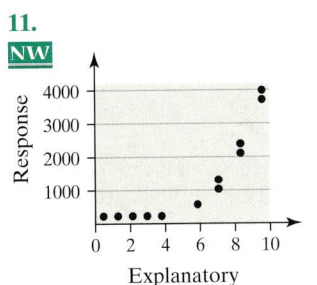

12.
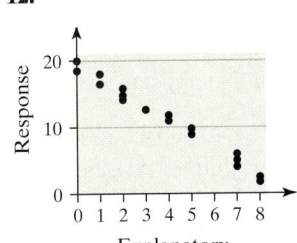

13.
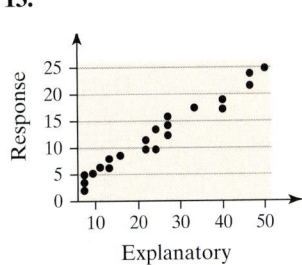

14.
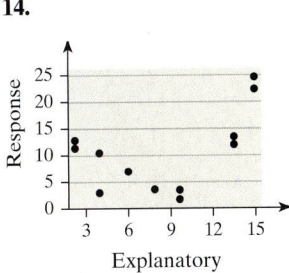

15. Match the linear correlation coefficient to the scatter diagram. The scales on the *x*- and *y*-axis are the same for each scatter
NW diagram.

(a) $r = 0.787$ (b) $r = 0.523$ (c) $r = 0.810$ (d) $r = 0.946$

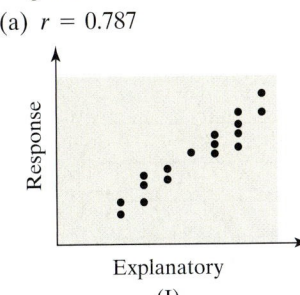
(I)

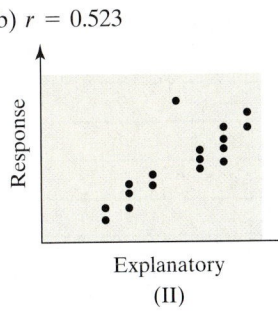
(II)

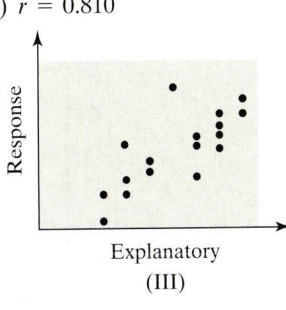
(III)

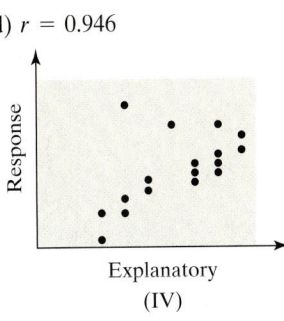
(IV)

16. Match the linear correlation coefficient to the scatter diagram. The scales on the *x*- and *y*-axis are the same for each scatter diagram.

(a) $r = -0.969$ (b) $r = -0.049$ (c) $r = -1$ (d) $r = -0.992$

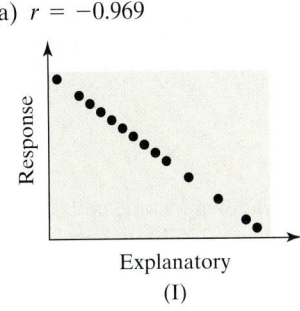
(I)

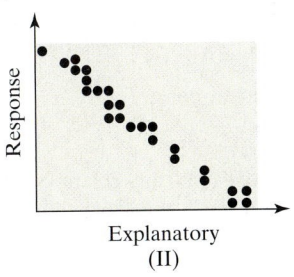
(II)

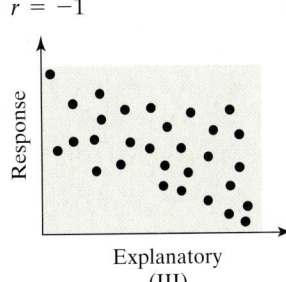
(III)

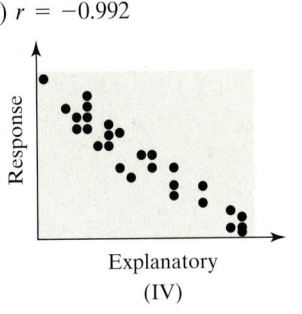
(IV)

17. **Does Education Pay?** The following scatter diagram drawn in MINITAB shows the relation between the percentage of the population of a state that has at least a bachelor's degree and the median income (in dollars) of the state for 2003.

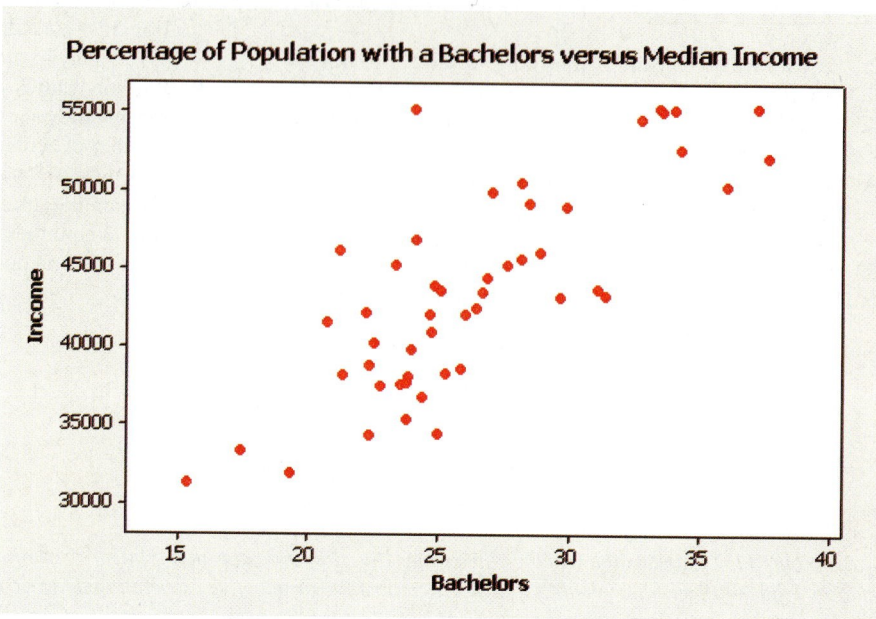

Source: U.S. Census Bureau

(a) Describe the relation that appears to exist between level of education and median income.

(b) One observation appears to stick out from the rest. Which one? This particular observation is for the state of Alaska. Can you think of any reasons why the state of Alaska might have a high median income, given the proportion of the population that has at least a bachelor's degree?

18. **Relation between Income and Birthrate?** The following scatter diagram drawn in Excel shows the relation between median income (in dollars) in a state and birthrate (births per 1000 women 15 to 44 years of age).

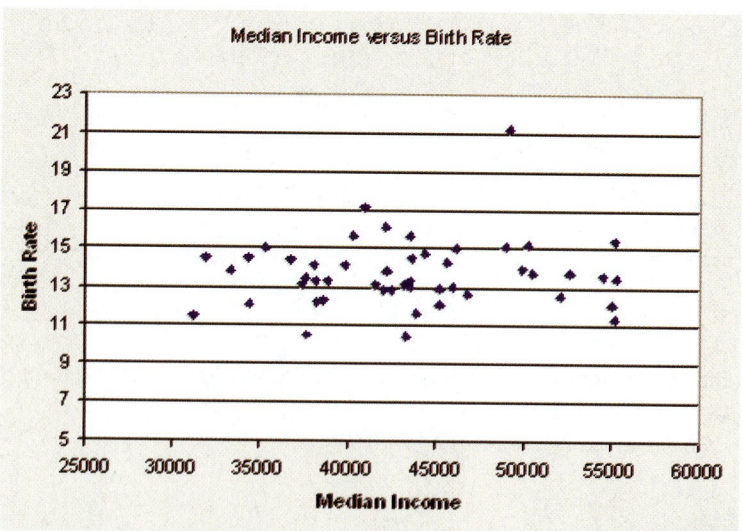

Source: U.S. Census Bureaus

(a) Does there appear to be any relation between median income and birth rate?

(b) One observation sticks out from the rest. Which one? This particular observation is for the state of Utah. Are there any explanations for this result?

In Problems 19–22, (a) draw a scatter diagram of the data, (b) by hand, compute the correlation coefficient, and (c) comment on the type of relation that appears to exist between x and y.

19.

x	2	4	8	8	9
y	1.4	1.8	2.1	2.3	2.6

20.

x	2	3	5	6	6
y	5.7	5.2	2.8	1.9	2.2

21.

x	1.2	1.8	2.3	3.5	4.1
y	8.4	7	7.3	4.5	2.4

22.

x	0	0.5	1.4	2.1	3.9	4.6
y	0.8	1.3	1.9	2.5	5.0	6.8

Applying the Concepts

23. Height versus Head Circumference A pediatrician wants
NW to determine the relation that may exist between a child's height and head circumference. She randomly selects 11 three-year-old children from her practice, measures their height and head circumference, and obtains the data shown in the table.

Height (inches)	Head Circumference (inches)	Height (inches)	Head Circumference (inches)
27.75	17.5	26.5	17.3
24.5	17.1	27	17.5
25.5	17.1	26.75	17.3
26	17.3	26.75	17.5
25	16.9	27.5	17.5
27.75	17.6		

Source: Denise Slucki, student at Joliet Junior College

(a) If the pediatrician wants to use height to predict head circumference, determine which variable is the explanatory variable and which is the response variable.
(b) Draw a scatter diagram.
(c) Compute the linear correlation coefficient between the height and head circumference of a child.
(d) Comment on the type of relation that appears to exist between the height and head circumference of a child on the basis of the scatter diagram and linear correlation coefficient.

24. Gestation Period versus Life Expectancy A researcher wants to know if the gestation period of an animal can be used to predict life expectancy. She collects the following data:

Animal	Gestation (or Incubation) Period (days)	Life Expectancy (years)
Cat	63	11
Chicken	22	7.5
Dog	63	11
Duck	28	10
Goat	151	12
Lion	108	10
Parakeet	18	8
Pig	115	10
Rabbit	31	7
Squirrel	44	9

Source: Time Almanac 2000

(a) Suppose the researcher wants to use the gestation period of an animal to predict its life expectancy. Determine which variable is the explanatory variable and which is the response variable.
(b) Draw a scatter diagram.
(c) Compute the linear correlation coefficient between gestation period and life expectancy.
(d) Comment on the type of relation that appears to exist between gestation period and life expectancy based on the scatter diagram and linear correlation coefficient.
(e) Remove the goat from the data set, and recompute the linear correlation coefficient between the gestation period and life expectancy. What effect did the removal of the data value have on the linear correlation coefficient? Provide a justification for this result.

25. Weight of a Car versus Miles per Gallon An engineer wanted to determine how the weight of a car affects gas mileage. The following data represent the weight of various domestic cars and their gas mileage in the city for the 2005 model year.

Car	Weight (pounds)	Miles per Gallon
Buick LeSabre	3565	20
Cadillac DeVille	3985	18
Chevrolet Corvette	3180	19
Chevrolet Monte Carlo	3340	21
Chrysler PT Cruiser	3100	21
Chrysler Sebring Sedan	3175	22
Dodge Neon	2580	27
Dodge Stratus Sedan	3175	22
Ford Focus	2655	26
Ford Mustang	3300	20
Lincoln LS	3680	20
Mercury Sable	3310	19
Pontiac Bonneville	3590	20
Pontiac Grand Am	3475	20
Pontiac Sunfire	2770	24
Saturn Ion	2690	26

Source: www.roadandtrack.com

(a) Determine which variable is the likely explanatory variable and which is the likely response variable.
(b) Draw a scatter diagram of the data.
(c) Compute the linear correlation coefficient between the weight of a car and its miles per gallon in the city.
(d) Comment on the type of relation that appears to exist between the weight of a car and its miles per gallon in the city based on the scatter diagram and the linear correlation coefficient.

26. Bone Length Research performed at NASA and led by Emily R. Morey-Holton measured the lengths of the right humerus and right tibia in 11 rats that were sent to space on Spacelab Life Sciences 2. The following data were collected.

Right Humerus (mm)	Right Tibia (mm)	Right Humerus (mm)	Right Tibia (mm)
24.8	36.05	25.9	37.38
24.59	35.57	26.11	37.96
24.59	35.57	26.63	37.46
24.29	34.58	26.31	37.75
23.81	34.2	26.84	38.5
24.87	34.73		

Source: NASA Life Sciences Data Archive

(a) Draw a scatter diagram, treating the length of the right humerus as the explanatory variable and the length of the right tibia as the response variable.
(b) Compute the linear correlation coefficient between the length of the right humerus and the length of the right tibia.
(c) Comment on the type of relation that appears to exist between the length of the right humerus and the length of the right tibia based on the scatter diagram and the linear correlation coefficient.
(d) Convert the data to inches (1 mm = 0.03937 inch), and recompute the linear correlation coefficient. What effect did the conversion from millimeters to inches have on the linear correlation coefficient?

27. Attending Class The following data represent the number of days absent and the final grade for a sample of college students in a general education course at a large midwestern state university.

# of absences	final grade
0	89.2
1	86.4
2	83.5
3	81.1
4	78.2
5	73.9
6	64.3
7	71.8
8	65.5
9	66.2

Source: College Teaching, Winter 2005, Vol. 53, Issue 1.

(a) The researcher wants to use the number of days absent to predict the final grade. Determine which variable is the explanatory variable and which is the response variable.
(b) Draw a scatter diagram of the data.
(c) Compute the linear correlation coefficient between the number of days absent and the final grade.
(d) Comment on the type of relation that appears to exist between the number of days absent and the final grade.
(e) Will going to class every day guarantee a passing grade? What other factors might need to be taken into account?

28. Antibiotics A study on antibiotic use among children in Manitoba, Canada, gave the following data for the number of prescriptions per 1000 children x years after 1995.

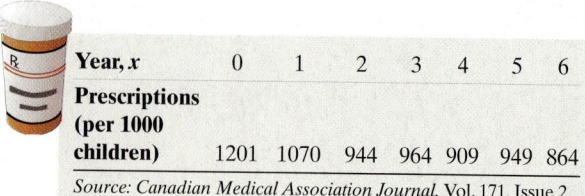

Year, x	0	1	2	3	4	5	6
Prescriptions (per 1000 children)	1201	1070	944	964	909	949	864

Source: Canadian Medical Association Journal, Vol. 171, Issue 2.

(a) Draw a scatter diagram of the data, treating year as the explanatory variable. What type of relation, if any, appears to exist between year and antibiotic prescriptions among children?

(b) Compute the linear correlation coefficient between year and antibiotic prescriptions among children.

(c) Comment on the type of relation that appears to exist between year and antibiotic prescriptions among children on the basis of the scatter diagram and the linear correlation coefficient.

29. **Age versus HDL Cholesterol** A doctor wanted to determine whether there was a relation between a male's age and his HDL (so-called good) cholesterol. He randomly selected 17 of his patients and determined their HDL cholesterol. He obtained the following data.

Age	HDL Cholesterol	Age	HDL Cholesterol
38	57	38	44
42	54	66	62
46	34	30	53
32	56	51	36
55	35	27	45
52	40	52	38
61	42	49	55
61	38	39	28
26	47		

Source: Data based on information obtained from the National Center for Health Statistics

(a) Draw a scatter diagram of the data, treating age as the explanatory variable. What type of relation, if any, appears to exist between age and HDL cholesterol?

(b) Compute the linear correlation coefficient between age and HDL cholesterol.

(c) Comment on the type of relation that appears to exist between age and HDL cholesterol on the basis of the scatter diagram and the linear correlation coefficient.

30. **Intensity of a Light Bulb** Cathy is conducting an experiment to measure the relation between a light bulb's intensity and the distance from the light source. She measures a 100-watt light bulb's intensity 1 meter from the bulb and at 0.1-meter intervals up to 2 meters from the bulb and obtains the following data.

Distance (meters)	Intensity	Distance (meters)	Intensity
1.0	0.29645	1.6	0.11450
1.1	0.25215	1.7	0.10243
1.2	0.20547	1.8	0.09231
1.3	0.17462	1.9	0.08321
1.4	0.15342	2.0	0.07342
1.5	0.13521		

(a) Draw a scatter diagram of the data, treating distance as the explanatory variable.

(b) Do you think that it is appropriate to compute the linear correlation coefficient between distance and intensity? Why?

31. **Does Size Matter?** Researchers wondered whether the size of a person's brain was related to the individual's mental capacity. They selected a sample of right-handed introductory psychology students who had SAT scores higher than 1350. The subjects took the Wechsler (1981) Adult Intelligence Scale-Revised to obtain their IQ scores. MRI scans were performed at the same facility for the subjects. The scans consisted of 18 horizontal MR images. The computer counted all pixels with nonzero gray scale in each of the 18 images, and the total count served as an index for brain size.

Gender	MRI Count	IQ	Gender	MRI Count	IQ
Female	816,932	133	Male	949,395	140
Female	951,545	137	Male	1,001,121	140
Female	991,305	138	Male	1,038,437	139
Female	833,868	132	Male	965,353	133
Female	856,472	140	Male	955,466	133
Female	852,244	132	Male	1,079,549	141
Female	790,619	135	Male	924,059	135
Female	866,662	130	Male	955,003	139
Female	857,782	133	Male	935,494	141
Female	948,066	133	Male	949,589	144

Source: Willerman, L., Schultz, R., Rutledge, J. N., and Bigler, E. (1991). "In Vivo Brain Size and Intelligence," *Intelligence,* 15, 223–228.

(a) Draw a scatter diagram, treating MRI count as the explanatory variable and IQ as the response variable. Comment on what you see.

(b) Compute the linear correlation coefficient between MRI count and IQ. Do you think that MRI count and IQ are linearly related?

(c) A lurking variable in the analysis is gender. Draw a scatter diagram, treating MRI count as the explanatory variable and IQ as the response variable, but use a different plotting symbol for each gender. For example, use a circle for males and a triangle for females. What do you notice?

(d) Compute the linear correlation coefficient between MRI count and IQ for females. Compute the linear correlation coefficient between MRI count and IQ for males. Do you believe that MRI count and IQ are linearly related? What is the moral?

32. Male versus Female Drivers The following data represent the number of licensed drivers in various age groups and the number of accidents within the age group by gender.

Age Group	Number of Male Licensed Drivers (000s)	Number of Crashes Involving a Male (000s)	Number of Female Licensed Drivers (000s)	Number of Crashes Involving a Female (000s)
16	816	244	764	178
17	1,198	233	1,115	175
18	1,342	243	1,212	164
19	1,454	229	1,333	145
20–24	7,866	951	7,394	618
25–29	9,356	899	8,946	595
30–34	10,121	875	9,871	571
35–39	10,521	901	10,439	566
40–44	9,776	692	9,752	455
45–49	8,754	667	8,710	390
50–54	6,840	390	6,763	247
55–59	5,341	290	5,258	165
60–64	4,565	218	4,486	133
65–69	4,234	191	4,231	121
70–74	3,604	167	3,749	104
75–79	2,563	118	2,716	77
80–84	1,400	61	1,516	45
≥85	767	34	767	20

Source: National Highway and Traffic Safety Institute

(a) On the same graph, draw a scatter diagram for both males and females. Be sure to use a different plotting symbol for each group. For example, use a square (□) or an M for males and a plus sign (+) or an F for females. Treat number of licensed drivers as the explanatory variable.
(b) Based on the scatter diagrams, do you think that insurance companies are justified in charging different insurance rates for males and females? Why?
(c) Compute the linear correlation coefficient between number of licensed drivers and number of crashes for males.
(d) Compute the linear correlation coefficient between number of licensed drivers and number of crashes for females.
(e) Which gender has the stronger linear relation between number of licensed drivers and number of crashes. Why?

33. Weight of a Car versus Miles per Gallon Suppose we add the Ford Taurus to the data in Problem 25. A Ford Taurus weighs 3305 pounds and gets 19 miles per gallon.
(a) Redraw the scatter diagram with the Taurus included.
(b) Recompute the linear correlation coefficient with the Taurus included.
(c) Compare the results of parts (a) and (b) with the results of Problem 25. Why are the results here reasonable?
(d) Now suppose we add the Toyota Prius to the data in Problem 25 (remove the Taurus). A Toyota Prius weighs 2890 pounds and gets 60 miles per gallon. Redraw the scatter diagram with the Prius included. What do you notice?

(e) Recompute the linear correlation coefficient with the Prius included. How did this new value affect your result?
(f) Why does this observation not follow the pattern of the data?

34. Gestation Period versus Life Expectancy Suppose we add humans to the data in Problem 24. Humans have a gestation period of 268 days and a life expectancy of 76.5 years.
(a) Redraw the scatter diagram with humans included.
(b) Recompute the linear correlation coefficient with humans included.
(c) Compare the results of (a) and (b) with the results of Problem 24. Provide a statement that explains the results.

35. Consider the following four data sets:

Data Set 1		Data Set 2		Data Set 3		Data Set 4	
x	y	x	y	x	y	x	y
10	8.04	10	9.14	10	7.46	8	6.58
8	6.95	8	8.14	8	6.77	8	5.76
13	7.58	13	8.74	13	12.74	8	7.71
9	8.81	9	8.77	9	7.11	8	8.84
11	8.33	11	9.26	11	7.81	8	8.47
14	9.96	14	8.10	14	8.84	8	7.04
6	7.24	6	6.13	6	6.08	8	5.25
4	4.26	4	3.10	4	5.39	8	5.56
12	10.84	12	9.13	12	8.15	8	7.91
7	4.82	7	7.26	7	6.42	8	6.89
5	5.68	5	4.47	5	5.73	19	12.50

Source: Anscombe, Frank, Graphs in statistical analysis, *American Statistician* 27 (1973):17–21.

(a) Compute the linear correlation coefficient for each data set.
(b) Draw a scatter diagram for each data set. Conclude that linear correlation coefficients and scatter diagrams must be used together in any statistical analysis of bivariate data.

36. The Best Predictor of the Winning Percentage The ultimate goal in any sport (besides having fun) is to win. One measure of how well a team does is the winning percentage. In baseball, a lot of effort goes in to figuring out the variable that best predicts of a team's winning percentage. The following data represent the winning percentages of teams in the National League along with potential explanatory variables. Which variable do you think is the best predictor of winning percentage? Why?

Team	Winning Percentage	Runs Scored	Home Runs	Team Batting Average	On-Base Percentage	Batting Average against Team	Team Earned-Run Average
Arizona	0.315	615	135	0.253	0.310	0.266	4.98
Atlanta	0.593	803	178	0.270	0.343	0.265	3.74
Chicago Cubs	0.549	789	235	0.268	0.328	0.247	3.81
Cincinnati	0.469	750	194	0.250	0.331	0.280	5.19
Colorado	0.420	833	202	0.275	0.345	0.290	5.54
Florida	0.512	718	148	0.264	0.329	0.256	4.10
Houston	0.568	803	187	0.267	0.342	0.258	4.05
Los Angeles	0.574	761	203	0.262	0.332	0.254	4.01
Milwaukee	0.416	634	135	0.248	0.321	0.259	4.24
Montreal	0.414	635	151	0.249	0.313	0.266	4.33
New York Mets	0.438	684	185	0.249	0.317	0.261	4.09
Philadelphia	0.531	840	215	0.267	0.345	0.264	4.45
Pittsburgh	0.447	680	142	0.260	0.321	0.267	4.29
San Diego	0.537	768	139	0.273	0.342	0.263	4.03
San Francisco	0.562	850	183	0.270	0.357	0.265	4.29
St. Louis	0.648	855	214	0.278	0.344	0.251	3.75

Source: espn.com

37. **Diversification** One basic theory of investing is diversification. The idea is that you want to have a basket of stocks that do not all "move in the same direction." In other words, if one investment goes down, you don't want a second investment in your portfolio that is also likely to go down. One hallmark of a good portfolio is a low correlation between investments. The following data represent the annual rates of return for various stocks. If you only wish to invest in two of the stocks, which two would you select if your goal is to have low correlation between the two investments? Which two would you select if your goal is to have one stock go up when the other goes down?

| Year | Rate of Return | | | | |
	Cisco Systems	Walt Disney	General Electric	Exxon Mobil	TECO Energy
1996	0.704	0.204	0.565	0.405	−0.012
1997	0.314	0.448	0.587	0.342	0.223
1998	1.50	−0.080	0.451	0.254	0.050
1999	1.31	−0.015	0.574	0.151	−0.303
2000	−0.286	−0.004	−0.055	0.127	0.849
2001	−0.527	−0.277	−0.151	−0.066	−0.150
2002	−0.277	−0.203	−0.377	−0.089	−0.369
2003	0.850	0.444	0.308	0.206	0.004
2004	−0.203	0.202	0.207	0.281	0.128

Source: Yahoo!Finance

38. **Lyme Disease versus Drownings** Lyme disease is an inflammatory disease that results in skin rash and flulike symptoms. It is transmitted through the bite of an infected deer tick. The following data represent the number of reported cases of Lyme disease and the number of drowning deaths for a rural county in the United States.

Month	J	F	M	A	M	J	J	A	S	O	N	D
Cases of Lyme Disease	3	2	2	4	5	15	22	13	6	5	4	1
Drowning Deaths	0	1	2	1	2	9	16	5	3	3	1	0

(a) Draw a scatter diagram of the data using cases of Lyme disease as the explanatory variable.
(b) Compute the correlation coefficient for the data.
(c) Based on your results from parts (a) and (b), what type of relation appears to exist between the number of reported cases of Lyme disease and drowning deaths? Do you believe that an increase in cases of Lyme disease causes an increase in drowning deaths?

39. **Television Stations and Life Expectancy** Based on data obtained from the *CIA World Factbook*, the linear correlation coefficient between number of television stations in a country and life expectancy of residents of the country is 0.599. What does this correlation imply? Do you believe that the more television stations a country has, the longer its population can expect to live? Why or why not?

40. **Caffeine and SIDS** A study on the relationship between caffeine consumption during pregnancy and sudden infant death syndrome (SIDS) showed that heavy caffeine consumption during pregnancy was associated with a signifi-

cant risk of SIDS. The study was later criticized on the claim that parental smoking was not properly assessed. Explain why this might be a concern.

41. **Influential** Consider the following set of data:

x	2.2	3.7	3.9	4.1	2.6	4.1	2.9	4.7
y	3.9	4.0	1.4	2.8	1.5	3.3	3.6	4.9

(a) Draw a scatter diagram of the data and compute the linear correlation coefficient.
(b) Draw a scatter diagram of the data and compute the linear correlation coefficient with the additional data point (10.4, 9.3). Comment on the effect the additional data point has on the linear correlation coefficient. Explain why correlations should always be reported with scatter diagrams.

42. **Faulty Use of Correlation** On the basis of the accompanying scatter diagram, explain what is wrong with the following statement: "Because the linear correlation coefficient between age and median income is 0.012, there is no relation between age and median income."

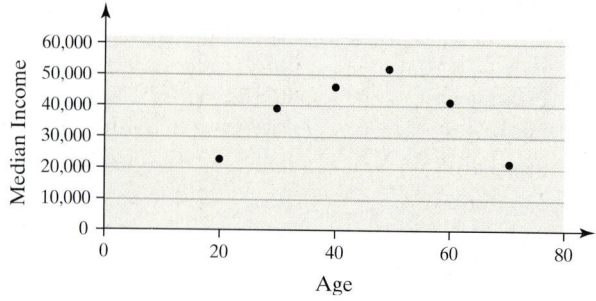

43. Name the Relation, Part I For each of the following statements, explain whether you think the variables will have positive correlation, negative correlation, or no correlation. Support your opinion.
(a) Number of children in the household under the age of 3 and expenditures on diapers
(b) Interest rates on car loans and number of cars sold
(c) Number of hours per week on the treadmill and cholesterol level
(d) Price of a Big Mac and number of McDonald's french fries sold in a week
(e) Shoe size and IQ

44. Name the Relation, Part II For each of the following statements, explain whether you think the variables will have positive correlation, negative correlation, or no correlation. Support your opinion.
(a) Number of cigarettes smoked by a pregnant woman each week and birth weight of her baby
(b) Annual salary and years of education
(c) Number of doctors on staff at a hospital and number of administrators on staff.
(d) Head circumference and IQ.
(e) Number of movie goers and movie ticket price

45. Transformations Consider the following data set:

x	5	6	7	7	8	8	8	8
y	4.2	5	5.2	5.9	6	6.2	6.1	6.9
x	9	9	10	10	11	11	12	12
y	7.2	8	8.3	7.4	8.4	7.8	8.5	9.5

(a) Draw a scatter diagram with the x-axis starting at 0 and ending at 30 and with the y-axis starting at 0 and ending at 20.
(b) Compute the linear correlation coefficient.
(c) Now multiply both x and y by 2.
(d) Draw a scatter diagram of the new data with the x-axis starting at 0 and ending at 30 and with the y-axis starting at 0 and ending at 20. Compare the scatter diagrams.
(e) Compute the linear correlation coefficient.
(f) Conclude that multiplying each value in the data set does not affect the correlation between the variables. Explain why this is the case.

46. Obesity In a study published in the *Journal of the American Medical Association* (May 16, 2001), researchers found that breast-feeding may help to prevent obesity in kids. In an interview, the head investigator stated. "It's not clear whether breast milk has obesity-preventing properties or the women who are breast-feeding are less likely to have fat kids because they are less likely to be fat themselves and may be more health conscious." Using this researcher's statement, explain what might be wrong with the conclusion that breast-feeding prevents obesity. Identify some lurking variables in the study.

47. How Well Will You Do in College? The College Board is a membership association composed of schools, colleges, universities, and other educational organizations. One of its better known programs is the administration of the SAT college entrance exam. In a recent study, the College Board wanted to learn what the best predictor of college grade-point average (GPA) was. The following correlations were obtained based on 48,039 students.

Correlation between College GPA and:	Correlation
SAT score combined with high school GPA	0.61
SAT verbal score	0.47
SAT math score	0.48
SAT combined verbal and math score	0.52
High school GPA	0.54

Source: The College Board

(a) Which variable is the best predictor of college GPA?
(b) Which variable is the worst predictor of college GPA?

48. Correlation Applet Load the correlation by eye applet.
(a) In the lower-left corner of the applet, add 10 points that line up with a positive slope so that the linear correlation between the points is about 0.8. Click "show r" to show the correlation.
(b) Add another point in the upper-right corner of the applet that roughly lines up with the 10 points you have in the lower-left corner. Comment on how the linear correlation coefficient changes.
(c) Drag the point in the upper-right corner straight down. Take note of the change in the linear correlation coefficient. Notice how a single point can have a substantial impact on the linear correlation coefficient.

49. Correlation Applet Load the correlation by eye applet. Add about 10 points that form an upside-down U. Certainly, there is a relation between x and y, but what is the value of the linear correlation coefficient? Conclude that a low linear correlation coefficient does not imply there is no relation between two variables; it means there may be no linear relation between two variables.

50. Correlation Applet Load the correlation by eye applet.
(a) Plot about 10 points that follow a linear trend and have a linear correlation coefficient that is close to 0.8.
(b) Clear the applet. Plot about 6 points vertically on top of each other on the left side of the applet. Add a seventh point to the right of the applet. Move the point until the linear correlation coefficient is close to 0.8.
(c) Clear the applet. Plot about 7 points in a U-shaped curve. Add an eighth point and move it around the applet until the linear correlation coefficient is close to 0.8.
(d) Conclude that a linear correlation coefficient can result from data that have many patterns and so you should always plot your data.

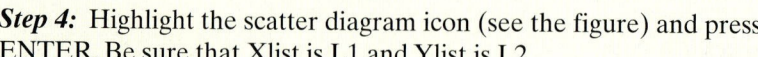

Technology Step-by-Step	**Drawing Scatter Diagrams and Determining the Correlation Coefficient**

TI-83/84 Plus   **Scatter Diagrams**

Step 1: Enter the explanatory variable in L1 and the response variable in L2.

Step 2: Press 2nd Y = to bring up the StatPlot menu. Select 1: Plot1.

Step 3: Turn Plot 1 ON by putting the cursor on the ON button and pressing ENTER.

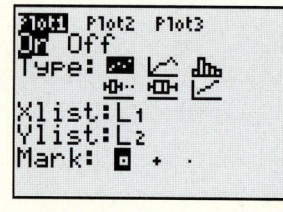

Step 4: Highlight the scatter diagram icon (see the figure) and press ENTER. Be sure that Xlist is L1 and Ylist is L2.

Step 5: Press ZOOM and select 9: ZoomStat.

Correlation Coefficient

Step 1: Turn the diagnostics on by selecting the catalog (2nd Ø). Scroll down and select DiagnosticOn. Hit ENTER to activate diagnostics.

Step 2: With the explanatory variable in L1 and the response variable in L2, press STAT, highlight CALC and select 4: LinReg (ax + b). With LinReg on the HOME screen press ENTER.

MINITAB   **Scatter Diagrams**

Step 1: Enter the explanatory variable in C1 and the response variable in C2. You may want to name the variables.

Step 2: Select the **Graph** menu and highlight **Plot** . . . .

Step 3: With the cursor in the Y column, select the response variable. With the cursor in the X column, select the explanatory variable. Click OK.

Correlation Coefficient

Step 1: With the explanatory variable in C1 and the response variable in C2, select the **Stat** menu and highlight **Basic Statistics**. Highlight **Correlation**.

Step 2: Select the variables whose correlation you wish to determine and click OK.

Excel   **Scatter Diagrams**

Step 1: Enter the explanatory variable in column A and the response variable in column B. Select the Chart Wizard icon.

Step 2: Follow the instructions in the Chart Wizard.

Correlation Coefficient

Step 1: Be sure the Data Analysis Tool Pak is activated by selecting the **Tools** menu and highlighting **Add-Ins** . . . . Check the box for the Analysis ToolPak and select OK.

Step 2: Select **Tools** and highlight **Data Analysis** . . . . Highlight **Correlation** and select OK.

Step 3: With the cursor in the Input Range, highlight the data. Select OK.

4.2 Least-Squares Regression

Preparing for This Section   Before getting started, review the following:

• Lines (Appendix B on CD, pp. B-1–B-5)

Objectives

1 **Find the least-squares regression line and use the line to make predictions**

2 **Interpret the slope and the *y*-intercept of the least-squares regression line**

3 **Compute the sum of squared residuals**

Once the scatter diagram and linear correlation coefficient indicate that a linear relation exists between two variables, we proceed to find a linear equation that describes the relation between the two variables. One way to obtain a line that describes the relation is to select two points from the data that appear to provide a good fit and find the equation of the line through these points.

| EXAMPLE 1 | **Finding an Equation That Describes Linearly Related Data** |

Problem: The data in Table 3 represent the club-head speed and the distance a golf ball travels for eight swings of the club. We determined that these data are linearly related in the last section.

! CAUTION

The method for obtaining an equation that describes the relation between two variables discussed in Example 1 is *not* the least-squares method. It is used to illustrate a point.

(a) Find a linear equation that relates club-head speed, x (the explanatory variable), and distance, y (the response variable), by selecting two points and finding the equation of the line containing the points.

(b) Graph the line on the scatter diagram.

(c) Use the equation to predict the distance a golf ball will travel if the club-head speed is 104 miles per hour.

Table 3

Club-Head Speed (mph) x	Distance (yards) y	(x, y)
100	257	(100, 257)
102	264	(102, 264)
103	274	(103, 274)
101	266	(101, 266)
105	277	(105, 277)
100	263	(100, 263)
99	258	(99, 258)
105	275	(105, 275)

Source: Paul Stephenson, student at Joliet Junior College

Approach:

(a) To answer part (a), we perform the following steps:

Step 1: Select two points from Table 3 so that a line drawn through the points appears to give a good fit. Call the points (x_1, y_1) and (x_2, y_2). Refer to Figure 1 for the scatter diagram.

In Other Words

A good fit means that the line drawn appears to describe the relation between the two variables well.

Step 2: Find the slope of the line containing these two points using $m = \dfrac{y_2 - y_1}{x_2 - x_1}$.

Step 3: Use the point–slope formula, $y - y_1 = m(x - x_1)$, to find the line through the points selected in Step 1. Express the line in the form $y = mx + b$, where m is the slope and b is the y-intercept.

(b) For part (b), draw a line through the points selected in Step 1 of part (a).

(c) Finally, for part (c), we let $x = 104$ in the equation found in part (a).

Solution

(a) **Step 1:** We will select $(x_1, y_1) = (99, 258)$ and $(x_2, y_2) = (105, 275)$, because a line drawn through these two points seems to give a good fit.

Step 2: $m = \dfrac{y_2 - y_1}{x_2 - x_1} = \dfrac{275 - 258}{105 - 99} = \dfrac{17}{6} = 2.8333$

Step 3: We use the point–slope form of a line to find the equation of the line.

$$y - y_1 = m(x - x_1)$$

$$y - 258 = 2.8333(x - 99) \qquad m = 2.8333, x_1 = 99, y_1 = 258$$

$$y - 258 = 2.8333x - 280.4967$$

$$y = 2.8333 - 22.4967 \tag{1}$$

The slope of the line is 2.8333 and the y-intercept is −22.4967.

(b) Figure 8 shows the scatter diagram along with the line drawn through the points $(99, 258)$ and $(105, 275)$.

Figure 8

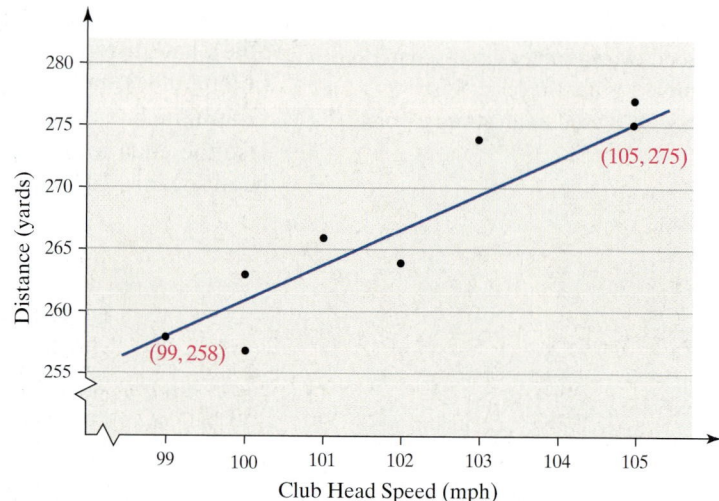

(c) We let $x = 104$ in equation (1) to predict the distance a golf ball travels when it is hit with a club-head speed of 104 miles per hour.

$$y = 2.8333(104) - 22.4967$$

$$= 272.2 \text{ yards}$$

We predict that a golf ball will travel 272.2 yards when it is hit with a club-head speed of 104 miles per hour.

Now Work Problems 11(a), 11(b), and 11(c).

1 ## Find the Least-Squares Regression Line and Use the Line to Make Predictions

Although the line that we found in Example 1 appears to describe the relation between club-head speed and distance well, is there a line that fits the data better? Is there a line that fits the data *best*?

Consider Figure 9. Each y-coordinate on the line corresponds to a predicted distance for a given club-head speed. For example, if club-head speed is 103 miles per hour, the predicted distance is $2.8333(103) - 22.4967 = 269.3$ yards. The observed distance for this club-head is speed is 274 yards. The difference between the observed value of y and the predicted value of y is the error or **residual**. For a swing speed of 103 mph the residual is

$$\text{Residual} = \text{observed } y - \text{predicted } y$$

$$= 274 - 269.3$$

$$= 4.7 \text{ yards}$$

The residual for a club-head speed of 103 miles per hour is labeled in Figure 9.

Figure 9

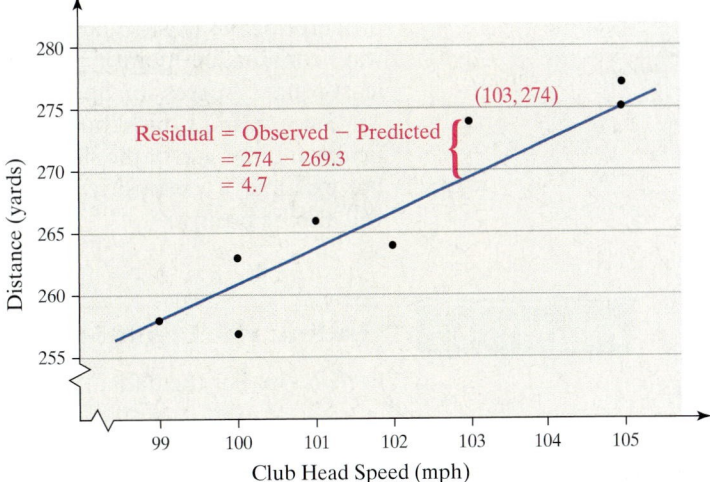

The line that *best* describes the relation between two variables is the one that minimizes the distance between the points and the line. The most popular technique for making the residuals as small as possible is the *method of least squares*, discovered by Adrien Marie Legendre.

Definition

Historical Note

Adrien Marie Legendre was born on September 18, 1752, into a wealthy family and was educated in mathematics and physics at the College Mazarin in Paris. From 1775 to 1780, he taught at Ecole Militaire. On March 30, 1783, Legendre was appointed an adjoint in the Académie des Sciences. On May 13, 1791, he became a member of the committee of the Académie des Sciences and was charged with the task of standardizing weights and measures. The committee worked to compute the length of the meter. During the revolution, Legendre lost his small fortune. In 1794, Legendre published *Eléments de géométrie*, which was the leading elementary text in geometry for around 100 years. In 1806, Legendre published a book on orbits, in which he developed the theory of least squares. He died on January 10, 1833.

Least-Squares Regression Criterion

The **least-squares regression line** is the one that minimizes the sum of the squared errors (or residuals). It is the line that minimizes the square of the vertical distance between the observed values of y and those predicted by the line, $\hat{y}$ (read "*y*-hat"). We represent this as

$$\text{Minimize } \sum \text{residuals}^2$$

The advantage of the least-squares criterion is that it allows for statistical inference on the predicted value and slope (Chapter 14). Another advantage of the least-squares criterion is explained by Legendre in his text *Nouvelles méthodes pour la determination des orbites des cométes*, published in 1806.

> Of all the principles that can be proposed for this purpose, I think there is none more general, more exact, or easier to apply, than that which we have used in this work; it consists of making the sum of squares of the errors a *minimum*. By this method, a kind of equilibrium is established among the errors which, since it prevents the extremes from dominating, is appropriate for revealing the state of the system which most nearly approaches the truth.

Equation of the Least-Squares Regression Line

The equation of the least-squares regression line is given by

$$\hat{y} = b_1 x + b_0$$

where

$$b_1 = r \cdot \frac{s_y}{s_x} \text{ is the \textbf{slope} of the least-squares regression line*} \qquad (2)$$

and

$$b_0 = \overline{y} - b_1 \overline{x} \text{ is the \textbf{y-intercept} of the least-squares regression line} \quad (3)$$

Note: $\overline{x}$ is the sample mean and s_x is the sample standard deviation of the explanatory variable x; $\overline{y}$ is the sample mean and s_y is the sample standard deviation of the response variable y.

*An equivalent formula is

$$b_1 = \frac{S_{xy}}{S_{xx}} = \frac{\sum x_i y_i - \dfrac{(\sum x_i)(\sum y_i)}{n}}{\sum x_i^2 - \dfrac{(\sum x_i)^2}{n}}$$

The notation $\hat{y}$ is used in the least-squares regression line to serve as a reminder that it is a predicted value of y for a given value of x. An interesting property of the least-squares regression line, $\hat{y} = b_1 x + b_0$, is that the line always contains the point $(\bar{x}, \bar{y})$. This property can be useful when drawing the least-squares regression line by hand.

Since s_y and s_x must both be positive, the sign of the linear correlation coefficient and the sign of the slope of the least-squares regression line are the same. For example, if r is positive, then the slope of the least-squares regression line will also be positive.

EXAMPLE 2

Historical Note

Sir Francis Galton was born on February 16, 1822. Galton came from a wealthy and well-known family. Charles Darwin was his first cousin. Galton studied medicine at Cambridge. After receiving a large inheritance, he left the medical field and traveled the world. He explored Africa from 1850 to 1852. In the 1860s, his study of meteorology led him to discover anticyclones. Influenced by Darwin, Galton always had an interest in genetics and heredity. He studied heredity through experiments with sweet peas. He noticed that the weight of the "children" of the "parent" peas reverted or *regressed* to the mean weight of all peas. Hence, the term *regression analysis*. Galton died January 17, 1911.

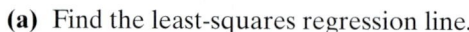

Finding the Least-Squares Regression Line

Problem: For the data in Table 3 on page 213,

(a) Find the least-squares regression line.

(b) Predict the distance a golf ball will travel when hit with a club-head speed of 103 miles per hour.

(c) Compute the residual for the prediction made in part (b).

(d) Draw the least-squares regression line on the scatter diagram of the data.

Approach

(a) From Example 2 in Section 4.1, we have the following:
$r = 0.9387, \bar{x} = 101.875, s_x = 2.2952, \bar{y} = 266.75,$ and $s_y = 7.74135.$
We substitute these values into Formula (2) to find the slope of the least-squares regression line. We use Formula (3) to find the intercept of the least-squares regression line.

(b) Substitute $x = 103$ into the least-squares regression line found in part (a) to find $\hat{y}$.

(c) The residual is the difference between the observed y and the predicted y. That is, residual $= y - \hat{y}$.

(d) To draw the least-squares regression line, select two values of x and use the equation to find the predicted values of y. Plot these points on the scatter diagram and draw a line through the points.

Solution

(a) Substituting $r = 0.9387, s_x = 2.2952,$ and $s_y = 7.74135$ into Formula (2), we obtain

$$b_1 = r \cdot \frac{s_y}{s_x} = 0.9387 \cdot \frac{7.74135}{2.2952} = 3.1661$$

We have that $\bar{x} = 101.875$ and $\bar{y} = 266.75$. Substituting these values into Formula (3), we obtain

$$b_0 = \bar{y} - b_1 \bar{x} = 266.75 - 3.1661(101.875) = -55.7964$$

The least-squares regression line is

$$\hat{y} = 3.1661x - 55.7964$$

(b) We let $x = 103$ in the equation $y = 3.1661x - 55.7964$ to predict the distance a golf ball hit with a club-head speed of 103 miles per hour will travel.

$$\hat{y} = 3.1661(103) - 55.7964$$

$$= 270.3 \text{ yards}$$

We predict that the distance the ball will travel is 270.3 yards.

CAUTION

Throughout the text, we will round the slope and y-intercept values to four decimal places. The predictions should be made to one more decimal place than the response variable.

(c) The actual distance the ball traveled is 274 yards. The residual is

$$\text{Residual} = \text{observed } y - \text{predicted } y$$

$$= y - \hat{y}$$

$$= 274 - 270.3$$

$$= 3.7 \text{ yards}$$

We underestimated the distance by 3.7 yards.

(d) Figure 10 shows the graph of the least-squares regression line drawn on the scatter diagram with the residual labeled.

Figure 10

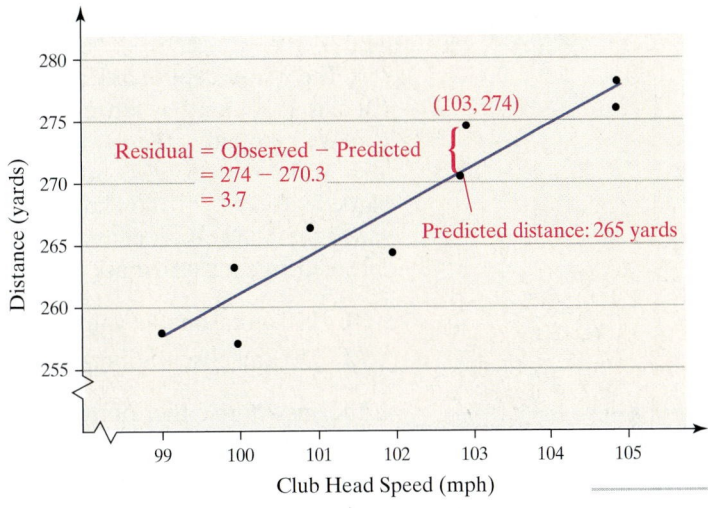

Notice that an underestimate results in a positive residual, while an overestimate results in a negative residual.

In Other Words

An underestimate means the residual is positive; an overestimate means the residual is negative. A residual of zero means the prediction is right on!

In-Class Activity: Paper Thin (Regression)

Each student, or small group of students, should receive a ruler with both inches and centimeters.

1. Use the ruler provided to measure the thickness of only one page of the text. Which unit of measurement did you use and why?

2. Grouping pages together (one page is one sheet), complete the following table:

No. of pages	25	50	75	150	200	225
Thickness						

3. Compute the least-squares regression line for your data.

4. Compare your data and your regression line to those around you. Did everyone get the same measurements and model? Explain why or why not.

5. Use your model to estimate the thickness of one page of the text. Is the value you obtained reasonable?

6. Use your model to estimate the thickness of 0 pages of the text. Is the value you obtained reasonable?

7. What, if anything, could you do to improve the model?

$x = 1.$

We can think of any point on the least-squares regression line as an estimate of the mean value of the response variable for a given value of the explanatory variable. For example, the mean distance a golf ball will travel when hit with a club-head speed of 103 miles per hour is 270.3 yards. In our experiment, when we hit the ball with a club-head speed of 103 miles per hour, the ball traveled 274 yards. So the distance the ball traveled was above the mean. Perhaps we hit the ball in the "sweet spot" of the club face or a breeze kicked up at our back.

② Interpret the Slope and *y*-Intercept of the Least-Squares Regression Line

The definition of the slope of a line is $\dfrac{\text{Rise}}{\text{Run}}$ or $\dfrac{\text{Change in } y}{\text{Change in } x}$. For a line whose slope is $\dfrac{2}{3}$, if x increases by 3, y will *increase* by 2. Or, if the slope of a line is $-4 = \dfrac{-4}{1}$, if x increases by 1, y will *decrease* by 4.

The *y*-intercept of any line is the point where the graph intersects the vertical axis. It is found by letting $x = 0$ in an equation and solving for y.

We found the regression equation in Example 2 to be $\hat{y} = 3.1661x - 55.7964$. So the slope of the line is 3.1661. We interpret this slope as follows: If the club-head speed increases by 1 mile per hour, the distance the ball travels increases by 3.1661 yards, on average. To interpret the *y*-intercept, we must first ask two questions:

1. Is 0 a reasonable value for the explanatory variable reasonable?
2. Do any observations near $x = 0$ exist in the data set?

CAUTION
Be careful when using the least-squares regression line to make predictions for values of the explanatory variable that are much larger or much smaller than those observed.

If the answer to either of these questions is no, we do not give an interpretation to the *y*-intercept. In the regression equation of Example 2, a swing speed of 0 miles per hour does not make sense, so an interpretation of the *y*-intercept is unreasonable. To interpret a *y*-intercept, we would say that it is the value of the response variable when the value of the explanatory variable is 0.

The second condition for interpreting the *y*-intercept is especially important because we should not use the regression model to make predictions **outside the scope of the model**. If this cannot be avoided, be cautious when using the regression model to make predictions for values of the explanatory variable that are much larger or much smaller than those observed, because we cannot be certain of the behavior of data for which we have no observations.

For example, it is inappropriate to use the line we determined in Example 2 to predict distance when club-head speed is 140 miles per hour. The highest observed club-head speed in our data set is 105 miles per hour. We cannot be certain that the linear relation between distance and club-head speed will continue. See Figure 11.

Figure 11

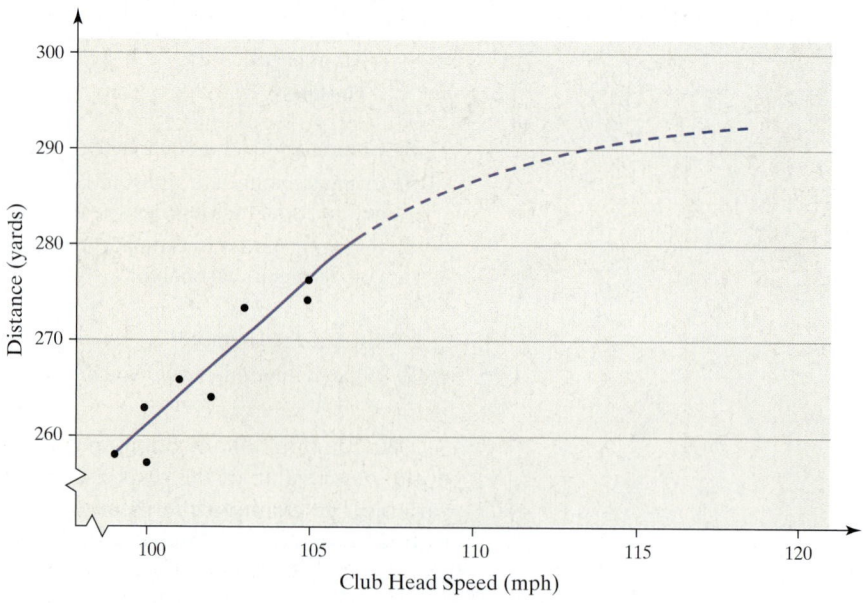

We have presented the procedure for determining the least-squares regression line by hand. In practice, however, a statistical spreadsheet or calculator with advanced statistical features is used to determine the least-squares regression line.

Finding the Least-Squares Regression Line Using Technology

Problem: Use a statistical spreadsheet or a graphing calculator with advanced statistical features to find the least-squares regression line of the data in Table 3.

Approach: Because technology plays a major role in obtaining the least-squares regression line, we will use a TI-84 Plus graphing calculator, MINITAB, and Excel to obtain the least-squares regression line. The steps for obtaining these lines are given in the Technology Step by Step on page 225.

Result: Figure 12(a) shows the output obtained from a TI-84 Plus graphing calculator, Figure 12(b) shows the output obtained from MINITAB with the slope and y-intercept highlighted, and Figure 12(c) shows partial output from Excel with the slope and y-intercept highlighted.

Figure 12

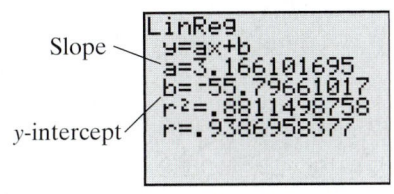

Slope

y-intercept

```
LinReg
y=ax+b
a=3.166101695
b=-55.79661017
r²=.8811498758
r=.9386958377
```

(a) TI-84 Plus output

```
The regression equation is
Distance (yards) = -55.8 + 3.17 Club Head Speed (mph)
Predictor                   Coef    SE Coef        T        P
Constant                  -55.80      48.37    -1.15    0.293
Club Head Speed (mph)     3.1661     0.4747     6.67    0.001

S = 2.88264      R - Sq = 88.1%    R - Sq(adj) = 86.1%
```

(b) MINITAB output

	Coefficients	Standard Error	t Stat	P-value
Intercept	−55.79661017	48.37134953	−1.153505344	0.29257431
Club Head Speed (mph)	3.166101695	0.47470539	6.669613957	0.00054983

(c) Excel output

Note: To get the linear correlation coefficient from MINITAB, use
$$r = \sqrt{\text{R} - \text{Sq}}. \text{ So } r = \sqrt{0.881} = 0.9386.$$

Now Work Problems 11(d) and 11(e).

③ Compute the Sum of Squared Residuals

Recall that the least-squares regression line is the line that minimizes the sum of the squared residuals. This means that the sum of the squared residuals, $\Sigma \text{ residuals}^2$, for the least-squares line will be smaller than for any other line that may describe the relation between the two variables. In particular, the sum of the squared residuals for the line obtained in Example 2 using the method of least squares will be smaller than the sum of the squared residuals for the line obtained in Example 1. It is worthwhile to verify this result.

EXAMPLE 4 **Comparing the Sum of Squared Residuals**

Problem: Compare the sum of squared residuals for the lines obtained in Examples 1 and 2.

Approach: We compute Σresiduals^2 using the predicted values of y, $\hat{y}$, for the lines obtained in Examples 1 and 2. This is best done by creating a table of values.

Solution: We create Table 4, which contains the value of the explanatory variable in column 1. Column 2 contains the corresponding response variable. Column 3 contains the predicted value using the equation obtained in Example 1, $\hat{y} = 2.8333x - 22.4967$. In column 4, we compute the residuals for each observation: residual = observed y − predicted y. For example, the first residual using the equation found in Example 1 is observed y − predicted y = 257 − 260.8 = −3.8. Column 5 contains the squares of the residuals obtained in column 4. Column 6 contains the predicted value using the least-squares regression equation obtained in Example 2: $\hat{y} = 3.1661x - 55.7964$. Column 7 represents the residuals for each observation and column 8 represents the squared residuals.

Table 4

Club-Head Speed (mph)	Distance (yards)	Example 1 ($\hat{y} = 2.8333x - 22.4967$)	Residual	Residual^2	Example 2 ($\hat{y} = 3.1661x - 55.7964$)	Residual	Residual^2
100	257	260.8	−3.8	14.44	260.8	−3.8	14.44
102	264	266.5	−2.5	6.25	267.2	−3.2	10.24
103	274	269.3	4.7	22.09	270.3	3.7	13.69
101	266	263.7	2.3	5.29	264.0	2.0	4.00
105	277	275.0	2.0	4.00	276.6	0.4	0.16
100	263	260.8	2.2	4.84	260.8	2.2	4.84
99	258	258.0	0.0	0.00	257.6	0.4	0.16
105	275	275.0	0.0	0.00	276.6	−1.6	2.56
				$\sum$ residual^2 = 56.91			$\sum$ residual^2 = 50.09

The sum of the squared residuals for the line found in Example 1 is 56.91; the sum of the squared residuals for the least-squares regression line is 50.09. Again, any line that describes the relation between distance and club-head speed will have a sum of squared residuals that is greater than 50.09.

Now Work Problems 11(f), (g) and (h)

We draw the graphs of the two lines obtained in Examples 1 and 2 on the same scatter diagram in Figure 13 to help the reader visualize the difference.

Figure 13

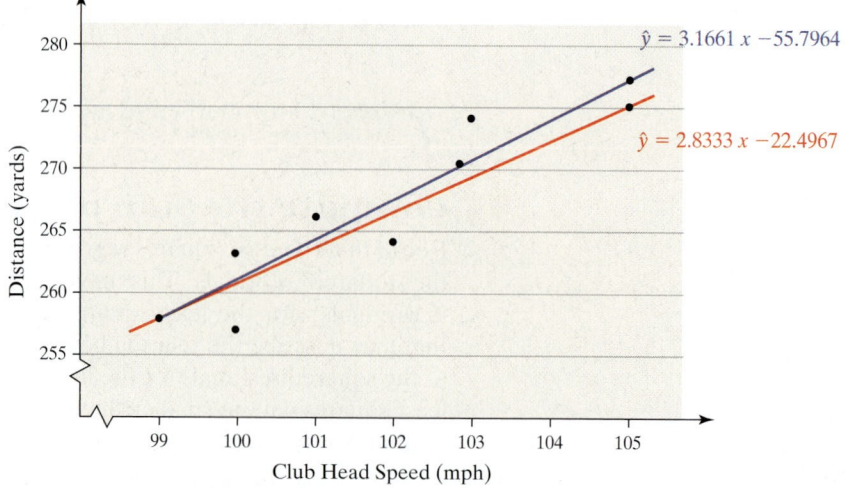

MAKING AN INFORMED DECISION

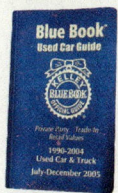

What Car Should I Buy?

You are still in the market to buy a car. As we all know, cars lose value over time. Therefore, another item to consider when purchasing a car is its depreciation rate. The higher the depreciation rate, the more value the car loses each year. Using the same three cars that you used in the Chapter 3 Decision, answer the following questions to determine depreciation rate.

1. Collect information regarding the three cars in your list by finding at least 12 cars of each car model that are for sale. Obtain the asking price and age of the car. Sources of data include your local newspaper classified ads or car Web sites, such as www.cars.com and www.vehix.com.

2. For each car type, draw a scatter diagram, treating age of the car as the explanatory variable and asking price as the response variable. Does the relation between the two variables appear linear?

3. The asking price of a car and the age of a car are related through the exponential equation $y = ab^x$, where y is the price of the car and x is the age of the car. The depreciation rate is $1 - b$. To estimate the values of a and b using least-squares regres-

sion, we need to transform the equation to a linear equation. This is accomplished by computing the logarithm of the asking price. For example, if the asking price of a car is $8000, compute $\log(8000) = 3.9031$. Compute the logarithm of each asking price for each car.

4. For each car model, draw a scatter diagram, treating age of the car as the explanatory variable and the logarithm of asking price as the response variable. Does the relation between age and the logarithm of asking price appear to be linear?

5. For each car model, find the least-squares regression line, treating age of the car as the explanatory variable and the logarithm of asking price as the response variable. The line will be $Y = \log a + (\log b)x = A + Bx$.

6. To determine a and b for the exponential equation $y = ab^x$, let $a = 10^A$ and $b = 10^B$. Graph the exponential equation on each scatter diagram from question 2.

7. For each exponential equation determined in question 6, the depreciation rate is $1 - b$. What are the depreciation rates for the three cars considered? Will this result affect your decision about which car to buy?

4.2 ASSESS YOUR UNDERSTANDING

Concepts and Vocabulary

1. Explain the least-squares regression criterion.
2. What is a residual? What does it mean when a residual is positive?
3. Explain the phrase *outside the scope of the model*. Why is it dangerous to make predictions outside the scope of the model?
4. If the linear correlation between two variables is negative, what can be said about the slope of the regression line?

5. In your own words, explain the meaning of Legendre's quote given on page 215.
6. *True or False*: The least-squares regression line always travels through the point $(\bar{x}, \bar{y})$.
7. In your own words explain what each point on the least-squares regression line represents.
8. If the linear correlation coefficient is 0, what is the equation of the least-squares regression line?

Skill Building

9. For the data set

x	0	2	3	5	6	6
y	5.8	5.7	5.2	2.8	1.9	2.2

(a) Draw a scatter diagram. Comment on the type of relation that appears to exist between x and y.
(b) Given that $\bar{x} = 3.667$, $s_x = 2.42212$, $\bar{y} = 3.933$, $s_y = 1.8239152$, and $r = -0.9476938$, determine the least-squares regression line.
(c) Graph the least-squares regression line on the scatter diagram drawn in part (a).

10. For the data set

x	2	4	8	8	9
y	1.4	1.8	2.1	2.3	2.6

(a) Draw a scatter diagram. Comment on the type of relation that appears to exist between x and y.
(b) Given $\bar{x} = 6.2$, $s_x = 3.03315$, $\bar{y} = 2.04$, that $s_y = 0.461519$, and $r = 0.957241$, determine the least-squares regression line.
(c) Graph the least-squares regression line on the scatter diagram drawn in part (a).

In Problems 11–16,
(a) Draw a scatter diagram treating x as the explanatory variable and y as the response variable.
(b) Select two points from the scatter diagram and find the equation of the line containing the points selected.
(c) Graph the line found in part (b) on the scatter diagram.
(d) Determine the least-squares regression line.
(e) Graph the least-squares regression line on the scatter diagram.
(f) Compute the sum of the squared residuals for the line found in part (b).
(g) Compute the sum of the squared residuals for the least-squares regression line found in part (d).
(h) Comment on the fit of the line found in part (b) versus the least-squares regression line found in part (d).

11.
NW

x	3	4	5	7	8
y	4	6	7	12	14

12.

x	3	5	7	9	11
y	0	2	3	6	9

13.

x	-2	-1	0	1	2
y	-4	0	1	4	5

14.

x	-2	-1	0	1	2
y	7	6	3	2	0

15.

x	20	30	40	50	60
y	100	95	91	83	70

16.

x	5	10	15	20	25
y	2	4	7	11	18

Applying the Concepts

Problems 17–22 use the results from Problems 23–28 in Section 4.1.

17. Height versus Head Circumference (Refer to Problem 23, Section 4.1) A pediatrician wants to determine the relation that exists between a child's height, x, and head circumference, y. She randomly selects 11 children from her practice, measures their height and head circumference, and obtains the following data.

Height, x (inches)	Head Circumference, y (inches)	Height, x (inches)	Head Circumference, y (inches)
27.75	17.5	26.5	17.3
24.5	17.1	27	17.5
25.5	17.1	26.75	17.3
26	17.3	26.75	17.5
25	16.9	27.5	17.5
27.75	17.6		

Source: Denise Slucki, student at Joliet Junior College

(a) Find the least-squares regression line, treating height as the explanatory variable and head circumference as the response variable.
(b) Interpret the slope and intercept, if appropriate.
(c) Use the regression equation to predict the head circumference of a child who is 25 inches tall.
(d) Compute the residual based on the observed head circumference of the 25-inch-tall child in the table. Is the head circumference of this child above average or below average?
(e) Draw the least-squares regression line on the scatter diagram of the data and label the residual from part (d).
(f) Notice that two children are 26.75 inches tall. One has a head circumference of 17.3 inches; the other has a head circumference of 17.5 inches. How can this be?
(g) Would it be reasonable to use the least-squares regression line to predict the head circumference of a child who was 32 inches tall? Why?

18. Gestation Period versus Life Expectancy (Refer to Problem 24, Section 4.1) The following data represent the gestation period , x, of various animals along with their life expectancy, y.

Animal	Gestation (or Incubation) Period (days), x	Life Expectancy (years), y
Cat	63	11
Chicken	22	7.5
Dog	63	11
Duck	28	10
Goat	151	12
Lion	108	10
Parakeet	18	8
Pig	115	10
Rabbit	31	7
Squirrel	44	9

Source: Time Almanac 2000

(a) Find the least-squares regression line, treating gestation period as the explanatory variable and life expectancy as the response variable.
(b) Interpret the slope and intercept, if appropriate.
(c) Suppose a new animal species has been discovered. After breeding the species in captivity, it is determined that the gestation period is 95 days. Use the least-squares regression line to predict the life expectancy of the animal.
(d) Use the regression equation to predict the life expectancy of a parakeet.
(e) Use the regression equation to predict the life expectancy of a rabbit.
(f) Compute the residual of the prediction made in part (e). Conclude that the least-squares regression

line sometimes provides accurate predictions (as in the case of the parakeet) and sometimes provides inaccurate predictions (as in the case of the rabbit). Unfortunately, when the value of the response variable is unknown, as in the case of the new animal species from part (c), we don't know the accuracy of the prediction.

19. **Weight of a Car versus Miles per Gallon** (Refer to Problem 25, Section 4.1) An engineer wants to determine how the weight of a car, y, affects gas mileage, y. The following data represent the weight of various domestic cars and their miles per gallon in the city for the 2005 model year.

Car	Weight (pounds), x	Miles per Gallon, y
Buick LeSabre	3565	20
Cadillac DeVille	3985	18
Chevrolet Corvette	3180	19
Chevrolet Monte Carlo	3340	21
Chrysler PT Cruiser	3100	21
Chrysler Sebring Sedan	3175	22
Dodge Neon	2580	27
Dodge Stratus Sedan	3175	22
Ford Focus	2655	26
Ford Mustang	3300	20
Lincoln LS	3680	20
Mercury Sable	3310	19
Pontiac Bonneville	3590	20
Pontiac Grand Am	3475	20
Pontiac Sunfire	2770	24
Saturn Ion	2690	26

Source: www.roadandtrack.com

(a) Find the least-squares regression line treating weight as the explanatory variable and miles per gallon as the response variable.
(b) Interpret the slope and intercept, if appropriate.
(c) Predict the miles per gallon of a Ford Mustang and compute the residual. Is the miles per gallon of a Mustang above average or below average for cars of this weight?
(d) Draw the least-squares regression line on the scatter diagram of the data and label the residual.
(e) Would it be reasonable to use the least-squares regression line to predict the miles per gallon of a Toyota Prius, a hybrid gas and electric car? Why or why not?

20. **Bone Length** (Refer to Problem 26, Section 4.1) Research performed at NASA and led by Emily R. Morey-Holton measured the lengths of the right humerus and right tibia in 11 rats that were sent to space on Space-lab Life Sciences 2. The following data were collected.

Right Humerus (mm)	Right Tibia (mm)	Right Humerus (mm)	Right Tibia (mm)
24.80	36.05	25.90	37.38
24.59	35.57	26.11	37.96
24.59	35.57	26.63	37.46
24.29	34.58	26.31	37.75
23.81	34.20	26.84	38.50
24.87	34.73		

Source: NASA Life Sciences Data Archive

(a) Find the least-squares regression line, treating the length of the right humerus, x, as the explanatory variable and the length of the right tibia, y, as the response variable.
(b) Interpret the slope and intercept, if appropriate.
(c) Determine the residual if the length of the right humerus is 26.11 mm and the actual length of the right tibia is 37.96 mm. Is the length of this tibia above or below average?
(d) Draw the least-squares regression line on the scatter diagram and label the residual from part (c).
(e) Suppose one of the rats sent to space experienced a broken right tibia due to a severe landing. The length of the right humerus is determined to be 25.31 mm. Use the least-squares regression line to estimate the length of the right tibia.

21. **Attending Class** (Refer to Problem 27, Section 4.1) The following data represent the number of days absent, x, and the final grade, y, for a sample of college students in a general education course at a large state university.

No. of absences, x	0	1	2	3	4	5	6	7	8	9
Final grade, y	89.2	86.4	83.5	81.1	78.2	73.9	64.3	71.8	65.5	66.2

Source: College Teaching, Winter 2005, Vol. 53, Issue 1.

(a) Find the least-squares regression line, treating number of absences as the explanatory variable and final grade as the response variable.
(b) Interpret the slope and intercept, if appropriate.
(c) Predict the final grade for a student who misses five class periods and compute the residual. Is the final grade above or below average for this number of absences?
(d) Draw the least-squares regression line on the scatter diagram of the data and label the residual.
(e) Would it be reasonable to use the least-squares regression line to predict the final grade for a student who has missed 15 class periods? Why or why not?

22. **Antibiotics** (Refer to Problem 28, Section 4.1) A study on antibiotic use among children in Manitoba, Canada, gave the following data for the number of prescriptions per 1000 children x years after 1995.

Year, x	0	1	2	3	4	5	6
Prescriptions (per 1000 children), y	1201	1070	944	964	909	949	864

Source: Canadian Medical Association Journal, Vol. 171, Issue 2.

(a) Find the least-squares regression line, treating year as the explanatory variable and prescriptions as the response variable.
(b) Interpret the slope and intercept, if appropriate.
(c) Predict the number of prescriptions per 1,000 children in Manitoba, Canada, in 2002 ($x = 7$).
(d) Draw the least-squares regression line on the scatter diagram of the data.
(e) Would it be reasonable to use the least-squares regression line to predict the number of prescriptions in Manitoba, Canada, in 2010? Why or why not?

23. **Does Size Matter?** Researchers wondered whether the size of a person's brain was related to the individual's mental capacity. They selected a sample of right-handed introductory psychology students who had SAT scores higher than 1350. The subjects were administered the Wechsler (1981) Adult Intelligence Scale-Revised to obtain their IQ scores. MRI scans, performed at the same facility, consisted of 18 horizontal MR images. The computer counted all pixels with nonzero gray scale in each of the 18 images, and the total count served as an index for brain size. The resulting data are presented in the table.

Gender	MRI Count, x	IQ, y	Gender	MRI Count, x	IQ, y
Female	816,932	133	Male	949,395	140
Female	951,545	137	Male	1,001,121	140
Female	991,305	138	Male	1,038,437	139
Female	833,868	132	Male	965,353	133
Female	856,472	140	Male	955,466	133
Female	852,244	132	Male	1,079,549	141
Female	790,619	135	Male	924,059	135
Female	866,662	130	Male	955,003	139
Female	857,782	133	Male	935,494	141
Female	948,066	133	Male	949,589	144

Source: Willerman, L., Schultz, R., Rutledge, J.N., and Bigler, E. (1991). "In Vivo Brain Size and Intelligence." Intelligence, 15, 223–228.

(a) Find the least-squares regression line treating MRI count as the explanatory variable and IQ as the response variable.
(b) What do you notice about the value of the slope? Why does this result seem reasonable based on the scatter diagram and linear correlation coefficient obtained in Problem 31 of Section 4.1?
(c) When there is no relation between the explanatory and response variable, we use the mean value of the response variable, $\bar{y}$, to predict. Predict the IQ of an individual whose MRI count is 1,000,000. Predict the IQ of an individual whose MRI count is 830,000.

24. **Male versus Female Drivers** The following data represent the number of licensed drivers in various age groups and the number of accidents within the age group by gender.

Age Group	Number of Male Licensed Drivers (000s)	Number of Crashes Involving a Male (000s)	Number of Female Licensed Drivers (000s)	Number of Crashes Involving a Female (000s)
16	816	244	764	178
17	1,198	233	1,115	175
18	1,342	243	1,212	164
19	1,454	229	1,333	145
20–24	7,866	951	7,394	618
25–29	9,356	899	8,946	595
30–34	10,121	875	9,871	571
35–39	10,521	901	10,439	566
40–44	9,776	692	9,752	455
45–49	8,754	667	8,710	390
50–54	6,840	390	6,763	247
55–59	5,341	290	5,258	165
60–64	4,565	218	4,486	133
65–69	4,234	191	4,231	121
70–74	3,604	167	3,749	104
75–79	2,563	118	2,716	77
80–84	1,400	61	1,516	45
≥85	767	34	767	20

Source: National Highway and Traffic Safety Institute

(a) Find the least-squares regression line for males, treating number of licensed drivers as the explanatory variable, x, and number of crashes, y, as the response variable. Repeat this procedure for females.

(b) Interpret the slope of the least-squares regression line for each gender, if appropriate. How might an insurance company use this information?

(c) Predict the number of accidents for males if there were 8700 thousand licensed drivers. Predict the number of accidents for females if there were 8700 thousand licensed drivers.

25. Mark Twain, in his book *Life on the Mississippi* (1884), makes the following observation:

Therefore, the Mississippi between Cairo and New Orleans was twelve hundred and fifteen miles long one hundred and seventy-six years ago. It was eleven hundred and eighty after the cut-off of 1722. It was one thousand and forty after the American Bend cut-off. It has lost sixty-seven miles since. Consequently its length is only nine hundred and seventy-three miles at present.

Now, if I wanted to be one of those ponderous scientific people, and "let on" to prove what had occurred in the remote past by what had occurred in a given time in the recent past, or what will occur in the far future by what has occurred in late years, what an opportunity is here! Geology never had such a chance, nor such exact data to argue from! Nor "development of species," either! Glacial epochs are great things, but they are vague—vague. Please observe:

In the space of one hundred and seventy-six years the Lower Mississippi has shortened itself two hundred and forty-two miles. That is an average of a trifle over one mile and a third per year. Therefore, any calm person, who is not blind or idiotic, can see that in the Old Oolitic Silurian Period, just a million years ago next November, the Lower Mississippi River was upwards of one million three hundred thousand miles long, and stuck out over the Gulf of Mexico like a fishing-rod. And by the same token any person can see that seven hundred and forty-two years from now the Lower Mississippi will be only a mile and three-quarters long, and Cairo and New Orleans will have joined their streets together, and be plodding comfortably along under a single mayor and a mutual board of aldermen. There is something fascinating about science. One gets such wholesale returns of conjecture out of such a trifling investment of fact.

Discuss how this relates to the material in this section.

26. **Regression Applet** Load the regression by eye applet. Create a scatter diagram with twelve points and a positive linear association. Try to choose the points so that the correlation is about 0.7.

(a) Draw a line that you believe describes the relation between the two variables well.

(b) Now click the Show Least-squares Line on the applet. Compare the sum of squared residuals for the line that you draw to the sum of squared residuals for the least-squares regression line. Repeat parts (a) and (b) as often as you like. Does your eyeballed line ever coincide with the least-squares regression line?

Technology Step-by-Step	**Determining the Least-Squares Regression Line**

TI-83/84 Plus Use the same steps that were followed to obtain the correlation coefficient.

MINITAB *Step 1:* With the explanatory variable in C1 and the response variable in C2, select the **Stat** menu and highlight **Regression**. Highlight **Regression** ....

Step 2: Select the explanatory (predictor) and response variables and click OK.

Excel *Step 1:* Be sure the Data Analysis Tool Pak is activated by selecting the **Tools** menu and highlighting **Add-Ins** .... Check the box for the Analysis ToolPak and select OK.

Step 2: Enter the explanatory variable in column A and the response variable in column B.

Step 3: Select the **Tools** menu and highlight **Data Analysis** ....

Step 4: Select the **Regression** option.

Step 5: With the cursor in the Y-range cell, highlight the column that contains the response variable. With the cursor in the X-range cell, highlight the column that contains the explanatory variable. Press OK.

4.3 Diagnostics on the Least-Squares Regression Line

Preparing for This Section Before getting started, review the following:

- Outliers (Section 3.4, pp. 170–172)

Objectives

1. **Compute and interpret the coefficient of determination**
2. **Perform residual analysis on a regression model**
3. **Identify influential observations**

In Section 4.2, we discussed the procedure for obtaining the least-squares regression line. In this section, we discuss additional characteristics of the least-squares regression line, along with graphical diagnostic tests that should be performed on every least-squares regression line.

1 Compute and Interpret the Coefficient of Determination

In Other Words

Just as a doctor performs tests in order to diagnose a patient, a researcher must use tests to diagnose the least-squares regression line.

Consider the club-head speed versus distance data introduced in Section 4.1. If we were asked to predict the distance of a randomly selected shot, what would be a good guess? Our best guess might be the average distance of all shots taken. Since we don't know this value, we would use the average distance from the sample data given in Table 1, $\overline{y} = 266.75$ yards.

Now suppose we were told this particular shot resulted from a swing with a club-head speed of 103 mph. We could use the least-squares regression line to adjust our guess to $\hat{y} = 3.1661(103) - 55.7964 = 270.3$ yards. Knowing the linear relation that exists between club-head speed and distance allows us to improve our estimate of the distance of the shot. In statistical terms, we say that some of the variation in distance is explained by the linear relation between club-head speed and distance.

The percentage of variation in distance that is explained by the least-squares regression line is called the *coefficient of determination*.

Definition

The **coefficient of determination**, R^2, measures the percentage of total variation in the response variable that is explained by the least-squares regression line.

In Other Words

The coefficient of determination is a measure of how well the least-squares regression line describes the relation between the explanatory and response variable. The closer R^2 is to 1, the better the line describes how changes in the explanatory variable affect the value of the response variable.

The coefficient of determination is a number between 0 and 1, inclusive. That is, $0 \le R^2 \le 1$. If $R^2 = 0$, the least-squares regression line has no explanatory value. If $R^2 = 1$, the least-squares regression line explains 100% of the variation in the response variable.

Consider Figure 14, where a horizontal line is drawn at $\overline{y} = 266.75$. This value represents the predicted distance of a shot without any knowledge of club-head speed. Armed with the additional information that the club-head speed is 103 miles per hour, we increased our guess to 270.3 yards. The difference between the predicted distance of 266.75 yards and the predicted distance of 270.3 yards is due to the fact that the club head-speed is 103 miles per hour. In other words, the difference between the prediction of $\hat{y} = 270.3$ and $\overline{y} = 266.75$ is explained by the linear relation between club-head speed and distance. The observed distance when club-head speed is 103 miles per hour is 274 yards (see Table 3 on page 213). The difference between our predicted value, $\hat{y} = 270.3$, and the actual value, $y = 274$, is due to factors (variables) other than the club-head speed and random error. The differences just discussed are called **deviations**.

Figure 14

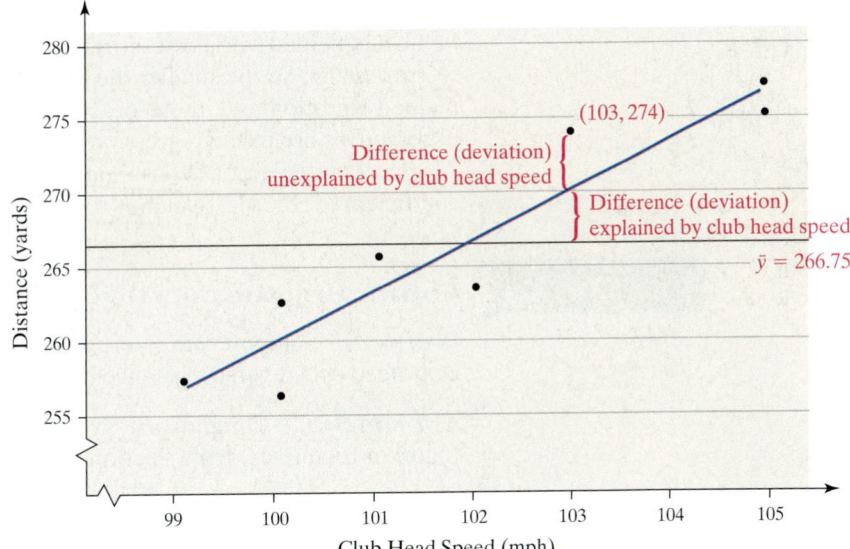

The deviation between the observed value of the response variable, y, and the mean value of the response variable, $\bar{y}$, is called the **total deviation**, so total deviation $= y - \bar{y}$. The deviation between the predicted value of the response variable, $\hat{y}$, and the mean value of the response variable, $\bar{y}$, is called the **explained deviation**, so explained deviation $= \hat{y} - \bar{y}$. Finally, the deviation between the observed value of the response variable, y, and the predicted value of the response variable, $\hat{y}$, is called the **unexplained deviation**, so unexplained deviation $= y - \hat{y}$. See Figure 15.

Figure 15

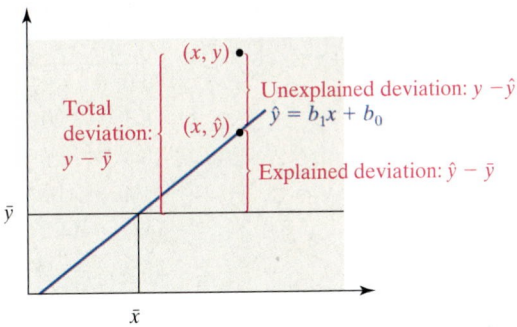

From the figure, it should be clear that

Total deviation = unexplained deviation + explained deviation

or

$$y - \bar{y} = (y - \hat{y}) + (\hat{y} - \bar{y})$$

Although beyond the scope of this text, it can be shown that

$$\sum (y - \bar{y})^2 = \sum (y - \hat{y})^2 + \sum (\hat{y} - \bar{y})^2$$

or

Total variation = unexplained variation + explained variation

Dividing both sides by total variation we obtain

$$1 = \frac{\text{unexplained variation}}{\text{total variation}} + \frac{\text{explained variation}}{\text{total variation}}$$

Subtracting $\dfrac{\text{unexplained variation}}{\text{total variation}}$ from both sides, we obtain

$$R^2 = \frac{\text{explained variation}}{\text{total variation}} = 1 - \frac{\text{unexplained variation}}{\text{total variation}}$$

Unexplained variation is found by summing the squares of the residuals, $\sum \text{residuals}^2$. So the smaller the sum of squared residuals, the smaller the unexplained variation and, therefore, the larger R^2 will be. Therefore, the closer the observed y's are to the regression line (the predicted y's), the larger R^2 will be.

The coefficient of determination, R^2, is the square of the linear correlation coefficient for the least-squares regression model. Written in symbols, $R^2 = (r)^2$.

EXAMPLE 1

Computing the Coefficient of Determination, R^2

Problem: Compute and interpret the coefficient of determination, R^2, for the club-head speed versus distance data shown in Table 2.

Approach: To compute R^2, we square the linear correlation coefficient, r, found in Example 2 from Section 4.1 on page 200.

Solution: $R^2 = (r^2) = 0.9387^2 = 0.8812 = 88.12\%$

Interpretation: 88.12% of the variation in distance is explained by the least-squares regression line, and 11.88% of the variation in distance is explained by other factors.

! CAUTION

Squaring the linear correlation coefficient to obtain the coefficient of determination works only for the least-squares linear regression model.

$$\hat{y} = b_0 + b_1 x.$$

The method does not work in general.

To help reinforce the concept of the coefficient of determination, consider the three data sets in Table 5.

Table 5

Data Set A		Data Set B		Data Set C	
x	y	x	y	x	y
3.6	8.9	3.1	8.9	2.8	8.9
8.3	15.0	9.4	15.0	8.1	15.0
0.5	4.8	1.2	4.8	3.0	4.8
1.4	6.0	1.0	6.0	8.3	6.0
8.2	14.9	9.0	14.9	8.2	14.9
5.9	11.9	5.0	11.9	1.4	11.9
4.3	9.8	3.4	9.8	1.0	9.8
8.3	15.0	7.4	15.0	7.9	15.0
0.3	4.7	0.1	4.7	5.9	4.7
6.8	13.0	7.5	13.0	5.0	13.0

Figure 16(a) represents the scatter diagram of data set A, Figure 16(b) represents the scatter diagram of data set B, and Figure 16(c) represents the scatter diagram of data set C.

Figure 16

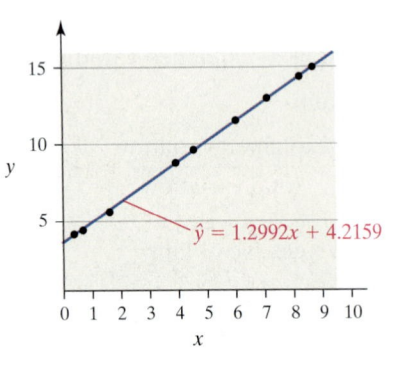

(a)

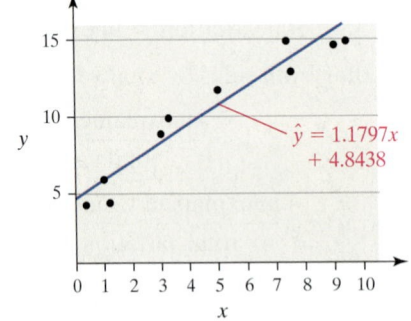

(b)

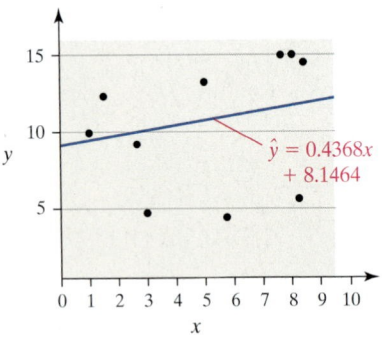

(c)

Notice that the y-values in each of the three data sets are the same. The variance of y is 17.49. If we look at the scatter diagram in Figure 16(a), we notice that almost 100% of the variability in y can be explained by the least-squares regression line, because the data almost lies perfectly on a straight line. In Figure 16(b), a high percentage of the variability in y can be explained by the least-squares regression line because the data have a strong linear relation. Higher x-values are associated with higher y-values. Finally, in Figure 16(c), a low percentage of the variability in y is explained by the least-squares regression line. If x increases, we cannot easily predict the change in y. If we compute the coefficient of determination, R^2, for the three data sets in Table 5, we obtain the following results:

Coefficient of determination for Data Set A: 99.99%

Coefficient of determination for Data Set B: 94.7%

Coefficient of determination for Data Set C: 9.4%

Notice that, as the explanatory ability of the line decreases, so does the coefficient of determination, R^2.

EXAMPLE 2

Figure 17

SUMMARY OUTPUT

Regression Statistics	
Multiple R	0.938695838
R Square	0.881149876
Adjusted R Square	0.861341522
Standard Error	2.882638465
Observations	8

Determining the Coefficient of Determination Using Technology

Problem: Determine the coefficient of determination, R^2, for the club-head speed versus distance data found in Example 2 from Section 4.1 using a statistical spreadsheet or graphing calculator with advanced statistical features.

Approach: We will use Excel to determine R^2. The steps for obtaining the coefficient of determination using Excel, MINITAB, and the TI-83/84 Plus graphing calculators are given in the Technology Step by Step on page 240.

Result: Figure 17 shows the results obtained from Excel. The coefficient of determination, R^2, is highlighted.

Now Work Problems 15 and 25(a).

 Perform Residual Analysis on a Regression Model

Recall that a residual is the difference between the observed value of y and the predicted value, $\hat{y}$. Residuals play an important role in determining the adequacy of the linear model. In fact, residuals can be used for the following purposes:

- To determine whether or not a linear model is appropriate to describe the relation between the explanatory and response variables
- To determine whether or not the variance of the residuals is constant
- To check for outliers

Let's see how the residuals can be used to verify the adequacy of the linear model.

Is a Linear Model Appropriate?   To answer this question, we make a scatter diagram with the residuals on the vertical axis and the explanatory variable on the horizontal axis.

> If a plot of the residuals against the explanatory variable shows a discernible pattern, such as a curve, then the response and explanatory variable may not be linearly related.

Figure 18 shows two residual plots. The residual plot in Figure 18(a) does not show any pattern, so a linear model is appropriate. However, the residual plot in Figure 18(b) shows an obvious pattern, which indicates that a linear model is inappropriate.

Figure 18

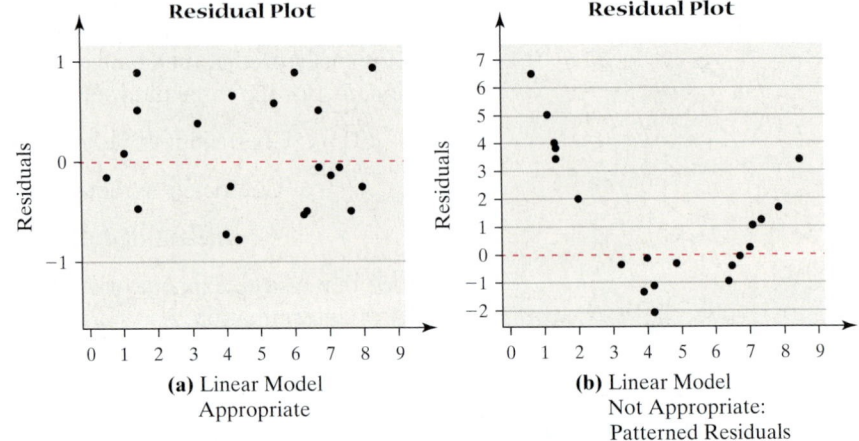

(a) Linear Model
Appropriate

(b) Linear Model
Not Appropriate:
Patterned Residuals

<hr />

EXAMPLE 3 **Is a Linear Model Appropriate?**

Problem: The data in Table 6 were collected by placing a temperature probe in a portable heater, removing the probe, and then recording temperature (in degrees Fahrenheit) over time (in seconds). Determine whether the relation between the temperature of the probe and time is linear.

Table 6

Time (seconds)	Temperature (°F)
1	164.5
3	163.2
5	162.0
7	160.5
9	158.6
11	156.8
13	155.1

Source: Michael Sullivan

Approach: We will find the least-squares regression line and determine the residuals. Plot the residuals against the explanatory variable, time. If an obvious pattern results, then we conclude that the linear model is not appropriate. If no pattern results, then a linear model is appropriate.

Solution: Enter the data into a statistical spreadsheet or a graphing calculator with advanced statistical features. Figure 19(a) shows a scatter diagram of the data. It appears that perhaps temperature and time are linearly related with a negative slope. In fact, the linear correlation coefficient between these two variables is −0.998. We use statistical software to compute the least-squares regression line and store the residuals. Figure 19(b) shows a plot of the residuals versus the explanatory variable, time. The upside-down U-shaped pattern in the plot indicates that the linear model is not appropriate. The predicted values overestimate the temperature for early and late times, whereas they underestimate the temperature for the middle times.

Figure 19

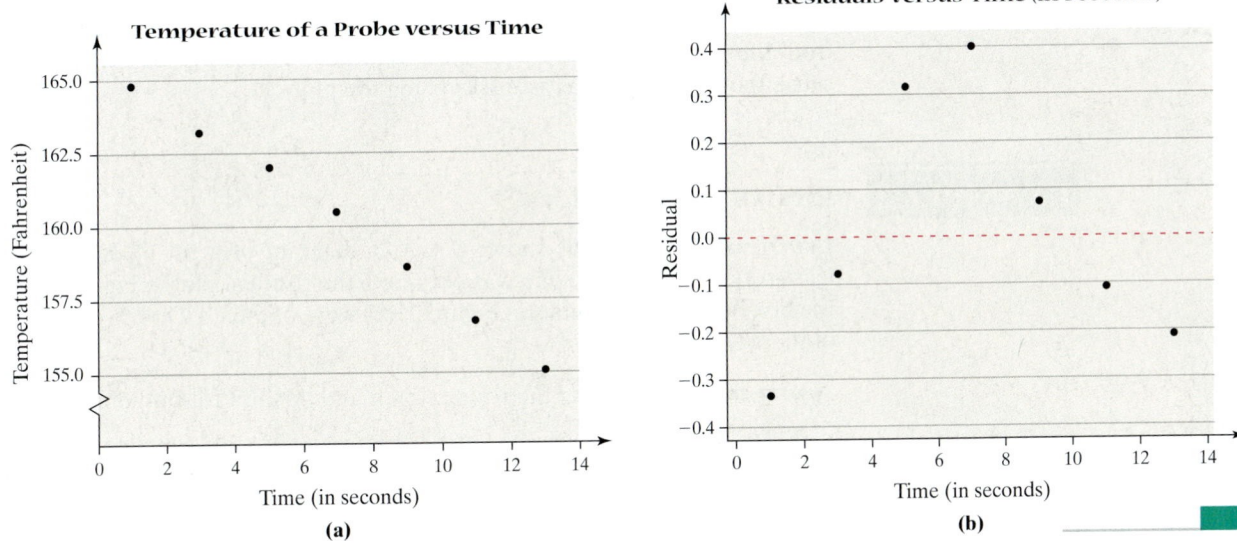

(a)

(b)

Now Work Problem 21.

Is the Variance of the Residuals Constant?   To determine whether the variance of the residuals is constant, we plot the residuals against the explanatory variable.

!**CAUTION**

If the model does not have constant error variance, statistical inference (the subject of Chapter 14) using the regression model is not reliable.

   If a plot of the residuals against the explanatory variable shows the spread of the residuals increasing or decreasing as the explanatory variable increases, then a strict requirement of the linear model is violated. This requirement is called constant error variance.*

EXAMPLE 4   **Constant Error Variance**

Figure 20 illustrates the idea behind constant error variance. Notice that in Figure 20(a) the data appear more spread out about the regression line for large x. From the residual plot in Figure 20(b), we see the absolute value of the residuals is larger for large x and smaller for the smaller x. This means that the predictions made using the regression equation will be less reliable for large x because there is more variability in y.

Figure 20

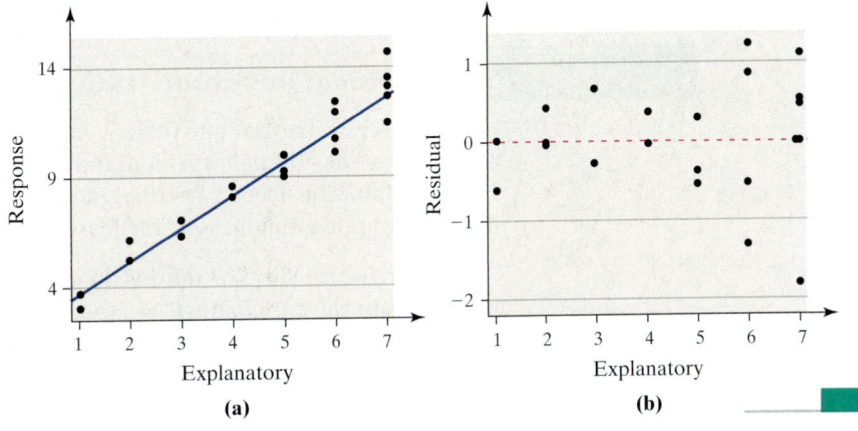

(a)

(b)

*The statistical term for constant error variance is **homoscedasticity**.

Are There Any Outliers?   In Section 3.4 we said that outliers are extreme observations. An **outlier** can also be thought of as an observation that does not fit the overall pattern of the data. To determine outliers, we can either construct a plot of residuals against the explanatory variable or draw a boxplot of the residuals.

We can find outliers from the residual plot because the residual will lie far from the rest of the plot. Outliers can be found from a boxplot of residuals because they will appear as asterisks (*) in the plot.

EXAMPLE 5   **Identifying Outliers**

Problem: Figure 21(a) shows a scatter diagram of a set of data. After the least-squares regression line was obtained, the residual plot in Figure 21(b) and the boxplot of the residuals in Figure 21(c) were obtained. Do the data have any outliers?

Approach: We look at the residual plot and boxplot for outliers.

Solution: We can see that the data do contain an outlier. We label the outlier in the scatter diagram given in Figure 21(a).

Figure 21

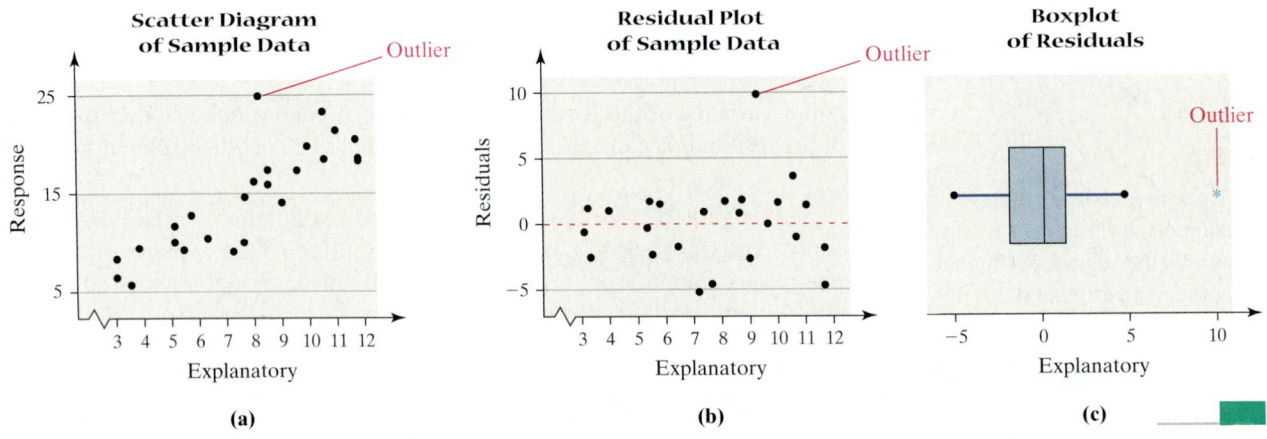

(a)   (b)   (c)

We will follow the practice of removing outliers only if they are the result of an error in recording, a miscalculation, or some other obvious blunder in the data-collection process (such as the observation not being from the population under study). If none of these are reasons for the outlier, it is recommended that a statistician be consulted.

EXAMPLE 6   **Graphical Residual Analysis**

Problem: The data in Table 4 on page 220 represent the club-head speed, the distance the golf ball traveled, and the residuals (column 7) for eight swings of a golf club. Construct a residual plot and boxplot of the residuals and comment on the appropriateness of the least-squares regression model.

Approach: We plot the residuals on the vertical axis and the corresponding values of the explanatory variable on the horizontal axis. We then look for any violations of the requirements of the regression model. We use a boxplot of the residuals to identify any outliers.

Solution: Figure 22 shows the residual plot. Figure 23 shows the boxplot.

Figure 22

Figure 23

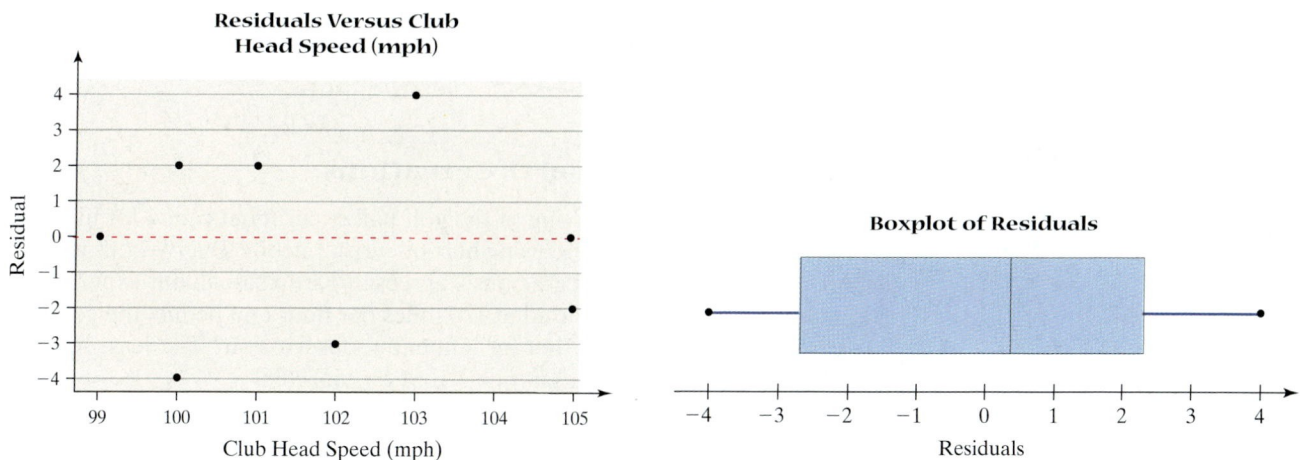

Because there is no discernible pattern in the residual plot, describing the relation between club-head speed and distance using a linear model is appropriate. The residuals display constant error variance, and there are no outliers in the boxplot of the residuals.

When outliers are identified, the first course of action is to determine whether the outlier occurred because of a data-entry or data-collection error. If either is the case, the data value should be removed. If there is no data-entry or data-collection error, the outlier should be removed only if there is reason to do so, such as the observation not being from the population being studied.

③ Identify Influential Observations

An **influential observation** is an observation that significantly affects the value of the slope and/or y-intercept of the least-squares regression line. How do we identify influential observations? We first remove the point that is believed to be influential from the data set, and then we recompute the regression line. If the slope or y-intercept changes significantly, we say that the removed point is influential. Consider the scatter diagram in Figure 24.

Figure 24

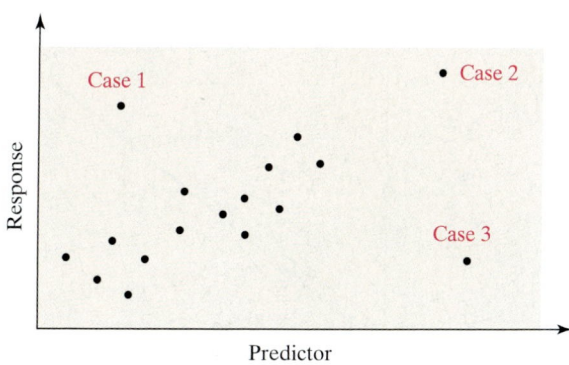

In Other Words

Draw a boxplot of the explanatory variable. If an outlier appears, then the observation corresponding to the outlier may be influential.

Case 1 would be an outlier because its Y value is large relative to its X value. Notice, however, that there are enough points to offset the effect Case 1 would have on the least-squares regression, so it is not influential. Case 2 is large relative to both its X value and Y value. It may or may not be influential because the point appears to lie along the linear pattern of the data. Case 3 is large relative to its X value, and its Y value is not consistent with the pattern of the data. This point would influence the values of the slope and y-intercept in the regression equation. We conclude the following:

Influential observations typically exist when the point is an outlier relative to its X value.

EXAMPLE 7 **Identifying Influential Observations**

Problem: Suppose our driving of the golf ball experiment called for nine trials, but the player that we were using hurt his wrist. Luckily, Tiger Woods is practicing on the same range and graciously agrees to participate in our experiment. His club-head speed is measured at 120 miles per hour and he hits the golf ball 305 yards. Is Tiger Woods's shot an influential observation? The least-squares regression line without Woods is $\hat{y} = 3.1661x - 55.7966$.

Approach: To determine whether Tiger Woods is influential, we recompute the least-squares regression line with Woods included. If the slope or y-intercept of the least-squares regression line changes substantially, we will conclude that Woods is influential.

Solution: Figure 25 shows partial output obtained from MINITAB with Woods included in the data set.

Figure 25   **Regression Analysis: Distance (yards) versus Club Head Speed (mph)**

```
The regression equation is
Distance (yards) = 39.5 + 2.23 Club Head Speed (mph)

Predictor                   Coef    SE Coef      T        P
Constant                   39.46      20.15    1.96    0.091
Club Head Speed (mph)     2.2287     0.1937   11.51    0.000

S = 3.51224    R - Sq = 95.0%      R - Sq(adj) = 94.3%
```

The least-squares regression line with Woods included in the data set is

$$\hat{y} = 2.2287x + 39.46$$

We graph the regression lines with Woods included (red) and with Woods excluded (blue) on the same scatter diagram. See Figure 26.

Figure 26

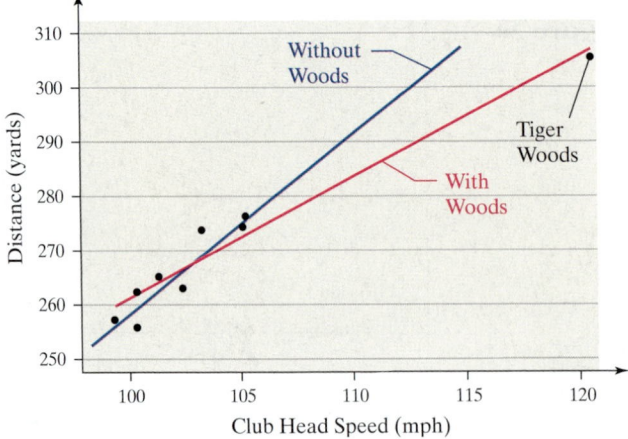

The inclusion of Woods in the data set causes the slope of the regression line to decrease (from 3.1661 to 2.1188) and the y-intercept to increase substantially (from -55.80 to $+39.46$). In other words, Wood's data pulls the least-squares regression line toward his observation. Wood's data is influential.

As with outliers, influential observations should be removed only if there is justification to do so. When an influential observation occurs in a data set and its removal is not warranted, two popular courses of action are to (1) collect more data so that additional points near the influential observation are obtained or (2) use techniques that reduce the influence of the influential observation. (These techniques are beyond the scope of this text.) In the case of Example 7, we are justified in removing Tiger Woods from the data set because our experiment called for the same player to swing the club each trial.

Now Work Problem 17.

4.3 ASSESS YOUR UNDERSTANDING

Concepts and Vocabulary

1. Suppose it is determined that $R^2 = 0.75$ when a linear regression is performed. Interpret this result.
2. Explain what is meant by total deviation, explained deviation, and unexplained deviation.
3. What does it mean if a plot of the residuals against the explanatory variable shows a discernible pattern (such as U-shaped)?
4. What is constant error variance? Why is it important?
5. What is an influential observation? If an influential observation is identified, what should be done?
6. Explain the difference between an outlier and an influential observation.

7. Explain how an influential observation affects the least-squares regression line.
8. Explain how a violation of the constant error variance assumption might affect the confidence placed on a predicted value.
9. If R^2 is 0.81, can you determine the value of r? If so, what is its value? If not, what else do you need to know?
10. Explain why it is important to do residual analysis when computing a least-squares regression line.

Skill Building

In Problems 11–14, analyze the residual plots and identify which, if any, of the conditions for an adequate linear model is not met.

11.   **12.**   **13.**   **14.**

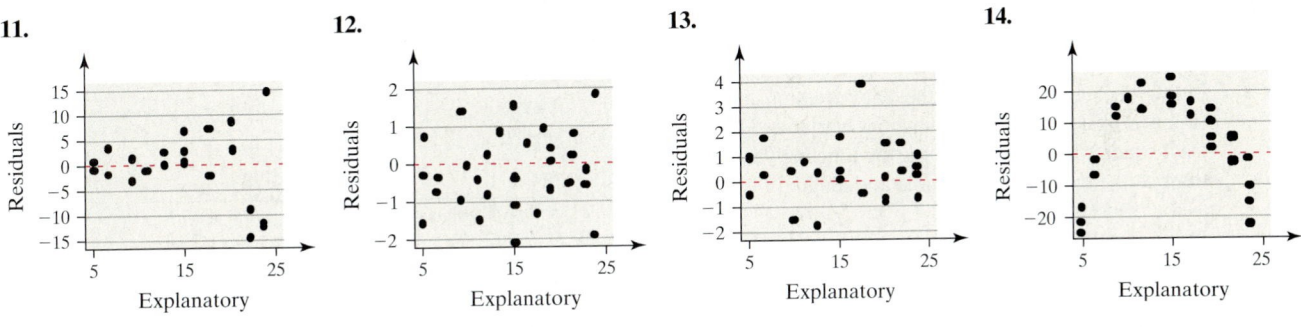

15. Match the coefficient of determination to the scatter diagram. The scales on the horizontal and vertical axis are the same for **NW** each scatter diagram.

(a) $R^2 = 0.58$ (b) $R^2 = 0.90$ (c) $R^2 = 1$ (d) $R^2 = 0.12$

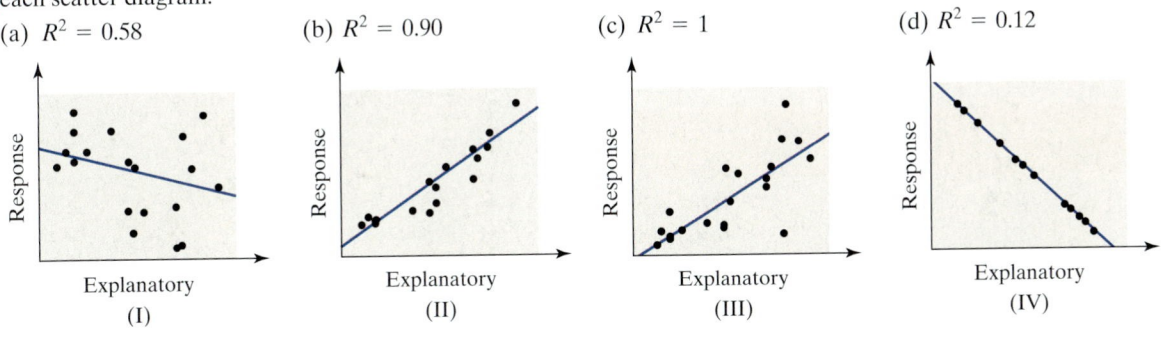

(I)   (II)   (III)   (IV)

16. Use the linear correlation coefficient given to determine the coefficient of determination, R^2. Interpret each R^2.

(a) $r = -0.32$ (b) $r = 0.13$ (c) $r = 0.40$ (d) $r = 0.93$

In Problems 17–20, a scatter diagram is given with one of the points drawn in blue. In addition, two least-squares regression lines are drawn: The line drawn in red is the least-squares regression line with the point in blue excluded. The line drawn in blue is the least-squares regression line with the point in blue included. On the basis of these graphs, do you think the point in blue is influential? Why?

17.
NW
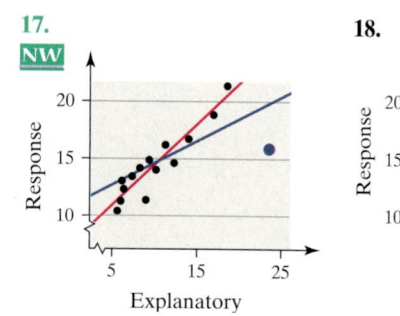

18.
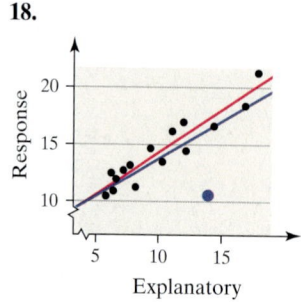

19.
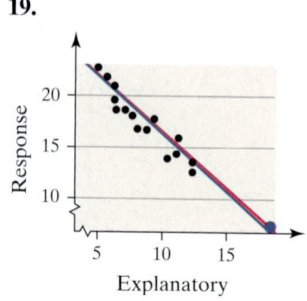

20.
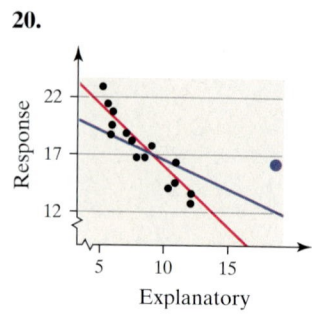

Applying the Concepts

21. The Other Old Faithful Perhaps you are familiar with the
NW famous Old Faithful geyser in Yellowstone National Park. Another Old Faithful geyser is located in Calistoga in California's Napa Valley. The following data represent the time between eruptions and the length of eruption for 11 randomly selected eruptions.

Time Between Eruptions, x	Length of Eruption, y	Time Between Eruptions, x	Length of Eruption, y
12.17	1.88	11.70	1.82
11.63	1.77	12.27	1.93
12.03	1.83	11.60	1.77
12.15	1.83	11.72	1.83
11.30	1.70		

Source: Ladonna Hansen, Park Curator

(a) Draw a scatter diagram of the data, treating time between eruptions as the explanatory variable. Find the least-squares regression, treating the time between eruptions as the explanatory variable and length of eruption as the response variable.
(b) Draw a residual plot. Do you think that time between eruptions and length of eruption are linearly related? Why or why not?
(c) The coefficient of determination is determined to be 83.0%. Interpret this result.

22. Concrete As concrete cures, it gains strength. The following data represent the 7-day and 28-day strength (in pounds per square inch) of a certain type of concrete.

7-Day Strength, x	28-Day Strength, y	7-Day Strength, x	28-Day Strength, y
2300	4070	2480	4120
3390	5220	3380	5020
2430	4640	2660	4890
2890	4620	2620	4190
3330	4850	3340	4630

(a) Draw a scatter diagram of the data, treating the 7-day strength as the explanatory variable. Find the least-squares regression, treating 7-day strength as the explanatory variable and 28-day strength as the response variable.
(b) Draw a residual plot. Do you think that 7-day strength and 28-day strength are linearly related? Why?
(c) The coefficient of determination, R^2, is determined to be 57.5%. Interpret this result.

23. Calories versus Sugar The following data represent the number of calories per serving and the number of grams of sugar per serving for a random sample of high-fiber cereals.

Calories, x	Sugar, y	Calories, x	Sugar, y
200	18	210	23
210	23	210	16
170	17	210	17
190	20	190	12
200	18	190	11
180	19	200	11

Source: Consumer Reports

(a) A scatter diagram with the least-squares regression line is shown. The least-squares regression equation is $\hat{y} = 0.0821x + 0.93$. Do you think that calories and sugar content are linearly related? Why?

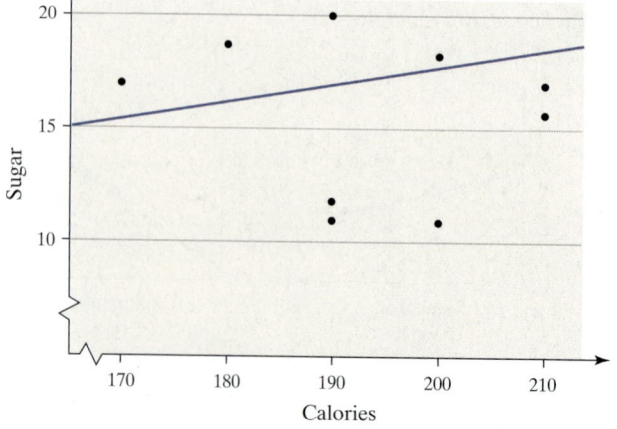

(b) The coefficient of determination, R^2, for these data is 6.8%. Interpret this result. Does this support your conclusion from part (a)? Why or why not?

(c) Suppose that we add Kellogg's All-Bran cereal, which has 80 calories and 6 grams of sugar per serving, to the data set. Draw a scatter diagram of the data with this cereal included. Do you think that Kellogg's All-Bran cereal is influential? *Hint:* The least-squares regression line with All-Bran included is $\hat{y} = 0.0933x - 1.288$. Why?

(d) The coefficient of determination, R^2, with Kellogg's All-Bran cereal included, is 42.1%. Interpret this result. Conclude that influential data values can cause the coefficient of determination to increase substantially, thereby increasing the apparent strength of the relation between two variables.

24. **Consumption versus Income** The following data represent the per capita disposable income (income after taxes) and per capita consumption in constant 2000 dollars in the United States for 1996–2004.

Year	Per Capita Disposable Income, x	Per Capita Consumption, y
1996	22,546	20,835
1997	23,065	21,365
1998	24,131	22,183
1999	24,564	23,050
2000	25,472	23,862
2001	25,698	24,216
2002	26,229	24,715
2003	26,570	25,270
2004	27,240	25,965

Source: Bureau of Economic Analysis

(a) A scatter diagram is shown. Do you think that per capita disposable income and per capita consumption are linearly related? Why or why not?

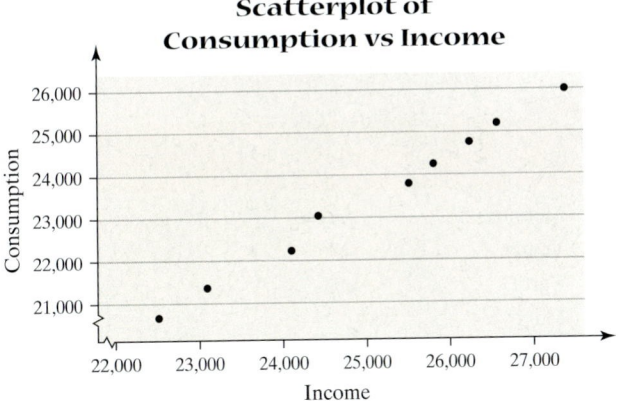

(b) The coefficient of determination, R^2, for the data is 99.4%. Interpret this result. Does this support your conclusion from part (a)? Why?

(c) A residual plot is shown below. Which, if any, of the requirements of the linear model are violated?

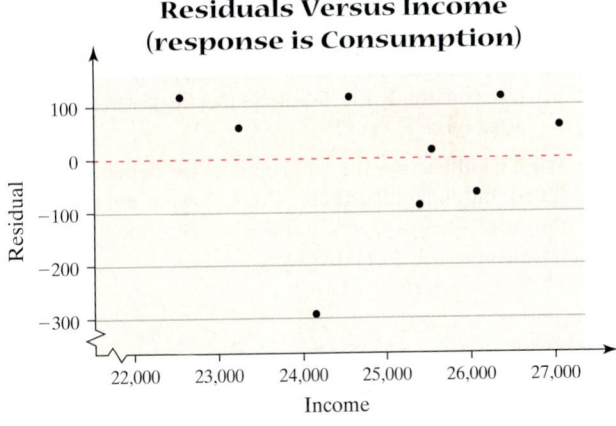

Problems 25–28, use the results from Problems 23–26, in Section 4.1 and Problems 17–20 in Section 4.2.

25. **Height versus Head Circumference** Use the results from Problem 23 in Section 4.1 and Problem 17 in Section 4.2 to

(a) compute the coefficient of determination, R^2.

(b) construct a residual plot to verify the requirements of the least-squares regression model.

(c) interpret the coefficient of determination and comment on the adequacy of the linear model.

26. **Gestation Period versus Life Expectancy** Use the results from Problem 24 in Section 4.1 and Problem 18 in Section 4.2 to

(a) compute the coefficient of determination, R^2.

(b) construct a residual plot to verify the requirements of the least-squares regression model.

(c) interpret the coefficient of determination and comment on the adequacy of the linear model.

27. **Weight of a Car versus Miles per Gallon** Use the results from Problem 25 in Section 4.1 and Problem 19 in Section 4.2 to

(a) compute the coefficient of determination, R^2.

(b) construct a residual plot to verify the requirements of the least-squares regression model.

(c) interpret the coefficient of determination and comment on the adequacy of the linear model.

28. **Bone Length** Use the results from Problem 26 in Section 4.1 and Problem 20 in Section 4.2 to

(a) compute the coefficient of determination, R^2.

(b) construct a residual plot to verify the requirements of the least-squares regression model.

(c) interpret the coefficient of determination and comment on the adequacy of the linear model.

29. Kepler's Law of Planetary Motion The time it takes for a planet to complete its orbit around the sun is called the planet's sidereal year. In 1618, Johann Kepler discovered that the sidereal year of a planet is related to the distance the planet is from the sun. The following data show the distances of the planets from the sun and their sidereal years.

Planet	Distance from Sun, x (millions of miles)	Sidereal Year, y
Mercury	36	0.24
Venus	67	0.62
Earth	93	1.00
Mars	142	1.88
Jupiter	483	11.9
Saturn	887	29.5
Uranus	1785	84.0
Neptune	2797	165.0
Pluto	3675	248.0

(a) Draw a scatter diagram of the data, treating distance from the sun as the explanatory variable.
(b) Compute the least-squares regression line.
(c) Plot the residuals against the distance from the sun.
(d) Do you think the least-squares regression line is a good model? Why?

30. Wind Chill Factor The wind chill factor depends on wind speed and air temperature. The following data represent the wind speed (in mph) and wind chill factor at an air temperature of 15° Fahrenheit.

Wind Speed, x (mph)	Wind Chill Factor, y	Wind Speed, x (mph)	Wind Chill Factor, y
5	12	25	−22
10	−3	30	−25
15	−11	35	−27
20	−17		

Source: Information Please Almanac.

(a) Draw a scatter diagram of the data treating wind speed as the explanatory variable.
(b) Compute the least-squares regression line.
(c) Plot the residuals against the wind speed.
(d) Do you think the least-squares regression line is a good model? Why?

31. Weight of a Car versus Miles per Gallon Suppose we add the Dodge Viper to the data in Problem 19 in Section 4.2. A Dodge Viper weighs 3425 pounds and gets 11 miles per gallon.

(a) Compute the coefficient of determination of the expanded data set. What effect does the addition of the Viper to the data set have on R^2?
(b) Is the point corresponding to the Dodge Viper influential? Is it an outlier?

32. Gestation Period versus Life Expectancy Suppose we add humans to the data in Problem 18 in Section 4.2. Humans have a gestation period of 268 days and a life expectancy of 76.5 years.

(a) Compute the coefficient of determination of the expanded data set. What effect does the addition of humans to the data set have on R^2?
(b) Is the point corresponding to humans influential? Is it an outlier?
(c) Remove humans from the data set. Suppose elephants are added to the data set (gestation period, 624 days; life expectancy, 35 years). Is the point corresponding to elephants influential? Is it an outlier?

33. Height versus Weight The following data represent the heights and weights of various professional baseball players.

Player	Height, x (inches)	Weight, y (pounds)
Alex Rodriguez	75	210
Derek Jeter	75	195
Greg Maddux	72	185
Randy Johnson	82	230
Al Leiter	75	220
Barry Bonds	74	210
Mike Bordick	71	175
Pete Harnisch	72	228
Randy Velarde	72	200
Jason Isringhausen	75	210

Source: Yahoo! Sports

(a) Draw a scatter diagram of the data, treating height as the explanatory variable and weight as the response variable.
(b) Compute the least-squares regression line and the correlation coefficient.
(c) Remove the value corresponding to Randy Johnson and recompute the least-squares regression line and the correlation coefficient. What effect does Randy Johnson have on the regression line and correlation coefficient?
(d) Do you think that Randy Johnson is an influential observation? Why?

34. Height versus Weight in the WNBA The following data represent the heights and weights of centers of the Eastern Conference in the Women's National Basketball Association.

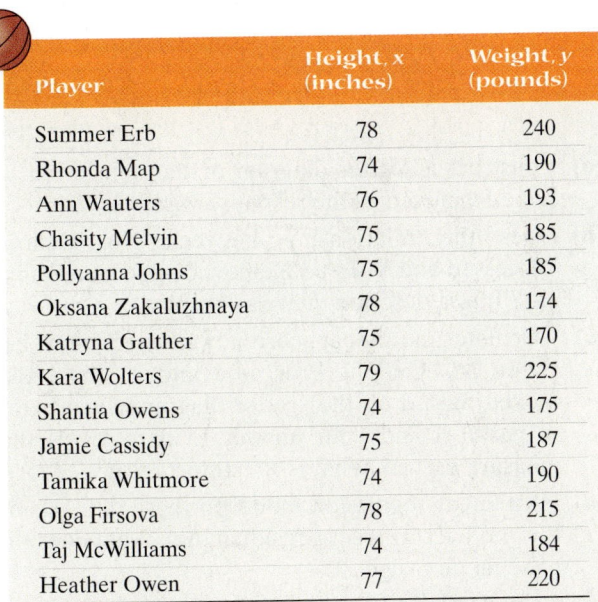

Player	Height, x (inches)	Weight, y (pounds)
Summer Erb	78	240
Rhonda Map	74	190
Ann Wauters	76	193
Chasity Melvin	75	185
Pollyanna Johns	75	185
Oksana Zakaluzhnaya	78	174
Katryna Galther	75	170
Kara Wolters	79	225
Shantia Owens	74	175
Jamie Cassidy	75	187
Tamika Whitmore	74	190
Olga Firsova	78	215
Taj McWilliams	74	184
Heather Owen	77	220

Source: Yahoo! Sports

(a) Draw a scatter diagram of the data treating height as the explanatory variable and weight as the response variable.
(b) Compute the least-squares regression line and the correlation coefficient.
(c) Remove the value corresponding to Oksana Zakaluzhnaya and recompute the least-squares regression line and the correlation coefficient.
(d) Do you think that Oksana Zakaluzhnaya is an influential observation? Why?

35. **Age versus Study Time** Professor Katula feels that there is a relation between the number of hours a statistics student studies each week and the student's age. She conducts a survey in which 26 statistics students are asked their age and the number of hours they study statistics each week. She obtains the following results:

Age, x	Number of Hours Studying, y	Age, x	Number of Hours Studying, y	Age, x	Number of Hours Studying, y
18	4.2	19	5.1	22	2.1
18	1.1	19	2.3	22	3.6
18	4.6	20	1.7	24	5.4
18	3.1	20	6.1	25	4.8
18	5.3	20	3.2	25	3.9
18	3.2	20	5.3	26	5.2
19	2.8	21	2.5	26	4.2
19	2.3	21	6.4	35	8.1
19	3.2	21	4.2		

(a) Draw a scatter diagram of the data. Comment on any potential influential observations.
(b) Find the least-squares regression line using all the data points.
(c) Find the least-squares regression line with the data point $(35, 8.1)$ removed.
(d) Draw each least-squares regression line on the scatter diagram obtained in part (a).
(e) Comment on the influence the point $(35, 8.1)$ has on the regression line.

36. **Regression Applet** Load the regression by eye applet. Create a scatter diagram with about 12 points clustered in the upper-left corner that have a strong, negative linear association.

(a) Click the "Show least-squares line" on the applet.
(b) Now add a thirteenth point in the bottom-right corner of the applet. The point should be near the least-squares regression line. How does the regression line change? Is this point influential?
(c) Drag the point in the bottom-right corner straight up. What makes this point influential?

(d) Clear the applet. Create a scatter diagram with 12 points that are spread out. The scatter diagram should have a strong, negative linear association. Click the "Show least-squares line".
(e) Now add a thirteenth point at the point near the mean of X and mean of Y. Does the least-square regression line change?
(f) Drag the thirteenth point straight up. Does the least-square regression line change much? Why or why not?

Consumer Reports — Fit to Drink

The taste, color, and clarity of the water coming out of home faucets have long concerned consumers. Recent reports of lead and parasite contamination have made unappetizing water a health, as well as an esthetic, concern. Water companies are struggling to contain cryptosporidium, a parasite that has caused outbreaks of illness that may be fatal to people with a weakened immune system. Even chlorination, which has rid drinking water of infectious organisms that once killed people by the thousands, is under suspicion as an indirect cause of miscarriages and cancer.

Concerns about water quality and taste have made home filtering increasingly popular. To find out how well they work, technicians at Consumer Reports tested 14 models to determine how well they filtered contaminants and whether they could improve the taste of our cabbage-soup testing mixture.

To test chloroform and lead removal, we added concentrated amounts of both to our water, along with calcium nitrate to increase water hardness. Every few days we analyzed the water to measure chloroform and lead content. The following table contains the lead measurements for one of the models tested.

No. Gallons Processed	% Lead Removed
25	85
26	87
73	86
75	88
123	90
126	87
175	92
177	94

(a) Construct a scatter diagram of the data using % Lead Removed as the response variable.

(b) Does the relationship between No. Gallons Processed and % Lead Removed appear to be linear? If not, describe the relationship.

(c) Calculate the linear correlation coefficient between No. Gallons Processed and % Lead Removed. Based on the scatter diagram constructed in part (a) and your answer to part (b), is this measure useful? What is R^2? Interpret R^2.

(d) Fit a linear regression model to these data. Using this model, construct a residual plot. What does the residual plot suggest?

(e) Using statistical software or a graphing calculator with advanced statistical features, fit a quadratic model to these data. Construct a residual plot. What is R^2? Which model appears to fit the data better?

(f) Given the nature of the variables being measured, describe the type of curve you would expect to show the true relationship between these variables (linear, quadratic, exponential, s-shaped). Support your position.

Note to Readers: In many cases, our test protocol and analytical methods are more complicated than described in these examples. The data and discussions have been modified to make the material more appropriate for the audience.

Technology Step-by-Step — Determining R^2 and Residual Plots

TI-83/84 Plus

The Coefficient of Determination, R^2

Use the same steps that were followed to obtain the correlation coefficient to obtain R^2. Diagnostics must be on.

Residual Plots

Step 1: Enter the raw data in L1 and L2. Obtain the least-squares regression line.

Step 2: Access STAT PLOTS. Select Plot1. Choose the scatter diagram icon and let XList be L1. Let YList be RESID by putting the cursor on List2, pressing 2nd STAT and choosing the list entitled RESID.

Step 3: Press ZOOM and select 9: ZoomStat.

MINITAB   **The Coefficient of Determination, R^2**

This is provided in the standard regression output.

Residual Plots

Follow the same steps as those used to obtain the regression output. (Section 4.2) Before selecting OK, click GRAPHS. In the cell that says "Residuals versus the variables" enter the name of the explanatory variable. Click OK.

Excel   **The Coefficient of Determination, R^2**

This is provided in the standard regression output.

Residual Plots

Follow the same steps as those used to obtain the regression output. (Section 4.2) Before selecting OK, click **Residual Plots**. Click OK.

Chapter 4 # Review

Summary

In this chapter, we introduced techniques that allow us to describe the relation between two quantitative variables. The first step in identifying the type of relation that might exist is to draw a scatter diagram. The explanatory variable is plotted on the horizontal axis and the corresponding response variable on the vertical axis. The scatter diagram can be used to discover whether the relation between the explanatory and the response variables is linear. In addition, for linear relations, we can judge whether the linear relation shows positive or negative association.

A numerical measure for the strength of linear relation between two quantitative variables is the linear correlation coefficient. It is a number between −1 and 1, inclusive. Values of the correlation coefficient near −1 are indicative of a negative linear relation between the two variables. Values of the correlation coefficient near +1 indicate a positive linear relation between the two variables. If the correlation coefficient is near 0, then there is little *linear* relation between the two variables.

Once a linear relation between the two variables has been discovered, we describe the relation by finding the least-

squares regression line. This line best describes the linear relation between the explanatory and the response variables. We can use the least-squares regression line to predict a value of the response variable for a given value of the explanatory variable.

The coefficient of determination, R^2, measures the percent of variation in the response variable that is explained by the least-squares regression line. It is a measure between 0 and 1 inclusive. The closer R^2 is to 1, the more explanatory value the line has. Whenever a least-squares regression line is obtained, certain diagnostics must be performed. These include verifying that the residuals have constant variance, verifying that the linear model is appropriate, and checking for outliers and influential observations.

One item worth mentioning again is that a researcher should never claim causation between two variables in a study unless the data are experimental. Observational data allow us to say that two variables might be associated, but we cannot claim causation.

Formulas

Correlation Coefficient

$$r = \frac{\sum \left(\frac{x_i - \bar{x}}{s_x} \right)\left(\frac{y_i - \bar{y}}{s_y} \right)}{n - 1}$$

Equation of the Least-Squares Regression Line

The equation of the least-squares regression line is given by

$$\hat{y} = b_1 x + b_0$$

where

$\hat{y}$ is the predicted value of the response variable,

$b_1 = r \cdot \dfrac{s_y}{s_x}$ is the slope of the least-squares regression line, and

$b_0 = \bar{y} - b_1 \bar{x}$ is the intercept of the least-squares regression line.

Coefficient of Determination, R^2

$$R^2 = \frac{\text{variation explained by explanatory variable}}{\text{total variation}}$$

$$= 1 - \frac{\text{unexplained variation}}{\text{total variation}}$$

$$= r^2 \text{ for the least-squares regression model } \hat{y} = b_1 x + b_0$$

Vocabulary

Bivariate data (p. 194)
Response variable (p. 195)
Explanatory variable (p. 195)
Predictor variable (p. 195)
Lurking variable (p. 195)
Scatter diagram (p. 195)
Positively associated (p. 197)

Negatively associated (p. 197)
Linear correlation coefficient (p. 198)
Correlation matrix (p. 201)
Residuals (p. 214)
Least-squares regression line (p. 215)
Slope (p. 215)
y-intercept (p. 215)

Outside the scope of the model (p. 218)
Coefficient of determination (p. 226)
Total deviation (p. 227)
Explained deviation (p. 227)
Unexplained deviation (p. 227)
Outlier (p. 232)
Influential observation (p. 233)

Objectives

Section	You should be able to . . .	Example	Review Exercises
4.1	1 Draw and interpret scatter diagrams (p. 195)	1, 3	1(a)–4(a), 9(a), 10(a), 20(a)–22(a)
	2 Understand the properties of the linear correlation coefficient (p. 197)		24
	3 Compute and interpret the linear correlation coefficient (p. 200)	2, 3	1(b)–4(b), 21(b)
4.2	1 Find the least-squares regression line and use the line to make predictions (p. 214)	2, 3	5(a)–8(a), 9(d)–10(d), 19(a), 20(b)
	2 Interpret the slope and y-intercept of the least-squares regression line (p. 218)	Page 218	5(c)–8(c)
	3 Compute the sum of squared residuals (p. 219)	4	9(g), 10(g)
4.3	1 Compute and interpret the coefficient of determination (p. 226)	1, 2	15(a)–18(a), 22(b)
	2 Perform residual analysis on a regression model (p. 229)	3 through 6	11–14, 15(b), (c) through–18(b), (c), 20(c) and (d), 22(c)
	3 Identify influential observations (p. 233)	7	15(d)–18(d), 19(b)

Review Exercises

1. **Engine Displacement versus Fuel Economy** The following data represent the size of a car's engine (in liters) versus its miles per gallon in the city for various 2005 domestic automobiles.

Car	Engine Displacement (liters), x	City MPG, y	Car	Engine Displacement (liters), x	City MPG, y
Buick Century	3.1	20	Ford Crown Victoria	4.6	18
Buick LeSabre	3.8	20	Ford Focus	2.0	26
Cadillac DeVille	4.6	18	Ford Mustang	3.8	20
Chevrolet Cavalier	2.2	25	Mercury Sable	3.0	19
Chevrolet Impala	3.8	21	Pontiac Grand Am	3.4	20
Chevrolet Malibu	2.2	24	Pontiac Sunfire	2.2	24
Chrysler Sebring Sedan	2.7	22	Saturn Ion	2.2	26
Dodge Magnum	3.5	21			

Source: www.roadandtrack.com

(a) Draw a scatter diagram treating engine displacement as the explanatory variable and miles per gallon as the response variable.
(b) Compute the linear correlation coefficient between engine displacement and miles per gallon.
(c) Based on the scatter diagram and the linear correlation coefficient, comment on the type of relation that appears to exist between the two variables.

2. **Temperature versus Cricket Chirps** Crickets make a chirping noise by sliding their wings rapidly over each other. Perhaps you have noticed that the number of chirps seems to increase with the temperature. The following data list the temperature (in degrees Fahrenheit) and the number of chirps per second for the striped ground cricket.

Temperature, x	Chirps per Second, y	Temperature, x	Chirps per Second, y
88.6	20.0	71.6	16.0
93.3	19.8	84.3	18.4
80.6	17.1	75.2	15.5
69.7	14.7	82.0	17.1
69.4	15.4	83.3	16.2
79.6	15.0	82.6	17.2
80.6	16.0	83.5	17.0
76.3	14.4		

Source: Pierce, George W. *The Songs of Insects.* Cambridge, MA Harvard University Press, 1949, pp. 12–21.

(a) Draw a scatter diagram treating temperature as the explanatory variable and chirps per second as the response variable.
(b) Compute the linear correlation coefficient between temperature and chirps per second.
(c) Based upon the scatter diagram and the linear correlation coefficient, comment on the type of relation that appears to exist between the two variables.

3. **Apartments**   The following data represent the square footage and rents for apartments in the Borough of Queens and Nassau County, New York.

Queens (New York City) Square Footage, x	Queens (New York City) Rent Per Month, y	Nassau County (Long Island) Square Footage, x	Nassau County (Long Island) Rent Per Month, y
500	650	1100	1875
588	1215	588	1075
1000	2000	1250	1775
688	1655	556	1050
825	1250	825	1300
460	1805	743	1475
1259	2700	660	1315
650	1200	975	1400
560	1250	1429	1900
1073	2350	800	1650
1452	3300	1906	4625
1305	3100	1077	1395

Source: apartments.com

(a) On the same graph, draw a scatter diagram for both Queens and Nassau County apartments, treating square footage as the explanatory variable. Be sure to use a different plotting symbol for each group.
(b) Compute the linear correlation coefficient between square footage and rent for each location.

(c) Given the scatter diagram and the linear correlation coefficient, comment on the type of relation that appears to exist between the two variables for each group.
(d) Does location appear to be a factor in rent?

4. **Boys versus Girls**   The following data represent the height (in inches) of boys and girls between the ages of 2 and 10 years.

Age	Boy Height, x	Girl Height, y	Age	Boy Height, x	Girl Height, y
2	36.1	39.0	6	49.8	43.7
2	34.2	38.6	7	43.2	50.5
2	31.1	33.6	7	47.9	47.7
3	36.3	41.3	8	51.4	44.0
3	39.5	40.9	8	48.3	62.1
4	41.5	43.2	8	50.9	44.8
4	38.6	39.8	9	52.2	50.9
5	45.6	50.5	9	51.3	55.6
5	44.8	38.3	10	55.6	61.4
5	44.6	43.9	10	59.5	50.8

Source: National Center for Health Statistics.

(a) On the same graph, draw a scatter diagram for both boys and girls, treating age as the explanatory variable. Be sure to use a different plotting symbol for each gender.
(b) Compute the linear correlation coefficient between age and height for each gender.
(c) Based on the scatter diagram and the linear correlation coefficient, comment on the type of relation that appears to exist between the age and height for each gender.
(d) Does gender appear to be a factor in determining height?

5. Using the data and results from Problem 1, do the following:
   (a) Find the least-squares regression line, treating engine displacement as the explanatory variable.
   (b) Draw the least-squares regression line on the scatter diagram.
   (c) Interpret the slope and y-intercept, if appropriate.
   (d) Predict the miles per gallon of a Ford Mustang whose engine displacement is 3.8 liters.
   (e) Compute the residual of the prediction found in part (d).
   (f) Is the miles per gallon above or below average for a Ford Mustang?

6. Using the data and results from Problem 2, do the following:
   (a) Find the least-squares regression line, treating temperature as the explanatory variable and chirps per second as the response variable.
   (b) Draw the least-squares regression line on the scatter diagram.
   (c) Interpret the slope and y-intercept, if appropriate.
   (d) Predict the chirps per second if it is 83.3°F.
   (e) Compute the residual of the prediction found in part (d).
   (f) Were chirps per second above or below average at 83.3°F?

7. Using the Queens data and results from Problem 3, do the following:
   (a) Find the least-squares regression line, treating square footage as the explanatory variable.
   (b) Draw the least-squares regression line on the scatter diagram.
   (c) Interpret the slope and y-intercept, if appropriate.
   (d) Predict the rent of an 825-square-foot apartment.
   (e) Compute the residual of the prediction found in part (d).
   (f) Is this apartment's rent above or below average?

8. Using the Boy Height data and results from Problem 4, do the following:
   (a) Find the least-squares regression line, treating age as the explanatory variable and height as the response variable.
   (b) Draw the least-squares regression line on the scatter diagram.
   (c) Interpret the slope and y-intercept, if appropriate.
   (d) Predict the height of a 6-year-old boy.
   (e) Compute the residual of the prediction found in part (d).
   (f) Is this boy's height above or below average?

In Problems 9 and 10, do the following:
(a) Draw a scatter diagram treating x as the predictor variable and y as the response variable.
(b) Select two points from the scatter diagram, and find the equation of the line containing the points selected.
(c) Graph the line found in part (b) on the scatter diagram.
(d) Determine the least-squares regression line.
(e) Graph the least-squares regression line on the scatter diagram.
(f) Compute the sum of the squared residuals for the line found in part (b).
(g) Compute the sum of the squared residuals for the least-squares regression line found in part (d).
(h) Comment on the fit of the line found in part (b) versus the least-squares regression line found in part (d).

9.

x	3	4	6	7	9
y	2.1	4.2	7.2	8.1	10.6

10.

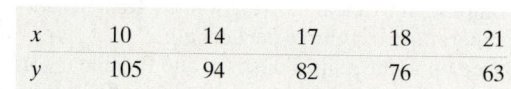

x	10	14	17	18	21
y	105	94	82	76	63

In Problems 11–14, a residual plot is given. From the residual plot, determine whether or not a linear model is appropriate. If not, state your reason.

11.
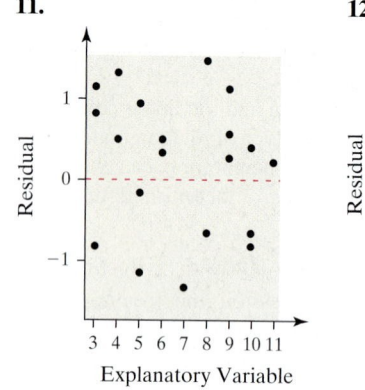

12.
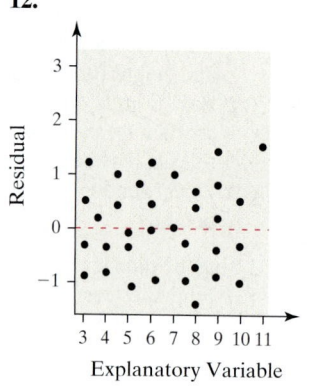

13.
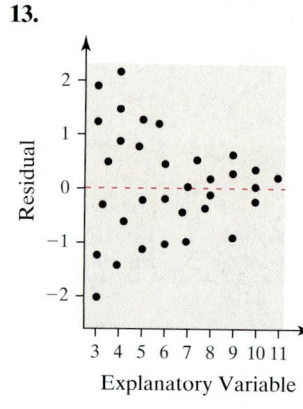

14.
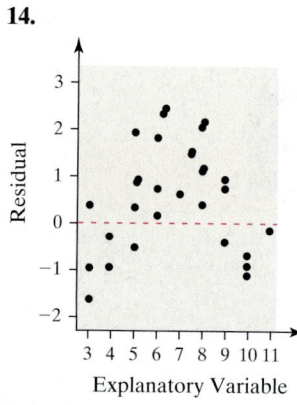

15. Use the results from Problems 1 and 5 for the following:
    (a) Compute and interpret R^2.
    (b) Plot the residuals against engine displacement.
    (c) For the graph obtained in part (b), is a linear model appropriate?
    (d) Are there any outliers or influential observations?
    (e) Add a Dodge Viper to the data set in Problem 1 (8.0-liter engine, 11 miles per gallon); then redo Problems 1, 5, and 15. Comment on the effect the Viper has on the regression model.

16. Use the results from Problems 2 and 6 for the following:
    (a) Compute and interpret R^2.
    (b) Plot the residuals against temperature.
    (c) For the graph obtained in part (b), is a linear model appropriate?
    (d) Are there any outliers or influential observations?

17. Use Queens data and the results from Problems 3 and 7 for the following:

(a) Compute and interpret R^2.
(b) Plot the residuals against square footage.
(c) For the graph obtained in part (b), is a linear model appropriate?
(d) Are there any outliers or influential observations?

18. Use the results from Problems 4 and 8 for the following:
    (a) Compute and interpret R^2.
    (b) Plot the residuals against age.
    (c) For the graph obtained in part (b), is a linear model appropriate?
    (d) Are there any outliers or influential observations?

19. **Apartment Rents** Using the Nassau County apartment rent data and the results of Problem 3, do the following:
    (a) Find the least-squares regression line, treating square footage as the explanatory variable.
    (b) Are there any outliers or influential observations?

20. **Depreciation** The following data represent the price of a random sample of used Chevy Camaros, by age.

Age, x (in years)	Price, y (in dollars)	Age, x (in years)	Price, y (in dollars)
2	$15,900	1	$20,365
5	$10,988	2	$16,463
2	$16,980	6	$10,824
5	$9,995	1	$19,995
4	$11,995	1	$18,650
5	$10,995	4	$10,488

Source: www.onlineauto.com

(a) Draw a scatter diagram treating the age of the car as the explanatory variable and price as the response variable.
(b) Find the least-squares regression line.
(c) Plot the residuals against the age of the car.
(d) Do you think that a linear model is appropriate for describing the relation between the age of the car and price? Why?

21. **200-Meter Dash** The following data represent the gold medal times, in seconds, for men and women in the 200-meter dash at the summer Olympics from 1948 to 2004.

Year	Time (Men)	Time (Women)	Year	Time (Men)	Time (Women)
1948	21.10	24.40	1980	20.19	22.03
1952	20.70	23.70	1984	19.80	21.81
1956	20.60	23.40	1988	19.75	21.34
1960	20.50	24.00	1992	20.01	21.81
1964	20.30	23.00	1996	19.32	22.12
1968	19.80	22.50	2000	20.09	21.84
1972	20.00	22.40	2004	19.79	22.05
1976	20.23	22.37			

Source: www.factmonster.com

(a) Draw a scatter diagram of the data using time for men as the explanatory variable and time for women as the response variable.
(b) Compute the correlation coefficient for the data.
(c) Based on your results from parts (a) and (b), what type of relation appears to exist between the gold medal time for men and the gold medal time for women in the 200-meter dash? Do you believe that the gold medal time for men causes the gold medal time for women?

22. **Population of Salem, Oregon** The following data represent the population of Salem, Oregon (in thousands of people), for 1990 to 2003.

Year	Population	Year	Population
1990	108	1997	124
1991	110	1998	127
1992	112	1999	129
1993	113	2000	137
1994	117	2001	139
1995	118	2002	141
1996	121	2003	143

Source: www.oregonlink.com

(a) A scatter diagram of the data is shown. Do you think that the year and the population of Salem, Oregon, are linearly related? Why or why not?

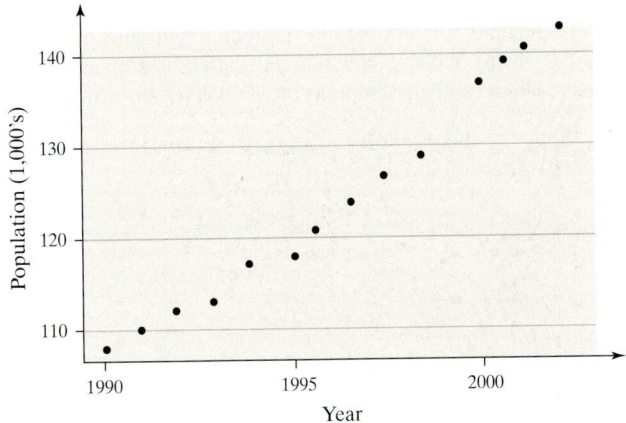

(b) The coefficient of determination, R^2, for the data is 97.9%. Interpret this result. Does this support your conclusion from part (a)? Why or why not?
(c) In the residual plot shown, Which, if any, of the requirements of the linear model are violated?

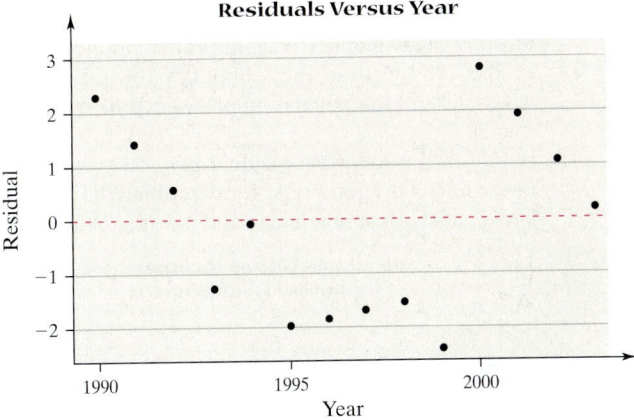

23. **Wine and Your Heart** The health benefits of moderate wine consumption are well documented. Researchers wanted to determine if alcohol consumption is positively related to heart-rate variability (HRV) in women with coronary heart disease (CHD). The purpose of the study was to shed some doubt on the heart–health benefits of wine. The researchers surveyed female patients who have recently been released

from the hospital after successful heart procedures such as by-pass surgery or angioplasty. A questionnaire evaluated self-reported consumption of individual alcoholic beverage types: beer, wine, and spirits. Other characteristics, such as age, body mass index, smoking habits, history of diabetes, menopausal status, and educational status, were also assessed. The researchers found that wine intake was associated with increased HRV. Based on this study, can we conclude that increased wine consumption in women with recent heart procedures causes an increase in heart-rate variability? Why?

24. List the seven properties of the linear correlation coefficient.

25. Analyzing a Newspaper Article In a newspaper article written in the *Chicago Tribune* on September 29, 2002, it was claimed that poorer school districts have shorter school days.

(a) The following scatter diagram was drawn using the data supplied in the article. In this scatter diagram, the response variable is length of the school day and the explanatory variable is percent of the population that is low income. The correlation between length and income is -0.461. Do you think that the scatter diagram and correlation coefficient support the position of the article?

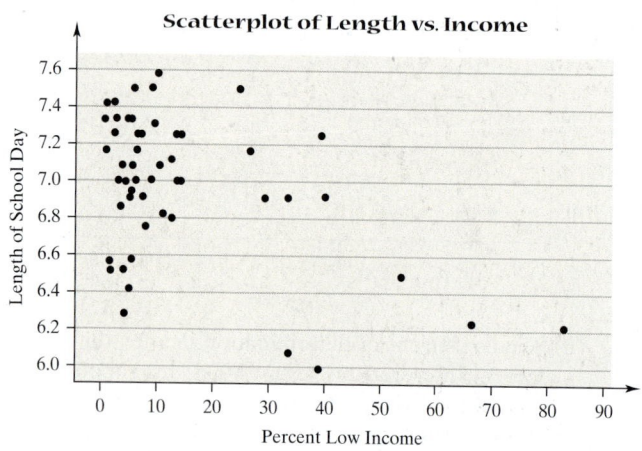

Scatterplot of Length vs. Income

(b) The least-squares regression line between length, y, and income, x, is $\hat{y} = -0.0102x + 7.11$. Interpret the slope of this regression line. Does it make sense to interpret the intercept? If so, interpret the intercept.

(c) Predict the length of the school day for a district in which 20% of the population is low income by letting $x = 20$.

(d) Based on the following residual plot, do you think a linear model is appropriate for describing the relation between length of school day and income? Why?

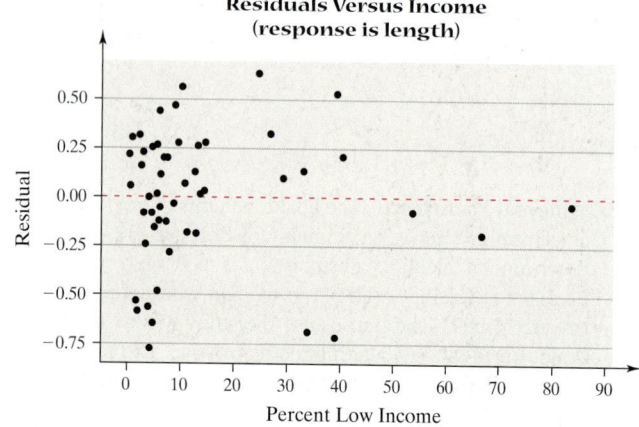

Residuals Versus Income
(response is length)

(e) Included in the output from MINITAB was a notification that three observations were influential. Based on the scatter diagram in part (a), which three observations do you think might be influential?

(f) We should not remove influential observations just because they are influential. What suggestions would you make to the author of the article to improve the study and therefore give more credibility to the conclusions made in the study?

(g) This same article included average Prairie State Achievement Examination (PSAE) scores for each district. The article implied that shorter school days result in lower PSAE scores. The correlation between PSAE score and length of school day is 0.517. A scatter diagram treating PSAE as the response variable is shown below. Do you believe that a longer school day is positively associated with a higher PSAE score?

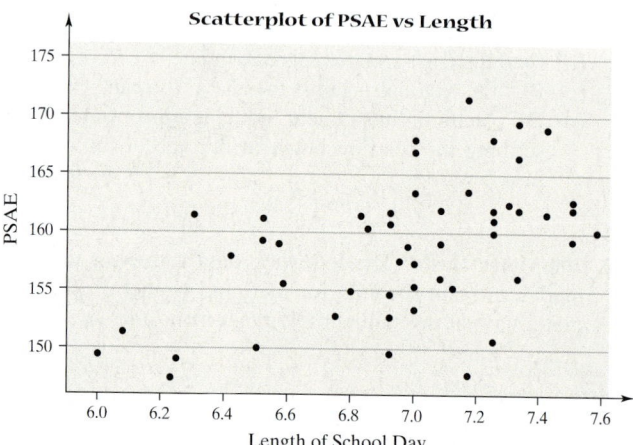

Scatterplot of PSAE vs Length

(h) The correlation between percentage of the population that is low income and PSAE score is -0.720. A scatter diagram treating PSAE score as the response variable is shown below.

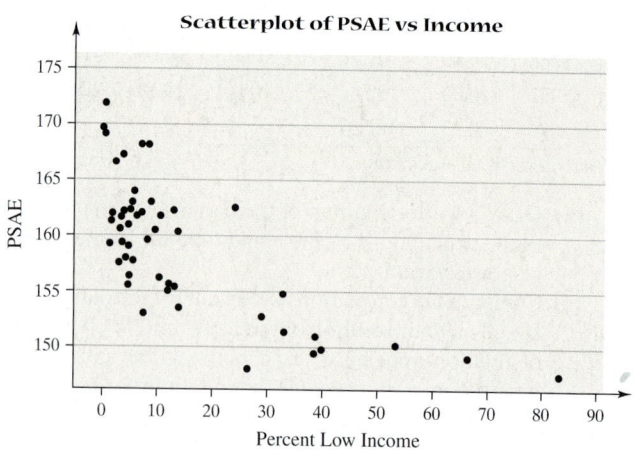

Scatterplot of PSAE vs Income

Do you believe that percentage of the population that is low income is negatively associated with PSAE score?

(i) Can you think of any lurking variables that are playing a role in this study?

Thomas Malthus, Population, and Subsistence

In 1798, Thomas Malthus published *An Essay on the Principle of Population* in which he stated

> "Population, when unchecked, increases in a geometrical ratio. Subsistence increases only in an arithmetical ratio. A slight acquaintance with numbers will show the immensity of the first power in comparison of the second."

What did he mean by this? Essentially he is claiming that population grows exponentially and food production increases linearly. The assumption is that the population grows unchecked. That is, without interference from man-made or natural disasters such as war, famine, disease, and so on. Malthus also stated,

> "By that law of our nature which makes food necessary to the life of man, the effects of these two unequal powers must be kept equal. This implies a strong and constantly operating check on population from the difficulty of subsistence. This difficulty must fall somewhere and must necessarily be severely felt by a large portion of mankind."

(a) Read the article *Food, Land, Population, and the American Economy* from Carrying Capacity Network. (*Source of article:* http://www.carryingcapacity.org/resources.html)

(b) Use the historic census data from the U.S. Census Bureau at www.census.gov for the U.S. for the years 1790-2000 to find an exponential model of the form $P = a \cdot b^t$ where t is the number of years after 1790. To do this, first take the logarithm of P. Now find the least-squares regression line between $\log P$ and t. The least-squares model is of the form $\log P = \log a + t \log b$. To find estimates of a and b, compute $10^{\log a}$ where $\log a$ is the intercept and $10^{\log b}$ is the slope of the least squares regression line. The quantity $b - 1$ is the growth rate as a decimal. For example, if $b = 1.01$, then the annual growth rate is $1.01 - 1 = 0.01$ or 1%.

(c) Multiply your model obtained in part (b) by 1.2 to obtain a function for the number of arable acres of land A required to have a diverse diet in the U.S. (as a function of time). [*Hint*: The model is $A = 1.2P$].

(d) Go the Statistical Abstract of the United States and look up the number of acres of farmland A in the United States for the most recent nine years for which data is available. Find the least squares regression line treating year as the explanatory variable.

(e) Determine when the U.S. will no longer be capable of providing a diverse diet for its entire population by setting the model in part (c) equal to the model in part (d) and solving for the year t.

(f) Does this result seem reasonable? Discuss some reasons why this prediction might be affected or changed. For example, rather than looking at the number of arable acres of land required to have a diverse diet, consider total food production. Compare the results using this variable as a measure of subsistence.

PART 3 Probability and Probability Distributions

CHAPTER 5
Probability

CHAPTER 6
Discrete Probability Distributions

CHAPTER 7
The Normal Probability Distribution

We now take a break from the statistical process. Why? Recall that we mentioned that inferential statistics uses methods that generalize results obtained from a sample to the population and measures their reliability. But how can we measure their reliability? It turns out that the methods we use to generalize results from a sample to a population are based on probability and probability models. Probability is a measure of the likelihood that something occurs.

Probability

CHAPTER 5

Outline

DECISIONS

Have you ever watched a sporting event on television in which the announcer cites an obscure statistic? Where do these numbers come from? Well, pretend that you are the statistician for your favorite sports team. Your job is to compile strange or obscure probabilities regarding your favorite team and a competing team. See the Decisions project on page 289.

 **Putting It All Together**

In Chapter 1, we learned the methods of collecting data. In Chapters 2 through 4, we learned how to summarize raw data using tables, graphs, and numbers. As far as the statistical process goes, we have discussed the collecting, organizing, and summarizing part of the process.

Before we can proceed with the analysis of data, we introduce probability, which forms the basis of inferential statistics. Why? Well, we can think of the probability of an outcome as the likelihood of observing that outcome. If something has a high likelihood of happening, it has a high probability (close to 1). If something has a small chance of happening, it has a low probability (close to 0). For example, in rolling a single die, it is unlikely that we would roll five straight sixes, so this result has a low probability. In fact, the probability of rolling five straight sixes is 0.0001286. So, if we were playing a game that entailed throwing a single die, and one of the players threw five sixes in a row, we would consider the player to be lucky (or a cheater) because it is such an unusual occurrence. Statisticians use probability in the same way. If something occurs that has a low probability, we investigate to find out "what's up."

5.1 Probability Rules

Preparing for This Section   Before getting started, review the following:

• Relative frequency (Section 2.1, p. 62)

Objectives

1 **Understand the rules of probabilities**

2 **Compute and interpret probabilities using the empirical method**

3 **Compute and interpret probabilities using the classical method**

4 **Use simulation to obtain data based on probabilities**

5 **Understand subjective probabilities**

Probability is a measure of the likelihood of a random phenomenon or chance behavior. Probability describes the long-term proportion with which a certain **outcome** will occur in situations with short-term uncertainty.

The long-term predictability of chance behavior is best understood through a simple experiment. Flip a coin 100 times and compute the proportion of heads observed after each toss of the coin. Suppose the first flip is tails, so the proportion of heads is $\frac{0}{1}$; the second flip is heads, so the proportion of heads is $\frac{1}{2}$; the third flip is heads, so the proportion of heads is $\frac{2}{3}$; and so on. Plot the proportion of heads versus the number of flips and obtain the graph in Figure 1(a). We repeat this experiment with the results shown in Figure 1(b).

Look at the graphs in Figures 1(a) and (b). Notice that in the short term (fewer flips of the coin) the observed proportion of heads is different and unpredictable for each experiment. As the number of flips of the coin increases, however, both graphs tend toward a proportion of 0.5. This is the basic premise of probability. Probability deals with experiments that yield random short-term results or **outcomes**, yet reveal long-term predictability. **The long-term proportion with which a certain outcome is observed is the probability of that outcome.** So, we say that the probability of observing a head is $\frac{1}{2}$ or 50% or 0.5, because as we flip the coin more times, the proportion of heads tends toward $\frac{1}{2}$. This phenomenon is referred to as the *Law of Large Numbers*.

In Other Words

Probability describes how likely it is that some event will happen. If we can look at the proportion of times an event has occurred over a long period of time (or over a large number of trials), we can be more certain of the likelihood of its occurrence.

Figure 1

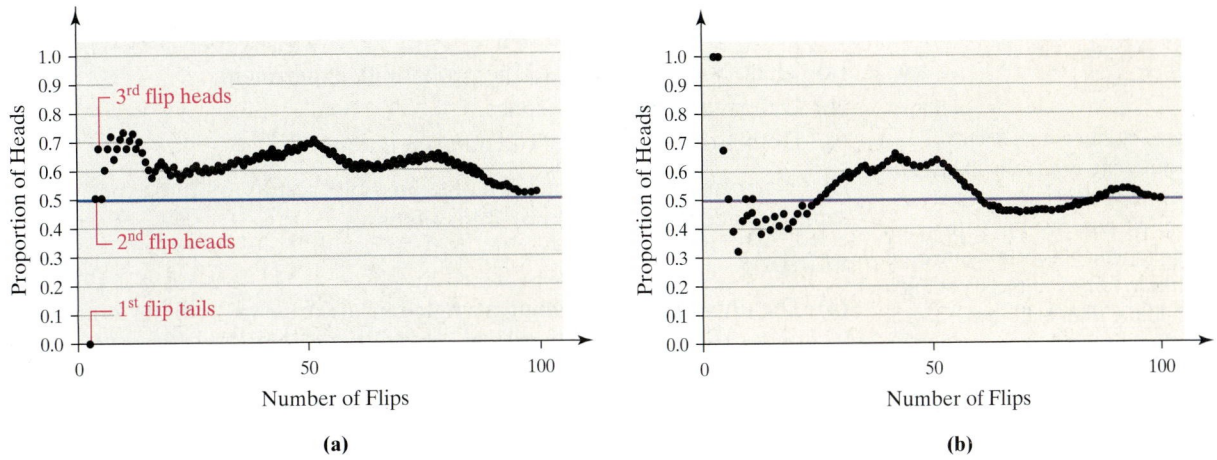

(a)

(b)

The Law of Large Numbers

As the number of repetitions of a probability experiment increases, the proportion with which a certain outcome is observed gets closer to the probability of the outcome.

The Law of Large Numbers is illustrated in Figure 1. For a few flips of the coin, the proportion of heads fluctuates wildly around 0.5, but as the number of flips increases, the proportion of heads settles down near 0.5. Jakob Bernoulli (a major contributor to the field of probability) believed that the Law of Large Numbers was common sense. This is evident in the following quote from his text *Ars Conjectandi*: "For even the most stupid of men, by some instinct of nature, by himself and without any instruction, is convinced that the more observations have been made, the less danger there is of wandering from one's goal."

In probability, an **experiment** is any process with uncertain results that can be repeated. The result of any single trial of the experiment is not known ahead of time. However, the results of the experiment over many trials produce regular patterns that enable us to predict with remarkable accuracy. For example, an insurance company cannot know ahead of time whether a particular 16-year-old driver will be involved in an accident over the course of a year. However, based on historical records, the company can be fairly certain that about three out of every ten 16-year-old male drivers will be involved in a traffic accident during the course of a year. Therefore, of the 816,000 16-year-old drivers (816,000 repetitions of the experiment), the insurance company is fairly confident that about 30%, or 244,800, of the drivers will be involved in an accident. This prediction forms the basis for establishing insurance rates for any particular 16-year-old male driver.

We now introduce some terminology that we will need to study probability.

Definitions

The **sample space**, S, of a probability experiment is the collection of all possible outcomes.

An **event** is any collection of outcomes from a probability experiment. An event may consist of one outcome or more than one outcome. We will denote events with one outcome, sometimes called *simple events*, e_i. In general, events are denoted using capital letters such as E.

In Other Words
An outcome is the result of one trial of a probability experiment. The sample space is a list of all possible results of a probability experiment.

We present an example to illustrate the definitions.

EXAMPLE 1

A *fair die* is one in which each possible outcome is equally likely. For example, rolling a 2 is just as likely as rolling a 5. We contrast this with a *loaded* die in which a certain outcome is more likely. For example, if rolling a 1 is more likely than rolling a 2, 3, 4, 5, or 6, the die is loaded.

Identifying Events and the Sample Space of a Probability Experiment

Problem: A probability experiment consists of rolling a single *fair* die.

(a) Identify the outcomes of the probability experiment.
(b) Determine the sample space.
(c) Define the event E = "roll an even number."

Approach: The outcomes are the possible results of the experiment. The sample space is a list of all possible outcomes.

Solution

(a) The outcomes from rolling a single fair die are e_1 = "rolling a one" = $\{1\}$, e_2 = "rolling a two" = $\{2\}$, e_3 = "rolling a three" = $\{3\}$, e_4 = "rolling a four" = $\{4\}$, e_5 = "rolling a five" = $\{5\}$, and e_6 = "rolling a six" = $\{6\}$.

(b) The set of all possible outcomes forms the sample space, $S = \{1, 2, 3, 4, 5, 6\}$. There are 6 outcomes in the sample space.

(c) The event E = "roll an even number" = $\{2, 4, 6\}$.

1 Understand the Rules of Probabilities

Probabilities have some rules that must be satisfied. In these rules, the notation $P(E)$ means "the probability of event E."

Rules of Probabilities

1. The probability of any event E, $P(E)$, must be greater than or equal to 0 and less than or equal to 1. If we let E denote any event, then $0 \leq P(E) \leq 1$.

2. The sum of the probabilities of all outcomes must equal 1. That is, if the sample space $S = \{e_1, e_2, \ldots, e_n\}$, then $P(e_1) + P(e_2) + \cdots + P(e_n) = 1$.

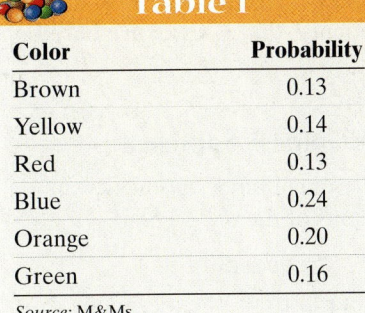

In Other Words
Rule 1 states that probabilities less than 0 or greater than 1 are not possible. Therefore, probabilities such as 1.32 or −0.3 are not possible. Rule 2 states when the probabilities of all outcomes are added, the sum must be 1.

A **probability model** lists the possible outcomes of a probability experiment and each outcome's probability. A probability model must satisfy rules 1 and 2 of the rules of probabilities.

EXAMPLE 2 **A Probability Model**

In a bag of M&M plain candies, the colors of the candies can be brown, yellow, red, blue, orange, or green. Suppose that a candy is randomly selected from a bag. Table 1 shows each color and the probability of drawing that color.

To verify this is a probability model we must show that rules 1 and 2 of the rules of probabilities are satisfied.

Each probability is greater than or equal to 0 and less than or equal to 1, so rule 1 is satisfied.

Because

$$0.13 + 0.14 + 0.13 + 0.24 + 0.20 + 0.16 = 1$$

rule 2 is also satisfied. The table is an example of a probability model.

Table 1

Color	Probability
Brown	0.13
Yellow	0.14
Red	0.13
Blue	0.24
Orange	0.20
Green	0.16

Source: M&Ms

If an event is **impossible**, the probability of the event is 0. If an event is a **certainty**, the probability of the event is 1.

Now Work Problem 9.

The closer a probability is to 1, the more likely the event will occur. The closer a probability is to 0, the less likely the event will occur. For example, an event with probability 0.8 is more likely to occur than an event with probability 0.75. An event with probability 0.8 will occur about 80 times out of 100 repetitions of the experiment, while an event with probability 0.75 will occur about 75 times out of 100.

Be careful of this interpretation. Just because an event has a probability of 0.75 does not mean that the event *must* occur 75 times out of 100. It simply means that we *expect* the number of occurrences to be close to 75 in 100 trials of the experiment. The more repetitions of the probability experiment, the closer the proportion with which the event occurs will be to 0.75 (the Law of Large Numbers).

One goal of this course is to learn how probabilities can be used to identify *unusual events*.

In Other Words
An unusual event is an event that is not likely to occur.

Definition   An **unusual event** is an event that has a low probability of occurring.

Typically, an event with a probability less than 5% is considered unusual, but this *cutoff point* is not set in stone. The researcher and the context of the problem determine the probability that separates unusual events from *not so unusual events*.

For example, suppose the probability of being wrongly convicted of a capital crime punishable by death is 3%. The probability is too high in light of the consequences (death for the wrongly convicted), so the event is not unusual (unlikely) enough. We would want this probability to be as close to zero as possible.

Now suppose that you are planning a picnic on a day for which there is a 3% chance of rain. In this context, you would consider "rain" an unusual (unlikely) event and proceed with the picnic plans.

The point is this: Selecting a probability that separates unusual events from not so unusual events is subjective and depends on the situation. Statisticians typically use cutoff points of 1%, 5%, and 10%. For many circumstances, any event that occurs with probability of 5% or less will be considered unusual.

Next, we introduce three methods for determining the probability of an event: (1) the empirical method, (2) the classical method, and (3) the subjective method.

! CAUTION

A probability of 0.05 should not always be used to separate unusual events from not so unusual events.

2   Compute and Interpret Probabilities Using the Empirical Method

Because probabilities deal with the long-term proportion with which a particular outcome is observed, it makes sense that we begin our discussion of determining probabilities using the idea of relative frequency. Probabilities computed in this manner rely on empirical evidence, that is, evidence based on the outcomes of a probability experiment.

> **Approximating Probabilities Using the Empirical Approach**
>
> The probability of an event E is approximately the number of times event E is observed divided by the number of repetitions of the experiment.
>
> $$P(E) \approx \text{relative frequency of } E = \frac{\text{frequency of } E}{\text{number of trials of experiment}} \quad \textbf{(1)}$$

The probability obtained using the empirical approach is approximate because different *runs* of the probability experiment lead to different outcomes and, therefore, different estimates of $P(E)$. Consider flipping a coin 20 times and recording the number of heads. Use the results of the experiment to estimate the probability of obtaining a head. Now repeat the experiment. Because the results of the second run of the experiment do not necessarily yield the same results, we cannot say the probability *equals* some proportion; rather we say the probability is *approximately* the proportion. As we increase the number of trials of a probability experiment, our estimate becomes more accurate (again, the Law of Large Numbers).

EXAMPLE 3   Using Relative Frequencies to Approximate Probabilities

A pit boss wanted to approximate the probability of rolling a seven using a pair of dice that have been in use for a while. To do this he rolls the dice 100 times and records 15 sevens. The approximate probability of rolling a seven is $\frac{15}{100} = 0.15$.

When we survey a random sample of individuals, the probabilities computed from the survey are approximate. In fact, we can think of a survey as a probability experiment, since the results of a survey are different each time the survey is conducted because different people are included.

EXAMPLE 4   Building a Probability Model from Survey Data

Problem: The data in Table 2 represent the results of a survey in which 200 people were asked their means of travel to work.

(a) Use the survey data to build a probability model for means of travel to work.
(b) Would it be unusual to randomly select an individual who walks to work?

Approach: To build a probability model, we estimate the probability of each outcome by determining the relative frequency of each outcome.

Table 2

Means of Travel	Frequency
Drive alone	153
Carpool	22
Public transportation	10
Walk	5
Other means	3
Work at home	7

Table 3

Means of Travel	Probability
Drive alone	0.765
Carpool	0.11
Public transportation	0.05
Walk	0.025
Other means	0.015
Work at home	0.035

Solution

(a) There are $153 + 22 + \cdots + 7 = 200$ individuals in the survey. The individuals can be thought of as trials of the probability experiment. The relative frequency for "drive alone" is $\dfrac{153}{200} = 0.765$. We compute the relative frequency of the other outcomes similarly and obtain the probability model in Table 3.

(b) The probability that an individual walks to work is 0.025. It is somewhat unusual to randomly choose a person who walks to work. ▬▬▬

Now Work Problem 37.

③ Compute and Interpret Probabilities Using the Classical Method

When using the empirical method, we obtain an approximate probability of an event by conducting a probability experiment.

The classical method of computing probabilities does not require that a probability experiment actually be performed. Rather, we use counting techniques to determine the probability of an event.

The classical method of computing probabilities requires *equally likely outcomes*. An experiment is said to have **equally likely outcomes** when each outcome has the same probability of occurring. For example, in throwing a fair die once, each of the six outcomes in the sample space, $\{1, 2, 3, 4, 5, 6\}$, has an equal chance of occurring. Contrast this situation with a loaded die in which a five or six is twice as likely to occur as a one, two, three, or four.

Computing Probability Using the Classical Method

If an experiment has n equally likely outcomes and if the number of ways that an event E can occur is m, then the probability of E, $P(E)$, is

$$P(E) = \frac{\text{Number of ways that } E \text{ can occur}}{\text{Number of possible outcomes}} = \frac{m}{n} \qquad (2)$$

So, if S is the sample space of this experiment,

$$P(E) = \frac{N(E)}{N(S)} \qquad (3)$$

where $N(E)$ is the number of outcomes in E, and $N(S)$ is the number of outcomes in the sample space.

EXAMPLE 5   **Computing Probabilities Using the Classical Method**

Problem: A pair of fair dice is rolled.

(a) Compute the probability of rolling a seven.
(b) Compute the probability of rolling "snake eyes"; that is, compute the probability of rolling a two.
(c) Comment on the likelihood of rolling a seven versus rolling a two.

Approach: To compute probabilities using the classical method, we count the number of outcomes in the sample space and count the number of ways the event can occur.

Solution

(a) In rolling a pair of fair dice, there are 36 equally likely outcomes in the sample space, as shown in Figure 2.

Figure 2

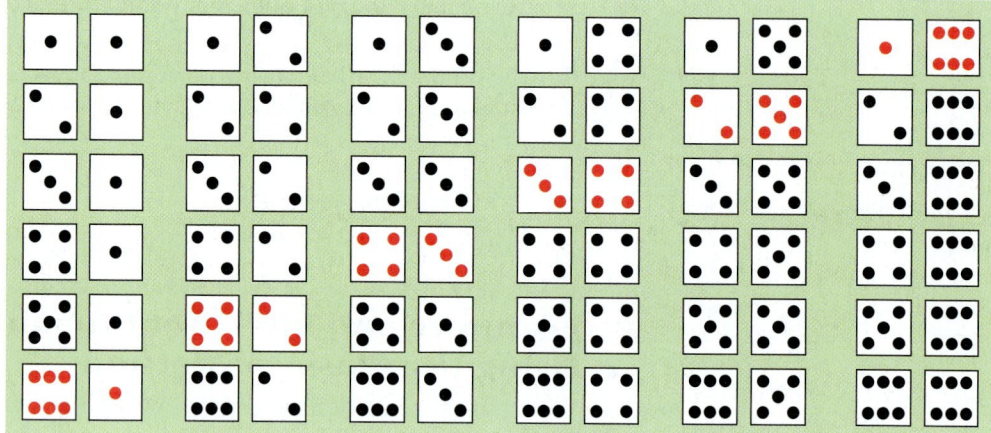

So $N(S) = 36$. The event E = "roll a seven" = $\{(1, 6),\ (2, 5),\ (3, 4),\ (4, 3),\ (5, 2),\ (6, 1)\}$ is composed of six outcomes, so $N(E) = 6$. Using Formula (3), the probability of rolling a seven is

$$P(E) = P(\text{roll a seven}) = \frac{N(E)}{N(S)} = \frac{6}{36} = \frac{1}{6}$$

(b) The event F = "roll a two" = $\{(1, 1)\}$ has one outcome, so $N(E) = 1$. Using Formula (3), the probability of rolling a two is

$$P(F) = P(\text{roll a two}) = \frac{N(E)}{N(S)} = \frac{1}{36}$$

(c) Since $P(\text{roll a seven}) = \dfrac{6}{36}$ and $P(\text{roll a two}) = \dfrac{1}{36}$, rolling a seven is six times as likely as rolling a two. In other words, in 36 rolls of the dice, we *expect* to observe about 6 sevens and only 1 two. ■

If we compare the empirical probability of rolling a seven, 0.15, obtained in Example 3 to the classical probability of rolling a seven, $\dfrac{1}{6} \approx 0.167$, obtained in Example 5(a), we see that they are not too far apart. In fact, if the dice are fair, we expect the relative frequency of sevens to get closer to 0.167 as we increase the number of rolls of the dice. That is, if the dice are fair, the empirical probability will get closer to the classical probability as the number of trials of the experiment increases. If the two probabilities do not get closer together, we may suspect that the dice are not fair.

In simple random sampling, each individual has the same chance of being selected. Therefore, we can use the classical method to compute the probability of obtaining a specific sample.

| EXAMPLE 6 | **Computing Probabilities Using Equally Likely Outcomes** |

Historical Note

Girolamo Cardano (in English Jerome Cardan) was born in Pavia, Italy, on September 24, 1501. He was an illegitimate child whose father was Fazio Cardano, a lawyer in Milan. Fazio was a part-time mathematician and taught Girolamo. In 1526, Cardano earned his medical degree. Shortly thereafter, his father died. Unable to maintain a medical practice, Cardano spent his inheritance and turned to gambling to help support himself. Cardano developed an understanding of probability that helped him to win. He wrote a booklet on probability. *Liber de Ludo Alaea*, which was not printed until 1663, 87 years after his death. The booklet is a practical guide to gambling, including cards, dice, and cheating. Eventually, Cardano became a lecturer of mathematics at the Piatti Foundation. This position allowed him to practice medicine and develop a favorable reputation as a doctor. In 1545, he published his greatest work, *Ars Magna*.

Problem: Sophia has three tickets to a concert. Yolanda, Michael, Kevin, and Marissa have all stated they would like to go to the concert with Sophia. To be fair, Sophia decides to randomly select the two people who can go to the concert with her.

(a) Determine the sample space of the experiment. In other words, list all possible simple random samples of size $n = 2$.
(b) Compute the probability of the event "Michael and Kevin attend the concert."
(c) Compute the probability of the event "Marissa attends the concert."

Approach: First, we determine the outcomes in the sample space by making a table. The probability of each event is the number of outcomes in the event divided by the number of outcomes in the sample space.

Solution

(a) The sample space is listed in Table 4.

Table 4		
Yolanda, Michael	Yolanda, Kevin	Yolanda, Marissa
Michael, Marissa	Kevin, Marissa	Michael, Kevin

(b) We have $N(S) = 6$, and there is one way the event "Michael and Kevin attend the concert" can occur. Therefore, the probability that Michael and Kevin attend the concert is $\frac{1}{6}$.

(c) We have $N(S) = 6$ and there are three ways the event "Marissa attends the concert" can occur. The probability that Marissa will attend is $\frac{3}{6} = 0.5 = 50\%$.

Now Work Problems 31 and 45.

| EXAMPLE 7 | **Comparing the Classical Method and Empirical Method** |

Problem: Suppose a survey is conducted in which 500 families with three children are asked to disclose the gender of their children. Based on the results, it was found that 180 of the families had two boys and one girl.

(a) Estimate the probability of having two boys and one girl in a three-child family using the empirical method.
(b) Compute the probability of having two boys and one girl in a three-child family using the classical method, assuming boys and girls are equally likely.

Approach To answer part (a), we determine the relative frequency of the event "two boys and one girl." To answer part (b), we must count the number of ways the event "two boys and one girl" can occur and divide this by the number of possible outcomes for this experiment.

Solution

(a) The empirical probability of the event $E =$ "two boys and one girl" is

$$P(E) \approx \text{ relative frequency of } E = \frac{180}{500} = 0.36 = 36\%$$

There is about a 36% probability that a family of three children will have two boys and one girl.

(b) To determine the sample space, we construct a **tree diagram** to list the equally likely outcomes of the experiment. We draw two branches corresponding to the two possible outcomes (boy or girl) for the first repetition of the experiment (the first child). For the second child, we draw four branches: two branches originate from the first boy and two branches originate from the first girl. This is repeated for the third child. See Figure 3, where B stands for boy and G stands for girl.

Figure 3

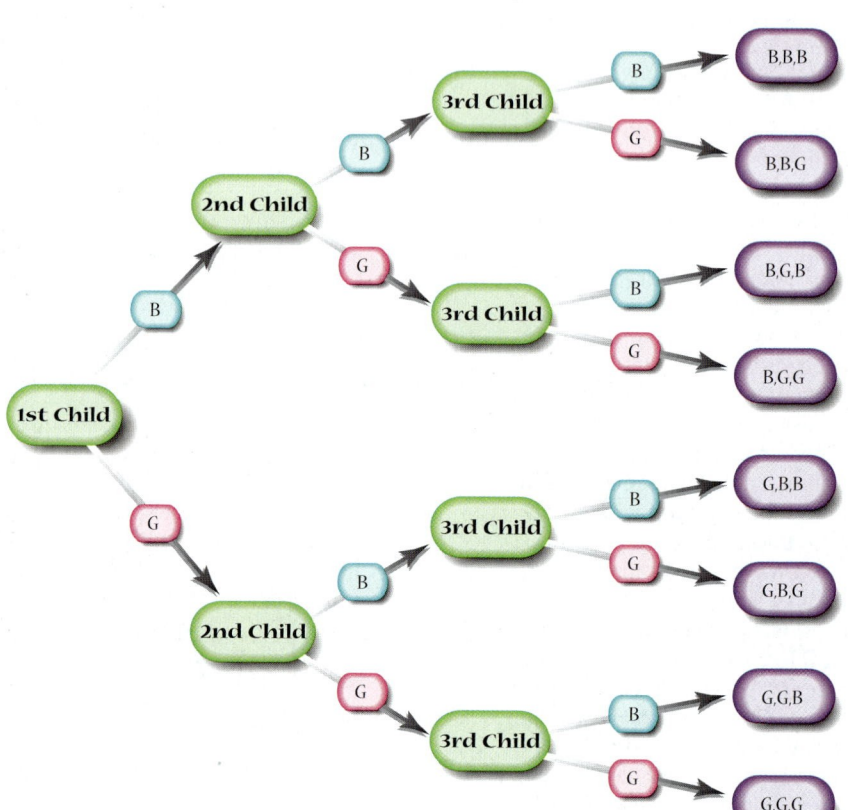

The sample space S of this experiment is found by following each branch to identify all the possible outcomes of the experiment:

$$S = \{BBB, BBG, BGB, BGG, GBB, GBG, GGB, GGG\}$$

So $N(S) = 8$.

For the event E = "two boys and a girl" = {BBG, BGB, GBB}, we have $N(E) = 3$. Since the outcomes are equally likely (for example, BBG is just as likely as BGB), the probability of E is

$$P(E) = \frac{N(E)}{N(S)} = \frac{3}{8} = 0.375 = 37.5\%$$

There is a 37.5% probability that a family of three children will have two boys and one girl.

In comparing the results of Examples 7(a) and 7(b), we notice that the two probabilities are slightly different. Empirical probabilities and classical probabilities often differ in value. As the number of repetitions in a probability experiment increases, the empirical probability should get closer to the classical probability. That is, the classical probability is the theoretical relative frequency of an event after a large number of trials of the probability experiment. However, it is also possible that the two probabilities differ because having a boy or girl are not equally likely events. (Maybe the probability of having a boy is 50.5% and the probability of having a girl is 49.5%.) If this is the case, the empirical probability will not get closer to the classical probability.

 ## Use Simulation to Obtain Data Based on Probabilities

Suppose we wanted to determine the probability of having a boy. Using classical methods, we assume that having a boy is just as likely as having a girl, so the probability of having a boy is 50%. We could also approximate this probability by looking in the *Statistical Abstract of the United States* under Vital Statistics and determining the number of boys and girls born for the most recent year for which data are available. For example, in 2002, there were 2,058,000 boys born and 1,964,000 girls. Based on empirical evidence, the probability of a boy is approximately $\dfrac{2,058,000}{(2,058,000 + 1,964,000)} = 0.512 = 51.2\%$.

However, instead of obtaining data from existing sources, we could simulate having babies by using a graphing calculator or statistical software to replicate the experiment as many times as we like.

EXAMPLE 8 | ### Simulating Probabilities

Problem

(a) Simulate the experiment of sampling 100 babies.

(b) Simulate the experiment of sampling 1000 babies.

Approach: To simulate probabilities, we use a random-number generator available in statistical software and most calculators. We assume the outcomes "have a boy" and "have a girl" are equally likely.

Solution

(a) We use MINITAB to perform the simulation. Set the seed in MINITAB to any value you wish, say 1204. Use the Integer Distribution* to generate random data that simulate having babies. If we agree to let 0 represent a boy and 1 represent a girl, we can approximate the probability of having a girl by summing the number of 1's (adding up the number of girls) and dividing by the number of repetitions of the experiment, 100. See Figure 4.

Figure 4

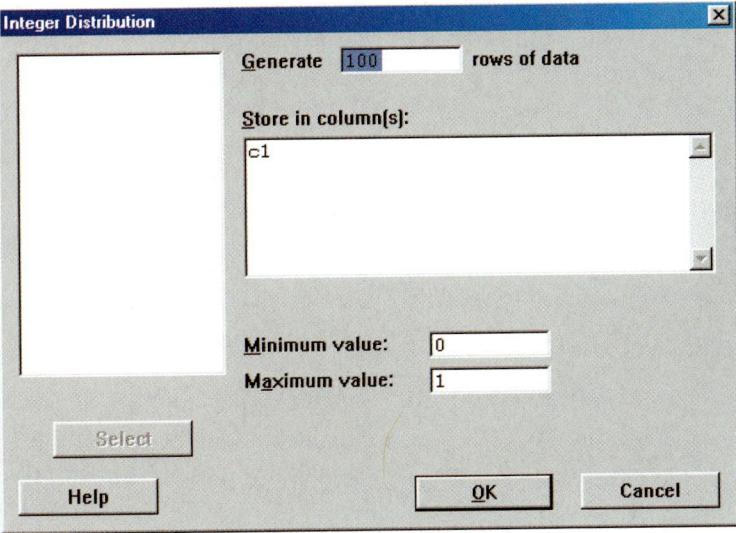

 **Historical Note**

Blaise Pascal was born on June 19, 1623, in Clermont, France. Pascal's father felt that Blaise should not be taught mathematics before age 15. Pascal couldn't resist studying mathematics on his own, and at the age of 12 started to teach himself geometry. In December 1639, the Pascal family moved to Rouen, where Pascal's father had been appointed as a tax collector. Between 1642 and 1645, Pascal worked on developing a calculator to help his father collect taxes. In correspondence with Fermat, he helped develop the theory of probability. This correspondence consisted of five letters written in the summer of 1654. They considered the dice problem and the problem of points. The dice problem deals with determining the expected number of times a pair of dice must be thrown before a pair of sixes is observed. The problem of points asks how to divide the stakes if a game of dice is incomplete. They solved the problem of points for a two-player game, but did not solve it for three or more players.

---

*The Integer Distribution involves a mathematical formula that uses a seed number to generate a sequence of equally likely random integers. Consult the technology manuals for setting the seed and generating sequences of integers.

Using MINITAB's Tally command, we can determine the number of 0s and 1s that MINITAB randomly generated. See Figure 5.

Figure 5

Summary Statistics for Discrete Variables

```
C1      Count      Percent
0          48        48.00
1          52        52.00
N=        100
```

Based on Figure 5, we approximate that there is a 48% probability of having a boy and a 52% probability of having a girl.

(b) Again, set the seed to 1204. Figure 6 shows the results of simulating the birth of 1000 babies.

Figure 6

Summary Statistics for Discrete Variables

```
C1      Count      Percent
0         501        50.10
1         499        49.90
N=       1000
```

We approximate that there is a 50.1% probability of having a boy and a 49.9% probability of having a girl.

Notice that more repetitions of the experiment (100 babies versus 1000 babies) result in a probability closer to $\frac{1}{2}$ for each gender. Notice also that our simulation results differ from the empirical probability of having a boy, namely 0.512. This serves as evidence against the belief that the probability of having a boy is $\frac{1}{2}$.

Now Work Problem 49.

5 Understand Subjective Probabilities

Suppose a sports reporter is asked what he thinks the chances are for the Boston Red Sox to return to the World Series. The sports reporter will likely process information about the Red Sox, their pitching staff, lead-off hitter, and so on, and then come up with an educated guess of the likelihood. The reporter may respond that there is a 20% chance the Red Sox return to the World Series. This forecast is a probability although it is not based on relative frequencies. We cannot, after all, repeat the experiment of playing a season under the same circumstances (same players, schedule, and so on) over and over. Nonetheless, the forecast of 20% does satisfy the criterion that a probability be between 0 and 1, inclusive. This forecast is known as a *subjective probability*.

Definition

A **subjective probability** of an outcome is a probability obtained on the basis of personal judgment.

It is important to understand that subjective probabilities are perfectly legitimate and are often the only method of assigning likelihood to an outcome. As another example, a financial reporter may ask an economist about the likelihood the economy will fall into recession next year. Again, we cannot conduct an experiment n times to obtain a relative frequency. The economist must use his or her knowledge of the current conditions of the economy and make an educated guess as to the likelihood of recession.

5.1 ASSESS YOUR UNDERSTANDING

Concepts and Vocabulary

1. Describe the difference between classical and empirical probability.
2. What is the probability of an event that is impossible? Suppose a probability is approximated to be zero based on empirical results. Does this mean the event is impossible?
3. In computing classical probabilities, all outcomes must be equally likely. Explain what this means.
4. What does it mean for an event to be unusual? Why should the cutoff for identifying unusual events not always be 0.05?

5. *True or False:* In a probability model, the sum of the probabilities of all outcomes must equal 1.
6. *True or False:* Probability is a measure of the likelihood of a random phenomenon or chance behavior.
7. In probability a(n) _____ is any process that can be repeated in which the results are uncertain.
8. A(n) _____ is any collection of outcomes from a probability experiment.

Skill Building

9. Verify that the following is a probability model. What do NW we call the outcome "blue"?

Color	Probability
red	0.3
green	0.15
blue	0
brown	0.15
yellow	0.2
orange	0.2

10. Verify that the following is a probability model. If the model represents the colors of M&Ms in a bag of plain M&Ms, explain what the model implies.

Color	Probability
red	0
green	0
blue	0
brown	0
yellow	1
orange	0

11. Why is the following not a probability model?

Color	Probability
red	0.3
green	−0.3
blue	0.2
brown	0.4
yellow	0.2
orange	0.2

12. Why is the following not a probability model?

Color	Probability
red	0.1
green	0.1
blue	0.1
brown	0.4
yellow	0.2
orange	0.3

13. Which of the following numbers could be the probability of an event?

$$0, 0.01, 0.35, -0.4, 1, 1.4$$

14. Which of the following numbers could be the probability of an event?

$$1.5, \frac{1}{2}, \frac{3}{4}, \frac{2}{3}, 0, -\frac{1}{4}$$

15. In five-card stud poker, a player is dealt five cards. The probability that the player is dealt two cards of the same value and three other cards of different value so that the player has a pair is 0.42. Explain what this probability means. If you play five-card stud 100 times, will you get a pair exactly 42 times? Why or why not?

16. In seven-card stud poker, a player is dealt seven cards. The probability that the player is dealt two cards of the same value and five other cards of different value so that the player has a pair is 0.48. Explain what this probability means. If you play seven-card stud 100 times, will you get a pair exactly 48 times? Why or why not?

17. Suppose you toss a coin 100 times and get 95 heads and 5 tails. Based on these results, what is the probability that the next flip results in a head?

18. Suppose you roll a die 100 times and get six 80 times. Based on these results, what is the probability the next roll results in six?

19. Bob is asked to construct a probability model for rolling a pair of fair dice. He lists the outcomes as 2, 3, 4, 5, 6, 7, 8, 9,

10, 11, 12. Because there are 11 outcomes, he reasoned, the probability of rolling a two must be $\frac{1}{11}$. What is wrong with Bob's reasoning?

20. **Blood Types** A person can have one of four blood types: A, B, AB, or O. If people are randomly selected, is the probability they have blood type A equal to $\frac{1}{4}$? Why?

21. If a person rolls a six-sided die and then flips a coin, describe the sample space of possible outcomes using 1, 2, 3, 4, 5, 6 for the die outcomes and H, T for the coin outcomes.

22. If a basketball player shoots three free throws, describe the sample space of possible outcomes using S for a made free throw and F for a missed free throw.

23. According to the U.S. Department of Education, the proportion of 3-year-olds that is enrolled in day care is 0.44. What is the probability a randomly selected three-year-old is enrolled in day care?

24. According to the American Veterinary Medical Association, the proportion of households owning a dog is 0.361. What is the probability a randomly selected household owns a dog?

For Problems 25–28, let the sample space be S = {1, 2, 3, 4, 5, 6, 7, 8, 9, 10}. Suppose the outcomes are equally likely.

25. Compute the probability of the event $E = \{1, 2, 3\}$.

26. Compute the probability of the event $F = \{3, 5, 9, 10\}$.

27. Compute the probability of the event $E =$ "an even number."

28. Compute the probability of the event $F =$ "an odd number."

Applying the Concepts

29. **Play Sports?** In a survey of 500 randomly selected high school students, it was determined that 288 played organized sports. What is the probability that a randomly selected high school student plays organized sports?

30. **Volunteer?** In a survey of 1100 female adults (18 years of age or older), it was determined that 341 volunteered at least once in the past year. What is the probability that a randomly selected adult female volunteered at least once in the past year?

31. **Planting Tulips** A bag of 100 tulip bulbs purchased from a nursery contains 40 red tulip bulbs, 35 yellow tulip bulbs, and 25 purple tulip bulbs.
    (a) What is the probability that a randomly selected tulip bulb is red?
    (b) What is the probability that a randomly selected tulip bulb is purple?

32. **Golf Balls** The local golf store sells an "onion bag" that contains 80 "experienced" golf balls. Suppose the bag contains 35 Titleists, 25 Maxflis, and 20 Top-Flites.
    (a) What is the probability that a randomly selected golf ball is a Titleist?
    (b) What is the probability that a randomly selected golf ball is a Top-Flite?

33. **Roulette** In the game of roulette, a wheel consists of 38 slots numbered 0, 00, 1, 2, . . . , 36. (See the photo.) To play the game, a metal ball is spun around the wheel and is allowed to fall into one of the numbered slots.

    (a) Determine the sample space.
    (b) Determine the probability that the metal ball falls into the slot marked 8. Interpret this probability.
    (c) Determine the probability that the metal ball lands in an odd slot. Interpret this probability.

34. **Birthdays** Exclude leap years from the following calculations and assume each birthday is equally likely:
    (a) Determine the probability that a randomly selected person has a birthday on the 1st day of a month. Interpret this probability.
    (b) Determine the probability that a randomly selected person has a birthday on the 31st day of a month. Interpret this probability.
    (c) Determine the probability that a randomly selected person was born in December. Interpret this probability.
    (d) Determine the probability that a randomly selected person has a birthday on November 8. Interpret this probability.
    (e) If you just met somebody and she asked you to guess her birthday, are you likely to be correct?
    (f) Do you think it is appropriate to use the methods of classical probability to compute the probability that a person is born in December?

35. **Genetics** A gene is composed of two alleles. An allele can be either dominant or recessive. Suppose a husband and wife, who are both carriers of the sickle-cell anemia allele but do not have the disease, decide to have a child. Because both parents are carriers of the disease, each has one dominant normal-cell allele and one recessive sickle-cell allele. Therefore, the genotype of each parent is Ss. Each parent contributes one allele to his or her offspring, with each allele being equally likely.
    (a) List the possible genotypes of their offspring.
    (b) What is the probability that the offspring will have sickle-cell anemia? In other words, what is the probability the offspring will have genotype ss? Interpret this probability.

(c) What is the probability that the offspring will not have sickle-cell anemia, but will be a carrier? In other words, what is the probability that the offspring will have one dominant normal-cell allele and one recessive sickle-cell allele? Interpret this probability.

36. **More Genetics** In Problem 35, we learned that for some diseases, such as sickle-cell anemia, an individual will get the disease only if he receives both recessive alleles. This is not always the case. For example, Huntington's disease only requires one dominant gene for an individual to contract the disease. Suppose a husband and wife, who both have a dominant Huntington's disease allele (S) and a normal recessive allele (s), decide to have a child.

(a) List the possible genotypes of their offspring.
(b) What is the probability that the offspring will not have Huntington's disease? In other words, what is the probability the offspring will have genotype ss? Interpret this probability.
(c) What is the probability that the offspring will have Huntington's disease?

37. **College Survey** In a national survey conducted by the Centers for Disease Control to determine college students' health-risk behaviors, college students were asked, "How often do you wear a seat belt when riding in a car driven by someone else?" The frequencies appear in the following table:

Response	Frequency
Never	125
Rarely	324
Sometimes	552
Most of the time	1257
Always	2518

(a) Construct a probability model for seat-belt use by a passenger.
(b) Would you consider it unusual to find a college student who never wears a seat belt when riding in a car driven by someone else? Why?

38. **College Survey** In a national survey conducted by the Centers for Disease Control to determine college students' health-risk behaviors, college students were asked, "How often do you wear a seat belt when driving a car?" The frequencies appear in the following table:

Response	Frequency
Never	118
Rarely	249
Sometimes	345
Most of the time	716
Always	3093

(a) Construct a probability model for seat-belt use by a driver.
(b) Is it unusual for a college student never to wear a seat belt when driving a car? Why?

39. **Larceny Theft** A police officer randomly selected 595 police records of larceny thefts. The following data represent the number of offenses for various types of larceny thefts.

Type of Larceny Theft	Number of Offenses
Pocket picking	5
Purse snatching	5
Shoplifting	118
From motor vehicles	197
Motor vehicle accessories	77
Bicycles	43
From buildings	105
From coin-operated machines	45

Source: U.S. Federal Bureau of Investigation

(a) Construct a probability model for type of larceny theft.
(b) Are purse-snatching larcenies unusual?
(c) Are larcenies from coin-operated machines unusual?

40. **Multiple Births** The following data represent the number of live multiple-delivery births (three or more babies) in 2002 for women 15 to 44 years old.

Age	Number of Multiple Births
15–19	93
20–24	511
25–29	1628
30–34	2832
35–39	1843
40–44	377

Source: National Vital Statistics Reports, Vol. 52, No. 10, December 17, 2003

(a) Construct a probability model for number of multiple births.
(b) In the sample space of all multiple births, are multiple births for 15- to 19-year-old mothers unusual?
(c) In the sample space of all multiple births, are multiple births for 40- to 44-year-old mothers unusual?

Problems 41–44, use the given table which lists six possible assignments of probabilities for tossing a coin twice, to answer the following questions.

Assignments	Sample Space			
	HH	**HT**	**TH**	**TT**
A	$\frac{1}{4}$	$\frac{1}{4}$	$\frac{1}{4}$	$\frac{1}{4}$
B	0	0	0	1
C	$\frac{3}{16}$	$\frac{5}{16}$	$\frac{5}{16}$	$\frac{3}{16}$
D	$\frac{1}{2}$	$\frac{1}{2}$	$-\frac{1}{2}$	$\frac{1}{2}$
E	$\frac{1}{4}$	$\frac{1}{4}$	$\frac{1}{4}$	$\frac{1}{8}$
F	$\frac{1}{9}$	$\frac{2}{9}$	$\frac{2}{9}$	$\frac{4}{9}$

41. Which of the assignments of probabilities are consistent with the definition of a probability model?

42. Which of the assignments of probabilities should be used if the coin is known to be fair?

43. Which of the assignments of probabilities should be used if the coin is known to always come up tails?

44. Which of the assignments of probabilities should be used if tails is twice as likely to occur as heads?

45. Going to Disney World John, Roberto, Clarice, Dominique, and Marco work for a publishing company. The company wants to send two employees to a statistics conference in Orlando. To be fair, the company decides that the two individuals who get to attend will have their names drawn from a hat. This is like obtaining a simple random sample of size 2.
  (a) Determine the sample space of the experiment. That is, list all possible simple random samples of size $n = 2$.
  (b) What is the probability that Clarice and Dominique attend the conference?
  (c) What is the probability that Clarice attends the conference?
  (d) What is the probability that John stays home?

46. Six Flags Six Flags over Mid-America in St. Louis has six roller coasters: The Screamin' Eagle, The Boss, River King Mine Train, Batman the Ride, Mr. Freeze, and Ninja. After a long day at the park, Ethan's parents tell him that he can ride two more coasters before leaving (but not the same one twice). Because he likes the rides equally, Ethan decides to randomly select the two coasters by drawing their names from his hat.
  (a) Determine the sample space of the experiment. That is, list all possible simple random samples of size $n = 2$.
  (b) What is the probability that Ethan will ride Batman and Mr. Freeze?
  (c) What is the probability that Ethan will ride the Screamin' Eagle?
  (d) What is the probability that Ethan will ride neither River King Mine Train nor Ninja?

47. Barry Bonds On October 5, 2001, Barry Bonds broke Mark McGwire's home-run record for a single season by hitting his 71st and 72nd home runs. Bonds went on to hit one more home run before the season ended, for a total of 73. Of the 73 home runs, 24 went to right field, 26 went to right center field, 11 went to center field, 10 went to left center field, and 2 went to left field. (*Source:* Baseball-almanac.com)
  (a) What is the probability that a randomly selected home run was hit to right field?
  (b) What is the probability that a randomly selected home run was hit to left field?
  (c) Was it unusual for Barry Bonds to hit a home run to left field? Explain.

48. Rolling a Die
  (a) Roll a single die 50 times, recording the result of each roll of the die. Use the results to approximate the probability of rolling a three.
  (b) Roll a single die 100 times, recording the result of each roll of the die. Use the results to approximate the probability of rolling a three.
  (c) Compare the results of (a) and (b) to the classical probability of rolling a three.

49. Simulation Use a graphing calculator or statistical software to simulate rolling a six-sided die 100 times, using an integer distribution with numbers one through six.
  (a) Use the results of the simulation to compute the probability of rolling a one.
  (b) Repeat the simulation. Compute the probability of rolling a one.
  (c) Simulate rolling a six-sided die 500 times. Compute the probability of rolling a one.
  (d) Which simulation resulted in the closest estimate to the probability that would be obtained using the classical method?

50. Classifying Probability Determine whether the following probabilities are computed using classical methods, empirical methods, or subjective methods.
  (a) The probability of having eight girls in an eight-child family is 0.390625%.
  (b) On the basis of a study of families with eight children, the probability of a family having eight girls is 0.54%.
  (c) According to a sports analyst, the probability that the Chicago Bears will win their next game is about 30%.
  (d) On the basis of clinical trials, the probability of efficacy of a new drug is 75%.

51. Checking for Loaded Dice You suspect a pair of dice to be loaded and conduct a probability experiment by rolling each die 200 times. The outcome of the experiment is listed in the following table:

Value of Die	Frequency	Value of Die	Frequency
1	105	4	49
2	47	5	51
3	44	6	104

Do you think the dice are loaded? Why?

52. Conduct a survey in your school by randomly asking 50 students whether they drive to school. Based on the results of the survey, approximate the probability that a randomly selected student drives to school.

53. In 2004, the median income of families in the United States was $57,500. What is the probability that a randomly selected family has an income greater than $57,500?

54. In 2004, 17% of Florida's population was 65 and over (*Source*: U.S. Census Bureau). What is the probability that a randomly selected Floridian is 65 or older?

55. The Probability Applet Load the long run probability applet on your computer.
  (a) Choose the "simulating the probability of a head with a fair coin" applet and simulate flipping a fair coin 10 times. What is the estimated probability of a head based on these 10 trials?

  (b) Reset the applet. Simulate flipping a fair coin 10 times a second time. What is the estimated probability of a head based on these 10 trials? Compare the results to part (a).
  (c) Reset the applet. Simulate flipping a fair coin 1000 times. What is the estimated probability of a head based on these 1000 trials? Compare the results to part (a).
  (d) Reset the applet. Simulate flipping a fair coin 1000 times. What is the estimated probability of a head based on these 1000 trials? Compare the results to part (c).
  (e) Choose the "simulating the probability of head with an unfair coin (P(H) = 0.2)" applet and simulate flipping a coin 1000 times. What is the estimated probability of a head based on these 1000 trials? If you did not know that the probability of heads was set to 0.2, what would you conclude about the coin? Why?

Technology Step-by-Step   Simulation

TI-83/84 Plus

Step 1: Set the seed by entering any number on the HOME screen. Press the STO ⇒ button, press the MATH button, highlight the PRB menu, and highlight 1:rand and hit ENTER. With the cursor on the HOME screen, hit ENTER.

Step 2: Press the MATH button and highlight the PRB menu. Highlight 5:randInt(and hit ENTER.

Step 3: After the randInt(on the HOME screen type 1,n, number of repetitions of experiment), where n is the number of equally likely outcomes. For example, to simulate rolling a single die 50 times, we type

```
randInt(1, 6, 50)
```

Step 4: Press the STO ⇒ button and then 2nd 1, and hit ENTER to store the data in L1.

Step 5: Draw a histogram of the data using the outcomes as classes. TRACE to obtain outcomes.

MINITAB

Step 1: Set the seed by selecting the **Calc** menu and highlighting **Set Base** .... Insert any seed you wish into the cell and click OK.

Step 2: Select the **Calc** menu, highlight **Random Data**, and then highlight **Integer**. To simulate rolling a single die 100 times, fill in the window as shown in Figure 4 on page 259.

Step 3: Select the **Stat** menu, highlight **Tables**, and then highlight **Tally** .... Enter C1 into the variables cell. Make sure that the Counts box is checked and click OK.

Excel

Step 1: With cell A1 selected, press the *fx* button.

Step 2: Highlight Math & Trig in the Function category window. Then highlight RANDBETWEEN in the Function Name: window. Click OK.

Step 3: To simulate rolling a die 50 times, enter 1 for the lower limit and 6 for the upper limit. Click OK.

Step 4: Copy the contents of cell A1 into cells A2 through A50.

5.2   The Addition Rule and Complements

Objectives

1   **Use the Addition Rule for disjoint events**
2   **Use the General Addition Rule**
3   **Compute the probability of an event using the Complement Rule**

1   **Use the Addition Rule for Disjoint Events**

Now we introduce some more rules for computing probabilities. However, before we present these rules, we must discuss *disjoint events.*

Definition

Two events are **disjoint** if they have no outcomes in common. Another name for disjoint events is **mutually exclusive** events.

In Other Words
Two events are disjoint if they cannot occur at the same time.

It is often helpful to draw pictures of events. Such pictures, called **Venn diagrams**, represent events as circles enclosed in a rectangle. The rectangle represents the sample space, and each circle represents an event. For example, suppose we randomly select chips from a bag. Each chip is labeled 0, 1, 2, 3, 4, 5, 6, 7, 8, 9. Let E represent the event "choose a number less than or equal to 2," and let F represent the event "choose a number greater than or equal to 8." Because E and F do not have any outcomes in common, they are disjoint. Figure 7 shows a Venn diagram of these disjoint events.

Figure 7

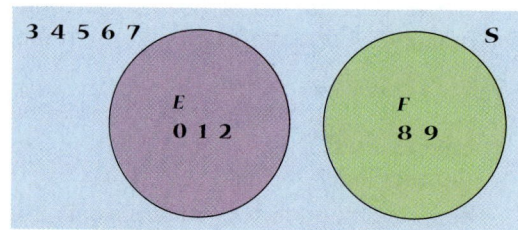

Notice that the outcomes in event E are inside circle E, and the outcomes in event F are inside the circle F. All outcomes that are in the sample space that are not in E nor in F are outside the circles, but inside the rectangle. From this diagram, we know that $P(E) = \dfrac{n(E)}{n(S)} = \dfrac{3}{10} = 0.3$ and $P(F) = \dfrac{n(F)}{n(S)} = \dfrac{2}{10} = 0.2$. In addition, $P(E \text{ or } F) = P(E) + P(F) = 0.3 + 0.2 = 0.5$. This result occurs because of the Addition Rule for Disjoint Events.

In Other Words
The Addition Rule for Disjoint Events states that if you have two events that have no outcomes in common, the probability that one or the other occurs is the sum of their probabilities.

Addition Rule for Disjoint Events

If E and F are disjoint (or mutually exclusive) events, then

$$P(E \text{ or } F) = P(E) + P(F)$$

The Addition Rule for Disjoint Events can be extended to more than two disjoint events. In general, if E, F, G, … each have no outcomes in common (they are pairwise disjoint), then

$$P(E \text{ or } F \text{ or } G \text{ or} \dots) = P(E) + P(F) + P(G) + \cdots$$

Let event G represent "the number is a 5 or 6." The Venn diagram in Figure 8 illustrates the Addition Rule for more than two disjoint events using the chip example. Notice that none of the events has any outcomes in common. So, from the Venn diagram, we can see $P(E) = \dfrac{n(E)}{n(S)} = \dfrac{3}{10} = 0.3$, $P(F) = \dfrac{n(F)}{n(S)} = \dfrac{2}{10} = 0.2$, and $P(G) = \dfrac{n(G)}{n(S)} = \dfrac{2}{10} = 0.2$. In addition, $P(E \text{ or } F \text{ or } G) = P(E) + P(F) + P(G) = 0.3 + 0.2 + 0.2 = 0.7$.

Figure 8

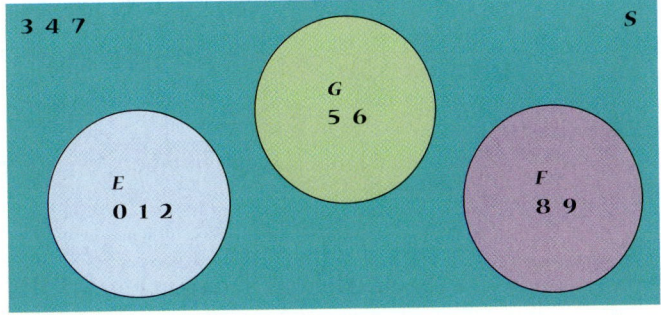

EXAMPLE 1   **Benford's Law and the Addition Rule for Disjoint Events**

Problem: Our number system consists of the digits 0, 1, 2, 3, 4, 5, 6, 7, 8, and 9. The first significant digit in any number must be 1, 2, 3, 4, 5, 6, 7, 8, or 9, because we do not write numbers such as 12 as 012, for example. Although we may think that each digit appears with equal frequency so that each digit has a $\dfrac{1}{9}$ probability of being the first significant digit, this in fact, is not true. In 1881, Simon Necomb discovered that digits do not occur with equal frequency. This same result was discovered again in 1938 by physicist Frank Benford. After studying lots and lots of data, he was able to assign probabilities of occurrence for each of the first digits, as shown in Table 5.

Table 5								
Digit	1	2	3	4	5	6	7	8	9
Probability	0.301	0.176	0.125	0.097	0.079	0.067	0.058	0.051	0.046

Source: The First Digit Phenomenon, T.P. Hill, *American Scientist*, July–August, 1998.

The probability model is now known as Benford's Law and plays a major role in identifying fraudulent data on tax returns and accounting books.

(a) Verify Benford's Law is a probability model.
(b) Use Benford's Law to determine the probability that a randomly selected first digit is 1 or 2.
(c) Use Benford's Law to determine the probability that a randomly selected first digit is at least 6.

Approach: For part (a), we need to verify that each probability is between 0 and 1 and that the sum of all probabilities equals 1. For parts (b) and (c), we use the Addition Rule for Disjoint Events.

Solution

(a) In looking at Table 5, we see that each probability is between 0 and 1. In addition, the sum of all the probabilities is 1.

$$0.301 + 0.176 + 0.125 + \cdots + 0.046 = 1$$

Because rules 1 and 2 are satisfied, Table 5 represents a probability model.

(b)
$$
\begin{aligned}
P(1 \text{ or } 2) &= P(1) + P(2) \\
&= 0.301 + 0.176 \\
&= 0.477
\end{aligned}
$$

If we looked at 100 numbers, we would expect about 48 to begin with 1 or 2.

(c)
$$
\begin{aligned}
P(\text{at least } 6) &= P(6 \text{ or } 7 \text{ or } 8 \text{ or } 9) \\
&= P(6) + P(7) + P(8) + P(9) \\
&= 0.067 + 0.058 + 0.051 + 0.046 \\
&= 0.222
\end{aligned}
$$

If we looked at 100 numbers, we would expect about 22 to begin with 6, 7, 8, or 9.

EXAMPLE 2   **A Deck of Cards and the Addition Rule for Disjoint Events**

Problem: Suppose a single card is selected from a standard 52-card deck, such as the one shown in Figure 9.

Figure 9

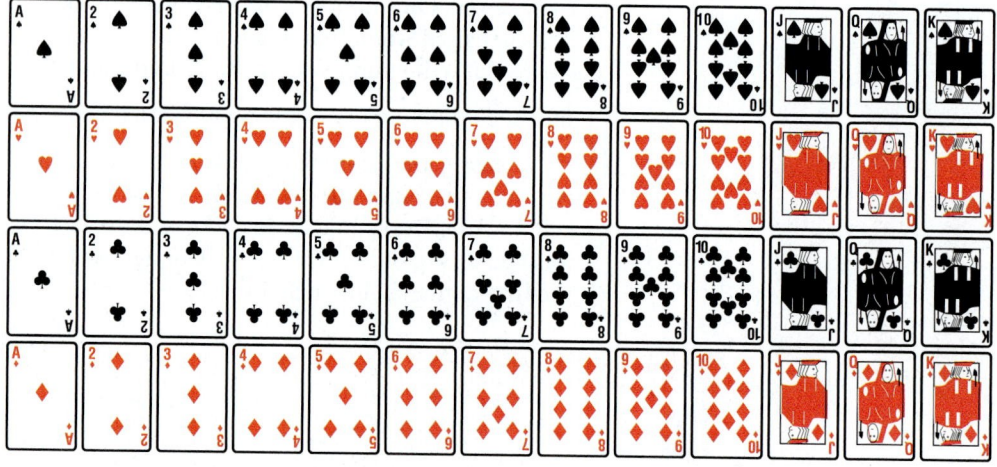

(a) Compute the probability of the event E = "drawing a king."
(b) Compute the probability of the event E = "drawing a king" or F = "drawing a queen."
(c) Compute the probability of the event E = "drawing a king" or F = "drawing a queen" or G = "drawing a jack."

Approach: We will use the classical method for computing the probabilities because the outcomes are equally likely and easy to count. We use the Addition Rule for Disjoint Events to compute the probabilities in parts (b) and (c) because the events are mutually exclusive. For example, you cannot simultaneously draw a king and a queen.

Solution: The sample space consists of the 52 cards in the deck, so $N(S) = 52$.

(a) A standard deck of cards has four kings, so $N(E) = 4$. Therefore,
$$P(\text{king}) = P(E) = \frac{N(E)}{N(S)} = \frac{4}{52} = \frac{1}{13}.$$

(b) A standard deck of cards also has four queens. Because events E and F are mutually exclusive (you cannot draw a king and queen simultaneously), we use the Addition Rule for Disjoint Events. So

$$P(\text{king or queen}) = P(E \text{ or } F)$$

$$= P(E) + P(F)$$

$$= \frac{4}{52} + \frac{4}{52} = \frac{8}{52} = \frac{2}{13}$$

(c) Because events E, F, and G are mutually exclusive, we use the Addition Rule for Disjoint Events extended to two or more disjoint events. So

$$P(\text{king or queen or jack}) = P(E \text{ or } F \text{ or } G)$$

$$= P(E) + P(F) + P(G)$$

$$= \frac{4}{52} + \frac{4}{52} + \frac{4}{52} = \frac{12}{52} = \frac{3}{13}$$

Now Work Problems 25(a)–(c).

2   **Use the General Addition Rule**

A question that you may be asking yourself is this: "What if I need to compute the probability of two events that are not disjoint?" This is a legitimate question and one that has an answer.

Consider the chip example. Suppose we are randomly selecting chips from a bag. Each chip is labeled 0, 1, 2, 3, 4, 5, 6, 7, 8, or 9. Let E represent the event "choose an odd number," and let F represent the event "choose a number less than or equal to 4." Because $E = \{1, 3, 5, 7, 9\}$ and $F = \{0, 1, 2, 3, 4\}$ have the outcomes 1 and 3 in common, the events are not disjoint. Figure 10 shows a Venn diagram of these events.

Figure 10

The overlapping region is E and F.

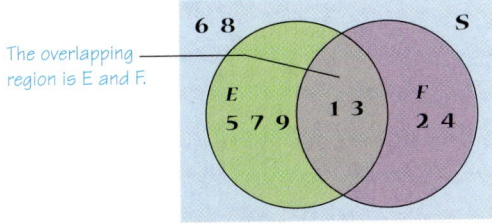

We can compute $P(E \text{ or } F)$ directly by counting because each outcome is equally likely. There are 7 outcomes in E or F and 10 outcomes in the sample space, so

$$P(E \text{ or } F) = \frac{N(E \text{ or } F)}{N(S)}$$

$$= \frac{7}{10}$$

If we attempt to compute $P(E \text{ or } F)$ using the Addition Rule for Disjoint Events, we obtain the following:

$$P(E \text{ or } F) = P(E) + P(F)$$

$$= \frac{5}{10} + \frac{4}{10}$$

$$= \frac{9}{10}$$

This result is incorrect because we counted the outcomes 1 and 3 twice: once for event E and once for event F. To avoid this double counting, we have to subtract the probability corresponding to the overlapping region, E and F. That is, we have to subtract $P(E \text{ and } F) = \dfrac{2}{10}$ from the result and obtain

$$P(E \text{ or } F) = P(E) + P(F) - P(E \text{ and } F)$$

$$= \frac{5}{10} + \frac{4}{10} - \frac{2}{10}$$

$$= \frac{7}{10}$$

which agrees with the result we obtained by counting. These results can be generalized in the following rule:

> **The General Addition Rule**
>
> For any two events E and F,
>
> $$P(E \text{ or } F) = P(E) + P(F) - P(E \text{ and } F)$$

EXAMPLE 3   **Computing Probabilities for Events That Are Not Disjoint**

Problem: Suppose a single card is selected from a standard 52-card deck. Compute the probability of the event E = "drawing a king" or H = "drawing a diamond."

Approach: The events are not disjoint because the outcome "king of diamonds" is in both events, so we use the General Addition Rule.

Solution
$$P(\text{king or diamond}) = P(\text{king}) + P(\text{diamond}) - P(\text{king of diamonds})$$

$$= \frac{4}{52} + \frac{13}{52} - \frac{1}{52}$$

$$= \frac{16}{52} = \frac{4}{13}$$

Now Work Problem 29.

Consider the data shown in Table 6, which represent the marital status of males and females 18 years old or older in the United States in 2003.

Table 6 is called a **contingency table** or **two-way table**, because it relates two categories of data. The **row variable** is marital status, because each row

Table 6			
	Males (in millions)	**Females** (in millions)	**Totals** (in millions)
Never married	28.6	23.3	51.9
Married	62.1	62.8	124.9
Widowed	2.7	11.3	14.0
Divorced	9.0	12.7	21.7
Totals (in millions)	102.4	110.1	212.5

Source: U.S. Census Bureau, *Current Population Reports*

in the table describes the marital status of each individual. The **column variable** is gender. Each box inside the table is called a **cell**. For example, the cell corresponding to married individuals who are male is in the second row, first column. Each cell contains the frequency of the category: There were 62.1 million married males in the United States in 2003. Put another way, in the United States in 2003, there were 62.1 million individuals who were male *and* married.

EXAMPLE 4   **Using the Addition Rule with Contingency Tables**

Problem: Using the data in Table 6, determine the following:

(a) Determine the probability that a randomly selected U.S. resident 18 years old or older is male.
(b) Determine the probability that a randomly selected U.S. resident 18 years old or older is widowed.
(c) Determine the probability that a randomly selected U.S. resident 18 years old or older is widowed or divorced.
(d) Determine the probability that a randomly selected U.S. resident 18 years old or older is male or widowed.

Approach: We first add up the entries in each row and column so that we get the total number of people in each category. We can then determine the probabilities using either the Addition Rule for Disjoint Events or the General Addition Rule.

Solution: Add the entries in each column. For example, in the male column, we find that there are $28.6 + 62.1 + 2.7 + 9.0 = 102.4$ million males 18 years old or older in the United States. Add the entries in each row. For example, in the never married row, we find there are $28.6 + 23.3 = 51.9$ million U.S. residents 18 years old or older who have never married. Adding the row totals or column totals, we find there are $102.4 + 110.1 = 51.9 + 124.9 + 14.0 + 21.7 = 212.5$ million U.S. residents 18 years old or older.

(a) There are 102.4 million males 18 years old or older and 212.5 million U.S. residents 18 years old or older. The probability that a randomly selected U.S. resident 18 years old or older is male is $\dfrac{102.4}{212.5} = 0.482$.

(b) There are 14.0 million U.S. residents 18 years old or older who are widowed. The probability that a randomly selected U.S. resident 18 years old or older is widowed is $\dfrac{14.0}{212.5} = 0.066$.

(c) The events widowed and divorced are disjoint. Do you see why? We use the Addition Rule for Disjoint Events.

$$P(\text{widowed or divorced}) = P(\text{widowed}) + P(\text{divorced})$$

$$= \frac{14.0}{212.5} + \frac{21.7}{212.5} = \frac{35.7}{212.5}$$

$$= 0.168$$

(d) The events male and widowed are not mutually exclusive. In fact, there are 2.7 million males who are widowed in the United States. Therefore, we use the General Addition Rule to compute $P(\text{male or widowed})$:

$$P(\text{male or widowed}) = P(\text{male}) + P(\text{widowed}) - P(\text{male and widowed})$$

$$= \frac{102.4}{212.5} + \frac{14.0}{212.5} - \frac{2.7}{212.5} = \frac{113.7}{212.5} = 0.535$$

Now Work Problem 37.

3   Compute the Probability of an Event Using the Complement Rule

Suppose the probability of an event E is known and we would like to determine the probability that E does not occur. This can easily be accomplished using the idea of *complements*.

Definition

Complement of an Event

Let S denote the sample space of a probability experiment and let E denote an event. The **complement of** E, denoted E^c, is all outcomes in the sample space S that are not outcomes in the event E.

In Other Words

The Complement Rule is used when you know the probability that some event will occur and you want to know the opposite: the chance it will not occur.

Because E and E^c are mutually exclusive,

$$P(E \text{ or } E^c) = P(E) + P(E^c) = P(S) = 1$$

Subtracting $P(E)$ from both sides, we obtain

$$P(E^c) = 1 - P(E)$$

We have the following result.

Complement Rule

If E represents any event and E^c represents the complement of E, then

$$P(E^c) = 1 - P(E)$$

Figure 11 illustrates the Complement Rule using a Venn diagram.

Figure 11

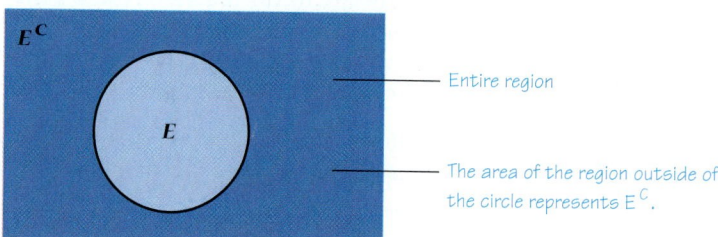

E^C

Entire region

E

The area of the region outside of the circle represents E^C.

EXAMPLE 5   **Computing Probabilities Using Complements**

Problem: According to the National Gambling Impact Study Commision, 52% of Americans have played state lotteries. What is the probability that a randomly selected American has not played a state lottery?

Approach: Not playing a state lottery is the complement of playing a state lottery. We compute the probability using the Complement Rule.

Solution

$$P(\text{not played state lottery}) = 1 - P(\text{played state lottery}) = 1 - 0.52 = 0.48$$

There is a 48% probability of randomly selecting an American who has not played a state lottery.

EXAMPLE 6   **Computing Probabilities Using Complements**

Problem: The data in Table 7 represent the income distribution of households in the United States in 2003.

Table 7

Annual Income	Number (in thousands)	Annual Income	Number (in thousands)
Less than $10,000	10,011	$50,000 to $74,999	20,191
$10,000 to $14,999	7,740	$75,000 to $99,999	12,313
$15,000 to $24,999	14,649	$100,000 to $149,999	10,719
$25,000 to $34,999	13,277	$150,000 to $199,999	3,372
$35,000 to $49,999	16,773	$200,000 or more	2,854

Source: U.S. Bureau of the Census

(a) Compute the probability that a randomly selected household earned $200,000 or more in 2003.

(b) Compute the probability that a randomly selected household earned less than $200,000 in 2003.

(c) Compute the probability that a randomly selected household earned at least $10,000 in 2003.

Approach: The probabilities will be determined by finding the relative frequency of each event. We have to find the total number of households in the United States in 2003.

Solution

(a) There were a total of $10,011 + 7,740 + \cdots + 2,854 = 111,899$ thousand households in the United States in 2003 and 2,854 thousand of them earned $200,000 or more. The probability that a randomly selected household in the United States earned $200,000 or more in 2003 is $\dfrac{2,854}{111,899} = 0.026$.

(b) We could compute the probability of randomly selecting a household that earned less than $200,000 in 2003 by adding the relative frequencies of each category less than $200,000, but it is easier to use complements. The complement of earning less than $200,000 is earning $200,000 or more. Therefore,

$$P(\text{less than } \$200,000) = 1 - P(\$200,000 \text{ or more})$$

$$= 1 - 0.026 = 0.974$$

There is a 97.4% probability of randomly selecting a household that earned less than $200,000 in 2003.

(c) The phrase *at least* means greater than or equal to. The complement of at least $10,000 is less than $10,000. In 2003, 10,011 thousand households earned less than $10,000. The probability of randomly selecting a household that earned at least $10,000 is

$$P(\text{at least } \$10,000) = 1 - P(\text{less than } \$10,000)$$

$$= 1 - \frac{10,011}{111,899} = 0.911$$

There is a 91.1% probability of randomly selecting a household that earned at least $10,000 in 2003.

Now Work Problems 25(d) and 27.

5.2 ASSESS YOUR UNDERSTANDING

Concepts and Vocabulary

1. What does it mean when two events are disjoint?

2. If E and F are disjoint events, then $P(E \text{ or } F) = $ _____.

3. If E and F are not disjoint events, then $P(E \text{ or } F) = $ _____.

4. What does it mean when two events are complements?

Skill Building

In Problems 5–12, a probability experiment is conducted in which the sample space of the experiment is $S = \{1, 2, 3, 4, 5, 6, 7, 8, 9, 10, 11, 12\}$. Let event $E = \{2, 3, 4, 5, 6, 7\}$, event $F = \{5, 6, 7, 8, 9\}$, event $G = \{9, 10, 11, 12\}$, and event $H = \{2, 3, 4\}$. Assume each outcome is equally likely.

5. List the outcomes in E and F. Are E and F mutually exclusive?

6. List the outcomes in F and G. Are F and G mutually exclusive?

7. List the outcomes in F or G. Now find $P(F \text{ or } G)$ by counting the number of outcomes in F or G. Determine $P(F \text{ or } G)$ using the General Addition Rule.

8. List the outcomes in E or H. Now find $P(E \text{ or } H)$ by counting the number of outcomes in E or H. Determine $P(E \text{ or } H)$ using the General Addition Rule.

9. List the outcomes in E and G. Are E and G mutually exclusive?

10. List the outcomes in F and H. Are F and H mutually exclusive?

11. List the outcomes in E^c. Find $P(E^c)$.

12. List the outcomes in F^c. Find $P(F^c)$.

In Problems 13–18, find the probability of the indicated event if $P(E) = 0.25$ and $P(F) = 0.45$.

13. Find $P(E \text{ or } F)$ if $P(E \text{ and } F) = 0.15$

14. Find $P(E \text{ and } F)$ if $P(E \text{ or } F) = 0.6$

15. Find $P(E \text{ or } F)$ if E and F are mutually exclusive

16. Find $P(E \text{ and } F)$ if E and F are mutually exclusive

17. Find $P(E^c)$

18. Find $P(F^c)$

19. If $P(E) = 0.60$, $P(E \text{ or } F) = 0.85$, and $P(E \text{ and } F) = 0.05$, find $P(F)$.

20. If $P(F) = 0.30$, $P(E \text{ or } F) = 0.65$, and $P(E \text{ and } F) = 0.15$, find $P(E)$.

In Problems 21–24, a golf ball is selected at random from a golf bag. If the golf bag contains 9 Titleists, 8 Maxflis, and 3 Top-Flites, find the probability of each event.

21. The golf ball is a Titleist or Maxfli.

22. The golf ball is a Maxfli or Top-Flite.

23. The golf ball is not a Titleist.

24. The golf ball is not a Top-Flite.

Applying the Concepts

25. **Weapon of Choice** The following probability model
NW shows the distribution of murders by type of weapon for murder cases from 1990 to 2002.

Weapon	Probability
Gun	0.668
Knife	0.125
Blunt object	0.048
Personal weapon	0.067
Strangulation	0.017
Fire	0.007
Other	0.068

Source: U.S. Federal Bureau of Investigation

(a) Verify this is a probability model.
(b) What is the probability a randomly selected murder resulted from a gun or knife? Interpret this probability.
(c) What is the probability a randomly selected murder resulted from a knife, blunt object, or strangulation? Interpret this probability.
(d) What is the probability a randomly selected murder resulted from a weapon other than a gun? Interpret this probability.
(e) Are murders by strangulation unusual?

26. **Doctorates Conferred** The following probability model shows the distribution of doctoral degrees from U.S. universities in 2003 by area of study.

Area of Study	Probability
Engineering	0.129
Physical sciences	0.101
Life sciences	0.206
Mathematics	0.024
Computer sciences	0.021
Social sciences	0.167
Humanities	0.133
Education	0.163
Professional and other fields	0.056

Source: U.S. National Science Foundation

(a) Verify this is a probability model.
(b) What is the probability a randomly selected doctoral candidate who earned a degree in 2003 studied physical science or life science? Interpret this probability.
(c) What is the probability a randomly selected doctoral candidate who earned a degree in 2003 studied physi-

cal science, life science, mathematics, or computer science? Interpret this probability.

(d) What is the probability a randomly selected doctoral candidate who earned a degree in 2003 did not study mathematics? Interpret this probability.

(e) Are doctoral degrees in mathematics unusual? Does this result surprise you?

27. Multiple Births The following data represent the number NW of live multiple-delivery births (three or more babies) in 2002 for women 15 to 54 years old:

Age	Number of Multiple Births
15–19	93
20–24	511
25–29	1628
30–34	2832
35–39	1843
40–44	377
45–54	117

Source: National Vital Statistics Report, Vol. 52, No. 10, December 17, 2003

(a) Determine the probability that a randomly selected multiple birth in 2002 for women 15 to 54 years old involved a mother 30 to 39 years old. Interpret this probability.

(b) Determine the probability that a randomly selected multiple birth in 2002 for women 15 to 54 years old involved a mother who was not 30 to 39 years old. Interpret this probability.

(c) Determine the probability that a randomly selected multiple birth in 2002 for women 15 to 54 years old involved a mother who was less than 45 years old. Interpret this probability.

(d) Determine the probability that a randomly selected multiple birth in 2002 for women 15 to 54 years old involved a mother who was at least 20 years old. Interpret this probability.

28. Housing The following probability model shows the distribution for the number of rooms in U.S. housing units.

Rooms	Probability
One	0.004
Two	0.013
Three	0.104
Four	0.209
Five	0.218
Six	0.190
Seven	0.123
Eight or more	0.139

Source: U.S. Bureau of the Census

(a) Verify this is a probability model.

(b) What is the probability that a randomly selected housing unit has four or more rooms? Interpret this probability.

(c) What is the probability that a randomly selected housing unit has fewer than eight rooms? Interpret this probability.

(d) What is the probability that a randomly selected housing unit has from four to six (inclusive) rooms? Interpret this probability.

(e) What is the probability that a randomly selected housing unit has at least two rooms? Interpret this probability.

29. A Deck of Cards A standard deck of cards contains 52 NW cards as shown in Figure 9. One card is randomly selected from the deck.

(a) Compute the probability of randomly selecting a heart or club from a deck of cards.

(b) Compute the probability of randomly selecting a heart or club or diamond from a deck of cards.

(c) Compute the probability of randomly selecting an ace or heart from a deck of cards.

30. A Deck of Cards A standard deck of cards contains 52 cards as shown in Figure 9. One card is randomly selected from the deck.

(a) Compute the probability of randomly selecting a two or three from a deck of cards.

(b) Compute the probability of randomly selecting a two or three or four from a deck of cards.

(c) Compute the probability of randomly selecting a two or club from a deck of cards.

31. Birthdays Exclude leap years from the following calculations:

(a) Compute the probability that a randomly selected person does not have a birthday on November 8.

(b) Compute the probability that a randomly selected person does not have a birthday on the 1st day of a month.

(c) Compute the probability that a randomly selected person does not have a birthday on the 31st day of a month.

(d) Compute the probability that a randomly selected person was not born in December.

32. Roulette In the game of roulette, a wheel consists of 38 slots numbered 0, 00, 1, 2, ... 36. The odd-numbered slots are red, and the even-numbered slots are black. The numbers 0 and 00 are green. To play the game, a metal ball is spun around the wheel and is allowed to fall into one of the numbered slots.

(a) What is the probability that the metal ball lands on green or red?

(b) What is the probability that the metal ball does not land on green?

33. Health Problems According to the Centers for Disease Control, the probability that a randomly selected citizen of the United States has hearing problems is 0.151. The probability that a randomly selected citizen of the United States has vision problems is 0.093. Can we compute the probability of randomly selecting a citizen of the United States who has hearing problems or vision problems by adding these probabilities? Why or why not?

34. **Visits to the Doctor** In a National Ambulatory Medical Care Survey administered by the Centers for Disease Control it was learned that the probability a randomly selected patient visited the doctor for a blood pressure check is 0.601. The probability a randomly selected patient visited the doctor for urinalysis is 0.128. Can we compute the probability of randomly selecting a patient who visited the doctor for a blood pressure check or urinalysis by adding these probabilities? Why or why not?

35. **Language Spoken at Home** According to the U.S. Census Bureau, the probability a randomly selected household speaks only English at home is 0.82. The probability a randomly selected household speaks only Spanish at home is 0.11.

    (a) What is the probability a randomly selected household speaks only English or only Spanish at home?
    (b) What is the probability a randomly selected household speaks a language other than only English or only Spanish at home?
    (c) What is the probability a randomly selected household speaks a language other than only English at home?
    (d) Can the probability that a randomly selected household speaks only Polish at home equal 0.08? Why or why not?

36. **Getting to Work** According to the U.S. Census Bureau the probability a randomly selected worker primarily drives a car to work is 0.82. The probability a randomly selected worker primarily takes public transportation to work is 0.053.

    (a) What is the probability a randomly selected worker primarily drives a car or takes public transportation to work?
    (b) What is the probability a randomly selected worker neither drives a car nor takes public transportation to work?
    (c) What is the probability a randomly selected worker does not drive a car to work?
    (d) Can the probability a randomly selected worker walks to work equal 0.15? Why or why not?

37. **Cigar Smoking** The data in the following table show the NW results of a national study of 137,243 U.S. men that investigated the association between cigar smoking and death from cancer. **Note:** Current cigar smoker means cigar smoker at time of death.

	Died from Cancer	Did Not Die from Cancer
Never smoked cigars	782	120,747
Former cigar smoker	91	7,757
Current cigar smoker	141	7,725

Source: Shapiro, Jacobs, and Thun. Cigar Smoking in Men and Risk of Death from Tobacco-Related Cancers, *Journal of the National Cancer Institute*, February 16, 2000.

    (a) If an individual is randomly selected from this study, what is the probability that he died from cancer?
    (b) If an individual is randomly selected from this study, what is the probability that he was a current cigar smoker?
    (c) If an individual is randomly selected from this study, what is the probability that he died from cancer and was a current cigar smoker?
    (d) If an individual is randomly selected from this study, what is the probability that he died from cancer or was a current cigar smoker?

38. **Civilian Labor Force** The following table represents the employment status and gender of the civilian labor force ages 16 to 24 (in millions).

	Male	Female
Employed	11.2	10.3
Unemployed	1.6	1.4

Source: U.S. Bureau of Labor Statistics, August 2002

    (a) What is the probability that a randomly selected 16- to 24-year-old individual from the civilian labor force is employed?
    (b) What is the probability that a randomly selected 16- to 24-year-old individual from the civilian labor force is male?
    (c) What is the probability that a randomly selected 16- to 24-year-old individual from the civilian labor force is employed and male?
    (d) What is the probability that a randomly selected 16- to 24-year-old individual from the civilian labor force is employed or male?

39. **Student Government Satisfaction Survey** The Committee on Student Life at a university conducted a survey of 375 undergraduate students regarding satisfaction with student government. Results of the survey are shown in the table by class rank.

	Freshman	Sophomore	Junior	Senior
Satisfied	57	49	64	61
Neutral	23	15	16	11
Not satisfied	21	18	14	26

    (a) If a survey participant is selected at random, what is the probability he or she is satisfied with student government?
    (b) If a survey participant is selected at random, what is the probability he or she is a junior?

(c) If a survey participant is selected at random, what is the probability he or she is satisfied and is a junior?

(d) If a survey participant is selected at random, what is the probability he or she is satisfied or is a junior?

40. The Placebo Effect   A company is testing a new medicine for migraine headaches. In the study, 150 women were given the new medicine and an additional 100 women were given a placebo. Each participant was directed to take the medicine when the first symptoms of a migraine occurred and then to record whether the headache went away within 45 minutes or lingered. The results are recorded in the following table:

	Headache Went Away	Headache Lingered
Given medicine	132	18
Given placebo	56	44

(a) If a study participant is selected at random, what is the probability she was given the placebo?

(b) If a study participant is selected at random, what is the probability her headache went away within 45 minutes?

(c) If a study participant is selected at random, what is the probability she was given the placebo and her headache went away within 45 minutes?

(d) If a study participant is selected at random, what is the probability she was given the placebo or her headache went away within 45 minutes?

41. Active Duty   The following table represents the number of active-duty military personnel by rank in the four major branches of the military as of February 28, 2005.

	Officers	Enlisted
Army	80,580	409,410
Navy	53,428	310,167
Air Force	73,331	159,164
Marine Corps	18,893	287,328

Source: U.S. Department of Defense

(a) If an active-duty military person is selected at random, what is the probability the individual is an officer?

(b) If an active-duty military person is selected at random, what is the probability the individual is in the Navy?

(c) If an active-duty military person is selected at random, what is the probability the individual is a Naval officer?

(d) If an active-duty military person is selected at random, what is the probability the individual is an officer or is in the Navy?

42. Driver Fatalities   The following data represent the number of driver fatalities in the United States in 2002 by age group for male and female drivers:

Age	Male	Female
Under 16	228	108
16–20	5696	2386
21–34	13,553	4148
35–54	14,395	5017
55–69	4937	1708
70 and over	3159	1529

Source: Traffic Safety Facts 2002, Federal Highway Administration, 2002.

(a) Determine the probability that a randomly selected driver fatality was male.

(b) Determine the probability that a randomly selected driver fatality was 16 to 20 years old.

(c) Determine the probability that a randomly selected driver fatality was a 16- to 20-year-old male.

(d) Determine the probability that a randomly selected driver fatality was male or 16 to 20 years old.

5.3   Independence and the Multiplication Rule

Objectives

1   Understand independence

2   Use the Multiplication Rule for independent events

3   Compute at-least probabilities

1   Understand Independence

The Addition Rule for Disjoint Events deals with probabilities involving the word *or*. That is, it is used for computing the probability of observing an outcome in event E or event F. We now describe a probability rule for computing the probability that E and F both occur.

Before we can present this rule, we must discuss the idea of *independent events*.

Definition

Two events E and F are **independent** if the occurrence of event E in a probability experiment does not affect the probability of event F. Two events are **dependent** if the occurrence of event E in a probability experiment affects the probability of event F.

To help you understand the idea of independence, we again look at a simple situation—flipping a coin. Suppose you flip a coin twice. Does the fact that you obtained a head on the first toss have any impact on the likelihood of obtaining a head on the second toss? Not unless you are a master coin flipper who can manipulate the outcome of a coin flip! For this reason, the outcome from the first flip is independent of the outcome from the second flip. Let's look at other examples.

EXAMPLE 1

Independent or Not?

(a) Suppose you flip a coin and roll a die. The events "obtain a head" and "roll a 5" are independent because the results of the coin flip do not impact the results of the die toss.

(b) Are the events "earned a bachelor's degree" and "earn more than $100,000 per year" independent? No, because knowing that an individual has a bachelor's degree affects the likelihood that the individual is earning more than $100,000 per year.

(c) Suppose two 24-year-old male drivers who live in the United States are randomly selected. The events "male 1 gets in a car accident during the year" and "male 2 gets in a car accident during the year" are independent because the males were randomly selected. This means what happens with one of the drivers has nothing to do with what happens to the other driver. ∎

In Other Words

In determining whether two events are independent, ask yourself whether the probability of one event is affected by the other event. For example, what is the probability that a 29-year-old male has high cholesterol? What is the probability that a 29-year-old male has high cholesterol, given that he eats fast food four times a week? Does the fact that the individual eats fast food four times a week change the likelihood that he has high cholesterol? If yes, the events are not independent.

In Example 1(c), we are able to conclude that the events "male 1 gets in an accident" and "male 2 gets in an accident" are independent because the individuals are randomly selected. By randomly selecting the individuals, it is reasonable to conclude that the individuals are not related in any way (related in the sense that they do not live in the same town, attend the same school, and so on). If the two individuals did have a common link between them (such as they both lived on the same city block), then knowing one of the males had a car accident may affect the likelihood that the other male had a car accident. After all, they could hit each other!

Now Work Problem 7.

Disjoint Events versus Independent Events

It is important that we understand that disjoint events and independent events are different concepts. Recall that two events are disjoint if they have no outcomes in common. In other words, two events are disjoint if knowing that one of the events occurs, we know the other event did not occur. Independence means that one event occurring does not affect the probability of the other event occurring. Therefore, knowing two events are disjoint means that the events are not independent.

CAUTION

Two events that are disjoint are not independent.

Consider the experiment of rolling a single die. Let E represent the event "roll an even number," and let F represent the event "roll an odd number." We can see that E and F are mutually exclusive because they have no outcomes in common. In addition, $P(E) = \dfrac{1}{2}$ and $P(F) = \dfrac{1}{2}$. However, if we are told that the roll of the die is going to be an even number, then what is the probability of event F? Because the outcome will be even, the probability of event F is now 0.

2 Use the Multiplication Rule for Independent Events

Suppose that you flip a fair coin twice. What is the probability that you obtain a head on both flips? Put another way, what is the probability that you obtain a head on the first flip *and* you obtain a head on the second flip? We can create a sample space that lists the outcomes of this experiment quite easily. In flipping a coin twice where H represents the outcome heads and T represents the outcome tails, we can obtain

$$S = \{HH, HT, TH, TT\}$$

There is one outcome with both heads. Because each outcome is equally likely, we have

$$P(1\text{heads on the 1}^{st}\text{ toss and heads on the 2}^{nd}\text{ toss}) = \frac{N(\text{heads on the 1}^{st}\text{ and heads on the 2}^{nd})}{N(S)}$$

$$= \frac{1}{4}$$

We may have intuitively been able to figure this out by recognizing $P(\text{head}) = \frac{1}{2}$ for each flip. So it seems reasonable that

$$P(\text{heads on the 1}^{st}\text{ toss and heads on the 2}^{nd}\text{ toss}) = P(\text{heads on 1}^{st}\text{ toss}) \cdot P(\text{heads on 2}^{nd}\text{ toss})$$

$$= \frac{1}{2} \cdot \frac{1}{2}$$

$$= \frac{1}{4}$$

Because both approaches result in the same answer, $\frac{1}{4}$, it seems reasonable to conjecture that $P(E \text{ and } F) = P(E) \cdot P(F)$.

> **Multiplication Rule for Independent Events**
>
> If E and F are independent events, then
> $$P(E \text{ and } F) = P(E) \cdot P(F)$$

EXAMPLE 2   Computing Probabilities of Independent Events

Problem: In the game of roulette, the wheel has slots numbered 0, 00, and 1 through 36. A metal ball is allowed to roll around a wheel until it falls into one of the numbered slots. You decide to play the game and place a bet on the number 17. What is the probability the ball will land in the slot numbered 17 two times in a row?

Approach: There are 38 outcomes in the sample space of the experiment. We use the classical method of computing probabilities because the outcomes are equally likely. In addition, we use the Multiplication Rule for Independent Events. The events "17 on first trial" and "17 on second trial" are independent because the ball does not remember it landed on 17 on the first trial, so this cannot affect the probability of landing on 17 on the second trial.

Solution: Because there are 38 possible outcomes to the experiment, the probability of the ball landing on 17 is $\frac{1}{38}$. Because the events "17 on first trial" and "17 on second trial" are independent, we have

$P(\text{ball lands in slot 17 in the first game and ball lands in slot 17 in the second game})$

$= P(\text{ball lands in slot 17 in the first game}) \cdot P(\text{ball lands in slot 17 in the second game})$

$= \dfrac{1}{38} \cdot \dfrac{1}{38} = \dfrac{1}{1,444} \approx 0.0006925$

It is very unlikely that the ball will land on 17 twice in a row. We expect the ball to land on 17 twice in a row about 7 times in 10,000 trials.

We can extend the Multiplication Rule for three or more independent events.

Multiplication Rule for *n* Independent Events

If events $E, F, G, \ldots$ are independent, then

$$P(E \text{ and } F \text{ and } G \text{ and } \ldots) = P(E) \cdot P(F) \cdot P(G) \ldots$$

EXAMPLE 3   **Life Expectancy**

Problem: The probability that a randomly selected male 24 years old will survive the year is 99.85% according to the National Vital Statistics Report, Vol. 47, No. 28. What is the probability that three randomly selected 24-year-old males will survive the year? What is the probability that 20 randomly selected 24-year-old males will survive the year?

Approach: We can safely assume that the outcomes of the probability experiment are independent, because there is no indication that the survival of one male affects the survival of the others. For example, if two of the males lived in the same house, a house fire could kill both males and we lose independence. (Knowledge that one male died in a house fire certainly affects the probability that the other died.) By randomly selecting the males, we minimize the chances that they are related in any way.

Solution:

$P(\text{all three males survive}) = P(1^{\text{st}} \text{ survives and } 2^{\text{nd}} \text{ survives and } 3^{\text{rd}} \text{ survives})$

$\qquad\qquad = P(1^{\text{st}} \text{ survives}) \cdot P(2^{\text{nd}} \text{ survives}) \cdot P(3^{\text{rd}} \text{ survives})$

$\qquad\qquad = (0.9985)(0.9985)(0.9985)$

$\qquad\qquad = 0.9955$

There is a 99.55% probability that all three males survive the year.

$P(\text{all 20 males survive}) = P(1^{\text{st}} \text{ survives and } 2^{\text{nd}} \text{ survives and} \ldots \text{and } 20^{\text{th}} \text{ survives})$

$\qquad\qquad \underset{\substack{\uparrow \\ \text{Independent events}}}{=} P(1^{\text{st}} \text{ survives}) \cdot P(2^{\text{nd}} \text{ survives}) \cdots \cdot P(20^{\text{th}} \text{ survives})$

$\qquad\qquad \underset{\substack{\text{Multiply 0.9985 by itself 20 times.}}}{=} (0.9985) \cdot (0.9985) \cdots \cdot (0.9985)$

$\qquad\qquad = (0.9985)^{20}$

$\qquad\qquad = 0.9704$

Now Work Problems 17(a) and (b).   There is a 97.04% probability that all 20 males survive the year.

3 **Compute At-least Probabilities**

We now present an example in which we compute at-least probabilities. These probabilities use the Complement Rule. The phrase *at least* means greater than or equal to. For example, a person must be at least 17 years old to see an R-rated movie.

EXAMPLE 4 **Computing At-Least Probabilities**

Problem: Compute the probability that at least 1 male out of 1000 aged 24 years will die during the course of the year if the probability that a randomly selected 24-year-old male survives the year is 0.9985.

Approach: The phrase *at-least* means greater than or equal, so we wish to know the probability that 1 or 2 or 3 or ... or 1000 males will die during the year. These events are mutually exclusive, so

$$P(1 \text{ or } 2 \text{ or } 3 \text{ or} \ldots \text{ or } 1000 \text{ die}) = P(1 \text{ dies}) + P(2 \text{ die}) + P(3 \text{ die}) + \cdots + P(1000 \text{ die})$$

Computing these probabilities is very time consuming. However, we notice that the complement of "at least one dying" is "none die." We use the Complement Rule to compute the probability.

Solution

$$
\begin{aligned}
P(\text{at least one dies}) &= 1 - P(\text{none die}) \\
&= 1 - P(\text{1st survives and 2nd survives and} \ldots \text{and 1000th survives}) \\
&= 1 - P(1^{\text{st}} \text{ survives}) \cdot P(2^{\text{nd}} \text{ survives}) \cdot \cdots \cdot P(1000^{\text{th}} \text{ survives}) \quad \textsf{\color{teal}{Independent events}} \\
&= 1 - (0.9985)^{1000} \\
&= 1 - 0.2229 \\
&= 0.7771 \\
&= 77.71\%
\end{aligned}
$$

There is a 77.71% probability that at least one 24-year-old male out of 1000 will die during the course of the year.

Now Work Problem 17(c).

Summary: Rules of Probability

1. The probability of any event must be between 0 and 1, inclusive. If we let E denote any event, then $0 \le P(E) \le 1$.
2. The sum of the probabilities of all outcomes must equal 1. That is, if the sample space $S = \{e_1, e_2, \ldots, e_n\}$, then $P(e_1) + P(e_2) + \cdots + P(e_n) = 1$.
3. If E and F are disjoint events, then $P(E \text{ or } F) = P(E) + P(F)$. If E and F are not disjoint events, then $P(E \text{ or } F) = P(E) + P(F) - P(E \text{ and } F)$.
4. If E represents any event and E^c represents the complement of E, then $P(E^c) = 1 - P(E)$.
5. If E and F are independent events, then $P(E \text{ and } F) = P(E) \cdot P(F)$.

Notice that *or* probabilities use the Addition Rule, whereas *and* probabilities use the Multiplication Rule. Accordingly, *or* probabilities imply addition, while *and* probabilities imply multiplication.

5.3 ASSESS YOUR UNDERSTANDING

Concepts and Vocabulary

1. Two events E and F are _____ if the occurrence of event E in a probability experiment does not affect the probability of event F.

2. The word *and* in probability implies that we use the _____ Rule.

3. The word *or* in probability implies that we use the _____ Rule.

4. *True or False:* When two events are disjoint, they are also independent.

5. If two events E and F are independent, $P(E \text{ and } F) =$ _____.

6. Suppose events E and F are disjoint. What is $P(E \text{ and } F)$?

Skill Building

7. Determine whether the events E and F are independent or dependent. Justify your answer.
   (a) E: It rains on June 30.
      F: It is cloudy on June 30.
   (b) E: Your car has a flat tire.
      F: The price of gasoline increases overnight.
   (c) E: You live at least 80 years.
      F: You smoke a pack of cigarettes every day of your life.

8. Determine whether the events A and B are independent or dependent. Justify your answer.
   (a) A: You earn an A on an exam.
      B: You study for an exam.
   (b) A: You are late for work.
      B: Your car has a flat tire.
   (c) A: You earn more than \$50,000 per year.
      B: You are born in the month of July.

9. Suppose events E and F are independent, $P(E) = 0.3$, and $P(F) = 0.6$. What is the $P(E \text{ and } F)$?

10. Suppose events E and F are independent, $P(E) = 0.7$, and $P(F) = 0.9$. What is the $P(E \text{ and } F)$?

Applying the Concepts

11. **Flipping a Coin** What is the probability of obtaining five heads in a row when flipping a coin? Interpret this probability.

12. **Rolling a Die** What is the probability of obtaining 4 ones in a row when rolling a fair, six-sided die? Interpret this probability.

13. **Southpaws** About 13% of the population is left-handed. If two people are randomly selected, what is the probability both are left-handed? What is the probability at least one is right-handed?

14. **Investing** Suppose your financial advisor has recommended two stocks, each of which has a 0.6 probability of increasing in value over the next year. Assuming the performance of one stock is independent of the other, what is the probability both stocks will rise over the next year? What is the probability at least one stock will not increase in value?

15. **False Positives** The ELISA is a test to determine whether the HIV antibody is present. The test is 99.5% effective. This means that the test will accurately come back negative if the HIV antibody is not present. The probability of a test coming back positive when the antibody is not present (a false positive) is 0.005. Suppose the ELISA is given to five randomly selected people who do not have the HIV antibody.
   (a) What is the probability that the ELISA comes back negative for all five people?
   (b) What is the probability that the ELISA comes back positive for at least one of the five people?

16. **Christmas Lights** Christmas lights are often designed with a series circuit. This means that when one light burns out, the entire string of lights goes black. Suppose the lights are designed so that the probability a bulb will last 2 years is 0.995. The success or failure of a bulb is independent of the success or failure of other bulbs.
   (a) What is the probability that in a string of 100 lights all 100 will last 2 years?
   (b) What is the probability at least one bulb will burn out in 2 years?

17. **Life Expectancy** The probability that a randomly selected 40-year-old male will live to be 41 years old is 0.99718 according to the National Vital Statistics Report, Vol. 48, No. 18.
   (a) What is the probability that two randomly selected 40-year-old males will live to be 41 years old?
   (b) What is the probability that five randomly selected 40-year-old males will live to be 41 years old?
   (c) What is the probability that at least one of five randomly selected 40-year-old males will not live to be 41 years old? Would it be unusual that at least one of five randomly selected 40-year-old males will not live to be 41 years old?

18. **Life Expectancy** The probability that a randomly selected 40-year-old female will live to be 41 years old is 0.99856 according to the National Vital Statistics Report, Vol. 48, No. 18.
   (a) What is the probability that two randomly selected 40-year-old females will live to be 41 years old?
   (b) What is the probability that five randomly selected 40-year-old females will live to be 41 years old?
   (c) What is the probability that at least one of five randomly selected 40-year-old females will not live to be 41 years old? Would it be unusual that at least one of five randomly selected 40-year-old females will not live to be 41 years old?

19. **Blood Types** Blood types can be classified as either Rh$^+$ or Rh$^-$. According to the *Information Please Almanac*, 99% of the Chinese population has Rh$^+$ blood.
    (a) What is the probability that two randomly selected Chinese people have Rh$^+$ blood?
    (b) What is the probability that six randomly selected Chinese people have Rh$^+$ blood?
    (c) What is the probability that at least one of six randomly selected Chinese people has Rh$^-$ blood? Would it be unusual that at least one of six randomly selected Chinese people has Rh$^-$ blood?

20. **Quality Control** Suppose a company selects two people who work independently inspecting two-by-four timbers. Their job is to identify low-quality timbers. Suppose the probability that an inspector does not identify a low-quality timber is 20%.
    (a) What is the probability that both inspectors do not identify a low-quality timber?
    (b) How many inspectors should be hired to keep the probability of not identifying a low-quality timber below 1%?

21. **Cold Streaks** Players in sports are said to have "hot streaks" and "cold streaks." For example, a batter in baseball might be considered to be in a slump or cold streak if he has made 10 outs in 10 consecutive at-bats. Suppose a hitter successfully reaches base 30% of the time he comes to the plate.
    (a) Find the probability that the hitter makes 10 outs in 10 consecutive at-bats, assuming that at-bats are independent events. *Hint:* The hitter makes an out 70% of the time.
    (b) Are cold streaks unusual?

22. **Hot Streaks** In a recent basketball game, a player who makes 65% of his free throws made eight consecutive free throws. Assuming free-throw shots are independent, determine whether this feat was unusual.

23. **Defense System** Suppose a satellite defense system is established in which four satellites acting independently have a 0.9 probability of detecting an incoming ballistic missile. What is the probability at least one of the four satellites detects an incoming ballistic missile? Would you feel safe with such a system?

24. **E.P.T. Pregnancy Tests** The packaging of an E.P.T. Pregnancy Test states that the test is "99% accurate at detecting typical pregnancy hormone levels." Assume the probability that a test will correctly identify a pregnancy is

0.99. Suppose 12 randomly selected pregnant women with typical hormone levels are each given the test.
    (a) What is the probability that all 12 tests will be positive?
    (b) What is the probability that at least one test will not be positive?

25. **Bowling** Suppose Ralph gets a strike when bowling 30% of the time.
    (a) What is the probability that Ralph gets two strikes in a row?
    (b) What is the probability that Ralph gets a turkey (three strikes in a row)?
    (c) When events are independent, their complements are independent as well. Use this result to determine the probability that Ralph gets a strike and then does not get a strike.

26. **NASCAR Fans** Among Americans who consider themselves auto racing fans, 59% identify NASCAR stock cars as their favorite type of racing. Suppose four auto racing fans are randomly selected.
    Source: ESPN / TNS Sports, reported in *USA Today*.
    (a) What is the probability that all four will identify NASCAR stock cars as their favorite type of racing?
    (b) What is the probability that at least one will not identify NASCAR stock cars as his or her favorite type of racing?
    (c) What is the probability that none will identify NASCAR stock cars as his or her favorite type of racing?
    (d) What is the probability that at least one will identify NASCAR stock cars as his or her favorite type of racing?

27. **Driving under the Influence** Among 21- to 25-year-olds, 29% say they have driven while under the influence of alcohol. Suppose three 21- to 25-year-olds are selected at random.
    Source: U.S. Department of Health and Human Services, reported in *USA Today*.
    (a) What is the probability that all three have driven while under the influence of alcohol?
    (b) What is the probability that at least one has not driven while under the influence of alcohol?
    (c) What is the probability that none of the three has driven while under the influence of alcohol?
    (d) What is the probability that at least one has driven while under the influence of alcohol?

5.4 Conditional Probability and the General Multiplication Rule

Objectives

1 Compute conditional probabilities

2 Compute probabilities using the General Multiplication Rule

1 Compute Conditional Probabilities

In the last section, we learned that when two events are independent the occurence of one event has no impact on the probability of the second event. For example, according to data from the Centers for Disease Control, 17% of adults have high cholesterol. So the probability that a randomly selected adult has high cholesterol is 0.17. Now, if we were told that this particular individual eats three Big Macs from McDonalds each week, do you think that the probability he has high cholesterol will change? Of course it will. The probability that the individual has high cholesterol will increase, given the fact that the individual eats three Big Macs each week. This is called a *conditional probability*.

Definition

Conditional Probability

The notation $P(F|E)$ is read "the probability of event F given event E." It is the probability that an event F occurs, given that the event E has occurred.

Let's look at an example.

EXAMPLE 1 **An Introduction to Conditional Probability**

Problem: Suppose that a single die is rolled. What is the probability that the die comes up 3? Now suppose the die is rolled a second time, but we are told the outcome will be an odd number. What is the probability that the die comes up 3?

Approach: We assume that the die is fair and compute the probabilities using equally likely outcomes.

Solution: In the first instance, there are six possibilities in the sample space, $S = \{1, 2, 3, 4, 5, 6\}$, so $P(3) = \frac{1}{6}$. In the second instance, there are three possibilities in the sample space, because the only possible outcomes are odd, so $S = \{1, 3, 5\}$. We express this probability symbolically as $P(3|\text{outcome is odd}) = \frac{1}{3}$, which is read "the probability of rolling a 3 given that the outcome is odd is one-third."

So conditional probabilities reduce the size of the sample space under consideration. Let's look at another example. The data in Table 8 represent the marital status of males and females 18 years old or older in the United States in 2003.

Table 8			
	Males (in millions)	**Females (in millions)**	**Totals (in millions)**
Never married	28.6	23.3	51.9
Married	62.1	62.8	124.9
Widowed	2.7	11.3	14.0
Divorced	9.0	12.7	21.7
Totals (in millions)	102.4	110.1	212.5

Source: U.S. Census Bureau, Current Population Reports

Suppose we want to know the probability that a randomly selected individual is widowed. This probability is found by dividing the number of widowed individuals by the total number of individuals who are 18 years old or older.

$$P(\text{widowed}) = \frac{14.0}{212.5}$$

$$= 0.066$$

Now suppose that we know the individual is female. Does this change the probability that she is widowed? Because the sample space now consists only of females, we can determine the probability that the individual is widowed, given that the individual is female, as follows:

$$P(\text{widowed}|\text{female}) = \frac{N(\text{widowed females})}{N(\text{females})}$$

$$= \frac{11.3}{110.1} = 0.103$$

So knowing that the individual is female increases the likelihood that the individual is widowed. The previous discussion leads to the following.

Conditional Probability Rule

If E and F are any two events, then

$$P(F|E) = \frac{P(E \text{ and } F)}{P(E)} = \frac{N(E \text{ and } F)}{N(E)} \tag{1}$$

The probability of event F occurring, given the occurrence of event E, is found by dividing the probability of E and F by the probability of E. Or the probability of event F occurring, given the occurrence of event E, is found by dividing the number of outcomes in E and F by the number of outcomes in E.

We used the second method for computing conditional probabilities in the widow example.

EXAMPLE 2   **Conditional Probabilities on Marital Status and Gender**

Problem: The data in Table 8 represent the marital status and gender of the residents of the United States aged 18 years old or older in 2003.

(a) Compute the probability that a randomly selected male has never married.
(b) Compute the probability that a randomly selected individual who has never married is male.

Approach

(a) We are given that the randomly selected person is male, so we concentrate on the male column. There are 102.4 million males and 28.6 million people who are male and never married, so $N(\text{male}) = 102.4$ million and $N(\text{male and never married}) = 28.6$ million. Compute the probability using the Conditional Probability Rule.

(b) We are given that the randomly selected person has never married, so we concentrate on the never married row. There are 51.9 million people who have never married and 28.6 million people who are male and have never married, so $N(\text{never married}) = 51.9$ million and $N(\text{male and never married}) = 28.6$ million. Compute the probability using the Conditional Probability Rule.

Solution

(a) Substituting into Formula (1), we obtain

$$P(\text{never married}|\text{male}) = \frac{N(\text{never married and male})}{N(\text{male})} = \frac{28.6}{102.4} \approx 0.279$$

There is a 27.9% probability the randomly selected individual has never married, given that he is male.

(b) Substituting into Formula (1), we obtain

$$P(\text{male}|\text{never married}) = \frac{N(\text{male and never married})}{N(\text{never married})} = \frac{28.6}{51.9} \approx 0.551$$

There is a 55.1% probability the randomly selected individual is male, given that he or she has never married.

What is the difference between the results of Example 2(a) and (b)? In Example 2(a), we found that 27.9% of males have never married, whereas in Example 2(b) we found that 55.1% of individuals who have never married are male. Do you see the difference?

Now Work Problem 17.

EXAMPLE 3 **Birth Weights of Preterm Babies**

Problem: In 2002, 11.96% of all births were preterm. (The gestation period of the pregnancy was less than 37 weeks.) Also in 2002, 0.23% of all births resulted in a preterm baby that weighed 8 pounds, 13 ounces or more. What is the probability that a randomly selected baby weighs 8 pounds, 13 ounces or more, given that the baby was preterm?

Approach: We want to know the probability that the baby weighs 8 pounds, 13 ounces or more, given the baby was preterm. We know that 0.23% of all babies weighed 8 pounds, 13 ounces or more and were preterm, so $P(\text{weighs 8 pounds, 13 ounces or more and preterm}) = 0.23\%$. We also know that 11.96% of all births were preterm, so $P(\text{preterm}) = 11.96\%$. We compute the probability by dividing the probability that a baby will weigh 8 pounds, 13 ounces or more *and* be preterm by the probability that a baby will be preterm.

Solution: $P(\text{weighs 8 pounds, 13 ounces or more}|\text{preterm})$

$$= \frac{P(\text{weighs 8 pounds, 13 ounces or more and preterm})}{P(\text{preterm})}$$

$$= \frac{0.23\%}{11.96\%} = \frac{0.0023}{0.1196} \approx 0.0192 = 1.92\%$$

There is a 1.92% probability that a randomly selected baby will weigh 8 pounds, 13 ounces or more, given the baby is preterm. It is unusual for preterm babies to weigh 8 pounds, 13 ounces or more.

Now Work Problem 13.

---

② Compute Probabilities Using the General Multiplication Rule

If we solve the Conditional Probability Rule for $P(E \text{ and } F)$, we obtain the General Multiplication Rule.

General Multiplication Rule

The probability that two events E and F both occur is

$$P(E \text{ and } F) = P(E) \cdot P(F|E)$$

In words, the probability of E and F is the probability of event E occurring times the probability of event F occurring, given the occurrence of event E.

| EXAMPLE 4 | **Using the General Multiplication Rule** |

Problem: The probability that a driver who is speeding gets pulled over is 0.8. The probability that a driver gets a ticket given he/she is pulled over is 0.9. What is the probability that a randomly selected driver who is speeding gets pulled over and gets a ticket?

Approach: Let E represent the event "driver who is speeding gets pulled over," and let F represent the event "driver gets a ticket." We use the General Multiplication Rule to compute $P(E \text{ and } F)$.

Solution: $P(\text{driver who is speeding gets pulled over and gets a ticket}) = P(E \text{ and } F) = P(E) \cdot P(F|E) = 0.8(0.9) = 0.72 = 72\%$. There is a 72% probability that a driver who is speeding gets pulled over and gets a ticket.

Now Work Problem 29.

| EXAMPLE 5 | **Acceptance Sampling** |

Problem: Suppose that a box of 100 circuits is sent to a manufacturing plant. Of the 100 circuits shipped, 5 are defective. The plant manager receiving the chips randomly selects 2 and tests them. If both chips work, she will accept the shipment. Otherwise, the shipment is rejected. What is the probability the plant manager discovers at least 1 defective circuit and rejects the shipment?

Approach: We wish to determine the probability that at least one of the tested circuits is defective. There are four possibilities in this probability experiment. None of the circuits are defective, the first is defective while the second is not, the first is not defective while the second is defective, or both circuits are defective. We cannot compute the probability that at least 1 is defective using the fact that there are four outcomes and three result in at least 1 defective, because the outcomes are not equally likely. We need a different approach. We could determine the probability of at least 1 defective by computing the probability the first is defective while the second is not, computing the probability the first is not defective while the second is defective, and computing the probability they both are defective and then adding these three probabilities together (because they are disjoint), but this is rather time consuming. It is easier to compute the probability that both circuits are not defective and use the Complement Rule to determine the probability of at least 1 defective.

Solution: We have 100 circuits and 5 of them are defective, so 95 circuits are not defective.

$$P(\text{at least 1 defective}) = 1 - P(\text{none defective})$$
$$= 1 - P(\text{1st not defective}) \cdot P(\text{2nd not defective} \mid \text{1st not defective})$$
$$= 1 - \left(\frac{95}{100}\right) \cdot \left(\frac{94}{99}\right)$$
$$= 1 - 0.9020$$
$$= 0.098$$

There is a 9.8% probability that the shipment will be not be accepted.

Now Work Problem 21.

Whenever a small random sample is taken from a large population, it is reasonable to compute probabilities of events assuming independence. Consider the following example.

EXAMPLE 6 Sickle-Cell Anemia

Problem: In a survey of 10,000 African Americans, it was determined that 27 had sickle-cell anemia.

(a) Suppose we randomly select 1 of the 10,000 African Americans surveyed. What is the probability that he or she will have sickle-cell anemia?
(b) If two individuals from this group are randomly selected, what is the probability that both have sickle-cell anemia?
(c) Compute the probability of randomly selecting two individuals from this group who have sickle-cell anemia, assuming independence.

Approach: We let the event E = "sickle-cell anemia," so $P(E)$ = number of African Americans who have sickle-cell anemia divided by the number in the survey. To answer part (b), we let E_1 = "first person has sickle-cell anemia" and E_2 = "second person has sickle-cell anemia," and then we compute $P(E_1 \text{ and } E_2) = P(E_1) \cdot P(E_2|E_1)$. To answer part (c), we use the Multiplication Rule for Independent Events.

Solution

(a) If one individual is selected, $P(E) = \dfrac{27}{10,000} = 0.0027 = 0.27\%$.

(b) Using the Multiplication Rule, we have

$$P(E_1 \text{ and } E_2) = P(E_1) \cdot P(E_2|E_1) = \frac{27}{10000} \cdot \frac{26}{9999} \approx 0.00000702$$

Notice $P(E_2|E_1) = \dfrac{26}{9999}$ because we are sampling without replacement, so after event E_1 occurs there is one less person with sickle cell anemia and one less person in the sample space.

(c) The assumption of independence means that the outcome of the first trial of the experiment does not affect the probability of the second trial. (It is like sampling with replacement.) Therefore, we assume $P(E_1) = P(E_2) = \dfrac{27}{10,000}$. Then

$$P(E_1 \text{ and } E_2) = P(E_1) \cdot P(E_2) = \frac{27}{10,000} \cdot \frac{27}{10,000} \approx 0.00000729$$

The probabilities in Examples 6(b) and 6(c) are extremely close in value. Based on these results, we infer the following principle:

> If small random samples are taken from large populations without replacement, it is reasonable to assume independence of the events. As a rule of thumb, if the sample size is less than 5% of the population size, we treat the events as independent.

For example, in Example 6, we can compute the probability of randomly selecting two African Americans who have sickle-cell anemia using independence because the sample size is less than 5% of the population size $\left(\dfrac{2}{10,000} = 0.0002 = 0.02\%\right)$.

Now Work Problem 35.

Historical Note

Andrei Nikolaevich Kolmogorov was born on April 25, 1903, in Tambov, Russia. His parents were not married. His mother's sister, Vera Yakovlena, raised Kolmogorov. In 1920, Kolmogorov enrolled in Moscow State University. He graduated from the university in 1925. That year he published eight papers, including his first on probability. In 1929, Kolmogorov received his doctorate. By this time he already had 18 publications. He became a professor at Moscow State University in 1931. Kolmogrov is quoted as saying, "The theory of probability as a mathematical discipline can and should be developed from axioms in exactly the same way as Geometry and Algebra." In addition to conducting research, Kolmogorov was interested in helping to educate gifted children. It did not bother him if the students did not become mathematicians; he simply wanted them to be happy. Andrei Kolmogorov died on October 20, 1987.

Conditional Probability and Independence

Two events are independent if the occurrence of event E in a probability experiment does not affect the probability of event F. We can now express independence using conditional probabilities.

Definition Two events E and F are independent if $P(E|F) = P(E)$.

So, the Multiplication Rule for Independent Events is a special case of the General Multiplication Rule.

MAKING AN INFORMED DECISION

Sports Probabilities

Have you ever watched a sporting event on television in which the announcer cites an obscure statistic? Where do these numbers come from? Well, pretend that you are the statistician for your favorite sports team. Your job is to compile strange probabilities regarding your favorite team and a competing team. For example, during the 2001 baseball season, the Boston Red Sox won 36% of the games they played on Wednesdays. As statisticians, we represent this as a conditional probability as follows: $P(\text{win}|\text{Wednesday}) = 0.36$. Suppose that Boston was playing the Seattle Mariners on a Wednesday and that Seattle won 68% of the games that it played on Wednesdays. From these statistics, we predict that Seattle will win the game. Other ideas for conditional probabilities include home versus road games, day versus night games, weather, and so on. For basketball, consider conditional probabilities such as the probability of winning if the team's leading scorer scores fewer than 12 points.

Use the statistics and probabilities that you compile to make a prediction about which team will win. Write an article that presents your predictions along with the supporting numerical facts. Maybe the article could include such "keys to the game" as "Our crack statistician has found that our football team wins 80% of its games when it holds opposing teams to less than 10 points." Repeat this exercise for at least five games. Following each game, determine whether the team you chose has won or lost. Compute your winning percentage for the games you predicted. Did you predict the winner in more than 50% of the games?

A great source for these obscure facts can be found at www.espn.com/. For baseball, a great site is www.mlb.com/. For basketball, go to www.nba.com/. For football, go to www.nfl.com/. For hockey, go to www.nhl.com/.

5.4 ASSESS YOUR UNDERSTANDING

Concepts and Vocabulary

1. The notation $P(F|E)$ means the probability of event _____ given event _____.

2. If $P(E) = 0.6$ and $P(E|F) = 0.34$, are events E and F independent?

Skill Building

3. Suppose that E and F are two events and that $P(E \text{ and } F) = 0.6$ and $P(E) = 0.8$. What is $P(F|E)$?

4. Suppose that E and F are two events and that $P(E \text{ and } F) = 0.21$ and $P(E) = 0.4$. What is $P(F|E)$?

5. Suppose that E and F are two events and that $N(E \text{ and } F) = 420$ and $N(E) = 740$. What is $P(F|E)$?

6. Suppose that E and F are two events and that $N(E \text{ and } F) = 380$ and $N(E) = 925$. What is $P(F|E)$?

7. Suppose that E and F are two events and that $P(E) = 0.8$ and $P(F|E) = 0.4$. What is $P(E \text{ and } F)$?

8. Suppose that E and F are two events and that $P(E) = 0.4$ and $P(F|E) = 0.6$. What is $P(E \text{ and } F)$?

9. According to the U.S. Census Bureau, the probability a randomly selected individual in the United States earns more than $75,000 per year is 18.4%. The probability a randomly selected individual in the United States earns more than $75,000 per year, given that the individual has earned a bachelor's degree, is 35.0%. Are the events "earn more than $75,000 per year" and "earned a bachelor's degree" independent?

10. The probability that a randomly selected individual in the United States 25 years and older has at least a bachelor's degree is 0.272. The probability that an individual in the United States 25 years and older has at least a bachelor's degree given that the individual is Hispanic, is 0.114. Are the events "bachelor's degree" and "Hispanic" independent? *Source: Educational Attainment in the United States, 2003.* U.S. Census Bureau, June 2004

Applying the Concepts

11. **Drawing a Card** Suppose a single card is selected from a standard 52-card deck. What is the probability that the card drawn is a club? Now suppose a single card is drawn from a standard 52-card deck, but we are told that the card is black. What is the probability that the card drawn is a club?

12. **Drawing a Card** Suppose a single card is selected from a standard 52-card deck. What is the probability that the card drawn is a king? Now suppose a single card is drawn from a standard 52-card deck, but we are told that the card is a heart. What is the probability that the card drawn is a king? Did the knowledge that the card is a heart change the probability that the card was a king? What is the term used to describe this result?

13. **Rainy Days** For the month of June in the city of Chicago, 37% of the days are cloudy. Also in the month of June in the city of Chicago, 21% of the days are cloudy and rainy. What is the probability that a randomly selected day in June will be rainy if it is cloudy?

14. **Cause of Death** According to the U.S. National Center for Health Statistics, in 2002, 0.2% of deaths in the United States were 25- to 34-year-olds whose cause of death was cancer. In addition, 1.97% of all those who died were 25 to 34 years old. What is the probability that a randomly selected death is the result of cancer if the individual is known to have been 25 to 34 years old?

15. **High School Dropout** According to the U.S. Census Bureau, 9.1% of high school dropouts are 16- to 17-year-olds. In addition, 5.8% of white high school dropouts are 16- to 17-year-olds. What is the probability a randomly selected dropout is white, given that he or she is 16 to 17 years old?

16. **Income by Region** According to the U.S Census Bureau, 19.1% of U.S. households are in the Northeast. In addition, 4.4% of U.S. households earn $75,000 per year or more and are located in the Northeast. Determine the probability that a randomly selected U.S. household earns more than $75,000 per year, given that the household is located in the Northeast.

17. **Health Insurance Coverage.** The following data represent, in thousands, the type of health insurance coverage of people by age in the year 2002.

	Age			
	<18	18–44	45–64	>64
Private health insurance	49,473	76,294	52,520	20,685
Government health insurance	19,662	11,922	9,227	32,813
No health insurance	8,531	25,678	9,106	258

Source: U.S. Census Bureau

(a) What is the probability that a randomly selected individual who is less than 18 years old has no health insurance?

(b) What is the probability that a randomly selected individual who has no health insurance is less than 18 years old?

18. **Cigar Smoking.** The data in the following table show the results of a national study of 137,243 U.S. men that investigated the association between cigar smoking and death from cancer. **Note**: Current cigar smoker means cigar smoker at time of death.

	Died from Cancer	Did Not Die from Cancer
Never smoked cigars	782	120,747
Former cigar smoker	91	7,757
Current cigar smoker	141	7,725

Source: Shapiro, Jacobs, and Thun. Cigar Smoking in Men and Risk of Death from Tobacco-Related Cancers, *Journal of the National Cancer Institute*, February 16, 2000

(a) What is the probability that a randomly selected individual from the study who died from cancer was a former cigar smoker?

(b) What is the probability that a randomly selected individual from the study who was a former cigar smoker died from cancer?

19. **Driver Fatalities** The following data represent the number of driver fatalities in the United States in 2002 by age for male and female drivers:

Age	Male	Female
Under 16	228	108
16–20	5696	2386
21–34	13,553	4148
35–54	14,395	5017
55–69	4937	1708
70 and over	3159	1529

Source: *Traffic Safety Facts 2002*. Federal Highway Administration, 2002

(a) What is the probability that a randomly selected driver fatality who was male was 16 to 20 years old?

(b) What is the probability that a randomly selected driver fatality who was 16 to 20 was male?

(c) Suppose you are a police officer called to the scene of a traffic accident with a fatality. The dispatcher states that the victim is 16 to 20 years old, but the gender is not known. Is the victim more likely to be male or female? Why?

20. **Marital Status** The following data, in thousands, represent the marital status of Americans 25 years old or older and their level of education in 2003.

	Did Not Graduate from High School	High School Graduate	Some College	College Graduate
Never married	4,333	8,637	7,403	8,321
Married, spouse present	14,787	35,376	28,892	34,693
Married, spouse absent	2,032	2,534	1,633	1,190
Separated	1,134	1,596	1,103	614
Widowed	4,582	5,155	2,487	1,746
Divorced	2,887	7,612	6,393	4,490

Source: Educational Attainment in the United States: 2003. U.S. Census Bureau, June 2004

(a) What is the probability that a randomly selected individual who has never married is a high school graduate?

(b) What is the probability that a randomly selected individual who is a high school graduate has never married?

---

21. **Acceptance Sampling** Suppose you just received a shipment of six televisions. Two of the televisions are defective. If two televisions are randomly selected, compute the probability that both televisions work. What is the probability at least one does not work?

22. **Committee** A committee consists of four women and three men. The committee will randomly select two people to attend a conference in Hawaii. Find the probability that both are women.

23. **Board Work** This past semester, I had a small business calculus section. The students in the class were Mike, Neta, Jinita, Kristin, and Dave. Suppose I randomly select two people to go to the board to work problems. What is the probability that Dave is the first person chosen to go to the board and Neta is the second?

24. **Party** My wife has organized a monthly neighborhood party. Five people are involved in the group: Yolanda (my wife), Lorrie, Laura, Kim, and Anne Marie. They decide to randomly select the first and second home that will host the party. What is the probability that my wife hosts the first party and Lorrie hosts the second? **Note:** Once a home has hosted, it cannot host again until all other homes have hosted.

25. **Playing a CD on the Random Setting** Suppose a compact disk (CD) you just purchased has 13 tracks. After listening to the CD, you decide that you like 5 of the songs. With the random feature on your CD player, each of the 13 songs is played once in random order. Find the probability that among the first two songs played

(a) You like both of them. Would this be unusual?

(b) You like neither of them.

(c) You like exactly one of them.

(d) Redo (a)–(c) if a song can be replayed before all 13 songs are played (if, for example, track 2 can play twice in a row).

26. **Packaging Error** Due to a manufacturing error, three cans of regular soda were accidentally filled with diet soda and placed into a 12-pack. Suppose that two cans are randomly selected from the case.

(a) Determine the probability that both contain diet soda.

(b) Determine the probability that both contain regular soda. Would this be unusual?

(c) Determine the probability that exactly one is diet and one is regular.

27. **Planting Tulips** A bag of 30 tulip bulbs purchased from a nursery contains 12 red tulip bulbs, 10 yellow tulip bulbs, and 8 purple tulip bulbs.

(a) What is the probability that two randomly selected tulip bulbs are both red?

(b) What is the probability that the first bulb selected is red and the second yellow?

(c) What is the probability that the first bulb selected is yellow and the second is red?

(d) What is the probability that one bulb is red and the other yellow?

28. **Golf Balls** The local golf store sells an "onion bag" that contains 35 "experienced" golf balls. Suppose the bag contains 20 Titleists, 8 Maxflis, and 7 Top-Flites.

(a) What is the probability that two randomly selected golf balls are both Titleists?

(b) What is the probability that the first ball selected is a Titleist and the second is a Maxfli?

(c) What is the probability that the first ball selected is a Maxfli and the second is a Titleist?

(d) What is the probability that one golf ball is a Titleist and the other is a Maxfli?

29. **Smokers** According to the National Center for Health Statistics, there is a 23.4% probability that a randomly selected resident of the United States aged 25 years or older is a smoker. In addition, there is a 21.7% probability that a randomly selected resident of the United States aged 25 years or older is female, given that he or she smokes. What is the probability that a randomly selected resident of the United States aged 25 years or older is female and smokes? Would it be unusual to randomly select a resident of the United States aged 25 years or older who is female and smokes?

30. Multiple Jobs According to the U.S. Bureau of Labor Statistics, there is a 5.84% probability that a randomly selected employed individual has more than one job (a multiple-job holder). Also, there is a 52.6% probability that a randomly selected employed individual is male, given that he has more than one job. What is the probability that a randomly selected employed individual is a multiple-job holder and male? Would it be unusual to randomly select an employed individual who is a multiple-job holder and male?

31. The Birthday Problem Determine the probability that at least 2 people in a room of 10 people share the same birthday, ignoring leap years and assuming each birth day is equally likely by answering the following questions:

(a) Compute the probability that 10 people have different birthdays. (*Hint:*The first person's birthday can occur 365 ways, the second person's birthday can occur 364 ways, because he or she cannot have the same birthday as the first person, the third person's birthday can occur 363 ways, because he or she cannot have the same birthday as the first or second person, and so on.)

(b) The complement of "10 people have different birthdays" is "at least 2 share a birthday." Use this information to compute the probability that at least 2 people out of 10 share the same birthday.

32. The Birthday Problem Using the procedure given in Problem 31, compute the probability that at least 2 people in a room of 23 people share the same birthday.

33. A Flush A flush in the card game of poker occurs if a player gets five cards that are all the same suit (clubs, diamonds, hearts, or spades). Answer the following questions to obtain the probability of being dealt a flush in five cards.

(a) We initially concentrate on one suit, say clubs. There are 13 clubs in a deck. Compute $P(\text{five clubs}) = P(\text{first card is clubs and second card is clubs and third card is clubs and fourth card is clubs and fifth card is clubs})$.

(b) A flush can occur if we get five clubs or five diamonds or five hearts or five spades. Compute $P(\text{five clubs or five diamonds or five hearts or five spades})$. Note the events are mutually exclusive.

34. A Royal Flush A royal flush in the game of poker occurs if the player gets the cards Ten, Jack, Queen, King, and Ace all in the same suit. Use the results of Problem 33 to compute the probability of being dealt a royal flush.

35. Independence in Small Samples from Large Populations Suppose a computer chip company has just shipped 10,000 computer chips to a computer company. Unfortunately, 50 of the chips are defective.

(a) Compute the probability that two randomly selected chips are defective using conditional probability.

(b) There are 50 defective chips out of 10,000 shipped. The probability that the first chip randomly selected is defective is $\frac{50}{10,000} = 0.005 = 0.5\%$. Compute the probability that two randomly selected chips are defective under the assumption of independent events. Compare your results to part (a). Conclude that, when small samples are taken from large populations without replacement, the assumption of independence does not significantly affect the probability.

36. Independence in Small Samples from Large Populations Suppose a poll is being conducted in the village of Lemont. The pollster identifies her target population as all residents of Lemont 18 years old or older. This population has 6494 people.

(a) Compute the probability that the first resident selected to participate in the poll is Roger Cummings and the second is Rick Whittingham. [These people are my neighbors in Lemont.]

(b) The probability that any particular resident of Lemont is the first person picked is $\frac{1}{6494}$. Compute the probability that Roger is selected first and Rick is selected second, assuming independence. Compare your results to part (a). Conclude that, when small samples are taken from large populations without replacement, the assumption of independence does not significantly affect the probability.

37. Independent? Refer to the contingency table in Problem 17 that relates age and health insurance coverage. Determine $P(<18 \text{ years old})$ and $P(<18 \text{ years old}|\text{no health insurance})$. Are the events "<18 years old" and "no health insurance" independent?

38. Independent? Refer to the contingency table in Problem 18 that relates cigar smoking and deaths from cancer. Determine $P(\text{died from cancer})$ and $P(\text{died from cancer}|\text{current cigar smoker})$. Are the events "died from cancer" and "current cigar smoker" independent?

39. Independent? Refer to the contingency table in Problem 19 that relates age of driving fatality to gender. Determine $P(\text{female})$ and $P(\text{female}|16–20)$. Are the events "female" and "16–20" independent?

40. Independent? Refer to the contingency table in Problem 20 that relates marital status and level of education. Determine $P(\text{divorced})$ and $P(\text{divorced}|\text{college graduate})$. Are the events "divorced" and "college graduate" independent?

Consumer Reports   His 'N' Hers Razor?

With so many men's and women's versions of different products, you might wonder how different they really are. To help answer this question, technicians at Consumers Union compared a new triple-edge razor for women with a leading double-edge razor for women and a leading triple-edge razor for men. We asked 30 women panelists to shave with the razors over a 4-week period, following a random statistical design.

After each shave, the panelists were asked to answer a series of questions related to the performance of the razor. One question involved rating the razor on a 5-point scale, with 1 being Poor and 5 being Excellent. The following table contains a summary of the results for this question.

Using the information in the table, answer the following questions:

(a) Calculate the probability that a randomly selected razor scored Very Good to Excellent.
(b) Calculate the probability that a randomly selected razor scored Poor.
(c) Calculate the probability of randomly selecting Razor B, given the score was Fair to Good.
(d) Calculate the probability of receiving an Excellent rating, given that Razor C was selected.
(e) Do you think that razor type and rating are independent?
(f) Which razor would you choose based on the information given? Support your decision.

Note to Readers: In many cases, our test protocol and analytical methods are more complicated than described in these examples. The data and discussions have been modified to make the material more appropriate for the audience.

Survey Results for Razor Study

Razor	Poor	Fair to Good	Very Good to Excellent
A	1	8	21
B	0	11	19
C	6	11	13

5.5 Counting Techniques

Objectives
1. Solve counting problems using the Multiplication Rule
2. Solve counting problems using permutations
3. Solve counting problems using combinations
4. Solve counting problems involving permutations with nondistinct items
5. Compute probabilities involving permutations and combinations

1 Solve Counting Problems Using the Multiplication Rule

Counting plays a major role in many diverse areas, including probability. In this section, we look at special types of counting problems and develop general formulas for solving them.

We begin with an example that demonstrates a general counting principle.

EXAMPLE 1   Counting the Number of Possible Meals

Problem: The fixed-price dinner at Mabenka Restaurant provides the following choices:

Appetizer:	soup or salad
Entrée:	baked chicken, broiled beef patty, baby beef liver, or roast beef au jus
Dessert:	ice cream or cheese cake

How many different meals can be ordered?

Approach: Ordering such a meal requires three separate decisions:

Choose an Appetizer	Choose an Entrée	Choose a Dessert
2 choices	4 choices	2 choices

We will draw a tree diagram that lists the possible meals that can be ordered.

Figure 12

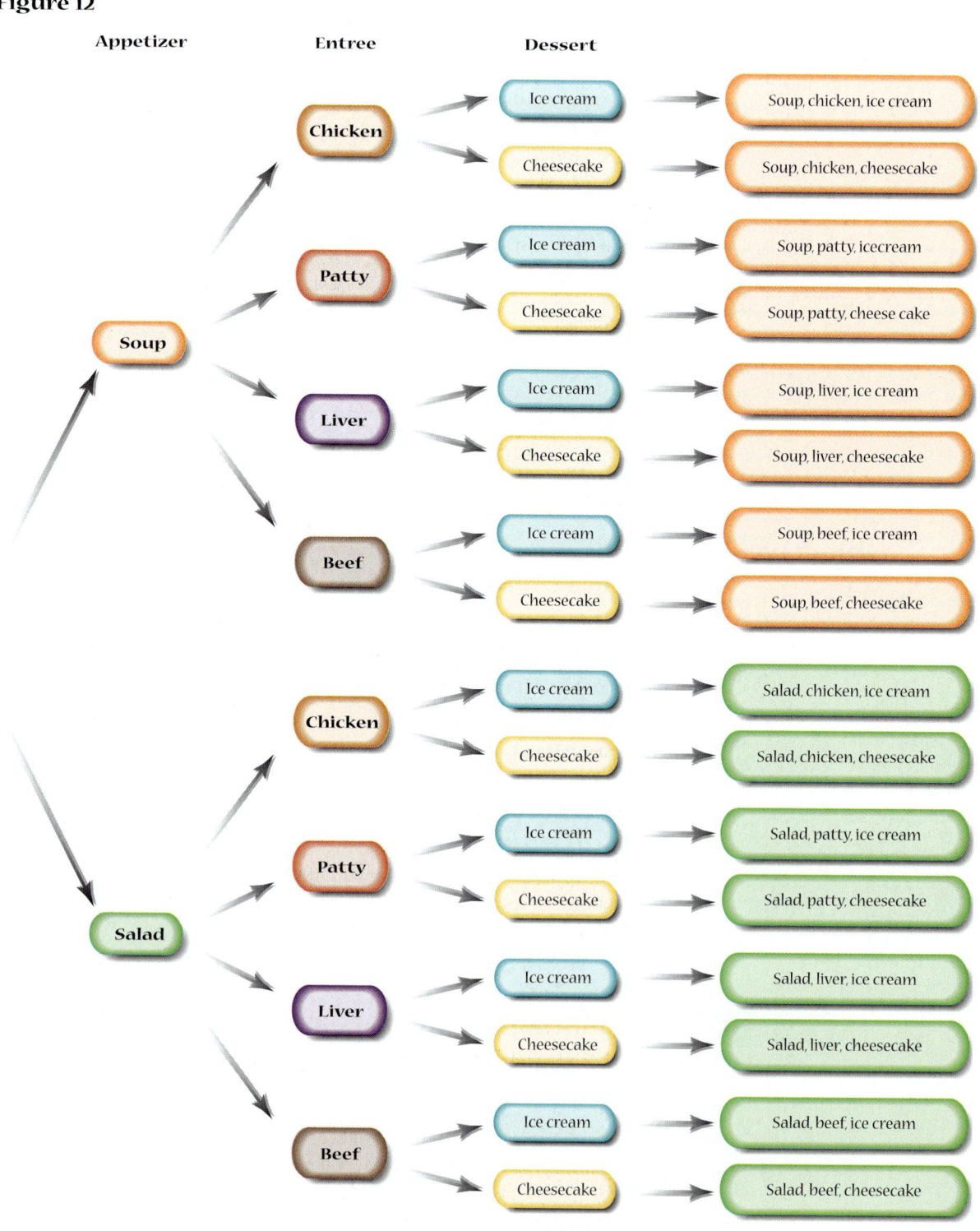

Solution: Look at the tree diagram in Figure 12. We see that, for each choice of appetizer, there are 4 choices of entrée, and that, for each of these $2 \cdot 4 = 8$ choices, there are 2 choices for dessert. A total of

$$2 \cdot 4 \cdot 2 = 16$$

different meals can be ordered.

Example 1 illustrates a general counting principle.

> ## Multiplication Rule of Counting
>
> If a task consists of a sequence of choices in which there are p selections for the first choice, q selections for the second choice, r selections for the third choice, and so on, then the task of making these selections can be done in
>
> $$p \cdot q \cdot r \cdot \ldots$$
>
> different ways.

EXAMPLE 2   Counting Airport Codes

Problem: The International Airline Transportation Association (IATA) assigns three-letter codes to represent airport locations. For example, the airport code for Fort Lauderdale is FLL. How many different airport codes are possible?

Approach: We are choosing 3 letters from 26 letters and arranging them in order. We notice that repetition of letters is allowed. We use the Multiplication Rule of Counting, recognizing there are 26 ways to choose the first letter, 26 ways to choose the second letter, and 26 ways to choose the third letter.

Solution: By the Multiplication Rule,

$$26 \cdot 26 \cdot 26 = 17{,}576$$

different airport codes are possible.

In Example 2, we were allowed to repeat a letter. For example, a valid airport code is FLL (Ft. Lauderdale International Airport), in which the letter L appears twice. In the next example, repetition is not allowed.

EXAMPLE 3   Counting without Repetition

Problem: Three members from a 14-member committee are to be randomly selected to serve as chair, vice-chair, and secretary. The first person selected is the chair; the second person selected, the vice-chair; and the third, the secretary. How many different committee structures are possible?

Approach: The task consists of making three selections. The first selection requires choosing from 14 members. Because a member cannot serve in more than one capacity, the second selection requires choosing from 13 members. The third selection requires choosing from 12 members. (Do you see why?) We use the Multiplication Rule to determine the number of possible committees.

Solution: By the Multiplication Rule,

$$14 \cdot 13 \cdot 12 = 2184$$

different committee structures are possible.

Now Work Problem 31.

The Factorial Symbol

We now introduce a special symbol that can assist us in representing certain types of counting problems.

Definition

If $n \geq 0$ is an integer, the **factorial symbol** $n!$ is defined as follows:

$$0! = 1 \qquad 1! = 1$$
$$n! = n(n-1)\cdots\cdots 3\cdot 2\cdot 1$$

USING TECHNOLOGY:

Your calculator has a factorial key. Use it to see how fast factorials increase in value. Find the value of 69!. What happens when you try to find 70!? In fact, 70! is larger than 10^{100} (a *googol*), the largest number most calculators can display.

For example, $2! = 2\cdot 1 = 2$, $3! = 3\cdot 2\cdot 1 = 6$, $4! = 4\cdot 3\cdot 2\cdot 1 = 24$, and so on. Table 9 lists the values of $n!$ for $0 \leq n \leq 6$.

Table 9						
n	0	1	2	3	4	5	6
$n!$	1	1	2	6	24	120	720

EXAMPLE 4   **The Traveling Salesman**

Problem: You have just been hired as a book representative for Prentice Hall. On your first day, you must travel to seven schools to introduce yourself. How many different routes are possible?

Approach: The seven schools are different. Let's call the schools A, B, C, D, E, F, and G. School A can be visited first, second, third, fourth, fifth, sixth or seventh. So we have seven choices for school A. We would then have six choices for school B, five choices for school C, and so on. We can use the Multiplication Rule and the factorial to find our solution.

Solution: $7\cdot 6\cdot 5\cdot 4\cdot 3\cdot 2\cdot 1 = 7! = 5040$ different routes are possible.

Now Work Problems 5 and 33.

2   Solve Counting Problems Using Permutations

Examples 3 and 4 illustrate a type of counting problem referred to as a *permutation*.

Definition

A **permutation** is an ordered arrangement in which r objects are chosen from n distinct (different) objects and repetition is not allowed. The symbol $_nP_r$ represents the number of permutations of r objects selected from n objects.

So we could represent the solution to the question posed in Example 3 as

$$_nP_r = {}_{14}P_3 = 14\cdot 13\cdot 12 = 2184$$

and the solution to Example 4 could be represented as

$$_7P_7 = 7\cdot 6\cdot 5\cdot 4\cdot 3\cdot 2\cdot 1 = 5040$$

To arrive at a formula for $_nP_r$, we note that there are n choices for the first selection, $n-1$ choices for the second selection, $n-2$ choices for the third selection, ..., and $n - (r - 1)$ choices for the rth selection. By the Multiplication Rule, we have

$$\begin{array}{cccc} & 1^{st} & 2^{nd} & 3^{rd} & & r^{th} \\ {}_nP_r & = & n \cdot (n-1) \cdot (n-2) \cdots [n-(r-1)] \\ & = & n \cdot (n-1) \cdot (n-2) \cdots (n-r+1) \end{array}$$

This formula for ${}_nP_r$ can be written in **factorial notation**:

$${}_nP_r = n \cdot (n-1) \cdot (n-2) \cdots (n-r+1)$$

$$= n \cdot (n-1) \cdot (n-2) \cdots (n-r+1) \cdot \frac{(n-r) \cdots 3 \cdot 2 \cdot 1}{(n-r) \cdots 3 \cdot 2 \cdot 1}$$

$$= \frac{n!}{(n-r)!}$$

We have the following result.

> **Number of Permutations of n Distinct Objects Taken r at a Time**
>
> The number of arrangements of r objects chosen from n objects, in which
>
> **1.** the n objects are distinct,
> **2.** once an object is used it cannot be repeated, and
> **3.** order is important,
>
> is given by the formula
>
> $${}_nP_r = \frac{n!}{(n-r)!} \qquad (1)$$

EXAMPLE 5 **Computing Permutations**

Problem: Evaluate: **(a)** ${}_7P_5$ **(b)** ${}_8P_2$ **(c)** ${}_5P_5$

Approach: To answer (a), we use Formula (1) with $n = 7$ and $r = 5$. To answer (b), we use Formula (1) with $n = 8$ and $r = 2$. To answer (c), we use Formula (1) with $n = 5$ and $r = 5$.

Solution

(a) ${}_7P_5 = \dfrac{7!}{(7-5)!} = \dfrac{7!}{2!} = \dfrac{7 \cdot 6 \cdot 5 \cdot 4 \cdot 3 \cdot 2!}{2!} = \underbrace{7 \cdot 6 \cdot 5 \cdot 4 \cdot 3}_{5 \text{ factors}} = 2520$

(b) ${}_8P_2 = \dfrac{8!}{(8-2)!} = \dfrac{8!}{6!} = \dfrac{8 \cdot 7 \cdot 6!}{6!} = \underbrace{8 \cdot 7}_{2 \text{ factors}} = 56$

(c) ${}_5P_5 = \dfrac{5!}{(5-5)!} = \dfrac{5!}{0!} = 5! = 5 \cdot 4 \cdot 3 \cdot 2 \cdot 1 = 120$

EXAMPLE 6 **Computing Permutations Using Technology**

Problem: Evaluate ${}_7P_5$ using a statistical spreadsheet or graphing calculator with advanced features.

Approach: We will use both Excel and a TI-84 Plus graphing calculator to evaluate ${}_7P_5$. The steps for computing permutations using Excel and the TI-83 or TI-84 Plus graphing calculators can be found in the Technology Step by Step on page 306.

Result: Figure 13(a) shows the result in Excel, and Figure 13(b) shows the result on a TI-84 Plus graphing calculator.

Figure 13

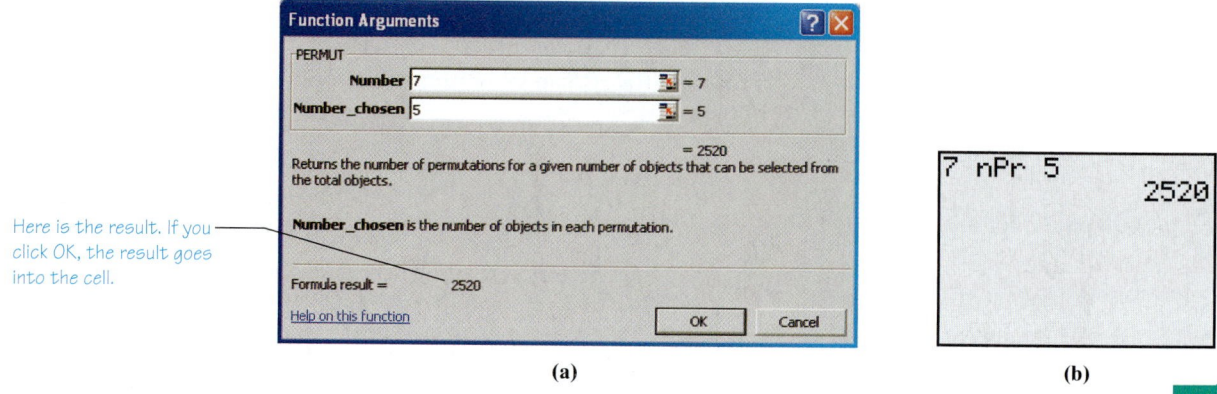

Here is the result. If you click OK, the result goes into the cell.

(a)

(b)

Now Work Problem 11.

Example 5(c) illustrates a general result.

$$_rP_r = r!$$

We need to recognize that, in a permutation, order matters. That is, if we wanted to permute the letters *ABC* by selecting them 3 at a time, the following arrangements are all different:

$$ABC, ACB, BAC, BCA, CAB, CBA$$

EXAMPLE 7 **Betting on the Trifecta**

Problem: In how many ways can horses in a 10-horse race finish first, second, and third?

Approach: The 10 horses are distinct. Once a horse crosses the finish line, that horse will not cross the finish line again, and, in a race, order is important. We have a permutation of 10 objects taken 3 at a time.

Solution: The top three horses can finish a 10-horse race in

$$_{10}P_3 = \frac{10!}{(10-3)!} = \frac{10!}{7!} = \frac{10 \cdot 9 \cdot 8 \cdot 7!}{7!} = \underset{3\ factors}{10 \cdot 9 \cdot 8} = 720 \text{ ways}$$

Now Work Problem 45.

③ Solve Counting Problems Using Combinations

In a permutation, order is important. For example, the arrangements *ABC*, *ACB*, *BAC*, *BCA*, *CAB*, and *CBA* are considered different arrangements of the letters *A*, *B*, and *C*. In many situations, though, order is unimportant. If order is unimportant, the six arrangements of the letters *A*, *B*, and *C* given above are not different. That is, we do not distinguish *ABC* from *BAC*. In the card game of poker, the order in which the cards are received does not matter. The *combination* of the cards is what matters.

Definition A **combination** is a collection, without regard to order, of n distinct objects without repetition. The symbol $_nC_r$ represents the number of combinations of n distinct objects taken r at a time.

EXAMPLE 8 **Listing Combinations**

Problem: Roger, Rick, Randy, and Jay are going to play golf. They will randomly select teams of two players each. List all possible team combinations. That is, list all the combinations of the four people Roger, Rick, Randy, and Jay taken two at a time. What is $_4C_2$?

Approach: We list the possible teams. We note that order is unimportant, so {Roger, Rick} is the same as {Rick, Roger}.

Solution: The list of all such teams (combinations) is

Roger, Rick; Roger, Randy; Roger, Jay; Rick, Randy; Rick, Jay; Randy, Jay

So

$$_4C_2 = 6$$

There are six ways of forming teams of two from a group of four players.

We can find a formula for $_nC_r$ by noting that the only difference between a permutation and a combination is that we disregard order in combinations. To determine $_nC_r$, we eliminate from the formula for $_nP_r$ the number of permutations that were rearrangements of a given set of r objects. In Example 8, for example, selecting {Roger, Rick} was the same as selecting {Rick, Roger}, so there were $2! = 2$ rearrangements of the two objects. This can be determined from the formula for $_nP_r$ by calculating $_rP_r = r!$. So, if we divide $_nP_r$ by $r!$, we will have the desired formula for $_nC_r$:

$$_nC_r = \frac{_nP_r}{r!} = \frac{n!}{r!(n-r)!}$$

We have the following result.

Number of Combinations of n Distinct Objects Taken r at a Time

The number of different arrangements of n objects using $r \leq n$ of them, in which

1. the n objects are distinct,
2. once an object is used, it cannot be repeated, and
3. order is not important

is given by the formula

$$_nC_r = \frac{n!}{r!(n-r)!} \tag{2}$$

Using Formula (2) to solve the problem presented in Example 8, we obtain

$$_4C_2 = \frac{4!}{2!(4-2)!} = \frac{4!}{2!2!} = \frac{4\cdot3\cdot2!}{2\cdot1\cdot2!} = \frac{12}{2} = 6$$

EXAMPLE 9   **Using Formula (2)**

Problem: Use Formula (2) to find the value of each expression.

(a) $_4C_1$ **(b)** $_6C_4$ **(c)** $_6C_2$

Approach: We use Formula (2): $_nC_r = \dfrac{n!}{r!(n-r)!}$

Solution

(a) $_4C_1 = \dfrac{4!}{1!(4-1)!} = \dfrac{4!}{1! \cdot 3!} = \dfrac{4 \cdot 3!}{1 \cdot 3!} = 4$ $n = 4, r = 1$

(b) $_6C_4 = \dfrac{6!}{4!(6-4)!} = \dfrac{6!}{4! \cdot 2!} = \dfrac{6 \cdot 5 \cdot 4!}{4! \cdot 2 \cdot 1} = \dfrac{30}{2} = 15$ $n = 6, r = 4$

(c) $_6C_2 = \dfrac{6!}{2!(6-2)!} = \dfrac{6!}{2!4!} = \dfrac{6 \cdot 5 \cdot 4!}{2 \cdot 1 \cdot 4!} = \dfrac{30}{2} = 15$ $n = 6, r = 2$

EXAMPLE 10   **Computing Combinations Using Technology**

Problem: Evaluate $_6C_4$ using a statistical spreadsheet or graphing calculator with advanced features.

Approach: We will use both Excel and a TI-84 Plus graphing calculator to evaluate $_6C_4$. The steps for computing combinations using Excel and the TI-83 or TI-84 Plus graphing calculators can be found in the Technology Step by Step on page 306.

Result: Figure 14(a) shows the result in Excel, and Figure 14(b) shows the result on a TI-84 Plus graphing calculator.

Figure 14

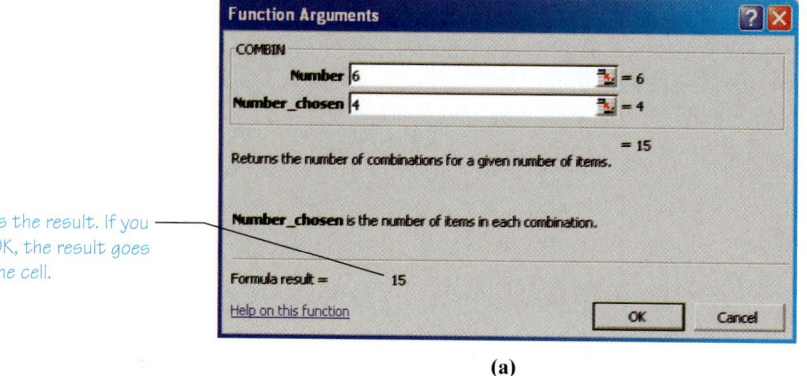

Here is the result. If you click OK, the result goes into the cell.

(a)

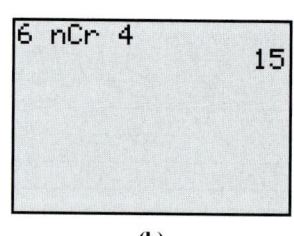

(b)

Notice in Example 9 that $_6C_4 = _6C_2$. This result can be generalized.

$$_nC_r = _nC_{n-r}$$

Now Work Problem 19.

EXAMPLE 11   **Simple Random Samples**

Problem: How many different simple random samples of size 4 can be obtained from a population whose size is 20?

Approach: The 20 individuals in the population are distinct. In addition, the order in which an individual is selected to be in the sample is unimportant. Thus, the number of simple random samples of size 4 from a population of size 20 is a combination of 20 objects taken 4 at a time.

Solution: Use Formula (2) with $n = 20$ and $r = 4$:

$$_{20}C_4 = \frac{20!}{4!(20-4)!} = \frac{20!}{4!16!} = \frac{20 \cdot 19 \cdot 18 \cdot 17 \cdot 16!}{4 \cdot 3 \cdot 2 \cdot 1 \cdot 16!} = \frac{116{,}280}{24} = 4845$$

There are 4845 different simple random samples of size 4 from a population whose size is 20.

> **Now Work Problem 51.**

④ Solve Counting Problems Involving Permutations with Nondistinct Items

Sometimes we wish to arrange objects in order, but some of the objects are not distinguishable.

EXAMPLE 12   **Forming Different Words**

Problem: How many distinguishable strings of letters can be formed by using all the letters in the word REARRANGE?

Approach: Each string formed will have nine letters: three R's, two A's, two E's, one N, and one G. To construct each word, we need to fill in nine positions with the nine letters:

$$\overline{1}\ \overline{2}\ \overline{3}\ \overline{4}\ \overline{5}\ \overline{6}\ \overline{7}\ \overline{8}\ \overline{9}$$

The process of forming a word consists of five tasks:

Step 1: Choose the positions for the three R's.
Step 2: Choose the positions for the two A's.
Step 3: Choose the positions for the two E's.
Step 4: Choose the position for the one N.
Step 5: Choose the position for the one G.

Task 1 can be done in $_9C_3$ ways. There then remain six positions to be filled, so Task 2 can be done in $_6C_2$ ways. There remain four positions to be filled, so Task 3 can be done in $_4C_2$ ways. There remain two positions to be filled, so Task 4 can be done in $_2C_1$ ways. The last position can be filled in $_1C_1$ way.

Solution: By the Multiplication Rule, the number of possible words that can be formed is

$$_9C_3 \cdot _6C_2 \cdot _4C_2 \cdot _2C_1 \cdot _1C_1 = \frac{9!}{6! \cdot 3!} \cdot \frac{6!}{4!2!} \cdot \frac{4!}{2! \cdot 2!} \cdot \frac{2!}{1! \cdot 1!} \cdot \frac{1!}{1!0!}$$

$$= \frac{9!}{3! \cdot 2! \cdot 2! \cdot 1! \cdot 1!}$$

$$= 15{,}120$$

The form of the answer to Example 12 is suggestive of a general result. Had the letters in REARRANGE each been different, there would have been $_9P_9 = 9!$ possible words formed. This is the numerator of the answer.

The presence of three R's, two A's, and two E's reduces the number of different words, as the entries in the denominator illustrate. We are led to the following result:

> ### Permutations with Nondistinct Items
>
> The number of permutations of n objects of which n_1 are of one kind, n_2 are of a second kind, ..., and n_k are of a kth kind is given by
>
> $$\frac{n!}{n_1! \cdot n_2! \cdot \cdots \cdot n_k!} \tag{3}$$
>
> where $n = n_1 + n_2 + \cdots + n_k$.

EXAMPLE 13 | ### Arranging Flags

Problem: How many different vertical arrangements are there of 10 flags if 5 are white, 3 are blue and 2 are red?

Approach: We seek the number of permutations of 10 objects, of which 5 are of one kind (white), 3 are of a second kind (blue), and 2 are of a third kind (red).

Solution: Using Formula (3), we find that there are

$$\frac{10!}{5! \cdot 3! \cdot 2!} = \frac{10 \cdot 9 \cdot 8 \cdot 7 \cdot 6 \cdot 5!}{5! \cdot 3! \cdot 2!} = 2520 \text{ different arrangements}$$

> **Now Work Problem 55.**

Summary

To summarize the differences between combinations and the various types of permutations, we present Table 10.

Table 10		
	Description	**Formula**
Combination	The selection of r objects from a set of n different objects when the order in which the objects is selected does not matter (so AB is the same as BA) and an object cannot be selected more than once (repetition is not allowed)	$_nC_r = \dfrac{n!}{r!(n-r)!}$
Permutation of Distinct Items with Replacement	The selection of r objects from a set of n different objects when the order in which the objects are selected matters (so AB is different from BA) and an object may be selected more than once (repetition is allowed)	n^r
Permutation of Distinct Items without Replacement	The selection of r objects from a set of n different objects when the order in which the objects are selected matters (so AB is different from BA) and an object cannot be selected more than once (repetition is not allowed)	$_nP_r = \dfrac{n!}{(n-r)!}$
Permutation of Nondistinct Items without Replacement	The number of ways n objects can be arranged (order matters) in which there are n_1 of one kind, n_2 of a second kind, ..., and n_k of a kth kind, where $n = n_1 + n_2 + \cdots + n_k$	$\dfrac{n!}{n_1! n_2! \cdots n_k!}$

 ## Compute Probabilities Involving Permutations and Combinations

The counting techniques presented in this section can be used to determine probabilities of certain events by using the classical method of computing probabilities. Recall that this method stated the probability of an event E is the number of ways event E can occur divided by the number of different possible outcomes of the experiment.

| EXAMPLE 14 | **Winning the Lottery** |

Problem: In the Illinois Lottery, an urn contains balls numbered 1 to 54. From this urn, 6 balls are randomly chosen without replacement. For a $1 bet, a player chooses two sets of 6 numbers. To win, all six numbers must match those chosen from the urn. The order in which the balls are selected does not matter. What is the probability of winning the lottery?

Approach: The probability of winning is given by the number of ways a ticket could win divided by the size of the sample space. Each ticket has two sets of six numbers, so there are two chances (for the two sets of numbers) of winning for each ticket. The size of the sample space S is the number of ways that 6 objects can be selected from 54 objects without replacement and without regard to order, so $N(S) = {}_{54}C_6$.

Solution: The size of the sample space is

$$N(S) = {}_{54}C_6 = \frac{54!}{6! \cdot (54-6)!} = \frac{54 \cdot 53 \cdot 52 \cdot 51 \cdot 50 \cdot 49 \cdot 48!}{6! \cdot 48!} = 25{,}827{,}165$$

Each ticket has two sets of 6 numbers, so a player has two chances of winning for each $1. If E is the event "winning ticket," then $N(E) = 2$. The probability of E is

$$P(E) = \frac{2}{25{,}827{,}165} \approx 0.000000077$$

There is about a 1 in 13,000,000 chance of winning the Illinois Lottery!   ▪

| EXAMPLE 15 | **Probabilities Involving Combinations** |

Problem: A shipment of 120 fasteners that contains 4 defective fasteners was sent to a manufacturing plant. The quality-control manager at the manufacturing plant randomly selects 5 fasteners and inspects them. What is the probability that exactly 1 fastener is defective?

Approach: The probability that exactly 1 fastener is defective is found by calculating the number of ways of selecting exactly 1 defective fastener in 5 fasteners and dividing this result by the number of ways of selecting 5 fasteners from 120 fasteners. To choose exactly 1 defective in the 5 requires choosing 1 defective from the 4 defectives and 4 nondefectives from the 116 nondefectives. The order in which the fasteners are selected does not matter, so we use combinations.

Solution: The number of ways of choosing 1 defective fastener from 4 defective fasteners is ${}_4C_1$. The number of ways of choosing 4 nondefective fasteners from 116 nondefectives is ${}_{116}C_4$. Using the Multiplication Rule, we find that the number of ways of choosing 1 defective and 4 nondefective fasteners is

$$({}_4C_1) \cdot ({}_{116}C_4) = 4 \cdot 7{,}160{,}245 = 28{,}640{,}980$$

The number of ways of selecting 5 fasteners from 120 fasteners is ${}_{120}C_5 = 190{,}578{,}024$. The probability of selecting exactly 1 defective fastener is

$$P(1 \text{ defective fastener}) = \frac{({}_4C_1)({}_{116}C_4)}{{}_{120}C_5} = \frac{28{,}640{,}980}{190{,}578{,}024} \approx 0.1503 = 15.03\%$$

There is a 15.03% probability of randomly selecting exactly 1 defective fastener.

Now Work Problem 61.

5.5 ASSESS YOUR UNDERSTANDING

Concepts and Vocabulary

1. A _____ is an ordered arrangement of r objects chosen from n distinct objects without repetition.

2. A _____ is an arrangement of r objects chosen from n distinct objects without repetition and without regard to order.

3. *True or False:* In a combination problem, order is not important.

4. Explain the difference between a combination and a permutation.

Skill Building

In Problems 5–10, find the value of each factorial.

5. 5!

6. 7!

7. 10!

8. 12!

9. 0!

10. 1!

In Problems 11–18, find the value of each permutation.

11. $_6P_2$

12. $_7P_2$

13. $_4P_4$

14. $_7P_7$

15. $_5P_0$

16. $_4P_0$

17. $_8P_3$

18. $_9P_4$

In Problems 19–26, find the value of each combination.

19. $_8C_3$

20. $_9C_2$

21. $_{10}C_2$

22. $_{12}C_3$

23. $_{52}C_1$

24. $_{40}C_{40}$

25. $_{48}C_3$

26. $_{30}C_4$

27. List all the permutations of five objects a, b, c, d, and e taken two at a time without repetition. What is $_5P_2$?

28. List all the permutations of four objects a, b, c, and d taken two at a time without repetition. What is $_4P_2$?

29. List all the combinations of five objects a, b, c, d, and e taken two at a time. What is $_5C_2$?

30. List all the combinations of four objects a, b, c, and d taken two at a time. What is $_4C_2$?

Applying the Concepts

31. **Clothing Options** A man has six shirts and four ties. Assuming that they all match, how many different shirt-and-tie combinations can he wear?

32. **Clothing Options** A woman has five blouses and three skirts. Assuming that they all match, how many different outfits can she wear?

33. **Arranging Songs on a CD** Suppose Dan is going to burn a compact disk (CD) that will contain 12 songs. In how many ways can Dan arrange the 12 songs on the CD?

34. **Arranging Students** In how many ways can 15 students be lined up?

35. **Traveling Salesman** A salesman must travel to eight cities to promote a new marketing campaign. How many different trips are possible if any route between cities is possible?

36. **Randomly Playing Songs** A certain compact disk player randomly plays each of 10 songs on a CD. Once a song is played, it is not repeated until all the songs on the CD have been played. In how many different ways can the CD player play the 10 songs?

37. **Stocks on the NYSE** Companies whose stocks are listed on the New York Stock Exchange (NYSE) have their company name represented by either one, two, or three letters (repetition of letters is allowed). What is the maximum number of companies that can be listed on the New York Stock Exchange?

38. **Stocks on the NASDAQ** Companies whose stocks are listed on the NASDAQ stock exchange have their company name represented by either four or five letters (repetition of letters is allowed). What is the maximum number of companies that can be listed on the NASDAQ?

39. **Garage Door Code** Outside a home, there is a keypad that can be used to open the garage if the correct four-digit code is entered.

(a) How many codes are possible?

(b) What is the probability of entering the correct code on the first try, assuming that the owner doesn't remember the code?

40. Social Security Numbers A Social Security number is used to identify each resident of the United States uniquely. The number is of the form xxx–xx–xxxx, where each x is a digit from 0 to 9.

(a) How many Social Security numbers can be formed?

(b) What is the probability of correctly guessing the Social Security number of the President of the United States?

41. Usernames Suppose a local area network requires eight letters for user names. Lower- and uppercase letters are considered the same. How many user names are possible for the local area network?

42. Passwords Suppose a local area network requires eight characters for a password. The first character must be a letter, but the remaining seven characters can be either a letter or a digit (0 through 9). Lower- and uppercase letters are considered the same. How many passwords are possible for the local area network?

43. Combination Locks A combination lock has 50 numbers on it. To open it, you turn counterclockwise to a number, then rotate clockwise to a second number, and then counterclockwise to the third number.

(a) How many different lock combinations are there?

(b) What is the probability of guessing a lock combination on the first try?

44. Forming License Plate Numbers How many different license plate numbers can be made by using one letter followed by five digits selected from the digits 0 through 9?

45. INDY 500 Suppose 40 cars start at the Indianapolis 500. In how many ways can the top 3 cars finish the race?

46. Betting on the Perfecta In how many ways can the top 2 horses finish in a 10-horse race?

47. Forming a Committee Four members from a 20-person committee are to be selected randomly to serve as chairperson, vice-chairperson, secretary, and treasurer. The first person selected is the chairperson; the second, the vice-chairperson; the third, the secretary; and the fourth, the treasurer. How many different leadership structures are possible?

48. Forming a Committee Four members from a 50-person committee are to be selected randomly to serve as chairperson, vice-chairperson, secretary, and treasurer. The first person selected is the chairperson; the second, the vice-chairperson; the third, the secretary; and the fourth, the treasurer. How many different leadership structures are possible?

49. Lottery Suppose a lottery exists where balls numbered 1 to 25 are placed in an urn. To win, you must match the four balls chosen in the correct order. How many possible outcomes are there for this game?

50. Forming a Committee In the U.S. Senate, there are 21 members on the Committee on Banking, Housing, and Urban Affairs. Nine of these 21 members are selected to be on the Subcommittee on Economic Policy. How many different committee structures are possible for this subcommittee?

51. Simple Random Sample How many different simple random samples of size 5 can be obtained from a population whose size is 50?

52. Simple Random Sample How many different simple random samples of size 7 can be obtained from a population whose size is 100?

53. Children A family has six children. If this family has exactly two boys, how many different birth and gender orders are possible?

54. Children A family has eight children. If this family has exactly three boys, how many different birth and gender orders are possible?

55. Forming Words How many different 10-letter words (real or imaginary) can be formed from the letters in the word STATISTICS?

56. Forming Words How many different nine-letter words (real or imaginary) can be formed from the letters in the word ECONOMICS?

57. Landscape Design A golf-course architect has four linden trees, five white birch trees, and two bald cypress trees to plant in a row along a fairway. In how many ways can the landscaper plant the trees in a row, assuming that the trees are evenly spaced?

58. Starting Lineup A baseball team consists of three outfielders, four infielders, a pitcher, and a catcher. Assuming that the outfielders and infielders are indistinguishable, how many batting orders are possible?

59. Little Lotto In the Illinois Lottery game Little Lotto, an urn contains balls numbered 1 to 30. From this urn, 5 balls are chosen randomly, without replacement. For a $1 bet, a player chooses one set of five numbers. To win, all five numbers must match those chosen from the urn. The order in which the balls are selected does not matter. What is the probability of winning Little Lotto with one ticket?

60. The Big Game In the Big Game, an urn contains balls numbered 1 to 50, and a second urn contains balls numbered 1 to 36. From the first urn, 5 balls are chosen randomly, without replacement. From the second urn, 1 ball is chosen randomly. For a $1 bet, a player chooses one set of five numbers to match the balls selected from the first urn and one number to match the ball selected from the second urn. To win, all six numbers must match, that is, the player must match the first 5 balls selected from the first urn *and* the single ball selected from the second urn. What is the probability of winning the Big Game with a single ticket?

61. Selecting a Jury The grade appeal process at a university requires that a jury be structured by selecting five individuals randomly from a pool of eight students and ten faculty.

(a) What is the probability of selecting a jury of all students?

(b) What is the probability of selecting a jury of all faculty?

(c) What is the probability of selecting a jury of two students and three faculty?

62. Selecting a Committee Suppose there are 55 Democrats and 45 Republicans in the U.S. Senate. A committee of seven senators is to be formed by selecting members of the Senate randomly.

(a) What is the probability that the committee is composed of all Democrats?

(b) What is the probability that the committee is composed of all Republicans?

(c) What is the probability that the committee is composed of three Democrats and four Republicans?

63. Acceptance Sampling Suppose a shipment of 120 electronic components contains 4 defective components. To determine whether the shipment should be accepted, a quality-control engineer randomly selects 4 of the components and tests them. If 1 or more of the components is defective, the shipment is rejected. What is the probability the shipment is rejected?

64. In the Dark A box containing twelve 40-watt light bulbs and eighteen 60-watt light bulbs is stored in your basement. Unfortunately, the box is stored in the dark and you need two 60-watt bulbs. What is the probability of randomly selecting two 60-watt bulbs from the box?

65. Randomly Playing Songs Suppose a compact disk (CD) you just purchased has 13 tracks. After listening to the CD, you decide that you like 5 of the songs. The random feature on your CD player will play each of the 13 songs once in a random order. Find the probability that among the first 4 songs played

(a) you like 2 of them;

(b) you like 3 of them;

(c) you like all 4 of them.

66. Packaging Error Through a manufacturing error, three cans marked "regular soda" were accidentally filled with diet soda and placed into a 12-pack. Suppose that three cans are randomly selected from the 12-pack.

(a) Determine the probability that exactly two contain diet soda.

(b) Determine the probability that exactly one contains diet soda.

(c) Determine the probability that all three contain diet soda.

67. Three of a Kind Suppose you are dealt 5 cards from a standard 52-card deck. Determine the probability of being dealt three of a kind (such as three aces or three kings) by answering the following questions:

(a) How many ways can 5 cards be selected from a 52-card deck?

(b) Each deck contains 4 two's, 4 three's, and so on. How many ways can three of the same card be selected from the deck?

(c) The remaining 2 cards must be different from the 3 chosen and different from each other. For example, if we drew three kings, the 4th card cannot be a king. After selecting the three of a kind, there are 12 different ranks of card remaining in the deck that can be chosen. If we have three kings, then we can choose two's, three's, and so on. Of the 12 ranks remaining, we choose 2 of them and then select one of the 4 cards in each of the two chosen ranks. How many ways can we select the remaining 2 cards?

(d) Use the General Multiplication Rule to compute the probability of obtaining three of a kind. That is, what is the probability of selecting three of a kind and two cards that are not like?

68. Two of a Kind Follow the outline presented in Problem 67 to determine the probability of being dealt exactly one pair.

69. Acceptance Sampling Suppose you have just received a shipment of 20 modems. Although you don't know this, 3 of the modems are defective. To determine whether you will accept the shipment, you randomly select 4 modems and test them. If all 4 modems work, you accept the shipment. Otherwise, the shipment is rejected. What is the probability of accepting the shipment?

70. Acceptance Sampling Suppose you have just received a shipment of 100 televisions. Although you don't know this, 6 are defective. To determine whether you will accept the shipment, you randomly select 5 televisions and test them. If all 5 televisions work, you accept the shipment; otherwise, the shipment is rejected. What is the probability of accepting the shipment?

Technology Step-by-Step

Factorials, Permutations, and Combinations

TI-83/84 Plus

Factorials

Step 1: To compute 7!, type 7 on the HOME screen.

Step 2: Press MATH, then highlight PRB, and then highlight 4 : ! Press ENTER. With 4! on the HOME screen, press ENTER again.

Permutations and Combinations

Step 1: To compute $_7P_3$, type 7 on the HOME screen.

Step 2: Press MATH, then highlight PRB, and then highlight 2 : $_nP_r$ and press ENTER.

Step 3: Type 3 on the HOME screen, and press ENTER.

Note: To compute $_7C_3$, select 3 : $_nC_r$ instead of 2 : $_nP_r$.

Excel **Factorials or Combinations**

Step 1: Select Insert and then Function ....

Step 2: Highlight Math & Trig in the Function category. For combinations, select COMBIN in the function name and fill in the appropriate cells. For a factorial, select FACT in the function name and fill in the appropriate cells.

Permutations

Step 1: Select Insert and then Function ....

Step 2: Highlight Statistical in the Function category. For permutations, select PERMUT in the function name and fill in the appropriate cells.

Chapter 5

Review

Summary

In this chapter, we introduced the concept of probability. Probability is a measure of the likelihood of a random phenomenon or chance behavior. Because we are measuring a random phenomenon, there is short-term uncertainty. However, this short-term uncertainty gives rise to long-term predictability.

Probabilities are numbers between zero and one, inclusive. The closer a probability is to one, the more likely the event is to occur. If an event has probability zero, it is said to be impossible. Events with probability one are said to be certain.

We introduced three methods for computing probabilities: (1) the empirical (or relative frequency) method, (2) the classical method, and (3) subjective probabilities. Empirical probabilities rely on the relative frequency with which an event happens. Classical probabilities require the events in the experiment to be equally likely. We count the number of ways an event can happen and divide this by the number of possible outcomes of the experiment. Empirical probabilities actually require that an experiment be performed, whereas classical probability does not. Subjective probabilities are probabilities based on the opinion of the individual providing the probability. They are educated guesses about the likelihood of an event occurring, but still represent a legitimate way of assigning probabilities.

We are also interested in probabilities of multiple outcomes. For example, we might be interested in the probability that either event E or event F happens. The Addition Rule is used to compute the probability of E *or* F; the Multiplication Rule is used to compute the probability that both E *and* F occur. Two events are mutually exclusive (or disjoint) if they do not have any outcomes in common. Two events E and F are independent if the probability of event E occurring does not affect the probability of event F. The complement of an event E, denoted E^c, is all the outcomes in the sample space that are not in E.

Finally, we introduced counting methods. The Multiplication Rule is used to count the number of ways a sequence of events can occur. Permutations are used to count the number of ways r distinct items can be arranged from a set of n items. Combinations are used to count the number of ways r distinct items can be selected from a set of n items without replacement and without regard to order. These counting techniques can be used to calculate probabilities using the classical method.

Formulas

Empirical Probability

$$P(E) \approx \frac{\text{frequency of } E}{\text{number of trials of experiment}}$$

Classical Probability

$$P(E) = \frac{\text{number of ways that } E \text{ can occur}}{\text{number of possible outcomes}} = \frac{N(E)}{N(S)}$$

Addition Rule for Disjoint Events

$$P(E \text{ or } F) = P(E) + P(F)$$

General Addition Rule

$$P(E \text{ or } F) = P(E) + P(F) - P(E \text{ and } F)$$

Probabilities of Complements

$$P(E^c) = 1 - P(E)$$

Multiplication Rule for Independent Events

$$P(E \text{ and } F) = P(E) \cdot P(F)$$

Multiplication Rule for n Independent Events

$$P(E \text{ and } F \text{ and } G \ldots) = P(E) \cdot P(F) \cdot P(G) \cdot \cdots$$

Conditional Probability Rule

$$P(F|E) = \frac{P(E \text{ and } F)}{P(E)} = \frac{N(E \text{ and } F)}{N(E)}$$

General Multiplication Rule

$$P(E \text{ and } F) = P(E) \cdot P(F|E)$$

Factorial Notation

$$n! = n \cdot (n-1) \cdot (n-2) \cdots \cdots 3 \cdot 2 \cdot 1$$

Combination

$$_nC_r = \frac{n!}{r!(n-r)!}$$

Permutation

$$_nP_r = \frac{n!}{(n-r)!}$$

Permutations with Nondistinct Items

$$\frac{n!}{n_1! \cdot n_2! \cdot \cdots \cdot n_k!}$$

Vocabulary

Probability (p. 251)
Outcome (p. 251)
Experiment (p. 252)
Sample space (p. 252)
Event (p. 252)
Probability model (p. 253)
Impossible event (p. 253)
Certainty (p. 253)
Unusual event (p. 253)

Equally likely outcomes (p. 255)
Tree diagram (p. 258)
Subjective probability (p. 260)
Disjoint (p. 266)
Mutually exclusive (p. 266)
Venn diagram (p. 266)
Contingency table (p. 271)
Row variable (p. 271)
Column variable (p. 271)

Cell (p. 271)
Complement (p. 272)
Independent (p. 278)
Dependent (p. 278)
Conditional probability (p. 284)
Factorial (p. 296)
Permutation (p. 296)
Combination (p. 299)

Objectives

Section	You should be able to . . .	Examples	Review Exercises
5.1	**1** Understand the rules of probabilities (p. 253)	2	1, 13(d), 18(a)
	2 Compute and interpret probabilities using the empirical method (p. 254)	3, 4, 7(a)	15(a), 16–18, 19(a) and (b), 20(a) and (b), 21(a), 39
	3 Compute and interpret probabilities using the classical method (p. 255)	5, 6, 7(b)	2–4, 13(a), 14
	4 Use simulation to obtain data based on probabilities (p. 259)	8	36
	5 Understand subjective probabilities (p. 260)		37
5.2	**1** Use the Addition Rule for disjoint events (p. 266)	1 and 2	3, 4, 7, 13(b) and (c), 18(b) and (c)
	2 Use the General Addition Rule (p. 269)	3 and 4	6, 19(d), 20(d)
	3 Compute the probability of an event using the Complement Rule (p. 272)	5 and 6	5, 15(b) and (e), 16, 18(d), 21(b), 21(d), 21(f), 22(c)
5.3	**1** Understand independence (p. 277)	1	9
	2 Use the Multiplication Rule for independent events (p. 279)	2 and 3	8, 21(c) and (e), 22(a) and (b), 23, 24
	3 Compute at-least probabilities (p. 281)	4	21(d), 21(f), 22(c)
5.4	**1** Compute conditional probabilities (p. 283)	1 through 3	11, 19(f), 20(e) and (f)
	2 Compute probabilities using the General Multiplication Rule (p. 286)	4 through 6	10, 15(c) and (d), 25, 26, 38
5.5	**1** Solve counting problems using the Multiplication Rule (p. 293)	1 through 4	27, 29
	2 Solve counting problems using permutations (p.296)	5 through 7	12(e) and (f), 28
	3 Solve counting problems using combinations (p. 298)	8 through 11	12(c) and (d), 31, 32
	4 Solve counting problems involving permutations with nondistinct items (p. 301)	12 and 13	30
	5 Compute probabilities involving permutations and combinations (p.302)	14 and 15	33–35

Review Exercises

1. (a) Which among the following numbers could be the probability of an event?

$$0, -0.01, 0.75, 0.41, 1.34$$

(b) Which among the following numbers could be the probability of an event?

$$\frac{2}{5}, \frac{1}{3}, -\frac{4}{7}, \frac{4}{3}, \frac{6}{7}$$

For Problems 2–5, let the sample space be S = {red, green, blue, orange, yellow}. Suppose the outcomes are equally likely.

2. Compute the probability of the event $E = \{yellow\}$.

3. Compute the probability of the event $F = \{green\ or\ orange\}$.

4. Compute the probability of the event $E = \{red\ or\ blue\ or\ yellow\}$.

5. Suppose that $E = \{yellow\}$. Compute the probability of E^c.

6. Suppose that $P(E) = 0.76$, $P(F) = 0.45$, and $P(E\ and\ F) = 0.32$. What is $P(E\ or\ F)$?

7. Suppose that $P(E) = 0.36$, $P(F) = 0.12$, and E and F are mutually exclusive. What is $P(E\ or\ F)$?

8. Suppose that events E and F are independent. In addition, $P(E) = 0.45$ and $P(F) = 0.2$. What is $P(E\ and\ F)$?

9. Suppose that $P(E) = 0.8$, $P(F) = 0.5$, and $P(E\ and\ F) = 0.24$. Are events E and F independent? Why?

10. Suppose that $P(E) = 0.59$ and $P(F|E) = 0.45$. What is $P(E\ and\ F)$?

11. Suppose that $P(E\ and\ F) = 0.35$ and $P(F) = 0.7$. What is $P(E|F)$?

12. Determine the value of each of the following:
    (a) $7!$
    (b) $0!$
    (c) $_9C_4$
    (d) $_{10}C_3$
    (e) $_9P_2$
    (f) $_{12}P_4$

13. **Roulette** In the game of roulette, a wheel consists of 38 slots, numbered 0, 00, 1, 2, ..., 36. (See the photo in Problem 33 from Section 5.1.) To play the game, a metal ball is spun around the wheel and is allowed to fall into one of the numbered slots. The slots numbered 0 and 00 are green, the odd numbers are red, and the even numbers are black.
    (a) Determine the probability that the metal ball falls into a green slot. Interpret this probability.
    (b) Determine the probability that the metal ball falls into a green or a red slot. Interpret this probability.
    (c) Determine the probability that the metal ball falls into 00 or a red slot. Interpret this probability.
    (d) Determine the probability that the metal ball falls into the number 31 and a black slot simultaneously. What term is used to describe this event?

14. **Craps** Craps is a dice game in which two fair dice are cast. If the roller shoots a 7 or 11 on the first roll, he or she wins. If the roller shoots a 2, 3, or 12 on the first roll, he or she loses.
    (a) Compute the probability that the shooter wins on the first roll. Interpret this probability.
    (b) Compute the probability that the shooter loses on the first roll. Interpret this probability.

15. **New Year's Holiday** Between 6:00 P.M. December 29, 2001, and 5:59 A.M. January 2, 2002, there were 575 traffic fatalities in the United States. Of these, 301 were alcohol related.
    (a) What is the probability that a randomly selected traffic fatality that happened between 6:00 P.M. December 29, 2001, and 5:59 A.M. January 2, 2002, was alcohol related?
    (b) What is the probability that a randomly selected traffic fatality that happened between 6:00 P.M. December 29, 2001, and 5:59 A.M. January 2, 2002, was not alcohol related?
    (c) What is the probability that two randomly selected traffic fatalities that happened between 6:00 P.M. December 29, 2001, and 5:59 A.M. January 2, 2002, were both alcohol related?
    (d) What is the probability that neither of two randomly selected traffic fatalities that happened between 6:00 P.M. December 29, 2001, and 5:59 A.M. January 2, 2002, were alcohol related?
    (e) What is the probability that of two randomly selected traffic fatalities that happened between 6:00 P.M. December 29, 2001, and 5:59 A.M. January 2, 2002, at least one was alcohol related?

16. **Cyclones** According to the National Hurricane Center, about 11% of tropical cyclones occur in the North Atlantic Ocean. What is the probability that a randomly selected cyclone occurs in the North Atlantic Ocean? What is the probability that a randomly selected cyclone occurs somewhere other than in the North Atlantic Ocean?

17. **Louisville Workers** The following data represent the distribution of class of workers in Louisville, Kentucky, in 2003.

Class of Worker	Number of Workers
Private wage and salary worker	99,402
Government worker	14,580
Self-employed worker	3,790
Unpaid family worker	164

Source: U.S. Census Bureau *American Community Survey*, 2003.

(a) Construct a probability model for Louisville workers.
(b) Is it unusual for a Louisville worker to be an unpaid family worker?
(c) Is it unusual for a Louisville worker to be self-employed?

18. **Girl Scout Cookies** The following probability model shows the distribution of the most popular selling Girl Scout Cookies®.

Cookie Type	Probability
Thin Mints	0.25
Samoas®/Caramel deLites	0.19
Peanut Butter Patties®/Tagalongs™	0.13
Peanut Butter Sandwich/Do-si-dos™	0.11
Shortbread/Trefoils	0.09
Other varieties	0.23

Source: www.girlscouts.org

(a) Verify this is a probability model.

(b) If a girl scout is selling cookies to people who randomly enter a shopping mall, what is the probability that the next box sold will be Peanut Butter Patties®/Tagalongs® or Peanut Butter Sandwich/Do-si-dos™?

(c) If a girl scout is selling cookies to people who randomly enter a shopping mall, what is the probability that the next box sold will be Thin Mints, Samoas®/Caramel deLites™, or Shortbread/Trefoils?

(d) What is the probability that the next box sold will not be Thin Mints?

19. Gestation Period versus Weight The following data represent the birth weights (in grams) of babies born in 2002, along with the period of gestation.

Birth Weight (in grams)	Period of Gestation		
	Preterm	Term	Postterm
Less than 1000	28,247	202	32
1000–1999	78,532	10,325	1,051
2000–2999	228,064	606,046	39,481
3000–3999	135,790	2,293,927	192,566
4000–4999	8,974	315,821	34,319
Over 5000	211	4,524	556

Source: National Vital Statistics Report, Vol. 52, No. 10, December 17, 2003

(a) What is the probability that a randomly selected baby born in 2002 was postterm?

(b) What is the probability that a randomly selected baby born in 2002 weighed 3000 to 3999 grams?

(c) What is the probability that a randomly selected baby born in 2002 weighed 3000 to 3999 grams and was postterm?

(d) What is the probability that a randomly selected baby born in 2002 weighed 3000 to 3999 grams or was postterm?

(e) What is the probability that a randomly selected baby born in 2002 weighed less than 1000 grams and was postterm? Is this event impossible?

(f) What is the probability that a randomly selected baby born in 2002 weighed 3000 to 3999 grams, given the baby was postterm?

(g) Are the events "postterm baby" and "weighs 3000 to 3999 grams" independent? Why?

20. Olympic Medals The following data represent the medal tallies of the top eight countries at the 2004 Summer Olympics in Athens.

Country	Gold	Silver	Bronze
United States	35	39	29
China	32	17	14
Russia	27	27	38
Australia	17	16	16
Japan	16	9	12
Germany	14	16	18
France	11	9	13
Italy	10	11	11

(a) If a medal is randomly selected from the top eight countries, what is the probability that it is gold?

(b) If a medal is randomly selected from the top eight countries, what is the probability that it was won by China?

(c) If a medal is randomly selected from the top eight countries, what is the probability that it is gold and was won by China?

(d) If a medal is randomly selected from the top eight countries, what is the probability that it is gold or was won by China?

(e) If a bronze medal is randomly selected from the top eight countries, what is the probability that it was won by Japan?

(f) If a medal that was won by Japan is randomly selected, what is the probability that it is bronze?

21. Better Business Bureau The Better Business Bureau reported that approximately 63% of consumer complaints in 2003 were filed online.

(a) If a consumer complaint from 2003 is randomly selected, what is the probability it was filed online?

(b) What is the probability that is was not filed online?

(c) If a random sample of five consumer complaints was selected, what is the probability that all five were filed online?

(d) If a random sample of five consumer complaints was selected, what is the probability that at least one was not filed online?

(e) If a random sample of five consumer complaints was selected, what is the probability that none was filed online?

(f) If a random sample of five consumer complaints was selected, what is the probability that at least one was filed online?

22. St. Louis Cardinals During the 2004 season, the St. Louis Cardinals won 64.8% of their games. Assume that the outcomes of the baseball games are independent and that the percentage of wins this season will be the same as in 2004, and answer the following questions:

(a) What is the probability that the Cardinals will win two games in a row?

(b) What is the probability that the Cardinals will win seven games in a row?

(c) What is the probability that the Cardinals will lose at least one of their next seven games?

23. Pick 3 For the Illinois Lottery's PICK 3 game, a player must match a sequence of three repeatable numbers, ranging from 0 to 9, in exact order (for example, 3–7–2). With a single ticket, what is the probability of matching the three winning numbers?

24. Pick 4 The Illinois Lottery's PICK 4 game is similar to PICK 3, except a player must match a sequence of four repeatable numbers, ranging from 0 to 9, in exact order (for example, 5–8–5–1). With a single ticket, what is the probability of matching the four winning numbers?

25. Acceptance Sampling Suppose you just received a shipment of 10 DVD players. One of the DVD players is defective. You will accept the shipment if two randomly selected DVD players work. What is the probability that you will accept the shipment?

26. Drawing Cards Suppose you draw 3 cards without replacement from a standard 52-card deck. What is the probability that all 3 cards are aces?

27. Forming License Plates A license plate is designed so that the first two characters are letters and the last four characters are digits (0 through 9). How many different license plates can be formed assuming letters and numbers can be used more then once?

28. Choosing a Seat If four students enter a classroom that has 10 vacant seats, in how many ways can they be seated?

29. Jumble In the game of Jumble, the letters of a word are scrambled. The player must form the correct word. In a recent game in a local newspaper, the Jumble "word" was LINCEY. How many different arrangements are there of the letters in this "word"?

30. Arranging Flags How many different vertical arrangements are there of 10 flags if 4 are white, 3 are blue, 2 are green, and 1 is red?

31. Simple Random Sampling How many different simple random samples of size 8 can be obtained from a population whose size is 55?

32. Forming Committees The U.S. Senate Appropriations Committee has 29 members. Suppose that a subcommittee is to be formed by randomly selecting 5 of the members of the Appropriations Committee. How many different committees could be formed?

33. Arizona's Fantasy 5 In one of Arizona's lotteries, balls are numbered 1 to 35. Five balls are selected randomly, without replacement. The order in which the balls are selected does not matter. To win, your numbers must match the five selected. Determine your probability of winning Arizona's Fantasy 5 with one ticket.

34. Pennsylvania's Cash 5 In one of Pennsylvania's lotteries, balls are numbered 1 to 39. Five balls are selected randomly, without replacement. The order in which the balls are selected does not matter. To win, your numbers must match the five selected. Determine your probability of winning Pennsylvania's Cash 5 with one ticket.

35. Packaging Error Because of a mistake in packaging, a case of 12 bottles of red wine contained 5 Merlot and 7 Cabernet, each without labels. All the bottles look alike and have an equal probability of being chosen. Three bottles are randomly selected.
(a) What is the probability that all three are Merlot?

(b) What is the probability that exactly two are Merlot?
(c) What is the probability that none is a Merlot?

36. Simulation Use a graphing calculator or statistical software to simulate the playing of the game of roulette, using an integer distribution with numbers 1 through 38. Repeat the simulation 100 times. Let the number 37 represent 0 and the number 38 represent 00. Use the results of the simulation to answer the following questions.
(a) What is the probability that the ball lands in the slot marked 7?
(b) What is the probability that the ball lands either in the slot marked 0 or in the one marked 00?

37. Explain what is meant by a subjective probability. List some examples of subjective probabilities.

38. Playing Five-Card Stud In the game of five-card stud, one card is dealt face down to each player and the remaining four cards are dealt face up. After two cards are dealt (one down and one up), the players bet. Players continue to bet after each additional card is dealt. Suppose three cards have been dealt to each of the five players at the table. You currently have three clubs in your hand, so you will attempt to get a flush (all cards in the same suit). Of the cards dealt, there are two clubs showing in other player's hands.
(a) How many clubs are in a standard 52-card deck?
(b) How many cards remain in the deck or are not known by you? Of this amount, how many are clubs?
(c) What is the probability you get dealt a club on the next card?
(d) What is the probability you get dealt two clubs in a row?
(e) Should you stay in the game?

39. Mark McGwire During the 1998 major league baseball season, Mark McGwire of the St. Louis Cardinals hit 70 home runs. Of the 70 home runs, 34 went to left field, 20 went to left center field, 13 went to center field, 3 went to right center field, and 0 went to right field Source:Miklasz, B., et al. *Celebrating 70: Mark McGwire's Historic Season*, Sporting News Publishing Co., 1998, pp. 179.
(a) What is the probability that a randomly selected home run was hit to left field? Interpret this probability.
(b) What is the probability that a randomly selected home run was hit to right field?
(c) Is it impossible for Mark McGwire to hit a homer to right field?

The Case of the Body in the Bag

The late spring morning broke along the banks of a river. Robert Donkin, an elderly retiree with fishing pole in hand, slipped through the underbrush that lined the river's banks. As he neared the shore, he saw a rather large canvas bag floating in the water, held by foliage that leaned over the river. The bag appeared to be stuffed, well-worn, and heavily stained. Upon closer inspection, Mr. Donkin observed what he believed to be hair floating through the bag's opening. Marking the spot of his discovery, the fisherman fetched the authorities.

Preliminary investigation at the scene revealed a body in the bag. Unfortunately, it was impossible to identify the corpse's sex or race immediately. Estimating age was also out of the question. Forensics was assigned the task of identifying the victim and estimating the cause and time of death. While waiting for the forensics analysis, you, as the detective in charge, have gathered the information shown on page 313 concerning victim–offender relationships from recent reports from the FBI.

Using the information contained in the tables, you are to develop a preliminary profile of the victim and offender by answering the following questions:

1. How likely is it that the offender is at least 18?
2. How likely is it that the offender is white?
3. How likely is it that the offender is male?
4. How likely is it that the offender is a white male?
5. How likely is it that the offender is either white or male?
6. How likely is it that the victim and the offender are from the same age category?
7. How likely is it that the victim and the offender are from different age categories?
8. How likely is it that the victim and the offender are of the same race?
9. How likely is it that the victim and the offender are of different races?
10. How likely is it that the victim and the offender are of the same sex?
11. How likely is it that the victim and the offender are of different sexes?
12. Without knowing the contents of the forensic team's report, what is your best prediction of the age, race, and sex of the victim? Explain your reasoning.
13. What is your best prediction as to the age, race, and sex of the offender? Explain your reasoning.

Soon after you finished this analysis, the preliminary forensics report was delivered to your desk. Although no identification had been made, the autopsy suggested that the cause of death was blunt-force trauma and that the body had been in the water at least two weeks. By using a variety of techniques, it was also determined that the victim was a white female with blonde hair. She was estimated as being in her mid-thirties, showed no signs of having had children, and was wearing no jewelry.

Based on this new information, you develop a new offender profile by answering the following questions:

14. How likely is it that the offender is at least 18?
15. How likely is it that the offender is white?
16. How likely is it that the offender is male?

Victim–Offender Relationship by Age

	Age of Offender		
Age of Victim	Less Than 18	At Least 18	Unknown
Less than 18	111	589	28
At least 18	231	5424	544
Unknown	4	76	17

Victim–Offender Relationship by Sex

	Sex of Offender		
Sex of Victim	Male	Female	Unknown
Male	4417	499	71
Female	1754	185	23
Unknown	49	7	19

Victim–Offender Relationship by Race

	Race of Offender			
Race of Victim	White	Black	Other	Unknown
White	307	501	44	41
Black	226	286	8	49
Other	47	26	122	4
Unknown	33	21	2	19

Sex of Victim by Race of Offender

	Race of Offender			
Sex of Victim	White	Black	Other	Unknown
Male	2163	2642	111	71
Female	1127	749	63	23
Unknown	33	21	2	19

Race of Victim by Sex of Offender

	Sex of Offender		
Race of Victim	Male	Female	Unknown
White	3199	363	41
Black	2793	305	49
Other	179	16	4
Unknown	49	7	19

Source: Uniform Crime Reports: Crime in the United States–2003.

17. How likely is it that the victim and the offender are from the same age category?
18. How likely is it that the victim and the offender are from different age categories?
19. How likely is it that the victim and the offender are of the same race?
20. How likely is it that the victim and the offender are of different races?
21. How likely is it that the victim and the offender are of the same sex?
22. How likely is it that the victim and the offender are of different sexes?
23. What is your best prediction of the age, race, and sex of the offender? Explain your reasoning.
24. Did your answers to the offender questions change once you knew the age, race, and sex of the victim? Explain.

Suppose that 45% of murder victims were known to be related to or acquainted with the offender, that 15% were murdered by an unrelated stranger, and that for 40% of victims relationship to their killer is unknown. Based on all the information available, complete your offender profile for this case.

Discrete Probability Distributions

Outline

DECISIONS

A woman who was shopping in Los Angeles had her purse stolen by a young, blonde female who was wearing a ponytail. Because there were no eyewitnesses and no real evidence, the prosecution used probability to make its case against the defendant. Your job is to play the role of both the prosecution and defense attorney to make probabilistic arguments both for and against the defendant. See the Decisions project on page 339.

●●● Putting It All Together

In Chapter 5, we discussed the idea of probability. The probability of an event is the long-term proportion with which the event is observed. That is, if we conduct an experiment 1000 times and observe the outcome 300 times, the probability of the outcome is 0.3. The more times we conduct the experiment, the more accurate the empirical probability. This is the Law of Large Numbers. We learned that we can use counting techniques to obtain theoretical probabilities provided that the outcomes in the experiment are equally likely. This is called classical probability.

We also learned that a probability model lists the possible outcomes to a probability experiment and each outcome's probability. A probability model must satisfy the rules of probability. In particular, all probabilities must be between 0 and 1, inclusive, and the sum of the probabilities must equal 1.

In this chapter, we introduce probability models for *random variables*. A random variable is a numerical measure of the outcome to a probability experiment. So, rather than listing specific outcomes to a probability experiment such as heads or tails, we might list the number of heads obtained in, say, three flips of a coin. In Section 6.1, we discuss random variables and describe the distribution of discrete random variables (shape, center, and spread). Then we discuss two specific discrete probability distributions: *the Binomial Probability Distribution* (Section 6.2) and *the Poisson Probability Distribution* (Section 6.3).

6.1 Discrete Random Variables

Preparing for This Section   Before getting started, review the following:

- Discrete versus continuous variables (Section 1.1, pp. 7–9)
- Relative frequency histograms for discrete data (Section 2.2, pp. 78–79)
- Mean (Section 3.1, pp. 121–124)
- Standard deviation (Section 3.2, pp. 143–145)
- Mean from grouped data (Section 3.3, pp. 157–158)
- Standard deviation from grouped data (Section 3.3, pp. 159–161)

Objectives

1 Distinguish between discrete and continuous random variables

2 Identify discrete probability distributions

3 Construct probability histograms

4 Compute and interpret the mean of a discrete random variable

5 Interpret the mean of a discrete random variable as an expected value

6 Compute the variance and standard deviation of a discrete random variable

1 Distinguish between Discrete and Continuous Random Variables

In Chapter 5, we presented the concept of an experiment and the outcomes of an experiment. Suppose we flip a coin two times. The possible outcomes of the experiment are {HH, HT, TH, TT}. Rather than being interested in the outcome, we might be interested in the number of heads. When experiments are conducted in a way such that the outcome is a numerical result, we say the outcome is a *random variable*.

Definition   A **random variable** is a numerical measure of the outcome of a probability experiment, so its value is determined by chance. Random variables are denoted using letters such as X.

So, in our coin flipping example, if the random variable X represents the number of heads in two flips of a coin the possible values of X are 0, 1, or 2.

We will follow the practice of using a capital letter to identify the random variable and a small letter to list the possible values of the random variable, that is, the sample space of the experiment. For example, if an experiment is conducted in which a single die is cast, then X represents the number of pips showing on the die and the possible values of X are $x = 1, 2, 3, 4, 5,$ or 6. As another example, suppose an experiment is conducted in which the time between arrivals of cars at a drive-through is measured. The random variable T might describe the time between arrivals, so the sample space of the experiment is $t > 0$.

There are two types of random variables, *discrete* and *continuous*.

Definitions

In Other Words

Discrete random variables typically result from counting, such as 0, 1, 2, 3, and so on. Continuous random variables are variables that result from measurement.

A **discrete random variable** has either a finite or countable number of values. The values of a discrete random variable can be plotted on a number line with space between each point. See Figure 1(a) on the next page.

A **continuous random variable** has infinitely many values. The values of a continuous random variable can be plotted on a line in an uninterrupted fashion. See Figure 1(b) on the next page.

Figure 1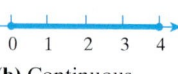

(a) Discrete Random Variable   (b) Continuous Random Variable

---

EXAMPLE 1 — Distinguishing between Discrete and Continuous Random Variables

(a) The number of A's earned in a section of statistics with 15 students enrolled is a discrete random variable because the value of the random variable results from counting. If we let the random variable X represent the number of A's, then the possible values of X are $x = 0, 1, 2, \ldots, 15$.

(b) The number of cars that travel through a McDonald's drive-through in the next hour is a discrete random variable because the value of the random variable results from counting. If we let the random variable X represent the number of cars through the drive-through in the next hour, the possible values of X are $x = 0, 1, 2, \ldots$.

(c) The speed of the next car that passes a state trooper is a continuous random variable because speed is measured. If we let the random variable S represent the speed of the next car, the possible values of S are all positive real numbers; that is, $s > 0$.

! **CAUTION**
Even though a radar gun may report the speed of a car as 37 miles per hour, it is actually any number greater than or equal to 36.5 mph and less than 37.5 mph. That is, $36.5 \leq s < 37.5$.

Now Work Problem 7.

In this chapter, we will concentrate on probabilities of discrete random variables. Probabilities for certain continuous random variables will be discussed in the next chapter.

2 Identify Discrete Probability Distributions

Because the value of a random variable is determined by chance, there are probabilities that correspond to the possible values of the random variable.

Definition
The **probability distribution** of a discrete random variable X provides the possible values of the random variable and their corresponding probabilities. A probability distribution can be in the form of a table, graph, or mathematical formula.

EXAMPLE 2 — A Discrete Probability Distribution

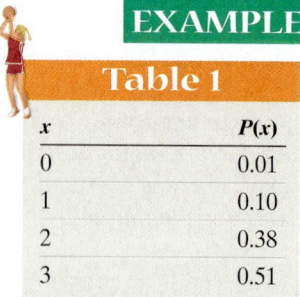

Table 1

x	$P(x)$
0	0.01
1	0.10
2	0.38
3	0.51

Suppose we ask a basketball player to shoot three free throws. Let the random variable X represent the number of shots made, so that $x = 0, 1, 2,$ or 3. Table 1 shows a probability distribution for the random variable X assuming the player historically makes 80% of her free throw attempts.

From the probability distribution in Table 1, we can see that the probability the player makes all three free-throw attempts is 0.51.

We will denote probabilities using the notation $P(x)$, where x is a specific value of the random variable. We read $P(x)$ as "the probability that the random variable X equals x." For example, $P(3) = 0.51$ is read "the probability that the random variable X equals 3 is 0.51."

Recall from Section 5.1 that probabilities must obey certain rules. We repeat the rules for a discrete probability distribution using the notation just introduced.

Rules for a Discrete Probability Distribution

Let $P(x)$ denote the probability that the random variable X equals x; then

1. $\sum P(x) = 1$
2. $0 \leq P(x) \leq 1$

Table 1 from Example 2 is a probability distribution because the sum of the probabilities equals 1 and each probability is between 0 and 1, inclusive. You are encouraged to verify this.

EXAMPLE 3 **Identifying Discrete Probability Distributions**

Problem: Which of the following is a discrete probability distribution?

(a)

x	$P(x)$
1	0.20
2	0.35
3	0.12
4	0.40
5	−0.07

(b)

x	$P(x)$
1	0.20
2	0.25
3	0.10
4	0.14
5	0.49

(c)

x	$P(x)$
1	0.20
2	0.25
3	0.10
4	0.14
5	0.31

Approach: In a discrete probability distribution, the sum of the probabilities must equal 1, and all probabilities must be greater than or equal to 0 and less than or equal to 1. We will verify these requirements for (a), (b), and (c).

Solution

(a) This is not a discrete probability distribution because $P(5) = -0.07$, which is less than 0.

(b) This is not a discrete probability distribution because

$$\sum P(x) = 0.2 + 0.25 + 0.10 + 0.14 + 0.49 = 1.18 \neq 1$$

(c) This is a discrete probability distribution because the sum of the probabilities equals 1, and each probability is greater than or equal to 0 and less than or equal to 1.

Now Work Problem 11.

Table 1 is a discrete probability distribution in table form. Probability distributions can also be represented through graphs or mathematical formulas. We discuss discrete probability distributions using graphs now and discuss probability distributions as mathematical formulas in the next section.

③ Construct Probability Histograms

A graphical depiction of a discrete probability distribution is typically done with a *probability histogram*.

Definition A **probability histogram** is a histogram in which the horizontal axis corresponds to the value of the random variable and the vertical axis represents the probability of each value of the random variable.

EXAMPLE 4 **Constructing a Probability Histogram**

Problem: Construct a probability histogram of the discrete probability distribution given in Table 1 from Example 2.

Approach: Probability histograms are constructed like relative frequency histograms, except that the vertical axis represents the probability of the random variable, rather than its relative frequency. Each rectangle is centered at the value of the discrete random variable.

Solution: Figure 2 presents the probability histogram.

In Other Words

A probability histogram is constructed the same way as a relative frequency histogram for discrete data. The only difference is that the vertical axis is a probability, rather than a relative frequency.

Figure 2

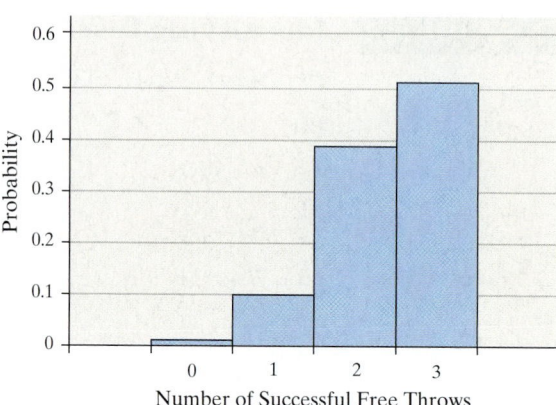

Notice that the area of each rectangle in the probability histogram equals the probability that the random variable assumes the particular value. For example, the area of the rectangle corresponding to the random variable $X = 2$ is $1 \cdot (0.38) = 0.38$, where 1 represents the width of the rectangle and 0.38 represents its height.

Probability histograms help us to determine the shape of the distribution. Recall that we describe distributions as skewed left, skewed right, or symmetric. For example, the probability histogram presented in Figure 2 is skewed left.

> **Now Work Problem 19(a) and (b).**

 Compute and Interpret the Mean of a Discrete Random Variable

Remember, when we describe the distribution of a variable, we describe its center, spread, and shape. We now introduce methods for identifying the center and spread of a discrete random variable. We will use the mean to describe the center of a random variable. The variance and standard deviation are used to describe the spread of a random variable.

To help see where the formula for computing the mean of a discrete random variable comes from, consider the following. One semester I had a small statistics class of 10 students. I asked them to disclose the number of people living in their household and obtained the following:

$$2, 4, 6, 6, 4, 4, 2, 3, 5, 5$$

What is the mean number of people in the 10 households? Of course, we could find the mean by adding the observations and dividing by 10. But we will take a different approach. Let the random variable X represent the number of people in the household and obtain the probability distribution in Table 2.

Table 2	
x	*P(x)*
2	$\frac{2}{10} = 0.2$
3	$\frac{1}{10} = 0.1$
4	$\frac{3}{10} = 0.3$
5	$\frac{2}{10} = 0.2$
6	$\frac{2}{10} = 0.2$

Now we compute the mean as follows:

$$\mu = \frac{\sum x_i}{N} = \frac{2 + 4 + 6 + 6 + 4 + 4 + 2 + 3 + 5 + 5}{10}$$

$$= \frac{\overbrace{2 + 2}^{2} + \overbrace{3}^{1} + \overbrace{4 + 4 + 4}^{3} + \overbrace{5 + 5}^{2} + \overbrace{6 + 6}^{2}}{10}$$

$$= \frac{2 \cdot 2 + 3 \cdot 1 + 4 \cdot 3 + 5 \cdot 2 + 6 \cdot 2}{10}$$

$$= 2 \cdot \frac{2}{10} + 3 \cdot \frac{1}{10} + 4 \cdot \frac{3}{10} + 5 \cdot \frac{2}{10} + 6 \cdot \frac{2}{10}$$

$$= 2 \cdot P(2) + 3 \cdot P(3) + 4 \cdot P(4) + 5 \cdot P(5) + 6 \cdot P(6)$$

$$= 2(0.2) + 3(0.1) + 4(0.3) + 5(0.2) + 6(0.2)$$

$$= 4.1$$

Based on the preceding computations, we conclude that the mean of a discrete random variable is found by multiplying each possible value of the random variable by its corresponding probability and adding these products.

In Other Words
To find the mean of a discrete random variable, multiply the value of each random variable by its probability. Then add these products.

The Mean of a Discrete Random Variable

The mean of a discrete random variable is given by the formula

$$\mu_X = \sum [x \cdot P(x)] \qquad \text{(1)}$$

where x is the value of the random variable and $P(x)$ is the probability of observing the random variable x.

EXAMPLE 5

Computing the Mean of a Discrete Random Variable

Problem: Compute the mean of the discrete random variable given in Table 1 from Example 2.

Approach: The mean of a discrete random variable is found by multiplying each value of the random variable by its probability and adding these products.

Solution: Refer to Table 3. The first two columns represent the discrete probability distribution. The third column represents $x \cdot P(x)$.

We substitute into Formula (1) to find the mean number of free throws made.

$$\mu_X = \sum [x \cdot P(x)] = 0(0.01) + 1(0.10) + 2(0.38) + 3(0.51) = 2.39 \approx 2.4$$

Table 3		
x	$P(x)$	$x \cdot P(x)$
0	0.01	$0 \cdot 0.01 = 0$
1	0.10	$1 \cdot 0.1 = 0.1$
2	0.38	0.76
3	0.51	1.53

We will follow the practice of rounding the mean, variance, and standard deviation to one more decimal place than the values of the random variable.

How to Interpret the Mean of a Discrete Random Variable

The mean of a discrete random variable can be thought of as the mean outcome of the probability experiment if we repeated the experiment many times. Consider the result of Example 5. If we repeated the experiment of shooting three free throws many times and recorded the number of free throws made, we would expect the average number of free throws made to be around 2.4.

In Other Words

We can think of the mean of a discrete random variable as the average outcome if the experiment is repeated many, many times.

Interpretation of the Mean of a Discrete Random Variable

Suppose an experiment is repeated n independent times and the value of the random variable X is recorded. As the number of repetitions of the experiment, n, increases, the mean value of the n trials will approach μ_X, the mean of the random variable X. In other words, let x_1 be the value of the random variable X after the first experiment, x_2 be the value of the random variable X after the second experiment, and so on. Then

$$\bar{x} = \frac{x_1 + x_2 + \cdots + x_n}{n}$$

The difference between $\bar{x}$ and μ_X gets closer to 0 as n increases.

EXAMPLE 6

Illustrating the Interpretation of the Mean of a Discrete Random Variable

Problem: The basketball player from Example 2 is asked to shoot three free throws 100 times. Compute the mean number of free throws made.

Approach: The player shoots three free throws and the number made is recorded. We repeat this experiment 99 more times and then compute the mean number of free throws made.

Solution: The results are presented in Table 4.

Table 4									
3	2	3	3	3	3	1	2	3	2
2	3	3	1	2	2	2	2	2	3
3	3	2	2	3	2	3	2	2	2
3	3	2	3	2	3	3	2	3	1
3	2	2	2	2	0	2	3	1	2
3	3	2	3	2	3	2	1	3	2
2	3	3	3	1	3	3	1	3	3
3	2	2	1	3	2	2	2	3	2
3	2	2	2	3	3	2	2	3	3
2	3	2	1	2	3	3	2	3	3

The first time the experiment was conducted, the player made all three free throws. The second time the experiment was conducted, the player made two out of three free throws. The hundredth time the experiment was conducted, the player made three out of three free throws. The mean number of free throws made was

$$\bar{x} = \frac{3 + 2 + 3 + \cdots + 3}{100} = 2.35$$

This is close to the mean of 2.4 (from Example 5). As the number of repetitions of the experiment increases, we expect $\bar{x}$ to get even closer to 2.4. ▪

Figure 3(a) and Figure 3(b) further demonstrate the interpretation of the mean of a discrete random variable. Figure 3(a) shows the mean number of free throws made versus the number of repetitions of the experiment for the data in Table 4. Figure 3(b) shows the mean number of free throws made versus the number of repetitions of the experiment when the same experiment of shooting three free throws 100 times is conducted a second time. In both plots the player starts off "hot," since the mean number of free throws made is above the theoretical level of 2.4. However, both graphs approach the theoretical mean of 2.4 as the number of repetitions of the experiment increases.

Figure 3

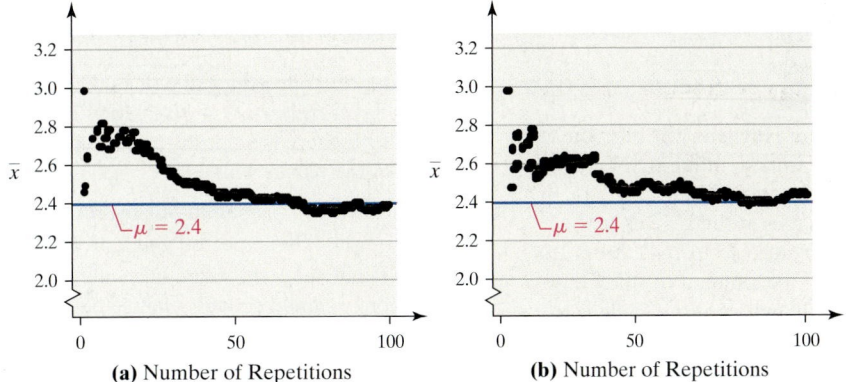

(a) Number of Repetitions   (b) Number of Repetitions

Now Work Problem 19(c).

5 ## Interpret the Mean of a Discrete Random Variable as an Expected Value

Because the mean of a random variable represents what we would expect to happen in the long run, the mean of a random variable is also called the **expected value**. The interpretation of expected value is the same as the interpretation of the mean of a discrete random variable.

EXAMPLE 7   ### Finding the Expected Value

In Other Words

The expected value of a discrete random variable is the mean of the discrete random variable.

Problem: A term life insurance policy will pay a beneficiary a certain sum of money upon the death of the policyholder. These policies have premiums that must be paid annually. Suppose a life insurance company sells a $250,000 one-year term life insurance policy to an 18-year-old male for $350. According to the *National Vital Statistics Report*, Vol. 47, No. 28, the probability the male will survive the year is 0.998789. Compute the expected value of this policy to the insurance company.

Approach: There are two possible outcomes to the experiment: survival or death. Let the random variable X represent the *payout* (money lost or gained), depending on survival or death of the insured. We assign probabilities to each of these random variables and substitute these values into Formula (1).

Solution

Step 1: We have $P(\text{survives}) = 0.998789$, so $P(\text{dies}) = 0.001211$. From the point of view of the insurance company, if the client survives the year, the insurance company makes $350. Therefore, we let $x = \$350$ if the client survives the year. If the client dies during the year, the insurance company must pay $250,000 to the client's beneficiary. However, the company still keeps the $350, so we let $x = \$350 - \$250{,}000 = -\$249{,}650$. The value is negative because it is money paid out by the insurance company. We have the probability distribution listed in Table 5.

Table 5

x	$P(x)$
$350 (survives)	0.998789
−$249,650 (dies)	0.001211

Step 2: Substituting into Formula (1), we obtain the expected value (from the point of view of the insurance company) of the policy.

$$E(X) = \mu_X = \sum xP(x) = \$350(0.998789) + (-\$249{,}650)(0.001211) = \$47.25$$

Interpretation: The company expects to make $47.25 for each 18-year-old male client it insures. The $47.25 profit of the insurance company is a long-term result. It does not make $47.25 on each person it insures, but rather the average profit per person insured is $47.25. Because this is a long-term result, the insurance "idea" will not work with only a few insured.

Now Work Problem 29.

Historical Note

Christiaan Huygens was born on April 14, 1629, into an influential Dutch family. He studied Law and Mathematics at the University of Leiden from 1645 to 1647. From 1647 to 1649, he continued to study Law and Mathematics at the College of Orange at Breda. Among his many great accomplishments, Huygens discovered the first moon of Saturn in 1655 and the shape of the rings of Saturn in 1656. While in Paris sharing his discoveries, he learned about probability through the correspondence of Fermat and Pascal. In 1657, Huygens published the first book on probability theory. In that text, Huygens introduced the idea of expected value.

In Other Words

The variance of a discrete random variable is a weighted average of the squared deviations where the weights are the probabilities.

In-Class Activity: Expected Value

Consider the following game of chance. A player pays $1 and rolls a pair of fair dice. If the player rolls a 2, 3, 4, 10, 11, or 12, the player loses the $1 bet. If the player rolls 5, 6, 8, or 9, there is a "push" and the player gets his or her dollar back. If the player rolls a 7, the player wins $1.

(a) Construct a probability distribution that describes the game.
(b) Compute the expected value of the game from the player's point of view.
(c) Actually play the game in a small group. Keep track of the results on paper (no money should actually change hands). Compute the mean earnings to the player from the game. Are the results close to what you expected? If not, why?

---

6 Compute the Variance and Standard Deviation of a Discrete Random Variable

We now introduce a method for computing the variance and standard deviation of a discrete random variable.

Variance and Standard Deviation of a Discrete Random Variable

The variance of a discrete random variable is given by

$$\sigma_X^2 = \sum [(x - \mu_X)^2 \cdot P(x)] \tag{2a}$$
$$= \sum [x^2 \cdot P(x)] - \mu_X^2 \tag{2b}$$

where x is the value of the random variable, μ_X is the mean of the random variable, and $P(x)$ is the probability of observing the random variable x.

To find the standard deviation of the discrete random variable, take the square root of the variance. That is, $\sigma_X = \sqrt{\sigma_X^2}$.

EXAMPLE 8   Computing the Variance and Standard Deviation of a Discrete Random Variable

Problem: Find the variance and standard deviation of the discrete random variable given in Table 1 from Example 2.

Approach: We will use Formula (2a) with the unrounded mean $\mu_X = 2.39$.

Solution: Refer to Table 6. The first two columns represent the discrete probability distribution. The third column represents $(x - \mu_X)^2 \cdot P(x)$. We sum the entries in the third column.

Table 6

x	$P(x)$	$(x - \mu_X)^2 \cdot P(x)$
0	0.01	$(0 - 2.39)^2 \cdot 0.01 = 0.057121$
1	0.10	$(1 - 2.39)^2 \cdot 0.10 = 0.19321$
2	0.38	$(2 - 2.39)^2 \cdot 0.38 = 0.057798$
3	0.51	$(3 - 2.39)^2 \cdot 0.51 = 0.189771$
		$\sum (x - \mu_X)^2 \cdot P(x) = 0.4979$

The variance of the discrete random variable X is

$$\sigma_X^2 = \sum (x - \mu_X)^2 \cdot P(x) = 0.4979 \approx 0.5$$

Approach: We will use Formula (2b) with the unrounded mean $\mu_X = 2.39$.

Solution: Refer to Table 7. The first two columns represent the discrete probability distribution. The third column represents $x^2 \cdot P(x)$.

Table 7

x	$P(x)$	$x^2 \cdot P(x)$
0	0.01	$0^2 \cdot 0.01 = 0$
1	0.10	$1^2 \cdot 0.10 = 0.10$
2	0.38	$2^2 \cdot 0.38 = 1.52$
3	0.51	$3^2 \cdot 0.51 = 4.59$
		$\sum x^2 \cdot P(x) = 6.21$

The variance of the discrete random variable X is

$$\sigma_X^2 = \sum [x^2 \cdot P(x)] - \mu_X^2 = 6.21 - 2.39^2 = 0.4979 \approx 0.5$$

The standard deviation of the discrete random variable is found by taking the square root of the variance.

$$\sigma_X = \sqrt{\sigma_X^2} = \sqrt{0.4979} \approx 0.7$$

Now Work Problems 19(d) and (e).

EXAMPLE 9   **Obtaining the Mean and Standard Deviation of a Discrete Random Variable Using Technology**

Figure 4

μ_X
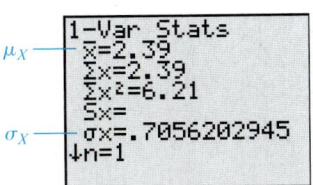
σ_X

Problem: Use a statistical spreadsheet or calculator to determine the mean and the standard deviation of the random variable whose distribution is given in Table 1.

Approach: We will use a TI-84 Plus graphing calculator to obtain the mean and standard deviation. The steps for determining the mean and standard deviation using a TI-83 or TI-84 Plus graphing calculator are given in the Technology Step by Step on page 327.

Result: Figure 4 shows the results from a TI-84 Plus graphing calculator. **Note:** The TI does not find s_X when the sum of L_2 is one.

6.1 ASSESS YOUR UNDERSTANDING

Concepts and Vocabulary

1. What is a random variable?
2. What is the difference between a discrete random variable and a continuous random variable? Provide your own examples of each.
3. What are the two requirements for a discrete probability distribution?
4. In your own words, provide an interpretation of the mean of a discrete random variable.
5. Suppose a baseball player historically hits 0.300. (This means that the player averages three hits in every 10 at-bats.) Suppose the player has zero hits in four at-bats in a game and enters the batter's box for the fifth time, whereupon the announcer declares that the player is "due for a hit." What is the flaw in the announcer's reasoning? If the player had four hits in the last four at-bats, is the player "due to make an out"?
6. A game is called a zero-sum game if the expected value of the game is zero. Explain what a game whose expected value is zero means.

Skill Building

In Problems 7–10, determine whether the random variable is discrete or continuous. In each case, state the possible values of the random variable.

7. (a) The number of light bulbs that burn out in the next
   NW week in a room of with 20 bulbs.
   (b) The time it takes to fly from New York City to Los Angeles.
   (c) The number of hits to a Web site in a day.
   (d) The amount of snow in Toronto during the winter.

8. (a) The time it takes for a light bulb to burn out.
   (b) The weight of a T-bone steak.
   (c) The number of free-throw attempts before the first shot is made.
   (d) In a random sample of 20 people, the number who are blood type A.

9. (a) The amount of rain in Seattle during April.
   (b) The number of fish caught during a fishing tournament.
   (c) The number of customers arriving at a bank between noon and 1:00 P.M.
   (d) The time required to download a file from the Internet.

10. (a) The number of defects in a roll of carpet.
    (b) The distance a baseball travels in the air after being hit.
    (c) The number of points scored during a basketball game.
    (d) The square footage of a house.

In Problems 11–16, determine whether the distribution is a discrete probability distribution. If not, state why.

NW **11.**

x	P(x)
0	0.2
1	0.2
2	0.2
3	0.2
4	0.2

12.

x	P(x)
0	0.1
1	0.5
2	0.05
3	0.25
4	0.1

13.

x	P(x)
10	0.1
20	0.23
30	0.22
40	0.6
50	−0.15

14.

x	P(x)
1	0
2	0
3	0
4	0
5	1

15.

x	P(x)
100	0.1
200	0.25
300	0.2
400	0.3
500	0.1

16.

x	P(x)
100	0.25
200	0.25
300	0.25
400	0.25
500	0.25

In Problems 17 and 18, determine the required value of the missing probability to make the distribution a discrete probability distribution.

17.

x	P(x)
3	0.4
4	?
5	0.1
6	0.2

18.

x	P(x)
0	0.30
1	0.15
2	?
3	0.20
4	0.15
5	0.05

Applying the Concepts

19. Parental Involvement In the following probability distribution, the random variable X represents the number of activities at least one parent of a K–5th grade student is involved in.

(a) Verify that this is a discrete probability distribution.
(b) Draw a probability histogram.
(c) Compute and interpret the mean of the random variable X.
(d) Compute the variance of the random variable X.
(e) Compute the standard deviation of the random variable X.
(f) What is the probability that a randomly selected student has at least one parent involved in three activities?
(g) What is the probability that a randomly selected student has at least one parent involved in three or four activities?

x	P(x)
0	0.035
1	0.074
2	0.197
3	0.320
4	0.374

Source: U.S. National Center for Education Statistics

20. Parental Involvement In the following probability distribution, the random variable X represents the number of activities at least one parent of a 6th–8th grade student is involved in.

(a) Verify that this is a discrete probability distribution.
(b) Draw a probability histogram.
(c) Compute and interpret the mean of the random variable X.
(d) Compute the variance of the random variable X.
(e) Compute the standard deviation of the random variable X.
(f) What is the probability that a randomly selected student has at least one parent involved in three activities?
(g) What is the probability that a randomly selected student has at least one parent involved in three or four activities?

x	P(x)
0	0.073
1	0.117
2	0.258
3	0.322
4	0.230

Source: U.S. National Center for Education Statistics

21. Ichiro's Hit Parade In the 2004 baseball season, Ichiro Suzuki of the Seattle Mariners set the record for most hits in a season with a total of 262 hits. In the following probability distribution, the random variable X represents the number of hits Ichiro obtained in a game.

x	P(x)
0	0.1677
1	0.3354
2	0.2857
3	0.1491
4	0.0373
5	0.0248

Source: Chicago Tribune

(a) Verify that this is a discrete probability distribution.
(b) Draw a probability histogram.
(c) Compute and interpret the mean of the random variable X.
(d) Compute the standard deviation of the random variable X.
(e) What is the probability that in a randomly selected game Ichiro got 2 hits?
(f) What is the probability that in a randomly selected game Ichiro got more than 1 hit?

22. Waiting in Line A Wendy's manager performed a study to determine a probability distribution for the number of people waiting in line X during lunch. The results were as follows:

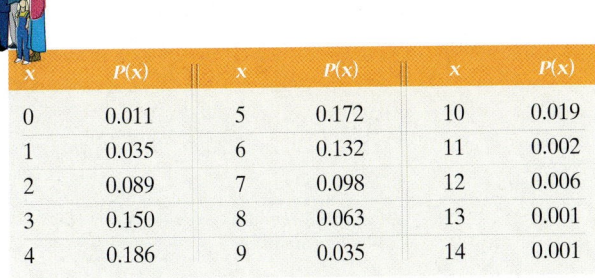

x	P(x)	x	P(x)	x	P(x)
0	0.011	5	0.172	10	0.019
1	0.035	6	0.132	11	0.002
2	0.089	7	0.098	12	0.006
3	0.150	8	0.063	13	0.001
4	0.186	9	0.035	14	0.001

(a) Verify that this is a discrete probability distribution.
(b) Draw a probability histogram.
(c) Compute and interpret the mean of the random variable X.
(d) Compute the variance of the random variable X.
(e) Compute the standard deviation of the random variable X.
(f) What is the probability that there are eight people waiting in line for lunch?
(g) What is the probability that there are 10 or more people waiting in line for lunch? Would this be unusual?

In Problems 23–26, (a) construct a discrete probability distribution for the random variable X [Hint: $P(x_i) = \dfrac{f_i}{N}$], (b) draw the probability histogram, (c) compute and interpret the mean of the random variable X, and (d) compute the standard deviation of the random variable X.

23. The World Series The following data represent the number of games played in each World Series from 1923 to 2004.

x (games played)	Frequency
4	15
5	15
6	18
7	33

Source: Information Please Almanac

24. Number of 5- to 9-Year-Old Girls The following data represent (in thousands) the number of 5- to 9-year-old females in the United States in 2000.

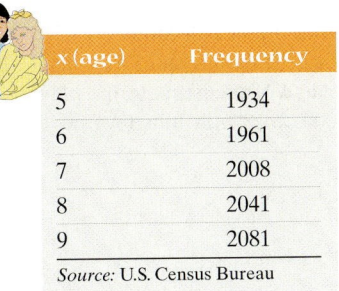

x (age)	Frequency
5	1934
6	1961
7	2008
8	2041
9	2081

Source: U.S. Census Bureau

25. Grade School Enrollment The following data represent (in thousands) the enrollment levels in grades 1 to 8 in the United States in 2000.

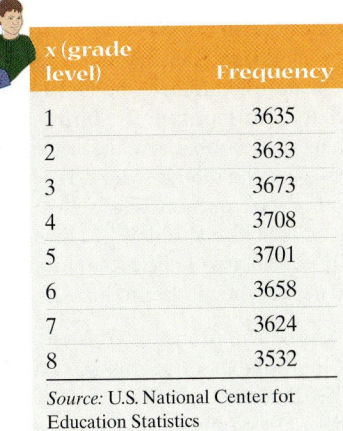

x (grade level)	Frequency
1	3635
2	3633
3	3673
4	3708
5	3701
6	3658
7	3624
8	3532

Source: U.S. National Center for Education Statistics

26. **High School Enrollment** The following data represent (in thousands) the enrollment levels in grades 9 to 12 in the United States in 2000.

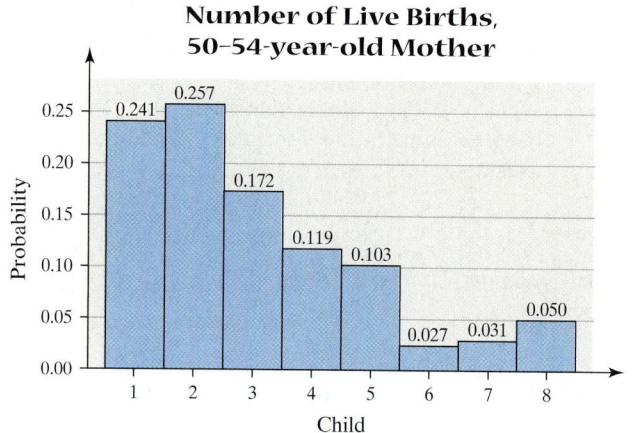

x (grade level)	Frequency
9	3958
10	3487
11	3080
12	2799

Source: U.S. National Center for Education Statistics

27. **Number of Births** The probability histogram that follows represents the number of live births by a mother 50 to 54 years old who had a live birth in 2002. The data are from the *National Vital Statistics Report*, Vol. 52, No. 10, December 17, 2003.

**Number of Live Births,
50–54-year-old Mother**

(Probability histogram: Child 1 = 0.241, 2 = 0.257, 3 = 0.172, 4 = 0.119, 5 = 0.103, 6 = 0.027, 7 = 0.031, 8 = 0.050)

(a) What is the probability that a randomly selected 50- to 54-year-old mother who had a live birth in 2002 has had her fourth live birth?

(b) What is the probability that a randomly selected 50- to 54-year-old mother who had a live birth in 2002 has had her fourth or fifth live birth?

(c) What is the probability that a randomly selected 50- to 54-year-old mother who had a live birth in 2002 has had her sixth or more live birth?

(d) If a 50- to 54-year-old mother who had a live birth in 2002 is randomly selected, how many live births would you expect the mother to have had?

28. **Rental Units** The probability histogram that follows represents the number of rooms in rented housing units in 2003. The data are from the U.S. Department of Housing and Urban Development.

(a) What is the probability that a randomly selected rental unit has five rooms?

(b) What is the probability that a randomly selected rental unit has five or six rooms?

Number of Rooms in Rental Unit

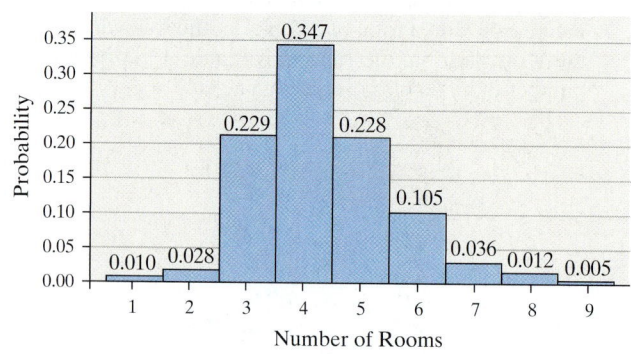

(Probability histogram: Number of Rooms 1 = 0.010, 2 = 0.028, 3 = 0.229, 4 = 0.347, 5 = 0.228, 6 = 0.105, 7 = 0.036, 8 = 0.012, 9 = 0.005)

(c) What is the probability that a randomly selected rental unit has seven or more rooms?

(d) If a rental unit is randomly selected, how many rooms would you expect the unit to have?

29. **Life Insurance** Suppose a life insurance company sells a $250,000 one-year term life insurance policy to a 20-year-old female for $200. According to the *National Vital Statistics Report*, Vol. 53, No 6, the probability that the female survives the year is 0.999546. Compute and interpret the expected value of this policy to the insurance company.

30. **Life Insurance** Suppose a life insurance company sells a $250,000 one-year term life insurance policy to a 20-year-old male for $350. According to the *National Vital Statistics Report*, Vol. 53, No. 6, the probability that the male survives the year is 0.998611. Compute and interpret the expected value of this policy to the insurance company.

31. **Investment** An investment counselor calls with a hot stock tip. He believes that if the economy remains strong the investment will result in a profit of $50,000. If the economy grows at a moderate pace, the investment will result in a profit of $10,000. However, if the economy goes into recession, the investment will result in a loss of $50,000. You contact an economist who believes there is a 20% probability the economy will remain strong, a 70% probability the economy will grow at a moderate pace, and a 10% probability the economy will slip into recession. What is the expected profit from this investment?

32. **Real Estate Investment** Shawn and Maddie purchase a foreclosed property for $50,000 and spend an additional $27,000 fixing up the property. They feel that they can resell the property for $120,000 with probability 0.15, $100,000 with probability 0.45, $80,000 with probability 0.25, and $60,000 with probability 0.15. Compute and interpret the expected profit for reselling the property.

33. **Roulette** In the game of roulette, a player can place a $5 bet on the number 17 and have a $\frac{1}{38}$ probability of winning. If the metal ball lands on 17, the player wins $175. Otherwise, the casino takes the player's $5. What is the expected value of the game to the player? If you played the game 1000 times, how much would you expect to lose?

34. Connecticut Lottery In the Cash Five Lottery in Connecticut, a player pays $1 for a single ticket with five numbers. Five Ping-Pong balls numbered 1 through 35 are randomly chosen from a bin without replacement. If all five numbers on a player's ticket match the five chosen, the player wins $100,000. The probability of this occurring is $\frac{1}{324,632}$. If four numbers match, the player wins $300. This occurs with probability $\frac{1}{2164}$. If three numbers match, the player wins $10. This occurs with probability $\frac{1}{75}$. Compute and interpret the expected value of the game from the player's point of view.

35. Powerball Powerball is a multi state lottery. The following probability distribution represents the cash prizes of Powerball with their corresponding probabilities.

x (cash prize, $)	P(x)
Grand prize	0.00000000684
200,000	0.00000028
10,000	0.000001711
100	0.000153996
7	0.004778961
4	0.007881463
3	0.01450116
0	0.9726824222

Source: www.powerball.com

(a) If the grand prize is $15,000,000, find and interpret the expected cash prize. If a ticket costs $1, what is your expected profit from one ticket?
(b) To the nearest million, how much should the grand prize be so that you can expect a profit? Assume nobody else wins so that you do not have to share the grand prize.
(c) Does the size of the grand prize affect your chance of winning? Explain.

36. SAT Test Penalty Some standardized tests, such as the SAT test, incorporate a penalty for wrong answers. For example, a multiple-choice question with five possible answers will have 1 point awarded for a correct answer and $\frac{1}{4}$ deducted point for an incorrect answer. Questions left blank are worth 0 points.
(a) Find the expected number of points received for a multiple-choice question with five possible answers when a student just guesses.
(b) Explain why there is a deduction for wrong answers.

37. Simulation Use the probability distribution from Problem 21 and a DISCRETE command for some statistical software to simulate 100 repetitions of the experiment (100 games). The number of hits is recorded. Approximate the mean and standard deviation of the random variable X based on the simulation. Repeat the simulation by performing 500 repetitions of the experiment. Approximate the mean and standard deviation of the random variable. Compare your results to the theoretical mean and standard deviation. What property is being illustrated?

38. Simulation Use the probability distribution from Problem 22 and a DISCRETE command for some statistical software to simulate 100 repetitions of the experiment. Approximate the mean and standard deviation of the random variable X based on the simulation. Repeat the simulation by performing 500 repetitions of the experiment. Approximate the mean and standard deviation of the random variable. Compare your results to the theoretical mean and standard deviation. What property is being illustrated?

Technology Step-by-Step   **Finding the Mean and Standard Deviation of a Discrete Random Variable Using Technology**

TI-83/84 Plus   **Step 1:** Enter the values of the random variable in L1 and their corresponding probabilities in L2.
Step 2: Press STAT, highlight CALC, and select 1: 1-Var Stats.
Step 3: With 1-Var Stats on the HOME screen, type L1 followed by a comma, followed by L2 as follows:

```
1-Var Stats L1, L2
```

Hit ENTER.

6.2 The Binomial Probability Distribution

Preparing for This Section Before getting started, review the following:

- Independence (Section 5.3, pp. 277–279)
- Combinations (Section 5.5, pp. 298–301)
- Multiplication Rule for Independent Events (Section 5.3, pp. 279–281)
- Addition Rule for Disjoint Events (Section 5.2, pp. 266–269)
- Complement Rule (Section 5.2, pp. 272–273)
- Empirical Rule (Section 3.2, pp. 145–146)

Objectives

1. **Determine whether a probability experiment is a binomial experiment**
2. **Compute probabilities of binomial experiments**
3. **Compute the mean and standard deviation of a binomial random variable**
4. **Construct binomial probability histograms**

1 Determine Whether a Probability Experiment Is a Binomial Experiment

In Section 6.1, we stated that probability distributions could be presented using tables, graphs, or mathematical formulas. In this section, we introduce a specific type of discrete probability distribution that can be presented using a formula, *the binomial probability distribution*.

The binomial probability distribution is a discrete probability distribution that describes probabilities for experiments in which there are two mutually exclusive (disjoint) outcomes. These two outcomes are generally referred to as *success* and *failure*. For example, a basketball player can either make a free throw (success) or miss (failure). A new surgical procedure can result in either life (success) or death (failure).

Experiments in which there are only two possible outcomes are referred to as *binomial experiments*, provided that certain criteria are met.

In Other Words

The prefix *bi* means two. This should help remind you that binomial experiments deal with situations in which there are only two outcomes: success and failure.

Criteria for a Binomial Probability Experiment

An experiment is said to be a **binomial experiment** provided

1. The experiment is performed a fixed number of times. Each repetition of the experiment is called a **trial**.
2. The trials are independent. This means the outcome of one trial will not affect the outcome of the other trials.
3. For each trial, there are two mutually exclusive (disjoint) outcomes: success or failure.
4. The probability of success is the same for each trial of the experiment.

Let the random variable X be the number of successes in n trials of a binomial experiment. Then X is called a **binomial random variable**. Before introducing the method for computing binomial probabilities, it is worthwhile to introduce some notation.

Notation Used in the Binomial Probability Distribution

- There are n independent trials of the experiment.
- Let p denote the probability of success so that $1 - p$ is the probability of failure.
- Let X denote the number of successes in n independent trials of the experiment. So $0 \le x \le n$.

EXAMPLE 1 Identifying Binomial Experiments

Problem: Determine which of the following probability experiments qualify as a binomial experiment. For those that are binomial experiments, identify the number of trials, probability of success, probability of failure, and possible values of the random variable X.

(a) An experiment in which a basketball player who historically makes 80% of his free throws is asked to shoot three free throws and the number of made free throws is recorded.

(b) The number of people with blood type O-negative based on a simple random sample of size 10 is recorded. According to the *Information Please Almanac*, 6% of the human population is blood type O-negative.

(c) A probability experiment in which three cards are drawn from a deck without replacement and the number of aces is recorded.

Approach: We determine whether or not the four conditions for a binomial experiment are satisfied.

1. The experiment is performed a fixed number of times.
2. The trials are independent.
3. There are only two possible outcomes of the experiment.
4. The probability of success for each trial is constant.

Solution

(a) This is a binomial experiment because

   1. There are $n = 3$ trials.
   2. The trials are independent.
   3. There are two possible outcomes: make or miss.
   4. The probability of success (make) is 0.8 and the probability of failure (miss) is 0.2. They are the same for each trial.

   The random variable X is the number of free throws made with $x = 0, 1, 2,$ or 3.

(b) This is a binomial experiment because

   1. There are 10 trials (the 10 randomly selected people).
   2. The trials are independent.*
   3. There are two possible outcomes: finding a person with blood type O-negative or not.
   4. The probability of success is 0.06 and the probability of failure is 0.94.

   The random variable X is the number of people with blood type O-negative with $x = 0, 1, 2, 3, \ldots, 10$.

(c) This is not a binomial experiment because the trials are not independent. The probability of an ace on the first trial is $\frac{4}{52}$. Because we are sampling without replacement, if an ace is selected on the first trial, the probability of an ace on the second trial is $\frac{3}{51}$. If an ace is not selected on the first trial, the probability of an ace on the second trial is $\frac{4}{51}$.

Now Work Problem 9.

*In sampling from large populations without replacement, the trials are assumed to be independent, provided that the sample size is small in relation to the size of the population. As a rule of thumb, if the sample size is less than 5% of the population size, the trials are assumed to be independent, although they are technically dependent. See Example 6 in Section 5.4.

> ⚠ **CAUTION**
>
> The probability of success, p, is always associated with the random variable X, the number of successes. So if X represents the number of 18-year-olds involved in an accident, then p represents the probability of an 18-year-old being involved in an accident.

It is worth mentioning that the word *success* does not necessarily imply that something positive has occurred. Success means that an outcome has occurred that corresponds with p, the probability of success. For example, a probability experiment might be to randomly select ten 18-year-old male drivers. We might let X denote the number who have been involved in an accident within the last year. In this case, a success would mean obtaining an 18-year-old male who was involved in an accident. This outcome is certainly not positive, but still represents a success as far as the experiment goes.

② Compute Probabilities of Binomial Experiments

We are now prepared to compute probabilities for a binomial random variable X. We present three methods for obtaining binomial probabilities: (1) the binomial probability distribution formula (2) a table of binomial probabilities, and (3) technology. We illustrate the binomial probability formula in Example 2.

EXAMPLE 2   **Constructing a Binomial Probability Distribution**

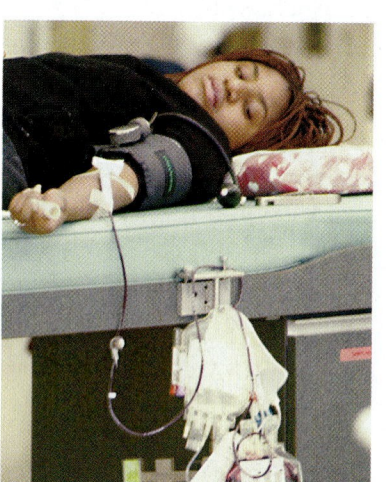

Problem: According to the *Information Please Almanac*, 6% of the human population is blood type O-negative. A simple random sample of size 4 is obtained, and the number of people X with blood type O-negative is recorded. Construct a probability distribution for the random variable X.

Approach: This is a binomial experiment with $n = 4$ trials. We define a success as selecting an individual with blood type O-negative. The probability of success, p, is 0.06, and X is the random variable representing the number of successes with $x = 0, 1, 2, 3,$ or 4.

Step 1: Construct a tree diagram listing the various outcomes of the experiment by listing each outcome as S (success) or F (failure).

Step 2: Compute the probabilities for each value of the random variable X.

Step 3: Construct the probability distribution.

Solution

Step 1: Figure 5 contains a tree diagram listing the 16 possible outcomes of the experiment.

Step 2: We now compute the probability for each possible value of the random variable X. We start with $P(0)$:

$$
\begin{aligned}
P(0) = P(FFFF) &= P(F) \cdot P(F) \cdot P(F) \cdot P(F) \quad \text{\color{teal}\small Multiplication Rule for Independent Events}\\
&= (0.94)(0.94)(0.94)(0.94)\\
&= (0.94)^4\\
&= 0.78075
\end{aligned}
$$

$$
\begin{aligned}
P(1) &= P(SFFF \text{ or } FSFF \text{ or } FFSF \text{ or } FFFS)\\
&= P(SFFF) + P(FSFF) + P(FFSF) + P(FFFS) \quad \text{\color{teal}\small Addition Rule for Disjoint Events}\\
&= (0.06)^1(0.94)^3 + (0.06)^1(0.94)^3 + (0.06)^1(0.94)^3 + (0.06)^1(0.94)^3 \quad \text{\color{teal}\small Multiplication Rule for Independent Events}\\
&= 4(0.06)^1(0.94)^3\\
&= 0.19934
\end{aligned}
$$

$$P(2) = P(SSFF \text{ or } SFSF \text{ or } SFFS \text{ or } FSSF \text{ or } FSFS \text{ or } FFSS)$$

$$= P(SSFF) + P(SFSF) + P(SFFS) + P(FSSF) + P(FSFS) + P(FFSS)$$

$$= (0.06)^2(0.94)^2 + (0.06)^2(0.94)^2 + (0.06)^2(0.94)^2 + (0.06)^2(0.94)^2 + (0.06)^2(0.94)^2 + (0.06)^2(0.94)^2$$

$$= 6(0.06)^2(0.94)^2$$

$$= 0.01909$$

Figure 5

1st Trial	2nd Trial	3rd Trial	4th Trial	Outcome	Number of Successes, X
			S	S,S,S,S	4
		S	F	S,S,S,F	3
	S	F	S	S,S,F,S	3
			F	S,S,F,F	2
S			S	S,F,S,S	3
	F	S	F	S,F,S,F	2
		F	S	S,F,F,S	2
			F	S,F,F,F	1
			S	F,S,S,S	3
	S	S	F	F,S,S,F	2
		F	S	F,S,F,S	2
F			F	F,S,F,F	1
			S	F,F,S,S	2
	F	S	F	F,F,S,F	1
		F	S	F,F,F,S	1
			F	F,F,F,F	0

Table 8

x	$P(x)$
0	0.78075
1	0.19934
2	0.01909
3	0.00081
4	0.00001

We compute $P(3)$ and $P(4)$ similarly and obtain $P(3) = 0.00081$ and $P(4) = 0.00001$. You are encouraged to verify these probabilities.

Step 3: We use these results and obtain the probability distribution in Table 8.

As we look back at the solution in Example 2, we note some interesting results. Consider the probability of obtaining $X = 1$ success:

$$P(1) = 4(0.06)^1 (0.94)^3$$

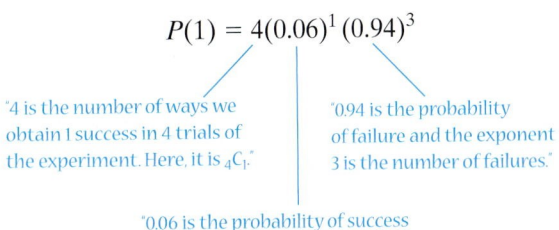

"4 is the number of ways we obtain 1 success in 4 trials of the experiment. Here, it is $_4C_1$."

"0.06 is the probability of success and the exponent 1 is the number of successes."

"0.94 is the probability of failure and the exponent 3 is the number of failures."

The coefficient 4 is the number of ways of obtaining one success in four trials. In general, the coefficient will be $_nC_x$, the number of ways of obtaining x successes in n trials. The second factor in the formula, $(0.06)^1$, is the probability of success, p, raised to the number of successes, x. The third factor in the formula, $(0.94)^3$, is the probability of failure, $1 - p$, raised to the number of failures, $n - x$. This formula holds for all binomial experiments, and we have the *binomial probability distribution function (pdf)*.

Binomial Probability Distribution Function

The probability of obtaining x successes in n independent trials of a binomial experiment, where the probability of success is p, is given by

$$P(x) = {_nC_x}\, p^x (1 - p)^{n-x}, \quad x = 0, 1, 2, \ldots, n \qquad (1)$$

!**CAUTION**

Before using the binomial probability distribution function, be sure the requirements for a binomial experiment are satisfied.

While reading probability problems, pay special attention to key phrases that translate into mathematical symbols. Table 9 lists various phrases and their corresponding mathematical equivalent.

Table 9	
Phrase	**Math Symbol**
"at least" or "no less than" or "greater than or equal to"	$\geq$
"more than" or "greater than"	$>$
"fewer than" or "less than"	$<$
"no more than" or "at most" or "less than or equal to"	$\leq$
"exactly or "equals" or "is"	$=$

EXAMPLE 3   Using the Binomial Probability Distribution Function

Problem: According to the Federal Communications Commission, 70% of all U.S. households have cable television.

(a) In a random sample of 15 households, what is the probability that exactly 10 have cable?

(b) In a random sample of 15 households, what is the probability that at least 13 have cable?

(c) In a random sample of 15 households, what is the probability that fewer than 13 have cable?

(d) In a random sample of 15 households, what is the probability that the number of households with cable is between 10 and 12, inclusive?

Approach: This is a binomial experiment with $n = 15$ independent trials with the probability of success, p, equal to 0.70. The possible values of the random variable X are $x = 0, 1, 2, \ldots, 15$. We use Formula (1) to compute the probabilities.

Solution

(a) $P(10) = {_{15}C_{10}}(0.70)^{10}(1 - 0.70)^{15-10} \qquad n = 15, x = 10, p = 0.70$

$$= \frac{15!}{10!(15 - 10)!}(0.70)^{10}(0.30)^5 \qquad {_nC_x} = \frac{n!}{x!(n - x)!}$$

$$= 3003(0.02825)(0.00243)$$

$$= 0.2061$$

Interpretation: The probability of getting exactly 10 households out of 15 with cable is 0.2061. In 100 trials of this experiment, we would expect about 21 trials to result in 10 households with cable.

(b) The phrase *at least* means greater than or equal to. The values of the random variable X greater than or equal to 13 are $X = 13, 14$, or 15.

$$P(X \geq 13) = P(13 \text{ or } 14 \text{ or } 15)$$
$$= P(13) + P(14) + P(15) \qquad \text{Addition Rule for Disjoint Events}$$
$$= {}_{15}C_{13}(0.70)^{13}(1 - 0.70)^{15-13} + {}_{15}C_{14}(0.70)^{14}(1 - 0.70)^{15-14} + {}_{15}C_{15}(0.70)^{15}(1 - 0.70)^{15-15}$$
$$= 0.0916 + 0.0305 + 0.0047$$
$$= 0.1268$$

Interpretation: There is a 0.1268 probability that in a random sample of 15 households at least 13 will have cable. In 100 trials of this experiment, we would expect about 13 trials to result in at least 13 households having cable.

(c) The values of the random variable X less than 13 are $X = 0, 1, 2, \ldots, 12$. Rather than compute $P(X \leq 12)$ directly by computing $P(0) + P(1) + \cdots + P(12)$, we can use the Complement Rule.

$$P(X < 13) = P(X \leq 12) = 1 - P(X \geq 13) = 1 - 0.1268 = 0.8732$$

Interpretation: There is a 0.8732 probability that in a random sample of 15 households, fewer than 13 will have cable. In 100 trials of this experiment, we expect about 87 trials to result in fewer than 13 households that have cable.

(d) The word *inclusive* means including, so we want to determine the probability that 10, 11, or 12 households have cable.

$$P(10 \leq X \leq 12) = P(10 \text{ or } 11 \text{ or } 12)$$
$$= P(10) + P(11) + P(12) \qquad \text{Addition Rule for Disjoint Events}$$
$$= {}_{15}C_{10}(0.70)^{10}(1 - 0.70)^{15-10} + {}_{15}C_{11}(0.70)^{11}(1 - 0.70)^{15-11} + {}_{15}C_{12}(0.70)^{12}(1 - 0.70)^{15-12}$$
$$= 0.2061 + 0.2186 + 0.1700$$
$$= 0.5947$$

Interpretation: The probability that the number of households with cable is between 10 and 12, inclusive, is 0.5947. In 100 trials of this experiment, we expect about 59 trials to result in 10 to 12 households having cable.

Obtaining Binomial Probabilities from Tables
Another method for obtaining probabilities is the binomial probability table. Table II in Appendix A gives probabilities for a binomial random variable X taking on a specific value such as $P(10)$ for select values of n and p. Table III in Appendix A gives cumulative probabilities of a binomial random variable X. This means Table III gives "less than or equal to" binomial probabilities such as $P(X \leq 6)$. We illustrate how to use Tables II and III in Example 4.

EXAMPLE 4   **Computing Binomial Probabilities Using the Binomial Table**

Problem: According to the National Endowment for the Arts, 20% of U.S. women attended a musical play in 2002.

(a) In a random sample of 15 U.S. women, what is the probability that exactly 5 have attended a musical play in 2002?
(b) In a random sample of 15 U.S. women, what is the probability that fewer than 7 attended a musical play in 2002?
(c) In a random sample of 15 U.S. women, what is the probability that 7 or more attended a musical play in 2002?

Approach: We use Tables II and III in Appendix A to obtain the probabilities.

Solution

(a) We have $n = 15$, $p = 0.20$, and $x = 5$. In Table II, we go to the section that contains $n = 15$ and the column that contains $p = 0.20$. Within the $n = 15$ section, we look for the row $x = 5$. The value at which the $x = 5$ row intersects with the $p = 0.20$ column is the probability we seek. See Figure 6. So $P(5) = 0.1032$.

Figure 6

n	x	0.01	0.05	0.10	0.15	0.20	0.25	0.30	0.35	0.40	0.45	0.50	0.55	0.60	0.65	0.70	0.75	0.80	0.85	0.85	0.90	0.95
15	0	0.8601	0.4633	0.2059	0.0874	0.0352	0.0134	0.0047	0.0016	0.0005	0.0001	0.0000	0.0000	0.0000	0.0000	0.0000	0.0000	0.0000	0.0000	0.0000	0.0000	0.0000
	1	0.1303	0.3658	0.3432	0.2312	0.1319	0.0668	0.0305	0.0126	0.0047	0.0015	0.0006	0.0001	0.0000	0.0000	0.0000	0.0000	0.0000	0.0000	0.0000	0.0000	0.0000
	2	0.0092	0.1348	0.2669	0.2856	0.2309	0.1559	0.0916	0.0476	0.0219	0.0090	0.0032	0.0010	0.0003	0.0001	0.0000	0.0000	0.0000	0.0000	0.0000	0.0000	0.0000
	3	0.0004	0.0307	0.1285	0.2184	0.2501	0.2252	0.1700	0.1110	0.0634	0.0318	0.0139	0.0052	0.0016	0.0004	0.0001	0.0000	0.0000	0.0000	0.0000	0.0000	0.0000
	4	0.0000	0.0049	0.0428	0.1156	0.1676	0.2252	0.2186	0.1782	0.1268	0.0780	0.0417	0.0191	0.0074	0.0024	0.0006	0.0001	0.0000	0.0000	0.0000	0.0000	0.0000
	5	0.0000	0.0006	0.0105	0.0449	0.1032	0.1651	0.2061	0.2123	0.1859	0.1404	0.0916	0.0515	0.0245	0.0096	0.0030	0.0007	0.0001	0.0000	0.0000	0.0000	0.0000
	6	0.0000	0.0000	0.0019	0.0132	0.0430	0.0917	0.1472	0.1906	0.2056	0.1914	0.1527	0.1048	0.0612	0.0298	0.0116	0.0034	0.0007	0.0001	0.0000	0.0000	0.0000
	7	1.0000	1.0000	…	0.9994	…	…	0.0500	…	0.7869	…	0.500	…	…	…	0.0500	…	…	…	…	…	…

Interpretation: There is a 0.1032 probability that in a random sample of 15 U.S. women, exactly 5 have attended a musical play in 2002. In 100 trials of this experiment, we expect about 10 trials to result in exactly 5 women who have attended a musical play in 2002.

(b) The values of the random variable X that are fewer than 7 are 0, 1, 2, 3, 4, 5, or 6. So, $P(X < 7) = P(X \leq 6)$. To compute $P(X \leq 6)$, we use the cumulative binomial table, Table III in Appendix A. The cumulative binomial table lists binomial probabilities less than or equal to a specified value. We have $n = 15$ and $p = 0.20$. In Table III, we go to the row that contains $n = 15$ and the column that contains $p = 0.20$. Within the $n = 15$ section, we look for the row $x = 6$. This row represents $P(X \leq 6)$. The value at which the $x = 6$ row intersects with the $p = 0.20$ column is the probability we seek. See Figure 7. So $P(X \leq 6) = 0.9819$.

Figure 7

n	x	0.01	0.05	0.10	0.15	0.20	0.25	0.30	0.35	0.40	0.45	0.50	0.55	0.60	0.65	0.70	0.75	0.80	0.85	0.85	0.90	0.95
15	0	0.8601	0.4633	0.2059	0.0874	0.0352	0.0134	0.0047	0.0016	0.0005	0.0001	0.0000	0.0000	0.0000	0.0000	0.0000	0.0000	0.0000	0.0000	0.0000	0.0000	0.0000
	1	0.9904	0.8290	0.5490	0.3188	0.1671	0.0802	0.0353	0.0142	0.0052	0.0017	0.0005	0.0001	0.0000	0.0000	0.0000	0.0000	0.0000	0.0000	0.0000	0.0000	0.0000
	2	0.9996	0.9538	0.8159	0.6042	0.3980	0.2361	0.1288	0.0617	0.0271	0.0107	0.0037	0.0011	0.0003	0.0001	0.0000	0.0000	0.0000	0.0000	0.0000	0.0000	0.0000
	3	1.0000	0.9945	0.9944	0.8227	0.6482	0.4613	0.2969	0.1727	0.0905	0.0424	0.0178	0.0083	0.0019	0.0005	0.0001	0.0000	0.0000	0.0000	0.0000	0.0000	0.0000
	4	1.0000	0.9994	0.9873	0.9388	0.8358	0.6886	0.5155	0.3519	0.2173	0.1204	0.0592	0.0255	0.0093	0.0028	0.0007	0.0001	0.0000	0.0000	0.0000	0.0000	0.0000
	5	1.0000	0.9999	0.9978	0.9964	0.9389	0.8516	0.7216	0.5843	0.4032	0.2608	0.1509	0.0789	0.0338	0.0124	0.0037	0.0008	0.0001	0.0000	0.0000	0.0000	0.0000
	6	1.0000	1.0000	0.9997	0.9964	0.9819	0.9434	0.9689	0.7546	0.6098	0.4522	0.3036	0.1618	0.0950	0.0422	0.0152	0.0042	0.0008	0.0001	0.0000	0.0000	0.0000
	7	1.0000	1.0000	1.0000	0.9994	0.9958	0.9627	0.9500	0.8868	0.7869	0.6535	0.500	0.3465	0.2131	0.1132	0.0500	0.0173	0.0042	0.0006	0.0000	0.0000	0.0000
	8	1.0000	1.0000	1.0000	0.9994	…	0.9627	0.0500	…	0.7869	…	0.500	…	…	…	0.0500	…	…	…	…	…	0.0000

Interpretation: There is a 0.9819 probability that in a random sample of 15 U.S. women, fewer than 7 have attended a musical play in 2002. In 100 trials of this experiment, we would expect about 98 trials to result in fewer than 7 women who have attended a musical play in 2002.

(c) To obtain $P(X \geq 7)$, we use the Complement Rule and the results of part (b) as follows:

$$P(X \geq 7) = 1 - P(X < 7)$$
$$= 1 - P(X \leq 6)$$
$$= 1 - 0.9819$$
$$= 0.0181$$

Interpretation: There is a 0.0181 probability that in a random sample of 15 U.S. women, at least 7 have attended a musical play in 2002. In 100 trials of this experiment, we expect about 2 trials to result in at least 7 women who have attended a musical play in 2002. Because this event only happens about 2 out of 100 times, we consider it to be unusual.

Obtaining Binomial Probabilities Using Technology
Statistical software and graphing calculators have the ability to compute binomial probabilities as well. We illustrate this approach to computing probabilities in the next example.

EXAMPLE 5

Obtaining Binomial Probabilities Using Technology

Problem: According to the National Endowment for the Arts, 20% of U.S. women attended a musical play in 2002.

(a) In a random sample of 15 U.S. women, what is the probability that exactly 5 have attended a musical play in 2002?
(b) In a random sample of 15 U.S. women, what is the probability that fewer than 7 attended a musical play in 2002?

Approach: Statistical software or graphing calculators with advanced statistical features have the ability to determine binomial probabilities. The steps for determining binomial probabilities using MINITAB, Excel, and the TI-83/84 Plus graphing calculators can be found in the Technology Step by Step on page 344.

Result: We will use Excel to determine the probability for part (a) and a TI-84 Plus to determine the probability for part (b).

(a) Using Excel's formula wizard, we obtain the results in Figure 8(a). So $P(5) = 0.1032$. This agrees with the results of Example 4(a).
(b) To compute probabilities such as $P(X < 7) = P(X \leq 6)$, it is best to use the **cumulative distribution function** (or **cdf**), which computes probabilities less than or equal to a specified value. Using a TI-84 Plus graphing calculator to compute $P(X \leq 6)$ with $n = 15$ and $p = 0.2$, we find $P(X \leq 6) = 0.9819$. See Figure 8(b).

Figure 8

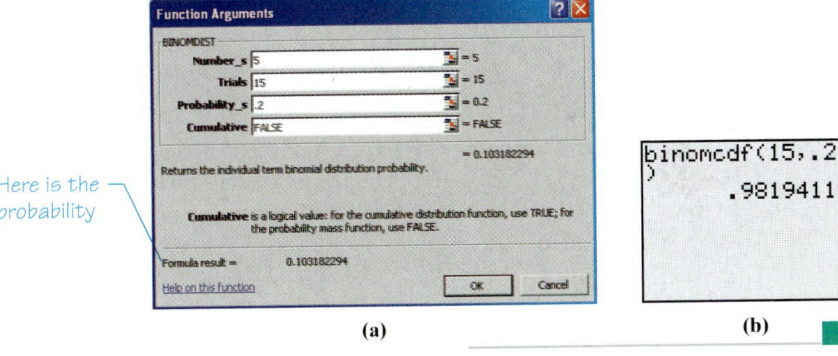

Here is the probability

(a)   (b)

Now Work Problem 35.

3 ## Compute the Mean and Standard Deviation of a Binomial Random Variable

We discussed finding the mean (or expected value) and standard deviation of a discrete random variable in Section 6.1. These formulas can be used to find the mean (or expected value) and standard deviation of a binomial random variable as well. However, there is a faster method.

In Other Words
The mean of a binomial random variable equals the number of trials of the experiment times the probability of success. It can be interpreted as the expected number of successes in n trials of the experiment.

Mean (or Expected Value) and Standard Deviation of a Binomial Random Variable
A binomial experiment with n independent trials and probability of success p has a mean and standard deviation given by the formulas

$$\mu_X = np \quad \text{and} \quad \sigma_X = \sqrt{np(1-p)} \tag{2}$$

EXAMPLE 6   **Finding the Mean and Standard Deviation of a Binomial Random Variable**

Problem: According to the Federal Communications Commission, 70% of all U.S. households had cable television in 2002. In a simple random sample of 300 households, determine the mean and standard deviation number of households that will have cable television.

Approach: This is a binomial experiment with $n = 300$ and $p = 0.70$. We can use Formula (2) to find the mean and standard deviation, respectively.

Solution: $\mu_X = np = 300(0.70) = 210$
and

$$\sigma_X = \sqrt{np(1-p)} = \sqrt{300(0.70)(1-0.70)} = \sqrt{63} = 7.9$$

Interpretation: We expect that in a random sample of 300 households 210 will have cable.

Now Work Problems 29(a), (b), and (c).

4   **Construct Binomial Probability Histograms**

Constructing binomial probability histograms is no different from constructing other probability histograms.

EXAMPLE 7   **Constructing Binomial Probability Histograms**

Problem

(a) Construct a binomial probability histogram with $n = 10$ and $p = 0.2$. Comment on the shape of the distribution.
(b) Construct a binomial probability histogram with $n = 10$ and $p = 0.5$. Comment on the shape of the distribution.
(c) Construct a binomial probability histogram with $n = 10$ and $p = 0.8$. Comment on the shape of the distribution.

Approach: To construct a binomial probability histogram, we will first obtain the probability distribution. We then construct the probability histogram of the probability distribution.

Solution

(a) We obtain the probability distribution with $n = 10$ and $p = 0.2$. See Table 10. Note in Table 10, $P(9) = 0.0000$. The probability is actually 0.000004096, but is written as 0.0000 to four significant digits. Same idea applies to $P(10)$. Figure 9 shows the corresponding probability histogram with the mean $\mu_X = 10(0.2) = 2$ labeled. The distribution is skewed right.

Table 10

x	$P(x)$
0	0.1074
1	0.2684
2	0.3020
3	0.2013
4	0.0881
5	0.0264
6	0.0055
7	0.0008
8	0.0001
9	0.0000
10	0.0000

Figure 9

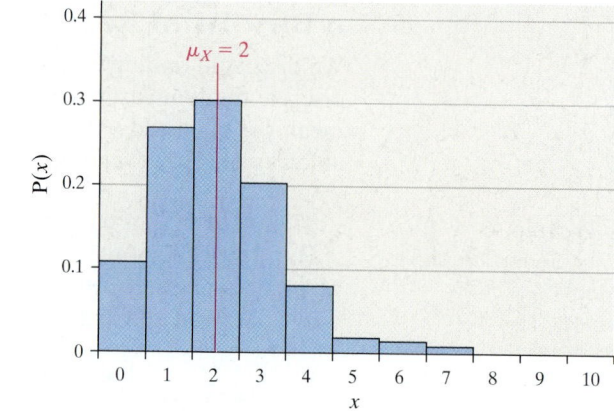

(b) We obtain the probability distribution with $n = 10$ and $p = 0.5$. See Table 11. Figure 10 shows the corresponding probability histogram with the mean $\mu_X = 10(0.5) = 5$ labeled. The distribution is symmetric and approximately bell shaped.

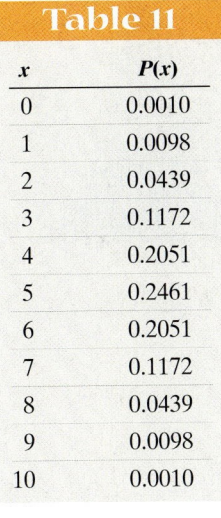

Table 11	
x	$P(x)$
0	0.0010
1	0.0098
2	0.0439
3	0.1172
4	0.2051
5	0.2461
6	0.2051
7	0.1172
8	0.0439
9	0.0098
10	0.0010

Figure 10

(c) We obtain the probability distribution with $n = 10$ and $p = 0.8$. See Table 12. Figure 11 shows the corresponding probability histogram with the mean $\mu_X = 10(0.8) = 8$ labeled. The distribution is skewed left.

Table 12	
x	$P(x)$
0	0.0000
1	0.0000
2	0.0001
3	0.0008
4	0.0055
5	0.0264
6	0.0881
7	0.2013
8	0.3020
9	0.2684
10	0.1074

Figure 11

Now Work Problem 29(d).

Based on the results of Example 7, we might conclude that the binomial probability distribution is skewed right if $p < 0.5$, symmetric and approximately bell shaped if $p = 0.5$, and skewed left if $p > 0.5$. Notice that Figure 9 ($p = 0.2$) and Figure 11 ($p = 0.8$) are mirror images.

The binomial probability distribution depends on the parameter p, and n, the number of trials. What role does n play in the shape of the distribution? To answer this question we compare the binomial probability histogram with $n = 10$ and $p = 0.2$. [see Figure 12(a)] to the binomial probability histogram with $n = 30$ and $p = 0.2$. [Figure 12(b)] and the binomial probability histogram with $n = 70$ and $p = 0.2$ [Figure 12(c)].

Figure 12(a) is skewed right. Figure 12(b) is slightly skewed right, and Figure 12(c) appears bell shaped.

Figure 12

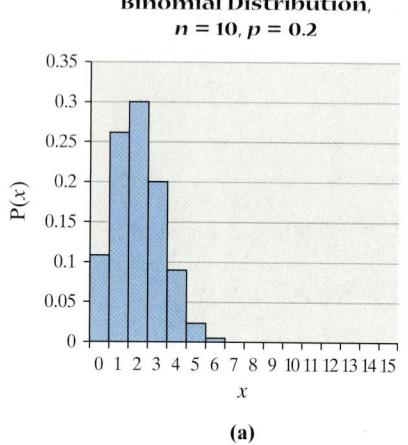

Binomial Distribution, $n = 10, p = 0.2$

(a)

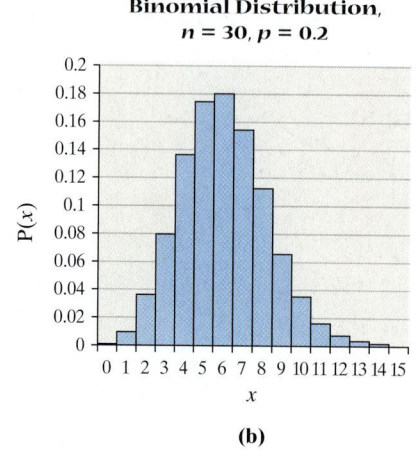

Binomial Distribution, $n = 30, p = 0.2$

(b)

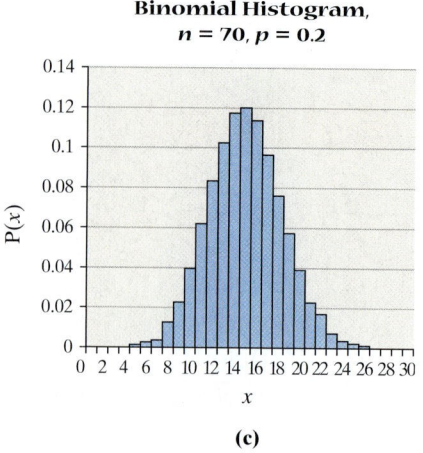

Binomial Histogram, $n = 70, p = 0.2$

(c)

We conclude the following:

> As the number of trials n in a binomial experiment increases, the probability distribution of the random variable X becomes bell shaped. As a rule of thumb, if $np(1 - p) \geq 10$,* the probability distribution will be approximately bell shaped.

This result allows us to use the Empirical Rule to identify unusual observations in a binomial experiment. Recall that the Empirical Rule states that in a bell-shaped distribution about 95% of all observations lie within two standard deviations of the mean. That is, about 95% of the observations lie between $\mu - 2\sigma$ and $\mu + 2\sigma$. Any observation that lies outside this interval may be considered unusual because the observation occurs less than 5% of the time.

EXAMPLE 8

Using the Mean, Standard Deviation, and Empirical Rule to Check for Unusual Results in a Binomial Experiment

Problem: According to the Federal Communications Commission, in 2002, 70% of all U.S. households had cable television. In a simple random sample of 300 households, 230 had cable. Is this result unusual?

Approach: Because $np(1 - p) = 300(0.70)(0.30) = 63 \geq 10$, the binomial probability distribution is approximately bell-shaped. Therefore, we can use the Empirical Rule to check for unusual observations. If the observation is less than $\mu - 2\sigma$ or greater than $\mu + 2\sigma$, we say it is unusual.

Solution: From Example 6, we have $\mu = 210$ and $\sigma = 7.9$.

$$\mu - 2\sigma = 210 - 2(7.9) = 210 - 15.8 = 194.2$$

and

$$\mu + 2\sigma = 210 + 2(7.9) = 210 + 15.8 = 225.8$$

*Ramsey, P. P., and P. H. Ramsey, Evaluating the Normal Approximation to the Binomial Test, *Journal of Educational Statistics* 13 (1998): 173–182.

Interpretation: Since any value less than 194.2 or greater than 225.8 is unusual, 230 is an unusual result. We should attempt to identify reasons for its value. It may be that the percentage of households that have cable has increased since 2002.

Now Work Problem 43.

MAKING AN INFORMED DECISION

Should We Convict?

A woman who was shopping in Los Angeles had her purse stolen by a young, blonde female who was wearing a ponytail. The blonde female got into a yellow car that was driven by a black male who had a mustache and a beard. The police located a blonde female named Janet Collins who wore her hair in a ponytail and had a friend who was a black male who had a mustache and beard and also drove a yellow car. The police arrested the two subjects.

Because there were no eyewitnesses and no real evidence, the prosecution used probability to make its case against the defendants. The following probabilities were presented by the prosecution for the known characteristics of the thieves.

Characteristic	Probability
Yellow car	$\frac{1}{10}$
Man with a mustache	$\frac{1}{4}$
Woman with a ponytail	$\frac{1}{10}$
Woman with blonde hair	$\frac{1}{3}$
Black man with beard	$\frac{1}{10}$
Interracial couple in car	$\frac{1}{1000}$

(a) Assuming that the characteristics listed are independent of each other, what is the probability that a randomly selected couple would have all these characteristics? That is, what is P ("yellow car" and "man with a mustache" and ... and "interracial couple in a car")?

(b) Would you convict the defendants based on this probability? Why?

(c) Now let n represent the number of couples in the Los Angeles area who could have committed the crime. Let p represent the probability a randomly selected couple has all six characteristics listed. Let the random variable X represent the number of couples who have all the characteristics listed in the table. Assuming that the random variable X follows the binomial probability function, we have

$$P(x) = {}_nC_x \cdot p^x (1 - p)^{n-x}, \quad x = 0, 1, 2, \ldots, n$$

Assuming that there were $n = 1{,}000{,}000$ couples in the Los Angeles area, what is the probability that more than one of them have the characteristics listed in the table? Does this result cause you to change your mind regarding the defendants' guilt?

(d) Now let's look at this case from a different point of view. We will compute the probability that more than one couple has the characteristics described, given that at least one couple has the characteristics.

$$P(X > 1 \mid X \ge 1) = \frac{P(X > 1 \text{ and } X \ge 1)}{P(X \ge 1)} \quad \text{Conditional Probability Rule}$$

$$= \frac{P(X > 1)}{P(X \ge 1)}$$

Compute this probability, assuming $n = 1{,}000{,}000$. Compute this probability again, but this time assume that $n = 2{,}000{,}000$. Do you think that the couple should be convicted "beyond all reasonable doubt"? Why?

6.2 ASSESS YOUR UNDERSTANDING

Concepts and Vocabulary

1. State the criteria for a binomial probability experiment.
2. What role does $_nC_x$ play in the binomial probability distribution function?
3. How can the Empirical Rule be used to identify unusual results in a binomial experiment? When can the Empirical Rule be used?
4. Describe how the value of n affects the shape of the binomial probability histogram.
5. Describe how the value of p affects the shape of the binomial probability histogram.
6. Explain what "success" means in a binomial experiment.

Skill Building

In Problems 7–16, determine which of the following probability experiments represents a binomial experiment. If the probability experiment is not a binomial experiment, state why.

7. A random sample of 15 college seniors is obtained, and the individuals selected are asked to state their ages.

8. A random sample of 30 cars in a used car lot is obtained, and their mileage recorded.

9. An experimental drug is administered to 100 randomly selected individuals, with the number of individuals responding favorably recorded.
NW

10. A poll of 1200 registered voters is conducted in which the respondents are asked whether they believe Congress should reform Social Security.

11. Three cards are selected from a standard 52-card deck without replacement. The number of aces selected is recorded.

12. Three cards are selected from a standard 52-card deck with replacement. The number of aces selected is recorded.

13. A basketball player who makes 80% of her free throws is asked to shoot free throws until she misses. The number of free-throw attempts is recorded.

14. A baseball player who reaches base safely 30% of the time is allowed to bat until he reaches base safely for the third time. The number of at-bats required is recorded.

15. An investor randomly purchases 10 stocks listed on the New York Stock Exchange. Historically, the probability that a stock listed on the NYSE will increase in value over the course of a year is 48%. The number of stocks that increase in value is recorded.

16. According to Nielsen Media Research, 70% of all U.S. households have cable television. In a small town of 40 households, a random sample of 10 households is asked whether they have cable television. The number of households with cable television is recorded.

In Problems 17–28, a binomial probability experiment is conducted with the given parameters. Compute the probability of x successes in the n independent trials of the experiment.

17. $n = 10, p = 0.4, x = 3$

18. $n = 15, p = 0.85, x = 12$

19. $n = 40, p = 0.99, x = 38$

20. $n = 50, p = 0.02, x = 3$

21. $n = 8, p = 0.35, x = 3$

22. $n = 20, p = 0.6, x = 17$

23. $n = 9, p = 0.2, x \leq 3$

24. $n = 10, p = 0.65, x < 5$

25. $n = 7, p = 0.5, x > 3$

26. $n = 20, p = 0.7, x \geq 12$

27. $n = 12, p = 0.35, x \leq 4$

28. $n = 11, p = 0.75, x \geq 8$

In Problems 29–34, (a) construct a binomial probability distribution with the given parameters; (b) compute the mean and standard deviation of the random variable using the methods of Section 6.1; (c) compute the mean and standard deviation, using the methods of this section; and (d) draw the probability histogram, comment on its shape, and label the mean on the histogram.

29. $n = 6, p = 0.3$
NW

30. $n = 8, p = 0.5$

31. $n = 9, p = 0.75$

32. $n = 10, p = 0.2$

33. $n = 10, p = 0.5$

34. $n = 9, p = 0.8$

Applying the Concepts

35. On-Time Flights According to American Airlines, its
NW flight 215 from Orlando to Los Angeles is on time 90% of
the time. Suppose 15 flights are randomly selected and the
number of on time flights is recorded.

(a) Explain why this is a binomial experiment.

(b) Find the probability that exactly 14 flights are on time.

(c) Find the probability that at least 14 flights are on
time.

(d) Find the probability that fewer than 14 flights are on
time.

(e) Find the probability that between 12 and 14 flights, inclusive, are on time.

36. Smokers According to the *Information Please Almanac*,
80% of adult smokers started smoking before turning 18
years old. Suppose 10 smokers 18 years old or older are
randomly selected and the number of smokers who started smoking before 18 is recorded.

(a) Explain why this is a binomial experiment.

(b) Find the probability that exactly 8 of them started
smoking before 18 years of age.

(c) Find the probability that at least 8 of them started
smoking before 18 years of age.

(d) Find the probability that fewer than 8 of them started
smoking before 18 years of age.

(e) Find the probability that between 7 and 9 of them, inclusive, started smoking before 18 years of age.

37. High-Speed Internet According to a report by the Commerce Department in the fall of 2004, 20% of U.S. households had some type of high-speed Internet connection.
Suppose 20 U.S. households are selected at random and
the number of households with high-speed Internet is
recorded.

(a) Find the probability that exactly 5 households have
high-speed Internet.

(b) Find the probability that at least 10 households have
high-speed Internet. Would this be unusual?

(c) Find the probability that fewer than 4 households
have high-speed Internet.

(d) Find the probability that between 2 and 5 households,
inclusive, have high-speed Internet.

38. Allergy Sufferers Clarinex-D is a medication whose purpose is to reduce the symptoms associated with a variety of
allergies. In clinical trials of Clarinex-D, 5% of the patients
in the study experienced insomnia as a side effect. Suppose
a random sample of 20 Clarinex-D users is obtained and
the number of patients who experienced insomnia is
recorded.

(a) Find the probability that exactly 3 experienced insomnia as a side effect.

(b) Find the probability that 3 or fewer experienced insomnia as a side effect.

(c) Find the probability that between 1 and 4 patients, inclusive, experienced insomnia as a side effect.

(d) Would it be unusual to find 4 or more patients who experienced insomnia as a side effect? Why?

39. Murder by Firearms According to the *Uniform Crime Report, 2003*, 66.9% of murders are committed with a
firearm. Suppose that 25 murders were randomly selected
and the number of murders committed with a firearm is
recorded.

(a) Find the probability that exactly 22 murders were
committed using a firearm.

(b) Find the probability that between 14 and 16 murders,
inclusive, were committed using a firearm.

(c) Would it be unusual if 22 or more murders were committed using a firearm? Why?

40. Migraine Sufferers Depakote is a medication whose purpose is to reduce the pain associated with migraine
headaches. In clinical trials of Depakote, 2% of the patients
in the study experienced weight gain as a side effect. Suppose a random sample of 30 Depakote users is obtained
and the number of patients who experienced weight gain is
recorded.

Source: Abbott Laboratories

(a) Find the probability that exactly 3 experienced weight
gain as a side effect.

(b) Find the probability that 3 or fewer experienced
weight gain as a side effect.

(c) Find the probability that 4 or more patients experienced weight gain as a side effect.

(d) Find the probability that between 1 and 4 patients, inclusive, experienced weight gain as a side effect.

41. Airline Satisfaction A CNN/USA Today/Gallup poll in
April 2005 reported that 75% of adult Americans were satisfied with the job the nation's major airlines were doing.
Suppose 10 adult Americans are selected at random and
the number who are satisfied is recorded.

(a) Find the probability that exactly 6 are satisfied with
the airlines.

(b) Find the probability that fewer than 7 are satisfied
with the airlines.

(c) Find the probability that 5 or more are satisfied with
the airlines.

(d) Find the probability that between 5 and 8, inclusive,
are satisfied with the airlines.

42. College Freshmen According to the Higher Education
Research Institute, 55% of college freshmen in 4-year colleges and universities during 2003 were female. Suppose
12 freshmen are randomly selected and the number of females is recorded.

(a) Find the probability that exactly 7 of them are
female.

(b) Find the probability that 5 or more are female.

(c) Find the probability that 8 or fewer are female.

(d) Find the probability that between 7 and 10, inclusive,
are female.

43. On-Time Flights According to American Airlines, its
NW flight 215 from Orlando to Los Angeles is on time 90% of
the time, Suppose 100 flights are randomly selected.

(a) Compute the mean and standard deviation of the random variable X, the number of on-time flights in 100 trials of the probability experiment.

(b) Interpret the mean.

(c) Would it be unusual to observe 80 on-time flights in a random sample of 100 flights from Orlando to Los Angeles? Why?

44. Smokers According to the *Information Please Almanac*, 80% of adult smokers started smoking before turning 18 years old.

(a) Compute the mean and standard deviation of the random variable X, the number of smokers who started before 18 in 200 trials of the probability experiment.

(b) Interpret the mean.

(c) Would it be unusual to observe 180 smokers who started smoking before turning 18 years old in a random sample of 200 adult smokers? Why?

45. High-Speed Internet According to a report by the Commerce Department in the fall of 2004, 20% of U.S. households had some type of high-speed Internet connection.

(a) Compute the mean and standard deviation of the random variable X, the number of U.S. households with a high-speed Internet connection in 100 households.

(b) Interpret the mean.

(c) Would it be unusual to observe 18 U.S. households that have a high-speed Internet connection in 100 households? Why?

46. Allergy Sufferers Clarinex-D is a medication whose purpose is to reduce the symptoms associated with a variety of allergies. In clinical trials of Clarinex-D, 5% of the patients in the study experienced insomnia as a side effect.

(a) If 240 users of Clarinex-D are randomly selected, how many would we expect to experience insomnia as a side effect?

(b) Would it be unusual to observe 20 patients experiencing insomnia as a side effect in 240 trials of the probability experiment? Why?

47. Murder by Firearms According to the *Uniform Crime Report, 2003*, 66.9% of murders are committed with a firearm.

(a) If 100 murders are randomly selected, how many would we expect to be committed with a firearm?

(b) Would it be unusual to observe 75 murders by firearm in a random sample of 100 murders? Why?

48. Migraine Sufferers Depakote is a medication whose purpose is to reduce the pain associated with migraine headaches. In clinical trials and extended studies of Depakote, 2% of the patients in the study experienced weight gain as a side effect. Would it be unusual to observe 16 patients who experience weight gain in a random sample of 600 patients who take the medication? Why?

49. Asthma Control Singulair is a medication whose purpose is to control asthma attacks. In clinical trials of Singulair, 18.4% of the patients in the study experienced headaches as a side effect. Would it be unusual to observe 86 patients who experience headaches in a random sample of 400 patients who use this medication? Why?

50. Simulation According to the U.S. National Center for Health Statistics, there is a 98% probability that a 20-year-old male will survive to age 30.

(a) Using statistical software, simulate taking 100 random samples of size 30 from this population.

(b) Using the results of the simulation, compute the probability that exactly 29 of the 30 males survive to age 30.

(c) Compute the probability that exactly 29 of the 30 males survive to age 30, using the binomial probability distribution. Compare the results with part (b).

(d) Using the results of the simulation, compute the probability that at most 27 of the 30 males survive to age 30.

(e) Compute the probability that at most 27 of the 30 males survive to age 30, using the binomial probability distribution. Compare the results with part (d).

(f) Compute the mean number of male survivors in the 100 simulations of the probability experiment. Is it close to the expected value?

(g) Compute the standard deviation of the number of male survivors in the 100 simulations of the probability experiment. Compare the result to the theoretical standard deviation of the probability distribution.

(h) Did the simulation yield any unusual results?

51. Probability Applet Load the binomial applet on your
APPLET computer.

(a) Set the probability of success, p, to 0.8 and the number of trials of the binomial experiment, n, to 10. Simulate shooting 10 free throws for $N = 1$. How many were made?

(b) Set the probability of success to 0.8 and the number of trials of the binomial experiment to 10. Simulate shooting 10 free throws $N = 1000$ times. Use the results of the simulation to estimate the probability of making 10 out of 10 free throws.

(c) Use the binomial probability formula to compute the probability of making 10 out of 10 free throws if the probability of success is 0.8.

(d) Use the results of the simulation to estimate the probability of making at least 8 out of 10 free throws.

(e) Use the binomial probability formula to compute the probability of making at least 8 out of 10 free throws.

(f) Determine the mean number of free throws made for the 1000 repetitions of the experiment. Is it close to the expected value?

52. Leisure Activity According to a 2002 survey by the National Endowment for the Arts, 60% of U.S. residents 18 and older attended a movie at least once in the previous year. Suppose you are performing a study and would like at least 12 people in the study to have attended a movie at least once in the past year.

(a) How many residents of the United States 18 years old or older do you expect to have to randomly select?

(b) How many residents of the United States 18 years old or older do you have to randomly select to have a 99% probability that the sample contains at least 12 who have attended a movie in the past year?

53. Educational Attainment According to the U.S. Census Bureau, in 2003 about 27% of residents of the United States 25 years old or older had earned at least a bachelor's degree. Suppose you are performing a study and would like at least 10 people in the study to have earned at least a bachelor's degree.

(a) How many residents of the United States 25 years old or older do you expect to randomly select?

(b) How many residents of the United States 25 years old or older do you have to randomly select to have probability 0.49 that the sample contains at least 10 who have earned at least a bachelor's degree?

54. Geometric Probability Distribution A probability distribution for the random variable X, the number of trials until a success is observed, is called the **geometric probability distribution**. It has the same requirements as the binomial distribution (see page 328), except that the number of trials is not fixed. Its probability distribution function (pdf) is

$$P(x) = p(1 - p)^{x-1}, \quad x = 1, 2, 3, \dots$$

where p is the probability of success.

(a) What is the probability that Shaquille O'Neal misses his first two free throws and makes the third? Over his career, he makes 53.6% of his free throws. That is, find $P(3)$.

(b) Construct a probability distribution for the random variable X, the number of free-throw attempts of Shaquille O'Neal before he makes a free throw. Construct the distribution for $x = 1, 2, 3, \dots, 10$. The probabilities are small for $x > 10$.

(c) Compute the mean of the distribution, using the formula presented in Section 6.1.

(d) Compare the mean obtained in part (c) with the value $\frac{1}{p}$. Conclude that the mean of a geometric probability distribution is $\mu_X = \frac{1}{p}$. How many free throws do we expect Shaq to take before we observe a made free throw?

55. Hypergeometric Probability Distribution Recall that, in sampling without replacement from a finite population, the trials are not independent. For example, suppose that a population has size 20 and that 8 of the people are female. The probability that the first person selected in the sample

is female will be $\frac{8}{20}$. If the person is not replaced, the probability that the second person is female will not be the same as with the first. So the requirements for a binomial experiment are violated. For large populations, this violation is so minor that it is overlooked. In cases where the trials are not independent, the random variable X has a **hypergeometric probability distribution**. The hypergeometric probability distribution is a distribution for the random variable X, the number of individuals in a sample of size n that have a particular attribute (such as female). The hypergeometric probability distribution function is

$$P(x) = \frac{(_{Np}C_x)(_{N(1-p)}C_{n-x})}{_NC_n}$$

where N is the size of the population, p is the proportion of the population with a certain attribute, x is the number of individuals from the population selected in the sample who have the attribute, and n is the number selected to be in the sample (so that $n - x$ is the number selected who do not have the attribute).

According to Nielsen Media Services, 75% of households have cable television. In a small town of $N = 500$ households, a random sample of $n = 20$ households is obtained.

(a) What is the probability exactly 15 of the households have cable?

(b) What is the probability between 15 and 17 of the households have cable?

(c) Verify the sample is less than 5% of the size of the population and use the binomial probability distribution to find approximate probabilities for parts (a) and (b). Compare the results.

56. Negative Binomial Probability Distribution The **negative binomial probability distribution** can be used to compute the probability of the random variable X, the number of trials necessary to observe r successes of a binomial experiment. The probability distribution function is given by

$$P(x) = (_{x-1}C_{r-1})p^r(1 - p)^{x-r}$$
$$x = r, r + 1, r + 2, \dots$$

Consider a roulette wheel. Remember, a roulette wheel has 2 green slots, 18 red slots, and 18 black slots.

(a) What is the probability that it will take $x = 1$ trial before observing $r = 1$ green?

(b) What is the probability that it will take $x = 20$ trials before observing $r = 2$ greens?

(c) What is the probability that it will take $x = 30$ trials before observing $r = 3$ greens?

(d) The expected number of trials before observing r successes is $\frac{r}{p}$. What is the expected number of trials before observing 3 greens?

Computing Binomial Probabilities via Technology

TI-83/84 Plus Computing *P*(*x*)

Step 1: Press 2nd VARS to access the probability distribution menu.

Step 2: Highlight 0: binompdf(for the TI-83 and A: binompdf(for the TI-84 and hit ENTER.

Step 3: With binompdf(on the HOME screen, type the number of trials *n*, the probability of success, *p*, and the number of successes, *x*. For example, with *n* = 10, *p* = 0.2, and *x* = 4, type

binompdf(10, 0.2, 4)

Then hit ENTER.

Computing *P*(*X* ≤ *x*)

Step 1: Press 2nd VARS to access the probability distribution menu.

Step 2: Highlight A: binomcdf(for the TI-83 and B: binomcdf(for the TI-84 and hit ENTER.

Step 3: With binomcdf(on the HOME screen, type the number of trials *n*, the probability of success, *p*, and the number of successes, *x*. For example, with *n* = 10, *p* = 0.2, and *x* = 4, type

binomcdf(10, 0.2, 4)

Then hit ENTER.

MINITAB Computing *P*(*x*)

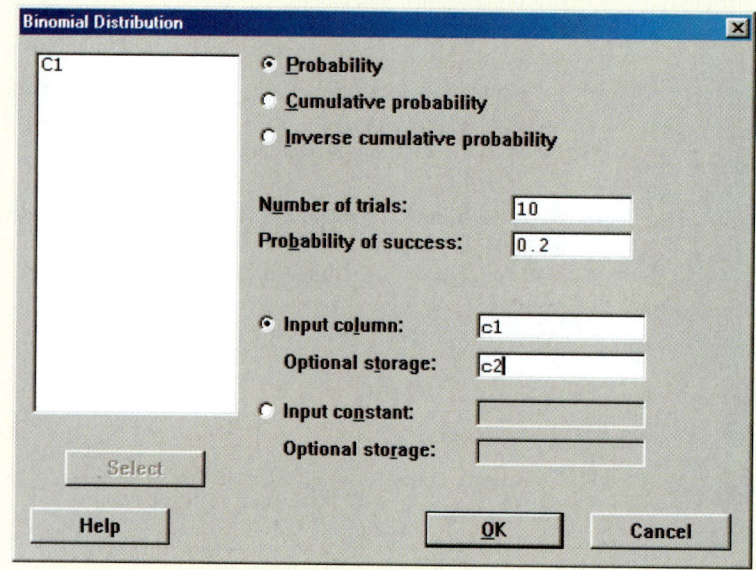

Step 1: Enter the possible values of the random variable *x* in C1. For example, with *n* = 10, *p* = 0.2, enter 0, 1, 2, ..., 10 into C1.

Step 2: Select the **CALC** menu, highlight **Probability Distributions**, then highlight **Binomial** ....

Step 3: Fill in the window as shown to the left. Click OK.

Computing *P*(*X* ≤ *x*)

Follow the same steps as those for computing *P*(*x*). In the window that comes up after selecting Binomial Distribution, select Cumulative probability instead of Probability.

Excel Computing *P*(*x*)

Step 1: Select the *fx* icon. Highlight Statistical in the Function category window. Highlight BINOMDIST in the Function name window.

Step 2: Fill in the window with the appropriate values. For example, if *x* = 5, *n* = 10, and *p* = 0.2, fill in the window as shown in Figure 8(a). Click OK.

Computing *P*(*X* ≤ *x*)

Follow the same steps as those presented for computing *P*(*x*). In the BINOMDIST window, type "TRUE" in the cumulative cell.

6.3 The Poisson Probability Distribution

Objectives
1. Understand when a probability experiment follows a Poisson process
2. Compute probabilities of a Poisson random variable
3. Find the mean and standard deviation of a Poisson random variable

1 Understand When a Probability Experiment Follows a Poisson Process

We now introduce another discrete probability model, the *Poisson Probability Distribution*, named after Siméon Denis Poisson. This probability distribution can be used to compute probabilities of experiments in which the random variable X counts the number of occurrences (successes) of a particular event within a specified interval (usually time or space). Consider the following example.

EXAMPLE 1   **Illustrating a Poisson Process**

A McDonald's manager knows from prior experience that cars arrive at the drive-through at the constant rate of two cars per minute between the hours of 12:00 noon and 1:00 P.M. The random variable X, the number of cars that arrive between 12:20 and 12:40, follows a Poisson process.

We now present the criteria for a random variable to follow a *Poisson process*.

Definition   A random variable X, the number of successes in a fixed interval, follows a **Poisson process** provided the following conditions are met.
1. The probability of two or more successes in any sufficiently small subinterval* is 0.
2. The probability of success is the same for any two intervals of equal length.
3. The number of successes in any interval is independent of the number of successes in any other interval provided the intervals are not overlapping.

In the McDonald's example, if we divide the time interval into a sufficiently small length (say, 1 second), it is impossible for more than one car to arrive. This would satisfy Part 1 of the definition. Part 2 of the definition is satisfied because the cars arrive at the constant rate of 2 cars per minute over the 1-hour interval. Part 3 is satisfied because the number of cars that arrive in any 1-minute interval (say between 12:23 P.M. and 12:24 P.M.) is independent of the number of cars that arrive in any other 1-minute interval (say between 12:35 P.M. and 12:36 P.M.).

2 Compute Probabilities of a Poisson Random Variable

If the random variable X follows a Poisson process, then we can use the following probability rule to compute Poisson probabilities.

---

*For example, the fixed interval might be any time between 0 and 5 minutes. A subinterval could be any time between 1 and 2 minutes.

In Other Words

Poisson probabilities are used to determine the probability of the number of successes in a fixed interval of time or space.

Poisson Probability Distribution Function

If X is the number of successes in an interval of fixed length t, the probability formula for X is

$$P(x) = \frac{(\lambda t)^x}{x!} e^{-\lambda t}, \quad x = 0, 1, 2, 3, \ldots \tag{1}$$

where λ (the Greek letter lambda) represents the average number of occurrences of the event in some interval of length 1 and $e \approx 2.71828$.

To clarify the roles of λ and t, let's revisit Example 1. In the McDonald's situation, $\lambda = 2$ cars per minute, while $t = 20$ minutes (the amount of time between 12:20 P.M. and 12:40 P.M.).

Now Work Problem 3.

EXAMPLE 2

Historical Note

Siméon Denis Poisson was born on June 21, 1781, in Pithiviers, France. He was educated by his father, who wanted him to become a doctor. However, one of his first patients died within a few hours of Poisson's treatment, so Poisson decided never to practice medicine again. After this mishap, Poisson enrolled in the Polytechnic school. While there, he drew the attention of Lagrange, Legendre, and Laplace. In 1837, Poisson published *Récherches sur la probabilité des jugements*, in which he first presented the probability distribution named after him.

Computing Probabilities of a Poisson Process

Problem: A McDonald's manager knows that cars arrive at the drive-through at the rate of 2 cars per minute between the hours of 12 noon and 1:00 P.M. She needs to determine and interpret the probability of the following events:

(a) Exactly 6 cars arrive between 12 noon and 12:05 P.M.
(b) Fewer than 6 cars arrive between 12 noon and 12:05 P.M.
(c) At least 6 cars arrive between 12 noon and 12:05 P.M.

Approach: The manager first needs a method that can be used to determine the probabilities. The cars arrive at a constant rate of 2 per minute over the time interval between 12:00 noon and 1:00 P.M. We know from Example 1 that the random variable X follows a Poisson process where $x = 0, 1, 2, \ldots$. The Poisson probability distribution function requires a value for λ and t. We are given that cars arrive at a rate of 2 per minute, so $\lambda = 2$. The interval of time we are interested in is 5 minutes, so $t = 5$.

Solution: We use the Poisson probability distribution function (1) to compute the probabilities.

(a) The probability that exactly six cars arrive between 12 noon and 12:05 P.M. is

$$P(6) = \frac{[2(5)]^6}{6!} e^{-2(5)} = \frac{1,000,000}{720}(0.0000454) \approx 0.0631$$

Interpretation: About 6 out of every 100 days, exactly 6 cars will arrive between 12:00 noon and 12:05 P.M.

(b) The probability that fewer than 6 cars arrive between 12:00 noon and 12:05 P.M. is

$$P(X < 6) = P(X \leq 5) = P(0) + P(1) + P(2) + P(3) + P(4) + P(5)$$

$$= \frac{[2(5)]^0}{0!} e^{-2(5)} + \frac{[2(5)]^1}{1!} e^{-2(5)} + \frac{[2(5)]^2}{2!} e^{-2(5)}$$

$$+ \frac{[2(5)]^3}{3!} e^{-2(5)} + \frac{[2(5)]^4}{4!} e^{-2(5)} + \frac{[2(5)]^5}{5!} e^{-2(5)}$$

$$= 0.00005 + 0.00045 + 0.00227 + 0.00757 + 0.01892 + 0.03783 = 0.0671$$

Interpretation: About 7 out of every 100 days, fewer than 6 cars will arrive between 12:00 noon and 12:05 P.M.

(c) The probability that at least 6 cars arrive between 12 noon and 12:05 P.M. is the complement of the probability that fewer than 6 cars arrive between 12 noon and 12:05 P.M. That is,

$$P(X \geq 6) = 1 - P(X < 6) = 1 - P(X \leq 5) = 1 - 0.0671 = 0.9329$$

Interpretation: About 93 out of every 100 days, at least six cars will arrive between 12:00 noon and 12:05 P.M.

Now Work Problem 11.

CAUTION
At-least probabilities must be computed using the Complement Rule for Poisson probabilities.

It is important to mention that at-least probabilities for a Poisson process must be found using the complement. This is because the random variable X in a Poisson process can be any integer greater than or equal to 0.

❸ Find the Mean and Standard Deviation of a Poisson Random Variable

If cars arrive at McDonald's at the rate of 2 per minute between 12:00 noon and 1:00 P.M., how many cars would you expect to arrive between the hours of 12:00 noon and 12:05 P.M.? Considering that 2 cars arrive every minute (on average) and we are observing the arrival of cars for 5 minutes, it might seem reasonable that we expect $2(5) = 10$ cars to arrive. Since the expected value of a random variable is the mean of the random variable, it seems reasonable that $\mu_X = \lambda t$ for interval t.

> **Mean and Standard Deviation of a Poisson Random Variable**
>
> A random variable X that follows a Poisson process with parameter λ has mean (or expected value) and standard deviation given by the formulas
>
> $$\mu_X = \lambda t \quad \text{and} \quad \sigma_X = \sqrt{\lambda t} = \sqrt{\mu_X}$$
>
> where t is the length of the interval.

Because $\mu_X = \lambda t$, we restate the Poisson probability distribution function in terms of its mean.

> **Poisson Probability Distribution Function**
>
> If X is the number of successes in an interval of fixed length and X follows a Poisson process with mean μ, the probability distribution function for X is
>
> $$P(x) = \frac{\mu^x}{x!}e^{-\mu}, \quad x = 0, 1, 2, 3 \ldots.$$

EXAMPLE 3   **Beetles and the Poisson Distribution**

Problem: A biologist performs an experiment in which 2000 Asian beetles are allowed to roam in an enclosed area of 1000 square feet. The area is divided into 200 subsections of 5 square feet each.

(a) Assuming that the beetles spread evenly throughout the enclosed area, how many beetles are expected to be within each subsection?
(b) What is the standard deviation of X, the number of beetles in a particular subsection?
(c) What is the probability of finding exactly 8 beetles in a particular subsection?
(d) Would it be unusual to find more than 16 beetles in a particular subsection?

Approach: If the beetles spread evenly throughout the enclosed region, we can model the distribution of the beetles using Poisson probabilities.

Solution:

(a) Assuming the beetles spread evenly throughout the enclosed area, we expect

$$\mu_X = \frac{2000 \text{ beetles}}{200 \text{ subsections}} = 10 \text{ beetles per subsection}$$

(b) Since $\mu_X = 10$, $\sigma_X = \sqrt{10}$.

(c) We use the expected value $\mu_X = 10$ in the Poisson probability distribution function to compute the probability of finding exactly 8 beetles in a subsection.

$$P(8) = \frac{10^8}{8!}e^{-10} \qquad P(x) = \frac{\mu^x}{x!}e^{-\mu}, \mu = 10, x = 8$$

$$\approx 0.1126$$

Figure 13

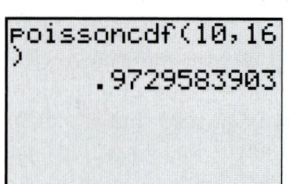

Interpretation: In 100 trials of this experiment, we expect to find 8 beetles in a particular subsection about 11 times.

(d) We compute $P(X > 16)$ using the Poisson probability distribution function, the Complement Rule, and a TI-84 Plus graphing calculator.

$$P(X > 16) = 1 - P(X \le 16)$$
$$= 1 - 0.9730 \qquad \text{See Figure 13.}$$
$$= 0.0270$$

Interpretation: According to the Poisson probability model, there will be more than 16 beetles in a subsection about 3 times in 100. To observe more than 16 beetles in a subsection is rather unusual.

6.3 ASSESS YOUR UNDERSTANDING

Concepts and Vocabulary

1. State the conditions required for a random variable X to follow a Poisson process.

2. Explain the role of λ and t in the Poisson probability formula.

Skill Building

In Problems 3–6, state the values of λ and t for each Poisson process.

3. The hits to a Web site occur at the rate of 10 per minute between 7:00 P.M. and 9:00 P.M. The random variable X is the number of hits to the Web site between 7:30 P.M. and 7:35 P.M.

4. The phone calls to a computer software help desk occur at the rate of 5 per minute between 3:00 P.M. and 4:00 P.M. The random variable X is the number of phone calls to the help desk between 3:10 P.M. and 3:20 P.M.

5. The flaws in a piece of timber occur at the rate of 0.07 per linear foot. The random variable X is the number of flaws in the next 20 linear feet of timber.

6. The potholes on a major highway in the city of Madison occur at the rate of 3.4 per mile. The random variable X is the number of potholes in 10 miles of randomly selected highway.

In Problems 7 and 8, the random variable X follows a Poisson process with the given mean.

7. Assuming $\mu = 5$, compute
   (a) $P(6)$
   (b) $P(X < 6)$
   (c) $P(X \ge 6)$
   (d) $P(2 \le X \le 4)$

8. Assuming $\mu = 7$, compute
   (a) $P(10)$
   (b) $P(X < 10)$
   (c) $P(X \ge 10)$
   (d) $P(7 \le X \le 9)$

In Problems 9 and 10, the random variable X follows a Poisson process with the given value of λ and t.

9. Assuming $\lambda = 0.07$ and $t = 10$, compute
   (a) $P(4)$
   (b) $P(X < 4)$
   (c) $P(X \ge 4)$
   (d) $P(4 \le X \le 6)$
   (e) μ_X and σ_X

10. Assuming $\lambda = 0.02$ and $t = 50$, compute
   (a) $P(2)$
   (b) $P(X < 2)$
   (c) $P(X \ge 2)$
   (d) $P(1 \le X \le 3)$
   (e) μ_X and σ_X

Applying the Concepts

11. **Hits to a Web Site** The number of hits to a Web site follows a Poisson process; hits occur at the rate of 1.4 per minute between 7:00 P.M. and 9:00 P.M. Compute the probability that the number of hits to the Web site between 7:30 P.M. and 7:35 P.M. is

(a) exactly seven. Interpret the result.
(b) fewer than seven. Interpret the result.
(c) at least seven. Interpret the result.

12. **Calls to the Help Desk** The phone calls to a computer software help desk occur at the rate of 2.1 per minute between 3:00 P.M. and 4:00 P.M. Compute the probability that the number of calls to the help desk between 3:10 P.M. and 3:15 P.M. is

(a) exactly eight. Interpret the result.
(b) fewer than eight. Interpret the result.
(c) at least eight. Interpret the result.

13. **Insect Fragments** The Food and Drug Administration sets a Food Defect Action Level (FDAL) for the various foreign substances that inevitably end up in the food we eat and liquids we drink. For example, the FDAL level for insect filth in peanut butter is 0.3 insect fragment (larvae, eggs, body parts, and so on) per gram. Suppose that a 5-gram sample of peanut butter contains 0.3 insect fragments per gram. Compute the probability that the number of insect fragments in a 5-gram sample of peanut butter is

(a) exactly two. Interpret the result.
(b) fewer than two. Interpret the result.
(c) at least two. Interpret the result.
(d) at least one. Interpret the result.
(e) Would it be unusual for a 5-gram sample of peanut butter to contain four or more insect fragments?

14. **Potholes** The potholes on a major highway in the city of Chicago occur at the rate of 3.4 per mile. Compute the probability that the number of potholes over 3 miles of randomly selected highway is

(a) exactly seven. Interpret the result.
(b) fewer than seven. Interpret the result.
(c) at least seven. Interpret the result.
(d) Would it be unusual for a randomly selected 3-mile stretch of highway in Chicago to contain more than 15 potholes?

15. **Airline Fatalities** According to the *Statistical Abstract of the United States*, airline fatalities occur at the rate of 0.04 death per 100 million miles (2002 data). Find the probability that, during the next 100 million miles of flight, there will be

(a) exactly zero deaths. Interpret the result.
(b) at least one death. Interpret the result.
(c) more than one death. Interpret the result.

16. **Traffic Fatalities** According to the *Statistical Abstract of the United States*, traffic fatalities occur at the rate of 1.5 deaths per 100 million miles (2002 data). Find the probability that, during the next 100 million vehicle miles, there will be

(a) exactly zero deaths. Interpret the result.
(b) at least one death. Interpret the result.

(c) more than one death. Interpret the result.
(d) Compare these results with the results in Problem 15. Given the choice, would you rather fly or drive?

17. **Florida Hurricanes** From 1900 to 2003 (104 years), Florida suffered 24 direct hits from major (category 3 to 5) hurricanes. Assuming this was typical and the number of hits per year follows a Poisson distribution, what is the probability Florida would be hit by at least three major hurricanes, as happened in 2004? Does this indicate that the 2004 hurricane season in Florida was unusual? (*Source:* National Hurricane Center)

18. **Police Dispatch** Officer Thompson of the Bay Ridge Police Department works the graveyard shift. He averages 4.5 calls per shift from his dispatcher. Assume the number of calls follows a Poisson distribution. Would it be unusual for Officer Thompson to get fewer than 2 calls in a shift?

19. **Wendy's Drive-Through** Cars arrive at Wendy's drive-through at a rate of 0.2 car per minute between the hours of 11:00 P.M. and 1:00 A.M. on Saturday evening. Wendy's begins an advertising blitz that touts its late-night service. After one week of advertising, Wendy's officials count the number of cars, X, arriving at Wendy's drive-through between the hours of 12:00 midnight and 12:30 A.M. at 200 of its restaurants. The results are shown in the following table:

x (number of cars arriving)	Frequency
1	4
2	5
3	13
4	23
5	25
6	28
7	25
8	27
9	21
10	15
11	5
12	3
13	2
14	2
15	0
16	2

(a) Construct a probability distribution for the random variable X, assuming it follows a Poisson process with $\lambda = 0.2$ and $t = 30$. This is the probability distribution of X before the advertising.
(b) Compute the expected number of restaurants that will have 0 arrivals, 1 arrival, and so on.
(c) Compare these results with the number of arrivals after the advertising. Does it appear the advertising was effective? Why?

20. **Quality Control** A builder ordered two hundred 8-foot grade A 2-by-4s for a construction job. To qualify as a grade A board, each 2-by-4 will have no knots and will average no more than 0.05 imperfection per linear foot. The following table lists the number of imperfections per 2-by-4 in the 200 ordered:

x (number of imperfections)	Frequency
0	124
1	51
2	20
3	5

(a) Construct a probability distribution for the random variable X, the number of imperfections per 8 feet of board, assuming it follows a Poisson process with $\lambda = 0.05$ and $t = 8$.

(b) Compute the expected number of 2-by-4s that will have 0 imperfections, 1 imperfection, and so on.

(c) Compare these results with the number of actual imperfections. Does it appear the 2-by-4s are of grade A quality? Why?

21. **Prussian Army** In 1898, Ladislaus von Bortkiewicz published *The Law of Small Numbers*, in which he demonstrated the power of the Poisson probability law. Before his publication, the law was used exclusively to approximate binomial probabilities. He demonstrated the law's power, using the number of Prussian cavalry soldiers who were kicked to death by their horses. The Prussian army monitored 10 cavalry corps for 20 years and recorded the number X of annual fatalities due to horse kicks for the 200 observations. The following table shows the data:

Number of Deaths, x	Number of Times x Deaths Were Observed
0	109
1	65
2	22
3	3
4	1

Adapted from *An Introduction to Mathematical Statistics* by Larsen et al. Prentice Hall, 2001

(a) Compute the proportion of years in which there were 0 deaths, 1 death, 2 deaths, 3 deaths, and 4 deaths.

(b) From the data in the table, what was the mean number of deaths per year?

(c) Use the mean number of deaths per year found in part (b) and the Poisson probability law to determine the theoretical proportion of years that 0 deaths should occur. Repeat this for 1, 2, 3, and 4 deaths.

(d) Compare the observed proportions to the theoretical proportions. Do you think the data can be modeled by the Poisson probability law?

22. **Simulation** Data from the National Center for Health Statistics show that, in 1995, spina bifida occurred at the rate of 28 per 100,000 live births. Let the random variable X represent the number of occurrences of spina bifida in a random sample of 100,000 live births.

(a) What is the expected number of children with spina bifida per 100,000 live births in any given year?

(b) Using statistical software such as MINITAB, simulate taking 200 random samples of 100,000 live births, assuming $\mu = 28$.

(c) Approximate the probability that fewer than 21 births per 100,000 result in spina bifida.

(d) In 1999, 20.83 births per 100,000 resulted in babies born with spina bifida. In light of the results of parts (b) and (c), is this an unusual occurrence? What might you conclude?

23. **Simulation** According to the National Center for Health Statistics, the common cold occurs at the rate of 23.8 colds per 100 people during the course of a year in the 18- to 24-year-old age group. Let the random variable X represent the number of 18- to 24-year-olds out of a sample of 500 who have had a common cold in the past year.

(a) What is the expected number of colds for every five hundred 18- to 24-year-olds?

(b) Using statistical software, such as MINITAB, simulate taking 100 random samples of size 500 from this population.

(c) Approximate the probability that at least 150 18- to 24-year-olds in a group of 500 will have experienced a common cold in the past year.

(d) Approximate the probability that fewer than 100 18- to 24-year-olds in a group of 500 will have experienced the common cold in the past year.

(e) Given the simulation, compute the mean number of 18- to 24-year-olds out of 500 who have experienced the common cold in the past year.

(f) Given the simulation, compute the standard deviation of the number of 18- to 24-year-olds out of 500 who have experienced the common cold in the past year.

(g) Compute the 5-number summary of the data. Did the simulation result in any unusual results?

24. **How Long Do I Have to Wait?** The number of hits to a Web site follows a Poisson process and occurs at the rate of 10 hits per minute between 7:00 P.M. and 9:00 P.M. How long should you expect to wait before the probability of at least one hit to the site is 95%?

Hint: $P(X \geq 1) = 1 - P(0)$

Consumer Reports | Quality Assurance in Customer Relations

The Customer Relations Department at Consumers Union (CU) receives thousands of letters and e-mails from customers each month. Some people write in asking how well a product performed during CU's testing, some people write in sharing their own experiences with their household products, and the remaining people write in for an array of other reasons.

To respond to each letter and e-mail that is received, Customer Relations recently upgraded its customer contact database. Although much of the process has been automated, it still requires employees to manually draft the responses. Given the current size of the department, each Customer Relations representative is required to draft approximately 300 responses each month.

As part of a quality assurance program, the Customer Relations manager would like to develop a plan that allows him to evaluate the performance of his employees. From past experience, he knows that the probability a new employee will write an initial draft of a response that contains errors is approximately 10%. The manager would like to know how many of the 300 responses he should sample to have a cost effective quality assurance program.

(a) Let X be a discrete random variable that represents the number of the $n = 300$ draft responses that contain errors. Describe the probability distribution for X. Be sure to include the name of the probability distribution, possible values for the random variable X, and values of the parameters.

(b) To be effective, suppose the manager would like to have a 95% probability of finding at least one draft document that contains an error. Assuming that the probability a draft document will have errors is known to be 10%, determine the appropriate sample size to satisfy the manager's requirements. *Hint*: We are required to find the number of draft documents that must be sampled so that the probability of finding at least one document containing an error is 95%. In other words, we have to determine n by solving: $P(x \geq 1) = 0.95$.

(c) Suppose the error rate is really 20%. What sample size will the manager have to review to have a 95% probability of finding one or more documents containing an error?

(d) Now let Y be a discrete random variable that represents the number of errors discovered in a single draft document. (It is possible for a single draft to contain more than one error.) The manager determined that errors occurred at the rate of 0.3 error per document. Describe the probability distribution for Y. Be sure to include the name of the probability distribution, possible values for the random variable Y, and values of the parameters.

(e) What is the probability that a document contains no errors? One error? At least two errors?

Note to Readers: In many cases, our test protocol and analytical methods are more complicated than described in these examples. The data and discussions have been modified to make the material more appropriate for the audience.

Technology Step-by-Step | **Computing Poisson Probabilities Via Technology**

TI-83/84 Plus | **Computing $P(x)$**

Step 1: Press 2nd VARS to access the probability distribution menu.

Step 2: Highlight B: poissonpdf (for the TI-83 and C: poissonpdf (for the TI-84 and hit ENTER.

Step 3: With poissonpdf (on the HOME screen, type the value of μ, followed by the number of successes, x. For example, with $\mu = 10$ and $x = 4$, type

$$\text{poissonpdf (10,4)}$$

Then hit ENTER.

Computing $P(X \leq x)$

Instead of selecting poissonpdf (, select poissoncdf (. Everything else is the same.

MINITAB | **Computing $P(x)$**

Step 1: Enter the desired values of the random variable X in C1.

Step 2: Select the **CALC** menu, highlight **Probability Distributions**, then highlight **Poisson** . . . .

Step 3: Select Probability, enter the mean, enter the input column as C1, and select C2 as the output column. Click OK.

Computing $P(X \le x)$

Follow the same steps as those followed for computing $P(x)$. In the window that comes up after selecting Poisson Distribution, select Cumulative probability instead of Probability.

Excel   ### Computing $P(x)$

Step 1: Enter the desired values of the random variable X in column A.

Step 2: With the cursor in cell B1, select the *fx* icon. Highlight Statistical in the Function category window. Highlight POISSON in the Function name window.

Step 3: In the cell labeled X, enter A1. In the cell labeled mean, enter the mean. In the cell labeled cumulative, type FALSE. Click OK.

Computing $P(X \le x)$

Follow the same steps as those presented for computing $P(x)$. In the POISSON window, type TRUE in the cumulative cell.

Chapter 6 # Review

Summary

In this chapter, we discussed discrete probability distributions. A random variable represents the numerical measurement of the outcome from a probability experiment. Discrete random variables have either a finite or a countable number of outcomes. The term *countable* means that the values result from counting. Probability distributions must satisfy the following two criteria: (1) All probabilities must be between 0 and 1, inclusive, and (2) the sum of all probabilities must equal 1. Discrete probability distributions can be presented in a table, graph, or mathematical formula.

The mean and standard deviation of a random variable describe the center and spread of the distribution. The mean of a random variable is also called its expected value.

We discussed two discrete probability distributions in particular, the binomial and Poisson. A probability experiment is considered a binomial experiment if there are n independent trials of the experiment with only two outcomes. The probability of success, p, is the same for each trial of the experiment. Special formulas exist for computing the mean and standard deviation of a binomial random variable.

The other probability distribution we discussed was the Poisson probability distribution. A Poisson process is one in which the following conditions are met: (1) The probability of two or more successes in any sufficiently small subinterval is 0. (2) The probability of success is the same for any two sufficiently short intervals of equal length. (3) Success or failure in any interval is independent of success or failure in any other interval. Special formulas exist for computing the mean and standard deviation of a random variable that follows a Poisson process. For the Poisson distribution, the mean and variance are equal.

Formulas

Mean (or Expected Value) of a Discrete Random Variable

$$\mu_X = E(X) = \sum x P(x)$$

Variance of a Discrete Random Variable

$$\sigma_X^2 = \sum (x - \mu_X)^2 \cdot P(x) = \sum [x^2 P(x)] - \mu_X^2$$

Binomial Probability Distribution Function

$$P(x) = {}_nC_x p^x (1 - p)^{n-x} \quad x = 0, 1, 2, \ldots$$

Mean of a Binomial Random Variable

$$\mu_X = np$$

Standard Deviation of a Binomial Random Variable

$$\sigma_X = \sqrt{np(1 - p)}$$

Poisson Probability Distribution Function

$$P(x) = \frac{(\lambda t)^x}{x!} e^{-\lambda t} = \frac{\mu^x}{x!} e^{-\mu} \quad x = 0, 1, 2, \ldots$$

Mean and Standard Deviation of a Poisson Random Variable

$$\mu_X = \lambda t \qquad \sigma_x = \sqrt{\lambda t}$$

Vocabulary

Random variable (p. 315)
Discrete random variable (p. 315)
Continuous random variable (p. 315)
Probability distribution (p. 316)

Probability histogram (p. 317)
Expected value (p. 321)
Trial (p. 328)
Binomial experiment (p. 328)

Binomial random variable (p. 328)
Poisson process (p. 345)

Objectives

Section	You should be able to . . .	Examples	Review Exercises
6.1	**1** Distinguish between discrete and continuous random variables (p. 315)	1	1, 2
	2 Identify discrete probability distributions (p. 316)	2 and 3	3, 4, 5(a), 6(a)
	3 Construct probability histograms (p. 317)	4	5(b), 6(b)
	4 Compute and interpret the mean of a discrete random variable (p. 318)	5, 6 and 9	5(c), 6(c), 15(b), 16(b)
	5 Interpret the mean of a discrete random variable as an expected value (p. 321)	7	7, 8
	6 Compute the variance and standard deviation of a discrete random variable (p. 322)	8 and 9	5(d), 6(d), 15(b), 16(b)
6.2	**1** Determine whether a probability experiment is a binomial experiment (p. 328)	1	9, 10
	2 Compute probabilities of binomial experiments (p. 330)	2 through 5	11(a)–(d), 12(a)–(d), 13(a)–(d), 14(a)–(d)
	3 Compute the mean and standard deviation of a binomial random variable (p. 335)	6	11(e), 12(e), 13(e), 14(e)
	4 Construct binomial probability histograms (p. 336)	7	15(d), 16(d)
6.3	**1** Understand when a probability experiment follows a Poisson process (p. 345)	1	24
	2 Compute probabilities of a Poisson random variable (p. 345)	2 and 3	17(a)–(d), 18(a)–(d), 19(a)–(c), 20–22
	3 Find the mean and standard deviation of a Poisson random variable (p. 347)	3	17(e), 18(e)

Review Exercises

In Problems 1 and 2, determine whether the random variable is discrete or continuous. In each case, state the possible values of the random variable.

1. (a) The number of inches of snow that falls in Buffalo during the winter season.
   (b) The number of days snow accumulates in Buffalo during the winter season.
   (c) The number of golf balls hit into the ocean on the famous 18th hole at Pebble Beach on a randomly selected Sunday.

2. (a) The miles per gallon of gasoline in a 2005 Toyota Sienna.
   (b) The number of children a randomly selected family has.
   (c) The number of goals scored by the Edmonton Oilers in a season.

In Problems 3 and 4, determine whether the distribution is a discrete probability distribution. If not, state why.

3. No,

x	$P(x)$
0	0.34
1	0.21
2	0.13
3	0.04
4	0.01

4.

x	$P(x)$
0	0.40
1	0.31
2	0.23
3	0.04
4	0.02

5. Stanley Cup   The Stanley Cup is a best-of-seven series to determine the champion of the National Hockey League. The following data represent the number of games played, X, in the Stanley Cup before a champion was determined from 1939 to 2004.

x	Frequency
4	20
5	16
6	17
7	13

Source: Information Please Almanac

   (a) Construct a probability model for the random variable X, the number of games in the Stanley Cup.
   (b) Draw a probability histogram.
   (c) Compute and interpret the mean of the random variable X.
   (d) Compute the standard deviation of the random variable X.

6. Property Crime   In 2003, 77% of crime was property crime, according to the National Crime Victimization Survey. Suppose that four crimes are randomly selected. Let the random variable X represent the number of property crimes.
   (a) Construct a probability model for the random variable X by constructing a tree diagram.
   (b) Draw a probability histogram.

(c) Compute and interpret the mean of the random variable X.

(d) Compute the standard deviation of the random variable X.

7. **Life Insurance** Suppose a life insurance company sells a $100,000 one-year term life insurance policy to a 35-year-old male for $200. According to the *National Vital Statistics Report*, Vol. 53, No. 6, the probability the male survives the year is 0.998592. Compute and interpret the expected value of this policy to the life insurance company.

8. **The Carnival** A carnival game is played as follows: You pay $2 to draw a card from an ordinary deck. If you draw an ace, you win $5. You win $3 for a face card and $10 for the seven of spades. If you pick anything else, you lose $2. On average, how much money can the operator expect to make per customer?

In Problems 9 and 10, determine which of the following probability experiments represents a binomial experiment. If the probability experiment is not a binomial experiment, state why.

9. According to the *Chronicle of Higher Education*, there is a 54% probability that a randomly selected incoming freshman will graduate from college within 6 years. Suppose 10 incoming male freshmen are randomly selected. After 6 years, each student is asked whether he or she graduated.

10. An experiment is conducted in which a single die is cast until a 3 comes up. The number of throws required is recorded.

11. **High Cholesterol** According to the National Center for Health Statistics, 8% of 20- to 34-year-old females have high serum cholesterol.

(a) In a random sample of 10 females 20- to 34 years old, find the probability that exactly 0 have high serum cholesterol. Interpret this result.

(b) In a random sample of 10 females 20- to 34 years old, find the probability that exactly 2 have high serum cholesterol. Interpret this result.

(c) In a random sample of 10 females 20- to 34 years old, find the probability that at least 2 have high serum cholesterol. Interpret this result.

(d) In a random sample of 10 females 20- to 34 years old, find the probability that exactly 9 will not have high serum cholesterol. Interpret this result.

(e) In a random sample of 250 females 20- to 34 years old, what is the expected number with high serum cholesterol? What is the standard deviation?

(f) If a random sample of 250 females 20- to 34 years old resulted in 12 of them having high serum cholesterol, would this be unusual? Why?

12. **Driving Age** According to a Gallup poll conducted December 17 to 19, 2004, 60% of U.S. women 18 years old or older stated that the minimum driving age should be 18 years or older.

(a) In a random sample of 15 U.S. women 18 years old or older, find the probability that exactly 10 believe the minimum driving age should be 18 years or older.

(b) In a random sample of 15 U.S. women 18 years old or older, find the probability that fewer than 5 believe the minimum driving age should be 18 years or older.

(c) In a random sample of 15 U.S. women 18 years old or older, find the probability that at least 5 believe the minimum driving age should be 18 years or older.

(d) In a random sample of 15 U.S. women 18 years old or older, find the probability that exactly 12 *do not* believe the minimum driving age should be 18 years or older.

(e) In a random sample of 200 U.S. women 18 years old or older, what is the expected number who believe the minimum driving age should be 18 years old or older? What is the standard deviation?

(f) If a random sample of 200 U.S. women 18 years old or older resulted in 110 who believed that the minimum driving age should be 18 years or older, would this be unusual? Why?

13. **Nielsen Ratings** Nielsen Media Research determines ratings for television programs by placing meters on 5000 televisions throughout the United States. The 2005 NCAA Basketball Championship broadcast resulted in a rating of 15.0, which means 15% of households were tuned in to the game.

(a) In a random sample of 20 households, find the probability that exactly 6 were tuned into the 2005 NCAA championship.

(b) In a random sample of 20 households, find the probability that fewer than 4 were tuned into the 2005 NCAA championship.

(c) In a random sample of 20 households, find the probability that at least 2 were tuned into the 2005 NCAA championship.

(d) In a random sample of 20 households, find the probability that exactly 17 were *not* tuned into the 2005 NCAA championship.

(e) In a random sample of 500 households, what is the expected number who were tuned in to the 2005 NCAA championship?

(f) If a random sample of 500 households resulted in 95 that were tuned into the game, would this be unusual? Why?

14. **Quit Smoking** The drug Zyban is meant to suppress the urge to smoke. In clinical trials, 35% of the study's participants experienced insomnia when taking 300 mg of Zyban per day.

(*Source:* GlaxoSmithKline)

(a) In a random sample of 25 users of Zyban, find the probability that exactly 8 will experience insomnia.

(b) In a random sample of 25 users of Zyban, find the probability that fewer than 4 will experience insomnia.

(c) In a random sample of 25 users of Zyban, find the probability that at least 5 will experience insomnia.

(d) In a random sample of 25 users of Zyban, find the probability that exactly 20 will not experience insomnia.

(e) In a random sample of 1000 users of Zyban, what is the expected number who experience insomnia? What is the standard deviation?

(f) If a random sample of 1000 users of Zyban results in 330 who experience insomnia, would this be unusual? Why?

In Problems 15 and 16, (a) construct a binomial probability distribution with the given parameters, (b) compute the mean and standard deviation of the random variable by using the methods of Section 6.1, (c) compute the mean and standard deviation by using the methods of Section 6.2, and (d) draw the probability histogram, comment on its shape, and label the mean on the histogram.

15. $n = 5$, $p = 0.2$

16. $n = 8$, $p = 0.75$

In Problems 17 and 18, the random variable X follows a Poisson process with the given value of λ and t.

17. Assuming that $\lambda = 0.05$ and $t = 8$, find
   (a) $P(3)$
   (b) $P(X < 3)$
   (c) $P(X \geq 3)$
   (d) $P(3 \leq X \leq 5)$
   (e) What are μ_X and σ_X?

18. Assuming that $\lambda = 0.15$ and $t = 10$, find
   (a) $P(2)$
   (b) $P(X < 2)$
   (c) $P(X \geq 2)$
   (d) $P(1 \leq X \leq 3)$
   (e) What are μ_X and σ_X?

19. Arrivals at the Bank   The number of cars that arrive at a bank's drive-through window between 3:00 P.M. and 6:00 P.M. on Friday follows a Poisson process at the rate of 0.41 car every minute. Compute the probability that the number of cars that arrive at the bank between 4:00 P.M. and 4:10 P.M. is
   (a) exactly four cars. Interpret this result.
   (b) fewer than four cars. Interpret this result.
   (c) at least four cars. Interpret this result.

20. Carpet Flaws   The mills at the Acme Carpet Company produce, on average, one flaw in every 500 yards of material they produce; the carpeting is sold in 100-yard rolls. If the number of flaws in a roll follows a Poisson distribution and the quality-control department rejects any roll with two or more flaws, what percent of the rolls is rejected?

21. Copier Maintenance   The student copy machine in the library requires a maintenance call an average of two times per month. Assuming the number of required maintenance calls follows a Poisson distribution, would it be unusual for the copy machine to require more than 4 maintenance calls in any given month?

22. Earthquakes   A great earthquake is an earthquake that measures 8.0 or higher on the Richter scale. From 1971 to 2004 (35 years), there have been 28 great earthquakes worldwide. The following data represent the observed number of years in which there were 0, 1, 2, 3, and 4 great earthquakes.

Number of Great Earthquakes in the Year, x	Number of Years in Which x Great Earthquakes Were Observed
0	17
1	11
2	5
3	1
4	1

Source: neic.usgs.gov

   (a) Compute the proportion of years in which there were 0, 1, 2, 3, and 4 great earthquakes.
   (b) What is the mean number of great earthquakes per year?
   (c) Use the mean number of great earthquakes per year found in part (b) and the Poisson probability law to determine the theoretical proportion of years that 0 great earthquakes should occur. Repeat this for 1, 2, 3, and 4 great earthquakes.
   (d) Compare the observed proportions to the theoretical proportions. Do you think the data can be modeled by the Poisson probability law?

23. State the condition required to use the Empirical Rule to check for unusual observations in a binomial experiment.

24. State the conditions required for a Poisson process.

25. In sampling without replacement, the assumption of independence required for a binomial experiment is violated. Under what circumstances can we sample without replacement and still use the binomial probability formula to approximate probabilities?

The Voyage of the St. Andrew

Throughout the picturesque valleys of mid-18th-century Germany echoed the song of the *Neuländer* (newlander). Their song enticed journeymen who struggled to feed their families with the dream and promise of colonial America. Traveling throughout the German countryside, the typical *Neuländer* sought to sign up several families from a village for immigration to a particular colony. By registering a group of neighbors, rather than isolated families, the agent increased the likelihood that his signees would not stray to the equally enticing proposals of a competitor. Additionally, by signing large groups, the *Neuländer* fattened his purse, to the tune of one to two florins a head.

Generally, the Germans who chose to undertake the hardship of a transAtlantic voyage were poor, yet the cost of such a voyage was high. Records from a 1753 voyage indicate that the cost of an adult fare (one freight) from Rotterdam to Boston was 7.5 pistole. Children between the ages of 4 and 13 were assessed at half the adult rate (one-half freight). Children under 4 were not charged. To get a sense of the expense involved, it has been estimated that the adult fare, 7.5 pistoles, is equivalent to approximately $2000! For a large family, the cost could easily be well beyond their means. Even though many immigrants did not have the necessary funds to purchase passage, they were determined to make the crossing. Years of indentured servitude for themselves and other family members were often the currency of last resort.

As a historian studying the influence of these German immigrants on colonial America, Hans Langenscheidt is interested in describing various demographic characteristics of these people. Unfortunately, accurate records are rare. In his research, he has discovered a partially reconstructed 1752 passenger list for a ship, the *St. Andrew*. This list contains the names of the head of families, a list of family members traveling with them, their parish of origin, and the number of freights each family purchased. Unfortunately, some of the data are missing for some of the families. Langenscheidt believes that the demographic parameters of this passenger list are likely to be similar to those of the other numerous voyages taken from Germany to America during the mid-eighteenth century. Assuming that he is correct, he believes that it is appropriate to create a discrete probability distribution for a number of demographic variables for this population of German immigrants. His distributions are presented next.

Probability Distribution of the Number of Families per Parish of German Immigrants on Board the 1752 *Voyage of the St. Andrew*	
Number of Families per Parish	**Probability**
1	0.706
2	0.176
3	0.000
4	0.059
5	0.000
6	0.059

Probability Distribution of the Known Number of Freights Purchased by the German Families on Board the 1752 *Voyage of the St. Andrew*	
Number of Freights	**Probability**
1.0	0.075
1.5	0.025
2.0	0.425
2.5	0.150
3.0	0.125
3.5	0.100
4.0	0.050
5.0	0.025
6.0	0.025

Probability Distribution of the Known Number of People in a Family for the Germans on Board the 1752 *Voyage of the St. Andrew*	
Number in Family	**Probability**
1	0.322
2	0.186
3	0.136
4	0.102
5	0.051
6	0.136
7	0.034
8	0.017
9	0.016

Source: Wilford W. Whitaker and Gary T. Horlacher, *Broad Bay Pioneers* (Rockport, Maine: Picton Press, 1998), 63–68. Distributions created from the partially reconstructed 1752 passenger list of the *St. Andrew* presented by Whitaker and Horlacher.

1. Using the information provided, describe, through histograms and numerical summaries such as the mean and standard deviation, each probability distribution.

2. Does it appear that, on average, the Neuländers were successful in signing more than one family from a parish? Does it seem likely that most of the families knew one another prior to undertaking the voyage? Explain your answers for both questions.

3. Using the mean number of freights purchased per family, estimate the average cost of the crossing for a family in pistoles and in U.S. dollars.

4. Is it appropriate to estimate the average cost of the voyage from the mean family size? Why or why not?

5. Langenscheidt came across a fragment of another ship's passenger list. This fragment listed information for six families. Of these six, five families purchased more than four freights. Using the information contained in the appropriate probability distribution for the *St. Andrew,* calculate the probability that at least five of six German immigrant families purchased more than four freights. Does it seem likely that these families came from a population similar to that of the Germans on board the *St. Andrew*? Explain.

6. Summarize your findings in a report. Discuss any assumptions made throughout this analysis. What are the consequences to your calculations and conclusions if your assumptions are subsequently determined to be invalid?

The Normal Probability Distribution

Outline

DECISIONS

You are interested in starting your own MENSA-type club. To qualify for the club, the potential member must have intelligence that is in the top 20% of all people. You must decide the baseline score that allows an individual to qualify. See the Decisions project on page 398.

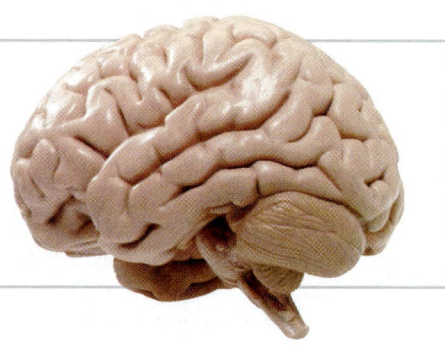

●●● Putting It All Together

In Chapter 6, we introduced discrete probability distributions and, in particular, the binomial and Poisson probability distributions. We computed probabilities for these discrete distributions using their probability distribution function.

However, we could also determine the probability of any discrete random variable from its probability histogram. For example, the figure shows the probability histogram for the binomial random variable X with $n = 5$ and $p = 0.35$.

From the probability histogram, we can see $P(1) \approx 0.31$. Notice that the width of each rectangle in the probability histogram is 1. Since the area of a rectangle equals height times width, we can think of $P(1)$ as the area of the rectangle corresponding to $X = 1$. Thinking of probability in this fashion makes the transition from computing discrete probabilities to continuous probabilities much easier.

In this chapter, we discuss two continuous distributions, *the uniform distribution* and *the normal distribution*. The greater part of the discussion will focus on the normal distribution, which has many uses and applications.

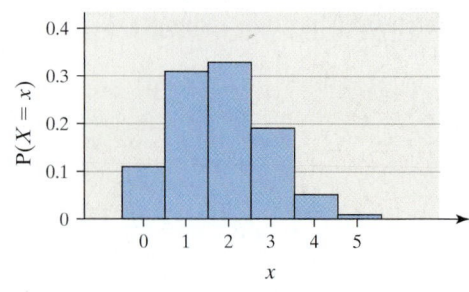

Binomial Probability Histogram;
$n = 5$, $p = 0.35$

7.1 Properties of the Normal Distribution

Preparing for This Section   Before getting started, review the following:

- Continuous variable (Section 1.1, p. 7)
- Rules for a discrete probability distribution (Section 6.1, p. 317)
- Z-score (Section 3.4, pp. 165–166)
- The Empirical Rule (Section 3.2, pp. 145–146)

Objectives

1. **Understand the uniform probability distribution**
2. **Graph a normal curve**
3. **State the properties of the normal curve**
4. **Understand the role of area in the normal density function**
5. **Understand the relation between a normal random variable and a standard normal random variable**

1 Understand the Uniform Probability Distribution

We illustrate a uniform distribution using an example. Using the uniform distribution makes it easy to see the relation between area and probability.

EXAMPLE 1   **Illustrating the Uniform Distribution**

Imagine that a friend of yours is always late. Let the random variable X represent the time from when you are supposed to meet your friend until he shows up. Further suppose that your friend could be on time ($x = 0$) or up to 30 minutes late ($x = 30$), with all 1-minute intervals of times between $x = 0$ and $x = 30$ equally likely. That is, your friend is just as likely to be from 3 to 4 minutes late as he is to be 25 to 26 minutes late. The random variable X can be any value in the interval from 0 to 30, that is, $0 \le x \le 30$. Because any two intervals of equal length between 0 and 30, inclusive, are equally likely, the random variable X is said to follow a **uniform probability distribution**.

When we compute probabilities for discrete random variables, we usually substitute the value of the random variable into a formula.

Things are not as easy for continuous random variables. Since there are an infinite number of possible outcomes for continuous random variables, it is reasonable that the probability of observing a particular value of a continuous random variable is zero. For example, the probability that your friend is exactly 12.9438823 minutes late is zero. This result is based on the fact that classical probability is found by dividing the number of ways an event can occur by the total number of possibilities. There is one way to observe 12.9438823, and there are an infinite number of possible values between 0 and 30, so we get a probability that is zero. To resolve this problem, we compute probabilities of continuous random variables over an interval of values. For example, we might compute the probability that your friend is between 10 and 15 minutes late. To find probabilities for continuous random variables, we use *probability density functions*.

Definition

A **probability density function** is an equation used to compute probabilities of continuous random variables that must satisfy the following two properties.

1. The total area under the graph of the equation over all possible values of the random variable must equal 1.

2. The height of the graph of the equation must be greater than or equal to 0 for all possible values of the random variable. That is, the graph of the equation must lie on or above the horizontal axis for all possible values of the random variable.

In Other Words
To find probabilities for continuous random variables, we do not use probability distribution functions (as we did for discrete random variables). Instead, we use probability density functions. The word *density* is used because it refers to the number of individuals per unit of area.

Property 1 is similar to the rule for discrete probability distributions that stated the sum of the probabilities must add up to 1. Property 2 is similar to the rule that stated all probabilities must be greater than or equal to 0.

Figure 1 illustrates the properties for the example about your friend who is always late. Since all possible values of the random variable between 0 and 30 are equally likely, the graph of the probability density function for uniform random variables is a rectangle. Because the random variable is any number between 0 and 30 inclusive, the width of the rectangle is 30. Since the area under the graph of the probability density function must equal 1 and the area of a rectangle equals height times width, the height of the rectangle must be $\frac{1}{30}$.

Figure 1
Uniform Density Function

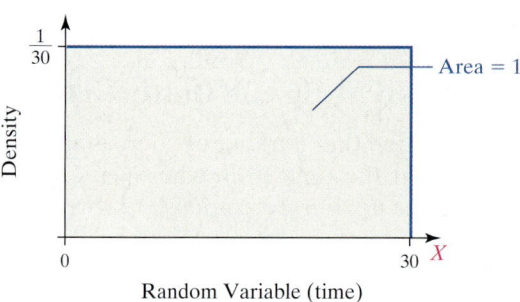

Random Variable (time)

A pressing question remains: How do we use density functions to find probabilities of continuous random variables?

The area under the graph of a density function over some interval represents the probability of observing a value of the random variable in that interval.

The following example illustrates this statement.

EXAMPLE 2 **Area as a Probability**

Problem: Refer to the situation presented in Example 1. What is the probability that your friend will be between 10 and 20 minutes late the next time you meet him?

Approach: Figure 1 presented the graph of the density function. We need to find the area under the graph between 10 and 20 minutes.

Solution: Figure 2 presents the graph of the density function with the area we wish to find shaded in green.

Figure 2

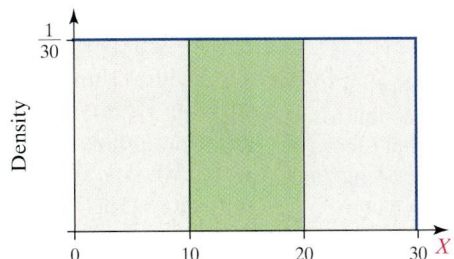

The width of the rectangle is 10 and its height is $\dfrac{1}{30}$. Therefore, the area between 10 and 20 is $10\left(\dfrac{1}{30}\right) = \dfrac{1}{3}$. The probability that your friend is between 10 and 20 minutes late is $\dfrac{1}{3}$.

Now Work Problem 13.

We introduced the uniform density function so that we could associate probability with area. We are now better prepared to discuss the most popular continuous distribution, the normal distribution.

② Graph a Normal Curve

Many continuous random variables, such as IQ scores, birth weights of babies, or weights of M&Ms, have relative frequency histograms that have a shape similar to Figure 3. Relative frequency histograms that have a shape similar to Figure 3 are said to have the shape of a **normal curve**.

Figure 3

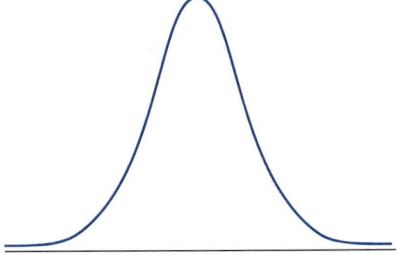

Definition   A continuous random variable is **normally distributed** or has a **normal probability distribution** if its relative frequency histogram of the random variable has the shape of a normal curve.

Figure 4 shows a normal curve, demonstrating the role μ and σ play in drawing the curve. Look back at Figure 5 on page 127 in Section 3.1. For any distribution, the mode represents the "high point" of the graph of the distribution. The median represents the point where 50% of the area under the distribution is to the left and 50% of the area under the distribution is to the right. The mean represents the balancing point of the graph of the distribution (See Figure 2 on page 123 in Section 3.1). For symmetric distributions, such as the normal distribution, the mean = median = mode. Because of this, the mean, μ, is the "high point" of the graph of the distribution.

The points at $x = \mu - \sigma$ and $x = \mu + \sigma$ are the *inflection points* on the normal curve. The **inflection points** are the points on the curve where the curvature

Figure 4

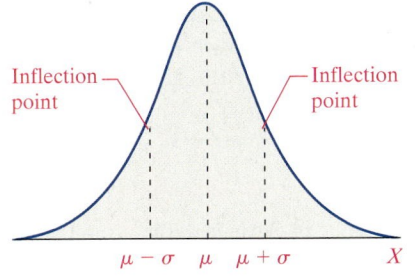

Inflection point — Inflection point

$\mu - \sigma \quad \mu \quad \mu + \sigma \qquad X$

Historical Note

Karl Pearson coined the phrase *normal curve*. He did not do this to imply that a distribution that is not normal is *abnormal*. Rather, Pearson wanted to avoid giving the name of the distribution a proper name, such as Gaussian (as in Carl Friedrich Gauss).

of the graph changes. To the left of $x = \mu - \sigma$ and to the right of $x = \mu + \sigma$, the curve is drawn upward ($\smile$ or $\diagdown$). In between $x = \mu - \sigma$ and $x = \mu + \sigma$, the curve is drawn downward ($\frown$).*

Figure 5 shows how changes in μ and σ change the position or shape of a normal curve. In Figure 5(a), two normal density curves are drawn with the location of the inflection points labeled. One density curve has $\mu = 0, \sigma = 1$, and the other has $\mu = 3, \sigma = 1$. We can see that increasing the mean from 0 to 3 caused the graph to shift three units to the right. In Figure 5(b), two normal density curves are drawn, again with the inflection points labeled. One density curve has $\mu = 0, \sigma = 1$, and the other has $\mu = 0, \sigma = 2$. We can see that increasing the standard deviation from 1 to 2 causes the graph to become flatter and more spread out.

Figure 5

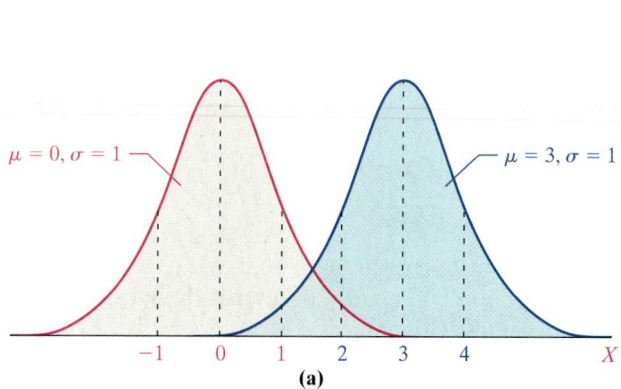

(a)

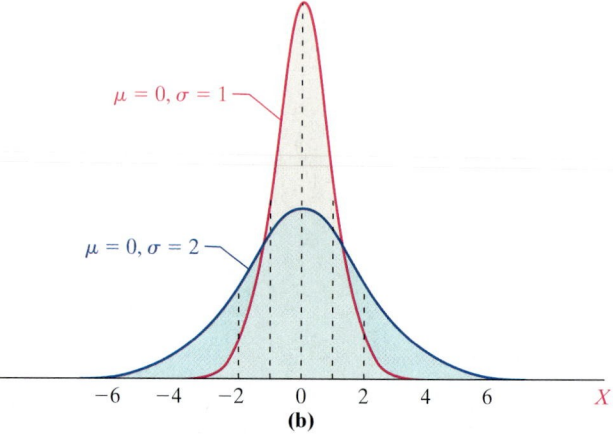

(b)

Now Work Problem 25.

③ State the Properties of the Normal Curve

The normal probability density function satisfies all the requirements that are necessary to have a probability distribution. We list the properties of the normal density curve next.

Historical Note

Abraham de Moivre was born in France on May 26, 1667. He is known as a great contributor to the areas of probability and trigonometry. In 1685, he moved to England. De Moivre was elected a fellow of the Royal Society in 1697. He was part of the commission to settle the dispute between Newton and Leibniz regarding who was the discoverer of calculus. He published *The Doctrine of Chance* in 1718. In 1733, he developed the equation that describes the normal curve. Unfortunately, de Moivre had a difficult time being accepted in English society (perhaps due to his accent) and was able to make only a meager living tutoring mathematics. An interesting piece of information regarding de Moivre; he correctly predicted the day of his death, Nov. 27, 1754.

Properties of the Normal Density Curve

1. It is symmetric about its mean, μ.
2. Because mean = median = mode, the highest point occurs at $x = \mu$.
3. It has inflection points at $\mu - \sigma$ and $\mu + \sigma$.
4. The area under the curve is 1.
5. The area under the curve to the right of μ equals the area under the curve to the left of μ, which equals $1/2$.
6. As x increases, without bound (gets larger and larger), the graph approaches, but never reaches, the horizontal axis. As x decreases without bound (gets larger and larger in the negative direction) the graph approaches, but never reaches, the horizontal axis.
7. The Empirical Rule: Approximately 68% of the area under the normal curve is between $x = \mu - \sigma$ and $x = \mu + \sigma$. Approximately 95% of the area under the normal curve is between $x = \mu - 2\sigma$ and $x = \mu + 2\sigma$. Approximately 99.7% of the area under the normal curve is between $x = \mu - 3\sigma$ and $x = \mu + 3\sigma$. See Figure 6.

*The vertical scale on the graph, which indicates density, is purposely omitted. The vertical scale, while important, will not play a role in any of the computations using this curve.

Figure 6

Normal Distribution

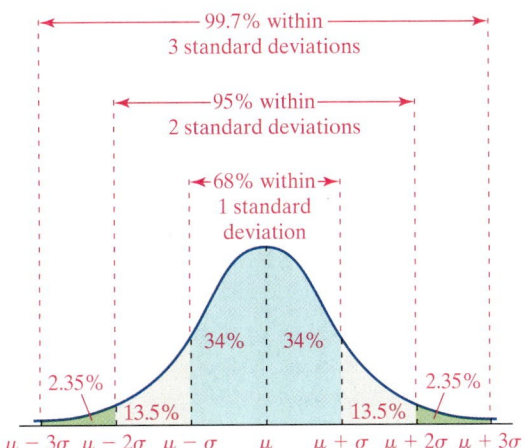

4 **Understand the Role of Area in the Normal Density Function**

Let's look at an example of a normally distributed random variable.

EXAMPLE 3 A Normal Random Variable

Problem: The relative frequency distribution given in Table 1 represents the heights of a pediatrician's 200 three-year-old female patients. The raw data indicate that the mean height of the patients is $\mu = 38.72$ inches with standard deviation $\sigma = 3.17$ inches.

(a) Draw a relative frequency histogram of the data. Comment on the shape of the distribution.

(b) Draw a normal curve with $\mu = 38.72$ inches and $\sigma = 3.17$ inches on the relative frequency histogram. Compare the area of the rectangle for heights between 40 and 40.9 inches to the area under the normal curve for heights between 40 and 40.9 inches.

Approach

(a) Draw the relative frequency histogram. If the histogram looks like Figure 4, we say that height is approximately normal. We say "approximately normal," rather than "normal," because the normal curve is an "idealized" description of the data and data rarely follows the curve exactly.

(b) Draw the normal curve on the histogram with the high point at μ and the inflection points at $\mu - \sigma$ and $\mu + \sigma$. Shade the rectangle corresponding to heights between 40 and 40.9 inches, and compare the area of the shaded region to the area under the normal curve between 40 and 40.9.

Solution

(a) Figure 7 shows the relative frequency distribution. The relative frequency histogram is symmetric and bell shaped.

Figure 7 Height of Three Year Old Females

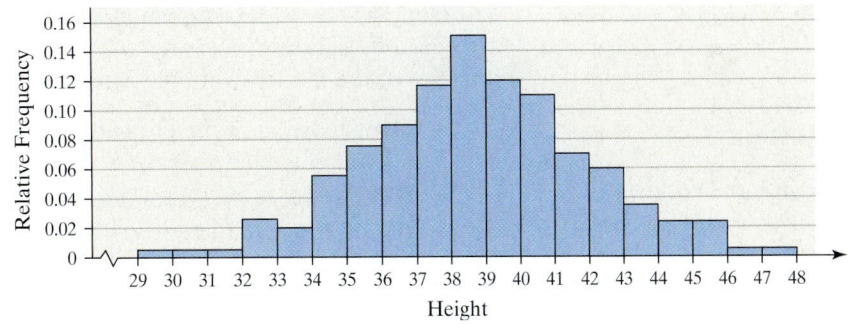

Table 1

Height (Inches)	Relative Frequency
29.0–29.9	0.005
30.0–30.9	0.005
31.0–31.9	0.005
32.0–32.9	0.025
33.0–33.9	0.02
34.0–34.9	0.055
35.0–35.9	0.075
36.0–36.9	0.09
37.0–37.9	0.115
38.0–38.9	0.15
39.0–39.9	0.12
40.0–40.9	0.11
41.0–41.9	0.07
42.0–42.9	0.06
43.0–43.9	0.035
44.0–44.9	0.025
45.0–45.9	0.025
46.0–46.9	0.005
47.0–47.9	0.005

CAUTION

It is rare for a continuous random variable to be exactly normal. Therefore, we usually say that a random variable is approximately normal if its histogram is bell shaped and symmetric.

(b) The normal curve with $\mu = 38.72$ and $\sigma = 3.17$ is superimposed on the relative frequency histogram in Figure 8. The figure demonstrates that the normal curve describes the heights of 3-year-old girls fairly well. We conclude that the heights of 3-year-old girls are approximately normal with $\mu = 38.72$ and $\sigma = 3.17$.

Figure 8 also shows the rectangle corresponding to heights between 40 and 40.9 inches. The area of this rectangle represents the proportion of 3-year-old females between 40 and 40.9 inches. Notice that the area of this shaded region is very close to the area under the normal curve for the same region, so we can use the area under the normal curve to approximate the proportion of 3-year-old females with heights between 40 and 40.9 inches!

Figure 8
Heights of 3-Year-Old Female Patients

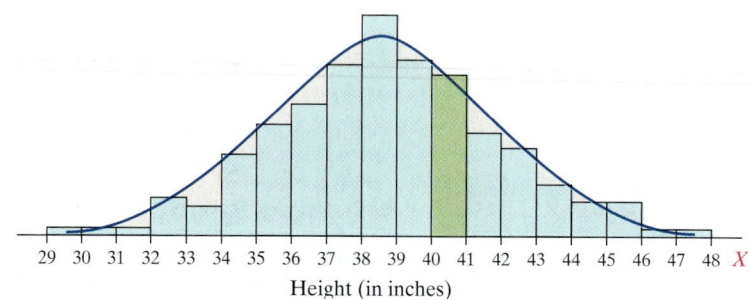

Height (in inches)

The normal curve drawn in Figure 8 is a *model*. In mathematics, a **model** is an equation, table, or graph that is used to describe reality. The normal distribution or normal curve is a model that is used to describe variables that are approximately normally distributed. For example, we saw in Example 3(b) that the normal curve drawn in Figure 8 does a good job of describing the observed distribution of heights of 3-year-old females.

The equation (model) that is used to determine the probability of continuous random variable is called a **probability density function** (or **pdf**). The **normal probability density function** is given by

$$y = \frac{1}{\sqrt{2\pi}} e^{\frac{-(x-\mu)^2}{2\sigma^2}}$$

where μ is the mean and σ is the standard deviation of the normal random variable. This equation represents the normal distribution. Don't feel threatened by this equation because we will not be using it in this course. Instead, we will use the normal distribution in graphical form by drawing the normal curve, as we did in Figure 5.

We now summarize the role area plays in the normal curve.

The Area under a Normal Curve

Suppose a random variable X is normally distributed with mean μ and standard deviation σ. The area under the normal curve for any interval of values of the random variable X represents either

- the proportion of the population with the characteristic described by the interval of values or

- the probability that a randomly selected individual from the population will have the characteristic described by the interval of values.

Historical Note

The normal probability distribution is often referred to as the Gaussian distribution in honor of Carl Gauss, the individual thought to have discovered the idea. However, it was actually Abraham de Moivre who first wrote down the equation of the normal distribution. Gauss was born in Brunswick, Germany, on April 30, 1777. Gauss, mathematical prowess was evident early in his life. At age 8 he was able to instantly add the first 100 integers. In 1799, Gauss earned his doctorate. The subject of his dissertation was the Fundamental Theorem of Algebra. In 1809, Gauss published a book on the mathematics of planetary orbits. In this book, he further developed the theory of least-squares regression by analyzing the errors. The analysis of these errors led to the discovery that errors follow a normal distribution. Gauss was considered to be "glacially cold" as a person and had troubled relationships with his family. Gauss died February 23, 1855.

EXAMPLE 4

Interpreting the Area under a Normal Curve

Problem: The serum total cholesterol for males 20 to 29 years old is known to be approximately normally distributed with mean $\mu = 180$ and $\sigma = 36.2$ based on data obtained from the National Health and Nutrition Examination Survey.

(a) Draw a normal curve with the parameters labeled.

(b) An individual with total cholesterol greater than 200 is considered to have high cholesterol. Shade the region under the normal curve to the right of $X = 200$.

(c) Suppose the area under the normal curve to the right of $X = 200$ is 0.2903. (You will learn how to find this area in Section 7.3.) Provide two interpretations of this result.

Approach: **(a)** Draw the normal curve with the mean $\mu = 180$ labeled at the high point and the inflection points at $\mu - \sigma = 180 - 36.2 = 143.8$ and $\mu + \sigma = 180 + 36.2 = 216.2$.

(b) Shade the region under the normal curve to the right of $X = 200$.

(c) The two interpretations of this shaded region are (1) the proportion of 20- to 29-year-old males who have high cholesterol and (2) the probability that a randomly selected 20- to 29-year-old male has high cholesterol.

Solution

(a) Figure 9 (a) shows the graph of the normal curve.

Figure 9

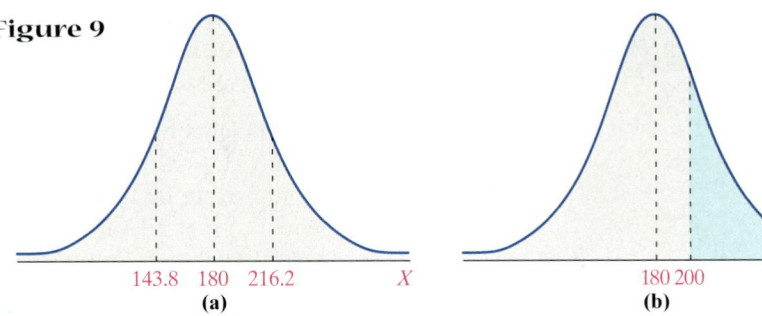

143.8 180 216.2   X   180 200   X
        (a)   **(b)**

(b) Figure 9 (b) shows the region under the normal curve to the right of $X = 200$ shaded.

(c) The two interpretations for the area of this shaded region are (1) the proportion of 20- to 29-year-old males that have high cholesterol is 0.2903 and (2) the probability that a randomly selected 20- to 29-year-old male has high cholesterol is 0.2903.

Now Work Problems 29 and 33.

5 ## Understand the Relation between a Normal Random Variable and a Standard Normal Random Variable

At this point, we know that a random variable X is approximately normally distributed if its relative frequency histogram has the shape of a normal curve. We use a normal random variable with mean μ and standard deviation σ to model the distribution of X. The area below the normal curve (model of X) represents the proportion of the population with a given characteristic or the probability that a randomly selected individual from the population will have a given characteristic.

The question now becomes, "How do I find the area under the normal curve?" Finding the area under a curve requires techniques introduced in calculus, which are beyond the scope of this text. An alternative would be to use a series of tables to find areas. However, this would result in an infinite number of tables being created for each possible mean and standard deviation!

A solution to the problem lies in the Z-score. Recall that the Z-score allows us to transform a random variable X with mean μ and standard deviation σ into a random variable Z with mean 0 and standard deviation 1.

In Other Words

The term "normal" refers to the shape of the distribution of a normal random variable.

Standardizing a Normal Random Variable

Suppose the random variable X is normally distributed with mean μ and standard deviation σ. Then the random variable

$$Z = \frac{X - \mu}{\sigma}$$

is normally distributed with mean $\mu = 0$ and standard deviation $\sigma = 1$. The random variable Z is said to have the **standard normal distribution**.

In Other Words

To find the area under any normal curve, we first find the Z-score of the normal random variable. Then use a table to find the area.

This result is powerful! We need only one table of areas corresponding to the standard normal distribution. If a normal random variable has mean different from 0 or standard deviation different from 1, we transform the normal random variable into a standard normal random variable Z, and then we use a table to find the area and, therefore, the probability.

We demonstrate the idea behind standardizing a normal random variable in the next example.

EXAMPLE 5   **Relation between a Normal Random Variable and a Standard Normal Random Variable**

Problem: The heights of a pediatrician's 200 three-year-old female patients are approximately normal with mean $\mu = 38.72$ inches and $\sigma = 3.17$ inches. We wish to demonstrate that the area under the normal curve between 35 and 38 inches is equal to the area under the standard normal curve between the Z-scores corresponding to heights of 35 and 38 inches.

Approach

Step 1: Draw a normal curve and shade the area representing the proportion of 3-year-old females between 35 and 38 inches tall.

Step 2: Standardize the random variable $X = 35$ and $X = 38$ using

$$Z = \frac{X - \mu}{\sigma}.$$

Step 3: Draw the standard normal curve with the standardized versions of $X = 35$ and $X = 38$ labeled. Shade the area that represents the proportion of 3-year-old females between 35 and 38 inches tall. Comment on the relation between the two shaded regions.

Solution

Step 1: Figure 10(a) shows the normal curve with mean $\mu = 38.72$ and $\sigma = 3.17$. The region between $X = 35$ and $X = 38$ is shaded.

Step 2: With $\mu = 38.72$ and $\sigma = 3.17$, the standardized version of $X = 35$ is

$$Z = \frac{X - \mu}{\sigma} = \frac{35 - 38.72}{3.17} = -1.17$$

The standardized version of $X = 38$ is

CAUTION

Recall that we round Z-scores to two decimal places.

$$Z = \frac{X - \mu}{\sigma} = \frac{38 - 38.72}{3.17} = -0.23$$

Step 3: Figure 10(b) shows the standard normal curve with the region between $Z = -1.17$ and $Z = -0.23$ shaded.

Figure 10

(a) Normal curve with $\mu = 38.72$ and
  $\sigma = 3.17$
(b) Standard normal curve

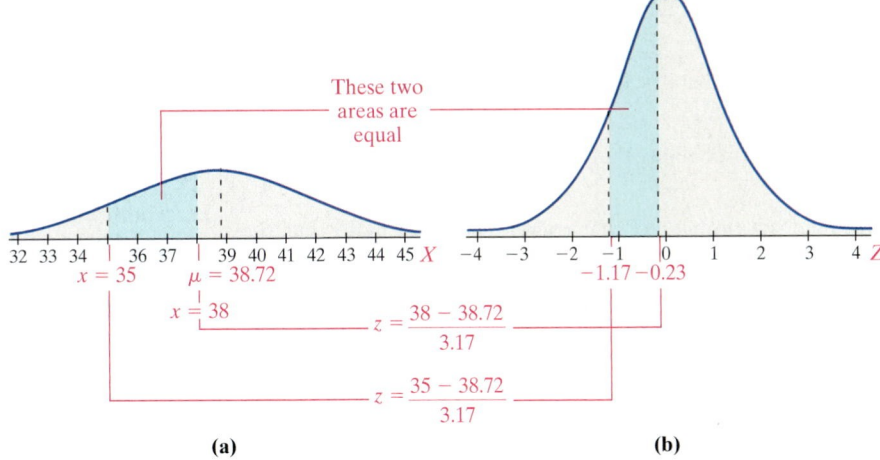

(a)   (b)

The area under the normal curve with $\mu = 38.72$ inches and $\sigma = 3.17$ inches bounded to the left by $X = 35$ and bounded to the right by $X = 38$ is equal to the area under the standard normal curve bounded to the left by $Z = -1.17$ and bounded to the right by $Z = -0.23$.

Now Work Problem 35.

7.1 ASSESS YOUR UNDERSTANDING

Concepts and Vocabulary

1. State the two characteristics of the graph of a probability density function.
2. To find the probabilities for continuous random variables, we do not use probability _____ functions, but instead we use probability _____ functions.
3. Provide two interpretations of the area under the graph of a probability density function.
4. Why do we standardize normal random variables to find the area under any normal curve?
5. The points at $x =$ _____ and $x =$ _____ are the inflection points on the normal curve.
6. As σ increases, the normal density curve becomes more spread out. Knowing the area under the density curve must be 1, what effect does increasing σ have on the height of the curve?

For Problems 7–12, determine whether the graph can represent a normal density function. If it cannot, explain why.

7.

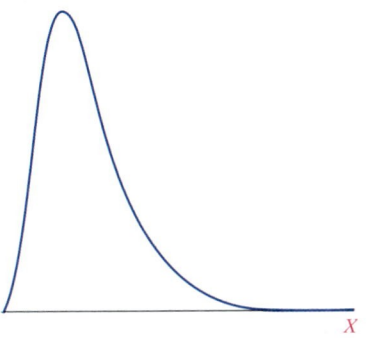

8.

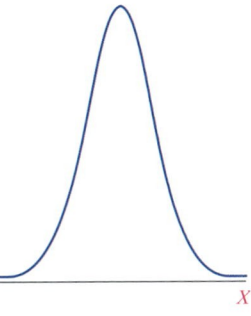

9.

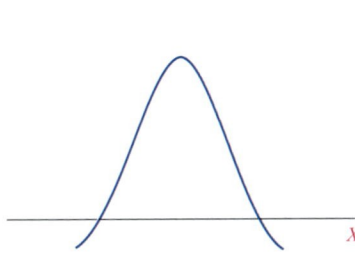

10.

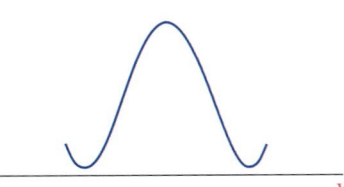

11.

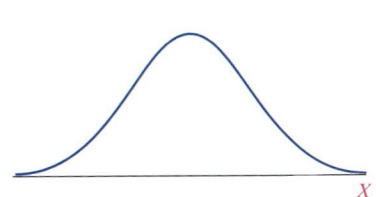

12.

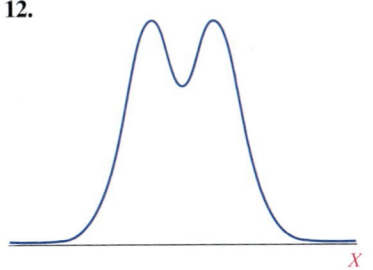

Skill Building

Problems 13–16 use the information presented in Examples 1 and 2.

13. Find the probability that your friend is between 5 and 10 minutes late.
NW

14. Find the probability that your friend is between 15 and 25 minutes late.

15. Find the probability that your friend is at least 20 minutes late.

16. Find the probability that your friend is no more than 5 minutes late.

17. **Uniform Distribution** The random-number generator on calculators randomly generates a number between 0 and 1. The random variable X, the number generated, follows a uniform probability distribution.

(a) Draw the graph of the uniform density function.
(b) What is the probability of generating a number between 0 and 0.2?

(c) What is the probability of generating a number between 0.25 and 0.6?
(d) What is the probability of generating a number greater than 0.95?
(e) Use your calculator or statistical software to randomly generate 200 numbers between 0 and 1. What proportion of the numbers are between 0 and 0.2? Compare the result with part (b).

18. **Uniform Distribution** Suppose the reaction time X (in minutes) of a certain chemical process follows a uniform probability distribution with $5 \le X \le 10$.

(a) Draw the graph of the density curve.
(b) What is the probability that the reaction time is between 6 and 8 minutes?
(c) What is the probability that the reaction time is between 5 and 8 minutes?
(d) What is the probability that the reaction time is less than 6 minutes?

In Problems 19–22, determine whether or not the histogram indicates that a normal distribution could be used as a model for the variable.

19. **Birth Weights** The following relative frequency histogram represents the birth weights (in grams) of babies whose term was 36 weeks.

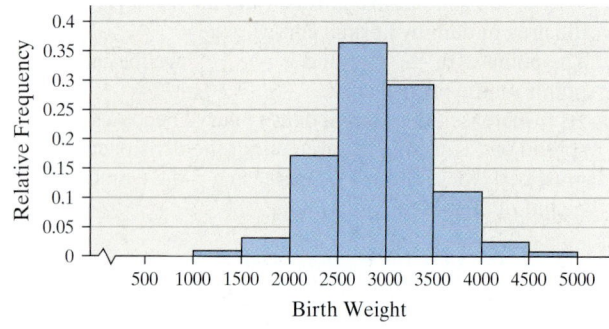

20. **Waiting in Line** The following relative frequency histogram represents the waiting time in line (in minutes) for the Demon Roller Coaster for 2000 randomly selected people on a Saturday afternoon in the summer.

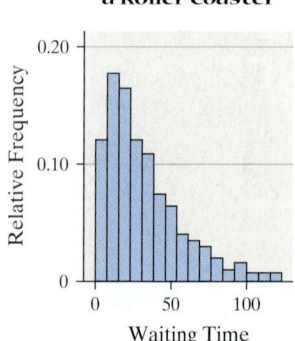

21. **Length of Phone Calls** The following relative frequency histogram represents the length of phone calls on my wife's cell phone during the month of September.

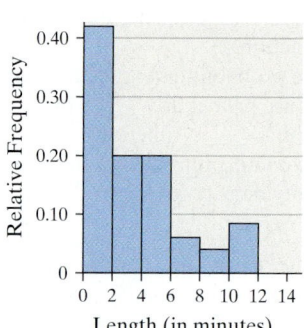

22. **Incubation Times** The following relative frequency histogram represents the incubation times of a random sample of Rhode Island Red Hens' eggs.

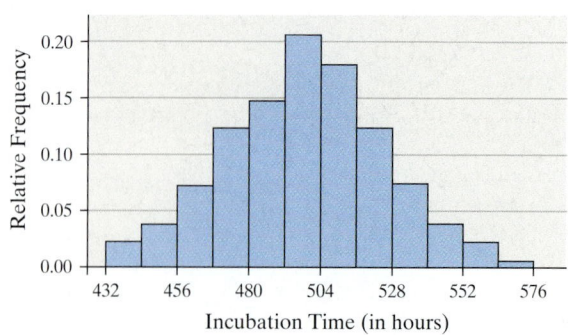

23. One graph in the following figure represents a normal distribution with mean $\mu = 10$ and standard deviation $\sigma = 3$. The other graph represents a normal distribution with mean $\mu = 10$ and standard deviation $\sigma = 2$. Determine which graph is which and explain how you know.

24. One graph in the following figure represents a normal distribution with mean $\mu = 8$ and standard deviation $\sigma = 2$. The other graph represents a normal distribution with mean $\mu = 14$ and standard deviation $\sigma = 2$. Determine which graph is which and explain how you know.

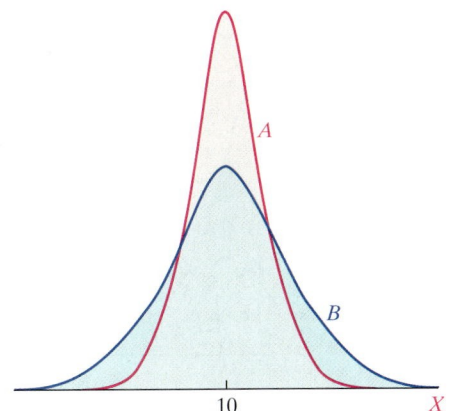

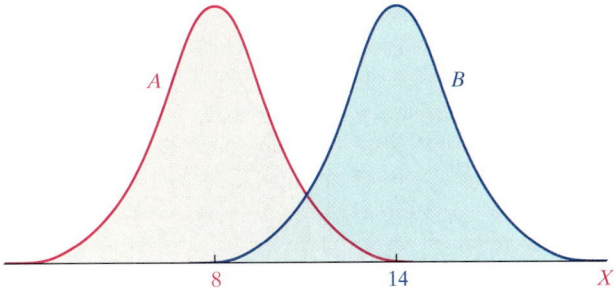

In Problems 25–28, the graph of a normal curve is given. Use the graph to identify the value of μ and σ.

25.
NW

26.

27.

28.

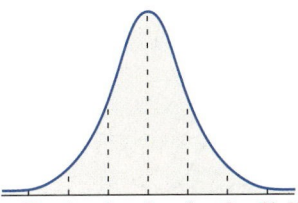

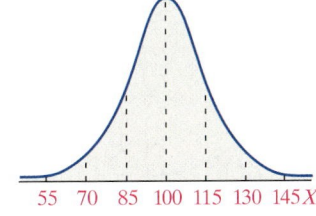

Applying the Concepts

29. Cell Phone Rates Suppose the monthly charge for cell phone plans in the United States is normally distributed with mean $\mu = \$62$ and standard deviation $\sigma = \$18$. (Source: based on information obtained from *Consumer Reports*)

(a) Draw a normal curve with the parameters labeled.
(b) Shade the region that represents the proportion of plans that charge less than $44.
(c) Suppose the area under the normal curve to the left of $X = \$44$ is 0.1587. Provide two interpretations of this result.

30. Refrigerators Suppose the life of refrigerators is normally distributed with mean $\mu = 14$ years and standard deviation $\sigma = 2.5$ years. (Source: based on information obtained from *Consumer Reports*)

(a) Draw a normal curve with the parameters labeled.
(b) Shade the region that represents the proportion of refrigerators that are kept for more than 17 years.
(c) Suppose the area under the normal curve to the right of $X = 17$ is 0.1151. Provide two interpretations of this result.

31. Birth Weights The birth weights of full-term babies are normally distributed with mean $\mu = 3400$ grams and $\sigma = 505$ grams. (Source: based on data obtained from the *National Vital Statistics Report*, Vol. 48, No. 3).

(a) Draw a normal curve with the parameters labeled.
(b) Shade the region that represents the proportion of full-term babies who weigh more than 4410 grams.
(c) Suppose the area under the normal curve to the right of $X = 4410$ is 0.0228. Provide two interpretations of this result.

32. Height of 10-Year-Old Males The heights of 10-year-old males are normally distributed with mean $\mu = 55.9$ inches and $\sigma = 5.7$ inches.

(a) Draw a normal curve with the parameters labeled.
(b) Shade the region that represents the proportion of 10-year-old males who are less than 46.5 inches tall.
(c) Suppose the area under the normal curve to the left of $X = 46.5$ is 0.0496. Provide two interpretations of this result.

33. Gestation Period The lengths of human pregnancy are **NW** normally distributed with $\mu = 266$ days and $\sigma = 16$ days.

(a) The following figure represents the normal curve with $\mu = 266$ days and $\sigma = 16$ days.

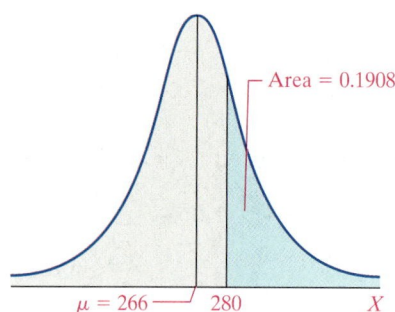

The area to the right of $X = 280$ is 0.1908. Provide two interpretations of this area.

(b) The following figure represents the normal curve with $\mu = 266$ days and $\sigma = 16$ days.

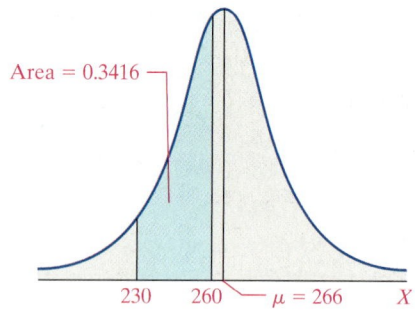

The area between $X = 230$ and $X = 260$ is 0.3416. Provide two interpretations of this area.

34. Miles per Gallon Elena conducts an experiment in which she fills up the gas tank on her Toyota Camry 40 times and records the miles per gallon for each fill-up. A histogram of the miles per gallon indicates that a normal distribution with mean of 24.6 miles per gallon and a standard deviation of 3.2 miles per gallon could be used to model the gas mileage for her car.

(a) The following figure represents the normal curve with $\mu = 24.6$ miles per gallon and $\sigma = 3.2$ miles per gallon.

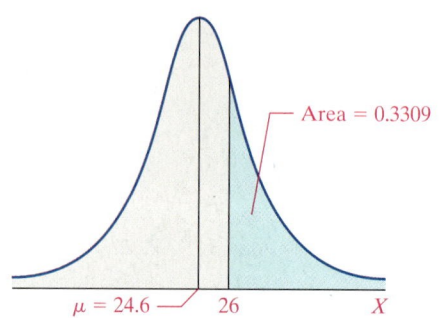

The area under the curve to the right of $X = 26$ is 0.3309. Provide two interpretations of this area.

(b) The following figure represents the normal curve with $\mu = 24.6$ miles per gallon and $\sigma = 3.2$ miles per gallon.

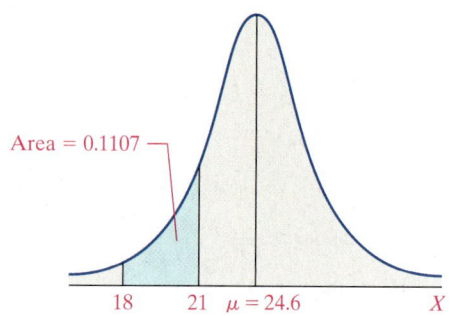

The area under the curve between $X = 18$ and $X = 21$ is 0.1107. Provide two interpretations of this area.

---

35. A random variable X is normally distributed with $\mu = 10$ **NW** and $\sigma = 3$.

(a) Compute $Z_1 = \dfrac{X_1 - \mu}{\sigma}$ for $X_1 = 8$.

(b) Compute $Z_2 = \dfrac{X_2 - \mu}{\sigma}$ for $X_2 = 12$.

(c) The area under the normal curve between $X_1 = 8$ and $X_2 = 12$ is 0.495. What is the area between Z_1 and Z_2?

36. A random variable X is normally distributed with $\mu = 25$ and $\sigma = 6$.

(a) Compute $Z_1 = \dfrac{X_1 - \mu}{\sigma}$ for $X_1 = 18$.

(b) Compute $Z_2 = \dfrac{X_2 - \mu}{\sigma}$ for $X_2 = 30$.

(c) The area under the normal curve between $X_1 = 18$ and $X_2 = 30$ is 0.6760. What is the area between Z_1 and Z_2?

37. Hitting a Pitching Wedge In the game of golf, distance control is just as important as how far a player hits the ball. Suppose Michael went to the driving range with his range finder and hit 75 golf balls with his pitching wedge and measured the distance each ball traveled (in yards). He obtained the following data:

100	97	101	101	103	100	99	100	100
104	100	101	98	100	99	99	97	101
104	99	101	101	101	100	96	99	99
98	94	98	107	98	100	98	103	100
98	94	104	104	98	101	99	97	103
102	101	101	100	95	104	99	102	95
99	102	103	97	101	102	96	102	99
96	108	103	100	95	101	103	105	100
94	99	95						

(a) Use MINITAB or some other statistical software to construct a relative frequency histogram. Comment on the shape of the distribution.

(b) Use MINITAB or some other statistical software to draw the normal density function on the relative frequency histogram.

(c) Do you think the normal density function accurately describes the distance Michael hits a pitching wedge? Why?

38. Heights of Five-Year-Old Females The following frequency distribution represents the heights (in inches) of eighty randomly selected 5-year-old females.

44.5	42.4	42.2	46.2	45.7	44.8	43.3	39.5
45.4	43.0	43.4	44.7	38.6	41.6	50.2	46.9
39.6	44.7	36.5	42.7	40.6	47.5	48.4	37.5
45.5	43.3	41.2	40.5	44.4	42.6	42.0	40.3
42.0	42.2	38.5	43.6	40.6	45.0	40.7	36.3
44.5	37.6	42.2	40.3	48.5	41.6	41.7	38.9
39.5	43.6	41.3	38.8	41.9	40.3	42.1	41.9
42.3	44.6	40.5	37.4	44.5	40.7	38.2	42.6
44.0	35.9	43.7	48.1	38.7	46.0	43.4	44.6
37.7	34.6	42.4	42.7	47.0	42.8	39.9	42.3

(a) Use MINITAB or some other statistical software to construct a relative frequency histogram. Comment on the shape of the distribution.

(b) Use MINITAB or some other statistical software to draw the normal density function on the relative frequency histogram.

(c) Do you think the normal density function accurately describes the heights of 5-year-old females? Why?

7.2 The Standard Normal Distribution

Preparing for This Section   Before getting started, review the following:

- The Complement Rule (Section 5.2, pp. 272–273)

Objectives

1 Find the area under the standard normal curve

2 Find Z-scores for a given area

3 Interpret the area under the standard normal curve as a probability

In Section 7.1, we introduced the normal distribution. We learned that, if X is a normally distributed random variable, we can use the area under the normal density function to obtain the proportion of a population, or the probability that a randomly selected individual from the population has a certain characteristic. To find the area under the normal curve, we first convert the random variable X to a standard normal random variable Z with mean $\mu = 0$ and standard deviation $\sigma = 1$ and find the area under the standard normal curve. This section discusses methods for finding the area under the standard normal curve.

Figure 11

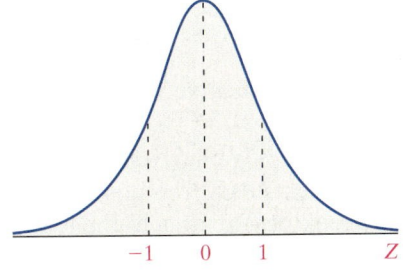

Properties of the Standard Normal Distribution

The standard normal distribution has a mean of 0 and a standard deviation of 1. The standard normal curve therefore, will have its high point located at 0 and inflection points located at −1 and +1. We use the random variable Z to represent a standard normal random variable. The graph of the standard normal curve is presented in Figure 11.

Although we stated the properties of normal curves in Section 7.1, it is worthwhile to restate them here in terms of the standard normal curve.

Properties of the Standard Normal Curve

1. It is symmetric about its mean, $\mu = 0$ and has standard deviation $\sigma = 1$.
2. The mean = median = mode = 0. Its highest point occurs at $\mu = 0$.
3. It has inflection points at $\mu - \sigma = 0 - 1 = -1$ and $\mu + \sigma = 0 + 1 = 1$.
4. The area under the curve is 1.
5. The area under the curve to the right of $\mu = 0$ equals the area under the curve to the left of $\mu = 0$, which equals $\frac{1}{2}$.
6. As Z increases, the graph approaches, but never equals, zero. As Z decreases, the graph approaches, but never equals, zero.
7. The Empirical Rule: Approximately 0.68 = 68% of the area under the standard normal curve is between -1 and 1. Approximately 0.95 = 95% of the area under the standard normal curve is between -2 and 2. Approximately 0.997 = 99.7% of the area under the standard normal curve is between -3 and 3. See Figure 12.

Figure 12

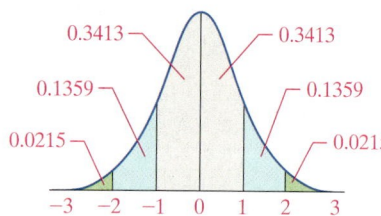

We now discuss the procedure for finding area under the standard normal curve.

Figure 13

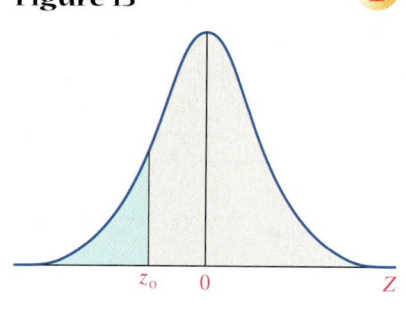

① Find the Area under the Standard Normal Curve

We discuss two methods for finding area under the standard normal curve. The first method uses a table of areas that has been constructed for various values of Z. The second method involves the use of statistical software or a calculator with advanced statistical features.

Table IV, which can be found in the inside back cover of the text or in Appendix A, gives areas under the standard normal curve for values to the left of a specified Z-score, z_0, as shown in Figure 13.

The shaded region represents the area under the standard normal curve to the left of $Z = z_0$. Whenever finding area under a normal curve, you should sketch a normal curve and shade the area you are finding.

EXAMPLE 1 Finding Area under the Standard Normal Curve to the Left of a Z-score

Problem: Find the area under the standard normal curve that lies to the left of $Z = 1.68$.

Approach

Step 1: Draw a standard normal curve with $Z = 1.68$ labeled, and shade the area under the curve to the left of $Z = 1.68$.

Step 2: The rows in Table IV represent the ones and tenths portion of Z, while the columns represent the hundredths portion. To find the area under the curve to the left of $Z = 1.68$, we need to split 1.68 as 1.6 and 0.08. Find the row that represents 1.6 and the column that represents 0.08 in Table IV. Identify where the row and column intersect. This value is the area.

Figure 14

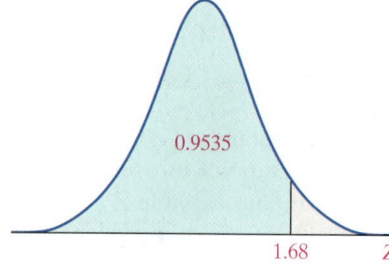

Solution

Step 1: Figure 14 shows the graph of the standard normal curve with $Z = 1.68$ labeled. The area left of $Z = 1.68$ is shaded.

Step 2: A portion of Table IV is presented in Figure 15. We have enclosed the row that represents 1.6 and the column that represents 0.08. The point where the row and column intersect is the area we are seeking. The area to the left of $Z = 1.68$ is 0.9535.

Figure 15

z	.00	.01	.02	.03	.04	.05	.06	.07	.08	.09
0.0	0.5000	0.5040	0.5080	0.5120	0.5160	0.5199	0.5239	0.5279	0.5319	0.5359
0.1	0.5398	0.5438	0.5478	0.5517	0.5557	0.5596	0.5636	0.5675	0.5714	0.5753
0.2	0.5793	0.5832	0.5871	0.5910	0.5948	0.5987	0.6026	0.6064	0.6103	0.6141
0.3	0.6179	0.6217	0.6255	0.6293	0.6331	0.6368	0.6406	0.6443	0.6480	0.6517
1.3	0.9032	0.9049	0.9066	0.9082	0.9099	0.9115	0.9131	0.9147	0.9162	0.9177
1.4	0.9192	0.9207	0.9222	0.9236	0.9251	0.9265	0.9279	0.9292	0.9306	0.9319
1.5	0.9332	0.9345	0.9357	0.9370	0.9382	0.9394	0.9406	0.9418	0.9429	0.9441
1.6	0.9452	0.9463	0.9474	0.9484	0.9495	0.9505	0.9515	0.9525	0.9535	0.9545
1.7	0.9554	0.9564	0.9573	0.9582	0.9591	0.9599	0.9608	0.9616	0.9625	0.9633
1.8	0.9641	0.9649	0.9656	0.9664	0.9671	0.9678	0.9686	0.9693	0.9699	0.9706

The area under a standard normal curve may also be determined using a statistical spreadsheet or graphing calculator with advanced statistical features.

EXAMPLE 2   **Finding the Area under a Standard Normal Curve Using Technology**

Problem: Find the area under the standard normal curve to the left of $Z = 1.68$ using a statistical spreadsheet or graphing calculator with advanced statistical features.

Approach: We will use MINITAB and a TI-84 Plus graphing calculator to find the area. The steps for determining the area under the standard normal curve for MINITAB, Excel, and the TI-83/84 Plus graphing calculators are given in the Technology Step by Step on page 384.

Result: Figure 16(a) shows the results from MINITAB, and Figure 16(b) shows the results from a TI-84 Plus graphing calculator.

Figure 16

Cumulative Distribution Function

Normal with mean = 0 and standard deviation = 1

```
   x        P( X <= x)
1.68         0.953521
```

```
normalcdf(-1E99,
1.68,0,1)
       .9535213678
```

(a)   (b)

Notice the output for MINITAB is titled Cumulative Distribution Function. Remember, the word cumulative means "less than or equal to," so MINITAB is giving the area under the standard normal curve for Z less than or equal to 1.68. The command required by the TI-84 Plus is normalcdf(. The cdf stands for Cumulative Distribution Function. The TI-graphing calculators require a left and right-bound. We use -1E99 for the left bound to obtain areas "less than or equal to" some value.

Now Work Problem 5.

Often, rather than being interested in the area under the standard normal curve to the left of $Z = z_0$, we are interested in obtaining the area under the standard normal curve to the right of $Z = z_0$. The solution to this type of problem

In Other Words

Area right = 1 − area left

uses the fact that the area under the entire standard normal curve is 1 and the Complement Rule. Therefore,

$$\left(\begin{array}{c}\text{Area under the normal curve}\\ \text{to the right of } z_0\end{array}\right) = 1 - \left(\begin{array}{c}\text{area to the left}\\ \text{of } z_0.\end{array}\right)$$

$$P(Z > z_0) \qquad\qquad = 1 - \qquad P(Z \le z_0)$$

EXAMPLE 3   Finding Area under the Standard Normal Curve to the Right of a Z-score

Problem: Find the area under the standard normal curve to the right of $Z = -0.46$.

Approach

Step 1: Draw a standard normal curve with $Z = -0.46$ labeled, and shade the area under the curve to the right of $Z = -0.46$.

Step 2: Find the row that represents −0.4 and the column that represents 0.06 in Table IV. Identify where the row and column intersect. This value is the area to the *left* of $Z = -0.46$.

Step 3: The area under the standard normal curve to the right of $Z = -0.46$ is 1 minus the area to the left of $Z = -0.46$.

Solution

Step 1: Figure 17 shows the graph of the standard normal curve with $Z = -0.46$ labeled. The area to the right of $Z = -0.46$ is shaded.

Step 2: A portion of Table IV is presented in Figure 18. We have enclosed the row that represents −0.4 and the column that represents 0.06. The point where the row and column intersect is the area to the left of $Z = -0.46$. The area to the left of $Z = -0.46$ is 0.3228.

Figure 17

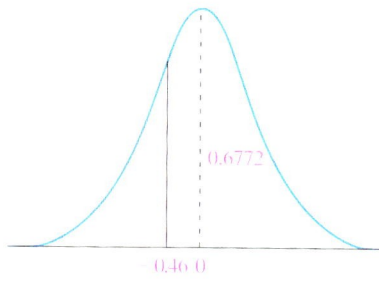

0.6772

−0.46 0

Figure 18

Z	.00	.01	.02	.03	.04	.05	.06	.07	.08	.09
−3.4	0.0003	0.0003	0.0003	0.0003	0.0003	0.0003	0.0003	0.0003	0.0003	0.0002
−3.3	0.0005	0.0005	0.0005	0.0004	0.0004	0.0004	0.0004	0.0004	0.0004	0.0003
−3.2	0.0007	0.0007	0.0006	0.0006	0.0006	0.0006	0.0006	0.0005	0.0005	0.0005
−3.1	0.0010	0.0009	0.0009	0.0009	0.0008	0.0008	0.0008	0.0008	0.0007	0.0007
−3.0	0.0013	0.0013	0.0013	0.0012	0.0012	0.0011	0.0011	0.0011	0.0010	0.0010
−0.5	0.3085	0.3050	0.3015	0.2981	0.2946	0.2912	0.2877	0.2843	0.2810	0.2776
−0.4	0.3446	0.3409	0.3372	0.3336	0.3300	0.3264	0.3228	0.3192	0.3156	0.3121
−0.3	0.3821	0.3783	0.3745	0.3707	0.3669	0.3632	0.3594	0.3557	0.3520	0.3483
−0.2	0.4207	0.4168	0.4129	0.4090	0.4052	0.4013	0.3974	0.3936	0.3897	0.3859
−0.1	0.4602	0.4562	0.4522	0.4483	0.4443	0.4404	0.4364	0.4325	0.4286	0.4247
−0.0	0.5000	0.4960	0.4920	0.4880	0.4840	0.4801	0.4761	0.4721	0.4681	0.4641

Step 3: The area under the standard normal curve to the right of $Z = -0.46$ is 1 minus the area to the left of $Z = -0.46$.

$$\begin{aligned}\text{Area right of } -0.46 &= 1 - (\text{area left of } -0.46)\\ &= 1 - 0.3228\\ &= 0.6772\end{aligned}$$

The area to the right of $Z = -0.46$ is 0.6772.

USING TECHNOLOGY

Statistical spreadsheets and graphing calculators can also find areas under the standard normal curve to the right of a Z-score.

Now Work Problem 7.

The next example presents a situation in which we are interested in the area between two Z-scores.

| **EXAMPLE 4** | **Find the Area under the Standard Normal Curve between Two Z-scores** |

Problem: Find the area under the standard normal curve between $Z = -1.35$ and $Z = 2.01$.

Approach

Step 1: Draw a standard normal curve with $Z = -1.35$ and $Z = 2.01$ labeled. Shade the area under the curve between $Z = -1.35$ and $Z = 2.01$.

Step 2: Find the area to the left of $Z = -1.35$. Find the area to the left of $Z = 2.01$.

Step 3: The area under the standard normal curve between $Z = -1.35$ and $Z = 2.01$ is the area to the left of $Z = 2.01$ minus the area to the left of $Z = -1.35$.

Solution

Step 1: Figure 19 shows the standard normal curve with the area between $Z = -1.35$ and $Z = 2.01$ shaded.

Step 2: Based upon Table IV, the area to the left of $Z = -1.35$ is 0.0885. The area to the left of $Z = 2.01$ is 0.9778.

Step 3: The area between $Z = -1.35$ and $Z = 2.01$ is

Figure 19

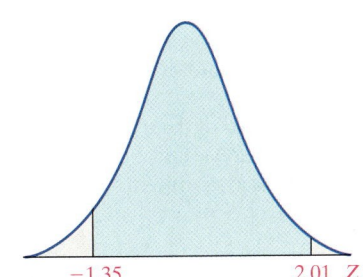

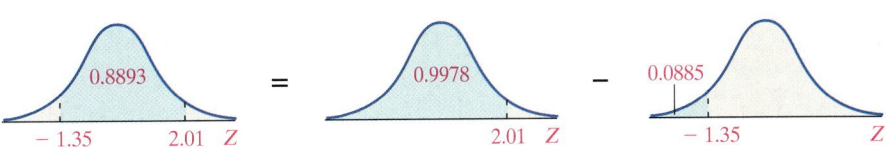

$$(\text{Area between } Z = -1.35 \text{ and } Z = 2.01) = (\text{area left of } Z = 2.01) - (\text{area left of } Z = -1.35)$$
$$= 0.9778 - 0.0885$$
$$= 0.8893$$

The area between $Z = -1.35$ and $Z = 2.01$ is 0.8893.

Now Work Problem 9.

We summarize the methods for obtaining area under the standard normal curve in Table 2.

Because the normal curve extends indefinitely in both directions on the Z-axis, there is no Z-value for which the area under the curve to the left of the Z-value is 1. For example, the area to the left of $Z = 10$ is less than 1, even though graphing calculators and statistical software state that the area is 1, because they can compute a limited number of decimal places. We will follow the practice of stating the area to the left of $Z = -3.90$ or to the right of $Z = 3.90$ as <0.0001. The area under the standard normal curve to the left of $Z = 3.90$ or to the right of $Z = -3.90$ will be stated as >0.9999.

CAUTION

State the area under the standard normal curve to the left of $Z = -3.90$ as <0.0001 (not 0). State the area under the standard normal curve to the left of $Z = 3.90$ as >0.9999 (not 1).

Table 2

Problem	Approach	Solution
Find the area to the left of z_0.	Shade the area to the left of z_0.	Use Table IV to find the row and column that correspond to z_0. The area is the value where the row and column intersect. Or use technology to find the area.
Find the area to the right of z_0.	Shade the area to the right of z_0.	Use Table IV to find the area left of z_0. The area to the right of z_0 is 1 minus the area to the left of z_0. Or use technology to find the area.
Find the area between z_0 and z_1.	Shade the area between z_0 and z_1.	Use Table IV to find the area to the left of z_0 and to the left of z_1. The area between z_0 and z_1 is (area to the left of z_1) − (area to the left of z_0). Or use technology to find the area.

2   Find Z-scores for a Given Area

Up to this point, we have found areas given the value of a Z-score. Often, we are interested in finding a Z-score that corresponds to a given area. The procedure to follow is the reverse of the procedure for finding areas given Z-scores.

EXAMPLE 5   **Finding a Z-score from a Specified Area to the Left**

Problem: Find the Z-score so that the area to the left of the Z-score is 0.32.

Approach

Step 1: Draw a standard normal curve with the area and corresponding un-known Z-score labeled.

Step 2: Look for the area in the table closest to 0.32.

Step 3: Find the Z-score that corresponds to the area closest to 0.32.

Solution

Step 1: Figure 20 shows the graph of the standard normal curve with the area of 0.32 labeled. We know z_0 must be less than 0. Do you know why?

Step 2: We refer to Table IV and look in the body of the table for an area clos-est to 0.32. The area closest to 0.32 is 0.3192. Figure 21 shows a partial represen-tation of Table IV with 0.3192 labeled.

Figure 20

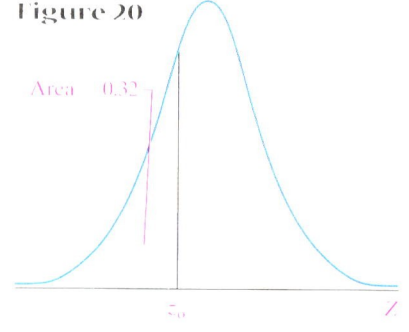

Figure 21

z	.00	.01	.02	.03	.04	.05	.06	.07	.08	.09
−3.4	0.0003	0.0003	0.0003	0.0003	0.0003	0.0003	0.0003	0.0003	0.0003	0.0002
−3.3	0.0005	0.0005	0.0005	0.0004	0.0004	0.0004	0.0004	0.0004	0.0004	0.0003
−3.2	0.0007	0.0007	0.0006	0.0006	0.0006	0.0006	0.0006	0.0005	0.0005	0.0005
−3.1	0.0010	0.0009	0.0009	0.0009	0.0008	0.0008	0.0008	0.0008	0.0007	0.0007
−3.0	0.0013	0.0013	0.0013	0.0012	0.0012	0.0011	0.0011	0.0011	0.0010	0.0010
−0.7	0.2420	0.2389	0.2358	0.2327	0.2296	0.2266	0.2236	0.2206	0.2177	0.2148
−0.6	0.2743	0.2709	0.2676	0.2643	0.2611	0.2578	0.2546	0.2514	0.2483	0.2451
−0.5	0.3085	0.3050	0.3015	0.2981	0.2946	0.2912	0.2877	0.2843	0.2810	0.2776
−0.4	0.3446	0.3409	0.3372	0.3336	0.3300	0.3264	0.3228	0.3192	0.3156	0.3121
−0.3	0.3821	0.3783	0.3745	0.3707	0.3669	0.3632	0.3594	0.3557	0.3520	0.3483
−0.2	0.4027	0.4168	0.4129	0.4090	0.4052	0.4013	0.3974	0.3936	0.3897	0.3859

Figure 22

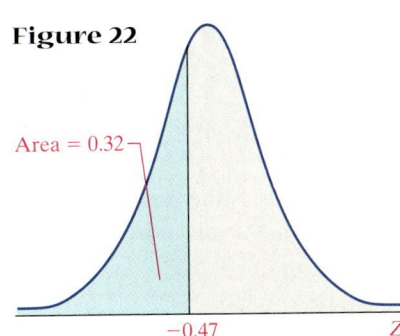

Area = 0.32

−0.47   Z

Step 3: From reading the table in Figure 21, we see that the approximate Z-score that corresponds to an area of 0.32 to its left is −0.47. So $z_0 = -0.47$. See Figure 22.

Statistical spreadsheets or graphing calculators with advanced statistical features can also be used to determine a Z-score corresponding to a specified area.

EXAMPLE 6   **Finding a Z-score from a Specified Area to the Left Using Technology**

Problem: Find the Z-score such that the area to the left of the Z-score is 0.32 using a statistical spreadsheet or graphing calculator with advanced statistical features.

Approach: We will use Excel and a TI-84 Plus graphing calculator to find the Z-score. The steps for determining the Z-score for MINITAB, Excel, and the TI-83/84 Plus graphing calculators are given in the Technology Step by Step on page 384.

Result: Figure 23(a) shows the results from Excel, and Figure 23(b) shows the results from a TI-84 Plus graphing calculator.

Figure 23

Here is the
Z-score

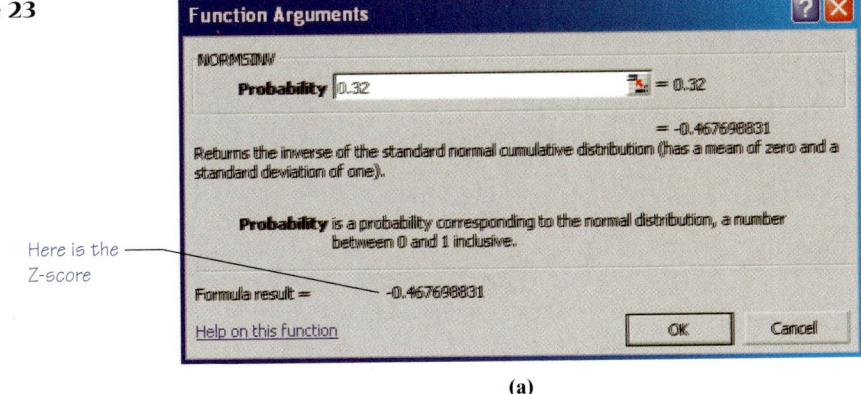

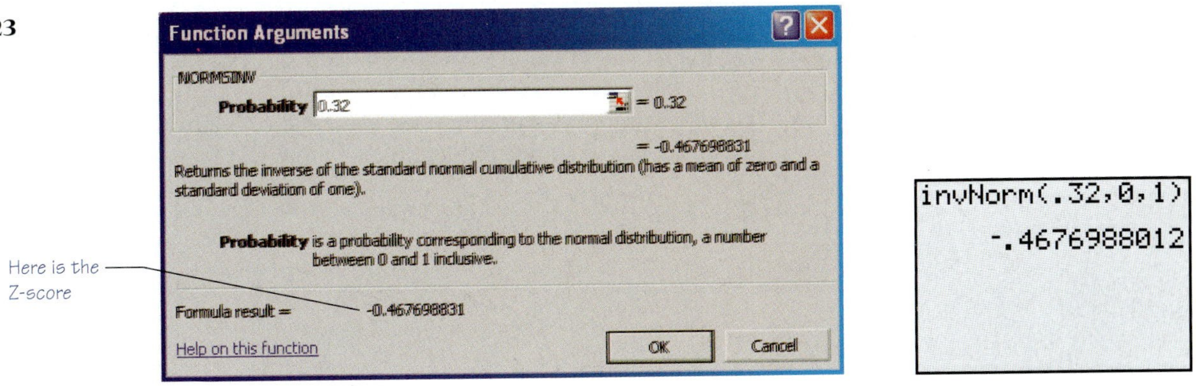

(a)   (b)

The Z-score that corresponds to an area of 0.32 to its left is −0.47, so $z_0 = -0.47$.

It is useful to remember that if the area to the left of the Z-score is less than 0.5 the Z-score must be less than 0. If the area to the left of the Z-score is greater than 0.5, the Z-score must be greater than 0.

Now Work Problem 15.

The next example deals with situations in which the area to the right of some unknown Z-score is given. The solution uses the fact that the area under the normal curve is 1.

EXAMPLE 7   Finding a Z-score from a Specified Area to the Right

Problem: Find the Z-score so that the area to the right of the Z-score is 0.4332.

Approach

Step 1: Draw a standard normal curve with the area and corresponding unknown Z-score labeled.

Step 2: Determine the area to the left of the unknown Z-score.

Step 3: Look for the area in the table closest to the area determined in Step 2 and record the Z-score that corresponds to the closest area.

Solution

Step 1: Figure 24 shows the standard normal curve with the area and unknown Z-score labeled.

Step 2: Since the area under the entire normal curve is 1, the area to the left of the unknown Z-score is 1 minus the area right of the unknown Z-score. Therefore,

$$\text{Area to the left} = 1 - \text{area to the right}$$
$$= 1 - 0.4332$$
$$= 0.5668$$

Step 3: We look in the body of Table IV for an area closest to 0.5668. See Figure 25. The area closest to 0.5668 is 0.5675.

> ⚠ **CAUTION**
> To find a Z-score given the area to the right, you must first determine the area to the left if you are using Table IV.

Figure 24

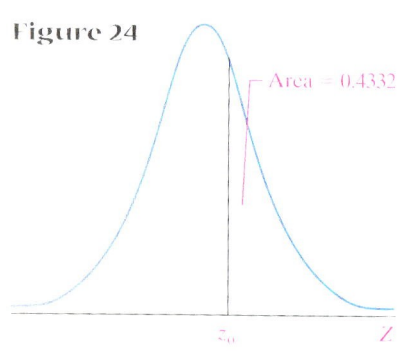

Area = 0.4332

Figure 25

z	.00	.01	.02	.03	.04	.05	.06	.07	.08	.09
0.0	0.5000	0.5040	0.5080	0.5120	0.5160	0.5199	0.5239	0.5279	0.5319	0.5359
0.1	0.5398	0.5438	0.5478	0.5517	0.5557	0.5596	0.5636	0.5675	0.5714	0.5753
0.2	0.5793	0.5832	0.5871	0.5910	0.5948	0.5987	0.6026	0.6064	0.6103	0.6141
0.3	0.6179	0.6217	0.6255	0.6293	0.6331	0.6368	0.6406	0.6443	0.6480	0.6517
0.4	0.6554	0.6591	0.6628	0.6664	0.6700	0.6736	0.6772	0.6808	0.6844	0.6879
0.5	0.6915	0.6950	0.6985	0.7019	0.7054	0.7088	0.7123	0.7157	0.7190	0.7224
0.6	0.7237	0.7291	0.7324	0.7357	0.7389	0.7422	0.7454	0.7486	0.7517	0.7549

The approximate Z-score that corresponds to a right tail area of 0.4332 is 0.17. Therefore, $z_0 = 0.17$. See Figure 26.

Figure 26

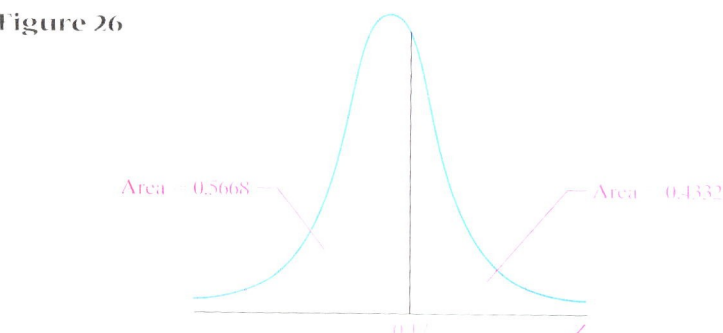

Area = 0.5668   Area = 0.4332

Now Work Problem 19.

In upcoming chapters, we will be interested in finding Z-scores that separate the middle area of the standard normal curve from the area in its tails.

EXAMPLE 8   Finding the Z-score from an Area in the Middle

Problem: Find the Z-score that divides the middle 90% of the area in the standard normal distribution from the area in the tails.

Approach

Step 1: Draw a standard normal curve with the middle $0.9 = 90\%$ of the area separated from the area of $5\% = 0.05$ in each of the two tails. Label the unknown Z-scores z_0 and z_1.

Step 2: Look in the body of Table IV to find the area closest to 0.05.

Step 3: Determine the Z-score in the left tail.

Step 4: The area to the right of z_1 is 0.05. Therefore, the area to the left of z_1 is 0.95. Look in Table IV for an area of 0.95 and find the corresponding Z-value.

Solution

Step 1: Figure 27 shows the standard normal curve with the middle 90% of the area separated from the area in the two tails.

Step 2: We look in the body of Table IV for an area closest to 0.05. See Figure 28.

Figure 27

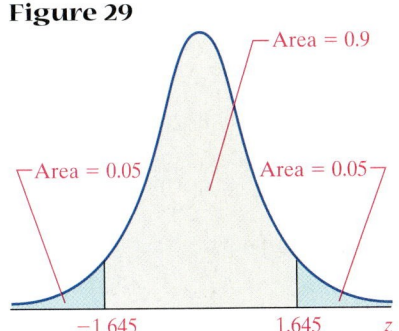

Area = 0.9
Area = 0.05   Area = 0.05
z_0 z_1 z

Figure 28

z	.00	.01	.02	.03	.04	.05	.06	.07	.08	.09
−3.4	0.0003	0.0003	0.0003	0.0003	0.0003	0.0003	0.0003	0.0003	0.0003	0.0002
−3.3	0.0005	0.0005	0.0005	0.0004	0.0004	0.0004	0.0004	0.0004	0.0004	0.0003
−3.2	0.0007	0.0007	0.0006	0.0006	0.0006	0.0006	0.0006	0.0005	0.0005	0.0005
−2.0	0.0228	0.0222	0.0217	0.0212	0.0207	0.0202	0.0197	0.0192	0.0188	0.0183
−1.9	0.0287	0.0281	0.0274	0.0268	0.0262	0.0256	0.0250	0.0244	0.0239	0.0233
−1.8	0.0359	0.0351	0.0344	0.0336	0.0329	0.0322	0.0314	0.0307	0.0301	0.0294
−1.7	0.0446	0.0436	0.0427	0.0418	0.0409	0.0401	0.0392	0.0384	0.0375	0.0367
−1.6	0.0548	0.0537	0.0526	0.0516	0.0505	0.0495	0.0485	0.0475	0.0465	0.0455
−1.5	0.0668	0.0655	0.0643	0.0630	0.0618	0.0606	0.0594	0.0582	0.0571	0.0559

Notice that 0.0495 and 0.0505 are equally close to 0.05. We agree to take the mean of the two Z-scores corresponding to the areas.

Step 3: The Z-score corresponding to an area of 0.0495 is -1.65. The Z-score corresponding to an area of 0.0505 is -1.64. Therefore, the approximate Z-score corresponding to an area of 0.05 to the left is

$$z_0 = \frac{-1.65 + (-1.64)}{2} = -1.645$$

Step 4: The area to the right of z_1 is 0.05. Therefore, the area to the left of $z_1 = 1 - 0.05 = 0.95$. In Table IV, we find an area of 0.9495 corresponding to $z = 1.64$ and an area of 0.9505 corresponding to $z = 1.65$. Consequently, the approximate Z-score corresponding to an area of 0.05 to the right is

$$z_1 = \frac{1.65 + 1.64}{2} = 1.645$$

See Figure 29.

Figure 29

Area = 0.9
Area = 0.05   Area = 0.05
−1.645   1.645 z

We could also obtain the solution to Example 8 using symmetry. Because the standard normal curve is symmetric about its mean, 0, the Z-score that corresponds to an area to the left of 0.05 will be the additive inverse (i.e. opposite) of the Z-score that corresponds to an area to the right of 0.05. Since the area to the left of $Z = -1.645$ is 0.05, the area to the right of $Z = 1.645$ is also 0.05.

Figure 30

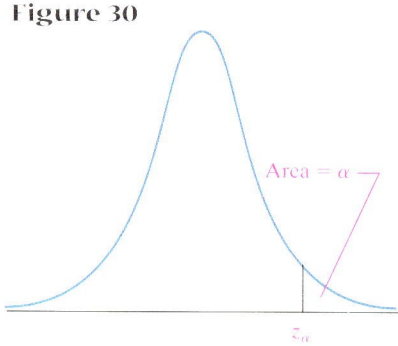

Area = α

z_α

Now Work Problem 23.

We are often interested in finding the Z-score that has a specified area to the right. For this reason, we have special notation to represent this situation.

The notation z_α (pronounced "z sub alpha") is the Z-score such that the area under the standard normal curve to the right of z_α is α. Figure 30 illustrates the notation.

EXAMPLE 9 Finding the Value of z_α

Figure 31

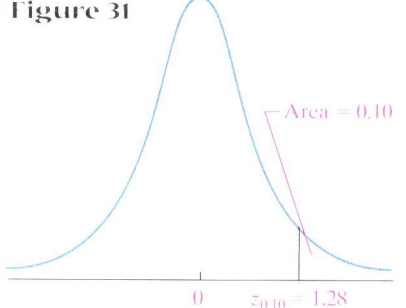

Area = 0.10

0    $z_{0.10} = 1.28$

Problem: Find the value of $z_{0.10}$.

Approach: We wish to find the Z-value such that the area under the standard normal curve to the right of the Z-value is 0.10.

Solution: The area to the right of the unknown Z-value is 0.10, so the area to the left of the Z-value is $1 - 0.10 = 0.90$. We look in Table IV for the area closest to 0.90. The area closest is 0.8997, which corresponds to a Z-value of 1.28. Therefore, $z_{0.10} = 1.28$. See Figure 31.

Now Work Problem 27.

3   Interpret the Area under the Standard Normal Curve as a Probability

Recall that the area under a normal curve can be interpreted either as a probability or as the proportion of the population with the given characteristic (as represented by an interval of numbers). When interpreting the area under the standard normal curve as a probability, we use the notation introduced in Chapter 6. For example, in Example 8, we found that the area under the standard normal curve to the left of $Z = -1.645$ is 0.05; therefore, the probability of randomly selecting a standard normal random variable that is less than -1.645 is 0.05. We write this statement with the notation $P(Z < -1.645) = 0.05$.

We will use the following notation to denote probabilities of a standard normal random variable, Z.

Notation for the Probability of a Standard Normal Random Variable

$P(a < Z < b)$	represents the probability a standard normal random variable is between a and b.
$P(Z > a)$	represents the probability a standard normal random variable is greater than a.
$P(Z < a)$	represents the probability a standard normal random variable is less than a.

EXAMPLE 10 Finding Probabilities of Standard Normal Random Variables

Problem: Evaluate $P(Z < 1.26)$.

Approach

Step 1: Draw a standard normal curve with the area we desire shaded.

Figure 32

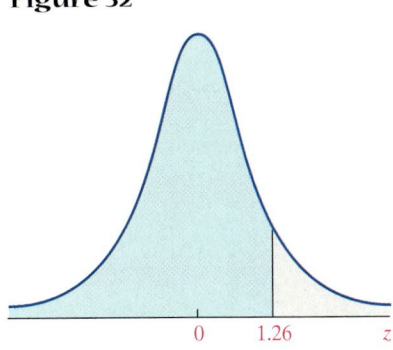

0   1.26   z

Step 2: Use Table IV to find the area of the shaded region. This area represents the probability.

Solution

Step 1: Figure 32 shows the standard normal curve with the area to the left of $Z = 1.26$ shaded.

Step 2: Using Table IV, we find the area under the standard normal curve to the left of $Z = 1.26$ is 0.8962. Because the area under the standard normal curve to the left of $Z = 1.26$ is 0.8962, we have $P(Z < 1.26) = 0.8962$.

Now Work Problem 33.

For any continuous random variable, the probability of observing a specific value of the random variable is 0. For example, for a standard normal random variable, $P(a) = 0$ for any value of a. This is because there is no area under the standard normal curve associated with a single value, so the probability must be 0. So, the following probabilities are equivalent.

$$P(a < Z < b) = P(a \leq Z < b) = P(a < Z \leq b) = P(a \leq Z \leq b)$$

For example, $P(Z < 1.26) = P(Z \leq 1.26) = 0.8962$.

7.2 ASSESS YOUR UNDERSTANDING

Concepts and Vocabulary

1. State the properties of the standard normal curve.
2. If the area under the standard normal curve to the left of $Z = 1.20$ is 0.8849, what is the area under the standard normal curve to the right of $Z = 1.20$?

3. *True or False:* The area under the standard normal curve to the left of $Z = 5.30$ is 1. Support your answer.
4. Explain why $P(Z < -1.30) = P(Z \leq -1.30)$.

Skill Building

In Problems 5–12, find the indicated areas. For each problem, be sure to draw a standard normal curve and shade the area that is to be found.

5. Determine the area under the standard normal curve that
NW lies to the left of
   (a) $Z = -2.45$
   (b) $Z = -0.43$
   (c) $Z = 1.35$
   (d) $Z = 3.49$

6. Determine the area under the standard normal curve that lies to the left of
   (a) $Z = -3.49$
   (b) $Z = -1.99$
   (c) $Z = 0.92$
   (d) $Z = 2.90$

7. Determine the area under the standard normal curve that
NW lies to the right of
   (a) $Z = -3.01$
   (b) $Z = -1.59$
   (c) $Z = 1.78$
   (d) $Z = 3.11$

8. Determine the area under the standard normal curve that lies to the right of
   (a) $Z = -3.49$
   (b) $Z = -0.55$
   (c) $Z = 2.23$
   (d) $Z = 3.45$

9. Determine the area under the standard normal curve that
NW lies between
   (a) $Z = -2.04$ and $Z = 2.04$
   (b) $Z = -0.55$ and $Z = 0$
   (c) $Z = -1.04$ and $Z = 2.76$

10. Determine the area under the standard normal curve that lies between
   (a) $Z = -2.55$ and $Z = 2.55$
   (b) $Z = -1.67$ and $Z = 0$
   (c) $Z = -3.03$ and $Z = 1.98$

11. Determine the area under the standard normal curve
  (a) to the left of $Z = -2$ or to the right of $Z = 2$
  (b) to the left of $Z = -1.56$ or to the right of $Z = 2.56$
  (c) to the left of $Z = -0.24$ or to the right of $Z = 1.20$

12. Determine the area under the standard normal curve
  (a) to the left of $Z = -2.94$ or to the right of $Z = 2.94$
  (b) to the left of $Z = -1.68$ or to the right of $Z = 3.05$
  (c) to the left of $Z = -0.88$ or to the right of $Z = 1.23$

In Problems 13 and 14, find the area of the shaded region for each standard normal curve.

13. (a)

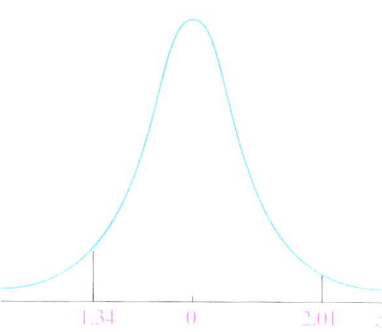

14. (a)

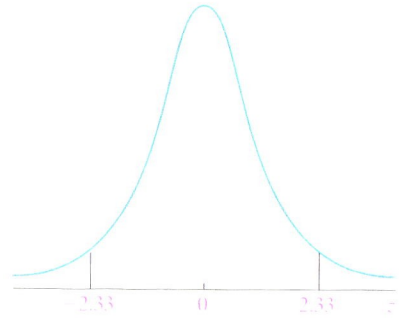

(b)

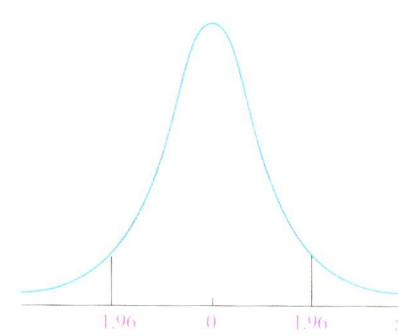

(b)

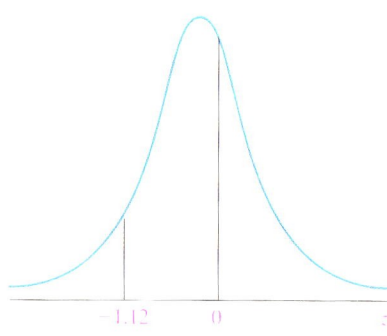

(c)

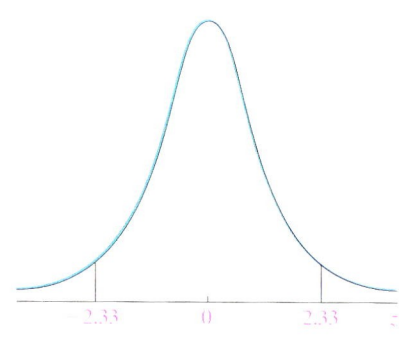

(c)

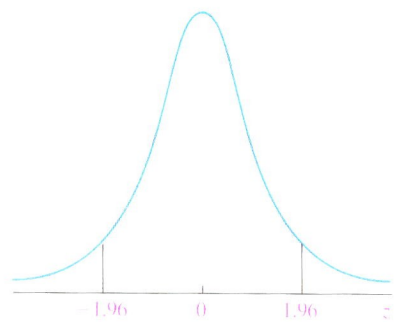

In Problems 15–26, find the indicated Z-score. Be sure to draw a standard normal curve that depicts the solution.

15. Find the Z-score such that the area under the standard normal curve to the left is 0.1.

16. Find the Z-score such that the area under the standard normal curve to the left is 0.2.

17. Find the Z-score such that the area under the standard normal curve to the left is 0.98.

18. Find the Z-score such that the area under the standard normal curve to the left is 0.85.

19. Find the Z-score such that the area under the standard normal curve to the right is 0.25.

20. Find the Z-score such that the area under the standard normal curve to the right is 0.35.

21. Find the Z-score such that the area under the standard normal curve to the right is 0.89.

22. Find the Z-score such that the area under the standard normal curve to the right is 0.75.

23. Find the Z-scores that separate the middle 80% of the dis-
NW tribution from the area in the tails of the standard normal
distribution.

24. Find the Z-scores that separate the middle 70% of the distribution from the area in the tails of the standard normal
distribution.

25. Find the Z-scores that separate the middle 99% of the
distribution from the area in the tails of the standard
normal distribution.

26. Find the Z-scores that separate the middle 94% of the distribution from the area in the tails of the standard normal
distribution.

In Problems 27–32, find the value of z_α.

27. $z_{0.05}$
NW

28. $z_{0.35}$

29. $z_{0.01}$

30. $z_{0.02}$

31. $z_{0.20}$

32. $z_{0.15}$

In Problems 33–44, find the indicated probability of the standard normal random variable Z.

33. $P(Z < 1.93)$

34. $P(Z < -0.61)$

35. $P(Z > -2.98)$

36. $P(Z > 0.92)$

37. $P(-1.20 \le Z < 2.34)$

38. $P(1.23 < Z \le 1.56)$

39. $P(Z \ge 1.84)$

40. $P(Z \ge -0.92)$

41. $P(Z \le 0.72)$

42. $P(Z \le -2.69)$

43. $P(Z < -2.56 \text{ or } Z > 1.39)$

44. $P(Z < -0.38 \text{ or } Z > 1.93)$

Applying the Concepts

45. **The Empirical Rule** The Empirical Rule states that
about 68% of the data in a bell-shaped distribution lies
within 1 standard deviation of the mean. This means
about 68% of the data lie between $Z = -1$ and $Z = 1$.
Verify this result. Verify that about 95% of the data lie
within 2 standard deviations of the mean. Finally, verify
that about 99.7% of the data lie within 3 standard deviations of the mean.

46. According to Table IV, the area under the standard normal curve to the left of $Z = -1.34$ is 0.0901. Without consulting Table IV, determine the area under the standard
normal curve to the right of $Z = 1.34$.

47. According to Table IV, the area under the standard normal curve to the left of $Z = -2.55$ is 0.0054. Without consulting Table IV, determine the area under the standard
normal curve to the right of $Z = 2.55$.

48. According to Table IV, the area under the standard normal
curve between $Z = -1.50$ and $Z = 0$ is 0.4332. Without
consulting Table IV, determine the area under the standard normal curve between $Z = 0$ and $Z = 1.50$.

49. According to Table IV, the area under the standard normal curve between $Z = -1.24$ and $Z = -0.53$ is 0.1906.
Without consulting Table IV, determine the area under
the standard normal curve between $Z = 0.53$ and
$Z = 1.24$.

50. (a) Suppose $P(Z < a) = 0.9938$; find a.
(b) Suppose $P(Z \ge a) = 0.4404$; find a.
(c) Suppose $P(-b < Z < b) = 0.8740$; find b.

Technology Step by Step | **The Standard Normal Distribution**

TI-83/84 Plus | **Finding Areas under the Standard Normal Curve**

Step 1: From the HOME screen, press 2nd VARS to access the DISTRibution menu.

Step 2: Select 2:normalcdf(

Step 3: With normalcdf(on the HOME screen, type *lowerbound, upperbound, 0, 1*). For example, to find the area left of $Z = 1.26$ under the standard normal curve, type

Normalcdf(-1E99, 1.26, 0, 1)

and hit ENTER.

Note: When there is no lowerbound, enter $-1E99$. When there is no upperbound, enter 1E99. The E shown is scientific notation; it is $\boxed{2^{nd}}$ $\boxed{,}$ on the keyboard.

Finding Z-scores Corresponding to an Area

Step 1: From the HOME screen, press 2nd VARS to access the DISTRibution menu.

Step 2: Select 3:invNorm(.

Step 3: With invNorm(on the HOME screen, type "*area left*", 0, 1). For example, to find the Z-score such that the area under the normal curve left of the Z-score is 0.79, type

InvNorm(0.79, 0 , 1)

and hit ENTER.

MINITAB | **Finding Areas under the Standard Normal Curve**

Step 1: MINITAB will find an area to the left of a specified Z-score. Select the **Calc** menu, highlight **Probability Distributions**, and highlight **Normal** . . . .

Step 2: Select **Cumulative Probability**. Set the mean to 0 and the standard deviation to 1. Select **Input Constant**, and enter the specified Z-score. Click OK.

Finding Z-scores Corresponding to an Area

Step 1: MINITAB will find the Z-score for an area to the left of an unknown Z-score. Select the **Calc** menu, highlight **Probability Distributions**, and highlight **Normal** . . . .

Step 2: Select **Inverse Cumulative Probability**. Set the mean to 0 and the standard deviation to 1. Select **Input Constant**, and enter the specified area. Click OK.

Excel | **Finding Areas under the Standard Normal Curve**

Step 1: Excel will find the area to the left of a specified Z-score. Select the *fx* button from the tool bar. In **Function Category:**, select "Statistical." In **Function Name:**, select "NormsDist." Click OK.

Step 2: Enter the specified Z-score. Click OK.

Finding Z-scores Corresponding to an Area

Step 1: Excel will find the Z-score for an area to the left of an unknown Z-score. Select the *fx* button from the tool bar. In **Function Category:**, select "Statistical." In **Function Name:**, select "NormsInv." Click OK.

Step 2: Enter the specified area. Click OK.

7.3 Applications of the Normal Distribution

Preparing for This Section   Before getting started, review the following:

- Percentiles (Section 3.4, pp. 167–169)

Objectives   ① **Find and interpret the area under a normal curve**
② **Find the value of a normal random variable**

① **Find and Interpret the Area under a Normal Curve**

Suppose that a random variable X is normally distributed with mean μ and standard deviation σ. The area below the normal curve represents a proportion or probability.

From the discussions in Section 7.1, we know that finding the area under a normal curve requires that we transform a normal random variable X with mean μ and standard deviation σ into a standard normal random variable Z with mean 0 and standard deviation 1. This is accomplished by letting

$Z = \dfrac{X - \mu}{\sigma}$ and using Table IV to find the area under the standard normal curve. This idea is illustrated in Figure 33.

Figure 33

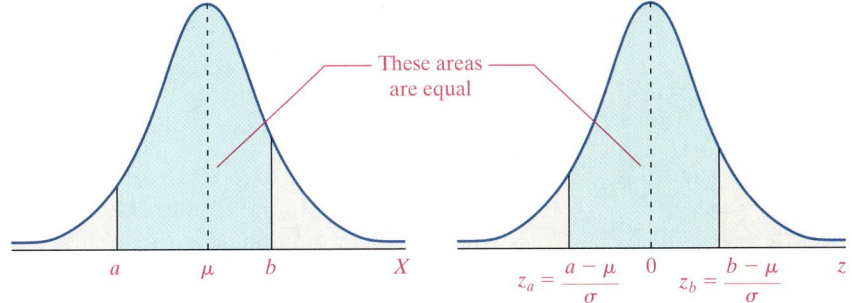

Now that we have the ability to find the area under a standard normal curve, we can find the area under any normal curve. We summarize the procedure next.

Finding the Area under Any Normal Curve

Step 1: Draw a normal curve and shade the desired area.

Step 2: Convert the values of X to Z-scores using $Z = \dfrac{X - \mu}{\sigma}$.

Step 3: Draw a standard normal curve and shade the area desired.

Step 4: Find the area under the standard normal curve. This area is equal to the area under the normal curve drawn in Step 1.

EXAMPLE 1   **Finding Area under a Normal Curve**

Problem: A pediatrician obtains the heights of her 200 three-year-old female patients. The heights are approximately normally distributed, with mean 38.72 inches and standard deviation 3.17 inches. Use the normal model to determine the proportion of the 3-year-old females that have a height less than 35 inches.

Approach: Follow Steps 1 through 4.

Solution

Step 1: Figure 34 shows the normal curve with the area to the left of 35 shaded.
Step 2: We convert $X = 35$ to a standard normal random variable Z.

$$Z = \frac{X - \mu}{\sigma} = \frac{35 - 38.72}{3.17} = -1.17$$

Step 3: Figure 35 shows the standard normal curve with the area to the left of $Z = -1.17$ shaded.

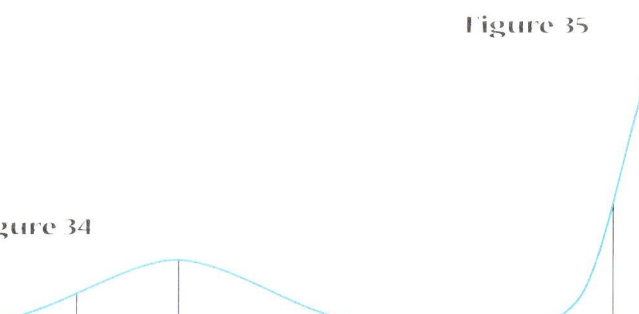

Figure 35

Figure 34

Table 3

Height (inches)	Relative Frequency
29.0–29.9	0.005
30.0–30.9	0.005
31.0–31.9	0.005
32.0–32.9	0.025
33.0–33.9	0.02
34.0–34.9	0.055
35.0–35.9	0.075
36.0–36.9	0.09
37.0–37.9	0.115
38.0–38.9	0.15
39.0–39.9	0.12
40.0–40.9	0.11
41.0–41.9	0.07
42.0–42.9	0.06
43.0–43.9	0.035
44.0–44.9	0.025
45.0–45.9	0.025
46.0–46.9	0.005
47.0–47.9	0.005

The area to the left of $Z = -1.17$ is equal to the area to the left of $X = 35$.
Step 4: Using Table IV, we find the area to the left of $Z = -1.17$ is 0.1210. The normal model indicates that the proportion of the pediatrician's 3-year-old females that are less than 35 inches tall is 0.1210.

According to the results of Example 1, the probability that a randomly selected 3-year-old female is shorter than 35 inches is 0.1210. If the normal curve is a good model for determining probabilities (or proportions), then about 12.1% of the 200 three-year-olds in Table 1 should be shorter than 35 inches. For convenience, the information provided in Table 1 is repeated in Table 3.

From the relative frequency distribution in Table 3, we determine that $0.005 + 0.005 + 0.005 + 0.025 + 0.02 + 0.055 = 0.115 = 11.5\%$ of the 3-year olds are less than 35 inches tall. The results based on the normal curve are in close agreement with the actual results. The normal curve accurately models the heights of 3-year-old females.

Because the area under the normal curve represents a proportion, we can also use the area under the normal curve to find percentile ranks of scores. Recall that the kth percentile divides the lower $k\%$ of a data set from the upper $(100 - k)\%$. In Example 1, 12% of the females have a height less than 35 inches, and 88% of the females have a height greater than 35 inches. Therefore, a child whose height is 35 inches is at the 12th percentile.

Statistical software and graphing calculators with advanced statistical features can also be used to find areas under any normal curve.

EXAMPLE 2 **Finding Area under a Normal Curve Using Technology**

Problem: Find the percentile rank of a 3-year old female whose height is 43 inches using a statistical spreadsheet or graphing calculator with advanced statistical features. From Example 1, we know the heights are approximately normally distributed with a mean of 38.72 inches and standard deviation of 3.17 inches.

Approach: We will use a TI-84 Plus graphing calculator to find the area. The steps for determining the area under the standard normal curve for MINITAB,

Figure 36

```
normalcdf(-1E99,
43,38.72,3.17)
        .9115172428
```

Excel, and the TI-83/84 Plus graphing calculators are given in the Technology Step by Step on page 393.

Result: Figure 36 shows the results from a TI-84 Plus graphing calculator. The area under the normal curve to the left of 43 is 0.91. Therefore, 91% of the heights are less than 43 inches and 9% of the heights are more than 43 inches. A child whose height is 43 inches is at the 91st percentile.

EXAMPLE 3 | **Finding the Probability of a Normal Random Variable**

Problem: For the pediatrician presented in Example 1, use the normal distribution to compute the probability that a randomly selected 3-year-old female is between 35 and 40 inches tall, inclusive. That is, find $P(35 \leq X \leq 40)$.

Approach: We follow the Steps 1 through 4 on page 385.

Solution

Step 1: Figure 37 shows the normal curve with the area between $X_1 = 35$ and $X_2 = 40$ shaded.

Step 2: Convert the values of $X_1 = 35$ and $X_2 = 40$ to Z-scores.

$$Z_1 = \frac{X_1 - \mu}{\sigma} = \frac{35 - 38.72}{3.17} = -1.17$$

$$Z_2 = \frac{X_2 - \mu}{\sigma} = \frac{40 - 38.72}{3.17} = 0.40$$

Step 3: Figure 38 shows the graph of the standard normal curve with the area between $Z_1 = -1.17$ and $Z_2 = 0.40$ shaded.

Figure 38

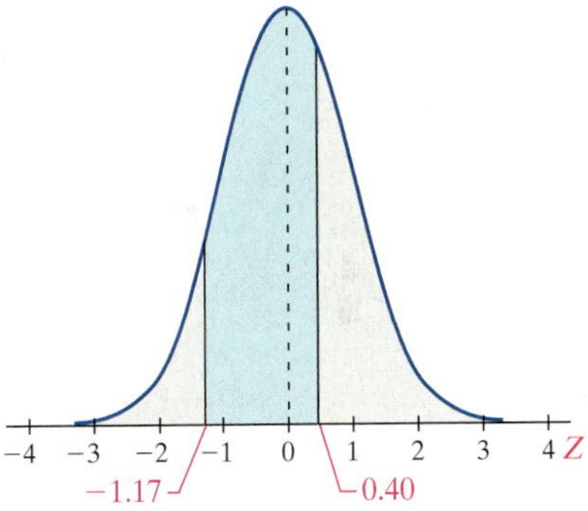

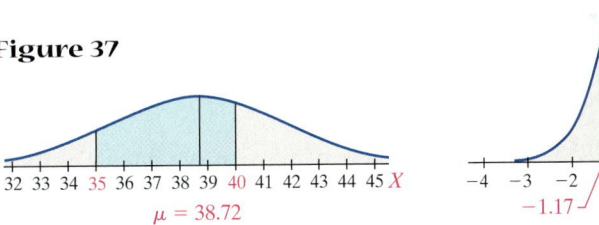

Figure 37

Step 4: Using Table IV, we find that the area to the left of $Z_2 = 0.40$ is 0.6554 and the area to the left of $Z_1 = -1.17$ is 0.1210. Therefore, the area between $Z_1 = -1.17$ and $Z_2 = 0.40$ is $0.6554 - 0.1210 = 0.5344$. We conclude that the probability a randomly selected 3-year-old female is between 35 and 40 inches tall is 0.5344. That is, $P(35 \leq X \leq 40) = P(-1.17 \leq Z \leq 0.40) = 0.5344$.

Now Work Problem 19.

In Other Words

The normal probability density function is used to model random variables that appear to be normal (such as girls' heights). A good model is one that yields results that are close to reality.

According to the relative frequency distribution in Table 3, the proportion of the 200 three-year-old females with heights between 35 inches and 40 inches is $0.075 + 0.09 + 0.115 + 0.15 + 0.12 = 0.55 = 55\%$. This is very close to the probability obtained in Example 3!

② Find the Value of a Normal Random Variable

Often, rather than being interested in the proportion or probability of a normal random variable, we are interested in calculating the value of a normal random variable required for the variable to correspond to a certain proportion or probability. For example, we might want to know the height of a 3-year-old girl at the 20th percentile. This means we want to know the height of a 3-year-old girl who is taller than 20% of all 3-year-old girls.

> **Procedure for Finding the Value of a Normal Random Variable Corresponding to a Specified Proportion, Probability, or Percentile**
>
> **Step 1:** Draw a normal curve and shade the area corresponding to the proportion, probability, or percentile.
> **Step 2:** Use Table IV to find the Z-score that corresponds to the shaded area.
> **Step 3:** Obtain the normal value from the fact that $X = \mu + Z\sigma$.*

EXAMPLE 4   **Finding the Value of a Normal Random Variable**

Problem: The heights of a pediatrician's 200 three-year-old females are approximately normally distributed with mean 38.72 inches and standard deviation 3.17 inches. Find the height of a 3-year-old female at the 20th percentile. That is, find the height of a 3-year-old female that separates the bottom 20% from the top 80%.

Approach: We follow Steps 1 through 3.

Solution:

Figure 39

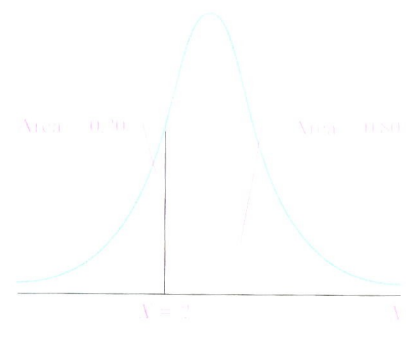

Step 1: Figure 39 shows the normal curve with the unknown value of X separating the bottom 20% of the distribution from the top 80% of the distribution.

Step 2: From Table IV, the area closest to 0.20 is 0.2005. The corresponding Z-score is -0.84.

Step 3: The height of a 3-year-old female that separates the bottom 20% of the data from the top 80% is computed as follows:

$$
\begin{aligned}
X &= \mu + Z\sigma \\
&= 38.72 + (-0.84)(3.17) \\
&= 36.1 \text{ inches}
\end{aligned}
$$

The height of a 3-year old female at the 20th percentile is 36.1 inches.

EXAMPLE 5   **Finding the Value of a Normal Random Variable Using Technology**

Problem: Use a statistical spreadsheet or a graphing calculator with advanced statistical features to verify the results of Example 4. That is, find the height for a 3-year-old female that is at the 20th percentile, assuming females' heights are approximately normally distributed with a mean of 38.72 inches and a standard deviation of 3.17 inches.

Approach: We will use MINITAB to find the height at the 20th percentile. The steps for determining the area under the standard normal curve for

$$
\begin{aligned}
*Z &= \frac{X - \mu}{\sigma} \\
Z\sigma &= X - \mu \\
X &= \mu + Z\sigma
\end{aligned}
$$

MINITAB, Excel, and the TI-83/84 Plus graphing calculators are given in the Technology Step by Step on page 393.

Result: Figure 40 shows the results obtained from MINITAB. The height of a three-year-old female at the 20th percentile is 36.1 inches.

Figure 40 **Inverse Cumulative Distribution Function**

```
Normal with mean = 38.7200 and standard deviation = 3.17000


P ( x <= x)        x
    0.2000    36.0521
```

Now Work Problem 27(a).

EXAMPLE 6 **Finding the Value of a Normal Random Variable**

Problem: The heights of a pediatrician's 200 three-year-old females are approximately normally distributed with mean 38.72 inches and standard deviation 3.17 inches. The pediatrician wishes to determine the heights that separate the middle 98% of the distribution from the bottom 1% and top 1%. In other words, find the 1st and 99th percentiles.

Approach: We follow Steps 1 through 3 given on page 388.

Solution:

Figure 41

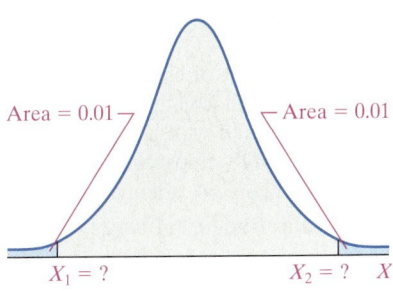

Step 1: Figure 41 shows the normal curve with the unknown values of X separating the bottom and top 1% of the distribution from the middle 98% of the distribution.

Step 2: First, we will find the Z-score that corresponds to the area 0.01 to the left. From Table IV, the area closest to 0.01 is 0.0099. The corresponding Z-score is -2.33. The Z-score that corresponds to the area 0.01 to the right is the Z-score that has the area 0.99 to the left. The area closest to 0.99 is 0.9901. The corresponding Z-score is 2.33.

Step 3: The height of a 3-year-old female that separates the bottom 1% of the distribution from the top 99% is

$$X_1 = \mu + Z\sigma$$
$$= 38.72 + (-2.33)(3.17)$$
$$= 31.3 \text{ inches}$$

The height of a 3-year-old female that separates the top 1% of the distribution from the bottom 99% is

$$X_2 = \mu + Z\sigma$$
$$= 38.72 + (2.33)(3.17)$$
$$= 46.1 \text{ inches}$$

A 3-year-old female whose height is less than 31.3 inches is in the bottom 1% of all 3-year-old females, and a 3-year-old female whose height is more than 46.1 inches is in the top 1% of all 3-year-old females. The pediatrician might use this information to identify those patients who have unusual heights.

Now Work Problem 27(b).

7.3 ASSESS YOUR UNDERSTANDING

Concepts and Vocabulary

1. Describe the procedure for finding the area under any normal curve.

2. Describe the procedure for finding the score corresponding to a probability.

Skill Building

In Problems 3–12, assume the random variable X is normally distributed with mean $\mu = 50$ and standard deviation $\sigma = 7$. Compute the following probabilities. Be sure to draw a normal curve with the area corresponding to the probability shaded.

3. $P(X > 35)$

4. $P(X > 65)$

5. $P(X \leq 45)$

6. $P(X \leq 58)$

7. $P(40 < X < 65)$

8. $P(56 < X < 68)$

9. $P(55 \leq X \leq 70)$

10. $P(40 \leq X \leq 49)$

11. $P(38 < X \leq 55)$

12. $P(56 \leq X < 66)$

In Problems 13–16, assume the random variable X, is normally distributed with mean $\mu = 50$ and standard deviation $\sigma = 7$. Find each indicated percentile for X.

13. The 9th percentile

14. The 90th percentile

15. The 81st percentile

16. The 38th percentile

Applying the Concepts

17. **Egg Incubation Times** The mean incubation time of fertilized chicken eggs kept at 100.5°F in a still-air incubator is 21 days. Suppose the incubation times are approximately normally distributed with a standard deviation of 1 day. Source: University of Illinois Extension.

    (a) What is the probability that a randomly selected fertilized chicken egg hatches in less than 20 days?
    (b) What is the probability that a randomly selected fertilized chicken egg takes over 22 days to hatch?
    (c) What is the probability that a randomly selected fertilized chicken egg hatches between 19 and 21 days?
    (d) Would it be unusual for an egg to hatch in less than 18 days?

18. **Medical Residents** In a 2003 study, the Accreditation Council for Graduate Medical Education found that medical residents' mean number of hours worked in a week is 81.7. Suppose the number of hours worked per week by medical residents is approximately normally distributed with a standard deviation of 6.9 hours.

    Source: www.medrecinst.com

    (a) What is the probability that a randomly selected medical resident works more than 80 hours in a week?
    (b) What is the probability that a randomly selected medical resident works more than 100 hours in a week?
    (c) What is the probability that a randomly selected medical resident works less than 60 hours in a week?

    (d) Would it be unusual for a medical resident to work less than 70 hours in a week?

19. **Chips Ahoy! Cookies** The number of chocolate chips in an 18-ounce bag of Chips Ahoy! chocolate chip cookies is approximately normally distributed with a mean of 1262 chips and standard deviation 118 chips according to a study by cadets of the U.S. Air Force Academy.

    Source: Brad Warner and Jim Rutledge, *Chance*, Vol. 12, No. 1, 1999, pp. 10–14.

    (a) What is the probability that a randomly selected 18-ounce bag of Chips Ahoy! cookies contains between 1000 and 1400 chocolate chips?
    (b) What is the probability that a randomly selected 18-ounce bag of Chips Ahoy! cookies contains fewer than 1000 chocolate chips?
    (c) What proportion of 18-ounce bags of Chip Ahoy! cookies contains more than 1200 chocolate chips?
    (d) What proportion of 18-ounce bags of Chip Ahoy! cookies contains fewer than 1125 chocolate chips?
    (e) What is the percentile rank of an 18-ounce bag of Chip Ahoy! cookies that contains 1475 chocolate chips?
    (f) What is the percentile rank of an 18-ounce bag of Chip Ahoy! cookies that contains 1050 chocolate chips?

20. **Earthquakes** The magnitude of earthquakes since 1900 that measure 0.1 or higher on the Richter scale in California is approximately normally distributed, with $\mu = 6.2$ and $\sigma = 0.5$, according to data obtained from the U.S. Geological Survey.

(a) What is the probability that a randomly selected earthquake in California has a magnitude of 6.0 or higher?

(b) What is the probability that a randomly selected earthquake in California has a magnitude less than 6.4?

(c) What is the probability that a randomly selected earthquake in California has a magnitude between 5.8 and 7.1?

(d) The great San Francisco Earthquake of 1906 had a magnitude of 8.25. Is an earthquake of this magnitude unusual in California?

(e) What is the percentile rank of a California earthquake that measures 6.8 on the Richter scale?

(f) What is the percentile rank of a California earthquake that measures 5.1 on the Richter scale?

21. **Hybrid Car** Introduced in the 2000 model year, the Honda Insight was the first hybrid automobile sold in the United States. The mean gas mileage for the model year 2005 Insight with an automatic transmission is 56 miles per gallon on the highway. Suppose the gasoline mileage of this automobile is approximately normally distributed with a standard deviation of 3.2 miles per gallon.

Source: www.fueleconomy.gov.

(a) What proportion of 2005 Honda Insights with automatic transmission gets over 60 miles per gallon on the highway?

(b) What proportion of 2005 Honda Insights with automatic transmission gets 50 miles per gallon or less on the highway?

(c) What proportion of 2005 Honda Insights with automatic transmission gets between 58 and 62 miles per gallon on the highway?

(d) What is the probability that a randomly selected 2005 Honda Insight with an automatic transmission gets less than 45 miles per gallon on the highway?

22. **Light Bulbs** General Electric manufactures a decorative Crystal Clear 60-watt light bulb that it advertises will last 1500 hours. Suppose the lifetimes of the light bulbs are approximately normally distributed with a mean of 1550 hours and a standard deviation of 57 hours.

(a) What proportion of the light bulbs will last less than the advertised time?

(b) What proportion of the light bulbs will last more than 1650 hours?

(c) What is the probability that a randomly selected GE Crystal Clear 60-watt light bulb lasts between 1625 and 1725 hours?

(d) What is the probability that a randomly selected GE Crystal Clear 60-watt light bulb lasts longer than 1400 hours?

23. **Heights of Females** As reported by the U.S. National Center for Health Statistics, the mean height of females 20 to 29 years old is $\mu = 64.1$ inches. If height is approximately normally distributed with $\sigma = 2.8$ inches, answer the following questions:

(a) What is the percentile rank of a 20- to 29-year-old female who is 60 inches tall?

(b) What is the percentile rank of a 20- to 29-year-old female who is 70 inches tall?

(c) What proportion of 20- to 29-year-old females are between 60 and 70 inches tall?

(d) Would it be unusual for a 20- to 29-year-old female to be taller than 70 inches?

24. **Gestation Period** The length of human pregnancies are approximately normally distributed with mean $\mu = 266$ days and standard deviation $\sigma = 16$ days.

(a) What percent of pregnancies lasts more than 270 days?

(b) What percent of pregnancies lasts less than 250 days?

(c) What percent of pregnancies lasts between 240 and 280 days?

(d) What is the probability that a randomly selected pregnancy lasts more than 280 days?

(e) What is the probability that a randomly selected pregnancy lasts no more than 245 days?

(f) A "very preterm" baby is one whose gestation period is less than 224 days. What proportion of births is "very preterm"?

25. **Manufacturing** Steel rods are manufactured with a mean length of 25 centimeter (cm). Because of variability in the manufacturing process, the lengths of the rods are approximately normally distributed with a standard deviation of 0.07 cm.

(a) What proportion of rods has a length less than 24.9 cm?

(b) Any rods that are shorter than 24.85 cm or longer than 25.15 cm are discarded. What proportion of rods will be discarded?

(c) Using the results of part (b), if 5000 rods are manufactured in a day, how many should the plant manager expect to discard?

(d) If an order comes in for 10,000 steel rods, how many rods should the plant manager manufacture if the order states that all rods must be between 24.9 cm and 25.1 cm?

26. **Manufacturing** Ball bearings are manufactured with a mean diameter of 5 millimeter (mm). Because of variability in the manufacturing process, the diameters of the ball bearings are approximately normally distributed with a standard deviation of 0.02 mm.

(a) What proportion of ball bearings has a diameter more than 5.03 mm?

(b) Any ball bearings that have a diameter less than 4.95 mm or greater than 5.05 mm are discarded. What proportion of ball bearings will be discarded?

(c) Using the results of part (b), if 30,000 ball bearings are manufactured in a day, how many should the plant manager expect to discard?

(d) If an order comes in for 50,000 ball bearings, how many bearings should the plant manager manufacture if the order states that all ball bearings must be between 4.97 mm and 5.03 mm?

27. **Egg Incubation Times** The mean of the incubation time
NW of fertilized chicken eggs kept at 100.5°F in a still-air incubator is 21 days. Suppose the incubation times are approx-

imately normally distributed with a standard deviation of 1 day.

Source: University of Illinois Extension.

(a) Determine the 17th percentile for incubation times of fertilized chicken eggs.

(b) Determine the incubation times that make up the middle 95% of fertilized chicken eggs?

28. **Medical Residents** In a 2003 study, the Accreditation Council for Graduate Medical Education found that medical residents' mean number of hours worked in a week is 81.7. Suppose the number of hours worked per week by medical residents is approximately normally distributed with a standard deviation of 6.9 hours.

Source: www.medrecinst.com

(a) Determine the 75th percentile for the number of hours worked in a week by medical residents.

(b) Determine the number of hours worked in a week that makes up the middle 80% of medical residents.

29. **Chips Ahoy! Cookies** The number of chocolate chips in an 18-ounce bag of Chips Ahoy! chocolate chip cookies is approximately normally distributed with a mean of 1262 chips and a standard deviation of 118 chips, according to a study by cadets of the U.S. Air Force Academy.

Source: Brad Warner and Jim Rutledge, *Chance*, Vol. 12, No. 1, 1999, pp. 10–14.

(a) Determine the 30th percentile for the number of chocolate chips in an 18-ounce bag of Chips Ahoy! cookies.

(b) Determine the number of chocolate chips in a bag of Chips Ahoy! that make up the middle 99% of bags.

30. **Earthquakes** The magnitude of earthquakes since 1900 that measure 0.1 or higher on the Richter scale in California is approximately normally distributed with $\mu = 6.2$ and $\sigma = 0.5$, according to data obtained from the U.S. Geological Survey.

(a) Determine the 40th percentile of the magnitude of earthquakes in California.

(b) Determine the magnitude of earthquakes that make up the middle 85% of magnitudes.

31. **Hybrid Car** Introduced in the 2000 model year, the Honda Insight was the first hybrid automobile sold in the United States. The mean was mileage for the model year 2005 Insight with an automatic transmission is 56 miles per gallon on the highway. Suppose the gasoline mileages of these automobiles are approximately normally distributed with standard deviation 3.2 miles per gallon.

Source: www.fueleconomy.gov.

(a) Determine the 97th percentile gasoline mileage for the model year 2005 Insight with an automatic transmission.

(b) Determine the mileage that makes up the middle 86% gasoline mileage of model year 2005 Insight with an automatic transmission.

32. **Speedy Lube** The time required for Speedy Lube to complete an oil change service on an automobile approximately follows a normal distribution, with a mean of 17 minutes and a standard deviation of 2.5 minutes.

(a) Speedy Lube guarantees customers that the service will take no longer than 20 minutes. If it does take longer, the customer will receive the service for half-price. What percent of customers receives the service for half price?

(b) If Speedy Lube does not want to give the discount to more than 3% of its customers, how long should it make the guaranteed time limit?

33. **Multiple Births** The following data represent the relative frequencies of live multiple-delivery births (three or more babies) in 2002 for women 15 to 44 years old.

Age	Relative Frequency
15–19	0.0128
20–24	0.0702
25–29	0.2235
30–34	0.3888
35–39	0.2530
40–44	0.0518

Source: National Vital Statistics Reports, Vol. 52, No. 10, December 17, 2003

Suppose the ages of multiple-birth mothers are approximately normally distributed with $\mu = 31.77$ years and standard deviation $\sigma = 5.19$ years.

(a) Compute the proportion of multiple-birth mothers in each class by finding the area under the normal curve.

(b) Compare the proportion to the actual proportions. Are you convinced that the ages of multiple-birth mothers are approximately normally distributed?

34. **Weather in Chicago** The following frequency distribution represents the daily high temperature in Chicago, November 16 to 30, for the years 1872 to 1999:

(a) Construct a relative frequency distribution.

(b) Draw a relative frequency histogram. Does the distribution of high temperatures appear to be normal?

(c) Compute the mean and standard deviation of high temperature.

(d) Use the information obtained in part (c) to compute the proportion of high temperatures in each class by finding the area under the normal curve.

(e) Are you convinced that high temperatures are approximately normally distributed?

Temperature	Frequency	Temperature	Frequency
5.0–9.9	1	40.0–44.9	375
10.0–14.9	10	45.0–49.9	281
15.0–19.9	15	50.0–54.9	233
20.0–24.9	40	55.0–59.9	160
25.0–29.9	95	60.0–64.9	101
30.0–34.9	217	65.0–69.9	21
35.0–39.9	325	70.0–74.9	1

Source: Chicago Tribune, November 27, 2000

More than 1 million skin cancers are expected to be diagnosed in the United States this year, almost half of all cancers diagnosed. The prevalence of skin cancer is attributable in part to a history of unprotected or under-protected sun exposure. Sunscreens have been shown to prevent certain types of lesions associated with skin cancer. They also protect skin against exposure to light that contributes to premature aging. As a result, sunscreen is now in moisturizers, lip balms, shampoos, hair-styling products, insect repellents, and makeup.

Consumer Reports tested 23 sunscreens and two moisturizers, all with a claimed sun-protection factor (SPF) of 15 or higher. SPF is defined as the degree to which a sunscreen protects the skin from UVB, the ultraviolet rays responsible for sunburn. (Some studies have shown that UVB, along with UVA, can increase the risk of skin cancers.) A person with untreated skin who can stay in the sun for 5 minutes before becoming sunburned should be able to stay in the sun for $15 \times 5 = 75$ minutes using a sunscreen rated at SPF15.

To test whether products met their SPF claims for UVB, we used a solar simulator (basically a sun lamp) to expose people to measured amounts of sunlight. First we determined the exposure time (in minutes) that caused each person's untreated skin to turn pink within 24 hours. Then we applied sunscreen to new areas of skin and made the same determination. To avoid potential sources of bias, samples of the sunscreens were applied to randomly assigned sites on the subjects' skin.

To determine the SPF rating of a sunscreen for a particular individual, the exposure time with sunscreen was divided by the exposure time without sunscreen.

The following table contains the mean and standard deviation of the SPF measurements for two particular sunscreens.

Product	Mean	Std Dev
A	15.5	1.5
B	14.7	1.2

(a) In designing this experiment, why is it important to obtain the exposure time without sunscreen first and then determine the exposure time with sunscreen for each person?

(b) Why is the random assignment of people and application sites to each treatment (the sunscreen) important?

(c) Calculate the percentage of SPF measurements that you expect to be less than 15, the advertised level of protection for each of the products. (Assume that the SPF ratings are approximately normal.)

(d) Calculate the percentage of SPF measurements that you expect to be greater than 17.5 for product A. Repeat this for product B.

(e) Calculate the percentage of SPF measurements that you expect to fall between 14.5 and 15.5 for product A. Repeat this for product B.

(f) Which product appears to be superior, A or B? Support your conclusion.

Note to Readers: *In many cases, our test protocol and analytical methods are more complicated than described in these examples. The data and discussions have been modified to make the material more appropriate for the audience.*

Technology Step by Step   **The Normal Distribution**

TI-83/84 Plus   **Finding Areas under the Normal Curve**

Step 1: From the HOME screen, press 2nd VARS to access the DISTRibution menu.

Step 2: Select `2:normalcdf(`

Step 3: With `normalcdf(` on the HOME screen, type *lowerbound, upperbound,* μ, σ). For example, to find the area to the left of $X = 35$ under the normal curve with $\mu = 40$ and $\sigma = 10$, type

```
Normalcdf(-1E99, 35, 40, 10)
```

and hit ENTER.

Note: When there is no lowerbound, enter −1E99. When there is no upperbound, enter 1E99. The E shown is scientific notation; it is 2nd , .

Finding Scores Corresponding to an Area

Step 1: From the HOME screen, press 2nd VARS to access the DISTRibution menu.

Step 2: Select 3:invNorm(

Step 3: With invNorm(on the HOME screen, type "*area*", μ, σ). For example, to find the score such that the area under the normal curve to the left of the score is 0.68 with $\mu = 40$ and $\sigma = 10$, type

$$InvNorm(0.68, 40, 10)$$

and hit ENTER.

MINITAB **Finding Areas under the Normal Curve**

Step 1: MINITAB will find an area to the left of a specified observation. Select the **Calc** menu, highlight **Probability Distributions**, and highlight **Normal** . . . .

Step 2: Select **Cumulative Probability**. Enter the mean, μ, and the standard deviation, σ. Select **Input Constant**, and enter the observation. Click OK.

Finding Z-Scores Corresponding to an Area

Step 1: MINITAB will find the score corresponding to a specified area. Select the **Calc** menu, highlight **Probability Distributions**, and highlight **Normal** . . . .

Step 2: Select **Inverse Cumulative Probability**. Enter the mean, μ, and the standard deviation, σ. Select **Input Constant**, and enter the area left of the unknown score. Click OK.

Excel **Finding Areas under the Normal Curve**

Step 1: Excel will find an area to the left of a specified observation. Select the *fx* button from the tool bar. In **Function Category**:, select "Statistical." In **Function Name**:, select NormDist. Click OK.

Step 2: Enter the specified observation, μ, and σ, and set **cumulative** to True. Click OK.

Finding Scores Corresponding to an Area

Step 1: Select the *fx* button from the tool bar. In **Function Category**:, select "Statistical." In **Function Name**:, select NormInv. Click OK.

Step 2: Enter the area left of the unknown score, μ, and σ. Click OK.

Assessing Normality

Preparing for This Section Before getting started, review the following:

- Shape of a distribution (Section 3.1, pp. 127–130)

Objectives ❶ **Draw normal probability plots to assess normality**

Suppose that we obtain a simple random sample from a population whose distribution is unknown. Many of the statistical tests that we perform on small data sets (sample size less than 30) require that the population from which the sample is drawn be normally distributed.

Up to this point, we have said that a random variable X is normally distributed, or at least approximately normal, provided the histogram of the data is symmetric and bell shaped. This method works well for large data sets, but the shape of a histogram drawn from a small sample of observations does not always accurately represent the shape of the population. For this reason, we need additional methods for assessing the normality of a random variable X when we are looking at a small set of sample data.

In Other Words

Normal probability plots are used to assess normality in small data sets.

1 Draw Normal Probability Plots to Assess Normality

A **normal probability plot** is a graph that plots observed data versus *normal scores*. A **normal score** is the expected Z-score of the data value if the distribution of the random variable is normal. The expected Z-score of an observed value depends on the number of observations in the data set.

To draw a normal probability plot requires the following steps:

> ### Drawing a Normal Probability Plot
>
> **Step 1:** Arrange the data in ascending order.
>
> **Step 2:** Compute $f_i = \dfrac{i - 0.375}{n + 0.25}$,* where i is the index (the position of the data value in the ordered list) and n is the number of observations. The expected proportion of observations less than or equal to the ith data value is f_i.
>
> **Step 3:** Find the Z-score corresponding to f_i from Table IV.
>
> **Step 4:** Plot the observed values on the horizontal axis and the corresponding expected Z-scores on the vertical axis.

Figure 42

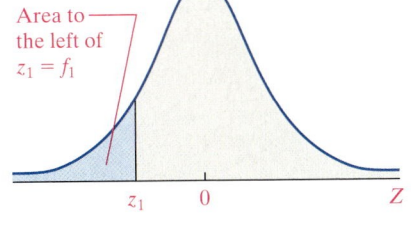

Area to the left of $z_1 = f_1$

The idea behind finding the expected Z-score is that, if the data come from a population that is normally distributed, we should be able to predict the area to the left of each data value. The value of f_i represents the expected area to the left of the ith observation when the data come from a population that is normally distributed. For example, f_1 is the expected area to the left of the smallest data value, f_2 is the expected area to the left of the second-smallest data value, and so on. Figure 42 illustrates the idea.

Once we determine each f_i, we find the Z-scores corresponding to f_1, f_2, and so on. The smallest observation in the data set will be the smallest expected Z-score, and the largest observation in the data set will be the largest expected Z-score. Also, because of the symmetry of the normal curve, the expected Z-scores are always paired as positive and negative values.

Normal random variables X and their Z-scores are linearly related $(X = \mu + Z\sigma)$, so a plot of observations of normal variables against their expected normal scores will be linear. We conclude the following:

> If sample data are taken from a population that is normally distributed, a normal probability plot of the observed values versus the expected Z-scores will be approximately linear.

Normal probability plots are typically drawn by using graphing calculators or statistical software. However, it is worthwhile to go through an example that demonstrates the procedure so that we can better understand the results supplied by software.

EXAMPLE 1

Constructing a Normal Probability Plot

Problem: The data in Table 4 represent the finishing time (in seconds) for six randomly selected races of a greyhound named Barbies Bomber in the 5/16-mile race at Greyhound Park in Dubuque, Iowa. Is there evidence to support the belief that the variable finishing time is normally distributed?

Approach: We follow Steps 1 through 4 listed above.

Table 4

31.35	32.52
32.06	31.26
31.91	32.37

Source: Greyhound Park, Dubuque, IA

*The derivation of this formula is beyond the scope of this text.

Solution

Step 1: The first column in Table 5 represents the index i. The second column represents the observed values in the data set, written in ascending order.

Step 2: The third column in Table 5 represents $f_i = \dfrac{i - 0.375}{n + 0.25}$ for each observation. This is the expected area under the normal curve to the left of the ith observation, assuming normality. For example, $i = 1$ corresponds to the finishing time of 31.26, and

$$f_1 = \frac{1 - 0.375}{6 + 0.25} = 0.1$$

So the area under the normal curve to the left of 31.26 is 0.1 if the sample data come from a population that is normally distributed.

Step 3: We use Table IV to find the Z-scores that correspond to f_i. The expected Z-scores are listed in the fourth column of Table 5. Look in Table IV for the area closest to $f_1 = 0.1$. The expected Z-score is -1.28. Notice that for each negative expected Z-score there is a corresponding positive expected Z-score, as a result of the symmetry of the normal curve.

Table 5

Index, i	Observed Value	f_i	Expected Z-score
1	31.26	$\dfrac{1 - 0.375}{6 + 0.25} = 0.1$	-1.28
2	31.35	$\dfrac{2 - 0.375}{6 + 0.25} = 0.26$	-0.64
3	31.91	0.42	-0.20
4	32.06	0.58	0.20
5	32.37	0.74	0.64
6	32.52	0.9	1.28

Step 4: We plot the actual observations on the horizontal axis and the expected Z-scores on the vertical axis. See Figure 43.

Figure 43
Normal Probability Plot

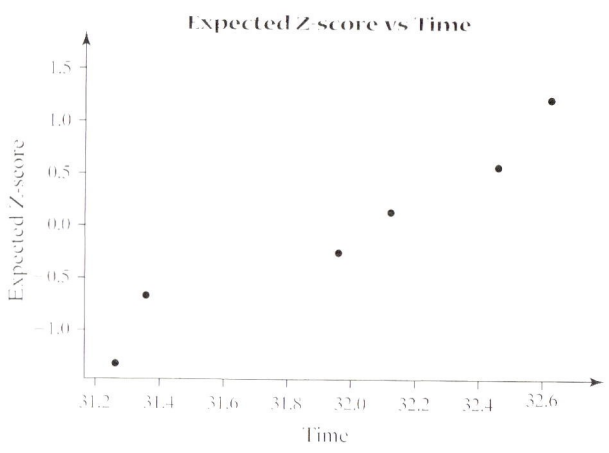

Although the normal probability plot in Figure 43 does show some curvature, it is roughly linear.* We conclude that the finishing time of Barbies Bomber in the $\frac{5}{16}$-mile race is approximately normally distributed.

*In fact, the correlation between the observed value and expected Z-score is 0.970.

Typically, normal probability plots are drawn by using either a graphing calculator with advanced statistical features or statistical software. Certain software, such as MINITAB, provides bounds that the data must lie within to support the belief that the sample data come from a population that is normally distributed.

EXAMPLE 2   **Assessing Normality Using Technology**

Problem: Using MINITAB or some other statistical software, draw a normal probability plot of the data in Table 4 and determine whether there is evidence to support the belief that the sample data come from a population that is normally distributed.

Approach: We will construct a normal probability plot using MINITAB. MINITAB provides curved *bounds* that can be used to assess normality. If the normal probability plot is roughly linear and all the data lie within the bounds provided by the software, we have reason to believe the data come from a population that is approximately normal. The steps for constructing normal probability plots using MINITAB, Excel, or the TI83/84 Plus graphing calculators can be found on page 401.

Solution: Figure 44 shows the normal probability plot. Notice that MINITAB gives area to the left of the expected Z-score, rather than the Z-score. For example, the area to the left of the expected Z-score of -1.28 is 0.10. MINITAB writes 0.10 as 10 percent.

Figure 44

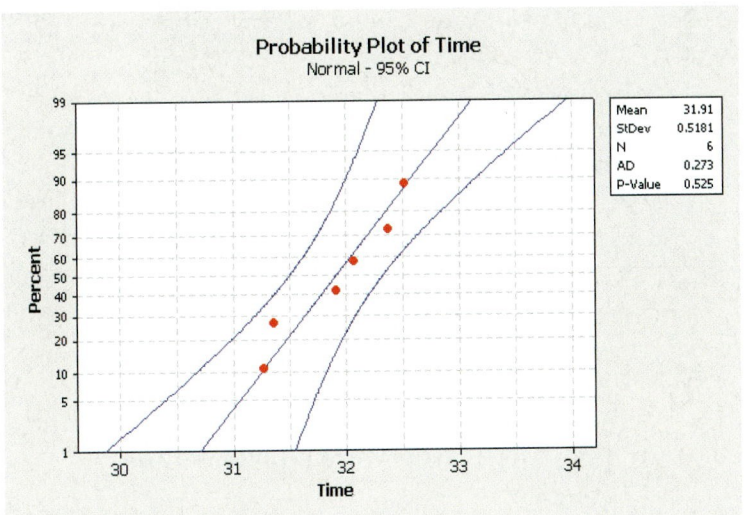

The normal probability plot is roughly linear, and all the data lie within the bounds provided by MINITAB. We conclude that the sample data could come from a population that is normally distributed.

Throughout the text, we will provide normal probability plots drawn with MINITAB so that assessing normality is straightforward.

EXAMPLE 3   **Assessing Normality**

Problem: The data in Table 6 represent the time spent waiting in line (in minutes) for the Demon Roller Coaster for 100 randomly selected riders. Is the random variable "waiting time" normally distributed?

7	3	5	107	8	37	16	41	7	25	22	19	1	40	1	29	93
33	76	14	8	9	45	15	81	94	10	115	18	0	18	11	60	34
30	6	21	0	86	6	11	1	1	3	9	79	41	2	9	6	19
4	3	2	7	18	0	93	68	6	94	16	13	24	6	12	121	30
35	39	9	15	53	9	47	5	55	64	51	80	26	24	12	0	
94	18	4	61	38	38	21	61	9	80	18	21	8	14	47	56	

Table 6

Approach: We will use MINITAB to draw a normal probability plot. If the normal probability plot is roughly linear and the data lie within the bounds provided by MINITAB, conclude that it is reasonable to believe that the sample data come from a population that follows a normal distribution.

Solution: Figure 45 shows a normal probability plot of the data drawn using MINITAB. Since the normal probability plot is not linear, the random variable "wait time" is not normally distributed. Figure 46 shows a histogram of the data in Table 6. The histogram indicates that the data are skewed right.

Figure 45

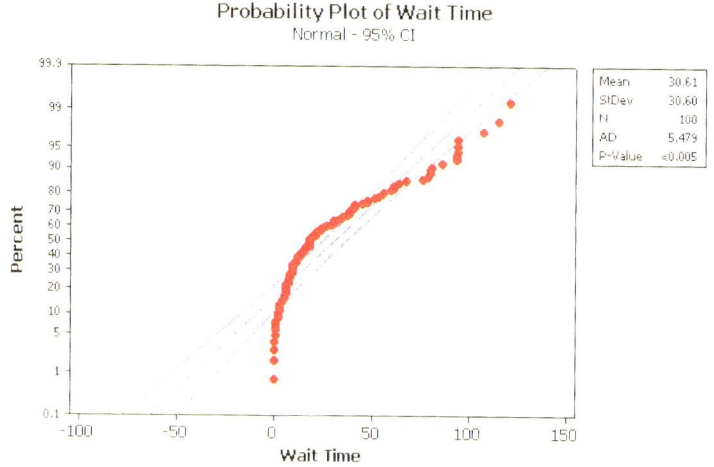

Figure 46

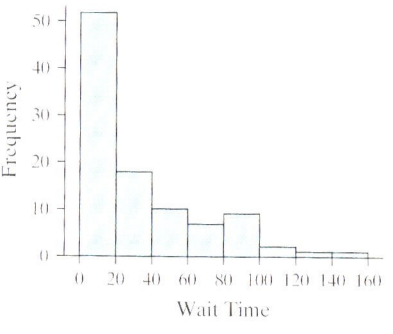

Now Work Problems 3 and 7

MAKING AN INFORMED DECISION

Join the Club

Suppose that you are interested in starting your own MENSA-type club. To qualify for the club, the potential member must have intelligence that is in the top 20% of all people. The problem that you face is that you do not have a baseline for measuring what qualifies as a top 20% score. To gather this data, you must obtain a random sample of at least 25 volunteers to take an online intelligence test. There are many online intelligence tests, but you need to make sure that the test will supply scored exams. One suggested site is www.queendom.com/tests/iq/classical_iq_r2_access.html.

Once you have obtained your sample of at least 25 test scores, answer the following questions.

(a) What is the mean test score? What is the standard deviation of the test scores?

(b) Do the sample data come from a population that is normally distributed? How do you know this?

(c) Assuming that the sample data come from a population that is normally distributed, determine the test score that would be required to join your club. That is, determine the test score that serves as a cutoff point for the top 20%. You can use this score to determine which potential members may join!

7.4 ASSESS YOUR UNDERSTANDING

Concepts and Vocabulary

1. Explain why normal probability plots should be linear if the data are normally distributed.

2. What does f_i represent?

Skill Building

In Problems 3–8, determine whether the normal probability plot indicates that the sample data could have come from a population that is normally distributed.

3.

NW

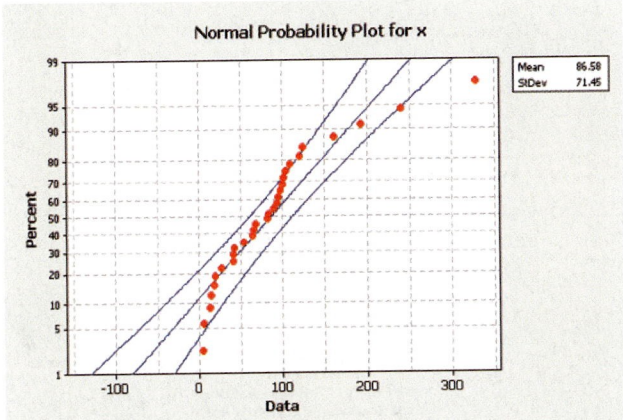

6.

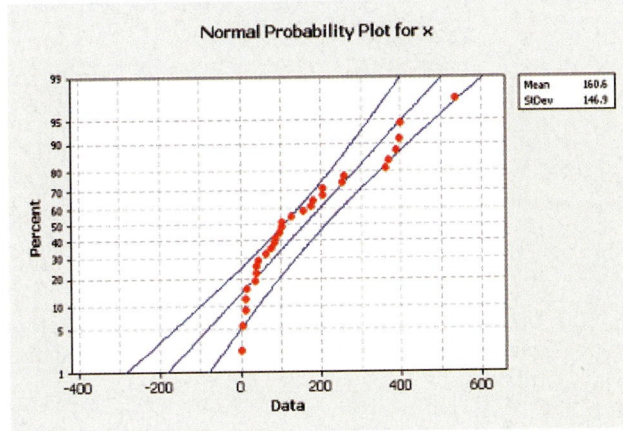

4.

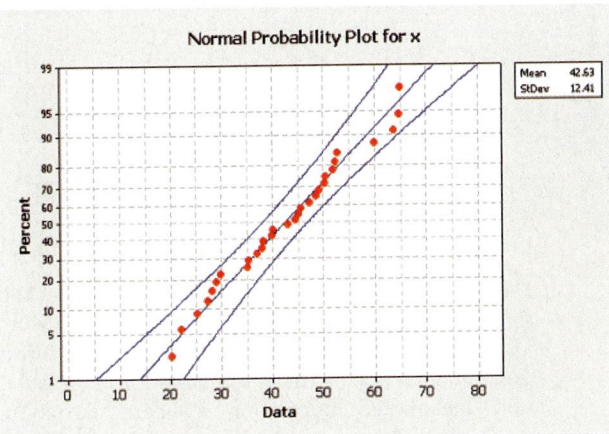

7.

NW

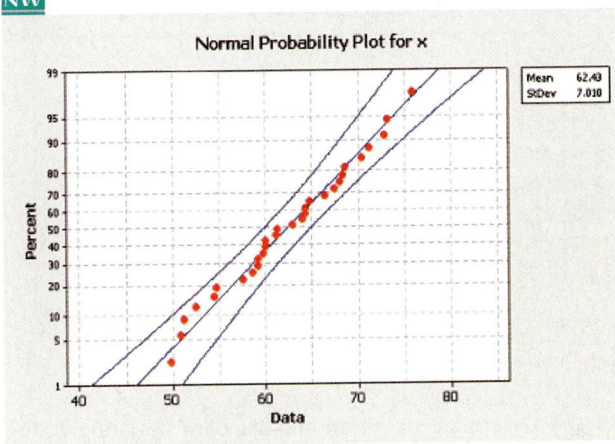

5.

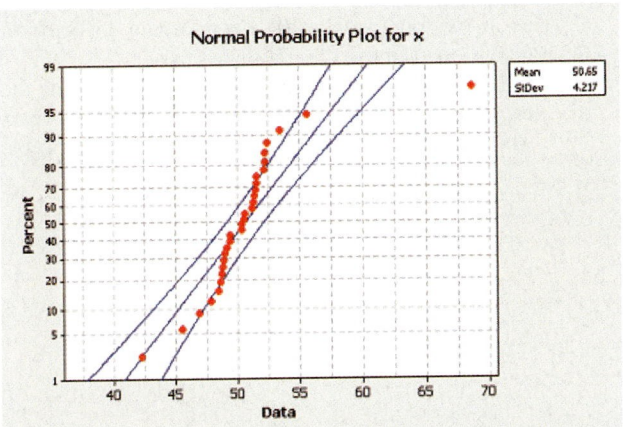

8.

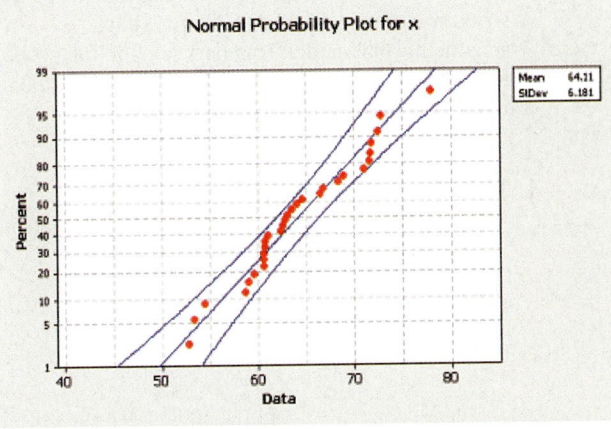

Applying the Concepts

9. **Chips per Bag** In a 1998 advertising campaign, Nabisco claimed that every 18-ounce bag of Chips Ahoy! cookies contained at least 1000 chocolate chips. Brad Warner and Jim Rutledge (*Chance*, Vol. 12, No. 1, 1999) tried to verify the claim. The following data represent the number of chips in an 18-ounce bag of Chips Ahoy! based on their study.

1087	1098	1103	1121	1132
1185	1191	1199	1200	1213
1239	1244	1247	1258	1269
1307	1325	1345	1356	1363
1135	1137	1143	1154	1166
1214	1215	1219	1219	1228
1270	1279	1293	1294	1295
1377	1402	1419	1440	1514

(a) Use the following normal probability plot to determine if the data could have come from a normal distribution.

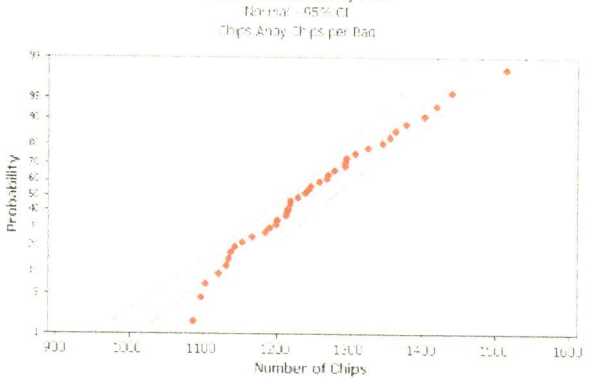

(b) Determine the mean and standard deviation of the sample data.

(c) Using the sample mean and sample standard deviation obtained in part (b) as estimates for the population mean and population standard deviation, respectively, draw a graph of a normal model for the distribution of chips in a bag a Chips Ahoy!

(d) Using the normal model from part (c), find the probability that an 18-ounce bag of Chips Ahoy! selected at random contains at least 1000 chips.

(e) Using the normal model from part (c), determine the proportion of 18-ounce bags of Chips Ahoy! that contains between 1200 and 1400 chips.

10. **Hours of TV** A random sample of college students aged 18 to 24 years was obtained, and the number of hours of television watched last week was recorded.

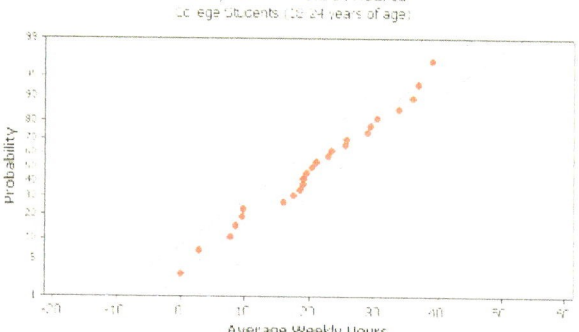

36.1	30.5	2.9	17.5	21.0
23.5	25.6	16.0	28.9	29.6
7.8	20.4	33.8	36.8	0.0
9.9	25.8	19.5	19.1	18.5
22.9	9.7	39.2	19.0	8.6

(a) Use the following normal probability plot to determine if the data could have come from a normal distribution.

(b) Determine the mean and standard deviation of the sample data.

(c) Using the sample mean and sample standard deviation obtained in part (b) as estimates for the population mean and population standard deviation, respectively, draw a graph of a normal model for the distribution of weekly hours of television watched.

(d) Using the normal model from part (c), find the probability that a college student aged 18 to 24 years, selected at random, watches between 20 and 35 hours of television each week.

(e) Using the normal model from part (c), determine the proportion of college students aged 18 to 24 years who watch more than 40 hours of television per week.

In Problems 11–14, use a normal probability plot to assess whether the sample data could have come from a population that is normally distributed.

11. O-Ring-Thickness A random sample of O-rings was obtained and the wall thickness (in inches) of each was recorded.

0.276	0.274	0.275	0.274	0.277
0.273	0.276	0.276	0.279	0.274
0.273	0.277	0.275	0.277	0.277
0.276	0.277	0.278	0.275	0.276

13. School Loans A random sample of 20 undergraduate students receiving student loans was obtained, and the amount of their loans for the 2004–2005 school year was recorded.

2,500	1,000	2,000	14,000	1,800
3,800	10,100	2,200	900	1,600
500	2,200	6,200	9,100	2,800
2,500	1,400	13,200	750	12,000

12. Customer Service A random sample of weekly work logs at an automobile repair station was obtained and the average number of customers per day was recorded.

26	24	22	25	23
24	25	23	25	22
21	26	24	23	24
25	24	25	24	25
26	21	22	24	24

14. Memphis Snowfall A random sample of 25 years between 1890 and 2005 was obtained, and the amount of snowfall, in inches, for Memphis was recorded.

Source: National Oceanic and Atmospheric Administration

24.0	7.9	1.5	0.0	0.3
0.4	8.1	4.3	0.0	0.5
3.6	2.9	0.4	2.6	0.1
16.6	1.4	23.8	25.1	1.6
12.2	14.8	0.4	3.7	4.2

Technology Step by Step	**Normal Probability Plots**

TI-83/84 Plus
Step 1: Enter the raw data into L1.
Step 2: Press 2nd Y = to access STAT PLOTS.
Step 3: Select 1:Plot1.
Step 4: Turn Plot1 ON by highlighting ON and pressing ENTER. Press the down-arrow key. Highlight the *normal probability plot* icon. It is the icon in the lower-right corner under Type:. Press ENTER to select this plot type. The Data List should be set at L1. The data axis should be the *x*-axis.
Step 5: Press ZOOM, and select 9:ZoomStat.

MINITAB
Step 1: Enter the raw data into C1.
Step 2: Select the **Graph** menu. Highlight **Probability Plot** ....
Step 3: In the Variables cell, enter the column that contains the raw data. Make sure Distribution is set to Normal. Click OK.

Excel
Step 1: Load the PHStat Add-in.
Step 2: Enter the raw data into column A.
Step 3: Select the **PHStat** menu. Highlight **Probability Distributions**, then highlight **Normal Probability Plot** ....
Step 4: With the cursor in the "Variable Cell Range:" cell, highlight the raw data. Enter a graph title, if desired. Click OK.

7.5 The Normal Approximation to the Binomial Probability Distribution

Preparing for This Section Before getting started, review the following:

- Binomial probability distribution (Section 6.2, pp. 328–339)

Objective

1 Approximate binomial probabilities using the normal distribution

1 Approximate Binomial Probabilities Using the Normal Distribution

In Section 6.2, we discussed the binomial probability distribution. A probability experiment is said to be a binomial experiment if the following conditions are met.

Criteria for a Binomial Probability Experiment

A probability experiment is said to be a binomial experiment if all the following are true:

1. The experiment is performed n independent times. Each repetition of the experiment is called a **trial**. Independence means that the outcome of one trial will not affect the outcome of the other trials.
2. For each trial, there are two mutually exclusive outcomes—success or failure.
3. The probability of success, p, is the same for each trial of the experiment.

The binomial probability formula can be used to compute probabilities of events in a binomial experiment. When there is a large number of trials of a binomial experiment, the binomial probability formula can be difficult to use. For example, suppose there are 500 trials of a binomial experiment and we wish to compute the probability of 400 or more successes. Using the binomial probability formula requires that we compute the following probabilities:

$$P(X \geq 400) = P(400) + P(401) + \cdots + P(500)$$

This would be time consuming to compute by hand! Fortunately, we have other means for approximating binomial probabilities, provided that certain conditions are met.

Recall, as the number of trials, n, in a binomial experiment increases, the probability histogram becomes more nearly symmetric and bell shaped (see page 338). We restate the conclusion here.

Historical Note

The normal approximation to the binomial was discovered by Abraham de Moivre in 1733. With the advance of computing technology, its importance has been diminished.

As the number of trials n in a binomial experiment increases, the probability distribution of the random variable X becomes more nearly symmetric and bell shaped. As a rule of thumb, if $np(1 - p) \geq 10$, the probability distribution will be approximately symmetric and bell shaped.

Because of this result, we might be inclined to think that binomial probabilities can be approximated by the area under the normal curve, provided that $np(1 - p) \geq 10$. This intuition is correct.

The Normal Approximation to the Binomial Probability Distribution

If $np(1 - p) \geq 10$, the binomial random variable X is approximately normally distributed with mean $\mu_X = np$ and standard deviation $\sigma_X = \sqrt{np(1 - p)}$.

Figure 47 shows a probability histogram for the binomial random variable X with $n = 40$ and $p = 0.5$ and a normal curve with $\mu_X = np = 40(0.5) = 20$ and standard deviation $\sigma_X = \sqrt{np(1 - p)} = \sqrt{40(0.5)(0.5)} = \sqrt{10}$. Notice that $np(1 - p) = 40(0.5)(1 - 0.5) = 10$.

Figure 47

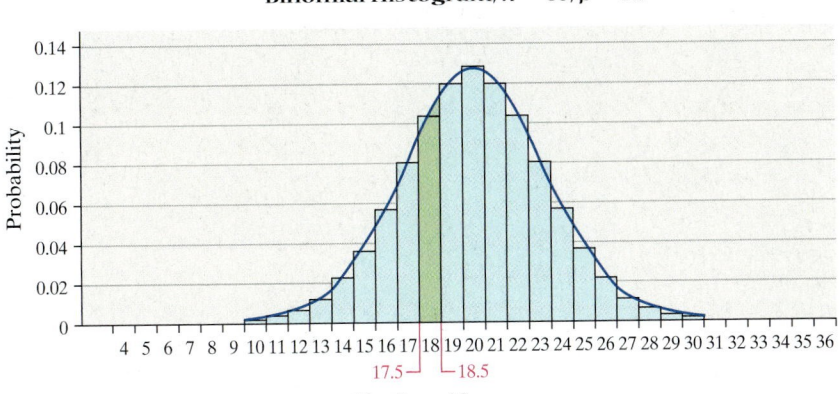

Binomial Histogram, $n = 40$, $p = 0.5$

Number of Successes, x

> **CAUTION**
> Don't forget about the correction for continuity. It is needed because we are using a continuous density function to approximate the probability of a discrete random variable.

We know from Section 6.2 that the area of the rectangle corresponding to $X = 18$ represents $P(18)$. The width of each rectangle is 1, so the rectangle extends from $X = 17.5$ to $X = 18.5$. The area under the normal curve from $X = 17.5$ to $X = 18.5$ is approximately equal to the area of the rectangle corresponding to $X = 18$. Therefore, the area under the normal curve between $X = 17.5$ and $X = 18.5$ is approximately equal to $P(18)$, where X is a binomial random variable with $n = 40$ and $p = 0.5$. We add and subtract 0.5 from $X = 18$ as a **correction for continuity**, because we are using a continuous density function to approximate a discrete probability.

Suppose we want to approximate $P(X \leq 18)$. Figure 48 illustrates the situation.

Figure 48

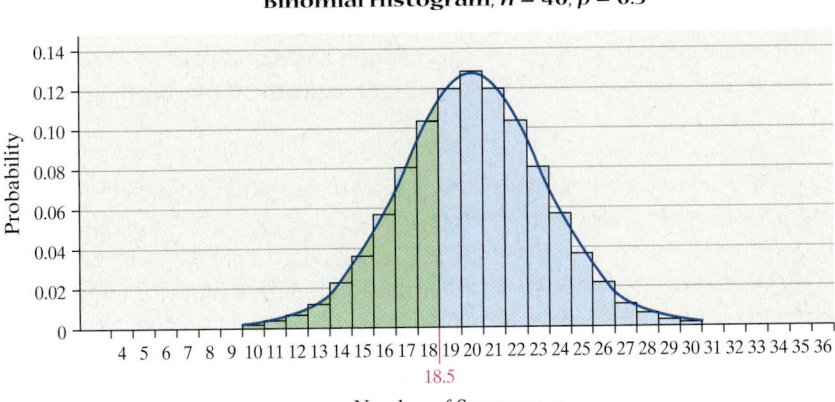

Binomial Histogram, $n = 40$, $p = 0.5$

Number of Successes, x

To approximate $P(X \leq 18)$, we compute the area under the normal curve for $X < 18.5$. Do you see why?

If we want to approximate $P(X \geq 18)$, we compute $P(X \geq 17.5)$. Do you see why? Table 7 summarizes how to use the correction for continuity.

Table 7

Exact Probability Using Binomial	Approximate Probability Using Normal	Graphical Depiction
$P(X = a)$	$P(a - 0.5 < X < a + 0.5)$	
$P(X \leq a)$	$P(X < a + 0.5)$	
$P(X \geq a)$	$P(X > a - 0.5)$	
$P(a \leq X \leq b)$	$P(a - 0.5 < X < b + 0.5)$	

A question remains, however. What do we do if the probability is of the form $P(X > a)$, $P(X < a)$, or $P(a < X < b)$? The solution is to rewrite the inequality in a form with $\leq$ or $\geq$. For example, $P(X > 4) = P(X \geq 5)$ and $P(X < 4) = P(X \leq 3)$ for binomial random variables, because the values of the random variables must be whole numbers.

EXAMPLE 1

The Normal Approximation to a Binomial Random Variable

Problem: According to the *Information Please Almanac*, 6% of the human population has blood type O-negative. What is the probability that, in a simple random sample of 500, fewer than 25 have blood type O-negative?

Approach

Step 1: We verify that this is a binomial experiment.

Step 2: Computing the probability by hand would be very tedious. Verify $np(1 - p) > 10$. Then we will know that the condition for using the normal distribution to approximate the binomial distribution is met.

Step 3: Approximate $P(X < 25) = P(X \leq 24)$ by using the normal approximation to the binomial distribution.

Solution

Step 1: There are 500 independent trials with each trial having a probability of success equal to 0.06. This is a binomial experiment.

Step 2: We verify $np(1 - p) \geq 10$.

$$np(1 - p) = 500(0.06)(0.94) = 28.2 \geq 10$$

Figure 49

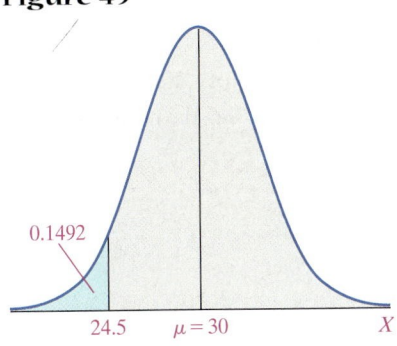

We can use the normal distribution to approximate the binomial distribution.

Step 3: We wish to know the probability that fewer than 25 people in the sample have blood type O-negative; that is, we wish to know $P(X < 25) = P(X \le 24)$. This is approximately equal to the area under the normal curve to the left of $X = 24.5$, with $\mu_X = np = 500(0.06) = 30$ and $\sigma_X = \sqrt{np(1 - p)} = \sqrt{500(0.06)(1 - 0.06)} = \sqrt{28.2} \approx 5.31$. See Figure 49. We convert $X = 24.5$ to a Z-score.

$$Z = \frac{24.5 - 30}{5.31} = -1.04$$

From Table IV, we find the area to the left of $Z = -1.04$ is 0.1492. Therefore, the approximate probability that fewer than 25 people will have blood type O-negative is $0.1492 = 14.92\%$.

Figure 50

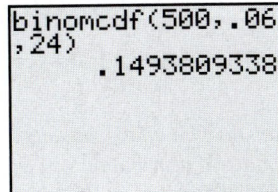

Using the *binomcdf(* command on a TI-84 + graphing calculator, we find that the exact probability is 0.1494. See Figure 50. The approximate result is close indeed!

Now Work Problem 21.

EXAMPLE 2

A Normal Approximation to the Binomial

Problem: According to the Federal Communications Commission, 70% of all U.S. households have cable television. Erica conducts a random sample of 1000 households in DuPage County and finds that 734 of them have cable.

(a) Assuming that 70% of households have cable, what is the probability of obtaining a random sample of at least 734 households with cable from a sample of size 1,000?
(b) Does the result from part (a) contradict the FCC information? Explain.

Approach: This is a binomial experiment with $n = 1000$ and $p = 0.70$. Erica needs to determine the probability of obtaining a random sample of at least 734 households with cable from a sample of size 1000, assuming 70% of households have cable. Computing this using the binomial probability formula would be difficult, so Erica will compute the probability using the normal approximation to the binomial, since $np(1 - p) = 1000(0.70)(0.30) = 210 \ge 10$. We approximate $P(X \ge 734)$ by computing the area under the standard normal curve to the right of $X = 733.5$ with $\mu_X = np = 1000(0.70) = 700$ and $\sigma_X = \sqrt{np(1 - p)} = \sqrt{1000(0.70)(1 - 0.70)} = \sqrt{210} \approx 14.491$.

Solution:

(a) Figure 51 shows the area we wish to compute. We convert $X = 733.5$ to a Z-score.

Figure 51

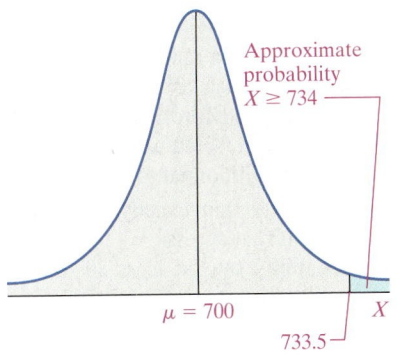

$$Z = \frac{733.5 - 700}{14.491} = 2.31$$

The area under the standard normal curve to the right of $Z = 2.31$ is $1 - 0.9896 = 0.0104$. There is a 1.04% probability of obtaining 734 or more households with cable from a sample of 1000 households, assuming that the percentage of households with cable is 70%.

(b) Yes. The result from part (a) means that about 1 sample in every 100 samples will have 734 or more households with cable if the true proportion is 0.7. Erica is not inclined to believe that her sample is one of the 1 in 100. She would rather believe that the proportion of households in DuPage County with cable is higher than 0.70.

Now Work Problem 27.

7.5 ASSESS YOUR UNDERSTANDING

Concepts and Vocabulary

1. List the conditions required for a binomial experiment.
2. Under what circumstances can the normal distribution be used to approximate binomial probabilities?
3. Why must we use a correction for continuity when using the normal distribution to approximate binomial probabilities?

4. *True or False:* Suppose X is a binomial random variable. To approximate $P(3 \leq X \leq 7)$ using the normal probability distribution, we compute $P(3.5 \leq X < 7.5)$.

Skill Building

In Problems 5–14, a discrete random variable is given. Assume the probability of the random variable will be approximated using the normal distribution. Describe the area under the normal curve that will be computed. For example, if we wish to compute the probability of finding at least five defective items in a shipment, we would approximate the probability by computing the area under the normal curve to the right of $X = 4.5$.

5. The probability that at least 40 households have a gas stove.
6. The probability of no more than 20 people who want to see *Roe v. Wade* overturned.
7. The probability that exactly eight defective parts are in the shipment.
8. The probability that exactly 12 students pass the course.
9. The probability that the number of people with blood type O-negative is between 18 and 24, inclusive.
10. The probability that the number of tornadoes that occur in the month of May is between 30 and 40, inclusive.

11. The probability that more than 20 people want to see the marriage tax penalty abolished.
12. The probability that fewer than 40 households have a pet.
13. The probability that more than 500 adult Americans support a bill proposing to extend daylight savings time.
14. The probability that fewer than 35 people support the privatization of Social Security.

In Problems 15–20, compute $P(x)$ using the binomial probability formula. Then determine whether the normal distribution can be used as an approximation for the binomial distribution. If so, approximate $P(x)$ and compare the result to the exact probability.

15. $n = 60$, $p = 0.4$, $X = 20$
16. $n = 80$, $p = 0.15$, $X = 18$
17. $n = 40$, $p = 0.25$, $X = 30$
18. $n = 100$, $p = 0.05$, $x = 50$
19. $n = 75$, $p = 0.75$, $X = 60$
20. $n = 85$, $p = 0.8$, $X = 70$

Applying the Concepts

21. **On-Time Flights** According to American Airlines, Flight 215 from Orlando to Los Angeles is on time 90% of the time. Suppose 150 flights are randomly selected. Use the normal approximation to the binomial to
    (a) approximate the probability that exactly 130 flights are on time.
    (b) approximate the probability that at least 130 flights are on time.
    (c) approximate the probability that fewer than 125 flights are on time.
    (d) approximate the probability that between 125 and 135 flights, inclusive, are on time.

22. **Smokers** According to *Information Please Almanac*, 80% of adult smokers started smoking before they were 18 years old. Suppose 100 smokers 18 years old or older are randomly selected. Use the normal approximation to the binomial to
    (a) approximate the probability that exactly 80 of them started smoking before they were 18 years old.
    (b) approximate the probability that at least 80 of them started smoking before they were 18 years old.
    (c) approximate the probability that fewer than 70 of them started smoking before they were 18 years old.
    (d) approximate the probability that between 70 and 90 of them, inclusive, started smoking before they were 18 years old.

23. **Migraine Sufferers** In clinical trials and extended studies of a medication whose purpose is to reduce the pain associated with migraine headaches, 2% of the patients in the study experienced weight gain as a side effect. Suppose a random sample of 600 users of this medication is obtained. Use the normal approximation to the binomial to
    (a) approximate the probability that exactly 20 will experience weight gain as a side effect.
    (b) approximate the probability that 20 or fewer will experience weight gain as a side effect.
    (c) approximate the probability that 22 or more patients will experience weight gain as a side effect.
    (d) approximate the probability that between 20 and 30 patients, inclusive, will experience weight gain as a side effect.

24. **High-Speed Internet** According to a report by the Commerce Department in the fall of 2004, 20% of U.S. households had some type of high-speed Internet connection. Suppose 80 U.S. households are selected at random. Use the normal approximation to the binomial to
    (a) approximate the probability that exactly 15 households have high-speed Internet access.
    (b) approximate the probability that at least 20 households have high-speed Internet access.

(c) approximate the probability that fewer than 10 households have high-speed Internet access.

(d) approximate the probability that between 12 and 18 households, inclusive, have high-speed Internet access.

25. Allergy Sufferers   Clarinex-D is a medication whose purpose is to reduce the symptoms associated with a variety of allergies. In clinical trials of Clarinex-D, 5% of the patients in the study experienced insomnia as a side effect. Suppose a random sample of 400 Clarinex-D users is obtained. Use the normal approximation to the binomial to

(a) approximate the probability that exactly 20 patients experienced insomnia as a side effect.

(b) approximate the probability that 15 or fewer patients experienced insomnia as a side effect.

(c) approximate the probability that 30 or more patients experienced insomnia as a side effect.

(d) approximate the probability that between 10 and 32 patients, inclusive, experienced insomnia as a side effect.

26. Murder by Firearms   According to the *Uniform Crime Report, 2003,* 66.9% of murders are committed with a firearm. Suppose that 50 murders are randomly selected. Use the normal approximation to the binomial to

(a) approximate the probability that exactly 40 murders are committed using a firearm.

(b) approximate the probability that at least 35 murders are committed using a firearm.

(c) approximate the probability that fewer than 25 murders are committed using a firearm.

(d) approximate the probability that between 30 and 35 murders, inclusive, are committed using a firearm.

27. Males Living at Home   According to the *Current Population Survey* (Internet release date: September 15, 2004), 55% of males between the ages of 18 and 24 years lived at home in 2003. (Unmarried college students living in a dorm are counted as living at home.) Suppose that a survey is administered at a community college to 200 randomly selected male students between the ages of 18 and 24 years and that 130 of them respond that they live at home.

(a) Approximate the probability that such a survey will result in at least 130 of the respondents living at home under the assumption that the true percentage is 55%.

(b) Does the result from part (a) contradict the results of the *Current Population Survey*? Explain.

28. Females Living at Home   According to the *Current Population Survey* (Internet release date: September 15, 2004), 46% of females between the ages of 18 and 24 years lived at home in 2003. (Unmarried college students living in a dorm are counted as living at home.) Suppose that a survey is administered at a community college to 200 randomly selected female students between the ages of 18 and 24 years and that 110 of them respond that they live at home.

(a) Approximate the probability that such a survey will result in at least 110 of the respondents living at home under the assumption that the true percentage is 46%.

(b) Does the result from part (a) contradict the results of the *Current Population Survey*? Explain.

29. Boys Are Preferred   In a Gallup poll conducted December 2–4, 2000, 42% of survey respondents said that, if they only had one child, they would prefer the child to be a boy. Suppose you conduct a survey of 150 randomly selected students on your campus and find that 80 of them would prefer a boy.

(a) Approximate the probability that, in a random sample of 150 students, at least 80 would prefer a boy, assuming the true percentage is 42%.

(b) Does this result contradict the Gallup poll? Explain.

30. Liars   According to a *USA Today* "Snapshot," 3% of Americans surveyed lie frequently. Suppose you conduct a survey of 500 college students and find that 20 of them lie frequently.

(a) Compute the probability that, in a random sample of 500 college students, at least 20 lie frequently, assuming the true percentage is 3%.

(b) Does this result contradict the *USA Today* "Snapshot"? Explain.

Chapter 7   **Review**

Summary

In this chapter, we introduced continuous random variables and the normal probability density function. A continuous random variable is said to be approximately normally distributed if a histogram of its values is symmetric and bell shaped. In addition, we can draw normal probability plots that are based on expected Z-scores. If these normal probability plots are approximately linear, we say the distribution of the random variable is approximately normal. The area under the normal density function can be used to find proportions or probabilities for normal random variables. Also, we can find the value of

a normal random variable that corresponds to a specific proportion, probability, or percentile.

If X is a binomial random variable with $np(1 - p) \geq 10$, we can use the area under the normal curve to approximate the probability of a binomial random variable. The parameters of the normal curve are $\mu_X = np$ and $\sigma_X = \sqrt{np(1 - p)}$, where n is the number of trials of the binomial experiment and p is the probability of success.

Formulas

Standardizing a Normal Random Variable

$$Z = \frac{X - \mu}{\sigma}$$

Finding the Score

$$X = \mu + Z\sigma$$

Vocabulary

Uniform probability distribution (p. 359)
Probability density function (p. 360)
Normal curve (p. 361)
Normal probability distribution (p. 361)

Inflection points (p. 361)
Normal probability density function (p. 364)
Standard normal distribution (p. 366)
Normal probability plot (p. 395)

Normal score (p. 395)
Trial (p. 402)
Correction for continuity (p. 403)

Objectives

Section	You should be able to . . .	Examples	Review Exercises
7.1	1 Understand the uniform probability distribution (p. 361)	1 and 2	37
	2 Graph a normal curve (p. 361)	Page 361	19–22
	3 State the properties of the normal curve (p. 362)	Page 362	38
	4 Understand the role of area in the normal density function (p. 363)	3 and 4	1, 2
	5 Understand the relation between a normal random variable and a standard normal random variable (p. 365)	5	3, 4
7.2	1 Find the area under the standard normal curve (p. 372)	1 through 4	5–8
	2 Find the Z-scores for a given area (p. 376)	5 through 9	13–18
	3 Interpret the area under standard normal curve as a probability (p. 380)	10	9–12
7.3	1 Find and interpret the area under a normal curve (p. 385)	1 through 3	19–22, 23(a)–(c), 24(a)–(c), 25(a)–(c), 26(a)–(d), 27, 28
	2 Find the value of a normal random variable (p. 388)	4 through 6	23(d), 24(d), (e), 25(d), (e), 26(e), (f)
7.4	1 Draw normal probability plots to assess normality (p. 395)	1 through 3	31–33
7.5	1 Approximate binomial probabilities using the normal distribution (p. 402)	1 and 2	29, 30

Review Exercise

1. Use the figure to answer the questions that follow:

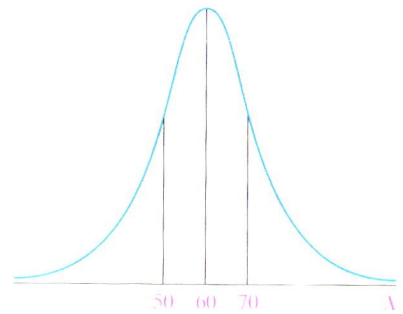

(a) What is μ?
(b) What is σ?
(c) Suppose the area under the normal curve to the right of $X = 75$ is 0.0668. Provide two interpretations for this area.
(d) Suppose the area under the normal curve between $X = 50$ and $X = 75$ is 0.7745. Provide two interpretations for this area.

2. Use the figure to answer the questions that follow:

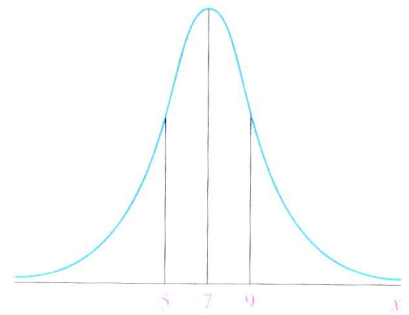

(a) What is μ?
(b) What is σ?
(c) Suppose the area under the normal curve to the left of $X = 10$ is 0.9332. Provide two interpretations for this area.
(d) Suppose the area under the normal curve between $X = 5$ and $X = 8$ is 0.5328. Provide two interpretations for this area.

3. A random variable X is approximately normally distributed with $\mu = 20$ and $\sigma = 4$.

 (a) Compute $Z_1 = \dfrac{X_1 - \mu}{\sigma}$ for $X_1 = 18$.

 (b) Compute $Z_2 = \dfrac{X_2 - \mu}{\sigma}$ for $X_2 = 21$.

 (c) The area under the normal curve between $X_1 = 18$ and $X_2 = 21$ is 0.2912. What is the area between Z_1 and Z_2?

4. A random variable X is approximately normally distributed with $\mu = 50$ and $\sigma = 8$.

 (a) Compute $Z_1 = \dfrac{X_1 - \mu}{\sigma}$ for $X_1 = 48$.

 (b) Compute $Z_2 = \dfrac{X_2 - \mu}{\sigma}$ for $X_2 = 60$.

 (c) The area under the normal curve between $X_1 = 48$ and $X_2 = 60$ is 0.4931. What is the area between Z_1 and Z_2?

In Problems 5–8, draw a standard normal curve and shade the area indicated. Then use Table IV to find the area under the normal curve.

5. The area to the left of $Z = -1.04$

6. The area to the right of $Z = 2.04$

7. The area between $Z = -0.34$ and $Z = 1.03$

8. The area between $Z = 1.93$ and $Z = 3.93$

In Problems 9–12, find the indicated probability of the standard normal random variable Z.

9. $P(Z < 1.19)$

10. $P(Z \geq 1.61)$

11. $P(-1.21 < Z \leq 2.28)$

12. $P(0.21 < Z < 1.69)$

13. Find the Z-score such that the area to the left of the Z-score is 0.84.

14. Find the Z-score such that the area to the right of the Z-score is 0.483.

15. Find the Z-scores that separate the middle 92% of the data from the area in the tails of the standard normal distribution.

16. Find the Z-scores that separate the middle 88% of the data from the area in the tails of the standard normal distribution.

17. Find the value of $z_{0.20}$

18. Find the value of $z_{0.04}$

In Problems 19–22, draw the normal curve with the parameters indicated. Then find the probability of the random variable X. Shade the area that represents the probability.

19. $\mu = 50, \sigma = 6, P(X > 55)$

20. $\mu = 30, \sigma = 5, P(X \leq 23)$

21. $\mu = 70, \sigma = 10, P(65 < X < 85)$

22. $\mu = 20, \sigma = 3, P(22 \leq X \leq 27)$

23. **Tire Wear** Suppose Dunlop Tire manufactures tires having the property that the mileage the tire lasts approximately follows a normal distribution with mean 70,000 miles and standard deviation 4400 miles.

 (a) What percent of the tires will last at least 75,000 miles?

 (b) Suppose Dunlop warrants the tires for 60,000 miles. What percent of the tires will last 60,000 miles or less?

 (c) What is the probability that a randomly selected Dunlop tire lasts between 65,000 and 80,000 miles?

 (d) Suppose that Dunlop wants to warrant no more than 2% of its tires. What mileage should the company advertise as its warranty mileage?

24. **Talk Time on a Cell Phone** Suppose the talk time in digital mode on a Motorola Timeport P8160 is approximately normally distributed with mean 324 minutes and standard deviation 24 minutes.

 (a) What proportion of the time will a fully charged battery last at least 300 minutes?

 (b) What proportion of the time will a fully charged battery last less than 340 minutes?

 (c) Suppose you charge the battery fully. What is the probability it will last between 310 and 350 minutes?

 (d) Determine the talk time that is in the top 20%.

 (e) Determine the talk time that makes up the middle 90% of talk time.

25. **Serum Cholesterol** As reported by the U.S. National Center for Health Statistics, the mean serum cholesterol of females 16 to 19 years old is $\mu = 171$. If serum cholesterol is approximately normally distributed with $\sigma = 39.8$, answer the following:

 (a) Determine the proportion of 16- to 19-year-old females with a serum cholesterol above 180.

 (b) Determine the proportion of 16- to 19-year-old females with a serum cholesterol between 150 and 200.

 (c) Suppose a 16- to 19-year-old female is randomly selected. Determine the probability her serum cholesterol is below 140.

 (d) Determine the serum cholesterol that divides the bottom 10% from the top 90% of all serum cholesterol levels of 16- to 19-year-old females.

 (e) According to the National Center for Health Statistics, the 25th percentile of serum cholesterol for 16- to 19-year-old females is 145. Is the 25th percentile on the normal curve close to the reported value of 145?

26. **Wechsler Intelligence Scale** The Wechsler Intelligence Scale for Children is approximately normally distributed with mean 100 and standard deviation 15.

 (a) What proportion of test takers will score above 125?

 (b) What proportion of test takers will score below 90?

 (c) What proportion of test takers will score between 110 and 140?

 (d) If a child is randomly selected, what is the probability that she scores above 150?

 (e) What intelligence score will place a child in the top 5% of all children?

 (f) If normal intelligence is defined as scoring in the middle 95% of all test takers, figure out the scores that differentiate normal intelligence from abnormal intelligence.

27. Major League Baseballs According to major league baseball rules, the ball must weigh between 5 and 5.25 ounces. (Source:www.baseball-almanac.com). Suppose a factory produces baseballs whose weights are approximately normally distributed with mean 5.11 ounces and standard deviation 0.062 ounces.

(a) What proportion of the baseballs produced by this factory are too heavy for use by major league baseball?

(b) What proportion of the baseballs produced by this factory are too light for use by the major league baseball?

(c) What proportion of the baseballs produced by this factory can be used by major league baseball?

28. Halogen Light Bulbs Feit Electric manufactures a Crystal Clear Halogen 60-watt light bulb that has an average life of 3000 hours. Suppose the lifetimes of the light bulbs are approximately normally distributed with standard deviation 183 hours.

(a) What proportion of the light bulbs will last more than 3300 hours?

(b) What proportion of the light bulbs will last less than 2500 hours?

(c) What is the probability that a randomly selected Feit Crystal Clear Halogen 60-watt light bulb lasts between 2900 and 3100 hours?

(d) What is the probability that a randomly selected Feit Crystal Clear Halogen 60-watt light bulb lasts less than 2600 hours?

(e) What is the percentile rank of a Feit Crystal Clear Halogen 60-watt light bulb that lasts 3350 hours?

(f) Would it be unusual for a Feit Crystal Clear Halogen 60-watt light bulb to last longer than 3400 hours?

29. High Cholesterol According to the National Center for Health Statistics, 8% of 20- to 34-year-old females have high serum cholesterol. Suppose you conduct a random sample of two hundred 20- to 34-year-old females.

(a) Verify that the conditions for using the normal distribution to approximate the binomial distribution are met.

(b) Approximate the probability that exactly 15 have high serum cholesterol. Interpret this result.

(c) Approximate the probability that more than 20 have high serum cholesterol. Interpret this result.

(d) Approximate the probability that at least 15 have high serum cholesterol. Interpret this result.

(e) Approximate the probability that fewer than 25 have high serum cholesterol. Interpret this result.

(f) Approximate the probability that between 15 and 25, inclusive, have high serum cholesterol. Interpret this result.

30. America Reads According to a Gallup poll conducted September 10–14, 1999, 56% of Americans 18 years old or older stated they had read at least six books (fiction and nonfiction) within the past year. Suppose you conduct a random sample of 250 Americans 18 years old or older.

(a) Verify that the conditions for using the normal distribution to approximate the binomial distribution are met.

(b) Approximate the probability that exactly 125 read at least six books within the past year. Interpret this result.

(c) Approximate the probability that fewer than 120 read at least six books within the past year. Interpret this result.

(d) Approximate the probability that at least 140 read at least six books within the past year. Interpret this result.

(e) Approximate the probability that between 100 and 120, inclusive, read at least six books within the past year. Interpret this result.

In Problems 31 and 32, a normal probability plot of a simple random sample of data from a population whose distribution is unknown was obtained. Given the normal probability plot, is there reason to believe the population is normally distributed?

31.

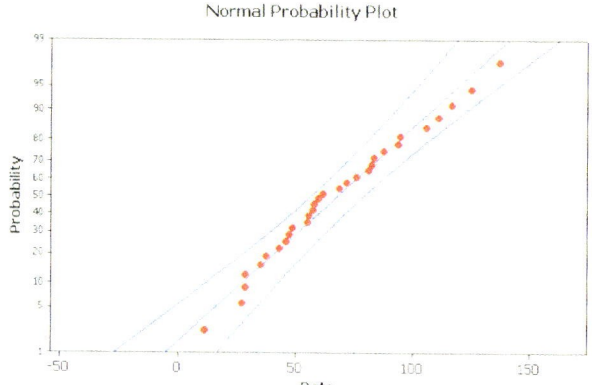

32.

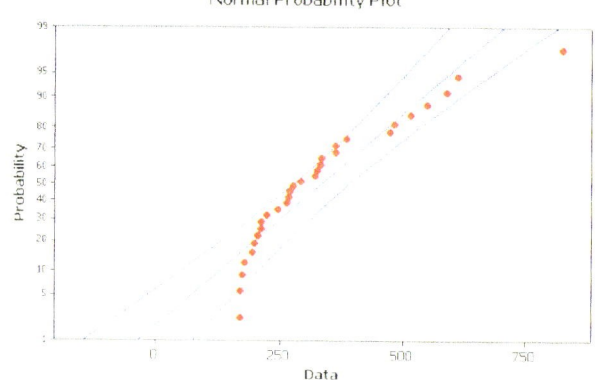

In Problems 33 and 34, assess the normality of the sample data.

33. Density of Earth In 1798, Henry Cavendish obtained 27 measurements of the density of Earth, using a torsion balance. The following data represent his estimates, represented as a multiple of the density of water.

5.50	5.57	5.42	5.61	5.53
5.47	4.88	5.62	5.63	4.07
5.29	5.34	5.26	5.44	5.55
5.34	5.30	5.36	5.79	5.29
5.10	5.86	5.58	5.27	5.85
5.65	5.39			

Source: Stigler, S.M. "Do robust estimators work with real data?" *Annals of Statistics,* 5 (1977), 1055–1078.

34. Life Expectancy The following data represent the life expectancy at birth in 2005 in a random sample of 20 countries.

75.84	77.11	75.91	80.39	78.92
74.23	78.62	77.79	77.83	80.10
79.95	76.58	76.84	79.14	77.62
79.21	78.35	77.71	80.40	78.19

Source: U.S. Census Bureau, International Database

35. Hybrid SUV As the first hybrid sports utility vehicle (SUV) with gasoline mileage certified by the Environmental Protection Agency, the Ford Escape HEV is the most fuel-efficient SUV in 2005. The mean mileage for the automatic four-wheel-drive Ford Escape HEV is 29 miles per gallon on the highway. (*Source:*www.fueleconomy.gov). Suppose the gasoline mileages of these SUVs are normally distributed with standard deviation 2 miles per gallon.

(a) What proportion of automatic four-wheel-drive 2005 Ford Escape HEVs gets over 25 miles per gallon on the highway?

(b) What proportion of automatic four-wheel-drive 2005 Ford Escape HEVs gets less than 30 miles per gallon on the highway?

(c) What is the probability that a randomly selected automatic four-wheel-drive 2005 Ford Escape HEV gets between 26 and 34 miles per gallon on the highway?

(d) What is the probability that a randomly selected automatic four-wheel-drive 2005 Ford Escape HEV gets over 35 miles per gallon on the highway?

(e) What is the percentile rank of an automatic four-wheel-drive 2005 Ford Escape HEV that gets 32 miles per gallon?

(f) What is the percentile rank of an automatic four-wheel-drive 2005 Ford Escape HEV that gets 25 miles per gallon?

36. Creative Thinking According to a *USA Today* "Snapshot," 20% of adults surveyed do their most creative thinking while driving. Suppose you conduct a survey of 250 adults and find that 30 do their most creative thinking while driving.

(a) Compute the probability that, in a random sample of 250 adults, 30 or fewer do their most creative thinking while driving.

(b) Does this result contradict the *USA Today* "Snapshot"? Explain.

37. Suppose a continuous random variable X is uniformly distributed with $0 \leq X \leq 20$.

(a) Draw a graph of the uniform density function.

(b) What is $P(0 \leq X \leq 5)$?

(c) What is $P(10 \leq X \leq 18)$?

38. List the properties of the standard normal curve.

39. Explain how to use a normal probability plot to assess normality.

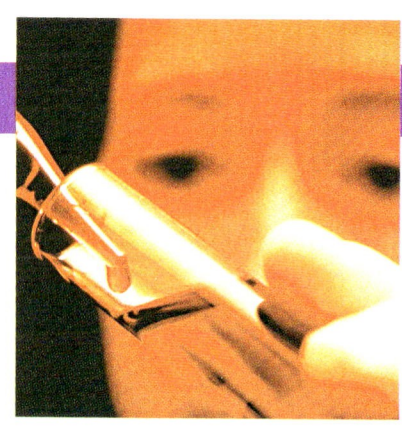

CASE STUDY

A Tale of Blood Chemistry and Health

Abby Tudor recently turned 40 years old. Her knees ache, and she often feels short of breath during exercise. She is experiencing fatigue and often feels that she is going in slow motion. Periodic dizziness plagues her during most days. According to the drugstore machine, her blood pressure is elevated. Her family has a history of cardiac disease, with both of her parents having experienced heart attacks. Additionally, an aunt on her mother's side has diabetes. Hypothyroidism also runs throughout her immediate family. Abby is approximately 20 pounds overweight. She has tried various diet and exercise programs in an attempt to lose weight. Her results have been disappointing.

With the advice of her physician, she scheduled an appointment for a full physical exam, including a complete blood workup. Her doctor is particularly interested in the level of various blood components that might shed some light on Abby's reported symptoms. Specifically, he wants to examine her white blood cell count, red blood cell count, hemoglobin, and hematocrit figures for indications of infection or anemia. Her serum glucose level will provide information concerning the possibility of the onset of diabetes. Cholesterol and triglyceride levels will provide insight into potential cardiac problems. Additionally, the possibility of hypothyroidism will be investigated by examining Abby's serum TSH level.

As instructed, two weeks before her doctor's appointment, she reported to her doctor's lab for a blood test. She confirmed for the lab technician that she had fasted for the 12 hours immediately preceding the exam.

During her physical, Abby's doctor went over the blood test report with her. He expressed concern over some of the results, but Abby was not convinced that she had a problem. Additional blood tests were not a viable option because it takes a good deal of time to get a sample of readings and they are expensive. At the time of her doctor's appointment, she was unwilling to accept the offered prescriptions. She chose to do a little research before committing herself to any drug regimen.

Her research revealed that many medical measurements, such as cholesterol, are normally distributed in healthy populations. Unfortunately, the lab report did not provide the means and standard deviations necessary for Abby to calculate the various probabilities of interest. However, the report did provide the appropriate reference intervals. Assuming that the reference intervals represent the range of values for each blood component for a healthy adult population, it is possible to estimate the various means and standard deviations for this population. Abby estimated each mean by taking the midpoint of its reference interval. Using the Range Rule of Thumb ($\sigma \approx$ range/4), standard deviations were estimated by dividing the reference interval range by 4. The following table lists Abby's blood test results, as well as the mean and standard deviation for a number of blood test components for the population of normal healthy adults.

For any blood component measurement that was below its population mean, Abby decided to calculate the probability that she would get a test value equal to the value obtained or less, given that she was a member of the healthy population. For example, her HDL cholesterol reading (42 mg/dL) was below the mean of the healthy population (92.5 mg/dL), so she calculated the following probability:

$$P(X \leq 42 \text{ mg/dL})$$

Blood Test Components for Healthy Adults and the Blood Test Results for Abby Tudor[*]

Abby's Blood Test Component	Unit	Standard Mean	Abby's Absolute Deviation	Result
White blood cell count	$10^3/\mu L$	7.25	1.625	5.3
Red blood cell count	$10^6/\mu L$	4.85	0.375	4.62
Hemoglobin	g/dL	14.75	1.125	14.6
Hematocrit	%	43.0	3.5	41.7
Glucose, serum	mg/dL	87.0	11.0	95.0
Creatine, serum	mg/dL	1.00	0.25	0.8
Sodium, serum	mEq/L	141.5	3.25	143.0
Potassium, serum	mEq/L	4.5	0.5	5.1
Chloride, serum	mEq/L	102.5	3.25	100.0
Carbon dioxide, total	mEq/L	26.0	3.0	25.0
Calcium, serum	mg/dL	9.55	0.525	10.1
Total cholesterol	mg/dL	149.5	24.75	253.0
Triglycerides	mg/dL	99.5	49.75	150.0
HDL cholesterol	mg/dL	92.5	28.75	42.0
LDL cholesterol	mg/dL	64.5	32.25	181.0
LDL/HDL ratio	Ratio	1.8	0.72	4.3
TSH, high sensitivity, serum	mcIU/mL	2.925	1.2875	3.15

[*] Population means and standard deviations were estimated from the reference intervals derived from an actual blood test report provided by TA LabCorp, Tampa, Florida. Means were estimated by taking the midpoints of the reference intervals. Standard deviations were estimated by dividing the reference interval ranges by 4. Test results attributed to Abby Tudor are actual results obtained from an anonymous patient.

Similarly, for any blood component measurement reading exceeding its population mean, Abby decided to calculate the probability that she would get a test value equal to the value obtained or more, given that she was a member of the healthy population. For example, her LDL cholesterol value (181 mg/dL) exceeds the mean of the healthy population (64.5 mg/dL), so she calculated the following probability:

$$P(X \geq 181 \text{ mg/dL})$$

To help her interpret the calculated probabilities, Abby decided to be concerned about only those blood components that had a probability less than 0.025. By choosing this figure, she is acknowledging that it is unlikely that she could have such an extreme blood component reading and still be part of the healthy population.

1. The reference interval for HDL cholesterol is 35 to 150 mg/dL. Use this information to confirm the mean and standard deviation provided for this blood component.
2. Using Abby's criteria and the means and standard deviations provided in her blood test report, determine which blood components should be a cause of concern for Abby. Write up a summary report of your findings. Be sure to include a discussion concening your assumptions and any limitations to your conclusions.

PART 4

Inference: From Samples to Population

In Chapter 1, we presented the following process of statistics:

Step 1: Identify a research objective.

Step 2: Collect the information needed to answer the questions posed in Step 1.

Step 3: Organize and summarize the information.

Step 4: Draw conclusions from the information.

The methods for conducting Steps 1 and 2 were discussed in Chapter 1. The methods for conducting Step 3 were discussed in Chapters 2 through 4. We took a break from the statistical process in Chapters 5 through 7 so that we could develop skills that allow us to tackle Step 4.

If the information (data) collected is from a population, we can use the summaries obtained in Step 3 to draw conclusions about the population being studied and the statistical process is over.

However, it is often difficult or impossible to gain access to populations, so the information obtained in Step 2 is often sample data. The sample data are used to make inferences about the population. For example, we might compute a sample mean from the information collected in Step 2 and use this information to draw conclusions regarding the population mean. The last part of this text discusses how sample data are used to draw conclusions about populations.

Sampling Distributions

Outline

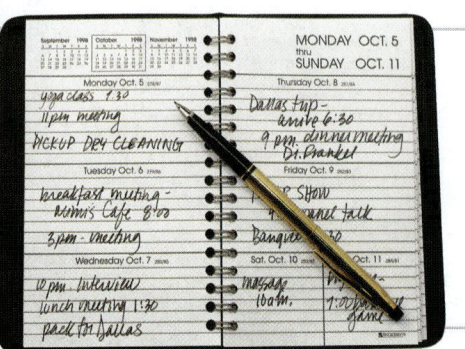

DECISIONS

The American Time Use Survey is a survey of adult Americans conducted by the Bureau of Labor Statistics. The purpose of the survey is to learn how Americans allocate their time during a day. As a reporter for the school newspaper, you wish to file a report that compares the typical student at your school to the rest of Americans. See the Decisions project on page 430.

●●● Putting It All Together

In Chapters 6 and 7, we learned about random variables and their probability distributions. A random variable is a numerical measure of the outcome to a probability experiment. A probability distribution provides a way to assign probabilities to the random variable. For discrete random variables, we discussed the binomial probability distribution and the Poisson probability distribution. We assigned probabilities using a formula. For continuous random variables, we discussed the normal probability distribution. To compute probabilities for a normal random variable, we found the area under a normal density curve.

In this chapter, we continue our discussion of probability distributions. Here statistics, such as $\bar{x}$, will be the random variable. Statistics are random variables because the value of a statistic varies from sample to sample. For this reason, statistics have probability distributions associated with them. For example, there is a probability distribution for the sample mean, sample variance, and so on. We use probability distributions to make probability statements regarding the statistic. So this chapter discusses the shape, center, and spread of statistics such as $\bar{x}$.

8.1 Distribution of the Sample Mean

Preparing for This Section   Before getting started, review the following:

- Simple random sampling (Section 1.2, pp. 16–19)
- The mean (Section 3.1, pp. 121–124)
- The standard deviation (Section 3.2, pp. 143–144)
- Applications of the normal distribution (Section 7.3, pp. 385–389)

Objectives **Understand the concept of a sampling distribution**

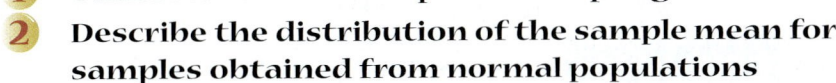 **Describe the distribution of the sample mean for samples obtained from normal populations**

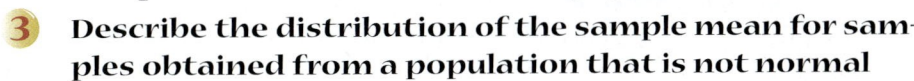 **Describe the distribution of the sample mean for samples obtained from a population that is not normal**

Suppose that the government wanted to estimate the mean income of all U.S. households. One approach the government could take is to literally survey each household in the United States to determine the population mean, μ. This would be a very expensive and time-consuming survey!

A second approach that the government could (and does) take is to survey a random sample of U.S. households and use the results of the survey to estimate the mean household income. This is done through the American Community Survey. The survey is administered to approximately 250,000 randomly selected households each month. Among the many questions on the survey, respondents are asked to report the income of each individual in the household. From this information, the federal government obtains a sample mean household income for U.S. households. For example, in 2003 the mean annual household income in the United States was estimated to be $\bar{x} = \$58,036$. The government might infer from this result that the mean annual household income of *all* U.S. households in 2003 was $\mu = \$58,036$. This type of statement is an example of **statistical inference** — using information from a sample to draw conclusions about a population.

The households that were administered the American Community Survey were determined by chance (random sampling). A second random sample of households would likely lead to a different sample mean such as $\bar{x} = \$58,132$, and a third random sample of households would likely lead to a third distinct sample mean such as $\bar{x} = \$58,095$. Because the households are selected by chance, the sample mean of household income is also determined by chance. We conclude from this that there is variability in our estimates. This variability leads to uncertainty as to whether our estimates are correct. Therefore, we need a way to assess the reliability of inferences made about a population based on sample data.

The measure of reliability is actually a statement of probability. Probability describes how likely an outcome is to occur. The goal of this chapter is to learn the distribution of statistics such as the sample mean so that our estimates are accompanied by statements that indicate the likelihood that our methods are accurate.

 ## Understand the Concept of a Sampling Distribution

In general, the sampling distribution of a statistic is a probability distribution for all possible values of the statistic computed from a sample of size n. The **sampling distribution of the sample mean** is the probability distribution of all possible values of the random variable $\bar{x}$ computed from a sample of size n from a population with mean μ and standard deviation σ.

The idea behind obtaining the sampling distribution of the mean is as follows:

Step 1: Obtain a simple random sample of size n.

Step 2: Compute the sample mean.

Step 3: Assuming that we are sampling from a finite population, repeat Steps 1 and 2 until all simple random samples of size n have been obtained. **Note:** Once a particular sample is obtained, it cannot be obtained a second time.

We present an example to illustrate the idea behind a sampling distribution.

EXAMPLE 1 **A Sampling Distribution**

Problem: One semester, Professor Goehl had a small statistics class of seven students. He asked them the ages of their cars and obtained the following data:

$$2, 4, 6, 8, 4, 3, 7$$

Construct a sampling distribution of the mean for samples of size $n = 2$. What is the probability of obtaining a sample mean between 4 and 6 years, inclusive that is, what is $P(4 \leq \overline{x} \leq 6)$?

Approach: We follow Steps 1 to 3 just listed to construct the probability distribution.

Solution: There are seven individuals in the population. We are selecting them two at a time without replacement. Therefore, there are $_7C_2 = 21$ samples of size $n = 2$. We list these 21 samples along with the sample means in Table 1.

	Table 1					
Sample	**Sample Mean**	**Sample**	**Sample Mean**	**Sample**	**Sample Mean**
2, 4	3	4, 8	6	6, 7	6.5
2, 6	4	4, 4	4	8, 4	6
2, 8	5	4, 3	3.5	8, 3	5.5
2, 4	3	4, 7	5.5	8, 7	7.5
2, 3	2.5	6, 8	7	4, 3	3.5
2, 7	4.5	6, 4	5	4, 7	5.5
4, 6	5	6, 3	4.5	3, 7	5

Table 2 displays the sampling distribution of the sample mean, $\overline{x}$.

	Table 2				
Sample Mean	**Frequency**	**Probability**	**Sample Mean**	**Frequency**	**Probability**
2.5	1	$\frac{1}{21}$	5.5	3	$\frac{3}{21}$
3	2	$\frac{2}{21}$	6	2	$\frac{2}{21}$
3.5	2	$\frac{2}{21}$	6.5	1	$\frac{1}{21}$
4	2	$\frac{2}{21}$	7	1	$\frac{1}{21}$
4.5	2	$\frac{2}{21}$	7.5	1	$\frac{1}{21}$
5	4	$\frac{4}{21}$			

From Table 2 we can compute

$$P(4 \le \overline{x} \le 6) = \frac{2}{21} + \frac{2}{21} + \frac{4}{21} + \frac{3}{21} + \frac{2}{21} = \frac{13}{21} = 0.619$$

If we took 10 simple random samples of size 2 from this population, about 6 of them would result in sample means between 4 and 6 years, inclusive.

The sample mean with the highest probability is $\overline{x} = 5$. This should not be surprising since the population mean of the data in Example 1 is $\mu = 4.9$, rounded to one decimal place. Figure 1 is a probability histogram of the sampling distribution for the sample mean given in Table 2.

Figure 1

Probability Distribution of the Sample Mean

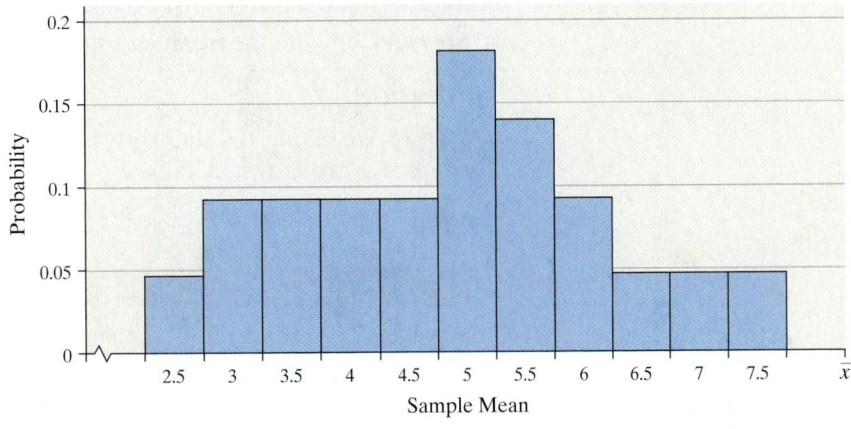

Sample Mean

Now Work Problem 31.

In-Class Activity-Sampling Distributions

Randomly select six students from the class to treat as a population. Choose a quantitative variable (such as pulse rate, age, or number of siblings) to use for this activity, and gather the data appropriately. Compute μ for the population. Divide the class into four groups and have one group list all samples of size $n = 2$, another group list all samples of size $n = 3$, and other groups list all samples of size $n = 4$ and $n = 5$. Each group should do the following:

(a) Compute the sample mean of each sample.
(b) Form the probability distribution for the sample mean.
(c) Draw a probability histogram of the probability distribution.
(d) Verify that $\mu_{\overline{x}} = \mu$.

Compare the spread in each probability distribution based on the probability histogram. What does this result imply about the standard deviation of the sample mean?

---

2 ## Describe the Distribution of the Sample Mean for Samples Obtained from Normal Populations

The point of Example 1 is to help you realize that statistics such as $\overline{x}$ are random variables and therefore have probability distributions associated with them. In practice, a single random sample of size n is obtained from a population. The probability distribution of the sample statistic (or sampling distribution) is determined from statistical theory. We will use simulation to help justify the result that statistical theory provides. We consider two possibilities. In the first case, we sample from a population that is known to be normally distributed. In the second case, we sample from a distribution that is not normally distributed.

Sampling Distribution of the Sample Mean: Population Normal

Problem: In Example 3 from Section 7.1, we learned that the height of 3-year-old females is approximately normally distributed with $\mu = 38.72$ inches and $\sigma = 3.17$ inches. Approximate the sampling distribution of $\bar{x}$ by taking 100 simple random samples of size $n = 5$.

Approach: Use MINITAB, Excel, or some other statistical software package to perform the simulation. We will perform the following steps:

Step 1: Obtain 100 simple random samples of size $n = 5$ from the population, using simulation.

Step 2: Compute the mean of each sample.

Step 3: Draw a histogram of the sample means.

Step 4: Compute the mean and standard deviation of the sample means.

Solution

Step 1: We obtain 100 simple random samples of size $n = 5$. All the samples of size $n = 5$ are shown in Table 3.

Sample	Sample of Size $n = 5$					Sample Mean
1	36.48	39.94	42.57	39.53	33.81	38.47
2	43.13	37.97	42.41	39.61	43.30	41.28
3	41.64	39.01	37.77	38.94	41.10	39.69
4	40.37	43.49	37.60	40.14	38.88	40.10
5	38.62	33.43	45.17	42.66	39.98	39.97
6	38.98	41.35	36.80	43.56	39.92	40.12
7	42.48	37.00	35.87	39.62	38.74	38.74
8	39.38	37.02	41.60	40.34	37.62	39.19
9	42.82	45.77	35.16	42.56	39.75	41.21
10	36.19	35.20	37.74	40.46	37.47	37.41
11	36.59	41.62	42.18	39.23	39.26	39.78
12	38.57	42.13	45.39	38.22	46.18	42.10
13	38.40	39.06	43.60	31.46	37.03	37.91
14	34.29	47.73	37.27	41.82	33.33	38.89
15	42.28	43.29	37.69	37.32	40.06	40.13
16	34.31	43.58	40.02	41.13	42.99	40.41
17	38.71	39.03	39.39	42.62	38.41	39.63
18	38.63	39.66	39.47	41.13	38.01	39.38
19	39.09	33.86	37.57	41.65	35.22	37.48
20	40.94	37.50	38.72	41.64	35.48	38.86
21	38.72	35.89	37.82	35.04	37.06	36.91
22	39.64	36.30	35.54	40.40	38.74	38.12
23	38.22	38.49	33.60	40.18	39.07	37.91
24	40.93	40.53	37.55	37.30	37.16	38.69
25	33.27	38.92	37.14	39.90	33.83	36.61
26	39.44	37.28	35.70	41.97	36.80	38.24
27	38.83	41.41	38.87	39.40	37.20	39.14
28	40.10	36.96	35.73	43.00	38.11	38.78

Table 3 (cont'd)

29	41.93	36.57	37.55	35.14	38.75	37.99
30	31.25	38.85	39.25	35.07	39.77	36.84
31	38.47	34.45	30.43	41.76	41.61	37.34
32	37.98	35.56	43.97	44.96	37.81	40.06
33	43.34	40.94	35.17	41.74	37.59	39.76
34	39.80	44.44	37.53	40.52	41.95	40.85
35	41.98	42.02	40.73	40.47	36.81	40.40
36	40.98	35.08	34.61	40.78	37.26	37.74
37	35.75	40.81	40.13	35.99	36.52	37.84
38	36.39	45.97	40.59	37.64	42.42	40.60
39	36.20	35.63	37.43	38.35	34.81	36.48
40	33.58	33.87	41.60	45.10	38.68	38.57
41	31.77	38.34	41.79	37.93	40.83	38.13
42	43.03	33.12	34.98	36.58	37.78	37.10
43	35.76	35.17	42.58	39.10	41.08	38.74
44	38.44	38.45	35.93	35.32	44.60	38.55
45	44.54	41.88	35.84	42.64	42.38	41.46
46	41.89	36.81	41.83	40.24	39.28	40.01
47	38.00	40.08	35.57	34.44	39.51	37.52
48	39.92	38.05	39.96	38.04	32.11	37.62
49	36.37	38.62	32.25	41.35	40.91	37.90
50	34.38	36.65	32.97	39.93	41.34	37.05
51	40.32	39.80	41.00	38.62	38.24	39.60
52	37.95	45.26	38.67	34.96	41.13	39.59
53	36.82	42.63	41.62	39.43	37.48	39.60
54	41.63	37.65	38.58	39.03	37.53	38.88
55	37.91	37.20	38.72	36.87	45.40	39.22
56	41.05	34.01	39.11	38.23	35.74	37.63
57	42.09	45.44	35.52	39.87	37.28	40.04
58	39.31	35.79	37.82	39.15	35.57	37.53
59	41.16	39.98	41.11	39.21	39.98	40.29
60	35.68	45.60	39.34	36.65	43.30	40.11
61	36.07	39.63	42.55	41.72	36.81	39.36
62	38.97	36.83	41.01	38.12	35.27	38.04
63	33.70	39.15	34.81	34.13	39.00	36.16
64	37.19	34.69	36.21	34.34	39.07	36.30
65	33.99	44.87	42.52	40.22	39.26	40.17
66	41.40	27.62	34.57	40.08	34.65	35.66
67	40.14	34.45	38.26	38.09	39.72	38.13
68	33.64	42.62	32.08	34.30	37.34	36.00
69	35.36	39.02	43.98	41.19	32.47	38.40
70	43.26	37.85	35.82	37.11	36.22	38.05
71	36.24	38.07	33.38	38.43	39.88	37.20
72	38.55	43.06	41.07	36.58	37.02	39.26
73	41.26	36.99	36.17	38.98	36.03	37.89
74	37.31	38.41	41.18	39.76	39.64	39.26
75	36.26	41.84	42.50	37.70	41.21	39.90
76	39.27	38.61	44.53	38.08	35.01	39.10

Table 3 (cont'd)						
77	39.14	40.83	39.83	37.78	36.51	38.82
78	42.53	43.41	41.01	33.71	39.47	40.03
79	45.34	32.61	33.81	39.03	40.32	38.22
80	36.31	35.55	37.12	38.74	40.80	37.70
81	31.40	41.80	40.15	42.53	37.62	38.70
82	41.01	39.02	39.68	36.61	38.44	38.95
83	34.15	36.19	35.98	36.02	36.32	35.73
84	31.50	37.61	43.29	39.82	38.78	38.20
85	43.26	34.01	41.18	40.23	39.28	39.59
86	41.76	41.40	39.02	38.20	39.42	39.96
87	37.06	35.95	39.98	40.00	43.36	39.27
88	41.01	37.56	36.95	39.71	37.97	38.64
89	34.97	38.36	36.30	38.48	34.24	36.47
90	38.38	38.94	40.96	36.13	35.98	38.08
91	39.41	30.78	37.66	37.31	42.04	37.44
92	39.83	35.88	30.20	45.07	40.06	38.21
93	36.25	39.56	34.53	40.69	37.03	37.61
94	45.64	40.66	44.51	40.50	39.43	42.15
95	37.63	44.77	38.31	36.53	38.41	39.13
96	39.78	33.34	43.42	43.63	38.77	39.79
97	41.48	37.39	38.62	43.83	34.26	39.12
98	37.68	40.66	38.93	40.94	37.54	39.15
99	39.72	32.61	32.62	40.35	38.65	36.79
100	39.25	41.06	41.17	38.30	38.24	39.60

Step 2: We compute the sample means for each of the 100 samples as shown in Table 3.

Step 3: We draw a histogram of the 100 sample means. See Figure 2.

Figure 2

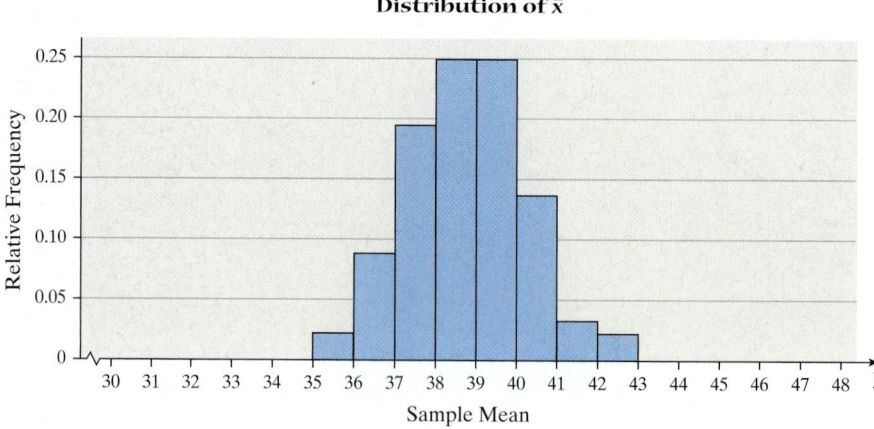

Step 4: The mean of the 100 sample means is 38.72 inches, and the standard deviation is 1.374 inches.

Look back at the histogram of the population data drawn in Figure 7 on page 363 from Section 7.1. Notice the center of the population distribution is the same as the center of the sampling distribution, but the spread of the population distribution is greater than that of the sampling distribution.

In Example 2 we were told that the data are approximately normal, with mean $\mu = 38.72$ inches and $\sigma = 3.17$ inches. The histogram in Figure 2 indicates that the distribution of sample means also appears to be normally distributed. In addition, the mean of the sample means is 38.72 inches, but the standard deviation is only 1.374 inches. We might conclude the following regarding the sampling distribution of $\bar{x}$.

1. *Shape*: It is normally distributed.
2. *Center*: It has mean equal to the mean of the population.
3. *Spread*: It has standard deviation less than the standard deviation of the population.

A question that we might ask is "What role does n, the sample size, play in the sampling distribution of $\bar{x}$?" Suppose the sample mean is computed for samples of size $n = 1$ through $n = 200$. That is, the sample mean is recomputed each time an additional individual is added to the sample. The sample mean is then plotted against the sample size in Figure 3.

Figure 3

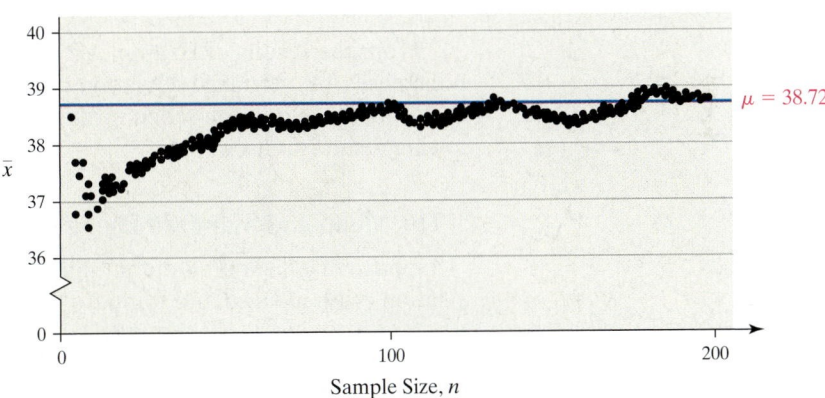

From the graph, we see that, as the sample size n increases, the sample mean gets closer to the population mean. This concept is known as the *Law of Large Numbers*.

The Law of Large Numbers

As additional observations are added to the sample, the difference between the sample mean, $\bar{x}$, and the population mean μ approaches zero.

In Other Words

As the sample size increases, the sample mean gets closer to the population mean.

So, according to the Law of Large Numbers, the more individuals we sample, the closer the sample mean gets to the population mean. This result implies that there is less variability in the distribution of the sample mean as the sample size increases. We demonstrate this result in the next example.

EXAMPLE 3   **The Impact of Sample Size on Sampling Variability**

Problem: Repeat the problem in Example 2 with a sample of size $n = 15$.

Approach: The approach will be identical to that presented in Example 2, except that we let $n = 15$ instead of $n = 5$.

Solution: Figure 4(a) shows the histogram of the sample means using the same scale as Figure 2. Compare this with the histogram in Figure 2. Notice the histogram in Figure 4(a) shows less dispersion than the histogram in Figure 2. This implies there is less variability in the distribution of $\bar{x}$ with $n = 15$.

We redraw the histogram in Figure 4(a) using a different scale in Figure 4(b). The histogram in Figure 4(b) is symmetric and mound shaped. This is an indication that the distribution of the sample mean is approximately normally distributed. The mean of the 100 sample means is 38.72 inches (just as in Example 2); however, the standard deviation is now 0.81 inches.

Figure 4

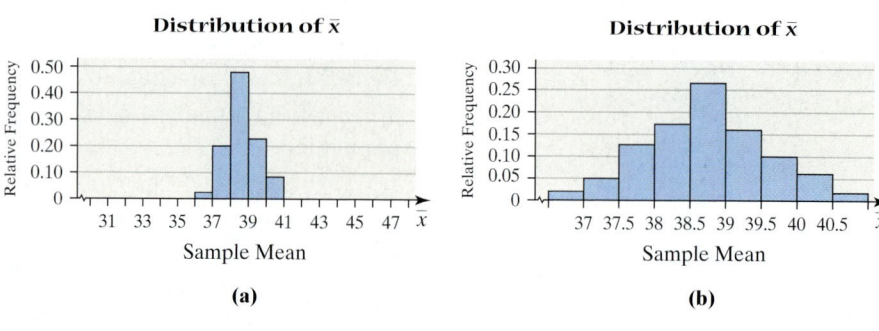

(a)

(b)

From the results of Examples 2 and 3, we conclude that, as the sample size n increases, the standard deviation of the distribution of $\bar{x}$ decreases. Although the proof is beyond the scope of this text, we should be convinced that the following result is reasonable.

The Mean and Standard Deviation of the Sampling Distribution of $\bar{x}$

Suppose that a simple random sample of size n is drawn from a large population* with mean μ and standard deviation σ. The sampling distribution of $\bar{x}$ will have mean $\mu_{\bar{x}} = \mu$ and standard deviation $\sigma_{\bar{x}} = \dfrac{\sigma}{\sqrt{n}}$. The standard deviation of the sampling distribution of $\bar{x}$ is called the **standard error of the mean** and is denoted $\sigma_{\bar{x}}$.

In Other Words

Regardless of the distribution of the population, the sampling distribution of $\bar{x}$ will have a mean equal to the mean of the population and a standard deviation equal to the standard deviation of the population divided by the square root of the sample size!

For the population presented in Example 2, if we draw a simple random sample of size $n = 5$, the sampling distribution $\bar{x}$ will have mean $\mu_{\bar{x}} = 38.72$ inches and standard deviation

$$\sigma_{\bar{x}} = \frac{\sigma}{\sqrt{n}} = \frac{3.17}{\sqrt{5}} \approx 1.418 \text{ inches}$$

Now Work Problem 11.

---

*Technically, we assume that we are drawing a simple random sample from an infinite population. For populations of finite size N, $\sigma_{\bar{x}} = \sqrt{\dfrac{N-n}{N-1}} \cdot \dfrac{\sigma}{\sqrt{n}}$. However, if the sample size is less than 5% of the population size ($n < 0.05N$), the effect of $\sqrt{\dfrac{N-n}{N-1}}$ (the finite population correction factor) can be ignored without affecting the results.

Now that we know how to determine the mean and standard deviation for any sampling distribution of $\bar{x}$, we can concentrate on the shape of the distribution. Refer back to Figures 2 and 4 from Examples 2 and 3. Recall that the population from which the sample was drawn was normal. The shapes of these histograms imply that the sampling distribution of $\bar{x}$ is also normal. This leads us to believe that if the population is normal the distribution of the sample mean is also normal.

Figure 5

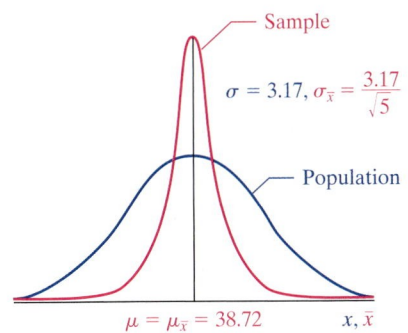

- Sample
- Population

$\sigma = 3.17, \sigma_{\bar{x}} = \dfrac{3.17}{\sqrt{5}}$

$\mu = \mu_{\bar{x}} = 38.72$ $x, \bar{x}$

The Shape of the Sampling Distribution of $\bar{x}$ If X Is Normal

If a random variable X is normally distributed, the distribution of the sample mean, $\bar{x}$, is normally distributed.

For example, the height of 3-year-old females is modeled by a normal random variable with mean $\mu = 38.72$ inches and standard deviation $\sigma = 3.17$ inches. The distribution of the sample mean, $\bar{x}$, the mean height of a simple random sample of $n = 5$ three-year-old females, is normal with mean $\mu_{\bar{x}} = 38.72$ inches and standard deviation $\sigma_{\bar{x}} = \dfrac{3.17}{\sqrt{5}}$ inches. See Figure 5.

EXAMPLE 4

Describing the Distribution of the Sample Mean

Problem: The height, X, of all 3-year-old females is approximately normally distributed with mean $\mu = 38.72$ inches and standard deviation $\sigma = 3.17$ inches. Compute the probability that a simple random sample of size $n = 10$ results in a sample mean greater than 40 inches. That is, compute $P(\bar{x} > 40)$.

Approach: The random variable X is normally distributed, so the sampling distribution of $\bar{x}$ will also be normally distributed. The mean of the sampling distribution is $\mu_{\bar{x}} = \mu$, and its standard deviation is $\sigma_{\bar{x}} = \dfrac{\sigma}{\sqrt{n}}$. We convert the random variable $\bar{x} = 40$ to a Z-score and then find the area under the standard normal curve to the right of this Z-score.

Figure 6

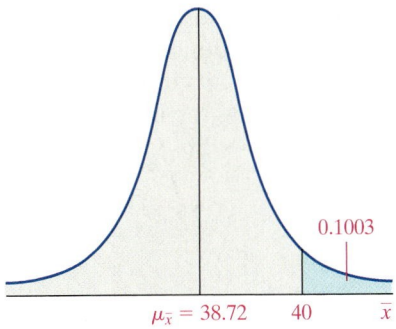

0.1003

$\mu_{\bar{x}} = 38.72$ 40 $\bar{x}$

Solution: The sample mean is normally distributed with mean $\mu_{\bar{x}} = 38.72$ inches and standard deviation $\sigma_{\bar{x}} = \dfrac{\sigma}{\sqrt{n}} = \dfrac{3.17}{\sqrt{10}} = 1.002$ inch.

Figure 6 displays the normal curve with the area we wish to compute shaded. We convert the random variable $\bar{x} = 40$ to a Z-score and obtain

$$Z = \frac{\bar{x} - \mu_{\bar{x}}}{\sigma_{\bar{x}}} = \frac{\bar{x} - \mu_{\bar{x}}}{\dfrac{\sigma}{\sqrt{n}}} = \frac{40 - 38.72}{1.002} = 1.28$$

The area to the right of $Z = 1.28$ is $1 - 0.8997 = 0.1003$.

Interpretation: The probability of obtaining a sample mean greater than 40 inches from a population whose mean is 38.72 inches is 0.1003. That is, $P(\bar{x} \geq 40) = 0.1003$. If we take 1,000 simple random samples of $n = 10$ three-year-olds from this population and if the population mean is 38.72 inches, about 100 of the samples will result in a mean height that is 40 inches or more.

Now Work Problem 19.

Describe the Distribution of the Sample Mean for Samples Obtained from a Population That Is Not Normal

What if the population from which the sample is drawn is not normal?

EXAMPLE 5 **Sampling from a Population That Is Not Normal**

Figure 7

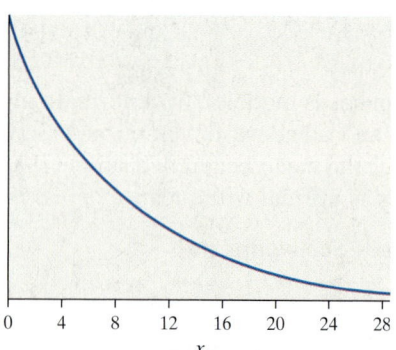

Problem: Figure 7 shows the graph of an *exponential density function* with mean and standard deviation equal to 10. The exponential distribution is used to model lifetimes of electronic components and to model the time required to serve a customer or repair a machine.

Clearly, the distribution of the population is not normal. Approximate the sampling distribution of $\overline{x}$ by obtaining, through simulation, 300 random samples of size (a) $n = 3$, (b) $n = 12$, and (c) $n = 30$ from the probability distribution.

Approach

Step 1: Use MINITAB, Excel, or some other statistical software to obtain 300 random samples for each sample size.

Step 2: Compute the sample mean of each of the 300 random samples.

Step 3: Draw a histogram of the 300 sample means.

Solution

Step 1: Using MINITAB, we obtain 300 random samples of size (a) $n = 3$, (b) $n = 12$, and (c) $n = 30$. For example, in the first random sample of size $n = 30$, we obtained the following results:

9.2	20.0	17.0	2.4	2.6	19.9	21.2	5.7	8.1	10.8
1.2	22.3	18.4	4.2	9.9	41.8	4.2	1.2	10.8	2.1
11.3	17.9	28.0	12.1	3.0	0.5	4.5	14.2	5.0	11.4

Step 2: We compute the mean of each of the 300 random samples, using MINITAB. For example, the sample mean of the first sample of size $n = 30$ is 11.36.

Step 3: Figure 8(a) displays the histogram of $\overline{x}$ that results from simulating 300 random samples of size $n = 3$ from an exponential distribution with $\mu = 10$ and $\sigma = 10$. Figure 8(b) displays the histogram of $\overline{x}$ that results from simulating 300 random samples of size $n = 12$, and Figure 8(c) displays the histogram of $\overline{x}$ that results from simulating 300 random samples of size $n = 30$.

Figure 8

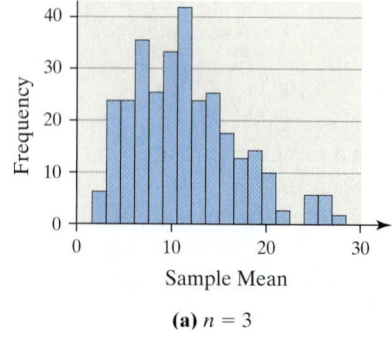

(a) $n = 3$

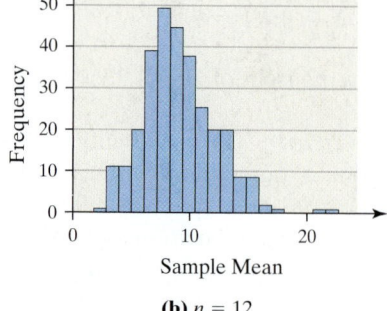

(b) $n = 12$

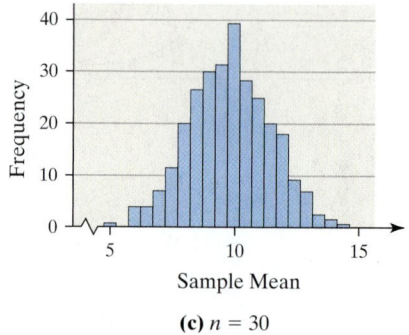

(c) $n = 30$

Notice that, as the sample size increases, the distribution of the sample mean becomes more normal, even though the population clearly is not normal!

We formally state the results of Example 5 as the *Central Limit Theorem*.

> **The Central Limit Theorem**
> Regardless of the shape of the population, the sampling distribution of $\bar{x}$ becomes approximately normal as the sample size n increases.

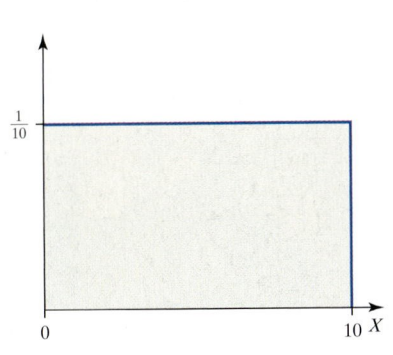

In Other Words

For any population, regardless of its shape, as the sample size increases, the shape of the distribution of the sample mean becomes more "normal."

So, if the random variable X is normally distributed, the sampling distribution of $\bar{x}$ will be normal. If the sample size is large enough, the sampling distribution of $\bar{x}$ will be approximately normal, *regardless of the shape of the distribution of X*. But how large does the sample size need to be before we can say that the sampling distribution of $\bar{x}$ is approximately normal? The answer depends on the shape of the distribution of the population. Distributions that are highly skewed will require a larger sample size for the distribution of $\bar{x}$ to become approximately normal.

For example, from Example 5 we see this right skewed distribution required a sample size of about 30 before the distribution of the sample mean is approximately normal. However, Figure 9(a) shows a uniform distribution for $0 \leq X \leq 10$. Figure 9(b) shows the distribution of the sample mean for $n = 3$. Figure 9(c) shows the distribution of the sample mean for $n = 12$, and Figure 9(d) shows the distribution of the sample mean for $n = 30$. Notice that even for $n = 3$ the distribution of the sample mean is approximately normal.

Figure 9

(a) Uniform Distribution

Distribution of $\bar{x}$

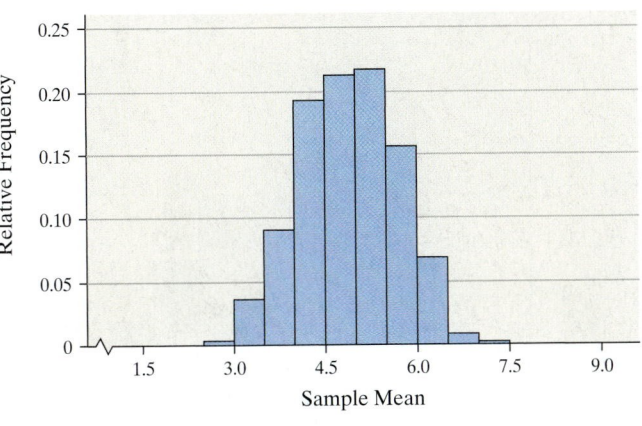

(b) Distribution of $\bar{x}$; $n = 3$

(c) Distribution of $\bar{x}$; $n = 12$

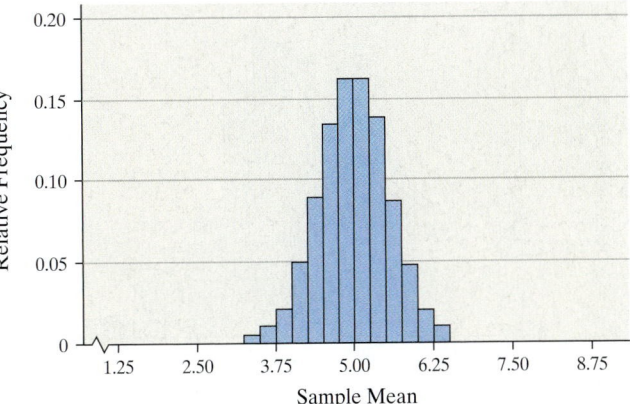

(d) Distribution of $\bar{x}$; $n = 30$

Table 4 shows the distribution of the cumulative number of children for 50-to 54-year-old mothers who had a live birth in 2002.

Table 4	
x (number of children)	*P(x)*
1	0.241
2	0.257
3	0.172
4	0.119
5	0.103
6	0.027
7	0.031
8	0.050

Source: U.S. Census Bureau

Figure 10(a) shows the probability histogram for this distribution. Figure 10(b) shows the distribution of the sample mean number of children for a random sample of $n = 3$ mothers. Figure 10(c) shows the distribution of the sample mean number of children for a random sample of $n = 12$ mothers, and Figure 10(d) shows the distribution of the sample mean for a random sample of $n = 30$ mothers. In this instance, the distribution of the sample mean is very close to normal for $n = 12$.

Figure 10a

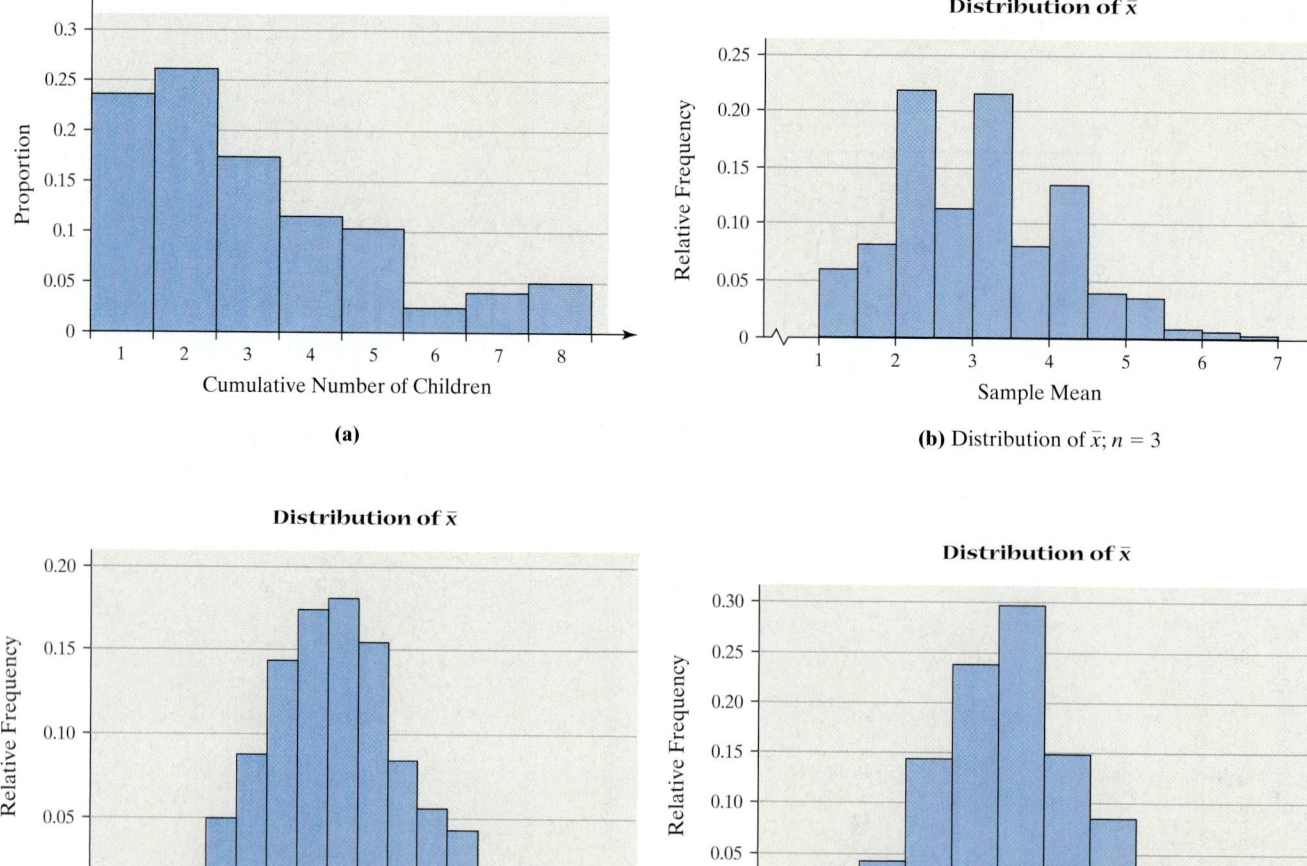

(a)

(b) Distribution of $\bar{x}$; $n = 3$

(c) Distribution of $\bar{x}$; $n = 12$

(d) Distribution of $\bar{x}$; $n = 30$

The results of Example 5 and Figures 9 and 10 confirm that the shape of the distribution of the population dictates the size of the sample required before the distribution of the sample mean can be called normal. With that said, so that we err on the side of caution, we will say that the distribution of the sample mean is approximately normal provided that the sample size is greater than or equal to 30 if the distribution of the population is unknown or not normal.

EXAMPLE 6   Applying the Central Limit Theorem

Problem: According to the U.S. Department of Agriculture, the mean calorie intake of males 20 to 39 years old is $\mu = 2716$, with standard deviation $\sigma = 72.8$. Suppose a nutritionist analyzes a simple random sample of $n = 35$ males between the ages of 20 and 39 years old and obtains a sample mean calorie intake of $\bar{x} = 2750$ calories. What is the probability that a random sample of 35 males between the ages of 20 and 39 years old would result in a sample mean of 2750 calories or higher? Are the results of the survey unusual? Why?

!**CAUTION**
The Central Limit Theorem has to do only with the *shape* of the distribution of the sample mean, not with its center and spread! The mean of the distribution of $\bar{x}$ is μ and the standard deviation of $\bar{x}$ is $\dfrac{\sigma}{\sqrt{n}}$, regardless of the size of the sample, n.

Approach

Step 1: We recognize that we are computing a probability regarding a sample mean, so we need to know the sampling distribution of $\bar{x}$. Because the population from which the sample is drawn is not known to be normal, the sample size must be greater than or equal to 30 to use the results of the Central Limit Theorem.

Step 2: Determine the mean and standard deviation of the sampling distribution of $\bar{x}$.

Step 3: Convert the sample mean to a Z-score.

Step 4: Use Table IV to find the area under the normal curve.

Solution

Step 1: Because the sample size is $n = 35$, the Central Limit Theorem states that the sampling distribution of $\bar{x}$ is approximately normal.

Step 2: The mean of the sampling distribution of $\bar{x}$ will equal the mean of the population, so $\mu_{\bar{x}} = 2716$. The standard deviation of the sampling distribution of $\bar{x}$ will equal the standard deviation of the population divided by the square root of the sample size, so $\sigma_{\bar{x}} = \dfrac{\sigma}{\sqrt{n}} = \dfrac{72.8}{\sqrt{35}} = 12.305$.

Step 3: We convert $\bar{x} = 2750$ to a Z-score.

$$Z = \frac{2750 - 2716}{\dfrac{72.8}{\sqrt{35}}} = 2.76$$

Step 4: We wish to know the probability that a random sample of $n = 35$ from a population whose mean is 2716 results in a sample mean of at least 2750. That is, we wish to know $P(\bar{x} \geq 2750)$. See Figure 11.

Figure 11

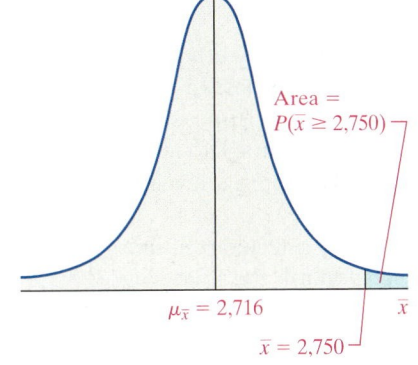

Area = $P(\bar{x} \geq 2750)$

$\mu_{\bar{x}} = 2,716$ $\bar{x}$

$\bar{x} = 2,750$

Historical Note

Pierre Simon Laplace was born on March 23, 1749 in Normandy, France. At age 16, Laplace attended Caen University, where he studied theology. While there, his mathematical talents were discovered, which led him to Paris, where he got a job as professor of mathematics at the École Militaire. In 1773, Laplace was elected to the Académie des Sciences. Laplace was not humble. It is reported that, in 1780, he stated that he was the best mathematician in Paris. In 1799, Laplace published the first two volumes of *Méchanique céleste*, in which he discusses methods for calculating the motion of the planets. On April 9, 1810, Laplace presented the Central Limit Theorem to the Academy.

This probability is represented by the area under the standard normal curve to the right of $Z = 2.76$.

$$P(\bar{x} \geq 2750) = P(Z \geq 2.76) = 1 - 0.9971 = 0.0029$$

Interpretation: If the population mean is 2716 calories, the probability that a random sample of 35 males between the ages of 20 and 39 will result in a sample mean calorie intake of 2750 calories or higher is 0.0029. This means that fewer than 1 sample in 100 will result in a sample mean of 2750 calories or higher if the population mean is 2716 calories. We can conclude one of two things based on this result.

1. The mean number of calories for males 20 to 39 years old is 2716, and we just happened to randomly select 35 individuals who, on average, consume more calories.

2. The mean number of calories consumed by 20- to 39-year-old males is higher than 2716 calories.

In statistical inference, we are inclined to accept the second possibility as the more reasonable choice. We recognize there is a possibility that our conclusion is incorrect.

Now Work Problem 25.

Summary: Shape, Center, and Spread of the Distribution of $\bar{x}$

Shape, Center, and Spread of Population	Distribution of the Sample Mean		
	Shape	**Center**	**Spread**
Normal with mean μ and standard deviation σ	Regardless of the sample size n, the shape of the distribution of the sample mean is normal	$\mu_{\bar{x}} = \mu$	$\sigma_{\bar{x}} = \dfrac{\sigma}{\sqrt{n}}$
Population is not normal with mean μ and standard deviation σ	As the sample size n increases, the distribution of the sample mean becomes approximately normal	$\mu_{\bar{x}} = \mu$	$\sigma_{\bar{x}} = \dfrac{\sigma}{\sqrt{n}}$

MAKING AN INFORMED DECISION

How Much Time Do You Spend in a Day . . . ?

The American Time Use Survey is a survey of adult Americans conducted by the Bureau of Labor Statistics. The purpose of the survey is to learn how Americans allocate their time in a day. As a reporter for the school newspaper, you wish to file a report that compares the typical student at your school to the rest of America.

For those Americans who are currently attending school, the mean amount of time spent in class in a day is 5.11 hours, and the mean amount of time spent studying and doing homework is 2.50 hours. The mean amount of time Americans spend watching television each day is 2.57 hours.

Conduct a survey of 35 randomly selected full-time students at your school in which you ask the following questions:

(a) On average, how much time do you spend attending class each day?

(b) On average, how much time do you spend studying and doing homework each day?

(c) On average, how much time do you spend watching television each day? If you do not watch television, write 0 hours.

1. For each question, describe the sampling distribution of the sample mean. Use the national norms as estimates for the population means for each variable. Use the sample standard deviation as an estimate of the population standard deviation.

2. Compute probabilities regarding the values of the statistics obtained from the survey. Are any of the results unusual?

Write an article for your newspaper reporting your findings.

8.1 ASSESS YOUR UNDERSTANDING

Concepts and Vocabulary

1. Explain what a sampling distribution is.
2. State the Central Limit Theorem.
3. The standard deviation of the sampling distribution of $\bar{x}$, denoted $\sigma_{\bar{x}}$, is called the _____ _____ of the _____.
4. As the sample size increases, the difference between the sample mean, $\bar{x}$, and the population mean, μ, approaches _____.
5. What are the mean and standard deviation of the sampling distribution of $\bar{x}$, regardless of the distribution of the population from which the sample was drawn?
6. If a random sample of size $n = 6$ is taken from a population, what is required to say that the sampling distribution of $\bar{x}$ is approximately normal?

7. To cut the standard error of the mean in half, the sample size must be increased by a factor of _____.
8. *True or False*: The mean and standard deviation of the distribution of $\bar{x}$ is $\mu_{\bar{x}} = \mu$ and $\sigma_{\bar{x}} = \dfrac{\sigma}{\sqrt{n}}$, respectively, even if the population is not normal.
9. Suppose a simple random sample of size $n = 10$ is obtained from a population that is normally distributed with $\mu = 30$ and $\sigma = 8$. What is the sampling distribution of $\bar{x}$?
10. Suppose a simple random sample of size $n = 40$ is obtained from a population with $\mu = 50$ and $\sigma = 4$. Does the population need to be normally distributed for the sampling distribution of $\bar{x}$ to be approximately normally distributed? Why? What is the sampling distribution of $\bar{x}$?

Skill Building

In Problems 11–14, determine $\mu_{\bar{x}}$ and $\sigma_{\bar{x}}$ from the given parameters of the population and the sample size.

11. $\mu = 80, \sigma = 14, n = 49$
    NW

12. $\mu = 64, \sigma = 18, n = 36$

13. $\mu = 52, \sigma = 10, n = 21$

14. $\mu = 27, \sigma = 6, n = 15$

15. Suppose a simple random sample of size $n = 49$ is obtained from a population with $\mu = 80$ and $\sigma = 14$.
    (a) Describe the sampling distribution of $\bar{x}$.
    (b) What is $P(\bar{x} > 83)$?
    (c) What is $P(\bar{x} \leq 75.8)$?
    (d) What is $P(78.3 < \bar{x} < 85.1)$?

16. Suppose a simple random sample of size $n = 36$ is obtained from a population with $\mu = 64$ and $\sigma = 18$.
    (a) Describe the sampling distribution of $\bar{x}$.
    (b) What is $P(\bar{x} < 62.6)$?
    (c) What is $P(\bar{x} \geq 68.7)$?
    (d) What is $P(59.8 < \bar{x} < 65.9)$?

17. Suppose a simple random sample of size $n = 12$ is obtained from a population with $\mu = 64$ and $\sigma = 17$.
    (a) What must be true regarding the distribution of the population in order to use the normal model to compute probabilities regarding the sample mean? Assuming this condition is true, describe the sampling distribution of $\bar{x}$.
    (b) Assuming the requirements described in part (a) are satisfied, determine $P(\bar{x} < 67.3)$?
    (c) Assuming the requirements described in part (a) are satisfied, determine $P(\bar{x} \geq 65.2)$?

18. Suppose a simple random sample of size $n = 20$ is obtained from a population with $\mu = 64$ and $\sigma = 17$.
    (a) What must be true regarding the distribution of the population in order to use the normal model to compute probabilities regarding the sample mean? Assuming this condition is true, describe the sampling distribution of $\bar{x}$.
    (b) Assuming the requirements described in part (a) are satisfied, determine $P(\bar{x} < 67.3)$?
    (c) Assuming the requirements described in part (a) are satisfied, determine $P(\bar{x} \geq 65.2)$.
    (d) Compare the results obtained in parts (b) and (c) with the results obtained in parts (b) and (c) in Problem 17. What effect does increasing the sample size have on the probabilities? Why do you think this is the case?

Applying the Concepts

19. **Gestation Period** The length of human pregnancies is approximately normally distributed with mean $\mu = 266$ days and standard deviation $\sigma = 16$ days.
    NW
    (a) What is the probability a randomly selected pregnancy lasts less than 260 days?
    (b) What is the probability that a random sample of 20 pregnancies has a mean gestation period of 260 days or less?
    (c) What is the probability that a random sample of 50 pregnancies has a mean gestation period of 260 days or less?

    (d) What might you conclude if a random sample of 50 pregnancies resulted in a mean gestation period of 260 days or less?
    (e) What is the probability a random sample of size 15 will have a mean gestation period within 10 days of the mean?

20. **Serum Cholesterol** As reported by the U.S. National Center for Health Statistics, the mean serum high-density-lipoprotein (HDL) cholesterol of females 20 to 29 years old is $\mu = 53$. If serum HDL cholesterol is normally distributed with $\sigma = 13.4$, answer the following questions:

   (a) What is the probability a randomly selected female 20 to 29 years old has a serum cholesterol above 60?

   (b) What is the probability a random sample of 15 female 20 to 29-year-olds has a mean serum cholesterol above 60?

   (c) What is the probability that a random sample of 20 female 20 to 29-year-olds has a mean serum cholesterol above 60?

   (d) What effect does increasing the sample size have on the probability? Provide an explanation for this result.

   (e) What might you conclude if a random sample of 20 female 20 to 29-year-olds has a mean serum cholesterol above 60?

21. **Old Faithful** The most famous geyser in the world, Old Faithful in Yellowstone National Park, has a mean time between eruptions of 85 minutes. If the interval of time between eruptions is normally distributed with standard deviation 21.25 minutes, answer the following questions: (*Source*: www.unmuseum.org)

   (a) What is the probability that a randomly selected time interval between eruptions is longer than 95 minutes?

   (b) What is the probability that a random sample of 20 time intervals between eruptions has a mean longer than 95 minutes?

   (c) What is the probability that a random sample of 30 time intervals between eruptions has a mean longer than 95 minutes?

   (d) What effect does increasing the sample size have on the probability? Provide an explanation for this result.

   (e) What might you conclude if a random sample of 30 time intervals between eruptions has a mean longer than 95 minutes?

22. **Medical Residents** In a 2003 study, the Accreditation Council for Graduate Medical Education found that medical residents work an average of 81.7 hours per week. Suppose the number of hours worked per week by medical residents is normally distributed with standard deviation 6.9 hours per week. (*Source*: www.medrecinst.com)

   (a) What is the probability that a randomly selected medical resident works less than 75 hours per week?

   (b) What is the probability that the mean number of hours worked per week by a random sample of five medical residents is less than 75 hours?

   (c) What is the probability that the mean number of hours worked per week by a random sample of eight medical resident is less than 75 hours?

   (d) What might you conclude if the mean number of hours worked per week by a random sample of eight medical residents is less than 75 hours?

23. **Rates of Return in Stocks** The S&P 500 is a collection of 500 stocks of publicly traded companies. Using data obtained from Yahoo!Finance, the monthly rates of return of the S&P 500 since 1950 are normally distributed. The mean rate of return is 0.007233 (0.7233%), and the standard deviation for rate of return is 0.04135 (4.135%).

   (a) What is the probability that a randomly selected month has a positive rate of return? That is, what is $P(x > 0)$?

   (b) Treating the next 12 months as a simple random sample, what is the probability the mean monthly rate of return will be positive? That is, with $n = 12$, what is $P(\bar{x} > 0)$?

   (c) Treating the next 24 months as a simple random sample, what is the probability the mean monthly rate of return will be positive?

   (d) Treating the next 36 months as a simple random sample, what is the probability the mean monthly rate of return will be positive?

   (e) Use the results of parts (b)–(d) to describe the likelihood of earning a positive rate of return on stocks as the investment time horizon increases.

24. **Gas Mileage** Based on tests of the Chevrolet Cobalt, engineers have found that the miles per gallon in highway driving are normally distributed, with a mean of 32 miles per gallon and a standard deviation 3.5 miles per gallon.

   (a) What is the probability that a randomly selected Cobalt gets more than 34 miles per gallon?

   (b) Suppose that 10 Cobalts are randomly selected and the miles per gallon for each car are recorded. What is the probability that the mean miles per gallon exceed 34 miles per gallon?

   (c) Suppose that 20 Cobalts are randomly selected and the miles per gallon for each car are recorded. What is the probability that the mean miles per gallon exceed 34 miles per gallon? Would this result be unusual?

25. **Oil Change** The shape of the distribution of the time required to get an oil change at a 10-minute oil-change facility is unknown. However, records indicate that the mean time for an oil change is 11.4 minutes and the standard deviation for oil-change time is 3.2 minutes.

   (a) To compute probabilities regarding the sample mean using the normal model, what size sample would be required?

   (b) What is the probability a random sample of $n = 40$ oil changes results in a sample mean time less than 10 minutes?

26. **Time Spent in the Drive-Through** The quality-control manager of a Long John Silver's restaurant wishes to analyze the length of time a car spends at the drive-through window waiting for an order. According to records obtained from the restaurants, it is determined that the mean time spent at the window is 59.3 seconds with a standard deviation of 13.1 seconds. The distribution of time at the window is skewed right (data based on information provided by Danica Williams, student at Joliet Junior College).

   (a) To obtain probabilities regarding a sample mean using the normal model, what size sample is required?

   (b) The quality-control manager wishes to use a new delivery system designed to get cars through the drive-through system faster. A random sample of 40 cars results in a sample mean time spent at the window of 56.8 seconds. What is the probability of obtaining a sample mean of 56.8 seconds or less assuming the population mean is 59.3 seconds? Do you think that the new system is effective?

27. Insect Fragments The Food and Drug Administration sets Food Defect Action Levels (FDALs) for some of the various foreign substances that inevitably end up in the food we eat and liquids we drink. For example, the FDAL for insect filth in peanut butter is 3 insect fragments (larvae, eggs, body parts, and so on) per 10 grams. A random sample of 50 ten-gram portions of peanut butter is obtained and results in a sample mean of $\bar{x} = 3.6$ insect fragments per ten-gram portion.

(a) Why is the sampling distribution of $\bar{x}$ approximately normal?

(b) What is the mean and standard deviation of the sampling distribution of $\bar{x}$?

[**Hint:** This is a Poisson process with $\mu = 3$ and $\sigma = \sqrt{3}$.]

(c) Suppose a simple random sample of $n = 50$ ten-gram samples of peanut butter results in a sample mean of 3.6 insect fragments. What is the probability a simple random sample of 50 ten-gram portions results in a mean of at least 3.6 insect fragments? Is this result unusual? What might we conclude?

28. Burger King's Drive-Through Suppose cars arrive at Burger King's drive-through at the rate of 20 cars every hour between 12:00 noon and 1:00 P.M. A random sample of 40 one-hour time periods between 12:00 noon and 1:00 P.M. is selected and has 22.1 as the mean number of cars arriving.

(a) Why is the sampling distribution of $\bar{x}$ approximately normal?

(b) What is the mean and standard deviation of the sampling distribution of $\bar{x}$?

[*Hint:* This is a Poisson process with $\mu = 20$ and $\sigma = \sqrt{20}$.]

(c) What is the probability that a simple random sample of 40 one-hour time periods results in a mean of at least 22.1 cars? Is this result unusual? What might we conclude?

29. Blows to the Head In a 2003 study of the long-term effects of concussions in football players, researchers at Virginia Tech concluded that college football players receive a mean of 50 strong blows to the head, each with an average of 40G (40 times the force of gravity). Assume the standard deviation is 16 strong blows to the head. What is the probability that a random sample of 60 college football players results in a mean of 45 or fewer strong blows to the head? Would this be unusual? (*Source*: Neuroscience for Kids, faculty.washington.edu/chudler/nfl.html).

30. Domestic Vacation Costs According to the AAA (American Automobile Association, April 20, 2005), a family of two adults and two children on vacation in the United States will pay an average of $247.02 per day for food and lodging with a standard deviation of $60.41 per day. Suppose a random sample of 50 families of two adults and two children is selected and monitored while on vacation in the United States. What is the probability that the average daily expenses for the sample are over $260.00 per day? Would this be unusual?

31. Sampling Distributions The following data represent the
NW ages of the winners of the Academy Award for Best Actor for the past 6 years (1999–2004).

2004: Jamie Foxx	37
2003: Sean Penn	43
2002: Adrien Brody	29
2001: Denzel Washington	47
2000: Russell Crowe	36
1999: Kevin Spacey	40

(a) Compute the population mean, μ.

(b) List all possible samples with size $n = 2$. There should be $_6C_2 = 15$ samples.

(c) Construct a sampling distribution for the mean by listing the sample means and their corresponding probabilities.

(d) Compute the mean of the sampling distribution.

(e) Compute the probability that the sample mean is within 3 years of the population mean age.

(f) Repeat parts (b)–(e) using samples of size $n = 3$. Comment on the effect of increasing the sample size.

32. Sampling Distributions The following data represent the running lengths (in minutes) of the winners of the Academy Award for Best Picture for the past 6 years (1999–2004).

2004: Million Dollar Baby	132
2003: The Lord of the Rings: The Return of the King	201
2002: Chicago	112
2001: A Beautiful Mind	134
2000: Gladiator	155
1999: American Beauty	120

(a) Compute the population mean, μ.

(b) List all possible samples with size $n = 2$. There should be $_6C_2 = 15$ samples.

(c) Construct a sampling distribution for the mean by listing the sample means and their corresponding probabilities.

(d) Compute the mean of the sampling distribution.

(e) Compute the probability that the sample mean is within 15 minutes of the population mean running times.

(f) Repeat parts (b)–(e) using samples of size $n = 3$. Comment on the effect of increasing the sample size.

33. Simulation Scores on the Stanford–Binet IQ test are normally distributed with $\mu = 100$ and $\sigma = 16$.

(a) Use MINITAB, Excel, or some other statistical software to obtain 500 random samples of size $n = 20$.

(b) Compute the sample mean for each of the 500 samples.

(c) Draw a histogram of the 500 sample means. Comment on its shape.

(d) What do you expect the mean and standard deviation of the sampling distribution of the mean to be?

(e) Compute the mean and standard deviation of the 500 sample means. Are they close to the expected values?

(f) Compute the probability that a random sample of 20 people results in a sample mean greater than 108.

(g) What proportion of the 500 random samples had a sample mean IQ greater than 108? Is this result close to the theoretical value obtained in part (f)?

34. Sampling Distribution Applet Load the sampling distribution applet on your computer. Set the applet so that the population is bell shaped. Take note of the mean and standard deviation.

(a) Obtain 1000 random samples of size $n = 5$. Describe the distribution of the sample mean based on the results of the applet. According to statistical theory, what is the distribution of the sample mean?

(b) Obtain 1000 random samples of size $n = 10$. Describe the distribution of the sample mean based on the results of the applet. According to statistical theory, what is the distribution of the sample mean?

(c) Obtain 1000 random samples of size $n = 30$. Describe the distribution of the sample mean based on the results of the applet. According to statistical theory, what is the distribution of the sample mean?

(d) Compare the results of parts (a)–(c). How are they the same? How are they different?

35. Sampling Distribution Applet Load the sampling distribution applet on your computer. Set the applet so that the population is skewed or draw your own skewed distribution. Take note of the mean and standard deviation.

(a) Obtain 1000 random samples of size $n = 5$. Describe the distribution of the sample mean based on the results of the applet.

(b) Obtain 1000 random samples of size $n = 10$. Describe the distribution of the sample mean based on the results of the applet.

(c) Obtain 1000 random samples of size $n = 50$. Describe the distribution of the sample mean based on the results of the applet. According to statistical theory, what is the distribution of the sample mean?

(d) Compare the results of parts (a)–(c). How are they the same? How are they different? What impact does the sample size have on the shape of the distribution of the sample mean?

8.2 Distribution of the Sample Proportion

Preparing for this Section   Before getting started, review the following:

- Applications of the Normal Distribution (Section 7.3, pp. 385–389)

Objectives

1. **Describe the sampling distribution of a sample proportion**
2. **Compute probabilities of a sample proportion**

1. ## Describe the Sampling Distribution of a Sample Proportion

Suppose we want to determine the proportion of households in a 100-house homeowners association that favor an increase in the annual assessments to pay for neighborhood improvements. One approach that we might take is to survey all households and determine which were in favor of higher assessments. If 65 of the 100 households favor the higher assessment, the population proportion, p, of households in favor of a higher assessment is

$$p = \frac{65}{100}$$

$$= 0.65$$

Of course, it is rare to gain access to all the individuals in a population. For this reason, we usually obtain estimates of population parameters such as p.

Definition

Suppose a random sample of size n is obtained from a population in which each individual either does or does not have a certain characteristic. The **sample proportion**, denoted $\hat{p}$ (read "p-hat"), is given by

$$\hat{p} = \frac{x}{n}$$

where x is the number of individuals in the sample with the specified characteristic.* The sample proportion is a statistic that estimates the population proportion, p.

*For those who studied Section 6.2 on binomial probabilities, x can be thought of as the number of successes in n trials of a binomial experiment.

EXAMPLE 1   **Computing a Sample Proportion**

Problem: Opinion Dynamics Corporation conducted a survey of 1000 adult Americans 18 years of age or older and asked, "Are you currently on some form of a low-carbohydrate diet?" Of the 1000 individuals surveyed, 150 indicated that they were on a low-carb diet. Find the sample proportion of individuals surveyed who are on a low-carb diet.

Approach: The sample proportion of individuals on a low-carb diet is found using the formula $\hat{p} = \dfrac{x}{n}$, where $x = 150$, the number of individuals in the survey with the characteristic "on a low-carb diet," and $n = 1000$.

Solution: Substituting $x = 150$ and $n = 1000$ into the formula $\hat{p} = \dfrac{x}{n}$, we have that $\hat{p} = \dfrac{150}{1000} = 0.15$. Opinion Dynamics Corporation estimates that 0.15 or 15% of adult Americans 18 years of age or older are on some form of low-carbohydrate diet.

If a second survey of 1000 American adults is conducted, it is likely the estimate of the proportion of Americans on a low-carbohydrate diet will be different because there will be different individuals in the sample. Because the value of $\hat{p}$ varies from sample to sample, it is a random variable and has a probability distribution.

To get a sense of the shape, center, and spread of the distribution of $\hat{p}$, we could repeat the exercise of obtaining simple random samples of 1000 adult Americans over and over. This would lead to a list of sample proportions. Each sample proportion would correspond to a simple random sample of 1000. A histogram of the sample proportions will give us a feel for the shape of the distribution of the sample proportion. The mean of the sample proportions will give us an idea of the center of the distribution of the sample proportion. The standard deviation of the sample proportions gives us an idea of the spread of the distribution of the sample proportions.

Rather than literally surveying 1000 adult Americans over and over again, we will use simulation to get an idea of the shape, center, and spread of the sampling distribution of the proportion.

EXAMPLE 2   **Using Simulation to Describe the Distribution of the Sample Proportion**

Problem: According to the Centers for Disease Control, 17% (or 0.17) of Americans have high cholestrol. Simulate obtaining 100 simple random samples of size (a) $n = 10$, (b) $n = 40$, and (c) $n = 80$. Describe the shape, center, and spread of the distribution for each sample size.

Approach: Use MINITAB, Excel, or some other statistical software package to conduct the simulation. We will perform the following steps:

Step 1: Obtain 100 simple random samples of size $n = 10$ from the population.
Step 2: Compute the sample proportion for each of the 100 samples.
Step 3: Draw a histogram of the sample proportions.
Step 4: Compute the mean and standard deviation of the sample proportions.

We then repeat these steps for samples of size $n = 40$ and $n = 80$.

Solution

Step 1: We simulate obtaining 100 simple random samples each of size $n = 10$ using MINITAB.

Step 2: The first sample of size $n = 10$ results in none of the individuals having high cholesterol, so $\hat{p} = \dfrac{0}{10} = 0$. The second sample results in two of the individuals having high cholesterol, so $\hat{p} = \dfrac{2}{10} = 0.2$. Table 5 shows the sample proportions for all 100 simple random samples of size $n = 10$.

Table 5

0.0	0.1	0.3	0.2	0.0	0.2	0.2	0.0	0.1	0.0
0.2	0.1	0.0	0.3	0.3	0.2	0.1	0.0	0.4	0.3
0.1	0.1	0.0	0.3	0.5	0.3	0.1	0.2	0.3	0.1
0.1	0.2	0.2	0.3	0.3	0.1	0.3	0.5	0.4	0.3
0.0	0.1	0.1	0.2	0.0	0.6	0.3	0.1	0.1	0.2
0.2	0.3	0.2	0.2	0.3	0.1	0.0	0.1	0.2	0.1
0.2	0.1	0.4	0.1	0.2	0.1	0.1	0.2	0.2	0.2
0.1	0.2	0.1	0.0	0.1	0.0	0.3	0.3	0.2	0.2
0.1	0.1	0.4	0.3	0.0	0.3	0.1	0.2	0.1	0.2
0.2	0.0	0.0	0.1	0.3	0.2	0.1	0.1	0.0	0.1

Step 3: Figure 12 shows a histogram of the 100 sample proportions. Notice that the shape of the distribution is skewed right.

Figure 12
Distribution of $\hat{p}$ with $n = 10$

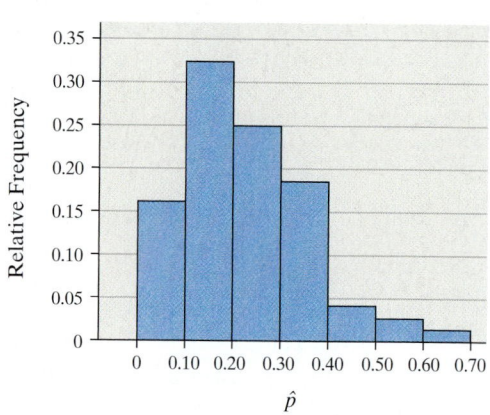

Step 4: The mean of the 100 sample proportions in Table 5 is 0.17. This is the same as the population proportion. The standard deviation of the 100 sample proportions in Table 5 is 0.1262.

We repeat Steps 1 through 4 for samples of size $n = 40$ and $n = 80$. Figure 13 shows the histogram for a sample of size $n = 40$. Notice that the shape of the distribution is skewed right (although not as skewed as the histogram with $n = 10$). The mean of the 100 sample proportions is 0.17 (the same as the population proportion), and the standard deviation is 0.0614 (less than the standard deviation for $n = 10$). Figure 14 shows the histogram for samples of size $n = 80$. Notice that the shape of the distribution is approximately normal. The mean of the 100 sample proportions for samples of size $n = 80$ is 0.17 (the same as the population proportion), and the standard deviation is 0.0408 (less than the standard deviation for $n = 40$).

We notice the following regarding the distribution of the sample proportion:

- *Shape*: As the size of the sample, n, increases, the shape of the distribution of the sample proportion becomes approximately normal.
- *Center*: The mean of the distribution of the sample proportion equals the population proportion, p.
- *Spread*: The standard deviation of the distribution of the sample proportion decreases as the sample size, n, increases.

Figure 13
Distribution of $\hat{p}$ with $n = 40$

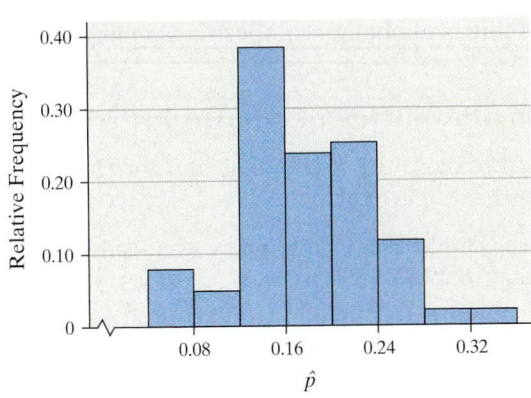

Figure 14
Distribution of $\hat{p}$ with $n = 80$

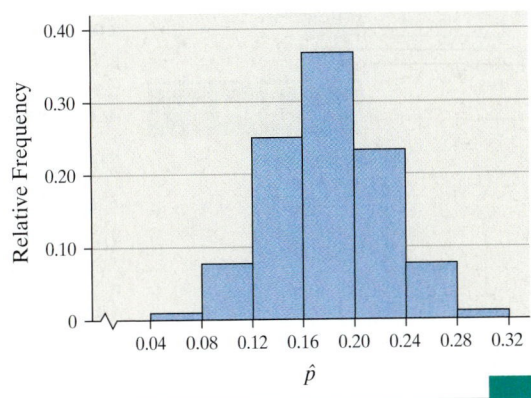

Although the proof is beyond the scope of this text, we should be convinced that the following result is reasonable.

> ### Sampling Distribution of $\hat{p}$
>
> For a simple random sample of size n such that $n \leq 0.05N$ (that is, the sample size is less than or equal to 5% of the population size)
>
> - The shape of the sampling distribution of $\hat{p}$ is approximately normal provided $np(1 - p) \geq 10$
> - The mean of the sampling distribution of $\hat{p}$ is $\mu_{\hat{p}} = p$
> - The standard deviation of the sampling distribution of $\hat{p}$ is
>
> $$\sigma_{\hat{p}} = \sqrt{\frac{p(1 - p)}{n}}$$

In Other Words

The reason that the sample size cannot be more than 5% of the population size is because the success or failure of identifying an individual in the population that has the specified characteristic should not be affected by earlier observations. For example, in a population of size 100 where 14 of the individuals have brown hair, the probability a randomly selected individual has brown hair is 14/100 = 0.14. The probability a second randomly selected student has brown hair is 13/99 = 0.13. The probability changes because the sampling is done without replacement.

The condition that the sample size is no more than 5% of the population size is needed so that result obtained from one individual in the survey is independent of the result obtained from any other individual in the survey. The condition for $np(1 - p)$ is at least 10 is needed for normality.

Also, regardless of whether $np(1 - p) \geq 10$ or not, the mean of the sampling distribution of $\hat{p}$ is p and the standard deviation of the sampling distribution of $\hat{p}$ is $\sqrt{\dfrac{p(1 - p)}{n}}$.

EXAMPLE 3   Describing the Distribution of the Sample Proportion

Problem: According to the Centers for Disease Control, 17% (or 0.17) of Americans have high cholesterol. Suppose we obtain a simple random sample of $n = 80$ Americans and determine which have high cholesterol. Describe the shape, center, and spread for the distribution of the sample proportion of Americans that have high cholesterol.

Approach: If the sample size is less than 5% of the population size and $np(1 - p)$ is at least 10, the sampling distribution of $\hat{p}$ is approximately normal, with mean $\mu_{\hat{p}} = p$ and standard deviation $\sigma_{\hat{p}} = \sqrt{\dfrac{p(1 - p)}{n}}$.

Solution: There are about 295 million people in the United States. The sample of $n = 80$ is certainly less than 5% of the population size. Also, $np(1 - p) = 80(0.17)(1 - 0.17) = 11.288 \geq 10$. The distribution of $\hat{p}$ is approximately normal, with mean $\mu_{\hat{p}} = 0.17$ and standard deviation

$$\sigma_{\hat{p}} = \sqrt{\frac{p(1 - p)}{n}} = \sqrt{\frac{0.17(1 - 0.17)}{80}} \approx 0.0420.$$

Now Work Problem 7.

Compute Probabilities of a Sample Proportion

Now that we can describe the distribution of the sample proportion, we can compute probabilities of obtaining a specific sample proportion.

EXAMPLE 4 **Compute Probabilities of a Sample Proportion**

Problem: According to the National Center for Health Statistics, 15% of all Americans have hearing trouble.

(a) In a random sample of 120 Americans, what is the probability at least 18% have hearing trouble?

(b) Would it be unusual if a random sample of 120 Americans results in 10 having hearing trouble?

Approach: First, we determine whether the distribution of the sampling distribution is approximately normal, with mean $\mu_{\hat{p}} = p$ and standard deviation $\sigma_{\hat{p}} = \sqrt{\dfrac{p(1 - p)}{n}}$, by verifying that the sample size is less than 5% of the population size and that $np(1 - p) \geq 10$. Then we can use the normal distribution to determine the probabilities.

Solution: There are approximately 295 million people in the United States. The sample size of $n = 120$ is definitely less than 5% of the population size. We are told that $p = 0.15$. Because $np(1 - p) = 120(0.15)(1 - 0.15) = 15.3 \geq 10$, the shape of the distribution of the sample proportion is approximately normal. The mean of the sample proportion $\hat{p}$ is $\mu_{\hat{p}} = 0.15$ and the standard deviation is $\sigma_{\hat{p}} = \sqrt{\dfrac{0.15(1 - 0.15)}{120}} \approx 0.0326$.

(a) We want to know the probability that a random sample of 120 Americans will result in a sample proportion of at least 18%, or 0.18. That is, we want to know $P(\hat{p} \geq 0.18)$. Figure 15(a) shows the normal curve with the area to the right of 0.18 shaded. To find this area, we convert $\hat{p} = 0.18$ to a standard normal random variable Z by subtracting the mean and dividing by the standard deviation. Don't forget that we round Z to two decimal places.

$$Z = \frac{\hat{p} - \mu_{\hat{p}}}{\sigma_{\hat{p}}} = \frac{0.18 - 0.15}{\sqrt{\dfrac{0.15(1 - 0.15)}{120}}} = 0.92$$

Figure 15(b) shows a standard normal curve with the area right of 0.92 shaded. Remember, the area to the right of $\hat{p} = 0.18$ is the same as the area to the right of $Z = 0.92$.

Figure 15

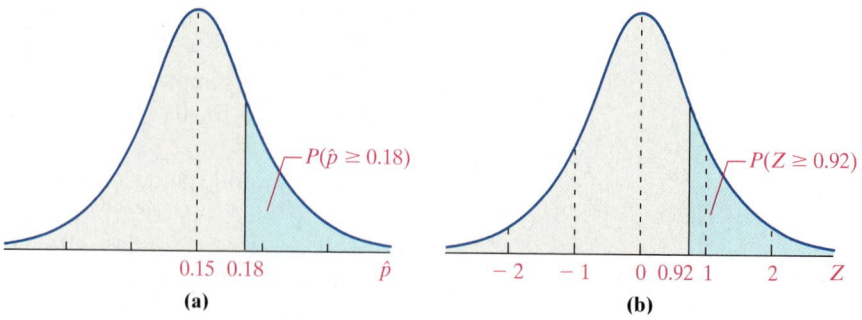

The area to the right of $Z = 0.92$ is 0.1788. Therefore,

$$P(\hat{p} \geq 0.18) = P(Z \geq 0.92) = 0.1788$$

Interpretation: The probability that a random sample of $n = 120$ Americans results in at least 18% having hearing trouble is 0.1788. This means that about 18 out of 100 random samples of size 120 will result in at least 18% having hearing trouble, even though the population proportion of Americans with hearing trouble is 0.15.

(b) A random sample of 120 Americans results in 10 having hearing trouble. The sample proportion of Americans with hearing trouble is $\hat{p} = \dfrac{10}{120} = 0.083$.

To determine whether a sample proportion 0.083 or less is unusual, we compute $P(\hat{p} \le 0.083)$ because if a sample proportion of 0.083 is unusual, then any sample proportion less than 0.083 is also unusual. Figure 16(a) shows the normal curve with the area to the left of 0.083 shaded. To find this area, we convert $\hat{p} = 0.083$ to a standard normal random variable Z.

$$Z = \frac{\hat{p} - \mu_{\hat{p}}}{\sigma_{\hat{p}}} = \frac{0.083 - 0.15}{\sqrt{\dfrac{0.15(1 - 0.15)}{120}}} = -2.06$$

Figure 16(b) shows a standard normal curve with the area left of -2.06 shaded. The area to the left of $\hat{p} = 0.083$ is the same as the area to the left of $Z = -2.06$.

Figure 16

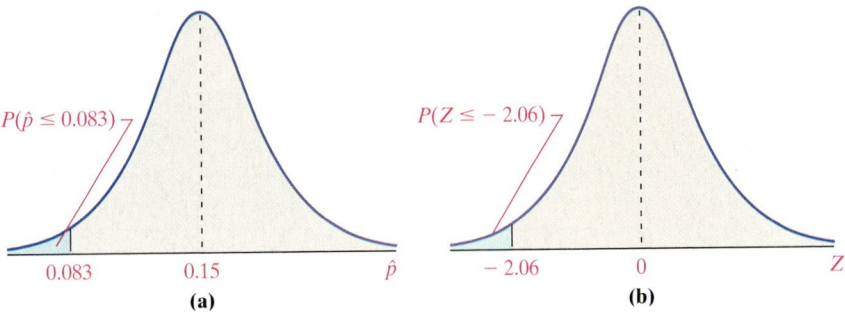

$P(\hat{p} \le 0.083)$

0.083   0.15 $\hat{p}$

(a)

$P(Z \le -2.06)$

-2.06 0 Z

(b)

The area to the left of $Z = -2.06$ is 0.0197. Therefore,

$$P(\hat{p} \le 0.083) = P(Z \le -2.06) = 0.0197$$

Interpretation: About 2 samples in 100 will result in a sample proportion of 0.083 or less from a population whose proportion is 0.15. We obtained a result that should only happen about 2 times in 100, so the results obtained are indeed unusual.

Now Work Problem 17.

8.2 ASSESS YOUR UNDERSTANDING

Concepts and Vocabulary

1. In a town of 500 households, 220 have a dog. The population proportion of dog owners in this town (expressed as a decimal) is $p =$ _____.

2. The _____ _____, denoted $\hat{p}$, is given by the formula $\hat{p} =$ _____, where x is the number of individuals with a specified characteristic in a sample of n individuals.

3. *True or False:* The population proportion and sample proportion always have the same value.

4. *True or False:* The mean of the sampling distribution of $\hat{p}$ is p.

5. Describe the circumstances under which the shape of the sampling distribution of $\hat{p}$ is approximately normal.

6. What happens to the standard deviation of $\hat{p}$ as the sample size increases? If the sample size is increased by a factor of 4, what happens to the standard deviation of $\hat{p}$?

Skill Building

In Problems 7–10, describe the sampling distribution of $\hat{p}$. Assume the size of the population is 25,000 for each problem.

7. $n = 500$, $p = 0.4$
NW

8. $n = 300$, $p = 0.7$

9. $n = 1000$, $p = 0.103$

10. $n = 1010$, $p = 0.84$

11. Suppose a simple random sample of size $n = 75$ is obtained from a population whose size is $N = 10,000$ and whose population proportion with a specified characteristic is $p = 0.8$.

(a) Describe the sampling distribution of $\hat{p}$.
(b) What is the probability of obtaining $x = 63$ or more individuals with the characteristic? That is, what is $P(\hat{p} \geq 0.84)$?
(c) What is the probability of obtaining $x = 51$ or fewer individuals with the characteristic? That is, what is $P(\hat{p} \leq 0.68)$?

12. Suppose a simple random sample of size $n = 200$ is obtained from a population whose size is $N = 25,000$ and whose population proportion with a specified characteristic is $p = 0.65$.

(a) Describe the sampling distribution of $\hat{p}$.
(b) What is the probability of obtaining $x = 136$ or more individuals with the characteristic? That is, what is $P(\hat{p} \geq 0.68)$?
(c) What is the probability of obtaining $x = 118$ or fewer individuals with the characteristic? That is, what is $P(\hat{p} \leq 0.59)$?

13. Suppose a simple random sample of size $n = 1000$ is obtained from a population whose size is $N = 1,000,000$ and whose population proportion with a specified characteristic is $p = 0.35$.

(a) Describe the sampling distribution of $\hat{p}$.
(b) What is the probability of obtaining $x = 390$ or more individuals with the characteristic?
(c) What is the probability of obtaining $x = 320$ or fewer individuals with the characteristic?

14. Suppose a simple random sample of size $n = 1460$ is obtained from a population whose size is $N = 1,500,000$ and whose population proportion with a specified characteristic is $p = 0.42$.

(a) Describe the sampling distribution of $\hat{p}$.
(b) What is the probability of obtaining $x = 657$ or more individuals with the characteristic?
(c) What is the probability of obtaining $x = 584$ or fewer individuals with the characteristic?

Applying the Concepts

15. Gardeners Suppose a simple random sample of size $n = 100$ households is obtained from a town with 5000 households. It is known that 30% of the households plant a garden in the spring.

(a) Describe the sampling distribution of $\hat{p}$.
(b) What is the probability that more than 37 households in the sample plant a garden? Is this result unusual?
(c) What is the probability that 18 or fewer households in the sample plant a garden? Is this result unusual?

16. Text Messaging A nationwide study in 2003 indicated that about 60% of college students with cell phones send and receive text messages with their phones. Suppose a simple random sample of $n = 1136$ college students with cell phones is obtained.
(*Source*: promomagazine.com)

(a) Describe the sampling distribution of $\hat{p}$, the sample proportion of college students with cell phones who send or receive text messages with their phones.
(b) What is the probability that 665 or fewer college students in the sample send and receive text messages with their cell phones? Is this result unusual?
(c) What is the probability that 725 or more college students in the sample send and receive text messages with their cell phone? Is this result unusual?

17. Credit Cards According to a *USA Today* "Snapshot,"
NW 26% of adults do not have any credit cards. Suppose a simple random sample of 500 adults is obtained.

(a) Describe the sampling distribution of $\hat{p}$, the sample proportion of adults who do not have a credit card.
(b) In a random sample of 500 adults, what is the probability that less than 24% have no credit cards?
(c) Would it be unusual if a random sample of 500 adults results in 150 or more having no credit cards?

18. Cell Phone Only According to a CNN report, 7% of the population do not have traditional phones and instead rely on only cell phones. Suppose a random sample of 750 telephone users is obtained.

(a) Describe the sampling distribution of $\hat{p}$, the sample proportion that are "cell-phone only."
(b) In a random sample of 750 telephone users, what is the probability that more than 8% are "cell-phone only"?
(c) Would it be unusual if a random sample of 750 adults results in 40 or fewer being "cell-phone only"?

19. **Phishing** A report released in May 2005 by First Data Corp. indicated that 43%, of adults had received a "phishing" contact (a bogus email that replicates an authentic site for the purpose of stealing personal information such as account numbers and passwords). Suppose a random sample of 800 adults is obtained.

    (a) In a random sample of 800 adults, what is the probability that no more than 40% have received a phishing contact?

    (b) Would it be unusual if a random sample of 800 adults resulted in 45% or more who had received a phishing contact?

20. **Second Homes** According to the National Association of Realtors, 23% of the roughly 8 million homes purchased in 2004 were considered investment properties. Suppose a random sample of 500 homes sold in 2004 is obtained.

    (a) In a random sample of 500 homes sold in 2004, what is the probability that at least 125 were purchased as investment properties?

    (b) Would it be unusual if a random sample of 500 homes sold in 2004 results in 20% or less being purchased as an investment property?

21. **Social Security Reform** A researcher studying public opinion of proposed Social Security changes obtains a simple random sample of 50 adult Americans and asks them whether or not they support the proposed changes. To say that the distribution of $\hat{p}$, the sample proportion who respond yes, is approximately normal, how many more adult Americans does the researcher need to sample if

    (a) 10% of all adult Americans support the changes?

    (b) 20% of all adult Americans support the changes?

22. **ADHD** A researcher studying ADHD among teenagers obtains a simple random sample of 100 teenagers aged 13 to 17 and asks them whether or not they have ever been prescribed medication for ADHD. To say that the distribution of $\hat{p}$, the sample proportion who respond no, is approximately normal, how many more teenagers aged 13 to 17 does the researcher need to sample if

    (a) 90% of all teenagers aged 13 to 17 have never been prescribed medication for ADHD?

    (b) 95% of all teenagers aged 13 to 17 have never been prescribed medication for ADHD?

23. **Simulation** The following exercise is meant to illustrate the normality of the distribution of the sample proportion, $\hat{p}$.

    (a) Using MINITAB or some other statistical spreadsheet, randomly generate 2000 samples of size 765 from a population with $p = 0.3$. Store the number of successes in a column called x.

    (b) Determine $\hat{p}$ for each of the 2000 samples by computing $\dfrac{x}{765}$. Store each $\hat{p}$ in a column called *phat*.

    (c) Draw a histogram of the 2000 estimates of p. Comment on the shape of the distribution.

    (d) Compute the mean and standard deviation of the sampling distribution of $\hat{p}$ in the simulation.

    (e) Compute the theoretical mean and standard deviation of the sampling distribution of $\hat{p}$. Compare the theoretical results to the results of the simulation. Are they close?

24. **The Sampling Distribution Applet** Load the sampling distribution applet on your computer. Set the applet so that the population is binary with probability of success equal to 0.2.

    (a) Obtain 1000 random samples of size $n = 5$. Describe the distribution of the sample proportion based on the results of the applet.

    (b) Obtain 1000 random samples of size $n = 30$. Describe the distribution of the sample proportion based on the results of the applet.

    (c) Obtain 1000 random samples of size $n = 100$. Describe the distribution of the sample proportion based on the results of the applet.

    (d) Compare the results of parts (a)–(c). How are they the same? How are they different?

25. **Finite Population Correction Factor** In this section, we assumed that the sample size was less than 5% of the size of the population. When sampling without replacement from a finite population in which $n > 0.05N$, the standard deviation of the distribution of $\hat{p}$ is given by

$$\sigma_{\hat{p}} = \sqrt{\frac{\hat{p}(1 - \hat{p})}{n - 1} \cdot \left(\frac{N - n}{N}\right)}$$

where N is the size of the population. Suppose a survey is conducted at a college having an enrollment of 6,502 students. The student council wants to estimate the percentage of students in favor of establishing a student union. In a random sample of 500 students, it was determined that 410 were in favor of establishing a student union.

    (a) Obtain the sample proportion, $\hat{p}$, of students surveyed who favor establishing a student union.

    (b) Calculate the standard deviation of the sampling distribution of $\hat{p}$.

Consumer Reports® | Tanning Salons

Medical groups have long warned about the dangers of indoor tanning. The American Medical Association and the American Academy of Dermatology unsuccessfully petitioned the Food and Drug Administration in 1994 to ban cosmetic-tanning equipment. Three years later, the Federal Trade Commission warned the public to beware of advertised claims that "unlike the sun, indoor tanning will not cause skin cancer or skin aging" or that you can "tan indoors with absolutely no harmful side effects."

In February 1999, still under pressure from the medical community, the FDA announced that current recommendations "may allow higher exposures" to UV radiation "than are necessary." The agency proposed reducing recommended exposures and requiring simpler wording on consumer warnings. But it has not yet implemented either of these changes. An FDA spokeswoman told us that "the agency decided to postpone amendment of its standard pending the results of ongoing research and discussions with other organizations."

To make matters worse, only about half the states have any rules for tanning parlors. In some of these states, the regulation is minimal and may not require licensing, inspections, training, record keeping, or parental consent for minors. Despite this, nearly 30 million Americans, including a growing number of teenage girls, are expected to visit a tanning salon in 2005.

In a recent survey of 296 indoor-tanning facilities around the country, to our knowledge the first nationwide survey of its kind, we found evidence of widespread failures to inform customers about the possible risks, including premature wrinkling and skin cancer, and to follow recommended safety procedures, such as wearing eye goggles. Many facilities made questionable claims about indoor tanning: that it's safer than sunlight, for example, and is well controlled.

(a) In designing this survey, why is it important to sample a large number of facilities? And why is it important to sample these facilities in multiple cities?

(b) Given the fact that there are over 150,000 tanning facilities in the United States, is the condition for independence of survey results satisfied? Why?

(c) Sixty-seven of the 296 tanning facilities surveyed stated that "tanning in a salon is the same as tanning in the sun with respect to causing skin cancer." Assuming that the true proportion is 25%, describe the sampling distribution of $\hat{p}$, the sample proportion of tanning facilities that state "tanning in a salon is the same as tanning in the sun with respect to causing skin cancer." Calculate the probability that less than 22.6% of randomly selected tanning salon facilities would state that tanning in a salon is the same as tanning in the sun with respect to causing skin cancer.

(d) Forty-two of the 296 tanning facilities surveyed stated "tanning in a salon does not cause wrinkled skin." Assuming that the true proportion is 18%, describe the sampling distribution of $\hat{p}$, the sample proportion of tanning facilities that state that "tanning in a salon does not cause wrinkled skin." Calculate the probability that at least 14.2% will state that tanning in a salon does not cause wrinkled skin. Would it be unusual for 50 or fewer facilities to state that tanning in a salon does not cause wrinkled skin?

Note to Readers: *In many cases, our test protocol and analytical methods are more complicated than described in this example. The data and discussion have been modified to make the material more appropriate for the audience.*

Chapter **8**   # Review

Summary

This chapter forms the bridge between probability and statistical inference. In Section 8.1, we discussed the distribution of the sample mean. We learned that the mean of the distribution of the sample mean equals the mean of the population $(\mu_{\bar{x}} = \mu)$ and that the standard deviation of the distribution of the sample mean is the standard deviation of the population divided by the square root of the sample size $\left(\sigma_{\bar{x}} = \dfrac{\sigma}{\sqrt{n}}\right)$. If the sample is obtained from a population that is known to be normally distributed, the shape of the distribution of the sample mean is also normal. If the sample is obtained from a population that is not normal, the shape of the distribution of the sample mean becomes approximately normal as the sample size increases. This result is known as the Central Limit Theorem.

In Section 8.2, we discussed the distribution of the sample proportion. We learned that the mean of the distribution of the sample proportion is the population proportion $(\mu_{\hat{p}} = p)$ and that the standard deviation of the distribution of the sample proportion is $\sigma_{\hat{p}} = \sqrt{\dfrac{p(1-p)}{n}}$. If $np(1-p) \geq 10$, then the shape of the distribution of $\hat{p}$ is approximately normal.

Formulas

Mean and Standard Deviation of the Sampling Distribution of $\bar{x}$

$$\mu_{\bar{x}} = \mu \quad \text{and} \quad \sigma_{\bar{x}} = \frac{\sigma}{\sqrt{n}}$$

Sample Proportion

$$\hat{p} = \frac{x}{n}$$

Mean and Standard Deviation of the Sampling Distribution of $\hat{p}$

$$\mu_{\hat{p}} = p \quad \text{and} \quad \sigma_{\hat{p}} = \sqrt{\frac{p(1-p)}{n}}$$

Standardizing a Normal Random Variable

$$Z = \frac{\bar{x} - \mu}{\frac{\sigma}{\sqrt{n}}} \quad \text{or} \quad Z = \frac{\hat{p} - p}{\sqrt{\frac{p(1-p)}{n}}}$$

Vocabulary

Statistical inference (p. 417)
Sampling distribution of the sample mean (p. 417)

Law of Large Numbers (p. 423)
Standard error of the mean (p. 424)
Central Limit Theorem (p. 427)

Sample proportion (p. 434)
Sampling distribution of $\hat{p}$ (p. 437)

Objectives

Section	You should be able to . . .	Example	Review Exercises
8.1	**1** Understand the concept of a sampling distribution (p. 417)	1	1
	2 Describe the distribution of the sample mean for samples obtained from normal populations (p. 419)	2 through 4	2, 4, 5, 6
	3 Describe the distribution of the sample mean for samples obtained from a population that is not normal (p. 426)	5 and 6	8, 9, 13, 14
8.2	**1** Describe the sampling distribution of a sample proportion (p. 434)	2 and 3	3, 4, 9(a)–12(a)
	2 Compute probabilities of a sample proportion (p. 438)	4	9(b), (c); 10(b), (c); 11(b), (c); 12(b), (c)

Review Exercises

1. In your own words, explain what a sampling distribution is.

2. Under what conditions is the sampling distribution of $\bar{x}$ normal?

3. Under what conditions is the sampling distribution of $\hat{p}$ approximately normal?

4. What is the mean and standard deviation of the sampling distribution of $\bar{x}$? What is the mean and standard deviation of the sampling distribution of $\hat{p}$?

5. **Energy Needs During Pregnancy** Suppose the total energy need during pregnancy is normally distributed with mean $\mu = 2600$ kcal/day and standard deviation $\sigma = 50$ kcal/day.
(*Source*: American Dietetic Association).
   (a) What is the probability that a randomly selected pregnant woman has an energy need of more than 2625 kcal/day? Is this result unusual?
   (b) Describe the sampling distribution of $\bar{x}$, the sample mean daily energy requirement for a random sample of 20 pregnant women.
   (c) What is the probability that a random sample of 20 pregnant women has a mean energy need of more than 2625 kcal/day? Is this result unusual?

6. **Battery Life** The charge life of a certain lithium ion battery for camcorders is normally distributed with mean 90 minutes and standard deviation 35 minutes.
   (a) What is the probability that a battery of this type, randomly selected, lasts more than 100 minutes on a single charge? Is this result unusual?
   (b) Describe the sampling distribution of $\bar{x}$, the sample mean charge life for a random sample of 10 such batteries.
   (c) What is the probability that a random sample of 10 such batteries has a mean charge life of more than 100 minutes? Is this result unusual?

7. **Copper Tubing** A machine at K&A Tube & Manufacturing Company produces a certain copper tubing component in a refrigeration unit. The tubing components produced by the manufacturer have a mean diameter of 0.75 inch with standard deviation 0.004 inch. The quality-control inspector takes a random sample of 30 components once each week and calculates the mean diameter of these components. If the mean is either less than 0.748 inch or greater than 0.752 inch, the inspector concludes that the machine needs an adjustment.
   (a) Describe the sampling distribution of $\bar{x}$, the sample mean diameter, for a random sample of 30 such components.
   (b) What is the probability that, based on a random sample of 30 such components, the inspector will conclude that the machine needs an adjustment?

8. **Filling Machines** A machine used for filling plastic bottles with a soft drink has a known standard deviation of $\sigma = 0.05$ liter. The target mean fill volume is $\mu = 2.0$ liter.
   (a) Describe the sampling distribution of $\bar{x}$, the sample mean fill volume, for a random sample of 45 such bottles.

(b) What is the probability that a random sample of 45 such bottles has a mean fill volume that is less than 1.995 liters?

(c) What is the probability that a random sample of 45 such bottles has a mean fill volume that is more than 2.015 liters? Is this result unusual? What might we conclude?

9. **Entrepreneurship** A Gallup survey in March 2005 indicated that 72% of 18- to 29-year-olds if given a choice, would prefer to start their own business rather than work for someone else. Suppose a random sample of 600 18- to 29-year-olds is obtained.

(a) Describe the sampling distribution of $\hat{p}$, the sample proportion of 18- to 29-year-olds who would prefer to start their own business.

(b) In a random sample of 600 18- to 29-year-olds, what is the probability that no more than 70% would prefer to start their own business?

(c) Would it be unusual if a random sample of 600 18- to 29-year-olds resulted in 450 or more who would prefer to start their own business?

10. **Smokers** According to the National Center for Health Statistics (2004), 22.4% of adults are smokers. Suppose a random sample of 300 adults is obtained.

(a) Describe the sampling distribution of $\hat{p}$, the sample proportion of adults who smoke.

(b) In a random sample of 300 adults, what is the probability that at least 50 are smokers?

(c) Would it be unusual if a random sample of 300 adults results in 18% or less being smokers?

11. **Advanced Degrees** According to the U.S. Census Bureau, roughly 9% of adults aged 25 years or older have an advanced degree. Suppose a random sample of 200 adults aged 25 years or older is obtained.

(a) Describe the sampling distribution of $\hat{p}$, the sample proportion of adults aged 25 years or older who have an advanced degree.

(b) In a random sample of 200 adults aged 25 years or older, what is the probability that no more than 6% have an advanced degree?

(c) Would it be unusual if a random sample of 200 adults aged 25 years or older results in 25 or more having an advanced degree?

12. **Peanut and Tree Nut Allergies** Peanut and tree nut allergies are considered to be the most serious food allergies. According to the National Institute of Allergy and Infectious Diseases, roughly 1% of Americans are allergic to peanuts or tree nuts. Suppose a random sample of 1500 Americans is obtained.

(a) Describe the sampling distribution of $\hat{p}$, the sample proportion of Americans allergic to peanuts or tree nuts.

(b) In a random sample of 1500 Americans, what is the probability that more than 1.5% are allergic to peanuts or tree nuts?

(c) Would it be unusual if a random sample of 1500 Americans results in fewer than 10 with peanut or tree nut allergies?

13. **Principals' Salaries** According to the *National Survey of Salaries and Wages in Public Schools*, the mean salary paid to public high school principals in 2004–2005 was $71,401. Assume the standard deviation is $26,145. What is the probability that a random sample of 100 public high school principals has an average salary under $65,000?

14. **Teaching Supplies** According to the National Education Association, public school teachers spend an average of $443 of their own money each year to meet the needs of their students. Assume the standard deviation is $175. What is the probability that a random sample of 50 public school teachers spends an average of more than $400 each year to meet the needs of their students?

CASE STUDY

Sampling Distribution of the Median

The exponential distribution is an example of a skewed distribution. It is defined by the density function

$$f(x) \geq \begin{cases} \lambda e^{\lambda x} = x \geq 0 \\ 0 \quad = x < 0 \end{cases}$$

with parameter λ. The mean of an exponential distribution is $\mu = \dfrac{1}{\lambda}$ and the median is $M = \dfrac{\ln 2}{\lambda}$.

The exponential distribution is used to model the failure rate of electrical components. For many electrical components, a plot of the failure rate over time appears to be shaped like a bathtub. The exponential distribution is used widely in reliability engineering for the constant hazard rate portion of the bathtub curve, where components spend most of their useful life. The portion of the curve prior to the constant rate is referred to as the burn-in period, and the portion after the constant rate is referred to as the wear-out period.

Another interesting property of the exponential distribution is the *memoryless property*. This says that the time to failure of a component does not depend on how long the component has been in use. For example, the probability that a component will fail in the next hour is the same whether the component has been in use for 10, 50, or 100 hours, and so on.

In our discussion of summary statistics, we have seen that the mean is used as a measure of the center for data sets whose distribution is (roughly) symmetric with no outliers. In the presence of outliers or skewness, we used the median as the measure of the center because it is resistant to extreme values. However, for skewed distributions, it is often difficult to determine which measure of center is most useful. In some cases, the mean, though not resistant, is easier to find than the median or is more intuitive to the reader within the context of the problem. Here we look at the sampling distributions of the mean and median for two different populations, one normal (symmetric) and one exponential (skewed).

1. Using a statistical spreadsheet, generate 250 samples (rows) of size $n = 80$ (columns) from a normal distribution with mean 50 and standard deviation 10.

2. Compute the mean for the first 10 values in each row and store the values in C81. You might label this column "n10mn" for normal distribution, sample size 10, mean.

3. Compute the median for the first 10 values in each row and store the values in C82. You might label this column 'n10med' for normal distribution, sample size 10, median.

4. Repeat parts 2 and 3 for samples of size 20 (C1–C20), 40 (C1–C40), and 80 (C1–C80), storing the results in consecutive columns, C83, C84, C85, and so on.

5. Compute the summary statistics for your data in columns C81 to C88.

6. How do the averages of your sample means compare to the actual population mean for the different sample sizes? How do the averages of your sample medians compare to the actual population median for the different sample sizes?

7. How does the standard deviation of sample means compare to the standard deviation of sample medians for different sample sizes?

8. Based on your results in parts 6 and 7, which measure of center seems more appropriate? Explain.

9. Construct histograms for each column of summary statistics. Describe the effect, if any, of increasing the sample size on the shape of the distribution of sample means and sample medians.

10. Using a statistical spreadsheet, generate 250 samples of size $n = 80$ from an exponential distribution with scale parameter 50. This could represent the distribution of failure times for a component with a failure rate of $\lambda = 0.02$.

Note: The threshold should be set to 0. Store the results in columns $101 - 180$.

11. Compute the mean for the first 10 values in each row and store the values in C181. You might label this column 'e10mn' for exponential distribution, sample size 10, mean.

12. Compute the median for the first 10 values in each row and store the values in C182. You might label this column 'e10med' for exponential distribution, sample size 10, median.

13. Repeat parts 11 and 12 for samples of size 20 (C101–C120), 40 (C101–C140), and 80 (C101–C180), storing the results in consecutive columns, C183, C184, C185, and so on.

14. Compute the summary statistics for your data in columns C181 to C188.

15. How do the averages of your sample means compare to the actual population mean for the different sample sizes? How do the averages of your sample medians compare to the actual population median for the different sample sizes?

16. How does the standard deviation of sample means compare to the standard deviation of sample medians for different sample sizes?

17. Based on your results in parts 15 and 16, which measure of center seems more appropriate? Explain?

18. Construct histograms for each column of summary statistics. Describe the effect, if any, of increasing the sample size on the shape of the distribution of sample means and sample medians.

19. Can the Central Limit Theorem be used to explain any of the results in part 18? Why or why not?

Estimating the Value of a Parameter Using Confidence Intervals

CHAPTER **9**

Outline

DECISIONS

One of the most difficult decisions that a person has to make while attending college is choosing a major field of study. This decision plays a large role in the career path of an individual. What major should you choose? The Decisions project on page 483 may help you decide.

●●● Putting It All Together

Chapters 1 through 7 laid the groundwork for the remainder of the course. These chapters dealt with data collection (Chapter 1), descriptive statistics (Chapters 2 through 4), and probability (Chapters 5 through 7). Chapter 8 formed a bridge between probability and statistical inference by giving us a model we can use to make probability statements about the sample mean and sample proportion.

We know from Section 8.1 that $\bar{x}$ is a random variable and so has a distribution associated with it. This distribution is called the sampling distribution of $\bar{x}$, the sample mean. The mean of this distribution is equal to the mean of the population, μ. The standard deviation of $\bar{x}$ is $\dfrac{\sigma}{\sqrt{n}}$. The shape of the distribution of $\bar{x}$ is normal if either the population is normal or approximately normal if the sample size is large. We learned in Section 8.2 that $\hat{p}$ is also a random variable whose mean is p and standard deviation is $\sqrt{\dfrac{p(1-p)}{n}}$. If $np(1-p) \geq 10$, the distribution of the random variable $\hat{p}$ is approximately normal.

We now discuss inferential statistics — the process of using information obtained in a sample and generalizing it to a population. We will study two areas of inferential statistics: (1) estimation—sample data are used to estimate the value of unknown parameters such as μ or p, and (2) hypothesis testing—claims regarding a characteristic of one or more populations are made and sample data are used to test the claims. In this chapter, we discuss estimation of an unknown parameter and in the next chapter, we will discuss hypothesis testing.

The information collected from a sample does not contain all the information in the population so we will assign probabilities to our estimates. These probabilities serve as a way of measuring what will happen if we estimate the value of the parameter many times and provide a measure of confidence in our results.

9.1   The Logic in Constructing Confidence Intervals about a Population Mean where the Population Standard Deviation Is Known

Preparing for This Section   Before getting started, review the following:

- Simple random sampling (Section 1.2, pp. 16–19)
- Sampling error (Section 1.4, p. 33)
- Parameter versus statistic (Section 3.1, p. 121)
- z_α notation (Section 7.2, p. 380)

- Normal probability plots (Section 7.4, pp. 394–398)
- Distribution of the sample mean (Section 8.1, pp. 417–430)

Objectives

1. **Compute a point estimate of the population mean**
2. **Construct and interpret a confidence interval about the population mean (assuming the population standard deviation is known)**
3. **Understand the role of margin of error in constructing a confidence interval**
4. **Determine the sample size necessary for estimating the population mean within a specified margin of error**

Our goal of this section is to estimate the value of an unknown population mean. We begin by introducing the approach to estimation by making the simplifying assumption that the population standard deviation σ is known. Granted, it is unlikely that we know the population standard deviation while not knowing the population mean. However, the assumption is made because it allows us to use the normal distribution to develop the technique of estimation. Put simply, this assumption liberates us from details and allows us to focus on the concept of estimation. We will drop this assumption in the next section.

1   Compute a Point Estimate of the Population Mean

The goal of statistical inference is to use information obtained from a sample and generalize the results to the population that is being studied. The first step in estimating the value of an unknown parameter such as μ or σ, is to obtain a random sample and use the data from the sample to obtain a *point estimate* of the parameter.

Definition   A **point estimate** is the value of a statistic that estimates the value of a parameter.

For example, the sample mean, $\bar{x}$, is a point estimate of the population mean, μ.

| EXAMPLE 1 | **Computing a Point Estimate** |

Problem: A two-lane highway with a posted speed limit of 45 miles per hour is located just outside a small 40-home subdivision. The residents of the neighborhood are concerned that the speed of cars on the highway is excessive and want an estimate of the population mean speed of the cars on the highway.

Approach: The Department of Transportation estimates that 2400 cars travel on the highway between the hours of 9 A.M. and 3 P.M. (nonpeak time). The residents decide to measure the speed of 12 cars on the highway during nonpeak times (congestion slows everyone down). The 12 cars will be selected using a simple random sample by determining 12 numbers between 1 and 2400 and measuring the speed of the cars using a radar gun. The sample mean speed of the 12 cars will be the point estimate of the population mean.

Solution: Table 1 shows the speeds (in miles per hour) of the 12 randomly selected cars. The sample mean is

Table 1

57.4	56.1	70.3	65.6
44.2	58.6	66.1	57.3
62.2	60.4	64.5	52.7

$$\bar{x} = \frac{57.4 + 44.2 + \cdots + 52.7}{12} = \frac{715.4}{12} = 59.62 \text{ miles per hour}$$

The point estimate of μ is 59.62 miles per hour.

Now Work Problem 25(a).

2 Construct and Interpret a Confidence Interval about the Population Mean

What do we do with the point estimate obtained in Example 1? Would you go to the police department and ask for the speed on this highway to be monitored based on these 12 observations? Or would you like to have more evidence? After all, we know that statistics such as $\bar{x}$ vary from sample to sample. So a different random sample of 12 cars might result in a different point estimate of the population mean, such as $\bar{x} = 54.39$ miles per hour. Assuming the method used in selecting the cars was done appropriately, both point estimates would be reasonably good guesses of the population mean. Because of this variability in the random variable $\bar{x}$, we are better off reporting a range (or interval) of values, with some measure of the likelihood that the interval includes the unknown population mean.

To help understand the idea of this interval, consider the following situation. Suppose you were asked to guess the mean age of the students in your statistics class. If the sample mean age based on a survey of five of the students is $\bar{x} = 24$, you might guess the mean age of *all* the students is 24 years. This would be a point estimate of μ, the mean age of all students in the class. As you look around the room, you realize that you did not survey everyone, so your guess is probably off. To account for this error, you might express your guess by producing a range of ages, such as 24 years old give or take 2 years (the *margin of error*). Mathematically, we write this as 24 ± 2. If asked how confident you are that the mean age is between 22 and 26, you might respond, "I am 80% confident that the mean age of all students in my statistics class is between 22 and 26 years." If asked to give a range for which your confidence increases to say, 90%, what do you think will happen to the interval? To have 90% confidence, you may need to increase your interval, say, to 20 to 28 years.

In Other Words

The symbol ± is read "plus or minus." It means "to add and subtract the quantity following the ± symbol."

In-Class Activity—Intro to Confidence Intervals (Tea Time)

When asked a question whose answer we are unsure of, we typically give an estimate along with some tolerance. For example, one might say the distance to a certain location is 5.5 miles, give or take (or within) a half-mile. The give-or-take amount reflects the certainty of our estimate.

What is the mean number gallons of tea consumed by individuals each year in the United States?

(a) Make a reasonable guess to answer the question. On what did you base your answer?

(b) Give an interval centered about your guess in which you are 99% confident the true value lies.

(c) Give an interval centered about your guess in which you are 90% confident the true value lies.

(d) Give an interval centered about your guess in which you are 50% confident the true value lies.

(e) How did the width of your interval change as you changed your level of confidence?

(f) Determine the mean number of gallons of tea consumed by individuals each year for your class.

(g) Obtain the true value from your instructor and compare it to your answer in parts (a) and (f). Discuss the comparison and the difficulties associated with estimating unknown quantities.

In statistics, we construct intervals for a population mean centered around a guess as well. The *guess* is the sample mean, the point estimate of the population mean.

Definition

In Other Words

A confidence interval is a range of numbers, such as 22–30. The level of confidence is the proportion of times that the interval contains the population mean if repeated samples were obtained.

- A **confidence interval** for an unknown parameter consists of an interval of numbers.
- The **level of confidence** represents the expected proportion of intervals that will contain the parameter if a large number of different samples is obtained. The level of confidence is denoted $(1 - \alpha) \cdot 100\%$.

For example, a 95% level of confidence ($\alpha = 0.05$) implies that, if 100 different confidence intervals are constructed, each based on a different sample from the same population, we will expect 95 of the intervals to include the parameter and 5 to not include the parameter.

Confidence interval estimates for the population mean are of the form

$$\text{Point estimate} \pm \text{margin of error}$$

The margin of error of a confidence interval estimate of a parameter depends on three factors:

1. *Level of confidence:* As the level of confidence increases, the margin of error also increases. This should feel logical, based on the discussion presented earlier regarding estimating the mean age of students in your statistics class.

2. *Sample size:* As the size of the random sample increases, the margin of error decreases. This is a consequence of the Law of Large Numbers, which states that as the sample size increases the difference between the statistic and parameter decreases.

3. *Standard deviation of the population:* The more spread there is in the population, the wider our interval will be for a given level of confidence.

Figure 1
Sampling distribution of $\bar{x}$

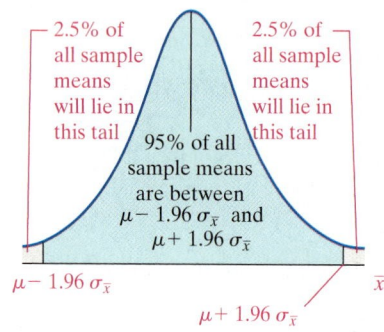

Now the question is "How do we construct an interval?" From Chapter 8, we know the following information about the distribution of the sample mean.

- The shape of the distribution of all possible sample means will be normal provided the population is normal or approximately normal if the sample size is large ($n \geq 30$).
- The mean of the distribution of sample means equals the mean of the population. That is, $\mu_{\bar{x}} = \mu$.
- The standard deviation of the sample mean equals the standard deviation of the population divided by the square root of the sample size. That is, $\sigma_{\bar{x}} = \dfrac{\sigma}{\sqrt{n}}$.

The sample mean is a point estimate of the population mean, so we expect the value of $\bar{x}$ to be close to μ, but we do not know how close. Because $\bar{x}$ is normally distributed, we know 95% of all sample means lie within 1.96 standard deviations of the population mean, μ, and 2.5% of the sample means lie in each tail. See Figure 1. That is, 95% of all sample means are in the interval

$$\mu - 1.96 \cdot \frac{\sigma}{\sqrt{n}} < \bar{x} < \mu + 1.96 \cdot \frac{\sigma}{\sqrt{n}}$$

With a little algebraic manipulation, we can rewrite this inequality with μ in the middle and obtain

$$\bar{x} - 1.96 \cdot \frac{\sigma}{\sqrt{n}} < \mu < \bar{x} + 1.96 \cdot \frac{\sigma}{\sqrt{n}} \tag{1}$$

This inequality states that 95% of *all* sample means will result in confidence interval estimates that contain the population mean. It is common to write the 95% confidence interval as

$$\bar{x} \pm 1.96 \cdot \frac{\sigma}{\sqrt{n}}$$

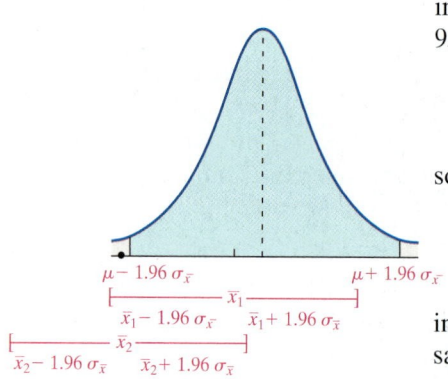

so that it is of the form

Point estimate $\pm$ margin of error

Perhaps you are wondering where the 1.96 comes from in a 95% confidence interval. For a 95% confidence interval, we are capturing the middle 95% of all sample means, so 2.5% of the sample means are in each tail. Remember that the notation z_α is used to represent the Z-value such that the area in the right tail is α. So $z_{0.025} = 1.96$ is the Z-value such that 2.5% of the area under the standard normal curve is to the right.

Let's look at an example that illustrates the concept of a 95% confidence interval.

EXAMPLE 2 **Constructing 95% Confidence Intervals Based on Twenty Samples**

Problem: It is known that scores on the Stanford–Binet IQ test are normally distributed with $\mu = 100$ and $\sigma = 16$. Use MINITAB, Excel, or some other statistical software to simulate obtaining 20 simple random samples of size $n = 15$. Use these 20 different samples to construct 95% confidence intervals for the population mean, μ.

Approach

Step 1: We will use MINITAB to obtain 20 simple random samples of size 15 from a population that is normally distributed with mean $\mu = 100$ and $\sigma = 16$. We then compute the sample means of each of the 20 samples.

Step 2: Construct 95% confidence intervals by computing

$$\bar{x} - 1.96 \cdot \frac{\sigma}{\sqrt{n}} = \bar{x} - 1.96 \cdot \frac{16}{\sqrt{15}}$$

and

$$\bar{x} + 1.96 \cdot \frac{\sigma}{\sqrt{n}} = \bar{x} + 1.96 \cdot \frac{16}{\sqrt{15}}$$

for each sample mean.

Solution

Step 1: Table 2 shows the 20 sample means obtained from MINITAB.

Table 2				
104.32	93.97	108.73	104.11	100.67
96.87	99.74	100.25	101.32	94.24
102.23	94.32	97.66	101.44	98.19
107.15	100.38	95.89	104.43	102.28

Step 2: We construct the 20 confidence intervals for each of the 20 sample means and present the results in Table 3.

Table 3				
Sample	**Sample Mean**	**Margin of Error** $1.96 \cdot \dfrac{\sigma}{\sqrt{n}} = 1.96 \cdot \dfrac{16}{\sqrt{15}} = 8.10$	**Lower Bound** $\bar{x} - 1.96 \cdot \dfrac{\sigma}{\sqrt{n}}$	**Upper Bound** $\bar{x} + 1.96 \cdot \dfrac{\sigma}{\sqrt{n}}$
1	104.32	8.10	$104.32 - 8.10 = 96.22$	$104.32 + 8.10 = 112.42$
2	93.97	8.10	$93.97 - 8.10 = 85.87$	$93.97 + 8.10 = 102.07$
3	108.73	8.10	$108.73 - 8.10 = 100.63$	$108.73 + 8.10 = 116.83$
4	104.11	8.10	96.01	112.21
5	100.67	8.10	92.57	108.77
6	96.87	8.10	88.77	104.97
7	99.74	8.10	91.64	107.84
8	100.25	8.10	92.15	108.35
9	101.32	8.10	93.22	109.42
10	94.24	8.10	86.14	102.34
11	102.23	8.10	94.13	110.33
12	94.32	8.10	86.22	102.42
13	97.66	8.10	89.56	105.76
14	101.44	8.10	93.34	109.54
15	98.19	8.10	90.09	106.29
16	107.15	8.10	99.05	115.25
17	100.38	8.10	92.28	108.48
18	95.89	8.10	87.79	103.99
19	104.43	8.10	96.33	112.53
20	102.28	8.10	94.18	110.38

From these results, we can see that 19 of the 20 (or 95%) samples resulted in 95% confidence intervals that contain the value of the population mean, 100. Notice that sample 3 results in a confidence interval that does not contain

Figure 2

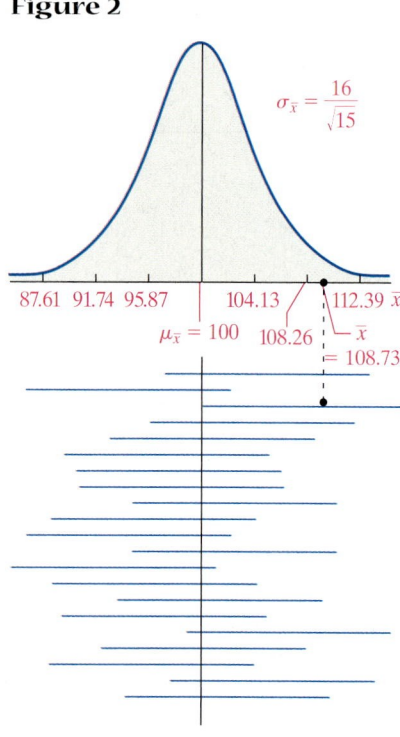

the population mean 100 (lower bound = 108.73 − 8.10 = 100.63; upper bound = 108.73 + 8.10 = 116.83).

Figure 2 presents the sampling distribution of $\bar{x}$ along with the 20 confidence intervals. Notice that the interval corresponding to $\bar{x} = 108.73$ does not intersect $\mu = 100$.

Note: In this particular simulation, exactly 95% of the samples resulted in intervals that contained μ. It is not always the case that exactly 95% of the sample means results in intervals that contain μ when performing this simulation. It could easily have been the case that all the intervals contained μ or that 17 out of the 20 contained μ. The interpretation of the confidence interval remains: in a 95% confidence interval, as the number of random samples of size n increases, the proportion of intervals that capture the population mean will approach 0.95. When we obtain only 20 samples, the exact proportion may be slightly different from 0.95.   ◄

We are not only interested in constructing 95% confidence intervals, so, we need a method for constructing any $(1 − \alpha) \cdot 100\%$ confidence interval. (Notice that if $\alpha = 0.05$ we are constructing a $(1 − 0.05) \cdot 100\% = 95\%$ confidence interval.)

We generalize Formula (1) given on page 450 by noting that $(1 − \alpha) \cdot 100\%$ of all sample means are in the interval

$$\mu − z_{\frac{\alpha}{2}} \cdot \frac{\sigma}{\sqrt{n}} < \bar{x} < \mu + z_{\frac{\alpha}{2}} \cdot \frac{\sigma}{\sqrt{n}}$$

as shown in Figure 3.

We rewrite this inequality with μ in the middle and obtain

$$\bar{x} − z_{\frac{\alpha}{2}} \cdot \frac{\sigma}{\sqrt{n}} < \mu < \bar{x} + z_{\frac{\alpha}{2}} \cdot \frac{\sigma}{\sqrt{n}}$$

So $(1 − \alpha) \cdot 100\%$ of the sample means will result in confidence intervals that contain the population mean. The sample means that are in the tails of the distribution in Figure 3 will not have confidence intervals that include the population mean.

Figure 3

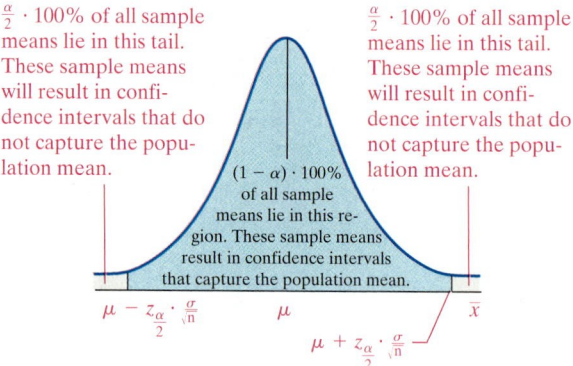

$\frac{\alpha}{2} \cdot 100\%$ of all sample means lie in this tail. These sample means will result in confidence intervals that do not capture the population mean.

$\frac{\alpha}{2} \cdot 100\%$ of all sample means lie in this tail. These sample means will result in confidence intervals that do not capture the population mean.

$(1 − \alpha) \cdot 100\%$ of all sample means lie in this region. These sample means result in confidence intervals that capture the population mean.

$\mu − z_{\frac{\alpha}{2}} \cdot \frac{\sigma}{\sqrt{n}}$ μ $\bar{x}$

$\mu + z_{\frac{\alpha}{2}} \cdot \frac{\sigma}{\sqrt{n}}$

The value $z_{\frac{\alpha}{2}}$ is called the **critical value** of the distribution. Table 4 shows some of the more common critical values used in the construction of confidence intervals.

Table 4		
Level of Confidence, $(1 − \alpha) \cdot 100\%$	Area in Each Tail, $\frac{\alpha}{2}$	Critical Value, $z_{\alpha/2}$
90%	0.05	1.645
95%	0.025	1.96
99%	0.005	2.575

A 95% confidence interval does *not* mean that there is a 95% probability that the interval contains μ. Remember, probability describes the likelihood of undetermined events. Therefore, it does not make sense to talk about the probability that the interval contains μ, since the population mean is a fixed value. Think of it this way: Suppose I flip a coin and obtain a head. If I ask you to determine the probability that the flip resulted in a head, it would not be 0.5, because the outcome has already been determined. Instead, the probability is 0 or 1. Confidence intervals work the same way. Because μ is already determined, we do not say that there is a 95% probability that the interval contains μ.

In Other Words

The interpretation of a confidence interval is this: We are (*insert level of confidence*) confident that the population mean is between (*lower bound*) and (*upper bound*). This is an abbreviated way of saying the method is correct $(1 - \alpha) \cdot 100\%$ of the time.

In a 95% confidence interval, $\alpha = 0.05$, so $z_{\alpha/2} = z_{0.05/2} = z_{0.025} = 1.96$. Any sample mean that lies within 1.96 standard deviations of the population mean will result in a confidence interval that contains μ, and any sample mean that is more than 1.96 standard deviations from the population mean will result in a confidence interval that does not contain μ. This is an extremely important point. **Whether a confidence interval contains μ depends solely on the sample mean, $\bar{x}$.** The population mean is a fixed value that either is or is not in the interval, and we do not know which of these possibilities is true for any computed confidence interval. However, if repeated samples are taken, we do know that $(1 - \alpha) \cdot 100\%$ of the confidence intervals will contain the population mean. This result leads to the following interpretation of a confidence interval.

Interpretation of a Confidence Interval

A $(1 - \alpha) \cdot 100\%$ confidence interval indicates that, if we obtained many simple random samples of size n from the population whose mean, μ, is unknown, then approximately $(1 - \alpha) \cdot 100\%$ of the intervals will contain μ.

For example, if we constructed a 90% confidence interval with a lower bound of 12 and an upper bound of 18, we would interpret the interval as follows: "We are 90% confident that the population mean, μ, is between 12 and 18."

Be sure that you understand that the level of confidence refers to the confidence in the method, not the specific interval. So a 90% confidence interval tells us that the method will result in an interval that contains the population mean, μ, 90% of the time. It does not tell us that there is a 90% probability that μ lies between 12 and 18.

We are now prepared to present a method for constructing a $(1 - \alpha) \cdot 100\%$ confidence interval about μ.

Constructing a $(1 - \alpha) \cdot 100\%$ Confidence Interval about μ, σ Known

Suppose a simple random sample of size n is taken from a population with unknown mean, μ, and known standard deviation σ. A $(1 - \alpha) \cdot 100\%$ confidence interval for μ is given by

$$\text{Lower bound: } \bar{x} - z_{\alpha/2} \cdot \frac{\sigma}{\sqrt{n}} \qquad \text{Upper bound: } \bar{x} + z_{\alpha/2} \cdot \frac{\sigma}{\sqrt{n}}$$

where $z_{\alpha/2}$ is the critical Z-value.

Note: The sample size must be large $(n \geq 30)$ or the population must be normally distributed. ◄

When constructing a confidence interval about the population mean, we must verify that the sample is big enough $(n \geq 30)$ or comes from a population that is normally distributed. Fortunately, the procedures for constructing confidence intervals presented in this section are **robust**, which means that minor departures from normality will not seriously affect the results. Nevertheless, it is important that the requirements for constructing confidence intervals are verified. We shall follow the practice of verifying the normality assumption for small sample sizes by drawing a normal probability plot and checking for outliers by drawing a boxplot.

Because the construction of the confidence interval with σ known uses Z-scores, it is sometimes referred to as constructing a **Z-interval**.

EXAMPLE 3   ### Constructing a Confidence Interval

Problem: Remember, we want to estimate the speed of cars on the highway outside a subdivision. In Example 1, we obtained a point estimate of 59.62 miles per hour for the population mean. We now wish to construct a 90% confidence interval about the population mean, μ. Assume the population standard deviation speed of vehicles on the highway is known to be 8 miles per hour, based on past studies.

Approach

Step 1: We are treating the cars as a simple random sample of size $n = 12$. We have to verify that the data come from a population that is normally distributed with no outliers. We will verify normality by constructing a normal probability plot. We check for outliers by drawing a boxplot.

Step 2: Since we are constructing a 90% confidence interval, $\alpha = 0.10$. Therefore, we have to identify the value of $z_{\alpha/2} = z_{0.1/2} = z_{0.05}$. This is the Z-score such that the area under the standard normal curve to the right of $z_{0.05}$ is 0.05.

Step 3: We compute the lower and upper bounds on the interval with $\overline{x} = 59.62$, $\sigma = 8$, and $n = 12$.

Step 4: We interpret the result by stating, "We are 90% confident that the mean speed of all cars traveling on the highway outside the subdivision is somewhere between *lower bound* and *upper bound*."

Solution

Step 1: A normal probability plot of the data in Table 1 is provided in Figure 4(a). A boxplot is presented in Figure 4(b).

Figure 4

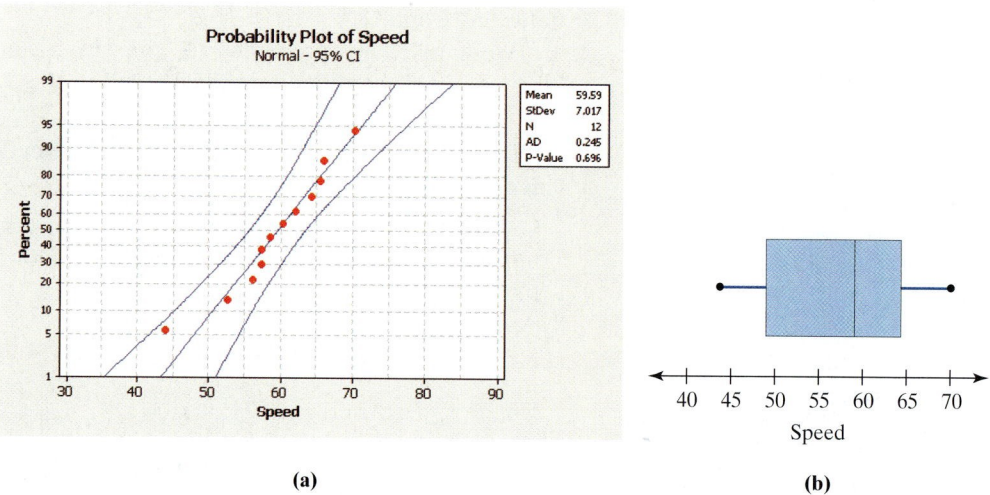

(a)   (b)

All the data values lie within the bounds on the normal probability plot, indicating that the data could come from a population that is normal. The boxplot does not show any outliers. The requirements for constructing a confidence interval are satisfied.

Step 2: We need to determine the value of $z_{0.05}$. We look in Table IV for an area equal to 0.95 (remember, the table gives areas left of the Z-scores). We have $z_{0.05} \doteq 1.645$.

Step 3: Substitute into the formulas for the lower and upper bound of the confidence interval.

$$\text{Lower bound:} \quad \overline{x} - z_{\alpha/2} \cdot \frac{\sigma}{\sqrt{n}} = 59.62 - 1.645 \cdot \frac{8}{\sqrt{12}} = 59.62 - 3.80 = 55.82$$

$$\text{Upper bound:} \quad \overline{x} + z_{\alpha/2} \cdot \frac{\sigma}{\sqrt{n}} = 59.62 + 1.645 \cdot \frac{8}{\sqrt{12}} = 59.62 + 3.80 = 63.42$$

Step 4: We are 90% confident that the mean speed of all cars traveling on the highway outside the subdivision is between 55.82 and 63.42 miles per hour.

Since the speed limit on the highway is 45 miles per hour, the residents now have some powerful evidence to bring to the police department. ▮

EXAMPLE 4 Constructing a Confidence Interval Using Technology

Problem: Construct a 90% confidence interval for the mean speed of all cars on the highway outside the subdivision using the sample data from Table 1.

Approach: We will use MINITAB to construct the confidence interval. The steps for constructing confidence intervals using the TI-83/84 graphing calculators, MINITAB, and Excel are given in the Technology Step by Step on page 465.

Result: Figure 5 shows the confidence interval obtained from MINITAB. The interval is reported in the form (*lower bound, upper bound*).

Figure 5 **One-Speed Z: Speed**

The assumed standard deviation = 8

Variable	N	Mean	StDev	SE Mean	90% CI
Speed	12	59.616	6.9563	2.3094	(55.8180, 63.4153)

USING TECHNOLOGY

Confidence intervals constructed by hand may differ from those using technology because of rounding.

Interpretation: We are 90% confident that the mean speed of all cars traveling on the highway outside the subdivision is between 55.82 and 63.42 miles per hour. ▮

Now Work Problems 25(b)–(e).

③ Understand the Role of the Margin of Error in Constructing a Confidence Interval

The width of the interval is determined by the *margin of error*.

Definition

The **margin of error**, E, in a $(1 - \alpha) \cdot 100\%$ confidence interval in which σ is known is given by

$$E = z_{\alpha/2} \cdot \frac{\sigma}{\sqrt{n}}$$

where n is the sample size.

In Other Words

The margin of error can be thought of as the "give or take" portion of the statement "The mean age of the class is 24, give or take 2 years."

Note: We require that the population from which the sample was drawn be normally distributed or the sample size n be greater than or equal to 30.

As we look at the formula for obtaining the margin of error, we see that its value depends on three quantities:

1. Level of confidence, $1 - \alpha$.
2. Standard deviation of the population, σ.
3. Sample size, n

We cannot control the standard deviation of the population, but we certainly can control the level of confidence and/or the sample size. Let's see how changing these values affects the margin of error.

EXAMPLE 5 Role of the Level of Confidence in the Margin of Error

Problem: For the problem of estimating the population mean speed of all cars on the highway outside the subdivision presented in Example 3, determine the effect on margin of error of increasing the level of confidence on the confidence interval to 99%.

Approach: With a 99% level of confidence, we have $\alpha = 0.01$. So, to compute the margin of error, E, we determine the value of $z_{\alpha/2} = z_{0.01/2} = z_{0.005}$. We then substitute this value into the formula for the margin of error with $\sigma = 8$ and $n = 12$.

Solution: After consulting Table IV, we determine $z_{0.005} = 2.575$. Substituting into the formula for margin of error, we obtain

$$E = z_{\alpha/2} \cdot \frac{\sigma}{\sqrt{n}} = 2.575 \cdot \frac{8}{\sqrt{12}} = 5.95$$

Notice the margin of error has increased from 3.80 to 5.95 when the level of confidence increases from 90% to 99%. If we want to be more confident that the sample results in an interval contains the population mean, we need to increase the width of the interval.

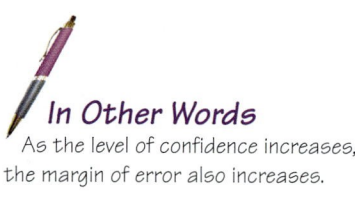

In Other Words
As the level of confidence increases, the margin of error also increases.

EXAMPLE 6 | **Role of Sample Size in the Margin of Error**

Problem: For the problem of estimating the population mean speed of all cars on the highway outside the subdivision presented in Example 3, determine the effect of increasing the sample size to $n = 48$ on the margin of error. Leave the level of confidence at 90%.

Approach: We compute the margin of error with $n = 48$ instead of $n = 12$.

Solution: Substituting $z_{0.05} = 1.645$, $\sigma = 8$, and $n = 48$ into the formula for computing margin of error, we obtain

$$E = z_{\alpha/2} \cdot \frac{\sigma}{\sqrt{n}} = 1.645 \cdot \frac{8}{\sqrt{48}} = 1.90$$

In Other Words
As the sample size increases, the margin of error decreases.

By increasing the sample size from $n = 12$ to $n = 48$, the margin of error, E, decreases from 3.80 to 1.90. Notice by quadrupling the sample size, the margin of error is cut in half.

This result should not be very surprising. Remember the Law of Large Numbers states that as the sample size n increases the sample mean approaches the value of the population mean. The smaller margin of error supports this idea.

Now Work Problem 17.

---

4 **Determine the Sample Size Necessary for Estimating the Population Mean within a Specified Margin of Error**

Suppose we want to know the number of cars that we should sample to estimate the speed of all cars traveling outside the subdivision within 2 miles per hour with 95% confidence. If we solve the formula for the margin of error, E, for n, we obtain a formula for determining sample size:

$$E = z_{\alpha/2} \cdot \frac{\sigma}{\sqrt{n}}$$

$$E\sqrt{n} = z_{\alpha/2} \cdot \sigma \qquad \text{Multiply both sides by}$$

$$\sqrt{n} = \frac{z_{\alpha/2} \cdot \sigma}{E} \qquad \text{Divide both sides by the margin of error, } E.$$

$$n = \left(\frac{z_{\alpha/2} \cdot \sigma}{E}\right)^2 \qquad \text{Square both sides.}$$

! **CAUTION**

Rounding *up* is different from rounding *off*. We round 5.32 *up* to 6 and *off* to 5.

Determining the Sample Size *n*

The sample size required to estimate the population mean, μ, with a level of confidence $(1 - \alpha) \cdot 100\%$ with a specified margin of error, E, is given by

$$n = \left(\frac{z_{\alpha/2} \cdot \sigma}{E} \right)^2$$

where *n* is rounded up to the nearest whole number.

When σ is unknown, it is common practice to conduct a preliminary survey to determine *s* and use it as an estimate of σ or use results from previous studies to obtain an estimate of σ. When using this approach, the size of the sample should be at least 30.

EXAMPLE 7   ### Determining Sample Size

Problem: We once again consider the problem of estimating the population mean speed of all cars traveling on the highway outside the subdivision. How large a sample is required to estimate the mean speed within 2 miles per hour with 90% confidence?

Approach: The sample size required can be obtained using the formula

$$n = \left(\frac{z_{\alpha/2} \cdot \sigma}{E} \right)^2$$

with $z_{\alpha/2} = z_{0.05} = 1.645$, $\sigma = 8$, and $E = 2$.

Solution: We substitute the values of z, σ, and E into the formula for determining sample size and obtain

$$n = \left(\frac{z_{\alpha/2} \cdot \sigma}{E} \right)^2 = \left(\frac{1.645 \cdot 8}{2} \right)^2 = 43.2964$$

! **CAUTION**

Don't forget to round up when determining sample size.

We round 43.2964 up to 44. A sample size of $n = 44$ results in an interval estimate of the population mean with a margin of error equal to 2 miles per hour with 90% confidence. If we obtain 100 samples of size $n = 44$, we expect about 90 intervals to capture the population mean, while 10 would not. ▇

Now Work Problem 37.

Some Final Thoughts

Many requirements must be satisfied while performing any type of statistical inference. It is worthwhile to list the requirements for constructing a confidence interval about μ with σ known in a single location for quick reference.

Requirements for Constructing a Confidence Interval about μ if σ Is Known

1. *The data obtained come from a simple random sample.* In Chapter 1, we introduced other sampling techniques, such as stratified, cluster, and systematic samples. The techniques introduced in this section apply only to samples obtained through simple random sampling. Although methods do exist for constructing confidence intervals when using the other sampling methods, they are beyond the scope of this text. If the

data are obtained from a suspect sampling method, such as voluntary response or convenience sampling, no methods exist for constructing confidence intervals. If data are collected in a flawed manner, any statistical inference performed on the data is useless!

2. *The data are obtained from a population that is normally distributed, or the sample size, n, is greater than or equal to 30.* When the sample size is small, we can use normal probability plots to help us judge whether the requirement of normality is satisfied. Remember, the techniques introduced in this section are robust. This means that minor departures from requirements will not have a severe effect on the results. However, we have to be aware that the sample mean is not resistant. Any outlier(s) in the data will affect the value of the sample mean and therefore affect the confidence interval. If the data contain outliers, we should proceed with caution when using the methods introduced in this section.

3. *The population standard deviation, σ, is assumed to be known.* However, it is unlikely that the population standard deviation is known when the population mean is not. We will drop this assumption in the next section.

9.1 ASSESS YOUR UNDERSTANDING

Concepts and Vocabulary

1. The construction of a confidence interval depends on three factors. What are they?
2. Why does the margin of error increase as the level of confidence increases?
3. Why does the margin of error decrease as the sample size n increases?
4. Suppose a confidence interval has a lower bound of 10 and an upper bound of 18. What is the sample mean of the data used to construct this interval? What is the margin of error?
5. A student constructs a 95% confidence interval for the mean age of students at his college. The lower bound is 21.4 years and the upper bound is 28.8 years. He interprets the interval as "There is a 95% probability that the mean age of a student is between 21.4 years and 28.8 years." What is wrong with this interpretation?
6. When determining the sample size required to estimate μ, we always round the value of n up to the next whole number. Explain why we do so.

Skill Building

In Problems 7–12, a simple random sample of size n < 30 has been obtained. From the normal probability plot and boxplot, judge whether a Z-interval should be constructed.

7.

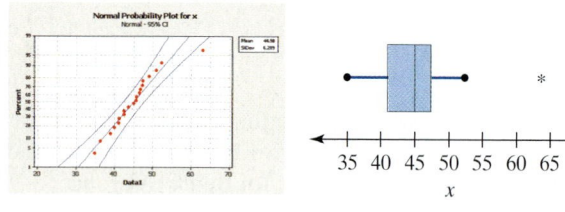

9.

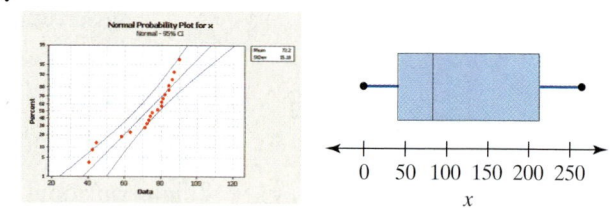

8.

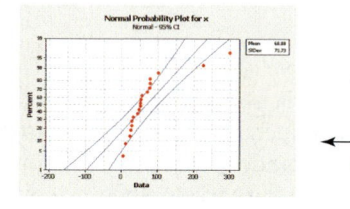

10.

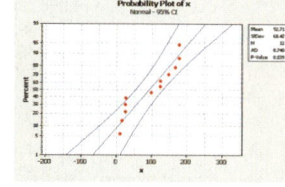

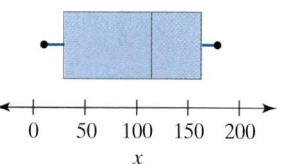

11.

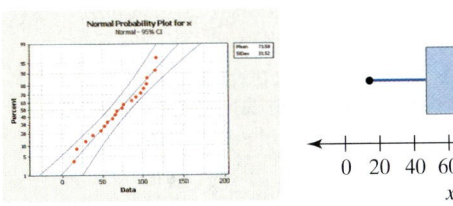

12.

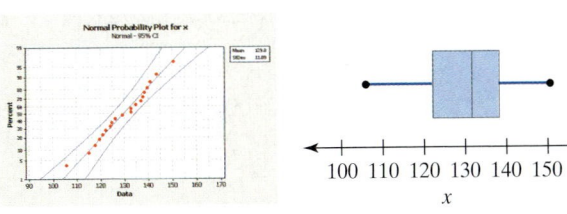

In Problems 13–16, compute the critical value $z_{\alpha/2}$ that corresponds to the given level of confidence.

13. 98%   **14.** 94%

15. 85%   **16.** 80%

17. A simple random sample of size n is drawn from a population whose population standard deviation, σ, is known to be 5.3. The sample mean, $\bar{x}$, is determined to be 34.2.

(a) Compute the 95% confidence interval about μ if the sample size, n, is 35.

(b) Compute the 95% confidence interval about μ if the sample size, n, is 50. How does increasing the sample size affect the margin of error, E?

(c) Compute the 99% confidence interval about μ if the sample size, n, is 35. Compare the results to those obtained in part (a). How does increasing the level of confidence affect the size of the margin of error, E?

(d) Can we compute a confidence interval about μ based on the information given if the sample size is $n = 15$? Why? If the sample size is $n = 15$, what must be true regarding the population from which the sample was drawn?

18. A simple random sample of size n is drawn from a population whose population standard deviation, σ, is known to be 3.8. The sample mean, $\bar{x}$, is determined to be 59.2.

(a) Compute the 90% confidence interval about μ if the sample size, n, is 45.

(b) Compute the 90% confidence interval about μ if the sample size, n, is 55. How does increasing the sample size affect the margin of error, E?

(c) Compute the 98% confidence interval about μ if the sample size, n, is 45. Compare the results to those obtained in part (a). How does increasing the level of confidence affect the size of the margin of error, E?

(d) Can we compute a confidence interval about μ based on the information given if the sample size is $n = 15$? Why? If the sample size is $n = 15$, what must be true regarding the population from which the sample was drawn?

19. A simple random sample of size n is drawn from a population that is normally distributed with population standard deviation, σ, known to be 13. The sample mean, $\bar{x}$, is found to be 108.

(a) Compute the 96% confidence interval about μ if the sample size, n, is 25.

(b) Compute the 96% confidence interval about μ if the sample size, n, is 10. How does decreasing the sample size affect the margin of error, E?

(c) Compute the 88% confidence interval about μ if the sample size, n, is 25. Compare the results to those obtained in part (a). How does decreasing the level of confidence affect the size of the margin of error, E?

(d) Could we have computed the confidence intervals in parts (a)–(c) if the population had not been normally distributed? Why?

(e) Suppose an analysis of the sample data revealed three outliers greater than the mean. How would this affect the confidence interval?

20. A simple random sample of size n is drawn from a population that is normally distributed with population standard deviation, σ, known to be 17. The sample mean, $\bar{x}$, is found to be 123.

(a) Compute the 94% confidence interval about μ if the sample size, n, is 20.

(b) Compute the 94% confidence interval about μ if the sample size, n, is 12. How does decreasing the sample size affect the margin of error, E?

(c) Compute the 85% confidence interval about μ if the sample size, n, is 20. Compare the results to those obtained in part (a). How does decreasing the level of confidence affect the size of the margin of error, E?

(d) Could we have computed the confidence intervals in parts (a)–(c) if the population had not been normally distributed? Why?

(e) Suppose an analysis of the sample data revealed one outlier greater than the mean. How would this affect the confidence interval?

Applying the Concepts

21. Time for Bed How much do Americans sleep each night? Based on a random sample of 1120 Americans 15 years of age or older, the mean amount of sleep per night is 8.17 hours according to the American Time Use Survey conducted by the Bureau of Labor Statistics. Assuming the population standard deviation for amount of sleep per night is 1.2 hours, construct and interpret a 95% confidence interval for the mean amount of sleep per night of Americans 15 years of age or older.

22. Hungry or Thirsty? How much time do Americans spend eating or drinking? Based on a random sample of 1120 Americans 15 years of age or older, the mean amount of time spent eating or drinking each day is 1.21 hours according to the American Time Use Survey conducted by the Bureau of Labor Statistics. Assuming the population standard deviation for amount of time spent eating or drinking is 0.65 hour, construct and interpret a 90% confidence interval for the mean amount of time spent eating or drinking each day by Americans 15 years of age or older.

23. **Time with Friends** A school social worker wishes to estimate the mean amount of time each week that high school students spend with friends. She obtains a random sample of 1175 high school students and finds that the mean weekly time spent with friends is 9.2 hours. Assuming the population standard deviation amount of time spent with friends is 6.7 hours, construct and interpret a 90% confidence interval for the mean time spent with friends each week.

24. **Travel Time to Work** An urban economist wishes to estimate the mean amount of time people spend traveling to work. He obtains a random sample of 50 individuals who are in the labor force and finds that the mean travel time is 24.2 minutes. Assuming the population standard deviation of travel time is 18.5 minutes, construct and interpret a 95% confidence interval for the mean travel time to work. **Note:** The standard deviation is large because many people work at home (travel time = 0 minutes) and many have commutes in excess of 1 hour. *Source:*Based on data obtained from the American Community Survey.

25. **Crash Test Results** The Insurance Institute for Highway
    NW Safety routinely conducts crash tests on vehicles to determine the cost of repairs. In four crashes of a Chevy Cavalier at 5 miles per hour, the institute found the cost of repairs to be $225, $462, $729, and $753. Treat these data as a simple random sample of four crashes, and answer the following questions, assuming that $\sigma = \$220$.

    (a) Use the data to compute a point estimate for the population mean cost of repairs on a Chevy Cavalier.
    (b) Because the sample size is small, we must verify that cost is approximately normally distributed and that the sample does not contain any outliers. The normal probability plot and boxplot are shown next. Are the conditions for constructing a Z-interval satisfied?

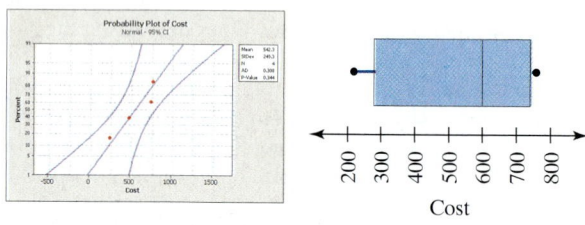

    (c) Construct a 95% confidence interval for the cost of repairs. Interpret this interval.
    (d) Construct a 90% confidence interval for the cost of repairs. Interpret this interval.
    (e) What happened to the width of the interval when the level of confidence decreased? Explain why this result is reasonable.

26. **Flight Time** A researcher for the FAA wants to estimate the average flight time (in minutes) from Albuquerque, New Mexico, to Dallas, Texas, for flights with American Airlines. He randomly selects nine flights between the two cities and obtains the data shown. Assume that $\sigma = 8$ minutes.

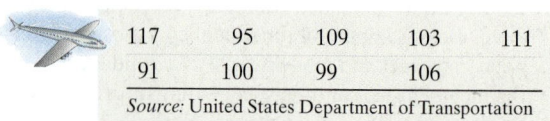

| 117 | 95 | 109 | 103 | 111 |
| 91 | 100 | 99 | 106 | |

Source: United States Department of Transportation

(a) Use the data to compute a point estimate for the population mean flight time between Albuquerque and Dallas on an American Airlines flight.
(b) Because the sample size is small, we must verify that flight time is approximately normally distributed and that the sample does not contain any outliers. The normal probability plot and boxplot are shown next. Are the conditions for constructing a Z-interval satisfied?

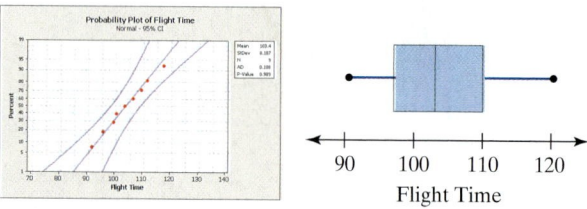

(c) Construct a 95% confidence interval for the flight time. Interpret this interval.
(d) Construct a 90% confidence interval for the flight time. Interpret this interval.
(e) What happened to the width of the interval when the level of confidence decreased? Explain why this result is reasonable.

27. **Investment Home Sales** A study by the National Association of Realtors reports that second-home sales accounted for more than a third of residential real estate transactions in 2004. According to the report, the average age of a person buying a second home as an investment rental property was 47 years. A real estate agent wants to estimate the average age of those buying investment property in his area. He randomly selects 15 of his clients who purchased an investment property and obtains the data shown. Assume that $\sigma = 7.9$.

FOR RENT	27	49	56	45	49
	35	55	40	46	48
	38	58	51	42	38

(a) Use the data to compute a point estimate for the population mean age of the real estate agent's clients who purchased investment property.
(b) Because the sample size is small, we must verify that client age is approximately normally distributed and that the sample does not contain any outliers. The normal probability plot and boxplot are shown next. Are the conditions for constructing a Z-interval satisfied?

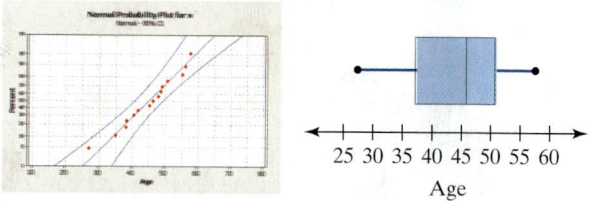

(c) Construct a 95% confidence interval for the mean age for all the real estate agent's clients who purchased investment property. Interpret this interval.
(d) Do the real estate agent's clients appear to differ in age from the general population? Why?

28. Vacation Home Sales A study by the National Association of Realtors reports that second-home sales accounted for more than a third of residential real estate transactions in 2004. According to the report, the average age of a person buying a second home as a vacation home was 55 years. A real estate agent wants to estimate the average age of those buying vacation homes in her area. She randomly selects 15 of her clients who purchased a vacation home and obtains the data shown. Assume that $\sigma = 4.7$.

47	51	51	47	47
35	55	54	37	40
52	49	51	41	48

(a) Use the data to compute a point estimate for the population mean age of the real estate agent's clients who purchased a vacation home.
(b) Because the sample size is small, we must verify that client age is approximately normally distributed and that the sample does not contain any outliers. The normal probability plot and boxplot are shown next. Are the conditions for constructing a Z-interval satisfied?

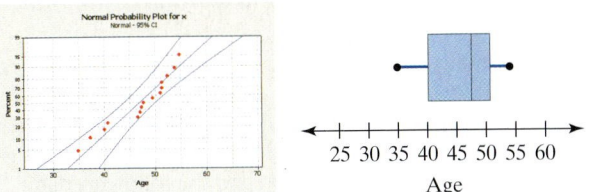

25 30 35 40 45 50 55 60
Age

(c) Construct a 95% confidence interval for the mean age for all the real estate agent's clients who purchased a vacation home. Interpret this interval.
(d) Do the real estate agent's clients appear to differ in age from the general population? Why?

29. Serum Cholesterol As reported by the U.S. National Center for Health Statistics, the mean serum high density lipoprotein (HDL) cholesterol of males 20 to 29 years old is $\mu = 47$. Dr. Paul Oswiecmiski wants to estimate the mean serum HDL cholesterol of his 20- to 29-year-old male patients. He randomly selects 15 of his 20- to 29-year-old patients and obtains the data shown. Assume that $\sigma = 12.5$.

48	61	46	35	48
34	64	42	29	71
64	47	56	27	53

Source: Dr. Paul Oswiecmiski

(a) Use the data to compute a point estimate for the population mean serum HDL cholesterol in Dr. Oswiecmiski's patients.
(b) A normal probability plot and boxplot indicate that the sample data come from a population that is approximately normally distributed with no outliers. Construct a 95% confidence interval for the mean serum HDL cholesterol for all Dr. Oswiecmiski's 20- to 29-year-old male patients. Interpret this interval.
(c) Do Dr. Oswiecmiski's patients appear to have a serum HDL different from that of the general population.
(d) If Dr. Oswiecmiski wanted a more precise confidence interval, what would you recommend?

30. Billing Process For a certain billing process, the number of days for customers to pay their bill from the date of invoice is approximately normally distributed, with mean $\mu = 47$ days and $\sigma = 11$ days. A random sample of 10 bills from the billing process during the month of June results in the following data:

55	45	45	42	65
58	35	36	34	60

(a) Use the data to compute a point estimate for the population mean number of days from invoice to pay the bill.
(b) Construct a 95% confidence interval for the mean number of days it takes customers to pay their bill. Interpret this interval.
(c) Is there evidence that June payments were typical?

31. Volume of Harley-Davidson Stock The volume of a stock is the number of shares traded on a given day. The following data, in millions (so that 3.78 represents 3,780,000 shares traded), represent the volume of Harley-Davidson stock traded for a random sample 40 trading days in 2004.

1.74	0.99	0.96	1.41	2.43	0.77	1.60	1.81	3.42	2.01	1.98
1.15	0.87	1.62	0.97	6.37	1.15	1.43	0.69	3.04	0.97	1.32
2.27	0.51	0.76	3.01	2.44	1.72	1.39	1.72	2.37	1.29	1.70
0.95	1.18	1.05	1.38	1.23	0.68	1.26				

Source: yahoo.finance.com.

The standard deviation of the number of shares traded in 2004 was $\sigma = 1.00$ million shares.

(a) Use the data to compute a point estimate for the population mean number of shares traded per day in 2004.
(b) Construct a 90% confidence interval for the population mean number of shares traded per day in 2004. Interpret the confidence interval.

(c) A second random sample of 40 days in 2004 resulted in the following data:

6.23	1.62	3.04	1.13	1.06	1.70	1.37	2.13	1.29	2.02	0.68
1.06	3.79	1.24	1.02	0.85	3.42	1.47	1.60	0.96	1.30	0.96
1.72	1.12	1.33	1.02	0.84	0.80	1.16	1.85	1.79	1.35	3.30
0.82	0.73	1.54	3.98	2.54	2.27	1.60				

Construct a 90% confidence interval for the population mean number of shares traded per day in 2004. Interpret the confidence interval.

(d) Explain why the confidence intervals obtained in parts (b) and (c) are different.

32. Volume of Google Stock The volume of a stock is the number of shares traded on a given day. The following data, in millions (so that 4.64 represents 4,640,000 shares traded), represent the volume of Google stock traded for a random sample of 35 trading days in 2004.

Google

4.64	7.57	7.49	5.54	21.16	13.36	3.11	3.55	15.27
2.60	22.31	11.19	11.90	4.57	19.83	13.89	6.60	14.85
2.03	2.50	3.91	10.64	6.26	7.57	3.93	9.14	8.77
5.52	7.03	14.41	16.75	8.47	5.84	8.57	5.90	

Source: yahoo.finance.com

The standard deviation of the number of shares traded in 2004 was $\sigma = 6.17$ million shares.

(a) Use the data to compute a point estimate for the population mean number of shares traded per day in 2004.

(b) Construct a 90% confidence interval for the population mean of the number of shares traded per day in 2004. Interpret the confidence interval.

(c) A second random sample of 35 days in 2004 resulted in the following data:

Google

12.41	2.93	13.89	3.93	11.90	14.85	2.60	6.89	5.84
3.61	2.68	5.24	3.79	7.39	10.67	7.06	6.26	4.60
22.31	7.57	3.62	11.06	4.35	7.54	12.37	9.14	7.57
7.67	8.77	15.28	7.63	3.55	4.27	32.76	7.49	

Construct a 90% confidence interval for the population mean for number of shares traded per day in 2004. Interpret the confidence interval.

(d) Explain why the confidence intervals obtained in parts (b) and (c) are different.

33. Dramas After watching a drama that seemed to last a long time, a student wondered how long the typical drama lasted. She obtained a random sample of 30 dramas and found the mean length of the movies to be 138.3 minutes. Assume the population standard deviation length of a drama is 27.3 minutes.

(a) An analysis of the data indicated that the distribution of lengths of dramas is skewed right. Why does the student have to have a large sample size?

(b) Construct and interpret a 99% confidence interval for the mean length of a drama.

34. What Is Wrong? The mean age of the 42 presidents of the United States on the day of inauguration is 54.8 years with a standard deviation of 6.2 years. A researcher constructed a 95% confidence interval for the mean age of presidents on inauguration day. He wrote that he was 95% confident the mean age of the president on inauguration day is between 53.0 and 56.7 years of age. What is wrong with the researcher's analysis?

35. Miles on a Saturn A researcher is interested in approximating the mean number of miles on 4-year-old Saturn SC1s. She finds a random sample of 33 such Saturn SC1s in the Chicago area and obtains the following results:

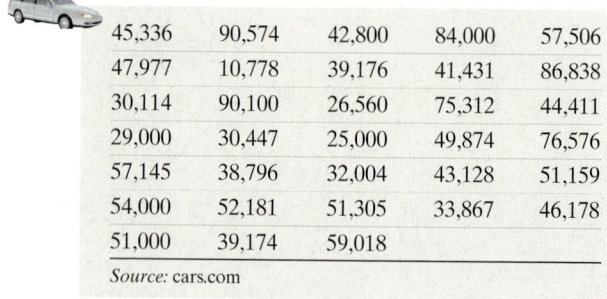

45,336	90,574	42,800	84,000	57,506
47,977	10,778	39,176	41,431	86,838
30,114	90,100	26,560	75,312	44,411
29,000	30,447	25,000	49,874	76,576
57,145	38,796	32,004	43,128	51,159
54,000	52,181	51,305	33,867	46,178
51,000	39,174	59,018		

Source: cars.com

(a) Obtain a point estimate of the population mean number of miles on a 4-year-old Saturn SC1.

(b) Construct and interpret a 99% confidence interval for the population mean for number of miles on a 4-year-old Saturn SC1. Assume that $\sigma = 19{,}700$.

(c) Construct and interpret a 95% confidence interval for the population mean for number of miles on a 4-year-old Saturn SC1. Assume that $\sigma = 19{,}700$.

(d) What effect does decreasing the level of confidence have on the interval?

(e) Do the confidence intervals computed in parts (b) and (c) represent an estimate for the population mean number of miles on Saturn SC1s in the United States? Why?

36. Miles on a Cavalier A researcher is interested in approximating the mean number of miles on 3-year-old Chevy Cavaliers. She finds a random sample of 35 Cavaliers in the Orlando, Florida, area and obtains the following results:

37,815	20,000	57,103	46,585	24,822
49,678	30,983	52,969	8,000	39,862
6,000	65,192	34,285	30,906	41,841
39,851	43,000	74,361	52,664	33,587
52,896	45,280	30,000	41,713	76,315
22,442	45,301	52,899	41,526	28,381
55,163	51,812	36,500	31,947	16,529

Source: cars.com

(a) Obtain a point estimate of the population mean number of miles on a 3-year-old Cavalier.

(b) Construct and interpret a 99% confidence interval for the population mean for number of miles on a 3-year-old Cavalier. Assume that $\sigma = 16{,}100$.

(c) Construct and interpret a 95% confidence interval for the population mean for number of miles on a 3-year-old Cavalier. Assume that $\sigma = 16{,}100$.

(d) What effect does decreasing the level of confidence have on the width of the interval?

(e) Do the confidence intervals computed in parts (b) and (c) represent an interval estimate for the population mean number of miles on Cavaliers in the United States? Why?

37. Sample Size Dr. Paul Oswiecmiski wants to estimate the mean serum HDL cholesterol of all 20- to 29-year-old females. How many subjects are needed to estimate the mean serum HDL cholesterol of all 20- to 29-year-old females within 2 points with 99% confidence, assuming that $\sigma = 13.4$? Suppose Dr. Oswiecmiski would be content with 95% confidence. How does the decrease in confidence affect the sample size required?

38. Sample Size Dr. Paul Oswiecmiski wants to estimate the mean serum HDL cholesterol of all 20- to 29-year-old males. How many subjects are needed to estimate the mean serum HDL cholesterol of all 20- to 29-year-old males within 1.5 points with 90% confidence, assuming that $\sigma = 12.5$? Suppose Dr. Oswiecmiski would prefer 98% confidence. How does the increase in confidence affect the sample size required?

39. How Much Do You Read? A Gallup poll conducted May 20–22, 2005 asked Americans how many books, either hardback or paperback, they read during the previous year. How many subjects are needed to estimate the number of books Americans read the previous year within one book with 95% confidence? Initial survey results indicate that $\sigma = 16.6$ books.

40. How Long Are You Online? A Gallup poll conducted January 23, 2003–February 10, 2003, asked American teens (aged 13 to 17) how much time they spent each week using the Internet. How many subjects are needed to estimate the time American teens spend on the Internet each week within 0.5 hour with 95% confidence? Initial survey results indicate that $\sigma = 6.6$ hours.

41. Miles on a Saturn A researcher wishes to estimate the mean number of miles on 4-year-old Saturn SC1s.

(a) How many cars should be in a sample to estimate the mean number of miles within 1000 miles with 90% confidence, assuming that $\sigma = 19{,}700$?

(b) How many cars should be in a sample to estimate the mean number of miles within 500 miles with 90% confidence, assuming that $\sigma = 19{,}700$?

(c) What effect does doubling the required accuracy have on the sample size? Why is this the expected result?

42. Miles on a Cavalier A researcher wishes to estimate the mean number of miles on 3-year-old Chevy Cavaliers.

(a) How many cars should be in a sample to estimate the mean number of miles within 2000 miles with 98% confidence, assuming that $\sigma = 16{,}100$?

(b) How many cars should be in a sample to estimate the mean number of miles within 1000 miles with 98% confidence, assuming that $\sigma = 16{,}100$?

(c) What effect does doubling the required accuracy have on the sample size? Why is this the expected result?

43. Simulation IQ scores as measured by the Stanford–Binet IQ test are normally distributed with $\mu = 100$ and $\sigma = 16$.

(a) Simulate obtaining 20 samples of size $n = 15$ from this population.

(b) Construct 95% confidence intervals for each of the 20 samples.

(c) How many of the intervals do you expect to include the population mean? How many actually contain the population mean?

44. Simulation Suppose the arrival of cars at Burger King's drive-through follows a Poisson process with $\mu = 4$ cars every 10 minutes.

(a) Simulate obtaining 30 samples of size $n = 40$ from this population.

(b) Construct 90% confidence intervals for each of the 30 samples. [*Hint:* $\sigma = \sqrt{\mu}$ in a Poisson process.]

(c) How many of the intervals do you expect to include the population mean? How many actually contain the population mean?

45. Effect of Nonnormal Data The *exponential probability distribution* is a probability distribution that can be used to model waiting time in line or the lifetime of electronic components. Its density function with $\mu = 5$ is shown in the accompanying figure. We can see that the distribution is skewed right.

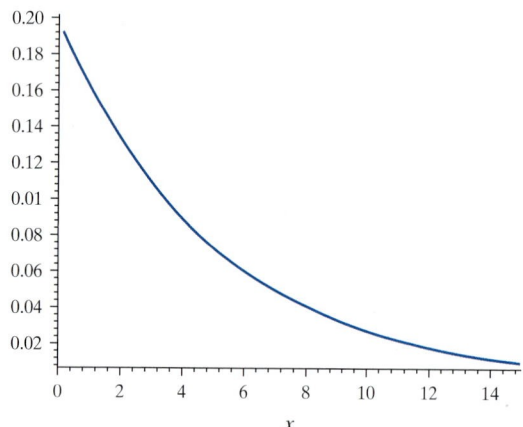

(a) Use MINITAB or some other statistical software to generate 100 random samples of size $n = 6$ from a population that follows the exponential probability distribution with $\mu = 5$. (It turns out that σ also equals 5.)

(b) Use the 100 samples to determine 100 95% confidence intervals with $\sigma = 5$.

(c) How many of the intervals do we expect to contain $\mu = 5$? How many of the 100 intervals contain $\mu = 5$?

(d) What are the consequences of not having data from a normal population when the sample size is small?

46. **Effect of Outliers** Suppose the following small data set represents a simple random sample from a population whose mean is 50 and standard deviation is 10.

43	63	53	50	58	44
53	53	52	41	50	43

(a) A normal probability plot indicates the data come from a population that is normally distributed with no outliers. Compute a 95% confidence interval for this data set, assuming $\sigma = 10$.

(b) Suppose the observation, 41, is inadvertently entered into the computer as 14. Verify that this observation is an outlier.

(c) Construct a 95% confidence interval on the data set with the outlier. What effect does the outlier have on the confidence interval?

(d) Consider the following data set, which represents a simple random sample of size 36 from a population whose mean is 50 and standard deviation is 10.

43	63	53	50	58	44
53	53	52	41	50	43
47	65	56	58	41	52
49	56	57	50	38	42
59	54	57	41	63	37
46	54	42	48	53	41

Verify that the sample mean for the large data set is the same as the sample mean for the small data set.

(e) Compute a 95% confidence interval for the large data set, assuming $\sigma = 10$. Compare the results to part (a). What effect does increasing the sample size have on the confidence interval?

(f) Suppose the last observation, 41, were inadvertently entered as 14. Verify that this observation is an outlier.

(g) Compute a 95% confidence interval for the large data set with the outlier, assuming $\sigma = 10$. Compare the results to part (e). What effect does an outlier have on a confidence interval when the data set is large?

47. By how many times does the sample size have to be increased to decrease the margin of error by a factor of $\dfrac{1}{2}$?

48. Suppose a certain population, A, has standard deviation $\sigma_A = 5$, and a second population, B, has standard deviation $\sigma_B = 10$. How many times larger than population A's sample size does population B's need to be to estimate margin of error μ with the same? [**Hint:** Compute n_B/n_A.]

49. **Resistance and Robustness** The following data sets represent simple random samples from a population whose mean is 100 and standard deviation is 15.

Data Set I				
106	122	91	127	88
74	77	108		

Data Set II				
106	122	91	127	88
74	77	108	87	88
111	86	113	115	97
122	99	86	83	102

Data Set III				
106	122	91	127	88
74	77	108	87	88
111	86	113	115	97
122	99	86	83	102
88	111	118	91	102
80	86	106	91	116

(a) Compute the sample mean of each data set.

(b) For each data set, construct a 95% confidence interval about the population mean with $\sigma = 15$.

(c) What impact does the sample size n have on the width of the interval?

For parts (d)–(e), suppose the data value 106 was accidentally recorded as 016.

(d) For each data set, construct a 95% confidence interval about the population mean with $\sigma = 15$ using the misentered data.

(e) Which intervals, if any, still capture the population mean, 100? What concept does this illustrate?

50. Confidence Interval Applet: The Role of Level of Confidence Load the confidence interval for a mean (the impact of a confidence level) applet.

(a) Set the shape to normal with mean = 50 and Std. Dev. = 10. Construct at least 1,000 confidence intervals with $n = 10$. For 95% confidence, what proportion of the intervals contain the population mean? What proportion did you expect to contain the population mean?

(b) Repeat part (a). Did the same proportion of intervals contain the population mean?

(c) For 99% confidence, what proportion of the intervals contain the population mean? What proportion did you expect to contain the population mean?

51. Confidence Interval Applet: The Role of Sample Size Load the confidence interval for a mean (the impact of a confidence level) applet.

(a) Set the shape to normal with mean = 50 and Std. Dev. = 10. Construct at least 1000 confidence intervals with $n = 10$. What proportion of the 95% confidence intervals contain the population mean? What proportion did you expect to contain the population mean?

(b) Repeat part (a). Did the same proportion of intervals contain the population mean?

(c) Set the shape to normal with mean = 50 and Std. Dev. = 10. Construct at least 1000 confidence intervals with $n = 50$. What proportion of the 95% confidence intervals contain the population mean? What proportion did you expect to contain the population mean? Does sample size have any impact on the proportion of intervals that capture the population mean?

(d) Compare the width of the intervals for the samples of size $n = 50$ obtained in part (c) to the width of the intervals for the samples of size $n = 10$ obtained in part (a). Which are wider? Why?

52. Confidence Interval Applet: The Role of Shape Load the confidence interval applet.

(a) Set the shape to normal with mean = 50 and Std. Dev. = 10. Construct at least 1000 95% confidence intervals with $n = 40$. How many of the intervals contain the population mean? How many did you expect to contain the population mean?

(b) Set the shape to right skewed with mean = 50 and Std. Dev. = 10. Construct at least 1000 95% confidence intervals with $n = 5$. How many of the intervals contain the population mean? How many did you expect to contain the population mean?

(c) What is the impact of nonnormal data on the confidence interval when the sample size is small?

Technology Step-by-Step	**Confidence Intervals about μ, σ Known**
TI-83/84 Plus	**Step 1:** If necessary, enter raw data in L1. **Step 2:** Press STAT, highlight TESTS, and select `7: ZInterval`. **Step 3:** If the data are raw, highlight DATA. Make sure `List1` is set to L1 and `Freq` to 1. If summary statistics are known, highlight STATS and enter the summary statistics. Following σ:, enter the population standard deviation. **Step 4:** Enter the confidence level following `C-Level:`. **Step 5:** Highlight `Calculate`; press ENTER.
MINITAB	**Step 1:** Enter raw data in column C1. **Step 2:** Select the **Stat** menu, highlight **Basic Statistics**, then highlight **1-Sample Z . . . .** **Step 3:** Enter C1 in the cell marked "Variables". Select Confidence Interval, and enter a confidence level. In the cell marked "Sigma", enter the value of σ. Click OK.
Excel	**Step 1:** If necessary, enter raw data in column A. **Step 2:** Load the PHStat Add-in. **Step 3:** Select the **PHStat menu**, highlight **Confidence Intervals . . .**, then highlight **Estimate for the mean, sigma known . . . .** **Step 4:** Enter the value of σ and the confidence level. If the summary statistics are known, click "Sample statistics known" and enter the sample size and sample mean. If summary statistics are unknown, click "Sample statistics unknown". With the cursor in the "Sample cell range" cell, highlight the data in column A. Click OK.

9.2 Confidence Intervals about a Population Mean in Practice Where the Population Standard Deviation Is Unknown

Preparing for This Section Before getting started, review the following:

- Simple random sampling (Section 1.2, pp. 16–19)
- Parameter versus statistic (Section 3.1, p. 121)
- Degrees of freedom (Section 3.2, p. 142)
- Normal probability plots (Section 7.4, pp. 394–398)
- Distribution of the sample mean (Section 8.1, pp. 417–430)

Objectives

1. **Know the properties of Student's *t*-distribution**
2. **Determine *t*-values**
3. **Construct and interpret a confidence interval about a population mean**

Now we drop the assumption that σ is known and learn how to construct confidence intervals about a population mean in practice. That is, we learn how to construct confidence intervals about a population mean under the more realistic scenario that the population standard deviation is unknown.

Know the Properties of Student's *t*-Distribution

In Section 9.1, we computed confidence intervals about a population mean, μ, under the assumption that the following conditions were met.

1. σ was known.
2. The population from which the sample was drawn followed a normal distribution or the sample size n was large ($n \geq 30$).
3. The sample was a simple random sample.

A $(1 - \alpha) \cdot 100\%$ confidence interval was then computed as

$$\bar{x} - z_{\alpha/2} \cdot \frac{\sigma}{\sqrt{n}} \quad \text{and} \quad \bar{x} + z_{\alpha/2} \cdot \frac{\sigma}{\sqrt{n}}$$

If σ is unknown, it seems reasonable to replace σ with s and proceed with the analysis. However, while it is true that $Z = \dfrac{\bar{x} - \mu}{\sigma/\sqrt{n}}$ is normally distributed with mean 0 and standard deviation 1, we cannot replace σ with s and say that $Z = \dfrac{\bar{x} - \mu}{s/\sqrt{n}}$ is normally distributed with mean 0 and standard deviation 1, because s itself is a random variable.

Instead, $\dfrac{\bar{x} - \mu}{s/\sqrt{n}}$ follows **Student's *t*-distribution**, developed by William Gosset. Gosset was in charge of conducting experiments at the Guinness brewery to identify the best barley variety. At the time, the only available distribution was the standard normal distribution, but Gosset always performed experiments with small data sets and he did not know σ. This led Gosset to determine the sampling distribution of $\dfrac{\bar{x} - \mu}{s/\sqrt{n}}$. He published his findings under the pen name *Student*.

Historical Note

William Sealey Gosset was born on June 13, 1876, in Canterbury, England. Gosset earned a degree in chemistry from New College in Oxford in 1899. He then got a job as a chemist for the Guinness Brewing Company in Dublin. Gosset, along with other chemists, was asked to find a way to make the best beer at the cheapest cost. This allowed him to concentrate on statistics. In 1904, Gosset wrote a paper on the brewing of beer that included a discussion of standard errors. In July, 1905, Gosset met with Karl Pearson to learn about the theory of standard errors. Over the next few years, he developed his *t*-distribution. The Guinness Brewery did not allow its employees to publish, so Gosset published his research using the pen name Student. Gosset died October 16, 1937.

Student's *t*-Distribution

Suppose a simple random sample of size *n* is taken from a population. If the population from which the sample is drawn follows a normal distribution, the distribution of

$$t = \frac{\overline{x} - \mu}{s/\sqrt{n}}$$

follows Student's *t*-distribution with $n - 1$ degrees of freedom,* where $\overline{x}$ is the sample mean and *s* is the sample standard deviation.

The interpretation of the *t*-statistic is the same as that of the *Z*-score. The *t*-statistic represents the number of sample standard errors $\overline{x}$ is from the population mean, μ. It turns out that the shape of the *t*-distribution depends on the sample size, *n*.

To help see how the *t*-distribution differs from the standard normal (or *z*-) distribution and the role the sample size *n* plays, we will go through the following simulation.

EXAMPLE 1

Comparing the Standard Normal Distribution to the *t*-Distribution Using Simulation

(a) Using a statistical spreadsheet such as MINITAB, obtain 1000 simple random samples of size $n = 5$ from a normal population with $\mu = 50$ and $\sigma = 10$.

(b) Calculate the sample mean and sample standard deviation for each sample.

(c) Compute $z = \dfrac{\overline{x} - \mu}{\sigma/\sqrt{n}}$ and $t = \dfrac{\overline{x} - \mu}{s/\sqrt{n}}$ for each sample.

(d) Draw histograms for both *z* and *t*.

Results   We use MINITAB to obtain the 1000 simple random samples and compute the 1000 sample means and sample standard deviations. We then compute $z = \dfrac{\overline{x} - \mu}{\sigma/\sqrt{n}} = \dfrac{\overline{x} - 50}{10/\sqrt{5}}$ and $t = \dfrac{\overline{x} - \mu}{s/\sqrt{n}} = \dfrac{\overline{x} - 50}{s/\sqrt{5}}$ for each of the 1000 samples. Figure 6(a) shows the histogram for *z*, and Figure 6(b) shows the histogram for *t*.

Figure 6

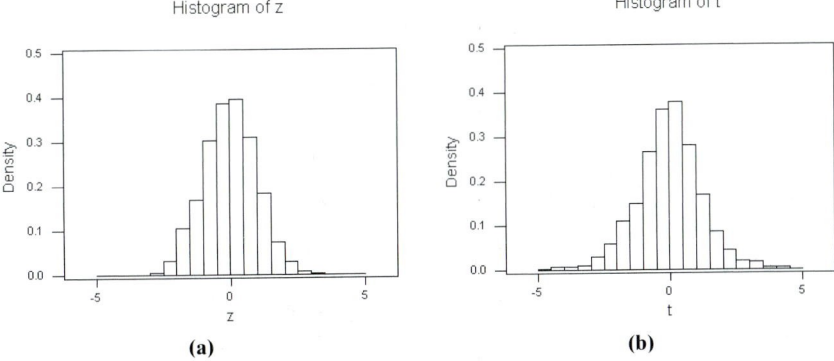

Histogram of z   Histogram of t

(a)   (b)

We notice that the histogram in Figure 6(a) is symmetric and bell shaped, with the center of the distribution at 0 and virtually all the rectangles between -3 and 3. In other words, *z* follows a standard normal distribution. The distribution of *t* is also symmetric and bell shaped and has its center at 0, but the

*The reader may wish to review the discussion of degrees of freedom in Section 3.2 on pp. 141–142.

distribution of t has longer tails (that is, t is more dispersed), so it is unlikely that t follows a standard normal distribution. The additional spread in the distribution of t can be attributed to the fact that we divide by $\dfrac{s}{\sqrt{n}}$ to find t instead of by $\dfrac{\sigma}{\sqrt{n}}$. Because the sample standard deviation is itself a random variable (rather than a constant such as σ), we have more dispersion in the distribution of t.

We now introduce the properties of the t-distribution.

Properties of the *t*-Distribution

1. The t-distribution is different for different degrees of freedom.

2. The t-distribution is centered at 0 and is symmetric about 0.

3. The area under the curve is 1. The area under the curve to the right of 0 equals the area under the curve to the left of 0 equals $\dfrac{1}{2}$.

4. As t increases without bound, the graph approaches, but never equals, zero. As t decreases without bound, the graph approaches, but never equals, zero.

5. The area in the tails of the t-distribution is a little greater than the area in the tails of the standard normal distribution, because we are using s as an estimate of σ, thereby introducing further variability into the t-statistic.

6. As the sample size n increases, the density curve of t gets closer to the standard normal density curve. This result occurs because, as the sample size n increases, the values of s get closer to the values of σ, by the Law of Large Numbers.

Figure 7

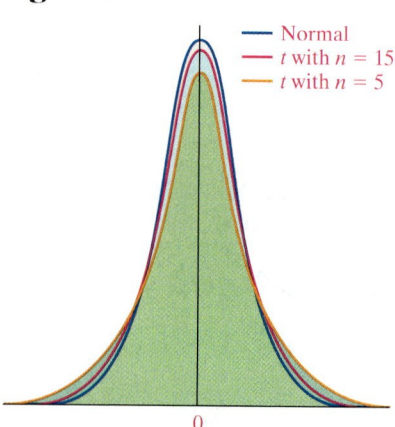

- Normal
- t with $n = 15$
- t with $n = 5$

0

In Figure 7, we show the t-distribution for the sample sizes $n = 5$ and $n = 15$. As a point of reference, we have also drawn the standard normal density curve.

Figure 8

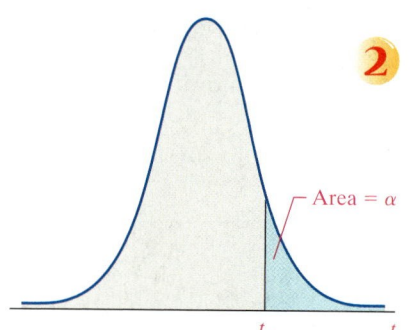

Area $= \alpha$

t_α t

2   Determine *t*-Values

Recall that the notation z_α is used to represent the z-score whose area under the normal curve to the right of z_α is α. Similarly, we let t_α represent the t-value such that the area under the t-distribution to the right of t_α is α. See Figure 8.

The shape of the t-distribution depends on the sample size, n. Therefore, the value of t_α depends not only on α, but also on the degrees of freedom, $n - 1$. In Table V in Appendix A, the far left column gives the degrees of freedom. The top row represents the area under the t-distribution to the right of some t-value.

EXAMPLE 2   Finding *t*-Values

Problem: Find the t-value such that the area under the t-distribution to the right of the t-value is 0.10, assuming 15 degrees of freedom. That is, find $t_{0.10}$ with 15 degrees of freedom.

Approach: We will perform the following steps.

Step 1: Draw a t-distribution with the unknown t-value labeled. Shade the area under the curve to the right of the t-value, as in Figure 8.

Step 2: Find the row in Table V that corresponds to 15 degrees of freedom and the column that corresponds to an area in the right tail of 0.10. Identify where the row and column intersect. This is the unknown t-value.

Figure 9

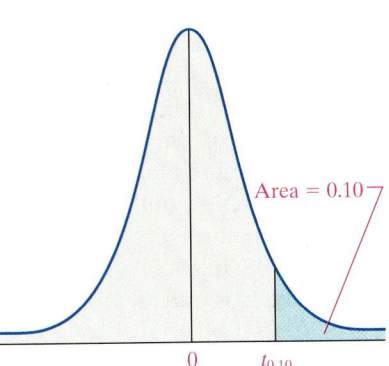

Area = 0.10

0        $t_{0.10}$

Solution

Step 1: Figure 9 shows the graph of the t-distribution with 15 degrees of freedom. The unknown value of t is labeled, and the area under the curve to the right of t is shaded.

Step 2: A portion of Table V is reproduced in Figure 10. We have enclosed the row that represents 15 degrees of freedom and the column that represents the area 0.10 in the right tail. The point where the row and column intersect is the t-value we are seeking. The value of $t_{0.10}$ with 15 degrees of freedom is 1.341; that is, the area under the t-distribution to the right of $t = 1.341$ with 15 degrees of freedom is 0.10.

Figure 10

| | | | | | Area in Right Tail | | | | | | | |
df	0.25	0.20	0.15	0.10	0.05	0.025	0.02	0.01	0.005	0.0025	0.001	0.0005
1	1.000	1.376	1.963	3.078	6.314	12.710	15.890	31.820	63.660	127.300	318.300	636.600
2	0.816	1.061	1.386	1.886	2.920	4.303	4.849	6.965	9.925	14.090	22.330	31.600
3	0.765	0.978	1.250	1.638	2.353	3.182	3.482	4.541	5.841	7.453	10.210	12.920
13	0.694	0.870	1.079	1.350	1.771	2.160	2.282	2.650	3.012	3.372	3.852	4.221
14	0.692	0.868	1.076	1.345	1.761	2.145	2.264	2.624	2.977	3.326	3.787	4.140
15	0.691	0.866	1.074	1.341	1.753	2.131	2.249	2.602	2.947	3.286	3.733	4.073
16	0.690	0.865	1.071	1.337	1.746	2.120	2.235	2.583	2.921	3.252	3.686	4.015

Now Work Problem 7(a).

USING TECHNOLOGY

The TI-84 Plus graphing calculator has an invT feature, which finds the value of t given an area left of the unknown t-value and the degrees of freedom.

If the degrees of freedom we desire are not available in Table V, we follow the practice of choosing the closest number of degrees of freedom available in the table. For example, if we have 43 degrees of freedom, we use 40 degrees of freedom from Table V. In addition, the last row of Table V provides the Z-values from the standard normal distribution. We use these values for situations where the degrees of freedom are more than 1000. This is acceptable because the t-distribution starts to behave like the standard normal distribution as n increases.

❸   Construct and Interpret a Confidence Interval about a Population Mean

The construction of confidence intervals about μ with σ unknown follows the same logic as the construction of confidence intervals with σ known. The only difference is that we use s in place of σ and t in place of z.

> **Constructing a $(1 - \alpha) \cdot 100\%$ Confidence Interval about μ, σ Unknown**
>
> Suppose a simple random sample of size n is taken from a population with unknown mean μ and unknown standard deviation σ. A $(1 - \alpha) \cdot 100\%$ confidence interval for μ is given by
>
> $$\text{Lower bound: } \bar{x} - t_{\alpha/2} \cdot \frac{s}{\sqrt{n}} \qquad \text{Upper bound: } \bar{x} + t_{\alpha/2} \cdot \frac{s}{\sqrt{n}} \quad \textbf{(1)}$$
>
> where $t_{\alpha/2}$ is computed with $n - 1$ degrees of freedom.
>
> **Note:** The interval is exact when the population is normally distributed. It is approximately correct for nonnormal populations, provided that n is large enough.

Notice that a confidence interval about μ with σ unknown can be computed for nonnormal populations even though Student's t-distribution required the population from which the sample was obtained to be normal. This is because the procedure for constructing the confidence interval is robust; that is, the procedure is accurate despite moderate departures from the normality requirement. Notice that we said the procedure is accurate for *moderate* departures from normality. If a small data set has outliers, the results are compromised, because neither the sample mean, $\overline{x}$, nor the sample standard deviation, s, is resistant to outliers. Sample data should always be inspected for serious departures from normality and for outliers. This is easily done with normal probability plots and boxplots.

Because this confidence interval uses the t-distribution, it is often referred to as the **t-interval**.

EXAMPLE 3   Constructing a Confidence Interval about a Population Mean

Table 5

64.0	33.4	45.8	56.0
51.5	29.2	63.7	

Source: Matt Gibson, instructor at Joliet West High School

Problem: An arborist is interesting in determining the mean diameter of mature white oak trees. The data in Table 5 represent the diameter (in centimeters) 1 meter above the base for a random sample of seven mature white oak trees in a forest preserve. Treat the sample as a simple random sample of all white oak trees in the forest preserve. Construct a 95% confidence interval for the mean diameter of a mature white oak tree 1 meter from the base. Interpret this interval.

Approach

Step 1: We draw a normal probability plot to verify the data come from a population that is normal and a boxplot to verify that there are no outliers.
Step 2: Compute $\overline{x}$ and s.
Step 3: Find the critical value $t_{\alpha/2}$ with $n - 1$ degrees of freedom.
Step 4: Compute the bounds on a $(1 - \alpha) \cdot 100\%$ confidence interval for μ, using the following:

$$\text{Lower bound: } \overline{x} - t_{\frac{\alpha}{2}} \cdot \frac{s}{\sqrt{n}} \qquad \text{Upper bound: } \overline{x} + t_{\frac{\alpha}{2}} \cdot \frac{s}{\sqrt{n}}$$

Step 5: Interpret the result.

Solution

Step 1: Because the sample size is small, we verify that the sample data come from a population that is normally distributed with no outliers. A normal probability plot of the data in Table 5 is shown in Figure 11(a). A boxplot is shown in Figure 11(b).

Figure 11

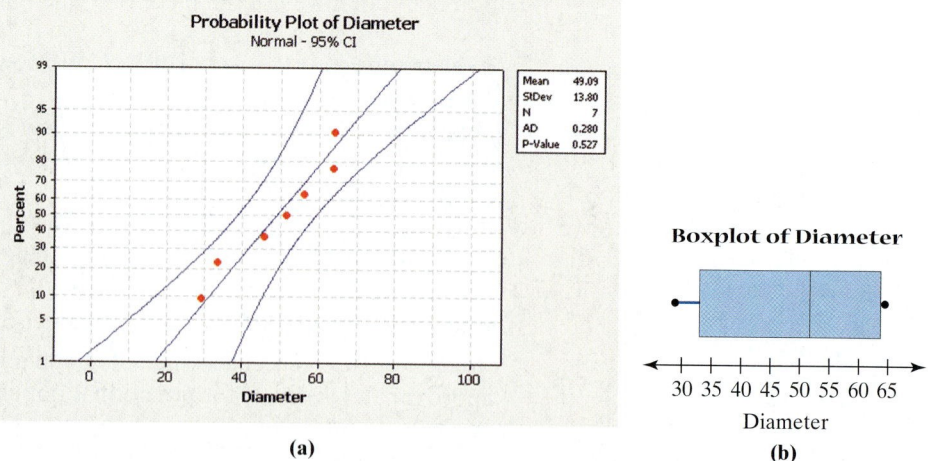

(a)

(b)

All the data lie within the bounds on the normal probability plot, indicating that the data could come from a population that is approximately normal. The box-plot does not display any outliers. The requirements for constructing the confidence interval are satisfied.

Step 2: We compute the sample mean and sample standard deviation using a calculator and find that $\bar{x} = 49.09$ centimeters and $s = 13.80$ centimeters.

Step 3: Because we wish to construct a 95% confidence interval, we have $\alpha = 0.05$. The sample size is $n = 7$. So we find $t_{\frac{0.05}{2}} = t_{0.025}$ with $7 - 1 = 6$ degrees of freedom. Referring to Table V, we find that $t_{0.025} = 2.447$.

Step 4: Using Formula (1), we find the lower and upper bounds:

$$\text{Lower bound:} \quad \bar{x} - t_{\frac{\alpha}{2}} \cdot \frac{s}{\sqrt{n}} = 49.09 - 2.447 \cdot \frac{13.80}{\sqrt{7}} = 49.09 - 12.76 = 36.33$$

$$\text{Upper bound:} \quad \bar{x} + t_{\frac{\alpha}{2}} \cdot \frac{s}{\sqrt{n}} = 49.09 + 2.447 \cdot \frac{13.80}{\sqrt{7}} = 49.09 + 12.76 = 61.85$$

Step 5: We are 95% confident that the mean diameter 1 meter above the base of mature white oak trees in the forest preserve is between 36.33 and 61.85 centimeters.

EXAMPLE 4	**Constructing a Confidence Interval about a Population Mean Using Technology**

Problem: Construct a 95% confidence interval for the mean diameter of a mature white oak tree 1 meter above the base using the sample data in Table 5. Interpret this interval.

Approach: We will use a TI-84 Plus graphing calculator to construct the confidence interval. The steps for constructing confidence intervals using the TI-83/84 Plus graphing calculators, MINITAB, and Excel are given in the Technology Step by Step on page 477.

Result: Figure 12 shows the results on a TI-84 Plus graphing calculator.

Figure 12   Lower Bound      Upper Bound

USING TECHNOLOGY
The results obtained from technology may differ from "by hand" intervals due to rounding error.

Interpretation: We are 95% confident that the mean diameter 1 meter above the base of mature white oak trees in the forest preserve is between 36.33 and 61.85 centimeters.

Now Work Problem 23.

EXAMPLE 5	**The Effect of Outliers**

Problem: The management of Disney World wanted to estimate the mean waiting time at the Dumbo ride. They randomly selected 15 riders and measured the amount of time (in minutes) the riders spent waiting in line. The results are in Table 6. Figure 13 shows a normal probability plot and boxplot for the data in the table. Figure 13(a) demonstrates that the sample data could have come from a population that is normally distributed except for the single outlier. Figure 13(b) shows the outlier as well. Determine a 95% confidence interval for the mean waiting time at the Dumbo ride both with and without the

Table 6				
30	29	34	41	37
16	18	30	24	25
29	16	80	19	30

outlier in the data set. Comment on the effect the outlier has on the confidence interval.

Figure 13

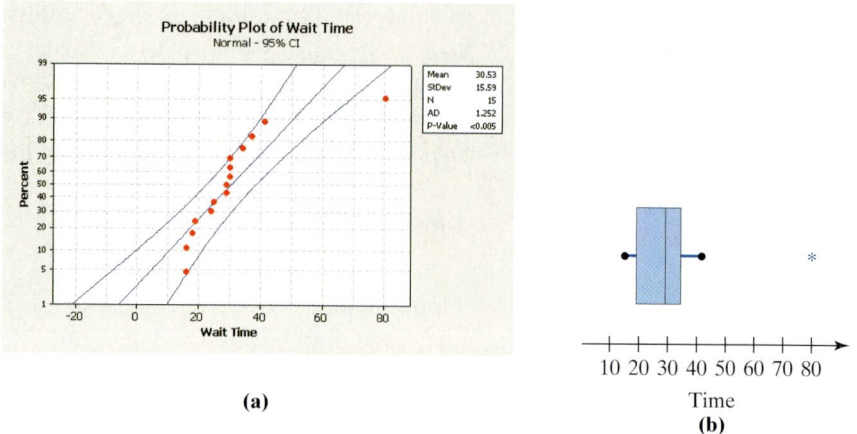

(a)

(b)

Approach: We will use MINITAB to construct the confidence intervals.

Solution: Figure 14(a) shows the confidence interval with the outlier included. Figure 14(b) shows the confidence interval with the outlier removed.

Figure 14   One-Sample T: Time(with outlier)

Variable	N	Mean	StDev	SE Mean	95.0% CI
Time	15	30.53	15.59	4.02	(21.90, 39.17)

(a)

One-Sample T: Time(without outlier)

Variable	N	Mean	StDev	SE Mean	95.0% CI
Time	14	27.00	7.75	2.07	(22.53, 31.47)

(b)

The 95% confidence interval with the outlier has a lower bound of 21.90 minutes and an upper bound of 39.17 minutes. The 95% confidence interval with the outlier removed has a lower bound of 22.53 minutes and an upper bound of 31.47. We notice a few things.

- With the outlier included, the sample mean is larger (i.e. drawn toward the outlier), because the sample mean is not resistant.
- With the outlier included, the sample standard deviation is larger, because the sample standard deviation is not resistant.
- With the outlier removed, the width of the interval decreased from $39.17 - 21.90 = 17.27$ minutes to $31.47 - 22.53 = 8.94$ minutes. The confidence interval is nearly twice as wide when the outlier is included.

What should we do if the requirements to compute a *t*-interval are not met? One option is to increase the sample size beyond 30 observations. The other option is to try to use *nonparametric procedures*. **Nonparametric procedures** (Chapter 15) typically do not require normality and the methods are resistant to outliers.

Now Work Problem 27.

In-Class Activity—Confidence Intervals

What is the mean number of rolls of a die before a 1 is observed? Either as an individual, or in a small group, roll a die until a 1 is observed. Repeat this process 30 times.

(a) Obtain a point estimate of the mean number of rolls of a die before a 1 is observed.

(b) The population standard deviation for the number of rolls before a 1 is observed is $\sqrt{30}$. Use this result to construct a 90% Z-interval for the mean number of rolls required before a 1 is observed.

(c) The population mean number of rolls before a 1 is observed is 6. Does your interval include 1? What proportion of the Z-intervals in the class include 1? How many did you expect to include 1?

(d) Construct a 90% t-interval for the mean number of rolls required before a 1 is observed.

(e) The population mean number of rolls before a 1 is observed is 6. Does your interval include 1? What proportion of the t-intervals in the class include 1? How many did you expect to include 1?

(f) Compare the Z-interval with the t-interval. Which has the smaller margin of error? Why?

9.2 ASSESS YOUR UNDERSTANDING

Concepts and Vocabulary

1. Explain the circumstances under which a Z-interval should be constructed. Under what circumstances should a t-interval be constructed? When can neither a Z- nor a t-interval be constructed?

2. Explain why the t-distribution has less spread as the number of degrees of freedom increases.

3. The procedures for constructing a t-interval are robust. Explain what this means.

4. Consider Figures 6(a) and (b). Why are the t-values more dispersed than the z-values?

5. Discuss the similarities and differences between the standard normal distribution and the t-distribution.

6. Explain what is meant by *degrees of freedom*.

Skill Building

7. **NW** **(a)** Find the t-value such that the area in the right tail is 0.10 with 25 degrees of freedom.

(b) Find the t-value such that the area in the right tail is 0.05 with 30 degrees of freedom.

(c) Find the t-value such that the area left of the t-value is 0.01 with 18 degrees of freedom. *[Hint: Use symmetry.]*

(d) Find the critical t-value that corresponds to 90% confidence. Assume 20 degrees of freedom.

8. **(a)** Find the t-value such that the area in the right tail is 0.02 with 19 degrees of freedom.

(b) Find the t-value such that the area in the right tail is 0.10 with 32 degrees of freedom.

(c) Find the t-value such that the area left of the t-value is 0.05 with 6 degrees of freedom. *[Hint: Use symmetry.]*

(d) Find the critical t-value that corresponds to 95% confidence. Assume 16 degrees of freedom.

9. A simple random sample of size n is drawn from a population that is normally distributed. The sample mean, $\bar{x}$, is found to be 108, and the sample standard deviation, s, is found to be 10.

(a) Construct a 96% confidence interval about μ if the sample size, n, is 25.

(b) Construct a 96% confidence interval about μ if the sample size, n, is 10. How does decreasing the sample size affect the margin of error, E?

(c) Construct a 90% confidence interval if the sample size, n, is 25. Compare the results to those obtained in part (a). How does decreasing the level of confidence affect the size of the margin of error, E?

(d) Could we have computed the confidence intervals in parts (a)–(c) if the population had not been normally distributed? Why?

10. A simple random sample of size n is drawn from a population that is normally distributed. The sample mean, $\bar{x}$, is found to be 50, and the sample standard deviation, s, is found to be 8.

(a) Construct a 98% confidence interval about μ if the sample size, n, is 20.

(b) Construct a 98% confidence interval about μ if the sample size, n, is 15. How does decreasing the sample size affect the margin of error, E?

(c) Construct a 95% confidence interval about μ if the sample size, n, is 20. Compare the results to those obtained in part (a). How does decreasing the level of confidence affect the size of the margin of error, E?

(d) Could we have computed the confidence intervals in parts (a)–(c) if the population had not been normally distributed? Why?

11. A simple random sample of size n is drawn. The sample mean, $\bar{x}$, is found to be 18.4, and the sample standard deviation, s, is found to be 4.5.

(a) Construct a 95% confidence interval about μ if the sample size, n, is 35.

(b) Construct a 95% confidence interval about μ if the sample size, n, is 50. How does increasing the sample size affect the margin of error, E?

(c) Construct a 99% confidence interval about μ if the sample size, n, is 35. Compare the results to those obtained in part (a). How does increasing the level of confidence affect the size of the margin of error, E?

(d) If the sample size is $n = 15$, what conditions must be satisfied to compute the confidence interval?

12. A simple random sample of size n is drawn. The sample mean, $\bar{x}$, is found to be 35.1, and the sample standard deviation, s, is found to be 8.7.

(a) Construct a 90% confidence interval about μ if the sample size, n, is 40.

(b) Construct a 90% confidence interval about μ if the sample size, n, is 100. How does increasing the sample size affect the margin of error, E?

(c) Construct a 98% confidence interval about μ if the sample size, n, is 40. Compare the results to those obtained in part (a). How does increasing the level of confidence affect the size of the margin of error, E?

(d) If the sample size is $n = 18$, what conditions must be satisfied to compute the confidence interval?

Applying the Concepts

13. How Much Do You Read? A Gallup poll conducted May 20–22, 2005, asked 1006 Americans, "During the past year, about how many books, either hardcover or paperback, did you read either all or part of the way through?" Results of the survey indicated that $\bar{x} = 13.4$ books and $s = 16.6$ books. Construct a 99% confidence interval for the mean number of books Americans read either all or part of during the preceding year. Interpret the interval.

14. How Much Do You Read? A Gallup poll conducted July 21–August 14, 1978, asked 1006 Americans, "During the past year, about how many books, either hardcover or paperback, did you read either all or part of the way through?" Results of the survey indicated that $\bar{x} = 18.8$ books and $s = 19.8$ books.

(a) Construct a 99% confidence interval for the mean number of books Americans read either all or part of during the preceding year. Interpret the interval.

(b) Compare these results to those of Problem 13. Were Americans reading more in 1978 than in 2005?

15. The SARS Epidemic Severe acute respiratory syndrome (or SARS) is a viral respiratory illness. It has the distinction of being the first new communicable disease of the 21st century. Researchers wanted to estimate the incubation period of patients with SARS. Based on interviews with 81 SARS patients, they found that the mean incubation period was 4.6 days with a standard deviation of 15.9 days. Based on this information, construct a 95% confidence interval for the mean incubation period of the SARS virus. Interpret the interval. (*Source:* Gabriel M. Leung et al., The Epidemiology of Severe Acute Respiratory Syndrome in the 2003 Hong Kong Epidemic: An Analysis of All 1755 Patients, *Annals of Internal Medicine*, 2004; 141:662–673.)

16. Tensile Strength Tensile strength is the amount of stress a material can withstand before it breaks. Researchers wanted to determine the tensile strength of a resin cement used in bonding crowns to teeth. The researchers bonded crowns to 72 extracted teeth. Using a tensile resistance test, they found the mean tensile strength of the resin cement to be 242.2 newtons (N) with a standard deviation of 70.6 newtons. Based on this information, construct a 90% confidence interval for the mean tensile strength of the resin cement. (*Source:* Simonides Consani et al., Effect of Cement Types on the Tensile Strength of Metallic Crowns Submitted to Thermocycling, *Brazilian Dental Journal*, Vol. 14, No. 3, 2003.)

17. How Much TV Do Teenagers Watch? A Gallup poll conducted January 17–February 6, 2005, asked 1028 teenagers aged 13 to 17, "Typically, how many hours per week do you spend watching TV?" Survey results indicate that $\bar{x} = 13$ hours and $s = 2.3$ hours. Construct a 95% confidence interval for the number of hours of TV teenagers watch each week. Interpret the interval.

18. Honda Accord Gas Mileage For a simple random sample of forty 2005 Honda Accords (4 cylinder, 2.4 liter, 5-speed automatic), the mean gas mileage was 23 miles per gallon with a standard deviation of 1.5 miles per gallon. Construct a 95% confidence interval for the mean gas mileage of similar 2005 Honda Accords. Interpret the interval.

19. Dividend Yields Many companies pay a portion of their profits to shareholders in the form of dividends. A stock's dividend yield is defined as the dividend paid by the company divided by its stock price. A stock analyst wants to estimate the mean dividend yield of financial stocks. He obtains a simple random sample of 11 financial stocks and obtains the following dividend yields. The data are in percent, so 3.17 represents 3.17%. Construct and interpret a 90% confidence interval for the dividend yield of financial stocks. An analysis of the data indicates that it is reasonable to conclude dividend yields follow a normal distribution. The data set has no outliers.

| 3.19 | 1.86 | 0 | 0.54 | 1.76 | 2.08 |
| 0.67 | 0.67 | 1.57 | 0 | 2.12 | |

Source: Morningstar

20. Tips A server at a restaurant wanted to estimate the mean tip percentage that she earns during dinner. She randomly selects 14 receipts from dinner, records the tip, and computes the tip rate. She obtains the following tip rates. The data are in percent, so 16.5 represents 16.5%. Construct and interpret a 95% confidence interval for the tip percent. An analysis of the data indicates that it is reasonable to conclude tip percent follows a normal distribution. The data has no outliers.

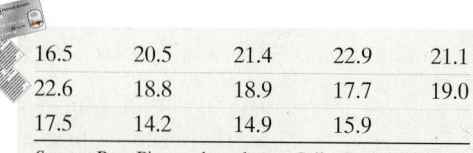

16.5	20.5	21.4	22.9	21.1
22.6	18.8	18.9	17.7	19.0
17.5	14.2	14.9	15.9	

Source: Pam Pimental, student at Joliet Junior College

21. Concentration of Dissolved Organic Carbon The following data represent the concentration of organic carbon (mg/L) collected from organic soil.

22.74	29.80	27.10	16.51	6.51
8.81	5.29	20.46	14.90	33.67
30.91	14.86	15.91	15.35	9.72
19.80	14.86	8.09	17.90	18.30
5.20	11.90	14.00	7.40	17.50
10.30	11.40	5.30	15.72	20.46
16.87	15.42	22.49		

Source: Lisa Emili, Ph.D. Candidate, University of Waterloo

Construct a 99% confidence interval for the mean concentration of dissolved organic carbon collected from organic soil. Interpret the interval. (**Note:** $\bar{x} = 15.92$ mg/L and $s = 7.38$ mg/L.)

22. Concentration of Dissolved Organic Carbon The following data represent the concentration of organic carbon (mg/L) collected from mineral soil.

8.50	3.91	9.29	21.00	10.89
10.30	11.56	7.00	3.99	3.79
5.50	4.71	7.66	11.72	11.80
8.05	10.72	21.82	22.62	10.74
3.02	7.45	11.33	7.11	9.60
12.57	12.89	9.81	17.99	21.40
8.37	7.92	17.90	7.31	16.92
4.60	8.50	4.80	4.90	9.10
7.90	11.72	4.85	11.97	7.85
9.11	8.79			

Source: Lisa Emili, Ph.D. candidate, University of Waterloo

(a) Construct a 99% confidence interval for the mean concentration of dissolved organic carbon collected from mineral soil. Interpret the interval. (**Note:** $\bar{x} = 10.03$ mg/L and $s = 4.98$ mg/L.)

(b) Compare the 99% confidence interval computed for organic soil (Problem 21) to the 99% confidence interval computed for mineral soil. Does there appear to be a difference between the concentration levels for the two soil types?

23. Crawling Babies The following data represent the age (in weeks) at which babies first crawl based on a survey of 12 mothers conducted by Essential Baby.

52	30	44	35
47	37	56	26
39	26	39	28

Source: www.essentialbaby.com

(a) Because the sample size is small, we must verify that the data come from a population that is normally distributed and that the sample size does not contain any outliers. The normal probability plot and boxplot are shown next. Are the conditions for constructing a confidence interval about the mean satisfied?

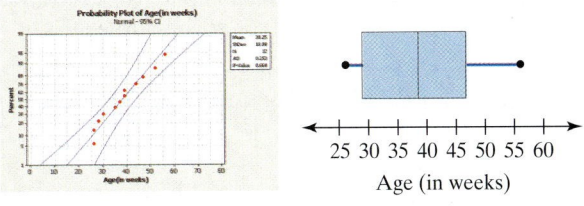

25 30 35 40 45 50 55 60
Age (in weeks)

(b) Construct a 95% confidence interval for the mean age at which a baby first crawls.

(c) What could be done to increase the accuracy of the interval without changing the level of confidence?

24. Battery Life The following data represent the battery life, in hours, for a random sample of 10 full charges on a fifth-generation iPod music player.

7.3	10.2	12.9	10.8	12.1
6.6	10.2	9.0	8.5	7.1

(a) Because the sample size is small, we must verify that the data come from a population that is normally distributed and that the sample size does not contain any outliers. The normal probability plot and boxplot are shown next. Are the conditions for constructing a confidence interval about the mean satisfied?

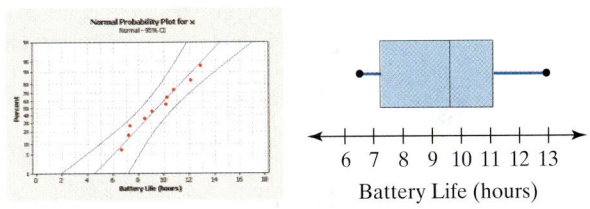

6 7 8 9 10 11 12 13
Battery Life (hours)

(b) Construct a 90% confidence interval for the mean number hours the battery will last on this player.

(c) Suppose you wanted more accuracy. What can be done to increase the accuracy of the interval without changing the level of confidence?

25. The Growing Season An agricultural researcher is interested in estimating the mean length of the growing season in the Chicago area. Treating the last 10 years as a simple random sample, he obtains the following data, which represent the number of days of the growing season.

155	162	146	144	168
188	215	180	166	151

Source: Midwest Climate Center

(a) Because the sample size is small, we must verify that the data come from a population that is normally distributed and that the sample size does not contain any outliers. The normal probability plot and boxplot are shown next. Are the conditions for constructing a confidence interval about the mean satisfied?

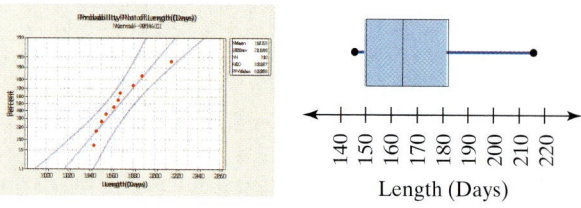

Length (Days)

(b) Construct a 95% confidence interval for the mean length of the growing season in the Chicago area.
(c) What could be done to increase the accuracy of the interval without changing the level of confidence?

26. **Disc Weight** When playing disc golf, the weight of the disc has a huge impact on disc performance. Lighter discs tend to fly farther, but are less stable. Heavier discs tend to give better control and are better suited for strong winds. The following data represent the weights, in grams, of a random sample of 12 discs produced by a certain manufacturer.

172	168	170	173	172	173
171	175	173	171	169	173

(a) Because the sample size is small, we must verify that the data come from a population that is normally distributed and that the sample size does not contain any outliers. The normal probability plot and boxplot are shown next. Are the conditions for constructing a confidence interval about the mean satisfied?

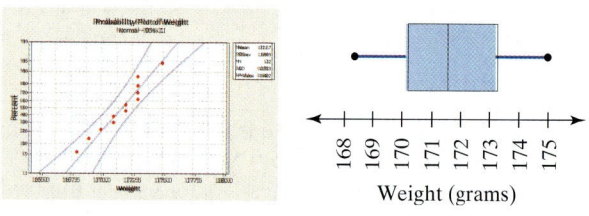

Weight (grams)

(b) Construct a 95% confidence interval for disc weight. Interpret the interval.

27. **Effect of Outliers** The following data represent the ask-
NW ing price of a simple random sample of homes for sale in Lexington, Kentucky in June 2005.

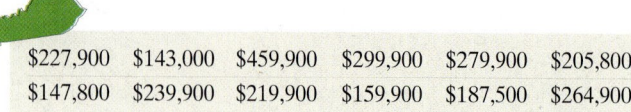

$227,900	$143,000	$459,900	$299,900	$279,900	$205,800
$147,800	$239,900	$219,900	$159,900	$187,500	$264,900

(a) Construct a boxplot to identify the outlier.
(b) Construct a 99% confidence interval with the outlier included.
(c) Construct a 99% confidence interval with the outlier removed.
(d) Comment on the effect the outlier has on the confidence interval.

28. **Effect of Outliers** The following data represent a simple random sample of ages of West Nile virus victims in the United States during 2004.

64	77	71	54	60	79
76	82	81	77	68	34

(a) Construct a boxplot to identify the outlier.
(b) Construct a 99% confidence interval with the outlier included.
(c) Construct a 99% confidence interval with the outlier removed.
(d) Comment on the effect the outlier has on the confidence interval.

29. **Simulation** IQ scores based on the Wechsler Intelligence Scale for Children (WISC) are known to be normally distributed with $\mu = 100$ and $\sigma = 15$.
(a) Simulate obtaining 20 samples of size $n = 15$ from this population.
(b) Obtain the sample mean and standard deviation for each of the 20 samples.
(c) Construct 95% t-intervals for each of the 20 samples.
(d) How many of the intervals do you expect to include the population mean? How many actually contain the population mean?

30. **Simulation** Suppose the arrival of cars at Burger King's drive-through follows a Poisson process with $\mu = 4$ cars every 10 minutes.
(a) Simulate obtaining 30 samples of size $n = 35$ from this population.
(b) Obtain the sample mean and standard deviation for each of the 30 samples.
(c) Construct 90% t-intervals for each of the 30 samples.
(d) How many of the intervals do you expect to include the population mean? How many actually contain the population mean?

31. **Confidence Interval Applet: The Role of Right Skewed Data** Load the confidence intervals for a mean (the impact of not knowing the standard deviation) applet.
(a) Set the shape to normal with mean = 100 and Std. Dev. = 15. Construct at least 1000 confidence intervals with $n = 10$. Compare the proportion of the 95% Z-intervals and t-intervals that contain the population mean. Is this what you would expect?
(b) Set the shape to skewed right with mean = 100 and Std. Dev. = 15. Construct at least 1000 confidence intervals with $n = 10$. Compare the proportion of the 95% Z-intervals and t-intervals that contain the population mean. Is this what you would expect?

Consumer Reports Tests Tires

Consumer Reports' specialized auto-test facility is the largest, most sophisticated consumer-based auto-testing facility in the world. Located on 327 acres in East Haddam, Connecticut, the facility is staffed by a team of experienced engineers and test personnel who buy and test more than 40 new cars, SUVs, minivans, and light trucks each year. For each vehicle, Consumer Reports conducts more than 46 individual tests, ranging from emergency handling, acceleration, and braking to fuel-economy measurements, noise-level evaluations, and bumper-impact tests.

Measurement	Distance
1	131.8
2	123.2
3	132.9
4	139.8
5	140.3
6	128.3
7	129.3
8	129.5
9	134.0

As part of testing vehicles, Consumer Reports tests tires. Our tire evaluations include dry and wet braking from 60 mph, braking on ice, snow traction on a flat surface, snow traction on a snow hill, emergency handling, routine cornering, ride comfort, rolling resistance, and noise. All the test data are recorded using an optical instrument that provides precise speed and distance measurements.

The following table contains the dry brake stopping distance (in feet) data for one brand of tires recently tested.

(a) A normal probability plot of the dry brake distance data is shown. What does this plot suggest about the distribution of the brake data?

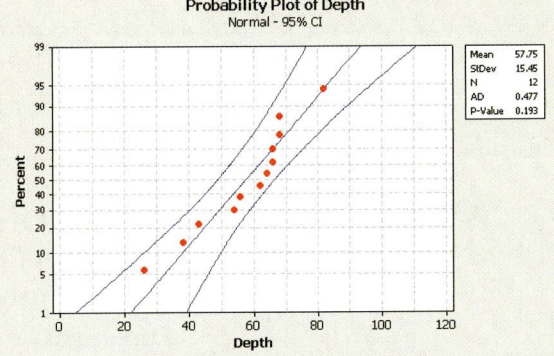

(b) Does the data set contain any outliers?

(c) Use these data to construct a 95% confidence interval for the mean dry braking distance for this brand of tires. Write a statement that explains the meaning of this confidence interval to the readers of *Consumer Reports* magazine.

Note to Readers: *In many cases, our test protocol and analytical methods are more complicated than described in these examples. The data and discussions have been modified to make the material more appropriate for the audience.*

Technology Step-by-Step	**Confidence Intervals about μ, σ Unknown**
TI-83/84 Plus	***Step 1:*** If necessary, enter raw data in L1.
	Step 2: Press STAT, highlight TESTS, and select `8: TInterval`.
	Step 3: If the data are raw, highlight DATA. Make sure `List1` is set to L1 and `Freq` to 1. If summary statistics are known, highlight STATS and enter the summary statistics.

Step 4: Enter the confidence level following `C-Level:`.

Step 5: Highlight `Calculate`; press ENTER.

MINITAB

Step 1: Enter raw data in column C1.

Step 2: Select the **Stat** menu, highlight **Basic Statistics**, then highlight **1-Sample t . . . .**

Step 3: Enter C1 in the cell marked "Variables". Select Confidence Interval, and enter a confidence level. Click OK.

Excel

Step 1: If necessary, enter raw data in column A.

Step 2: Load the PHStat Add-in.

Step 3: Select the **PHStat menu**, highlight **Confidence Intervals . . . ,** then highlight **Estimate for the mean, sigma unknown . . . .**

Step 4: Enter the confidence level. If the summary statistics are known, click "Sample statistics known" and enter the sample size, sample mean, and sample standard deviation. If summary statistics are unknown, click "Sample statistics unknown". With the cursor in the "Sample cell range" cell, highlight the data in column A. Click OK.

9.3 Confidence Intervals about a Population Proportion

Preparing for This Section Before getting started, review the following:

- Distribution of the sample proportion (Section 8.2, pp. 434–439)

Objectives

1 **Obtain a point estimate for the population proportion**

2 **Construct and interpret a confidence interval for the population proportion**

3 **Determine the sample size necessary for estimating a population proportion within a specified margin of error**

Probably the most frequently reported confidence interval is one involving the proportion of a population. Researchers are often interested in estimating the proportion of the population that has a certain characteristic. For example, in a poll conducted by the Gallup Organization in early December, 2004, a random sample of 1004 Americans resulted in 84% of the respondents stating that they are satisfied with the way things are going in their personal lives. The poll had a margin of error of ±3% with 95% confidence. Based on discussions in Sections 9.1 and 9.2, we know that this means the pollsters are 95% confident the true proportion of Americans who are satisfied with the way things are going in their personal lives is between 81% and 87% (84% ± 3%).

In this section, we will discuss the techniques for estimating the population proportion population, *p*. In addition, we present methods for determining the sample size required to estimate the population proportion.

/ **In Other Words**

The symbol ± is read "plus or minus." It means "to add and subtract the quantity following the ± symbol."

1 **Obtain a Point Estimate for the Population Proportion**

A point estimate is an unbiased estimator of the parameter. The point estimate for the population proportion is $\hat{p} = \dfrac{x}{n}$, where *x* is the number of individuals in the sample with the specified characteristic and *n* is the sample size.

| **EXAMPLE 1** | **Obtaining a Point Estimate of a Population Proportion** |

Problem: In a poll conducted March 17–21, 2005, by the Pew Research Center for the People and the Press, a simple random sample of 1505 American adults was asked whether they were in favor of tighter enforcement of government rules on TV content during hours when children are most likely to be watching. Of the 1505 adults, 1129 responded yes. Obtain a point estimate for the population proportion of Americans who are in favor of tighter enforcement of government rules on TV content during hours when children are most likely to be watching.

Approach: The point estimate of the population proportion is $\hat{p} = \dfrac{x}{n}$, where $x = 1129$ and $n = 1505$.

Solution: Substituting into the formula, we obtain $\hat{p} = \dfrac{x}{n} = \dfrac{1129}{1505} = 0.750 = 75.0\%$. We estimate that 75% of adult Americans are in favor of tighter enforcement of government rules on TV content during hours when children are most likely to be watching.

> **Now Work Problem 11(a).**

② Construct and Interpret a Confidence Interval for the Population Proportion

The point estimate obtained in Example 1 is a single estimate of the unknown parameter, p. We recognize it is unlikely that the sample proportion will equal the population proportion. Therefore, rather than reporting the value of the sample proportion alone, it is preferred to report an interval about the sample proportion.

In Section 8.2, we discussed the sampling distribution of $\hat{p}$. For convenience, we repeat the sampling distribution of $\hat{p}$ here.

Sampling Distribution of $\hat{p}$

For a simple random sample of size n where $n \leq 0.05N$ (that is, the sample size is no more than 5% of the population size), the sampling distribution of $\hat{p}$ is approximately normal with mean $\mu_{\hat{p}} = p$ and standard deviation $\sigma_{\hat{p}} = \sqrt{\dfrac{p(1-p)}{n}}$, provided that $np(1-p) \geq 10$.

We use the sampling distribution of $\hat{p}$ to construct a confidence interval for the population proportion p.

Constructing a $(1 - \alpha) \cdot 100\%$ Confidence Interval for a Population Proportion

Suppose a simple random sample of size n is taken from a population. A $(1 - \alpha) \cdot 100\%$ confidence interval for p is given by the following quantities:

$$\text{Lower bound:} \quad \hat{p} - z_{\alpha/2} \cdot \sqrt{\dfrac{\hat{p}(1-\hat{p})}{n}}$$

$$\text{Upper bound:} \quad \hat{p} + z_{\alpha/2} \cdot \sqrt{\dfrac{\hat{p}(1-\hat{p})}{n}} \tag{1}$$

Note: It must be the case that $n\hat{p}(1 - \hat{p}) \geq 10$ and $n \leq 0.05N$ to construct this interval.

Notice we use $\hat{p}$ in place of p in the standard deviation. This is because p is unknown and $\hat{p}$ is the best point estimate of p.

EXAMPLE 2

Constructing a Confidence Interval for a Population Proportion

Problem: In a poll conducted March 17–21, 2005, by the Pew Research Center for the People and the Press, a simple random sample of 1505 American adults was asked whether they were in favor of tighter enforcement of government rules on TV content during hours when children are most likely to be watching. Of the 1505 adults, 1129 responded yes. Obtain a 95% confidence interval for the proportion of Americans who are in favor of tighter enforcement of government rules on TV content during hours when children are most likely to be watching.

Approach

Step 1: Compute the value of $\hat{p}$.

Step 2: We can compute a 95% confidence interval about p provided that $n\hat{p}(1 - \hat{p}) \geq 10$ and $n \leq 0.05N$.

Step 3: Determine the critical value, $z_{\alpha/2}$.

Step 4: Determine the lower and upper bounds of the confidence interval.

Step 5: Interpret the result.

Solution:

Step 1: From Example 1 we have that $\hat{p} = 0.750$.

Step 2: $n\hat{p}(1 - \hat{p}) = 1505(0.750)(1 - 0.750) = 282.1875 \geq 10$. There are over 180 million adult Americans, so our sample size is definitely less than 5% of the population size. We can proceed to construct the confidence interval.

Step 3: Because we want a 95% confidence interval, we have $\alpha = 0.05$, so $z_{\alpha/2} = z_{0.05/2} = z_{0.025} = 1.96$.

Step 4: Substituting into Formula (1) with $n = 1505$, we obtain the lower and upper bounds of the confidence interval:

Lower bound:

$$\hat{p} - z_{\alpha/2} \cdot \sqrt{\frac{\hat{p}(1 - \hat{p})}{n}} = 0.750 - 1.96 \cdot \sqrt{\frac{0.750(1 - 0.750)}{1505}}$$

$$= 0.750 - 0.022 = 0.728$$

Upper bound:

$$\hat{p} + z_{\alpha/2} \cdot \sqrt{\frac{\hat{p}(1 - \hat{p})}{n}} = 0.750 + 1.96 \cdot \sqrt{\frac{0.750(1 - 0.750)}{1505}}$$

$$= 0.750 + 0.022 = 0.772$$

Step 5: We are 95% confident that the proportion of adult Americans who are in favor of tighter enforcement of government rules on TV content during hours when children are most likely to be watching is between 0.728 and 0.772.

EXAMPLE 3

Constructing a Confidence Interval about a Population Proportion Using Technology

Problem: In a poll conducted March 17–21, 2005, by the Pew Research Center for the People and the Press, a simple random sample of 1505 American adults was asked whether they were in favor of tighter enforcement of government rules on TV content during hours when children are most likely to be watching. Of the 1505 adults, 1129 responded yes. Obtain a 95% confidence interval for the proportion of Americans who are in favor of tighter enforcement of government rules on TV content during hours when children are most likely to be watching.

Approach: We will use MINITAB to construct the confidence interval. The steps for constructing confidence intervals using MINITAB, Excel, and the TI-83/84 graphing calculators are given in the Technology Step by Step on page 486.

Result: Figure 15 shows the results obtained from MINITAB.

Figure 15   **Test and CI for One Proportion**

```
Test of p = 0.5 vs p not = 0.5

Sample    X      N    Sample p       95% CI         Z-Value P-Value
1       1129   1505   0.750166 (0.728294,0.772038)  19.41   0.000
```

The lower bound is 0.728 and the upper bound is 0.772.

Interpretation: We are 95% confident that the proportion of adult Americans who are in favor of tighter enforcement of government rules on TV content during hours when children are most likely to be watching is between 0.728 and 0.772.

USING TECHNOLOGY

The results obtained using technology may differ from those obtained by hand due to rounding.

It is important to remember the correct interpretation of a confidence interval. The statement "95% confident" means that, if 100 samples of size 1505 were taken, about 95 of the intervals will contain the parameter p and about 5 will not. Unfortunately, we cannot know whether the interval we computed in Examples 2 and 3 is one of the 95 intervals that contains p or one of the 5 that does not contain p.

Often, polls will report their results by giving the value of $\hat{p}$ obtained from the sample data along with the margin of error, rather than reporting a confidence interval. In Examples 2 and 3, the Pew Research Center might say, "In a survey conducted March 17–21, 2005, 75% of adult Americans are in favor of tighter enforcement of government rules on TV content during hours when children are most likely to be watching. The survey results have a margin of error of 2.2%.

CAUTION

Beware of surveys that do not report a margin of error. Survey results should also report sample size, sampling technique, and the population that was being studied.

> **Now Work Problems 11(b), (c), and (d).**

In-Class Activity—Confidence Intervals (M&M's)

Mars, Inc. reports that the proportion of Plain M&M's it produces that are blue is 0.24. However, you suspect that this is not the case.

(a) Each student should obtain a 1-pound bag of plain M&M's candies and bring it to class unopened.

(b) In class, count the number of candies in your bag for each of the six colors (red, brown, blue, yellow, green, and orange).

(c) Obtain a point estimate for the population proportion of blue candies.

(d) Verify the requirements for constructing a confidence interval about $\hat{p}$ are satisfied. If the conditions are not met, what could you do to rectify the situation?

(e) Construct a 95% confidence interval for the proportion of blue candies. Interpret the interval. Does your interval contain 0.24?

(f) Compare your interval with the rest of the class. What proportion of intervals do not contain 0.24? Is this result what you would expect?

③ Determine the Sample Size Necessary for Estimating a Population Proportion within a Specified Margin of Error

In Section 9.1, we introduced a method for determining the sample size n required to estimate the sample mean within a certain margin of error with a specified level of confidence. The formula was obtained by solving the margin of error, $E = z_{\alpha/2} \cdot \dfrac{\sigma}{\sqrt{n}}$, for n. We can follow the same approach to determine sample size when estimating a population proportion. In Formula (1) on page 479, we notice that the margin of error, E, is given by $E = z_{\alpha/2} \cdot \sqrt{\dfrac{\hat{p}(1 - \hat{p})}{n}}$. We solve the margin of error for n and obtain $n = \hat{p}(1 - \hat{p})\left(\dfrac{z_{\alpha/2}}{E}\right)^2$

The problem with this formula is that it depends on $\hat{p}$, and $\hat{p} = \dfrac{x}{n}$ depends on the sample size, n, which is what we are trying to determine in the first place! How do we resolve this issue? There are two possibilities: (1) We could use an estimate of p based on a pilot study or an earlier study, or (2) we could let $\hat{p} = 0.5$. When $\hat{p} = 0.5$ the maximum value of $\hat{p}(1 - \hat{p}) = 0.25$ is obtained as illustrated in Figure 16. Using the maximum value gives the largest possible value of n for a given level of confidence and a given margin of error.

Figure 16

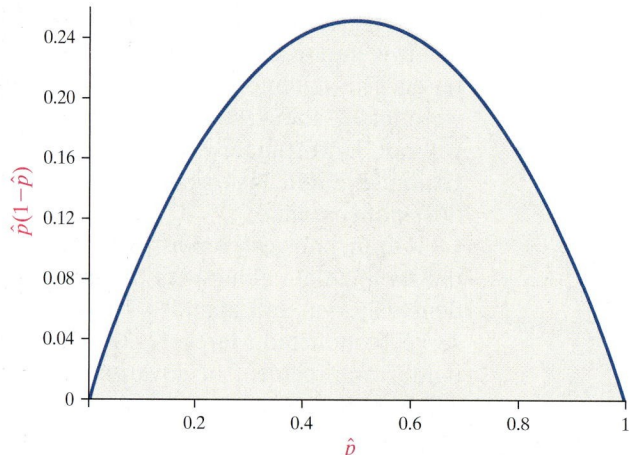

The disadvantage of the second option is that it could lead to a larger sample size than is necessary. Because of the time and expense of sampling, it is desirable to avoid too large a sample.

Sample Size Needed for Estimating the Population Proportion p

The sample size required to obtain a $(1 - \alpha) \cdot 100\%$ confidence interval for p with a margin of error E is given by

$$n = \hat{p}(1 - \hat{p})\left(\frac{z_{\alpha/2}}{E}\right)^2 \qquad (2)$$

(rounded up to the next integer), where $\hat{p}$ is a prior estimate of p. If a prior estimate of p is unavailable, the sample size required is

$$n = 0.25\left(\frac{z_{\alpha/2}}{E}\right)^2 \qquad (3)$$

rounded up to the next integer. The margin of error should always be expressed as a decimal when using Formulas (2) and (3).

In Other Words

There are two formulas for determining sample size when estimating the population proportion. Formula (2) requires a prior estimate of p; Formula (3) does not.

EXAMPLE 4 Determining Sample Size

Problem: A sociologist wishes to estimate the percentage of the U.S. population living in poverty. What size sample should be obtained if she wishes the estimate to be within 2 percentage points with 99% confidence if

(a) she uses the 2003 estimate of 12.7% obtained from the American Community Survey.

(b) she does not use any prior estimates.

Approach: In both cases, we have $E = 0.02$ (2% = 0.02) and $z_{\alpha/2} = z_{0.01/2} = z_{0.005} = 2.575$. To answer part (a), we let $\hat{p} = 0.127$ in Formula (2). To answer part (b), we use Formula (3).

Solution

(a) Substituting $E = 0.02$, $z_{0.005} = 2.575$, and $\hat{p} = 0.127$ into Formula (2), we obtain

$$n = \hat{p}(1 - \hat{p})\left(\frac{z_{\alpha/2}}{E}\right)^2 = 0.127(1 - 0.127)\left(\frac{2.575}{0.02}\right)^2 = 1837.9$$

We round this value up to 1838, so she must survey 1838 randomly selected residents of the United States.

(b) Substituting $E = 0.02$ and $z_{0.005} = 2.575$ into Formula (3), we obtain

$$n = 0.25\left(\frac{z_{\alpha/2}}{E}\right)^2 = 0.25\left(\frac{2.575}{0.02}\right)^2 = 4144.1$$

We round this value up to 4145, so she must survey 4145 randomly selected residents of the United States.

We can see the effect of not having a prior estimate of p: The required sample size more than doubled!

CAUTION
We always round up when determining sample size.

> Now Work Problem 17.

MAKING AN INFORMED DECISION

Help Wanted

What's Your Major?

One of the most difficult decisions a person has to make while attending college is choosing a major field of study. This decision plays a huge role in the career path of the individual. The purpose of this project is to help you make a more informed decision regarding your major.

Randomly select people who have recently graduated in the major that you have chosen or randomly select individuals who have recently secured a job in a career that you are considering. If you have not yet chosen a major, select one that interests you. Create a survey that will allow you to make a more informed decision regarding your major field of study. Some suggested sample questions follow:

(a) What was your major in college?
(b) On average, how many hours do you work?

(c) Approximately how many weeks did it take you to find your job?
(d) What was your starting annual salary?
(e) Are you satisfied or dissatisfied with your career?
(f) Do you believe you have job security?

Administer the survey. Be sure to explain the purpose of the survey and make it look professional. Because it is important that we obtain accurate results, make sure that the surveys are completed anonymously. We will treat the randomly selected individuals as a simple random sample.

Estimate the unknown parameter for each question asked. For example, in question (b), estimate the population mean number of hours worked. In question (e), estimate the proportion of individuals satisfied with their career. Then construct 95% confidence intervals about each parameter. Write a report detailing your findings.

9.3 ASSESS YOUR UNDERSTANDING

Concepts and Vocabulary

1. What is the best point estimate of a population proportion?

2. What are the requirements that must be satisfied to construct a confidence interval about a population proportion?

3. When determining the sample size required to obtain an estimate for a population proportion, is the researcher better off using a prior estimate of p or no prior estimate of p? Why? List some pros and cons for each scenario.

4. Explain why you should be wary of surveys that do not report a margin of error.

Skill Building

In Problems 5–10, construct a confidence interval of the population proportion at the given level of confidence.

5. $x = 30, n = 150$, 90% confidence

6. $x = 80, n = 200$, 98% confidence

7. $x = 120, n = 500$, 99% confidence

8. $x = 400, n = 1200$, 95% confidence

9. $x = 860, n = 1100$, 94% confidence

10. $x = 540, n = 900$, 96% confidence

Applying the Concepts

11. **Lipitor** **NW** The drug Lipitor is meant to lower cholesterol levels. In a clinical trial of 863 patients who received 10-mg doses of Lipitor daily, 47 reported a headache as a side effect.
    (a) Obtain a point estimate for the population proportion of Lipitor users who will experience a headache as a side effect.
    (b) Verify that the requirements for constructing a confidence interval about $\hat{p}$ are satisfied.
    (c) Construct a 90% confidence interval for the population proportion of Lipitor users who will report a headache as a side effect.
    (d) Interpret the confidence interval.

12. **Pepcid** A study of 74 patients with ulcers was conducted in which they were prescribed 40 mg of Pepcid. After 8 weeks, 58 reported ulcer healing.
    (a) Obtain a point estimate for the proportion of patients with ulcers receiving Pepcid who will report ulcer healing.
    (b) Verify that the requirements for constructing a confidence interval about $\hat{p}$ are satisfied.
    (c) Construct a 99% confidence interval for the proportion of patients with ulcers receiving Pepcid who will report ulcer healing.
    (d) Interpret the confidence interval.

13. **Defense Spending** In a February 2005 poll conducted by the Gallup Organization, 302 of 1008 randomly selected adults aged 18 or older stated that they believe the United States is spending too little on national defense.
    (a) Obtain a point estimate for the proportion of adults aged 18 or older who feel the United States is spending too little on national defense.
    (b) Verify that the requirements for constructing a confidence interval about $\hat{p}$ are satisfied.
    (c) Construct a 98% confidence interval for the proportion of adults aged 18 or older who believe the United States is spending too little on national defense. Interpret the interval.

14. **Partial Birth Abortions** In an October 2003 poll conducted by the Gallup Organization, 684 of 1006 randomly selected adults aged 18 years old or older stated they think the government should make partial birth abortions illegal, except in cases necessary to save the life of the mother.
    (a) Obtain a point estimate for the proportion of adults aged 18 or older who think the government should make partial birth abortions illegal, except in cases necessary to save the life of the mother.
    (b) Verify that the requirements for constructing a confidence interval about $\hat{p}$ are satisfied.
    (c) Construct a 98% confidence interval for the proportion of adults aged 18 or older who think the government should make partial birth abortions illegal, except in cases necessary to save the life of the mother. Interpret the interval.

15. **Packer Fans** In a Harris Poll conducted October 20–25, 2004, 381 of 2114 randomly selected adults who follow professional football said the Green Bay Packers were their favorite team.
    (a) Verify that the requirements for constructing a confidence interval about $\hat{p}$ are satisfied.
    (b) Construct a 90% confidence interval for the proportion of adults who follow professional football who say the Green Bay Packers is their favorite team. Interpret this interval.
    (c) Construct a 99% confidence interval for the proportion of adults who follow professional football who say the Green Bay Packers is their favorite team. Interpret this interval.
    (d) What is the effect of increasing the level of confidence on the width of the interval?

16. **Taking It Easy** In a Gallup poll conducted December 11–14, 2003, 455 of 1011 randomly selected adults aged 18 and older said they had too little time for relaxing or doing nothing.
    (a) Verify that the requirements for constructing a confidence interval about $\hat{p}$ are satisfied.
    (b) Construct a 92% confidence interval for the proportion of adults aged 18 and older who say they have too

little time for relaxing or doing nothing. Interpret this interval.

(c) Construct a 96% confidence interval for the proportion of adults aged 18 and older who say they have too little time for relaxing or doing nothing. Interpret this interval.

(d) What is the effect of increasing the level of confidence on the width of the interval?

17. High-Speed Internet Access A researcher wishes to estimate the proportion of adults who have high-speed Internet access. What size sample should be obtained if she wishes the estimate to be within 0.03 with 99% confidence if

(a) she uses a 2004 estimate of 0.44 obtained from a Harris poll?

(b) she does not use any prior estimates?

18. Home Ownership An urban economist wishes to estimate the proportion of Americans who own their house. What size sample should be obtained if he wishes the estimate to be within 0.02 with 90% confidence if

(a) he uses an estimate of 0.675 from the fourth quarter of 2000 obtained from the U.S. Census Bureau?

(b) he does not use any prior estimates?

19. Affirmative Action A sociologist wishes to conduct a poll to estimate the percentage of Americans who favor affirmative action programs for women and minorities for admission to colleges and universities. What sample size should be obtained if she wishes the estimate to be within 4 percentage points with 90% confidence if

(a) she uses a 2003 estimate of 55% obtained from a Gallup Youth Survey?

(b) she does not use any prior estimates?

20. Affirmative Action A sociologist wishes to conduct a poll to estimate the percentage of Americans who judge that affirmative action programs should require businesses to hire a specific number or quota of minorities and women. What size sample should be obtained if she wishes the estimate to be within 4 percentage points with 90% confidence if

(a) she uses a 1999 estimate of 37% obtained from a Time/CNN poll?

(b) she does not use any prior estimates?

21. A Penny for Your Thoughts A researcher for the U.S. Department of the Treasury wishes to estimate the percentage of Americans who support abolishing the penny. What size sample should be obtained if he wishes the estimate to be within 2 percentage points with 98% confidence if

(a) he uses a June 2004 estimate of 23% obtained from a Harris Poll?

(b) he does not use any prior estimate?

22. Credit Card Debt A school administrator is concerned about the amount of credit card debt college students have. She wishes to conduct a poll to estimate the percentage of full-time college students who have credit card debt of $2000 or more. What size sample should be obtained if

she wishes the estimate to be within 2.5 percentage points with 94% confidence if

(a) a pilot study indicates the percentage is 34%?

(b) no prior estimates are used?

23. Death Penalty In a Harris Poll conducted in July 2000, 64% of the people polled answered yes to the following question: "Do you believe in capital punishment, that is the death penalty, or are you opposed to it?" The margin of error in the poll was $\pm 3\% = \pm 0.03$, and the estimate was made with 95% confidence. How many people were surveyed?

24. Own a Gun? In a Harris Poll conducted in May 2000, 39% of the people polled answered yes to the following question: "Do you happen to have in your home or garage any guns or revolvers?" The margin of error in the poll was $\pm 3\% = \pm 0.03$, and the estimate was made with 95% confidence. How many people were surveyed?

25. 2004 Presidential Election The Gallup Organization conducted a poll of 2014 likely voters just prior to the 2004 presidential election. The results of the survey indicated that George W. Bush would receive 49% of the popular vote and John Kerry would receive 47% of the popular vote. The margin of error was reported to be 3%. The Gallup Organization reported that the race was too close to call. Use the concept of a confidence interval to explain what this means.

26. Confidence Interval Applet: The Role of Level of Confidence Load the confidence interval for proportions applet.

(a) Construct at least 1,000 confidence intervals for the population proportion with $n = 100$, $p = 0.3$. What proportion of the 95% confidence intervals contain the population proportion? What proportion did you expect to contain the population proportion?

(b) Repeat part (a). Did the same proportion of intervals contain the population proportion?

(c) Construct at least 1,000 confidence intervals for the population proportion with $n = 100$, $p = 0.3$. What proportion of the 99% confidence intervals contain the population proportion 0.3? What proportion did you expect to contain the population proportion?

27. Confidence Interval Applet: The Role of Sample Size Load the confidence interval for proportions applet.

(a) Construct at least 1,000 confidence intervals for the population proportion with $n = 10$, $p = 0.3$. What proportion of the 95% confidence intervals contain the population proportion 0.3?

(b) Construct at least 1,000 confidence intervals for the population proportion with $n = 40$, $p = 0.3$. What proportions of the 95% confidence intervals contain the population proportion 0.3?

(c) Construct at least 1,000 confidence intervals for the population proportion with $n = 100$, $p = 0.3$. What proportion of the 95% confidence intervals contain the population proportion, 0.3?

(d) What happens to the proportion of intervals that capture the population proportion as the sample size, n, increases.

Confidence Intervals about p

TI-83/84 Plus **Step 1:** Press STAT, highlight TESTS, and select `A:1-PropZInt..`

Step 2: Enter the values of x and n.

Step 3: Enter the confidence level following `C-Level:`

Step 4: Highlight `Calculate`; press ENTER.

MINITAB **Step 1:** If you have raw data, enter the data in column C1.

Step 2: Select the **Stat** menu, highlight **Basic Statistics**, then highlight **1 Proportion . . . .**

Step 3: Enter C1 in the cell marked "Sample in Columns" if you have raw data. If you have summary statistics, click "Summarized data" and enter the number of trials, n, and the number of successes, x.

Step 4: Click the Options button. Select a Confidence Level. Click "Use test based on a normal distribution" (provided that the assumptions stated are satisfied). Click OK twice.

Excel **Step 1:** Load the PHStat Add-in.

Step 2: Select the **PHStat menu**, highlight **Confidence Intervals . . . ,** then highlight **Estimate for the proportion . . . .**

Step 3: Enter the confidence level. Enter the sample size, n, and the number of successes, x. Click OK.

9.4 Confidence Intervals about a Population Standard Deviation

Objectives

 Find critical values for the chi-square distribution

 Construct and interpret confidence intervals about the population variance and standard deviation

In this section, we discuss methods for estimating a population variance or standard deviation. Just as we discovered the sampling distribution of $\bar{x}$ and $\hat{p}$, we must find s^2, the point estimate of σ^2, and the sampling distribution of s^2. We then construct intervals about the point estimate of σ^2, using this sampling distribution.

Why might we be interested in obtaining estimates of σ^2? Many production processes not only require accuracy on average (the mean); they also require consistency. Consider a filling machine (such as a coffee machine) that consistently over- and underfills cups, but, on average, fills correctly. Certainly, customers are not happy if the machine underfills their cups, and they might be dissatisfied even if it overfills, because of spilling. A machine that consistently delivers the correct amount of liquid is desired.

As another example, consider a mutual fund that claims an average rate of return of 12% per year over the past 20 years. An investor might prefer consistent year-to-year returns near 12% to returns that fluctuated wildly yet resulted in a mean return of 12%. Both of these situations illustrate the importance of measuring variability, the topic of this section.

 Find Critical Values for the Chi-Square Distribution

We begin by exploring the sampling distribution of s^2 through a simulation. Suppose we obtain 2000 samples of size $n = 10$ from a population that is known to be normally distributed with mean 100 and standard deviation 15. We then perform the following steps:

Step 1: Compute the sample variance of each of the 2000 samples.

Step 2: Compute $\dfrac{(n-1)s^2}{\sigma^2} = \dfrac{9s^2}{15^2}$ for each sample.

Step 3: Draw a histogram of these values as shown in Figure 17.

Figure 17

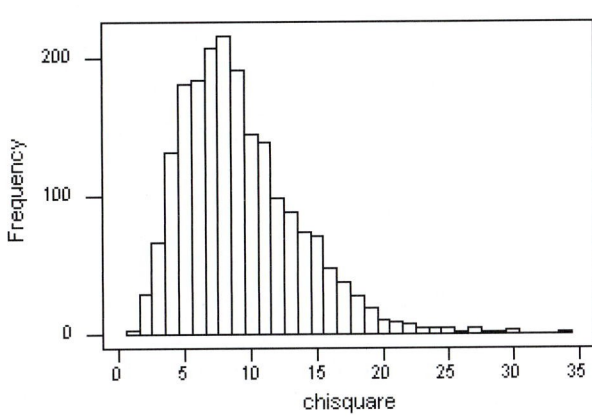

Chi-Square Distribution

The histogram represents the sampling distribution of $\dfrac{(n-1)s^2}{\sigma^2}$. We can see that the histogram does not imply that the sampling distribution of $\dfrac{(n-1)s^2}{\sigma^2}$ is normal. Rather, the histogram appears to be skewed right as indicated by the extended right tail of the histogram.

Chi-Square Distribution

If a simple random sample of size n is obtained from a normally distributed population with mean μ and standard deviation σ, then

$$\chi^2 = \frac{(n-1)s^2}{\sigma^2}$$

has a **chi-square distribution** with $n-1$ degrees of freedom.

The symbol χ^2, chi-square, is pronounced "kigh-square" (to rhyme with "sky-square"). We can find critical values of the chi-square distribution in Table VI in Appendix A of the text. Before discussing how to read Table VI, we introduce characteristics of the chi-square distribution.

Figure 18

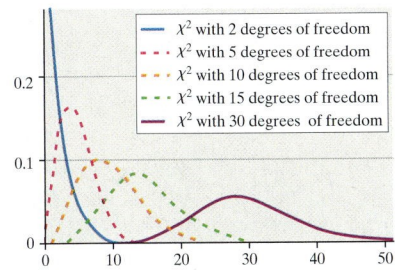

Characteristics of the Chi-Square Distribution

1. It is not symmetric.
2. The shape of the chi-square distribution depends on the degrees of freedom, just as Student's t-distribution.
3. As the number of degrees of freedom increases, the chi-square distribution becomes more nearly symmetric. See Figure 18.
4. The values of χ^2 are nonnegative; that is, values of χ^2 are always greater than or equal to 0.

Because the χ^2 distribution is not symmetric, we cannot construct a confidence interval for σ^2 by computing "point estimate ± margin of error." Instead, we must determine the left and right bounds by using different critical values.

Table VI is structured similarly to Table V for the t-distribution. The left column represents the degrees of freedom, and the top row represents the area under the chi-square distribution to the right of the critical value. We use the notation χ^2_α to denote the critical χ^2-value such that the area under the chi-square distribution to the right of χ^2_α is α.

EXAMPLE 1   Finding Critical Values for the Chi-Square Distribution

Problem: Find the critical values that separate the middle 90% of the chi-square distribution from the 5% area in each tail, assuming 15 degrees of freedom.

Approach: We perform the following steps to obtain the critical values.

Step 1: Draw a chi-square distribution with the critical values and areas labeled.
Step 2: Use Table VI to find the critical values.

Solution

Step 1: Figure 19 shows the chi-square distribution with 15 degrees of freedom and the unknown critical values labeled. The area to the right of the right critical value is 0.05. We denote this critical value $\chi^2_{0.05}$. The area to the right of the left critical value is $1 - 0.05 = 0.95$. We denote this critical value $\chi^2_{0.95}$.

Figure 19

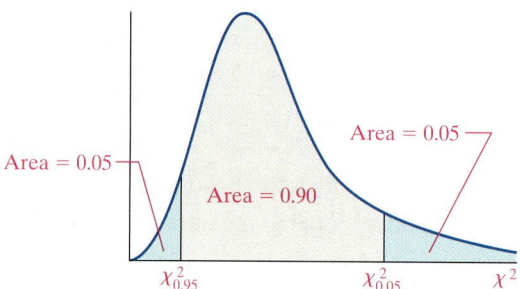

Step 2: Figure 20 shows a partial representation of Table VI. The row containing 15 degrees of freedom is boxed. The columns corresponding to an area to the right of 0.95 and 0.05 are also boxed. The critical values are $\chi^2_{0.95} = 7.261$ and $\chi^2_{0.05} = 24.996$.

Figure 20

Degrees of Freedom	Area to the Right of the Critical Value									
	0.995	0.99	0.975	0.95	0.90	0.10	0.05	0.025	0.01	0.005
1	—	—	0.001	0.004	0.016	2.706	3.841	5.024	6.635	7.879
2	0.010	0.020	0.051	0.103	0.211	4.605	5.991	7.378	9.210	10.597
3	0.072	0.115	0.216	0.352	0.584	6.251	7.815	9.348	11.345	12.838
12	3.074	3.571	4.404	5.226	6.304	18.549	21.026	23.337	26.217	28.299
13	3.565	4.107	5.009	5.892	7.042	19.812	22.362	24.736	27.688	29.819
14	4.075	4.660	5.629	6.571	7.790	21.064	23.685	26.119	29.141	31.319
15	4.601	5.229	6.262	7.261	8.547	22.307	24.996	27.488	30.578	32.801
16	5.142	5.812	6.908	7.962	9.312	23.542	26.296	28.845	32.000	34.267
17	5.697	6.408	7.564	8.672	10.085	24.769	27.587	30.191	33.409	35.718
18	6.365	7.215	8.231	9.288	10.265	25.200	28.601	31.595	24.265	36.456

In studying Table VI, we notice that the degrees of freedom are numbered 1 to 30 inclusive, then 40, 50, 60, ..., 100. If the number of degrees of freedom is not found in the table, we follow the practice of choosing the degrees of freedom closest to that desired. If the degrees of freedom are exactly between two values, find the mean of the values. For example, to find the critical value corresponding to 75 degrees of freedom, compute the mean of the critical values corresponding to 70 and 80 degrees of freedom.

Now Work Problem 5.

② Construct and Interpret Confidence Intervals about the Population Variance and Standard Deviation

The sample variance, s^2, is the best point estimate of the population variance, σ^2. We use the sample standard deviation as the point estimate of σ.*

We now must develop a method for constructing a confidence interval for the population variance. Suppose that we take a simple random sample of size n from a population that is normally distributed with mean μ and standard deviation σ; then $\chi^2 = \dfrac{(n-1)s^2}{\sigma^2}$ follows a chi-square distribution with $n-1$ degrees of freedom. Therefore, $(1-\alpha) \cdot 100\%$ of the values of χ^2 will lie between $\chi^2_{1-\alpha/2}$ and $\chi^2_{\alpha/2}$. Figure 21 illustrates the situation.

Figure 21

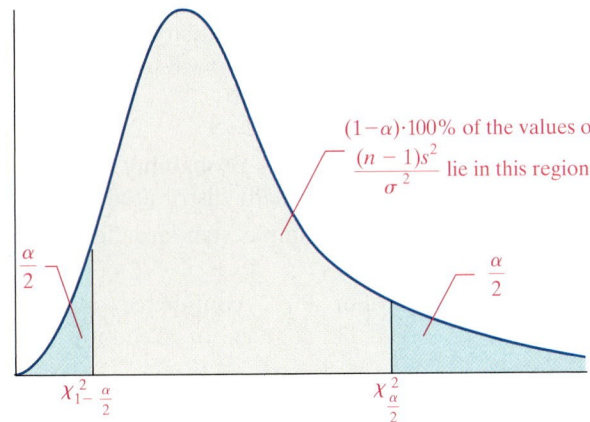

$(1-\alpha)\cdot 100\%$ of the values of $\dfrac{(n-1)s^2}{\sigma^2}$ lie in this region

$\dfrac{\alpha}{2}$

$\dfrac{\alpha}{2}$

$\chi^2_{1-\frac{\alpha}{2}}$

$\chi^2_{\frac{\alpha}{2}}$

We say that $(1-\alpha) \cdot 100\%$ of the values of $\dfrac{(n-1)s^2}{\sigma^2}$ lie within the interval defined by the inequality

$$\chi^2_{1-\alpha/2} < \frac{(n-1)s^2}{\sigma^2} < \chi^2_{\alpha/2}$$

If we rewrite this inequality with σ^2 in the center, we have the formula for a $(1-\alpha) \cdot 100\%$ confidence interval about σ^2:

$$\frac{(n-1)s^2}{\chi^2_{\alpha/2}} < \sigma^2 < \frac{(n-1)s^2}{\chi^2_{1-\alpha/2}}$$

! CAUTION
A confidence interval about the population variance or standard deviation is not of the form "point estimate ± margin of error" because the sampling distribution of the sample variance is not symmetric.

*Although the sample standard deviation is a biased estimator of σ, it is common practice to use s as the estimator of σ.

> ### A $(1 - \alpha) \cdot 100\%$ Confidence Interval about σ^2
>
> If a simple random sample of size n is taken from a normal population with mean μ and standard deviation σ, then a $(1 - \alpha) \cdot 100\%$ confidence interval about σ^2 is given by
>
> $$\text{Lower bound:} \quad \frac{(n-1)s^2}{\chi^2_{\alpha/2}} \qquad \text{Upper bound:} \quad \frac{(n-1)s^2}{\chi^2_{1-\alpha/2}} \qquad \textbf{(1)}$$

EXAMPLE 2

Constructing a Confidence Interval about a Population Variance and Standard Deviation

Table 7

$41,844	$41,500	$39,995
$36,995	$40,990	$37,995
$41,995	$38,900	$42,995
$36,995	$43,995	$35,950

Source: cars.com

Problem: Table 7 shows the sale price of 12 randomly selected 3-year-old Chevy Corvettes. Construct a 90% confidence interval about the population variance and standard deviation price of a 3-year-old Chevy Corvette.

Approach: We perform the following steps to construct the confidence interval:

Step 1: Check to see whether it is reasonable to conclude that the data come from a normally distributed population.

Step 2: Compute the sample variance.

Step 3: Determine the critical values.

Step 4: Use Formula (1) to construct the confidence interval for the population variance.

Step 5: Compute the square root of the lower bound and upper bound to obtain the confidence interval for the population standard deviation.

Solution

Step 1: A normal probability plot and boxplot indicate that the price of Corvettes is normally distributed with no outliers.

Step 2: The sample standard deviation is $2615.19. The sample variance is $(\$2615.19)^2$.

Step 3: For 90% confidence, we have $\alpha = 0.10$. Using Table VI with $12 - 1 = 11$ degrees of freedom, we find the left critical value to be $\chi^2_{1-\alpha/2} = \chi^2_{1-0.10/2} = \chi^2_{0.95} = 4.575$. We find the right critical value to be $\chi^2_{\alpha/2} = \chi^2_{0.10/2} = \chi^2_{0.05} = 19.675$. See Figure 22.

Figure 22

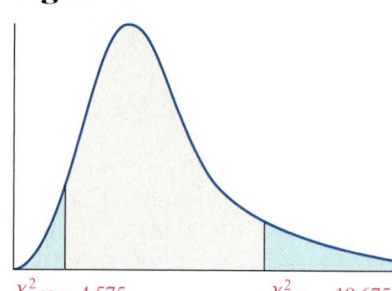

$\chi^2_{0.95} = 4.575$ $\chi^2_{0.05} = 19.675$

Step 4: Substitute these values into Formula (1), yielding

$$\text{Lower bound:} \quad \frac{(n-1)s^2}{\chi^2_{\alpha/2}} = \frac{11(2615.19)^2}{19.675} = 3,823,705.52$$

$$\text{Upper bound:} \quad \frac{(n-1)s^2}{\chi^2_{1-\alpha/2}} = \frac{11(2615.19)^2}{4.575} = 16,444,023.19$$

Step 5: Taking the square root of each part,* we obtain

Lower bound: $1955   Upper bound: $4055

We are 90% confident that the population standard deviation of the price of a 3-year-old Chevy Corvette is between $1955 and $4055.

USING TECHNOLOGY

Of the three statistical packages we have been discussing in this text, only MINITAB constructs confidence intervals about the population standard deviation. See the Technology Step by Step on page 492.

---

*Be sure to take the square root of the *unrounded* values.

The formula presented for constructing the confidence interval about a population variance requires that the data come from a normal distribution. Confidence intervals for the population mean and population proportion are robust (not sensitive to moderate departures from normality), but the intervals computed for population variance are not robust. Therefore, it is vital that the requirement of normality be verified before proceeding.

> Now Work Problem 11.

9.4 ASSESS YOUR UNDERSTANDING

Concepts and Vocabulary

1. State the characteristics of the chi-square distribution.
2. How must the population be distributed to construct a confidence interval about a population standard deviation?

3. Explain how to obtain a confidence interval for the population standard deviation.
4. *True or False:* Confidence intervals are always of the form "*point estimate ± margin of error*".

Skill Building

In Problems 5–8, find the critical values $\chi^2_{1-\alpha/2}$ and $\chi^2_{\alpha/2}$ for the given level of confidence and sample size.

5. 90% confidence; $n = 20$
NW
6. 95% confidence, $n = 25$
7. 98% confidence, $n = 23$
8. 99% confidence, $n = 14$

9. A simple random sample of size n is drawn from a population that is known to be normally distributed. The sample variance, s^2, is determined to be 12.6.
   (a) Construct a 90% confidence interval for σ^2 if the sample size, n, is 20.
   (b) Construct a 90% confidence interval for σ^2 if the sample size, n, is 30. How does increasing the sample size affect the width of the interval?

(c) Construct a 98% confidence interval for σ^2 if the sample size, n, is 20. Compare the results with those obtained in part (a). How does increasing the level of confidence affect the confidence interval?

10. A simple random sample of size n is drawn from a population that is known to be normally distributed. The sample variance, s^2, is determined to be 19.8.
   (a) Construct a 95% confidence interval for σ^2 if the sample size, n, is 10.
   (b) Construct a 95% confidence interval for σ^2 if the sample size, n, is 25. How does increasing the sample size affect the width of the interval?
   (c) Construct a 99% confidence interval for σ^2 if the sample size, n, is 10. Compare the results with those obtained in part (a). How does increasing the level of confidence affect the confidence interval?

Applying the Concepts

11. **Crawling Babies** The following data represent the age (in weeks) at which babies first crawl based on a survey of 12 mothers conducted by Essential Baby.

NW

52	30	44	35
47	37	56	26
39	26	39	28

Source: www.essentialbaby.com

In Problem 23 from Section 9.2, it was verified that the data are normally distributed and that $s = 10.00$ weeks. Construct a 95% confidence interval for the population standard deviation of the age (in weeks) at which babies first crawl. Interpret the interval.

12. **Battery Life** The following data represent the battery life, in hours, for a random sample of 10 full charges on a fifth-generation iPod music player.

7.3	10.2	12.9	10.8	12.1
6.6	10.2	9.0	8.5	7.1

In Problem 24 from Section 9.2, it was verified that the data are normally distributed and that $s = 2.14$ hours. Construct a 95% confidence interval for the population standard deviation of the battery life (in hours). Interpret the interval.

13. The Growing Season An agricultural researcher is interested in estimating the variation in the length of the growing season in the Chicago area. Treating the last 10 years as a simple random sample, he obtains the following data, which represent the number of days of the growing season.

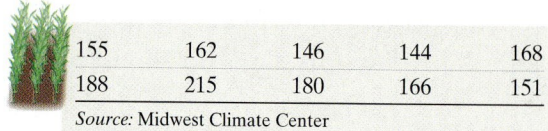

155	162	146	144	168
188	215	180	166	151

Source: Midwest Climate Center

In Problem 25 from Section 9.2, it was verified that the data are normally distributed and that $s = 21.88$ days. Construct a 99% confidence interval for the population standard deviation of the number of days in the growing season. Interpret the interval.

14. Disc Weight When playing disc golf, the weight of the disc has a huge impact on disc performance. Lighter discs tend to fly farther, but are less stable. Heavier discs tend to give better control and are better suited for strong winds. The following data represent the weights, in grams, of a random sample of 12 distance drivers produced by a certain manufacturer.

172	168	170	173	172	173
171	175	173	171	196	173

In Problem 26 from Section 9.2, it was verified that the data are normally distributed and that $s = 1.97$ grams. Construct a 99% confidence interval for the population standard deviation of the disc weight (in grams). Interpret the interval.

15. Peanuts A jar of peanuts is supposed to have 16 ounces of peanuts. The filling machine inevitably experiences fluctuations in filling, so a quality-control manager randomly samples 12 jars of peanuts from the storage facility and measures their contents. She obtains the following data:

15.94	15.74	16.21	15.36	15.84	15.84
15.52	16.16	15.78	15.51	16.28	16.53

(a) Verify that the data are normally distributed by constructing a normal probability plot.

(b) Determine the sample standard deviation.
(c) Construct a 90% confidence interval for the population standard deviation of the number of ounces of peanuts.
(d) Suppose the quality control manager wants the machine to have a population standard deviation below 0.20 ounce. Does the confidence interval validate this desire?

16. Investment Risk Investors not only desire a high return on their money, but they would also like the rate of return to be stable from month to month. An investment manager invests with the goal of reducing volatility (month-to-month fluctuations in the rate of return). The following data represent the rate of return (in percent) for his mutual fund for the past 12 months.

13.8	15.9	10.0	12.4	11.3	6.6
9.6	12.4	10.3	8.7	14.9	6.7

(a) Verify that the data are normally distributed by constructing a normal probability plot.
(b) Determine the sample standard deviation.
(c) Construct a 95% confidence interval for the population standard deviation of the rate of return.
(d) Suppose the investment manager wants to have a population standard deviation for the rate of return below 6%. Does the confidence interval validate this desire?

17. Critical Values Sir R. A. Fisher, a famous statistician, showed that the critical values of a chi-square distribution can be approximated by the standard normal distribution

$$\chi_k^2 = \frac{(z_k + \sqrt{2v - 1})^2}{2}$$

where v is the degrees of freedom and z_k is the Z-score such that the area under the standard normal curve to the right of z_k is k. Use Fisher's approximation to find $\chi_{0.975}^2$ and $\chi_{0.025}^2$ with 100 degrees of freedom. Compare the results with those found in Table VI.

Technology Step-by-Step | **Confidence Intervals about σ**

TI-83/84 Plus The TI-83/84 Plus do not construct confidence intervals about σ.

MINITAB *Step 1:* Enter raw data in column C1.
Step 2: Select the **Stat** menu, highlight **Basic Statistics**, then highlight **Graphical Summary** . . . .
Step 3: Enter C1 in the cell marked "Variables."
Step 4: Enter the Confidence level desired. Click OK. The confidence interval for sigma is reported in the output.

Excel Excel does not construct confidence intervals about σ.

9.5 Putting It All Together: Which Procedure Do I Use?

Objective   ❶ Determine the appropriate confidence interval to construct

❶ **Determine the Appropriate Confidence Interval to Construct**

Perhaps the most difficult aspect of constructing a confidence interval is determining which type of confidence interval to construct. To assist in your decision making, we present Figure 23.

Figure 23

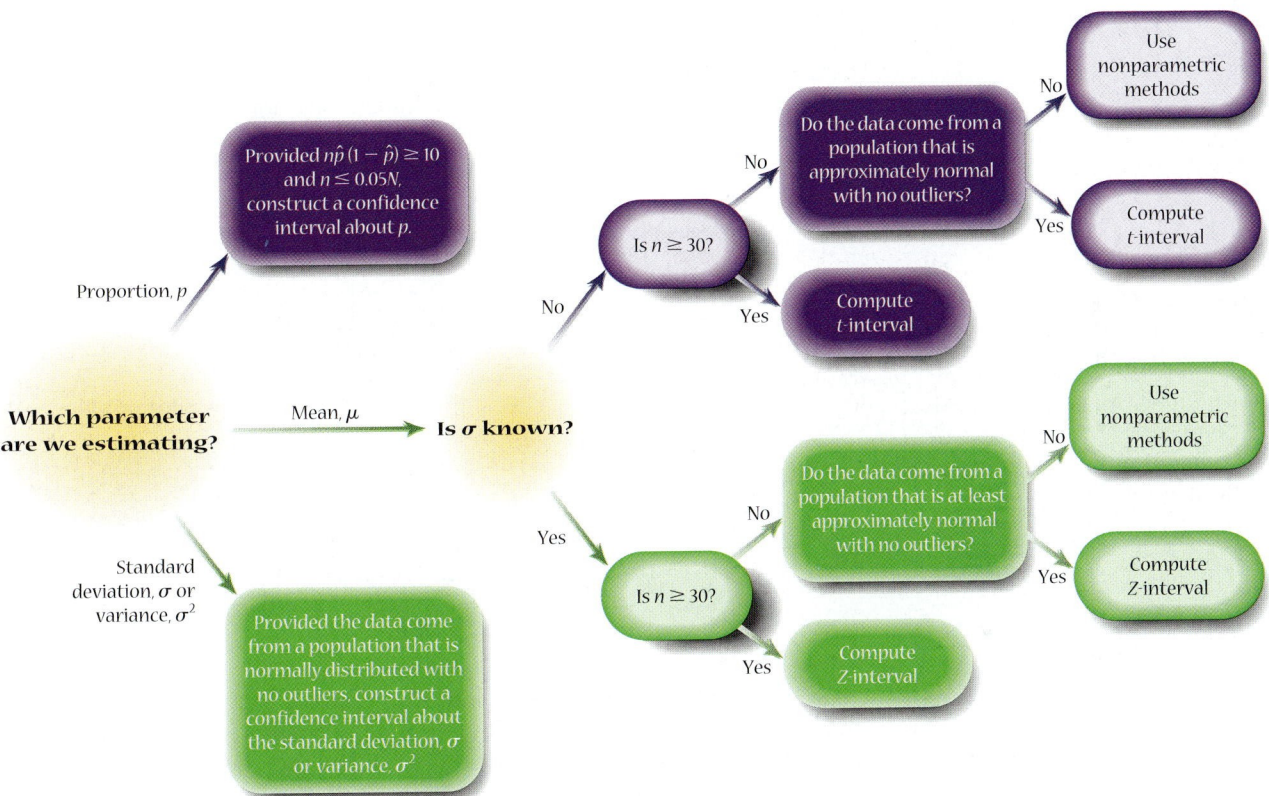

From the flow chart, the first step is to determine which parameter we wish to estimate. If we are estimating a proportion, standard deviation, or variance we verify the requirements to construct the interval and proceed. However, if we are estimating a mean than we have another decision to make: we must determine whether the population standard deviation is known, or not. If the population standard deviation is not known, we construct a *t*-interval (provided the sample size is large or the sample data come from a population that is normally distributed with no outliers). If the population standard deviation is known, we construct a *Z*-interval (again provided the sample size is large or the sample data come from a population that is normally distributed with no outliers).

EXAMPLE 1   **Constructing a Confidence Interval: Which Method Do I Use?**

Problem: Robert wishes to estimate the mean number of miles that his Buick Rendezvous can be driven on a full tank of gas. He fills up his car with regular unleaded gasoline from the same gas station 10 times and records the number of miles that he drives until his low-tank indicator light comes on. He obtains the data shown in Table 8. Construct a 95% confidence interval for the mean number of miles driven on a full tank of gas.

Table 8				
323.9	326.8	370.6	398.8	417.5
450.7	368.8	423.8	382.7	343.1

Approach: We will follow the flow chart given in Figure 23.

Solution: The problem is asking us to construct a confidence interval for the *mean* number of miles driven. So we wish to estimate the population mean, μ. No information is given about the value of the population standard deviation, so σ is unknown. Because the sample size is small, we need to verify that the data come from a population that is normally distributed with no outliers. Figure 24 shows a normal probability plot and boxplot.

Figure 24

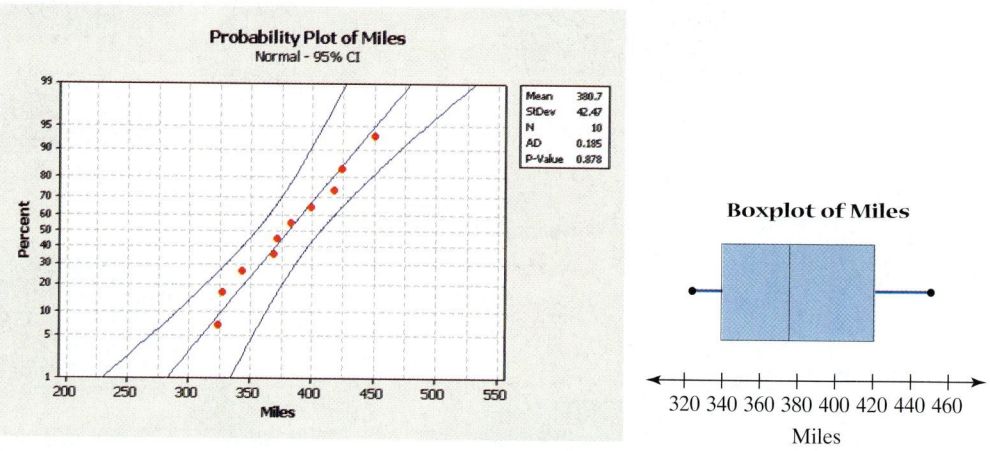

The normal probability plot indicates that the sample data come from a population that is normally distributed, and the boxplot indicates that there are no outliers, so we construct a *t*-interval. Based on the sample data from Table 8, we have that $n = 10$, $\bar{x} = 380.67$, and $s = 42.47$. For a 95% confidence interval with $n - 1 = 10 - 1 = 9$ degrees of freedom, we have that $t_{\frac{\alpha}{2}} = t_{\frac{0.05}{2}} = t_{0.025} = 2.262$.

$$\text{Lower bound:} \quad \bar{x} - t_{\frac{\alpha}{2}} \cdot \frac{s}{\sqrt{n}} = 380.67 - 2.262 \cdot \frac{42.47}{\sqrt{10}} = 350.29$$

$$\text{Upper bound:} \quad \bar{x} - t_{\frac{\alpha}{2}} \cdot \frac{s}{\sqrt{n}} = 380.67 + 2.262 \cdot \frac{42.47}{\sqrt{10}} = 411.05$$

Robert is 95% confident that the mean of the miles he can drive on a full tank of gas is between 350.29 miles and 411.05 miles.

9.5 ASSESS YOUR UNDERSTANDING

Concepts and Vocabulary

1. State the circumstances under which we construct a *t*-interval. What are the circumstances under which we construct a *Z*-interval?

2. What are the requirements that must be satisfied before we can construct a confidence interval about a population proportion?

Skill Building

In Problems 3–12, construct the appropriate confidence interval.

3. A simple random sample of size $n = 14$ is drawn from a population that is normally distributed with $\sigma = 20$. The sample mean is found to be $\bar{x} = 60$. Construct a 95% confidence interval about the population mean.

4. A simple random sample of size $n = 22$ is drawn from a population that is normally distributed with $\sigma = 37$. The sample mean is found to be $\bar{x} = 122.5$. Construct a 90% confidence interval about the population mean.

5. A simple random sample of size $n = 300$ individuals who are currently employed is asked if they work at home at least once per week. Of the 300 employed individuals surveyed, 35 responded that they did work at home at least once per week. Construct a 99% confidence interval about the population proportion of employed individuals who work at home at least once per week.

6. A simple random sample of size $n = 785$ adults was asked if they follow college football. Of the 785 surveyed, 275 responded that they did follow college football. Construct a 95% confidence interval about the population proportion of adults who follow college football.

7. A simple random sample of size $n = 12$ is drawn from a population that is normally distributed. The sample mean is found to be $\bar{x} = 45$ and the sample standard deviation is found to be $s = 14$. Construct a 90% confidence interval about the population mean.

8. A simple random sample of size $n = 17$ is drawn from a population that is normally distributed. The sample mean is found to be $\bar{x} = 3.25$, and the sample standard deviation is found to be $s = 1.17$. Construct a 95% confidence interval about the population mean.

9. A simple random sample of size $n = 40$ is drawn from a population. The sample mean is found to be $\bar{x} = 120.5$, and the sample standard deviation is found to be $s = 12.9$. Construct a 99% confidence interval about the population mean.

10. A simple random sample of size $n = 210$ is drawn from a population. The sample mean is found to be $\bar{x} = 20.1$, and the sample standard deviation is found to be $s = 3.2$. Construct a 90% confidence interval about the population mean.

11. A simple random sample of size $n = 12$ is drawn from a population that is normally distributed. The sample variance is found to be $s^2 = 23.7$. Construct a 90% confidence interval about the population variance.

12. A simple random sample of size $n = 25$ is drawn from a population that is normally distributed. The sample variance is found to be $s^2 = 3.97$. Construct a 95% confidence interval about the population standard deviation.

Applying the Concepts

13. **Aggravated Assault** In a random sample of 40 felons convicted of aggravated assault, it was determined that the mean length of sentencing was 54 months, with a standard deviation of 8 months. Construct and interpret a 95% confidence interval for the mean length of sentence for an aggravated assault conviction. *Source:* Based on data obtained from the U.S. Department of Justice.

14. **Click It** In a February 2005 Harris Poll, 769 of 1010 randomly selected adults said that they always wear their seatbelt. Construct and interpret a 98% confidence interval for the proportion of adults who always wear their seatbelt.

15. **Estate Tax Returns** In a random sample of 100 estate tax returns that was audited by the Internal Revenue Service, it was determined that the mean amount of additional tax owed was $3137. Assuming the population standard deviation of the additional amount owed is $2694, construct and interpret a 90% confidence interval for the mean additional amount of tax owed for estate tax returns.

16. **Muzzle Velocity** Fifty rounds of a new type of ammunition were fired from a test weapon, and the muzzle velocity of the projectile was measured. The sample had a mean muzzle velocity of $\bar{x} = 863$ meters per second, with $s = 2.7$ meters per second. Construct and interpret a 99% confidence interval for the mean muzzle velocity.

17. **Worried about Retirement?** In a survey of 1010 adult Americans, the Gallup Organization asked, "Are you worried or not worried about having enough money for retirement?" Of the 1010 surveyed, 606 stated that they were worried about having enough money for retirement. Construct a 90% confidence interval for the proportion of adult Americans who are worried about having enough money for retirement.

18. **Theme Park Spending** In a random sample of 40 visitors to a certain theme park, it was determined that the mean amount of money spent per person at the park (including ticket price) was $93.43 per day. Assuming the population standard deviation of the amount spent per person is $15, construct and interpret a 95% confidence interval for the mean amount spent daily per person at the theme park.

In Problems 19–24, construct a 95% Z-interval or a 95% t-interval about the population mean. If neither can be constructed, state the reason why. For convenience, a normal probability plot and boxplot are given.

19. **Height of Males** The heights of 20- to 29-year-old males are known to have population standard deviation $\sigma = 2.9$ inches. A simple random sample of $n = 15$ males 20 to 29 years old results in the following data:

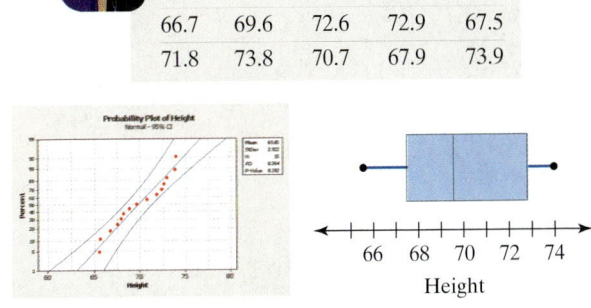

65.5	72.3	68.2	65.6	68.8
66.7	69.6	72.6	72.9	67.5
71.8	73.8	70.7	67.9	73.9

20. Gestation Period The gestation period of humans has population standard deviation $\sigma = 16$ days. A simple random sample of $n = 12$ live births results in the following data:

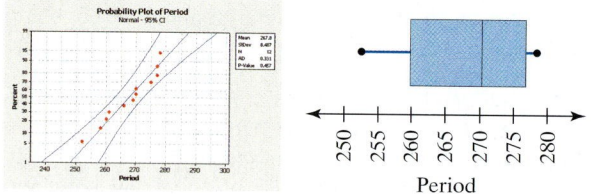

| 266 | 270 | 277 | 278 | 258 | 275 |
| 261 | 260 | 270 | 269 | 252 | 277 |

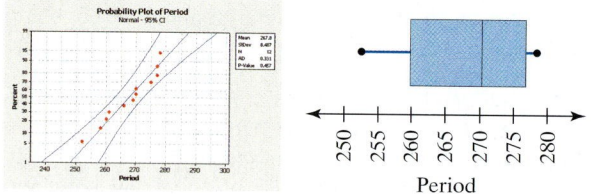

Period

21. Officer Friendly A police officer hides behind a billboard to catch speeders. The following data represent the number of minutes he waits before first observing a car that is exceeding the speed limit by more than 10 miles per hour on 10 randomly selected days:

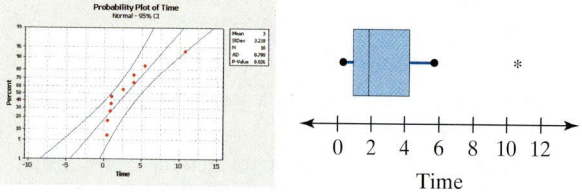

| 1.0 | 5.4 | 0.8 | 10.7 | 0.5 |
| 0.9 | 3.9 | 0.4 | 2.5 | 3.9 |

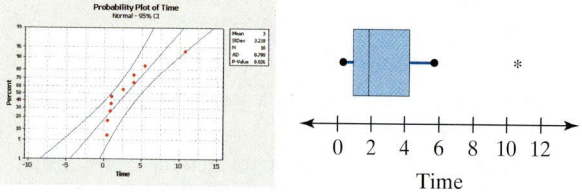

Time

22. M&Ms A quality-control engineer wanted to estimate the mean weight (in grams) of a random sample of 12 plain M&M candies.

0.87	0.88	0.82	0.90
0.84	0.84	0.91	0.94
0.86	0.86	0.88	0.87

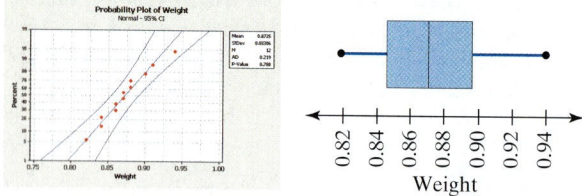

Weight

23. Pulse Fifteen randomly selected women were asked to work on a StairMaster for 3 minutes. After the 3 minutes, their pulses were measured and the following data were obtained:

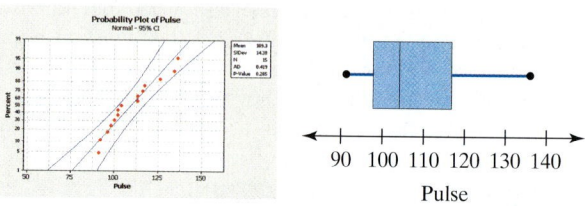

117	102	98	100	116
113	91	92	96	136
134	126	104	113	102

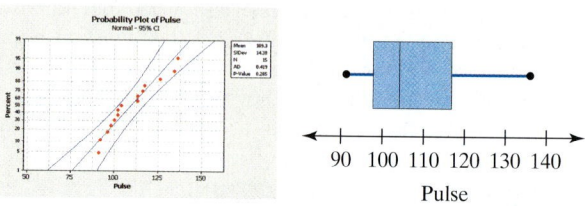

Pulse

24. Law Grads' Pay A random sample of recent graduates of law school was obtained in which the graduates were asked to report their starting salary. The data, based on results reported by the National Association for Law Placement, are as follows:

75,000	49,000	79,000	81,000	38,000
36,500	39,000	41,500	131,000	45,500
92,000	62,500	68,000	37,500	39,500

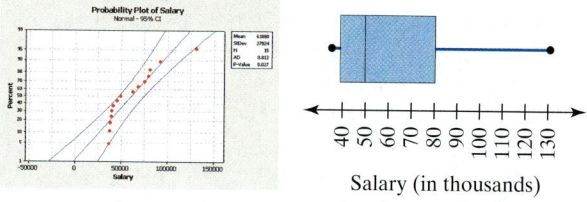

Salary (in thousands)

Review

Summary

In this chapter, we discussed estimation methods. We estimated the values of the parameters μ, p, and σ. We started by estimating the population mean under the assumption that the population standard deviation was known. This assumption allowed us to construct a confidence interval about μ by utilizing the standard normal distribution. To construct the interval, we required either that the population from which the sample was drawn be normal or that the sample size, n, be greater than or equal to 30. Also, the sampling method had to be simple random sampling. With these requirements satisfied, the $(1 - \alpha) \cdot 100\%$ confidence interval about μ is $\bar{x} \pm z_{\alpha/2} \cdot \dfrac{\sigma}{\sqrt{n}}$. We have $(1 - \alpha) \cdot 100\%$ confidence that the unknown value of μ lies within the interval.

In Section 9.2, we dropped the assumption that the population standard deviation be known. With σ unknown, the sampling distribution of $t = \dfrac{\bar{x} - \mu}{s/\sqrt{n}}$ follows Student's t-distribution with $n - 1$ degrees of freedom. We use the t-distribution to construct the confidence interval about μ. To construct this interval, either the population from which the sample was drawn must be normal or the sample size must be large. Also, the sampling method must be simple random sampling. With these requirements satisfied, the $(1 - \alpha) \cdot 100\%$ confidence interval about μ is $\bar{x} \pm t_{\alpha/2} \cdot \dfrac{s}{\sqrt{n}}$, where $t_{\alpha/2}$ has $n - 1$ degrees of freedom. This means that the procedure results in an interval that contains μ, the population mean, $(1 - \alpha) \cdot 100\%$ of the time.

In Section 9.3, a confidence interval regarding the population proportion, p, was constructed. This confidence interval is constructed about the binomial parameter p. Provided that the sample is obtained by simple random sampling, $n\hat{p}(1 - \hat{p}) \geq 10$, and the sample is less than 5% of the population size the $(1 - \alpha) \cdot 100\%$ confidence interval about p is $\hat{p} \pm z_{\alpha/2} \cdot \sqrt{\dfrac{\hat{p}(1 - \hat{p})}{n}}$. We have $(1 - \alpha) \cdot 100\%$ confidence that the unknown value of p lies within the interval.

Finally, we introduced a method for estimating the population standard deviation. To perform this estimation, the population from which the sample is drawn must be normal, and the sampling method must be simple random sampling. If these requirements are satisfied, then $\chi^2 = \dfrac{(n - 1)s^2}{\sigma^2}$ follows the chi-square distribution with $n - 1$ degrees of freedom. The $(1 - \alpha) \cdot 100\%$ confidence interval about σ^2 is $\dfrac{(n - 1)s^2}{\chi^2_{\alpha/2}} < \sigma^2 < \dfrac{(n - 1)s^2}{\chi^2_{1-\alpha/2}}$. To construct the $(1 - \alpha) \cdot 100\%$ confidence interval about σ, we take the square root of each part of the inequality and obtain $\sqrt{\dfrac{(n - 1)s^2}{\chi^2_{\alpha/2}}} < \sigma < \sqrt{\dfrac{(n - 1)s^2}{\chi^2_{1-\alpha/2}}}$. We have $(1 - \alpha) \cdot 100\%$ confidence that the unknown value of σ lies within the inteval.

Formulas

Confidence Intervals

- A $(1 - \alpha) \cdot 100\%$ confidence interval about μ with σ known is $\bar{x} \pm z_{\alpha/2} \cdot \dfrac{\sigma}{\sqrt{n}}$, provided that the population from which the sample was drawn is normal or that the sample size is large ($n \geq 30$).

- A $(1 - \alpha) \cdot 100\%$ confidence interval about μ with σ unknown is $\bar{x} \pm t_{\alpha/2} \cdot \dfrac{s}{\sqrt{n}}$, where $t_{\alpha/2}$ has $n - 1$ degrees of freedom, provided that the population from which the sample was drawn is normal or that the sample size is large ($n \geq 30$).

- A $(1 - \alpha) \cdot 100\%$ confidence interval about p is $\hat{p} \pm z_{\alpha/2} \cdot \sqrt{\dfrac{\hat{p}(1 - \hat{p})}{n}}$, provided that $n\hat{p}(1 - \hat{p}) \geq 10$ and $n \leq 0.05\,N$.

- A $(1 - \alpha) \cdot 100\%$ confidence interval about σ^2 is $\dfrac{(n - 1)s^2}{\chi^2_{\alpha/2}} < \sigma^2 < \dfrac{(n - 1)s^2}{\chi^2_{1-\alpha/2}}$, where χ^2 has $n - 1$ degrees of freedom, provided that the population from which the sample was drawn is normal.

Sample Size

- To estimate the population mean within a margin of error E at a $(1 - \alpha) \cdot 100\%$ level of confidence requires a sample of size $n = \left(\dfrac{z_{\alpha/2} \cdot \sigma}{E}\right)^2$ (rounded up to the next integer).

- To estimate the population proportion within a margin of error E at a $(1 - \alpha) \cdot 100\%$ level of confidence requires a sample of size $n = \hat{p}(1 - \hat{p})\left(\dfrac{z_{\alpha/2}}{E}\right)^2$ (rounded up to the next integer), where $\hat{p}$ is a prior estimate of the population proportion.

- To estimate the population proportion within a margin of error E at a $(1 - \alpha) \cdot 100\%$ level of confidence requires a sample of size $n = 0.25\left(\dfrac{z_{\alpha/2}}{E}\right)^2$ (rounded up to the next integer) when no prior estimate is available.

Vocabulary

Point estimate (p. 447)
Confidence interval (p. 449)
Level of confidence (p. 449)
Critical value (p. 452)

Robust (p. 453)
Z-interval (p. 453)
Margin of error (p. 455)
Student's t-distribution (p. 466)

t-interval (p. 470)
Chi-square distribution (p. 487)

Objectives

Section	You should be able to . . .	Examples	Review Exercises
9.1	**1** Compute a point estimate of the population mean (p. 447)	1	10(a), 11(a), 12(b), 15(a), 16(a)
	2 Construct and interpret a confidence interval about a population mean assuming the population standard deviation is known (p. 448)	2–4	5, 7(b), 7(c), 8(b), 8(c), 11(b), 11(c), 12(c), 12(d)
	3 Understand the role of margin of error in constructing a confidence interval (p. 455)	5 and 6	
	4 Determine the sample size necessary for estimating the population mean within a specified margin of error (p. 456)	7	7(d), 8(d)
9.2	**1** Know the properties of Student's t-distribution (p. 466)	1	23–25
	2 Determine t-values (p. 468)	2	1, 2
	3 Construct and interpret a confidence interval about a population mean (p. 469)	3–5	6, 9(b), 10(b), 13(b), 13(c), 14(b), 14(c), 15(c), 15(d), 16(c), 16(d), 17, 18
9.3	**1** Obtain a point estimate for the population proportion (p. 478)	1	19(a), 20(a)
	2 Construct and interpret a confidence interval for the population proportion (p. 479)	2 and 3	19(b), 20(b)
	3 Determine the sample size necessary for estimating a population proportion within a specified margin of error (p. 482)	4	19(c), 19(d), 20(c), 20(d)
9.4	**1** Find critical values for the chi-square distribution (p. 486)	1	3, 4
	2 Construct and interpret confidence intervals about the population variance and standard deviation (p. 489)	2	6(d), 15(e), 16(e), 21, 22
9.5	**1** Determine the appropriate confidence interval to construct (p. 493)	1	5–22

Review Exercises

In Problems 1 and 2, find the critical t-value for constructing a confidence interval about a population mean at the given level of confidence for the given sample size, n.

1. 99% confidence; $n = 18$

2. 90% confidence; $n = 27$

In Problems 3 and 4, find the critical values $\chi^2_{1-\alpha/2}$ and $\chi^2_{\alpha/2}$ required to construct a confidence interval about σ for the given level of confidence and sample size.

3. 95% confidence; $n = 22$

4. 99% confidence; $n = 12$

5. A simple random sample of size n is drawn from a population that is known to be normally distributed. The sample mean, $\bar{x}$, is determined to be 54.8.
   (a) Construct the 90% confidence interval about the population mean if the population standard deviation, σ, is known to be 10.5 and the sample size, n, is 20.
   (b) Construct the 90% confidence interval about the population mean if the population standard deviation, σ,

is known to be 10.5 and the sample size, n, is 30. How does increasing the sample size affect the width of the interval?
   (c) Construct the 99% confidence interval about the population mean if the population standard deviation, σ, is known to be 10.5 and the sample size, n, is 20. Compare the results to those obtained in part (a). How does increasing the level of confidence affect the confidence interval?

6. A simple random sample of size n is drawn from a population that is known to be normally distributed. The sample mean, $\bar{x}$, is determined to be 104.3 and the sample standard deviation, s, is determined to be 15.9.
   (a) Construct the 90% confidence interval about the population mean if the sample size, n, is 15.
   (b) Construct the 90% confidence interval about the population mean if the sample size, n, is 25. How does increasing the sample size affect the width of the interval?

(c) Construct the 95% confidence interval about the population mean if the sample size, n, is 15. Compare the results to those obtained in part (a). How does increasing the level of confidence affect the confidence interval?

(d) Construct the 90% confidence interval about the population standard deviation if the sample size, n, is 15.

7. Tire Wear Suppose Michelin wishes to estimate the mean mileage for its HydroEdge tire. In a random sample of 40 tires, the sample mean mileage was $\bar{x} = 100{,}294$.

(a) Why can we say that the sampling distribution of $\bar{x}$ is approximately normal?

(b) Construct a 90% confidence interval for the mean mileage for all HydroEdge tires, assuming that $\sigma = 4600$ miles. Interpret this interval.

(c) Construct a 95% confidence interval for the mean mileage for all HydroEdge tires, assuming that $\sigma = 4600$ miles. Interpret this interval.

(d) How many tires would Michelin require to estimate the mean mileage for all HydroEdge tires within 1500 miles with 99% confidence?

8. Talk Time on a Cell Phone Suppose Motorola wishes to estimate the mean talk time for its V505 camera phone before the battery must be recharged. In a random sample of 35 phones, the sample mean talk time was 325 minutes.

(a) Why can we say that the sampling distribution of $\bar{x}$ is approximately normal?

(b) Construct a 94% confidence interval for the mean talk time for all Motorola V505 camera phones, assuming that $\sigma = 31$ minutes. Interpret this interval.

(c) Construct a 98% confidence interval for the mean talk time for all Motorola V505 camera phones, assuming that $\sigma = 31$ minutes. Interpret this interval.

(d) How many phones would Motorola need to test to estimate the mean talk time for all V505 camera phones within 5 minutes with 95% confidence?

9. Watching TV The Gallup Organization surveyed 1028 teenagers between the ages of 13 and 17 years and asked, "Typically, how many hours per week do you spend watching TV?" The sample mean number of hours was 13, and the sample median number of hours was 7. The sample standard deviation was 5.8 hours.

(a) Based on the sample mean and median, describe the shape of the distribution of the population. Why is a large sample size necessary to construct a confidence interval about the mean using the methods presented in this chapter?

(b) Construct and interpret a 90% confidence for the mean number of hours that 13- to 17-year-olds watch television per week.

10. Bottling Soda A soft-drink manufacturer is worried that its filling machine is underfilling the bottles that are supposed to have 19.2 ounces. The quality-control manager takes a random sample of 20 bottles from the production line and obtains the following data:

18.9	18.9	19.0	18.9	19.2	19.0	19.0	19.2	19.0	19.3
18.9	19.1	18.7	18.5	19.0	19.3	18.8	19.1	19.5	19.0

(a) Use the data to compute a point estimate for the population mean for the content of the bottles produced by the manufacturer.

(b) A normal probability plot and boxplot indicate that the sample data come from a population that is normally distributed with no outliers. Construct a 95% confidence interval for the mean content for all bottles produced by the manufacturer. Interpret this interval.

(c) Does the sample indicate that the bottling process is operating properly? Why?

11. Working for a Living The following data represent the number of hours worked per week for a random sample of 15 adults in the United States and Canada, based on data from an August 2004 study by the Gallup Organization.

USA				
70	26	14	41	53
49	40	46	34	33
66	38	46	35	44

Canada				
49	22	42	35	36
63	39	40	35	40
31	68	31	53	28

(a) Obtain a point estimate for the population mean number of hours worked each week for each county.

(b) A normal probability plot and boxplot indicate that the U.S. sample data come from a population that is normally distributed with no outliers. Construct a 99% confidence interval for the population mean number of hours worked each week for the United States, assuming that $\sigma = 12.8$ hours.

(c) A normal probability plot and boxplot indicate that the Canadian sample data come from a population that is normally distributed with no outliers. Construct a 99% confidence interval for the population mean number of hours worked each week for Canada, assuming that $\sigma = 10.8$ hours.

(d) Does it appear to be the case that Americans work more hours per week than Canadians? Why?

12. Math Achievement The following data represent the mathematics achievement test scores for a random sample of 15 male and 15 female students who had just completed high school in the United States, according to data obtained from the International Association for the Evaluation of Education Achievement study.

Male				
488	350	547	488	474
471	443	385	477	452
418	388	441	463	412

Female				
433	389	520	479	454
563	411	458	398	337
418	492	442	494	514

(a) Verify that the scores for each gender are normally distributed with no outliers.
(b) Obtain a point estimate for the population mean score of each gender.
(c) Construct a 95% confidence interval for the population mean achievement score of males, assuming that $\sigma = 64.8$.
(d) Construct a 95% confidence interval for the population mean achievement score of females, assuming that $\sigma = 56.9$.
(e) Does there appear to be any difference between the scores of males and those of females? Why?

13. Family Size A random sample of 60 married couples who have been married 7 years was asked the number of children they have. The results of the survey are as follows:

0	0	0	3	3	3	1	3	2	2	3	1
3	2	4	0	3	3	3	1	0	2	3	3
1	4	2	3	1	3	3	5	0	2	3	0
4	4	2	2	3	2	2	2	2	3	4	3
2	2	1	4	3	2	4	2	1	2	3	2

Note: $\bar{x} = 2.27$, $s = 1.22$.
(a) What is the shape of the distribution of $\bar{x}$? Why?
(b) Compute a 95% confidence interval for the mean number of children of all couples who have been married 7 years. Interpret this interval.
(c) Compute a 99% confidence interval for the mean number of children of all couples who have been married 7 years. Interpret this interval.

14. Waiting in Line The following data represent the number of cars that arrive at McDonald's drive-through between 11:50 A.M. and 12:00 noon for a random sample of Wednesdays. **Note:** $\bar{x} = 4.08$, $s = 2.12$.

1	7	3	8	2	3	8	2	6	3
6	5	6	4	3	4	3	8	1	2
5	3	6	3	3	4	3	2	1	2
4	4	9	3	5	2	3	5	5	5
2	5	6	1	7	1	5	3	8	4

(a) What is the shape of the distribution of $\bar{x}$? Why?
(b) Compute a 90% confidence interval for the mean number of cars waiting in line between 11:50 A.M. and 12:00 noon on Wednesdays. Interpret this interval.
(c) Compute a 95% confidence interval for the mean number of cars waiting in line between 11:50 A.M. and 12:00 noon on Wednesdays. Interpret this interval.

15. Blood Plasma In a study of aerobic activity, the blood plasma volume (in liters) of 12 women was measured, and the following data were obtained.

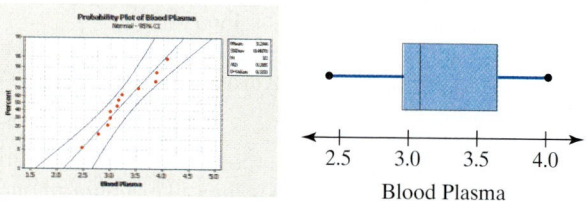

3.15	2.99	2.77	3.12	2.45	3.85
2.99	3.87	4.06	2.94	3.53	3.21

Source: Journal of Applied Physiology 65, 6 (December 1988), p. 361, from Hogg and Tanis Probability and Statistical Inference, 6/e

(a) Use the data to compute a point estimate for the population mean and population standard deviation.
(b) Because the sample size is small, we must verify that blood plasma volume is normally distributed and that the data do not contain any outliers. The figures show the normal probability plot and boxplot. Are the conditions for constructing a confidence interval about μ satisfied?

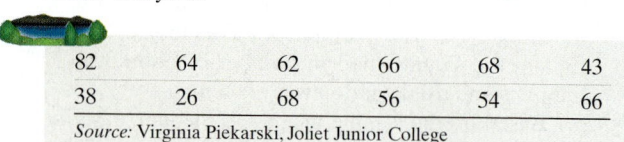

(c) Construct a 95% confidence interval for the mean blood plasma volume for all women. Interpret this interval.
(d) Construct a 99% confidence interval for the mean blood plasma volume for all women. Interpret this interval.
(e) Construct a 99% confidence interval for the population standard deviation of blood plasma volume. Interpret this interval.

16. Water Clarity The campus at Joliet Junior College has a lake. A Secchi disk is used to measure the water clarity of the lake's water by lowering the dish into the water and measuring the distance below water until the disk is no longer visible. The following measurements (in inches) were taken on the lake at various points in time over the course of a year.

82	64	62	66	68	43
38	26	68	56	54	66

Source: Virginia Piekarski, Joliet Junior College

(a) Use the data to compute a point estimate for the population mean and population standard deviation.
(b) Because the sample size is small, we must verify that the data are normally distributed and do not contain any outliers. The figures show the normal probability plot and boxplot. Are the conditions for constructing a confidence interval about μ satisfied?

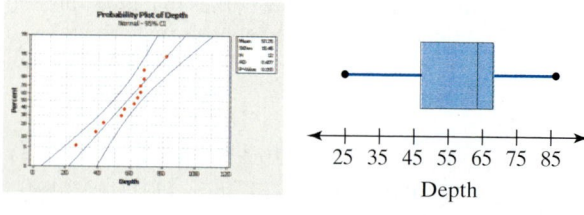

(c) Construct a 95% confidence interval for the mean Secchi disk measurement. Interpret this interval.

(d) Construct a 99% confidence interval for the mean Secchi disk measurement. Interpret this interval.

(e) Construct a 95% confidence interval for the population standard deviation Secchi disk measurement. Interpret this interval.

17. Working for a Living Redo Problems 11(b)–(d), assuming that σ is unknown.

18. Math Achievement Redo Problems 12(c)–(e), assuming that σ is unknown.

19. Hypertension In a random sample of 678 adult males 20 to 34 years of age, it was determined that 58 of them have hypertension (high blood pressure). *Source:* The Centers for Disease Control.

(a) Obtain a point estimate for the proportion of adult males 20 to 34 years of age who have hypertension.

(b) Construct a 95% confidence interval for the proportion of adult males 20 to 34 years of age who have hypertension. Interpret the confidence interval.

(c) Suppose you wish to conduct your own study to determine the proportion of adult males 20 to 34 years old who have hypertension. What sample size would be needed for the estimate to be within 3 percentage points with 95% confidence if you use the point estimate obtained in part (a)?

(d) Suppose you wish to conduct your own study to determine the proportion of adult males 20 to 34 years old who have hypertension. What sample size would be needed for the estimate to be within 3 percentage points with 95% confidence if you don't have a prior estimate?

20. Carbon Monoxide From a random sample of 1201 Americans, it was discovered that 1139 of them lived in neighborhoods with acceptable levels of carbon monoxide. *Source:* The Environmental Protection Agency.

(a) Obtain a point estimate for the proportion of Americans who live in neighborhoods with acceptable levels of carbon monoxide.

(b) Construct a 99% confidence interval for the proportion of Americans who live in neighborhoods with acceptable levels of carbon monoxide.

(c) Suppose you wish to conduct your own study to determine the proportion of Americans who live in neighborhoods with acceptable levels of carbon monoxide. What sample size would be needed for the estimate to be within 1.5 percentage points with 90% confidence if you use the estimate obtained in part (a)?

(d) Suppose you wish to conduct your own study to determine the proportion of Americans who live in neighborhoods with acceptable levels of carbon monoxide. What sample size would be needed for the estimate to be within 1.5 percentage points with 90% confidence if you don't have a prior estimate?

21. Working for a Living Use the data in Problem 11 to answer these questions.

(a) Compute the 95% confidence interval for the population standard deviation of hours worked each week for the United States.

(b) Compute the 95% confidence interval for the population standard deviation of hours worked each week for Canada.

(c) Does it appear to be the case that the number of hours worked per week for Canadians is less dispersed than the number of hours worked each week for Americans?

22. Math Achievement Use the data in Problem 12 to answer these questions:

(a) Compute the 95% confidence interval for the population standard deviation test score of males.

(b) Compute the 95% confidence interval for the population standard deviation test score of females.

(c) Does it appear to be the case that the test scores of males are more dispersed than the test scores of females?

23. The area under the t-distribution with 18 degrees of freedom to the right of $t = 1.56$ is 0.0681. What is the area under the t-distribution with 18 degrees of freedom to the left of $t = -1.56$? Why?

24. Which is larger, the area under the t-distribution with 10 degrees of freedom to the right of $t = 2.32$ or the area under the standard normal distribution to the right of $z = 2.32$? Why?

25. State the properties of Student's t-distribution.

The Search for a Fire-Safe Cigarette

In the United States, approximately 1000 deaths and 3000 serious injuries result annually from fires caused by dropped or discarded cigarettes. The dollar value of the property lost in these fires is staggering. Over one-fifth of all deadly fires have cigarettes as their ignition source! Cigarettes are the single largest cause of these life-consuming blazes.

The National Fire Protection Agency describes the cigarette fire mortality rate among young children as moderate, and that for children between the ages of 10 and 17 is considered low. For adults, the cigarette fire mortality rate steadily increases with age, peaking above age 85. The elderly smoker who uses sedating medications or alcohol is particularly imperiled *(Source:The Fire-Safe Cigarette: the Search for a Standard.* www.burnfoundation.org/firesafecig.html).

Recently, the U.S. National Institute of Standards and Technology (NIST) conducted a study to assess the risk of fire from a dropped or discarded cigarette. Two types of cigarettes, designated 529 and 531, were chosen for their experiment. Both cigarettes were purchased in the same geographical region and were advertised as being 100 mm long and 25 mm in circumference. Each brand used expanded, flue-cured tobacco. Cigarette 529 was wrapped with low-air-permeability paper; cigarette 531 was wrapped with paper of conventional air permeability.

The study's response variables were binary (for example, full-length burn or not and substrate ignition or not), so a binomial probability distribution was used to perform the analysis. Specifically, the NIST study relied on confidence intervals of binomial (population) proportions. The report noted some limitations concerning the use of these confidence intervals. Specifically, the text mentioned that the distribution of a binomial proportion is approximately normal if the sample size is sufficiently large. Furthermore, the sample size necessary for the appropriate use of the normal approximation depends directly on the value of the proportion. That is, if the proportion is near zero or one, a very large sample size is needed for the normal approximation to work well. For outcomes when the normal approximation was not appropriate, the researchers used other, more computationally intensive methods.

One phase of the study dealt with the relative ignition strengths of cigarettes 529 and 531. The number of successful full-length burns was counted for each brand of cigarettes for different numbers of original filter paper substrate layers. The results for tests conducted in 1993 and 2000 are presented in the following table:

	Relative Ignition Strengths of Cigarettes 529 and 531 (Number of Full-Length Burns/Number of Trials)*					
	1993			**2000**		
Cigarette	**3 Layers**	**10 Layers**	**15 Layers**	**3 Layers**	**10 Layers**	**15 Layers**
529	$\frac{22}{32}$	$\frac{0}{32}$	$\frac{0}{32}$	$\frac{18}{48}$	$\frac{0}{48}$	
531	$\frac{32}{32}$	$\frac{30}{32}$	$\frac{28}{32}$		$\frac{44}{48}$	$\frac{38}{48}$

Where appropriate, for each sample year, calculate the 95% confidence interval for the proportion of full-length burns for the various combinations of cigarette and number of original filter paper substrate layers. State conclusions based on your findings concerning the relative ignition strengths of cigarettes 529 and 531. If you could repeat this experiment, how large a sample would you need to ensure that the use of the normal distribution was appropriate for all the various combinations?

Another phase of the study addressed the ignition propensities of conventionally wrapped and banded ("modified product") cigarettes. The Mock-up Ignition Method, developed under the Fire-Safe Cigarette Act of 1990, was followed during this test. The mock-up method measures whether a cigarette transfers enough heat to a fabric/foam substrate (simulated piece of furniture) to cause ignition. Substrate materials were formed by placing a fabric on a piece of nonfire-retardant foam. To ensure adequate contact between layers, a metal rim was placed on top of the substrate. An ignition was declared when the char advanced 10 mm away from the burning cigarette. Lit cigarettes of each type were exposed to four different substrate materials, with the number of ignitions being recorded for each scenario. The results of this experiment are presented in the following table:

Ignition Propensities of Conventional and Banded Cigarettes of the Same Brand, Measured with of the Mock-up Ignition Method (Number of Ignitions/Number of Trials)[*]				
	Substrate			
Cigarette	**Duck #10**	**Duck #6**	**Duck #4**	**Ivory Cotton**
Conventional	64/64	64/64	12/64	64/64
Banded	24/64	32/64	2/64	38/64

Where appropriate, calculate the 95% confidence interval for the proportion of each substrate that was ignited by conventional and banded cigarettes. State conclusions based on your findings regarding the ignition propensities of conventional and banded cigarettes. If you could repeat this experiment, how large a sample would you need to ensure that the use of the normal distribution was appropriate for each scenario?

Write a final report detailing your findings. Include in this document a discussion concerning any assumptions made.

*Richard G. Gann *et al. Relative Ignition Propensity of Test Market Cigarettes*. National Institute of Standards and Technology, Technical Note 1436, January 2001. Web address: www.bfrl.nist.gov/pdf/cigaretterpt.pdf. The original data contained in this report were doubled to ensure that some of the binomial proportions would satisfy the normal approximation requirements.

Testing Claims Regarding a Parameter

Outline

DECISIONS

Many of the products we buy have labels that indicate the net weight of the contents. For example, a candy bar wrapper might state that the net weight of the candy bar is 4 ounces. Should we believe the manufacturer? See the Decisions project on page 537.

●●● Putting It All Together

In Chapter 9, we mentioned that there are two areas of inferential statistics: (1) estimation and (2) hypothesis testing. We have already discussed procedures for estimating the population mean, the population proportion and the population standard deviation.

We now focus our attention on hypothesis testing. Hypothesis testing is used to test claims regarding a characteristic of one or more populations. In this chapter, we will test claims regarding a single population parameter. The claims that we test regard the population mean, the population proportion, and the population standard deviation.

10.1 The Language of Hypothesis Testing

Preparing for This Section   Before getting started, review the following:

- Simple random sampling (Section 1.2, pp.16–19)
- Sampling Distribution of $\bar{x}$ (Section 8.1, pp.417–430)
- Parameter versus statistic (Section 3.1, p. 121)
- Table 9 (Section 6.2, p. 332)

Objectives

1. **Determine the null and alternative hypotheses from a claim**
2. **Understand Type I and Type II errors**
3. **State conclusions to hypothesis tests**

Let's begin with an example that introduces the idea behind hypothesis testing.

EXAMPLE 1   **Illustrating Hypothesis Testing**

Problem: According to the National Center for Chronic Disease Prevention and Health Promotion, 73.8% of females between the ages of 18 and 29 exercise. Kathleen believes that more women between the ages of 18 and 29 are now exercising, so she obtains a simple random sample of 1000 women and finds that 750 of them are exercising. Is this evidence that the percent of women between the ages of 18 and 29 who are exercising has increased? What if Kathleen's sample resulted in 920 women exercising?

Approach: Here is the situation Kathleen faces. If 73.8% of 18- to 29-year-old females exercise, she would expect 738 of the 1000 respondents in the sample to exercise. The questions that Kathleen wants to answer are "How likely is it to obtain a sample of 750 out of 1000 women exercising from a population when the percentage of women who exercise is 73.8%? How likely is a sample that has 920 women exercising?"

Solution: The result of 750 women who exercise is close to what we would expect, so Kathleen is not inclined to believe that the percentage of women exercising has increased. However, the likelihood of obtaining a sample of 920 women who exercise is extremely low if the actual percentage of women who exercise is 73.8%. For the case of obtaining a sample of 920 women who exercise, Kathleen can conclude one of two things: either the proportion of women who exercise is 73.8% and her sample just happens to include a lot of women who exercise, or the proportion of women who exercise has increased. Provided the sampling was performed in a correct fashion, Kathleen is more inclined to believe that the percentage of women who exercise has increased.

---

1. ## Determine the Null and Alternative Hypotheses from a Claim

Example 1 presents the basic premise behind hypothesis testing: A claim is made, information is collected, and this information is used to test the claim. The steps in conducting a hypothesis test are presented next.

> **Steps in Hypothesis Testing**
> 1. A claim is made.
> 2. Evidence (sample data) is collected to test the claim.
> 3. The data are analyzed to assess the plausibility of the claim.

In this section, we introduce the language of hypothesis testing. Sections 10.2 to 10.5 discuss the formal process of testing a hypothesis.

Definition

A **hypothesis** is a statement or claim regarding a characteristic of one or more populations.

In Other Words
In this chapter, we will focus on hypothesis testing regarding a single population parameter. The characteristic we test regards the population mean, proportion, or standard deviation.

In this chapter, we look at hypotheses regarding a single population parameter. The following are examples of claims regarding a characteristic of a single population.

(A) According to a Gallup poll conducted in 1995, 74% of Americans felt that men were more aggressive than women. A researcher claims the percentage of Americans that feel men are more aggressive than women is different today (a claim regarding a population proportion).

(B) The packaging on a light bulb states that the bulb will last 500 hours under normal use. A consumer advocate would like to know if the mean lifetime of a bulb is less than 500 hours (a claim regarding the population mean).

(C) The standard deviation of the rate of return for a certain class of mutual funds is 0.08. A mutual fund manager claims the standard deviation of the rate of return for his fund is less than 0.08 (a claim regarding the population standard deviation).

CAUTION
If population data are available, there is no need for inferential statistics.

We test these types of claims using sample data because it is usually impossible or impractical to gain access to the entire population. The procedure (or process) that we use to test these claims is called *hypothesis testing*.

Definition

Hypothesis testing is a procedure, based on sample evidence and probability, used to test claims regarding a characteristic of one or more populations.

Because a claim can either be true or false, hypothesis testing is based on two types of hypotheses.

Definitions

The **null hypothesis**, denoted H_0 (read "H-naught"), is a statement to be tested. The null hypothesis is assumed true until evidence indicates otherwise. In this chapter, it will be a statement regarding the value of a population parameter.

The **alternative hypothesis**, denoted H_1 (read "H-one"), is a claim to be tested. We are trying to find evidence for the alternative hypothesis. In this chapter, it will be a claim regarding the value of a population parameter.

In Other Words
The null hypothesis is a statement of status quo or no difference and always contains a statement of equality. The null hypothesis is assumed to be true until we have evidence to the contrary. The claim that we seek evidence for always becomes the alternative hypothesis.

For the light bulb manufacturer in Situation B, the consumer advocate wishes to test the claim that the mean lifetime of the bulb is less than 500 hours. Because we are trying to obtain evidence for this claim, it is expressed as the alternative hypothesis using the notation $H_1: \mu < 500$. The statement made by the manufacturer is that the bulb lasts 500 hours on average. We express the statement to be tested using the notation $H_0: \mu = 500$.

In this chapter, there are three ways to set up the null and alternative hypotheses.

1. Equal hypothesis versus not equal hypothesis **(two-tailed test)**
   H_0: parameter = some value
   H_1: parameter ≠ some value

2. Equal versus less than **(left-tailed test)**
   H_0: parameter = some value
   H_1: parameter < some value

3. Equal versus greater than **(right-tailed test)**
   H_0: parameter = some value
   H_1: parameter > some value

Left- and right-tailed tests are referred to as **one-tailed tests**. Notice that in the left-tailed test the direction of the inequality sign in the alternative hypothesis points to the left ($<$), while in the right-tailed test the direction of the inequality sign in the alternative hypothesis points to the right ($>$). Notice that in all three tests the null hypothesis contains a statement of equality. The statement of equality comes from existing information.

Refer to the three claims made on page 506. In Situation A, the null hypothesis is H_0: $p = 0.74$. This is a statement of *status quo* or no difference. The Latin phrase status quo means the existing state or condition. It means that American opinions have not changed from 1995. In Situation B, the null hypothesis is H_0: $\mu = 500$. This is a statement of no difference between the population mean and the lifetime stated on the label. In Situation C, the null hypothesis is H_0: $\sigma = 0.08$. This is a statement of no difference between the population standard deviation rate of return of the manager's mutual fund and all mutual funds.

The wording of the claim, which is dictated by the researcher before any data are collected, determines the structure of the alternative hypothesis (two-tailed, left-tailed, or right-tailed). For example, the label on a can of soda states that the can contains 12 ounces of liquid. A consumer advocate would be concerned only if the mean contents are less than 12 ounces, so the alternative hypothesis is H_1: $\mu < 12$. However, a quality-control engineer for the soda manufacturer would be concerned if there is too little or too much soda in the can, so the alternative hypothesis would be H_1: $\mu \neq 12$. In both cases, however, the null hypothesis is a statement of no difference between the manufacturer's assertion on the label and the actual mean contents of the can. So the null hypothesis is H_0: $\mu = 12$.

EXAMPLE 2   **Forming Hypotheses**

Problem: For each of the following claims, determine the null and alternative hypotheses. State whether the test is two-tailed, left-tailed, or right-tailed.

(a) The Medco pharmaceutical company has just developed a new antibiotic for children. Among the competing antibiotics, 2% of children who take the drug experience headaches as a side effect. A researcher for the Food and Drug Administration wishes to test the claim that the percentage of children taking the new antibiotic who experience headaches as a side effect is more than 2%.

(b) The Blue Book value of a used 3-year-old Chevy Corvette is $37,500. Grant wishes to test the claim that the mean price of a used 3-year-old Chevy Corvette in the Miami metropolitan area is different from $37,500.

(c) The standard deviation of the contents in a 64-ounce bottle of detergent using an old filling machine was known to be 0.23 ounce. The company purchased a new filling machine and wants to test the claim that the standard deviation for the new filling machine is less than 0.23 ounce.

Approach: In each case, we must first identify the parameter about which the claim is made and the status quo. We then assess the direction of the claim (greater than, less than, or not equal to) to help us form the alternative hypothesis.

In Other Words

Structuring the null and alternative hypothesis:

1. Identify the parameter in the claim.

2. Determine the status quo value of the parameter.

3. Identify the claim that we want evidence for to determine the alternative hypothesis.

Solution

(a) The claim is regarding a population proportion, p. If the new drug is no different from current drugs on the market, the proportion of individuals taking the new drug who experience a headache will be 0.02, so the null hypothesis is H_0: $p = 0.02$. The phrase "more than" is represented symbolically as $>$, so the claim is $p > 0.02$. Therefore, the alternative hypothesis is H_1: $p > 0.02$. This is a right-tailed test because the alternative hypothesis contains a $>$ symbol.

In Other Words

Look for key phrases in the claim. For example, "more than" means $>$; "different from" means $\neq$; "less than" means $<$; and so on. See Table 9 on page 332 for a list of key phrases and the symbols they translate into.

(b) The claim is regarding a population mean, μ. If the mean price of a 3-year-old Corvette in Grant's neighborhood is no different from the Blue Book price, the population mean in Grant's neighborhood will be $37,500, so the null hypothesis is $H_0: \mu = 37,500$. Grant wishes to test the claim that the mean price is different from $37,500, so that the alternative hypothesis is $H_1: \mu \neq 37,500$. This is a two-tailed test because the alternative hypothesis contains a $\neq$ symbol.

(c) The claim is regarding a population standard deviation, σ. If the new machine is no different from the old machine, the standard deviation of the amount in the bottles filled by the new machine will be 0.23 ounce, so the null hypothesis is $H_0: \sigma = 0.23$. The phrase "less than" is represented symbolically by $<$, so the claim is $\sigma < 0.23$ ounce and the alternative hypothesis is $H_1: \sigma < 0.23$. This is a left-tailed test because the alternative hypothesis contains a $<$ symbol.

Now Work Problem 17(a).

2   Understand Type I and Type II Errors

As stated earlier, we use sample data to determine whether to reject or not reject the null hypothesis. Because the decision to reject or not reject the null hypothesis is based on incomplete (sample) information, there is always the possibility of making an incorrect decision. In fact, there are four possible outcomes from hypothesis testing.

In Other Words

When you are testing a hypothesis, there is always the possibility that your conclusion will be wrong. To make matters worse, you won't know whether you are wrong or not! Don't fret, however, we have tools to help manage these incorrect conclusions.

Four Outcomes from Hypothesis Testing

1. We reject the null hypothesis when the alternative hypothesis is true. This decision would be correct.
2. We do not reject the null hypothesis when the null hypothesis is true. This decision would be correct.
3. We reject the null hypothesis when the null hypothesis is true. This decision would be incorrect. This type of error is called a **Type I error**.
4. We do not reject the null hypothesis when the alternative hypothesis is true. This decision would be incorrect. This type of error is called a **Type II error**.

Figure 1 illustrates the two types of errors that can be made in hypothesis testing.

Figure 1

Conclusion		Reality	
		H_o Is True	H_1 Is True
	Do Not Reject H_o	Correct Conclusion	Type II Error
	Reject H_o	Type I Error	Correct Conclusion

We illustrate the idea of Type I and Type II errors by looking at hypothesis testing from the point of view of a criminal trial. In any trial, the defendant is assumed to be innocent. (We give the defendant the benefit of the doubt.) The district attorney must present evidence proving that the defendant is guilty.

Because we are seeking evidence for guilt, it becomes the alternative hypothesis. Innocence is assumed, so this is the null hypothesis. The hypotheses for a trial are written

H_0: the defendant is innocent
H_1: the defendant is guilty

The trial is the process whereby information (sample data) is obtained. The jury then deliberates about the evidence (the data analysis). Finally, the jury either convicts the defendant (rejects the null hypothesis) or declares the defendant not guilty (fails to reject the null hypothesis).

Note that the defendant is never declared innocent. That is, we never conclude that the null hypothesis is true. Using this analogy, the two correct decisions are to conclude that an innocent person is not guilty or conclude that a guilty person is guilty. The two incorrect decisions are to convict an innocent person (a Type I error) or to let a guilty person go free (a Type II error). It is helpful to think in this way when trying to remember the difference between a Type I and a Type II error.

In Other Words

A Type I error is like putting an innocent person in jail. A Type II error is like letting a guilty person go free.

EXAMPLE 3   **Type I and Type II Errors**

Problem: The Medco pharmaceutical company has just developed a new antibiotic. Among the competing antibiotics, 2% of children who take the drug experience headaches as a side effect. A researcher for the Food and Drug Administration claims that the percentage of children taking the new antibiotic who experience a headache as a side effect is more than 2%. To test this claim, we conduct a hypothesis test with H_0: $p = 0.02$ and H_1: $p > 0.02$. Provide statements explaining what it would mean to make (a) a Type I error and (b) a Type II error.

Approach: A Type I error occurs if the null hypothesis is rejected when, in fact, the null hypothesis is true. A Type II error occurs if the null hypothesis is not rejected when, in fact, the alternative hypothesis is true.

Solution

(a) We make a Type I error if the sample evidence leads us to believe that $p > 0.02$ (that is, we reject the null hypothesis) when, in fact, the proportion of children who experience a headache is not greater than 0.02.

(b) We make a Type II error if we do not reject the null hypothesis that the proportion of children experiencing a headache is equal to 0.02 when, in fact, the proportion of children who experience a headache is more than 0.02. For example, the sample evidence led the researcher to believe $p = 0.02$ when in fact the true proportion is some value larger than 0.2.

Now Work Problems 17(b) and 17(c).

Understand the Probability of Making a Type I or Type II Error

Recall that we never know whether a confidence interval contains the unknown parameter. We only know the likelihood that a confidence interval captures the parameter. Similarly, we never know whether the outcome of a hypothesis test results in an error or not. However, just as we place a level of confidence in the construction of a confidence interval, we can determine the probability of making errors. The following notation is commonplace:

$$\alpha = P(\text{Type I error}) = P(\text{rejecting } H_0 \text{ when } H_0 \text{ is true})$$

$$\beta = P(\text{Type II error}) = P(\text{not rejecting } H_0 \text{ when } H_1 \text{ is true})$$

The symbol β is the Greek letter beta (pronounced "BAY tah"). The probability of making a Type I error, α, is chosen by the researcher *before* the sample data are collected. This probability is referred to as the *level of significance*.

Definition

> The **level of significance**, α, is the probability of making a Type I error.

The choice of the level of significance depends on the consequences of making a Type I error. If the consequences are severe, the level of significance should be small (say, $\alpha = 0.01$). However, if the consequences of making a Type I error are not severe, a higher level of significance can be chosen (say $\alpha = 0.05$ or $\alpha = 0.10$).

Why is the level of significance not always set at $\alpha = 0.01$? By reducing the probability of making a Type I error, you increase the probability of making a Type II error, β. Using our court analogy, a jury is instructed that the prosecution must provide proof of guilt "beyond all reasonable doubt." This implies that we are choosing to make α small so that the probability we will send an innocent person to jail is very small. The consequence of the small α, however, is a large β, which means many guilty defendants will go free. For now, we are content to recognize the inverse relation between α and β (as one goes up the other goes down).

In Other Words

As the probability of a Type I error increases, the probability of a Type II error decreases, and vice versa.

③ State Conclusions to Hypothesis Tests

Once the decision to reject or not reject the null hypothesis is made, the researcher must state his or her conclusion. It is important to recognize that we never *accept* the null hypothesis. Again, the court system analogy helps to illustrate the idea. The null hypothesis is H_0: innocent. When the evidence presented to the jury is not enough to convict beyond all reasonable doubt, the jury comes back with a verdict of not guilty.

! **CAUTION**

We never *accept* the null hypothesis, because, without having access to the entire population, we don't know the exact value of the parameter stated in the null. Rather, we say that we do not reject the null hypothesis. This is just like the court system. We never declare a defendant innocent, but rather say the defendant is not guilty.

Notice that the verdict does not state that the null hypothesis of innocence is true; it states that there is not enough evidence to conclude guilt. This is a huge difference. Being told that you are not guilty is very different from being told that you are innocent!

So sample evidence can never prove the null hypothesis to be true. When we do not reject the null hypothesis, we are saying that the evidence indicates that the null hypothesis *could* be true.

EXAMPLE 4 | **Stating the Conclusion**

Problem: The Medco pharmaceutical company has just developed a new antibiotic. Among the competing antibiotics, 2% of children who take the drug experience a headache as a side effect. A researcher for the Food and Drug Administration claims that the percentage of children taking the new antibiotic who experience a headache as a side effect is more than 2%. From Example 2(a), we know the null hypothesis is H_0: $p = 0.02$ and the alternative hypothesis is H_1: $p > 0.02$.

(a) Suppose the sample evidence indicates that the null hypothesis is rejected. State the conclusion.

(b) Suppose the sample evidence indicates that the null hypothesis is not rejected. State the conclusion.

Approach: When the null hypothesis is rejected, we say that there is sufficient evidence to support the claim. When the null hypothesis is not rejected, we say that there is not sufficient evidence to support the claim. We never say that the null hypothesis is true!

Solution

(a) The claim is that the percentage of children taking the new antibiotic who experience a headache as a side effect is more than 2%. Because the null hypothesis ($p = 0.02$) is rejected, we conclude there is sufficient evidence to support the claim that the proportion of children who experience a headache as a side effect is more than 0.02.

(b) Because the null hypothesis is not rejected, we conclude that there is not sufficient evidence to support the claim that the proportion of children who experience a headache as a side effect is more than 0.02.

Now Work Problem 25.

10.1 ASSESS YOUR UNDERSTANDING

Concepts and Vocabulary

1. Explain what it means to make a Type I error. Explain what it means to make a Type II error.
2. Suppose the consequences of making a Type I error are severe. Would you choose the level of significance, α, to equal 0.01, 0.05, or 0.10? Why?
3. What happens to the probability of making a Type II error, β, as the level of significance, α, decreases? Why is this result intuitive?
4. If a hypothesis is tested at the $\alpha = 0.05$ level of significance, what is the probability of making a Type I error?
5. The following is a quotation from Sir Ronald A. Fisher, a famous statistician.

   "For the logical fallacy of believing that a hypothesis has been proved true, merely because it is not contradicted by the available facts, has no more right to insinuate it-

self in statistics than in other kinds of scientific reasoning .... It would, therefore, add greatly to the clarity with which the tests of significance are regarded if it were generally understood that tests of significance, when used accurately, are capable of rejecting or invalidating hypotheses, in so far as they are contradicted by the data: but that they are never capable of establishing them as certainly true ...."

   In your own words, explain what this quotation means.
6. In your own words, explain the difference between "beyond all reasonable doubt" and "beyond all doubt."
7. *True or False:* Sample evidence can prove that a null hypothesis is true.
8. *True or False:* Type I and Type II errors are independent events.

Skill Building

In Problems 9–14, a null and alternative hypothesis is given. Determine whether the hypothesis test is left-tailed, right-tailed, or two-tailed. What parameter is being tested?

9. $H_0: \mu = 5$
   $H_1: \mu > 5$

10. $H_0: p = 0.2$
    $H_1: p < 0.2$

11. $H_0: \sigma = 4.2$
    $H_1: \sigma \neq 4.2$

12. $H_0: p = 0.76$
    $H_1: p > 0.76$

13. $H_0: \mu = 120$
    $H_1: \mu < 120$

14. $H_0: \sigma = 7.8$
    $H_1: \sigma \neq 7.8$

For each claim in Problems 15–22, (a) determine the null and alternative hypotheses, (b) explain what it would mean to make a Type I error, and (c) explain what it would mean to make a Type II error.

15. **Teenage Mothers** According to the U.S. Census Bureau, 11.8% of registered births in the United States in 2000 were to teenage mothers. A sociologist claims that this percentage has decreased since then.

16. **Charitable Contributions** According to *Giving and Volunteering in the United States, 2001 Edition*, the mean charitable contribution per household in the United States in 2000 was $1623. A researcher claims that the level of giving has changed since then.

17. **Single-Family Home Price** According to the Federal
    NW Housing Finance Board, the mean price of a single-family

home in October 2003 was $243,756. A real estate broker believes that because of changes in interest rates, as well as other economic factors, the mean price has increased since then.

18. **Fair Packaging and Labeling** Federal law requires that a jar of peanut butter that is labeled as containing 32 ounces must contain at least 32 ounces. A consumer advocate feels that a certain peanut butter manufacturer is shorting customers by underfilling the jars so that the mean content is less than the 32 ounces stated on the label.

19. Valve Pressure The standard deviation in the pressure required to open a certain sprung-type valve is known to be $\sigma = 0.7$ psi. Due to changes in the manufacturing process, the quality-control manager feels that the pressure variability has been reduced.

20. Overweight According to the Centers for Disease Control and Prevention, 16% of children aged 6 to 11 years are overweight. A school nurse thinks that the percentage of 6- to 11-year-olds who are overweight is higher in her school district.

21. Cell Phone Service According to the *Statistical Abstract of the United States,* the mean monthly cell phone bill was $49.91 in 2003. A researcher suspects that the mean monthly cell phone bill is different today.

22. SAT Math Scores In 2004, the standard deviation SAT math score for all students taking the exam was 114. A teacher believes that, due to changes to the SAT Reasoning Test in 2005, the standard deviation of SAT math scores will increase.

In Problems 23–34, state the conclusion based on the results of the test.

23. For the claim made in Problem 15, suppose the null hypothesis is rejected.

24. For the claim made in Problem 16, suppose the null hypothesis is not rejected.

25. For the claim made in Problem 17, suppose the null hypothesis is not rejected.

26. For the claim made in Problem 18, suppose the null hypothesis is rejected.

27. For the claim made in Problem 19, suppose the null hypothesis is not rejected.

28. For the claim made in Problem 20, suppose the null hypothesis is not rejected.

29. For the claim made in Problem 21, suppose the null hypothesis is rejected.

30. For the claim made in Problem 22, suppose the null hypothesis is not rejected.

31. For the claim made in Problem 15, suppose the null hypothesis is not rejected.

32. For the claim made in Problem 16, suppose the null hypothesis is rejected.

33. For the claim made in Problem 17, suppose the null hypothesis is rejected.

34. For the claim made in Problem 18, suppose the null hypothesis is not rejected.

Applying the Concepts

35. Fruits and Vegetables According to the *Statistical Abstract of the United States*, the mean consumption of fruits in 2003 was 98.4 pounds. A dietician believes more people are becoming health conscious and that fruit consumption has risen since then.
(a) Determine the null and alternative hypotheses.
(b) Suppose sample data indicate that the null hypothesis should be rejected. State the conclusion of the researcher.
(c) Suppose, in fact, the mean consumption of fruits is 98.4 pounds. Was a Type I or Type II error committed? If we tested this hypothesis at the $\alpha = 0.05$ level of significance, what is the probability of committing a Type I error?

36. Test Preparation The mean score on the SAT Math Reasoning exam is 518. A test preparation company claims that the mean scores of students who take its course are higher than the mean of 518.
(a) Determine the null and alternative hypotheses.
(b) Suppose sample data indicate that the null hypothesis should not be rejected. State the conclusion of the company.
(c) Suppose, in fact, the mean score of students taking the preparatory course is 522. Was a Type I or Type II error committed? If we tested this hypothesis at the $\alpha = 0.01$ level, what is the probability of committing a Type I error?

(d) If we wanted to decrease the probability of making a Type II error, would we need to increase or decrease the level of significance?

37. Marijuana Use According to the Centers for Disease Control and Prevention, in 2001, 10.2% of high school students had tried marijuana for the first time before the age of 13. The Drug Abuse and Resistance Education (DARE) program underwent several major changes to keep up with technology and issues facing students in the 21st century. After the changes, a school resource officer (SRO) claims that the proportion of high school students who have tried marijuana for the first time before the age of 13 has decreased from the 2001 level.
(a) Determine the null and alternative hypotheses.
(b) Suppose sample data indicate that the null hypothesis should not be rejected. State the conclusion of the SRO.
(c) Suppose, in fact, the proportion of high school students who have tried marijuana for the first time before the age of 13 was 9.5%. Was a Type I or Type II error committed?

38. Internet Use According to the *Statistical Abstract of the United States*, in 2000, 10.7% of Americans over 65 years of age used the Internet. A researcher believes the proportion of Americans over 65 years of age who use the Internet is higher than 10.7% today.
(a) Determine the null and alternative hypotheses.

(b) Suppose sample data indicate that the null hypothesis should be rejected. State the conclusion of the researcher.

(c) Suppose, in fact, the percentage of Americans over 65 years of age who use the Internet is still 10.7%. Was a Type I or Type II error committed?

39. Consumer Reports The following is an excerpt from a *Consumer Reports* article from February 2001.

> The Platinum Gasaver makes some impressive claims. The device, $188 for two, is guaranteed to increase gas mileage by 22% says the manufacturer, National Fuelsaver. Also, the company quotes "the government" as concluding, "Independent testing shows greater fuel savings with *Gasaver* than the 22 percent claimed by the developer." Readers have told us they want to know more about it.
>
> The Environmental Protection Agency (EPA), after its lab tests of the Platinum Gasaver, concluded in 1991, "Users of the device would not be expected to realize either an emission or fuel economy benefit." The Federal Trade Commission says, "No government agency endorses gas-saving products for cars."

Determine the null and alternative hypotheses that the EPA used in to draw the conclusion stated in the second paragraph.

40. Prolong Engine Treatment The manufacturer of Prolong Engine Treatment claims that if you add one 12-ounce bottle of their $20 product your engine will be protected from excessive wear. An infomercial claims that a woman drove 4 hours without oil, thanks to Prolong. *Consumer Reports* magazine tested engines in which they added Prolong to the motor oil, ran the engines, drained the oil, and then determined the time until the engines seized.

(a) Determine the null and alternative hypotheses *Consumer Reports* will test.

(b) Both engines took exactly 13 minutes to seize. What conclusion might *Consumer Reports* draw based on this evidence?

41. Refer to the claim made in Problem 18. Researchers must choose the level of significance based on the consequences of making a Type I error. In your opinion, is a Type I error or Type II error more serious? Why? On the basis of your answer, decide on a level of significance, α. Be sure to support your opinion.

10.2 **Testing Claims about a Population Mean Assuming the Population Standard Deviation Is Known**

Preparing for This Section Before getting started, review the following:

- Using probabilities to identify unusual events (Section 5.1, p. 253)
- z_α Notation (Section 7.2, p. 380)
- Sampling distribution of $\bar{x}$ (Section 8.1, pp. 417–430)
- Computing normal probabilities (Section 7.3, pp. 385–389)

Objectives

1 Understand the logic of hypothesis testing

2 Test a claim about a population mean with σ known using the classical approach

3 Test a claim about a population mean with σ known using *P*-values

4 Test a claim about a population mean with σ known using confidence intervals

5 Understand the difference between statistical significance and practical significance

1 Understand the Logic of Hypothesis Testing

Now that we know the language of hypothesis testing, we are ready to present the methods for conducting a hypothesis test.

In this section, we present three approaches to testing a claim about a population mean, μ. As we did for confidence intervals, we begin by assuming that we know the value of the population standard deviation, σ. The assumption is made because it allows us to use the normal model to test claims regarding the population mean. Because the normal model is fairly easy to use, we can concentrate on the techniques of hypothesis testing without getting bogged down with other details. The assumption that σ is known will be dropped in the next section.

To test a claim regarding the population mean assuming the population standard deviation is known, two requirements must be satisfied.

- A simple random sample is obtained.
- The population from which the sample is drawn is normally distributed or the sample size is large ($n \geq 30$).

If these requirements are met, then the distribution of $\bar{x}$ is normal with mean μ and standard deviation $\dfrac{\sigma}{\sqrt{n}}$.

The first method that we use in testing a claim regarding a population mean is referred to as the classical (traditional) approach, the second method is the *P*-value approach, and the third method uses confidence intervals. Your instructor may choose to cover one, two, or all three approaches to hypothesis testing. Do not be alarmed if one or two of the approaches are not covered by your instructor.

Let's lay out a scenario that will be used to help understand both the classical approach to hypothesis testing and the *P*-value approach. Suppose a consumer advocate claims a manufacturer of potato chips is underfilling its bags. The bag states the contents weigh 12.5 ounces. In hypothesis testing we assume that the manufacturer is "not guilty," which means we assume that the population mean contents of the bags of chips is $\mu = 12.5$ ounces. We are looking for evidence to support the claim that the manufacturer is underfilling the bags. To test the claim, we have the following hypotheses:

$$H_0: \mu = 12.5 \quad \text{versus} \quad H_1: \mu < 12.5$$

Suppose the consumer advocate gathers evidence by obtaining a simple random sample of $n = 36$ bags of chips, weighing the contents, and obtaining a sample mean of 12.45 ounces. Does this result provide sufficient evidence to reject the null hypothesis? What is sufficient or *statistically significant* evidence?

Definition

When observed results are unlikely under the assumption that the null hypothesis is true, we say the result is **statistically significant**. When results are found to be statistically significant, we reject the null hypothesis.

Figure 2

Before we can test the claim, we need to know the distribution of the sample mean, since a different sample of 36 bags of chips will likely result in a different sample mean. Since the sample size is large, the Central Limit Theorem says that the shape of the distribution of the sample mean is approximately normal. Regardless of the size of the sample, the mean of the distribution of the sample mean is $\mu_{\bar{x}} = \mu = 12.5$ ounces because we assume the statement in the null hypothesis to be true, until we have evidence to the contrary. Suppose the population standard deviation is known to be 0.12 ounce; then the standard deviation of the distribution of the sample mean is $\sigma_{\bar{x}} = \dfrac{\sigma}{\sqrt{n}} = \dfrac{0.12}{\sqrt{36}} = 0.02$ ounce. Figure 2 shows the sampling distribution of the sample mean for our potato chip example.

12.48   12.5   12.52 $\bar{x}$

Now that we have a model that describes the distribution of the sample mean, we can look at two approaches to testing the claim that the potato chip company is underfilling the bags.

The Logic of the Classical Approach

One criterion we may use for sufficient evidence is to reject the null hypothesis if the sample mean is too many standard deviations below the hypothesized (or status quo) population mean of 12.5 ounces. For example, our criterion might be to reject the null hypothesis if the sample mean is more than 2 standard deviations below the assumed mean of 12.5 ounces.

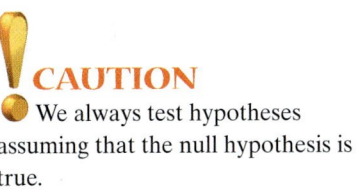

CAUTION

We always test hypotheses assuming that the null hypothesis is true.

Recall that $Z = \dfrac{\overline{x} - \mu}{\sigma/\sqrt{n}}$ represents the number of standard deviations that $\overline{x}$ is from the population mean, μ. Our simple random sample of 36 bags results in sample mean weight of $\overline{x} = 12.45$ ounces, so under the assumption that the null hypothesis is true, we have

$$ Z = \frac{\overline{x} - \mu}{\sigma/\sqrt{n}} = \frac{12.45 - 12.5}{0.12/\sqrt{36}} = -2.5 $$

The sample mean is 2.5 standard deviations below the hypothesized mean. Because the sample mean is more than 2 standard deviations (that is, "too far") below the hypothesized population mean, we will reject the null hypothesis and conclude that there is sufficient evidence to refute the claim that the bag has 12.5 ounces of potato chips. This conclusion will lead the consumer advocate to wage a "truth in advertising campaign" against the potato chip manufacturer.

Why does it make sense to reject the null hypothesis if the sample mean is more than two standard deviations away from the hypothesized mean? The area under the standard normal curve to the left of $Z = -2$ is 0.0228, as shown in Figure 3.

Figure 4 shows that, if the null hypothesis is true, 97.72% of all sample means will be 12.46 ounces or more and only 2.28% of the sample means will be less than 12.46 ounces, as indicated by the region shaded green in Figure 4. Remember, the 12.46 comes from the fact that 12.46 is 2 standard deviations below the hypothesized mean of 12.5 ounces. If a sample mean lies in the green region we are inclined to believe that it came from a population whose mean is less than 12.5, rather than believe that the population mean equals 12.5 and our sample just happened to result in an unusual outcome (a bunch of underfilled bags).

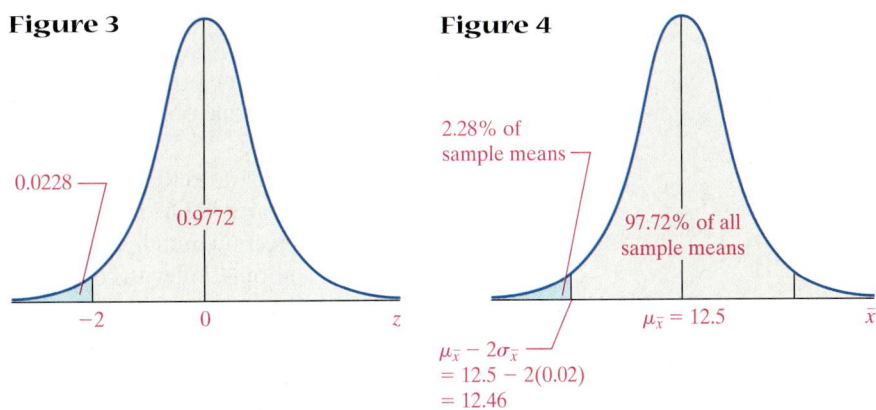

Figure 3

0.0228

0.9772

−2   0 z

Figure 4

2.28% of sample means

97.72% of all sample means

$\mu_{\overline{x}} = 12.5$ $\overline{x}$

$\mu_{\overline{x}} - 2\sigma_{\overline{x}}$
$= 12.5 - 2(0.02)$
$= 12.46$

Notice that our criterion for rejecting the null hypothesis will lead to making a Type I error (rejecting a true null hypothesis) 2.28% of the time. That is, the probability of making a Type I error is 2.28%.

The previous discussion leads to the following premise of hypothesis testing using the classical approach:

> If the sample mean is too many standard deviations from the mean stated in the null hypothesis, we reject the null hypothesis.

The Logic of the *P*-value Approach

A second criterion we may use for sufficient evidence to support the claim that the manufacturer is underfilling the bags is to compute how likely it is to obtain a sample mean of 12.45 ounces or less from a population whose mean is 12.5 ounces. If a sample mean of 12.45 or less is unlikely (or unusual), we have evidence against the null hypothesis. If the sample mean of 12.45 is not unlikely (not unusual), we do not have sufficient evidence against the null hypothesis.

We can compute the probability of obtaining a sample mean of 12.45 or less from a population whose mean is 12.5 using the normal model. Figure 5 shows the area that represents $P(\bar{x} < 12.45)$.

Figure 5

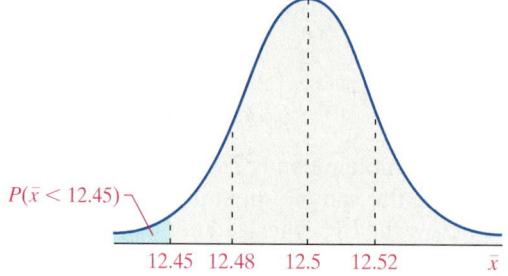

Because

$$Z = \frac{\bar{x} - \mu}{\sigma_{\bar{x}}} = \frac{12.45 - 12.5}{0.02} = -2.5$$

we compute

$$P(\bar{x} \leq 12.45) = P(Z \leq -2.5) = 0.0062$$

The probability of obtaining a sample mean of 12.45 ounces or less from a population whose mean is 12.5 ounces is 0.0062. This means that less than 1 sample in 100 will give a mean as low or lower than the one we obtained *if* the population mean really is 12.5 ounces. Because these results are so unusual, we take this as evidence against the statement in the null hypothesis.

This discussion leads to the following premise of testing a hypothesis using the *P*-value approach:

> Assuming H_0 is true, if the probability of getting a sample mean as extreme or more extreme than the one obtained is small, we reject the null hypothesis.

Figure 6 further illustrates the situation for both the classical and *P*-value approach. The sample mean of 12.45 is too far from the assumed population mean of 12.5. Therefore, we reject the null hypothesis that $\mu = 12.5$ and conclude that the sample came from a population with some population mean less than 12.5 ounces, as indicated by the distribution in blue. We don't know what the population mean weight of the bags is, but we have evidence that it is less than 12.5 ounces.

Figure 6

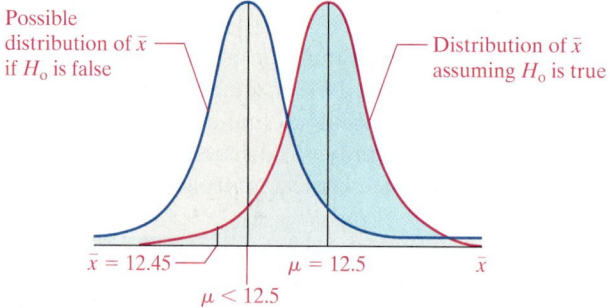

2 Test a Claim about a Population Mean with σ Known Using the Classical Approach

We now formalize the procedure for testing a claim regarding the population mean when the population standard deviation, σ, is known using the classical approach.

Testing a Claim Regarding the Population Mean with σ Known Using the Classical Approach

If a claim is made regarding the population mean with σ known, we can use the steps that follow to test the claim as long as the following two requirements are satisfied.

1. the sample is obtained using simple random sampling,
2. the sample has no outliers and the population from which the sample is drawn is normally distributed or the sample size, n, is large ($n \geq 30$).

Step 1: A claim is made regarding the population mean. The claim is used to determine the null and alternative hypotheses. Again, the hypotheses can be structured in one of three ways:

Two-Tailed	Left-Tailed	Right-Tailed
$H_0: \mu = \mu_0$	$H_0: \mu = \mu_0$	$H_0: \mu = \mu_0$
$H_1: \mu \neq \mu_0$	$H_1: \mu < \mu_0$	$H_1: \mu > \mu_0$

Note: μ_0 is the assumed or status quo value of the population mean.

Step 2: Select a level of significance α based on the seriousness of making a Type I error.

Step 3: Provided the population from which the sample is drawn is normal or the sample size is large ($n \geq 30$) and the population standard deviation σ, is known, the distribution of the sample mean, $\overline{x}$, is normal with mean μ_0 and standard deviation $\dfrac{\sigma}{\sqrt{n}}$. Therefore,

$$z_0 = \frac{\overline{x} - \mu_0}{\sigma/\sqrt{n}}$$

represents the number of standard deviations the sample mean is from the assumed mean, μ_0. This value is called the **test statistic**.

Step 4: The level of significance is used to determine the *critical value*. The **critical value** represents the maximum number of standard deviations the sample mean can be from μ_0 before the null hypothesis is rejected. For example, the critical value in the left-tailed test is $-z_\alpha$. The shaded region(s) represents the *critical (or rejection) region(s)*. The **critical region** or **rejection region** is the set of all values such that the null hypothesis is rejected.

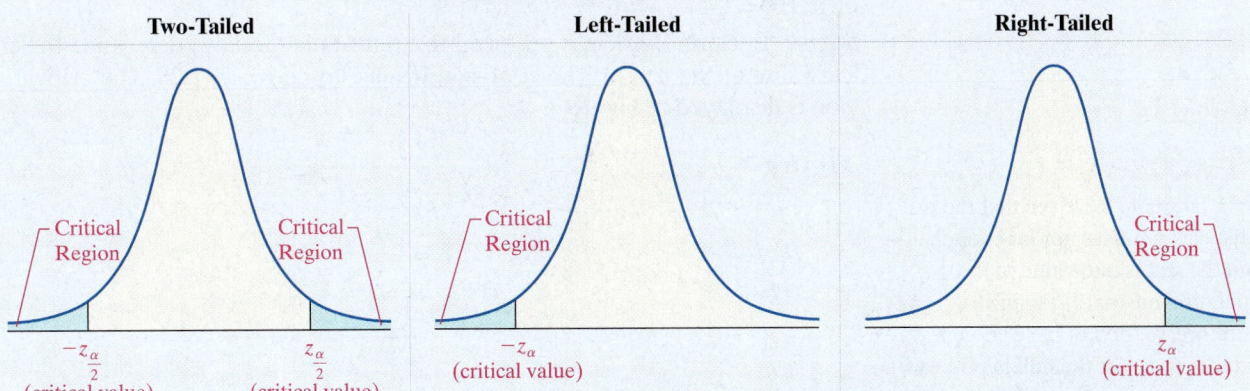

Step 5: Compare the critical value with the test statistic:

Two-Tailed	Left-Tailed	Right-Tailed
If $z_0 < -z_{\alpha/2}$ or $z_0 > z_{\alpha/2}$ reject the null hypothesis	If $z_0 < -z_\alpha$ reject the null hypothesis	If $z_0 > z_\alpha$ reject the null hypothesis

The comparison of the test statistic and critical value is called the **decision rule**.

Step 6: State the conclusion.

The procedure is **robust**, which means that minor departures from normality will not adversely affect the results of the test. However, for small samples, if the data have outliers, the procedure should not be used.

For small samples, we will verify that the data come from a population that is normal by constructing normal probability plots (to assess normality) and boxplots (to determine whether there are outliers). If the normal probability plot indicates that the data do not come from a population that is normally distributed or the boxplot reveals outliers, nonparametric tests should be performed, as discussed in Chapter 15, Section 3.

EXAMPLE 1 | **The Classical Approach of Hypothesis Testing:**
Right-Tailed, Large Sample

Problem: According to the U.S. Federal Highway Administration, the mean number of miles driven annually is 12,200. Patricia believes that residents of the state of Montana drive more than the national average. She obtains a simple random sample of 35 drivers from a list of registered drivers in the state of Montana. The mean number of miles driven for the 35 drivers is 12,895.9. Assuming $\sigma = 3800$ miles, test Patricia's claim at the $\alpha = 0.1$ level of significance.

Approach: Because the sample size is large, we can proceed to Steps 1 through 6.

Solution

Step 1: Patricia claims people in Montana are driving more than 12,200 miles annually. This claim can be written $\mu > 12,200$. This is a right-tailed test and we have

$$H_0: \mu = 12,200 \quad \text{versus} \quad H_1: \mu > 12,200$$

Step 2: The level of significance is $\alpha = 0.1$.

Step 3: Patricia found the sample mean, $\bar{x}$, to be 12,895.9 miles. The test statistic is

$$z_0 = \frac{\bar{x} - \mu_0}{\sigma/\sqrt{n}} = \frac{12,895.9 - 12,200}{3800/\sqrt{35}} = 1.08$$

The sample mean of 12,895.9 miles is 1.08 standard deviations above the mean of 12,200.

Step 4: Because Patricia is performing a right-tailed test, we determine the critical value at the $\alpha = 0.1$ level of significance to be $z_{0.1} = 1.28$. The critical region is displayed in Figure 7.

Figure 7

!**CAUTION**

In Example 1, we see that the test statistic $z_0 = 1.08$ is not far enough from the status quo value of the population mean, 12,200 miles. Therefore, we do not have enough evidence to reject the null hypothesis. However, this does not mean we are accepting the null hypothesis that the mean number of miles driven in Montana is 12,200. We are saying that we don't have enough evidence to say it is greater than 12,200 miles. Be sure you understand the difference between these two comments.

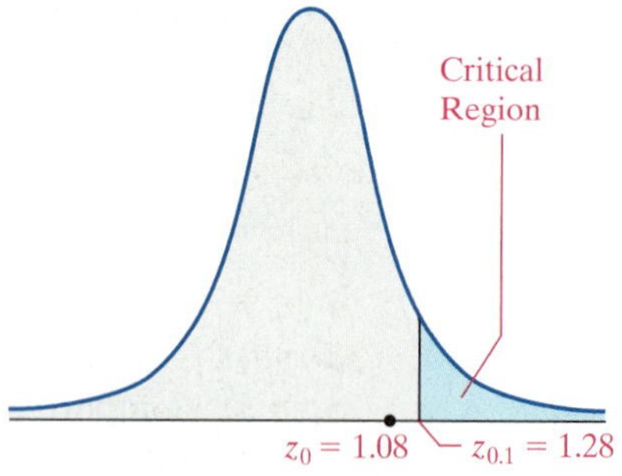

Step 5: Because the test statistic $z_0 = 1.08$ is less than the critical value $z_{0.1} = 1.28$, we do not reject the null hypothesis. That is, the value of the test statistic does not fall within the critical region, so we do not reject H_0. We label the test statistic in Figure 7.

Step 6: There is not sufficient evidence at the $\alpha = 0.1$ level of significance to support Patricia's claim that residents of the state of Montana drive more than the national average of 12,200 miles.

NW **Now Work Problem 19 Using the Classical Approach.**

Now let's look at a two-tailed hypothesis test.

EXAMPLE 2

The Classical Approach of Hypothesis Testing: Two-Tailed, Small Sample

Table 1

94.25	38.94	79.15	56.78
70.07	115.59	77.56	37.01
55.00	76.05	27.29	52.48

Problem: According to CTIA–The Wireless Association, the mean monthly cell phone bill in 2004 was $50.64. A market researcher believes that the mean monthly cell phone bill is different today, but is not sure whether bills have declined because of technological advances or increased due to additional use. To test the claim, the researcher phones a simple random sample of 12 cell phone subscribers and obtains the data in Table 1.

Assuming $\sigma = \$18.49$, use these data to test the claim that the mean monthly cell phone bills is different from $50.64 at the $\alpha = 0.05$ level of significance.

Approach: Because the sample size, n, is less than 30, we must verify that the data come from a population that is approximately normal with no outliers. We will construct a normal probability plot and boxplot to verify these requirements. We then proceed to follow Steps 1 through 6.

Solution: Figure 8 displays the normal probability plot and boxplot.

Figure 8

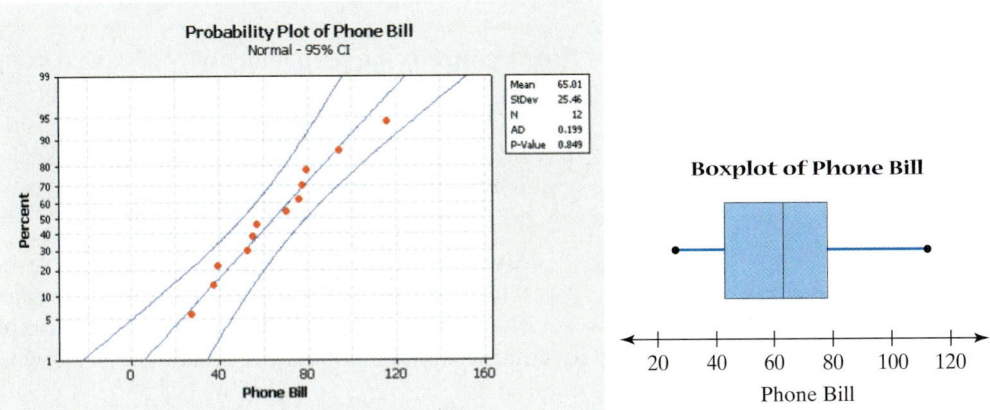

The normal probability plot indicates that the data could come from a population that is normal. The boxplot does not show any outliers.

Step 1: The claim is that the mean cell phone bill is different from $50.64. The claim can be written $\mu \neq 50.64$. This is a two-tailed test and we have

$$H_0: \mu = 50.64 \quad \text{versus} \quad H_1: \mu \neq 50.64$$

Step 2: The level of significance is $\alpha = 0.05$.

Step 3: From the data in Table 1, the sample mean is computed to be $65.014. The test statistic is

$$z_0 = \frac{\bar{x} - \mu_0}{\sigma/\sqrt{n}} = \frac{65.014 - 50.64}{18.49/\sqrt{12}} = 2.69$$

The sample mean of $65.014 is 2.69 standard deviations above the assumed population mean of $50.64.

Step 4: Because this is a two-tailed test, we determine the critical values at the $\alpha = 0.05$ level of significance to be $-z_{0.05/2} = -1.96$ and $z_{0.05/2} = 1.96$. The critical regions are displayed in Figure 9.

Figure 9

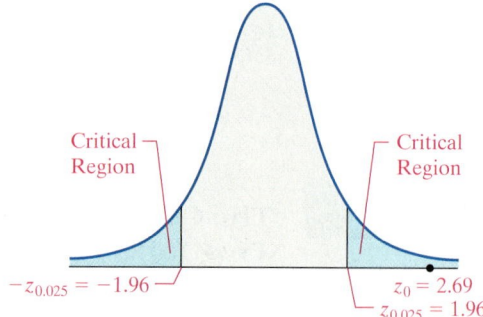

Critical Region   Critical Region

$-z_{0.025} = -1.96$ $z_0 = 2.69$
$z_{0.025} = 1.96$

Step 5: Because the test statistic, $z_0 = 2.69$, is greater than the critical value $z_{0.025} = 1.96$, we reject the null hypothesis. That is, the value of the test statistic falls within the critical region, so we reject H_0. We label this point in Figure 9.

Step 6: There is sufficient evidence at the $\alpha = 0.05$ level of significance to support the claim that the mean monthly cell phone bill is different from the mean amount in 2004, $50.64.

NW **Now Work Problem 25 Using the Classical Approach.**

③ Test a Claim about a Population Mean with σ Known Using *P*-Values

Now let's look at testing hypotheses with *P*-values.

Definition

A ***P*-value** is the probability of observing a sample statistic as extreme or more extreme than the one observed under the assumption that the null hypothesis is true. Put another way, the *P*-value is the likelihood or probability that a sample will result in a sample mean such as the one obtained if the null hypothesis is true.

In Other Words

The smaller the *P*-value, the greater the evidence against the null hypothesis.

A small *P*-value implies that the sample mean is unlikely if the null hypothesis is true and would be considered as evidence against the null hypothesis.

The following procedures can be used to compute *P*-values when testing a hypothesis about a population mean with σ known.

Testing a Claim Regarding the Population Mean Using *P*-Values

If a claim is made regarding the population mean with σ known, we can use the steps that follow to compute the *P*-value, provided that the following two requirements are satisfied.

1. the sample is obtained using simple random sampling, and
2. the sample has no outliers and the population from which the sample is drawn is normally distributed or the sample size, n, is large ($n \geq 30$).

Step 1: A claim is made regarding the population mean. The claim is used to determine the null and alternative hypotheses. The hypotheses can be structured in one of three ways:

Two-Tailed	Left-Tailed	Right-Tailed
$H_0: \mu = \mu_0$	$H_0: \mu = \mu_0$	$H_0: \mu = \mu_0$
$H_1: \mu \neq \mu_0$	$H_1: \mu < \mu_0$	$H_1: \mu > \mu_0$

Note: μ_0 is the assumed value of the population mean.

Step 2: Decide on a level of significance, α, depending on the seriousness of making a Type I error.

Step 3: Compute the test statistic, $z_0 = \dfrac{\bar{x} - \mu_0}{\sigma/\sqrt{n}}$.

Step 4: Determine the *P*-value.

Two-Tailed	Left-Tailed	Right-Tailed

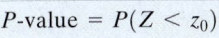

Two-Tailed

$P\text{-value} = P(Z < -|z_0| \text{ or } Z > |z_0|)$
$= 2P(Z > |z_0|)$

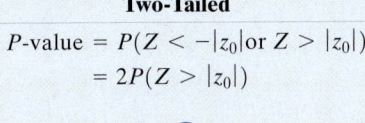

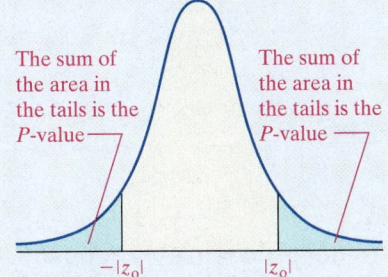

Left-Tailed

$P\text{-value} = P(Z < z_0)$

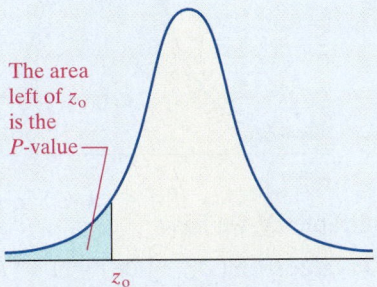

Right-Tailed

$P\text{-value} = P(Z > z_0)$

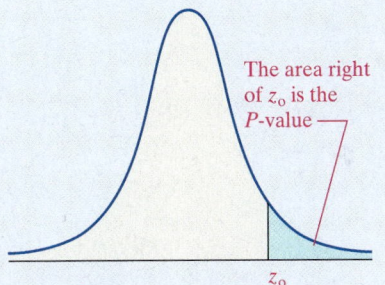

Interpretation: The *P*-value is the probability of obtaining a sample mean that is more than $|z_0|$ standard deviations from the hypothesized mean, μ_0, assuming H_0 is true.

Interpretation: The *P*-value is the probability of obtaining a sample mean of $\bar{x}$ or smaller under the assumption that H_0 is true. In other words, it is the probability of obtaining a sample mean that is more than z_0 standard deviations to the left of μ_0.

Interpretation: The *P*-value is the probability of obtaining a sample mean of $\bar{x}$ or larger under the assumption that H_0 is true. In other words, it is the probability of obtaining a sample mean that is more than z_0 standard deviations to the right of μ_0.

Step 5: Reject the null hypothesis if the *P*-value is less than the level of significance, α. The comparison of the *P*-value and the level of significance is called the **decision rule**.

Step 6: State the conclusion.

EXAMPLE 3

The *P*-value Approach of Hypothesis Testing: Right-Tailed, Large Sample

Problem: According to the U.S. Federal Highway Administration, the mean number of miles driven annually is 12,200. Patricia believes that residents of the state of Montana drive more than the national average. She obtains a simple random sample of 35 drivers from a list of registered drivers in the state of Montana. The mean number of miles driven for the 35 drivers is 12,895.9. Assuming $\sigma = 3800$ miles, test Patricia's claim at the $\alpha = 0.1$ level of significance using the *P*-value approach.

Approach: Because the sample size is large, we can proceed to Steps 1 through 6.

Solution

Step 1: Patricia claims people in Montana are driving more than 12,200 miles annually. This claim can be written $\mu > 12{,}200$. This is a right-tailed test and we have

$$H_0: \mu = 12{,}200 \quad \text{versus} \quad H_1: \mu > 12{,}200$$

Step 2: The level of significance is $\alpha = 0.1$.

Step 3: Patricia found the sample mean, $\bar{x}$, to be 12,895.9 miles. The test statistic is

$$z_0 = \frac{\bar{x} - \mu_0}{\sigma/\sqrt{n}} = \frac{12{,}895.9 - 12{,}200}{3800/\sqrt{35}} = 1.08$$

The sample mean of 12,895.9 miles is 1.08 standard deviations above the mean of 12,200.

In Other Words

For this test (a right-tailed test), the *P*-value is the area under the standard normal curve to the right of $z_0 = 1.08$.

Step 4: Because Patricia is performing a right-tailed test,

$$P\text{-value} = P(Z > z_0) = P(Z > 1.08)$$

We have to determine the area under the standard normal curve to the right of $z_0 = 1.08$ as shown in Figure 10.

Figure 10

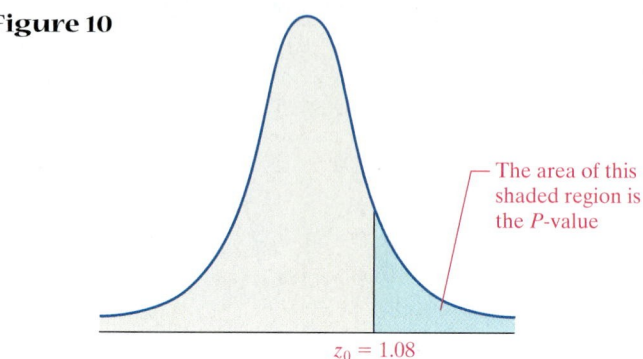

The area of this shaded region is the *P*-value

$z_0 = 1.08$

Using Table IV, we have

$$P\text{-value} = P(Z > 1.08) = 1 - P(Z \leq 1.08) = 1 - 0.8599 = 0.1401$$

The probability of obtaining a sample mean of 12,895.9 miles or higher from a population whose mean is 12,200 miles is 0.1401. So about 14 samples in 100 will result in a sample mean of 12,895.9 miles or higher from a population whose mean is 12,200 miles.

Step 5: A probability of an event of 0.1401 is not unusual. The decision rule that we use is this: If the *P*-value is less than the level of significance, α, we reject the null hypothesis. Because $0.1401 > 0.1$, we do not reject the null hypothesis.

In Other Words

If *P*-value $< \alpha$, then reject the null hypothesis.

Step 6: There is not sufficient evidence at the $\alpha = 0.1$ level of significance to support the claim that residents of the state of Montana drive more than the national average of 12,200 miles.

> **NW** Now Work Problem 19 Using the *P*-Value Approach.

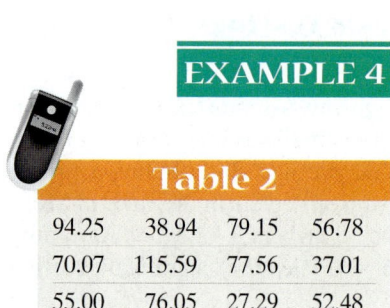

EXAMPLE 4 | **The *P*-Value Approach of Hypothesis Testing: Two-Tailed, Small Sample**

Table 2

94.25	38.94	79.15	56.78
70.07	115.59	77.56	37.01
55.00	76.05	27.29	52.48

Problem: According to CTIA–The Wireless Association, the mean monthly cell phone bill in 2004 was $50.64. A market researcher believes that the mean monthly cell phone bill is different today, but is not sure whether bills have declined because of technological advances or increased due to additional use. To test the claim, the researcher phones a simple random sample of 12 cell phone subscribers and obtains the data in Table 2.

Assuming $\sigma = \$18.49$, use these data to test the claim that the mean monthly cell phone bill is different from $50.64 at the $\alpha = 0.05$ level of significance. Use the *P*-value approach.

Approach: Because the sample size, n, is less than 30, we must verify that the data come from a population that is approximately normal with no outliers. We will construct a normal probability plot and boxplot to verify these requirements. We then proceed to follow Steps 1 through 6.

Solution: Figure 8 on page 519 displays the normal probability plot and boxplot. The normal probability plot indicates that the data could come from a population that is normal. The boxplot does not show any outliers.

Step 1: The claim is that the mean of the cell phone bill is different from $50.64. The claim can be written $\mu \neq 50.64$. This is a two-tailed test and we have

$$H_0: \mu = 50.64 \quad \text{versus} \quad H_1: \mu \neq 50.64$$

Step 2: The level of significance is $\alpha = 0.05$.

Step 3: From the data in Table 2, the sample mean is computed to be $65.014 and $n = 12$. We assume $\sigma = 18.49. The test statistic is

$$z_0 = \frac{\bar{x} - \mu_0}{\sigma/\sqrt{n}} = \frac{65.014 - 50.64}{18.49/\sqrt{12}} = 2.69$$

The sample mean of $65.014 is 2.69 standard deviations above the assumed population mean of $50.64.

Step 4: Because we are performing a two-tailed test,

$$P\text{-value} = 2P(Z > |z_0|) = 2P(Z > 2.69)$$

We need to determine the area under the standard normal curve to the right of $Z = 2.69$ and to the left of $Z = -2.69$, as shown in Figure 11.

In Other Words

To find the P-value for a two-tailed test, first determine whether the test statistic, z_0, is positive or negative. If z_0 is negative, determine the area under the standard normal curve to the left of z_0 and then multiply this area by 2. If z_0 is positive, find the area under the standard normal curve to the right of z_0 and double this value.

Figure 11

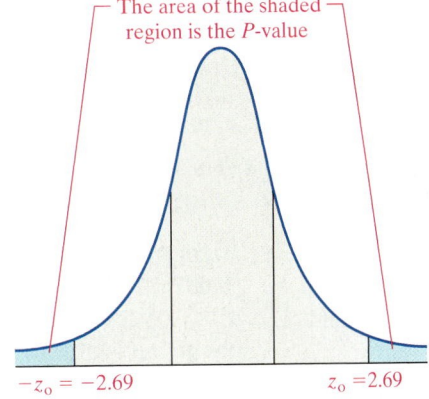

The area of the shaded region is the *P*-value

$-z_0 = -2.69$ $z_0 = 2.69$

Using Table IV, we have

$$
\begin{aligned}
P\text{-value} = P(Z < -2.69 \text{ or } Z > 2.69) &= P(Z < -2.69) + P(Z > 2.69) \\
&= 2P(Z > 2.69) \\
&= 2[1 - P(Z \leq 2.69)] \\
&= 2(1 - 0.9964) \\
&= 2(0.0036) \\
&= 0.0072
\end{aligned}
$$

The probability of obtaining a sample mean that is more than 2.69 standard deviations from the status quo population mean of $50.64 is 0.0072. This means less than 1 sample in 100 will result in a sample mean such as the one we obtained if the statement in the null hypothesis is true.

Step 5: Because the *P*-value is less than the level of significance $(0.0072 < 0.05)$, we reject the null hypothesis.

Step 6: There is sufficient evidence at the $\alpha = 0.05$ level of significance to support the claim that the mean monthly cell phone bill is different from $50.64, the mean amount in 2004.

Now Work Problem 25 Using the *P*-Value Approach.

EXAMPLE 5   Testing a Claim about a Population Mean Using Technology

Problem: According to CTIA–The Wireless Association, the mean monthly cell phone bill in 2004 was $50.64. A market researcher believes that the mean monthly cell phone bill is different today, but is not sure whether bills have declined because of technological advances, or increased due to additional use. To test the claim, the researcher phones a simple random sample of 12 cell phone subscribers and obtains the data in Table 2.

Assuming $\sigma = \$18.49$, use these data to test the claim that the mean monthly cell phone bill is different from $50.64 at the $\alpha = 0.05$ level of significance.

Approach: We will use MINITAB to test the claim. The steps for testing claims about a population mean with σ known using the TI-83/84 Plus graphing calculators, MINITAB, and Excel are given in the Technology Step by Step on page 530.

Result: Figure 12 shows the results obtained from MINITAB.

Figure 12

One-Sample Z: Cell Phone Bill

```
Test of mu = 50.64 vs not = 50.64
The assumed standard deviation = 18.49

Variable       N    Mean    StDev  SE Mean       95% CI            Z      P
Cell Phone
Bill          12  65.0142  25.4587  5.3376  (54.5527, 75.4757)  2.69   0.007
```

USING TECHNOLOGY

The P-value obtained from technology may be slightly different from the P-value obtained by hand because of rounding.

The P-value is highlighted. From MINITAB, we have P-value = 0.007.

Interpretation: Because the P-value is less than the level of significance $(0.007 < 0.05)$, we reject the null hypothesis. There is sufficient evidence at the $\alpha = 0.05$ level of significance to support the claim that the mean monthly cell phone bill is different from $50.64, the mean amount in 2004.

One advantage of using P-values over the classical approach in hypothesis testing is that P-values provide information regarding the strength of the evidence. In Example 2, we rejected the null hypothesis, but did not learn anything about the strength of the evidence against the null hypothesis. Example 4 tested the same claim using P-values. The P-value was 0.0072. This result not only led us to reject the null hypothesis, but also indicates the strength of the evidence against the null hypothesis: Less than 1 sample in 100 would give us the sample mean that we got if the null hypothesis, H_0: $\mu = 50.64$, were true.

Another advantage of P-values is that they are interpreted the same way, regardless of the type of hypothesis test being performed. If P-value $< \alpha$, reject the null hypothesis.

④ Test a Claim about a Population Mean with σ Known Using Confidence Intervals

Recall that the level of confidence in a confidence interval is a probability that represents the percentage of intervals that will contain μ if repeated samples are obtained. The level of confidence is denoted $(1 - \alpha) \cdot 100\%$. We can use confidence intervals to test H_0: $\mu = \mu_0$ versus H_1: $\mu \neq \mu_0$ using the following criterion.

> When testing H_0: $\mu = \mu_0$ versus H_1: $\mu \neq \mu_0$, if a $(1 - \alpha) \cdot 100\%$ confidence interval contains μ_0, we do not reject the null hypothesis. However, if the confidence interval does not contain μ_0, we have sufficient evidence that supports the claim stated in the alternative hypothesis and conclude $\mu \neq \mu_0$ at the level of significance, α.

EXAMPLE 6   **Testing a Claim about a Population Mean Using a Confidence Interval**

Problem: Test the claim presented in Examples 2 and 4 at the $\alpha = 0.05$ level of significance by constructing a 95% confidence interval about μ, the population mean monthly cell phone bill.

Approach: We construct the 95% confidence interval using the data in Table 1 or Table 2. If the interval contains the status quo mean of $50.64, we do not reject the null hypothesis.

Solution: We use the formula on page 453 to find the lower and upper bounds with $\bar{x} = \$65.014$, $\sigma = \$18.49$, and $n = 12$.

$$\text{Lower bound:}\quad \bar{x} - z_{\alpha/2} \cdot \frac{\sigma}{\sqrt{n}} = \$65.014 - 1.96 \cdot \frac{18.49}{\sqrt{12}} = \$65.014 - \$10.462 = \$54.552$$

$$\text{Upper bound:}\quad \bar{x} + z_{\alpha/2} \cdot \frac{\sigma}{\sqrt{n}} = \$65.014 + 1.96 \cdot \frac{18.49}{\sqrt{12}} = \$65.014 + \$10.462 = \$75.476$$

We are 95% confident the mean monthly cell phone bill is between $54.552 and $75.476. Because the mean stated in the null hypothesis, $H_0: \mu = 50.64$, is not included in this interval, we reject the null hypothesis. There is sufficient evidence at the $\alpha = 0.05$ level of significance to support the claim that the mean monthly cell phone bill is different from $50.64.

Now Work Problem 35.

5   Understand the Difference between Statistical Significance and Practical Significance

In Other Words
Results are statistically significant if the difference between the observed result and the statement made in the null hypothesis is unlikely to occur due to chance alone.

When a large sample size is used in a hypothesis test, the results could be statistically significant even though the difference between the sample statistic and mean stated in the null hypothesis may have no *practical significance*.

Definition   **Practical significance** refers to the idea that small differences between the statistic and parameter stated in the null hypothesis are statistically significant, while the difference is not large enough to cause concern or be considered important.

EXAMPLE 7   **Statistical versus Practical Significance**

Problem: According to the American Community Survey, the mean travel time to work in Dallas, Texas, in 2003 was 23.6 minutes. Suppose the Department of Transportation in Dallas just reprogrammed all the traffic lights in an attempt to reduce travel time. To test the claim that travel times in Dallas have decreased as a result of the reprogramming, they obtain a sample of 2500 commuters, record their travel time to work, and obtain a sample mean of 23.3 minutes. Assuming that the population standard deviation travel time to work is known to be 8.4 minutes, test the claim that travel times in Dallas have decreased as a result of the reprogramming at the $\alpha = 0.05$ level of significance.

Approach: We will use both the classical approach and *P*-value approach to test the claim.

Solution

Step 1: The claim is that the mean travel time to work has decreased from 23.6 minutes. This claim can be written $\mu < 23.6$. This is a left-tailed test and we have

$$H_0: \mu = 23.6 \quad \text{versus} \quad H_1: \mu < 23.6$$

Step 2: The level of significance is $\alpha = 0.05$.

Step 3: The test statistic is

$$z_0 = \frac{\overline{x} - \mu_0}{\dfrac{\sigma}{\sqrt{n}}} = \frac{23.3 - 23.6}{\dfrac{8.4}{\sqrt{2500}}} = -1.79$$

Classical Approach

Step 4: This is a left-tailed test. With $\alpha = 0.05$, the critical value is $-z_{0.05} = -1.645$.

Step 5: Because the test statistic is less than the critical value (the critical value falls in the critical region), we reject the null hypothesis.

P-value Approach

Step 4: Because this is a left-tailed test, the P-value is P-value $= P(Z < z_0) = P(Z < -1.79) = 0.0367$

Step 5: Because the P-value is less than the level of significance $(0.0367 < 0.05)$, we reject the null hypothesis.

Step 6: There is sufficient evidence at the $\alpha = 0.05$ level of significance to support the claim that the mean travel time to work has decreased.

While the difference between 23.3 and 23.6 is statistically significant, it really has no practical meaning. After all, is 0.3 minutes (18 seconds) really going to make anyone feel better about their commute to work?

CAUTION!

Beware of studies with large sample sizes that claim statistical significance because the differences may not have any practical meaning.

The reason that the results from Example 7 were statistically significant had to do with the large sample size. The moral of the story is this:

> Large sample sizes can lead to results that are statistically significant, while the difference between the statistic and parameter in the null hypothesis is not enough to be considered practically significant.

10.2 ASSESS YOUR UNDERSTANDING

Concepts and Vocabulary

1. State the requirements that must be satisfied to test a claim regarding a population mean with σ known.
2. Determine the critical value for a right-tailed test regarding a population mean with σ known at the $\alpha = 0.01$ level of significance.
3. Determine the critical value for a two-tailed test regarding a population mean with σ known at the $\alpha = 0.05$ level of significance.
4. The procedures for testing a claim regarding a population mean with σ known are robust. What does this mean?
5. Explain what a P-value is. What is the criterion for rejecting the null hypothesis using the P-value approach?
6. Suppose that we are testing the hypotheses $H_0: \mu = \mu_0$ versus $H_1: \mu < \mu_0$ and we find the P-value to be 0.23.

   Explain what this means. Would you reject H_0? Why?
7. Suppose that we are testing the hypotheses $H_0: \mu = \mu_0$ versus $H_1: \mu \neq \mu_0$ and we find the P-value to be 0.02. Explain what this means. Would you reject H_0? Why?
8. Discuss the advantages and disadvantages of using the classical approach to hypothesis testing. Discuss the advantages and disadvantages of using the P-value approach to hypothesis testing.
9. In your own words, explain the difference between *statistical significance* and *practical significance*.
10. *True or False*: To test $H_0: \mu = \mu_0$ versus $H_1: \mu \neq \mu_0$ using a 5% level of significance, we could construct a 95% confidence interval.

Skill Building

11. To test H_0: $\mu = 50$ versus H_1: $\mu < 50$, a random sample of size $n = 24$ is obtained from a population that is known to be normally distributed with $\sigma = 12$.

(a) If the sample mean is determined to be $\bar{x} = 47.1$, compute the test statistic.

(b) If the researcher decides to test this hypothesis at the $\alpha = 0.05$ level of significance, determine the critical value.

(c) Draw a normal curve that depicts the critical region.

(d) Will the researcher reject the null hypothesis? Why?

12. To test H_0: $\mu = 40$ versus H_1: $\mu > 40$, a random sample of size $n = 25$ is obtained from a population that is known to be normally distributed with $\sigma = 6$.

(a) If the sample mean is determined to be $\bar{x} = 42.3$, compute the test statistic.

(b) If the researcher decides to test this hypothesis at the $\alpha = 0.1$ level of significance, determine the critical value.

(c) Draw a normal curve that depicts the critical region.

(d) Will the researcher reject the null hypothesis? Why?

13. To test H_0: $\mu = 100$ versus H_1: $\mu \neq 100$, a random sample of size $n = 23$ is obtained from a population that is known to be normally distributed with $\sigma = 7$.

(a) If the sample mean is determined to be $\bar{x} = 104.8$, compute the test statistic.

(b) If the researcher decides to test this hypothesis at the $\alpha = 0.01$ level of significance, determine the critical values.

(c) Draw a normal curve that depicts the critical regions.

(d) Will the researcher reject the null hypothesis? Why?

14. To test H_0: $\mu = 80$ versus H_1: $\mu < 80$, a random sample of size $n = 22$ is obtained from a population that is known to be normally distributed with $\sigma = 11$.

(a) If the sample mean is determined to be $\bar{x} = 76.9$, compute the test statistic.

(b) If the researcher decides to test this hypothesis at the $\alpha = 0.02$ level of significance, determine the critical value.

(c) Draw a normal curve that depicts the critical region.

(d) Will the researcher reject the null hypothesis? Why?

15. To test H_0: $\mu = 20$ versus H_1: $\mu < 20$, a random sample of size $n = 18$ is obtained from a population that is known to be normally distributed with $\sigma = 3$.

(a) If the sample mean is determined to be $\bar{x} = 18.3$, compute and interpret the P-value.

(b) If the researcher decides to test this hypothesis at the $\alpha = 0.05$ level of significance, will the researcher reject the null hypothesis? Why?

16. To test H_0: $\mu = 4.5$ versus H_1: $\mu > 4.5$, a random sample of size $n = 13$ is obtained from a population that is known to be normally distributed with $\sigma = 1.2$.

(a) If the sample mean is determined to be $\bar{x} = 4.9$, compute and interpret the P-value.

(b) If the researcher decides to test this hypothesis at the $\alpha = 0.1$ level of significance, will the researcher reject the null hypothesis? Why?

17. To test H_0: $\mu = 105$ versus H_1: $\mu \neq 105$, a random sample of size $n = 35$ is obtained from a population whose standard deviation is known to be $\sigma = 12$.

(a) Does the population need to be normally distributed to compute the P-value?

(b) If the sample mean is determined to be $\bar{x} = 101.2$, compute and interpret the P-value.

(c) If the researcher decides to test this hypothesis at the $\alpha = 0.02$ level of significance, will the researcher reject the null hypothesis? Why?

18. To test H_0: $\mu = 45$ versus H_1: $\mu \neq 45$, a random sample of size $n = 40$ is obtained from a population whose standard deviation is known to be $\sigma = 8$.

(a) Does the population need to be normally distributed to compute the P-value?

(b) If the sample mean is determined to be $\bar{x} = 48.3$, compute and interpret the P-value.

(c) If the researcher decides to test this hypothesis at the $\alpha = 0.05$ level of significance, will the researcher reject the null hypothesis? Why?

Applying the Concepts

19. **Are Women Getting Taller?** A researcher claims that the [NW] average height of a woman aged 20 years or older is greater than the 1994 mean height of 63.7 inches, on the basis of data obtained from the Centers for Disease Control and Prevention's, *Advance Data Report*, No. 347. She obtains a simple random sample of 45 women and finds the sample mean height to be 63.9 inches. Assume that the population standard deviation is 3.5 inches. Test the researcher's claim using either the classical approach or the P-value approach at the $\alpha = 0.05$ level of significance.

20. **ATM Withdrawals** The manufacturer of a certain type of ATM machine reports that the mean ATM withdrawal is $60. The manager of a convenience store with an ATM machine thinks that mean withdrawal from his machine is less than this amount. He obtains a simple random sample of 35 withdrawals over the past year and finds the sample mean to be $52. Assume that the population standard deviation is $13. Test the manager's claim using either the classical approach or the P-value approach at the $\alpha = 0.05$ level of significance.

21. **SAT Exam Scores** A school administrator claims that students whose first language learned is not English score worse on the verbal portion of the SAT exam than students whose first language is English. The mean SAT verbal score of students whose first language is English is 515, on the basis of data obtained from the College Board. Suppose a simple random sample of 20 students whose first language learned was not English results in a sample mean SAT verbal score of 458. SAT verbal scores are normally distributed with a population standard deviation of 112.

(a) Why is it necessary for SAT verbal scores to be normally distributed to test the claim using the methods of this section?

(b) Test the researcher's claim using the classical approach or the P-value approach at the $\alpha = 0.10$ level of significance.

22. **SAT Exam Scores** A school administrator claims that students whose first language learned is not English score differently on the math portion of the SAT exam than students whose first language is English. The mean SAT math score of students whose first language is English is 516, on the basis of data obtained from the College Board. Suppose a simple random sample of 20 students whose first language learned was not English results in a sample mean SAT math score of 522. SAT math scores are normally distributed with a population standard deviation of 114.

    (a) Why is it necessary for SAT math scores to be normally distributed to test the claim using the methods of this section?

    (b) Test the researcher's claim using the classical approach or the *P*-value approach at the $\alpha = 0.10$ level of significance.

23. **Acid Rain** In 1990, the mean pH level of the rain in Pierce County, Washington, was 5.03. A biologist claims that the acidity of rain has increased. (This would mean that the pH level of the rain has decreased.) From a random sample of 19 rain dates in 2004, she obtains the following data:

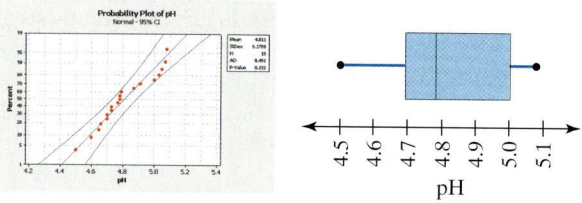

5.08	4.66	4.70	4.87
4.78	5.00	4.50	4.73
4.79	4.65	4.91	5.07
5.03	4.78	4.77	4.60
4.73	5.05	4.70	

Source: National Atmospheric Deposition Program

    (a) Because the sample size is small, she must verify that pH level is normally distributed and the sample does not contain any outliers. The normal probability plot and boxplot are shown next. Are the conditions for testing the hypothesis satisfied?

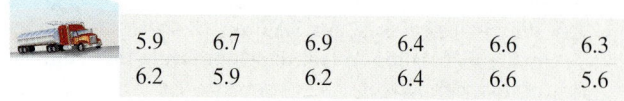

    (b) Test the hypothesis, assuming that $\sigma = 0.2$ at the $\alpha = 0.01$ level of significance.

24. **Fuel Catalyst** To improve fuel efficiency and reduce pollution, the owner of a trucking fleet decides to install a new fuel catalyst in all his semitrucks. He feels that the catalyst will help to increase the number of miles per gallon. Before the installation, his trucks had a mean gas mileage of 5.6 miles per gallon. A random sample of 12 trucks after the installation gave the following gas mileages:

5.9	6.7	6.9	6.4	6.6	6.3
6.2	5.9	6.2	6.4	6.6	5.6

    (a) Because the sample size is small, he must verify that mileage is normally distributed and the sample does

not contain any outliers. The normal probability plot and boxplot are shown. Are the conditions for testing the hypothesis satisfied?

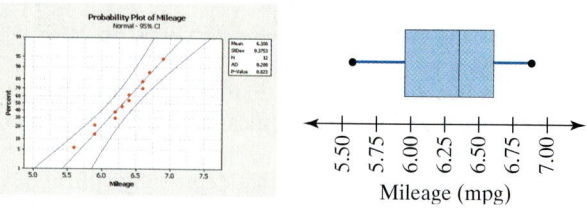

    (b) Test the hypothesis, assuming that $\sigma = 0.5$ mile per gallon at the 0.05 level of significance.

25. **Filling Bottles** A certain brand of apple juice is supposed to have 64 ounces of juice. Because the punishment for underfilling bottles is severe, the target mean amount of juice is 64.05 ounces. However, the filling machine is not precise, and the exact amount of juice varies from bottle to bottle. The quality-control manager wishes to verify that the mean amount of juice in each bottle is 64.05 ounces so that she can be sure that the machine is not over- or underfilling. She randomly samples 22 bottles of juice and measures the content and obtains the following data:

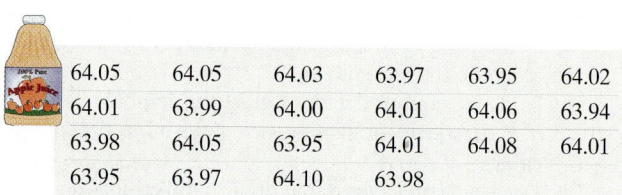

64.05	64.05	64.03	63.97	63.95	64.02
64.01	63.99	64.00	64.01	64.06	63.94
63.98	64.05	63.95	64.01	64.08	64.01
63.95	63.97	64.10	63.98		

    (a) Because the sample size is small, she must verify that the amount of juice is normally distributed and the sample does not contain any outliers. The normal probability plot and boxplot are shown. Are the conditions for testing the hypothesis satisfied?

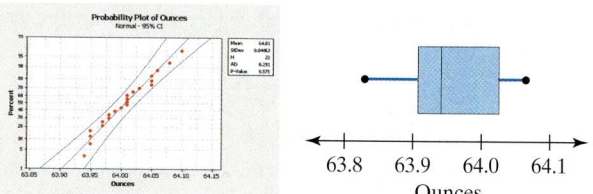

    (b) Test the hypothesis, assuming that $\sigma = 0.06$ at the $\alpha = 0.01$ level of significance.

    (c) Should the assembly line be shut down so that the machine can be recalibrated?

    (d) Explain why a level of significance of $\alpha = 0.01$ might be more reasonable than $\alpha = 0.1$. [**Hint:** Consider the consequences of incorrectly rejecting the null hypothesis.]

26. **SAT Reasoning Test** In 2005, in response to various criticisms, the College Board implemented changes to the SAT test. In particular, the math portion eliminated quantitative comparisons and expanded the topics covered. A school administrator believed that the new math portion is more difficult and will result in lower scores on the math portion compared to the 2004 average score of 516. A ran-

dom sample of 25 students taking the new SAT test resulted in the following scores on the math portion.

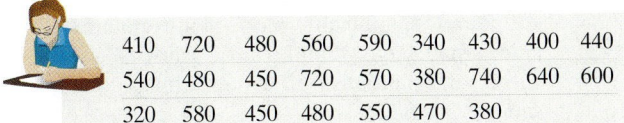

410	720	480	560	590	340	430	400	440
540	480	450	720	570	380	740	640	600
320	580	450	480	550	470	380		

(a) Because the sample size is small, he must verify that the new math scores are normally distributed and the sample does not contain any outliers. The normal probability plot and boxplot are shown. Are the conditions for testing the hypothesis satisfied?

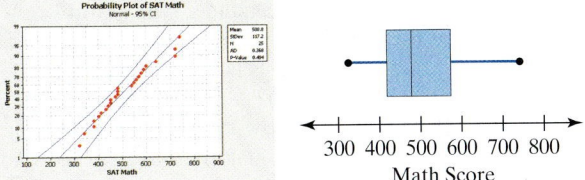

(b) Test the hypothesis, assuming $\sigma = 114$, at the $\alpha = 0.1$ level of significance.

27. Are Cars Younger? Suppose you have just been hired by Ford Motor Company. Management asks you to test the claim that cars are younger today versus 1995. According to the Nationwide Personal Transportation Survey conducted by the U.S. Department of Transportation, the mean age of a car in 1995 was 8.33 years. Based on a random sample of 18 automobile owners, you obtain the ages as shown in the following table:

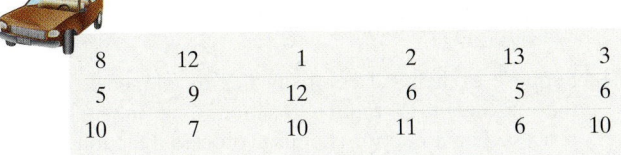

8	12	1	2	13	3
5	9	12	6	5	6
10	7	10	11	6	10

(a) Because the sample size is small, you must verify that the data are normally distributed with no outliers by drawing a normal probability plot and boxplot. Based on the following graphs, can you perform a hypothesis test?

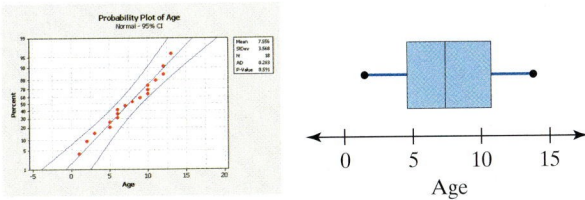

(b) Test the hypothesis, assuming that $\sigma = 3.8$ years at the $\alpha = 0.1$ level of significance.

28. It's a Hot One! Recently, a friend of mine claimed that the summer of 2000 in Houston, Texas, was hotter than usual. To test his claim, I went to AccuWeather.com and randomly selected 12 days in the summer of 2000. I then recorded the departure from normal, with positive values indicating above-normal temperatures and negative val-

ues indicating below-normal temperatures, as shown in the following:

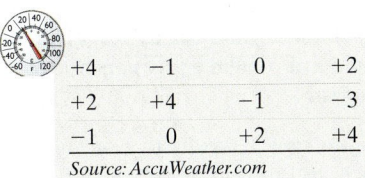

+4	−1	0	+2
+2	+4	−1	−3
−1	0	+2	+4

Source: AccuWeather.com

(a) Because the sample size is small, I must verify that the temperature departure is normally distributed and the sample does not contain any outliers. The normal probability plot and boxplot are shown. Are the conditions for testing the hypothesis satisfied?

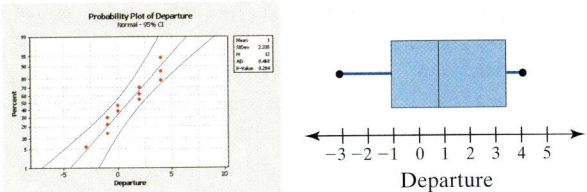

(b) Test the hypothesis, assuming that $\sigma = 1.8$ at the $\alpha = 0.05$ level of significance.

29. Farm Size In 1990, the average farm size in Kansas was 694 acres, according to data obtained from the U.S. Department of Agriculture. A researcher claims that farm sizes are larger now due to consolidation of farms. She obtains a random sample of 40 farms and determines the mean size to be 731 acres. Assume that $\sigma = 212$ acres. Test the researcher's claim at the $\alpha = 0.05$ level of significance.

30. Oil Output An energy official claims that the oil output per well in the United States has declined from the 1998 level of 11.1 barrels per day. He randomly samples 50 wells throughout the United States and determines the mean output to be 10.7 barrels per day. Assume that $\sigma = 1.3$ barrels. Test the researcher's claim at the $\alpha = 0.05$ level of significance.

31. Volume of Dell Computer Stock The average daily volume of Dell Computer stock in 2000 was $\mu = 31.8$ million shares, with a standard deviation of $\sigma = 14.8$ million shares, according to Yahoo!Finance. A stock analyst claims that the stock volume in 2004 is different from the 2000 level. Based on a random sample of 35 trading days in 2004, he finds the sample mean to be 23.5 million shares. Test the analyst's claim at the $\alpha = 0.05$ level of significance.

32. Volume of Motorola Stock The average daily volume of Motorola stock in 2000 was $\mu = 11.4$ million shares, with a standard deviation of $\sigma = 8.3$ million shares, according to Yahoo!Finance. A stock analyst claims the stock volume in 2004 is different from the 2000 level. Based on a random sample of 35 trading days in 2004, he finds the sample mean to be 13.3 million shares. Test the analyst's claim at the $\alpha = 0.05$ level of significance.

33. Using Confidence Intervals to Test Hypotheses Test the claim made in Problem 25 by constructing a 99% confidence interval.

34. Using Confidence Intervals to Test Hypotheses Test the claim made in Problem 26 by constructing a 95% confidence interval.

35. Using Confidence Intervals to Test Hypotheses Test the
NW claim made in Problem 31 by constructing a 95% confidence interval.

36. Using Confidence Intervals to Test Hypotheses Test the claim made in Problem 32 by constructing a 95% confidence interval.

37. Statistical Significance versus Practical Significance A math teacher claims that she has developed a review course that increases the scores of students on the math portion of the SAT exam. Based on data from the College Board, SAT scores are normally distributed with $\mu = 514$ and $\sigma = 113$. The teacher obtains a random sample of 1800 students, puts them through the review class, and finds that the mean SAT math score of the 1800 students is 518.

(a) State the null and alternative hypotheses.

(b) Test the claim at the $\alpha = 0.10$ level of significance. Is a mean SAT math score of 518 significantly higher than 514?

(c) Do you think that a mean SAT math score of 518 versus 514 will affect the decision of a school admissions administrator? In other words, does the increase in the score have any practical significance?

(d) Test the claim at the $\alpha = 0.10$ level of significance with $n = 400$ students. Assume that the sample mean is still 518. Is a sample mean of 518 significantly more than 514? Conclude that large sample sizes cause P-values to shrink substantially, all other things being the same.

38. Statistical Significance versus Practical Significance The manufacturer of a daily dietary supplement claims that their product will help people lose weight. The company obtains a random sample of 950 adult males aged 20 to 74 who take the supplement and finds their mean weight loss after eight weeks to be 0.9 pounds. Assume the population standard deviation weight loss is $\sigma = 7.2$ pounds.

(a) State the null and alternative hypotheses.

(b) Test the claim at the $\alpha = 0.1$ level of significance. Is a mean weight loss of 0.9 pound significant?

(c) Do you think that a mean weight loss of 0.9 pounds is worth the expense and commitment of a daily dietary supplement? In other words, does the weight loss have any practical significance?

(d) Test the claim at the $\alpha = 0.1$ level of significance with $n = 40$ subjects. Assume that the sample mean weight loss is still 0.9 pounds. Is a sample mean weight loss of 0.9 pounds significantly more than 0 pounds? Conclude that large sample sizes cause P-values to shrink substantially, all other things being the same.

39. Simulation Simulate drawing 50 simple random samples of size $n = 20$ from a population that is normally distributed with mean 80 and standard deviation 7.

(a) Test the null hypothesis H_0: $\mu = 80$ versus the alternative hypothesis H_1: $\mu \neq 80$.

(b) Suppose we were testing this hypothesis at the $\alpha = 0.1$ level of significance. How many of the 50 samples would you expect to result in a Type I error?

(c) Count the number of samples that lead to a rejection of the null hypothesis. Is it close to the expected value determined in part (b)?

(d) Describe how we know that a rejection of the null hypothesis results in making a Type I error in this situation.

40. Simulation Simulate drawing 40 simple random samples of size $n = 35$ from a population that is exponentially distributed with mean 8 and standard deviation $\sqrt{8}$.

(a) Test the null hypothesis H_0: $\mu = 8$ versus the alternative hypothesis H_1: $\mu \neq 8$.

(b) Suppose we were testing this hypothesis at the $\alpha = 0.05$ level of significance. How many of the 40 samples would you expect to result in a Type I error?

(c) Count the number of samples that lead to a rejection of the null hypothesis. Is it close to the expected value determined in part (b)?

(d) Describe how we know that a rejection of the null hypothesis results in making a Type I error in this situation.

41. Suppose a chemical company has developed a catalyst that is meant to reduce reaction time in a chemical process. For a certain chemical process, reaction time is known to be 150 seconds. The researchers conducted an experiment with the catalyst 40 times and measured reaction time. The researchers reported that the catalysts reduced reaction time with a P-value of 0.02.

(a) Identify the null and alternative hypotheses.

(b) Explain what this result means. Do you believe that the catalyst is effective?

Technology Step-by-Step   **Hypothesis Tests Regarding μ, σ Known**

TI-83/84 Plus   *Step 1:* If necessary, enter raw data in L1.

Step 2: Press STAT, highlight TESTS, and select `1:Z-Test`.

Step 3: If the data are raw, highlight DATA; make sure that List1 is set to L1 and Freq is set to 1. If summary statistics are known, highlight STATS and enter the summary statistics. Following σ, enter the population standard deviation. For the value of μ_0, enter the value of the mean stated in the null hypothesis.

Step 4: Select the direction of the alternative hypothesis.

Step 5: Highlight Calculate and press ENTER. The TI-83/84 Plus gives the *P*-value.

MINITAB   **Step 1:** Enter raw data in column C1.

Step 2: Select the **Stat** menu, highlight **Basic Statistics**, and then highlight **1-Sample Z** ....

Step 3: Click Options. In the cell marked "Alternative," select the appropriate direction for the alternative hypothesis. Click OK.

Step 4: Enter C1 in the cell marked "Variables." In the cell labeled "Test Mean", enter the value of the mean stated in the null hypothesis. In the cell labeled "standard deviation," enter the value of σ. Click OK.

Excel   **Step 1:** If necessary, enter raw data in column A.

Step 2: Load the PHStat Add-in.

Step 3: Select the **PHStat** menu, highlight **One Sample Tests . . .**, and then highlight **Z Test for the mean, sigma known . . . .**

Step 4: Enter the value of the null hypothesis, the level of significance, α, and the value of σ. If the summary statistics are known, click "Sample statistics known" and enter the sample size and sample mean. If summary statistics are unknown, click "Sample statistics unknown." With the cursor in the "Sample cell range" cell, highlight the data in column A. Click the option corresponding to the desired test [two-tail, upper (right) tail, or lower (left) tail]. Click OK.

10.3 Testing Claims about a Population Mean in Practice

Preparing for This Section   Before getting started, review the following:

- Sampling distribution of $\bar{x}$ (Section 8.1, pp. 417–430)
- The *t*-distribution (Section 9.2, pp. 466–469)
- Using probabilities to identify unusual events (Section 5.1, p. 253)

Objectives   **1**   **Test a claim about a population mean with σ unknown**

In Other Words
When σ is known, use z; when σ is unknown use t.

In Section 10.2, we assumed that the population standard deviation, σ, was known when testing claims regarding the population mean. We now introduce procedures for testing claims regarding a population mean when σ is not known. The only difference from the situation where σ is known is that we must use the *t*-distribution rather than the *z*-distribution.

We do not replace σ with s and say that $z = \dfrac{\bar{x} - \mu}{s/\sqrt{n}}$ is normally distributed

with mean 0 and standard deviation 1. Instead, $t = \dfrac{\bar{x} - \mu}{s/\sqrt{n}}$ follows **Student's *t*-**

distribution with $n - 1$ degrees of freedom. Let's review the properties of the *t*-distribution.

Properties of the *t*-Distribution

1. The *t*-distribution is different for different degrees of freedom.

2. The *t*-distribution is centered at 0 and is symmetric about 0.

3. The area under the curve is 1. Because of the symmetry, the area under the curve to the right of 0 equals the area under the curve to the left of 0, which equals $\frac{1}{2}$.

4. As t increases without bound, the graph approaches, but never equals, zero. As t decreases without bound, the graph approaches, but never equals, zero.

5. The area in the tails of the t-distribution is a little greater than the area in the tails of the standard normal distribution, because using s as an estimate of σ introduces more variability to the t-statistic.

6. As the sample size n increases, the density curve of t gets closer to the standard normal density curve. This result occurs because, as the sample size n increases, the values of s get closer to the values of σ, by the Law of Large Numbers.

❶ Testing a Claim about a Population Mean with σ Unknown

From Section 10.2, we know that there are two approaches (besides using confidence intervals) we can use to test claims regarding a population mean, the classical approach and the P-value approach. We will present both methods here. Your instructor may choose one or both approaches.

Both the classical approach and the P-value approach to testing a claim about μ with σ unknown follow the exact same logic as testing a claim about μ with σ known. The only difference is that we use Student's t-distribution, rather than the normal distribution.

Testing a Claim Regarding a Population Mean with σ Unknown

If a claim is made regarding the population mean with σ unknown, we can use the following steps to test the claim, provided that

1. the sample is obtained using simple random sampling
2. the sample has no outliers and the population from which the sample is drawn is normally distributed or the sample size, n, is large ($n \geq 30$).

Step 1: A claim is made regarding the population mean. The claim is used to determine the null and alternative hypotheses. The hypotheses can be structured in one of three ways:

Two-Tailed	Left-Tailed	Right-Tailed
$H_0: \mu = \mu_0$	$H_0: \mu = \mu_0$	$H_0: \mu = \mu_0$
$H_1: \mu \neq \mu_0$	$H_1: \mu < \mu_0$	$H_1: \mu > \mu_0$

Note: μ_0 is the assumed value of the population mean.

Step 2: Select a level of significance α, depending on the seriousness of making a Type I error.

Step 3: Compute the test statistic

$$t_0 = \frac{\bar{x} - \mu_0}{s/\sqrt{n}}$$

which follows Student's t-distribution with $n - 1$ degrees of freedom.

Classical Approach

Step 4: Use Table V to determine the critical value using $n - 1$ degrees of freedom.

P-value Approach

Step 4: Use Table V to estimate the P-value using $n - 1$ degrees of freedom.

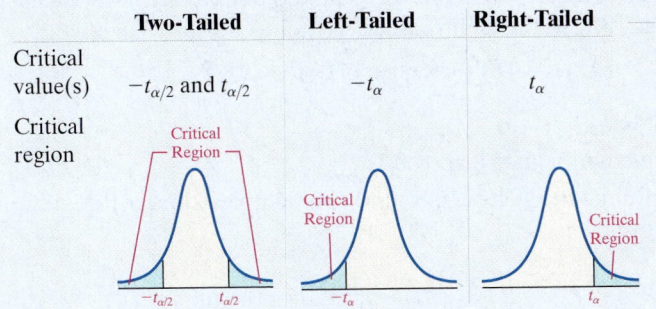

	Two-Tailed	**Left-Tailed**	**Right-Tailed**
Critical value(s)	$-t_{\alpha/2}$ and $t_{\alpha/2}$	$-t_{\alpha}$	t_{α}

Step 5: Compare the critical value with the test statistic.

Step 5: If the P-value $< \alpha$, reject the null hypothesis.

Two-Tailed	**Left-Tailed**	**Right-Tailed**
If $t_0 < -t_{\alpha/2}$ or $t_0 > t_{\alpha/2}$, reject the null hypothesis	If $t_0 < -t_{\alpha}$, reject the null hypothesis	If $t_0 > t_{\alpha}$, reject the null hypothesis

Step 6: State the conclusion.

Notice that the procedure just presented requires either that the population from which the sample was drawn be normal or that the sample size be large ($n \geq 30$). The procedure is robust, so minor departures from normality will not adversely affect the results of the test. However, if the data include outliers, the procedure should not be used.

Just as we did for hypothesis tests with σ known, we will verify these assumptions by constructing normal probability plots (to assess normality) and boxplots (to discover whether there are outliers). If the normal probability plot indicates that the data do not come from a normal population or if the boxplot reveals outliers, nonparametric tests should be performed, as discussed in Section 15.3.

Before we look at a couple of examples, it is important to understand that we cannot find exact P-values using the t-distribution table (Table V) because the table provides t-values only for certain areas. However, we can use the table to calculate lower and upper bounds on the P-value. To find exact P-values, we use statistical software or a graphing calculator with advanced statistical features.

In Other Words

When σ is unknown, exact P-values can be found using technology.

EXAMPLE 1   **Testing a Claim about a Population Mean, Large Sample**

Problem: According to the Centers for Disease Control, the mean number of cigarettes smoked per day by individuals who are daily smokers is 18.1. A researcher claims that retired adults smoke less than the general population of daily smokers. To test this claim, she obtains a random sample of 40 retired adults who are current smokers and records the number of cigarettes smoked on a randomly selected day. The data result in a sample mean of 16.8 cigarettes and a standard deviation of 4.7 cigarettes. Is there sufficient evidence at the $\alpha = 0.1$ level of significance to support the claim that retired adults who are daily smokers smoke less than the general population of daily smokers?

Approach: Because the sample size is large, we can follow the steps to testing a claim about a population mean given on pages 532–533.

Solution

Step 1: The researcher claims that retired adults smoke less than the general population. The mean number of cigarettes smoked per day by individuals who are daily smokers is 18.1, so the claim can be written $\mu < 18.1$. We have

$$H_0: \mu = 18.1 \quad \text{versus} \quad H_1: \mu < 18.1$$

This is a left-tailed test.

Step 2: The level of significance is $\alpha = 0.1$.

Step 3: The sample mean is $\bar{x} = 16.8$, and the sample standard deviation is $s = 4.7$. The test statistic is

$$t_0 = \frac{\bar{x} - \mu_0}{s/\sqrt{n}} = \frac{16.8 - 18.1}{4.7/\sqrt{40}} = -1.749$$

Classical Approach

Step 4: Because this is a left-tailed test, we determine the critical t-value at the $\alpha = 0.1$ level of significance with $n - 1 = 40 - 1 = 39$ degrees of freedom to be $-t_{0.1} = -1.304$. The critical region is displayed in Figure 13.

Figure 13

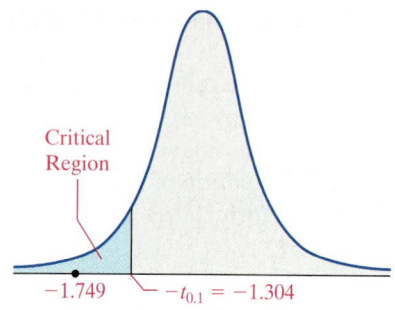

Step 5: Because the test statistic $t_0 = -1.749$ is less than the critical value $-t_{0.1} = -1.304$, the researcher rejects the null hypothesis. We label this point in Figure 13.

P-value Approach

Step 4: Because this is a left-tailed test, the P-value is the area under the t-distribution with $40 - 1 = 39$ degrees of freedom to the left of the test statistic, $t_0 = -1.749$, as shown in Figure 14(a). That is, P-value $= P(t < t_0) = P(t < -1.749)$, with 39 degrees of freedom.

Because of the symmetry of the t-distribution, the area under the distribution to the left of -1.749 equals the area under the distribution to the right of 1.749. So P-value $= P(t < -1.749) = P(t > 1.749)$. See Figure 14(b).

Using Table V, we find the row that corresponds to 39 degrees of freedom. The value 1.749 lies between 1.685 and 2.023. The value of 1.685 has an area under the t-distribution of 0.05 to the right, with 39 degrees of freedom. The area under the t-distribution with 39 degrees of freedom to the right of 2.023 is 0.025. See Figure 15.

Because 1.749 is between 1.685 and 2.023, the P-value is between 0.025 and 0.05. So

$$0.025 < P\text{-value} < 0.05$$

Figure 14

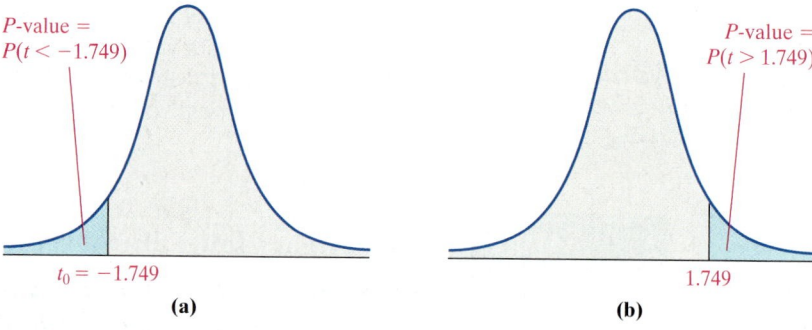

Step 5: Because the P-value is less than the level of significance $\alpha = 0.1$, we reject the null hypothesis.

Step 6: There is sufficient evidence to support the researcher's claim that retired adults smoke less than the general population of daily smokers at the $\alpha = 0.1$ level of significance.

Figure 15

					Area in Right Tail							
df	0.25	0.20	0.15	0.10	0.05	0.025	0.02	0.01	0.005	0.0025	0.001	0.0005
1	1.000	1.376	1.963	3.078	6.314	12.706	15.894	31.821	63.657	127.321	318.289	636.558
2	0.816	1.061	1.386	1.886	2.920	4.303	4.849	6.965	9.925	14.089	22.328	31.600
3	0.765	0.978	1.250	1.638	2.353	3.182	3.482	4.541	5.841	7.453	10.214	12.924
37	0.681	0.851	1.051	1.305	1.687	2.026	2.129	2.431	2.715	2.985	3.326	3.574
38	0.681	0.851	1.051	1.304	1.686	2.024	2.127	2.429	2.712	2.980	3.339	3.566
39	0.641	0.851	1.050	1.304	1.685	2.023	2.125	2.426	2.704	2.976	3.313	3.558
40	0.681	0.851	1.050	1.303	1.654	2.021	2.123	2.423	2.706	2.971	3.307	3.551
50	0.679	0.849	1.047	1.299	1.676	2.009	2.109	2.403	2.678	2.937	3.261	3.496
60	0.679	0.848	1.045	1.296	1.671	2.000	2.099	2.390	2.660	2.915	3.232	3.460

Obtaining the approximate P-value in Example 1 was somewhat challenging. With the aid of technology, we can find the exact P-value quite painlessly.

EXAMPLE 2   **Testing a Claim about a Population Mean Using Technology**

Problem: Obtain an exact P-value for the problem in Example 1 using statistical software or a graphing calculator with advanced statistical features.

Approach: We will use a TI-84 Plus graphing calculator to obtain the P-value. The steps for testing claims about a population mean with σ unknown using the TI-83/84 Plus graphing calculators, MINITAB, and Excel are given in the Technology Step by Step on page 544.

Result: Figure 16(a) shows the results from the TI-84 Plus using the Calculate option. Figure 16(b) shows the results using the Draw option. The P-value is 0.044. Notice the P-value is between 0.025 and 0.05. This agrees with the results in Example 1.

Figure 16

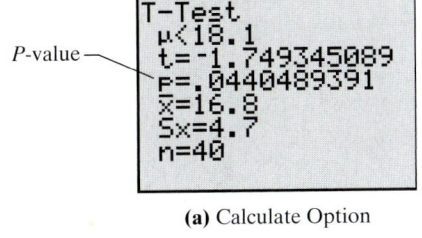

P-value

(a) Calculate Option

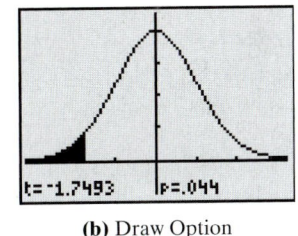

(b) Draw Option

Now Work Problem 13.

EXAMPLE 3   **Testing a Claim about a Population Mean, Small Sample**

Problem: The "fun size" of a Snickers bar is supposed to weigh 20 grams. Because the punishment for selling candy bars that weigh less than 20 grams is so severe, the manufacturer calibrates the machine so that the mean weight is 20.1 grams. The quality-control engineer at M&M–Mars, the company that manufactures Snickers bars, is concerned that the candy does not weigh 20.1 grams. She obtains a random sample of 11 candy bars, weighs them, and obtains the data in Table 3. Test the claim that the Snickers do not have a mean weight of 20.1 grams. Because shutting down the plant is very expensive, she decides to test the claim at the $\alpha = 0.01$ level of significance.

Table 3		
19.68	20.66	19.56
19.98	20.65	19.61
20.55	20.36	21.02
21.5	19.74	

Source: Michael Carlisle, student at Joliet Junior College

Approach: Before we can perform the hypothesis test, we must verify that the data come from a population that is normally distributed with no outliers. We will construct a normal probability plot and boxplot to verify these requirements. We then proceed to follow Steps 1 through 6.

Solution: Figure 17 displays the normal probability plot and boxplot.

Figure 17

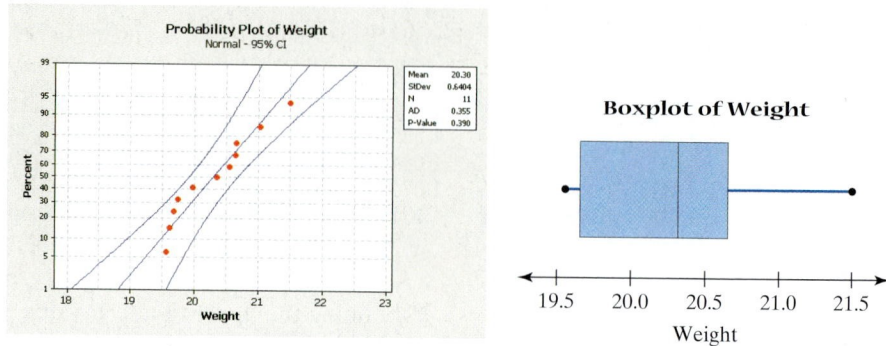

The normal probability plot indicates that the data come from a population that is approximately normal. The boxplot does not show any outliers. We can proceed with the hypothesis test.

Step 1: The quality-control engineer wishes to test the claim that the Snickers do not have a mean weight of 20.1 grams. The claim can be written $\mu \neq 20.1$. We have

$$H_0: \mu = 20.1 \quad \text{versus} \quad H_1: \mu \neq 20.1$$

This is a two-tailed test.

Step 2: The level of significance is $\alpha = 0.01$.

Step 3: From the data in Table 3, the sample mean is $\overline{x} = 20.3$, and the sample standard deviation is $s = 0.64$. The test statistic is

$$t_0 = \frac{\overline{x} - \mu_0}{s/\sqrt{n}} = \frac{20.3 - 20.1}{0.64/\sqrt{11}} = 1.036$$

Classical Approach

Step 4: Because this is a two-tailed test, we determine the critical t-values at the $\alpha = 0.01$ level of significance with $n - 1 = 11 - 1 = 10$ degrees of freedom to be $-t_{0.01/2} = -t_{0.005} = -3.169$ and $t_{0.01/2} = t_{0.005} = 3.169$. The critical regions are displayed in Figure 18.

Figure 18

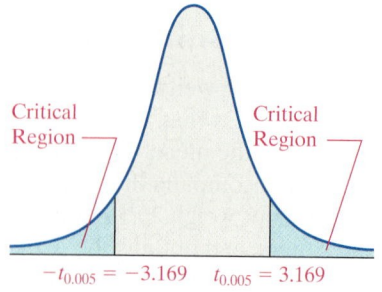

P-value Approach

Step 4: Because this is a two-tailed test, the P-value is the area under the t-distribution with $11 - 1 = 10$ degrees of freedom to the left of the test statistic $-t_0 = -1.036$ and to the right of $t_0 = 1.036$, as shown in Figure 19. That is, P-value $= P(t < -1.036) + P(t > 1.036) = 2P(t > 1.036)$, with 10 degrees of freedom.

Using Table V, we find the row that corresponds to 10 degrees of freedom. The value 1.036 lies between 0.879 and 1.093. The value of 0.879 has an area under the t-distribution of 0.20 to the right. The area under the t-distribution with 10 degrees of freedom to the right of 1.093 is 0.015.

Because 1.036 is between 0.879 and 1.093, the P-value is between 2(0.20) and 2(0.15). So

$$0.30 < P\text{-value} < 0.40$$

Using MINITAB, we find that the exact P-value is 0.323.

Figure 19

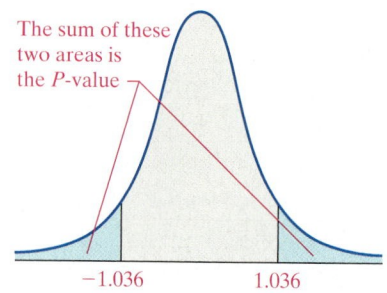

Step 5: The test statistic is $t_0 = -1.036$. Because the test statistic is between the critical values $-t_{0.005} = -3.169$ and $t_{0.005} = 3.169$, the quality-control engineer does not reject the null hypothesis.

Step 5: Because the P-value is greater than the level of significance $(0.323 > 0.01)$, the quality-control engineer does not reject the null hypothesis.

Step 6: There is not sufficient evidence to support the claim that the Snickers do not have a mean weight of 20.1 grams at the $\alpha = 0.01$ level of significance.

Now Work Problem 21.

Sections 10.2 and 10.3 discussed performing hypothesis tests about a population mean. The main criterion for choosing which test to use is whether the population standard deviation, σ, is known. Provided that the population from which the sample is drawn is normal or that the sample size is large,

- if σ is known, use the z-test procedures from Section 10.2;
- if σ is unknown, use the t-test procedures from Section 10.3.

In Section 10.4, we will discuss testing hypotheses about a population proportion; Section 10.5 will present methods for testing hypotheses about a population variance or standard deviation.

In-Class Activity: Stringing Them Along (Part I)

How skilled are people at estimating the length of a piece of rope? Do you think that they will tend to overestimate its length? Underestimate? Or are you not sure?

1. Look at the piece of rope that your instructor is holding and estimate the length of the rope in inches.
2. Using the null hypothesis $H_0: \mu = \mu_0$, where μ_0 represents the actual length of the rope, select an appropriate alternative hypothesis and a level of significance based on your responses to the questions posed at the beginning of the activity.
3. Obtain the actual length of the rope from your instructor. Combine the data for the entire class and test the hypothesis formed in part (b). What did you conclude?

[**Note:** Save the class data for use in another activity]

MAKING AN INFORMED DECISION

What Does It Really Weigh?

Many consumer products that we purchase have labels that describe the net weight of the contents. For example, the net weight of a candy bar might be listed as 4 ounces. Choose any consumer product that reports the net weight of the contents on the packaging.

(a) Obtain a random sample of size 8 or more of the consumer product. We will treat the random purchases as a simple random sample. Weigh the contents without the packaging.
(b) If your sample size is less than 30, verify that the population from which the sample was drawn is normal and that the sample does not contain any outliers.
(c) As the consumer, you are concerned only with situations in which you are getting ripped off. Determine the null and alternative hypotheses from the point of view of the consumer.
(d) Test the claim that the consumer is getting ripped off at the $\alpha = 0.05$ level of significance. Are you getting ripped off? What makes you say so?
(e) Suppose you are the quality-control manager. How would you structure the alternative hypothesis? Test this claim at the $\alpha = 0.05$ level of significance. Is there anything wrong with the manufacturing process? What makes you say so?

10.3 ASSESS YOUR UNDERSTANDING

Concepts and Vocabulary

1. State the requirements that must be satisfied to test a claim about a population mean with σ unknown.

2. Determine the critical value for a right-tailed test of a population mean with σ unknown at the $\alpha = 0.01$ level of significance with 15 degrees of freedom.

3. Determine the critical value for a two-tailed test of a population mean with σ unknown at the $\alpha = 0.05$ level of significance with 12 degrees of freedom.

4. Determine the critical value for a left-tailed test of a population mean with σ unknown at the $\alpha = 0.05$ level of significance with 19 degrees of freedom.

Skill Building

5. To test $H_0: \mu = 50$ versus $H_1: \mu < 50$, a simple random sample of size $n = 24$ is obtained from a population that is known to be normally distributed.
   (a) If $\bar{x} = 47.1$ and $s = 10.3$, compute the test statistic.
   (b) If the researcher decides to test this hypothesis at the $\alpha = 0.05$ level of significance, determine the critical value.
   (c) Draw a t-distribution that depicts the critical region.
   (d) Will the researcher reject the null hypothesis? Why?

6. To test $H_0: \mu = 40$ versus $H_1: \mu > 40$, a simple random sample of size $n = 25$ is obtained from a population that is known to be normally distributed.
   (a) If $\bar{x} = 42.3$ and $s = 4.3$, compute the test statistic.
   (b) If the researcher decides to test this hypothesis at the $\alpha = 0.1$ level of significance, determine the critical value.
   (c) Draw a t-distribution that depicts the critical region.
   (d) Will the researcher reject the null hypothesis? Why?

7. To test $H_0: \mu = 100$ versus $H_1: \mu \neq 100$, a simple random sample of size $n = 23$ is obtained from a population that is known to be normally distributed.
   (a) If $\bar{x} = 104.8$ and $s = 9.2$, compute the test statistic.
   (b) If the researcher decides to test this hypothesis at the $\alpha = 0.01$ level of significance, determine the critical values.
   (c) Draw a t-distribution that depicts the critical region.
   (d) Will the researcher reject the null hypothesis? Why?

8. To test $H_0: \mu = 80$ versus $H_1: \mu < 80$, a simple random sample of size $n = 22$ is obtained from a population that is known to be normally distributed.
   (a) If $\bar{x} = 76.9$ and $s = 8.5$, compute the test statistic.
   (b) If the researcher decides to test this hypothesis at the $\alpha = 0.02$ level of significance, determine the critical value.
   (c) Draw a t-distribution that depicts the critical region.
   (d) Will the researcher reject the null hypothesis? Why?

9. To test $H_0: \mu = 20$ versus $H_1: \mu < 20$, a simple random sample of size $n = 18$ is obtained from a population that is known to be normally distributed.
   (a) If $\bar{x} = 18.3$ and $s = 4.3$, compute the test statistic.
   (b) Draw a t-distribution with the area that represents the P-value shaded.
   (c) Approximate and interpret the P-value.
   (d) If the researcher decides to test this hypothesis at the $\alpha = 0.05$ level of significance, will the researcher reject the null hypothesis? Why?

10. To test $H_0: \mu = 4.5$ versus $H_1: \mu > 4.5$, a simple random sample of size $n = 13$ is obtained from a population that is known to be normally distributed.
    (a) If $\bar{x} = 4.9$ and $s = 1.3$, compute the test statistic.
    (b) Draw a t-distribution with the area that represents the P-value shaded.
    (c) Approximate and interpret the P-value.
    (d) If the researcher decides to test this hypothesis at the $\alpha = 0.1$ level of significance, will the researcher reject the null hypothesis? Why?

11. To test $H_0: \mu = 105$ versus $H_1: \mu \neq 105$, a simple random sample of size $n = 35$ is obtained.
    (a) Does the population have to be normally distributed to test this hypothesis by using the methods presented in this section?
    (b) If $\bar{x} = 101.9$ and $s = 5.9$, compute the test statistic.
    (c) Draw a t-distribution with the area that represents the P-value shaded.
    (d) Determine and interpret the P-value.
    (e) If the researcher decides to test this hypothesis at the $\alpha = 0.01$ level of significance, will the researcher reject the null hypothesis? Why?

12. To test $H_0: \mu = 45$ versus $H_1: \mu \neq 45$, a simple random sample of size $n = 40$ is obtained.
    (a) Does the population have to be normally distributed to test this hypothesis by using the methods presented in this section?
    (b) If $\bar{x} = 48.3$ and $s = 8.5$, compute the test statistic.
    (c) Draw a t-distribution with the area that represents the P-value shaded.
    (d) Determine and interpret the P-value.
    (e) If the researcher decides to test this hypothesis at the $\alpha = 0.01$ level of significance, will the researcher reject the null hypothesis? Why?

Applying the Concepts

13. Effects of Alcohol on the Brain In a study published in **NW** the *American Journal of Psychiatry* (157:737–744, May 2000), researchers wanted to measure the effect of alcohol on the development of the hippocampal region in adolescents. The hippocampus is the portion of the brain responsible for long-term memory storage. The researchers randomly selected 12 adolescents with alcohol use disorders. They wanted to test the claim that the hippocampal volumes in the alcoholic adolescents were less than the normal volume of 9.02 cm^3. An analysis of the sample data revealed that the hippocampal volume is approximately normal with $\bar{x} = 8.10$ and $s = 0.7$. Test the researchers' claim at the $\alpha = 0.01$ level of significance.

14. Effects of Plastic Resin Para-nonylphenol is found in polyvinyl chloride (PVC) used in the food processing and packaging industries. Researchers wanted to determine the effect this substance had on the organ weight of first-generation mice when both parents were exposed to $50\ \mu g/L$ of para-nonylphenol in drinking water for 4 weeks. After 4 weeks, the mice were bred. After 100 days, the offspring of the exposed parents were sacrificed and the kidney weights were determined. The mean weight of the 12 offspring was found to be 396.9 mg with a standard deviation of 45.4 mg. Is there significant evidence to support the claim that the kidney weight of the offspring whose parents were exposed to $50\ \mu g/L$ of para-nonylphenol in drinking water for 4 weeks is greater than 355.7 mg, the mean weight of kidneys in normal 100-day old mice at the $\alpha = 0.05$ level of significance?

(*Source:* Vendula Kyselova et al., Effects of p-nonylphenol and resveratrol on body and organ weight and in vivo fertility of outbred CD-1 mice, *Reproductive Biology and Endocrinology* 2003).

15. Got Milk? The U.S. Food and Drug Administration recommends that individuals consume 1000 mg of calcium daily. After an advertising campaign aimed at male teenagers, the International Dairy Foods Association (IDFA) claims that male teenagers consume more than the recommended daily amount of calcium. To support this claim, the IDFA obtained a random sample of 50 male teenagers and found that the mean amount of calcium consumed was 1081 mg, with a standard deviation of 426 mg. Is there significant evidence to support the claim of the IDFA at the $\alpha = 0.05$ level of significance?

16. Too Much Salt? A nutritionist claims that children under the age of 10 years are consuming more than the U.S. Food and Drug Administration's recommended daily allowance of sodium, which is 2400 mg. To test this claim, she obtains a random sample of 75 children under the age of 10 and measures their daily consumption of sodium. The mean amount of sodium consumed was determined to be 2993 mg, with a standard deviation of 1489 mg. Is there significant evidence to support the claim of the nutritionist at the $\alpha = 0.05$ level of significance?

17. Normal Temperature Carl Reinhold August Wunderlich said that the mean temperature of humans is 98.6°F. Researchers Philip Mackowiak, Steven Wasserman, and Myron Levine [*JAMA*, Sept. 23–30 1992; 268(12):1578–80] thought that the mean temperature of humans is less than 98.6°F. They measured the temperature of 148 healthy adults 1 to 4 times daily for 3 days, obtaining 700 measurements. The sample data resulted in a sample mean of 98.2°F and a sample standard deviation of 0.7°F.

(a) Using the classical approach, judge whether there is evidence to support the researchers' claim at the $\alpha = 0.01$ level of significance.

(b) Determine and interpret the *P*-value.

18. Normal Temperature Carl Reinhold August Wunderlich said that the mean temperature of humans is 98.6°F. Researchers Philip Mackowiak, Steven Wasserman, and Myron Levine [*JAMA*, Sept. 23–30 1992; 268(12):1578–80] measured the temperatures of 26 females 1 to 4 times daily for 3 days to get a total of 123 measurements. The sample data yielded a sample mean of 98.4°F and a sample standard deviation of 0.7°F.

(a) Using the classical approach, judge whether there is evidence to support the claim that the normal temperature of women is less than 98.6°F at the $\alpha = 0.01$ level of significance.

(b) Determine and interpret the *P*-value.

19. Age of Death-Row Inmates In 2002, the mean age of an inmate on death row was 40.7 years, according to data obtained from the U.S. Department of Justice. A sociologist wants to test the claim that the mean age of a death-row inmate has changed since then. She randomly selects 32 death-row inmates and finds that their mean age is 38.9, with a standard deviation of 9.6.

(a) Test the sociologist's claim at the $\alpha = 0.05$ level of significance.

(b) Test the sociologist's claim by constructing a 95% confidence interval.

20. Energy Consumption In 2001, the mean household expenditure for energy was $1493, according to data obtained from the U.S. Energy Information Administration. An economist wanted to know whether this amount has changed significantly from its 2001 level. In a random sample of 35 households, he found the mean expenditure (in 2001 dollars) for energy during the most recent year to be $1618, with standard deviation $321.

(a) Test the economist's claim that the mean expenditure has changed significantly from the 2001 level at the $\alpha = 0.05$ level of significance.

(b) Test the economist's claim by constructing a 95% confidence interval.

21. Conforming Golf Balls The United States Golf Association tion requires that golf balls have a diameter that is 1.68 inches. An engineer for the USGA wishes to discover whether Maxfli XS golf balls have a mean diameter different from 1.68 inches. A random sample of Maxfli XS golf balls was selected. Their diameters are shown in the table.

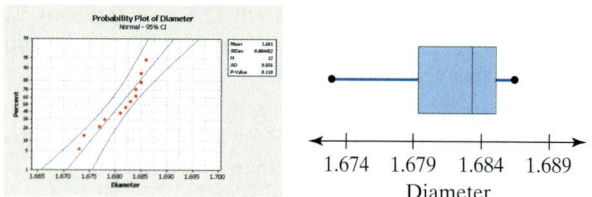

1.683	1.677	1.681
1.685	1.678	1.686
1.684	1.684	1.673
1.685	1.682	1.674

Source: Michael McCraith, Joliet Junior College

(a) Because the sample size is small, he must verify that the diameter is normally distributed and the sample does not contain any outliers. The normal probability plot and boxplot are shown. Are the conditions for testing the hypothesis satisfied?

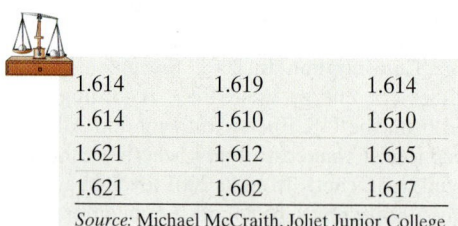

(b) Using the classical approach, test the claim that the golf balls have a mean diameter that is different from 1.68 inches at the $\alpha = 0.05$ level of significance.
(c) Determine and interpret the *P*-value.

22. Conforming Golf Balls The USGA requires that golf balls have a weight that is less than 1.62 ounces. An engineer for the USGA wants to test the claim that Maxfli XS golf balls have a mean weight less than 1.62 ounces. He obtains a random sample of 12 Maxfli XS golf balls. Their weights are in the table.

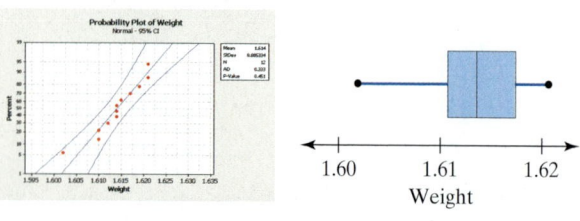

1.614	1.619	1.614
1.614	1.610	1.610
1.621	1.612	1.615
1.621	1.602	1.617

Source: Michael McCraith, Joliet Junior College

(a) Because the sample size is small, he must verify that weight is normally distributed and that the sample does not contain any outliers. The normal probability plot and boxplot are shown. Are the conditions for testing the hypothesis satisfied?

(b) Decide whether the golf balls meet Maxfli's standard at the $\alpha = 0.1$ level of significance.

23. Waiting in Line The mean waiting time at the drive-through of a fast-food restaurant from the time an order is placed to the time the order is received is 84.3 seconds. A manager devises a new drive-through system that he believes will decrease wait time. To test this claim, he initiates the new system at his restaurant and measures the wait time for 10 randomly selected orders. The wait times are provided in the table.

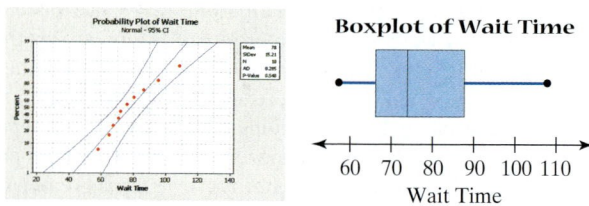

108.5	67.4	58.0	75.9	65.1
80.4	95.5	86.3	70.9	72.0

(a) Because the sample size is small, the manager must verify that wait time is normally distributed and the sample does not contain any outliers. The normal probability plot and boxplot are shown. Are the conditions for testing the claim satisfied?

(b) Test the manager's claim at the $\alpha = 0.1$ level of significance.

24. Calibrating a pH Meter An engineer wants to measure the bias in a pH meter. She uses the meter to measure the pH in 14 neutral substances (pH = 7.0) and obtains the data shown in the table.

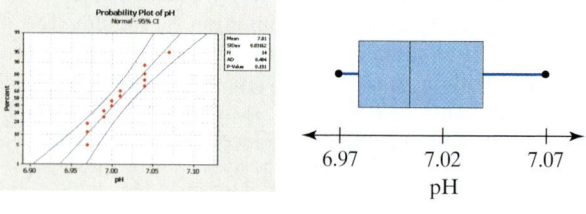

7.01	7.04	6.97	7.00	6.99	6.97	7.04
7.04	7.01	7.00	6.99	7.04	7.07	6.97

(a) Because the sample size is small, she must verify that pH is normally distributed and the sample does not contain any outliers. The normal probability plot and boxplot are shown. Are the conditions for testing the hypothesis satisfied?

(b) Is there sufficient evidence to support the claim that the pH meter is not correctly calibrated at the $\alpha = 0.05$ level of significance?

25. P/E Ratio A stock analyst believes that the price-to-earnings (P/E) ratio of companies listed on the Standard and Poor's 500 (S&P 500) Index is less than its December 1, 2000, level of 22.0, in response to economic uncertainty. The P/E ratio is the price an investor is willing to pay for $1 of earnings. For example, a P/E of 23 means the investor pays $23 for each $1 of earnings. A higher P/E is an indication of investor optimism. Lower P/Es are generally assigned to companies with lower earnings growth. To test his claim, he randomly samples 14 companies listed on the S&P 500 and calculates their P/E ratios. He obtains the following data:

Company	P/E Ratio	Company	P/E Ratio
Boeing	25.1	Dow Chemical	13.7
General Motors	7.8	Citigroup	18.5
Halliburton	35.8	Merck and Co.	26.8
Norfolk Southern	25.6	Sara Lee	11.9
Agilent Technologies	22.5	Harley-Davidson	37.4
Old Kent Financial	20.0	Circuit City	14.5
Cendent	15.0	Minnesota Mining and Manufacturing	23.8

Source: Checkfree Corporation

(a) Verify that P/E ratios are normally distributed, and check for outliers by drawing a normal probability plot and boxplot.
(b) Test the analyst's claim at the $\alpha = 0.05$ level of significance.
(c) Determine and interpret the *P*-value.

26. Enlarged Prostate Benign prostatic hyperplasia is a common cause of urinary outflow obstruction in aging males. The efficacy of Cardura (doxazosin mesylate) was measured in clinical trials of 173 patients with benign prostatic hyperplasia. Researchers wanted to discover whether Cardura significantly increased the urinary flow rate. It was found that an average increase of 0.8 mL/sec was obtained. This was said to be significant, with a *P*-value less than 0.01. State the null and alternative hypotheses of the researchers and interpret the *P*-value.

27. Systolic Blood Pressure of Surgical Patients A nursing student maintained that the mean systolic blood pressure of her male patients on the surgical floor was less than 130 mm Hg. She randomly selected 19 male surgical patients and collected the systolic blood pressures shown in the table.

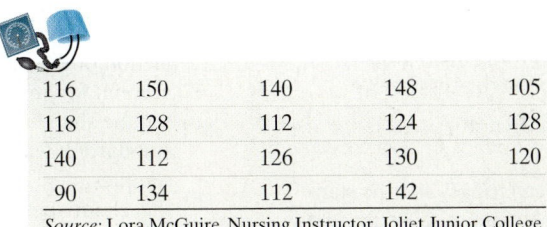

116	150	140	148	105
118	128	112	124	128
140	112	126	130	120
90	134	112	142	

Source: Lora McGuire, Nursing Instructor, Joliet Junior College

(a) Because the sample size is small, she must verify that the systolic blood pressure is normally distributed and the sample does not contain any outliers. The normal probability plot and boxplot are shown. Are the conditions for testing the hypothesis satisfied?

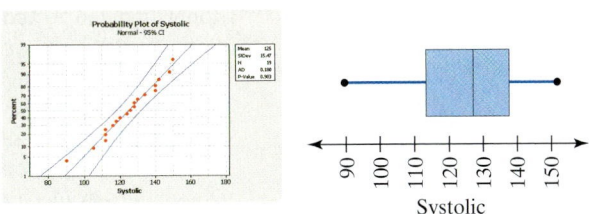

Systolic

(b) The student enters the data into MINITAB and obtains the following results:

T-Test of the Mean

```
Test of mu = 130.00 vs mu < 130.00

Variable           N      Mean    StDev   SE Mean      T      P
Systolic          19       125    15.47      3.55  -1.41  0.088
```

What are the null and alternative hypotheses? Identify the *P*-value. Will the nursing student reject the null hypothesis at the $\alpha = 0.05$ level of significance? State her conclusion.

28. Temperature of Surgical Patients A nursing student suspects that the mean temperature of surgical patients is above the normal temperature, 98.2°F. (See Problem 17.) She randomly selects 32 surgical patients and obtains the temperatures shown in the table.

97.4	98.6	98.2	98.2	98.4	98.6
99.8	97.7	97.8	98.9	97.8	97.8
96.7	97.8	98.3	98.5	98.0	98.7
96.8	98.6	98.4	97.4	99.1	98.7
97.8	99.2	99.2	98.1	98.6	98.4
99.2	98.6				

Source: Lora McGuire, Nursing Instructor, Joliet Junior College

(a) What are the null and alternative hypotheses of the student?

(b) The student enters the data into MINITAB and obtains the following results:

T-Test of the Mean

```
Test of mu = 98.200 vs mu > 98.200

Variable          N     Mean    StDev    SE Mean      T       P
Temperat         32   98.291    0.689      0.122    0.74    0.23
```

What is the *P*-value of the test? State the nursing student's conclusion.

---

29. Soybean Yield The mean yield per acre of soybeans on farms in the United States in 2003 was 33.5 bushels, according to data obtained from the U.S. Department of Agriculture. A farmer in Iowa claimed the yield was higher than the reported mean. He randomly sampled 35 acres on his farm and determined the mean yield to be 37.1 bushels, with a standard deviation of 2.5 bushels. He computed the *P*-value to be less than 0.0001 and concluded that the U.S. Department of Agriculture was wrong. Why should his conclusions be looked on with skepticism?

30. Significance Test Applet Load the hypothesis tests for a mean applet.

(a) Set the shape to normal, the mean to 100, and the standard deviation to 15. These parameters describe the distribution of IQ scores. Obtain 1000 simple random samples of size $n = 10$ from this population, and test the claim that the mean is different from 100. How many samples led to a rejection of the null hypothesis if $\alpha = 0.05$? How many would we expect to lead to a rejection of the null hypothesis? For this level of significance, what is the probability of a Type I error?

(b) Set the shape to normal, the mean to 100, and the standard deviation to 15. These parameters describe the distribution of IQ scores. Obtain 1000 simple random samples of size $n = 30$ from this population, and test the claim that the mean is different from 100. How many samples led to a rejection of the null hypothesis if $\alpha = 0.05$? How many would we expect to lead to a rejection of the null hypothesis? For this level of significance, what is the probability of a Type I error?

(c) Compare the results of parts (a) and (b). Did the sample size have any impact on the number of samples that incorrectly rejected the null hypothesis?

31. Significance Test Applet: Violating Assumptions Load the hypothesis tests for a mean applet.

(a) Set the shape to right skewed, the mean to 50, and the standard deviation to 10. Obtain 1000 simple random samples of size $n = 8$ from this population, and test the claim that the mean is different from 50. How many of the samples led to a rejection of the null hypothesis if $\alpha = 0.05$? How many would we expect to lead to a rejection of the null hypothesis if $\alpha = 0.05$? What might account for any discrepancies?

(b) Set the shape to right skewed, the mean to 50, and the standard deviation to 10. Obtain 1000 simple random samples of size $n = 40$ from this population, and test the claim that the mean is different from 50. How many of the samples led to a rejection of the null hypothesis if $\alpha = 0.05$? How many would we expect to lead to a rejection of the null hypothesis if $\alpha = 0.05$?

32. Simulation Simulate drawing 40 simple random samples of size $n = 20$ from a population that is normally distribution with mean 50 and standard deviation 10.

(a) Test the null hypothesis $H_0: \mu = 50$ versus the alternative hypothesis $H_1: \mu \neq 50$ for each of the 40 samples using a *t*-test.

(b) Suppose we were testing this hypothesis at the $\alpha = 0.05$ level of significance. How many of the 40 samples would you expect to result in a Type I error?

(c) Count the number of samples that lead to a rejection of the null hypothesis. Is it close to the expected value determined in part (b)?

(d) Describe why we know a rejection of the null hypothesis results in making a Type I error in this situation.

Consumer Reports® | **Eyeglass Lenses**

Eyeglasses are part medical device and part fashion statement, a marriage that has always made them a tough buy. Aside from the thousands of different frames the consumer has to choose from, various lens materials and coatings can add to the durability, and the cost, of a pair of eyeglasses. One manufacturer even goes so far as to claim that its lenses are "the most scratch-resistant plastic lenses ever made." With a claim like that, we had to test the lenses.

One test involved tumbling the lenses in a drum containing scrub pads of grit of varying size and hardness. Afterward, readings of the lenses' haze were taken on a spectrometer to determine how scratched they had become. To evaluate their scratch resistance, we measured the difference between the haze reading before and after tumbling.

The photo illustrates the difference between an uncoated lens (on the left) and the manufacturer's "scratch-resistant" lens (on the right).

The following table contains the haze measurements both before and after the scratch resistance test for this manufacturer. Haze difference is measured by subtracting the before score from the after score. In other words, haze difference is computed as After–Before.

(a) Suppose it is known that the closest competitor to the manufacturer's lens has a mean haze difference of 1.0. Do the data support the manufacturer's scratch resistance claim?

(b) Write the null and alternative hypotheses, letting μ_{hdiff} represent the mean haze difference for the manufacturer's lens.

(c) We used MINITAB to perform a one-sample t-test. The results are shown below.

Using the MINITAB output, answer the following questions:

1. What is the value of the test statistic?

2. What is the P-value of the test?

3. What is the conclusion of this test? Write a paragraph for the readers of *Consumer Reports* magazine that explains your findings.

Note to Readers: In many cases, our test protocol and analytical methods are more complicated than described in these examples. The data and discussions have been modified to make the material more appropriate for the audience.

Before	After	Difference
0.18	0.72	0.54
0.16	0.85	0.69
0.20	0.71	0.51
0.17	0.42	0.25
0.21	0.76	0.55
0.21	0.51	0.30

One-Sample T: Difference

```
Test of mu = 1 vs mu < 1

Variable        N            Mean      StDev     SE Mean
Difference      6          0.4733     0.1665      0.0680

Variable      95.0%     Upper Bound        T           P
Difference              0.6103        -7.75       0.000
```

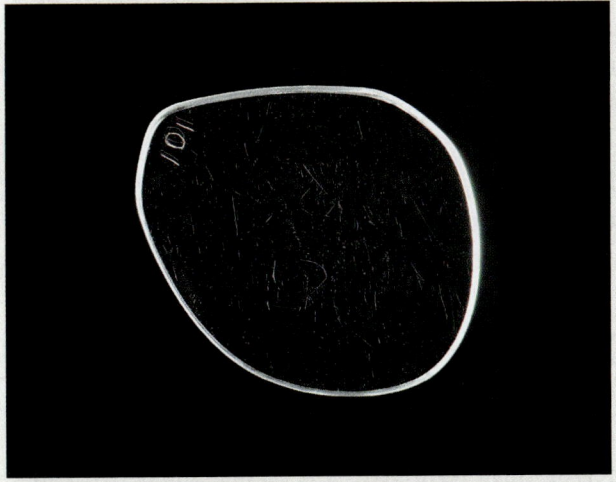
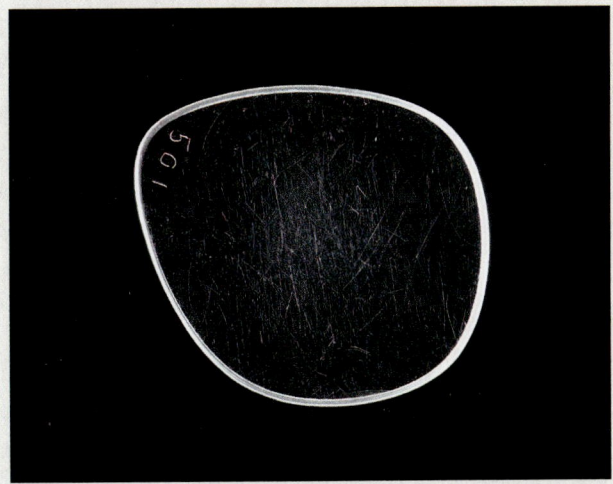

Technology Step-by-Step	**Hypothesis Tests Regarding μ, σ Unknown**
TI-83/84 Plus	*Step 1:* If necessary, enter raw data in L1.
	Step 2: Press STAT, highlight TESTS, and select `2:T-Test`.
	Step 3: If the data are raw, highlight DATA; make sure that List1 is set to L1 and Freq is set to 1. If summary statistics are known, highlight STATS and enter the summary statistics. For the value of μ_0, enter the value of the mean stated in the null hypothesis.
	Step 4: Select the direction of the alternative hypothesis.
	Step 5: Highlight **Calculate** and press ENTER. The TI-83/84 gives the *P*-value.
MINITAB	*Step 1:* Enter raw data in column C1.
	Step 2: Select the **Stat** menu, highlight **Basic Statistics**, then highlight **1-Sample t . . . .**
	Step 3: Enter C1 in the cell marked "samples in columns." Enter the value of the mean stated in the null hypothesis in the cell marked "Test mean:" Click Options. In the cell marked "Alternative," select the direction of the alternative hypothesis. Click OK twice.
Excel	*Step 1:* If necessary, enter raw data in column A.
	Step 2: Load the PHStat Add-in.
	Step 3: Select the **PHStat menu**, highlight **One Sample Tests . . . ,** and then highlight **t Test for the mean, sigma known . . . .**
	Step 4: Enter the value of the null hypothesis and the level of significance, α. If the summary statistics are known, click "Sample statistics known" and enter the sample size, sample mean, and sample standard deviation. If summary statistics are unknown, click "Sample statistics unknown." With the cursor in the "Sample cell range" cell, highlight the data in column A. Click the option corresponding to the desired test [two-tail, upper (right) tail, or lower (left) tail]. Click OK.

10.4 Testing Claims about a Population Proportion

Preparing for This Section Before getting started, review the following:

- Binomial probability distribution (Section 6.2, pp. 328–339)
- Distribution of the sample proportion (Section 8.2, pp. 434–439)

Objectives

  1   **Test a claim about a population proportion using the normal model**

  2   **Test a claim about a population proportion using the binomial probability distribution**

  1   **Test a Claim about a Population Proportion Using the Normal Model**

Recall that the best point estimate of p, the proportion of the population with a certain characteristic, is given by

$$\hat{p} = \frac{x}{n}$$

where x is the number of individuals in the sample with the specified characteristic and n is the sample size. Recall from Section 8.2 that the sampling distribution of $\hat{p}$ is approximately normal, with mean $\mu_{\hat{p}} = p$ and standard deviation $\sigma_{\hat{p}} = \sqrt{\dfrac{p(1-p)}{n}}$, provided that the following requirements are satisfied.

- The sample is a simple random sample
- $np(1-p) \geq 10$

- $n \leq 0.05N$ (that is the sample size is no more than 5% of the population size).

Testing a hypothesis about the population proportion, p, follows the same logic as the testing of hypotheses about a population mean with σ known. The only difference is that the **test statistic** is

$$z_0 = \frac{\hat{p} - p_0}{\sqrt{\dfrac{p_0(1 - p_0)}{n}}}$$

where p_0 is the value of the population proportion stated in the null hypothesis.

The careful reader will notice that we are using p_0 in computing the standard error rather than $\hat{p}$ (as we did in computing confidence intervals about p). This is because, when we test a hypothesis, the null hypothesis is always assumed true. Therefore, we are assuming that the population proportion is p_0.

> ⚠ **CAUTION**
>
> When determining the standard error for the sampling distribution of $\hat{p}$ for hypothesis testing, use the assumed value of the population proportion, p_0.

Testing a Claim Regarding a Population Proportion, p

If a claim is made regarding the population proportion, we can use the following steps to test the claim, provided that

1. the sample is obtained by simple random sampling
2. $np_0(1 - p_0) \geq 10$ with $n \leq 0.05N$ (the sample size, n, is no more than 5% of the population size, N).

Step 1: A claim is made regarding the population proportion. The claim is used to determine the null and alternative hypotheses. The hypotheses can be structured in one of three ways:

Two-Tailed	Left-Tailed	Right-Tailed
$H_0: p = p_0$	$H_0: p = p_0$	$H_0: p = p_0$
$H_1: p \neq p_0$	$H_1: p < p_0$	$H_1: p > p_0$

Note: p_0 is the assumed value of the population proportion.

Step 2: Select a level of significance α, depending on the seriousness of making a Type I error.

Step 3: Compute the test statistic

$$z_0 = \frac{\hat{p} - p_0}{\sqrt{\dfrac{p_0(1 - p_0)}{n}}}$$

Classical Approach

Step 4: Use Table IV to determine the critical value.

P-value Approach

Step 4: Use Table IV to determine the *P*-value.

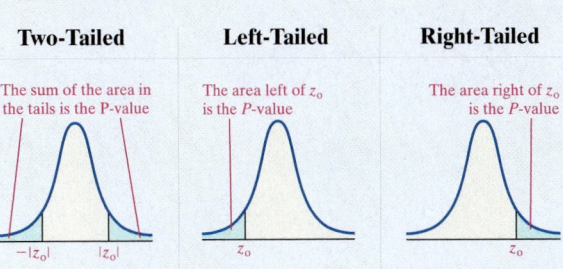

Step 5: Compare the critical value with the test statistic. | **Step 5:** If P-value $< \alpha$, reject the null hypothesis.

Two-Tailed	Left-Tailed	Right-Tailed
If $z_0 < -z_{\alpha/2}$ or $z_0 > z_{\alpha/2}$ reject the null hypothesis	If $z_0 < -z_\alpha$ reject the null hypothesis	If $z_0 > z_\alpha$ reject the null hypothesis

Step 6: State the conclusion.

EXAMPLE 1 Testing a Claim about a Population Proportion: Right-Tailed Test

Problem: In 2004, 65% of adult Americans thought that the death penalty was morally acceptable. In a poll conducted by the Gallup Organization May 2–5, 2005, a simple random sample of 1005 adult Americans resulted in 704 respondents stating that they believe the death penalty was morally acceptable when asked, "Do you believe the death penalty is morally acceptable or morally wrong?" The choices "morally acceptable" and "morally wrong" were randomly interchanged for each interview. Is there significant evidence to indicate that the proportion of adult Americans who believe that the death penalty is morally acceptable has increased from the level reported in 2004 at the $\alpha = 0.05$ level of significance?

Approach: We must verify the requirements to perform the hypothesis test; that is, the sample must be a simple random sample, $np_0(1 - p_0) \geq 10$, and the sample size cannot be more than 5% of the population size. Then we follow Steps 1 through 6 listed on pages 545–546).

> **! CAUTION**
> Always verify the requirements before conducting a hypothesis test.

Solution: The claim is that the proportion of adult Americans who believe that the death penalty is morally acceptable has increased since 2004. The claim can be written $p > 0.65$. The sample is a simple random sample. Also, $np_0(1 - p_0) = (1005)(0.65)(1 - 0.65) = 228.6 > 10$. There are over 200 million adult Americans, so the sample size is less than 5% of the population size. The requirements are satisfied, so we now proceed to Steps 1 through 6.

Step 1: The claim is that $p > 0.65$. We have

$$H_0: p = 0.65 \quad \text{versus} \quad H_1: p > 0.65$$

Step 2: The level of significance is $\alpha = 0.05$.

Step 3: The assumed value of the population proportion is $p_0 = 0.65$. The point estimate of the population proportion is $\hat{p} = \dfrac{x}{n} = \dfrac{704}{1005} = 0.70$. The test statistic is

$$z_0 = \frac{\hat{p} - p_0}{\sqrt{\dfrac{p_0(1 - p_0)}{n}}} = \frac{0.70 - 0.65}{\sqrt{\dfrac{0.65(1 - 0.65)}{1005}}} = 3.32$$

Classical Approach

Step 4: Because this is a right-tailed test, we determine the critical value at the $\alpha = 0.05$ level of significance to be $z_{0.05} = 1.645$. The critical region is displayed in Figure 20.

Figure 20

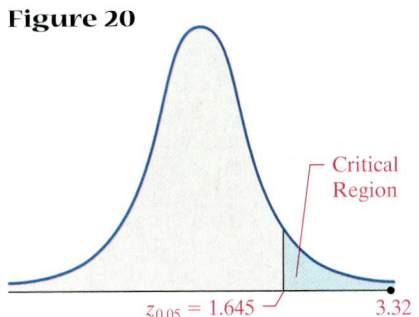

Step 5: The test statistic is $z_0 = 3.32$. We label this point in Figure 20. Because the test statistic is greater than the critical value $(3.32 > 1.645)$, we reject the null hypothesis.

P-value Approach

Step 4: Because this is a right-tailed test, the P-value is the area under the standard normal distribution to the right of the test statistic, $z_0 = 3.32$, as shown in Figure 21. That is, P-value $= P(Z > z_0) = P(Z > 3.32) = 1 - P(Z \le 3.32) = 1 - 0.9995 = 0.0005$.

Figure 21

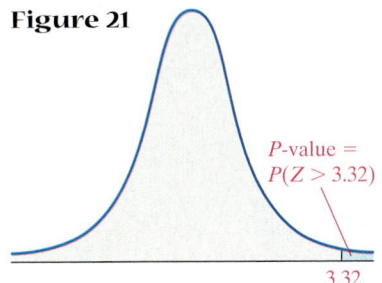

Step 5: Because the P-value is less than the level of significance $\alpha = 0.05$ $(0.0005 < 0.05)$, we reject the null hypothesis.

Step 6: There is sufficient evidence to support the claim that the proportion of adult Americans who believe that the death penalty is morally acceptable has increased since 2004 at the $\alpha = 0.05$ level of significance.

Now Work Problem 9.

EXAMPLE 2　**Testing a Claim about a Population Proportion, Two-Tailed Test**

Problem: The drug Prevnar is a vaccine meant to prevent meningitis. (It also helps control ear infections.) It is typically administered to infants. In clinical trials, the vaccine was administered to 710 randomly sampled infants between 12 and 15 months of age. Of the 710 infants, 121 experienced a loss of appetite. Is there significant evidence to conclude that the proportion of infants who receive Prevnar and experience loss of appetite is different from 0.135, the proportion of children who experience a loss of appetite with competing medications at the $\alpha = 0.01$ level of significance?

Approach: We must verify the requirements to perform the hypothesis test: the sample must be a simple random sample with $np_0(1 - p_0) \ge 10$, and the sample size cannot be more than 5% of the population. Then we follow Steps 1 through 6, as laid out previously.

Solution: We are testing the claim that the proportion of infants who experience a loss of appetite is different from 0.135; that is, $p \ne 0.135$. The sample is a simple random sample. Also, $np_0(1 - p_0) = (710)(0.135)(1 - 0.135) = 82.9 > 10$. Because there are about 1 million babies between 12 and 15 months of age, the sample size is less than 5% of the population size. The requirements are satisfied, so we now proceed to follow Steps 1 through 6.

Step 1: We are testing the claim that $p \neq 0.135$. This is a two-tailed test.

$$H_0: p = 0.135 \quad \text{versus} \quad H_1: p \neq 0.135$$

Step 2: The level of significance is $\alpha = 0.01$.

Step 3: The assumed value of the population proportion is $p_0 = 0.135$. The point estimate of the population proportion is $\hat{p} = \dfrac{x}{n} = \dfrac{121}{710} = 0.170$. The test statistic is

$$z_0 = \frac{\hat{p} - p_0}{\sqrt{\dfrac{p_0(1 - p_0)}{n}}} = \frac{0.170 - 0.135}{\sqrt{\dfrac{0.135(1 - 0.135)}{710}}} = 2.73$$

Classical Approach

Step 4: Because this is a two-tailed test, we determine the critical values at the $\alpha = 0.01$ level of significance to be $-z_{0.01/2} = -z_{0.005} = -2.575$ and $z_{0.01/2} = z_{0.005} = 2.575$. The critical regions are displayed in Figure 22.

Figure 22

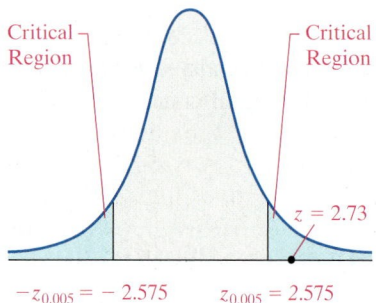

$$-z_{0.005} = -2.575 \qquad z_{0.005} = 2.575$$

Step 5: The test statistic is $z_0 = 2.73$. We label this point in Figure 22. Because the test statistic is greater than the critical value $(2.73 > 2.575)$, we reject the null hypothesis.

P-value Approach

Step 4: Because this is a two-tailed test, the P-value is the area under the standard normal distribution to the left of $-z_0 = -2.73$ and to the right of $z_0 = 2.73$ as shown in Figure 23. That is, P-value $= P(Z < -|z_0|) + P(Z > |z_0|) = 2P(Z < -2.73) = 2(0.0032) = 0.0064$.

Figure 23

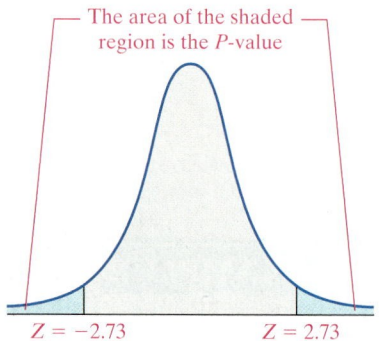

The area of the shaded region is the P-value

$Z = -2.73 \qquad Z = 2.73$

Step 5: Because the P-value is less than the level of significance $\alpha = 0.01$ $(0.0064 < 0.01)$, we reject the null hypothesis.

Step 6: There is sufficient evidence to support the claim that the proportion of infants who experienced a loss of appetite when receiving Prevnar is different from 0.135 at the $\alpha = 0.01$ level of significance.

Now Work Problem 15.

EXAMPLE 3   **Testing a Claim Regarding a Population Proportion Using Technology**

Problem: Test the claim presented in Example 2 by obtaining the P-value using statistical software or a graphing calculator with advanced statistical features.

Approach: We will use MINITAB to obtain the P-value. The steps for testing claims about a population proportion using the TI-83/84 Plus graphing calculator, MINITAB, and Excel are given in the Technology Step by Step on page 552.

Result: Figure 24 shows the results using MINITAB. The P-value is 0.006. Because the P-value is less than the level of significance $(0.006 < 0.01)$, we reject the null hypothesis.

Figure 24   **Test and CI for One Proportion**

```
Test of p = 0.135 vs p not = 0.135

Sample    X    N    Sample          99% CI          Z-Value    P-Value
1        121  710   0.170423   (0.134075, 0.206770)    2.76      0.006
```

In-Class Activity (Hypothesis Testing): Taste the Rainbow

The advertising campaign for Skittles® Brand candy in 2005 said to "Taste the Rainbow!". While the original candies do not have all the colors of the rainbow, they do come in red, orange, yellow, green, and purple (violet). But are the proportions of each color the same? If so, the proportion of each color would be $p = 0.2$.

(a) Obtain a 1-pound bag of Skittles (original flavor).

(b) Select one of the five original colors. Count the total number of candies in your bag, as well as the number for the color you selected.

(c) Is $np_0(1 - p_0) \geq 10$? If not, what could you do?

(d) Test the claim that the proportion of the selected color is different from $p = 0.2$.

(e) Compare your results to others in the class. Did everyone arrive at the same conclusion for the same color? What about different colors?

 Test a Claim about a Population Proportion Using the Binomial Probability Distribution

For the sampling distribution of $\hat{p}$ to be approximately normal, we require that $np(1 - p)$ be at least 10. What if this requirement is not satisfied? In Section 6.2, we used the binomial probability formula to identify unusual events. We stated that an event was unusual if the probability of observing the event was less than 0.05. This criterion is based on the P-value approach to testing hypotheses; the probability that we computed was the P-value. We use this same approach to test hypotheses regarding a population proportion for small samples.

EXAMPLE 4   **Hypothesis Test for a Population Proportion: Small Sample Size**

Problem: According to the U.S. Department of Agriculture, 48.9% of males between 20 and 39 years of age consume the minimum daily requirement of calcium. After an aggressive "Got Milk" advertising campaign, the USDA conducts a survey of 35 randomly selected males between the ages of 20 and 39 and finds that 21 of them consume the recommended daily allowance of calcium. At the $\alpha = 0.10$ level of significance, is there evidence to conclude that the percentage of males between the ages of 20 and 39 who consume the recommended daily allowance of calcium has increased?

Approach: We use the following steps:

Step 1: Determine the null and alternative hypotheses.

Step 2: Check whether $np_0(1 - p_0)$ is greater than or equal to 10, where p_0 is the proportion stated in the null hypothesis. If it is, then the sampling distribution of $\hat{p}$ is approximately normal and we can use the steps presented on pages 545–546. Otherwise, we use Steps 3 and 4, presented next.

Step 3: Compute the P-value. For right-tailed tests, the P-value is the probability of obtaining x or more successes. For left-tailed tests, the P-value is the probability of obtaining x or fewer successes.* The P-value is always computed with

---

*We won't address P-values for two-tailed hypothesis tests. For those who are interested, the P-value is 2 times the probability of obtaining x or more successes if $\hat{p} > p$ and 2 times the probability of obtaining x or fewer successes if $\hat{p} < p$.

the proportion given in the null hypothesis. Remember, we assume that the null is true until we have evidence to the contrary.

Step 4: If the P-value is less than the level of significance, α, reject the null hypothesis.

Solution

Step 1: The status quo or no change proportion of 20- to 39-year-old males who consume the minimum daily requirement of calcium is 0.489. We wish to know whether or not the advertising campaign helped to increase this proportion. Therefore,

$$H_0: p = 0.489 \quad \text{and} \quad H_1: p > 0.489$$

Step 2: From the null hypothesis, we have $p_0 = 0.489$. There were $n = 35$ individuals surveyed, so $np_0(1 - p_0) = 35(0.489)(1 - 0.489) = 8.75$. Because $np_0(1 - p_0) < 10$, the sampling distribution of $\hat{p}$ is not approximately normal.

Step 3: Let the random variable X represent the number of individuals who consume the daily requirement of calcium. We have $x = 21$ successes in $n = 35$ trials, so $\hat{p} = \dfrac{21}{35} = 0.6$. We want to judge whether the larger proportion is due to an increase in the population proportion or to sampling error. We obtained $x = 21$ successes in the survey and this is a right tailed test, so the P-value is $P(X \geq 21)$.

$$P\text{-value} = P(X \geq 21) = 1 - P(X < 21) = 1 - P(X \leq 20)$$

We will compute this P-value by using statistical software or a graphing calculator with advanced statistical features, with $n = 35$ and $p = 0.489$. Figure 25 shows the results using a TI-84 Plus graphing calculator.

The P-value is 0.1261. MINITAB will compute exact P-values using this approach as well.

Step 4: The P-value is greater than the level of significance $(0.1261 > 0.10)$, so we do not reject H_0. There is insufficient evidence (at the $\alpha = 0.1$ level of significance) to conclude that the proportion of 20- to 39-year-old males who consume the recommended daily allowance of calcium has increased.

Figure 25

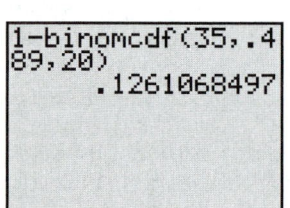

Now Work Problem 21.

10.4 ASSESS YOUR UNDERSTANDING

Concepts and Vocabulary

1. State the assumptions required to test a hypothesis about a population proportion.

2. A poll conducted by CNN, *USA Today*, and Gallup reported the following results: "According to the most recent CNN/*USA Today*/Gallup poll, conducted June 28–July 1, a majority of Americans (52%) approve of the job Bush is doing as president, ..." The poll results were obtained by conducting simple random sample of 1014 adults aged 18 years old or older, with a margin of error of ±3 percentage points. State what is wrong with the conclusions presented by the pollsters.

Skill Building

In Problems 3–8, test the hypothesis, using (a) the classical approach and then (b) the P-value approach. Be sure to verify the requirements of the test.

3. $H_0: p = 0.3$ versus $H_1: p > 0.3$
   $n = 200; x = 75; \alpha = 0.05$

4. $H_0: p = 0.6$ versus $H_1: p < 0.6$
   $n = 250; x = 124; \alpha = 0.01$

5. $H_0: p = 0.55$ versus $H_1: p < 0.55$
   $n = 150; x = 78; \alpha = 0.1$

6. $H_0: p = 0.25$ versus $H_1: p < 0.25$
   $n = 400; x = 96; \alpha = 0.1$

7. $H_0: p = 0.9$ versus $H_1: p \neq 0.9$
   $n = 500; x = 440; \alpha = 0.05$

8. $H_0: p = 0.4$ versus $H_1: p \neq 0.4$
   $n = 1000; x = 420; \alpha = 0.01$

Applying the Concepts

9. Lipitor The drug Lipitor is meant to reduce total choles-
NW terol and LDL cholesterol. In clinical trials, 19 out of 863
patients taking 10 mg of Lipitor daily complained of flulike symptoms. Suppose that it is known that 1.9% of patients taking competing drugs complain of flulike
symptoms. Is there sufficient evidence to support the claim
that more than 1.9% of Lipitor users experience flulike
symptoms as a side effect at the $\alpha = 0.01$ level of significance?

10. Nexium Nexium is a drug that can be used to reduce the
acid produced by the body and heal damage to the esoph-
agus due to acid reflux. Suppose the manufacturer of Nexium claims that more than 94% of patients taking Nexium
are healed within 8 weeks. In clinical trials, 213 of 224 patients suffering from acid reflux disease were healed after
8 weeks. Test the manufacturer's claim at the $\alpha = 0.01$
level of significance.

11. Americans Reading In a Gallup Poll conducted May
20–22, 2005, 835 of 1006 adults aged 18 or older said they
had read at least one book during the previous year. In
December 1990, 81% of adults aged 18 or older had
read at least one book during the previous year. Is there
sufficient evidence to support the claim that the percent
of adults who have read at least one book in the last
year is different from 1990 at the $\alpha = 0.05$ significance
level?

12. Haunted Houses In September 1996, 33% of adult Americans believed in haunted houses. In a Gallup Poll conducted June 6–8, 2005, 370 of 1002 adult Americans aged
18 or older believed in haunted houses. Is there sufficient
evidence to support the claim that the proportion of adult
Americans who believe in haunted houses has increased
at the $\alpha = 0.05$ level of significance?

13. Eating Together In December 2001, 38% of adults with
children under the age of 18 reported that their family ate
dinner together 7 nights a week. Suppose in a recent poll,
337 of 1122 adults with children under the age of 18 reported that their family ate dinner together 7 nights a
week. Is there sufficient evidence that the proportion of
families with children under the age of 18 who eat dinner
together 7 nights a week has decreased at the $\alpha = 0.05$
significance level?

14. Critical Job Skills In August 2003, 56% of employed
adults in the United States reported that basic mathematical skills were critical or very important to their job.
The supervisor of the job placement office at a 4-year
college thinks this percentage has increased due to increased use of technology in the workplace. He takes a
random sample of 480 employed adults and finds that
297 of them feel that basic mathematical skills are critical or very important to their job. Is there sufficient evidence to conclude that the percentage of employed
adults who feel basic mathematical skills are critical or
very important to their job has increased at the $\alpha = 0.05$
level of significance?

15. Chance for Promotion In October 1998, 30% of em-
NW ployed adults were satisfied with their chances for promotion. A human resource manager wants to determine if
this percentage has changed significantly since then. She
randomly selects 280 employed adults and finds that 112
of them are completely satisfied with their chances for
promotion. Is there sufficient evidence to conclude that
the percentage of employed adults satisfied with their
chances for promotion is significantly different from the
percentage in 1998, at the $\alpha = 0.1$ level of significance?

16. Living Alone? In 2000, 58% of females aged 15 years of
age and older lived alone, according to the U.S. Census
Bureau. A sociologist tests whether this percentage is different today by conducting a random sample of 500 females aged 15 years of age and older and finds that 285 are
living alone. Is there sufficient evidence at the $\alpha = 0.1$
level of significance to support belief in a change?

17. Confidence in Schools In 1995, 40% of adults aged 18
years or older reported that they had "a great deal" of
confidence in the public schools. On June 1, 2005, the
Gallup Organization released results of a poll in which
372 of 1004 adults aged 18 years or older stated that they
had "a great deal" of confidence in the public schools. Is
there evidence at the $\alpha = 0.05$ level of significance to
conclude that the proportion of adults aged 18 years or
older having "a great deal" of confidence in the public
schools is significantly lower in 2005 than the 1995 proportion?

18. Pathological Gamblers Pathological gambling is an im-
pulse-control disorder. The American Psychiatric Association lists 10 characteristics that indicate the disorder in its
DSM-IV manual. The National Gambling Impact Study
Commission randomly selected 2417 adults and found
that 35 were pathological gamblers. Is there evidence to
support the claim that more than 1% of the adult population are pathological gamblers at the $\alpha = 0.05$ level of significance?

19. Talk to the Animals In a survey conducted by the American Animal Hospital Association, 37% of respondents
stated that they talk to their pets on the answering machine or telephone. A veterinarian found this result hard
to believe, so he randomly selected 150 pet owners and
discovered that 54 of them spoke to their pet on the answering machine or telephone. Test the veterinarian's
claim that less than 37% of pet owners speak to their pets
on the answering machine or telephone, at the $\alpha = 0.05$
level of significance.

20. Eating Salad According to a survey conducted by the Association for Dressings and Sauces (this is an actual association!), 85% of American adults eat salad at least once a
week. A nutritionist suspects that the percentage is higher
than this. She conducts a survey of 200 American adults
and finds that 171 of them eat salad at least once a week.
Is there sufficient evidence to support the nutritionist's
claim at the $\alpha = 0.10$ level of significance?

21. Small-Sample Hypothesis Test In 1997, 4% of mothers **NW** smoked more than 21 cigarettes during their pregnancy. An obstetrician believes that the percentage of mothers who smoke 21 cigarettes or more is less than 4% today. She randomly selects 120 pregnant mothers and finds that 3 of them smoked 21 or more cigarettes during pregnancy. Test the researcher's claim at the $\alpha = 0.05$ level of significance.

22. Small-Sample Hypothesis Test According to the United States Census Bureau, in 2000, 3.2% of Americans worked at home. An economist believes that the percentage of Americans working at home has increased since then. He randomly selects 150 working Americans and finds that 8 of them work at home. Test the economist's claim at the $\alpha = 0.05$ level of significance.

23. Small-Sample Hypothesis Test According to the U.S. Census Bureau, in 2000, 9.6% of Californians had a travel time to work of more than 60 minutes. An urban economist believes that the percentage of Californians who have a travel time to work of more than 60 minutes has increased since then. She randomly selects 80 working Californians and finds that 13 of them have a travel time to work that is more than 60 minutes. Test the urban economist's claim at the $\alpha = 0.1$ level of significance.

24. Small-Sample Hypothesis Test According to the U.S. Census Bureau, in 2000, 1.5% of males living in Colorado were teachers. A researcher believes that this percentage has decreased since then. She randomly selects 500 males living in Colorado and finds that 3 of them are teachers. Test the researcher's claim at the $\alpha = 0.1$ level of significance.

25. Statistics in the Media One of the more popular statistics reported in the media is the president's job approval rating. The approval rating is reported as the proportion of Americans who approve of the job that the sitting president is doing and is typically based on a random sample of about 1100 Americans.

(a) This proportion tends to fluctuate from week to week. Name some reasons for the fluctuation in the statistic.

(b) A recent article had the headline "Bush Ratings Show Decline." This headline was written because an April poll showed President Bush's approval rating to be 0.48 (48%). A poll in June based on 1100 randomly selected Americans showed that 506 approved of the job Bush was doing. Do the results of the June poll indicate that the proportion of Americans who approve of the job Bush is doing is significantly less than April's level? Explain.

26. Statistics in the Media In May 2002, 71% (0.71) of Americans favored the death penalty for a person convicted of murder. In May 2005, 1005 adult Americans were asked by the Gallup Organization, "Are you in favor of the death penalty for a person convicted of murder?" Of the 1005 adults surveyed, 744 responded yes. The headline in the article reporting the survey results stated, "Americans' Views of Death Penalty More Positive This Year." Use a test of significance to support or refute this headline.

Technology Step-by-Step | **Hypothesis Tests Regarding a Population Proportion**

TI-83/84 Plus

Step 1: Press STAT, highlight TESTS, and select `5:1-PropZTest`.

Step 2: For the value of p_0, enter the "status quo" value of the population proportion.

Step 3: Enter the number of successes, x, and the sample size, n.

Step 4: Select the direction of the alternative hypothesis.

Step 5: Highlight Calculate or Draw, and press ENTER. The TI-83 or TI-84 gives the P-value.

MINITAB

Step 1: Select the **Stat** menu, highlight **Basic Statistics**, then highlight **1-Proportion**.

Step 2: Select "Summarized data."

Step 3: Enter the number of trials, n, and the number of successes, x.

Step 4: Click Options. Enter the "status quo" value of the population proportion in the cell "Test proportion." Enter the direction of the alternative hypothesis. If $np_0(1 - p_0) \geq 10$, check the box marked "Use test and interval based on normal distribution." Click OK twice.

Excel

Step 1: Load the PHStat Add-in.

Step 2: Select the **PHStat menu**, highlight **One Sample Tests . . .**, then highlight **Z Test for proportion**.

Step 3: Enter the value of the null hypothesis, the level of significance, α, the number of successes, x, and the number of trials, n. Click the option corresponding to the desired test [two-tail, upper (right) tail, or lower (left) tail]. Click OK.

10.5 Testing Claims about a Population Standard Deviation

Preparing for This Section Before getting started, review the following:

* Confidence intervals about a population standard
  deviation (Section 9.4, pp. 486–491)

Objectives   1 Test a claim about a population standard deviation.

In this section, we discuss methods for testing claims regarding a population variance or standard deviation.

Why might we be interested in testing claims regarding σ^2 or σ? Many production processes not only require accuracy on average (the mean), they also require consistency. Consider a filling machine (such as a coffee machine) that consistently over- and underfills cups, but, on average, fills correctly. Certainly, customers are not happy if the machine underfills their cups, and they might be dissatisfied even if it overfills, because of spilling. A machine that *consistently* delivers the correct amount of liquid is desired. As another example, consider a mutual fund that invests in the stock market. An investor would prefer consistent year-to-year returns near 12%, rather than returns that fluctuate wildly, yet result in a mean return of 12%. Both these situations illustrate the importance of measuring variability. The main idea is that the standard deviation and the variance are measures of consistency. The less consistent the values of a variable are, the higher the standard deviation of the variable.

We begin by reviewing the chi-square distribution.

Chi-Square Distribution

If a simple random sample of size n is obtained from a normally distributed population with mean μ and standard deviation σ, then

$$\chi^2 = \frac{(n-1)s^2}{\sigma^2}$$

where s^2 is a sample variance has a chi-square distribution with $n-1$ degrees of freedom.

Remember, the symbol χ^2 is pronounced "kigh-square." You can find critical values of the chi-square distribution in Table VI in Appendix A of the text.

Figure 26
Chi-Square Distributions

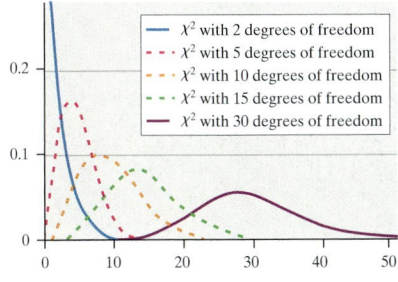

Characteristics of the Chi-Square Distribution

1. It is not symmetric.
2. The shape of the chi-square distribution depends on the degrees of freedom, just as with Student's t-distribution.
3. As the number of degrees of freedom increases, the chi-square distribution becomes more nearly symmetric, as is illustrated in Figure 26.
4. The values of χ^2 are nonnegative: that is, the values of χ^2 are greater than or equal to 0.

Recall that Table VI is structured similarly to Table V (for the t-distribution). The left column represents the degrees of freedom, and the top row represents the area under the chi-square distribution to the right of the critical value. We use the notation χ^2_α to denote the critical χ^2-value for which the area under the chi-square distribution to the right of χ^2_α is α. For example, the area under the chi-square distribution to the right of $\chi^2_{0.10}$ is 0.10.

Testing a Claim about a Population Standard Deviation

We now present the steps for testing a claim about a population variance or standard deviation.

Testing a Claim about a Population Variance or Standard Deviation

If a claim is made about the population variance or standard deviation, we can use the following steps to test the claim, provided that

1. the sample is obtained using simple random sampling
2. the population is normally distributed.

Step 1: A claim is made regarding the population variance or standard deviation. The claim is used to determine the null and alternative hypotheses. We present the three cases for a claim regarding a population standard deviation.

Two-Tailed	Left-Tailed	Right-Tailed
$H_0: \sigma = \sigma_0$	$H_0: \sigma = \sigma_0$	$H_0: \sigma = \sigma_0$
$H_1: \sigma \neq \sigma_0$	$H_1: \sigma < \sigma_0$	$H_1: \sigma > \sigma_0$

Note: σ_0 is the assumed value of the population standard deviation.

Step 2: Select a level of significance α based on the seriousness of making a Type I error.

Step 3: Compute the test statistic

$$\chi_0^2 = \frac{(n-1)s^2}{\sigma_0^2}$$

Classical Approach

Step 4: Use Table VI to determine the critical value using $n-1$ degrees of freedom.

Two-Tailed

Critical Region

$\chi_{1-\frac{\alpha}{2}}^2$ $\chi_{\frac{\alpha}{2}}^2$

(critical value) (critical value)

Left-tailed

Critical Region

$\chi_{1-\alpha}^2$

(critical value)

Right-tailed

Critical Region

χ^2

(critical value)

Step 5: Compare the critical value with the test statistic.

Two-Tailed	Left-Tailed	Right-Tailed
If $\chi_0^2 < \chi_{1-\alpha/2}^2$ or $\chi_0^2 > \chi_{\alpha/2}^2$, reject the null hypothesis.	If $\chi_0^2 < \chi_{1-\alpha}^2$, reject the null hypothesis.	If $\chi_0^2 > \chi_\alpha^2$, reject the null hypothesis.

P-value Approach

Step 4: Use Table VI to determination approximate P-value for a left-tailed or right-tailed test by determining the area under the chi-square distribution with $n-1$ degrees of freedom to the left (for a left-tailed test) or right (for a right-tailed test) of the test statistic. For two-tailed tests, we recommend the classical approach.

Left-Tailed

P-value = $P(\chi^2 < \chi_0^2)$

χ_0^2

Right-Tailed

P-value = $P(\chi^2 > \chi_0^2)$

χ_0^2

Step 5: If P-value $< \alpha$, reject the null hypothesis.

Step 6: State the conclusion.

!**CAUTION**

● The procedures in this section are not robust.

The methods presented for testing a hypothesis about a population variance or standard deviation are not robust. Therefore, if analysis of the data indicates that the variable does not come from a popultion that is normally distributed, the procedures presented in this section are not valid. Again, always be sure to verify that the requirements are satisfied before proceeding with the test!

EXAMPLE 1	**Testing a Claim about a Population Standard Deviation,**

Testing a Claim about a Population Standard Deviation, Left-Tailed Test

Table 4		
19.68	20.66	19.56
19.98	20.65	19.61
20.55	20.36	21.02
21.50	19.74	

Source: Michael Carlisle, student at Joliet Junior College

Problem: In Example 2 from Section 10.3, the quality-control engineer for M&M–Mars tested the claim that the mean weight of fun-size Snickers was not 20.1 grams. Suppose that the standard deviation of the weight of the candy was 0.75 gram before a machine recalibration. The quality-control engineer claims that the recalibration resulted in the standard deviation weight being less than 0.75 gram. Test this claim at the $\alpha = 0.05$ level of significance. For convenience, the data are reproduced in Table 4.

Approach: Before we can perform the hypothesis test, we must verify that the data come from a population that is normally distributed. We will construct a normal probability plot to verify this requirement. We then proceed to follow Steps 1 through 6.

Solution: A normal probability plot was drawn for these data in Figure 17. The plot indicates that the weight of the candy comes from a population that is normally distributed.

Step 1: We need evidence to conclude that the population standard deviation is less than 0.75 gram. This can be written $\sigma < 0.75$. We have

$$H_0: \sigma = 0.75 \quad \text{versus} \quad H_1: \sigma < 0.75$$

!**CAUTION**

● The test statistic requires s^2 and σ^2, so be careful if the problem gives s or σ.

This is a left-tailed test.

Step 2: The level of significance is $\alpha = 0.05$.

Step 3: From the data in Table 4, the sample standard deviation is computed to be $s = 0.6404$. The test statistic is

$$\chi_0^2 = \frac{(n-1)s^2}{\sigma_0^2} = \frac{(11-1)(0.6404)^2}{0.75^2} = 7.291$$

Classical Approach

Step 4: Because this is a left-tailed test, we determine the critical χ^2-value at the $\alpha = 0.05$ level of significance with $n - 1 = 11 - 1 = 10$ degrees of freedom to be $\chi_{1-0.05}^2 = \chi_{0.95}^2 = 3.940$. The critical region is displayed in Figure 27.

Figure 27

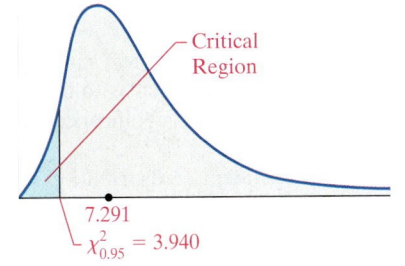

Critical Region

7.291

$\chi_{0.95}^2 = 3.940$

P-value Approach

Step 4: Because this is a left-tailed test, the P-value is the area under the χ^2-distribution with $11 - 1 = 10$ degrees of freedom to the left of the test statistic, $\chi_0^2 = 7.291$, as shown in Figure 28. That is, P-value $= P(\chi^2 < \chi_0^2) = P(\chi^2 < 7.291)$, with 10 degrees of freedom.

Using Table VI, we find the row that corresponds to 10 degrees of freedom. The value 7.291 is greater than 4.865. The value of 4.865 has an area under the χ^2-distribution of 0.10 to the left. So the P-value is greater than 0.10.

Figure 28

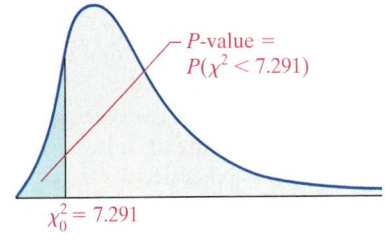

P-value $= P(\chi^2 < 7.291)$

$\chi_0^2 = 7.291$

Step 5: Because the test statistic $\chi_0^2 = 7.291$ is greater than the critical value $\chi_{0.95}^2 = 3.940$, the quality-control engineer does not reject the null hypothesis. We label this point in Figure 27.

Step 5: Because the P-value is greater than the level of significance $\alpha = 0.05$, we do not reject the null hypothesis.

Step 6: There is not sufficient evidence at the $\alpha = 0.05$ level of significance to support the claim that the standard deviation of the weight of fun-size Snickers is less than 0.75 gram.

Figure 29

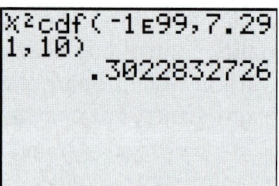

Neither MINITAB, Excel, nor the TI-83/84 Plus graphing calculators currently have procedures for testing claims regarding a population variance or standard deviation. However, we can find exact P-values by finding the area under the chi-square distribution. For example, Figure 29 shows the results of finding the area under the chi-square distribution to the left of $\chi_0^2 = 7.291$ with 10 degrees of freedom using a TI-84 Plus graphing calculator. The exact P-value is 0.302.

Now Work Problem 9.

10.5 ASSESS YOUR UNDERSTANDING

Concepts and Vocabulary

1. State the requirements to test a claim regarding a population standard deviation.

2. Determine the critical value for a right-tailed test of a population standard deviation with 18 degrees of freedom at the $\alpha = 0.05$ level of significance.

Skill Building

3. To test $H_0: \sigma = 50$ versus $H_1: \sigma < 50$, a random sample of size $n = 24$ is obtained from a population that is known to be normally distributed.
   (a) If the sample standard deviation is determined to be $s = 47.2$, compute the test statistic.
   (b) If the researcher decides to test this hypothesis at the $\alpha = 0.05$ level of significance, determine the critical value.
   (c) Draw a chi-square distribution and depict the critical region.
   (d) Will the researcher reject the null hypothesis? Why?

4. To test $H_0: \sigma = 35$ versus $H_1: \sigma > 35$, a random sample of size $n = 15$ is obtained from a population that is known to be normally distributed.
   (a) If the sample standard deviation is determined to be $s = 37.4$, compute the test statistic.
   (b) If the researcher decides to test this hypothesis at the $\alpha = 0.01$ level of significance, determine the critical value.
   (c) Draw a chi-square distribution and depict the critical region.
   (d) Will the researcher reject the null hypothesis? Why?

5. To test $H_0: \sigma = 1.8$ versus $H_1: \sigma > 1.8$, a random sample of size $n = 18$ is obtained from a population that is known to be normally distributed.
   (a) If the sample standard deviation is determined to be $s = 2.4$, compute the test statistic.
   (b) If the researcher decides to test this hypothesis at the $\alpha = 0.10$ level of significance, determine the critical value.

   (c) Draw a chi-square distribution and depict the critical region.
   (d) Will the researcher reject the null hypothesis? Why?

6. To test $H_0: \sigma = 0.35$ versus $H_1: \sigma < 0.35$, a random sample of size $n = 41$ is obtained from a population that is known to be normally distributed.
   (a) If the sample standard deviation is determined to be $s = 0.23$, compute the test statistic.
   (b) If the researcher decides to test this hypothesis at the $\alpha = 0.01$ level of significance, determine the critical value.
   (c) Draw a chi-square distribution and depict the critical region.
   (d) Will the researcher reject the null hypothesis? Why?

7. To test $H_0: \sigma = 4.3$ versus $H_1: \sigma \neq 4.3$, a random sample of size $n = 12$ is obtained from a population that is known to be normally distributed.
   (a) If the sample standard deviation is determined to be $s = 4.8$, compute the test statistic.
   (b) If the researcher decides to test this hypothesis at the $\alpha = 0.05$ level of significance, determine the critical values.
   (c) Draw a chi-square distribution and depict the critical regions.
   (d) Will the researcher reject the null hypothesis? Why?

8. To test $H_0: \sigma = 1.2$ versus $H_1: \sigma \neq 1.2$, a random sample of size $n = 22$ is obtained from a population that is known to be normally distributed.
   (a) If the sample standard deviation is determined to be $s = 0.8$, compute the test statistic.

(b) If the researcher decides to test this hypothesis at the $\alpha = 0.10$ level of significance, determine the critical values.

(c) Draw a chi-square distribution and depict the critical regions.

(d) Will the researcher reject the null hypothesis? Why?

Applying the Concepts

9. **Mutual Fund Risk** One measure of the risk of a mutual
NW fund is the standard deviation of its rate of return. Suppose a mutual fund qualifies as having moderate risk if the standard deviation of its monthly rate of return is less than 4%. A mutual-fund manager claims that his fund has moderate risk. A mutual-fund rating agency does not believe this claim and randomly selects 25 months and determines the rate of return for the fund. The standard deviation of the rate of return is computed to be 3.01%. Is there sufficient evidence to support the claim that the fund has moderate risk at the $\alpha = 0.05$ level of significance? A normal probability plot indicates that the monthly rates of return are normally distributed.

10. **Filling Machine** A machine fills bottles with 64 fluid ounces of liquid. The quality-control manager determines that the fill levels are normally distributed with a mean of 64 ounces and a standard deviation of 0.42 ounces. He has an engineer recalibrate the machine in an attempt to lower the standard deviation. After the recalibration, the quality-control manager randomly selects 19 bottles from the line and determines that the standard deviation is 0.38 ounces. Is there sufficient evidence for the manager to conclude that the standard deviation has decreased at the $\alpha = 0.01$ level of significance?

11. **Pump Design** The piston diameter of a certain hand pump is 0.5 inch. The quality-control manager determines that the diameters are normally distributed, with a mean of 0.5 inch and a standard deviation of 0.004 inch. The machine that controls the piston diameter is recalibrated in an attempt to lower the standard deviation. After recalibration, the quality-control manager randomly selects 25 pistons from the production line and determines that the standard deviation is 0.0025 inch. Is there significant evidence for the manager to conclude that the standard deviation has decreased at the $\alpha = 0.01$ level of significance?

12. **Counting Carbs** The manufacturer of processed deli meats claims that the standard deviation of the number of carbohydrates in its smoked turkey breast is 0.5 gram per 2-ounce serving. A dietician does not believe this claim and randomly selects eighteen 2-ounce servings of the smoked turkey breast and determines the number of carbohydrates per serving. The standard deviation of the number of carbs is computed to be 0.62 gram per serving. Is there sufficient evidence to indicate that the standard deviation is not 0.5 gram per serving at the $\alpha = 0.05$ level of significance? A normal probability plot indicates that the number of carbohydrates per serving is normally distributed.

13. **Acid Rain** In Problem 23 from Section 10.2, we tested a claim regarding the mean amount of acidity in rain in Pierce County, Washington. To test the claim, we verified that the data are normally distributed and assumed that $\sigma = 0.2$. Test this assumption at the $\alpha = 0.05$ level of significance, given that $s = 0.1708$.

5.08	4.66	4.70	4.87
4.78	5.00	4.50	4.73
4.79	4.65	4.91	5.07
5.03	4.78	4.77	4.60
4.73	5.05	4.70	

Source: National Atmospheric Deposition Program

14. **Fuel Catalyst** In Problem 24 from Section 10.2, we tested a claim regarding the mean number of miles per gallon for semitrucks. To test this claim, we verified that the data are normally distributed and assumed that $\sigma = 0.5$ mile per gallon. Test this assumption at the $\alpha = 0.05$ level of significance, given that $s = 0.375$ mile per gallon.

5.9	6.7	6.9	6.4	6.6	6.3
6.2	5.9	6.2	6.4	6.6	5.6

15. **Conforming Golf Balls** In Problem 21 from Section 10.3, we verified that Maxfli XS golf balls conformed to the USGA requirements for mean diameter. Suppose the USGA also requires that the standard deviation of the diameter of a golf ball must be less than 0.004 inch. Determine whether these balls conform to this requirement at the $\alpha = 0.01$ level of significance. (**Note:** In Problem 21 from Section 10.3, we verified that the data are normally distributed.)

1.683	1.677	1.681
1.685	1.678	1.686
1.684	1.684	1.673
1.685	1.682	1.674

Source: Michael McCraith, Joliet Junior College

16. **Conforming Golf Balls** In Problem 22 from Section 10.3, we verified that Maxfli XS golf balls conformed to USGA requirements for mean weight. Suppose the USGA also requires that the standard deviation of the weight of a golf ball must be less than 0.005 ounce. Determine whether these balls conform to this requirement at the $\alpha = 0.01$ level of significance. (**Note:** In Problem 22 from Section 10.3, we verified that the data are normally distributed.)

1.614	1.619	1.614
1.614	1.610	1.610
1.621	1.612	1.615
1.621	1.602	1.617

Source: Michael McCraith, Joliet Junior College

17. **An Inconsistent Player** We often hear sports announcers say, "I wonder which player will show up to play today." This is the announcer's way of saying that the player is inconsistent, that his or her performance varies dramatically from game to game. Suppose the standard deviation of the number of points scored by shooting guards in the NBA is 8.3. A random sample of 25 games played by Allen Iverson of the Philadelphia 76ers results in a sample standard deviation of 6.7 points. Assuming that a normal probability plot indicates that the points scored are approximately normally distributed, test the claim that Allen Iverson is more consistent than other shooting guards in the NBA at the $\alpha = 0.10$ level of significance.

18. **Waiting in Line** One aspect of Queuing theory is to consider waiting time in lines. Suppose a fast-food chain is trying to determine whether it should switch from having four cash registers with four separate lines to four cash registers with a single line. It has been determined that the mean wait time in both lines is equal. However, the chain is uncertain about which line has less variability in wait time. From experience, the chain knows that the wait times in the four separate lines are normally distributed with $\sigma = 1.2$ minutes. In a study, the chain reconfigured five restaurants to have a single line and measured the wait times for 50 randomly selected customers. The sample standard deviation was determined to be $s = 0.84$ minute. Test the claim that the variability in wait time is less for a single line than for multiple lines at the $\alpha = 0.05$ level of significance.

19. **Heights of Baseball Players** Data obtained from the National Center for Health Statistics show that men between the ages of 20 and 29 have a mean height of 69.3 inches, with a standard deviation of 2.9 inches. A baseball analyst claims that the standard deviation of heights of major-

league baseball players is less than 2.9 inches. The heights (in inches) of 20 randomly selected players are shown in the table.

72	74	71	72	76
70	77	75	72	72
77	72	75	70	73
73	75	73	74	74

Source: espn.com

(a) Verify that the data are normally distributed by drawing a normal probability plot.
(b) Compute the sample standard deviation.
(c) Test the claim at the $\alpha = 0.01$ level of significance.

20. **NCAA Softball** NCAA rules require the circumference of a softball to be 12 ± 0.125 inches. A softball manufacturer bidding on an NCAA contract is shown to meet the requirements for mean circumference. Suppose the NCAA also requires that the standard deviation of the softball circumferences not exceed 0.05 inch. A representative from the NCAA believes the manufacturer does not meet this requirement. She collects a random sample of 20 softballs from the production line and finds that $s = 0.09$ inch. Is there enough evidence to support the representative's claim at the $\alpha = 0.05$ level of significance?

21. **P-Values** Determine the exact *P*-value of the hypothesis test in Problem 9.

22. **P-Values** Determine the exact *P*-value of the hypothesis test in Problem 10.

10.6 Putting It All Together: Which Method Do I Use?

Objective   **1** Determine the appropriate hypothesis test to perform

1 ### Determine the Appropriate Hypothesis Test to Perform

Perhaps the most difficult aspect of testing claims is determining which hypothesis test to conduct. To assist in the decision making, we present Figure 30 which shows which approach to take in testing claims for the three parameters discussed in this chapter.

Figure 30

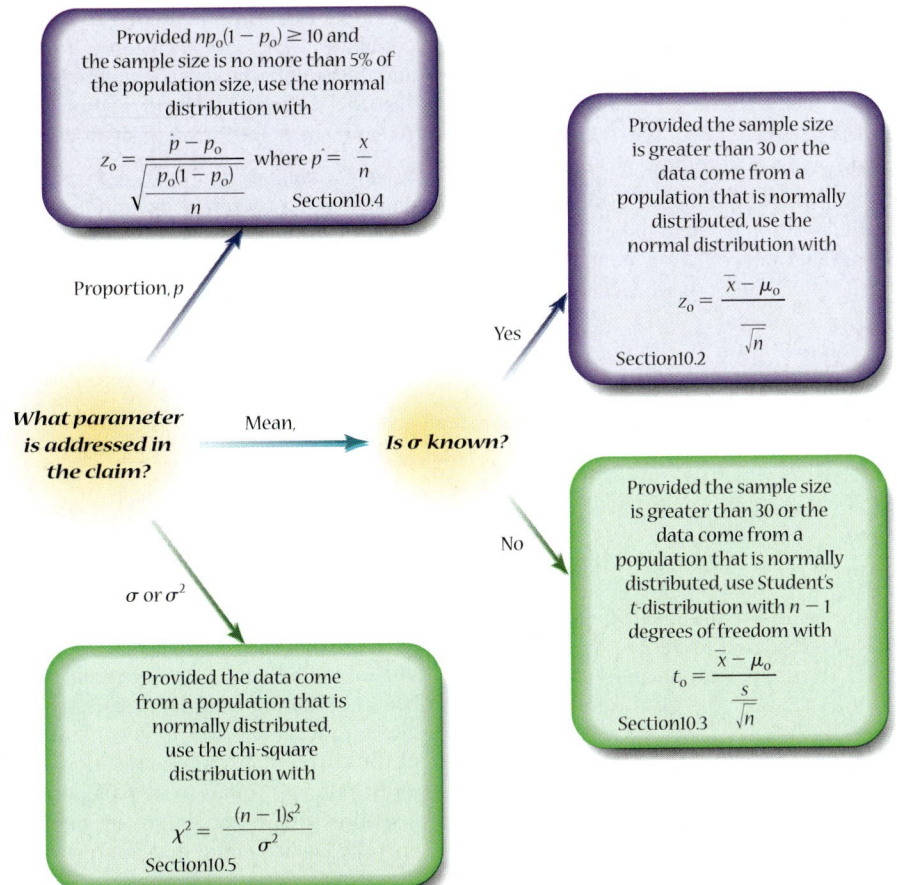

Provided $np_0(1 - p_0) \geq 10$ and the sample size is no more than 5% of the population size, use the normal distribution with

$$z_0 = \frac{\hat{p} - p_0}{\sqrt{\dfrac{p_0(1 - p_0)}{n}}} \text{ where } \hat{p} = \frac{x}{n}$$

Section 10.4

Provided the sample size is greater than 30 or the data come from a population that is normally distributed, use the normal distribution with

$$z_0 = \frac{\bar{x} - \mu_0}{\frac{\sigma}{\sqrt{n}}}$$

Section 10.2

What parameter is addressed in the claim?   Mean,   **Is σ known?**   Yes

Proportion, p

No

σ or σ^2

Provided the sample size is greater than 30 or the data come from a population that is normally distributed, use Student's t-distribution with $n - 1$ degrees of freedom with

$$t_0 = \frac{\bar{x} - \mu_0}{\frac{s}{\sqrt{n}}}$$

Section 10.3

Provided the data come from a population that is normally distributed, use the chi-square distribution with

$$\chi^2 = \frac{(n - 1)s^2}{\sigma^2}$$

Section 10.5

10.6 ASSESS YOUR UNDERSTANDING

Concepts and Vocabulary

1. What are the requirements that must be satisfied to test a claim about a population mean? When do we use the normal model to test a claim about a population mean? When do we use Student's t-distribution to test a claim about a population mean?

2. What are the requirements that must be satisfied before we can test a claim about a population proportion?

Skill Building

In Problems 3–12, test each claim.

3. A simple random sample of size $n = 14$ is drawn from a population that is normally distributed with $\sigma = 20$. The sample mean is found to be $\bar{x} = 60$. Test the claim that the population mean is less than 70 at the $\alpha = 0.1$ level of significance.

4. A simple random sample of size $n = 19$ is drawn from a population that is normally distributed. The sample mean is found to be $\bar{x} = 0.8$, and the sample standard deviation is found to be $s = 0.4$. Test the claim that the population mean is less than 1.0 at the $\alpha = 0.01$ level of significance.

5. A simple random sample of size $n = 200$ individuals with a valid driver's license is asked if they drive an American-made automobile. Of the 200 individuals surveyed, 115 responded that they drive an American-made automobile. Test the claim that more than half of those with a valid driver's license drive an American-made automobile at the $\alpha = 0.05$ level of significance.

6. A simple random sample of size $n = 20$ is drawn from a population that is normally distributed. The sample variance is found to be $s^2 = 49.3$. Test the claim that the population variance is less than 95 at the $\alpha = 0.1$ level of significance.

7. A simple random sample of size $n = 15$ is drawn from a population that is normally distributed. The sample mean is found to be $\bar{x} = 23.8$, and the sample standard deviation is found to be $s = 6.3$. Test the claim that the population mean is different from 25 at the $\alpha = 0.01$ level of significance.

8. A simple random sample of size $n = 25$ is drawn from a population that is normally distributed with $\sigma = 7$. The sample mean is found to be $\bar{x} = 53.2$. Test the claim that the population mean is different from 55 at the $\alpha = 0.05$ level of significance.

9. A simple random sample of size $n = 16$ is drawn from a population that is normally distributed. The sample variance is found to be $s^2 = 13.7$. Test the claim that the population variance is greater than 10 at the $\alpha = 0.05$ level of significance.

10. A simple random sample of size $n = 65$ is drawn from a population. The sample mean is found to be $\bar{x} = 583.1$, and the sample standard deviation is found to be $s = 114.9$. Test the claim that the population mean is different from 600 at the $\alpha = 0.1$ level of significance.

11. A simple random sample of size $n = 40$ is drawn from a population. The sample mean is found to be $\bar{x} = 108.5$, and the sample standard deviation is found to be $s = 17.9$. Test the claim that the population mean is greater than 100 at the $\alpha = 0.05$ level of significance.

12. A simple random sample of size $n = 320$ adults was asked their favorite ice cream flavor. Of the 320 individuals surveyed, 58 responded that they preferred mint chocolate chip. Test the claim that less than 25% of adults prefer mint chocolate chip ice cream at the $\alpha = 0.01$ level of significance.

Applying the Concepts

13. **Family Size** In 1985, the mean for the ideal number of children for a family was considered to be 2.5. A Gallup Poll of 1006 adults aged 18 or older conducted February 16–17, 2004, reported the mean for the ideal number of children to be 2.6.

    (a) Assuming $\sigma = 1.2$, is there sufficient evidence to support the claim that the mean ideal number of children has changed since 1985 at the $\alpha = 0.1$ level of significance?

    (b) Do the results have any practical significance? Explain.

14. **Americans Online** In spring 2002, 66% of adults in the United States aged 18 years or older had Internet access. A Harris Interactive poll in February and May of 2005 found that 1496 of 2022 adults surveyed had Internet access. Is this enough evidence to support the claim that more adults had Internet access in 2005 than in 2002 at the $\alpha = 0.05$ level of significance?

15. **Yield Strength** A manufacturer of high-strength, low-alloy steel beams requires that the standard deviation of yield strength not exceed 7000 psi. The quality-control manager selected a sample of 20 steel beams and measured their yield strength. The standard deviation of the sample was 7500 psi. Assume yield strengths are normally distributed. Test the claim that the standard deviation of yield strength exceeds 7000 psi at the $\alpha = 0.01$ level of significance.

16. **Tattoos** In 2001, 23% of American university undergraduate students had at least one tattoo. A health practitioner suspects that the percent has changed since then. She obtains a random sample of 1026 university undergraduates and finds that 254 have at least one tattoo. Is this sufficient evidence to support the practitioner's claim at the $\alpha = 0.1$ level of significance?

17. **Mortgage Rates** In 2001, the mean contract interest rate for a conventional 30-year first loan for the purchase of a single-family home was 6.3 percent, according to the U.S. Federal Housing Board. A real estate agent believes that interest rates are lower today and obtains a random sample of 41 recent 30-year conventional loans. The mean interest rate was found to be 6.05 percent, with a standard deviation of 1.75 percent. Is this enough evidence to support the agent's claim that interest rates are lower at the $\alpha = 0.05$ level of significance?

18. **Pharmaceuticals** A pharmaceutical company manufactures a 200-mg pain reliever. Company specifications include that the standard deviation of the amount of the active ingredient must not exceed 5 mg. The quality-control manager selects a random sample of 30 tablets from a certain batch and finds that the sample standard deviation is 7.3 mg. Assume that the amount of the active ingredient is normally distributed. Test the claim that the standard deviation of the amount of the active ingredient is greater than 5 mg at the $\alpha = 0.05$ level of significance.

19. **Auto Insurance** According to the Insurance Information Institute, the mean expenditure for auto insurance in the United States was $774 for 2002. An insurance salesman believes that mean expenditure for auto insurance is different now. He obtains a random sample of 35 auto insurance policies and determines the mean expenditure to be $735 with a standard deviation of $48.31. Is there enough evidence to support the claim that the mean expenditure for auto insurance is different from the 2002 amount at the $\alpha = 0.01$ level of significance?

20. Toner Cartridge The manufacturer of a toner cartridge claims the mean number of printouts is 10,000 for each cartridge. A consumer advocate believes the actual mean number of printouts is lower. He selects a random sample of 14 such cartridges and obtains the following number of printouts:

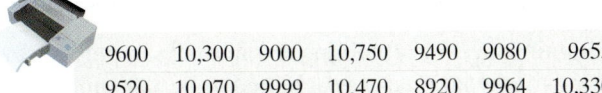

9600	10,300	9000	10,750	9490	9080	9655
9520	10,070	9999	10,470	8920	9964	10,330

(a) Because the sample size is small, he must verify that the number of printouts is normally distributed and the sample does not contain any outliers. The normal probability plot and boxplot are shown. Are the conditions for testing the hypothesis satisfied?

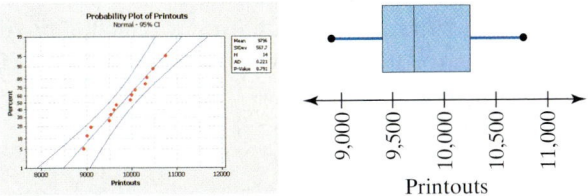

Printouts

(b) Is there enough evidence to support the advocate's claim at the $\alpha = 0.05$ level of significance?

21. Vehicle Emission Inspection A certain vehicle emission inspection station claims that the mean wait time for customers is less than 8 minutes. A local resident is skeptical and collects a random sample of 49 wait times for customers at the testing station. He finds that the sample mean is 7.34 minutes, with a standard deviation of 3.2 minutes. Is there enough evidence to support the testing station's claim at the $\alpha = 0.01$ level of significance?

22. Lights Out With a previous contractor, the mean time to replace a streetlight was 3.2 days. A city councilwoman thinks that the new contractor is not getting the streetlights replaced as quickly. She selects a random sample of 12 streetlight service calls and obtains the following times to replacement (in days).

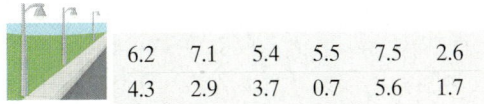

6.2	7.1	5.4	5.5	7.5	2.6
4.3	2.9	3.7	0.7	5.6	1.7

(a) Because the sample size is small, she must verify that replacement time is normally distributed and the sample does not contain any outliers. The normal probability plot and boxplot are shown below. Are the conditions for testing the hypothesis satisfied.

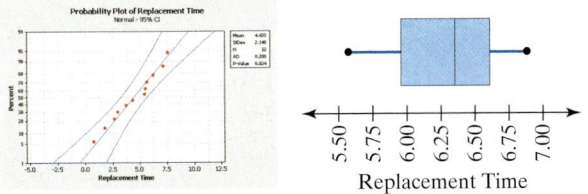

Replacement Time

(b) Is there enough evidence to support the councilwoman's claim at the $\alpha = 0.05$ level of significance?

10.7 # The Probability of a Type II Error and the Power of the Test

Preparing for This Section   Before getting started, review the following:

• Computing normal probabilities (Section 7.3, pp. 385–389)

Objectives   **1** **Determine the probability of making a Type II error**
             2 **Compute the power of the test**

1   ## Determine the Probability of Making a Type II Error

Recall, from Section 10.1, that a Type II error is made if we fail to reject the null hypothesis when the alternative hypothesis is true. The probability of making a Type II error is called β. It can be shown that the probability of making a Type II error; the probability of making a Type I error, α; and the sample size, n, are related. Once two of the three values are chosen, the third value can be determined when a specific value of the parameter in the alternative hypothesis is chosen. Usually researchers choose a level of α and the sample size n. To minimize the value of β for a given level of α, a researcher should choose n to be as large as time, money, and other resources allow.

In this section we present the procedure for computing the probability of making a Type II error for tests regarding the population mean when the population standard deviation is known. We will not find β for any other hypothesis tests.

In Example 1 from Section 10.2 we tested

$$H_0: \mu = 12{,}200 \text{ versus } H_1: \mu > 12{,}200$$

at the $\alpha = 0.10$ level of significance by obtaining a random sample of size $n = 35$. The population standard deviation σ was assumed to be 3800 miles. The alternative hypothesis is true if the true population mean is greater than 12,200 miles, and a Type II error is committed whenever H_0 is not rejected and $\mu > 12{,}200$. So β is

$$P(\text{Type II Error}) = P(\text{do not reject } H_0 \text{ given } H_1 \text{ is true})$$

To compute the probability of making a Type II error we need to know the criterion for not rejecting H_0. Since in this example $\alpha = 0.10$, we would not reject H_0 if $Z_0 < z_{0.1} = 1.28$. We can write the criterion for not rejecting the null hypothesis in terms of a sample mean. Using the Z-score, we will not reject H_0 if

$$Z_0 = \frac{\bar{x} - 12{,}200}{\dfrac{3800}{\sqrt{35}}} < 1.28$$

Solving the inequality for $\bar{x}$, we would not reject H_0 if

$$\bar{x} < 12{,}200 + 1.28\left(\frac{3800}{\sqrt{35}}\right) = 13{,}022$$

Figure 31 shows the "reject" and "do not reject" regions.

To compute the probability of making a Type II error, we also need a specific value of the population mean that is in the alternative hypothesis. For example, suppose the true population mean is 12,700 miles. Then

$$P(\text{Type II error}) = P(\text{do not reject } H_0 \text{ given } H_1 \text{ is true})$$

$$= P(\text{sample mean is less than 13,022 given that } \mu = 12{,}700)$$

$$= P\left(Z < \frac{13{,}022 - 12{,}700}{\dfrac{3800}{\sqrt{35}}}\right)$$

$$= P(Z < 0.50)$$

$$= 0.6915$$

The probability of making a Type II error is 0.6915 if the population mean is 12,700 miles. Figure 32 helps to illustrate the concept.

Figure 31

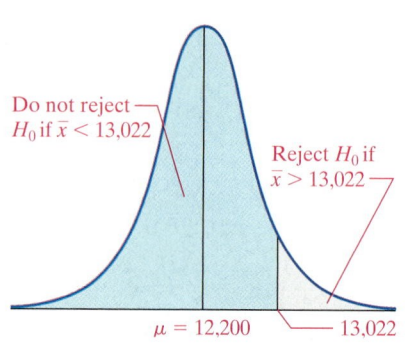

Do not reject H_0 if $\bar{x} < 13{,}022$

Reject H_0 if $\bar{x} > 13{,}022$

$\mu = 12{,}200$ 13,022

Figure 32

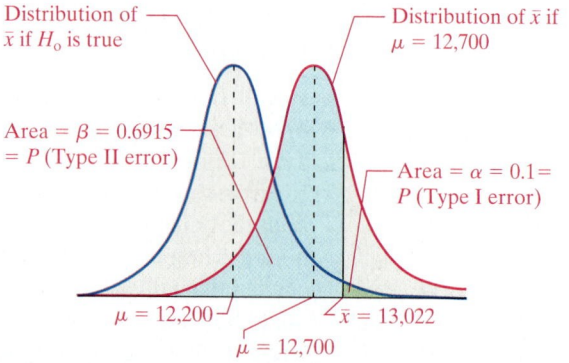

Distribution of $\bar{x}$ if H_0 is true

Distribution of $\bar{x}$ if $\mu = 12{,}700$

Area $= \beta = 0.6915$ $= P$ (Type II error)

Area $= \alpha = 0.1 = P$ (Type I error)

$\mu = 12{,}200$ $\bar{x} = 13{,}022$

$\mu = 12{,}700$

There are an infinite number of values for the mean that are greater than 12,200, and each has its own Type II error probability.

We summarize the procedure for computing the probability of making a Type II error with the steps that follow.

Probability of a Type II Error

Step 1: Determine the sample mean that separates the rejection region from the nonrejection region.

Two-Tailed Test $(H_1: \mu \neq \mu_0)$	**Right-Tailed Test** $(H_1: \mu > \mu_0)$	**Left-Tailed Test** $(H_1: \mu < \mu_0)$
$\bar{x}_L = \mu_0 - z_{\alpha/2} \cdot \dfrac{\sigma}{\sqrt{n}}$	$\bar{x} = \mu_0 + z_\alpha \cdot \dfrac{\sigma}{\sqrt{n}}$	$\bar{x} = \mu_0 - z_\alpha \cdot \dfrac{\sigma}{\sqrt{n}}$
$\bar{x}_U = \mu_0 + z_{\alpha/2} \cdot \dfrac{\sigma}{\sqrt{n}}$		

Note: z is the critical value, μ_0 is the assumed value of the population mean, σ is the population standard deviation, and n is the sample size.

Step 2: Draw a normal curve whose mean is a particular value from the alternative hypothesis, with the sample mean(s) found in Step 1 labeled.

Step 3:

Two-Tailed Test $(H_1: \mu \neq \mu_0)$	**Right-Tailed Test** $(H_1: \mu > \mu_0)$	**Left-Tailed Test** $(H_1: \mu < \mu_0)$
Find the area under the normal curve drawn in Step 2 between $\bar{x}_L$ and $\bar{x}_U$ found in Step 1. This area represents β, the probability of not rejecting the null hypothesis when the alternative hypothesis is true.	Find the area under the normal curve drawn in Step 2 to the left of the sample mean, $\bar{x}$, found in Step 1. This area represents β, the probability of not rejecting the null hypothesis when the alternative hypothesis is true.	Find the area under the normal curve drawn in Step 2 to the right of the sample mean, $\bar{x}$, found in Step 1. This area represents β, the probability of not rejecting the null hypothesis when the alternative hypothesis is true.

Let's look at an example that uses these steps.

EXAMPLE 1   ## Computing the Probability of a Type II Error

Problem: In Example 1 from Section 10.2, we tested $H_0: \mu = 12,200$ versus $H_1: \mu > 12,200$ for a random sample of size $n = 35$, with the population standard deviation, σ, assumed to be 3800 miles at the $\alpha = 0.1$ level of significance. Compute the probability of a Type II error, given that the population mean is $\mu = 12,500$.

Approach: We follow Steps 1 through 3.

Solution

Step 1: We first determine $z_{0.1}$ to be 1.28. We let $Z = 1.28$, $\mu_0 = 12,200$, $\sigma = 3800$, and $n = 35$ to find the sample mean that separates the rejection region from the nonrejection region:

$$1.28 = \frac{\bar{x} - 12,200}{3800/\sqrt{35}}$$

Now solve for the sample mean, $\bar{x}$:

$$1.28 \cdot \frac{3800}{\sqrt{35}} = \bar{x} - 12,200 \qquad \text{Multiply both sides by } \frac{3800}{\sqrt{35}}.$$

$$\bar{x} = 12,200 + 1.28 \cdot \frac{3800}{\sqrt{35}} \approx 13,022 \qquad \text{Add 12,200 to both sides and simplify.}$$

For any sample mean less than 13,022, we do not reject the null hypothesis.

Step 2: Figure 33 shows the normal curve with $\mu = 12{,}500$. The sample mean found in Step 1, $\bar{x} = 13{,}022$, is also labeled.

Figure 33

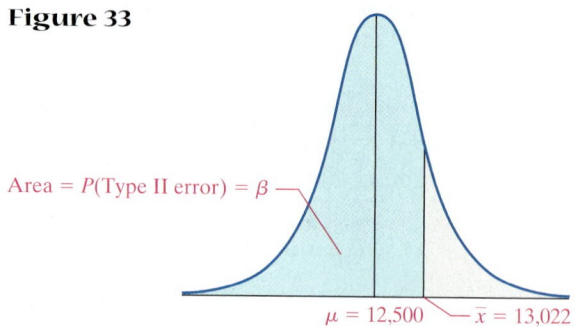

Area = P(Type II error) = β

$\mu = 12{,}500$ $\bar{x} = 13{,}022$

Step 3: Because this is a right-tailed test, the area to the left of $\bar{x} = 13{,}022$ represents β, the probability of a Type II error.

$$P(\text{Type II error}) = P(\bar{x} < 13{,}022 \text{ given that } \mu = 12{,}500)$$
$$= P\left(Z < \frac{13{,}022 - 12{,}500}{3800/\sqrt{35}}\right) = P(Z < 0.81) = 0.7910$$

There is a 0.7910 probability of making a Type II error if the true population mean is 12,500.

Now Work Problem 13.

Notice that, as the population mean gets closer to the hypothesized mean, the probability of making a Type II error increases.

2 Compute The Power of the Test

The probability of rejecting the null hypothesis when the alternative hypothesis is true is $1 - \beta$. The value of $1 - \beta$ is referred to as the **power of the test**. The higher the power of the test is, the more likely the test will reject the null when the alternative hypothesis is true.

EXAMPLE 2 **Computing the Power of the Test**

Problem: Compute the power of the test for the situation in Example 1.

Approach: The power of the test is $1 - \beta$. We computed β in Example 1 to be 0.7910 when the true population mean is 12,500.

In Other Words
The power of the test is the probability that you will correctly reject H_0.

Solution: The power of the test is $1 - 0.7910 = 0.2090$. There is a 0.2090 probability of rejecting the null hypothesis if the true population mean is 12,500 miles.

In Table 5, we show the probability of making a Type II error and the power of the test for a number of different values of the population mean greater than 12,200. Notice that the closer the population mean gets to the hypothesized value of 12,200, the larger the probability of a Type II error becomes. This should not be surprising. When the true population mean is close to the population mean stated in the null hypothesis, there is a high probability that we will not reject the null hypothesis and will make a Type II error.

We can use the values of μ and their corresponding powers to construct a *power curve*. A **power curve** is a graph that shows the power of the test against values of the population mean that make the null hypothesis false. Figure 34 shows the power curve for the values in Table 5.

	Table 5	
Value of μ	**Probability of Making a Type II Error, β**	**Power, $1 - \beta$**
12,201	0.8997	0.1003
12,500	0.7910	0.2090
12,700	0.6915	0.3085
13,000	0.5120	0.4880
13,500	0.2296	0.7704
14,000	0.0643	0.9357

Figure 34

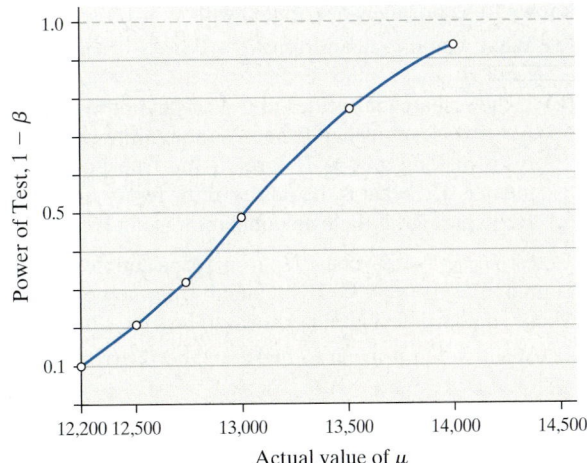

Figure 34 shows that, as the true mean gets farther from the hypothesized mean, the power of the test approaches 1.

10.7 ASSESS YOUR UNDERSTANDING

Concepts and Vocabulary

1. Explain what it means to make a Type II error.
2. Explain the term *power of the test*.
3. What happens to the power of the test as the true mean gets closer to the hypothesized value of the mean? Why is this result reasonable?

4. What effect would increasing the sample size have on the power of the test, assuming all else remains unchanged?

Skill Building

5. To test H_0: $\mu = 50$ versus H_1: $\mu < 50$, a simple random sample of size $n = 24$ is obtained from a population that is known to be normally distributed with $\sigma = 12$.
   (a) What would it mean to make a Type II error for this test?
   (b) If the researcher decides to test this hypothesis at the $\alpha = 0.05$ level of significance, compute the probability of making a Type II error if the true population mean is 48. What is the power of the test?
   (c) Redo part (b) if the true population mean is 49.2.

6. To test H_0: $\mu = 40$ versus H_1: $\mu > 40$, a simple random sample of size $n = 25$ is obtained from a population that is known to be normally distributed with $\sigma = 6$.
   (a) What would it mean to make a Type II error for this test?
   (b) If the researcher decides to test this hypothesis at the $\alpha = 0.05$ level of significance, compute the probability of making a Type II error if the true population mean is 42. What is the power of the test?
   (c) Redo part (b) if the true population mean is 40.6.

7. To test H_0: $\mu = 100$ versus H_1: $\mu \neq 100$, a simple random sample of size $n = 23$ is obtained from a population that is known to be normally distributed with $\sigma = 7$.
   (a) What would it mean to make a Type II error for this test?

   (b) If the researcher decides to test this hypothesis at the $\alpha = 0.01$ level of significance, compute the probability of making a Type II error if the true population mean is 103. What is the power of the test?
   (c) Redo part (b) if the true population mean is 101.4.

8. To test H_0: $\mu = 80$ versus H_1: $\mu < 80$, a simple random sample of size $n = 22$ is obtained from a population that is known to be normally distributed with $\sigma = 11$.
   (a) What would it mean to make a Type II error for this test?
   (b) If the researcher decides to test this hypothesis at the $\alpha = 0.01$ level of significance, compute the probability of making a Type II error if the true population mean is 79. What is the power of the test?
   (c) Redo part (b) if the true population mean is 77.5.

9. To test H_0: $\mu = 20$ versus H_1: $\mu < 20$, a simple random sample of size $n = 18$ is obtained from a population that is known to be normally distributed with $\sigma = 3$.
   (a) What would it mean to make a Type II error for this test?
   (b) If the researcher decides to test this hypothesis at the $\alpha = 0.05$ level of significance, compute the probability of making a Type II error if the true population mean is 17.4. What is the power of the test?
   (c) Redo part (b) if the true population mean is 19.2.

10. To test $H_0: \mu = 4.5$ versus $H_1: \mu > 4.5$, a simple random sample of size $n = 13$ is obtained from a population that is known to be normally distributed with $\sigma = 1.2$.

(a) What would it mean to make a Type II error for this test?

(b) If the researcher decides to test this hypothesis at the $\alpha = 0.05$ level of significance, compute the probability of making a Type II error if the true population mean is 4.7. What is the power of the test?

(c) Redo part (b) if the true population mean is 5.4.

11. To test $H_0: \mu = 105$ versus $H_1: \mu \neq 105$, a simple random sample of size $n = 35$ is obtained from a population whose standard deviation is known to be $\sigma = 12$.

(a) What would it mean to make a Type II error for this test?

(b) If the researcher decides to test this hypothesis at the $\alpha = 0.1$ level of significance, compute the probability of making a Type II error if the true population mean is 101.8. What is the power of the test?

(c) Redo part (b) if the true population mean is 103.1.

12. To test $H_0: \mu = 45$ versus $H_1: \mu \neq 45$, a simple random sample of size $n = 40$ is obtained from a population whose standard deviation is known to be $\sigma = 8$.

(a) What would it mean to make a Type II error for this test?

(b) If the researcher decides to test this hypothesis at the $\alpha = 0.1$ level of significance, compute the probability of making a Type II error if the true population mean is 46.2. What is the power of the test?

(c) Redo part (b) if the true population mean is 48.3.

Applying the Concepts

13. Are Women Getting Taller? A researcher claims that the average height of a woman aged 20 years or older is greater than the 1994 mean height of 63.7 inches, on the basis of data obtained from the Centers for Disease Control and Prevention *Advance Data Report*, No. 347. She obtains a simple random sample of 45 women. Assume that the population standard deviation is 6.5 inches.

(a) What would it mean to make a Type II error for this test?

(b) If the researcher decides to test this hypothesis at the $\alpha = 0.05$ level of significance, compute the probability of making a Type II error if the true population mean is 64 inches. What is the power of the test?

(c) Redo part (b) if the true population mean is 66 inches.

14. ATM Withdrawals The manufacturer of a certain type of ATM machine reports that the mean ATM withdrawal is $60. The manager of a convenience store with an ATM machine thinks that mean withdrawals from his machine are less than this amount. He obtains a simple random sample of 35 withdrawals over the past year. Assume that the population standard deviation is $13.

(a) What would it mean to make a Type II error for this test?

(b) If the manager decides to test this hypothesis at the $\alpha = 0.1$ level of significance, compute the probability of making a Type II error if the true population mean is $57. What is the power of the test?

(c) Redo part (b) if the true population mean is $54.

15. SAT Exam Scores A school administrator claims that students whose first language learned is not English score worse on the verbal portion of the SAT exam than students whose first language is English. The mean SAT verbal score of students whose first language is English is 515, on the basis of data obtained from the College Board. Suppose the administrator obtains a simple random sample of 20 students whose first language learned was not English. SAT verbal scores are normally distributed with a population standard deviation of 112.

(a) What would it mean to make a Type II error for this test?

(b) If the administrator decides to test this hypothesis at the $\alpha = 0.01$ level of significance, compute the probability of making a Type II error if the true population mean is 500. What is the power of the test?

(c) Redo part (b) if the true population mean is 480.

16. SAT Exam Scores A school administrator claims that students whose first language learned is not English score differently on the math portion of the SAT exam than students whose first language is English. The mean SAT math score of students whose first language is English is 516, on the basis of data obtained from the College Board. Suppose the administrator obtains a simple random sample of 20 students whose first language learned was not English. SAT math scores are normally distributed with a population standard deviation of 114.

(a) What would it mean to make a Type II error for this test?

(b) If the administrator decides to test this hypothesis at the $\alpha = 0.05$ level of significance, compute the probability of making a Type II error if the true population mean is 505. What is the power of the test?

(c) Redo part (b) if the true population mean is 475.

17. Effect of α Redo Problem 5(b) with $\alpha = 0.01$. What effect does lowering α have on the power of the test?

18. Effect of α Redo Problem 6(b) with $\alpha = 0.1$. What effect does increasing α have on the power of the test?

19. Probability of a Type II Error in Tests on Proportions Suppose we wish to test $H_0: p = 0.75$ versus $H_1: p < 0.75$. In this problem, we compute the power of the test for proportions.

(a) Suppose the researcher decides to test this hypothesis at the $\alpha = 0.05$ level of significance by conducting a simple random sample of $n = 500$ people. Compute the probability of making a Type II error if the true population proportion is 0.70. (**Hint:** The hypothesized mean is $p_0 = 0.75$, and the standard deviation is $\sqrt{\dfrac{p_0(1 - p_0)}{n}}$.)

(b) What is the power of the test?

20. Probability of a Type II Error in Tests on Proportions Suppose we wish to test H_0: $p = 0.35$ versus H_1: $p > 0.35$. In this problem, we compute the power of the test for proportions.

(a) Suppose the researcher decides to test this hypothesis at the $\alpha = 0.05$ level of significance by conducting a simple random sample of $n = 700$ people. Compute the probability of making a Type II error if the true population proportion is 0.39. **Hint:** The hypothesized mean is $p_0 = 0.35$ and the standard deviation is $\sqrt{\dfrac{p_0(1 - p_0)}{n}}$.

(b) What is the power of the test?

Chapter 10   Review

Summary

In this chapter, we discussed testing claims using hypothesis testing. A claim is made regarding a population parameter, which leads to a null and an alternative hypothesis. The null hypothesis is assumed true. Given sample data, we either reject or do not reject the null hypothesis. In performing a hypothesis test, there is always the possibility of making a Type I error (rejecting the null hypothesis when it is true) or of making a Type II error (not rejecting the null hypothesis when it is false). The probability of making a Type I error is equal to the level of significance, α, of the test.

We discussed four types of hypothesis tests in this chapter. First, we performed tests about a population mean with σ known. Second, we performed tests about a population mean with σ unknown.

In both of these cases, we required that either the sample size be large ($n \geq 30$) or the population be approximately normal with no outliers. For small sample sizes, we verified the normality of the data by using normal probability plots. Boxplots were used to check for outliers.

The third test we performed regarded claims about a population proportion. To perform these tests, we required a large sample size so that $np(1 - p) \geq 10$, yet the sample size could be no more than 5% of the population size.

Finally, we performed tests regarding a population standard deviation (or variance). These tests require that the population from which the sample is drawn be normally distributed. This test is not robust, so deviation from the requirement of normality was not allowed.

All four hypothesis tests were performed by using classical methods and the P-value approach. The P-value approach to testing hypotheses has appeal, because the rejection rule is always to reject the null hypothesis if the P-value is less than the level of significance, α.

We concluded the section by learning how to compute the probability of making a Type II error and the power of the test. The power of the test is the probability of rejecting the null hypothesis when the alternative hypothesis is true. The closer the true value of the population mean is to the hypothesized value of the population mean, the lower the power of the test.

Formulas

Test Statistics

- $z_0 = \dfrac{\bar{x} - \mu_0}{\sigma/\sqrt{n}}$ follows the standard normal distribution if the population from which the sample was drawn is normal or if the sample size is large ($n \geq 30$).

- $t_0 = \dfrac{\bar{x} - \mu_0}{s/\sqrt{n}}$ follows Student's t-distribution with $n - 1$ degrees of freedom if the population from which the sample was drawn is normal or if the sample size is large ($n \geq 30$).

- $z_0 = \dfrac{\hat{p} - p_0}{\sqrt{\dfrac{p_0(1 - p_0)}{n}}}$ follows the standard normal distribution if $np_0(1 - p_0) \geq 10$ and $n \leq 0.05N$.

- $\chi_0^2 = \dfrac{(n - 1)s^2}{\sigma_0^2}$ follows the χ^2-distribution with $n - 1$ degrees of freedom if the population from which the sample was drawn is normal.

Type I and Type II Errors
- $\alpha = P(\text{Type I error}) = P(\text{rejecting } H_0 \text{ when } H_0 \text{ is true})$
- $\beta = P(\text{Type II error}) = P(\text{not rejecting } H_0 \text{ when } H_1 \text{ is true})$

Vocabulary

Hypothesis (p. 506)	Type I error (p. 508)	Decision rule (p. 517, 521)
Hypothesis testing (p. 506)	Type II error (p. 508)	Robust (p. 518)
Null hypothesis (p. 506)	Level of significance (p. 510)	*P*-value (p. 520)
Alternative hypothesis (p. 506)	Statistically significant (p. 514)	Practical significance (p. 525)
Two-tailed test (p. 506)	Test statistic (p. 517)	Power of the test (p. 564)
Left-tailed test (p. 506)	Critical value (p. 517)	Power curve (p. 564)
Right-tailed test (p. 506)	Critical region (p. 517)	

Objectives

Section	You should be able to . . .	Examples	Review Exercises
10.1	**1** Determine the null and alternative hypothesis from a claim (p. 505)	2	1, 2
	2 Understand Type I and Type II errors (p. 508)	3	1–4, 17(b), (c); 18(b), (c)
	3 State conclusions to hypothesis tests (p. 510)	4	1–2
10.2	**1** Understand the logic of hypothesis testing (p. 513)	pp. 515–518	30
	2 Test a claim about a population mean with σ known using the classical approach (p. 516)	1 and 2	5, 6, 15, 16, 19, 20
	3 Test a claim about a population mean with σ known using *P*-values (p. 520)	3–5	5, 6, 15, 16, 19, 20
	4 Test a claim about a population mean with σ known using confidence intervals (p. 524)	6	6(f)
	5 Understand the difference between statistical significance and practical significance (p. 525)	7	29
10.3	**1** Test a claim about a population mean with σ unknown (p. 532)	1–3	7, 8, 13, 14, 17(a), 18(a)
10.4	**1** Test a claim about a population proportion using the normal model (p. 544)	1–3	9, 10, 21, 22
	2 Test a claim about a population proportion using the binomial probability distribution (p. 549)	4	27, 28
10.5	**1** Test a claim about a population standard deviation (p. 554)	1	11, 12, 23, 24
10.6	**1** Determine the appropriate hypothesis test to perform (p. 558)	p. 561	5–24
10.7	**1** Determine the probability of making a Type II error (p. 561)	1	25(a), 26(a)
	2 Compute the power of the test (p. 564)	2	25(b), 26(b)

Review Exercises

For each claim in Problems 1 and 2, (a) determine the null and alternative hypotheses, (b) explain what it would mean to make a Type I error, (c) explain what it would mean to make a Type II error, (d) state the conclusion that would be drawn if the null hypothesis is not rejected, and (e) state the conclusion that would be reached if the null hypothesis is rejected.

1. **Credit Card Debt**   According to the *Statistical Abstract of the United States*, the mean outstanding credit card debt per cardholder was $4277 in 2000. A consumer credit counselor believes that the mean outstanding credit card debt per cardholder is now more than the 2000 amount.

2. **Downloading Music**   According to a study by Ipsos-Reid, 61% of Internet users aged 18 to 24 years old had downloaded music from the Internet by the end of 2000. A researcher believes that the percentage is now higher than 61%.

3. Suppose a test is conducted at the $\alpha = 0.05$ level of significance. What is the probability of a Type I error?

4. Suppose β is computed to be 0.113. What is the probability of a Type II error? What is the power of the test? How would you interpret the power of the test?

5. To test $H_0: \mu = 30$ versus $H_1: \mu < 30$, a simple random sample of size $n = 12$ is obtained from a population that is known to be normally distributed with $\sigma = 4.5$.
   (a) If the sample mean is determined to be $\bar{x} = 28.6$, compute the test statistic.
   (b) If the researcher decides to test this hypothesis at the $\alpha = 0.05$ level of significance, determine the critical value.
   (c) Draw a normal curve that depicts the rejection region.
   (d) Will the researcher reject the null hypothesis? Why?
   (e) What is the *P*-value?

6. To test $H_0: \mu = 65$ versus $H_1: \mu \neq 65$, a simple random sample of size $n = 23$ is obtained from a population that is known to be normally distributed with $\sigma = 12.3$.
   (a) If the sample mean is determined to be $\bar{x} = 70.6$, compute the test statistic.

(b) If the researcher decides to test this hypothesis at the $\alpha = 0.1$ level of significance, determine the critical values.

(c) Draw a normal curve that depicts the rejection region.

(d) Will the researcher reject the null hypothesis? Why?

(e) What is the P-value?

(f) Test the claim by constructing a 90% confidence interval.

7. To test $H_0: \mu = 8$ versus $H_1: \mu \neq 8$, a simple random sample of size $n = 15$ is obtained from a population that is known to be normally distributed.

(a) If $\bar{x} = 7.3$ and $s = 1.8$, compute the test statistic.

(b) If the researcher decides to test this hypothesis at the $\alpha = 0.02$ level of significance, determine the critical values.

(c) Draw a t-distribution that depicts the rejection region.

(d) Will the researcher reject the null hypothesis? Why?

(e) Determine the P-value.

8. To test $H_0: \mu = 3.9$ versus $H_1: \mu < 3.9$, a simple random sample of size $n = 25$ is obtained from a population that is known to be normally distributed.

(a) If $\bar{x} = 3.5$ and $s = 0.9$, compute the test statistic.

(b) If the researcher decides to test this hypothesis at the $\alpha = 0.05$ level of significance, determine the critical value.

(c) Draw a t-distribution that depicts the rejection region.

(d) Will the researcher reject the null hypothesis? Why?

(e) Determine the P-value.

In Problems 9 and 10, test the hypothesis at the $\alpha = 0.05$ level of significance, using (a) the classical approach and (b) the P-value approach. Be sure to verify the requirements of the test.

9. $H_0: p = 0.6$ versus $H_1: p > 0.6$
$n = 250; x = 165; \alpha = 0.05$

10. $H_0: p = 0.35$ versus $H_1: p \neq 0.35$
$n = 420; x = 138; \alpha = 0.01$

11. To test $H_0: \sigma = 5.2$ versus $H_1: \sigma \neq 5.2$, a simple random sample of size $n = 18$ is obtained from a population that is known to be normally distributed.

(a) If the sample standard deviation is determined to be $s = 4.9$, compute the test statistic.

(b) If the researcher decides to test this hypothesis at the $\alpha = 0.05$ level of significance, determine the critical values.

(c) Draw a chi-square distribution that depicts the rejection region.

(d) Will the researcher reject the null hypothesis? Why?

12. To test $H_0: \sigma = 15.7$ versus $H_1: \sigma > 15.7$, a simple random sample of size $n = 25$ is obtained from a population that is known to be normally distributed.

(a) If the sample standard deviation is determined to be $s = 16.5$, compute the test statistic.

(b) If the researcher decides to test this hypothesis at the $\alpha = 0.1$ level of significance, determine the critical value.

(c) Draw a chi-square distribution that depicts the rejection region.

(d) Will the researcher reject the null hypothesis? Why?

13. Linear Rotary Bearing A linear rotary bearing is designed so that the distance between the retaining rings is 0.875 inch. The quality-control manager suspects that the manufacturing process needs to be recalibrated and that the mean distance between the retaining rings is longer than 0.875 inch. In a random sample of 36 bearings, he finds the sample mean distance between the retaining rings to be 0.876 inch with standard deviation 0.005 inch. Test the quality-control manager's claim at the $\alpha = 0.05$ level of significance.

14. Education Pays The U.S. Census Bureau reported that the mean annual salary in 2002 was $51,194 for an individual whose highest degree was a bachelor's. A government economist believes that the mean annual salary for individuals whose highest degree is a bachelor's is different today. She obtains a random sample of 300 employed adults whose highest degree is a bachelor's and determines the mean annual salary to be $55,988 with a standard deviation of $26,855 (both in 2002 dollars). Test the economist's claim at the $\alpha = 0.05$ level of significance.

15. SAT Math Scores A mathematics instructor wanted to know whether or not use of a calculator improves SAT math scores. In 2000, the SAT math scores of students who used a calculator once or twice weekly were normally distributed, with a mean of 474 and a standard deviation 103. In a random sample of 50 students who use a calculator every day, the mean score was 539. Is there significant evidence to support the claim that students who use a calculator "frequently" score better on the SAT math portion than those who use a calculator "infrequently"? Test at the $\alpha = 0.01$ level of significance.

16. Birth Weight An obstetrician wants to determine whether a new diet significantly increases the birth weight of babies. In 2002, birth weights of full-term babies (gestation period of 37 to 41 weeks) were normally distributed, with mean 7.53 pounds and standard deviation 1.15 pounds, according to the *National Vital Statistics Report*, Vol. 48, No. 3. The obstetrician randomly selects 50 recently pregnant mothers and persuades them to partake of this new diet. The obstetrician then records the birth weights of the babies and obtains a mean of 7.79 pounds. Is there sufficient evidence to support the claim that the new diet increases the birth weights of newborns at the $\alpha = 0.01$ level of significance?

17. High Cholesterol A nutritionist maintains that 20- to 39-year-old males consume too much cholesterol. The USDA-recommended daily allowance of cholesterol is 300 mg. In a survey conducted by the U.S. Department of Agriculture of 404 20- to 39-year-old males, it was determined the mean daily cholesterol intake was 326 milligrams, with standard deviation 342 milligrams.

(a) Is there evidence to support the nutritionist's claim at the $\alpha = 0.05$ level of significance?

(b) What would it mean for the nutritionist to make a Type I error? A Type II error?

(c) What is the probability the nutritionist will make a Type I error?

18. Sodium A nutritionist claims that 20- to 39-year-old females consume too much sodium. The USDA-recommended daily allowance of sodium is 2400 mg. In a survey conducted by the U.S. Department of Agriculture of 257 20- to 39-year-old females, it was determined the mean daily sodium intake was 2919 milligrams and the standard deviation was 906 milligrams.

(a) Is there evidence to support the nutritionist's claim at the $\alpha = 0.10$ level of significance?

(b) What would it mean for the nutritionist to make a Type I error? A Type II error?

(c) What is the probability the nutritionist will make a Type I error?

19. Acid Rain In 1990, the mean pH level of the rain in Barnstable County, Massachussetts, was 4.61. A biologist fears that the acidity of rain has increased (in other words that the pH level of the rain has decreased). She draws a random sample of 25 rain dates in 2004 and obtains the following data:

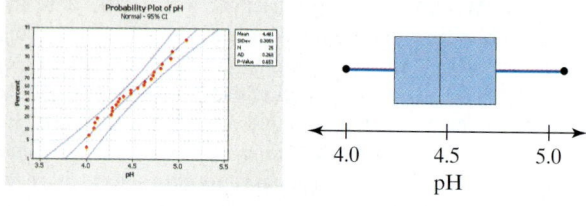

4.80	4.27	4.09	4.55	5.08	4.34	4.08
4.36	4.82	4.70	4.40	4.73	4.62	4.48
4.28	4.28	4.48	4.72	4.12	4.00	4.93
4.91	4.32	4.63	4.03			

Source: National Atmospheric Deposition Program

(a) Because the sample size is small, she must verify that the pH level is normally distributed and the sample does not contain any outliers. The normal probability plot and boxplot are shown. Are the conditions for testing the hypothesis satisfied?

(b) Test the claim, assuming $\sigma = 0.26$, at the $\alpha = 0.01$ level of significance.

20. Hemoglobin A medical researcher maintains that the mean hemoglobin reading of surgical patients is different from 14.0 grams per deciliter. She randomly selects nine surgical patients and obtains the following data:

| 14.6 | 12.8 | 8.9 | 9.0 | 9.9 |
| 10.7 | 13.0 | 12.0 | 13.0 | |

Source: Lora McGuire, Nursing Instructor, Joliet Junior College

(a) Because the sample size is small, she must verify that hemoglobin is normally distributed and the sample does not contain any outliers. The normal probability plot and boxplot are shown. Are the conditions for testing the hypothesis satisfied?

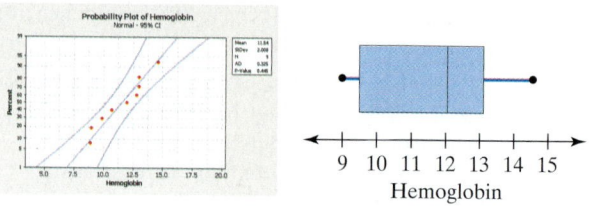

(b) Test the claim, assuming that $\sigma = 2.001$ at the $\alpha = 0.01$ level of significance.

21. Tuberculosis According to the Centers for Disease Control, 56% of all tuberculosis cases in 1999 were of foreign-born residents of the United States. A researcher believes that this proportion has increased from its 1999 level. She obtains a simple random sample of 300 tuberculosis cases in the United States and determines that 170 of them are foreign-born. Is there sufficient evidence to support the claim that the percentage of cases of tuberculosis of foreign-born residents has increased at the $\alpha = 0.01$ level of significance?

22. Phone Purchases In 1997, 39.4% of females had ordered merchandise or services by phone in the last 12 months. A market research analyst feels that the percentage of females ordering merchandise or services by phone has declined from the 1997 level because of Internet purchases. She obtains a random sample of 500 females and determines that 191 of them have ordered merchandise or services by phone in the last 12 months. Test the researcher's claim that the percentage of females ordering merchandise or services by phone has decreased from its 1997 proportion at the $\alpha = 0.10$ level of significance.

23. SAT Math Scores A researcher believes there is more variability in SAT math scores for students living in low-income households than for students living in high-income households. According to the College Board, SAT math scores of students living in households with a household income above \$50,000 per year are normally distributed, with a standard deviation of 104. In a random sample of 23 students who live in households with an income below \$30,000 per year, it was determined their SAT math scores had a standard deviation of 112. Test the researcher's claim that students in low-income households have a higher standard deviation on their SATs than do students in high-income households at the $\alpha = 0.10$ level of significance.

24. Birth Weights   An obstetrician maintains that preterm babies (gestation period less than 37 weeks) have a higher variability in birth weight than do full-term babies (gestation period 37 to 41 weeks). According to the *National Vital Statistics Report*, Vol. 48, No. 3, the birth weights of full-term babies are normally distributed, with standard deviation 505.6 grams. A random sample of 41 preterm babies results in a standard deviation equal to 840 grams. Test the researcher's claim that the variability in the birth weight of preterm babies is more than the variability in birth weight of full-term babies, at the $\alpha = 0.01$ level of significance.

25. SAT Math Scores   Refer to Problem 15. Suppose the true mean SAT math score of students who use a calculator every day is 510.
(a) What is the probability of a Type II error?
(b) What is the power of the test?

26. Birth Weight   Refer to Problem 16. Suppose the true mean weight of babies whose mothers partake of the new diet is 7.59 pounds.
(a) What is the probability of a Type II error?
(b) What is the power of the test?

27. Teen Prayer   In 1995, 40% of adolescents stated they prayed daily. A researcher wanted to know whether this percentage has risen since then. He surveys 40 adolescents and finds that 18 pray on a daily basis. Is this evidence to support the claim that the proportion of adolescents who pray daily has increased at the $\alpha = 0.05$ level of significance?

28. Leisure Activities   In 1999, 27% of Americans stated that reading was their favorite leisure-time activity. A researcher wants to know whether this percentage has declined since then. He surveys 35 Americans and finds that 4 of them consider reading to be their favorite leisure-time activity. Is this evidence to support the claim that the proportion of Americans who consider reading to be their favorite leisure-time activity has decreased at the $\alpha = 0.05$ level of significance?

29. A New Teaching Method   A large university has a college algebra enrollment of 5000 students each semester. Because of space limitations, the university decides to offer its college algebra courses in a self-study format in which students learn independently but have access to tutors and other help in a lab setting. Historically, students in traditional college algebra scored 73.2 points on the final exam and the coordinator of this course is concerned that test scores are going to decrease in the new format. At the end of the first semester using the new delivery system, 3851 students took the final exams and had a mean score of 72.8 and a standard deviation of 12.3. Treating these students as a simple random sample of all students, determine whether or not the scores decreased significantly at the $\alpha = 0.05$ level of significance. Do you think that the decrease in scores has any practical significance?

30. In your own words, explain the procedure for testing a claim about a population mean when assuming the population standard deviation is known.

CASE STUDY

How Old Is Stonehenge?

Approximately 8 miles north of Salisbury, Wiltshire, England, stands a large circular stone monument surrounded by an earthwork. This prehistoric structure is known throughout the world as Stonehenge. Its name is derived from the Old English word *hengen*, referring to something hung up. In the case of the monument, this name refers to the large horizontal lintel stones. The monument consists of an outer ring of sarsen stones, surrounding two inner circles of bluestones. The first and third circles are adorned with the familiar stone lintels. The entire structure is surrounded by a ditch and bank. Just inside the bank are 56 pits, named the Aubrey Holes, after their discoverer. These holes appear to have been filled shortly after their excavation.

Recently, it has been discovered that a number of the stone alignments are associated with important solar and lunar risings and settings, suggesting that the site served as some sort of massive astronomical calendar. If this conclusion is accurate, it seems likely that the monument might have been used as a temple for sky worshipers.

Corinn Dillion is interested in dating the construction of the structure. Excavations at the site uncovered a number of unshed antlers, antler tines, and animal bones. Carbon 14 dating methods were used to estimate the ages of the Stonehenge artifacts. Carbon 14 is one of three carbon isotopes found in Earth's atmosphere. Carbon 12 makes up 99% of all the carbon dioxide in the air. Virtually all the remaining 1% is composed of carbon 13. By far, the rarest form of carbon isotope found in the atmosphere is carbon 14.

The ratio of carbon 14 to carbon 12 remains constant in living organisms. However, once the organism dies, the amount of carbon-14 in the remains of the organism begins to decline, because it is radioactive, with a half-life of 5730 years (the "Cambridge half-life"). So the decay of carbon-14 into ordinary nitrogen makes possible a reliable estimate about the time of death of the organism. The counted carbon 14 decay events can be modeled by the normal distribution.

Dillion's team used two different carbon 14 dating methods to arrive at age estimates for the numerous Stonehenge artifacts. The liquid scintillation counting (LSC) method utilizes benzene, acetylene, ethanol, methanol, or a similar chemical. Unlike the LSC method, the accelerator mass spectrometry (AMS) technique offers direct carbon 14 isotope counting. The AMS method's greatest advantage is that it requires only milligram-sized samples for testing. The AMS method was used only on recovered artifacts that were of extremely small size.

Stonehenge's main ditch was dug in a series of segments. Excavations at the base of the ditch uncovered a number of antlers, which bore signs of heavy use. These antlers could have been used by the builders as picks or rakes. The fact that no primary silt was discovered beneath the antlers suggests that they were buried in the ditch shortly after its completion. Another researcher, Phillip Corbin, using an archeological markings approach, had previously claimed that the mean date for the construction of the ditch was 2950 B.C. A sample of nine age estimates from unshed antlers excavated from the ditch produced a mean of 3033.1 B.C., with standard deviation 66.9 years. Assume that the ages are normally distributed with no obvious outliers. At an $\alpha = 0.05$ significance level, is there any reason to dispute Corbin's claim?

Four animal bone samples were discovered in the ditch terminals. These bones bore signs of attempts at artificial preservation and might have been in use for a substantial period of time before being placed at Stonehenge. When dated, these bones had a mean age of 3187.5 B.C. and standard deviation of 67.4 years. Assume that the ages are normally distributed with no obvious outliers.

Use an $\alpha = 0.05$ significance level to test the claim that the population mean age of the site is different from 2950 B.C.

In the center of the monument are two concentric circles of igneous rock pillars, called bluestones. The construction of these circles was never completed. These circles are known as the Bluestone Circle and the Bluestone Horseshoe. The stones in these two formations were transported to the site from the Prescelly Mountains in Pembrokeshire, southwest Wales. Excavation at the center of the monument revealed an antler, an antler tine, and an animal bone. Each artifact was submitted for dating. It was determined that this sample of three artifacts had a mean age of 2193.3 B.C, with a standard deviation of 104.1 years. Assume that the ages are normally distributed with no obvious outliers. Use an $\alpha = 0.05$ significance level to test the claim that the population mean age of the Bluestone formations is different from Corbin's declared mean age of the ditch, that is, 2950 B.C.

Finally, three additional antler samples were uncovered at the Y and Z holes. These holes are part of a formation of concentric circles, 11 meters and 3.7 meters, respectively, outside of the Sarsen Circle. The sample mean age of these antlers is 1671.7 B.C. with a standard deviation of 99.7 years. Assume that the ages are normally distributed with no obvious outliers. Use an $\alpha = 0.05$ significance level to test whether the population mean age of the Y and Z holes is different from Corbin's stated mean age of the ditch, that is, 2950 B.C.

From your analysis, does it appear that the mean ages of the artifacts from the ditch, the ditch terminals, the Bluestones, and the Y and Z holes dated by Dillion are consistent with Corbin's claimed mean age of 2950 B.C. for construction of the ditch? Can you use the results from your hypothesis tests to infer the likely construction order of the various Stonehenge structures? Explain.

Using Dillion's data, construct a 95% confidence interval for the population mean ages of the various sites. Do these confidence intervals support Corbin's claim? Can you use these confidence intervals to infer the likely construction order of the various Stonehenge structures? Explain.

Which statistical technique, hypothesis testing or confidence intervals, is more useful in assessing the age and likely construction order of the Stonehenge structures? Explain.

Discuss the limitations and assumptions of your analysis. Is there any additional information that you would like to have before publishing your findings? Would another statistical procedure be more useful in analyzing these data? If so, which one? Explain. Write a report to Corinn Dillion detailing your analysis.

Source: This fictional account is based on information obtained from *Archaeometry and Stonehenge* (www.eng_h.gov.uk/stoneh). The means and standard deviations used throughout this case study were constructed by calculating the statistics from the midpoint of the calibrated date range supplied for each artifact.

Inferences on Two Samples

CHAPTER 11

Outline

DECISIONS

Suppose that you have just received an inheritance of $10,000 and decide that the money should be invested, rather than blown on frivolous items. You have decided that you will invest the money in one of two types of mutual funds. But which type? See the Decisions project on page 595.

●●● Putting It All Together

In Chapters 9 and 10 we discussed inferences regarding a single population parameter. The inferential methods presented in these chapters will be modified slightly in this chapter so that we can compare two population parameters.

The first two sections of this chapter deal with testing for the difference of two population means. The methods presented in this chapter can be used to determine whether a certain treatment results in significantly different sample statistics. From a design-of-experiments point of view, the methods presented in Section 11.1 are used to handle matched-pairs designs (Section 1.5, pages 43–44) with a quantitative-response variable. For example, we might want to know whether married couples have similar IQs. To test this theory, we could randomly select 20 married couples and determine the difference in their IQs.

Section 11.2 presents inferential methods used to handle completely randomized designs when there is a single treatment that has two levels (Section 1.5, pages 42–43) and the response variable is quantitative. For example, we might randomly divide 100 volunteers who have the common cold into two groups, a control group and an experimental group. The control group would receive a placebo and the experimental group would receive a predetermined amount of some experimental drug. The response variable might be the time until the cold symptoms go away.

Section 11.3 discusses the difference between two population proportions. Again, we can use a completely randomized design to compare two population proportions. However, rather than having a quantitative response variable, we would have a binomial response variable; that is, either the experimental unit has a characteristic or it does not.

Finally, Section 11.4 presents a discussion for comparing two population standard deviations.

Inference about Two Means: Dependent Samples

Preparing for This Section   Before getting started, review the following:

- Matched-pairs design (Section 1.5, pp. 43–44)
- Confidence intervals about μ, σ unknown (Section 9.2, pp. 466–473)
- Hypothesis tests about μ, σ unknown (Section 10.3, pp. 531–537)
- Type I and Type II errors (Section 10.1, pp. 508–509)

Objectives

1 Distinguish between independent and dependent sampling

2 Test claims made regarding matched-pairs data

3 Construct and interpret confidence intervals about the population mean difference of matched-pairs data

1 ## Distinguish between Independent and Dependent Sampling

In Other Words

If the individuals in two samples are somehow related (husband–wife, siblings, similar characteristics, or even the same person), the sampling is dependent.

To perform inference on the difference of two population means, we must first determine whether the data come from an *independent* or *dependent* sample. A sampling method is **independent** when the individuals selected for one sample do not dictate which individuals are to be in a second sample. A sampling method is **dependent** when the individuals selected to be in one sample are used to determine the individuals to be in the second sample. For example, if we are conducting a study that compares the IQs of husbands and wives, once a husband is selected to be in the study, his wife is automatically matched with him. Dependent samples are often referred to as **matched-pairs** samples.

EXAMPLE 1 ### Distinguishing between Independent and Dependent Sampling

Problem: For each of the following experiments, determine whether the sampling method is independent or dependent.

(a) Researcher Steven J. Sperber, MD, and his associates wanted to determine the effectiveness of a new medication* in the treatment of discomfort associated with the common cold. They randomly divided 430 subjects into two groups: Group 1 received the new medication and Group 2 received a placebo. The goal of the study was to determine whether the mean of the symptom assessment scores of the individuals receiving the new medication (Group 1) was less than that of the placebo group (Group 2).

(b) In an experiment conducted in a biology class, Professor Andy Neill measured the time required for 12 students to catch a falling meter stick using their dominant hand and nondominant hand. The goal of the study was to determine whether the reaction time in an individual's dominant hand is different from the reaction time in the nondominant hand.

Approach: We must determine whether the individuals in one group were used to determine the individuals in the other group. If so, the sampling method is dependent. If not, the sampling method is independent.

Solution

(a) The sampling method is independent because the individuals in Group 1 were not used to determine which individuals are in Group 2.

---

*The medication was a combination of pseudoephedrine and acetaminophen. The study is published in the *Archives of Family Medicine* 9(2000): 979–985.

(b) The sampling method is dependent because the individuals are related. The measurements for the dominant and nondominant hand are on the same individual.

Now Work Problem 5.

In this section, we will discuss inference on the difference of two means for dependent sampling. Section 11.2 addresses inference when the sampling is independent.

② Test Claims Made Regarding Matched-Pairs Data

Inference on matched-pairs data is very similar to inference regarding a population mean when the population standard deviation is unknown. Recall that if the population from which the sample was drawn is normally distributed or the sample size is large ($n \geq 30$), we said that

$$t = \frac{\overline{x} - \mu}{\dfrac{s}{\sqrt{n}}}$$

follows Student's t-distribution with $n - 1$ degrees of freedom.

When analyzing matched-pairs data, we compute the difference in each matched pair and then perform inference on the differenced data using the methods of Section 9.2 or 10.3.

In Other Words

Statistical inference methods on matched-pairs data use the same methods as inference on a single population mean with σ unknown, except that the differences are analyzed.

Testing a Claim Regarding the Difference of Two Means Using a Matched-Pairs Design

If a claim is made regarding the mean difference of matched-pairs data, we can use the following steps to test the claim, provided that

1. the sample is obtained using simple random sampling;
2. the sample data are matched pairs;
3. the differences are normally distributed with no outliers or the sample size, n, is large ($n \geq 30$).

Step 1: A claim is made regarding the mean difference from matched-pairs data. The claim is used to determine the null and alternative hypotheses. The hypotheses can be structured in one of three ways, where μ_d is the population mean difference of the matched-pairs data.

Two-Tailed	Left-Tailed	Right-Tailed
$H_0: \mu_d = 0$	$H_0: \mu_d = 0$	$H_0: \mu_d = 0$
$H_1: \mu_d \neq 0$	$H_1: \mu_d < 0$	$H_1: \mu_d > 0$

Step 2: Select a level of significance α, depending on the seriousness of making a Type I error.

Step 3: Compute the test statistic

$$t_0 = \frac{\overline{d} - 0}{\dfrac{s_d}{\sqrt{n}}} = \frac{\overline{d}}{\dfrac{s_d}{\sqrt{n}}}$$

which approximately follows Student's t-distribution with $n - 1$ degrees of freedom. The values of $\overline{d}$ and s_d are the mean and standard deviation of the differenced data.

Classical Approach

Step 4: Use Table V to determine the critical value using $n - 1$ degrees of freedom.

	Two-Tailed	Left-Tailed	Right-Tailed
Critical value(s)	$-t_{\alpha/2}$ and $t_{\alpha/2}$	$-t_\alpha$	t_α
Critical region(s)	Critical Region Critical Region	Critical Region	Critical Region

$-t_{\alpha/2}$ $t_{\alpha/2}$ $-t_\alpha$ t_α

Step 5: Compare the critical value with the test statistic.

Two-Tailed	Left-Tailed	Right-Tailed
If $t_0 < -t_{\alpha/2}$ or $t_0 > t_{\alpha/2}$, reject the null hypothesis	If $t_0 < -t_\alpha$, reject the null hypothesis	If $t_0 > t_\alpha$, reject the null hypothesis

P-value Approach

Step 4: Use Table V to estimate the P-value using $n - 1$ degrees of freedom.

	Two-Tailed	Left-Tailed	Right-Tailed
	The sum of the area in the tails is the P-value	The area left of t_0 is the P-value	The area right of t_0 is the P-value

$-|t_0|$ $|t_0|$ t_0 t_0

Step 5: If P-value $< \alpha$, reject the null hypothesis.

Step 6: State the conclusion.

The procedures just presented are **robust**, which means that minor departures from normality will not adversely affect the results of the test. If the data have outliers, however, the procedure should not be used.

We will verify the assumption that the differenced data come from a population that is normally distributed by constructing normal probability plots. We use boxplots to determine whether there are outliers. If the normal probability plot indicates that the differenced data are not normally distributed or the boxplot reveals outliers, nonparametric tests should be performed as discussed in Section 15.4.

EXAMPLE 2   **Testing a Claim Regarding Matched-Pairs Data**

Problem: Professor Andy Neill measured the time (in seconds) required to catch a falling meter stick for 12 randomly selected students' dominant hand and nondominant hand. Professor Neill claims that the reaction time in an individual's dominant hand is less than the reaction time in their nondominant hand. Test the claim at the $\alpha = 0.05$ level of significance. The data obtained are presented in Table 1

	Table 1		
Student	**Dominant Hand, X_i**		**Nondominant Hand, Y_i**
1	0.177		0.179
2	0.210		0.202
3	0.186		0.208
4	0.189		0.184
5	0.198		0.215
6	0.194		0.193
7	0.160		0.194
8	0.163		0.160
9	0.166		0.209
10	0.152		0.164
11	0.190		0.210
12	0.172		0.197

Source: Professor Andy Neill, Joliet Junior College

Approach: This is a matched-pairs design because the variable is measured on the same subject for both the dominant and nondominant hand, the treatment in this experiment. We compute the difference between the dominant time and the nondominant time. So, for the first student we compute $X_1 - Y_1$, for the second student we compute $X_2 - Y_2$, and so on. If the reaction time in the dominant hand is less than the reaction time in the nondominant hand, we would expect the values of $X_i - Y_i$ to be negative. Before we perform the hypothesis test, we must verify that the differences come from a population that is approximately normally distributed with no outliers because the sample size is small. We will construct a normal probability plot and boxplot of the differenced data to verify these requirements. We then proceed to follow Steps 1 through 6.

Solution: We compute the differences as $d_i = X_i - Y_i$ = time of dominant hand for ith student minus time of nondominant hand for ith student. We expect these differences to be negative, so we wish to test the claim that $\mu_d < 0$. Table 2 displays the differences.

Table 2			
Student	**Dominant Hand, X_i**	**Nondominant Hand, Y_i**	**Difference, d_i**
1	0.177	0.179	$0.177 - 0.179 = -0.002$
2	0.210	0.202	$0.210 - 0.202 = 0.008$
3	0.186	0.208	-0.022
4	0.189	0.184	0.005
5	0.198	0.215	-0.017
6	0.194	0.193	0.001
7	0.160	0.194	-0.034
8	0.163	0.160	0.003
9	0.166	0.209	-0.043
10	0.152	0.164	-0.012
11	0.190	0.210	-0.020
12	0.172	0.197	-0.025

$$\sum d_i = -0.158$$

! CAUTION

The way that we define the difference determines the direction of the alternative hypothesis in one-tailed tests. In Example 1, we expect $X_i < Y_i$, so the difference $X_i - Y_i$ is expected to be negative. Therefore, the alternative hypothesis is $H_1: \mu_d < 0$, and we have a left-tailed test. However, if we computed the differences as $Y_i - X_i$, we'd expect the differences to be positive, and we have a right-tailed test!

We compute the mean and standard deviation of the differences and obtain $\bar{d} = -0.0132$ rounded to four decimal places and $s_d = 0.0164$ rounded to four decimal places. We must verify that the data come from a population that is approximately normal with no outliers. Figure 1 shows the normal probability plot and boxplot of the differenced data.

Figure 1

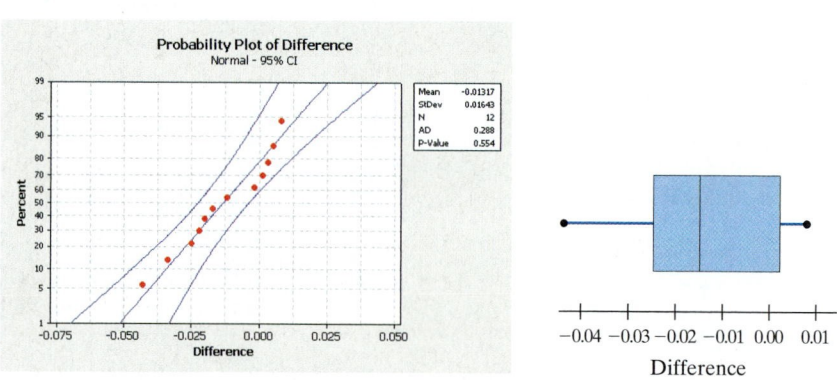

The normal probability plot is roughly linear and the boxplot does not indicate any outliers. We can proceed with the hypothesis test.

Step 1: Professor Neill claims that the reaction time in the dominant hand is less than the reaction time in the nondominant hand. We express this claim as $\mu_d < 0$. We have

$$H_0: \mu_d = 0 \quad \text{versus} \quad H_1: \mu_d < 0$$

This test is left tailed.

Step 2: The level of significance is $\alpha = 0.05$.

Step 3: The sample mean is $\overline{d} = -0.0132$ second, and the sample standard deviation is $s_d = 0.0164$ second. The test statistic is

$$t_0 = \frac{\overline{d}}{\frac{s_d}{\sqrt{n}}} = \frac{-0.0132}{\frac{0.0164}{\sqrt{12}}} = -2.788$$

Classical Approach

Step 4: Because this is a left-tailed test, we determine the critical t-value at the $\alpha = 0.05$ level of significance with $n - 1 = 12 - 1 = 11$ degrees of freedom to be $-t_{0.05} = -1.796$. The critical region is displayed in Figure 2.

Figure 2

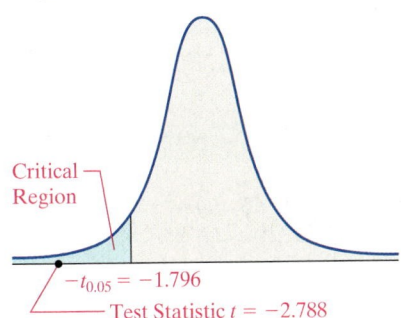

Critical Region

$-t_{0.05} = -1.796$

Test Statistic $t = -2.788$

Step 5: Because the test statistic $t_0 = -2.788$ is less than the critical value $-t_{0.05} = -1.796$, Professor Neill rejects the null hypothesis. We label this point in Figure 2.

P-value Approach

Step 4: Because this is a left-tailed test, the P-value is the area under the t-distribution with $12 - 1 = 11$ degrees of freedom to the left of the test statistic, $t_0 = -2.788$, as shown in Figure 3(a). That is, P-value $= P(t < t_0) = P(t < -2.788)$, with 11 degrees of freedom.

Because of the symmetry of the t-distribution, the area under the t-distribution to the left of -2.788 equals the area under the t-distribution to the right of 2.788. So P-value $= P(t_0 < -2.788) = P(t_0 > 2.788)$. See Figure 3(b).

Using Table V, we find the row that corresponds to 11 degrees of freedom. The value 2.788 lies between 2.718 and 3.106. The value of 2.718 has an area under the t-distribution with 11 degrees of freedom of 0.01 to the right. The value of 3.106 has an area under the t-distribution with 11 degrees of freedom of 0.005 to the right.

Because 2.788 is between 2.718 and 3.106, the P-value is between 0.005 and 0.01. So $0.005 < P\text{-value} < 0.01$.

Figure 3

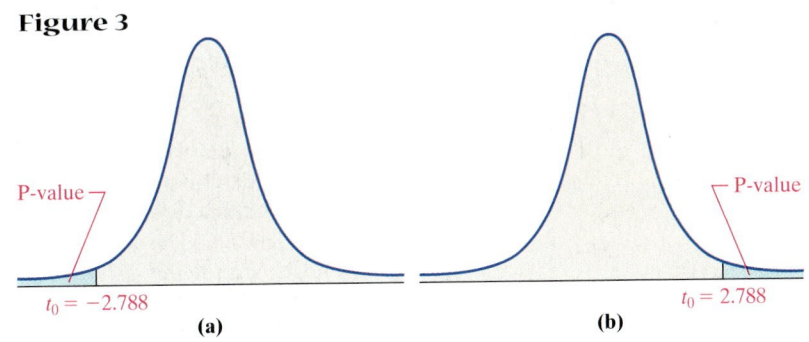

P-value

$t_0 = -2.788$

(a)

P-value

$t_0 = 2.788$

(b)

Step 5: Because the P-value is less than the level of significance $\alpha = 0.05$, we reject the null hypothesis.

Step 6: There is sufficient evidence to support Professor Neill's claim that the reaction time in the dominant hand is less than the reaction time in the nondominant hand at the $\alpha = 0.05$ level of significance.

 **Testing a Claim Regarding Matched-Pairs Data Using Technology**

Problem: Obtain an exact *P*-value for the problem in Example 2 using statistical software or a graphing calculator with advanced statistical features.

Approach: We will use MINITAB to obtain the *P*-value. The steps for testing claims regarding matched-pairs data using the TI-83/84 Plus graphing calculators, MINITAB, and Excel are given in the Technology Step-by-Step on page 587.

Result: Figure 4 shows the results obtained from MINITAB. Notice the *P*-value is 0.009.

Figure 4   **Paired T-Test and Confidence Interval**

```
Paired T for Dominant — Non-dominant

                N        Mean       StDev      SE Mean
Dominant        12       0.17975    0.01752    0.00506
Non-domi        12       0.19292    0.01799    0.00519
Difference      12      -0.01317    0.01643    0.00474


95% CI for mean difference: (-0.02361, -0.00273)
T-Test of mean difference = 0 (vs < 0): T-Value = -2.78 P-Value = 0.009
```

Interpretation: There is a 0.009 probability of obtaining a sample mean difference of -0.0132 or less from a population whose mean difference is 0. The results that we obtained would happen about 1 time in 100 samples if the null hypothesis were true. At the $\alpha = 0.05$ level of significance, there is sufficient evidence to reject the null hypothesis because the *P*-value is less than α. We conclude that the reaction time in the dominant hand is faster than the reaction time in the nondominant hand.

In-Class Activity: Measuring Reaction Time

We just saw that reaction time is different between your dominant and nondominant hand. Let's try to reproduce the results of Example 2. We can estimate reaction time by measuring reaction distance, in particular, the distance a ruler will fall before being caught. Would you expect this distance to be larger or smaller with your dominant hand?

(a) Pair up with another student in the class.

(b) Randomly determine whether to work first with the dominant hand or the nondominant hand. Why is this step important?

(c) One student holds his hand perpendicular to the floor so that the thumb and forefinger are roughly 5 cm apart. The other student holds a ruler (with centimeters) vertically just above the first student's fingers. The second student releases the ruler so that it falls vertically between the first student's thumb and forefinger (without hinting at when it will be dropped), and the first student grasps the ruler as quickly as possible. Record the number of centimeters required to grasp the ruler.

(d) Repeat the process for the other hand of the first student and then for both hands of the second student.

(e) Compute the difference in each student's reaction distance by computing distance$_{\text{dom. hand}}$ − distance$_{\text{nondom. hand}}$

(f) Combine your team data with the rest of the class and conduct an appropriate hypothesis test to answer the question.

Now Work Problem 13(a).

Construct and Interpret Confidence Intervals about the Population Mean Difference of Matched-Pairs Data

We can also create a confidence interval for the population mean difference, μ_d, using the sample mean difference, $\overline{d}$, the sample standard deviation difference, s_d, the sample size, and $t_{\alpha/2}$. Remember, a confidence interval about a population mean is given in the following form:

Point estimate ± margin of error

Based on the preceding formula, we compute the confidence interval about μ_d as follows:

> **Confidence Interval for Matched-Pairs Data**
>
> A $(1 - \alpha) \cdot 100\%$ confidence interval for μ_d is given by
>
> $$\text{Lower bound: } \overline{d} - t_{\alpha/2} \cdot \frac{s_d}{\sqrt{n}} \qquad \text{Upper bound: } \overline{d} + t_{\alpha/2} \cdot \frac{s_d}{\sqrt{n}} \quad (1)$$
>
> The critical value $t_{\alpha/2}$ is determined using $n - 1$ degrees of freedom. The values of $\overline{d}$ and s_d are the mean and standard deviation of the differenced data.
>
> **Note:** The interval is exact when the population is normally distributed and approximately correct for nonnormal populations, provided that n is large.

EXAMPLE 4   **Constructing a Confidence Interval for Matched-Pairs Data**

Problem: Using the data from Table 2, construct a 95% confidence interval estimate of the mean difference, μ_d.

Approach

Step 1: Compute the differenced data. Because the sample size is small, we must verify that the differenced data come from a population that is approximately normal with no outliers.

Step 2: Compute the sample mean difference, $\overline{d}$, and the sample standard deviation difference, s_d.

Step 3: Determine the critical value, $t_{\alpha/2}$, with $\alpha = 0.05$ and $n - 1$ degrees of freedom.

Step 4: Use Formula (1) to determine the lower and upper bounds.

Step 5: Interpret the results.

Solution

Step 1: We computed the differenced data and verified that they come from a population that is approximately normally distributed with no outliers in Example 2.

Step 2: We computed the sample mean difference, $\overline{d}$, to be -0.0132 and the sample standard deviation of the difference, s_d, to be 0.0164 in Example 2.

Step 3: Using Table V with $\alpha = 0.05$ and $12 - 1 = 11$ degrees of freedom, we find $t_{\alpha/2} = t_{0.025} = 2.201$.

Step 4: Substituting into Formula (1), we find

$$\text{Lower bound: } \overline{d} - t_{\alpha/2} \cdot \frac{s_d}{\sqrt{n}} = -0.0132 - 2.201 \cdot \frac{0.0164}{\sqrt{12}} = -0.0236$$

$$\text{Upper bound: } \overline{d} + t_{\alpha/2} \cdot \frac{s_d}{\sqrt{n}} = -0.0132 + 2.201 \cdot \frac{0.0164}{\sqrt{12}} = -0.0028$$

Step 5: We are 95% confident that the mean difference between the dominant hand's reaction time and the nondominant hand's reaction time is between

−0.0236 and −0.0028 second. In other words, we are 95% confident that the dominant hand has a mean reaction time that is somewhere between 0.0028 second and 0.0236 second faster than the nondominant hand. Notice that the confidence interval does not contain zero. This evidence supports the claim that the reaction time of a person's dominant hand is different from the reaction time of the nondominant hand.

Now Work Problem 13(b).

We can see that the results of Example 4 agree with the 95% confidence interval determined by MINITAB in Figure 4.

11.1 ASSESS YOUR UNDERSTANDING

Concepts and Vocabulary

1. A sampling method is _____ when the individuals selected for one sample do not dictate which individuals are in a second sample.

2. A sampling method is _____ when the individuals selected to be in one sample are used to determine the individuals in the second sample.

3. Suppose a researcher claims the mean from population 1 is less than the mean from population 2 in matched-pairs data. How would you define μ_d? How would you determine d_i?

4. What are the requirements to test a claim regarding the difference of two means with dependent sampling?

Skill Building

In Problems 5–10, determine whether the sampling is dependent or independent.

5. A sociologist wishes to compare the annual salaries of married couples. She obtains a random sample of 50 married couples in which both spouses work and determines each spouse's annual salary.

6. A researcher wishes to determine the effects of alcohol on people's reaction times to a stimulus. She randomly divides 100 people aged 21 or older into two groups. Group 1 is asked to drink 3 ounces of alcohol, while group 2 drinks a placebo. Both drinks taste the same, so the individuals in the study do not know which group they belong to. Thirty minutes after consuming the drink, the subjects in each group perform a series of tests meant to measure reaction time.

7. An educator wants to determine whether a new curriculum significantly improves standardized test scores for third-grade students. She randomly divides 80 third-graders into two groups. Group 1 is taught using the new curriculum, while group 2 is taught using the traditional curriculum. At the end of the school year, both groups are given the standardized test and the mean scores are compared.

8. A psychologist wants to know whether subjects respond faster to a go/no go stimulus or a choice stimulus. With the go/no go stimulus, subjects must respond to a particular stimulus by pressing a button and disregard other stimuli. In the choice stimulus, the subjects respond differently depending on the stimulus. She randomly selects 20 subjects and each subject is presented a series of go/no go stimuli and choice stimuli. The mean reaction time to each stimulus is compared.

9. A study was conducted by researchers designed "to determine the genetic and nongenetic factors to structural brain abnormalities on schizophrenia." The researchers examined the brains of 29 patients diagnosed with schizophrenia and compared them with 29 healthy patients. The whole-brain volumes of the two groups were compared.

   Source: William F.C. Baare et al., Volumes of Brain Structures in Twins Discordant for Schizophrenia, *Archives of General Psychiatry* **58**: (2000) 33–40.

10. An agricultural researcher wanted to determine whether there were any significant differences in the plowing method used on crop yield. He divided a parcel of land that had uniform soil quality into 30 subplots. He then randomly selected 15 of the plots to be chisel plowed and 15 plots to be fall plowed. He recorded the crop yield at the end of the growing season to determine whether there was a significant difference in the mean crop yield.

In Problems 11 and 12, assume that the paired data came from a population that is normally distributed.

11.

Observation	1	2	3	4	5	6	7
X_i	7.6	7.6	7.4	5.7	8.3	6.6	5.6
Y_i	8.1	6.6	10.7	9.4	7.8	9.0	8.5

(a) Determine $d_i = X_i - Y_i$ for each pair of data.
(b) Compute $\bar{d}$ and s_d.
(c) Test the claim that $\mu_d < 0$ at the $\alpha = 0.05$ level of significance.
(d) Compute a 95% confidence interval about the population mean difference $\mu_d = \mu_1 - \mu_2$.

12.

Observation	1	2	3	4	5	6	7	8
X_i		19.4	18.3	22.1	20.7	19.2	11.8	20.1	18.6
Y_i		19.8	16.8	21.1	22.0	21.5	18.7	15.0	23.9

(a) Determine $d_i = X_i - Y_i$ for each pair of data.
(b) Compute $\bar{d}$ and s_d.
(c) Test the claim that $\mu_d \neq 0$ at the $\alpha = 0.01$ level of significance.
(d) Compute a 99% confidence interval about the population mean difference $\mu_d = \mu_1 - \mu_2$.

Applying the Concepts

13. Muzzle Velocity The following data represent the muzzle velocity (in feet per second) of rounds fired from a 155-mm gun. For each round, two measurements of the velocity were recorded using two different measuring devices, with the following data obtained:

Observation	1	2	3	4	5	6	7	8	9	10	11	12
A	793.8	793.1	792.4	794.0	791.4	792.4	791.7	792.3	789.6	794.4	790.9	793.5
B	793.2	793.3	792.6	793.8	791.6	791.6	791.6	792.4	788.5	794.7	791.3	793.5

Source: Christenson, Ronald and Blackwood, Larry. Tests for Precision and Accuracy of Multiple Measuring Devices. *Technometrics*, Nov. 93, Vol 35, Issue 4, pp. 411–421.

(a) Why are these matched-pairs data?
(b) Test the claim that there is a difference in the measurement of the muzzle velocity between device A and device B at the $\alpha = 0.01$ level of significance. **Note:** A normal probability plot and boxplot of the data indicate that the differences are approximately normally distributed with no outliers.
(c) Construct a 99% confidence interval about the population mean difference. Interpret your results.
(d) Draw a boxplot of the differenced data. Does this visual evidence support the results obtained in part (b)?

14. Reaction Time In an experiment conducted online at the University of Mississippi, study participants are asked to react to a stimulus. In one experiment, the participant must press a key on seeing a blue screen. Reaction time (in seconds) to press the key is measured. The same person is then asked to press a key on seeing a red screen, again with reaction time measured. The results for six randomly sampled study participants are as follows:

Participant Number:	1	2	3	4	5	6
Reaction time to blue	0.582	0.481	0.841	0.267	0.685	0.450
Reaction time to red	0.408	0.407	0.542	0.402	0.456	0.533

Source: PsychExperiments at the University of Mississippi

(a) Why are these matched-pairs data?
(b) Test the claim that the reaction time to the blue stimulus is different from the reaction time to the red stimulus at the $\alpha = 0.01$ level of significance. **Note:** A normal probability plot and boxplot of the data indicate that the differences are approximately normally distributed with no outliers.
(c) Construct a 98% confidence interval about the population mean difference. Interpret your results.
(d) Draw a boxplot of the differenced data. Does this visual evidence support the results obtained in part (b)?

15. Secchi Disk A Secchi disk is an 8-inch-diameter weighted disk that is painted black and white and attached to a rope. The disk is lowered into water and the depth (in inches) at which it is no longer visible is recorded. The measurement is an indication of water clarity. A environmental biologist is interested in determining whether the water clarity of the lake at Joliet Junior College is improving. She takes measurements at the same location on the same dates during the course of a year and repeats the measurements on the same dates 5 years later. She obtains the following results:

Observation:	1	2	3	4	5	6	7	8
Date:	5/11	6/7	6/24	7/8	7/27	8/31	9/30	10/12
Initial depth	38	58	65	74	56	36	56	52
Depth 5 years later	52	60	72	72	54	48	58	60

Source: Virginia Piekarski, Joliet Junior College

(a) Why is it important to take the measurements on the same date?
(b) Test the claim that the clarity of the lake is improving at the $\alpha = 0.05$ level of significance. **Note:** A normal probability plot and boxplot of the data indicate that the differences are approximately normally distributed with no outliers.
(c) Construct a 95% confidence interval about the population mean difference. Interpret your results.
(d) Draw a boxplot of the differenced data. Does this visual evidence support the results obtained in part (b)?

16. **Rat's Hemoglobin.** Hemoglobin helps the red blood cells transport oxygen and remove carbon dioxide. Researchers at NASA wanted to determine the effects of space flight on a rat's hemoglobin. The following data represent the hemoglobin (in grams per deciliter) at lift-off minus 3 days (H-L3) and immediately upon the return (H-R0) for 12 randomly selected rats sent to space on the Spacelab Sciences 1 flight.

Rat No.:	1	2	3	4	5	6	7	8	9	10	11	12
H-L3	15.2	16.1	15.3	16.4	15.7	14.7	14.3	14.5	15.2	16.1	15.1	15.8
H-R0	15.8	16.5	16.7	15.7	16.9	13.1	16.4	16.5	16.0	16.8	17.6	16.9

Source: NASA Life Sciences Data Archive

(a) Test the claim that the hemoglobin levels at lift-off minus 3 days are less than the hemoglobin levels upon return at the $\alpha = 0.05$ level of significance. **Note:** A normal probability plot and boxplot of the data indicate that the differences are approximately normally distributed with no outliers.
(b) Construct a 90% confidence interval about the population mean difference. Interpret your results.
(c) Draw a boxplot of the differenced data. Does this visual evidence support the results obtained in part (a)?

17. **Getting Taller?** To test the claim that sons are taller than their fathers, a researcher randomly selects 13 fathers who have adult male children. She records the height of both the father and son in inches and obtains the following data.

	1	2	3	4	5	6	7	8	9	10	11	12	13
Height of father	70.3	67.1	70.9	66.8	72.8	70.4	71.8	70.1	69.9	70.8	70.2	70.4	72.4
Height of son	74.1	69.2	66.9	69.2	68.9	70.2	70.4	69.3	75.8	72.3	69.2	68.6	73.9

Source: Anna Behounek, student at Joliet Junior College

Test the claim that sons are taller than their fathers at the $\alpha = 0.1$ level of significance. **Note:** A normal probability plot and boxplot of the data indicate that the differences are approximately normally distributed with no outliers.

18. **Waiting in Line** A quality-control manager at an amusement park feels that the amount of time that people spend waiting in line for the American Eagle roller coaster is too long. To determine if a new loading/unloading procedure is effective in reducing wait time in line, he measures the amount of time (in minutes) people are waiting in line on 7 days. After implementing the new procedure, he again measures the amount of time (in minutes) people are waiting in line on 7 days and obtains the following data. To make a reasonable comparison, he chooses days when the weather conditions are similar.

Day:	Mon (2 P.M.)	Tues (2 P.M.)	Wed (2 P.M.)	Thurs (2 P.M.)	Fri (2 P.M.)	Sat (11 A.M.)	Sat (4 P.M.)	Sun (12 noon)	Sun (4 P.M.)
Wait time before new procedure	11.6	25.9	20.0	38.2	57.3	32.1	81.8	57.1	62.8
Wait time after new procedure	10.7	28.3	19.2	35.9	59.2	31.8	75.3	54.9	62.0

Test the claim that the new loading/unloading procedure is effective in reducing wait time at the $\alpha = 0.05$ level of significance. **Note:** A normal probability plot and boxplot of the data indicate that the differences are approximately normally distributed with no outliers.

19. **Hardness Testing** The manufacturer of hardness testing equipment uses steel-ball indenters to penetrate metal that is being tested. However, the manufacturer thinks it would be better to use a diamond indenter so that they can test all types of metal. Because of the differences between the two types of indenters, it is suspected that the two methods will produce different hardness readings. The metal specimens to be tested are large enough so that two indentions can be made. Therefore, the manufacturer wants to use both indenters on each specimen and compare the readings.

	Specimen								
	1	2	3	4	5	6	7	8	9
Steel ball	50	57	61	71	68	54	65	51	53
Diamond	52	56	61	74	69	55	68	51	56

Test the claim that the two indenters result in different measurements at the $\alpha = 0.05$ level of significance. **Note:** A normal probability plot and boxplot of the data indicate that the differences are approximately normally distributed with no outliers.

20. **Car Rentals** The following data represent the daily rental for a compact automobile charged by two car rental companies, Thrifty and Hertz, in 10 locations.

City	Thrifty	Hertz	City	Thrifty	Hertz
Chicago	21.81	18.99	Seattle	21.96	22.99
Los Angeles	29.89	48.99	Pittsburgh	20.90	19.99
Houston	17.90	19.99	Phoenix	47.75	36.99
Orlando	27.98	35.99	New Orleans	33.81	26.99
Boston	24.61	25.60	Minneapolis	33.49	20.99

Source: Yahoo!Travel

Test the claim that Thrifty is less expensive than Hertz at the $\alpha = 0.1$ level of significance. **Note:** A normal probability plot and boxplot of the data indicate that the differences are approximately normally distributed with no outliers.

21. **DUI Simulator** To illustrate the effects of driving under the influence (DUI) of alcohol, a police officer brought a DUI simulator to a local high school. Student reaction time in an emergency was measured with unimpaired vision and also while wearing a pair of special goggles to simulate the effects of alcohol on vision. For a random sample of nine teenagers, the time (in seconds) required to bring the vehicle to a stop from a speed of 60 miles per hour was recorded. Construct and interpret a 95% confidence interval for the mean difference in braking time with impaired vision and normal vision where the differences are computed as "impaired minus normal." **Note:** A normal probability plot and boxplot of the data indicate that the differences are approximately normally distributed with no outliers.

	Subject								
	1	**2**	**3**	**4**	**5**	**6**	**7**	**8**	**9**
Normal	4.47	4.24	4.58	4.65	4.31	4.80	4.55	5.00	4.79
Impaired	5.77	5.67	5.51	5.32	5.83	5.49	5.23	5.61	5.63

22. **Braking Distance** An automotive researcher wanted to estimate the difference in distance required to come to a complete stop while traveling 40 miles per hour on wet versus dry pavement. Because car type plays a role, the researcher used eight different cars with the same driver and tires. The braking distance (in feet) on both wet and dry pavement is shown in the following table. Construct a 95% confidence interval for the mean difference in braking distance on wet versus dry pavement where the differences are computed as "wet minus dry." Interpret the interval. **Note:** A normal probability plot and boxplot of the data indicate that the differences are approximately normally distributed with no outliers.

Car No.:	1	2	3	4	5	6	7	8
Wet braking distance	106.9	100.9	108.8	111.8	105.0	105.6	110.6	107.9
Dry braking distance	71.8	68.8	74.1	73.4	75.9	75.2	75.7	81.0

23. **Does Octane Matter?** Octane is a measure of how much the fuel can be compressed before it spontaneously ignites. Some people believe that higher-octane fuels result in better gas mileage for their car. To test this claim, a researcher randomly selected 11 individuals (and their cars) to participate in the study. Each participant received 10 gallons of gas and drove his car on a closed course that simulated both city and highway driving. The number of miles driven until the car ran out of gas was recorded. A coin flip was used to determine whether the car was filled up with 87-octane or 92-octane fuel first, and the driver did not know which type of fuel was in the tank. The results are in the following table:

Driver:	1	2	3	4	5	6	7	8	9	10	11
Miles on 87 octane	234	257	243	215	114	287	315	229	192	204	547
Miles on 92 octane	237	238	229	224	119	297	351	241	186	209	562

(a) Why is it important that the matching be done by driver and car?
(b) Why is it important to conduct the study on a closed track?
(c) The normal probability plots for miles on 87 octane and miles on 92 octane are shown. Are either of these variables normally distributed?

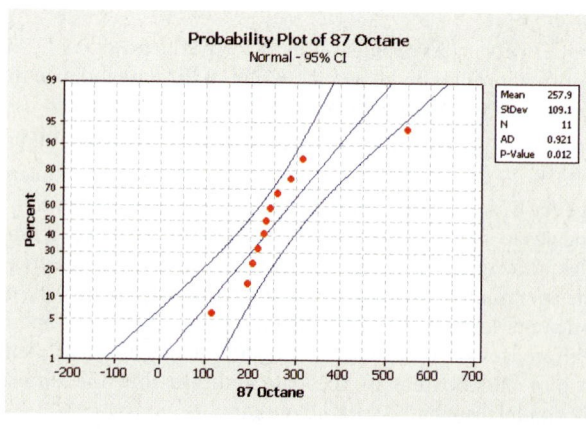

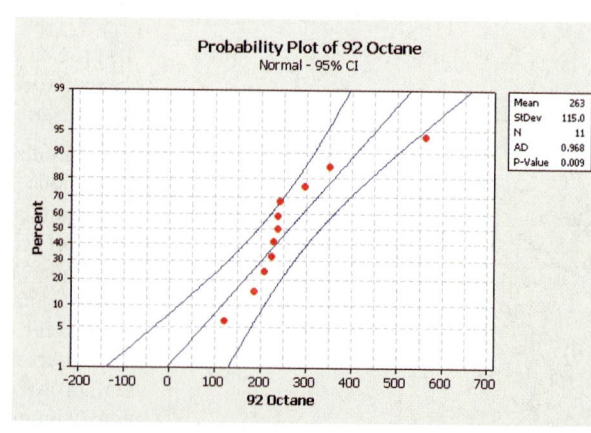

(d) The differences are computed as 92 octane minus 87 octane. The normal probability plot of the differences is shown. Is there reason to believe that the differences are normally distributed? Conclude that the differences can be normally distributed even though the original data are not.

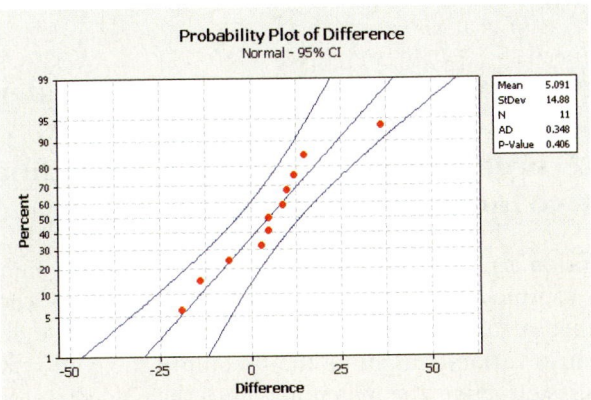

(e) The researchers used MINITAB to test the claim that the mileage from 92 octane is greater than the mileage from 87 octane. The results are as follows:

Paired T-Test and CI: 92 Octane, 87 Octane
```
Paired T for 92 Octane - 87 Octane

                 N      Mean     StDev    SE Mean
92 Octane       11   263.000   115.041     34.686
87 Octane       11   257.909   109.138     32.906
Difference      11   5.09091   14.87585    4.48524

95% lower bound for mean difference: -3.03841
T-Test of mean difference = 0 (vs > 0): T-Value = 1.14  P-Value = 0.141
```

What do you conclude regarding the claim? Why?

| **Technology Step-by-Step** | **Two-sample *t*-tests, Dependent Sampling** |

TI-83/84 Plus

Hypothesis Tests

Step 1: If necessary, enter raw data in L1 and L2. Let L3 = L1 − L2 (or L2 − L1), depending on how the alternative hypothesis was defined.

Step 2: Press STAT, highlight TESTS, and select 2: T-Test.

Step 3: If the data are raw, highlight Data, making sure that List is set to L3 with frequency set to 1. If summary statistics are known, highlight Stats and enter the summary statistics.

Step 4: Highlight the appropriate relation between μ_1 and μ_2 in the alternative hypothesis.

Step 5: Highlight Calculate or Draw and press ENTER. Calculate gives the test statistic and *P*-value. Draw will draw the *t*-distribution with the *P*-value shaded.

Confidence Intervals

Follow the same steps as those given for hypothesis tests, except select 8: TInterval. Also, select a confidence level (such as 95% = 0.95).

MINITAB

Step 1: Enter raw data in columns C1 and C2.

Step 2: Select the **Stat** menu, highlight **Basic Statistics**, and then highlight **Paired-t . . . .**

Step 3: Enter C1 in the cell marked "First Sample" and enter C2 in the cell marked "Second Sample." Under OPTIONS, select the direction of the alternative hypothesis and select a confidence level. Click OK.

Excel

Step 1: Enter raw data in columns A and B.

Step 2: Select the **Tools** menu and highlight **Data Analysis . . . .**

Step 3: Select "*t*-test: Paired Two-Sample for Means." With the cursor in the "Variable 1 Range" cell, highlight the data in column A. With the cursor in the "Variable 2 Range" cell, highlight the data in column B. Enter the hypothesized difference in the means (usually 0) and a value for alpha. Click OK.

11.2 Inference about Two Means: Independent Samples

Preparing for This Section Before getting started, review the following:

- The Completely Randomized Design (Section 1.5, pp. 42–43)
- Confidence intervals about μ, σ unknown (Section 9.2, pp. 466–473)
- Hypothesis tests about μ, σ unknown (Section 10.3, pp. 531–537)
- Type I and Type II errors (Section 10.1, pp. 508–509)

Objectives

1. **Test claims regarding the difference of two independent means**

2. **Construct and interpret confidence intervals regarding the difference of two independent means**

We now turn our attention to inferential methods for comparing means from two independent samples. For example, suppose we wish to know whether a new experimental drug relieves symptoms attributable to the common cold. The response variable might be the time until the cold symptoms go away. If the drug is effective, the mean time until the cold symptoms go away should be less for individuals taking the drug than for those not taking the drug. If we let μ_1 represent the mean time until cold symptoms go away for the individuals taking the drug, and μ_2 represent the mean time until cold symptoms go away for individuals taking a placebo, the null and alternative hypotheses will be

$$H_0: \mu_1 = \mu_2 \quad \text{versus} \quad H_1: \mu_1 < \mu_2$$

or, equivalently,

$$H_0: \mu_1 - \mu_2 = 0 \quad \text{versus} \quad H_1: \mu_1 - \mu_2 < 0$$

To test this claim, we might randomly divide 500 volunteers who have a common cold into two groups: an experimental group (Group 1) and a control group (Group 2). The control group will receive a placebo and the experimental group will receive a predetermined amount of the experimental drug. Next, determine the time until the cold symptoms go away. Compute $\bar{x}_1$, the sample mean time until cold symptoms go away in the experimental group, and $\bar{x}_2$, the sample mean time until cold symptoms go away in the control group. Now, we determine whether the difference in the sample means, $\bar{x}_1 - \bar{x}_2$, is significantly different from 0, the assumed difference stated in the null hypothesis. To do this, we need to know the sampling distribution of $\bar{x}_1 - \bar{x}_2$.

It is unreasonable to expect to know information regarding σ_1 and σ_2 without knowing information regarding the population means. Therefore, we must develop a sampling distribution for the difference of two means when the population standard deviations are unknown.

The comparison of two means with unequal (and unknown) population variances is called the Behrens–Fisher problem. While an exact method for performing inference on the equality of two means with unequal population standard deviations does not exist, an approximate solution is available. The approach that we use is known as **Welch's approximate** *t*, in honor of English statistician Bernard Lewis Welch (1911–1989).

Sampling Distribution of the Difference of Two Means: Independent Samples with Population Standard Deviations Unknown (Welch's *t*)

Suppose a simple random sample of size n_1 is taken from a population with unknown mean μ_1 and unknown standard deviation σ_1. In addition, a simple random sample of size n_2 is taken from a second population with unknown mean μ_2 and unknown standard deviation σ_2. If the two populations are normally distributed or the sample sizes are sufficiently large ($n_1 \geq 30$ and $n_2 \geq 30$), then

$$t = \frac{(\overline{x}_1 - \overline{x}_2) - (\mu_1 - \mu_2)}{\sqrt{\dfrac{s_1^2}{n_1} + \dfrac{s_2^2}{n_2}}} \qquad \textbf{(1)}$$

approximately follows Student's t-distribution with the smaller of $n_1 - 1$ or $n_2 - 1$ degrees of freedom, where $\overline{x}_1$ is the sample mean and s_1 is the sample standard deviation from population 1, and $\overline{x}_2$ is the sample mean and s_2 is the sample standard deviation from population 2.

1 ## Test Claims Regarding the Difference of Two Independent Means

Now that we know the approximate sampling distribution of $\overline{x}_1 - \overline{x}_2$ we can introduce a procedure that can be used to test claims regarding two population means.

Testing a Claim Regarding the Difference of Two Means

If a claim is made regarding two population means, μ_1 and μ_2, with unknown population standard deviations, we can use the following steps to test the claim, provided that

1. the samples are obtained using simple random sampling;
2. the samples are independent;
3. the populations from which the samples are drawn are normally distributed or the sample sizes are large ($n_1 \geq 30$ and $n_2 \geq 30$).

Step 1: A claim is made regarding two population means. The claim is used to determine the null and alternative hypotheses. The hypotheses are structured in one of three ways:

Two-Tailed	Left-Tailed	Right-Tailed
$H_0\colon \mu_1 = \mu_2$	$H_0\colon \mu_1 = \mu_2$	$H_0\colon \mu_1 = \mu_2$
$H_1\colon \mu_1 \neq \mu_2$	$H_1\colon \mu_1 < \mu_2$	$H_1\colon \mu_1 > \mu_2$

Note: μ_1 is the population mean for population 1, and μ_2 is the population mean for population 2.

Step 2: Select a level of significance α, depending on the seriousness of making a Type I error.

Step 3: Compute the test statistic

$$t_0 = \frac{(\overline{x}_1 - \overline{x}_2) - (\mu_1 - \mu_2)}{\sqrt{\dfrac{s_1^2}{n_1} + \dfrac{s_2^2}{n_2}}}$$

which approximately follows Student's t-distribution.

Classical Approach

Step 4: Use Table V to determine the critical value using the smaller of $n_1 - 1$ or $n_2 - 1$ degrees of freedom.

	Two-Tailed	Left-Tailed	Right-Tailed
Critical Value	$-t_{\alpha/2}$ and $t_{\alpha/2}$	$-t_\alpha$	t_α

Step 5: Compare the critical value with the test statistic.

Two-Tailed	Left-Tailed	Right-Tailed
If $t_0 < -t_{\alpha/2}$ or $t_0 > t_{\alpha/2}$ reject the null hypothesis	If $t_0 < -t_\alpha$, reject the null hypothesis	If $t_0 > t_\alpha$ reject the null hypothesis

P-value Approach

Step 4: Use Table V to estimate the P-value using the the smaller of $n_1 - 1$ or $n_2 - 1$ degrees of freedom.

Two-tailed	Left-tailed	Right-tailed

Step 5: If P-value $< \alpha$, reject the null hypothesis.

Step 6: State the conclusion.

The procedure just presented is robust, so minor departures from normality will not adversely affect the results of the test. If the data have outliers, however, the procedure should not be used.

We will verify these requirements by constructing normal probability plots (to assess normality) and boxplots (to determine whether there are outliers). If the normal probability plot indicates that the data came from a population that is not normally distributed or the boxplot reveals outliers, nonparametric tests should be performed as discussed in, Section 15.5.

EXAMPLE 1 **Testing a Claim Regarding Two Means**

Problem: In the Spacelab Life Sciences 2 payload, 14 male rats were sent to space. Upon their return, the red blood cell mass (in milliliters) of the rats was determined. A control group of 14 male rats was held under the same conditions (except for space flight) as the space rats, and their red blood cell mass was also determined when the space rats returned. The project, led by Dr. Paul X. Callahan, resulted in the data listed in Table 3. Test the claim that the flight animals have a different red blood cell mass from the control animals at the $\alpha = 0.05$ level of significance.

Note that this experiment is a completely randomized design with two levels of treatment: flight and control.

Table 3							
Flight				**Control**			
8.59	8.64	7.43	7.21	8.65	6.99	8.40	9.66
6.87	7.89	9.79	6.85	7.62	7.44	8.55	8.70
7.00	8.80	9.30	8.03	7.33	8.58	9.88	9.94
6.39	7.54			7.14	9.14		

Source: NASA Life Sciences Data Archive

Approach: We verify that each sample comes from a population that is approximately normal with no outliers by drawing normal probability plots and boxplots. The boxplots will be drawn on the same graph so that we can visually compare the two samples. We then follow Steps 1 through 6, listed on page 589.

Solution: Figure 5 shows normal probability plots of the data, which indicate that the data could come from populations that are normal. On the basis of the boxplots, it seems that there is not much difference in the red blood cell mass of the two samples, although the flight group might have a slightly lower red blood cell mass. We have to determine if this difference is significant or due to chance.

Figure 5

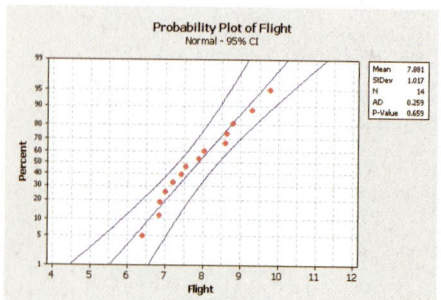
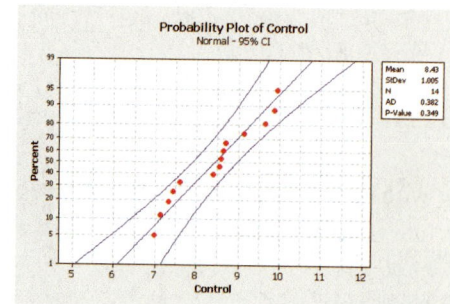
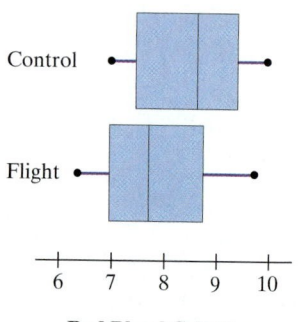

Red Blood Cell Mass

Step 1: The claim is that the flight animals have a different red blood cell mass from the control animals. Let μ_1 represent the mean red blood cell mass of the flight animals and μ_2 represent the mean red blood cell mass of the control animals. Then the claim can be expressed as $\mu_1 \neq \mu_2$, and we have the hypotheses

$$H_0: \mu_1 = \mu_2 \qquad\qquad H_0: \mu_1 - \mu_2 = 0$$
$$\text{versus} \qquad\text{or}\qquad \text{versus}$$
$$H_1: \mu_1 \neq \mu_2 \qquad\qquad H_1: \mu_1 - \mu_2 \neq 0$$

Step 2: The level of significance $\alpha = 0.05$.

Step 3: The sample statistics for the data in Table 3 are shown in Table 4. The test statistic is

Table 4

	Flight Animals	Control Animals
Sample size	$n_1 = 14$	$n_2 = 14$
Sample mean	$\bar{x}_1 = 7.881$	$\bar{x}_2 = 8.430$
Sample standard deviation	$s_1 = 1.017$	$s_2 = 1.005$

$$t_0 = \frac{(\bar{x}_1 - \bar{x}_2) - (\mu_1 - \mu_2)}{\sqrt{\dfrac{s_1^2}{n_1} + \dfrac{s_2^2}{n_2}}} = \frac{(7.881 - 8.430) - 0}{\sqrt{\dfrac{1.017^2}{14} + \dfrac{1.005^2}{14}}}$$

$$= \frac{-0.549}{0.3821288115} = -1.437$$

Classical Approach

Step 4: This is a two-tailed test with $\alpha = 0.05$. Since the sample sizes of the experimental group and control group are both 14, we have $n_1 - 1 = 14 - 1 = 13$ degrees of freedom. The critical values are $t_{\frac{\alpha}{2}} = t_{\frac{0.05}{2}} = t_{0.025} = 2.160$ and $-t_{0.025} = -2.160$.

The critical region is displayed in Figure 6.

Figure 6

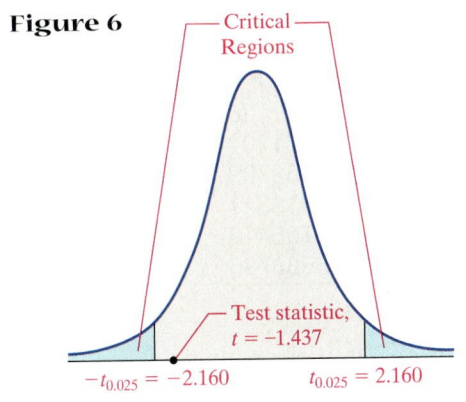

Critical Regions

Test statistic, $t = -1.437$

$-t_{0.025} = -2.160$ $t_{0.025} = 2.160$

P-value Approach

Step 4: Because this is a two-tailed test, the P-value is the area under the t-distribution to the left of $t_0 = -1.437$ plus the area under the t-distribution to the right of $t_0 = 1.437$. See Figure 7.

Figure 7   The sum of the area in the tails is the P-value

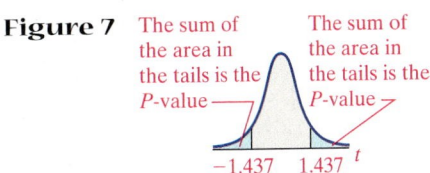

The sum of the area in the tails is the P-value

-1.437 1.437

Since the sample size of the experimental group and control group are both 14, we have $n_1 - 1 = 14 - 1 = 13$ degrees of freedom. Because of symmetry, we use Table V to estimate the area under the t-distribution to the right of $t_0 = 1.437$ and double it.

$$P\text{-value} = P(t_0 < -1.437 \text{ or } t_0 > 1.437)$$
$$= 2P(t_0 > 1.437)$$

Using Table V, we find the row that corresponds to 13 degrees of freedom. The value 1.437 lies between 1.350 and 1.771. The area under the t-distribution with 13 degrees of freedom to the right of 1.350 is 0.10. The area under the t-distribution with 13 degrees of freedom to the right of 1.771 is 0.05. After doubling these values, we have

$$0.10 < P\text{-value} < 0.20$$

Step 5: Because the test statistic does not lie within a critical region, we do not reject the null hypothesis.

Step 5: Because P-value $> \alpha$, we do not reject the null hypothesis.

Step 6: There is not sufficient evidence to conclude that the flight animals have a different red blood cell mass from the control animals at the $\alpha = 0.05$ level of significance.

The degrees of freedom used to determine the critical value(s) presented in Example 1 are conservative. Results that are more accurate can be obtained by using the following degrees of freedom:

$$df = \frac{\left(\dfrac{s_1^2}{n_1} + \dfrac{s_2^2}{n_2}\right)^2}{\dfrac{\left(\dfrac{s_1^2}{n_1}\right)^2}{n_1 - 1} + \dfrac{\left(\dfrac{s_2^2}{n_2}\right)^2}{n_2 - 1}}$$

(2)

When using Formula (2) to compute degrees of freedom, round down to the nearest integer to use Table V. For hand inference, it is recommended that we use the smaller of $n_1 - 1$ or $n_2 - 1$ as the degrees of freedom to ease computation. However, computer software will use Formula (2) when computing the degrees of freedom for increased precision in determining the P-value.

EXAMPLE 2   Testing a Claim Regarding Two Means Using Technology

Problem: Obtain an exact P-value for the problem in Example 1 using statistical software or a graphing calculator with advanced statistical features.

Approach: We will use Excel to obtain the P-value. The steps for testing claims regarding two means using the TI-83/84 Plus graphing calculator, MINITAB, and Excel are given in the Technology Step-by-Step on page 601.

Result: Figure 8 shows the results obtained from Excel. The P-value is 0.1627.

Figure 8

t-Test: *Two-Sample Assuming Unequal Variances*

	Flight	Control
Mean	7.880714286	8.43
Variance	1.035207143	1.010969231
Observations	14	14
Hypothesized Mean Difference	0	
df	26	
t Stat	−1.436781704	
P(T<=t) one-tail	0.081352709	
t Critical one-tail	1.705616341	
P(T<=t) two-tail	0.162705419	
t Critical two-tail	2.055530786	

Interpretation: There is a 0.1627 probability of obtaining a sample mean difference that is $|t_0| = |-1.437| = 1.44$ sample standard deviations from the hypothesized mean difference of 0. There is not sufficient evidence to conclude that the flight animals have a different red blood cell mass from the control animals at the $\alpha = 0.05$ level of significance.

Notice that the degrees of freedom in the technology solution are 26* versus 13 in the conservative solution done by hand in Example 1. With the lower degrees of freedom, the critical t is larger (2.160 with 13 degrees of freedom versus 2.056 with approximately 26 degrees of freedom). The larger critical value increases the number of standard deviations the difference in the sample means must be from the hypothesized mean difference before the null hypothesis is rejected. Therefore, in using the smaller of $n_1 - 1$ or $n_2 - 1$ degrees of

CAUTION
The degrees of freedom in by-hand solutions will not equal the degrees of freedom in technology solutions unless you use Formula (2) to compute degrees of freedom.

*Actually, the degrees of freedom are 25.996, but Excel rounded to 26.

freedom, we need more substantial evidence to reject the null hypothesis. This requirement decreases the probability of a Type I error (rejecting the null when the null is true) below the actual level of α chosen by the researcher. This is what we mean when we say that the method of using the lesser of $n_1 - 1$ and $n_2 - 1$ as a proxy for degrees of freedom is conservative compared with using Formula (2).

Now Work Problem 13(a) and (b)

In-Class Activity: Stringing Them Along (Part II)

Do you believe that it is easier to estimate the length of a longer rope or shorter rope by sight?

(a) Look at the piece of rope that your instructor is holding and estimate the length of the rope in inches. Now look at the second piece of rope and estimate its length in inches. After your instructor tells you the actual length of each piece of rope, compute the absolute value of the difference between your estimate and the actual length. That is, compute |actual−estimate| for each student's estimate.

(b) Assuming that estimates are just as good for long ropes as they are for short ropes, the null hypothesis is $H_0: \mu_{longer} = \mu_{shorter}$ (where μ represents the population mean absolute difference between the actual length and the estimated length). Based on your answer to the question posed at the beginning of the activity, select an appropriate alternative hypothesis and a level of significance.

(c) Combine your data with the rest of the class and obtain the length of the rope from your instructor. Conduct the hypothesis test you outlined in part (b). What did you conclude?

(d) Could this test have been done as a matched-pairs test? Explain.

[**Note:** Save the class data for use in another activity]

---

❷ Construct and Interpret Confidence Intervals Regarding the Difference of Two Independent Means

Constructing a confidence interval about the difference of two means is an extension of the results presented in Section 9.2.

Constructing a $(1 - \alpha) \cdot 100\%$ Confidence Interval about the Difference of Two Means

Suppose a simple random sample of size n_1 is taken from a population with unknown mean μ_1 and unknown standard deviation σ_1. Also, a simple random sample of size n_2 is taken from a population with unknown mean μ_2 and unknown standard deviation σ_2. If the two populations are normally distributed or the sample sizes are sufficiently large ($n_1 \geq 30$ and $n_2 \geq 30$), a $(1 - \alpha) \cdot 100\%$ confidence interval about $\mu_1 - \mu_2$ is given by

$$\text{Lower bound:} \quad (\bar{x}_1 - \bar{x}_2) - t_{\alpha/2} \cdot \sqrt{\frac{s_1^2}{n_1} + \frac{s_2^2}{n_2}}$$

and

$$\text{Upper bound:} \quad (\bar{x}_1 - \bar{x}_2) + t_{\alpha/2} \cdot \sqrt{\frac{s_1^2}{n_1} + \frac{s_2^2}{n_2}} \quad \text{(3)}$$

where $t_{\alpha/2}$ is computed using the smaller of $n_1 - 1$ or $n_2 - 1$ degrees of freedom or Formula (2).

EXAMPLE 3 ### Constructing a Confidence Interval about the Difference of Two Means

Problem: Construct a 95% confidence interval about $\mu_1 - \mu_2$ using the data presented in Table 3.

Approach: A normal probability plot and boxplot (Figure 5) indicate that the data are approximately normal with no outliers. We compute the confidence interval with $\alpha = 0.05$ using Formula (3).

Solution: We have already found the sample statistics in Example 1. In addition, we found $t_{\alpha/2} = t_{0.025}$ with 13 degrees of freedom to be 2.160. Substituting into Formula (3), we obtain the following results:

Lower bound:

$$(\bar{x}_1 - \bar{x}_2) - t_{\alpha/2} \cdot \sqrt{\frac{s_1^2}{n_1} + \frac{s_2^2}{n_2}} = (7.881 - 8.430) - 2.160 \cdot \sqrt{\frac{1.017^2}{14} + \frac{1.005^2}{14}}$$

$$= -0.549 - 0.825 = -1.374$$

Upper bound:

$$(\bar{x}_1 - \bar{x}_2) + t_{\alpha/2} \cdot \sqrt{\frac{s_1^2}{n_1} + \frac{s_2^2}{n_2}} = (7.881 - 8.430) + 2.160 \cdot \sqrt{\frac{1.017^2}{14} + \frac{1.005^2}{14}}$$

$$= -0.549 + 0.825 = 0.276$$

Interpretation: We are 95% confident that the mean difference between the red blood cell mass of the flight animals and control animals is between -1.374 and 0.276 mL. Because the confidence interval contains zero, there is not sufficient evidence to support the claim that there is a difference in the red blood cell mass of the flight group and the control group.

Now Work Problems 13(c) and (d).

! CAUTION
We would use the pooled two-sample *t*-test when the two samples come from populations that have the same variance. *Pooling* refers to finding a weighted average of the two sample variances from the independent samples. It is difficult to verify that two sample variances might be equal, so we will always use Welch's *t* when comparing two means.

What about the Pooled Two-Sample t-tests?
Perhaps you noticed that statistical software and graphing calculators with advanced statistical features provide an option for two types of two-sample *t* tests: one that assumes equal population variances (pooling) and one that does not assume equal population variances. Welch's *t*-statistic does not assume that the population variances are equal and can be used whether the population variances are equal or not. The test that assumes equal population variances is referred to as the *pooled t-statistic*.

The **pooled t-statistic** is computed by finding a weighted average of the sample variances and uses this average in the computation of the test statistic. The advantage of this test statistic is that it exactly follows Student's *t*-distribution with $n_1 + n_2 - 2$ degrees of freedom.

The disadvantage of the test statistic is that it requires that the population variances be equal. How is this requirement to be verified? While a test for determining the equality of variances does exist (*F*-test, Section 11.4), the test *requires* that each population be normally distributed. However, the *F*-test is not robust. Any minor departures from normality will make the results of the *F*-test unreliable. It has been recommended by many statisticians* that a preliminary *F*-test to check the requirement of equality of variance not be performed. In fact, George Box once said, "To make preliminary tests on variances is rather like putting to sea in a rowing boat to find out whether conditions are sufficiently calm for an ocean liner to leave port!"

*Moser and Stevens, Homogeneity of Variance in the Two-Sample Means Test, *American Statistician*, Vol. 46, No. 1.

Because the formal F-test for testing the equality of variances is so volatile, we are content to use Welch's t. This test is more conservative than the pooled t. The price that must be paid for the conservative approach is that the probability of a Type II error is higher in Welch's t than in the pooled t when the population variances are equal. However, the two tests typically provide the same conclusion, even if the assumption of equal population standard deviations seems reasonable.

MAKING AN INFORMED DECISION

Where Should I Invest?

Suppose that you have just received an inheritance of $10,000 and you decide that you should invest the money rather than blow it on frivolous items. You have decided that you will invest the money in one of two types of mutual funds. The first type you are considering follows a *large-value* approach to investing. This means that the mutual fund invests only in large, established companies that are considered to be a good bargain. The second type of mutual fund you are considering follows a *large-growth* approach to investing. This means that the mutual fund invests in large companies that are experiencing solid revenue growth.

To make an informed decision, you decide to research the rate of return of the past 3 years for each of these types of mutual funds. The mutual fund must have a Morningstar rating of four or five stars.

The Morningstar mutual-fund rating system ranks mutual funds, using one to five stars. The stars divide the mutual-fund performance into quintiles; that is, a mutual fund with a one-star rating is in the bottom 20% of mutu-

al funds in its category, a mutual fund with a two-star rating has an investment performance between the 21st and 40th percentile, and so on. These data can be found at www.morningstar.com or screen.yahoo.com/funds.html.

(a) Obtain a simple random sample of at least 15 mutual funds for each investment category. Determine the 3-year rate of return for each fund.

(b) Verify that the 3-year rates of return come from a population that is normally distributed. Also, verify that the data have no outliers. If the data do not come from a population that is normally distributed, you will have to increase the sample size so that the Central Limit Theorem can be used.

(c) Construct a boxplot for the rate of return of each fund category using the same scale. Which investment category, if any, seems superior?

(d) Obtain a 95% confidence interval for the difference between the mean rates of return. Interpret the interval.

(e) Write a report that details which investment category seems to be superior.

11.2 ASSESS YOUR UNDERSTANDING

Concepts and Vocabulary

1. What are the requirements that need to be satisfied to test a hypothesis regarding the difference of two means with σ unknown?

2. Explain why using the smaller of $n_1 - 1$ or $n_2 - 1$ degrees of freedom to determine the critical t instead of Formula (2) is conservative.

Skill Building*

In Problems 3–8, assume that the populations are normally distributed.

3. (a) Test the claim that $\mu_1 \neq \mu_2$ at the $\alpha = 0.05$ level of significance for the given sample data.
   (b) Construct a 95% confidence interval about $\mu_1 - \mu_2$.

	Population 1	Population 2
n	15	15
$\overline{x}$	15.3	14.2
s	3.2	3.5

4. (a) Test the claim that $\mu_1 \neq \mu_2$ at the $\alpha = 0.05$ level of significance for the given sample data.
   (b) Construct a 95% confidence interval about $\mu_1 - \mu_2$.

	Population 1	Population 2
n	20	20
$\overline{x}$	111	104
s	8.6	9.2

5. (a) Test the claim that $\mu_1 > \mu_2$ at the $\alpha = 0.1$ level of significance for the given sample data.

(b) Construct a 90% confidence interval about $\mu_1 - \mu_2$.

	Population 1	Population 2
n	25	18
$\bar{x}$	50.2	42.0
s	6.4	9.9

7. (a) Test the claim that $\mu_1 < \mu_2$ at the $\alpha = 0.02$ level of significance for the given sample data.

(b) Construct a 90% confidence interval about $\mu_1 - \mu_2$.

	Population 1	Population 2
n	32	25
$\bar{x}$	103.4	114.2
s	12.3	13.2

6. (a) Test the claim that $\mu_1 < \mu_2$ at the $\alpha = 0.05$ level of significance for the given sample data.

(b) Construct a 95% confidence interval about $\mu_1 - \mu_2$.

	Population 1	Population 2
n	40	32
$\bar{x}$	94.2	115.2
s	15.9	23.0

8. (a) Test the claim that $\mu_1 > \mu_2$ at the $\alpha = 0.05$ level of significance for the given sample data.

(b) Construct a 95% confidence interval about $\mu_1 - \mu_2$.

	Population 1	Population 2
n	23	13
$\bar{x}$	43.1	41.0
s	4.5	5.1

Applying the Concepts

9. Treating Bipolar Mania In a study published in the *Archives of General Psychiatry* entitled "Efficacy of Olanzapine in Acute Bipolar Mania" (Vol. 57, No. 9, pp. 841–849), researchers conducted a randomized, double-blind study to measure the effects of the drug olanzapine on patients diagnosed with bipolar disorder. A total of 115 patients with a DSM-IV diagnosis of bipolar disorder were randomly divided into two groups. Group 1 ($n_1 = 55$) received 5 to 20 mg per day of olanzapine, while Group 2 ($n_2 = 60$) received a placebo. The effectiveness of the drug was measured using the Young–Mania Rating Scale total score with the net improvement in the score recorded. The results are presented in the table.

	Experimental Group	Control Group
n	55	60
Mean improvement	14.8	8.1
Sample standard deviation	12.5	12.7

(a) Test the claim that the experimental group experienced a larger mean improvement than the control group at the $\alpha = 0.01$ level of significance.

(b) Construct a 95% confidence interval about $\mu_1 - \mu_2$ and interpret the results.

10. Hormone Replacement Therapy Coronary heart disease is the leading cause of death among older women. In observational studies, the number of deaths due to coronary heart disease has been reduced in postmenopausal women who take hormone replacement therapy. Low levels of serum high-density lipoprotein (HDL) cholesterol are considered to be one of the risk factors predictive of death from coronary heart disease. Researchers at the Washington University School of Medicine claimed that serum HDL increases when patients participate in hormone replacement therapy. The researchers randomly divided 59 sedentary women 75 years of age or older into two groups. The 30 patients in Group 1 (the experimental group) took hormone replacement pills for 9 months. The 29 patients in Group 2 (the control group) took a placebo for 9 months. At the conclusion of the treatment, the patient's serum HDL was recorded. The experiment was double-blind. The following results were obtained, where the means and standard deviations are in milligrams per deciliter (mg/dL).

	Experimental Group	Control Group
Sample size	$n_1 = 30$	$n_2 = 29$
Mean increase in HDL	$\bar{x}_1 = 8.1$	$\bar{x}_2 = 2.4$
Sample standard deviation	$s_1 = 10.5$	$s_2 = 4.3$

Source: Ellen F. Binder et al., Effects of Hormone Replacement Therapy on Serum Lipids in Elderly Women, *Annals of Internal Medicine* 134 (May 2001): pp. 754–760.

(a) What type of experimental design is this? What are the treatments? How many levels does the factor have?

(b) Test the claim that the experimental group had a larger mean increase in serum HDL levels than the control group at the $\alpha = 0.01$ level of significance (serum HDL is normally distributed).

(c) Construct a 95% confidence interval about $\mu_1 - \mu_2$ and interpret the results.

*The confidence intervals in the back of the text were computed using the smaller of $n_1 - 1$ or $n_2 - 1$ degrees of freedom. These intervals will be wider than those obtained using technology.

11. **Walking in the Airport, Part I** Do people walk faster in the airport when they are departing (getting on a plane) or when they are arriving (getting off a plane)? Researcher Seth B. Young measured the walking speed of travelers in San Francisco International Airport and Cleveland Hopkins International Airport. His findings are summarized in the table.

Direction of Travels	Departure	Arrival
Mean speed (feet per minute)	260	269
Standard deviation (feet per minute)	53	34
Sample size	35	35

Source: Young, Seth B., Evaluation of Pedestrian Walking Speeds in Airport Terminals, *Transportation Research Record*, Paper 99-0824.

(a) Is this an observational study or a designed experiment? Why?

(b) Explain why it is reasonable to use Welch's *t*-test.

(c) Test the claim that individuals walk at different speeds depending on whether they are departing or arriving at the $\alpha = 0.05$ level of significance.

(d) Construct a 95% confidence interval about $\mu_{arrival} - \mu_{departure}$. Interpret the interval.

12. **Walking in the Airport, Part II** Do business travelers walk at a different pace than leisure travelers? Researcher Seth B. Young measured the walking speed of business and leisure travelers in San Francisco International Airport and Cleveland Hopkins International Airport. His findings are summarized in the table.

Type of Traveler	Business	Leisure
Mean speed (feet per minute)	272	261
Standard deviation (feet per minute)	43	47
Sample size	20	20

Source: Young, Seth B., Evaluation of Pedestrian Walking Speeds in Airport Terminals, *Transportation Research Record*, Paper 99-0824.

(a) Is this an observational study or a designed experiment? Why?

(b) What must be true regarding the populations to use Welch's *t*-test to compare the means?

(c) Assuming the requirements listed in part (b) are satisfied, test the claim that business travelers walk at a different speed from leisure travelers at the $\alpha = 0.05$ level of significance.

(d) Construct a 95% confidence interval about $\mu_{business} - \mu_{leisure}$. Interpret the interval.

13. **Concrete Strength** An engineer wanted to know whether the strength of two different concrete mix designs differed significantly. He randomly selected 9 cylinders, measuring 6 inches in diameter and 12 inches in height, into which mixture 67-0-301 was poured. After 28 days, he measured the strength (in pounds per square inch) of the cylinder. He also randomly selected 10 cylinders of mixture 67-0-400 and performed the same test. The results are as follows:

Mixture 67-0-301			Mixture 67-0-400			
3960	4090	3100	4070	4890	5020	4330
3830	3200	3780	4640	5220	4190	3730
4080	4040	2940	4120	4620		

(a) Is it reasonable to use Welch's *t*-test? Why?

Note: Normal probability plots indicate that the data are approximately normal and boxplots indicate that there are no outliers.

(b) Test the claim that mixture 67-0-400 is stronger than mixture 67-0-301 at the $\alpha = 0.05$ level of significance.

(c) Construct a 90% confidence interval about $\mu_{400} - \mu_{301}$ and interpret the results.

(d) Draw boxplots of each data set using the same scale. Does this visual evidence support the results obtained in part (b)?

14. **Measuring Reaction Time** Researchers at the University of Mississippi wanted to determine whether the reaction time (in seconds) of males differed from that of females to a go/no go stimulus. The researchers randomly selected 20 females and 15 males to participate in the study. The go/no go stimulus required the student to respond to a particular stimulus and not to respond to other stimuli. The results are as follows:

Female Students				Male Students		
0.588	0.652	0.442	0.293	0.375	0.256	0.427
0.340	0.636	0.391	0.367	0.654	0.563	0.405
0.377	0.646	0.403	0.377	0.374	0.465	0.402
0.380	0.403	0.617	0.434	0.373	0.488	0.337
0.443	0.481	0.613	0.274	0.224	0.477	0.655

Source: PsychExperiments at the University of Mississippi. The results are as follows:

(a) Is it reasonable to use Welch's *t*-test? Why?

Note: Normal probability plots indicate that the data are approximately normal and boxplots indicate that there are no outliers.

(b) Test the claim that there is no difference in the reaction time of males and females at the $\alpha = 0.05$ level of significance.

(c) Construct a 90% confidence interval about $\mu_f - \mu_m$ and interpret the results.

(d) Draw boxplots of each data set using the same scale. Does this visual evidence support the results obtained in part (b)?

15. Bacteria in Hospital Carpeting Researchers wanted to determine if carpeted rooms contained more bacteria than uncarpeted rooms. To determine the amount of bacteria in a room, researchers pumped the air from the room over a Petri dish at the rate of 1 cubic foot per minute for eight carpeted rooms and eight uncarpeted rooms. Colonies of bacteria were allowed to form in the 16 Petri dishes. The results are presented in the table. A normal probability plot and boxplot indicate that the data are approximately normally distributed with no outliers.

Carpeted Rooms (bacteria/cubic foot)		Uncarpeted Rooms (bacteria/cubic foot)	
11.8	10.8	12.1	12.0
8.2	10.1	8.3	11.1
7.1	14.6	3.8	10.1
13.0	14.0	7.2	13.7

Source: Walter, William G., and Stober, Angie, Microbial Air Sampling in a Carpeted Hospital. *Journal of Environmental Health*, **30** (1968), p. 405.

(a) Test the claim that carpeted rooms have more bacteria than uncarpeted rooms at the $\alpha = 0.05$ level of significance.

(b) Construct a 95% confidence interval about $\mu_{\text{carpet}} - \mu_{\text{uncarpeted}}$ and interpret the results.

16. Visual versus Textual Learners Researchers wanted to know whether there was a difference in comprehension among students learning a computer program based on the style of the text. They randomly divided 36 students into two groups of 18 each. The researchers verified that the 36 students were similar in terms of educational level, age, and so on. Group 1 individuals learned the software using a visual manual (*multimodal instruction*), while Group 2 individuals learned the software using a textual manual (*unimodal instruction*). The following data represent scores the students received on an exam given to them after they studied from the manuals.

Visual Manual		Textual Manual	
51.08	60.35	64.55	56.54
57.03	76.60	57.60	39.91
44.85	70.77	68.59	65.31
75.21	70.15	50.75	51.95
56.87	47.60	49.63	49.07
75.28	46.59	43.58	48.83
57.07	81.23	57.40	72.40
80.30	67.30	49.48	42.01
52.20	60.82	49.57	61.16

Source: Mark Gellevij et al., Multimodal Versus Unimodal Instruction in a Complex Learning Context, *Journal of Experimental Education*, 2002, 70(3), pp. 215–239.

(a) What type of experimental design is this?

(b) What are the treatments?

(c) A normal probability plot and boxplot indicate it is reasonable to use Welch's *t*-test. Test the claim that there is a difference in test scores at the $\alpha = 0.05$ level of significance.

(d) Construct a 95% confidence interval about $\mu_{\text{visual}} - \mu_{\text{textual}}$ and interpret the results.

17. Does the Designated Hitter Help? In baseball, the American League allows a designated hitter (DH) to bat for the pitcher, who is typically a weak hitter. In the National League, the pitcher must bat. The common belief is that this results in American League teams scoring more runs. In interleague play, when American League teams visit National League teams, the American League pitcher must bat. So, if the DH does result in more runs, we would expect that American League teams will score fewer runs when visiting National League parks. To test this claim, a random sample of runs scored by American League teams with and without their DH is given in the following table. Test the claim that the designated hitter results in more runs scored at the $\alpha = 0.05$ level of significance. **Note:** $\bar{x}_{\text{NL}} = 4.3$, $s_{\text{NL}} = 2.6$, $\bar{x}_{\text{AL}} = 6.0$, $s_{\text{AL}} = 3.5$.

National League Park (without DH)					American League Park (with DH)				
1	5	5	4	7	6	2	3	6	8
2	6	2	9	2	1	3	7	6	4
8	8	2	10	4	4	12	5	6	13
4	3	4	1	9	6	9	5	6	7
3	5	1	3	3	4	3	2	5	5
3	5	2	7	2	6	14	14	7	0

Source: espn.com

18. **Rhythm & Blues versus Alternative** A music industry producer wondered whether there is a difference in lengths (in seconds) of rhythm & blues songs versus alternative songs. He obtained a random sample of each music category and documented song lengths. The results are in the following table. Test the claim that the length of rhythm & blues songs is different from the length of alternative songs at the $\alpha = 0.1$ level of significance. **Note:** $\bar{x}_{RB} = 242.7$, $s_{RB} = 26.9$, $\bar{x}_{ALT} = 238.3$, $s_{ALT} = 28.9$.

Rhythm & Blues (in seconds)					Alternative (in seconds)				
267	244	233	293	231	246	279	226	255	249
224	271	246	258	255	225	216	197	216	232
281	256	236	231	224	256	307	237	216	187
203	258	237	228	217	258	253	223	264	255
205	217	227	211	235	227	274	192	213	272
241	211	257	321	264	226	251	202	278	216

Source: www.yahoo.com/music

19. **Kids and Leisure** Young children require a lot of time. This time commitment cuts into a parent's leisure time. A sociologist wanted to estimate the difference in the amount of daily leisure time (in hours) of adults who do not have children under the age of 18 years and the amount of daily leisure time (in hours) of adults who have children under the age of 18 years. A random sample of 40 adults with no children under the age of 18 years results in a mean daily leisure time of 5.62 hours with a standard deviation of 2.43 hours. A random sample of 40 adults with children under the age of 18 years results in a mean daily leisure time of 4.10 hours with a standard deviation of 1.82 hours. Construct and interpret a 90% confidence interval for the mean difference in leisure time between adults with no children and adults with children.

Source: American Time Use Survey

20. **Aluminum Bottles** The aluminum bottle, first introduced in 1991 by CCL Container for mainly personal and household items such as lotions, has become popular with beverage manufacturers. Besides being lightweight and requiring less packaging, the aluminum bottle is reported to cool faster and stay cold longer than typical glass bottles. A small brewery tests this claim and obtains the following information regarding the time (in minutes) required to chill a bottle of beer from room temperature (75°F) to serving temperature (45°F). Construct and interpret a 90% confidence interval for the mean difference in cooling time for clear glass versus aluminum.

	Clear Glass	Aluminum
Sample Size	$n_1 = 42$	$n_2 = 35$
Mean time to chill	133.8	92.4
Sample standard deviation	9.9	7.3

21. **Comparing Step Pulses** A physical therapist wanted to know whether the mean step pulse of men was less than the mean step pulse of women. She randomly selected 51 men and 70 women to participate in the study. Each subject was required to step up and down onto a 6-inch platform for 3 minutes. The pulse of each subject (in beats per minute) was then recorded. After the data were entered into MINITAB, the following results were obtained.

Two Sample T-Test and Confidence Interval

```
Two sample T for Men vs Women

          N    Mean   StDev   SE Mean
Men      51   112.3    11.3       1.6
Women    70   118.3    14.2       1.7

95% CI for mu Men − mu Women: (−10.7, −1.5)
T-Test mu Men = mu Women (vs <):T = −2.61 P = 0.0051  DF = 118
```

(a) State the null and alternative hypotheses.
(b) Identify the P-value and state the researcher's conclusion if the level of significance was $\alpha = 0.01$.
(c) What is the 95% confidence interval for the mean difference in pulse rates of men versus women? Interpret this interval.

22. **Comparing Flexibility** A physical therapist claims that women are more flexible than men. She measures the flexibility of 31 randomly selected women and 45 randomly selected men by determining the number of inches subjects could reach while sitting on the floor with their legs straight out and back perpendicular to the ground. The more flexible an individual is, the higher the measured flexibility will be. After entering the data into MINITAB, she obtained the following results:

Two Sample T-Test and Confidence Interval

```
Two sample T for Men vs Women

          N    Mean   StDev   SE Mean
Men      45   18.64    3.29      0.49
Women    31   20.99    2.07      0.37

95% CI for mu Men − mu Women: (−3.58, −1.12)
T-Test mu Men = mu Women (vs <):T = −3.82 P = 0.0001 DF = 73
```

(a) State the null and alternative hypotheses.
(b) Identify the P-value and state the researcher's conclusion if the level of significance was $\alpha = 0.01$.
(c) What is the 95% confidence interval for the mean difference in flexibility of men versus women? Interpret this interval.

Consumer Reports® The High Cost of Convenience

Consumer Reports was interested in comparing a name-brand paper towel with a new version packaged in a box. The towels in the box, which cost nearly twice as much as the traditional roll, are marketed for their convenience. Given the difference in cost, one might wonder if the boxed version performs better than the traditional roll. To help answer this question, technicians at Consumers Union subjected both types of towels to five physical tests: absorption time in water, absorption time in oil, absorption capacity in water, absorption capacity in oil, and wet strength. For brevity, we will discuss only the results of the absorption time in water test.

The absorption time in water was defined as the amount of time necessary for a single sheet to absorb a predetermined amount of water. To compare the absorption times of the two types of towels, we tested six randomly selected sheets of paper towels. To avoid potential sources of bias, the individual sheets were taken from different samples of the products, and the tests were conducted in a randomly chosen order.

The claim being tested is that the water absorption time for the boxed version is less than the water absorption time for the traditional roll.

(a) Write the null and alternative hypotheses, letting μ_{box} represent the mean absorption time for the boxed version and μ_{roll} represent the mean absorption time for the roll version.

(b) Normal probability plots of the water absorption times for the two products are shown next. Based

on the normal probability plots, is it reasonable to conduct a two-sample hypothesis test?

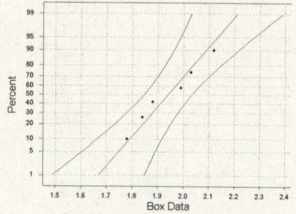
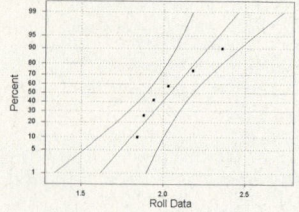

(c) A boxplot of the water absorption times for the two products follows:

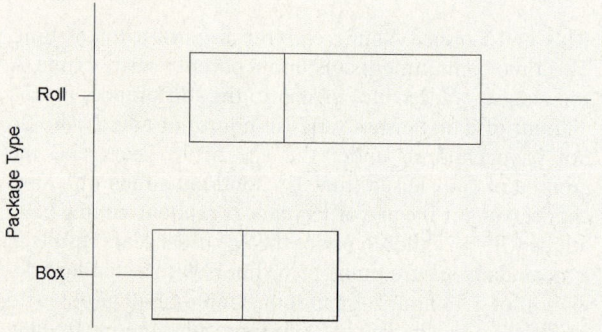

Does the data set have any outliers? Based on the boxplots, do you think that the absorption times for the boxed version are lower than the absorption times for the roll?

(d) To test the claim, we used MINITAB to perform a two-sample *t*-test. The results are as shown. Using the MINITAB output, determine the value of the test statistic. What is the *P*-value of the test? Although they are not discussed here, the other phys-ical tests provided similar results. Write an article that states your conclusion and any recommendations that you would make regarding the purchase of the two products.

```
Two-Sample T-Test and CI: Absorption Time In Water, CU
Two-sample    T    for Absorption    Time    In Water

CU-Text       N              Mean    StDev    SE Mean
Box           6            0.9717   0.0538      0.022
Roll          6            1.0200   0.0942      0.038

Difference = mu (Box) - mu (Roll)
Estimate for difference: -0.0483
95% upper bound for difference: 0.0320
T-Test of difference = 0 (vs <): T-Value = -1.09 P-Value = 0.150 DF = 10
Both use Pooled StDev = 0.0767
```

***Note to Readers:** In many cases, our test protocol and analytical methods are more complicated than described in these examples. The data and discussions have been modified to make the material more appropriate for the audience.*

| Technology Step-by-Step | **Two-sample *t*-tests, Independent Sampling** |

TI-83/84 Plus

Hypothesis Tests

Step 1: If necessary, enter raw data in L1 and L2.

Step 2: Press STAT, highlight TESTS, and select `4:2-SampTTest ....`

Step 3: If the data are raw, highlight Data, making sure that List1 is set to L1 and List2 is set to L2, with frequencies set to 1. If summary statistics are known, highlight Stats and enter the summary statistics.

Step 4: Highlight the appropriate relation between μ_1 and μ_2 in the alternative hypothesis. Set Pooled to NO.

Step 5: Highlight Calculate or Draw and press ENTER. Calculate gives the test statistic and *P*-value. Draw will draw the *t*-distribution with the *P*-value shaded.

Confidence Intervals

Follow the steps given for hypothesis tests, except select `0::2-SampTInt`. Also, select a confidence level (such as 95% = 0.95).

MINITAB

Step 1: Enter raw data in columns C1 and C2.

Step 2: Select the **Stat** menu, highlight **Basic Statistics**, then highlight **2-Sample t . . . .**

Step 3: Select "Samples in different columns." Enter C1 in the cell marked "First" and enter C2 in the cell marked "Second." Select the direction of the alternative hypothesis and select a confidence level. Click OK.

Excel

Step 1: Enter raw data in columns A and B.

Step 2: Select the **Tools menu** and highlight **Data Analysis . . . .**

Step 3: Select "*t*-test: Two-Sample Assuming Unequal Variances." With the cursor in the "Variable 1 Range" cell, highlight the data in column A. Enter the hypothesized difference in the means (usually 0) and a value for alpha. Click OK.

11.3   Inference about Two Population Proportions

Preparing for This Section   Before getting started, review the following:

- Confidence intervals about a population proportion (Section 9.3, pp. 478–483)

- Hypothesis tests about a population proportion (Section 10.4, pp. 544–550)

Objectives

1. **Test claims regarding two population proportions**
2. **Construct and interpret confidence intervals for the difference between two population proportions**
3. **Determine the sample size necessary for estimating the difference between two population proportions within a specified margin of error**

In Sections 9.3 and 10.4, we discussed inference regarding a single population proportion. We will now discuss inferential methods for comparing two population proportions. For example, in clinical trials of the drug Nasonex, a drug that is meant to relieve allergy symptoms, 26% of patients receiving 200 micrograms (μg) of Nasonex reported a headache as a side effect while 22% of patients receiving a placebo reported a headache as a side effect. Researchers want to determine whether the proportion of patients receiving the treatment and complaining of headaches is significantly higher than the proportion of patients receiving the placebo and complaining of headaches.

To conduct inference about two population proportions, we must first determine the sampling distribution of the difference of two proportions. Recall that the point estimate of a population proportion, p, is given by $\hat{p} = \dfrac{x}{n}$, where x is the number of the n individuals in the sample that have a specific characteristic. In addition, we recall that the sampling distribution of $\hat{p}$ is approximately normal with mean $\mu_{\hat{p}} = p$ and standard deviation $\sigma_{\hat{p}} = \sqrt{\dfrac{p(1-p)}{n}}$, provided that $np(1-p) \geq 10$, so

$$z = \frac{\hat{p} - p}{\sqrt{\dfrac{p(1-p)}{n}}}$$

is approximately normal with mean 0 and standard deviation 1. Using this information along with the idea of independent sampling from two populations, we obtain the sampling distribution of the difference between two proportions:

Sampling Distribution of the Difference between Two Proportions

Suppose a simple random sample of size n_1 is taken from a population where x_1 of the individuals have a specified characteristic, and a simple random sample of size n_2 is independently taken from a different population where x_2 of the individuals have a specified characteristic. The sampling distribution of $\hat{p}_1 - \hat{p}_2$ where $\hat{p}_1 = \dfrac{x_1}{n_1}$ and $\hat{p}_2 = \dfrac{x_2}{n_2}$, is approximately normal, with mean $\mu_{\hat{p}_1 - \hat{p}_2} = p_1 - p_2$ and standard deviation

$$\sigma_{\hat{p}_1 - \hat{p}_2} = \sqrt{\frac{p_1(1-p_1)}{n_1} + \frac{p_2(1-p_2)}{n_2}}, \quad \text{provided that } n_1\hat{p}_1(1-\hat{p}_1) \geq 10$$

and $n_2\hat{p}_2(1-\hat{p}_2) \geq 10$. The standardized version of $\hat{p}_1 - \hat{p}_2$ is then written as

$$z = \frac{(\hat{p}_1 - \hat{p}_2) - (p_1 - p_2)}{\sqrt{\dfrac{p_1(1-p_1)}{n_1} + \dfrac{p_2(1-p_2)}{n_2}}}$$

which has an approximate standard normal distribution.

---

1 Test Claims Regarding Two Population Proportions

Now that we know the approximate sampling distribution of $\hat{p}_1 - \hat{p}_2$, we can introduce a procedure that can be used to test claims regarding two population proportions. We first consider the test statistic. Following the discussion for comparing two means, it seems reasonable that the test statistic for the difference of two population proportions would be

$$z = \frac{(\hat{p}_1 - \hat{p}_2) - (p_1 - p_2)}{\sqrt{\dfrac{p_1(1-p_1)}{n_1} + \dfrac{p_2(1-p_2)}{n_2}}} \tag{1}$$

When comparing two population proportions, the null hypothesis will always be H_0: $p_1 = p_2$. Because the null hypothesis is assumed to be true, the test assumes $p_1 = p_2$, so $p_1 - p_2 = 0$. We also assume both p_1 and p_2 equal p, where p is the common population proportion. If we substitute this value of p into Equation (1), we obtain

$$z = \frac{(\hat{p}_1 - \hat{p}_2) - (p_1 - p_2)}{\sqrt{\dfrac{p_1(1 - p_1)}{n_1} + \dfrac{p_2(1 - p_2)}{n_2}}} = \frac{\hat{p}_1 - \hat{p}_2 - 0}{\sqrt{\dfrac{p(1 - p)}{n_1} + \dfrac{p(1 - p)}{n_2}}} = \frac{\hat{p}_1 - \hat{p}_2}{\sqrt{p(1 - p)}\sqrt{\dfrac{1}{n_1} + \dfrac{1}{n_2}}} \qquad (2)$$

We need a point estimate of p because it is unknown. The best point estimate of p is called the **pooled estimate of p**, denoted $\hat{p}$, where

$$\hat{p} = \frac{x_1 + x_2}{n_1 + n_2}$$

Substituting the pooled estimate of p into Equation (2), we obtain

$$z = \frac{\hat{p}_1 - \hat{p}_2}{\sigma_{\hat{p}_1 - \hat{p}_2}} = \frac{\hat{p}_1 - \hat{p}_2}{\sqrt{\hat{p}(1 - \hat{p})}\sqrt{\dfrac{1}{n_1} + \dfrac{1}{n_2}}}$$

This test statistic will be used to test claims regarding two population proportions. We now present the steps necessary to test claims regarding two population proportions.

Hypothesis Test Regarding the Difference between Two Population Proportions

If a claim is made regarding the two population proportions, p_1 and p_2, we can use the steps that follow to test the claim, provided that

1. the samples are independently obtained using simple random sampling,
2. $n_1\hat{p}_1(1 - \hat{p}_1) \geq 10$ and $n_2\hat{p}_2(1 - \hat{p}_2) \geq 10$,
3. $n_1 \leq 0.05N_1$ and $n_2 \leq 0.05N_2$ (the sample size is no more than 5% of the population size); this requirement ensures the independence necessary for a binomial experiment.

Step 1: A claim is made regarding two population proportions. The claim is used to determine the null and alternative hypotheses. The hypotheses can be structured in one of three ways

Two-Tailed	Left-Tailed	Right-Tailed
H_0: $p_1 = p_2$	H_0: $p_1 = p_2$	H_0: $p_1 = p_2$
H_1: $p_1 \neq p_2$	H_1: $p_1 < p_2$	H_1: $p_1 > p_2$

Note: p_1 is the population proportion for population 1, and p_2 is the population proportion for population 2.

Step 2: Select a level of significance α, depending on the seriousness of making a Type I error.

Step 3: Compute the test statistic

$$z_0 = \frac{(\hat{p}_1 - \hat{p}_2)}{\sqrt{\hat{p}(1 - \hat{p})}\sqrt{\dfrac{1}{n_1} + \dfrac{1}{n_2}}}$$

where $\hat{p} = \dfrac{x_1 + x_2}{n_1 + n_2}$.

Classical Approach

Step 4: Use Table IV to determine the critical value.

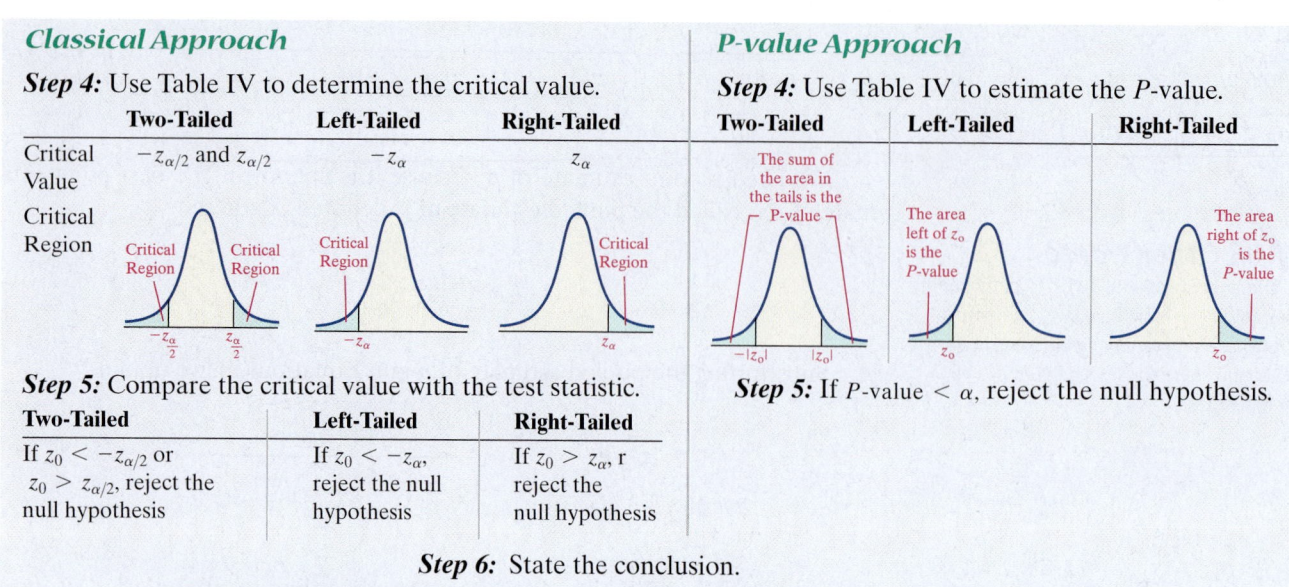

	Two-Tailed	Left-Tailed	Right-Tailed
Critical Value	$-z_{\alpha/2}$ and $z_{\alpha/2}$	$-z_{\alpha}$	z_{α}

Step 5: Compare the critical value with the test statistic.

Two-Tailed	Left-Tailed	Right-Tailed
If $z_0 < -z_{\alpha/2}$ or $z_0 > z_{\alpha/2}$, reject the null hypothesis	If $z_0 < -z_{\alpha}$, reject the null hypothesis	If $z_0 > z_{\alpha}$, r reject the null hypothesis

P-value Approach

Step 4: Use Table IV to estimate the P-value.

Two-Tailed	Left-Tailed	Right-Tailed
The sum of the area in the tails is the P-value	The area left of z_0 is the P-value	The area right of z_0 is the P-value

Step 5: If P-value $< \alpha$, reject the null hypothesis.

Step 6: State the conclusion.

EXAMPLE 1   Testing a Claim Regarding Two Population Proportions

Problem: In clinical trials of Nasonex, 3774 adult and adolescent allergy patients (patients 12 years and older) were randomly divided into two groups. The patients in Group 1 (experimental group) received 200 mcg of Nasonex, while the patients in Group 2 (control group) received a placebo. Of the 2103 patients in the experimental group, 547 reported headaches as a side effect. Of the 1671 patients in the control group, 368 reported headaches as a side effect. Is there significant evidence to support the claim that the proportion of Nasonex users that experienced headaches as a side effect is greater than the proportion in the control group at the $\alpha = 0.05$ level of significance?

Approach: We must verify the requirements to perform the hypothesis test. That is, the sample must be a simple random sample and $n_1\hat{p}_1(1 - \hat{p}_1) \geq 10$ and $n_2\hat{p}_2(1 - \hat{p}_2) \geq 10$. In addition, the sample size cannot be more than 5% of the population size. Then we follow the preceding Steps 1 through 6.

Solution: First we verify that the requirements are satisfied.

1. The samples are independently obtained using simple random sampling.
2. We have $x_1 = 547$, $n_1 = 2103$, $x_2 = 368$, and $n_2 = 1671$, so

$$\hat{p}_1 = \frac{x_1}{n_1} = \frac{547}{2103} = 0.26 \text{ and } \hat{p}_2 = \frac{x_2}{n_2} = \frac{368}{1671} = 0.22. \text{ Therefore,}$$

$$n_1\hat{p}_1(1 - \hat{p}_1) = 2103(0.26)(1 - 0.26) = 404.6172 \geq 10$$
$$n_2\hat{p}_2(1 - \hat{p}_2) = 1671(0.22)(1 - 0.22) = 286.7436 \geq 10$$

3. More than 10 million Americans 12 years old or older are allergy sufferers, so the sample sizes are less than 5% of the population size.

All three requirements are satisfied, so we now proceed to follow Steps 1 through 6.

Step 1: The claim is that the proportion of patients taking Nasonex who experience a headache is greater than the proportion of patients taking the placebo who experience a headache. Letting p_1 represent the population proportion of patients taking Nasonex who experience a headache and p_2 represent the population proportion of patients taking the placebo who experience a headache, we can express the claim as $p_1 > p_2$. This is a right-tailed hypothesis with

$$H_0: p_1 = p_2 \quad \text{versus} \quad H_1: p_1 > p_2$$

or, equivalently,

$$H_0: p_1 - p_2 = 0 \quad \text{versus} \quad H_1: p_1 - p_2 > 0$$

Step 2: The level of significance is $\alpha = 0.05$.

Step 3: From verifying requirement 2, we have that $\hat{p}_1 = 0.26$ and $\hat{p}_2 = 0.22$. To find the test statistic, we first compute the pooled estimate of $\hat{p}$:

$$\hat{p} = \frac{x_1 + x_2}{n_1 + n_2} = \frac{547 + 368}{2103 + 1671} = 0.242$$

The test statistic is

$$z_0 = \frac{(\hat{p}_1 - \hat{p}_2)}{\sqrt{\hat{p}(1 - \hat{p})}\sqrt{\dfrac{1}{n_1} + \dfrac{1}{n_2}}} = \frac{0.26 - 0.22}{\sqrt{0.242(1 - 0.242)}\sqrt{\dfrac{1}{2103} + \dfrac{1}{1671}}} = \frac{0.04}{0.0140357419} = 2.85$$

Classical Approach

Step 4: This is a right-tailed test with $\alpha = 0.05$. The critical value is $z_{0.05} = 1.645$. The critical region is displayed in Figure 9.

Figure 9

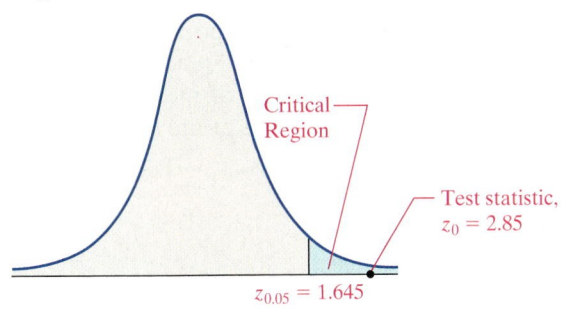

Step 5: Because $z_0 > z_{0.05}$ (the test statistic lies within critical region), we reject the null hypothesis.

P-value Approach

Step 4: Because this is a right-tailed test, the P-value is the area under the standard normal distribution to the right of $Z_0 = 2.85$. See Figure 10.

Figure 10

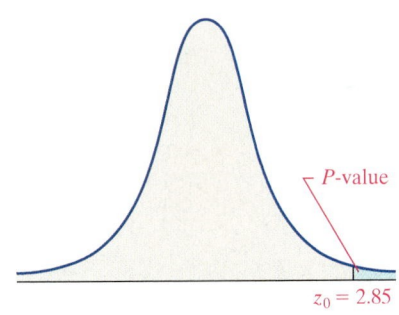

$$P\text{-value} = P(Z_0 > 2.85)$$
$$= 0.0022$$

Step 5: Because P-value $< \alpha$ (0.0022 $<$ 0.05), we reject the the null hypothesis.

Step 6: There is sufficient evidence at the $\alpha = 0.05$ level of significance to support the claim that the proportion of individuals 12 years and older taking 200 mcg of Nasonex who experience headaches is greater than the proportion of individuals 12 years and older taking a placebo who experience headaches.

> **⚠ CAUTION**
> In any statistical study, be sure to consider practical significance. Many statistically significant results can be produced simply by increasing the sample size.

In looking back at the results of Example 1, we notice that the proportion of individuals taking 200 mcg of Nasonex who experience headaches is *statistically significantly* greater than the proportion of individuals 12 years and older taking a placebo who experience headaches. However, we need to ask ourselves a pressing question. Would you not take an allergy medication because 26% of patients experienced a headache taking the medication versus 22% who experienced a headache taking a placebo? Most people would be willing to accept the additional risk of a headache to relieve their allergy symptoms. While the difference of 4% is statistically significant, it does not have any *practical significance*.

Testing a Claim Regarding the Difference of Two Population Proportions Using Technology

Problem: Obtain the exact P-value for the problem in Example 1 using statistical software or a graphing calculator with advanced statistical features.

Approach: We will use MINITAB to obtain the P-value. The steps for testing claims regarding two proportions using the TI-83/84 Plus graphing calculator, MINITAB, and Excel are given in the Technology Step-by-Step on page 612.

Result: Figure 11 shows the results obtained from MINITAB. The P-value is 0.004.

The P-value obtained from MINITAB differs from the P-value obtained by-hand in Example 1 because of rounding.

Figure 11

Test and CI for Two Proportions

```
Sample      X       N    Sample p
1         547    2103    0.260105
2         368    1671    0.220227


Difference = p (1) − p (2)
Estimate for difference: 0.0398772
90% CI for difference: (0.0169504, 0.0628040)
Test for difference = 0 (vs > 0): Z = 2.86 P-Value = 0.002
```

Interpretation: There is a 0.004 probability of obtaining a difference in sample proportions that is 2.68 standard deviations from the hypothesized mean difference of 0. There is sufficient evidence to support the claim that the proportion of individuals 12 years and older taking 200 mcg of Nasonex who experience headaches is greater than the proportion of individuals 12 years and older taking a placebo who experience headaches at the $\alpha = 0.05$ level of significance.

Now Work Problem 13(a).

 ## Construct and Interpret Confidence Intervals for the Difference between Two Population Proportions

The sampling distribution of the difference of two proportions, $\hat{p}_1 - \hat{p}_2$, can also be used to construct confidence intervals for the difference of two proportions.

Constructing a $(1 - \alpha) \cdot 100\%$ Confidence Interval for the Difference between Two Population Proportions

To construct a $(1 - \alpha) \cdot 100\%$ confidence interval for the difference between two population proportions, the following requirements must be satisfied.

1. the samples are obtained independently, using simple random sampling,
2. $n_1 \hat{p}_1 (1 - \hat{p}_1) \geq 10$ and $n_2 \hat{p}_2 (1 - \hat{p}_2) \geq 10$,
3. $n_1 \leq 0.05 N_1$ and $n_2 \leq 0.05 N_2$ (the sample size is no more than 5% of the population size); this ensures the independence necessary for a binomial experiment.

Provided that these requirements are met, a $(1 - \alpha) \cdot 100\%$ confidence interval for $p_1 - p_2$ is given by

Lower bound: $(\hat{p}_1 - \hat{p}_2) - z_{\alpha/2} \cdot \sqrt{\dfrac{\hat{p}_1(1 - \hat{p}_1)}{n_1} + \dfrac{\hat{p}_2(1 - \hat{p}_2)}{n_2}}$

$$\text{Upper bound:} \quad (\hat{p}_1 - \hat{p}_2) + z_{\alpha/2} \cdot \sqrt{\frac{\hat{p}_1(1 - \hat{p}_1)}{n_1} + \frac{\hat{p}_2(1 - \hat{p}_2)}{n_2}} \quad \textbf{(3)}$$

Notice that we do not pool the sample proportions. This is because we are not making any assumptions regarding their equality, as we did in hypothesis testing.

EXAMPLE 3   **Constructing a Confidence Interval for the Difference between Two Population Proportions**

Problem: In clinical trials of Nasonex, 750 randomly selected pediatric patients (ages 3 to 11 years old) were randomly divided into two groups. The patients in Group 1 (experimental group) received 100 mcg of Nasonex, while the patients in Group 2 (control group) received a placebo. Of the 374 patients in the experimental group, 64 reported headaches as a side effect. Of the 376 patients in the control group, 68 reported headaches as a side effect. Construct a 90% confidence interval for the difference between the two population proportions, $p_1 - p_2$.

Approach: We can compute a 90% confidence interval about $p_1 - p_2$, provided that the requirements stated above are satisfied. We then construct the interval by using Formula (3).

Solution

Step 1: We have to verify the requirements for constructing a confidence interval about the difference between two population proportions. (1) The samples were randomly divided into two groups. (2) For the experimental group (Group 1), we have $n_1 = 374$ and $x_1 = 64$, so $\hat{p}_1 = \frac{x_1}{n_1} = \frac{64}{374} = 0.171$. For the control group (Group 2), we have $n_2 = 376$ and $x_2 = 68$, so that $\hat{p}_2 = \frac{x_2}{n_2} = \frac{68}{376} = 0.181$. Therefore,

$$n_1\hat{p}_1(1 - \hat{p}_1) = 374(0.171)(1 - 0.171) = 53.02 \geq 10$$
$$n_2\hat{p}_2(1 - \hat{p}_2) = 376(0.181)(1 - 0.181) = 55.74 \geq 10$$

(3) The samples were independently obtained and the sample sizes are less than 5% of the population size. (There are over 20 million children between the ages of 3 and 11 in the United States.)

Step 2: Because we want a 90% confidence interval, we have $\alpha = 0.10$, so $z_{\alpha/2} = z_{0.05} = 1.645$.

Step 3: Substituting into Formula (3) with $\hat{p}_1 = 0.171$, $n_1 = 374$, $\hat{p}_2 = 0.181$, and $n_2 = 376$, we obtain the lower and upper bounds on the confidence interval:

Lower bound:
$$(\hat{p}_1 - \hat{p}_2) - z_{\alpha/2} \cdot \sqrt{\frac{\hat{p}_1(1 - \hat{p}_1)}{n_1} + \frac{\hat{p}_2(1 - \hat{p}_2)}{n_2}}$$
$$= (0.171 - 0.181) - 1.645 \cdot \sqrt{\frac{0.171(1 - 0.171)}{374} + \frac{0.181(1 - 0.181)}{376}}$$
$$= -0.010 - 0.046 = -0.056$$

Upper bound:
$$(\hat{p}_1 - \hat{p}_2) + z_{\alpha/2} \cdot \sqrt{\frac{\hat{p}_1(1 - \hat{p}_1)}{n_1} + \frac{\hat{p}_2(1 - \hat{p}_2)}{n_2}}$$
$$= (0.171 - 0.181) + 1.645 \cdot \sqrt{\frac{0.171(1 - 0.171)}{374} + \frac{0.181(1 - 0.181)}{376}}$$
$$= -0.010 + 0.046 = 0.036$$

USING TECHNOLOGY

Graphing calculators with advanced statistical features and statistical spreadsheets can be used to construct confidence intervals about the difference between two population proportions. Figure 12 shows the results using a TI-84 Plus graphing calculator.

Figure 12

```
2-PropZInt
(-.0555,.03601)
p̂1=.1711229947
p̂2=.1808510638
n1=374
n2=376
```

Based on the results of the study, we are 90% confident that the difference between the proportion of headaches in the experimental group and the control group is between -0.056 and 0.036. Because the confidence interval contains 0, there is no evidence to support the claim that the proportion of 3 to 11 year old patients complaining of headaches who receive Nasonex is different from those who do not receive Nasonex at the $\alpha = 0.1$ level of significance.

Now Work Problem 13(b).

③ Determine the Sample Size Necessary for Estimating the Difference between Two Population Proportions within a Specified Margin of Error

In Section 9.3, we introduced a method for determining the sample size n required to estimate a single population proportion within a specified margin of error, E, with a specified level of confidence. This formula was obtained by solving the margin of error, $E = z_{\alpha/2} \cdot \sqrt{\dfrac{\hat{p}(1 - \hat{p})}{n}}$, for n. We can follow the same approach to determine the sample size when we want to estimate two population proportions. Notice that the margin of error, E, in Formula (3) is given by

$$E = z_{\alpha/2} \cdot \sqrt{\frac{\hat{p}_1(1 - \hat{p}_1)}{n_1} + \frac{\hat{p}_2(1 - \hat{p}_2)}{n_2}}.$$ Assuming that $n_1 = n_2 = n$, we can solve this expression for $n = n_1 = n_2$ and obtain the following result:

CAUTION
When doing sample size calculations, always round up.

> ### Sample Size for Estimating $p_1 - p_2$
> The sample size required to obtain a $(1 - \alpha) \cdot 100\%$ confidence interval with a margin of error, E, is given by
> $$n = n_1 = n_2 = [\hat{p}_1(1 - \hat{p}_1) + \hat{p}_2(1 - \hat{p}_2)]\left(\frac{z_{\alpha/2}}{E}\right)^2 \qquad (4)$$
> rounded up to the next integer, if prior estimates of P_1 and P_2, $\hat{p}_1$ and $\hat{p}_2$, are available. If prior estimates of P_1 and P_2 are unavailable, the sample size is
> $$n = n_1 = n_2 = 0.5\left(\frac{z_{\alpha/2}}{E}\right)^2 \qquad (5)$$
> rounded up to the next integer. The margin of error should always be expressed as a decimal when using Formulas (4) and (5).

EXAMPLE 4 Determining Sample Size

Problem: A nutritionist wishes to estimate the difference between the proportion of males and females who consume the USDA's recommended daily intake of calcium. What sample size should be obtained if she wishes the estimate to be within 3 percentage points with 95% confidence, assuming that

(a) she uses the results of the USDA's 1994–1996 Diet and Health Knowledge Survey, according to which 51.1% of males and 75.2% of females consume the USDA's recommended daily intake of calcium,

(b) she does not use any prior estimates?

Approach: We have $E = 0.03$ and $z_{\alpha/2} = z_{0.05/2} = z_{0.025} = 1.96$. To answer part (a), we let $\hat{p}_1 = 0.511$ (for males) and $\hat{p}_2 = 0.752$ (for females) in Formula (4). To answer part (b), we use Formula (5).

Solution

(a) Substituting $E = 0.03$, $z_{0.025} = 1.96$, $\hat{p}_1 = 0.511$, and $\hat{p}_2 = 0.752$ into Formula (4), we obtain

$$n_1 = n_2 = [\hat{p}_1(1 - \hat{p}_1) + \hat{p}_2(1 - \hat{p}_2)]\left(\frac{z_{\alpha/2}}{E}\right)^2 = [0.511(1 - 0.511) + 0.752(1 - 0.752)]\left(\frac{1.96}{0.03}\right)^2$$

$$= 1862.6$$

We round this value up to 1863. The nutritionist must survey 1863 randomly selected males and 1863 randomly selected females.

(b) Substituting $E = 0.03$ and $z_{0.025} = 1.96$ into Formula (5), we obtain

$$n_1 = n_2 = 0.5\left(\frac{z_{\alpha/2}}{E}\right)^2 = 0.5\left(\frac{1.96}{0.03}\right)^2 = 2134.2$$

We round this value up to 2135. The nutritionist must survey 2135 randomly selected males and 2135 randomly selected females.

Now Work Problem 23.

In Other Words

If possible, obtain a prior estimate of $\hat{p}$ when doing sample size computations.

We can see that having prior estimates of the population proportions reduces the number of individuals that need to be surveyed.

11.3 ASSESS YOUR UNDERSTANDING

Concepts and Vocabulary

1. Explain why we determine a pooled estimate of the population proportion when testing claims regarding the difference of two proportions, but do not pool when constructing confidence intervals about the difference of two proportions.

2. State the requirements that must be satisfied to test a claim regarding two population proportions.

Skill Building

In Problems 3–6, test the claim at the $\alpha = 0.05$ level of significance by determining (a) the null and alternative hypotheses, (b) the test statistic, (c) the critical value, and (d) the P-value. Assume the samples were obtained independently using simple random sampling.

3. Claim $p_1 > p_2$. Sample data:
$x_1 = 368$, $n_1 = 541$, $x_2 = 351$, $n_2 = 593$

4. Claim $p_1 < p_2$. Sample data:
$x_1 = 109$, $n_1 = 475$, $x_2 = 78$, $n_2 = 325$

5. Claim $p_1 \neq p_2$. Sample data:
$x_1 = 28$, $n_1 = 254$, $x_2 = 36$, $n_2 = 301$

6. Claim $p_1 \neq p_2$. Sample data:
$x_1 = 804$, $n_1 = 874$, $x_2 = 902$, $n_2 = 954$

In Problems 7–10, construct a confidence interval for $p_1 - p_2$ at the given level of confidence.

7. $x_1 = 368$, $n_1 = 541$, $x_2 = 421$, $n_2 = 593$, 90% confidence

8. $x_1 = 109$, $n_1 = 475$, $x_2 = 78$, $n_2 = 325$, 99% confidence

9. $x_1 = 28$, $n_1 = 254$, $x_2 = 36$, $n_2 = 301$, 95% confidence

10. $x_1 = 804$, $n_1 = 874$, $x_2 = 892$, $n_2 = 954$, 95% confidence

Applying the Concepts

11. Prevnar The drug Prevnar is a vaccine meant to prevent certain types of bacterial meningitis. It is typically administered to infants starting around 2 months of age. In randomized, double-blind clinical trials of Prevnar, infants were randomly divided into two groups. Subjects in Group 1 received Prevnar while subjects in Group 2 received a control vaccine. After the first dose, 107 of 710 subjects in the experimental group (Group 1) experienced fever as a side effect. After the first dose, 67 of 611 of the subjects in the control group (Group 2) experienced fever as a side effect.

(a) Test the claim that a higher proportion of subjects in Group 1 experienced fever as a side effect than subjects in Group 2 at the $\alpha = 0.05$ level of significance.

(b) Construct a 90% confidence interval for the difference between the two population proportions, $p_1 - p_2$.

12. Prevnar The drug Prevnar is a vaccine meant to prevent certain types of bacterial meningitis. It is typically administered to infants starting around 2 months of age. In randomized, double-blind clinical trials of Prevnar, infants were randomly divided into two groups. Subjects in Group 1 received Prevnar, while subjects in Group 2 received a control vaccine. After the second dose, 137 of 452 subjects in the experimental group (Group 1) experienced drowsiness as a side effect. After the second dose, 31 of 99 subjects in the control group (Group 2) experienced drowsiness as a side effect.

(a) Test the claim that a different proportion of subjects in Group 1 experienced drowsiness as a side effect than subjects in Group 2 at the $\alpha = 0.05$ level of significance.

(b) Construct a 99% confidence interval for the difference between the two population proportions, $p_1 - p_2$.

13. Too Much Cholesterol in Your Diet? A nutritionist NW claims that the proportion of individuals who have at most an eighth-grade education and consume more than the USDA's recommended daily allowance of 300 mg of cholesterol is higher than the proportion of individuals who have at least some college and consume too much cholesterol. In interviews with 320 individuals who have at most an eighth-grade education, she determined that 114 of them consumed too much cholesterol. In interviews with 350 individuals with at least some college, she determined that 112 of them consumed too much cholesterol per day. (Based on data obtained from the USDA's Diet and Health Knowledge Survey.)

(a) Test the claim that the proportion of individuals with at most an eighth-grade education who consume too much cholesterol is higher than the proportion of individuals who have at least some college and consume too much cholesterol at the $\alpha = 0.1$ level of significance.

(b) Construct a 95% confidence interval for the difference between the two population proportions, $p_8 - p_c$.

14. Consumption of Saturated Fat A nutritionist claims that the proportion of females who consume too much saturated fat is lower than the proportion of males who consume too much saturated fat. In interviews with 513 randomly selected females, she determines that 300 consume too much saturated fat. In interviews with 564 randomly selected males, she determines that 391 consume too much saturated fat, based on data obtained from the USDA's Diet and Health Knowledge Survey.

(a) Test the claim that a lower proportion of females than males consume too much saturated fat at the $\alpha = 0.05$ level of significance.

(b) Construct a 95% confidence interval for the difference between the two population proportions, $p_f - p_m$.

15. Percentage of Americans Who Smoke on the Decline? On November 3–5, 2005, the Gallup Organization surveyed 1028 adults and found that 226 of them had smoked at least one cigarette in the past week. In 1990, Gallup also asked 1028 adults the same question and determined that 278 adults had smoked at least one cigarette in the past week. Can we say the proportion of American adults who smoked at least one cigarette in the past week has decreased since 1990? Why?

16. Life on Mars? On March 19–21, 1999, the Gallup Organization surveyed 535 adults aged 18 years old or older and asked, "Do you think there is life of some form on other planets in the universe or not?" Of the 535 individuals surveyed, 326 responded yes. When the same question was asked on September 3–5, 1996, 385 of the 535 individuals surveyed responded yes.

(a) Test the claim that the proportion of adults who believe that there is life on other planets has decreased since September, 1996, at the $\alpha = 0.10$ level of significance.

(b) Construct a 90% confidence interval for the difference between the two population proportions, $p_{1996} - p_{1999}$.

17. Salk Vaccine On April 12, 1955, Dr. Jonas Salk released the results of clinical trials for his vaccine to prevent polio. In these clinical trials, 400,000 children were randomly divided in two groups. The subjects in Group 1 (the experimental group) were given the vaccine, while the subjects in Group 2 (the control group) were given a placebo. Of the 200,000 children in the experimental group, 33 developed polio. Of the 200,000 children in the control group, 115 developed polio.

(a) Test the claim that the proportion of subjects in the experimental group who contracted polio is less than the proportion of subjects in the control group who contracted polio at the $\alpha = 0.01$ level of significance.

(b) Construct a 90% confidence interval for the difference between the two population proportions, $p_1 - p_2$.

18. Mind over Body On May 10–14, 2001, the Gallup Organization surveyed 1002 adult Americans and asked them if they believed in psychic or spiritual healing. Of the 1002 individuals surveyed, 551 said yes. When the same question was asked on June 6–8, 2005, 541 of the 1002 individuals surveyed responded yes.

(a) Test the claim that the proportion of adult Americans who believe in psychic or spiritual healing has changed since May 2001 at the $\alpha = 0.05$ level of significance.

(b) Construct a 90% confidence interval for the difference between the two population proportions, $p_{2001} - p_{2005}$.

19. Side Effects   In clinical trials of the allergy medicine Clarinex (5 mg), it was reported that 50 out of 1655 individuals in the Clarinex group and 31 out of 1652 individuals in the placebo group experienced dry mouth as a side effect of their respective treatments.

(*Source:*www.clarinex.com).

(a) Test the claim that the proportion of individuals experiencing dry mouth is greater for those taking Clarinex than for those taking a placebo at the $\alpha = 0.05$ level of significance.

(b) Is the difference between the groups practically significant?

20. Practical versus Statistical Significance   Suppose that in clinical trials for treatment of a skin disorder 642 of 2105 patients receiving the current standard treatment were cured of the disorder and 697 of 2115 patients receiving a new proposed treatment were cured of the disorder.

(a) Test the claim that the new procedure cures a higher percentage of patients at the $\alpha = 0.05$ level of significance.

(b) Do you think that the difference in success rates is practically significant? What factors might influence your decision?

21. Own a Gun?   In October 2004, the Gallup Organization surveyed 1134 American adults and found that 431 owned a gun. In February 1999, the Gallup Organization had surveyed 1134 American adults and found that 408 owned a gun. Suppose that a newspaper article has a headline that reads, "Percentage of American Gun Owners on the Rise." Is this an accurate headline? Why?

22. Accupril   Accupril, a medication supplied by Pfizer Pharmaceuticals, is meant to control hypertension. In clinical trials of Accupril, 2142 subjects were divided into two groups. The 1563 subjects in Group 1 (the experimental group) received Accupril. The 579 subjects in Group 2 (the control group) received a placebo. Of the 1563 subjects in the experimental group, 61 experienced dizziness as a side effect. Of the 579 subjects in the control group, 15 experienced dizziness as a side effect. To test the claim that the proportion experiencing dizziness in the experimental group is greater than that in the control group, the

researchers entered the data into MINITAB statistical software and obtained the following results:

```
Test and Confidence Interval for Two Proportions
Sample      X       N   Sample p
1          61     1563   0.039028
2          15      579   0.025907


Estimate for p(1) - p(2): 0.0131208
95% CI for p(1) - p(2): (-0.00299150, 0.0292330)
Test for p(1) - p(2) = 0 (vs > 0): Z = 1.46 P-Value = 0.072
```

What conclusion can be drawn from the clinical trials?

23. Determining Sample Size   A physical therapist wants to determine the difference in the proportion of men and women who participate in regular sustained physical activity. What sample size should be obtained if she wishes the estimate to be within three percentage points with 95% confidence, assuming that

(a) she uses the 1998 estimates of 21.9% male and 19.7% female from the U.S. National Center for Chronic Disease Prevention and Health Promotion?

(b) she does not use any prior estimates?

24. Determining Sample Size   An educator wants to determine the difference between the proportion of males and females who have completed 4 or more years of college. What sample size should be obtained if she wishes the estimate to be within two percentage points with 90% confidence, assuming that

(a) she uses the 1999 estimates of 27.5% male and 23.1% female from the U.S. Census Bureau?

(b) she does not use any prior estimates?

25. Treating Apnea   An individual with apnea stops breathing while sleeping. In a randomized, double-blind study to evaluate the efficacy of caffeine citrate for treatment of apnea, infants with six or more episodes of apnea were administered 10 mg/kg of caffeine citrate intravenously, followed by 2.5 mg/kg/day for up to 10 days or a placebo. (Caffeine Citrate for the Treatment of Apnea in Prematurity: A Double-blind, Placebo Controlled Study, *Pharmacotherapy*, June 2000, 20(6):644–652). With successful treatment defined as a 50% or more reduction in apnea episodes, it was determined that caffeine citrate was significantly more effective than the placebo in reducing apnea episodes by at least 50% in 6 days with the *P*-value less than 0.05. State any conclusions, using this *P*-value.

Technology Step-by-Step	**Inference about Two Population Proportions**

TI-83/84 Plus

Hypothesis Tests

Step 1: Press STAT, highlight TESTS, and select `6:2-PropZTest ....`

Step 2: Enter the values of x_1, n_1, x_2, and n_2.

Step 3: Highlight the appropriate relation between p_1 and p_2 in the alternative hypothesis.

Step 4: Highlight Calculate or Draw and press ENTER. Calculate gives the test statistic and *P*-value. Draw will draw the *Z*-distribution with the *P*-value shaded.

Confidence Intervals

Follow the same steps given for hypothesis tests, except select `B:2-PropZInt ....` Also, select a confidence level (such as 95% = 0.95).

MINITAB

Step 1: Select the **Stat** menu, highlight **Basic Statistics**, then highlight **2 Proportions . . . .**

Step 2: Click the Summarized Data Option. Enter the number of trials, n_1, and the number of successes, x_1. Enter the number of trials, n_2, and the number of successes, x_2.

Step 3: Under OPTIONS, select the direction of the alternative hypothesis and select a confidence level. Click OK.

Excel

Step 1: Load the PHStat add-in.

Step 2: Select the **PHStat** menu. Highlight **Two-Sample Tests** and then highlight **Z Test for Differences in Two Proportions . . . .**

Step 3: Enter the hypothesized difference in the two proportions (ususally 0). Enter the level of significance, α. For the Population 1 sample, enter the number of successes, x_1, and the sample size, n_1. For the Population 2 sample, enter the number of successes, x_2, and the sample size, n_2. Select the appropriate test option. Click OK.

11.4 Inference about Two Population Standard Deviations

Objectives

  ① **Find critical values of the *F*-distribution**

  ② **Test claims regarding two population standard deviations**

① Find Critical Values of the *F*-distribution

In this section, we discuss methods for comparing two population standard deviations (or variances). For example, we might be interested in testing the claim that the rate of return for Cisco Systems stock is more volatile than General Electric (GE) stock. If Cisco Systems stock is more volatile than GE, the standard deviation rate of return on Cisco Systems would be higher than the standard deviation rate of return on GE. To test claims regarding population standard deviations, we use Fisher's *F*-distribution, named in honor of Sir Ronald A. Fisher. Certain requirements must be satisfied to test claims regarding two population standard deviations.

> **Requirements for Testing Claims Regarding Two Population Standard Deviations**
>
> **1.** The samples are independent simple random samples.
> **2.** The populations from which the samples are drawn are normally distributed.

The second requirement is critical. The procedures introduced in this section are **not robust**, so any departures from normality will adversely affect the results of the test and make them unreliable. One reason for this is that the standard deviation is not a good measure of spread in nonsymmetric distributions because the standard deviation is not resistant to extreme values. Populations that are skewed will have extreme values that inflate the value of the standard deviation. Therefore, whenever performing inference regarding two population standard deviations, we must verify the requirement of the normality of the population using normal probability plots.

We use the following notation when describing the two populations.

Notation Used when Comparing Two Population Standard Deviations

σ_1^2: Variance for population 1
σ_2^2: Variance for population 2
s_1^2: Sample variance for population 1
s_2^2: Sample variance for population 2
n_1: Sample size for population 1
n_2: Sample size for population 2

Before we can perform statistical inference regarding two population standard deviations, we want to know the sampling distribution of the test statistic. In comparing two population standard deviations, the test statistic follows Fisher's F-distribution.

Fisher's F-distribution

If $\sigma_1^2 = \sigma_2^2$ and s_1^2 and s_2^2 are sample variances from independent simple random samples of size n_1 and n_2, respectively, drawn from normal populations, then

$$F = \frac{s_1^2}{s_2^2}$$

follows the F-distribution with $n_1 - 1$ degrees of freedom in the numerator and $n_2 - 1$ degrees of freedom in the denominator.

We can find critical values of the F-distribution using Table VII in Appendix A. Before discussing how to read Table VII, we present characteristics of the F-distribution.

Characteristics of the F-distribution

1. The F-distribution is not symmetric. It is skewed right.
2. The shape of the F-distribution depends on the degrees of freedom in the numerator and denominator. See Figure 13. This is similar to the χ^2 distribution and Student's t-distribution, whose shapes depend on their degrees of freedom.
3. The total area under the curve is 1.
4. The values of F are always greater than or equal to zero.

Figure 13
F-Distributions

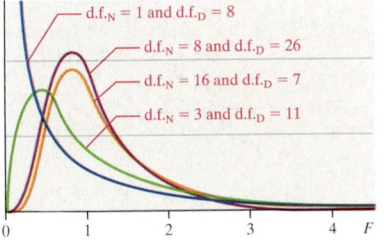

Table VII is structured differently from the t- or χ^2 distribution. The row across the top provides the degrees of freedom in the numerator, while the column on the left provides the degrees of freedom in the denominator. Corresponding to the degrees of freedom in the numerator and denominator, we have various areas (0.1, 0.05, 0.025, 0.01, 0.001) in the right tail of the F-distribution. The body of the table provides the critical F values. We use the notation

$$F_{\alpha, n_1 - 1, n_2 - 1}$$

where $n_1 - 1$ is the degrees of freedom in the numerator, $n_2 - 1$ is the degrees of freedom in the denominator, and α is the area to the right of $F_{\alpha, n_1 - 1, n_2 - 1}$. To determine the critical value that has an area of α to the left, we use the following property:

$$F_{1-\alpha, n_1 - 1, n_2 - 1} = \frac{1}{F_{\alpha, n_2 - 1, n_1 - 1}}$$

So, to find the critical F that has an area of $\alpha = 0.05$ to the left with 12 degrees of freedom in the numerator and 20 degrees of freedom in the denominator, find the critical F that has an area of $\alpha = 0.05$ to the right with 20 degrees of freedom in the numerator and 12 degrees of freedom in the denominator and compute its reciprocal.

EXAMPLE 1 · Finding Critical Values for the F-distribution

Problem: Find the critical F-values

(a) for a right-tailed test with $\alpha = 0.05$, degrees of freedom in the numerator $= 10$, and degrees of freedom in the denominator $= 7$.

(b) for a two-tailed test with $\alpha = 0.05$, degrees of freedom in the numerator $= 15$, and degrees of freedom in the denominator $= 20$.

Approach: We will perform the following steps to obtain the critical values.

Step 1: Draw an F-distribution with the critical value and area labeled.

Step 2: Use Table VII to find the critical value.

Solution

(a) Step 1: Figure 14 shows the F-distribution with 10 degrees of freedom in the numerator and 7 degrees of freedom in the denominator. The area to the right of the unknown critical value is 0.05. We denote this critical value $F_{0.05, 10, 7}$.

Step 2: Figure 15 shows a partial representation of Table VII. We box the column corresponding to 10 degrees of freedom (the numerator degrees of freedom) and the row corresponding to 7 degrees of freedom (the denominator degrees of freedom) with $\alpha = 0.05$. The critical value is $F_{0.05, 10, 7} = 3.64$.

Figure 14

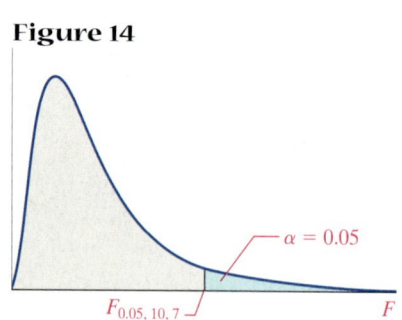

$\alpha = 0.05$

$F_{0.05,\,10,\,7}$ F

Figure 15

	Area to the Right of Critical Value	Degrees of Freedom in the Numerator							
		9	10	15	20	30	60	120	1000
	0.100	59.86	60.19	61.22	61.74	62.26	62.79	63.06	63.30
	0.050	240.54	241.88	245.95	248.01	250.10	252.20	253.25	254.19
1	0.025	963.28	968.63	984.87	993.10	1001.4	1009.8	1014	1017.7
	0.010	6022.5	6055.8	6157.3	6208.7	6260.6	6313	6339.4	6362.7
	0.001	602284	605621	615764	620908	626099	631337	633972	636301
	0.100	2.72	2.70	2.63	2.59	2.56	2.51	2.49	2.47
	0.050	3.68	3.64	3.51	3.44	3.38	3.30	3.27	3.23
7	0.025	4.82	4.76	4.57	4.47	4.36	4.25	4.20	4.15
	0.010	6.72	6.62	6.31	6.16	5.99	5.82	5.74	5.66
	0.001	14.33	14.08	13.32	12.93	12.53	12.12	11.91	11.72
	0.100	2.56	2.54	2.46	2.42	2.38	2.34	2.32	2.30
	0.050	3.39	3.35	3.22	3.15	3.08	3.01	2.97	2.93
8	0.025	4.36	4.30	4.10	4.00	3.89	3.78	3.73	3.68
	0.010	5.91	5.81	5.52	5.36	5.20	5.03	4.95	4.87
	0.001	11.77	11.54	10.84	10.48	10.11	9.73	9.53	9.36

(b) ***Step 1:*** Figure 16 shows the F-distribution with 15 degrees of freedom in the numerator and 20 degrees of freedom in the denominator. The area to the right of the right critical value is $\frac{\alpha}{2} = 0.025$. We denote this critical value $F_{0.025,15,20}$. The area to the right of the left critical value is $1 - \frac{\alpha}{2} = 0.975$. We denote this critical value $F_{0.975,15,20}$.

Figure 16

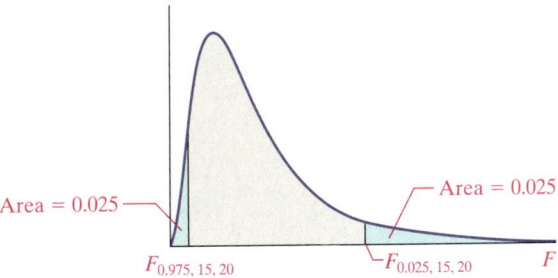

Area = 0.025

Area = 0.025

$F_{0.975,\,15,\,20}$ $F_{0.025,\,15,\,20}$ F

Step 2: We refer to Table VII. To find the right critical value, identify the column that represents 15 degrees of freedom (the numerator degrees of freedom) and the row that represents 20 degrees of freedom (the denominator degrees of freedom) with $\frac{\alpha}{2} = 0.025$. The right critical value is $F_{0.025,15,20} = 2.57$. To find the left critical value, we use the fact that $F_{1-\alpha,n_1-1,n_2-1} = \dfrac{1}{F_{\alpha,n_2-1,n_1-1}}$. Therefore, we use Table VII and find that $F_{0.025,20,15} = 2.76$. So the left critical value is $F_{0.975,15,20} = \dfrac{1}{F_{0.025,20,15}} = \dfrac{1}{2.76} = 0.36$.

In studying Table VII, we notice that some values of the degrees of freedom for either the numerator or denominator are not in the table. If the number of degrees of freedom is not found in the table, we follow the practice of choosing the degrees of freedom closest to that desired. If the degrees of freedom is exactly between two values, find the mean of the values. For example, to find the critical value corresponding to 35 degrees of freedom in the numerator, compute the mean of the critical values corresponding to 30 and 40 degrees of freedom in the numerator.

> **Now Work Problem 3.**

2 ## Test Claims Regarding Two Population Standard Deviations

Now that we know the approximate sampling distribution of $\dfrac{s_1^2}{s_2^2}$ and how to find critical values in the F-distribution, we can introduce a procedure that can be used to test claims regarding two population standard deviations (or variances).

> ### Test Claims Regarding Two Population Standard Deviations
>
> If a claim is made regarding two population standard deviations, σ_1 and σ_2, we can use the following steps to test the claim provided:
>
> 1. The sample is obtained using simple random sampling.
> 2. The sample data are independent.
> 3. The populations from which the samples are drawn are normally distributed.

Step 1: A claim is made regarding two population standard deviations. The claim is used to determine the null and alternative hypotheses. The hypotheses can be structured in one of three ways:

Two-Tailed	Left-Tailed	Right-Tailed
$H_0: \sigma_1 = \sigma_2$	$H_0: \sigma_1 = \sigma_2$	$H_0: \sigma_1 = \sigma_2$
$H_1: \sigma_1 \neq \sigma_2$	$H_1: \sigma_1 < \sigma_2$	$H_1: \sigma_1 > \sigma_2$

Note: σ_1 is the population standard deviation for population 1, σ_2 is the population standard deviation for population 2.

Step 2: Select a level of significance α, depending on the seriousness of making a Type I error.

Step 3: Compute the test statistic

$$F_0 = \frac{s_1^2}{s_2^2}$$

which follows Fisher's F-distribution with $n_1 - 1$ degrees of freedom in the numerator and $n_2 - 1$ degrees of freedom in the denominator.

Classical Approach

Step 4: Use Table VII to determine the critical value(s) using $n_1 - 1$ degrees of freedom in the numerator and $n_2 - 1$ degrees of freedom in the denominator. The shaded regions represent the critical region.

P-value Approach

Step 4: Use technology to determine the P-value.

	Two-Tailed	Left-Tailed	Right-Tailed
Critical value(s)	$F_{1-\alpha/2,n_1-1,n_2-1}$ and $F_{\alpha/2,n_1-1,n_2-1}$	$F_{1-\alpha,n_1-1,n_2-1}$	F_{α,n_1-1,n_2-1}

Step 5: Compare the critical value with the test statistic.

Step 5: If P-value $< \alpha$, reject the null hypothesis.

Two-Tailed	Left-Tailed	Right-Tailed
If $F_0 < F_{1-\alpha/2,n_1-1,n_2-1}$ or $F_0 > F_{\alpha/2,n_1-1,n_2-1}$, reject the null hypothesis	If $F_0 < F_{1-\alpha,n_1-1,n_2-1}$, reject the null hypothesis	If $F_0 > F_{\alpha,n_1-1,n_2-1}$, reject the null hypothesis

Step 6: State the conclusion.

Because the procedures just presented are **not robust**, minor departures from normality will adversely affect the results of the test. Therefore, the test should be used only when the requirement of normality has been verified. We will verify this requirement by constructing normal probability plots.

EXAMPLE 2 Testing a Claim Regarding Two Population Standard Deviations

⚠ CAUTION
The test for equality of population standard deviations is not robust. Thus, any departures from normality make the results of the inference useless..

Problem: An investor believes that Cisco Systems is a more volatile stock than General Electric. The volatility of a stock is measured by the standard deviation rate of return on the stock. The data in Table 5 represent the monthly rate of return between 1990 and 2005 for 10 randomly selected months for Cisco Systems stock and 14 randomly selected months for General Electric stock.

Table 5

Monthly Rate of Return for Cisco Systems Stock (%)		Monthly Rate of Return for General Electric Stock (%)	
1.93	31.11	1.15	−1.92
11.64	4.87	4.68	−2.32
23.13	−5.56	3.23	−5.19
5.18	8.91	0.98	4.80
−7.60	9.72	−0.55	5.29
		−5.88	9.00
		−3.26	−0.60

Source: Yahoo! Finance

Test the investor's claim that Cisco Systems stock is more volatile than General Electric stock at the $\alpha = 0.05$ level of significance.

Approach: First we verify that both variables are normally distributed by constructing normal probability plots. We then follow Steps 1 through 6 to test the claim.

Solution: Figure 17(a) shows the normal probability plot for Cisco Systems, while Figure 17(b) shows the normal probability plot for General Electric.

Figure 17

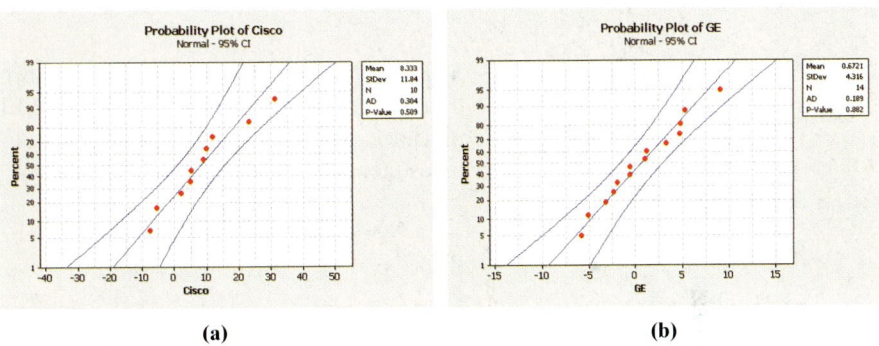

(a)   (b)

Both normal probability plots are linear, so the data come from a population that is normally distributed. We now follow Steps 1 through 6.

Step 1: The investor claims that Cisco Systems is a more volatile stock than General Electric. The standard deviation of the rate of return is a measure of a stock price's volatility. If we let σ_1 represent the standard deviation of the rate of return for Cisco Systems and σ_2 represent the standard deviation of the rate of return for General Electric, this claim can be written $\sigma_1 > \sigma_2$. We have

$$H_0: \sigma_1 = \sigma_2 \quad \text{versus} \quad H_1: \sigma_1 > \sigma_2$$

This is a right-tailed test.

Step 2: The level of significance is $\alpha = 0.05$.

Step 3: We compute the sample standard deviation of the rate of return for Cisco Systems, s_1, to be 11.84% and the sample standard deviation of the rate of return for General Electric, s_2, to be 4.32%. Because the data are normally distributed, the test statistic is

$$F_0 = \frac{s_1^2}{s_2^2} = \frac{11.84^2}{4.32^2} = 7.51$$

Classical Approach

Step 4: Because this is a right-tailed test, we determine the critical value at the $\alpha = 0.05$ level of significance with $n_1 - 1 = 10 - 1 = 9$ degrees of freedom in the numerator and $n_2 - 1 = 14 - 1 = 13$ degrees of freedom in the denominator and find it to be $F_{0.05,9,13} \approx 2.80$ (using 12 degrees of freedom in the denominator, because that is closest). The critical region is displayed in Figure 18.

Figure 18

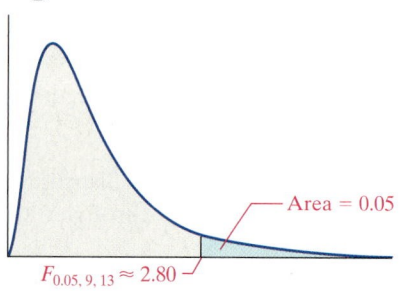

Area = 0.05

$F_{0.05, 9, 13} \approx 2.80$

Step 5: Because the test statistic $F_0 = 7.51$ is greater than the critical value $F_{0.05,9,13} \approx 2.80$, we reject the null hypothesis.

P-value Approach Using Technology

Step 4: We will use Excel to obtain the P-value. The steps for testing claims comparing two population standard deviations using the TI-83/84 Plus graphing calculators, MINITAB, and Excel are given in the Technology Step-by-Step on page 623. Figure 19. shows the results obtained from Excel.

Figure 19

F-Test Two-Sample for Variances

	Cisco	GE
Mean	8.333	0.67214286
Variance	140.082668	18.6297566
Observations	10	14
df	9	13
F	7.51929673	
P(F<=t) one-tail	0.00069363	
F Critical one-tail	2.71435852	

The P-value is 0.0007.

Step 5: Because the P-value is less than the level of significance $\alpha = 0.05$, we reject the null hypothesis.

Step 6: There is sufficient evidence to support the investor's claim that Cisco Systems stock is more volatile than General Electric stock at the $\alpha = 0.05$ level of significance. Investors would use this result to demand a higher rate of return on investments in Cisco Systems.

Now Work Problem 19.

EXAMPLE 3 **Testing a Claim Regarding Two Population Standard Deviations**

Problem: In the Spacelab Life Sciences 2 payload, 14 male rats were sent to space. Upon their return, the red blood cell mass (in milliliters) of the rats was determined. A control group of 14 male rats was held under the same conditions (except for spaceflight) as the space rats, and their red blood cell mass was likewise determined when the space rats returned. The project was led by Dr. Paul X. Callahan. The data in Table 6 were obtained.

Test the claim that the standard deviation of the red blood cell mass in the flight animals is different from the standard deviation of the red blood cell mass in the control animals at the $\alpha = 0.05$ level of significance.

Approach: We have already verified that the data are normally distributed, in Example 1 from Section 11.2. See Figure 5.

Step 1: The claim is that the standard deviation of the red blood cell mass in the flight animals is different from the standard deviation of the red blood cell mass in the control animals. If we let σ_1 represent the standard deviation of the red blood cell mass for the flight animals and σ_2 represent the standard deviation of the red blood cell mass for the control animals, this claim can be written $\sigma_1 \neq \sigma_2$. We have

$$H_0: \sigma_1 = \sigma_2 \quad \text{versus} \quad H_1: \sigma_1 \neq \sigma_2$$

Table 6

Flight		Control	
8.59	8.64	8.65	6.99
6.87	7.89	7.62	7.44
7.00	8.80	7.33	8.58
6.39	7.54	7.14	9.14
7.43	7.21	8.40	9.66
9.79	6.85	8.55	8.70
9.30	8.03	9.88	9.94

Source: NASA, Life Sciences Data Archive

This is a two-tailed test.

Step 2: The level of significance is $\alpha = 0.05$.

Step 3: In Example 1 from Section 11.2, we computed the sample standard deviation of the red blood cell mass for the experimental group, s_1, to be 1.017 and the sample standard deviation of the red blood cell mass for the control group, s_2, to be 1.005. Because the data are normally distributed, the test statistic is

$$F_0 = \frac{s_1^2}{s_2^2} = \frac{1.017^2}{1.005^2} = 1.024$$

Classical Approach

Step 4: Because this is a two-tailed test, we determine the critical value at the $\alpha = 0.05$ level of significance with $n_1 - 1 = 14 - 1 = 13$ degrees of freedom in the numerator and $n_2 - 1 = 14 - 1 = 13$ degrees of freedom in the denominator and find them to be $F_{0.025,13,13} \approx 3.28$ (using 12 degrees of freedom in the denominator, because that is closest) and

$$F_{0.975,13,13} = \frac{1}{F_{0.025,13,13}} = \frac{1}{3.28} = 0.30.$$

The critical regions are displayed in Figure 20.

Figure 20

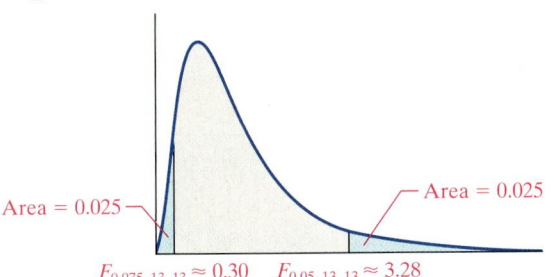

Area = 0.025

Area = 0.025

$F_{0.975, 13, 13} \approx 0.30$ $F_{0.05, 13, 13} \approx 3.28$

Step 5: Because the test statistic $F_0 = 1.024$ lies between the critical values 0.30 and 3.28; we do not reject the null hypothesis.

P-value Approach Using Technology

Step 4: We will use a TI-84 graphing calculator to obtain the P-value. The steps for testing claims comparing two population standard deviations using the TI-83/84 Plus graphing calculators, MINITAB, and Excel are given in the Technology Step-by-Step on page 623. Figure 21 shows the result obtained from the TI-84 Plus using the DRAW option.

Figure 21

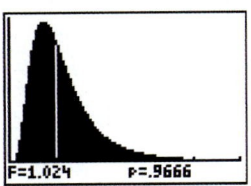

F=1.024   P=.9666

The P-value is 0.9666.

Step 5: Because the P-value is greater than the level of significance $\alpha = 0.05$, we do not reject the null hypothesis.

Step 6: There is not sufficient evidence to support the claim that the standard deviation of the red blood cell mass in the experimental group is significantly different from the red blood cell mass in the control group at the $\alpha = 0.05$ level of significance.

In-Class Activity: Stringing Them Along (Part III)

Do you believe there is more variability in the error (estimate minus actual) in the length of a longer rope or shorter rope? Continue the activity on page 593 by doing the following.

(a) Assuming that the errors in the longer and shorter ropes have equal variability, the null hypothesis is $H_0: \sigma_{\text{longer}} = \sigma_{\text{shorter}}$ (where σ represents the population standard deviation of absolute differences between the actual length and the estimated length). Based on your answer to the question posed at the beginning of the activity, select an appropriate alternative hypothesis and a level of significance.

(b) Retrieve the class data for Stringing Them Along (Part II) on page 593.

(c) Conduct the hypothesis test you outlined in part (a). What did you conclude?

(d) Are the two sets of data normally distributed? If not, how will that affect your results?

11.4 ASSESS YOUR UNDERSTANDING

Concepts and Vocabulary

1. Is the test for comparing two population standard deviations robust? Why is the assumption of normality so important to perform the test presented in this section?

2. State the assumptions that must be satisfied to compare two population standard deviations.

Skill Building

3. Find the critical value for a right-tailed test with $\alpha = 0.05$, NW degrees of freedom in the numerator = 9, and degrees of freedom in the denominator = 10.

4. Find the critical value for a right-tailed test with $\alpha = 0.01$, degrees of freedom in the numerator = 20, and degrees of freedom in the denominator = 25.

5. Find the critical values for a two-tailed test with $\alpha = 0.05$, degrees of freedom in the numerator = 6, and degrees of freedom in the denominator = 8.

6. Find the critical values for a two-tailed test with $\alpha = 0.02$, degrees of freedom in the numerator = 5, and degrees of freedom in the denominator = 7.

7. Find the critical value for a left-tailed test with $\alpha = 0.10$, degrees of freedom in the numerator = 25, and degrees of freedom in the denominator = 20.

8. Find the critical value for a left-tailed test with $\alpha = 0.01$, degrees of freedom in the numerator = 15, and degrees of freedom in the denominator = 20.

9. Find the critical value for a right-tailed test with $\alpha = 0.05$, degrees of freedom in the numerator = 45, and degrees of freedom in the denominator = 15.

10. Find the critical value for a right-tailed test with $\alpha = 0.01$, degrees of freedom in the numerator = 55, and degrees of freedom in the denominator = 50.

In Problems 11–16, assume the populations are normally distributed.

11. Test the claim that $\sigma_1 \neq \sigma_2$ at the $\alpha = 0.05$ level of significance for the given sample data.

	Population 1	Population 2
n	16	16
s	3.2	3.5

12. Test the claim that $\sigma_1 \neq \sigma_2$ at the $\alpha = 0.1$ level of significance for the given sample data.

	Population 1	Population 2
n	21	21
s	8.6	9.2

13. Test the claim that $\sigma_1 > \sigma_2$ at the $\alpha = 0.01$ level of significance for the given sample data.

	Population 1	Population 2
n	26	19
s	9.9	6.4

14. Test the claim that $\sigma_1 < \sigma_2$ at the $\alpha = 0.05$ level of significance for the given sample data.

	Population 1	Population 2
n	21	26
s	15.9	23.0

15. Test the claim that $\sigma_1 < \sigma_2$ at the $\alpha = 0.1$ level of significance for the given sample data.

	Population 1	Population 2
n	51	26
s	8.3	13.2

16. Test the claim that $\sigma_1 > \sigma_2$ at the $\alpha = 0.05$ level of significance for the given sample data.

	Population 1	Sample for Population 2
n	23	13
s	7.5	5.1

Applying the Concepts

17. **Treating Bipolar Mania** In a study published in the *Archives of General Psychiatry* entitled Efficacy of Olanzapine in Acute Bipolar Mania (Vol. 57, No. 9, pp. 841–849), researchers conducted a randomized, double-blind study to measure the effects of the drug olanzapine on patients diagnosed with bipolar disorder. One hundred fifteen patients with a DSM-IV diagnosis of bipolar disorder were randomly divided into two groups. Group 1 ($n = 55$) received 5 to 20 mg per day of olanzapine, while Group 2 ($n = 60$) received a placebo. The effectiveness of the drug was measured by the Young–Mania Rating Scale total score, with the net improvement in the score recorded. The results are presented in the following table:

	Treatment Group	Control Group
n	55	60
Mean improvement	14.8	8.1
Sample standard deviation	12.5	12.7

Assuming that the data are normally distributed, test the claim that the standard deviation in the treatment group is different from the standard deviation in the control group at the $\alpha = 0.05$ level of significance.

18. **Treating the Common Cold** Researcher Steven J. Sperber, MD, and his associates wanted to determine the effectiveness of pseudoephedrine and acetaminophen in the treatment of discomfort associated with the common cold. They published their results in the *Archives of Family Medicine 2000* 9: 979–985. In their study, they randomly divided 430 subjects into two groups: Group 1 ($n = 216$) received pseudoephedrine and acetaminophen; Group 2 ($n = 214$) received a placebo. The goal of the study was to discover whether the mean symptom assessment score of the individuals receiving the treatment (Group 1) was less than that of the control group (Group 2). In Group 1, the mean reduction in the symptom assessment scores was 1.30, with a standard deviation of 0.88; in Group 2, participants had a mean reduction in the symptom assessment score of 0.93, with a standard deviation of 0.88. Assuming that the data are normally distributed, test the claim that the standard deviation of the symptom assessment score of the subjects in Group 1 is different from the standard deviation of the symptom assessment score for the subjects in Group 2 at the $\alpha = 0.05$ level of significance.

19. **Vitamin A Supplements in Low-Birth-Weight Babies** Low-birth-weight babies are at increased risk of respiratory infections in the first few months of life and have low liver stores of vitamin A. In a randomized, double-blind experiment, 130 low-birth-weight babies were randomly divided into two groups. Subjects in Group 1 (the treatment group, $n_1 = 65$) were given 25,000 IU of vitamin A on study days 1, 4, and 8, where study day 1 was between 36 and 60 hours after delivery. Subjects in Group 2 (the control group, $n_2 = 65$) were given a placebo. The treatment group had a mean serum retinol concentration 45.77 μg/dL, with a standard deviation of 17.07 μg/dL. The control group had a mean serum retinol concentration of 12.88 μg/dl, with a standard deviation of 6.48 μg/dl. Test the claim that the treatment group had a higher standard deviation for serum retinol concentration than did the control group at the $\alpha = 0.01$ level of significance. It is known that serum retinol concentration is normally distributed.

20. **SAT Test Scores** A researcher wants to know if students who do not plan to apply for financial aid had more variability on the SAT I math test than those who do plan to do so. She obtains a random sample of 35 students who do not plan to apply for financial aid and a random sample of 38 students who do plan to apply for financial aid and obtains the following results:

Do Not Plan to Apply for Financial Aid	Plan to Apply for Financial Aid
$n_1 = 35$	$n_2 = 38$
$s_1 = 123.1$	$s_2 = 119.4$

Test the claim that students who do not plan to apply for financial aid have a higher standard deviation on the SAT I math exam than do students who do plan to apply for financial aid at the $\alpha = 0.01$ level of significance. SAT I math exam scores are known to be normally distributed.

21. **Waiting Time in Line** McDonald's executives want to experiment with redesigning its restaurants so that the customers form one line leading to four registers to place orders, rather than four lines leading to four separate registers. They redesign 30 randomly selected restaurants with the single line. In addition, they randomly select 30 restaurants with the four-line configuration to participate in the study. At each restaurant, an employee monitors the wait time (in minutes) of randomly selected patrons. The following data are collected.

Single Line			
1.2	2.1	1.7	2.8
1.9	2.1	2.4	3.0
2.1	2.3	1.9	2.9
2.7	2.0	2.9	2.3
2.8	1.1	3.1	1.8

Multiple Lines			
1.1	1.6	2.9	4.6
3.8	2.9	2.8	2.9
4.3	2.3	2.0	2.6
1.3	1.3	2.0	2.7
2.0	3.2	2.3	0.9

(a) Test the claim that the standard deviation for wait time in the single line is less than the standard deviation for wait time in the multiple lines at the $\alpha = 0.05$ level of significance. **Note:** Normal probability plots indicate the data are normally distributed.

(b) Draw boxplots of each data set to confirm the results of part (a) visually.

22. **Filling Machines** A quality-control engineer wants to find out whether or not a new machine that fills bottles with liquid has less variability than the machine currently in use. The engineer calibrates each machine to fill bottles with 16 ounces of a liquid. After running each machine for 5 hours, she randomly selects 15 filled bottles from each machine and measures their contents. She obtains the following results:

	Old Machine	
16.01	16.04	15.96
16.00	16.07	15.89
16.04	16.05	15.91
16.10	16.01	16.00
15.92	16.16	15.92

	New Machine	
16.02	15.96	16.05
15.95	15.99	16.02
16.00	15.97	16.03
16.06	16.05	15.94
16.08	15.96	15.95

(a) Test the claim that the standard deviation of fill in the new machine is less than the standard deviation of fill in the old machine at the $\alpha = 0.05$ level of significance. **Note:** Normal probability plots indicate the data are normally distributed.

(b) Draw boxplots of each data set to confirm the results of part (a) visually.

23. Systolic Blood Pressure A nurse was interested in discovering whether men have more variability in their systolic blood pressure than women. She randomly selects 20 males and 17 females from the surgical floor of her hospital and records their systolic blood pressures. The data are as follows:

	Females		
140	124	126	106
126	132	126	140
136	120	110	122
118	130	118	114
140			

Source: Lora McGuire, nursing instructor, Joliet Junior College.

	Males		
112	134	118	112
140	105	134	128
148	95	124	140
120	130	150	112
126	116	128	142

(a) Given the following normal probability plots, comment on whether the requirement of normality required for the F-test is satisfied and justify your answer.

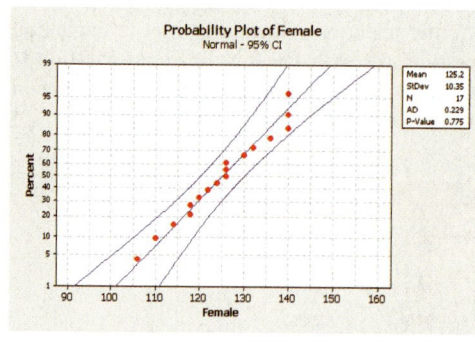

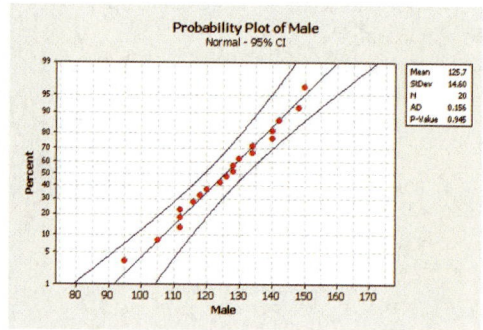

(b) The nurse enters the data into MINITAB to test the claim that males have a higher standard deviation of systolic blood pressure than females at the $\alpha = 0.05$ level of significance. She obtains a P-value of 0.0844. What should the nurse conclude?

(c) Compute the test statistic, and use Table VII in the back of the text to verify that the P-value computed is correct.

24. Measuring Reaction Time Researchers at the University of Mississippi wanted to discover whether the standard deviation for reaction time to a go/no go stimulus of males differed from that of females. The researchers randomly selected 20 females and 15 males to participate in the study. The go/no go stimulus required the student to respond to a particular stimulus and not to respond to other stimuli. The data are as follows:

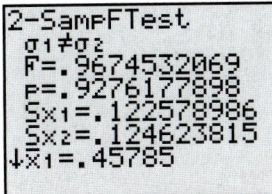

Female Students						
0.588	0.403	0.293	0.377	0.613	0.377	0.391
0.367	0.442	0.274	0.434	0.403	0.636	0.481
0.652	0.443	0.380	0.646	0.340	0.617	

Male Students						
0.375	0.477	0.374	0.465	0.402	0.337	0.655
0.488	0.427	0.373	0.224	0.654	0.563	0.405
0.256						

Source: PsychExperiments at the University of Mississippi (www.olemiss.edu/psychexps) are as follows:

```
2-SampFTest
  σ1≠σ2
  F=.9674532069
  P=.9276177898
  Sx1=.122578986
  Sx2=.124623815
↓x̄1=.45785
```

Normal probability plots indicate the requirement of normality is satisfied. The output to the left is from a TI-84 Plus.

(a) Use the results to test the claim that there is a difference between the standard deviations in reaction time of males and females at the $\alpha = 0.05$ level of significance.

(b) Draw boxplots of each data set, using the same scale. Does this visual evidence support the results obtained in part (a)?

Technology Step-by-Step | **Comparing Two Population Standard Deviations**

TI-83/84 Plus
Step 1: If necessary, enter raw data in L1 and L2.
Step 2: Press STAT, highlight TESTS, and select `D:2-SampleFTest`.
Step 3: If the data are raw, highlight DATA and make sure List1 is set to L1 and List2 is set to L2, with frequencies set to 1. If summary statistics are known, highlight STATS and enter the summary statistics.
Step 4: Highlight the appropriate relation between σ_1 and σ_2 in the alternative hypothesis.
Step 5: Highlight Calculate or Draw and press ENTER. Calculate gives the test statistic and *P*-value. Draw draws the *F*-distribution with the *P*-value shaded.

MINITAB
Step 1: Enter data in column C1 and subscripts (usually 1 and 2) in C2 so as to identify the population of the data in C1.
Step 2: Select the **Stat** menu, highlight **ANOVA**, then highlight **Homogeneity of Variance** . . . .
Step 3: Enter C1 in the cell marked "Response" and enter C2 in the cell marked "Factors." Select the confidence level. Click OK.

Excel
Step 1: Enter raw data in columns A and B.
Step 2: Select the **Tools** menu; highlight **Data Analysis** . . . .
Step 3: Select "*F*-test Two-Sample for Variances." With the cursor in the "Variable 1 Range" cell, highlight the data in column A. With the cursor in the "Variable 2 Range" cell and highlight the data in column B. Enter a value for alpha. Click OK.

Chapter **11** **Review**

Summary

This chapter discussed performing statistical inference by comparing two population parameters. We began with a discussion regarding the comparison of two population means. To determine the method to use, we must know whether the sampling was dependent or independent. A sampling method is independent when the choice of individuals for one sample does not dictate which individuals will be in a second sample. A sampling method is dependent when the individuals selected for one sample are used to determine the individuals in the second sample. For dependent sampling, we use the paired *t*-test to perform statistical inference. For independent sampling, we use Welch's two-sample *t*. For both tests, the population must be normally distributed or the sample sizes must be large.

Section 11.3 dealt with statistical inference for comparing two population proportions. To perform these tests, $n\hat{p}(1 - \hat{p})$ must be greater than or equal to 10 for each population; and each sample size could be no more than 5% of the population size.then distribution of $\hat{p}_1 - \hat{p}_2$ is approximately normal, with mean $p_1 - p_2$ and standard deviation

$$\sqrt{\frac{p_1(1 - p_1)}{n_1} + \frac{p_2(1 - p_2)}{n_2}}.$$

We wrapped up the chapter with a discussion on inference for comparing two population standard deviations. We use the F-test, but require that both populations be normally distributed. This test is not robust, so if the data show any departures from normality, the test should not be used.

To help determine which test to use, we include the flow chart in Figure 22.

Figure 22

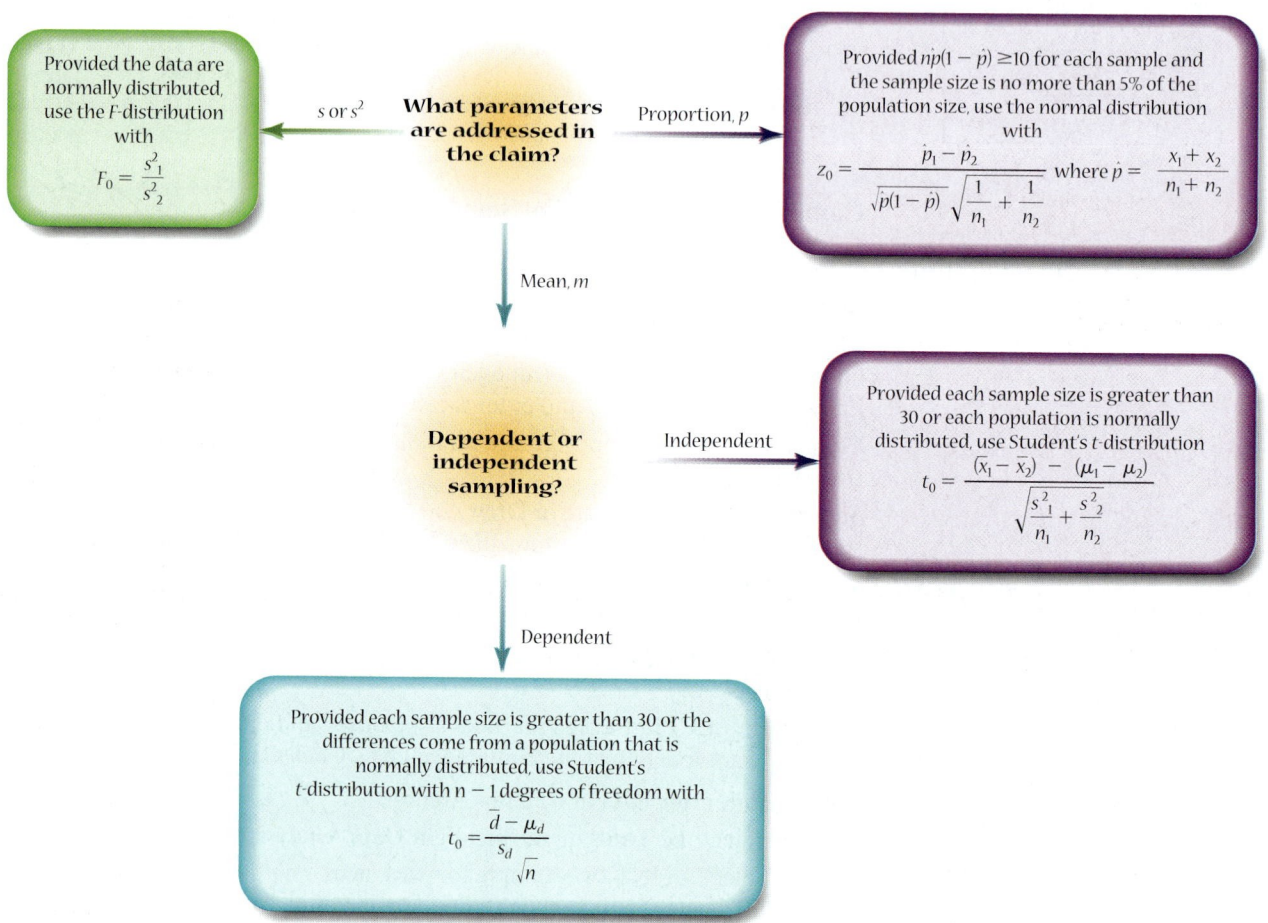

Formulas

- Test statistic for matched-pairs data:

$$t_0 = \frac{\bar{d}}{\frac{s_d}{\sqrt{n}}}.$$

where $\bar{d}$ is the mean and s_d is the standard deviation of the differenced data

- Confidence interval for matched-pairs data:

Lower bound: $\bar{d} - t_{\alpha/2} \cdot \dfrac{s_d}{\sqrt{n}}$

Upper bound: $\bar{d} + t_{\alpha/2} \cdot \dfrac{s_d}{\sqrt{n}}$

- Test statistic comparing two means (independent sampling):

$$t_0 = \frac{(\bar{x}_1 - \bar{x}_2) - (\mu_1 - \mu_2)}{\sqrt{\frac{s_1^2}{n_1} + \frac{s_2^2}{n_2}}}$$

- Confidence interval for the difference of two means (independent samples):

Lower bound: $(\bar{x}_1 - \bar{x}_2) - t_{\alpha/2} \cdot \sqrt{\dfrac{s_1^2}{n_1} + \dfrac{s_2^2}{n_2}}$

Upper bound: $(\bar{x}_1 - \bar{x}_2) + t_{\alpha/2} \cdot \sqrt{\dfrac{s_1^2}{n_1} + \dfrac{s_2^2}{n_2}}$

- Test statistic comparing two population proportions:

$$z_0 = \frac{\hat{p}_1 - \hat{p}_2}{\sqrt{\hat{p}(1 - \hat{p})}\sqrt{\dfrac{1}{n_1} + \dfrac{1}{n_2}}}$$

where $\hat{p} = \dfrac{x_1 + x_2}{n_1 + n_2}$.

- Confidence interval for the difference of two proportions:

Lower bound: $(\hat{p}_1 - \hat{p}_2) - z_{\alpha/2} \cdot \sqrt{\dfrac{\hat{p}_1(1 - \hat{p}_1)}{n_1} + \dfrac{\hat{p}_2(1 - \hat{p}_2)}{n_2}}$

Upper bound: $(\hat{p}_1 - \hat{p}_2) + z_{\alpha/2} \cdot \sqrt{\dfrac{\hat{p}_1(1 - \hat{p}_1)}{n_1} + \dfrac{\hat{p}_2(1 - \hat{p}_2)}{n_2}}$

- Sample size for estimating $p_1 - p_2$

$$n = n_1 = n_2 = [\hat{p}_1(1 - \hat{p}_1) + \hat{p}_2(1 - \hat{p}_2)]\left(\frac{z_{\alpha/2}}{E}\right)^2$$

$$n = n_1 = n_2 = 0.5\left(\frac{z_{\alpha/2}}{E}\right)^2$$

- Finding a critical F for the left tail:

$$F_{1-\alpha,n_1-1,n_2-1} = \frac{1}{F_{\alpha,n_2-1,n_1-1}}$$

- Test statistic for comparing two population standard deviations:

$$F_0 = \frac{s_1^2}{s_2^2}$$

Vocabulary

Dependent sampling (p. 575)
Independent sampling (p. 575)
Matched pairs (p. 575)

robust (p. 577)
Welch's approximate t (p. 588)
Pooled t-statistic (p. 594)

Pooled estimate of p (p. 603)
F-distribution (p. 613)

Objectives

Section	You should be able to ...	Examples	Review Exercises
11.1	**1** Distinguish between independent and dependent sampling (p. 575)	1	1–4, 15(a), 16(a), 17(a), 18(a)
	2 Test claims regarding matched-pairs data (p. 576)	2 and 3	7(c), 8(c), 15(b), 18(b)
	3 Construct and interpret confidence intervals about the population mean difference of matched-pairs data (p. 581)	4	18(c), 25
11.2	**1** Test claims regarding the difference of two independent means (p. 589)	1 and 2	9(a), 10(a), 11(a), 12(a), 16(b), 17(b)
	2 Construct and interpret confidence intervals regarding the difference of two independent means (p. 593)	3	26, 27
11.3	**1** Test claims regarding two population proportions (p. 602)	1 and 2	13, 14, 19(a), 20(a)
	2 Construct and interpret confidence intervals for the difference between two population proportions (p. 606)	3	19(b), 20(b)
	3 Determine the sample size necessary for estimating the difference between two population proportions within a specified margin of error (p. 608)	4	21, 22
11.4	**1** Find critical values of the F-distribution (p. 612)	1	5, 6
	2 Test claims regarding two population standard deviations (p. 615)	2 and 3	23, 24

Review Exercises

In Problems 1–4, determine if the sampling is dependent or independent.

1. A researcher wants to know if the mean length of stay in for-profit hospitals is different from the mean length of stay in not-for-profit hospitals. He randomly selected 20 individuals in the for-profit hospital and matched them with 20 individuals in the not-for-profit hospital by diagnosis.

2. An urban economist believes that commute times to work in the South are less than commute times to work in the Midwest. He randomly selects 40 employed individuals in the South and 45 employed individuals in the Midwest and determines their commute times.

3. A stock analyst wants to know if there is a difference between the mean rate of return from energy stocks and that from financial stocks. He randomly selects 13 energy stocks and computes the rate of return for the past year. He randomly selects 13 financial stocks and computes the rate of return for the past year.

4. A prison warden wants to know if men receive longer sentences for crimes than women. He randomly samples 30 men and matches them with 30 women by type of crime committed and records their lengths of sentence.

5. (a) Find the critical F-value for a right-tailed test with $\alpha = 0.05$, degrees of freedom in the numerator $= 8$, and degrees of freedom in the denominator $= 9$.
  (b) Find the critical F-value for a two-tailed test with $\alpha = 0.05$, degrees of freedom in the numerator $= 10$, and degrees of freedom in the denominator $= 5$.

6. (a) Find the critical F-value for a right-tailed test with $\alpha = 0.1$, degrees of freedom in the numerator $= 20$, and degrees of freedom in the denominator $= 12$.
  (b) Find the critical F-value for a left-tailed test with $\alpha = 0.1$, degrees of freedom in the numerator $= 12$, and degrees of freedom in the denominator $= 20$.

In Problems 7 and 8, assume that the paired data came from a population that is normally distributed.

7.

Observation	1	2	3	4	5	6
X_i	34.2	32.1	39.5	41.8	45.1	38.4
Y_i	34.9	31.5	39.5	41.9	45.5	38.8

  (a) Compute $d_i = X_i - Y_i$ for each pair of data.
  (b) Compute $\bar{d}$ and s_d.
  (c) Test the claim that $\mu_d < 0$ at the $\alpha = 0.05$ level of significance.
  (d) Compute a 98% confidence interval about the population mean difference μ_d.

8.

Observation	1	2	3	4	5	6	7
X_i	18.5	21.8	19.4	22.9	18.3	20.2	23.1
Y_i	18.3	22.3	19.2	22.3	18.9	20.7	23.9

  (a) Compute $d_i = X_i - Y_i$ for each pair of data.
  (b) Compute $\bar{d}$ and s_d.
  (c) Test the claim that $\mu_d \neq 0$ at the $\alpha = 0.01$ level of significance.
  (d) Compute a 95% confidence interval about the population mean difference μ_d.

In Problems 9–12, assume that the populations are normally distributed and that independent sampling occurred.

9.

	Population 1	Population 2
n	13	8
$\bar{x}$	32.4	28.2
s	4.5	3.8

  (a) Test the claim that $\mu_1 \neq \mu_2$ at the $\alpha = 0.1$ level of significance for the given sample data.
  (b) Construct a 90% confidence interval about $\mu_1 - \mu_2$.
  (c) Test the claim that $\sigma_1 \neq \sigma_2$ at the $\alpha = 0.05$ level of significance for the given sample data.

11.

	Population 1	Population 2
n	45	41
$\bar{x}$	48.2	45.2
s	8.4	10.3

  (a) Test the claim that $\mu_1 > \mu_2$ at the $\alpha = 0.01$ level of significance for the given sample data.
  (b) Construct a 90% confidence interval about $\mu_1 - \mu_2$.
  (c) Test the claim that $\sigma_1 < \sigma_2$ at the $\alpha = 0.01$ level of significance for the given sample data.

10.

	Population 1	Population 2
n	24	27
$\bar{x}$	104.2	110.4
s	12.3	8.7

  (a) Test the claim that $\mu_1 \neq \mu_2$ at the $\alpha = 0.1$ level of significance for the given sample data.
  (b) Construct a 95% confidence interval about $\mu_1 - \mu_2$.
  (c) Test the claim that $\sigma_1 > \sigma_2$ at the $\alpha = 0.1$ level of significance for the given sample data.

12.

	Population 1	Population 2
n	13	8
$\bar{x}$	96.6	98.3
s	3.2	2.5

  (a) Test the claim that $\mu_1 < \mu_2$ at the $\alpha = 0.05$ level of significance for the given sample data.
  (b) Construct a 99% confidence interval about $\mu_1 - \mu_2$.
  (c) Test the claim that $\sigma_1 \neq \sigma_2$ at the $\alpha = 0.05$ level of significance for the given sample data.

In Problems 13 and 14, test the claim at the $\alpha = 0.05$ level of significance by (a) determining the null and alternative hypotheses, (b) computing the test statistic, (c) computing the critical value, and (d) computing the P-value. Assume that the samples were obtained independently by simple random sampling.

13. Claim $p_1 \neq p_2$. Sample data: $x_1 = 451$, $n_1 = 555$, $x_2 = 510$, $n_2 = 600$

14. Claim $p_1 < p_2$. Sample data: $x_1 = 156$, $n_1 = 650$, $x_2 = 138$, $n_2 = 540$

15. Height versus Arm Span A statistics student claimed that an individual's arm span is equal to the individual's height. To test this claim, the student used a random sample of 10 students and obtained the following data.

Student:	1	2	3	4	5
Height (inches)	59.5	69	77	59.5	74.5
Arm span (inches)	62	65.5	76	63	74

Student:	6	7	8	9	10
Height (inches)	63	61.5	67.5	73	69
Arm span (inches)	66	61	69	70	71

Source: John Climent, Cecil Community College

(a) Is the sampling method dependent or independent? Why?

(b) Test the claim that an individual's height and arm span are the same at the $\alpha = 0.05$ level of significance. **Note:** A normal probability plot indicates that the data and differenced data are normally distributed. A boxplot indicates that the data and differenced data have no outliers.

16. Acid Rain A researcher wants to know whether the acidity of rain (pH) near Houston, Texas, is significantly different from that near Chicago, Illinois. He randomly selects 12 rain dates in Texas and 14 rain dates in Illinois and obtains the following data:

Texas

4.69	5.10	5.22	4.46	4.93	4.65
5.22	4.76	4.25	5.14	4.11	4.71

Illinois

4.40	4.69	4.22	4.64	4.54	4.35	4.69
4.40	4.75	4.63	4.45	4.49	4.36	4.52

Source: National Atmospheric Deposition Program

(a) Is the sampling method dependent or independent? Why?

(b) Test the claim that there is no difference between the acidity of rain in Houston and that in Chicago at the $\alpha = 0.05$ level of significance.

(c) Draw boxplots of each data set, using the same scale. Does this visual evidence support the results obtained in part (b)?

17. McDonald's versus Wendy's A student wanted to determine whether the wait time in the drive-through at McDonald's differed from that at Wendy's. She used a random sample of 30 cars at McDonald's and 27 cars at Wendy's and obtained these results:

Wait Time at McDonald's Drive-Through (in seconds)				
151.09	227.38	111.84	131.21	128.75
191.60	126.91	137.90	195.44	246.59
141.78	127.35	121.21	101.03	95.09
122.06	122.62	100.04	71.37	153.34
140.44	126.62	116.72	131.69	100.94
115.66	147.28	81.43	86.31	156.34

Wait Time at Wendy's Drive-Through (in seconds)				
281.90	71.02	204.29	128.59	133.56
187.53	199.86	190.91	110.55	110.64
196.84	233.65	171.01	182.54	183.79
284.48	363.34	270.82	390.50	471.62
123.66	174.43	385.90	386.71	155.53
203.62	119.61			

Source: Catherine M. Simmons, student at Joliet Junior College

(a) Is the sampling method dependent or independent?

(b) Test the claim that there is a difference in wait times at each restaurant's drive-through at the $\alpha = 0.1$ level of significance.

(c) Draw boxplots of each data set using the same scale. Does this visual evidence support the results obtained in part (b)?

18. Gastric Bypass Laparoscopic gastric bypass surgery reduces the size of the stomach and can cause malabsorption of nutrients, leading to weight loss according to webmd.com. Researchers wanted to test the claim that body mass index (BMI) reduction is significantly greater than zero 10 months after surgery. They selected 110 patients who elected to have laparoscopic gastric bypass surgery and measured their BMI prior to surgery. After 10 months they again measured the BMI. The mean reduction in BMI in the 110 patients was found to be 12.61 with a standard deviation of 4.90 [*Source:*St. Peter, Shawn D., et al., Impact of Advanced Age on Weight Loss and Health Benefits after Laparoscopic Gastric Bypass, *Archives of Surgery,* Vol. 140, No. 2, February, 2005].

(a) What type of experimental design is this? Why?

(b) Test the claim that the BMI reduction is greater than zero at the $\alpha = 0.05$ level of significance.

(c) Construct a 90% confidence interval for the mean BMI reduction after laparoscopic gastric bypass surgery. Write a sentence that a doctor could read to his or her patients that explains the expected BMI reduction 10 months after the surgery.

19. Treatment for Osteoporosis Osteoporosis is a condition in which people experience decreased bone mass and an increase in the risk of bone fracture. Actonel is a drug that helps combat osteoporosis in postmenopausal women. In clinical trials, 1374 postmenopausal women were randomly divided into experimental and control groups. The subjects in the experimental group were administered 5 mg of Actonel, while the subjects in the control group were administered a placebo. The number of women who experienced a bone fracture over the course of 1 year was recorded. Of the 696 women in the experimental group, 27 experienced a fracture during the course of the year. Of the 678 women in the control group, 49 experienced a fracture during the course of the year.

(a) Test the claim that a lower proportion of women in the experimental group experienced a bone fracture than the women in the control group at the $\alpha = 0.01$ level of significance.

(b) Construct a 95% confidence interval for the difference between the two population proportions, $p_{exp} - p_{control}$.

(c) What type of experimental design is this? What is the treatment? How many levels does it have?

20. Zoloft Zoloft is a drug that is used to treat obsessive–compulsive disorder (OCD). In randomized, double-blind clinical trials, 926 patients diagnosed with OCD were randomly divided into two groups. Subjects in Group 1 (experimental group) received 200 mg per day of Zoloft, while subjects in Group 2 (control group) received a placebo. Of the 553 subjects in the experimental group, 77 experienced dry mouth as a side effect. Of the 373 subjects in the control group, 34 experienced dry mouth as a side effect.

(a) Test the claim that a higher proportion of the subjects in the experimental group experienced dry mouth than did the subjects in the control group at the $\alpha = 0.05$ level of significance.

(b) Construct a 90% confidence interval for the difference between the two population proportions, $p_1 - p_2$.

21. Determining Sample Size A nutritionist wants to estimate the difference between the percentage of men and women who have high cholesterol. What sample size should be obtained if she wishes the estimate to be within 2 percentage points with 90% confidence, assuming

(a) that she uses the 1994 estimates of 18.8% male and 20.5% female from the National Center for Health Statistics?

(b) that she does not use any prior estimates?

22. Determining Sample Size A researcher wants to estimate the difference between the percentage of individuals without a high school diploma who smoke and the percentage of individuals with bachelor's degrees who smoke. What sample size should be obtained if she wishes the estimate to be within 4 percentage points with 95% confidence, assuming

(a) that she uses the 1999 estimates of 32.2% of those without a high school diploma and 11.1% of those with a bachelor's degree, from the National Center for Health Statistics?

(b) that she does not use any prior estimates?

23. Wait Time Using the data from Problem 17, test the claim that the standard deviation of wait time at Wendy's is more than that at McDonald's at the $\alpha = 0.05$ level of significance.

24. Acid Rain Using the data from Problem 16, test the claim that the standard deviation of acidity in the rain near Houston, Texas, is different from that near Chicago, Illinois, at the $\alpha = 0.05$ level of significance.

25. Height versus Arm Span Construct and interpret a 95% confidence interval about the population mean difference between height and arm span using the data from Problem 15. What does the interval lead us to conclude regarding any differences between height and arm span?

26. Acid Rain Construct and interpret a 90% confidence interval about $\mu_T - \mu_I$ using the data from Problem 16.

27. McDonald's versus Wendy's Construct and interpret a 95% confidence interval about $\mu_M - \mu_W$ using the data from Problem 17. How might a marketing executive with McDonald's use this information?

28. Explain when the matched-pairs t should be used instead of Welch's t in comparing two population means. What are some advantages in designing a matched-pairs experiment versus using Welch's t?

CASE STUDY

Control in the Design of an Experiment

Dr. Penelope Nicholls is interested in exploring a possible connection between high plasma homocysteine (a toxic amino acid created by the body as it metabolizes protein) levels and cardiac hypertrophy (enlargement of the heart) in humans. Because there are many complex relationships among human characteristics, it will be difficult to answer her research question because there is a significant risk that confounding factors will cloud her inferences. She wants to be sure that any differences in cardiac hypertrophy are due to high plasma homocysteine levels and not to other factors. Consequently, she needs to design her experiment carefully so that she controls lurking variables to the extent possible. Therefore she decides to design a two-sample experiment with independent sampling: one of the groups will be the experimental group, the other a control group. Knowing that many factors can affect the degree of cardiac hypertrophy (the response variable), Dr. Nicholls controls these factors by randomly assigning the experimental units to the experimental or control group. She hopes the randomization will result in both groups having similar characteristics.

In her preliminary literature review, Dr. Nicholls uncovered an article in which the authors hypothesized that there might be a relationship between high plasma homocysteine levels in patients with end-stage renal disease (ESRD) and cardiac hypertrophy. She has asked you, as her assistant, to review this article.

Upon reading the article, you discover that the authors used a nonrandom process to select a control and an ESRD group. The researchers enlisted 75 stable ESRD patients into their study. All these patients were on hemodialysis for between 6 and 312 months. The control group subjects were chosen so as to eliminate any intergroup differences in terms of mean blood pressure (BP) and gender. In an effort to minimize situational contaminants, all physical and biochemical measurements were made after an overnight fast. The results of the clinical characteristics and the biochemical findings for the control and ESRD groups are reproduced in the following tables:

Clinical Characteristics (Mean ± Standard Deviation)

Parameters	Controls ($n = 57$)	ESRD Subjects ($n = 75$)
Age (years)	49.2 ± 14.7	57.3 ± 15.1
Sex (M/F ratio)	1.4 ± 0.50	1.4 ± 0.50
Body surface area (m^2)	1.85 ± 0.25	1.67 ± 0.20
Body mass index (kg/m^2)	26.0 ± 4.70	23.7 ± 3.90
Systolic BP (mmHg)	145.0 ± 15.5	148.8 ± 29.7
Diastolic BP (mmHg)	85.0 ± 14.8	80.2 ± 14.3
Mean BP (mmHg)	104.5 ± 14.2	103.6 ± 17.4
Pulse pressure (mmHg)	59.4 ± 15.5	68.6 ± 24.3
Heart rate (beats/min)	63.0 ± 8.0	70.0 ± 9.0

Biological Findings (Mean ± Standard Deviation)

Parameters	Controls ($n = 57$)	ESRD Subjects ($n = 75$)
Total cholesterol (mmol/L)	5.28 ± 1.04	4.91 ± 1.06
HDL cholesterol (mmol/L)	1.38 ± 0.39	1.07 ± 0.38
Triglycerides (mmol/L)	1.39 ± 0.63	1.90 ± 1.02
Serum albumin (g/L)	44.7 ± 2.60	39.9 ± 3.00
Plasma fibrinogen (g/L)	3.21 ± 0.78	4.75 ± 1.04
Plasma creatinine (mmol/L)	0.10 ± 0.01	0.90 ± 0.13
Blood urea (mmol/L)	6.10 ± 1.20	24.3 ± 2.00
Calcium (mmol/L)	2.46 ± 0.08	2.45 ± 0.12
Phosphates (mmol/L)	1.03 ± 0.21	1.88 ± 0.38

Source: Jacques Blacher et al., Association between Plasma Homocysteine Concentrations and Cardiac Hypertrophy in End-Stage Renal Disease. *Journal of Nephrology* 12, 4 (July–August, 1999):248–255. Article available at http://www.sin_italia.org/jnonline/vol12n4/blacher/blacher.htm.

Which type of sampling method, independent or dependent, was used in this experiment? Explain.

Using the appropriate hypothesis-testing procedure, determine whether the control and ESRD groups have equivalent population means for each of the various clinical and biochemical parameters. Dr. Nicholls requires that you indicate those parameters that have *P*-values less than 0.05 and those less than 0.01.

Detail any assumptions and the rationale behind making them that you made while carrying out your analysis. Is there any additional information that you would like to have? Explain. Are there any additional statistical procedures that you think might be useful for analyzing these data? Explain.

Based on your findings, does it appear that the control and ESRD groups have similar initial clinical characteristics and biochemical findings? Does it appear that the authors of this article were successful in reducing the likelihood that a confounding effect would obscure their results?

Even though Dr. Nicholls does not wish to restrict her research to patients with end-stage renal disease, how might the information presented for this research assist her in designing her own experiment?

Submit a written report to Dr. Nicholls that outlines all your findings and recommendations.

Inference on Categorical Data

Outline

DECISIONS

Are there benefits to attending college? If so, what are they? See the Decisions project on page 661.

●●● Putting It All Together

In Chapters 9–11, we introduced statistical methods that can be used to test claims regarding a population parameter such as μ or σ.

Often, however, rather than being interested in testing a claim regarding a parameter of a probability distribution, we are interested in testing a claim regarding the entire probability distribution. For example, we might wish to test the claim that the distribution of colors in a bag of plain M&M candies is 13% brown, 14% yellow, 13% red, 20% orange, 24% blue, and 16% green. We introduce methods for testing claims such as this in Section 12.1.

Remember the correlation coefficient? It is a measure of the strength of a linear relation between two quan-titative variables. In Section 12.2, we introduce methods that can be used to get a feel for the strength of a relation between two qualitative variables for population data.

In Section 12.3, we discuss a method that can be used to determine whether two variables are independent based on a sample. If they are not independent, the value of one variable impacts the value of the other variable, so the variables are somehow related. We conclude Section 12.3 by introducing tests for homogeneity. This procedure is used to compare proportions from two or more populations. It is an extension of the two-sample z-test for proportions discussed in Section 11.3.

12.1 Goodness-of-Fit Test

Preparing for This Section Before getting started, review the following:

- Chi-square distribution (Section 9.4, pp. 486–489)
- Expected value (Section 6.1, pp. 321–322)
- Mean of a binomial random variable (Section 6.2, pp. 335–336)
- Mutually exclusive (Section 5.2, pp. 266–269)

Objectives **1** Perform a goodness-of-fit test

Figure 1
Chi-square distributions

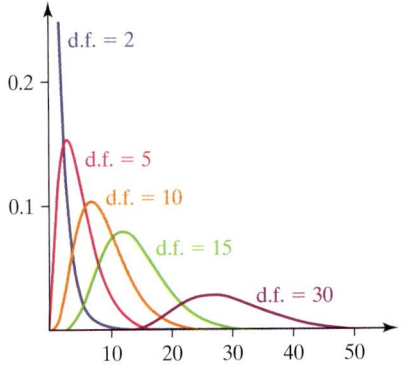

1 ## Perform a Goodness-of-Fit Test

In this section, we present a procedure that can be used to test claims regarding a probability distribution. For example, we might want to test the claim that the distribution of plain M&M candies in a bag is 13% brown, 14% yellow, 13% red, 20% orange, 24% blue, and 16% green. Or we might want to test the claim that the number of hits a player gets in his next four at-bats follows a binomial distribution with $n = 4$ and $p = 0.298$.

The methods used to test these types of claims use the chi-square distribution, introduced in Section 9.4. Recall the following properties of the chi-square distribution.

> **Characteristics of the Chi-Square Distribution**
>
> 1. It is not symmetric.
> 2. The shape of the chi-square distribution depends on the degrees of freedom, just like Student's *t*-distribution.
> 3. As the number of degrees of freedom increases, the chi-square distribution becomes more symmetric, as illustrated in Figure 1.
> 4. The values of χ^2 are nonnegative. That is, the values of χ^2 are greater than or equal to 0.

The critical values of the chi-square distribution can be found in Table VI.

Definition A **goodness-of-fit test** is an inferential procedure used to determine whether a frequency distribution follows a claimed distribution.

As an example, we might want to test the claim that a die is fair. This would mean the probability of each outcome is $\frac{1}{6}$ when a die is cast. We express this claim as

$$H_0: p_1 = p_2 = p_3 = p_4 = p_5 = p_6 = \frac{1}{6}$$

Here's another example: According to the U.S. Census Bureau in 2000, 19.0% of the population of the United States resided in the Northeast; 22.9% resided in the Midwest, 35.6% resided in the South, and 22.5% resided in the West. We might want to test the claim that the distribution of U.S. residents is the same today as it was in 2000. Remember, the null hypothesis is a statement of "no change," so for this claim, the null hypothesis is

H_0: The distribution of residents in the United States is the same today as it was in 2000.

The idea behind testing these types of claims is to compare the actual number of observations for each category of data with the number of observations we would expect if the null hypothesis were true. If a significant difference between the observed counts and expected counts exists, we have evidence against the null hypothesis.

The method for obtaining the expected counts is an extension of the expected value of a binomial random variable. Recall that the mean (and therefore expected value) of a binomial random variable with n independent trials and probability of success, p, is given by $E = \mu = np$.

In Other Words

The expected count for each category is the number of trials of the experiment times the probability of success in the category.

Expected Counts

Suppose there are n independent trials of an experiment with $k \geq 3$ mutually exclusive possible outcomes. Let p_1 represent the probability of observing the first outcome and E_1 represent the expected count of the first outcome, p_2 represent the probability of observing the second outcome and E_2 represent the expected count of the second outcome, and so on. The expected counts for each possible outcome are given by

$$E_i = \mu_i = np_i \quad \text{for} \quad i = 1, 2, \ldots, k$$

EXAMPLE 1   Finding Expected Counts

Problem: An urban economist wishes to determine whether the distribution of residents in the United States is the same today as it was in 2000. That year, 19.0% of the population of the United States resided in the Northeast, 22.9% resided in the Midwest, 35.6% resided in the South, and 22.5% resided in the West (based on data obtained from the Census Bureau). If the economist randomly selects 1500 households in the United States, compute the expected number of households in each region, assuming that the distribution of households did not change from 2000.

Approach

Step 1: Determine the probabilities for each outcome.

Step 2: There are $n = 1500$ trials (the 1500 households surveyed) of the experiment. We expect $np_{\text{northeast}}$ of the households surveyed to reside in the Northeast, np_{midwest} of the households to reside in the Midwest, and so on.

Solution

Step 1: The probabilities are the relative frequencies from the 2000 distribution: $p_{\text{northeast}} = 0.190$, $p_{\text{midwest}} = 0.229$, $p_{\text{south}} = 0.356$, and $p_{\text{west}} = 0.225$.

Step 2: The expected counts for each location within the United States are as follows:

Expected count of Northeast: $np_{\text{northeast}} = 1500(0.190) = 285$

Expected count of Midwest: $np_{\text{midwest}} = 1500(0.229) = 343.5$

Expected count of South: $np_{\text{south}} = 1500(0.356) = 534$

Expected count of West: $np_{\text{west}} = 1500(0.225) = 337.5$

Of the 1500 households surveyed, the economist expects 285 households in the Northeast, 343.5 households in the Midwest, 534 households in the South, and 337.5 households in the West if the distribution of residents of the United States is the same today as it was in 2000.

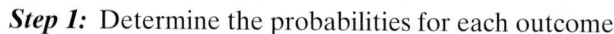

Now Work Problem 5.

To test a claim, we compare the observed counts with the expected counts. If the observed counts are significantly different from the expected counts, we have evidence against the null hypothesis. We need a test statistic and sampling distribution to test the claim.

Historical Note

The goodness-of-fit test was invented by Karl Pearson (the Pearson of correlation coefficient fame). Pearson believed that statistics should be done by determining the distribution of a random variable. Such a determination could be made only by looking at large numbers of data. This philosophy caused Pearson to "butt heads" with Ronald Fisher, because Fisher believed in analyzing small samples.

CAUTION
Goodness-of-fit tests are used to test a claim regarding the distribution of a variable based on a single population. If you wish to compare two or more populations, you must use the tests for homogeneity presented in Section 12.3.

Test Statistic for Goodness-of-Fit Tests

Let O_i represent the observed counts of category i, E_i represent the expected counts of category i, k represent the number of categories, and n represent the number of independent trials of an experiment. Then the formula

$$\chi^2 = \sum \frac{(O_i - E_i)^2}{E_i} \qquad i = 1, 2, \ldots, k$$

approximately follows the chi-square distribution with $k - 1$ degrees of freedom, provided that

1. all expected frequencies are greater than or equal to 1 (all $E_i \geq 1$) and
2. No more than 20% of the expected frequencies are less than 5.

Note: $E_i = np_i$ for $i = 1, 2, \ldots, k$.

From Example 1, there were $k = 4$ categories (Northeast, Midwest, South, and West). For the Northeast, the expected frequency, E, is 295.

Now that we know the distribution of goodness-of-fit tests, we can present a method for testing claims regarding the distribution of a random variable.

The Goodness-of-Fit Test

If a claim is made regarding a distribution, we can use the steps that follow to test the claim.

Step 1: The claim made regarding a distribution is used to determine the null and alternative hypotheses:

H_0: The random variable follows the claimed distribution.

H_1: The random variable does not follow the claimed distribution.

Step 2: Decide on a level of significance, α, depending on the seriousness of making a Type I error.

Step 3:

(a) Calculate the expected counts for each of the k categories. The expected counts are $E_i = np_i$ for $i = 1, 2, \ldots, k$, where n is the number of trials and p_i is the probability of the ith category, assuming that the null hypothesis is true.

(b) Verify that the requirements for the goodness-of-fit test are satisfied.
1. All expected counts are greater than or equal to 1 (all $E_i \geq 1$).
2. No more than 20% of the expected counts are less than 5.

(c) Compute the **test statistic:**

$$\chi_0^2 = \sum \frac{(O_i - E_i)^2}{E_i}$$

Note: O_i is the observed count for the ith category.

CAUTION
If requirements in Step 3(b) are not satisfied, one option is to combine two of the low-frequency categories into a single category.

Classical Approach

Step 4: Determine the critical value. All goodness-of-fit tests are right-tailed tests, so the critical value is χ_α^2 with $k - 1$ degrees of freedom. See Figure 2.

P-value Approach

Step 4: Use Table VI to obtain an approximate P-value by determining the area under the chi-square distribution with $k - 1$ degrees of freedom to the right of the test statistic. See Figure 3.

(*continued on next page*)

Classical Approach

Figure 2

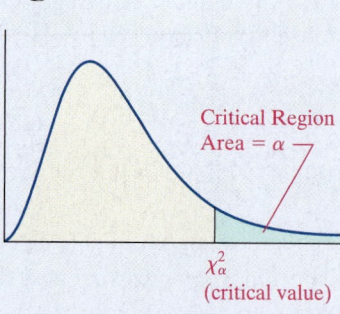

Step 5: Compare the critical value to the test statistic. If $\chi_0^2 > \chi_\alpha^2$, reject the null hypothesis.

P-value Approach

Figure 3

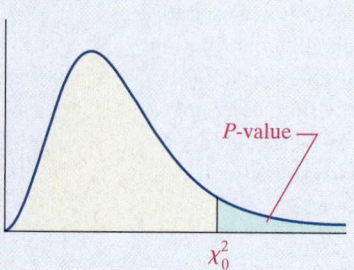

Step 5: If P-value $< \alpha$, reject the null hypothesis.

Step 6: State the conclusion.

EXAMPLE 2

Testing a Claim Using the Goodness-of-Fit Test

Problem: An urban economist wishes to test the claim that the distribution of U.S. residents in the United States is different today than it was in 2000. In 2000, 19.0% of the population of the United States resided in the Northeast, 22.9% resided in the Midwest, 35.6% resided in the South, and 22.5% resided in the West (based on data obtained from the Census Bureau). The economist randomly selects 1500 households in the United States and obtains the frequency distribution shown in Table 1.

Test the claim that the distribution of residents in the United States is different today from the distribution in 2000 at the $\alpha = 0.05$ level of significance.

Approach: We follow Steps 1 through 6 on pages 635–636.

Solution

Step 1: The claim is that the distribution of residents is different today from the distribution in 2000. This claim will be the null hypothesis.

H_0: The distribution of residents of the United States is the same today as it was in 2000.

H_1: The distribution of residents of the United States is different today from what it was in 2000.

Step 2: The level of significance is $\alpha = 0.05$.

Step 3:

(a) The expected counts were computed in Example 1. For convenience, we show the observed and expected counts in Table 2.

Table 1

Region	Frequency
Northeast	274
Midwest	303
South	564
West	359

In Other Words

Remember, the null hypothesis is always a statement of "no change." Therefore, the null hypothesis is that there is no change in the distribution from 2000.

Table 2		
Region	**Observed Counts**	**Expected Counts**
Northeast	274	285
Midwest	303	343.5
South	564	534
West	359	337.5

(b) Since all expected counts are greater than or equal to 5, the requirements for the goodness-of-fit test are satisfied.

(c) The test statistic is

$$\chi_0^2 = \sum \frac{(O_i - E_i)^2}{E_i} = \frac{(274 - 285)^2}{285} + \frac{(303 - 343.5)^2}{343.5} + \frac{(564 - 534)^2}{534} + \frac{(359 - 337.5)^2}{337.5}$$
$$= 8.255$$

Classical Approach

Step 4: There are $k = 4$ categories, so we find the critical value using $4 - 1 = 3$ degrees of freedom. The critical value is $\chi_{0.05}^2 = 7.815$. See Figure 4.

Figure 4

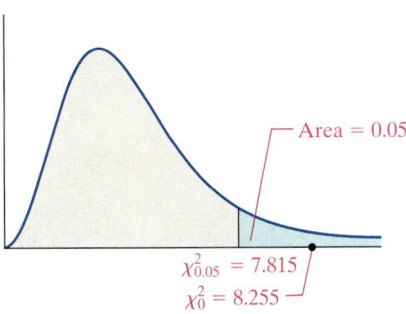

Area = 0.05

$\chi_{0.05}^2 = 7.815$
$\chi_0^2 = 8.255$

Step 5: Because the test statistic, 8.255, is greater than the critical value, 7.815, we reject the null hypothesis.

P-value Approach

Step 4: There are $k = 4$ categories. The *P*-value is the area under the chi-square distribution with $4 - 1 = 3$ degrees of freedom to the right of $\chi_0^2 = 8.255$, as shown in Figure 5.

Figure 5

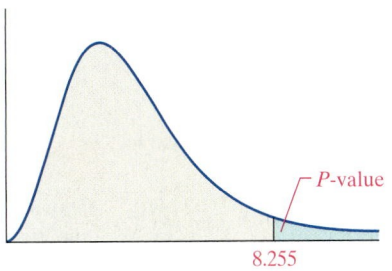

P-value

8.255

Using Table VI, we find the row that corresponds to 3 degrees of freedom. The value of 8.255 lies between 7.815 and 9.348. The value of 7.815 has an area under the chi-square distribution of 0.05 to the right. The area under the chi-square distribution with 3 degrees of freedom to the right of 9.348 is 0.025. Because 8.255 is between 7.815 and 9.348, the *P*-value is between 0.025 and 0.05. So $0.025 < P\text{-value} < 0.05$.

Step 5: Because the *P*-value is less than the level of significance, $\alpha = 0.05$, we reject the null hypothesis.

Step 6: There is sufficient evidence at the $\alpha = 0.05$ level of significance to support the claim that the distribution of United States residents today is different from the distribution in 2000.

If we compare the observed and expected counts, we notice that the Northeast and Midwest regions of the United States have observed counts lower than expected, while the South and West regions of the United States have observed counts higher than expected. So we might conclude that residents of the United States are moving to southern and western locations.

Now Work Problem 11.

In the next example, each of the k categories is equally likely.

EXAMPLE 3 **Testing a Claim Using a Goodness-of-Fit Test**

Problem: An obstetrician wants to know whether or not the proportion of children born each day of the week is the same. She randomly selects 500 birth records and obtains the data shown in Table 3 (based on data obtained from *Vital Statistics of the United States*, 2002).

Is there reason to believe that the day on which a child is born occurs with equal frequency at the $\alpha = 0.01$ level of significance?

Table 3

Day of Week	Frequency
Sunday	57
Monday	78
Tuesday	74
Wednesday	76
Thursday	71
Friday	81
Saturday	63

Approach We follow Steps 1 through 6 presented on pages 633–634

Solution:

Step 1: We will assume that the day on which a child is born occurs with equal frequency. If we let 1 represent Sunday, 2 represent Monday, and so on, we can express this claim as $p_1 = p_2 = \cdots = p_7 = \dfrac{1}{7}$. Thus, we have

$$H_0: p_1 = p_2 = p_3 = p_4 = p_5 = p_6 = p_7 = \frac{1}{7}$$

H_1: At least one of the proportions is different from the others.

Step 2: The level of significance is $\alpha = 0.01$.

Step 3:

(a) The expected counts for each category (day of the week), assuming the null hypothesis is true, are shown in Table 4.

Table 4

Day of the Week	Frequency (observed count)	Assumed Probability	Expected Count
Sunday	57	$\dfrac{1}{7}$	$500\left(\dfrac{1}{7}\right) = \dfrac{500}{7}$
Monday	78	$\dfrac{1}{7}$	$\dfrac{500}{7}$
Tuesday	74	$\dfrac{1}{7}$	$\dfrac{500}{7}$
Wednesday	76	$\dfrac{1}{7}$	$\dfrac{500}{7}$
Thursday	71	$\dfrac{1}{7}$	$\dfrac{500}{7}$
Friday	81	$\dfrac{1}{7}$	$\dfrac{500}{7}$
Saturday	63	$\dfrac{1}{7}$	$\dfrac{500}{7}$

(b) Since all expected counts are greater than or equal to $5\left(\dfrac{500}{7} \approx 71.4\right)$, the requirements for the goodness-of-fit test are satisfied.

(c) The test statistic is

$$\chi_0^2 = \frac{(57 - 500/7)^2}{500/7} + \frac{(78 - 500/7)^2}{500/7} + \frac{(74 - 500/7)^2}{500/7} + \frac{(76 - 500/7)^2}{500/7} +$$

$$\frac{(71 - 500/7)^2}{500/7} + \frac{(81 - 500/7)^2}{500/7} + \frac{(63 - 500/7)^2}{500/7} = 6.184$$

Classical Approach

Step 4: There are $k = 7$ categories, so we find the critical value using $7 - 1 = 6$ degrees of freedom. The critical value is $\chi_{0.01}^2 = 16.812$. See Figure 6.

P-value Approach

Step 4: There are $k = 7$ categories. The P-value is the area under the chi-square distribution with $7 - 1 = 6$ degrees of freedom to the right of $\chi_0^2 = 6.184$, as shown in Figure 7.

Classical Approach (cont'd)

Figure 6

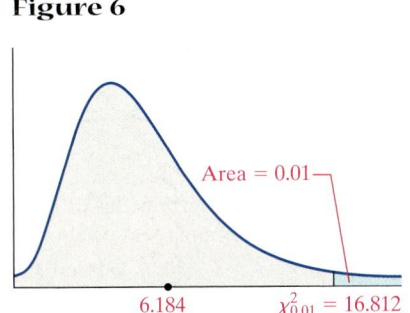

Area = 0.01

6.184 $\chi^2_{0.01} = 16.812$

Step 5: Because the test statistic, 6.184, is less than the critical value, 16.812, we do not reject the null hypothesis.

P-value Approach (cont'd)

Figure 7

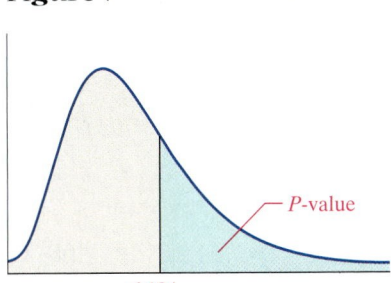

P-value

6.184

Using Table VI, we find the row that corresponds to 6 degrees of freedom. The value of 6.184 is less than 10.546, which has an area under the chi-square distribution of 0.10 to the right. The P-value is greater than 0.1. So P-value > 0.10.

Step 5: Because the P-value is greater than the level of significance, $\alpha = 0.01$, we do not reject the null hypothesis.

Step 6: There is not sufficient evidence at the $\alpha = 0.01$ level of significance to reject the belief that the day of the week on which a child is born occurs with equal frequency.

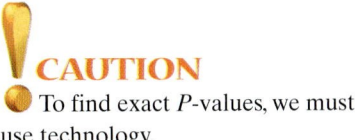

Now Work Problem 17.

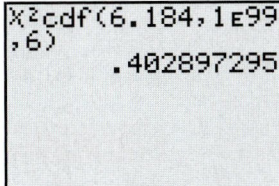

! CAUTION

To find exact P-values, we must use technology.

Finding Exact *P*-values Using Technology

We cannot obtain exact P-values by hand, but statistical software will provide P-values based on the value of the test statistic and the degrees of freedom. We can compute the P-value of the hypothesis tested in Example 3, using the χ^2-cdf command on a TI-84 Plus graphing calculator. See Figure 8. The area under the chi-square distribution to the right of 6.184 with 6 degrees of freedom is 0.403, so the P-value is 0.403. See Figure 9. The P-value is greater than the level of significance, so we do not reject the null hypothesis.

Figure 8

```
χ²cdf(6.184,1E99
,6)
        .402897295
```

Figure 9

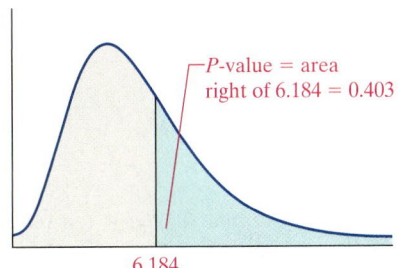

P-value = area right of 6.184 = 0.403

6.184

In-Class Activity: If the Candy Fits—Eat It

Researchers can sometimes help determine or refute the authorship of a text or play by looking at the distribution of word lengths and sentence lengths. On a tastier note, could you determine the type of M&Ms candy just by knowing how many of each color are in the bag?

(continued on next page)

In-Class Activity (continued)

(a) Each student should purchase a bag of M&Ms that are either plain, peanut, almond, crispy, or peanut butter. Make sure the bag is large enough to obtain a decent sample size.

(b) At home, record the number of each color from your bag (ignore broken candies).

(c) In class, give your data to your partner and see if he or she can determine which type of candy you purchased while you try to determine his or her type.
[**Note:** Actual distributions can be found at us.mms.com/us/about/products].

(d) Compare your results to the rest of the class. Were there types that were harder to distinguish? Why do you think this was the case?

12.1 ASSESS YOUR UNDERSTANDING

Concepts and Vocabulary

1. Why is goodness-of-fit a good choice for the title of the procedures used in this section?
2. Explain why chi-square goodness-of-fit tests are always right tailed.
3. State the requirements to perform a goodness-of-fit test.
4. Suppose the expected count of a category is less than 1. What can be done to the categories so that a goodness-of-fit test can be performed?

Skill Building

In Problems 5 and 6, determine the expected counts for each outcome.

5. | $n = 500$ | | | | |
| --- | --- | --- | --- | --- |
| p_i | 0.2 | 0.1 | 0.45 | 0.25 |
| Expected counts | | | | |

6. | $n = 700$ | | | | |
| --- | --- | --- | --- | --- |
| p_i | 0.15 | 0.3 | 0.35 | 0.20 |
| Expected Counts | | | | |

In Problems 7–10, determine (a) the χ^2 test statistic, (b) the degrees of freedom, and (c) the critical value using $\alpha = 0.05$, and (d) test the hypothesis at the $\alpha = 0.05$ level of significance.

7. H_0: $p_A = p_B = p_C = p_D = \dfrac{1}{4}$.

H_1: At least one of the proportions is different from the others.

Outcome	A	B	C	D
Observed	30	20	28	22
Expected	25	25	25	25

8. H_0: $p_A = p_B = p_C = p_D = p_E = \dfrac{1}{5}$,

H_1: At least one of the proportions is different from the others.

Outcome	A	B	C	D	E
Observed	38	45	41	33	43
Expected	40	40	40	40	40

9. H_0: The random variable X is binomial with $n = 4, p = 0.8$.

H_1: The random variable X is not binomial with $n = 4, p = 0.8$.

X	0	1	2	3	4
Observed	1	38	132	440	389
Expected	1.6	25.6	153.6	409.6	409.6

10. H_0: The random variable X is binomial with $n = 4, p = 0.3$.

H_1: The random variable X is not binomial with $n = 4, p = 0.8$.

X	0	1	2	3	4
Observed	260	400	280	50	10
Expected	240.1	411.6	264.6	75.6	8.1

Applying the Concepts

11. Plain M&Ms According to the manufacturer of M&Ms, NW 13% of the plain M&Ms in a bag should be brown, 14% yellow, 13% red, 20% orange, 24% blue, and 16% green. A student wanted to determine whether a randomly selected bag of plain M&Ms had contents that followed this distribution. He counted the number of M&Ms that were each color and obtained the results shown in the table. Test the claim that plain M&Ms follow the distribution stated by M&M/Mars at the $\alpha = 0.05$ level of significance.

Color	Frequency
Brown	61
Yellow	64
Red	54
Blue	61
Orange	96
Green	64

12. Peanut M&Ms According to the manufacturer of M&Ms, 12% of the peanut M&Ms in a bag should be brown, 15% yellow, 12% red, 23% blue, 23% orange, and 15% green. A student wanted to determine whether a randomly selected bag of peanut M&Ms had contents that followed this distribution. He counted the number of M&Ms that were each color and obtained the results shown in the table. Test the claim that peanut M&Ms follow the distribution stated by M&M/Mars at the $\alpha = 0.05$ level of significance.

Color	Frequency	Color	Frequency
Brown	53	Blue	96
Yellow	66	Orange	88
Red	38	Green	59

13. Benford's Law, Part I Our number system consists of the digits 0, 1, 2, 3, 4, 5, 6, 7, 8, and 9. The first significant digit in any number must be 1, 2, 3, 4, 5, 6, 7, 8, or 9 because we do not write numbers such as 12 as 012. Although we may think that each digit appears with equal frequency so that each digit has a $\frac{1}{9}$ probability of being the first significant digit, this is not true. In 1881, Simon Newcomb discovered that digits do not occur with equal frequency. This same result was discovered again in 1938 by physicist Frank Benford. After studying much data, he was able to assign probabilities of occurrence to the first digit in a number as shown.

Digit	1	2	3	4	5	6	7	8	9
Probability	0.301	0.176	0.125	0.097	0.079	0.067	0.058	0.051	0.046

Source: T.P. Hill, The First Digit Phenomenon, *American Scientist*, July–August, 1998.

The probability distribution is now known as Benford's Law and plays a major role in identifying fraudulent data on tax returns and accounting books. For example, the following distribution represents the first digits in 200 allegedly fraudulent checks written to a bogus company by an employee attempting to embezzle funds from his employer.

First digit of fraudulent check	1	2	3	4	5	6	7	8	9
Frequency	36	32	28	26	23	17	15	16	7

Source: State of Arizona vs. Wayne James Nelson

(a) Because these data are meant to prove that someone is guilty of fraud, what would be an appropriate level of significance when performing a goodness-of-fit test?

(b) Using the level of significance chosen in part (a), test the claim that the first digits in the allegedly fraudulent checks obey Benford's law.

(c) Based on the results of part (b), do you think that the employee is guilty of embezzlement?

14. Benford's Law, Part II Refer to Problem 13. The following distribution lists the first digit of the surface area (in square miles) of 335 rivers. Is there evidence at the $\alpha = 0.05$ level of significance to support the claim that the distribution follows Benford's law?

First digit of surface area	1	2	3	4	5	6	7	8	9
Frequency	104	55	36	38	24	29	18	14	17

Source: Eric W. Weisstein, Benford's Law, From *MathWorld*—A Wolfram Web Resource.

15. Are Some Months Busier than Others? A researcher wants to know whether the distribution of birth month is uniform. The following data, based on results obtained from *Vital Statistics of the United States*, 2005, Volume 1, represent the distribution of months in which 500 randomly selected children were born:

Month	Frequency	Month	Frequency
Jan.	40	July	45
Feb.	38	Aug.	44
March	41	Sep.	44
April	40	Oct.	43
May	42	Nov.	39
June	41	Dec.	43

Is there reason to believe that each birth month occurs with equal frequency at the $\alpha = 0.05$ level of significance?

16. Bicycle Deaths A researcher wanted to determine whether bicycle deaths were uniformly distributed over the days of the week. She randomly selected 200 deaths that involved a bicycle, recorded the day of the week on which the death occurred, and obtained the following results (the data are based on information obtained from the Insurance Institute for Highway Safety).

Day of the Week	Frequency	Day of the Week	Frequency
Sunday	16	Thursday	34
Monday	35	Friday	41
Tuesday	16	Saturday	30
Wednesday	28		

Is there reason to believe that bicycle fatalities occur with equal frequency with respect to day of the week at the $\alpha = 0.05$ level of significance?

17. Pedestrian Deaths A researcher wanted to determine
NW whether pedestrian deaths were uniformly distributed over the days of the week. She randomly selected 300 pedestrian deaths, recorded the day of the week on which the death occurred, and obtained the following results (the data are based on information obtained from the Insurance Institute for Highway Safety).

Day of the Week	Frequency	Day of the Week	Frequency
Sunday	39	Thursday	41
Monday	40	Friday	49
Tuesday	30	Saturday	61
Wednesday	40		

Test the belief that the day of the week on which a fatality happens involving a pedestrian occurs with equal frequency at the $\alpha = 0.05$ level of significance.

18. Is the Die Loaded? A player in a craps game suspects that one of the dice being used in the game is loaded. A loaded die is one in which not all the possibilities (1, 2, 3, 4, 5, and 6) are equally likely. The player throws the die 400 times, records the outcome after each throw, and obtains the following results:

Outcome	Frequency	Outcome	Frequency
1	62	4	62
2	76	5	57
3	76	6	67

(a) Is there evidence to support the claim that the die is loaded at the $\alpha = 0.01$ level of significance?
(b) Why do you think the player might test the claim at the $\alpha = 0.01$ level of significance, rather than, say, the $\alpha = 0.1$ level of significance?

19. Home Schooling A school social worker wants to determine if the grade distribution of home-schooled children is different in her district than nationally. The U.S. National Center for Education Statistics provided her with the following data, which represent the relative frequency of home-schooled children by grade level.

Grade	Relative Frequency
K	0.076
1–3	0.253
4–5	0.158
6–8	0.235
9–12	0.278

She obtains a sample of 25 home-schooled children within her district that yields the following data:

Grade	Frequency
K	6
1–3	9
4–5	3
6–8	4
9–12	3

(a) Because of the low cell counts, combine cells into three categories K–3, 4–8, and 9–12.
(b) Test the claim that the grade distribution of home-schooled children is different in her district than nationally at the $\alpha = 0.05$ level of significance.

20. Golden Benford The Fibonacci sequence is a famous sequence of numbers whose elements commonly occur in nature. The terms in the Fibonacci sequence are 1, 1, 2, 3, 5, 8, 13, 21,...The ratio of consecutive terms approaches the "golden ratio", $\Phi = \dfrac{1 + \sqrt{5}}{2}$. If we examine the first digit of the first 85 terms in the Fibonacci sequence, the distribution of digits is as shown in the following table:

Digit	1	2	3	4	5	6	7	8	9
Frequency	25	16	11	7	7	5	4	6	4

Is there evidence to support the belief that the first digit of the Fibonacci numbers follow the Benford distribution (shown in Problem 13) at the $\alpha = 0.05$ level of significance?

21. **Testing the Random-Number Generator** Statistical spreadsheets and graphing calculators with advanced statistical features have random-number generators that create random numbers conforming to a specified distribution.

   (a) Use a random-number generator to create a list of 500 randomly selected integers numbered 1 to 5.
   (b) What proportion of the numbers generated should be 1? 2? 3? 4? 5?
   (c) Test the claim that the random-number generator is generating random integers between 1 and 5 with equal likelihood by performing a chi-square goodness-of-fit test at the $\alpha = 0.01$ level of significance.

22. **Testing the Random-Number Generator** Statistical spreadsheets and graphing calculators with advanced statistical features have random-number generators that create random numbers conforming to a specified distribution.

   (a) Use a random-number generator to create a list of 500 trials of a binomial experiment with $n = 5$ and $p = 0.2$.
   (b) What proportion of the numbers generated should be 0? 1? 2? 3? 4? 5?
   (c) Test the claim that the random-number generator is generating random outcomes of a binomial experiment with $n = 5$ and $p = 0.2$ by performing a chi-square goodness-of-fit test at the $\alpha = 0.01$ level of significance.

In Section 10.4, we tested claims regarding a population proportion using a z-test. However, we can also use the chi-square goodness-of-fit test to test claims with $k = 2$ possible outcomes. In Problems 23 and 24, we test claims with the use of both methods.

23. **Low Birth Weight** According to the U.S. Census Bureau, 7.1% of all babies born to nonsmoking mothers are of low birth weight ($<5\,\text{lb}, 8\,\text{oz}$). An obstetrician wanted to know whether mothers between the ages of 35 and 39 years give birth to a higher percentage of low-birth-weight babies. She randomly selected 160 births for which the mother was 35 to 39 years old and found 15 low-birth-weight babies.

   (a) If the proportion of low-birth-weight babies for mothers in this age group is 0.071, compute the expected number of low-birth-weight births to 35 to 39 year old mothers. What is the expected number of births to mothers 35 to 39 years old that are not low birth weight?
   (b) Test the obstetrician's claim at the $\alpha = 0.05$ level of significance using the chi-square goodness-of-fit test.
   (c) Test the claim by using the approach presented in Section 10.4.

24. **Living Alone?** In 2000, 25.8% of Americans 15 years of age or older lived alone, according to the Census Bureau. A sociologist claims that this percentage is greater today, conducts a random sample of 400 Americans 15 years of age or older, and finds that 164 are living alone.

   (a) If the proportion of Americans aged 15 years or older living alone is 0.258, compute the following expected numbers: Americans 15 years of age or older who live alone; Americans 15 years of age or older who do not live alone?
   (b) Test the sociologist's claim at the $\alpha = 0.05$ level of significance using the goodness-of-fit test.
   (c) Test the claim by using the approach presented in Section 10.4.

25. Using the results of Problem 7, compute $\Sigma(O - E)$. Explain why this result is reasonable.

12.2 Contingency Tables and Association

Preparing for This Section   Before getting started, review the following:

- Side-by-side bar graphs (Section 2.1, pp. 64–65)
- Conditional probability (Section 5.4, pp. 283–286)

Objectives

 Compute the marginal distribution of a variable

 Use the conditional distribution to identify association among categorical data

As we saw in Section 12.1, data, whether qualitative or quantitative, can be organized into categories. For example, a person might be categorized as a male or as a female. A person might also be categorized as a 20- to 29-year-old.

Consider the data (measured in thousands) in Table 5, which represent the employment status and level of education of all U.S. residents 25 years old or older in 2004. By definition, an individual is unemployed if he or she is actively seeking work but is unable to find work. An individual is considered not to be in the labor force if he or she is not employed and is not actively seeking employment.

Table 5

Employment Status	Level of Education			
	Did Not Finish High School	High School Graduate	Some College	Four or More Years of College
Employed	11,537	35,857	32,977	39,293
Unemployed	1,109	2,069	1,461	1,098
Not in the labor force	15,509	21,554	13,117	11,462

Source: Bureau of Labor Statistics

Table 5 is referred to as a **contingency table** or a **two-way table**, because it relates two categories of data. The **row variable** is employment status, because each row in the table describes the employment status of a group. The **column variable** is level of education. Each box inside the table is referred to as a **cell**. For example, the cell corresponding to employed individuals who are high school graduates is in the first row, second column. Each cell contains the frequency of the category: There were 11,537 thousand employed individuals who did not finish high school in 2004.

The data presented in Table 5 describe two characteristics regarding the population of U.S. residents who are 25 years or older: their employment status and their level of education. Because the data represent the entire population, it does not make sense to perform any type of statistical inference. However, we still might want to investigate whether the two variables are associated. For example, are individuals who have a higher level of education more likely to be employed?

1 ## Compute the Marginal Distribution of a Variable

The first step in analyzing data in a contingency table is to look at the distribution of each variable separately. To determine the distribution of either the column variable or the row variable, we create a *marginal distribution*.

Definition

A **marginal distribution** of a variable is a frequency or relative frequency distribution of either the row or column variable in the contingency table.

In Other Words

The distributions are called marginal distributions because they appear in the right and the bottom margin of the contingency table.

A marginal distribution removes the effect of either the row variable or the column variable in the contingency table.

To create a marginal distribution for a variable, we calculate the row and column totals for each category of the variable. The row totals for each category are found by adding up the cell entries in each column of that row. The row totals represent the distribution of the row variable. The column totals for each category are found by adding up the cell entries in each row of that column. The column totals represent the distribution of the column variable.

EXAMPLE 1

Determining Frequency Marginal Distributions

Problem: Find the frequency marginal distributions for employment status and level of education from the data in Table 5.

Approach: The row total for the category "Employed" is found by adding the number of employed individuals who did not finish high school, the number of employed individuals who finished high school, and so on. We repeat this process for each category of employment status.

The column total for the category "Did not finish high school" is found by adding the number of employed individuals who did not finish high school, the number of unemployed individuals who did not finish high school, and the number of individuals not in the labor force who did not finish high school. We repeat this process for each level of education.

Solution: In Table 6, the blue entries represent the marginal distribution of the row variable, employment status. For example, there were $11,537 + 35,857 + 32,977 + 39,293 = 119,664$ thousand employed individuals in 2004. The red entries represent the marginal distribution of the column variable, level of education.

Table 6

| Employment Status | Level of Education | | | | |
	Did Not Finish High School	High School Graduate	Some College	Four or More Years of College	Totals
Employed	11,537	35,857	32,977	39,293	119,664
Unemployed	1,109	2,069	1,461	1,098	5,737
Not in the Labor Force	15,509	21,554	13,117	11,462	61,642
Totals	28,155	59,480	47,555	51,853	187,043

The marginal distribution for employment status removes the effect of level of education; the marginal distribution for level of education removes the effect of employment status. From the marginal distribution of employment status, we can see that there were about twice as many employed Americans as there were Americans not in the labor force (119,664 thousand versus 61,642 thousand) in 2004. From the marginal distribution of level of education, we can see that 59,480 thousand Americans were high school graduates. We can also learn from the table that there were 187,043 thousand U.S. residents 25 years old or older.

Now Work Problem 5(a).

We can use the row and column totals obtained in Example 1 to calculate the relative frequency marginal distribution for level of education and employment status.

EXAMPLE 2   **Determining Relative Frequency Marginal Distributions**

Problem: Determine the relative frequency marginal distribution for level of education and employment status from the data in Table 6.

Approach: The relative frequency marginal distribution for the row variable, employment status, is found by dividing the row total for each employment status by the table total, 187,043. The relative frequency marginal distribution for the column variable, level of education, is found by dividing the column total for each level of education by the table total.

Solution: Table 7 represents the relative frequency marginal distribution for each variable.

Table 7					
	Did Not Finish High School	High School Graduate	Some College	Four or More Years of College	Totals
Employed	11,537	35,857	32,977	39,293	$\dfrac{119,664}{187,043} = 0.640$
Unemployed	1,109	2,069	1,461	1,098	$\dfrac{5,737}{187,043} = 0.031$
Not in the labor force	15,509	21,554	13,117	11,462	$\dfrac{61,642}{187,043} = 0.330$
Totals	$\dfrac{28,155}{187,043} = 0.151$	$\dfrac{59,480}{187,043} = 0.318$	$\dfrac{47,555}{187,043} = 0.254$	$\dfrac{51,853}{187,043} = 0.277$	1

From Table 7, we learn that 15.1% of U.S. residents 25 years old or older did not graduate from high school. We also learn that 64.0% of U.S. residents 25 years old or older were employed in 2004.

Now Work Problems 5(b), (c), and (d).

2 · Use the Conditional Distribution to Identify Association among Categorical Data

As we look at the information contained in Tables 6 and 7; we might ask whether there is a relation between employment status and level of education. In other words, what role does the level of education play in the employment status of an individual? If level of education does not play any role, we would expect the relative frequencies for employment status at each level of education to be close to the relative frequency marginal distribution for employment status given in blue in Table 7. So we would expect 64.0% of individuals who did not finish high school to be employed, 64.0% of individuals who finished high school to be employed, 64.0% of individuals with some college to be employed, and 64.0% of individuals with four or more years of college to be employed. If the relative frequencies for the various levels of education are different, we might associate this difference with the level of education.

The marginal distributions in Tables 6 and 7 allow us to see the distribution of either the row variable, employment status, or the column variable, level of education, but we do not get a sense of association from these tables.

When describing any association between two categories of data, we must use relative frequencies instead of frequencies, because frequencies are difficult to compare when there are different numbers of observations for the categories of a variable.

> ! **CAUTION**
> For the data in Table 7, row or column totals might not sum exactly to 1, due to rounding.

> ! **CAUTION**
> To describe the association between two categorical variables, relative frequencies must be used, because there are different numbers of observations for the categories.

EXAMPLE 3 Comparing Two Categories of a Variable

Problem:

(a) What proportion of the individuals who did not finish high school is employed?
(b) What proportion of high school graduates is employed?
(c) What proportion of the individuals who finished some college is employed?
(d) What proportion of the individuals who attended four or more years of college is employed?

Approach: For the first question, we are asking, "of the individuals who did not finish high school, what proportion are employed?" To determine the proportion of employed individuals who did not finish high school, we divide the number of employed individuals who did not finish high school by the number of people who did not finish high school. We repeat this process for the remaining three questions.

Solution:

(a) In 2004, 28,155 thousand individuals 25 years old or older did not finish high school. (See Table 6.) Of this number, 11,537 thousand were employed. Therefore, $\frac{11,537}{28,155} = 0.410$ represents the proportion of employed individuals who did not finish high school.

(b) In 2004, 59,480 thousand individuals 25 years old or older were high school graduates. Of this number, 35,857 thousand were employed. Therefore, $\frac{35,857}{59,480} = 0.603$ represents the proportion of employed individuals who graduated from high school.

(c) In 2004, 47,555 thousand individuals 25 years old or older had some college. Of this number, 32,977 thousand were employed. Therefore, $\frac{32,977}{47,555} = 0.693$ were employed.

(d) In 2004, 51,853 thousand individuals 25 years old or older had four or more years of college. Of this number, 39,293 thousand were employed. Therefore, $\frac{39,293}{51,853} = 0.758$ were employed. From these relative frequencies, we can see that, as the level of education increases, the proportion of individuals employed increases.

The results in Example 3 are only partial. In general, we create a relative frequency distribution for each employment status of individuals who did not finish high school, a second relative frequency distribution for each employment status of individuals who are high school graduates, and so on. These relative frequency distributions are called *conditional distributions*.

In Other Words

Since we are finding the conditional distribution of employment status by level of education, the level of education is the *explanatory variable* and the employment status is the *response variable*.

Definition

> A **conditional distribution** lists the relative frequency of each category of a variable, given a specific value of the other variable in the contingency table.

An example should help solidify our understanding of the definition.

EXAMPLE 4   **Constructing a Conditional Distribution**

Problem: Find the conditional distribution of the variable employment status by level of education for the data in Table 5.

Approach: We first compute the relative frequency for each employment status, given that the individual did not finish high school. We next compute the relative frequency for each employment status, given that the individual is a high school graduate. Then we compute the relative frequency for each employment status, given that the individual had some college. Finally, we compute the relative frequency for each employment status, given that the individual had at least four years of college. This is the same approach taken in Example 3 for employed individuals.

Solution: We start with the individuals who did not finish high school. We use the number of individuals who did not finish high school (28,155 thousand) as the denominator in computing the relative frequencies.

We then compute the relative frequency for each employment status, given that the individual is a high school graduate. In this case, since we know the individual is a high school graduate, we use the number of individuals who graduated from high school as the denominator in computing relative frequencies.

Next, compute the relative frequency for each employment status, given that the individual had some college and then given that the individual had four or more years of college. We obtain Table 8.

	Level of Education			
Employment Status	Did Not Finish High School	High School Graduate	Some College	Four or More Years of College
Employed	$\frac{11{,}537}{28{,}155} = 0.410$	$\frac{35{,}857}{59{,}480} = 0.603$	$\frac{32{,}977}{47{,}555} = 0.693$	$\frac{39{,}293}{51{,}853} = 0.758$
Unemployed	$\frac{1{,}109}{28{,}155} = 0.039$	$\frac{2{,}069}{59{,}480} = 0.035$	$\frac{1{,}461}{47{,}555} = 0.031$	$\frac{1{,}098}{51{,}853} = 0.021$
Not in the labor force	$\frac{15{,}509}{28{,}155} = 0.551$	$\frac{21{,}554}{59{,}480} = 0.362$	$\frac{13{,}117}{47{,}555} = 0.276$	$\frac{11{,}462}{51{,}853} = 0.221$
Totals	1	1	1	1

Table 8

Notice that the "Employed" row in Table 8 shows the results from Example 3. From the conditional distributions by level of education, associations become apparent. As the amount of schooling increases, the proportion employed within each category also increases. As the amount of schooling increases, the proportion unemployed and the proportion not in the labor force decrease.

Information about individuals' level of education provides some insight into their employment status. For example, suppose we are asked to predict the number of employed individuals out of 100. We might predict that about 64 of the 100 are employed, because the employment rate for the entire United States is 64.0%. (See Table 7.) However, if we are given the additional information that all the 100 individuals had completed four or more years of college, we would increase our guess to about 76. Do you see why? The additional information about the individuals' level of education allows us to adjust our predictions.

Now Work Problem 5(e).

We could also construct a conditional distribution by employment status. The procedure is the same, except the distribution uses the rows instead of the columns. In this case, the explanatory variable is employment status and the response variable is level of education.

As is usually the case, a graph can provide a powerful depiction of the data. A bar graph of the conditional distribution is recommended.

EXAMPLE 5 **Drawing a Bar Graph of a Conditional Distribution**

Problem: Using the results of Example 4, draw a bar graph that represents the conditional distribution of employment status by level of education.

Approach: We will draw three bars, side by side, for each employment status. The horizontal axis will represent the level of education and the vertical axis will represent the relative frequency.

Solution: Figure 10 contains the bar graph: The blue bars represent the proportion employed for each level of education, the green bars represent the proportion unemployed for each level of education, and the purple bars represent the proportion not in the labor force for each level of education.

Figure 10

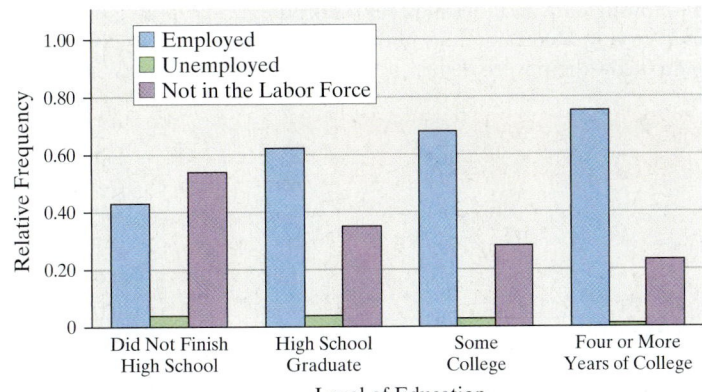

Employment Status by Level of Education

The graph contains the same information as that obtained from Table 8. The proportion employed increases as the level of education increases. We also see that the proportion of individuals unemployed or not in the labor force decreases as the level of education increases. _____

Now Work Problems 5(f) and (g).

The methods presented in this section for identifying the association between two categorical variables are different from the methods for measuring association between two quantitative variables. The measure of association in this section is based on whether there are differences in the relative frequencies of the response variable (employment status) for the different categories of the explanatory variable (level of education). If differences exist, we might attribute these differences to the explanatory variable.

In addition, because the data in Example 4 are observational, we do not make any statements regarding causation. Level of education is not said to be a cause of employment status, because a controlled experiment was not conducted. We can never make statements about causation when the data are observational.

Finally, we should note that the data that we analyzed were population data. Therefore, we do not perform any inferential statistics on the data. Any differences that exist in the marginal distributions are differences in the populations. For example, we can state that a higher level of education is associated with a higher level of employment with 100% certainty, because the analysis was performed on population data. If the data were sample data, we would not have this level of confidence.

CAUTION

The data in this section are population data, so there is no need to perform inference. We check for association in sample data in Section 12.3.

12.2 ASSESS YOUR UNDERSTANDING

Concepts and Vocabulary

1. What is meant by a marginal distribution? What is meant by a conditional distribution?

2. Refer to Table 5. Is constructing a conditional distribution by level of education different from constructing a conditional distribution by employment status? If they are different, explain the difference.

3. Explain why we use the term *association* rather than *correlation* when describing the relation between two variables in this section.

4. Explain why we use relative frequencies when describing the association between two categorical variables.

Applying the Concepts

5. **Medicaid** The following data, in thousands, represent the ages of persons receiving Medicaid and whether their income places them below the poverty level.

	Age			
Poverty Level	<18	18-44	45-64	≥65
Below poverty level	7571	3693	1718	1025
Above poverty level	9694	4745	2282	2258

Source: Statistical Abstract, 2004

(a) Construct a frequency marginal distribution.
(b) Construct a relative frequency marginal distribution.
(c) What percentage of people receiving Medicaid are below the poverty level?
(d) What percentage of people receiving Medicaid are less than 18 years old?
(e) Construct a conditional distribution by poverty level.
(f) Draw a bar graph of the conditional distribution.
(g) What, if any, association exists between age and poverty level? Discuss the association.

6. **SAT Verbal Scores** The following data represent the scores received on the 2000 SAT I: Reasoning Test–Verbal, by gender.

	Gender	
SAT Verbal Score	Male	Female
200–299	18,262	21,205
300–399	78,550	97,482
400–499	198,858	242,430
500–599	208,578	238,870
600–699	119,747	123,584
700–800	36,277	35,166

Source: College Board

(a) Construct a frequency marginal distribution.
(b) Construct a relative frequency marginal distribution.
(c) What percentage of the students taking the SAT had verbal scores of 500–599?
(d) What percentage of the students taking the SAT verbal is female?
(e) Construct a conditional distribution by gender.
(f) Draw a bar graph of the conditional distribution.
(g) What, if any, association exists between SAT verbal score and gender? Discuss the association.

7. **Driver Fatalities** The following data represent the number of driver fatalities in 2000, by age group, for male and female drivers.

	Gender	
Age	Male	Female
16–24	4901	1601
25–34	3693	1003
35–44	3493	1143
45–54	3032	979
55–64	1866	679
65–69	594	228
>69	2105	1100
Unknown	17	4

Source: National Highway Traffic Institute

(a) Construct a frequency marginal distribution.
(b) Construct a relative frequency marginal distribution.
(c) What percentage of the drivers killed was 25 to 34?
(d) What percentage of the drivers killed was female?
(e) Construct a conditional distribution by gender.
(f) Draw a bar graph of the conditional distribution found in part (e).
(g) What, if any, association exists between age and gender? Discuss the association.

8. **Marital Status** The following data (in thousands) represent the marital status of Americans 25 years old or older and their level of education in 2003.

	Level of Education			
Marital Status	Did Not Graduate High School	High School Graduate	Some College	College Graduate
Never married	4,333	8,637	7,432	8,293
Married, spouse present	14,787	35,376	29,006	34,693
Married, spouse absent	894	938	535	574
Separated	1,134	1,596	1,098	614
Widowed	4,582	5,155	2,487	1,746
Divorced	2,887	7,612	6,415	4,490

Source: U.S. Census Bureau

(a) Construct a frequency marginal distribution.
(b) Construct a relative frequency marginal distribution.
(c) What percentage of Americans 25 years old or older has never married?
(d) What percentage of Americans 25 years old or older has graduated from high school?
(e) Construct a conditional distribution by level of education.
(f) Draw a bar graph of the conditional distribution found in part (e).
(g) What, if any, association exists between level of education and marital status? Discuss the association.

9. **Cause of Death** Based on information provided by the *National Vital Statistics Reports*, Vol. 53, No. 15, February 28, 2005, the following data represent the cause of death for various age groups in 2003.

Age (years)	Cause of Death			
	Accidents and Adverse Effects	Malignant Neoplasms	Diseases of Heart	Cerebrovascular Diseases
1–4	1,679	383	186	59
5–14	2,561	1,060	252	83
15–24	14,966	1,628	1,083	204
25–44	27,844	19,041	16,283	3,004
45–64	23,669	144,936	101,713	15,971
65 and older	33,976	387,475	564,204	138,397

(a) Construct a frequency marginal distribution.
(b) Construct a relative frequency marginal distribution.
(c) What percentage of the deaths was due to accidents and adverse effects?
(d) What percentage of the deaths was of people 15 to 24 years old?
(e) Construct a conditional distribution by age.
(f) Draw a bar graph of the conditional distribution.
(g) What, if any, association exists between age and cause of death? Discuss the association.

10. **Recreation** The following data, in thousands, represent the number of participants in various recreational activities by age for 2002.

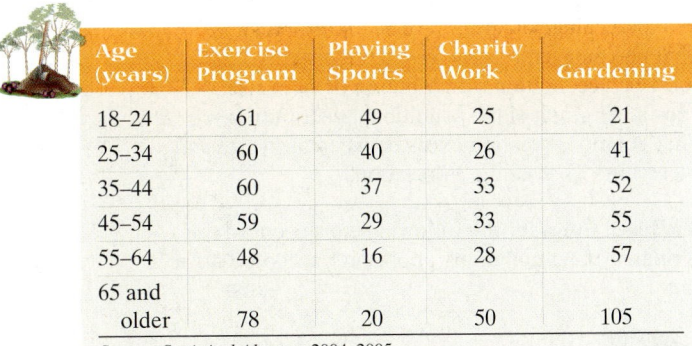

Age (years)	Exercise Program	Playing Sports	Charity Work	Gardening
18–24	61	49	25	21
25–34	60	40	26	41
35–44	60	37	33	52
45–54	59	29	33	55
55–64	48	16	28	57
65 and older	78	20	50	105

Source: Statistical Abstract, 2004–2005

(a) Construct a frequency marginal distribution.
(b) Construct a relative frequency marginal distribution.
(c) What percentage of these participants was in an exercise program?
(d) What percentage of these participants was 35 to 44 years old?
(e) Construct a conditional distribution by age.
(f) Draw a bar graph of the conditional distribution.
(g) What if any association exists between age and recreational activity? Discuss the association.

11. Abortion The following data represent the number of abortions in thousands, completed in a year, by age and year:

| Age (years) | Year | | |
	1990	1995	2000
≤19	369	274	244
20–24	532	441	430
≥25	713	645	640

Source: World Almanac, 2000

(a) Construct a frequency marginal distribution.
(b) Construct a relative frequency marginal distribution.
(c) What percentage of the abortions were performed in 1990?
(d) What percentage of the women having abortions were 19 years old or younger?
(e) Construct a conditional distribution by year.
(f) Draw a bar graph of the conditional distribution.
(g) What, if any, association exists between age and year? Discuss the association.

12. Farm Size The data in the table represent the size of a farm and the tenure of the operator (a *full owner* owns all the acreage he or she farms; a *part owner* owns some of the land and rents the rest; a *tenant* leases the land he or she farms).

| Size (acres) | Tenure | | |
	Full Owner	Part Owner	Tenant
Under 50	639	64	41
50–179	487	131	41
180–499	203	153	33
500–999	54	91	17
1000 or more	46	112	18

Source: Statistical Abstract of the United States, 2004–2005

(a) Construct a frequency marginal distribution.
(b) Construct a relative frequency marginal distribution.
(c) What percentage of farms have 50 to 179 acres?
(d) What percentage of farms have a full owner?
(e) Construct a conditional distribution by farm size.
(f) Draw a bar graph of the conditional distribution.
(g) What, if any, association exists between farm size and ownership? Discuss the association.

13. Gender Bias? Consider the following contingency table, which relates the number of applicants accepted to a college.

	Accepted	Denied
Male	98	522
Female	90	200

(a) Construct a conditional distribution by gender.
(b) What proportion of males was accepted? What proportion of females was accepted?
(c) What might you conclude about the admittance policies of the school?

A lurking variable is the type of school applied to. This particular college has two programs of study: business and social work. The following table shows applications by type of school.

| | Business School | |
	Accepted	Denied
Male	90	510
Female	10	60

| | Social Work School | |
	Accepted	Denied
Male	8	12
Female	80	140

(d) What proportion of males who applied to the business school was accepted? What proportion of females who applied to the business school was accepted?
(e) What proportion of males who applied to the social work school was accepted? What proportion of females who applied to the social work school was accepted?
(f) Notice that for both the business school and social work school a higher proportion of males was accepted, yet for the whole university, the university appears to admit females at a rate double that of males! This result illustrates **Simpson's paradox**. When the lurking variable type of school is considered, the apparent gender bias goes away. Explain carefully how the bias disappears when type of school is considered.

14. An Unhealthy Hospital The following contingency table relates the number of patients admitted for surgery to Palos Community Hospital and General Hospital along with their survival status for the month of July.

	PCH	GH
Survived	150	230
Died	20	55

(a) Construct a conditional distribution by hospital.
(b) What proportion of surgeries at Palos Community Hospital resulted in death? What proportion of surgeries at General Hospital resulted in death?
(c) What might you conclude about General Hospital?

A lurking variable is type of hospital. General Hospital is a trauma center, so emergency or high-risk patients (such as gunshot wounds) are directed to General Hospital. Palos Community Hospital typically does planned procedures. The following table below shows survival by type of surgical procedure.

| | Planned | |
	PCH	GH
Survived	138	160
Died	12	13

| | Emergency | |
	PCH	GH
Survived	12	70
Died	8	42

(d) What proportion of planned surgeries at PCH had patients that survived? What proportion of planned surgeries at GH had patients that survived?

(e) What proportion of emergency surgeries at PCH had patients that survived? What proportion of emergency surgeries at GH had patients that survived?

(f) Notice that for both the planned and emergency surgeries a higher proportion of survivors were at GH, yet for all surgeries PCH has a higher proportion of survivors! This result illustrates Simpson's paradox. When the lurking variable type of surgery is considered, the apparent superiority of Palos Community Hospital goes away. Explain carefully how the bias disappears when type of surgery is considered.

12.3 Tests for Independence and the Homogeneity of Proportions

Preparing for This Section   Before getting started, review the following:

- The language of hypothesis tests (Section 10.1, pp. 505–511)
- Independent events (Section 5.3, pp. 277–280)
- Mean of a binomial random variable (Section 6.2, pp. 335–336)
- Testing a hypothesis about two population proportions (Section 11.3, pp. 601–606)

Objectives

 Perform a test for independence

 Perform a test for homogeneity of proportions

In Section 12.2, we discussed the association between two variables in which we had data for an entire population. Under this situation, we could say two variables are associated by examining the relative frequency marginal distributions. If differences existed among the relative frequency marginal distributions, we could claim association among the variables. However, if we have sample data, we must perform statistical inference to determine whether any differences we have obtained are significant. In this section, we develop methods for performing statistical inference on two categorical variables to determine whether there is any association between the two variables. We call the method the *chi-square test for independence*.

 Perform a Test for Independence

We begin with a definition.

Definition

The **chi-square test for independence** is used to determine whether there is an association between a row variable and column variable in a contingency table constructed from sample data. The null hypothesis is that the variables are not associated; in other words, they are independent. The alternative hypothesis is that the variables are associated, or dependent.

In Other Words

In a chi-square independence test, the null hypothesis is always

H_0: The variables are independent

The alternative hypothesis is always

H_1: The variables are not independent

The idea behind testing these types of claims is to compare actual counts to the counts we would expect if the null hypothesis were true (if the variables are independent). If a significant difference between the actual counts and expected counts exists, we take this as evidence against the null hypothesis.

The method for obtaining the expected counts requires that we compute the number of observations expected within each cell under the assumption that the null hypothesis is true. Recall, if two events E and F are independent, then $P(E \text{ and } F) = P(E) \cdot P(F)$. We can use the Multiplication Rule for independent events to obtain the expected proportion of observations within each cell under the assumption of independence. We then multiply this result by n, the sample size, to obtain the expected count within each cell.* We present an example to introduce the method for obtaining expected counts.

*Recall that the expected value of a binomial random variable for n independent trials of a binomial experiment with probability of success p is given by $E = \mu = np$.

EXAMPLE 1 | **Determining the Expected Counts in a Test for Independence**

Problem: Blood type is classified as A, B, AB, or O. In addition, blood can be classified as Rh^+ or Rh^-. In a survey of 500 randomly selected individuals, a phlebotomist obtained the results shown in Table 9.

Table 9				
	Blood Type			
Rh-status	**A**	**B**	**AB**	**O**
Rh^+	176	28	22	198
Rh^-	30	12	4	30

Compute the expected counts within each cell, assuming that Rh-status and blood type are independent.

Approach

Step 1: Compute the row and column totals.

Step 2: Compute the relative marginal frequencies for the row variable and column variable.

Step 3: Use the Multiplication Rule for Independent Events to compute the proportion of observations within each cell under the assumption of independence.

Step 4: Multiply the proportions by 500, the sample size, to obtain the expected counts within each cell.

Solution

Step 1: The row totals (blue) and column totals (red) are presented in Table 10.

Table 10					
	A	**B**	**AB**	**O**	**Row Totals**
Rh^+	176	28	22	198	424
Rh^-	30	12	4	30	76
Column totals	206	40	26	228	500

Step 2: The relative frequencies for the row variable (Rh-status) and column variable (blood type) are presented in Table 11.

Table 11					
	A	**B**	**AB**	**O**	**Relative Frequency**
Rh^+	176	28	22	198	$\frac{424}{500} = 0.848$
Rh^-	30	12	4	30	$\frac{76}{500} = 0.152$
Relative Frequency	$\frac{206}{500} = 0.412$	$\frac{40}{500} = 0.08$	$\frac{26}{500} = 0.052$	$\frac{228}{500} = 0.456$	1

Step 3: Assuming blood type and Rh-status are independent, we use the Multiplication Rule for Independent Events to compute the proportion of observations we would expect in each cell. For example, the proportion of individuals who are Rh^+ and of blood type A would be

$$\left(\begin{array}{c} \text{Proportion } Rh^+ \text{ and} \\ \text{blood type A} \end{array} \right) = (\text{proportion } Rh^+) \cdot (\text{proportion blood type A})$$

$$= (0.848)(0.412)$$

$$= 0.349376$$

Table 12 contains the expected proportion in each cell, under the assumption of independence.

Table 12				
	A	**B**	**AB**	**O**
Rh^+	0.349376	0.06784	0.044096	0.386688
Rh^-	0.062624	0.01216	0.007904	0.069312

Step 4: We multiply the expected proportions in Table 12 by 500, the sample size, to obtain the expected counts under the assumption of independence. The results are presented in Table 13.

Table 13				
	A	**B**	**AB**	**O**
Rh^+	500(0.349376) = 174.688	500(0.06784) = 33.92	500(0.044096) = 22.048	193.344
Rh^-	31.312	6.08	3.952	34.656

If blood type and Rh-status are independent, we would expect a random sample of 500 individuals to contain about 175 who are of blood type A and are Rh^+.

The technique used in Example 1 to find the expected counts might seem rather tedious. It certainly would be more pleasant if we could determine a shortcut formula that could be used to obtain the expected counts. Let's consider the expected count for blood type A and Rh^+. This expected count was obtained by multiplying the proportion of individuals who were of blood type A, the proportion of individuals who were Rh^+, and the number of individuals in the sample. That is,

$$\text{Expected count} = (\text{proportion } Rh^+)(\text{proportion blood type A})(\text{sample size})$$

$$= \frac{424}{500} \cdot \frac{206}{500} \cdot 500$$

$$= \frac{424 \cdot 206}{500} \qquad \textit{Cancel the 500s.}$$

$$= \frac{(\text{row total for } Rh^+)(\text{column total for blood type A})}{\text{table total}}$$

This leads to the following general result:

> ### Expected Frequencies in a Chi-Square Test for Independence
>
> To find the expected frequencies in a cell when performing a chi-square independence test, multiply the row total of the row containing the cell by the column total of the column containing the cell and divide this result by the table total. That is,
>
> $$\text{Expected frequency} = \frac{(\text{row total})(\text{column total})}{\text{table total}} \qquad \textbf{(1)}$$

For example, to calculate the expected frequency for Rh^+, blood type A, we compute

$$\text{Expected frequency} = \frac{(\text{row total})(\text{column total})}{\text{table total}} = \frac{(424)(206)}{500} = 174.688$$

Now Work Problem 7(a).

This result agrees with the result obtained in Table 13.

To test a claim regarding the independence of two variables, we compare the actual (observed) counts to those expected. If the observed counts are significantly different from the expected counts, we take this as evidence against the null hypothesis. We need a test statistic and sampling distribution to test the claim.

> ### Test Statistic for the Test of Independence
>
> Let O_i represent the observed number of counts in the ith cell and E_i represent the expected number of counts in the ith cell. Then
>
> $$\chi^2 = \sum \frac{(O_i - E_i)^2}{E_i}$$
>
> approximately follows the chi-square distribution with $(r - 1)(c - 1)$ degrees of freedom, where r is the number of rows and c is the number of columns in the contingency table, provided that (1) all expected frequencies are greater than or equal to 1 and (2) no more than 20% of the expected frequencies are less than 5.

From Example 1, there were $r = 2$ rows and $c = 4$ columns.

We now present a method for testing claims regarding the association between two variables in a contingency table.

> ### Chi-Square Test for Independence
>
> If a claim is made regarding the association between (or independence of) two variables in a contingency table, we can use the steps that follow to test the claim.
>
> **Step 1:** A claim is made regarding the independence of the data.
>
> H_0: The row variable and column variable are independent.
>
> H_1: The row variable and column variable are dependent.
>
> **Step 2:** Choose a level of significance, α, depending on the seriousness of making a Type I error.
>
> **Step 3:**
>
> **(a)** Calculate the expected frequencies (counts) for each cell in the contingency table using Formula (1).
>
> **(b)** Verify that the requirements for the goodness-of-fit test are satisfied:
>
> **1.** All expected frequencies are greater than or equal to 1 (all $E_i \geq 1$).
>
> **2.** No more than 20% of the expected frequencies are less than 5.

(c) Compute the **test statistic**:

$$\chi_0^2 = \sum \frac{(O_i - E_i)^2}{E_i}$$

Note: O_i is the observed frequency for the ith category.

Classical Approach

Step 4: Determine the critical value. All chi-square tests for independence are right-tailed tests, so the critical value is χ_α^2 with $(r-1)(c-1)$ degrees of freedom, where r is the number of rows and c is the number of columns in the contingency table. See Figure 11.

Figure 11

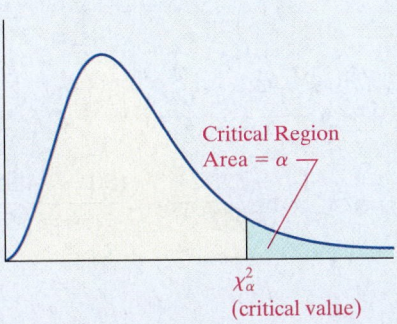

Step 5: Compare the critical value to the test statistic. If $\chi_0^2 > \chi_\alpha^2$, reject the null hypothesis.

P-value Approach

Step 4: Use Table VI to determine an approximate P-value by determining the area under the chi-square distribution with $(r-1)(c-1)$ degrees of freedom, where r is the number of rows and c is the number of columns in the contingency table to the right of the test statistic. See Figure 12.

Figure 12

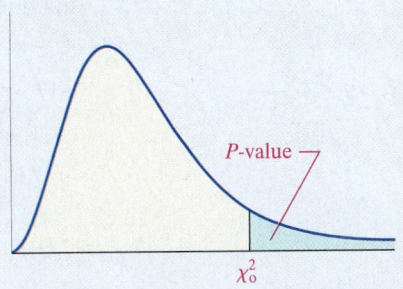

Step 5: If P-value $< \alpha$, reject the null hypothesis.

Step 6: State the conclusion.

EXAMPLE 2   **Performing a Chi-Square Test for Independence**

Problem: Blood is classified as A, B, AB, or O. In addition, blood can be classified as Rh$^+$ or Rh$^-$. In a survey of 500 randomly selected individuals, a phlebotomist obtained the results shown in Table 14.

Table 14

Rh-status	Blood Type			
	A	**B**	**AB**	**O**
Rh$^+$	176	28	22	198
Rh$^-$	30	12	4	30

Test whether blood type and Rh-status are independent at the $\alpha = 0.05$ level of significance.

Approach: We follow Steps 1 through 6 just given.

Solution

Step 1: We wish to determine whether blood type and Rh-status are independent. We state the hypotheses as follows:

H_0: Blood type and Rh-status are independent (or not related).
H_1: Blood type and Rh-status are dependent (or somehow related).

Step 2: The level of significance is $\alpha = 0.05$.

Step 3:

(a) The expected frequencies were computed in Example 1. Table 15 shows the observed frequencies, with the expected frequencies in parentheses.

	Table 15			
	Observed (and Expected) Frequencies			
	A	**B**	**AB**	**O**
Rh$^+$	176	28	22	198
	(174.688)	(33.92)	(22.048)	(193.344)
Rh$^-$	30	12	4	30
	(31.312)	(6.08)	(3.952)	(34.656)

(b) Since all expected frequencies are greater than or equal to 1 and only 1 out of 8 (=12.5%) expected frequencies is less than 5, the requirements for the goodness-of-fit test are satisfied.

(c) The test statistic is

$$\chi_0^2 = \frac{(176 - 174.688)^2}{174.688} + \frac{(28 - 33.92)^2}{33.92} + \frac{(22 - 22.048)^2}{22.048} + \frac{(198 - 193.344)^2}{193.344}$$

$$+ \frac{(30 - 31.312)^2}{31.312} + \frac{(12 - 6.08)^2}{6.08} + \frac{(4 - 3.952)^2}{3.952} + \frac{(30 - 34.656)^2}{34.656}$$

$$= 7.601$$

Classical Approach

Step 4: There are $r = 2$ rows and $c = 4$ columns, so we find the critical value using $(r - 1)(c - 1) = (2 - 1)(4 - 1) = 3$ degrees of freedom. The critical value is $\chi_{0.05}^2 = 7.815$. See Figure 13.

Figure 13

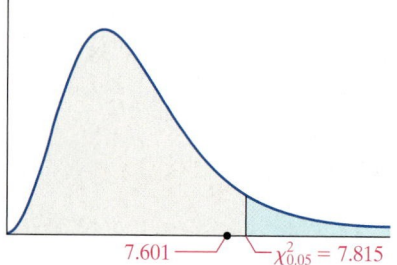

7.601 $\chi_{0.05}^2 = 7.815$

Step 5: Because the test statistic, 7.601, is less than the critical value, 7.815, we do not reject the null hypothesis.

P-value Approach

Step 4: There are $r = 2$ rows and $c = 4$ columns, so we find the P-value using $(r - 1)(c - 1) = (2 - 1)(4 - 1) = 3$ degrees of freedom. The P-value is the area under the chi-square distribution with 3 degrees of freedom to the right of $\chi_0^2 = 7.601$, as shown in Figure 14.

Figure 14

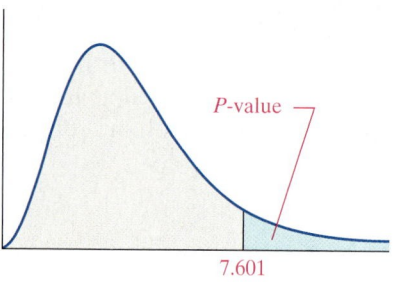

P-value

7.601

Using Table VI, we find the row that corresponds to 3 degrees of freedom. The value of 7.601 lies between 6.251 and 7.815. The area under the chi-square distribution with 3 degrees of freedom to the right of 6.251 is 0.10. The value of 7.815 has an area under the chi-square distribution of 0.05 to the right. Because 7.601 is between 6.251 and 7.815, the P-value is between 0.05 and 0.10. So $0.05 < P\text{-value} < 0.10$.

The exact P-value using the TI-84 Plus's χ_{cdf}^2 command is 0.055.

Step 5: Because the P-value is greater than the level of significance, $\alpha = 0.05$, we do not reject the null hypothesis.

Step 6: There is not enough evidence, at the $\alpha = 0.05$ level of significance, to support the belief that Rh-status and blood type are dependent. We conclude that Rh-status and blood type are not related.

Now Work Problems 7(b), (c), (d), and (e).

While we did not reject the null hypothesis that Rh-status and blood type are independent at the $\alpha = 0.05$ level of signficance, the closeness of the test statistic to the critical value could be considered suggestive of some type of relation between Rh-status and blood type.

To see the relation between Rh-status and blood type, we draw bar graphs of the conditional distributions of Rh-status by blood type. Recall that a conditional distribution lists the relative frequency of each category of a variable, given a specific value of the other variable in a contingency table. For example, we can calculate the relative frequency of Rh^+ and Rh^-, given that an individual is of blood type A. We repeat this for the remaining blood types.

| **EXAMPLE 3** | **Constructing a Conditional Distribution and Bar Graph** |

Problem: Find the conditional distribution of the variable Rh-status by blood type for the data in Table 14. Then draw a bar graph that represents the conditional distribution of Rh-status by blood type.

Approach: First, compute the relative frequency for Rh-status, given that the individual is blood type A. Then compute the relative frequency for Rh-status, given that the individual is blood type B, and so on. We will draw two bars, side by side, for each blood type. The horizontal axis represents the blood type, and the vertical axis represents the relative frequency of Rh-status for each blood type.

Solution: We start with the individuals who are blood type A. The relative frequency with which we observe an individual who is Rh^+, given that the individual is blood type A, is $\frac{176}{206} = 0.854$. Therefore, 85.4% of individuals who are blood type A are Rh^+. The relative frequency with which we observe an individual who is Rh^-, given that the individual is blood type A, is $\frac{30}{206} = 0.146$. We now proceed to compute the relative frequency for each Rh-status, given that the individual is blood type B. The relative frequency with which we observe an individual who is Rh^+, given that the individual is blood type B, is $\frac{28}{40} = 0.70$.

Repeat this process for individuals who are Rh^-. Finally, compute the relative frequency for each Rh-status, given that the individual is blood type AB and then given that the individual is blood type O. We obtain Table 16.

	Table 16			
	A	**B**	**AB**	**O**
Rh$^+$	$\frac{176}{206} = 0.854$	$\frac{28}{40} = 0.700$	$\frac{22}{26} = 0.846$	$\frac{198}{228} = 0.868$
Rh$^-$	$\frac{30}{206} = 0.146$	$\frac{12}{40} = 0.300$	$\frac{4}{26} = 0.154$	$\frac{30}{228} = 0.132$

From the conditional distributions by blood type, the association between blood type and Rh-status should be apparent. The proportion of individuals who are Rh^+ is less for those individuals who are of blood type B than for the other blood types. However, the difference is not significant at the $\alpha = 0.05$ level of significance.

Figure 15 contains the bar graph of the conditional distribution. The blue bars represent the proportion of individuals with Rh^+ blood for each blood type, and the green bars represent the proportion of individuals with Rh^- blood for each blood type.

Figure 15

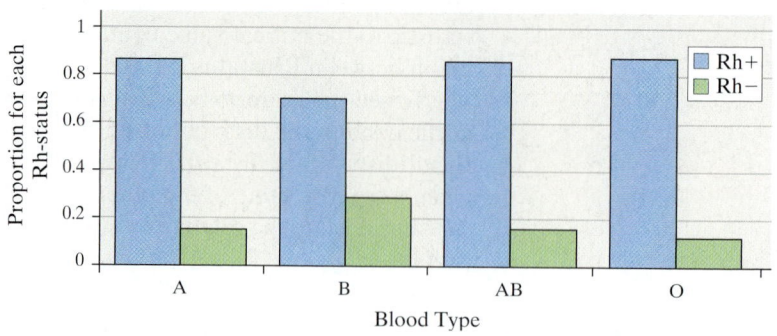

Proportion of Each Rh-status by Blood Type

Now Work Problem 7(f).

EXAMPLE 4 **Chi-square Test for Independence Using Technology**

Problem: Using the data presented in Table 14 from Example 2, test whether blood type and Rh-status are independent at the $\alpha = 0.05$ level of significance using statistical software or a graphing calculator with advanced statistical features.

Approach: We will use MINITAB to test the claim. The steps for testing for independence using the TI-83/84 Plus graphing calculators, MINITAB, and Excel are given in the Technology Step-by-Step on page 668.

Result: Figure 16 shows the results obtained from MINITAB.

Figure 16

Chi-Square Test

Expected counts are printed below observed counts

	C1	C2	C3	C4	Total
1	176	28	22	198	424
	174.69	33.92	22.05	193.34	
2	30	12	4	30	76
	31.31	6.08	3.95	34.66	
Total	206	40	26	228	500

Chi-Sq = 0.010 + 1.033 + 0.000 + 0.112 +
         0.055 + 5.764 + 0.001 + 0.626 = 7.601

DF = 3, P-Value = 0.055
1 cells with expected counts less than 5.0

Interpretation: Because the *P*-value is greater than the level of significance $(0.055 > 0.05)$, we do not reject the null hypothesis. There is not sufficient evidence at the $\alpha = 0.05$ level of significance to support the claim that blood type and Rh-status are independent.

2 **Perform a Test for Homogeneity of Proportions**

The chi-square test for independence is a test regarding a sample from a *single* population. Each individual in this single population is classified in two ways (such as blood type and Rh-status). We now discuss a second type of chi-square test, which can be used to compare the population proportions from *different* populations. The test is an extension of the two-sample *z*-test introduced in Section 11.3, where we compared two population proportions.

In Other Words
The chi-square test for homogeneity of proportions is used to compare proportions from two or more populations.

Definition
In a **chi-square test for homogeneity of proportions**, we test the claim that different populations have the same proportion of individuals with some characteristic.

For example, we might look at the proportion of individuals who experience headaches as a side effect for a placebo group (Group 1), for an experimental group that receives 50 mg per day of a medication (Group 2), and for an experimental group that receives 100 mg per day of the medication (Group 3). Under this circumstance, we might claim that the proportion of individuals in each group who experience a headache as a side effect is the same in each population, so our null hypothesis would be

$$H_0: p_1 = p_2 = p_3$$

versus the alternative,

H_1: At least one of the population proportions is different from the others.

The procedures for performing a test of homogeneity are identical to those for a test of independence.

EXAMPLE 5   A Test for Homogeneity of Proportions

Problem: Zocor is a drug manufactured by Merck and Co. that is meant to reduce the level of LDL (bad) cholesterol, while increasing the level of HDL (good) cholesterol. In clinical trials of the drug, patients were randomly divided into three groups. Group 1 received Zocor, Group 2 received a placebo, and Group 3 received cholestyramine, a cholesterol-lowering drug currently available. Table 17 contains the number of patients in each group who did and did not experience abdominal pain as a side effect.

Table 17

	Group 1 (Zocor)	Group 2 (Placebo)	Group 3 (Cholestyramine)
Number of people who experienced abdominal pain	51	5	16
Number of people who did not experience abdominal pain	1532	152	163

Source: Merck and Co.

Is there evidence to indicate that the proportion of subjects in each group who experienced abdominal pain is different at the $\alpha = 0.01$ level of significance?

Approach: We will follow Steps 1 through 6 on pages 654–655.

Solution

Step 1: For the sake of the test, we assume that the proportion of subjects in each group who experienced abdominal pain are equal. We state the hypotheses as follows:

   $H_0: p_1 = p_2 = p_3$
   H_1: At least one of the proportions is different from the others.

Here, p_1, p_2, and p_3 are the proportions in groups 1, 2, and 3, respectively.

Step 2: The level of significance is $\alpha = 0.01$.

Step 3:

(a) The expected frequency of subjects who experienced abdominal pain in Group 1 is computed by multiplying the row total of individuals who experienced abdominal pain by the column total of number of individuals in Group 1 and dividing this result by the total number of subjects in the study. There were $51 + 5 + 16 = 72$ subjects who experienced abdominal pain and $51 + 1532 = 1583$ subjects in Group 1. There were a total of $51 + 5 + 16 + 1532 + 152 + 163 = 1919$ subjects in the study. The number of subjects expected to experience abdominal pain in Group 1 is

$$E = \frac{72 \cdot 1583}{1919} = 59.393$$

Table 18 contains the row and column totals along with the observed frequencies. The expected frequencies are in parentheses.

Table 18

| | Observed (and Expected) Frequencies | | | |
	Group 1 (Zocor)	Group 2 (Placebo)	Group 3 (Cholestyramine)	Row Totals
Number of people who experienced abdominal pain	51 (59.393)	5 (5.891)	16 (6.716)	72
Number of people who did not experience abdominal pain	1532 (1523.607)	152 (151.109)	163 (172.284)	1847
Column totals	1583	157	179	1919

(b) All the expected frequencies are greater than 5. The requirements have been satisfied.

(c) The test statistic is

$$\chi_0^2 = \frac{(51 - 59.393)^2}{59.393} + \frac{(5 - 5.891)^2}{5.891} + \frac{(16 - 6.716)^2}{6.716} +$$

$$\frac{(1532 - 1523.607)^2}{1523.607} + \frac{(152 - 151.109)^2}{151.109} + \frac{(163 - 172.284)^2}{172.284} = 14.707$$

Classical Approach

Step 4: There are $r = 2$ rows and $c = 3$ columns, so we find the critical value using $(2 - 1)(3 - 1) = 2$ degrees of freedom. The critical value is $\chi_{0.01}^2 = 9.210$. See Figure 17.

Figure 17

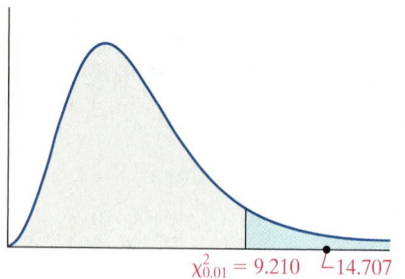

$\chi_{0.01}^2 = 9.210$ 14.707

Step 5: Because the test statistic, 14.707, is greater than the critical value, 9.210, we reject the null hypothesis.

P-value Approach

Step 4: There are $r = 2$ rows and $c = 3$ columns, so we find the P-value using $(2 - 1)(3 - 1) = 2$ degrees of freedom. The P-value is the area under the chi-square distribution with 2 degrees of freedom to the right of $\chi_0^2 = 14.707$, as shown in Figure 18.

Figure 18

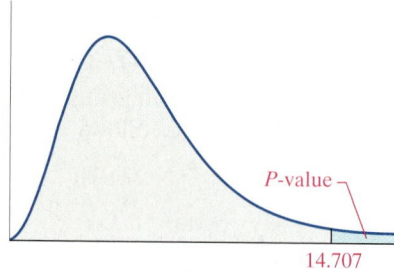

P-value

14.707

Using Table VI, we find the row that corresponds to 2 degrees of freedom. The area under the chi-square distribution with 2 degrees of freedom to the right of 10.597 is 0.005. Because 14.707 is to the right of 10.597, the P-value is less than 0.005. So P-value < 0.005.

Step 5: Because the P-value is less than the level of significance, $\alpha = 0.01$, we reject the null hypothesis.

!**CAUTION**
If we reject the null hypothesis in a chi-square test for homogeneity, we are saying that there is sufficient evidence for us to believe that at least one proportion is different from the others. However, it does not tell us which proportions differ.

Step 6: There is sufficient evidence at the $\alpha = 0.01$ level of significance to reject the claim that the proportion of subjects in each group who experience abdominal pain are equal. We conclude that at least one of the three groups experiences abdominal pain at a rate different from the other two groups. ▪

Figure 19 shows the bar graph, with the blue bars representing the proportion of individuals who experienced abdominal pain for each group and the green bars representing the proportion of individuals who did not experience abdominal pain for each group.

Figure 19

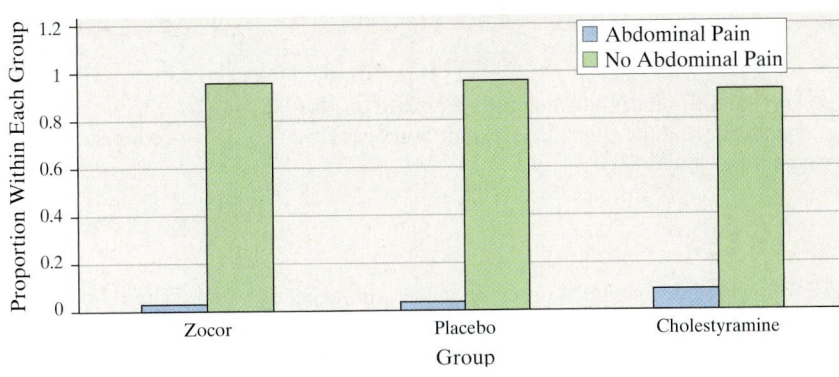

From the graph, it is apparent that a higher proportion of patients taking cholestyramine experience abdominal pain as a side effect.

Now Work Problem 15.

Recall that the requirements for performing a chi-square test are that all expected frequencies are greater than 1 and that at most 20% of the expected frequencies can be less than 5. If these requirements are not satisfied, the researcher has one of two options: (1) combine two columns (or rows) to increase the expected frequencies or (2) increase the sample size.

MAKING AN INFORMED DECISION

Benefits of College

Are there benefits to attending college? If so, what are they? In this project, we will identify some of the perks that a college education provides. Obtain a random sample of at least 50 people aged 21 years or older and administer the following survey:

Please answer the following questions:

1. What is the highest level of education you have attained?
   _____ Have not completed high school
   _____ High school graduate
   _____ College graduate

2. What is your employment status?
   _____ Employed
   _____ Unemployed, but actively seeking work
   _____ Unemployed, but not actively seeking work

3. If you are employed, what is your annual income?
   _____ Less than $20,000
   _____ $20,000–$39,999
   _____ $40,000–$60,000
   _____ More than $60,000

4. If you are employed, which statement best describes the level of satisfaction you have with your career? Answer this question only if you are employed.
   _____ Satisfied—I enjoy my job and am happy with my career.
   _____ Somewhat satisfied—Work is work, but I am not unhappy with my career.
   _____ Somewhat dissatisfied—I do not enjoy my work, but I also have no intention of leaving.
   _____ Dissatisfied—Going to work is painful. I would quit tomorrow if I could.

(continued on next page)

(a) Use the results of the survey to create a contingency table for each of the following categories:
- Level of education/employment status
- Level of education/annual income
- Level of education/job satisfaction
- Annual income/job satisfaction

(b) Perform a chi-square test for independence on each contingency table from part (a).

(c) Draw bar graphs for each contingency table from part (a).

(d) Write a report that details your findings.

12.3 ASSESS YOUR UNDERSTANDING

Concepts and Vocabulary

1. Explain the differences between the chi-square test for independence and the chi-square test for homogeneity. What are the similarities?

2. Why does the test for homogeneity follow the same procedures as the test for independence?

Skill Building

3. The following table contains observed values and expected values in parentheses for two categorical variables, X and Y, where variable X has three categories and variable Y has two categories:

	X_1	X_2	X_3
Y_1	34	43	52
	(36.26)	(44.63)	(48.11)
Y_2	18	21	17
	(15.74)	(19.37)	(20.89)

(a) Compute the value of the chi-square test statistic.
(b) Test the claim that X and Y are independent at the $\alpha = 0.05$ level of significance.
(c) What is the P-value?

4. The following table contains observed values and expected values in parentheses for two categorical variables, X and Y, where variable X has three categories and variable Y has two categories:

	X_1	X_2	X_3
Y_1	87	74	34
	(75.12)	(80.43)	(39.46)
Y_2	12	32	18
	(23.88)	(25.57)	(12.54)

(a) Compute the value of the chi-square test statistic.
(b) Test the claim that X and Y are independent at the $\alpha = 0.05$ level of significance.
(c) What is the P-value?

5. The following table contains the number of successes and failures for three categories of a variable.

	Category 1	Category 2	Category 3
Success	76	84	69
Failure	44	41	49

Test the claim that the proportions are equal for each category at the $\alpha = 0.01$ level of significance. What is the P-value?

6. The following table contains the number of successes and failures for three categories of a variable.

	Category 1	Category 2	Category 3
Success	204	199	214
Failure	96	121	98

Test the claim that the proportions are equal for each category at the $\alpha = 0.01$ level of significance. What is the P-value?

Applying the Concepts

7. **Family Structure and Sexual Activity** A sociologist wants to discover whether the sexual activity of females between the ages of 15 and 19 years of age and family structure are associated. She randomly selects 380 females between the ages of 15 and 19 years of age and asks each to disclose her family structure at age 14 and whether she has had sexual intercourse. The results are shown in the table on the next page. Data are based on information obtained from the U.S. National Center for Health Statistics.

(a) Compute the expected values of each cell under the assumption of independence.
(b) Verify that the requirements for performing a chi-square test of independence are satisfied.
(c) Compute the chi-square test statistic.
(d) Test whether family structure and sexual activity of 15- to 19-year-old females are independent at the $\alpha = 0.05$ level of significance.

Sexual Activity	Family Structure			
	Both Biological Adoptive Parents	Single Parent	Parent and Stepparent	Nonparental Guardian
Had sexual intercourse	64	59	44	32
Did not have sexual intercourse	86	41	36	18

(e) Compare the observed frequencies with the expected frequencies. Which cell contributed most to the test statistic? Was the expected frequency greater than or less than the observed frequency? What does this information tell you?

(f) Construct a conditional distribution by family structure and draw a bar graph. Does this evidence support your conclusion in part (d)?

(g) Compute the *P*-value for this test by finding the area under the chi-square distribution to the right of the test statistic.

---

8. Prenatal Care An obstetrician wants to learn whether the amount of prenatal care and the wantedness of the pregnancy are associated. He randomly selects 939 women who had recently given birth and asks them to disclose whether their pregnancy was intended, unintended, or mistimed. In addition, they were to disclose when they started receiving prenatal care, if ever. The results of the survey are as follows:

Wantedness of Pregnancy	Months Pregnant before Prenatal Care Began		
	Less Than 3 Months	3 to 5 Months	More Than 5 Months (or never)
Intended	593	26	33
Unintended	64	8	11
Mistimed	169	19	16

(a) Compute the expected values of each cell under the assumption of independence.

(b) Verify that the requirements for performing a chi-square test of independence are satisfied.

(c) Compute the chi-square test statistic.

(d) Test whether prenatal care and the wantedness of pregnancy are independent at the $\alpha = 0.05$ level of significance.

(e) Compare the observed frequencies with the expected frequencies. Which cell contributed most to the test statistic? Was the expected frequency greater than or less than the observed frequency? What does this information tell you?

(f) Construct a conditional distribution by wantedness of the pregnancy and draw a bar graph. Does this evidence support your conclusion in part (d)?

---

9. Education versus Area of Country An urban economist wants to determine whether the region of the United States a resident lives in is associated with level of education. He randomly selects 1804 residents of the United States and asks them to disclose the region of the United States in which they reside and their level of education. He obtains the data in the following table:

Area of Country	Level of Education			
	Not a High School Graduate	High School Graduate	Some College	Bachelor's Degree or Higher
Northeast	52	123	70	94
Midwest	123	146	102	96
South	119	204	148	144
West	62	106	111	104

Source: U.S. Census Bureau

(a) Test whether level of education and region of the United States are independent at the $\alpha = 0.05$ level of significance.

(b) Compare the observed frequencies with the expected frequencies. Which cell contributed most to the test statistic? Was the expected frequency greater than or less than the observed frequency? What does this information tell you?

(c) Construct a conditional distribution by level of education and draw a bar graph. Does this evidence support your conclusion in part (a)?

10. Profile of Smokers The following data represent the smoking status by level of education for residents of the United States 18 years old or older from a random sample of 1054 residents.

Number of Years of Education	Smoking Status		
	Current	Former	Never
<12	178	88	208
12	137	69	143
13–15	44	25	44
16 or more	34	33	51

Source: National Health Interview Survey

(a) Test whether smoking status and level of education are independent at the $\alpha = 0.05$ level of significance.
(b) Compute the P-value for this test by finding the area under the chi-square distribution to the right of the test statistic.
(c) Construct a conditional distribution by number of years of education, and draw a bar graph. Does this evidence support your conclusion in part (a)?

11. Legalization of Marijuana On May 14, 2001, the Supreme Court, by a vote of 8 to 0, struck down state laws that legalized marijuana for medicinal purposes. The Gallup Organization later conducted surveys of randomly selected Americans 18 years old or older and asked whether they support the limited use of marijuana when prescribed by physicians to relieve pain and suffering. The results of the survey, by age group, are as follows:

Opinion	Age		
	18–29 Years Old	30–49 Years Old	50 Years or Older
For	172	313	258
Against	52	103	119

(a) Test whether age and opinion regarding the legalization of marijuana are independent at the $\alpha = 0.05$ level of significance.
(b) Compute the P-value for this test by finding the area under the chi-square distribution to the right of the test statistic.
(c) Construct a conditional distribution by age and draw a bar graph. Does this evidence support your conclusion in part (a)?

12. Pro Life or Pro Choice A recent Gallup Organization Poll asked male and female Americans whether they were pro life or pro choice when it comes to abortion issues. The results of the survey are as follows:

Gender	Pro Life/Pro Choice	
	Pro Life	Pro Choice
Men	196	199
Women	239	249

(a) Test whether an individual's opinion regarding abortion is independent of gender at the $\alpha = 0.1$ level of significance.
(b) Compute the P-value for this test by finding the area under the chi-square distribution to the right of the test statistic.
(c) Construct a conditional distribution by gender and draw a bar graph. Does this evidence support your conclusion in part (a)?

13. Delinquencies A delinquency offense is an act committed by a juvenile for which an adult could be prosecuted in a criminal court. The following data represent the number of various types of delinquencies by gender in a random sample of 750 delinquencies.

Gender	Delinquency			
	Person	Property	Drugs	Public Order
Female	24	85	7	28
Male	97	367	39	103

(a) Test whether gender is independent of type of delinquency at the $\alpha = 0.05$ level of significance.
(b) Compute the P-value for this test by finding the area under the chi-square distribution to the right of the test statistic.
(c) Construct a conditional distribution by type of delinquency and draw a bar graph. Does this evidence support your conclusion in part (a)?

14. Visits to the Emergency Room The following data represent the gender and age of 489 randomly selected patients who visited the emergency room with an injury-related emergency.

Gender	Age (years)					
	Under 15	15–24	25–44	45–64	65–74	75 and Older
Male	66	56	92	37	8	11
Female	44	39	66	34	12	24

(a) Test the claim that the age of an individual visiting the emergency room is independent of gender at the $\alpha = 0.1$ level of significance.
(b) Compute the P-value for this test by finding the area under the chi-square distribution to the right of the test statistic.
(c) Construct a conditional distribution by age and draw a bar graph. Does this evidence support your conclusion in part (a)?

15. Smoked Lately? Suppose a researcher wants to investigate whether the proportion of smokers within different age groups is the same. He divides the American population into four age groups: 18 to 29 years old, 30 to 49 years old, 50 to 64 years old, and 65 years or older. Within each age group, he surveys 80 individuals and asks, "Have you smoked at least one cigarette in the past week?" The results of the survey are as follows:

Smoking Status	Age (years)			
	18–29	30–49	50–64	65 or Older
Smoked at least one cigarette in past week	24	21	23	12
Did not smoke at least one cigarette in past week	56	59	57	68

Source: Gallup Organization

(a) Is there evidence to indicate that the proportion of individuals within each age group who have smoked at least one cigarette in the past week is different at the $\alpha = 0.05$ level of significance?

(b) Compute the *P*-value for this test by finding the area under the chi-square distribution to the right of the test statistic.

(c) Construct a conditional distribution by age and draw a bar graph. Does this evidence support your conclusion in part (a)?

16. **Are You Satisfied?** Suppose an economist wants to gauge the level of satisfaction of Americans. He randomly samples 150 people 18 years old or older from four geographic regions of the United States: East, South, Midwest, and West. He asks the individuals selected, "Are you satisfied or dissatisfied with the way things are going in the United States at this time?" The following table gives results of the survey.

Satisfaction	Region			
	East	South	Midwest	West
Satisfied	77	84	93	83
Dissatisfied	73	66	57	67

Source: Gallup Organization

(a) Test whether the proportions of Americans who are satisfied with the way things are going in the United States for each region of the country are equal at the $\alpha = 0.1$ level of significance.

(b) Compute the *P*-value for this test by finding the area under the chi-square distribution to the right of the test statistic.

(c) Construct a conditional distribution by region of the country and draw a bar graph. Does this evidence support your conclusion in part (a)?

17. **Celebrex** Celebrex is a drug manufactured by Pfizer, Inc., used to relieve symptoms associated with osteoarthritis and rheumatoid arthritis in adults. It is considered to be one of the nonsteroidal anti-inflammatory drugs. These types of drugs are known to be associated with gastrointestinal toxicity, such as bleeding, ulceration, and perforation of the stomach, small intestine, or large intestine. In clinical trials, researchers wanted to learn whether the proportion of subjects taking Celebrex who experienced these side effects differed significantly from that in other treatment groups. The following data were collected (Naproxen is a nonsteroidal anti-inflammatory drug that is also used in the treatment of arthritis).

Side Effect	Placebo	Treatment			
		Celebrex (50 mg per day)	Celebrex (100 mg per day)	Celebrex (200 mg per day)	Naproxen (500 mg per day)
Experienced ulcers	5	8	7	13	34
Did not experience ulcers	212	225	220	208	176

Source: Pfizer, Inc.

(a) Test whether the proportion of subjects within each treatment group is the same at the $\alpha = 0.01$ level of significance.

(b) Compute the *P*-value for this test by finding the area under the chi-square distribution to the right of the test statistic.

(c) Construct a conditional distribution by treatment and draw a bar graph. Does this evidence support your conclusion in part (a)?

18. **Celebrex** Celebrex is a drug manufactured by Pfizer, Inc., used to relieve symptoms associated with osteoarthritis and rheumatoid arthritis in adults. In clinical trials of the medication, some subjects reported dizziness as a side effect. The researchers wanted to discover whether the proportion of subjects taking Celebrex who reported dizziness as a side effect differed significantly from that for other treatment groups. The following data were collected.

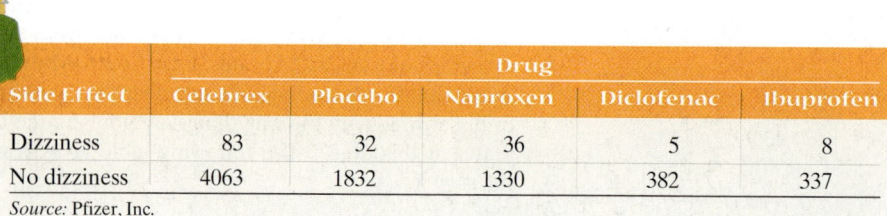

Side Effect	Drug				
	Celebrex	Placebo	Naproxen	Diclofenac	Ibuprofen
Dizziness	83	32	36	5	8
No dizziness	4063	1832	1330	382	337

Source: Pfizer, Inc.

(a) Test whether the proportion of subjects within each treatment group who experienced dizziness are the same at the $\alpha = 0.01$ level of significance.

(b) Compute the *P*-value for this test by finding the area under the chi-square distribution to the right of the test statistic.

(c) Construct a conditional distribution by treatment and draw a bar graph. Does this evidence support your conclusion in part (a)?

---

19. **Dropping a Course** A survey of 52 randomly selected students who dropped a course in the current semester was conducted at a community college. The goal of the survey was to learn why students drop courses. The following data were collected: "Personal" drop reasons include financial, transportation, family issues, health issues, and lack of child care. "Course" drop reasons include reducing one's load, being unprepared for the course, the course was not what was expected, dissatisfaction with teaching, and not getting the desired grade. "Work" drop reasons include an increase in hours, a change in shift, and obtaining full-time employment. "Career" drop reasons include not needing the course and a change of plans. The results of the survey are as follows:

(a) Construct a contingency table for the two variables.

(b) Test the claim that gender is independent of drop reason at the $\alpha = 0.1$ level of significance.

(c) Compute the *P*-value for this test by finding the area under the chi-square distribution to the right of the test statistic.

(d) Construct a conditional distribution by drop reason and draw a bar graph. Does this evidence support your conclusion in part (a)?

20. **Political Affiliation** A political scientist wanted to learn whether there is any association between the education level of a registered voter and his or her political party affiliation. He randomly selected 46 registered voters and obtained the following data:

Gender	Drop Reason	Gender	Drop Reason
Male	Personal	Male	Work
Female	Personal	Male	Work
Male	Work	Female	Course
Male	Personal	Male	Work
Male	Course	Female	Course
Male	Course	Female	Course
Female	Course	Female	Course
Female	Course	Male	Work
Male	Course	Male	Personal
Female	Course	Male	Course
Male	Personal	Female	Course
Male	Work	Female	Course
Male	Work	Male	Course
Male	Course	Female	Course
Male	Course	Male	Work
Male	Work	Male	Course
Female	Personal	Female	Work
Male	Course	Male	Personal
Female	Work	Male	Work
Male	Work	Female	Course
Male	Work	Male	Course
Female	Course	Male	Personal
Female	Personal	Female	Course
Female	Personal	Female	Work
Female	Personal	Male	Work

Education	Political Party	Education	Political Party
Grade school	Democrat	High school	Democrat
College	Republican	College	Republican
High school	Democrat	College	Republican
High school	Republican	Grade school	Democrat
High school	Democrat	High school	Republican
Grade school	Democrat	High school	Democrat
College	Republican	High school	Democrat
Grade school	Democrat	College	Republican
High school	Democrat	High school	Republican
High school	Democrat	Grade school	Democrat
Grade school	Democrat	High school	Democrat
College	Republican	College	Democrat
Grade school	Democrat	College	Republican
College	Democrat	High school	Republican
College	Democrat	College	Democrat
Grade school	Republican	College	Democrat
College	Republican	High school	Democrat
Grade school	Republican	College	Republican
College	Republican	College	Democrat
High school	Democrat	High school	Republican
College	Democrat	College	Republican
College	Republican	High school	Republican
College	Democrat	College	Democrat

(a) Construct a contingency table for the two variables.
(b) Test the claim that level of education is independent of political affiliation at the $\alpha = 0.1$ level of significance.

(c) Compute the P-value for this test by finding the area under the chi-square distribution to the right of the test statistic.
(d) Construct a conditional distribution by level of education and draw a bar graph. Does this evidence support your conclusion in part (a)?

In Problem 21, we demonstrate that the z-test for comparing two population proportions is equivalent to the chi-square test for homogeneity when there are two possible outcomes.

21. **Percentage of Americans Who Smoke on the Decline?**
On November 3, 2004, the Gallup Organization surveyed 1007 adults and found that 222 of them had smoked at least one cigarette in the past week. In 1998, they also asked 1007 adults the same question and determined that 282 adults had smoked at least one cigarette in the past week. The results are presented in the following table:

Smoking Status	Year	
	1998	2004
Smoked	282	222
Did not smoke	725	785

(a) Compute the expected number of adult Americans who have smoked at least one cigarette in the past week and the expected number who have not, assuming $p_{1990} = p_{2004}$.
(b) Compute the chi-square test statistic.
(c) Test the researchers' claim at the $\alpha = 0.05$ level of significance using the chi-square test.
(d) Compute the Z-test statistic. Now compute z^2. Compare z^2 with the chi-square test statistic. Conclude that $z^2 = \chi^2$.

Consumer Reports Dirty Birds?

Hungry for a cheap, low-fat alternative to beef, Americans are eating more chicken than ever. Although precise figures are impossible to obtain, the U.S. Centers for Disease Control and Prevention reported that the number of cases of outbreaks of illness caused by chicken rose threefold between 1988 and 1992. Salmonella bacteria were the most common cause of these outbreaks.

In a study for *Consumer Reports*, we purchased 1000 fresh, whole broiler chickens at grocery stores in 36 cities across the United States over a 5-week period. Our shoppers packed the birds in coolers and shipped them overnight to the lab. There, tests were conducted to determine the presence of salmonella and campylobacter, another chicken-related bug. The results of the study for salmonella were as follows:

Brand	Salmonella		Total
	Present	Absent	
A	8	192	200
B	17	183	200
C	27	173	200
D	14	186	200
E	20	180	200
Total	86	914	1000

Assuming that the chickens represent a random sample from each brand included in the study, use the information presented in the table to answer the following:

(a) Calculate the proportion of incidence for each brand shown in the table.

(b) Compute a 95% confidence interval for the incidence of salmonella for brand C.

(c) Using a chi-square test of homogeneity of proportions, is there evidence that the five brands have the same incidence rate for salmonella?

(d) Brands A and D are major competitors in the same market. The manufacturer of brand A claims to have improved its cleanliness and claims that it is substantially cleaner than brand D. Is there evidence to support this contention?

(e) Write a paragraph for the readers of *Consumer Reports* magazine that explains your conclusions.

Note to Readers: In many cases, our test protocol and analytical methods are more complicated than described in these examples. The data and discussions have been modified to make the material more appropriate for the audience.

Technology Step-by-Step	Chi-Square Tests

TI-83/84 Plus

Step 1: Access the MATRX menu. Highlight the EDIT menu, and select `1: [A]`.

Step 2: Enter the number of rows and columns of the matrix.

Step 3: Enter the cell entries for the observed matrix, and press 2nd QUIT. Repeat Steps 1–3 for the expected values, but enter the expected frequencies in matrix B.

Step 4: Press STAT, highlight the TESTS menu, and select `C: χ²-Test`

Step 5: With the cursor after the **Observed:**, enter matrix [A] by accessing the MATRX menu, highlighting NAMES, and selecting `1: [A]`.

Step 6: With the cursor after the **Expected:**, enter matrix [B] by accessing the MATRX menu, highlighting NAMES, and selecting `2: [B]`.

Step 7: Highlight **Calculate** or **Draw**, and press ENTER.

MINITAB

Step 1: Enter the data into the MINITAB spreadsheet.

Step 2: Select the **Stat** menu, highlight **Tables**, and select **Chi-Square Test . . . .**

Step 3: Select the columns that contain the data, and press OK.

Excel

Step 1: Enter the observed frequencies in the spreadsheet.

Step 2: Compute the expected frequencies and enter them in a different location in the spreadsheet.

Step 3: Select *fx* from the tool bar. Select **Statistical** for the function category and highlight **CHITEST** in the function name.

Step 4: With the cursor in the actual cell, highlight the observed data. With the cursor in the expected cell, highlight the expected frequencies. Click OK. The output provided is the *P*-value.

Note: This test can also be performed by using the PHStat add-in. See the Excel Technology Manual.

Chapter **12** # Review

Summary

In this chapter, we introduced chi-square methods. The first chi-square method involved tests for goodness-of-fit. We used the chi-square distribution to test the claim that a random variable followed a certain distribution. This is done by comparing values expected based on the distribution of the random variable to observed values.

The next section of the chapter introduced methods that allow us to describe any association that might exist between two qualitative variables. This is done through contingency tables. Both marginal and conditional distributions allow us to describe the effect one variable might have on the other variable in the study. We also construct bar graphs to see the association between the two variables in the study.

Finally, we introduced chi-square methods that allowed us to perform tests for independence and homogeneity. In a test for independence, the researcher obtains random data for two variables and tests whether the variables are associated. The

null hypothesis in these tests is always that the variables are not associated (independent). The test statistic compares the values expected if the variables were independent to those observed. If the expected and observed values differ significantly, we reject the null hypothesis and conclude that there is evidence to support the belief that the variables are not independent (they are associated). We draw bar graphs of the marginal distributions to help us see the association, if any.

The last chi-square test was the test for homogeneity of proportions. This test is similar to the test for independence, except we are testing that the proportion of individuals in the study with a certain characteristic are equal ($p_1 = p_2 = \cdots = p_k$). To perform this test, we take random samples of a predetermined size from each group under consideration (a random sample of size n_1 for group 1, a random sample of size n_2 for group 2, and so on).

Formulas

Expected Counts In a Goodness of Fit Test

$E_i = \mu_i = np_i$ for $i = 1, 2, \ldots, k$

Chi-Square Test Statistic

$\chi^2 = \sum \dfrac{(O_i - E_i)^2}{E_i}$ $i = 1, 2, \ldots, k$

Expected Frequencies In a Test for Independence

$$\text{Expected frequency} = \dfrac{(\text{row total})(\text{column total})}{\text{table total}}$$

Vocabulary

Goodness-of-fit test (p. 631)
Expected counts (p. 632)
Contingency (or two-way) table (p. 642)
Row variable (p. 642)

Column variable (p. 642)
Cell (p. 642)
Marginal distribution (p. 642)
Conditional distribution (p. 645)

Chi-square independence test (p. 651)
Chi-square test for homogeneity of proportions (p. 658)

Objectives

Section	You should be able to . . .	Examples	Review Exercises
12.1	**1** Perform a goodness-of-fit test (p. 631)	2 and 3	1–4
12.2	**1** Compute the marginal distribution of a variable (p. 642)	1 and 2	5(a), (b)–8(a), (b)
	2 Use the conditional distribution to identify association among categorical data (p. 644)	3–5	5(e)–8(e)
12.3	**1** Perform a test for independence (p. 651)	1–4	9–12
	2 Perform a test for homogeneity of proportions (p. 658)	5	13, 14, 15

Review Exercises

1. Roulette Wheel A pit boss suspects that a roulette wheel is out of balance. A roulette wheel has 18 black slots, 18 red slots, and 2 green slots. The pit boss spins the wheel 500 times and records the following frequencies:

Outcome	Frequency
Black	233
Red	237
Green	30

Test the claim that the wheel is out of balance at the $\alpha = 0.05$ level of significance.

2. Fair Dice? A pit boss is concerned that a pair of dice being used in a craps game is not fair. The distribution of the expected sum of two fair dice is as follows:

Sum of Two Dice	Probability	Sum of Two Dice	Probability
2	$\frac{1}{36}$	8	$\frac{5}{36}$
3	$\frac{2}{36}$	9	$\frac{4}{36}$
4	$\frac{3}{36}$	10	$\frac{3}{36}$
5	$\frac{4}{36}$	11	$\frac{2}{36}$
6	$\frac{5}{36}$	12	$\frac{1}{36}$
7	$\frac{6}{36}$		

The pit boss rolls the dice 400 times and records the sum of the dice. The following are the results:

Sum of Two Dice	Frequency	Sum of Two Dice	Frequency
2	16	8	59
3	23	9	45
4	31	10	34
5	41	11	19
6	62	12	11
7	59		

Test the claim that the dice are fair at the $\alpha = 0.01$ level of significance.

3. Educational Attainment A researcher wanted to test the claim that the distribution of educational attainment of Americans today is different from the distribution in 1994. The distribution of educational attainment in 1994 is as follows:

Education	Relative Frequency
Not a high school graduate	0.191
High school graduate	0.344
Some college	0.174
Associate's degree	0.070
Bachelor's degree	0.147
Advanced degree	0.075

Source: Statistical Abstract of the United States

To test the claim, the researcher randomly selects 500 Americans, learns their levels of education, and obtains the following data:

Education	Frequency
Not a high school graduate	89
High school graduate	152
Some college	83
Associate's degree	39
Bachelor's degree	93
Advanced degree	44

At the $\alpha = 0.1$ level of significance, test the claim that the distribution of educational attainment today is different from that of educational attainment in 1994.

4. School Violence A school administrator is concerned that the distribution of school violence has changed and has become more violent. The distribution of school crime for 1992 is as follows:

Crime	Theft	Violent	Serious Violent
Relative frequency	0.619	0.314	0.067

Source: National Center for Educational Statistics

To test the claim, the researcher randomly selects 800 school crimes, finds out whether they were theft, violent, or serious violent crimes, and obtains the following data:

Crime	Theft	Violent	Serious Violent
Frequency	421	311	68

At the $\alpha = 0.05$ level of significance, test the claim that the distribution of school crime today is different from the 1992 distribution.

5. Police Officers The Federal Bureau of Investigation publishes the *Uniform Crime Report*, which contains information regarding various crimes throughout the United States. The following data represent the number of law enforcement officers feloniously killed by type of firearm and by region of the country for the years 1994 to 2003.

	Region			
Firearm	Northeast	Midwest	South	West
Handgun	43	79	182	86
Rifle	7	21	45	20
Shotgun	3	5	20	6

(a) Construct a frequency marginal distribution.
(b) Construct a relative frequency marginal distribution.
(c) What percentage of law enforcement officers were killed by a handgun?
(d) What percentage of law enforcement officers killed were in the Northeast?
(e) Construct a conditional distribution by region of the United States.
(f) Draw a bar graph of the conditional distribution.
(g) What, if any, association exists between gun type and region of the United States? Discuss the association.

6. Educational Attainment versus Age The following data (in thousands) represent the educational attainment and age of residents of the United States in 2003.

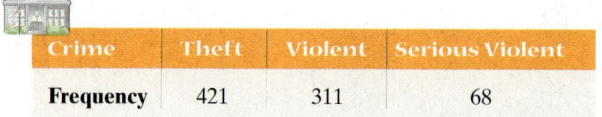

	Level of Education			
Age (years)	Not a High School Graduate	High School Graduate	Some College	Four or More Years of College
25–34	5062	11,380	10,988	11,773
35–44	5289	13,883	11,944	12,958
45–54	4426	12,432	11,104	12,231
55–64	4055	9,179	6,768	7,398
At Least 65	9786	12,396	6,128	5,942

Source: U.S. Census Bureau, Current Population Report

(a) Construct a frequency marginal distribution.
(b) Construct a relative frequency marginal distribution.
(c) What percentage of adults 25 years old or older are high school graduates?
(d) What percentage of adults 25 years old or older are at least 65 years of age?
(e) Construct a conditional distribution by level of education.
(f) Draw a bar graph of the conditional distribution.
(g) What association, if any, exists between level of education and age? Discuss the association.

7. **Child Fatalities** The following data represent the number of fatalities involving children under the age of 13 and a motor vehicle, by type of fatality and gender in 2003.

Type of Fatality	Gender	
	Male	Female
Passenger in vehicle	588	568
Pedestrian	215	114
Bicycle	79	14

Source: U.S. Department of Transportation

(a) Construct a frequency marginal distribution.
(b) Construct a relative frequency marginal distribution.
(c) What percentage of children under the age of 13 killed by a motor vehicle was male?
(d) What percentage of children under the age of 13 was killed in accidents involving motor vehicles while riding a bicycle?
(e) Construct a conditional distribution by gender.
(f) Draw a bar graph of the conditional distribution.
(g) What association, if any, exists between type of fatality and gender? Discuss the association.

8. **September 11** The following data represent the number of murder victims of the September 11, 2001, terrorist attack by location and gender.

Gender	World Trade Center	Pentagon	Somerset County, PA
Male	2128	108	20
Female	621	71	20

Source: U.S. Federal Bureau of Investigation

(a) Construct a frequency marginal distribution.
(b) Construct a relative frequency marginal distribution.
(c) What percentage of September 11th victims were female?
(d) What percentage of September 11th victims were at the World Trade Center?
(e) Construct a conditional distribution by gender.
(f) Draw a bar graph of the conditional distribution.

9. **Evolution or Creation?** The Gallup Organization conducted a poll of 1016 randomly selected Americans aged 18 years old or older in February 2001 and 1017 randomly selected Americans aged 18 years old or older in June 1993 and asked them the following question:

> Which of the following statements comes closest to your views on the origin and development of human beings? (1) Human beings have developed over millions of years from less advanced forms of life, but God guided this process. (2) Human beings have developed over millions of years from less advanced forms of life, and God had no part in this process. (3) God created human beings pretty much in their present form at one time within the last 10,000 years or so.

The results of the survey are as follows:

Date	Belief			
	Humans Developed, with God Guiding	Humans Developed, but God Had No Part in Process	God Created Humans in Present Form	Other/No Opinion
February 2001	376	122	457	61
June 1993	356	112	478	71

(a) Compute the expected values of each cell under the assumption of independence.
(b) Verify that the requirements for performing a chi-square test of independence are satisfied.
(c) Compute the chi-square test statistic.
(d) Test the claim that people's opinion regarding human origin is independent of the date the question is asked at the $\alpha = 0.05$ level of significance.
(e) Compare the observed frequencies with the expected frequencies. Which cell contributed most to the test statistic? Was the expected frequency greater than or less than the observed frequency? What does this information tell you?
(f) Construct a conditional distribution by date and draw a bar graph. Does this evidence support your conclusion in part (d)?
(g) Compute the P-value for this test by finding the area under the chi-square distribution to the right of the test statistic.

10. Caesarean Sections An obstetrician wanted to discover whether the method of delivering a baby was independent of race. The following data represent the race of the mother and the method of delivery for 365 randomly selected births.

Race	Delivery Method	
	Vaginal	Caesarean
White	242	63
Black	47	13

Source: Statistical Abstract of the United States

(a) Compute the expected values of each cell under the assumption of independence.
(b) Verify that the requirements to perform a chi-square test of independence are satisfied.
(c) Compute the chi-square test statistic.
(d) Test the claim that method of delivery is independent of the race of the mother at the $\alpha = 0.05$ level of significance.
(e) Compare the observed frequencies with the expected frequencies. Which cell contributed most to the test statistic? Was the expected frequency greater than or less than the observed frequency? What does this information tell you?
(f) Construct a conditional distribution by method of delivery and draw a bar graph. Does this evidence support your conclusion in part (d)?
(g) Compute the *P*-value for this test by finding the area under the chi-square distribution to the right of the test statistic.

11. Race versus Region of the United States A sociologist wanted to determine whether the locations in which individuals live are independent of their races. He randomly selects 2712 U.S. residents and asks them to disclose their race and the location in which they live. He obtains the following data:

Race	Region			
	Northeast	Midwest	South	West
White	421	520	656	400
Black	56	57	158	28
American Indian, Eskimo, Aleut	2	4	6	9
Asian or Pacific Islander	13	8	11	40
Hispanic	38	17	68	101
Other	17	8	24	50

Source: Statistical Abstract of the United States

At the $\alpha = 0.01$ level of significance, test the claim that race is independent of the location in which a resident lives within the United States.

12. Marital Status and Gender A sociologist wanted to determine whether marital status and gender were independent. He randomly sampled 201 residents of the United States who were 18 years old or older and asked them to disclose their gender and marital status. The following data were collected.

Marital Status	Gender	
	Male	Female
Never married	26	22
Married	59	60
Divorced	9	11
Widowed	3	11

At the $\alpha = 0.05$ level of significance, test the claim that gender is independent of marital status.

13. The Common Cold A doctor wanted to determine whether the proportion of Americans who have had symptoms associated with the common cold is the same for all four regions of the United States. He randomly sampled 120 individuals from each region of the United States and obtained the following data:

Symptoms	Region			
	Northeast	Midwest	South	West
Symptoms within last year	26	31	23	35
No symptoms within last year	94	89	97	85

Source: National Center for Health Statistics

At the $\alpha = 0.05$ level of significance, test the claim that the proportion of Americans who have had symptoms associated with the common cold is the same for all four regions of the United States.

14. Hardworking? In a *Newsweek* poll conducted on June 24, 1999, 750 randomly selected adults were asked, "Do you believe Americans today are as willing to work hard at their jobs to get ahead as they were in the past, or are not as willing to work hard to get ahead?" The results of the survey are as follows:

Response	Age (years)		
	18–29	30–49	50 and Older
As willing	58	75	70
Not as willing	192	175	180

Source: pollingreport.com

At the $\alpha = 0.05$ level of significance, test the claim that the proportion of Americans who believe that Americans are willing to work just as hard today is the same for all three age groups.

15. Night Owls For a Gallup poll conducted on June 6–8, 2005, 1279 randomly selected adults aged 18 and older were asked, "Are you a morning person or a night person?" The results of the survey are as follows:

Response	Age			
	18–29	30–49	50–64	65 or Older
Morning person	97	177	210	210
Night person	234	147	107	97

At the $\alpha = 0.05$ level of significance, test the claim that the proportion of adults aged 18 years or older who are "morning people" are the same for all four age groups.

16. Simpson's Paradox in Baseball The following tables summarize the number of hits and at-bats for Dustan Mohr (Minnesota Twins/San Francisco Giants) and Darin Erstad (Anaheim Angels) during the 2003 MLB season when accounting for runners in scoring position.

Source:
ite.pubs.informs.org/Vol5No1/KvamSokol/index.php

Runner(s) in Scoring Position	Mohr	Erstad
Hits	19	9
At-bats	97	50

No Runners in Scoring Position	Mohr	Erstad
Hits	68	56
At-bats	251	208

(a) Which player had the better batting average (hits divided by at-bats) in each situation?
(b) Which player had the better batting average overall (total hits divided by total at-bats)?
(c) Explain why your answer to part (b) differs from part (a).

CASE STUDY

Feeling Lucky? Well, Are You?

In fiscal year (FY) 2000–01 (October 2000–September 2001), the Florida Lottery generated $2,360,561,022 in total sales. Over that period, the state spent $32,630,000 on advertising to promote its various games. Rand Advertising is interested in gaining access to this lucrative market. You have been assigned the task of preparing a report on the lottery sales structure for three of Florida's online (non-scratch-off ticket) games: Fantasy Five, Mega Money, and Lotto. Your findings will become part of a proposal by Rand to the Florida Lottery.

In Fantasy 5, a player picks five numbers from 1 to 36, at one dollar per play. Drawings are held 7 days a week. If there is no jackpot winner for a drawing, the money allocated for the top prize "rolls down" to the next prize tier (4 of 5). In FY 2000–01, Fantasy 5 generated $191,614,355 in sales. The prize structure for this game is as follows:

Prize Structure for Florida's Fantasy 5 Online Game

Match	Estimated Prize Amount per Winner	Probability
5 of 5	$100,000	$\dfrac{1}{376{,}992}$
4 of 5	$100	$\dfrac{1}{2432}$
3 of 5	$10	$\dfrac{1}{81}$
2 of 5	Free ticket	$\dfrac{1}{8}$

The Mega Money game produced $108,841,978 in sales for FY 2000–01. At one dollar per ticket, players pick 4 numbers from 1 to 32 and one MEGABALL number from 1 to 32. Drawings are held on Tuesdays and Fridays. If there is no jackpot winner for a drawing, the money allocated for the top prize rolls over to the next drawing, adding to the total of the next jackpot. The following is the prize structure for Mega Money.

Prize Structure for Florida's Mega Money Online Game

Match	Estimated Prize Amount per Winner	Probability
4 of 4 + MEGABALL	$200,000	$\dfrac{1}{1{,}150{,}720}$
4 of 4	$550	$\dfrac{1}{37{,}120}$
3 of 4 + MEGABALL	$350	$\dfrac{1}{10{,}274}$
3 of 4	$50	$\dfrac{1}{331}$
2 of 4 + MEGABALL	$25	$\dfrac{1}{507}$
2 of 4	$1	$\dfrac{1}{16}$

In terms of sales, Lotto is Florida's most lucrative lottery product, with $845,565,515 in FY 2000–01 sales. Players pick 6 numbers from 1 to 53, at one dollar per play. Drawings are held on Wednesdays and Saturdays. Like Mega Money, if there is no jackpot winner, the top prize is rolled over to the next drawing. Because of the game design, it is much more difficult to win the Lotto jackpot. The difficulty in winning the game leads to numerous jackpot rollovers, creating, at times, jackpots exceeding $50,000,000! The propensity to roll over makes it difficult to determine an estimated prize payout per winner. However, the odds structure is shown in the table to the right.

Odds Structure for Florida's Lotto Online Game

Match	Probability
6 of 6	$\dfrac{1}{22{,}957{,}480}$
5 of 6	$\dfrac{1}{81{,}410}$
4 of 6	$\dfrac{1}{1416}$
3 of 6	$\dfrac{1}{71}$

To conduct your study, you have obtained the sales figures for each of the three games by district sales office for the week of September 17–23, 2001. These data are as follows:

Number of Tickets Sold for Florida Lottery Games by Sales District for the Week of September 17–23, 2001

District	Florida Lottery Game		
	Fantasy 5	Mega Money	Lotto
1	87,030	27,221	270,256
3	186,780	56,822	934,451
4	267,955	94,123	1,004,651
5	232,451	61,019	707,934
6	727,390	262,840	2,746,438
8	462,874	135,321	1,563,491
9	353,532	139,421	1,492,549
10	377,627	140,285	1,433,077
11	470,993	153,591	1,743,370
12	588,154	185,946	1,852,986
13	1,283,042	474,067	2,883,453

Use an $\alpha = 0.05$ significance level to test the claim that the numbers of tickets sold for each lottery game and sales district are independent. Construct a bar graph that represents the conditional distribution of game by sales district. Does this graphical evidence support your conclusion regarding the relationship between the type of game and the sales district? Explain.

Additionally, you are interested in the daily sales structure for the various districts. The following are the numbers of tickets sold each day for the week of September 17 for the three games in District 1.

Number of Tickets Sold for Florida Lottery Games by Day of the Week (September 17–September 23, 2001) for District 1

Day	Florida Lottery Game		
	Fantasy 5	Mega Money	Lotto
Mon. 9/17	12,794	2,082	13,093
Tue. 9/18	11,564	9,983	18,200
Wed. 9/19	12,562	1,040	60,746
Thu. 9/20	11,408	1,677	18,054
Fri. 9/21	15,299	10,439	38,513
Sat. 9/22	15,684	1,502	115,686
Sun. 9/23	7,728	498	5,964

Use an $\alpha = 0.05$ significance level to test the claim that the proportion of Fantasy 5 sales is the same for each day of the week. Perform a similar test for Mega Money and Lotto.

Write a report detailing your assumptions, analyses, findings, and conclusions.

Comparing Three or More Means

CHAPTER 13

Outline

DECISIONS

Remember the inheritance you received in Chapter 11? Suppose you have decided that you now want to try investing in individual stocks rather than mutual funds. Which type of stocks should you invest in to earn the highest rate of return? See the Decisions project on page 723.

●●● Putting It All Together

Do you remember the progression for comparing proportions? Chapters 9 and 10 discussed inference about a single proportion, Chapter 11 discussed inference about two proportions, and Chapter 12 presented a discussion about inference for three or more proportions (homogeneity of proportions).

We had this same progression of topics for inference on means. In Chapters 9 and 10, we discussed inferential techniques about a population mean. In Chapter 11, we discussed inferential techniques for comparing two means. In this chapter, we continue to expand our horizons by learning inferential techniques for comparing three or more means. Just as we needed to use a different distribution to compare multiple proportions (the chi-square distribution), we use a different distribution for comparing three or more means. Although the choice of distribution initially may seem strange, once the logic of the procedure is understood, the choice of distribution makes sense.

13.1 Comparing Three or More Means (One-Way Analysis of Variance)

Preparing for This Section   Before getting started, review the following:

- Completely randomized design (Section 1.5, pp. 42–43)
- Nature of hypothesis testing (Section 10.1, pp. 505–511)
- Comparing two population means (Section 11.2, pp. 588–593)

- Normal probability plots (Section 7.4, pp. 396–400)
- *F*-distribution (Section 11.4, pp. 612–615)
- Boxplots (Section 3.5, pp. 177–180)

Objectives

1. **Verify the requirements to perform a one-way ANOVA**
2. **Test a claim regarding three or more means using one-way ANOVA**

In Section 11.2, we compared two population means. Just as we extended the concept of comparing two population proportions (Section 11.3) to comparing three or more population proportions (Tests for Homogeneity of Proportions, Section 12.3), we now extend the concept of comparing two population means to comparing three or more population means. The procedure for doing this is called *Analysis of Variance*, or *ANOVA* for short.

Definition   **Analysis of Variance (ANOVA)** is an inferential method that is used to test the equality of three or more population means.

In Other Words

In ANOVA, the null hypothesis is always that the means of the different populations are equal. The alternative hypothesis is always that the mean of at least one population is different from the others.

For example, a family doctor might claim the mean HDL (so-called good) levels of cholesterol of males in the age groups 20 to 29 years old, 40 to 49 years old, and 60 to 69 years old are different. To test this claim, we assume the mean HDL cholesterol of each age group is the same. If we call the 20- to 29-year-olds population 1, 40- to 49-year-olds population 2, and 60 to 69-year-olds population 3, our null hypothesis would be

$$H_0: \mu_1 = \mu_2 = \mu_3$$

versus the alternative hypothesis,

H_1: At least one of the population means is different from the others

As another example, a medical researcher might want to compare the effect different levels of an experimental drug have on hair growth. The researcher might randomly divide a group of subjects into three different treatment groups. Group 1 might receive a placebo once a day, group 2 might receive 50 mg of the experimental drug once a day, and group 3 might receive 100 mg of the experimental drug once a day. The researcher then compares the mean numbers of new hair follicles for each of the three treatment groups. The three different treatment groups correspond to three different populations.

It is tempting to test the null hypothesis $H_0: \mu_1 = \mu_2 = \mu_3$ by comparing the population means two at a time using the techniques introduced in Section 11.2. If we proceeded this way, we would need to test three different hypotheses:

$$H_0: \mu_1 = \mu_2 \quad \text{and} \quad H_0: \mu_1 = \mu_3 \quad \text{and} \quad H_0: \mu_2 = \mu_3$$

CAUTION!

Do not test $H_0: \mu_1 = \mu_2 = \mu_3$ by conducting three separate hypothesis tests, because the probability of making a Type I error will be much higher than α.

Each hypothesis would have a probability of Type I error (rejecting the null hypothesis when it is true) of α. If we used an $\alpha = 0.05$ level of significance, each hypothesis would have a 95% probability of rejecting the null hypothesis when the alternative hypothesis is true (i.e. a 95% probability of making a correct decision). The probability that all three hypotheses correctly reject the null hypothesis is $0.95^3 = 0.86$ (assuming the tests are independent). There is a $1 - 0.95^3 = 1 - 0.86 = 0.14$, or 14%, probability that at least one hypothesis will lead to an incorrect rejection of H_0. A 14% probability of a Type I error is much higher than the desired 5% probability. As the number of populations that are to be compared increases, the probability of making a Type I error using multiple t-tests for a given value of α also increases.

To address this problem, Sir Ronald A. Fisher (1890–1962) introduced the method of analysis of variance. Although it seems strange to name a procedure that is used to compare *means* analysis of *variance*, the justification for the name will become clear as we see the procedure in action.

The procedure used in this section is called *one-way* analysis of variance because there is only one factor that distinguishes the various populations in the study. For example, in comparing the mean HDL cholesterol levels of males, the only factor that distinguishes the three groups is age. In comparing the effectiveness of the hair growth drug, the only factor that distinguishes the three groups is the amount of the experimental drug received (0 mg, 50 mg, or 100 mg).

In performing one-way analysis of variance, care must be taken so that the subjects are similar in all characteristics except for the level of the treatment. Fisher stated that this is easiest to accomplish through randomization, which neutralizes the effect of the uncontrolled variables. In the hair growth example, the subjects should be similar in terms of eating habits, age, and so on, by randomly assigning the subjects to the three groups.

> **! CAUTION**
> It is vital that individuals be randomly assigned to treatments.

① Verify the Requirements to Perform a One-Way ANOVA

To perform a one-way ANOVA test, certain requirements must be satisfied.

> **Requirements of a One-Way ANOVA Test**
>
> 1. There are k simple random samples from k populations.
> 2. The k samples are independent of each other; that is, the subjects in one group cannot be related in any way to subjects in a second group.
> 3. The populations are normally distributed.
> 4. The populations have the same variance; that is, each treatment group has population variance σ^2.

Suppose we are testing a claim regarding $k = 3$ population means so that the null hypothesis is

$$H_0: \mu_1 = \mu_2 = \mu_3$$

and the alternative hypothesis is

$$H_1: \text{at least one of the population means is different from the others}$$

Figure 1(a) shows the distribution of each population if the null hypothesis is true, and Figure 1(b) shows what the distribution of each population might look like if the alternative hypothesis is true.

Figure 1

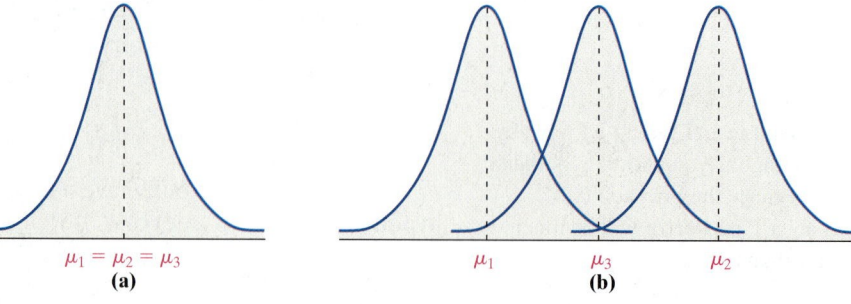

$\mu_1 = \mu_2 = \mu_3$
(a)

$\mu_1 \quad \mu_3 \quad \mu_2$
(b)

In Other Words

Try to design experiments that use ANOVA so that each treatment group is the same size.

The methods of one-way ANOVA are **robust**, so small departures from the requirement of normality will not significantly affect the results of the procedure. In addition, the requirement of equal population variances does not need to be strictly adhered to, especially if the sample size for each treatment group is the same. Therefore, it is worthwhile to design an experiment in which the samples from the populations are roughly equal in size.

We can verify the requirement of normality by constructing normal probability plots. The requirement of equal population variances is more difficult to verify. However, a general rule of thumb is as follows:

> **Verifying the Requirement of Equal Population Variances**
>
> The one-way ANOVA procedures may be used provided that the largest sample standard deviation is no more than two times larger than the smallest sample standard deviation.

EXAMPLE 1   **Testing the Requirements of One-Way ANOVA**

Problem: Researcher Jelodar Gholamali wanted to determine the effectiveness of various treatments on glucose levels of diabetic rats. He randomly assigned diabetic albino rats into four treatment groups. Group 1 rats served as a control group and were fed a regular diet. Group 2 rats were served a regular diet supplemented with a herb, fenugreek. Group 3 rats were served a regular diet supplemented with garlic. Group 4 rats were served a regular diet supplemented with onion. The basis for the study is that Persian folklore states that diets supplemented with fenugreek, garlic, or onion help to treat diabetes. After 15 days of treatment, the blood glucose was measured in milligrams per deciliter (mg/dL). The results presented in Table 1 are based on the results published in the article. Verify that the requirements to perform one-way ANOVA are satisfied. *Source:*Jelodar, G. A., et.al., Effect of Fenugreek, Onion and Garlic on Blood Glucose and Histopathology of Pancreas of Alloxan-Induced Diabetic Rats, *Indian Journal of Medical Science*, 2005;59:64–69.

Table 1			
Control	**Fenugreek**	**Garlic**	**Onion**
288.1	229.1	177.4	299.7
296.8	240.7	202.2	258.3
267.8	239.4	163.1	286.8
256.7	207.7	184.7	244.0
292.1	225.7	197.9	267.1
282.9	230.8	164.6	297.1
260.3	206.6	193.9	249.9
283.8	213.3	158.1	265.1

Approach: We must verify the previously listed requirements.

Solution

1. The rats were randomly assigned to each treatment group.
2. None of the subjects selected is related in any way, so the samples are independent.
3. Figure 2 shows the normal probability plots for all four treatment groups. All the normal probability plots are roughly linear, so we conclude that the sample data come from populations that are approximately normally distributed.

Figure 2

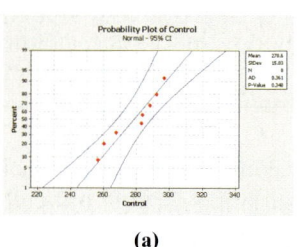

(a)

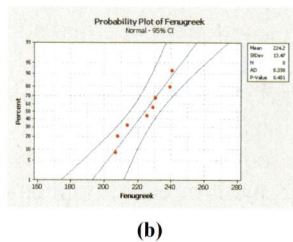

(b)

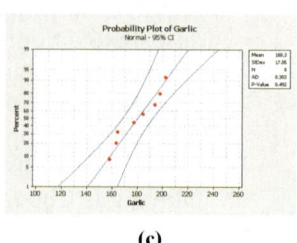

(c)

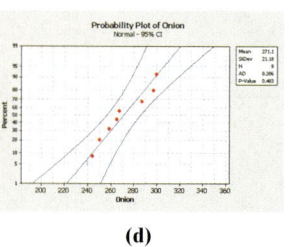

(d)

4. The sample standard deviations for each sample are computed using MINITAB and presented in Figure 3. The largest standard deviation is 21.18 mg/dL, and the smallest standard deviation is 13.49 mg/dL. Because the largest standard deviation is not more than two times larger than the smallest standard deviation $(2 \cdot 13.49 = 26.98 > 21.18)$, the requirement of equal population variances is satisfied.

Figure 3

Descriptive Statistics: Control, Fenugreek, Garlic, Onion

Variable	N	N*	Mean	SE Mean	StDev	Minimum	Q1	Median	Q3	Maximum
Control	8	0	278.56	5.31	15.03	256.70	262.18	283.35	291.10	296.80
Fenugreek	8	0	224.16	4.77	13.49	206.60	209.10	227.40	237.25	240.70
Garlic	8	0	180.24	6.03	17.06	158.10	163.48	181.05	196.90	202.20
Onion	8	0	271.00	7.49	21.18	244.00	252.06	266.10	294.53	299.70

Because all four requirements are satisfied, we can perform a one-way ANOVA.

2 Test a Claim Regarding Three or More Means Using One-Way ANOVA

The computations in performing any analysis of variance test are tedious, so virtually all researchers use statistical software to conduct the test. When using software, it is easiest to use the P-value approach. The nice thing about P-values is that the decision rule is always the same, regardless of the type of hypothesis being tested.

> **Decision Rule in the One-Way ANOVA Test**
> If the P-value is less than the level of significance, α, reject the null hypothesis.

EXAMPLE 2 **Performing One-Way ANOVA Using Technology**

Problem: The researcher in Example 1 wishes to determine if there is a difference in the mean glucose among the four treatment groups. Test this claim at the $\alpha = 0.05$ level of significance.

Approach: We will use MINITAB, Excel, and a TI-84 Plus graphing calculator to test the hypothesis. If the P-value is less than the level of significance, we reject the null hypothesis.

Result: The claim is that there is a difference in the mean glucose among the four treatment groups. For the sake of the test, we assume that the mean glucose among the four treatment groups is the same. So the null hypothesis is

$$H_0: \mu_{\text{control}} = \mu_{\text{fenugreek}} = \mu_{\text{garlic}} = \mu_{\text{onion}}$$

versus the alternative hypothesis

$$H_1: \text{at least one of the population means is different from the others}$$

Figure 4(a) shows the output from MINITAB, Figure 4(b) shows the output from Excel, and Figure 4(c) shows the output from a TI-84 Plus graphing calculator.

Figure 4

One-Way ANOVA: Control, Fenugreek, Garlic, Onion

```
Source      DF        SS        MS        F        P
Factor       3     50091     16697    58.21    0.000
Error       28      8032       287
Total       31     58122

S = 16.94    R - Sq = 86.18%    R - Sq (adj) = 84.70%
```

```
                              Individual 95% CIs For Mean
                              Based on Pooled StDev
Level       N   Mean   StDev  --+---------+---------+---------+-------
Control     8 278.56   15.03                                (---*--)
Fenugreek   8 224.16   13.49                     (--*---)
Garlic      8 180.24   17.06  (---*---)
Onion       8 271.00   21.18                              (--*---)
                             --+---------+---------+---------+-------
                             175       210       245       280
```

```
Pooled St Dev = 16.94
```

(a) MINITAB Output

Anova: Single Factor

SUMMARY

Groups	Count	Sum	Average	Variance
Control	8	2228.5	278.5625	225.7713
Fenugreek	8	1793.3	224.1625	181.9884
Garlic	8	1441.9	180.2375	291.0341
Onion	8	2168	271	448.58

ANOVA

Source of Variation	SS	df	MS	F	P-value	F crit
Between Groups	50090.69	3	16696.9	58.2091	3.74E-12	2.946685
Within Groups	8031.616	28	286.8434			
Total	58122.31	31				

(b) Excel Output

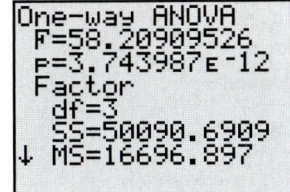
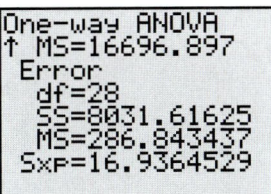

(c) TI–84 Plus Output

Notice that MINITAB indicates that the *P*-value is 0.000. This does not mean that the *P*-value is 0, but instead means that the *P*-value is less than 0.0001. The output of Excel and the TI-84 Plus confirm this by indicating the *P*-value is 3.7×10^{-12}. Because the *P*-value is less than the level of significance, we reject the null hypothesis. There is sufficient evidence to support the claim that at least one of the population means for glucose levels is different from the others.

Whenever you perform analysis of variance, it is always a good idea to present visual evidence that supports the conclusions of the test. Side-by-side boxplots are a great way to help see the results of the ANOVA procedure. Figure 5 shows the side-by-side boxplots of the data presented in Table 1. The boxplots support the ANOVA results from Example 2.

Figure 5

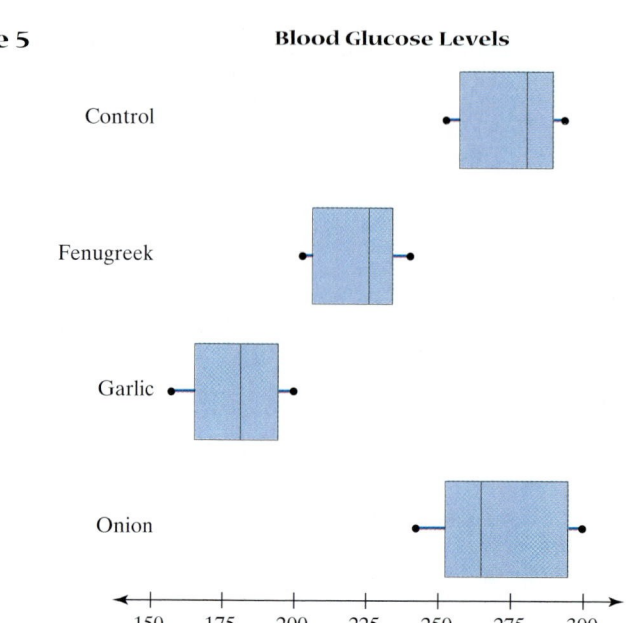

Blood Glucose Levels

Now Work Problems 11(a)–(d).

A Conceptual Understanding of One-Way ANOVA

Look again at Figure 4. You may have noticed that each of the three outputs included an F-value ($F = 58.21$). We now illustrate the idea behind the F-test statistic. Remember, in testing any hypothesis, the null hypothesis is assumed to be true until the evidence indicates otherwise. In testing the claim regarding k population means, we assume that $\mu_1 = \mu_2 = \cdots = \mu_k = \mu$. That is, we assume that all k samples come from the same normal population whose mean is μ and variance is σ^2. Table 2 shows the statistics that result by sampling from each of the k populations.

Table 2			
Population	**Sample Size**	**Sample Mean**	**Sample Standard Deviation**
1	n_1	$\overline{x}_1$	s_1
2	n_2	$\overline{x}_2$	s_2
3	n_3	$\overline{x}_3$	s_3
$\vdots$	$\vdots$	$\vdots$	$\vdots$
k	n_k	$\overline{x}_k$	s_k

The computation of the F test statistic requires that we understand *mean squares*. A **mean square** is an average (mean) of squared values. For example, any variance is a mean square. The F-test statistic is the ratio of two mean squares.

If the null hypothesis is true, then each treatment group comes from the same population whose mean is μ and whose variance is σ^2. The sample mean of the entire set of data (all treatment groups combined) is a good estimate of μ. We will call this sample mean $\overline{x}$. Similarly, the sample mean of Sample 1 (or treatment 1) will be $\overline{x}_1$, the sample mean of Sample 2 (or treatment 2) will be $\overline{x}_2$, and so on.

Finding an estimate of σ^2 is somewhat more complicated. One approach is to estimate σ^2 by computing a measure of variation in sample means from one treatment group to the next, weighted by the sample size of the corresponding treatment group. We call this value the **mean square due to treatment**, denoted MST. It is computed as follows:

$$\text{MST} = \frac{n_1(\bar{x}_1 - \bar{x})^2 + n_2(\bar{x}_2 - \bar{x})^2 + \cdots + n_k(\bar{x}_k - \bar{x})^2}{k - 1} \quad \text{(1)}$$

If the null hypothesis is true, then MST is an unbiased estimator of σ^2, the variance of the population.

A second approach to estimating σ^2 is to compute the sample variance for each sample (or treatment), and then to find a weighted average of the sample variances. We call this the **mean square due to error**, denoted MSE. The mean square due to error is an unbiased estimator of σ^2 whether or not the null hypothesis is true. It is computed as follows:

$$\text{MSE} = \frac{(n_1 - 1)s_1^2 + (n_2 - 1)s_2^2 + \cdots + (n_k - 1)s_k^2}{n - k} \quad \text{(2)}$$

where n is the total size of the sample. In other words, $n = n_1 + n_2 + \cdots + n_k$.

The F-test statistic is the ratio of the two estimates of σ^2.

$$F = \frac{\text{mean square due to treatment}}{\text{mean square due to error}} = \frac{\text{MST}}{\text{MSE}}$$

If the null hypothesis is true, both MST and MSE provide unbiased estimates for σ^2, the population variance. So, if the null hypothesis is true, we would expect the F-test statistic to be close to 1. However, if the null hypothesis is not true, at least one of the sample means from a treatment group will be "far away" from $\bar{x}$, the sample mean of the entire data set. This will cause MST to be large relative to MSE, which ultimately leads to an F-test statistic substantially larger than 1.

We now present the steps to be used in the computation of the F-test statistic.

Computing the *F*-Test Statistic

Step 1: Compute the sample mean of the combined data set by adding up all the observations and dividing by the number of observations. Call this value $\bar{x}$.

Step 2: Find the sample mean for each sample (or treatment). Let $\bar{x}_1$ represent the sample mean of sample 1, $\bar{x}_2$ represent the sample mean of sample 2, and so on.

Step 3: Find the sample variance for each sample (or treatment). Let s_1^2 represent the sample variance for sample 1, s_2^2 represent the sample variance for sample 2, and so on.

Step 4: Compute the mean square due to treatment, MST.

Step 5: Compute the mean square due to error, MSE.

Step 6: Compute the F-test statistic:

$$F = \frac{\text{mean square due to treatment}}{\text{mean square due to error}} = \frac{\text{MST}}{\text{MSE}}$$

EXAMPLE 3   **Computing the F-Test Statistic**

Problem: Compute the F-test statistic for the data presented in Example 1.

Approach: We follow Steps 1–6 just presented.

Solution

Step 1: Compute the mean of the entire data set.

$$\bar{x} = \frac{288.1 + 296.8 + \cdots + 249.9 + 265.1}{32} = \frac{7633.30}{32} = 238.54$$

Step 2: Find the sample mean of each treatment. Call the control group population 1, the fenugreek group population 2, the garlic group population 3, and the onion group population 4. Then

$$\bar{x}_1 = \frac{288.1 + 296.8 + \cdots + 283.8}{8} = 278.56 \qquad \bar{x}_2 = \frac{229.1 + 240.7 + \cdots + 213.3}{8} = 224.16$$

$$\bar{x}_3 = \frac{177.4 + 202.2 + \cdots + 158.1}{8} = 180.24 \qquad \bar{x}_4 = \frac{299.7 + 258.3 + \cdots + 265.1}{8} = 271.00$$

Step 3: Find the sample variance for each treatment group.

$$s_1^2 = \frac{(288.1 - 278.56)^2 + (296.8 - 278.56)^2 + \cdots + (283.8 - 278.56)^2}{8 - 1} = 225.77$$

$$s_2^2 = \frac{(229.1 - 224.16)^2 + (240.7 - 224.16)^2 + \cdots + (213.3 - 224.16)^2}{8 - 1} = 181.99$$

$$s_3^2 = \frac{(177.4 - 180.24)^2 + (202.2 - 180.24)^2 + \cdots + (158.1 - 180.24)^2}{8 - 1} = 291.03$$

$$s_4^2 = \frac{(299.7 - 271)^2 + (258.3 - 271)^2 + \cdots + (265.1 - 271)^2}{8 - 1} = 448.58$$

Step 4: Compute MST.

$$\text{MST} = \frac{8(278.56 - 238.54)^2 + 8(224.16 - 238.54)^2 + 8(180.24 - 238.54)^2 + 8(271 - 238.54)^2}{4 - 1}$$

$$= \frac{50,087.4112}{3}$$

$$= 16,695.80$$

Step 5: Compute MSE.

$$\text{MSE} = \frac{(8 - 1)225.77 + (8 - 1)181.99 + (8 - 1)291.03 + (8 - 1)448.48}{32 - 4}$$

$$= \frac{8030.89}{28}$$

$$= 286.82$$

Step 6: Compute the F-test statistic.

$$F = \frac{\text{mean square due to treatment}}{\text{mean square due to error}} = \frac{\text{MST}}{\text{MSE}} = \frac{16,695.80}{286.84} = 58.21$$

This is the same result provided by the statistical software and graphing calculator in Example 2.

Looking back at the formula for the mean square due to treatment, we notice that if the null hypothesis is true the overall mean, $\bar{x}$, should be close to the means computed from each sample (treatment), $\bar{x}_1$, $\bar{x}_2$, and so on. If one or more of the means computed from each sample is substantially different from the overall mean, MST will be large, which in turn makes the F-statistic large. In Example 3, we can see that the control group has a sample mean much larger than the overall mean ($\bar{x}_1 = 278.56$ versus $\bar{x} = 238.54$), and the garlic group has a sample mean much smaller than the overall mean ($\bar{x}_3 = 180.24$ versus $\bar{x} = 238.54$).

The results of the computations that lead to the F-test statistic are presented in an **ANOVA table**, the form of which is shown in Table 3.

	Table 3				
Source of Variation	**Sum of Squares**	**Degrees of Freedom**	**Mean Squares**	**F-Test Statistic**
Treatment	50,087.41	$k - 1 = 4 - 1 = 3$	16,695.80	58.21
Error	8030.89	$n - k = 32 - 4 = 28$	286.82	
Total	58,118.3	$n - 1 = 32 - 1 = 31$		

Figure 6

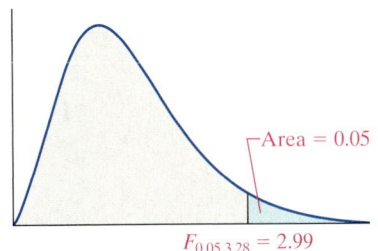

Area = 0.05

$F_{0.05, 3, 28} = 2.99$

In Other Words

If we reject the null hypothesis when doing ANOVA, we are rejecting the assumption that the population means are all equal. However, the test doesn't tell us which means differ.

Notice that the "sum of squares treatment" is the numerator of the computation for the mean square due to treatment. The "sum of squares error" is the numerator of the computation for the mean square due to error. Each entry in the mean square column is the sum of squares divided by the corresponding degrees of freedom. In addition,

Sum of squares total = sum of squares treatment + sum of squares error

This result is a consequence of the requirement of independence of the observations within the groups.

If we were conducting this ANOVA test by hand, we would compare the F-test statistic with a critical F-value. The critical F-value is the F-value whose area in the right tail is α with $k - 1$ degrees of freedom in the numerator and $n - k$ degrees of freedom in the denominator. The critical F-value for the claim made in Example 1 at the $\alpha = 0.05$ level of significance is $F_{0.05, 3.28} \approx 2.99$. Because the F-test statistic, 58.21, is greater than the critical F, reject the null hypothesis. See Figure 6.

Suppose the null hypothesis of equal population means is rejected. This conclusion tells us that at least one of the population means is different from the others, but we don't know which one. We can determine which population means differ using Tukey's tests discussed in the next section.

13.1 ASSESS YOUR UNDERSTANDING

Concepts and Vocabulary

1. The acronym ANOVA stands for _____ _____ _____.
2. What are the requirements to perform a one-way ANOVA? Is the test robust?
3. What is the mean square due to treatment estimate of σ^2?

What is the mean square due to error estimate of σ^2?
4. Why does a large value of the F statistic provide evidence against the null hypothesis $H_0: \mu_1 = \mu_2 = \cdots = \mu_k$?

Skill Building

In Problems 5 and 6, fill in the ANOVA table.

5.

Source of Variation	Sum of Squares	Degrees of Freedom	Mean Squares	F-Test Statistic
Treatment	387	2		
Error	8042	27		
Total				

6.

Source of Variation	Sum of Squares	Degrees of Freedom	Mean Squares	F-Test Statistic
Treatment	2814	3		
Error	4915	36		
Total				

In Problems 7 and 8, determine the F-test statistic based on the given summary statistics. $\left[\textit{Hint:}\ \bar{x} = \dfrac{\Sigma n_i \bar{x}_i}{\Sigma n_i}.\right]$

7.

Population	Sample Size	Sample Mean	Sample Variance
1	10	40	48
2	10	42	31
3	10	44	25

8.

Population	Sample Size	Sample Mean	Sample Variance
1	15	105	34
2	15	110	40
3	15	108	30
4	15	90	38

9. The following data represent a simple random sample of $n = 4$ from three populations that are known to be normally distributed. Verify that the F-test statistic is 2.04.

Sample 1	Sample 2	Sample 3
28	22	25
23	25	24
30	17	19
27	23	30

10. The following data represent a simple random sample of $n = 5$ from three populations that are known to be normally distributed. Verify that the F-test statistic is 2.599.

Sample 1	Sample 2	Sample 3
73	67	72
82	77	80
82	66	87
81	67	77
97	83	96

Applying the Concepts

11. Corn Production The data in the table represent the number of corn plants in randomly sampled rows (a 17-foot by 5-inch strip) for various types of plot. An agricultural researcher wants to know whether the mean numbers of plants for each plot type are equal.

Plot Type	Number of Plants					
Sludge plot	25	27	33	30	28	27
Spring disk	32	30	33	35	34	34
No till	30	26	29	32	25	29

Source: Andrew Dieter and Brad Schmidgall. Joliet Junior College

(a) Write the null and alternative hypotheses.
(b) State the requirements that must be satisfied to use the one-way ANOVA procedure.
(c) Use the following partial MINITAB output to test the researcher's claim at the $\alpha = 0.05$ level of significance.

One-way ANOVA: Sludge Plot, Spring Disk, No Till

```
Source    DF      SS      MS      F      P
Factor     2   84.11   42.06   7.10  0.007
Error     15   88.83    5.92
Total     17  172.94
```

(d) Shown are side-by-side boxplots of each type of plot. Do these boxplots support the results obtained in part (c)?

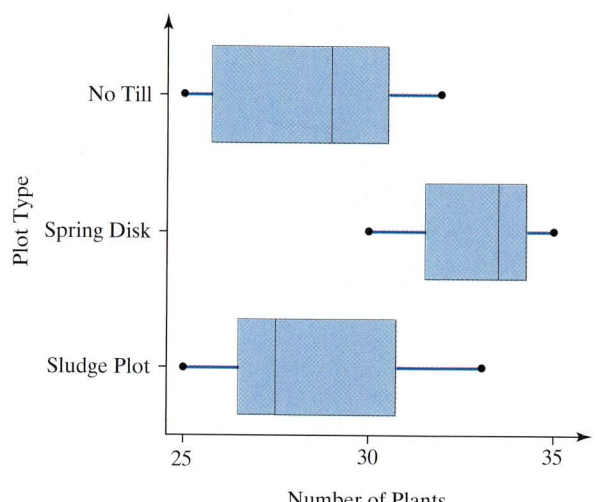

(e) Verify that the F-test statistic is 7.10.

12. Soybean Yield The data in the table represent the number of pods on a random sample of soybean plants for various plot types. An agricultural researcher wants to test the claim that the mean numbers of pods for each plot type are equal.

Plot Type	Pods								
Liberty	32	31	36	35	41	34	39	37	38
No till	34	30	31	27	40	33	37	42	39
Chisel plowed	34	37	24	23	32	33	27	34	30

Source: Andrew Dieter and Brad Schmidgall, Joliet Junior College

(a) Write the null and alternative hypotheses.
(b) State the requirements that must be satisfied to use the one-way ANOVA procedure.
(c) Use the following MINITAB output to test the researcher's claim at the $\alpha = 0.05$ level of significance.

One-way ANOVA: Liberty, No Till Chisel Plowed

Source	DF	SS	MS	F	P
Factor	2	149.0	74.5	3.77	0.038
Error	24	474.7	19.8		
Total	26	623.6			

(d) Shown are side-by-side boxplots of each type of plot. Do these boxplots support the results obtained in part (c)?

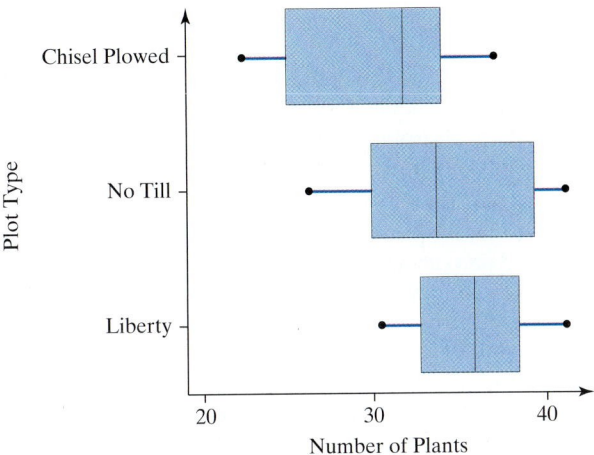

(e) Verify that the F-test statistic is 3.77.
(f) Based on the boxplots, which type of plot appears to have a significantly different mean number of plants?

13. **Births by Day of Week** An obstetrician knew that there were more live births during the week than on weekends. She wanted to determine whether the mean number of births was the same for each of the five days of the week. She randomly selected eight dates for each of the five days of the week and obtained the following data:

Monday	Tuesday	Wednesday	Thursday	Friday
10,456	11,621	11,084	11,171	11,545
10,023	11,944	11,570	11,745	12,321
10,691	11,045	11,346	12,023	11,749
10,283	12,927	11,875	12,433	12,192
10,265	12,577	12,193	12,132	12,422
11,189	11,753	11,593	11,903	11,627
11,198	12,509	11,216	11,233	11,624
11,465	13,521	11,818	12,543	12,543

Source: National Center for Health Statistics

(a) Write the null and alternative hypotheses.
(b) State the requirements that must be satisfied to use the one-way ANOVA procedure.
(c) Use the following MINITAB output to test the researcher's claim at the $\alpha = 0.01$ level of significance:

One-way ANOVA: Mon, Tues, Wed, Thurs, Fri

Source	DF	SS	MS	F	P
Factor	4	11507633	2876908	9.80	0.000
Error	35	10270781	293451		
Total	39	21778414			

(d) Shown are side-by-side boxplots of each type of plot. Do these boxplots support the results obtained in part (c)?

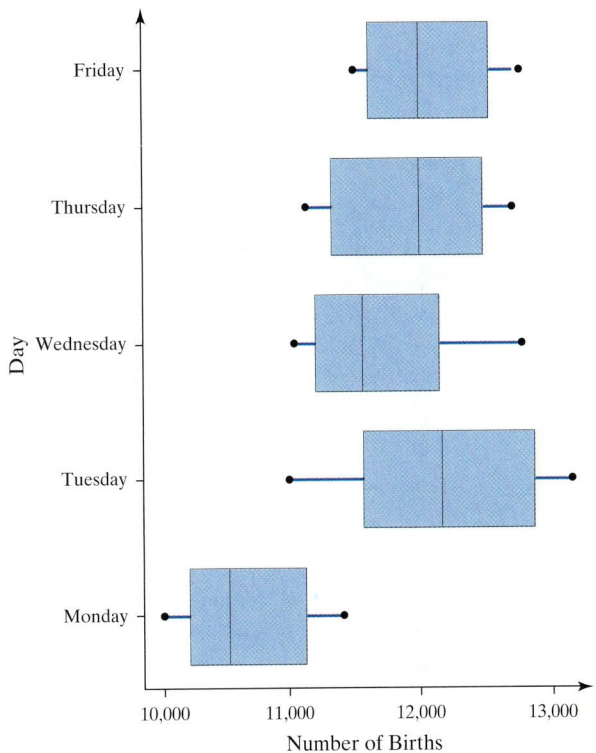

(e) Verify that the F-test statistic is 9.80.
(f) Based on the boxplots, which day of the week appears to have a significantly different number of births?

14. **Punkin Chunkin** The World Championship Punkin Chunkin contest is held every fall in Millsboro, Delaware. Contestants build devices meant to hurl 8 to 10-pound pumpkins across a field. One class of entry is the air cannon, which must use compressed air to fire a pumpkin. The following data represent a simple random sample of distances that pumpkins have traveled (in feet) for the years 2001 to 2004. Is there evidence to conclude that the

mean distance that a pumpkin is fired is different for the various years?

2001	2002	2003	2004
3494.70	3881.54	3895.85	4224.00
3360.20	3232.74	4434.28	4065.60
3911.02	3696.19	3448.56	3967.95
3124.40	3414.54	3665.89	3596.09
3718.77	3816.64	3539.66	3970.00
3453.12	3631.78	3591.16	3877.68

Source: www.punkinchunkin.com/main.htm

(a) Write the null and alternative hypotheses.
(b) State the requirements that must be satisfied to use the one-way ANOVA procedure.
(c) Use the following MINITAB output to test the claim at the $\alpha = 0.05$ level of significance.

One-way ANOVA: 2001, 2002, 2003, 2004

```
Source    DF      SS      MS      F      P
Factor     3  659242  219747   2.82  0.065
Error     20 1559456   77973
Total     23 2218698
```

(d) Shown are side-by-side boxplots of each type of plot. Do these boxplots support the results obtained in part (c)?

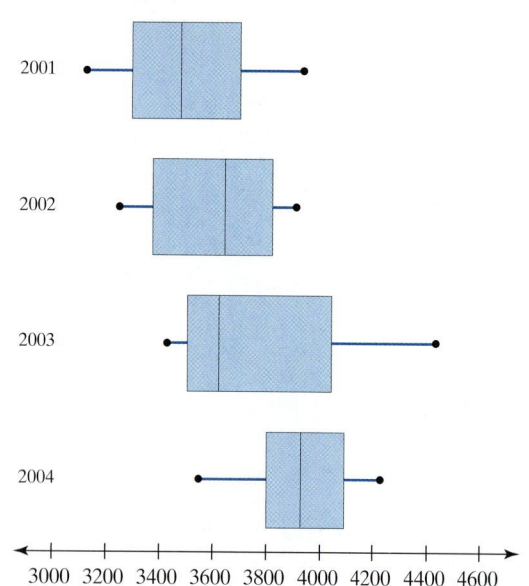

Boxplot of 2001, 2002, 2003, 2004

(e) Verify that the F-test statistic is 2.82.

15. **Rates of Return** A stock analyst wondered whether the mean rate of return of financial, energy, and utility stocks differed over the past 5 years. He obtained a simple random sample of eight companies from each of the three

sectors and obtained the 5-year rates of return shown in the following table (in percent):

Financial	Energy	Utilities
10.76	12.72	11.88
15.05	13.91	5.86
17.01	6.43	13.46
5.07	11.19	9.90
19.50	18.79	3.95
8.16	20.73	3.44
10.38	9.60	7.11
6.75	17.40	15.70

Source: Morningstar.com

(a) State the null and alternative hypothesis.
(b) Verify that the requirements to use the one-way ANOVA procedure are satisfied. Normal probability plots indicate that the sample data come from normal populations.
(c) Test the claim that the mean rates of return are different at the $\alpha = 0.05$ level of significance.
(d) Draw boxplots of the three sectors to support the results obtained in part (c).

16. **Reaction Time** In an online psychology experiment sponsored by the University of Mississippi, researchers asked study participants to respond to various stimuli. Participants were randomly assigned to one of three groups. Subjects in group 1 were in the simple group. They were required to respond as quickly as possible after a stimulus was presented. Subjects in group 2 were in the go/no-go group. These subjects were required to respond to a particular stimulus while disregarding other stimuli. Finally, subjects in group 3 were in the choice group. They needed to respond differently, depending on the stimuli presented. Depending on the type of whistle sound, the subject must press a certain button. The reaction time (in seconds) for each stimulus is presented in the table.

Simple	Go/No Go	Choice
0.430	0.588	0.561
0.498	0.375	0.498
0.480	0.409	0.519
0.376	0.613	0.538
0.402	0.481	0.464
0.329	0.355	0.725

Source: PsychExperiments; The University of Mississippi; www.olemiss.edu/psychexps/

The researcher wants to test the claim that the mean reaction times for each stimulus are equal.

(a) State the null and alternative hypotheses.
(b) Verify that the requirements to use the one-way ANOVA procedure are satisfied. Normal probability plots indicate that the sample data come from a normal population.
(c) Test the claim that the mean reaction times to the three stimuli are the same at the $\alpha = 0.05$ level of significance.
(d) Draw boxplots of the three stimuli to support the analytic results obtained in part (c).

17. **Crash Data** The Insurance Institute for Highway Safety conducts experiments in which cars are crashed into a fixed barrier at 40 mph. In the Institute's 40-mph offset test, 40% of the total width of each vehicle strikes a barrier on the driver's side. The barrier's deformable face is made of aluminum honeycomb, which makes the forces in the test similar to those involved in a frontal offset crash between two vehicles of the same weight, each going just less than 40 mph. Suppose you are in the market to buy a new family car. You want to know whether the mean chest compression resulting from this offset crash is the same for large family cars, passenger vans, and midsize utility vehicles. The following data were collected from the institute's study.

Large Family Cars	Chest Compression (mm)	Passenger Vans	Chest Compression (mm)	Midsize Utility Vehicles	Chest Compression (mm)
Hyundai XG350	33	Toyota Sienna	29	Honda Pilot	31
Ford Taurus	28	Honda Odyssey	28	Toyota 4Runner	36
Buick LeSabre	28	Ford Freestar	27	Mitsubishi Endeavor	35
Chevrolet Impala	26	Mazda MPV	30	Nissan Murano	29
Chrysler 300	34	Chevrolet Uplander	26	Ford Explorer	29
Pontiac Grand Prix	34	Nissan Quest	33	Jeep Liberty	36
Toyota Avalon	31	Kia Sedona	21	Buick Randezvous	29

Source: Insurance Institute for Highway Safety

The researcher wants to test the claim that the means for chest compression for each class of vehicle are equal.

(a) State the null and alternative hypotheses.
(b) Verify that the requirements to use the one-way ANOVA procedure are satisfied. Normal probability plots indicate that the sample data come from normal populations.
(c) Test the claim that the mean chest compression for each vehicle type is the same at the $\alpha = 0.01$ level of significance.
(d) Draw boxplots of the three types of vehicle to support the analytic results obtained in part (c).

18. **Crash Data** The Insurance Institute for Highway Safety conducts experiments in which cars are crashed into a fixed barrier at 40 mph. In the Institute's 40-mph offset test, 40% of the total width of each vehicle strikes a barrier on the driver's side. The barrier's deformable face is made of aluminum honeycomb, which makes the forces in the test similar to those involved in a frontal offset crash between two vehicles of the same weight, each going just less than 40 mph. Suppose you are in the market to buy a new family car. You want to know if the mean head injury resulting from this offset crash is the same for large family cars, passenger vans, and midsize utility vehicles. The following data were collected from the institute's study.

Large Family Cars	Head Injury (hic)	Passenger Vans	Head Injury (hic)	Midsize Utility Vehicles	Head Injury (hic)
Hyundai XG350	264	Toyota Sienna	148	Honda Pilot	202
Ford Taurus	170	Honda Odyssey	238	Toyota 4Runner	216
Buick LeSabre	409	Ford Freestar	340	Mitsubishi Endeavor	186
Chevrolet Impala	204	Mazda MPV	693	Nissan Murano	517
Chrysler 300	149	Chevrolet Uplander	550	Ford Explorer	202
Pontiac Grand Prix	627	Nissan Quest	470	Kia Sorento	552
Toyota Avalon	166	Kia Sedona	332	Chevy Trailblazer	386

Source: Insurance Institute for Highway Safety

The researcher wants to test the claim that the means for head injury for each class of vehicle are equal.

(a) State the null and alternative hypotheses.

(b) Verify that the requirements to use the one-way ANOVA procedure are satisfied. Normal probability plots indicate that the sample data come from normal populations.

(c) Test the claim that the mean head injury for each vehicle type is the same at the $\alpha = 0.01$ level of significance.

(d) Draw boxplots of the three vehicle types to support the analytic results obtained in part (c).

19. pH in Rain An environmentalist wanted to determine if the mean acidity of rain differed among Alaska, Florida, and Texas. He randomly selected six rain dates at each of the three locations and obtained the data in the following table.

Alaska	Florida	Texas
5.41	4.87	5.46
5.39	5.18	6.29
4.90	4.40	5.57
5.14	5.12	5.15
4.80	4.89	5.45
5.24	5.06	5.30

Source: National Atmospheric Deposition Program

(a) State the null and alternative hypothesis.

(b) Verify that the requirements to use the one-way ANOVA procedure are satisfied. Normal probability plots indicate that the sample data come from a normal population.

(c) Test the claim that the mean pHs in the rainwater are different at the $\alpha = 0.05$ level of significance.

(d) Draw boxplots of the pH in rain for the three states to support the results obtained in part (c).

20. Lower Your Cholesterol Researchers Francisco Fuentes and his colleagues wanted to determine the most effective diet for reducing LDL cholesterol, the so-called "bad" cholesterol, among three diets: (1) a saturated-fat diet, (2) the Mediterranean diet, and (3) the U.S. National Cholesterol Education Program or NCEP-1 Diet. The participants in the study were shown to have the same levels of LDL cholesterol before the study. Participants were randomly assigned to one of the three treatment groups. Individuals in group 1 received the saturated fat diet, which is 15% protein, 47% carbohydrates, 38% fat (20% saturated fat, 12% monounsaturated fat, and 6% polyunsaturated fat). Individuals in group 2 received the Mediterranean diet, which is 47% carbohydrates, 38% fat (<10% saturated fat, 22% monounsaturated fat, and 6% polyunsaturated fat). Individuals in group 3 received the NCEP-1 Diet (<10% saturated fat, 12% monounsaturated fat, and 6% polyunsaturated fat). After 28 days, their LDL cho-

lesterol levels were recorded. The data in the following table are based on this study.

Saturated Fat	Mediterranean	NCEP-1
245	56	125
123	78	100
166	101	140
104	158	151
196	145	138
300	118	268
140	145	75
240	211	71
218	131	184
173	125	116
223	160	144
177	130	101
193	83	135
224	263	144
149	150	130

(a) State the null and alternative hypothesis.

(b) Verify that the requirements to use the one-way ANOVA procedure are satisfied. Normal probability plots indicate that the sample data come from normal populations.

(c) Test the claim that the mean LDL cholesterol levels are different at the $\alpha = 0.05$ level of significance.

(d) Draw boxplots of the three LDL cholesterol levels for the three groups to support the analytic results obtained in part (c).

21. Concrete Strength An engineer wants to know if the mean strengths of three different concrete mix designs differ significantly. He randomly selects 9 cylinders that measure 6 inches in diameter and 12 inches in height in which mixture 67-0-301 is poured, 9 cylinders of mixture 67-0-400, and 9 cylinders of mixture 67-0-353. After 28 days, he measures the strength (in pounds per square inch) of the cylinders. The results are presented in the following table:

Mixture 67-0-301		Mixture 67-0-400		Mixture 67-0-353	
3960	4090	4070	4120	4150	3820
4040	3830	4330	4640	3820	3750
3780	3940	4620	4190	4010	3990
3890	4080	3730	3850	4150	4320
3990		4890		4190	

(a) State the null and alternative hypotheses.

(b) Explain why we cannot use one-way ANOVA to test these hypotheses.

22. **Analyzing Journal Article Results** Researchers (Brian G. Feagan et al., Erythropoietin with Iron Supplementation to Prevent Allogeneic Blood Transfusion in Total Hip Joint Arthroplasty, *Annals of Internal Medicine, Vol. 133, No. 11*) wanted to determine whether epoetin alfa was effective in increasing the hemoglobin concentration in patients undergoing hip arthroplasty. The researchers screened patients for eligibility by performing a complete medical history and physical of the patients. Once eligible patients were identified, the researchers used a computer-generated schedule to assign the patients to the high-dose epoetin group, low-dose epoetin group, or placebo group. The study was double-blind. Based on an analysis of variance, it was determined that there were significant differences in the increase in hemoglobin concentration in the three groups with a *P*-value less than 0.001. The mean increase in hemoglobin in the high-dose epoetin group was 19.5 g/L, the mean increase in hemoglobin in the low-dose epoetin group was 17.2 g/L, and mean increase in hemoglobin in the placebo group was 1.2 g/L.

(a) Why do you think it was necessary to screen patients for eligibility?

(b) Why was a computer-generated schedule used to assign patients to the various treatment groups?

(c) What does it mean for a study to be double-blind? Why do you think the researchers desired a double-blind study?

(d) Interpret the reported *P*-value.

Technology Step-by-Step **ANOVA**

TI-83/84 Plus **Step 1:** Enter the raw data into L1, L2, L3, and so on, for each population or treatment.

Step 2: Press STAT, highlight TESTS, and select F:ANOVA(.

Step 3: Enter the list names for each sample or treatment after ANOVA(. For example, if there are three treatments in L1, L2, and L3, enter

ANOVA(L1,L2,L3)

Press ENTER.

MINITAB **Step 1:** Enter the raw data into C1, C2, C3, and so on, for each sample or treatment.

Step 2: Select **Stat**, then highlight **ANOVA**, and select **One-way** (**Unstacked**).

Step 3: Enter the column names in the cell marked "Responses." Click OK.

Excel **Step 1:** Enter the raw data in columns A, B, C, and so on, for each sample or treatment.

Step 2: Be sure the Data Analysis Tool Pak is activated. This is done by selecting the **Tools** menu and highlighting **Add-Ins** .... Check the box for the Analysis ToolPak and select OK. Select **Tools**, then highlight **Data Analysis**. Select ANOVA: Single Factor and click OK.

Step 3: With the cursor in the "Input Range:" cell, highlight the data. Click OK.

13.2 Post Hoc Tests on One-Way Analysis of Variance

Preparing for This Section Before getting started, review the following:

- Using Confidence Intervals to Test Hypotheses (Section 9.2, p. 466–473)

Objective **①** **Perform the Tukey Test**

Suppose we perform a one-way ANOVA and the results lead us to conclude that at least one population mean is different from the others. To determine which means differ significantly, we make additional comparisons between means. The procedures for making these comparisons are called **multiple comparison methods.**

Suppose we used a one-way ANOVA to test the null hypothesis $H_0: \mu_1 = \mu_2 = \mu_3 = \mu_4$ and the sample data led us to reject the null hypothesis. At this point, we probably would want to know which means differ. For example, we might suspect that μ_1 is different from μ_2, which would lead us to test $H_0: \mu_1 = \mu_2$ versus $H_1: \mu_1 \neq \mu_2$. Or we might suspect that μ_1 and μ_2 are

In Other Words
Post hoc is Latin for after this. Post hoc tests are performed after rejecting the null hypothesis that three or more population means are equal.

equal, but different from μ_3, so we would test $H_0: \mu_1 + \mu_2 = 2\mu_3$ versus the alternative $H_0: \mu_1 + \mu_2 \neq 2\mu_3$.

The researcher typically has some idea as to the type of comparisons that are of interest. In most situations, and in this text, we will only be concerned with comparing pairs of means as in $H_0: \mu_1 = \mu_2$ versus $H_1: \mu_1 \neq \mu_2$. Many different procedures can be used to compare two means. In this text, we present the method introduced by John Tukey (the same Tukey of box-plot fame).

① Perform the Tukey Test

The Tukey test is also known as the *Honestly Significant Difference Test* or the *Wholly Significant Difference Test*. It is designed to compare pairs of means after the null hypothesis of equal means has been rejected. That is, it tests $H_0: \mu_i = \mu_j$ versus $H_1: \mu_i \neq \mu_j$ for all means where $i \neq j$. The goal of the test is to determine which population means differ significantly.

The computation of the test statistic for Tukey's test follows the same logic as the test for comparing two means from independent sampling. Suppose that we wish to test $H_0: \mu_1 = \mu_2$ versus $H_1: \mu_1 \neq \mu_2$. This is equivalent to testing $H_0: \mu_2 - \mu_1 = 0$ versus $H_1: \mu_2 - \mu_1 \neq 0$. The test statistic is based on the sample mean difference, $\bar{x}_2 - \bar{x}_1$, and the standard error of the sample mean difference. The standard error is not the same as the standard error used in comparing two means from independent samples. Instead, the standard error is

$$SE = \sqrt{\frac{s^2}{2} \cdot \left(\frac{1}{n_1} + \frac{1}{n_2} \right)}$$

where s^2 is the mean square error estimate (MSE) of σ^2 from the one-way ANOVA, n_1 is the sample size from population 1, and n_2 is the sample size from population 2.

Test Statistic for Tukey's Test

The test statistic for Tukey's test when testing $H_0: \mu_1 = \mu_2$ versus $H_1: \mu_1 \neq \mu_2$ is given by

$$q = \frac{(\bar{x}_2 - \bar{x}_1) - (\mu_2 - \mu_1)}{\sqrt{\frac{s^2}{2} \cdot \left(\frac{1}{n_1} + \frac{1}{n_2} \right)}} = \frac{\bar{x}_2 - \bar{x}_1}{\sqrt{\frac{s^2}{2} \cdot \left(\frac{1}{n_1} + \frac{1}{n_2} \right)}}$$

where $\bar{x}_2 > \bar{x}_1$

s^2 is the mean square error estimate of σ^2 (MSE) from ANOVA

n_1 is the sample size from population 1

n_2 is the sample size from population 2

Because we assume that the null hypothesis is true, until we have evidence to the contrary, $\mu_2 - \mu_1 = 0$ in the computation of the test statistic.

The question now becomes, "What distribution does this test statistic follow?" The q-test statistic follows a distribution called the **Studentized range distribution**. The shape of the distribution depends on the error degrees of freedom, ν, and the total number of means being compared, k. We will compare the test statistic, q, to a critical value from this distribution, $q_{\alpha, \nu, k}$, where α is the level of significance of the test and therefore the probability of making a Type I error (rejecting the null hypothesis when the null hypothesis is true). The level of significance α is called the **experimentwise error rate** or the **familywise error rate**.

Critical Value for Tukey's Test

The critical value for Tukey's test using a familywise error rate α is given by

$$q_{\alpha,\nu,k}$$

where

> ν is the degrees of freedom due to error (the degrees of freedom due to error is the total number of subjects sampled size minus the number of means being compared, or $n - k$)
>
> k is the total number of means being compared

We can determine the critical value from the Studentized range distribution by referring to Table VIII. We have provided tables for $\alpha = 0.01$ and $\alpha = 0.05$.

EXAMPLE 1 **Finding the Critical Value from the Studentized Range Distribution**

Problem: Find the critical value from the Studentized range distribution with $\nu = 7$ degrees of freedom and $k = 3$ degrees of freedom, with a familywise error rate $\alpha = 0.05$.

Approach: We look in Table VIII with $\alpha = 0.05$ and where the row corresponding to $\nu = 7$ intersects with the column corresponding to $k = 3$. The value in the cell is the critical value.

Solution: See Figure 7. The critical value is $q_{0.05,7,3} = 4.165$.

Figure 7 $\alpha = 0.05$

ν	k(or p):2	3	4	5	6	7	8	9	10
1	17.97	26.98	32.82	37.08	40.41	43.12	45.40	47.36	49.07
2	6.085	8.331	9.798	10.88	11.74	12.44	13.03	13.54	13.99
3	4.501	5.910	6.825	7.502	8.037	8.478	8.853	9.177	9.462
4	3.927	5.040	5.757	6.287	6.707	7.053	7.347	7.602	7.826
5	3.635	4.602	5.218	5.673	6.033	6.330	6.582	6.802	6.995
6	3.461	4.339	4.896	5.305	5.628	5.895	6.122	6.319	6.493
7	3.344	4.165	4.681	5.060	5.359	5.606	5.815	5.998	6.158
8	3.261	4.041	4.529	4.886	5.167	5.399	5.597	5.767	5.918
9	3.199	3.949	4.415	4.756	5.024	5.244	5.432	5.595	5.739
10	3.151	3.877	4.327	4.654	4.912	5.124	5.305	5.461	5.599

Now Work Problem 3.

Now that we know the distribution of the test statistic q and how to find the critical value, we can introduce a procedure that can be used to make multiple comparisons using Tukey's test.

Tukey's Test

After rejecting the null hypothesis $H_0: \mu_1 = \mu_2 = \cdots = \mu_k$, the following steps can be used to compare pairs of means for significant differences, provided that

1. There are k simple random samples from k populations.
2. The k samples are independent of each other.
3. The populations are normally distributed.
4. The populations have the same variance.

Step 1: Arrange the sample means in ascending order.

Step 2: Compute the pairwise differences, $\bar{x}_i - \bar{x}_j$, where $\bar{x}_i > \bar{x}_j$.

Step 3: Compute the test statistic, $q = \dfrac{\bar{x}_i - \bar{x}_j}{\sqrt{\dfrac{s^2}{2} \cdot \left(\dfrac{1}{n_i} + \dfrac{1}{n_j}\right)}}$, for each pairwise difference.

Step 4: Determine the critical value, $q_{\alpha,\nu,k}$, where α is the level of significance (the familywise error rate).

Step 5: If $q \geq q_{\alpha,\nu,k}$, reject the null hypothesis that $H_0: \mu_i = \mu_j$ and conclude that the means are significantly different.

Step 6: Compare all pairwise differences to identify which means are considered equal.

Let's go through an example to illustrate Tukey's test.

EXAMPLE 2   Performing Tukey's Test

Problem: In Example 1 from Section 13.1, we rejected the null hypothesis $H_0: \mu_{\text{control}} = \mu_{\text{fenugreek}} = \mu_{\text{garlic}} = \mu_{\text{onion}}$. Use Tukey's test to determine which pairwise means differ using a familywise error rate of $\alpha = 0.05$.

Approach: We will follow Steps 1–6 to determine which pairwise means differ.

Solution

Step 1: The means for each category are listed in Table 4. The means, in ascending order, are $\bar{x}_G = 180.24$, $\bar{x}_F = 224.16$, $\bar{x}_O = 271.00$, $\bar{x}_C = 278.56$.

Step 2: We now compute the pairwise differences, $\bar{x}_i - \bar{x}_j$, where $\bar{x}_i > \bar{x}_j$. That is, subtract the smaller sample mean from the larger sample mean, as shown in the second column of Table 5. It is helpful to write the differences in descending order.

Table 4

Control	Fenugreek	Garlic	Onion
278.56	224.16	180.24	271.00

Table 5

Comparison	Difference, $\bar{x}_i - \bar{x}_j$	Test Statistic, q	Critical Value	Decision
Control vs. garlic (control − garlic)	$278.56 - 180.24 = 98.32$	$q = \dfrac{\bar{x}_C - \bar{x}_G}{\sqrt{\dfrac{s^2}{2} \cdot \left(\dfrac{1}{n_F} + \dfrac{1}{n_G}\right)}} = \dfrac{98.32}{\sqrt{\dfrac{287}{2} \cdot \left(\dfrac{1}{8} + \dfrac{1}{8}\right)}} = 16.415$	3.845	Reject $H_0: \mu_C = \mu_G$ since $q > q_{0.05,28,4}$
Garlic vs. onion (onion − garlic)	$271.00 - 180.24 = 90.76$	$q = \dfrac{\bar{x}_G - \bar{x}_O}{\sqrt{\dfrac{s^2}{2} \cdot \left(\dfrac{1}{n_C} + \dfrac{1}{n_O}\right)}} = \dfrac{90.76}{\sqrt{\dfrac{287}{2} \cdot \left(\dfrac{1}{8} + \dfrac{1}{8}\right)}} = 15.153$	3.845	Reject $H_0: \mu_G = \mu_O$ since $q > q_{0.05,28,4}$
Control vs. fenugreek (control − fenugreek)	$278.56 - 224.16 = 54.4$	$q = \dfrac{\bar{x}_C - \bar{x}_F}{\sqrt{\dfrac{s^2}{2} \cdot \left(\dfrac{1}{n_C} + \dfrac{1}{n_F}\right)}} = \dfrac{54.4}{\sqrt{\dfrac{287}{2} \cdot \left(\dfrac{1}{8} + \dfrac{1}{8}\right)}} = 9.082$	3.845	Reject $H_0: \mu_C = \mu_F$ since $q > q_{0.05,28,4}$
Fenugreek vs. onion (onion − fenugreek)	$271.00 - 224.16 = 46.84$	$q = \dfrac{\bar{x}_F - \bar{x}_O}{\sqrt{\dfrac{s^2}{2} \cdot \left(\dfrac{1}{n_F} + \dfrac{1}{n_O}\right)}} = \dfrac{46.84}{\sqrt{\dfrac{287}{2} \cdot \left(\dfrac{1}{8} + \dfrac{1}{8}\right)}} = 7.820$	3.845	Reject $H_0: \mu_F = \mu_O$ since $q > q_{0.05,28,4}$
Fenugreek vs. garlic (fenugreek − garlic)	$224.16 - 180.24 = 43.92$	$q = \dfrac{\bar{x}_F - \bar{x}_G}{\sqrt{\dfrac{s^2}{2} \cdot \left(\dfrac{1}{n_F} + \dfrac{1}{n_G}\right)}} = \dfrac{43.92}{\sqrt{\dfrac{287}{2} \cdot \left(\dfrac{1}{8} + \dfrac{1}{8}\right)}} = 7.333$	3.845	Reject $H_0: \mu_F = \mu_G$ since $q > q_{0.05,28,4}$
Control vs onion (control − onion)	$278.56 - 271.00 = 7.56$	$q = \dfrac{\bar{x}_C - \bar{x}_O}{\sqrt{\dfrac{s^2}{2} \cdot \left(\dfrac{1}{n_C} + \dfrac{1}{n_O}\right)}} = \dfrac{7.56}{\sqrt{\dfrac{287}{2} \cdot \left(\dfrac{1}{8} + \dfrac{1}{8}\right)}} = 1.262$	3.845	Do not reject $H_0: \mu_C = \mu_O$ since $q < q_{0.05,28,4}$

Step 3: Compute the test statistic q for each pairwise difference. From the ANOVA table in Figure 4, we have the mean square error $= 287$. This is the value of s^2 in the computation of the test statistic. The test statistic for each pairwise difference is shown in the third column of Table 5.

Step 4: Find the critical value using an $\alpha = 0.05$ familywise error rate with $\nu = n - k = 32 - 4 = 28$ and $k = 4$. From Table VIII, the critical value is $q_{0.05,28,3} = 3.845$. Since $q_{0.05,28,4}$ is not in Table VIII (because it does not have an entry for $\nu = 28$), we use the entry corresponding to the value closest to ν, which is 30. The critical value is placed in the fourth column of Table 5.

Step 5: If $q \geq q_{0.05,28,4}$, we reject the null hypothesis that the pairwise means are equal. We state our decisions in the fifth column of Table 5. By writing the differences in descending order, once we "do not reject H_0", we can stop because we won't reject smaller differences either!

Step 6: The conclusions of Tukey's test are

$$\mu_C = \mu_O \neq \mu_F \neq \mu_G$$

It is helpful to use lines to indicate which sample means are not significantly different, as follows:

$$\underline{\mu_C \ \mu_O} \ \ \mu_F \ \ \mu_G$$

The means listed over a common line are said to be equal. So we have evidence that indicates a diet rich in garlic is most helpful in reducing blood sugar levels. Fenugreek reduces blood sugar significantly more than a diet supplemented with onion. And onion does not significantly change blood sugar from a normal diet. The bottom line—eat garlic to reduce blood sugar. ▪

EXAMPLE 3   Tukey's Test Using Technology

Problem: Use MINITAB to obtain the results of Tukey's test for the blood sugar data from Example 1 from Section 13.1.

Approach: The steps for obtaining the results of Tukey's test using MINITAB are given in the Technology Step by Step on page 700.

Result: The results of Tukey's test using MINITAB are presented in Figure 8. Instead of reporting P-values to test each hypothesis that the pairwise means

Figure 8   **Tukey 95% Simultaneous Confidence Intervals**
All Pairwise Comparisons

```
Individual confidence level = 98.92%

Control subtracted from:

            Lower  Center   Upper  +---------+---------+---------+---------
Fenugreek -77.51 -54.40 -31.29          (---*---)
Garlic    -121.44 -98.33 -75.21 (---*--)
Onion     -30.68  -7.56  15.55                      (---*---)
                                +---------+---------+---------+---------
                              -120       -60        0        60


Fenugreek subtracted from:

           Lower  Center   Upper  +---------+---------+---------+---------
Garlic    -67.04 -43.92 -20.81          (---*---)
Onion      23.72  46.84  69.95                           (---*---)
                                +---------+---------+---------+---------
                              -120       -60        0        60


Garlic subtracted from:

           Lower  Center   Upper  +---------+---------+---------+---------
Onion      67.65  90.76 113.88                               (---*---)
                                +---------+---------+---------+---------
                              -120       -60        0        60
```

are equal, MINITAB reports confidence intervals. Recall, we can use confidence intervals to test $H_0: \mu_1 = \mu_2$ versus $H_1: \mu_1 \neq \mu_2$ (or, equivalently, $H_0: \mu_1 - \mu_2 = 0$ versus $H_1: \mu_1 - \mu_2 \neq 0$) by determining whether the interval contains 0. If 0 is contained in the interval, we do not reject the null hypothesis, but if the interval does not contain 0, we reject the null hypothesis at the α level of significance.

If we reject the null hypothesis, $H_0: \mu_1 = \mu_2$, we can look at the bounds of the confidence interval to determine which mean (μ_1 or μ_2) is higher. For example, if the interval contains only positive numbers, then $\bar{x}_1 - \bar{x}_2$ is positive, so we would conclude that there is evidence that $\mu_1 > \mu_2$. Similarly, if the confidence interval contains only negative numbers, we have evidence that $\mu_1 < \mu_2$.

If we look at the first group of confidence intervals, we are comparing means from the control group and the other three groups. From the output, the lower bound on the confidence interval fenugreek–control is -77.51 mg/dL and the upper bound is -31.29 mg/dL. Because the interval does not contain 0, we reject the null hypothesis that the mean blood sugar of subjects consuming fenugreek equals the mean blood sugar of subjects in the control group. In addition, because fenugreek–control is negative, we know that $\mu_C > \mu_F$. Next, we compare the control group with the garlic group. The lower bound is -121.44 mg/dL and the upper bound is -75.21 mg/dL. Because the interval does not contain 0, we reject the null hypothesis of equal means. We read the rest of the output in the same fashion.

Now Work Problem 11.

Some Cautions Regarding Tukey's Test

Sometimes the results of Tukey's test are ambiguous. Suppose that the null hypothesis $H_0: \mu_1 = \mu_2 = \mu_3 = \mu_4$ is rejected. Also suppose that the results of Tukey's test indicate the following:

$$\underline{\mu_1 \ \mu_2 \ \mu_3} \ \mu_4$$

It appears that populations 1, 2, and 3 have a population mean that is different from populations 3 and 4. Clearly, this is impossible. Do you see why? The population mean from treatment 3 can not equal the population mean from treatment 1 and 2 while at the same time equal the population mean from treatment 4, whose mean is different from population 1 and 2! This basically means that at least one Type II error has been committed by Tukey's test.

This result allows us to say that $\mu_1 = \mu_2 \neq \mu_4$, but we can not tell how μ_3 is related to the other population means. A solution to this problem is to increase the sample size so that the test is more powerful.

It can also happen that the one-way ANOVA rejects $H_0: \mu_1 = \mu_2 = \cdots = \mu_k$, but the Tukey test does not detect any pairwise differences. This result occurs because one-way ANOVA is more powerful than Tukey's test. Again, the solution is to increase the sample size.

13.2 ASSESS YOUR UNDERSTANDING

Concepts and Vocabulary

1. Explain the purpose of multiple comparison tests. When do we use them?

2. What is the familywise error rate?

Skill Building

3. (a) Find the critical value from the Studentized range distribution for $\alpha = 0.05$, $\nu = 10$, $k = 3$.
   (b) Find the critical value from the Studentized range distribution for $\alpha = 0.05$, $\nu = 24$, $k = 5$.
   (c) Find the critical value from the Studentized range distribution for $\alpha = 0.05$, $\nu = 32$, $k = 8$.

   (d) Find the critical value from the Studentized range distribution for $H_0: \mu_1 = \mu_2 = \mu_3 = \mu_4 = \mu_5$, with $n = 65$ at $\alpha = 0.05$.

4. (a) Find the critical value from the Studentized range distribution for $\alpha = 0.05$, $\nu = 12$, $k = 3$.
   (b) Find the critical value from the Studentized range distribution for $\alpha = 0.05$, $\nu = 20$, $k = 4$.

(c) Find the critical value from the Studentized range distribution for $\alpha = 0.05$, $\nu = 42$, $k = 5$.

(d) Find the critical value from the Studentized range distribution for $H_0: \mu_1 = \mu_2 = \mu_3 = \mu_4$, with $n = 34$ at $\alpha = 0.05$.

5. Suppose that there is sufficient evidence to reject $H_0: \mu_1 = \mu_2 = \mu_3$ using a one-way ANOVA. The mean square error from ANOVA is determined to be 26.2. The sample means are $\bar{x}_1 = 9.5$, $\bar{x}_2 = 9.1$, $\bar{x}_3 = 18.1$, with $n_1 = n_2 = n_3 = 5$. Use Tukey's test to determine which pairwise means are significantly different using a familywise error rate of $\alpha = 0.05$.

6. Suppose that there is sufficient evidence to reject $H_0: \mu_1 = \mu_2 = \mu_3 = \mu_4$ using a one-way ANOVA. The mean square error from ANOVA is determined to be 26.2. The sample means are $\bar{x}_1 = 42.6$, $\bar{x}_2 = 49.1$, $\bar{x}_3 = 46.8$, $\bar{x}_4 = 63.7$, with $n_1 = n_2 = n_3 = n_4 = 6$. Use Tukey's test to determine which pairwise means are significantly different using a familywise error rate of $\alpha = 0.05$.

7. Suppose the following data are taken from three different populations that are known to be normally distributed with equal population variances based on independent simple random samples.

Sample 1	Sample 2	Sample 3
35.4	42.0	43.3
35.0	39.4	48.6
39.2	33.4	42.0
44.8	35.1	53.9
36.9	32.4	46.8
28.9	22.0	51.7

(a) Test the claim that each sample comes from a population with the same mean at the $\alpha = 0.05$ level of significance. That is, test $H_0: \mu_1 = \mu_2 = \mu_3$.

(b) If you rejected the null hypothesis in part (a), use Tukey's test to determine which pairwise means differ using a familywise error rate of $\alpha = 0.05$.

(c) Draw boxplots of each set of sample data to support your results of parts (a) and (b).

8. Suppose the following data are taken from four different populations that are known to be normally distributed with equal population variances based on independent simple random samples.

Sample 1	Sample 2	Sample 3	Sample 4
110	138	98	130
85	140	100	116
83	130	94	157
95	115	110	137
103	101	104	144
105	130	118	124
107	123	102	139

(a) Test the claim that each sample comes from a population with the same mean at the $\alpha = 0.05$ level of significance. That is, test $H_0: \mu_1 = \mu_2 = \mu_3 = \mu_4$.

(b) If you rejected the null hypothesis in part (a), use Tukey's test to determine which pairwise means differ using a familywise error rate of $\alpha = 0.05$.

(c) Draw boxplots of each set of sample data to support your results for parts (a) and (b).

9. **Corn Production** (*See Problem 11 from Section 13.1*) An agricultural researcher wanted to know whether the mean number of plants for each plot type differed. A one-way analysis of variance was performed to test $H_0: \mu_{SP} = \mu_{SD} = \mu_{NT}$. The null hypothesis was rejected with a P-value of 0.007. The researcher proceeded to conduct Tukey's test to determine which means differed using MINITAB. The results are presented next. Which pairwise means differ? Which plot type would you recommend?

Tukey 95% Simultaneous Confidence Intervals
All Pairwise Comparisons

```
Individual confidence level = 97.97%

Sludge Plot subtracted from:

Spring   Lower  Center  Upper  ------+---------+---------+---------+---
Disk     1.021   4.667  8.313                        (------*-------)
No Till -3.479   0.167  3.813              (------*-------)
                                ------+---------+---------+---------+---
                                    -5.0       0.0       5.0      10.0

Spring Disk subtracted from:

         Lower   Center   Upper  ------+---------+---------+---------+---
No Till -8.146  -4.500  -0.854  (------*------)
                                ------+---------+---------+---------+---
                                    -5.0       0.0       5.0      10.0
```

10. **Soybean Yield** (*See Problem 12 from Section 13.1*) An agricultural researcher wanted to determine if the mean soybean yield differed for different plot types. A one-way analysis of variance was performed to test $H_0: \mu_{Lib} = \mu_{NoTill} = \mu_{Chis}$. The null hypothesis was rejected with a P-value of 0.038. The researcher proceeded to conduct Tukey's test to determine which pairwise means differed using MINITAB. The results are presented next. Which pairwise means differ? Which plot type would you recommend?

Tukey 95% Simultaneous Confidence Intervals
All Pairwise Comparisons

```
Individual confidence level = 98.02%

Liberty subtracted from:

              Lower   Center  Upper  --------+---------+---------+---------+-
No Till      -6.344  -1.111   4.122              (--------*--------)
Chisel                                       (--------*--------)
Plowed     -10.677  -5.444  -0.212  --------+---------+---------+---------+-
                                         -6.0       0.0       6.0      12.0

No Till subtracted from:

              Lower   Center  Upper  --------+---------+---------+---------+-
Chisel                                       (--------*-------)
Plowed      -9.566  -4.333   0.900  --------+---------+---------+---------+-
                                         -6.0       0.0       6.0      12.0
```

Applying the Concepts

11. pH in Rain In Problem 19 from Section 13.1, we rejected
NW H_0: $\mu_{AK} = \mu_{FL} = \mu_{TX}$ and concluded that the mean pH in
rain falling in the three states differ. Use Tukey's test to
determine which pairwise means differ using a familywise
error rate of $\alpha = 0.05$.

12. Lower Your Cholesterol In Problem 20 from Section 13.1,
we rejected H_0: $\mu_{SF} = \mu_{MED} = \mu_{NCEP}$ and concluded that
there is at least one treatment that results in a mean LDL
cholesterol that differs from the rest. Use Tukey's test to
determine which pairwise means differ using a familywise
error rate of $\alpha = 0.05$.

13. Attention Deficit–Hyperactivity Disorder Researchers
William E. Pelham et. al. studied the impact that varying
levels of Ritalin and Adderall had on a child's ability to
follow rules when the child is diagnosed with attention
deficit–hyperactivity disorder (ADHD). They randomly
assigned children to one of five treatment groups: placebo, 10 mg Ritalin, 17.5 mg of Ritalin, 7.5 mg of Adderall,
or 12.5 mg of Adderall twice a day. They recorded a score
that indicated the child's ability to follow rules, with a
higher score indicating a higher ability to follow rules.
The following data are based on their study, A Comparison of Ritalin and Adderall: Efficacy and Time-course in
Children with Attention Deficit/Hyperactivity Disorder,
Pediatrics, Vol. 103, No. 4.

Placebo	Ritalin 10 mg	Ritalin 17.5 mg	Adderall 7.5 mg	Adderall 12.5 mg
47	96	37	57	84
20	75	83	71	105
34	57	76	85	83
60	44	43	92	92
28	83	87	75	74
72	69	62	55	92
25	57	99	73	64

(a) Test the null hypothesis that the mean score for each
treatment is the same at the $\alpha = 0.05$ level of significance. [**Note:** The requirements for a one-way
ANOVA are satisfied].

(b) If the null hypothesis is rejected in part (a), use
Tukey's test to determine which pairwise means differ using a familywise error rate of $\alpha = 0.05$.

(c) Draw boxplots of the five treatment levels to support
the analytic results obtained in parts (a) and (b).

14. Nutrition Researchers Sharon Peterson and Madeleine
Sigman-Grant wanted to compare the overall nutrient
intake of American children (ages 2 to 19) who exclusively use lean meats, mixed meats, or higher-fat meats.
The data given represent the daily consumption of calcium (in mg) for a random sample of eight children in each
category and are based on the results presented in their
article," Impact of Adopting Lower-fat Food Choices on

Nutrient Intake of American Children, " *Pediatrics,* Vol.
100, No. 3.

Lean Meats	Mixed Meats	Higher-Fat Meats
844.2	897.7	843.4
745.0	908.1	862.2
773.1	948.8	790.5
823.6	836.6	876.5
812.0	871.6	790.8
758.9	945.9	847.2
810.7	859.4	772.0
790.6	920.2	851.3

(a) Test the null hypothesis that the mean calcium for
each category is the same at the $\alpha = 0.05$ level of significance.[**Note:** The requirements for a one-way
ANOVA are satisfied].

(b) If the null hypothesis is rejected in part (a), use Tukey's
test to determine which pairwise means differ using a
familywise error rate of $\alpha = 0.05$.

(c) Draw boxplots of the three treatment levels to support the analytic results obtained in parts (a) and (b).

15. Heart Rate of Smokers A researcher wants to determine
the impact that smoking has on resting heart rate. She randomly selects seven individuals from three categories: (1)
nonsmokers, (2) light smokers (fewer than 10 cigarettes
per day), and (3) heavy smokers (more than 10 cigarettes
per day) and obtains the following heart rate data (beats
per minute):

Nonsmokers	Light Smokers	Heavy Smokers
56	78	77
53	62	86
53	70	65
65	73	83
70	67	79
58	75	80
51	65	77

(a) Test the null hypothesis that the mean resting heart
rate for each category is the same at the $\alpha = 0.05$
level of significance.[**Note:** The requirements for a
one-way ANOVA are satisfied].

(b) If the null hypothesis is rejected in part (a), use Tukey's
test to determine which pairwise means differ using a
familywise error rate of $\alpha = 0.05$.

(c) Draw boxplots of the three treatment levels to support the results obtained in parts (a) and (b).

16. Price to Earnings Ratios One measure of the value of a
stock is its price to earnings ratio (or P/E ratio). It is the

ratio of the price of a stock per share to the earnings per share and can be thought of as the price an investor is willing to pay for $1 of earnings in a company. A stock analyst wants to know whether the P/E ratio for three industry categories differs significantly. The following data represent a simple random sample of companies from three categories: (1) financial, (2) food, and (3) leisure goods.

Financial	Food	Leisure Goods
8.83	19.75	14.10
12.75	17.87	10.12
13.48	15.18	15.57
14.42	22.84	13.48
10.06	15.60	11.27

Source: Yahoo! Finance

(a) Test the null hypothesis that the mean P/E ratio for each category is the same at the $\alpha = 0.05$ level of significance. [**Note:** The requirements for a one-way ANOVA are satisfied].

(b) If the null hypothesis is rejected in part (a), use Tukey's test to determine which pairwise means differ using a familywise error rate of $\alpha = 0.05$.

(c) Draw boxplots of the three categories to support the analytic results obtained in parts (a) and (b).

17. Got Milk? Researchers Sharon Peterson and Madeleine Sigman-Grant wanted to compare the overall nutrient intake of American children (ages 2 to 19) who exclusively use skim milk instead of 1%, 2%, or whole milk. The researchers combined children who consumed 1% or 2% milk into a "mixed milk" category. The following data represent the daily calcium intake (in milligrams) for a random sample of eight children in each category and is based on the results presented in their article, Impact of Adopting Lower-fat Food Choices on Nutrient Intake of American Children, *Pediatrics*, Vol. 100, No. 3.

Skim Milk	Mixed Milks	Whole Milk
857	1006	879
853	991	938
865	1015	841
904	1035	818
916	1024	870
886	1013	874
854	1065	881
856	1002	836

(a) Is there sufficient evidence to support the belief that at least one of the means is different from the others at the $\alpha = 0.05$ level of significance? [**Note:** The requirements for a one-way ANOVA are satisfied].

(b) If the null hypothesis is rejected in part (a), use Tukey's test to determine which pairwise means differ using a familywise error rate of $\alpha = 0.05$.

(c) Draw boxplots of the three categories to support the analytic results obtained in parts (a) and (b).

18. The **comparisonwise error rate**, denoted α_c, is the probability of making a Type I error when comparing two means. It is related to the familywise error rate, α, through the formula $1 - \alpha = (1 - \alpha_c)^k$, where k is the number of means being compared.

(a) If the familywise error rate is $\alpha = 0.05$ and there are $k = 3$ means being compared, what is the comparisonwise error rate?

(b) If the familywise error rate is $\alpha = 0.05$ and there are $k = 5$ means being compared, what is the comparisonwise error rate?

(c) Based on the results of parts (a) and (b), what happens to the comparisonwise error rate as the number of means compared increases?

Consumer Reports Tea Antioxidants: An Unexpected Side Benefit to a Cheap and Tasty Beverage

Antioxidant compounds are thought to be beneficial to human health. In a study for *Consumer Reports*, 15 brands of tea were tested for total antioxidant content. The antioxidant content was measured as oxygen radical absorbance capacity (ORAC) per 8-oz serving at an outside laboratory. Among the 15 brands, 4 were bagged black, 3 were bagged green, 2 were instant, and 6 were ready-to-drink teas.

The data for the 6 ready-to-drink teas are given in the table. To avoid potential sources of bias, measurements were taken from three different lots of each brand, and the tests were conducted in a randomly chosen order. A control was also run repeatedly through-out the test period to determine if there was a time effect, but the data are not included here.

Brand	ORAC	Brand	ORAC
1	427.4	4	484.3
1	481.1	4	617.3
1	428.3	4	624.9
2	667.7	5	664.1
2	737.6	5	623.1
2	630.9	5	647.5
3	715.2	6	606.0
3	724.1	6	793.9
3	686.9	6	668.7

MINITAB was used to perform a one-way analysis of variance of the data. The results are shown next:

```
Analysis of Variance for ORAC
Source DF      SS     MS      F       P
Brand    5  147257  29451   8.80   0.001
Error   12   40180   3348
Total   17  187437
                       Individual 95% CIs For Mean
                       Based on Pooled StDev
Level  N   Mean   StDev  ----------+---------+---------+-------
1      3  445.60  30.75  (-----*-----)
2      3  678.73  54.20                    (-----*-----)
3      3  708.73  19.42                       (-----*-----)
4      3  575.50  79.07             (-----*-----)
5      3  644.90  20.62                 (-----*-----)
6      3  689.53  95.67                    (-----*-----)
                       ----------+---------+---------+-------
Pooled StDev = 57.87             480       600       720
```

Using the MINITAB output, answer the following questions:

(a) What are the null and alternative hypotheses?

(b) Do there appear to be any outliers or other unusual observations? What must be true regarding the distribution of the response variable, ORAC, to perform a one-way analysis of variance?

(c) Is there a statistical difference among brands? What is the significance level of the test?

(d) Which brands, if any, appear to have similar amounts of ORAC?

(e) Use Tukey's test to determine which pairwise means differ using a familywise error rate of $\alpha = 0.05$.

(f) Write a paragraph for the readers of *Consumer Reports* magazine that explains your findings. Which brand of tea would you recommend? Why?

Note to Readers: In many cases, our test protocol and analytical methods are more complicated than described in these examples. The data and discussions have been modified to make the material more appropriate for the audience.

Technology Step-by-Step	**Tukey's Test**
MINITAB	**Step 1:** Enter the raw data in C1, C2, and so on, for each population or treatment.

Step 2: Select **Stat**, then highlight **ANOVA**, and then select **One-Way (Unstacked).**

Step 3: Enter the column names in the cell marked "Responses". Click "Comparisons . . . ". Check the box labeled "Tukey's, family error rate:". Enter the familywise error rate in the box. For $\alpha = 0.05$, enter 5. Click OK twice.

13.3 The Randomized Complete Block Design

Preparing for This Section Before getting started, review the following:

- Randomized Block Design (Section 1.5, pp. 44–46)
- Inference on Two Means: Dependent Samples (Section 11.1, pp. 575–582)

Objectives

1 **Conduct analysis of variance on the randomized complete block design**

2 **Perform the Tukey test**

In the completely randomized design, a single **factor** is manipulated and fixed at different levels. The experimental units are then randomly assigned to one of the factor levels or treatments. If there are only two levels of the factor (or two treatments), we can analyze the data using the methods introduced in Section 11.2. If there are three or more treatments, we analyze the data using the methods of one-way analysis of variance.

Remember, in designing an experiment, the researcher identifies the various factors that may affect the value of the response variable. The researcher can then deal with the factors in one of three ways:

1. Control their levels so they remain fixed throughout the experiment
2. Manipulate or set them at fixed levels
3. Randomize so that any effects not identified or uncontrollable are minimized

In the completely randomized design, the researcher manipulates a single factor and fixes it at two or more levels and then randomly assigns experimental units to a treatment. The researcher may have the ability to control certain factors in this design setting so that they remain fixed. The remaining factors are dealt with by randomly assigning the experimental units to treatment groups.

The completely randomized design is not always sufficient, because the researcher may be aware of additional factors that cannot be fixed at a single level throughout the experiment.

One such experimental design that captures more information (and therefore reduces experimental error) is the randomized block design, first introduced in Section 1.5, where we discussed the impact fertilizer might have on crop yield. We designed an experiment in which three levels of fertilizer (the three treatments) were each analyzed on two different plant types (the blocks).

To help further understand the purpose of the completely randomized block design, consider the following example.

> ✒ **In Other Words**
>
> A block is a method for controlling experimental error. Blocks should form a homogeneous group. For example, if age is thought to explain some of the variability in the response variable, we can remove age from the experimental error by forming blocks of experimental units with the same age. Gender is another common block.

EXAMPLE 1 **Illustrating the Randomized Complete Block Design**

Suppose that a researcher is interested in determining the tread wear on four different tires after 5000 miles of highway driving. The researcher has four of each brand of tire and four cars along with four drivers. For simplicity, we call the tire types A, B, C, and D. The researcher wishes to test $H_0: \mu_A = \mu_B = \mu_C = \mu_D$ versus the alternative that at least one of the means differs, where μ_i represents the mean tread loss for each tire type.

The researcher must first determine the variables that affect tire wear. Certainly, the type of car, the driver, tire inflation, and road conditions are a few. For the sake of time, the researcher decides to use all four cars in the experiment rather than one. He decides to set the inflation of each tire at 32 pounds and drive the cars on the same oval track. Each car is assigned one driver.

One possible experimental design in this setting would be to randomly assign the tires to the cars. Tire 1A would be assigned number 1, tire 2A would be assigned number 2, and so on, until we reach tire 4D, which would be assigned number 16. We can then number the cells as shown in Table 6(a). Now use a random number generator to assign the tires to the cars. If the first number generated is a 15, we would place tire 1A on car 4 on the back left position. If the next random number is 12, tire 2A is placed on car 3 on the back right position. We continue until all the tires are assigned to cars. Let X_{1A} represent the tread loss for the first tire A, let X_{2A} represent the tread loss for the second tire A, and so on. Under these circumstances, we might end up with the design shown in Table 6(b).

Table 6

	Front Left	Front Right	Back Left	Back Right
Car 1	1	2	3	4
Car 2	5	6	7	8
Car 3	9	10	11	12
Car 4	13	14	15	16

(a)

	Front Left	Front Right	Back Left	Back Right
Car 1	X_{4C}	X_{4B}	X_{3D}	X_{4D}
Car 2	X_{2B}	X_{3A}	X_{3B}	X_{2A}
Car 3	X_{4A}	X_{1D}	X_{2C}	X_{1B}
Car 4	X_{2D}	X_{3C}	X_{1A}	X_{1C}

(b)

A quick look at Table 6(b) indicates that tire A is not on car 1, tire C is not on car 2, tire D is not on car 2, and tire B is not on car 4. Therefore, any impact that the car might have on tire wear is not completely accounted for. That is, tire wear variability might be confounded with car variability.

To resolve this issue, we might require that each car must have all four tire types on it. This would result in a randomized complete block design. We use the word *complete* because each block (car) gets every treatment (tire type). We use the word *randomized* because the order in which the treatment is applied within each block is random. To obtain the randomness, we randomly assign one treatment (tire brand) to each block (car). For example, we have four treatments (tire brand), so we randomly select an integer between 1 and 4 for the tire number and a second random integer between 1 and 4 for the tire brand. For example, if a random-number generator results in a 2 and then a 1, we place tire 2 of brand A on car 1. Continuing to place tires on car 1, we repeat the process, but we cannot select brand A again because each brand must be represented on car 1. By randomly assigning each treatment to each block, we might end up with the design shown in Table 7*.

---

*Table 6 reveals another flaw. Notice that tire B is never in the front-left location. In all likelihood the position of the tire on the car also affects tire wear. We could account for this by rotating the tires every 1250 miles, or we could require that each treatment (tire) must also be placed in each position on the car, so one car must have tire A in the front left, a second car must have tire A in the front right, and so on. If we impose this second requirement, we would have a **Latin Square design**.

Table 7				
	Front Left	**Front Right**	**Back Left**	**Back Right**
Car 1	X_{2A}	X_{3C}	X_{2D}	X_{3B}
Car 2	X_{1A}	X_{2B}	X_{1D}	X_{2C}
Car 3	X_{4D}	X_{4A}	X_{1B}	X_{4C}
Car 4	X_{1C}	X_{4B}	X_{3A}	X_{3D}

In looking at Table 7, you should notice that horizontal lines are used in the table to help the reader see how the blocks are formed. Using this randomized complete block design, the null hypothesis is

$$H_0: \mu_A = \mu_B = \mu_C = \mu_D$$

versus

$$H_1: \text{at least one of the means is different}$$

One specific type of randomized complete block design is the matched-pairs design introduced in Section 1.5. This type of experimental design requires dependent sampling in which the experimental units are somehow related, such as in husband/wife, the same experimental unit; or some other characteristic that relates the individuals. The related experimental units form the blocks in this experimental design. We learned how to analyze data obtained from this experimental design in Section 11.1 using the paired t-test. For example, in Section 11.1, Example 2, we analyzed matched-pairs data where the blocks were the students.

The methods used in this section can also be used when we want to compare the means resulting from an observational study in which there are three or more categories whose means we are interested in comparing (the categories act like the treatment in an experiment) along with blocks. For example, suppose we obtain a simple random sample of individuals and asked them to disclose their number of years of education. We might also wonder if the mean number of years of education differs by age group, where the groups are segmented as 25 to 34 years of age, 35 to 49 years of age, and 50 years and older. Since gender or ethnicity might be considered factors that contribute to variability in the number of years of education, we might use gender or ethnicity as a block.

> ! **CAUTION!**
> When we block, we are not interested in determining whether the block is significant. We only want to remove experimental error to reduce the mean square error.

1 ## Conduct Analysis of Variance on the Randomized Complete Block Design

The following requirements must be satisfied to analyze data from a randomized complete block design.

Requirements for Analyzing Data from a Randomized Complete Block Design

1. The response variable for each of the k populations is normally distributed.
2. The response variable for each of the k populations has the same variance; that is, each treatment group has population variance σ^2.*

*There are a number of tests for comparing variances. MINITAB compares variances using two different tests: Bartlett's test and Levene's test. Levene's test will be conducted provided there are no missing observations (we do not cover analysis when there are missing observations because it is beyond the scope of the text). Levene's test is considered to be more robust than Bartlett's test, but neither is very reliable. Therefore, the best approach is to design your experiment so that the sample size for each factor is the same, because analysis of variance is not affected too much when we have equal sample sizes even if there is inequality in the variances.

As in one-way ANOVA, we will verify the requirement of equal population variances by comparing the sample standard deviations. The procedures can be used provided the largest sample standard deviation is no more than two times larger than the smallest sample standard deviation.

Because the computations in performing the analysis of the randomized complete block design are tedious, we will use statistical software to conduct the analysis. We use the P-value approach to test the hypothesis. Again, the decision rule will be to reject the null hypothesis if the P-value is less than the level of significance, α.

EXAMPLE 2 **Analyzing the Randomized Complete Block Design**

Problem: A researcher is interested in the effect of four different diets on the weight gain of newborn rats. To account for genetic differences among the rats, the researcher decides that she should randomly assign rats from the same mother to each treatment so that a set of four rats constitutes a block. The researcher identifies 5 mother rats that had 4 offspring, for a total of 20 rats in the study. For each block (sibling rats), the researcher randomly assigns the experimental units (rats) to a treatment using a random-number generator by generating integers from 1 to 4. After 6 months, she records the weight gain of the 20 rats (in grams) and obtains the data in Table 8. Is there sufficient evidence to conclude that the weight gains of the rats on the four diets differ at the $\alpha = 0.05$ level of significance?

Table 8				
	Diet 1	**Diet 2**	**Diet 3**	**Diet 4**
Block 1	15	14.6	17.7	10.4
Block 2	17.9	17.4	16.0	12.2
Block 3	17.5	14.8	14.2	14.8
Block 4	16.3	17.3	14.4	12.0
Block 5	15.4	19.3	18.8	14.3

Approach: This is a randomized complete block design. We will analyze the results of the experiment using MINITAB to determine if the weight gains differ significantly.

Solution: We wish to test

$$H_0: \mu_1 = \mu_2 = \mu_3 = \mu_4 \quad \text{vs.} \quad H_1: \text{at least one of the means is different}$$

A normal probability plot for the data from each of the four diets indicates that the requirement of normality has been satisfied. The diet with the highest standard deviation, diet 2 with $s_2 = 2.018$, is not more than two times the diet with the lowest standard deviation, diet 1 with $s_1 = 1.268$, so we can proceed with the analysis.

We enter the data into MINITAB in three columns. The first column represents the block, the second column the diet, and the third column the response variable, weight gain. So, for the rat in block 1 that received diet 1, we enter a 1 in column C1, a 1 in C2, and 15 in C3. For the rat in block 1 that received diet 2, we enter 1 in C1, 2 in C2, and 14.6 in C3, and so on.

We then analyze the data by selecting two-way ANOVA and choose the row factor to be the block and the column factor to be the treatment, diet. Be sure to check the box marked "Additive Model." Also check the box for "Display Means" so that we can perform multiple comparisons if we reject the null hypothesis. We obtain the results in Figure 9.

In Other Words!
Remember, the smaller the *P*-value is, the stronger the evidence against the null hypothesis.

Figure 9

Two-way ANOVA: Weight Gain versus Block, Treatment
```
Analysis of Variance for Weight G
Source        DF      SS      MS      F       P
Block          4   14.71    3.68   1.21   0.359
Treatment      3   51.87   17.29   5.67   0.012
Error         12   36.62    3.05
Total         19  103.21


                        Individual 95% CI
Block      Mean     --+---------+---------+---------+---------
1          14.42    (----------*----------)
2          15.87           (----------*----------)
3          15.32         (----------*----------)
4          15.00      (----------*----------)
5          16.95                (----------*----------)
                     --+---------+---------+---------+---------
                     12.80    14.40    16.00    17.60


                        Individual 95% CI
Treatment  Mean     -----+---------+---------+---------+------
1          16.42                    (-------*--------)
2          16.68                    (-------*--------)
3          16.22                  (-------*--------)
4          12.74    (--------*-------)
                    -----+---------+---------+---------+------
                    12.00    14.00    16.00    18.00
```

We are not interested in whether the weight gains among the blocks are equal, so we won't bother looking at the *P*-value for blocks. We are interested in knowing whether the treatment appears to have resulted in different weight gains. We take the *P*-value of 0.012 due to treatment as evidence against the null hypothesis of equal weight gain among treatments. Therefore, we conclude that the treatment, diet, does result in significantly different weight gain. We do not know which diets result in different mean weight gains, however.

Now Work Problem 7(a) and (b).

2 Perform the Tukey Test

Once the null hypothesis of equal population means is rejected, we proceed to determine which means differ significantly. This is called a *post hoc* (after this) test, and the method that we use to conduct the analysis is Tukey's test. The steps are identical to those presented in Section 13.2. The critical value is $q_{\alpha,\nu,k}$ using a familywise error rate of α with $\nu = (r - 1)(c - 1) =$ the error degrees of freedom (r is the number of blocks and c is the number of treatments) and k is the number of means being tested.

EXAMPLE 3   **Multiple Comparisons Using Tukey's Test**

Problem: Use Tukey's test to determine which pairwise means differ for the data presented in Example 2, with using a familywise error rate of $\alpha = 0.05$, using MINITAB.

Approach: The steps for conducting Tukey's test using MINITAB are given in the Technology Step by Step on page 711.

Result: The results of Tukey's test using MINITAB are presented in Figure 10. MINITAB allows the user to perform pairwise comparisons using either confidence intervals or *P*-values. In the output given next, Figure 10(a) shows the pairwise comparisons using 95% confidence intervals. The first sets of confidence intervals compare diet 1 to diets 2 through 4. We can see that the interval diet 1–diet 4 does not include 0. Therefore, we reject the null hypothesis that the mean weight gain from diet 1 equals the mean weight gain from diet 4. In addition, the interval contains negative numbers, so we conclude the weight gain from diet 1 is less than that from diet 4. Continuing the comparisons leads us to conclude that diet 4 results in more weight gain than the other three diets.

If we look at Figure 10(b), we see that the *P*-values comparing diet 4 with the remaining diets are all small (less than 0.05). Therefore, we conclude that the weight gain from diet 4 is significantly different from the weight gain from the other three diets.

Figure 10

```
Tukey 95.0% Simultaneous Confidence Intervals
Response Variable Gain
All Pairwise Comparisons among Levels of Diet

Diet = 1 subtracted from:

Diet  Lower   Center    Upper   -----+---------+---------+---------+-
2    -3.021   0.260    3.5413                    (----------*----------)
3    -3.481  -0.200    3.0813                 (----------*----------)
4    -6.961  -3.680   -0.3987       (----------*----------)
                                -----+---------+---------+---------+-
                                   -6.0      -3.0      0.0       3.0


Diet = 2 subtracted from:

Diet  Lower   Center    Upper   -----+---------+---------+---------+-
3    -3.741  -0.460    2.8213                 (---------*----------)
4    -7.221  -3.940   -0.6587       (----------*----------)
                                -----+---------+---------+---------+-
                                   -6.0      -3.0      0.0       3.0


Diet = 3 subtracted from:

Diet  Lower   Center    Upper   -----+---------+---------+---------+-
4    -6.761  -3.480   -0.1987       (----------*----------)
                                -----+---------+---------+---------+-
                                   -6.0      -3.0      0.0       3.0
```

(a)

```
Tukey Sumultaneous Tests
Response Variable Gain
All Pairwise Comparisons among Levels of Diet

Diet = 1 subtracted from:

Level  Difference     SE of                Adjusted
Diet    of Means   Difference   T-Value    P-Value
2         0.260       1.105      0.235      0.9952
3        -0.200       1.105     -0.181      0.9978
4        -3.680       1.105     -3.331      0.0266


Diet = 2 subtracted from:

Level  Difference     SE of                Adjusted
Diet    of Means   Difference   T-Value    P-Value
3        -0.460       1.105     -0.416      0.9746
4        -3.940       1.105     -3.566      0.0176


Diet = 3 subtracted from:

Level  Difference     SE of                Adjusted
Diet    of Means   Difference   T-Value    P-Value
4        -3.480       1.105     -3.150      0.0365
```

(b)

Now Work Problem 7(c) and (d).

In-Class Activity: Flight of Fancy: Part I

How does the design affect the flight of a paper airplane? Explore this concept by comparing the flight distances of three different designs. Since the type of paper could have an impact, use the same type of paper (for example—newspaper, brown shipping paper, typing paper, or other) and block by gender.

(a) Create a table to record your flight data based on gender (block) and plane design.
(b) Search the Internet (for example, using the Google search engine) for different designs of paper airplanes.
(c) Discuss the model being used and other variables that might affect flight distance. How can you account for these other variables?
(d) Fly each plane the same number of times and record the flight distance in your table.
(e) Was there a significant difference in flight distance for the different types of designs?
(f) If you answered yes to the previous question, conduct Tukey's test to determine which differences in flight distance are significant using the familywise error rate $\alpha = 0.05$.

13.3 ASSESS YOUR UNDERSTANDING

Concepts and Vocabulary

1. How does the completely randomized design differ from a randomized complete block design?
2. What is blocking? Why might a researcher want to block?
3. What does *randomized* mean in the randomized complete block design? What does *complete* mean in the randomized complete block design?
4. What requirements must be satisfied to analyze a randomized complete block design?
5. Name the three ways that a researcher can deal with explanatory variables in a designed experiment.
6. How is a matched-pairs design related to the randomized complete block design?

Skill Building

7. Given the following ANOVA output, answer the questions that follow:

Source	DF	SS	MS	F	P
Block	4	768.27	192.067	10.96	0.002
Treatment	2	278.53	139.267	7.95	0.013
Error	8	140.13	17.517		
Total	14	1186.93			

(a) The researcher wishes to test $H_0: \mu_1 = \mu_2 = \mu_3$ against H_1: at least one of the means is different. Based on the ANOVA table, what should the researcher conclude?
(b) What is the mean square error?
(c) The following output represents the results of Tukey's test. What should the researcher conclude?

```
Tukey Simultaneous Tests
Response Variable Response
All Pairwise Comparisons among Levels
of Treatment
Treatment = 1 subtracted from:

Treat-   Difference   SE of                Adjusted
ment     of Means   Difference  T-Value    P-Value
2          -1.000      2.647    -0.3778     0.9251
3           8.600      2.647     3.2489     0.0282


Treatment = 2 subtracted from:

Treat-   Difference   SE of                Adjusted
ment     of Means   Difference  T-Value    P-Value
3           9.600      2.647     3.627      0.0165
```

8. Given the following ANOVA output, answer the questions that follow:

Source	DF	SS	MS	F	P
Block	6	1637.81	272.968	7.91	0.001
Treatment	2	707.43	353.714	10.25	0.003
Error	12	413.90	34.492		
Total	20	2759.14			

(a) The researcher wishes to test $H_0: \mu_1 = \mu_2 = \mu_3$ against H_1: at least one of the means is different. Based on the ANOVA table, what should the researcher conclude?
(b) What is the mean square error?
(c) The following output represents the results of Tukey's test. What should the researcher conclude?

```
Tukey 95% Simultaneous Confidence Intervals
Response Variable Response
All Pairwise Comparisons among Levels
of Treatment
Treatment = 1 subtracted from:

Treatment Lower Center Upper  -----+---------+---------+---------+-
2         -3.511  4.857 13.23  (-----------*-----------)
3          5.631 14.000 22.37              (-----------*-----------)
                                -----+---------+---------+---------+-
                                   0.0      7.0      14.0     21.0

Treatment = 2 subtracted from:

Treatment Lower Center Upper  -----+---------+---------+---------+-
3          0.7743 9.143 17.51         (-----------*-----------)
                                -----+---------+---------+---------+-
                                   0.0      7.0      14.0     21.0
```

9. Given the following ANOVA output, answer the questions that follow:

Analysis of Variance for Response

Source	DF	SS	MS	F	P
Block	8	2105.436	263.179	278.66	0.000
Treatment	3	6.393	2.131	2.26	0.108
Error	24	22.667	0.944		
Total	35	2134.496			

(a) The researcher wishes to test $H_0: \mu_1 = \mu_2 = \mu_3 = \mu_4$ against H_1: at least one of the means is different. Based on the ANOVA table, what should the researcher conclude?
(b) What is the mean square error?
(c) Explain why it is not necessary to use Tukey's test on these data.

10. Given the following ANOVA output, answer the questions that follow:

Analysis of Variance for Response

Source	DF	SS	MS	F	P
Block	6	1712.37	285.39	134.20	0.000
Treatment	3	2.27	0.76	0.36	0.786
Error	18	38.28	2.13		
Total	27	1752.91			

(a) The researcher wishes to test $H_0: \mu_1 = \mu_2 = \mu_3 = \mu_4$ against H_1: at least one of the means is different. Based on the ANOVA table, what should the researcher conclude?
(b) What is the mean square error?
(c) Explain why it is not necessary to use Tukey's test on these data.

In Problems 11 and 12, assume that the data come from populations that are normally distributed with the same variance.

11.

Block	Treatment 1	Treatment 2	Treatment 3
1	9.7	8.4	8.8
2	10.4	8.9	8.5
3	10.5	9.3	9.0
4	10.7	10.5	9.3
5	11.1	10.7	10.3

(a) Test $H_0: \mu_1 = \mu_2 = \mu_3$ against H_1: at least one of the means is different, where μ_1 is the mean for treatment 1, and so on, at the $\alpha = 0.05$ level of significance.
(b) If the null hypothesis from part (a) was rejected, use Tukey's test to determine which pairwise means differ using a familywise error rate of $\alpha = 0.05$.
(c) Draw boxplots of the data for each treatment using the same scale to support the analytical results obtained in parts (a) and (b).

12.

Block	Treatment 1	Treatment 2	Treatment 3
1	15.8	15.0	15.3
2	16.0	15.8	17.2
3	21.6	18.3	21.5
4	21.6	20.8	21.3
5	22.5	21.5	23.5
6	17.5	16.2	16.8

(a) Test $H_0: \mu_1 = \mu_2 = \mu_3$ against H_1: at least one of the means is different, where μ_1 is the mean for treatment 1, and so on, at the $\alpha = 0.05$ level of significance.
(b) If the null hypothesis from part (a) was rejected, use Tukey's test to determine which pairwise means differ using a familywise error rate of $\alpha = 0.05$.
(c) Draw boxplots of the data for each treatment using the same scale to support the analytical results obtained in parts (a) and (b).

Applying the Concepts

13. **Octane** An automotive engineer wanted to determine whether the octane of gasoline used in a car increases gas mileage. He recognized that car and driver are variables that affect gas mileage. He selected six different brands of car and assigned a driver to each car, so he blocked by car type and driver. For each car (and driver), the researcher randomly selected a number from 1 to 3, with 1 representing 87-octane gasoline, 2 representing 89-octane gasoline, and 3 representing 92-octane gasoline. Then 5 gallons of the gasoline selected was placed in the car. The car was driven around a closed track at 40 miles per hour until the car ran out of gas. The number of miles driven was recorded and then divided by 5 to obtain the miles per gallon. He obtained the following results:

	87 Octane	89 Octane	92 Octane
Chevrolet Impala	28.3	28.4	28.7
Chrysler 300M	27.1	26.9	27.2
Ford Taurus	26.4	26.1	26.8
Lincoln LS	26.1	26.4	27.3
Toyota Camry	28.4	28.9	29.1
Volvo S60	25.3	25.1	25.8

(a) Normal probability plots for each treatment indicate that the requirement of normality has been satisfied. Verify that the requirement of equal population variances for each treatment has been satisfied.
(b) Is there evidence that the mean miles per gallon are different among the three octane levels at the $\alpha = 0.05$ level of significance?
(c) If the null hypothesis from part (b) was rejected, use Tukey's test to determine which pairwise means differ using a familywise error rate of $\alpha = 0.05$.
(d) Based on your results for part (c), what do you conclude?

14. Healing Rate A medical researcher wanted to determine the effectiveness of coagulants on the healing rate of a razor cut on lab mice. Because healing rates of mice vary from mouse to mouse, the researcher decided to block by mouse. First, the researcher gave each mouse a local anesthesia and then made a 5-mm incision that was 2 mm deep on each mouse. He randomly selected one of the three treatments and recorded the time that the wound stopped bleeding. He repeated this process two more times on each mouse and obtained the results shown.

Mouse	No Drug	Experimental Drug 1	Experimental Drug 2
1	3.2	3.4	3.4
2	4.8	4.4	3.4
3	6.6	5.9	5.4
4	6.5	6.3	5.2
5	6.4	6.3	6.1

(a) Normal probability plots for each treatment indicate that the requirement of normality has been satisfied. Verify that the requirement of equal population variances for each treatment has been satisfied.
(b) Is there evidence that the mean heal time is different among the three treatments at the $\alpha = 0.05$ level of significance?
(c) If the null hypothesis from part (b) was rejected, use Tukey's test to determine which pairwise means differ using a family-wise error rate of $\alpha = 0.05$.
(d) Based on your results for part (c), what do you conclude?

15. Crash Tests The Insurance Institute for Highway Safety regularly tests cars for various safety factors. In one such test, the institute tests the bumpers in 5-mile per hour (mph) crashes. The following data represent the cost of repair (in dollars) after four different 5-mph crashes on small utility vehicles. The institute blocks by location of crash, and the treatment is car model.

	Jeep Cherokee	Saturn VUE	Toyota RAV4	Hyundai Santa Fe
Front into flat barrier	652	416	489	539
Rear into flat barrier	824	556	1897	1504
Front into angle barrier	1448	1179	1151	1578
Rear into pole	1553	1335	2377	1988

Source: Insurance Institute for Highway Safety

(a) Normal probability plots for each treatment indicate that the requirement of normality has been satisfied. Verify that the requirement of equal population variances for each treatment has been satisfied.
(b) Is there evidence that the mean cost of repair is different among the four SUVs at the $\alpha = 0.05$ level of significance?
(c) If the null hypothesis from part (b) was rejected, use Tukey's test to determine which pairwise means differ using a family-wise error rate of $\alpha = 0.05$.

16. Lodging A travel agent wanted to know whether the price of Marriott, Hyatt, and Sheraton Hotels differed significantly. She knew that location of the hotel is a factor in determining price, so she blocked each hotel by location. After randomly selecting six cities and obtaining the room rate for each hotel, she obtained the following data:

	Marriott	Hyatt	Sheraton
Chicago	179	139.40	150
Los Angeles	169	161.50	161
Houston	163	187	189
Boston	189	179.10	169
Denver	179	168	112
Orlando	147	159	147

Source: Expedia.com

(a) Normal probability plots for each treatment indicate that the requirement of normality has been satisfied. Verify that the requirement of equal population variances for each treatment has been satisfied.
(b) Is there evidence that the mean cost of the room is different among the three hotel chains at the $\alpha = 0.05$ level of significance?
(c) If the null hypothesis from part (b) was rejected, use Tukey's test to determine which pairwise means differ using a family-wise error rate of $\alpha = 0.05$.

17. Rats in Space Researchers at NASA wanted to determine the effects of space flight on a rat's daily consumption of water. The following data represent the water consumption (in milliliters per day) at lift-off minus 1, return plus 1, and 1 month after return for six rats sent to space on the Spacelab Sciences 1 flight.

Concrete	2 Days	7 Days	28 Days
1	2830	3505	4470
2	3295	3430	4740
3	2710	3670	5115
4	2855	3355	4880
5	2980	3985	4445
6	3065	3630	4080
7	3765	4570	5390

Source: "Measurement-Error-Model Collinearities," *Technometrics*, Vol. 34 454–464.

(a) Normal probability plots for each treatment indicate that the requirement of normality has been satisfied. Verify that the requirement of equal population variances for each treatment has been satisfied.

(b) Is there evidence that the water consumption is different for the three days at the $\alpha = 0.05$ level of significance?

(c) If the null hypothesis from part (b) was rejected, use Tukey's test to determine which pairwise means differ using a family-wise error rate of $\alpha = 0.05$.

18. Concrete Strength Researchers Olivia Carrillo-Gamboa and Richard Gunst presented the following data in their article, "Measurement-Error-Model Collinearities." The data represent the compressive strength (in pounds per square inch) of a random sample of concrete 2 days, 7 days, and 28 days after pouring.

Rat	Lift-Off Minus 1	Return Plus 1	Return Plus 1 Month
1	18.5	32	30
2	17.5	18	34
3	28.0	31	39
4	28.5	29	44
5	31.0	48	54
6	22.5	25	32

Source: NASA Life Sciences Data Archive

(a) Normal probability plots for each treatment indicate that the requirement of normality has been satisfied. Verify that the requirement of equal population variances for each treatment has been satisfied.

(b) Is there evidence that the mean strength is different among the three days at the $\alpha = 0.05$ level of significance?

(c) If the null hypothesis from part (b) was rejected, use Tukey's test to determine which pairwise means differ using a familywise error rate of $\alpha = 0.05$.

19. Waiting in Line A quality-control manager at an amusement park feels that the amount of time that people spend waiting in line for the American Eagle roller coaster is too long. To determine if a new loading/unloading procedure is effective in reducing wait time in line, he measures the amount of time (in minutes) people are waiting in line for seven days. After implementing the new procedure, he again measures the amount of time (in minutes) people are waiting in line for seven days and obtains the following data. To make a reasonable comparison, he chooses days when weather conditions are alike.

Day	Mon	Tues	Wed	Thurs	Fri	Sat	Sun
Wait time before new procedure	11.6	25.9	20.0	38.2	57.3	32.1	81.8
Wait time after new procedure	10.7	28.3	19.2	35.9	59.2	31.8	75.3

Treat each day as a block and the wait times before and after the procedure as the treatment.

(a) Using the methods introduced in this section, determine whether there is sufficient evidence to conclude that the two loading procedures are resulting in different measurements of the wait time at the $\alpha = 0.05$ level of significance.

(b) Using the methods introduced in Section 11.1, determine whether there is sufficient evidence to conclude that the two loading procedures are resulting in different measurements of the wait time at the $\alpha = 0.05$ level of significance.

(c) Compare the *P*-values of both approaches. Can you conclude that the method presented in this section is a generalization of the matched-pair *t*-test?

Randomized Complete Block Design

MINITAB   **Conducting the Analysis of Variance**

Step 1: In C1, enter the block of the response variable; in C2, enter the treatment of the response variable; and in C3, enter the value of the response variable.

Step 2: Select the **Stat** menu. Highlight **ANOVA** and select **Two-way. . . .**

Step 3: Enter C3 in the box marked "Response Variable." Enter C1 in the box marked "row factor," and enter C2 in the box marked "column factor." Check the "Additive Model" box and check the "Display means" box. Click OK.

Tukey's Test

Step 1: With the data entered into the MINITAB spreadsheet as specified in Step 1 above, select the **Stat** menu. Highlight **ANOVA** and select **General Linear Model. . . .**

Step 2: Enter the name of the response variable in the box marked "Responses." Enter "C1 C2," in the box marked "Model," where C1 is the block of the response variable and C2 is the treatment. Enter C1 in the box marked "Random factors."

Step 3: Click "Comparisons . . .". Click the "Pairwise comparisons" radio button. In the box marked "Terms," enter C2, the treatment level. Check the box marked "Tukey." Check the box marked "Confidence Interval," if you want confidence intervals. Check the box marked "Test" if you want to conduct hypothesis tests of the pairwise differences.

Step 4: Click OK twice.

13.4   Two-Way Analysis of Variance

Objectives

1. **Analyze a two-way ANOVA design**
2. **Draw interaction plots**
3. **Perform the Tukey test**

> **CAUTION**
>
> The randomized complete block design is a design that varies a single factor. We are not interested in determining whether the block is significant! If we want to analyze the effect of fixing two different factors at different levels, we use the methods of two-way ANOVA.

One-way analysis of variance is used to compare k population means. In this design, there is one factor set at k levels. For example, we randomly assign 40 males between the ages of 20 and 29 years to $k = 3$ different treatment groups. Group 1 receives 5 mg per day of an experimental drug, group 2 receives 7.5 mg per day of an experimental drug, and group 3 receives a placebo. The expectation is that the 40 males are alike in all ways except the treatment. The experimental design that allows for this type of analysis is the completely randomized design.

The methods of one-way ANOVA can also be used in analyzing observational data. For example, a researcher might wonder whether the mean of the hemoglobin levels varies with age. In this circumstance, the factor is age, such as 20 to 29 years old, 30 to 39 years old, and 40 to 49 years old. In this case, the factor has $k = 3$ levels. Of course, in observational studies, there is never any talk of causation when it comes to the relation between the factor (age) and response variable (hemoglobin). In both the experimental design and the observational study, we should notice that the factor is a categorical (qualitative) variable that classifies an individual into one of k groups. The categorical variable in the treatment experiment is the amount of the drug. The categorical variable in the hemoglobin study is age group.

An improvement on analysis involving one-way ANOVA is the randomized complete block design in which experimental units are organized through a technique called *blocking*. For example, if the researcher attempting to determine the effectiveness of a drug feels that genetics may explain some of the variation in the response variable, he may select families that have three male siblings and block by sibling. For example, one sibling is randomly assigned to group 1, one sibling is randomly assigned to group 2, and the third sibling to Group 3. It is important to recognize that we are still only interested in determining whether the factor, drug, affects the response variable. We block so that we reduce experimental error.

 ## Analyze a Two-Way Analysis of Variance Design

We now present analysis in which two factors can be used to explain variability in the response variable. We deal with the two factors by fixing them at different levels. Again, remember that there are three ways to deal with factors: (1) fix them at a single level, (2) manipulate them by fixing them at different levels, and (3) randomize so that their effect on the response variable is minimized. In both the completely randomized design and the randomized complete block design, we manipulated one factor to see how varying it affected the response variable. In this section, we manipulate two factors.

Throughout the section, we will call the two factors factor A and factor B. If factor A has two levels and factor B has two levels, we have a **2 × 2 factorial design**, (read "two by two factorial design"). For example, suppose we want to determine the impact an experimental drug has on hemoglobin levels in humans. We might have two levels in the first factor (placebo or 5 mg per day of the drug) and two levels in the second factor (male or female). In this way, our design will look like Figure 11.

Figure 11

		Factor B	
		Placebo	**5 mg per day**
Factor A	**Male**	Hemoglobin of males receiving the placebo	Hemoglobin of males receiving 5 mg per day
	Female	Hemoglobin of females receiving the placebo	Hemoglobin of females receiving 5 mg per day

In general, the two-factor design looks as presented in Figure 12, where there are n replications for each combination of factor A and factor B. Factor A has a levels and factor B has b levels.

Figure 12

			Factor B		
		1	**2**	. . .	**b**
Factor A	**1**	n observations of the response variable with level 1 of factor A and level 1 of factor B	n observations of the response variable with level 1 of factor A and level 2 of factor B	. . .	n observations of the response variable with level 1 of factor A and level b of factor B
	2	n observations of the response variable with level 2 of factor A and level 1 of factor B	n observations of the response variable with level 2 of factor A and level 2 of factor B	. . .	n observations of the response variable with level 2 of factor A and level b of factor B
	⋮	⋮	⋮	⋮	⋮
	a	n observations of the response variable with level a of factor A and level 1 of Factor B	n observations of the response variable with level a of factor A and level 2 of Factor B	. . .	n observations of the response variable with level a of factor A and level b of Factor B

So, if there are three levels of factor A and two levels of factor B, we have a 3×2 factorial design as shown in Figure 13(a). If there are three levels of factor A and three levels of factor B, we have a 3×3 factorial design as shown in Figure 13(b).

Figure 13

Factor B

		1	**2**
	1	n observations of the response variable with level 1 of factor A and level 1 of factor B	n observations of the response variable with level 1 of factor A and level 2 of factor B
Factor A	**2**	n observations of the response variable with level 2 of factor A and level 1 of factor B	n observations of the response variable with level 2 of factor A and level 2 of factor B
	3	n observations of the response variable with level 3 of factor A and level 1 of factor B	n observations of the response variable with level 3 of factor A and level 2 of factor B

(a)

Factor B

		1	**2**	**3**
	1	n observations of the response variable with level 1 of factor A and level 1 of factor B	n observations of the response variable with level 1 of factor A and level 2 of factor B	n observations of the response variable with level 1 of factor A and level 3 of factor B
Factor A	2	n observations of the response variable with level 2 of factor A and level 1 of factor B	n observations of the response variable with level 2 of factor A and level 2 of factor B	n observations of the response variable with level 2 of factor A and level 3 of factor B
	3	n observations of the response variable with level 3 of factor A and level 1 of factor B	n observations of the response variable with level 3 of factor A and level 2 of factor B	n observations of the response variable with level 3 of factor A and level 3 of factor B

(b)

Each of the n observations of the response variable for the different levels of the factors exists within a **cell**. In a 2×2 factorial design, there are four cells with n observations within each cell. In a 2×3 factorial design, there are six cells with n observations within each cell. In a 3×3 factorial design, there are nine cells with n observations within each cell.

EXAMPLE 1   **A 2 $\times$ 3 Factorial Design**

High-density lipoprotein (HDL) cholesterol is the so-called good cholesterol because it helps to reduce the amount of bad cholesterol in your system. A pharmaceutical company developed a drug that is meant to increase HDL levels in patients. They obtained volunteers whose HDL cholesterol was roughly the same and randomly assigned them to one of three groups. Group 1 was a placebo group that received a sugar tablet. Group 2 received 5 mg of the experimental drug, and group 3 received 10 mg of the experimental drug. In addition to the three treatment groups, the researchers considered age to be a factor in the analysis, so they divided patients into an 18- to 39-year-old category and a 40 and older category. The design results in a 2×3 factorial design. Table 9 shows the increase in HDL cholesterol (in mg/dL) for each patient in the study after 10 weeks. Each cell has $n = 3$ replications.

Table 9

| | Drug Treatment | | |
	Placebo	5 mg	10 mg
	4	9	14
18–39 years	3	5	12
	−1	6	10
	3	3	10
40 or older	2	6	8
	0	7	7

Age labels the rows (18–39 years and 40 or older).

Now Work Problem 17(a).

In the factorial design, all levels of factor A are combined with all levels of factor B, so we say that the factors are **crossed**. The effect of factor A is the change in the response variable that results from changing the level of factor A. The effect of factor B is the change in the response variable that results from changing the level of factor B. Together these two effects are called the **main effects**. If changes in the level of factor A result in different changes in the value of the response variable for the different levels of factor B, we say that there is an **interaction effect** between the factors. An example will help clarify the idea of interaction effect.

EXAMPLE 2   Illustrating Main Effects and Interaction Effect

Suppose we have two factors, A and B, each fixed at two levels, high and low. The value of the response variable at each level is presented in Table 10. For example, the response variables when both factor A and B are set to low are 5 and 3.

Table 10

| | | Factor B | |
		Low	High
	Low	5	8
Factor A		3	6
	High	8	17
		4	18

The main effect of factor A can be thought of as the difference between the mean value of the response variable when factor A is set at low and the mean value of the response variable when factor A is set at high. That is,

$$\text{Main effect of factor A} = \frac{8 + 4 + 17 + 18}{4} - \frac{5 + 3 + 8 + 6}{4} = 6.25 \text{ units}$$

Increasing factor A from low level to high level causes the value of the response variable to increase by 6.25 units, on average.

The main effect of factor B is

$$\text{Main effect of factor B} = \frac{8 + 6 + 17 + 18}{4} - \frac{5 + 3 + 8 + 4}{4} = 7.25 \text{ units}$$

Increasing factor B from low level to high level causes the value of the response variable to increase by 7.25 units, on average.

To see the interaction effect between the factors, we look at one level of one of the factors and then see how changing the other factor affects the value of the response variable. For example, if we look only at the low level of factor B, we can see the effect of changing factor A from low to high is

$$\frac{8 + 4}{2} - \frac{5 + 3}{2} = 2 \text{ units}$$

This indicates that if we are at the low level of factor B then increasing factor A from low to high causes the response variable to increase by 2 units, on average. If we look only at the high level of factor B, we can see the effect of changing factor A from low to high is

$$\frac{17 + 18}{2} - \frac{8 + 6}{2} = 10.5 \text{ units}$$

CAUTION!
If interaction between two factors exists, then interpret the main effects with extreme caution.

This indicates that if we are at the high level of factor B then increasing factor A from low to high causes the response variable to increase by 10.5, on average.

The main effect of factor A is 6.25 units, but this is really misleading, because if we are at the low level of factor B, then the mean increase in the response variable is only 2 units, but if we are at the high level of factor B, the mean increase in the response variable is 10.5 units. So the increase in the value of the response variable depends on the level of factor B. For this reason, we say there is an interaction effect between factors A and B.

The moral: If interaction between two factors exists, it is extremely dangerous to look at main effects because they are misleading.

Now Work Problem 9(a).

Now let's learn to test hypotheses that involve a factorial design. The requirements for the model are as follows:

Requirements for the Two-Way Analysis of Variance

1. The populations from which the samples are drawn must be normal.
2. The samples are independent.
3. The populations all have the same variance.

We can check the normality requirement through a normal probability plot of the data within each cell. The requirement for equal variances can be checked by verifying that the largest sample standard deviation is no more than two times the smallest sample standard deviation.

Although not a requirement to conduct two-way ANOVA, we also assume that there are an equal number of observations for each combination of the factors because the power of the test is increased. That is, it reduces the likelihood of making a Type II error.

In a two-way ANOVA, we will be testing three separate hypotheses. The three hypotheses go with the three types of effects presented earlier. The first hypothesis deals with the significance of any interaction effect. So, we have

Hypotheses Regarding Interaction Effect

H_0: there is no interaction between the factors
H_1: there is interaction between the factors

Hypotheses Regarding Main Effects

H_0: there is no effect of factor A on the response variable
H_1: there is an effect of factor A on the response variable

H_0: there is no effect of factor B on the response variable
H_1: there is an effect of factor B on the response variable

> Whenever conducting a two-way ANOVA, we always first test the hypothesis regarding interaction effect. If the null hypothesis of no interaction is rejected, we do not interpret the results of the hypotheses involving the main effects. This is because the interaction clouds the interpretation of the main effects.

The decision rule for each of the above hypotheses will be the same as always: if the P-value is less than the level of significance, we reject the null hypothesis in favor of the alternative. We will obtain the P-value from the output of statistical software. Let's look at an example.

EXAMPLE 3   Examining a Two-Way ANOVA

Problem: In Example 1, we presented data for an experimental drug that was meant to increase HDL cholesterol. The data are presented in Table 11.

Table 11

Age		Drug Treatment		
		Placebo	5 mg	10 mg
18–39 years		4	9	14
		3	5	12
		−1	6	10
40 or older		3	3	10
		2	6	8
		0	7	7

(a) HDL cholesterol levels are known to have a distribution that is approximately normal. Verify that the largest sample standard deviation of a cell is no more than two times the smallest sample standard deviation of a cell.

(b) Use MINITAB to test whether there is an interaction effect between the drug treatment and gender.

(c) If the null hypothesis of no interaction is not rejected, determine whether there is sufficient evidence to conclude that the mean increase in HDL cholesterol is different among each drug treatment group. If the null hypothesis of no interaction is not rejected, determine whether there is sufficient evidence to conclude that the mean increase in HDL cholesterol is different for each age group.

Approach: We use MINITAB to obtain the two-way analysis of variance. If the P-value corresponding to each hypothesis is small (say, less than $\alpha = 0.05$), we reject the null hypothesis in favor of the alternative hypothesis.

Solution

(a) Obtain the descriptive statistics from MINITAB shown in Figure 14. The largest sample standard deviation, 2.65 mg/dL, is not twice as large as the smallest sample standard deviation, 1.528 mg/dL, so the requirement of equal population variances has been satisfied.

Figure 14   **Descriptive Statistics: P/18–39, 5/18–39, 10/18–39, P/40, 5/40, 10/40**

Variable	N	Mean	Median	TrMean	StDev	SE Mean
P/18-39	3	2.00	3.00	2.00	2.65	1.53
5/18-39	3	6.67	6.00	6.67	2.08	1.20
10/18-39	3	12.00	12.00	12.00	2.00	1.15
P/40	3	1.667	2.000	1.667	1.528	0.882
5/40	3	5.33	6.00	5.33	2.08	1.20
10/40	3	8.333	8.000	8.333	1.528	0.882

Variable	Minimum	Maximum	Q1	Q3
P/18-39	−1.00	4.00	−1.00	4.00
5/18-39	5.00	9.00	5.00	9.00
10/18-39	10.00	14.00	10.00	14.00
P/40	0.000	3.000	0.000	3.000
5/40	3.00	7.00	3.00	7.00
10/40	7.000	10.000	7.000	10.000

(b) We first test the hypotheses

H_0: there is no interaction between drug treatment and age

versus

H_1: there is an interaction between drug treatment and age

Enter the data into MINITAB. Let column 1 represent the drug treatment; so if the observation is from the placebo group, we enter a 1, if the observation is from 5 mg, we enter a 2, and so on. We let column 2 represent the age and enter 1 for 18- to 39-year-olds and enter 2 for 40 or older. Column 3 gets the value of the response variable, increase in HDL cholesterol. The results of the two-way analysis of variance are presented in Figure 15.

Figure 15

Two-way ANOVA: HDL versus Drug, Age

Analysis of Variance for HDL

Source	DF	SS	MS	F	P
Drug	2	208.33	104.17	25.68	0.000
Age	1	14.22	14.22	3.51	0.086
Interaction	2	8.78	4.39	1.08	0.370
Error	12	48.67	4.06		
Total	17	280.00			

The P-value for the interaction effect is 0.370. Because the P-value is large, we do not reject the null hypothesis and conclude that there is no interaction effect.

(c) We now test the hypotheses regarding the main effects. The hypothesis regarding the main effect, drug treatment, is

$$H_0: \mu_{\text{placebo}} = \mu_{5\text{ mg}} = \mu_{10\text{ mg}}$$

versus

H_1: at least one of the means differs

The hypothesis regarding the main effect, age, is

$$H_0: \mu_{18\text{-}39} = \mu_{40\text{ or older}} \quad \text{versus} \quad H_1: \text{the means differ}$$

The P-value for the main effect, drug treatment, is given as 0.000. The small value of the P-value is taken as evidence against the null hypothesis. We conclude that at least one of the mean increases in HDL is different for the different levels of the drug treatment. Therefore, the level of the drug treatment is a significant contributor to explaining the increase in HDL cholesterol.

! CAUTION!

Although we do not reject the null hypothesis that the mean increase in HDL is the same at the different age levels, there is some evidence that age plays a role in HDL levels.

The *P*-value for the main effect, age, is given as 0.086. Because this is greater than the level of significance for the test ($\alpha = 0.05$), we do not reject the null hypothesis and conclude that the mean increase in HDL cholesterol does not change with age.

Now Work Problems 17(b) and (c).

2 Draw Interaction Plots

As is always the case in statistics, we like to support results graphically whenever possible. There is a way to graphically represent the role interaction plays in any factorial design. We call these plots **interaction plots**.

Constructing Interaction Plots

Step 1: Compute the mean value of the response variable within each cell. In addition, compute the row mean value of the response variable and the column mean value of the response variable with each level of each factor.

Step 2: In a Cartesian plane, label the horizontal axis for each level of factor A. The vertical axis will represent the mean value of the response variable. For each level of factor A, plot the mean value of the response variable for each level of factor B. Draw straight lines connecting the points for the common level of factor B. You should have as many lines as there are levels of factor B. The more difference there is in the slopes of the two lines, the stronger the evidence of interaction.

An example should help you to understand the steps given.

EXAMPLE 4   **Drawing an Interaction Plot**

Problem: Draw an interaction plot for the data from Example 3.

Approach: We will follow Steps 1 and 2.

Solution

Step 1: We compute the mean value of the response variable for each cell as shown in Table 12.

Table 12

		Placebo	Drug Treatment 5 mg	10 mg
Age	18–39 years	$\frac{4 + 3 + (-1)}{3} = 2$	$\frac{9 + 5 + 6}{3} = 6.7$	$\frac{14 + 12 + 10}{3} = 12$
	40 or older	$\frac{3 + 2 + 0}{3} = 1.7$	$\frac{3 + 6 + 7}{3} = 5.3$	$\frac{10 + 8 + 7}{3} = 8.3$

Figure 16

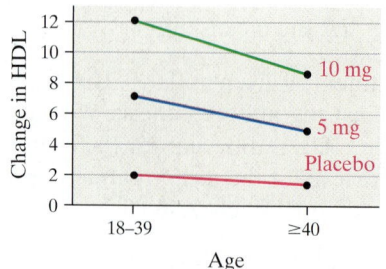

Step 2: We draw a Cartesian plane with the horizontal axis labeled Age with the levels 18-39 years and 40 or older indicated. The vertical axis is labeled as the response. For each age group, we plot the corresponding mean values of the response variable for drug. Then connect the points and label each line with its corresponding drug treatment. See Figure 16.

When interpreting interaction plots, we look at the level of parallelism between the lines. The more parallel the lines seem to be the stronger the visual evidence is of no interaction. From Figure 16, we see that the lines are roughly parallel, so our conclusion of no interaction is verified. Another bit of information that can be learned from the interaction plot is the impact age has on the response variable, change in HDL. Notice that the change in HDL is decreasing as the age increases. From this, we can see that the decrease in HDL cholesterol is largest for the 10-mg treatment and smallest for the placebo treatment.

EXAMPLE 5   Drawing an Interaction Plot Using Technology

Problem: Draw an interaction plot for the data in Example 3 using a statistical package.

Approach: We will use MINITAB to draw the interaction plot. The steps to follow are given in the Technology Step by Step on page 727.

Result: Figure 17 shows interaction plots from MINITAB using the data in Example 3.

Figure 17

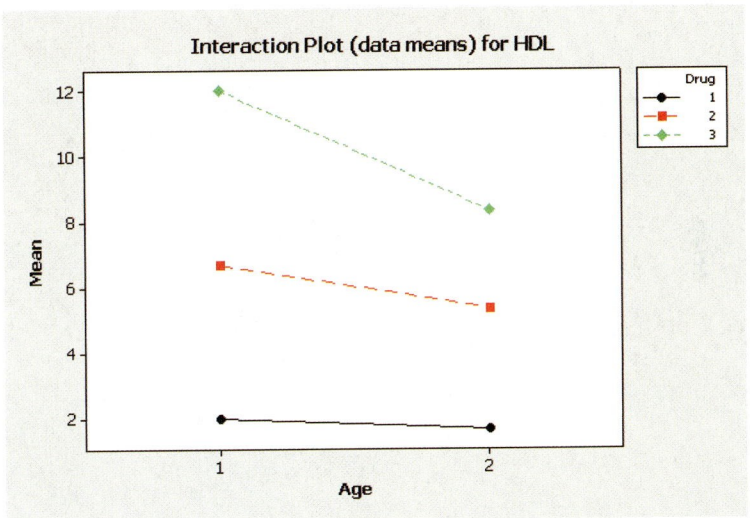

Now Work Problem 17(d).

Figure 18 gives some interaction plots and the level of interaction indicated by the graph. In the graphs, A_1, A_2, and A_3 represent the three levels of treatment for Factor A, while B_1, B_2, and B_3 represent the three levels of treatment for Factor B. Remember, though, that the interaction plots are solely meant to help us visualize the interaction (just as boxplots allow us to visualize differences among treatments in one-way ANOVA). They are not meant to be used to test for interaction. We can only test for interaction by looking at the F-test statistic and its P-value for interaction in the analysis of variance.

Figure 18

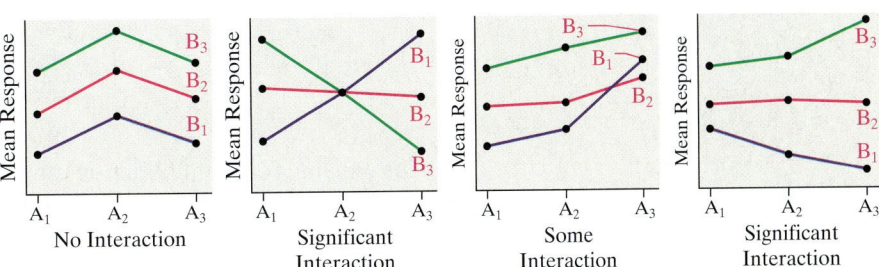

EXAMPLE 6 ## Analyzing a Two-by-Two Factorial Design

Problem: Suppose that an educational psychologist wants to conduct an experiment to determine whether varying the conditions in which learning and testing take place affects test results. She randomly selects 20 students who are unfamiliar with the Battle at Gettysburg from the Civil War. After presenting the material to the students and allowing them ample time to study, the students are given an exam. Ten of the students receive the lecture in a large lecture hall and ten receive the lecture in a small classroom (both lectures have the same instructor). Of the students in the large lecture hall, five take the exam in the same large lecture hall, while five take the exam in a small classroom. Of the students in the small classroom, five take the exam in a large lecture hall and five take the exam in the same small classroom that they received the lecture in. The experimental design is, therefore, a 2 by 2 factorial design with 5 replications in each cell. The scores for each student are presented in Table 13.

Table 13		Lecture Administered	
		Large Lecture Hall	Classroom
Exam Given	Large lecture hall	70	63
		63	46
		95	47
		84	71
		72	66
	Classroom	51	76
		73	67
		65	80
		72	90
		43	79

Preliminary analysis indicates that the requirements to conduct a two-way ANOVA are satisfied. Test whether there is an interaction effect between the lecture room and exam. Draw an interaction plot to confirm your results.

Approach: We use MINITAB to perform the two-way analysis of variance and to draw the interaction plot.

Solution: Figure 19 shows the analysis of variance from MINITAB.

Figure 19

Two-way ANOVA: Score versus Exam, Lecture

```
Analysis of Variance for Score
Source         DF      SS      MS       F      P
Exam            1      18      18    0.13  0.718
Lecture         1       0       0    0.00  0.954
Interaction     1    1602    1602   11.97  0.003
Error          16    2142     134
```

The *P*-value for the interaction effect is 0.003, indicating that there is a significant interaction effect between lecture location and exam location. Therefore, we will not test the main effects. Figure 20 shows the interaction plot from MINITAB, which confirms the analytic results from the ANOVA.

Figure 20

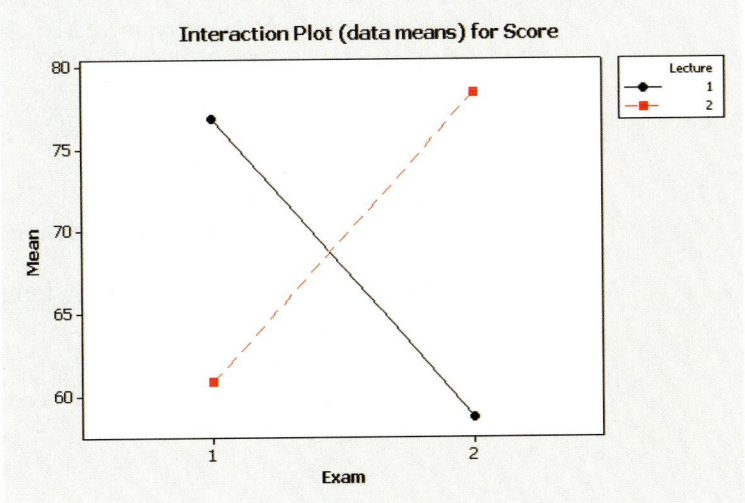

In looking at the interaction plot, it appears that students perform best if they take their exams in the same classroom where their lecture was given. The two high means are the means for (large lecture, large exam) and (classroom lecture, classroom exam). In other words, where a student performs best on the exam seems to depend on where the student attended lecture.

③ Perform the Tukey Test

Once the null hypothesis of equal population means for either factor is rejected, we proceed to determine which means differ significantly. Again, the method that we use to conduct the analysis is Tukey's test. The steps are identical to those presented for one-way ANOVA. However, the critical value and the estimate of the standard error are a little different. The critical value is $q_{\alpha,\nu,k}$, using a familywise error rate of α with $\nu = N - ab$, where N is the total number of observations (or replications), a is the number of levels for factor A, and b is the number of levels for factor B; k is the number of means being tested for the factor. The standard error is

$$\text{SE} = \sqrt{\frac{\text{MSE}}{m}}$$

where m is the number of levels for the factor times the number of observations within each cell.

EXAMPLE 7   **Multiple Comparisons Using Tukey's Test**

Problem: Use statistical software to perform Tukey's test to determine which pairwise means differ for the data presented in Example 3 using a familywise error rate of $\alpha = 0.05$.

Approach: We will use MINITAB to perform Tukey's test. The steps for performing this test can be found in the Technology Step by Step on page 727.

Result: The results of Tukey's test using MINITAB are presented in Figure 21. MINITAB allows the user to perform pairwise comparisons using either confidence intervals or P-values. In the output shown, Figure 21 gives the pairwise comparisons using 95% confidence intervals. For example, the confidence interval for drug level 2–drug level 1 (5 mg–placebo) does not contain 0 and is positive (lower bound: 1.067 mg/dL; upper bound: 7.266 mg/dL). Therefore, the mean increase in HDL cholesterol resulting from 5 mg of the drug is significantly greater than the mean increase due to the placebo. The confidence interval for "Drug Level 3 – Drug Level 1" (10 mg – Placebo) does not contain 0 and is positive. This indicates that the mean increase in HDL cholesterol resulting from 10 mg of the drug is also greater than the mean increase due to the placebo. Moreover if we look at "Drug Level 3 – Drug Level 2", we see the interval also does not contain 0, so the mean increase in HDL level with the 10 mg drug is higher than the increase with 5 mg of the drug.

Figure 21

Tukey 95% Simultaneous Confidence Intervals
Response Variable HDL
All Pairwise Comparisons among Levels of Age

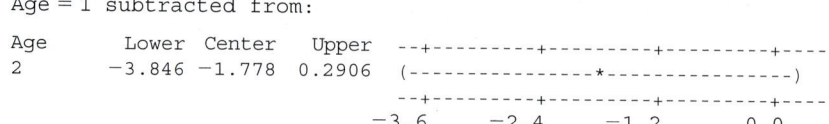

```
Age = 1 subtracted from:

Age      Lower  Center   Upper   --+---------+---------+---------+----
2       -3.846 -1.778  0.2906         (----------------*----------------)
                                 --+---------+---------+---------+----
                               -3.6      -2.4      -1.2       0.0
```

Tukey 95% Simultaneous Confidence Intervals
Response Variable HDL
All Pairwise Comparisons among Levels of Drug

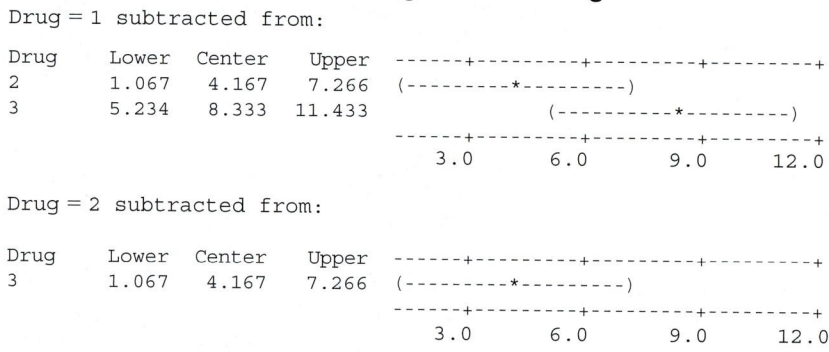

```
Drug = 1 subtracted from:

Drug    Lower  Center   Upper   ------+---------+---------+---------+
2       1.067  4.167    7.266   (---------*---------)
3       5.234  8.333   11.433              (----------*---------)
                                ------+---------+---------+---------+
                                    3.0       6.0       9.0      12.0


Drug = 2 subtracted from:

Drug    Lower  Center   Upper   ------+---------+---------+---------+
3       1.067  4.167    7.266   (---------*---------)
                                ------+---------+---------+---------+
                                    3.0       6.0       9.0      12.0
```

Now Work Problem 17(e).

In-Class Activity: Flight of Fancy: Part II

How does the type of paper affect the flight of a paper airplane? In Part I of this activity, we looked at how the design of the plane affected flight distance. While we blocked on gender. Now, we also want to know whether type of paper plays a role in paper airplane flight. Use three different types of paper (for example, newspaper, brown shipping paper, typing paper) and perform a two-way ANOVA.

(a) Determine if there is significant interaction between paper type and plane design.

(b) Draw an interaction plot of the data to support the results.

(c) If there is not significant interaction, determine if there is significant difference in the means for the three plane designs. If there is not significant interaction, determine if there is significant difference in the means for the paper types.

(d) If there is significant difference in the means for the three plane designs, use Tukey's test to determine which pairwise means differ using a familywise error rate of $\alpha = 0.05$. If there is significant difference in the means for the paper type, use Tukey's test to determine which pairwise means differ using a familywise error rate of $\alpha = 0.05$.

(e) Check the requirements for the model. Are they all met? If not, how will that affect your conclusions?

MAKING AN INFORMED DECISION

Where Should I Invest? Part II

Remember that inheritance you received in Chapter 11? Suppose you have decided that you now want to try investing in individual stocks rather than mutual funds. You are told by a friend who is a finance major that there are three broad stock types: large-capitalization stocks, midcapitalization stocks, and small-capitalization stocks. Large-capitalization stocks are companies that have a market value in excess of $10 billion. Small-capitalization stocks are companies with a market value below $2 billion. Mid-caps fall in between. Your finance friend also tells you that there are two investment styles: value and growth. Value stocks are companies thought to be undervalued relative to their peers, while growth stocks are companies thought to have a higher growth rate than their peers.

(a) Obtain a random sample of five stocks within each category. Determine the 1- and 5-year rates of return for each stock.

(b) Verify that the rates of return are normally distributed. Also verify that two times the smallest standard deviation is less than the largest standard deviation. If any requirements are not satisfied, increase the sample size until the requirements become satisfied.

(c) Perform a one-way ANOVA with stock type as the factor. Is there any difference in the mean rate of return among the three stock types? If so, perform a Tukey test to determine which mean rates of return differ.

(d) Perform a two-way ANOVA with stock type and investment style as the factors. Is there any interaction between stock type and investment style? If not, do the means of either factor differ? If the means do differ, perform a Tukey test to determine which mean rates of return differ.

(e) Write a report that details which investment category seems to be best.

13.4 ASSESS YOUR UNDERSTANDING

Concepts and Vocabulary

1. Explain the differences between the completely randomized design, randomized complete block design, and factorial design.

2. Explain what an interaction effect is. Why is it dangerous to analyze main effects if there is an interaction effect?

3. What is an interaction plot? Why are they useful?

4. How many treatments are there in a 2×3 factorial design? How many levels does each treatment have?

Skill Building

In Problems 5–8, determine whether the interaction plot suggests that significant interaction exists among the factors.

5.
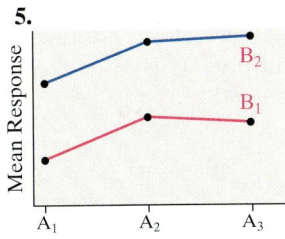

6.
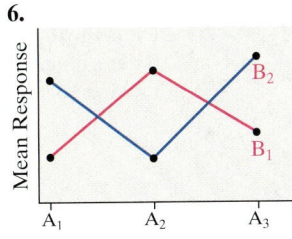

7.
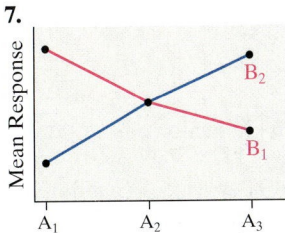

8.
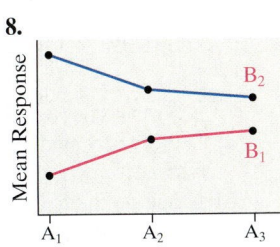

In Problems 9 and 10, (a) use the same analysis as that presented in Example 2 to conjecture whether interaction exists between factor A and factor B. (b) Draw an interaction plot to verify your conjecture from part (a).

9.
NW

		Factor B	
		Low	High
	Low	12	6
		8	7
Factor A	High	20	1
		14	3

10.

		Factor B	
		Low	High
	Low	104	143
Factor A		96	121
	High	84	55
		76	43

11. Given the following ANOVA output, answer the questions that follow.

Analysis of Variance for Response

Source	DF	SS	MS	F	P
Factor A	1	531.2	531.2	11.73	0.003
Factor B	2	3018.0	1509.0	33.33	0.000
Interaction	2	16.3	8.2	0.18	0.836
Error	18	814.9	45.3		

(a) Is there any evidence of an interaction effect? Why or why not?
(b) Based on the P-value, is there evidence of a difference in the means from factor A? Based on the P-value, is there evidence of a difference in the means from factor B?
(c) What is the mean square error?

12. Given the following ANOVA output, answer the questions that follow.

Analysis of Variance for Response

Source	DF	SS	MS	F	P
Factor A	2	156	78	0.39	0.679
Factor B	2	132	66	0.33	0.720
Interaction	4	311	78	0.39	0.813
Error	27	5354	198		
Total	35	5952			

(a) Is there any evidence of an interaction effect? Why or why not?
(b) Based on the P-value, is there evidence of a difference in the means from factor A? Based on the P-value, is there evidence of a difference in the means from Factor B?
(c) What is the mean square error?

13. Given the following ANOVA output, answer the questions that follow.

Analysis of Variance for Response

Source	DF	SS	MS	F	P
Factor A	2	2269.8	1134.9	35.63	0.000
Factor B	2	115.2	57.6	1.81	0.183
Interaction	4	1694.8	423.7	13.30	0.000
Error	27	860.0	31.9		
Total	35	4939.8			

(a) Is there any evidence of an interaction effect? Why or why not?
(b) Based on the P-value, is there evidence of a difference in the means from factor A? Based on the P-value, is there evidence of a difference in the means from factor B?
(c) What is the mean square error?

14. Given the following ANOVA output, answer the questions that follow.

Analysis of Variance for Response

Source	DF	SS	MS	F	P
Factor A	2	1209.7	604.8	8.32	0.002
Factor B	2	577.6	288.8	3.97	0.031
Interaction	4	1474.9	368.7	5.07	0.004
Error	27	1962.2	72.7		
Total	35	5224.3			

(a) Is there any evidence of an interaction effect? Why or why not?
(b) Based on the P-value, is there evidence of a difference in the means from factor A? Based on the P-value, is there evidence of a difference in the means from Factor B?
(c) What is the mean square error?

In Problems 15 and 16, assume that the data come from populations that are normally distributed with the same variance.

15.

		Factor B		
		Level 1	**Level 2**	**Level 3**
		64.9	53.3	51.9
		59.3	44.2	61.8
	Level 1	32.9	46.2	68.5
		59.4	43.3	65.8
Factor A		50.7	38.4	53.1
	Level 2	59.3	50.7	57.9
		58.1	40.2	55.7
		33.5	52.6	74.4

(a) Determine whether or not there is significant interaction between factor A and factor B.
(b) If there is no significant interaction, determine if there is a significant difference in the means for the two levels of factor A. If there is no significant interaction, determine if there is a significant difference in the means for the three levels of factor B.
(c) Draw an interaction plot of the data to support the results of parts (a) and (b).
(d) If there is a significant difference in the means for the two levels of factor A, use Tukey's test to determine which pairwise means differ using a familywise error rate of $\alpha = 0.05$. If there is a significant difference in the means for the three levels of factor B, use Tukey's test to determine which pairwise means differ using a familywise error rate of $\alpha = 0.05$.

16.

		Factor B		
		Level 1	**Level 2**	**Level 3**
		104	111	95
	Level 1	82	104	84
		81	74	104
		111	109	106
Factor A	**Level 2**	112	115	110
		82	97	99
		112	108	92
	Level 3	108	99	129
		117	104	120

(a) Determine whether or not there is significant interaction between factor A and factor B.
(b) If there is no significant interaction, determine whether there is a significant difference in the means for the three levels of factor A. If there is no significant interaction, determine whether there is a significant difference in the means for the three levels of Factor B.
(c) Draw an interaction plot of the data. Does this plot support the results of parts (a) and (b)?
(d) If there is a significant difference in the means for the three levels of factor A, use Tukey's test to determine which pairwise means differ using a familywise error rate of $\alpha = 0.05$. If there is a significant difference in the means for the three levels of factor B, use Tukey's test to determine which pairwise means differ using a familywise error rate of $\alpha = 0.05$.

Applying the Concepts

17. Cholesterol Levels NW A family physician wanted to know if age and gender were factors that explained levels of serum cholesterol (in mg/dL) in her adult patients. She randomly selects two patients for each category of data and obtains the following results: based on data obtained from the

		Age (years)		
		18–34	35–54	55 and older
Gender	Female	180, 192	205, 226	218, 231
	Male	175, 193	213, 222	203, 185

Source: National Center for Health Statistics

Serum cholesterols are known to be approximately normally distributed and the population variances are equal.
(a) What type of factorial design is this? How many replications are there within each cell?
(b) Determine if there is significant interaction between age and gender.
(c) If there is not significant interaction, determine whether there is significant difference in the means for the three age groups. If there is not significant interaction, determine whether there is significant difference in the means for the genders.
(d) Draw an interaction plot of the data to support the results of parts (b) and (c).
(e) If there is significant difference in the means for the three age groups, use Tukey's test to determine which pairwise means differ using a familywise error rate of $\alpha = 0.05$. If there is significant difference in the means for gender, use Tukey's test to determine which pairwise means differ using a familywise error rate of $\alpha = 0.05$.

18. Reaction Time In an online psychology experiment sponsored by the University of Mississippi, researchers asked study participants to respond to various stimuli. Participants were randomly assigned to one of three treatment groups. Subjects in group 1 were in the simple group. They were required to respond as quickly as possible after a stimulus was presented. Subjects in group 2 were in the go/no-go group. These subjects were required to respond to a particular stimulus while disregarding other stimuli. Finally, subjects in group 3 were in the choice group. They needed to respond differently, depending on the stimuli presented. The researcher felt that age may be a factor in determining reaction time, so she organized the experimental units by age and obtained the following data:

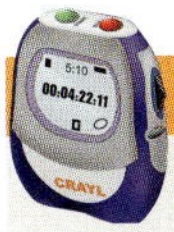

		Stimulus		
		Simple	Go/No-Go	Choice
		0.384	0.338	0.586
	18–24	0.248	0.495	0.509
		0.191	0.631	0.364
Age		0.203	0.485	0.626
	25–34	0.331	0.389	0.858
		0.438	0.629	0.529
		0.494	0.585	0.520
	35 and older	0.467	0.782	0.854
		0.302	0.529	0.700

Source: PsychExperiments; University of Mississippi; psychexps.olemiss.edu/

(a) What type of factorial design is this? How many replications are there within each cell?
(b) Normal probability plots indicate that it is reasonable to believe that the data come from populations that are normally distributed. Verify the requirement of equal population variances.
(c) Determine if there is significant interaction between stimulus and age.
(d) If there is not significant interaction, determine whether there is significant difference in the means for the three types of stimulus. If there is not significant interaction, determine whether there is significant difference in the means for the three categories of age.
(e) Draw an interaction plot of the data to support the results of parts (c) and (d).

19. Concrete Strength An engineer wants to know if the mean strengths of three different concrete mix designs differ significantly. He also suspects that slump may be a predictor of concrete strength. Slump is a measure of the uniformity of the concrete, with a higher slump indicating a less uniform mixture. The following data represent the

28-day strength (in pounds per square inch) of three different mixtures with three different slumps.

		Mixture 67-0-301	Mixture 67-0-400	Mixture 67-0-353
		3960	4815	4595
	3.75	4005	4595	4145
		3445	4185	4585
		4010	4070	3855
Slump	4	3415	4545	3675
		3710	4175	4010
		3290	4020	3875
	5	3390	4355	3700
		3740	3935	3350

(a) Normal probability plots indicate that it is reasonable to believe that the data come from populations that are normally distributed. Verify the requirement of equal population variances.
(b) Determine whether there is significant interaction between mixture type and slump.
(c) If there is not significant interaction, determine whether there is significant difference in the means for the three types of mixture. If there is not significant interaction, determine whether there is significant difference in the means for the slumps.
(d) Draw an interaction plot of the data to support the results of parts (b) and (c).
(e) If there is significant difference in the means for the three mixture types, use Tukey's test to determine which pairwise means differ using a familywise error rate of $\alpha = 0.05$. If there is significant difference in the means for the slumps, use Tukey's test to determine which pairwise means differ using a familywise error rate of $\alpha = 0.05$.

20. **Diet and Birth Weight** An obstetrician wanted to determine the impact that three experimental diets had on the birth weights of pregnant mothers. She randomly selected 27 pregnant mothers in the first trimester of whom 9 were 20 to 29 years old, 9 were 30 to 39 years old, and 9 were 40 or older. For each age group, she randomly assigned the mothers to one of the three diet treatments. After delivery she measured the birth weight (in grams) of the babies and obtained the following data:

		Birth Weight (grams)		
		Diet 1	Diet 2	Diet 3
		4473	3961	3667
	20–29 years	3878	3557	3139
		3936	3321	3356
		3886	3330	2762
Age	30–39 years	4147	3644	3551
		3693	2811	3272
		3878	2937	2781
	40 or older	4002	3228	3138
		3382	2732	3435

(a) Birth weights are known to be approximately normally distributed. Verify the requirement of equal population variances.
(b) Determine whether there is significant interaction between age and diet.
(c) If there is no significant interaction, determine whether there is significant difference in the means for the three age groups. If there is no significant interaction, determine whether there is significant difference in the means for the diets.
(d) Draw an interaction plot of the data to support the results of parts (b) and (c).
(e) If there is significant difference in the means for the three age groups, use Tukey's test to determine which pairwise means differ using a familywise error rate of $\alpha = 0.05$. If there is significant difference in the means for the diets, use Tukey's test to determine which pairwise means differ using a familywise error rate of $\alpha = 0.05$.

21. **Oil Changes** The following data represent the cost of an oil change in three different geographic regions for two types of service centers. A specialty chain is an oil change facility that specializes in oil changes, while a general service station provides a wide array of services in addition to oil changes.

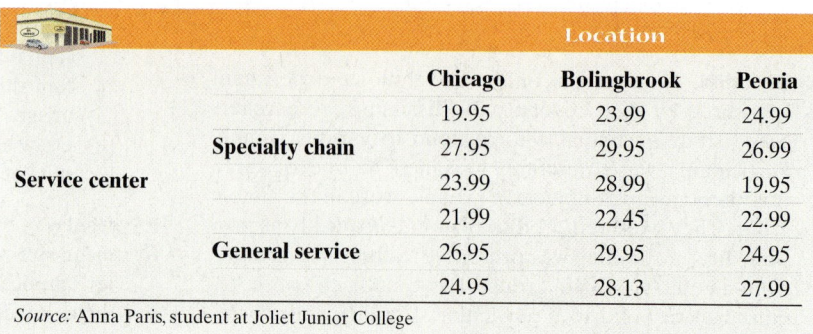

		Location		
		Chicago	Bolingbrook	Peoria
		19.95	23.99	24.99
	Specialty chain	27.95	29.95	26.99
		23.99	28.99	19.95
Service center		21.99	22.45	22.99
	General service	26.95	29.95	24.95
		24.95	28.13	27.99

Source: Anna Paris, student at Joliet Junior College

(a) The prices of oil changes are approximately normally distributed. Verify the requirement of equal population variances.

(b) Determine if there is significant interaction between location and service center type.

(c) If there is not significant interaction, determine whether there is significant difference in the means for the three locations. If there is no significant interaction, determine whether there is significant difference in the means for the two service center types.

(d) Draw an interaction plot of the data to support the results of parts (b) and (c).

(e) If there is significant difference in the means for the three locations, use Tukey's test to determine which pairwise means differ using a familywise error rate of $\alpha = 0.05$. If there is significant difference in the means for the service center type, use Tukey's test to determine which pairwise means differ using a familywise error rate of $\alpha = 0.05$.

Technology Step-by-Step

MINITAB

Two-Way ANOVA

Obtaining Two-Way ANOVA

Step 1: In C1 enter the level of factor A (that is for level 1 enter a 1, and so on), in C2 enter the level of factor B, and in C3 enter the value of the response variable.

Step 2: Select the **Stat** menu. Highlight **ANOVA** and select **Two-way . . . .**

Step 3: Enter C3 in the box marked "Response Variable." Enter C1 in the box marked "row factor" and enter C2 in the box marked "column factor." Do not check the "Additive Model" box and check the "Display means" box. Click OK. We want to display means so that we can use them for Tukey's test.

Interaction Plots

Step 1: Enter the data as described in Step 1 above. Select the **Stat** menu. Highlight **ANOVA** and select **Interaction Effects Plot . . . .**

Step 2: Enter C3 in the box marked "Response Variable." Enter C1 and C2 in the box marked "Factors." Click OK.

Tukey's Test

Step 1: With the data entered into the MINITAB spreadsheet as specified in Step 1 above, select the **Stat** menu. Highlight **ANOVA** and select **General Linear Model . . . .**

Step 2: Enter the name of the response variable in the box marked "Responses." Enter C1, C2, and C1*C2 in the box marked "Model," where C1 is Factor A and C2 is Factor B.

Step 3: Click "Comparisons . . . ". Click the "Pairwise comparisons" radio button. In the box marked "Terms," enter C1* and C2. Check the box marked "Tukey." Check the box marked "Confidence Interval," if you want confidence intervals. Check the box marked "Test" if you want to conduct hypothesis tests of the pairwise differences.

Step 4: Click OK twice.

Main Effects Plots

Step 1: Enter the data as described in Step 1 above. Select the **Stat** menu. Highlight **ANOVA** and select **Main Effects Plot . . . .**

Step 2: Enter C3 in the box marked "Response Variable." Enter C1 and C2 in the box marked "Factors." Click OK.

Chapter 13 **Review**

Summary

We began the chapter with a discussion of one-way analysis of variance (ANOVA). One-way ANOVA is used to compare k means for equality when there is a single factor that has k levels. To perform any ANOVA test, the sample data must come from a population that is normally distributed. The samples must be obtained independently, and the largest sample standard deviation can be no more than two times the smallest standard deviation. The ANOVA procedures are robust, so minor departures from the normality requirement do not seriously affect the results.

If the null hypothesis, $H_0: \mu_1 = \mu_2 = \cdots = \mu_k$, is rejected in a one-way ANOVA, at least one of the population means is different from the others. To determine which means differ, we use Tukey's test, which compares each pair of sample means for significant differences. That is, it tests $H_0: \mu_i = \mu_j$ versus $H_1: \mu_i \neq \mu_j$ for $i \neq j$.

A different ANOVA procedure is used for analyzing experiments designed as randomized complete block designs. Blocking describes a way of organizing experimental units according to some common characteristic for which the response variable is expected to be similar, such as gender. By blocking, we reduce experimental error.

When data are analyzed from a randomized complete block design, we are not interested in whether the block is significant or not. Once again, if the null hypothesis of equal population means is rejected, we can use Tukey's test to compare each pair of means.

Finally, we analyzed data in which there are two factors. Factor A can be set at a levels and factor B can be set at b levels. When there are two factors, we use a two-way ANOVA procedure to test $H_0: \mu_1 = \mu_2 = \cdots = \mu_k$ versus H_1: at least one of the means differs. When performing a two-way ANOVA, we first look for a significant interaction effect. This means the changes in factor A result in different changes in the value of the response variable for different levels of factor B. If interaction exists, we do not consider main effects.

An example of a main effect would be to see how changes in the levels of factor B affect changes in the response variable. If no interaction exists, we then look for significant main effects. If the main effects indicate that at least one of the means differs, we use Tukey's test to determine which means differ.

Vocabulary

Analysis of Variance (p.677)
Mean square (p.682)
Mean square due to treatment (p.683)
Mean square due to error (p.683)
ANOVA table (p.685)
Multiple comparison methods (p.691)

Studentized range distribution (p.692)
Experimentwise error rate (p.692)
Family wise error rate (p.692)
Factor (p.701)
2×2 factorial design (p.712)
Cell (p.713)

Crossed (p.714)
Main effects (p.714)
Interaction effects (p.714)
Interaction plots (p.718)

Formulas

Test Statistic for One-Way ANOVA

$$F = \frac{\text{mean square due to treatment}}{\text{mean square due to error}} = \frac{\text{MST}}{\text{MSE}}, \text{ where}$$

$$\text{MST} = \frac{n_1(\bar{x}_1 - \bar{x})^2 + n_2(\bar{x}_2 - \bar{x})^2 + \cdots + n_k(\bar{x}_k - \bar{x})^2}{k - 1}$$

$$\text{MSE} = \frac{(n_1 - 1)s_1^2 + (n_2 - 1)s_2^2 + \cdots + (n_k - 1)s_k^2}{n - k}$$

Test Statistic for Tukey's Test after One-Way ANOVA

$$q = \frac{(\bar{x}_2 - \bar{x}_1) - (\mu_2 - \mu_1)}{\sqrt{\frac{s^2}{2} \cdot \left(\frac{1}{n_1} + \frac{1}{n_2}\right)}} = \frac{\bar{x}_2 - \bar{x}_1}{\sqrt{\frac{s^2}{2} \cdot \left(\frac{1}{n_1} + \frac{1}{n_2}\right)}}$$

Objectives

Section	You should be able to . . .	Examples	Review Exercises
13.1	1 Verify the requirements to perform a one-way ANOVA (p. 678)	1	3(b), 4(b)
	2 Test a claim regarding three or more means using one-way ANOVA (p. 680)	2 and 3	3(c), 4(c), 6(a)
13.2	1 Perform the Tukey test (p. 692)	1–3	2, 3(f), 5, 6(b)
13.3	1 Conduct analysis of variance on the randomized complete block design (p. 703)	2	7(a), 8(a)
	2 Perform the Tukey test (p. 705)	3	7(b), 8(b)
13.4	1 Analyze a two-way ANOVA design (p. 712)	1–3 and 6	9(b), (c), 10(b), (c)
	2 Draw interaction plots (p. 718)	4–6	9(d), 10(d)
	3 Perform the Tukey test (p. 721)	7	9(e), 10(e)

Exercises

1. (a) Find the critical value from the Studentized range distribution for $\alpha = 0.05$, $\nu = 16$, $k = 7$.
   (b) Find the critical value from the Studentized range distribution for $\alpha = 0.05$, $\nu = 30$, $k = 6$.
   (c) Find the critical value from the Studentized Range Distribution for $\alpha = 0.01$, $\nu = 42$, $k = 4$. (**Note:** We use $q_{0.05,40,8}$ since $q_{0.05,42,8}$ is not in the table.)
   (d) Find the critical value from the Studentized range distribution for H_0: $\mu_1 = \mu_2 = \mu_3 = \mu_4 = \mu_5 = \mu_6$, with $n = 46$, $\alpha = 0.05$.

2. Suppose that there is sufficient evidence to reject H_0: $\mu_1 = \mu_2 = \mu_3 = \mu_4$ using a one-way ANOVA. The mean square error from ANOVA is determined to be 373. The sample means are $\bar{x}_1 = 78$, $\bar{x}_2 = 64$, $\bar{x}_3 = 70$, $\bar{x}_4 = 47$ with $n_1 = n_2 = n_3 = n_4 = 31$. Use Tukey's test to determine which pairwise means are significantly different using a familywise error rate of $\alpha = 0.05$.

3. **Soil Testing**   A researcher took water samples in a forest with a stream running through it. The samples were collected over the course of a year. Each sample was analyzed for the concentration of dissolved organic carbon (mg/L) in it, with the results presented in the following table:

Organic				Mineral					Surface				
22.74	14.90	17.90	11.40	8.50	5.50	3.02	7.31	4.85	10.83	11.94	15.76	15.03	15.77
29.80	14.86	18.30	5.30	3.91	4.71	7.45	16.92	11.97	8.74	12.40	10.96	14.53	6.40
27.10	15.91	5.2	15.72	9.29	7.66	11.33	4.60	7.85	9.20	10.30	19.78	12.04	14.19
16.51	15.35	11.90	20.46	21.00	11.72	7.11	8.50	9.11	8.12	10.48	20.56	16.82	13.89
6.51	9.72	14.00	16.87	10.89	11.80	17.99	4.80	8.79	7.60	12.88	17.93	10.70	8.69
8.81	19.80	7.40	15.42	10.30	8.05	21.40	4.90	9.60	6.30	19.01	14.28	16.00	10.46
5.29	14.86	17.50	22.49	11.56	10.72	8.37	9.10	12.57	6.68	19.19	13.11	20.70	16.23
20.46	8.09	10.30		7.00	21.82	7.92	7.90	12.89	7.34	13.14	12.27	13.70	15.43
				3.99	22.62	17.90	11.72	9.81	9.52	12.51	16.47	16.12	
				3.79	10.74								

Source: Lisa Emili, PhD Candidate, Department of Geography and Wetlands Research Centre, University of Waterloo

Each sample was then categorized according to the type of water collected. Water was collected from streams (surface water), groundwater was collected from organic soil, and groundwater was collected from mineral soil. The researcher wanted to test the claim that the mean concentration of dissolved organic carbon was the same for each collection area.
(a) State the null and alternative hypotheses.
(b) State the requirements that must be satisfied to use the one-way ANOVA procedure.

(c) Use the following MINITAB output to test the researcher's claim at the $\alpha = 0.01$ level of significance.

One-way ANOVA: Organic, Mineral, Surface

Source	DF	SS	MS	F	P
Factor	2	471.4	235.7	9.43	0.000
Error	119	2973.8	25.0		
Total	121	3445.1			

(d) Shown are side-by-side boxplots of each type of plot. Do the boxplots support the analytic results?

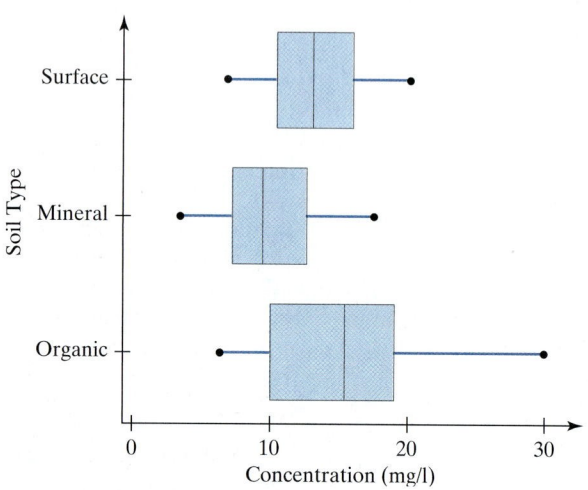

(e) Verify that the F-test statistic is 9.43.

(f) The following display represents the results of Tukey's test obtained from MINITAB. Which pairwise means are different?

```
Tukey 95% Simultaneous Confidence Intervals
All Pairwise Comparisons

Individual confidence level = 98.09%

Organic subtracted from:

          Lower   Center   Upper  --+---------+---------+---------+-------
Mineral  -7.588  -4.840  -2.092  (-------*-------)
Surface  -4.606  -1.821   0.964              (-------*-------)
                                  --+---------+---------+---------+-------
                                  -7.0      -3.5       0.0       3.5


Mineral subtracted from:

          Lower   Center   Upper  --+---------+---------+---------+-------
Surface   0.527   3.019   5.510                         (-------*------)
                                  --+---------+---------+---------+-------
                                  -7.0      -3.5       0.0       3.5
```

4. **Crash Data**   The Insurance Institute for Highway Safety conducts experiments in which cars are crashed into a fixed barrier at 40 mph. In the Institute's 40-mph offset test, 40% of the total width of each vehicle strikes a barrier on the driver's side. The barrier's deformable face is made of aluminum honeycomb, which makes the forces in the test similar to those involved in a frontal offset crash between two vehicles of the same weight, each going just less than 40 mph. Suppose you are in the market to buy a new family car. You want to know if the mean femur force in kilonewtons (kN)

on the left leg resulting from this offset crash is the same for large family cars, passenger vans, and midsize utility vehicles. The following data were collected from the institute's study.

Large Family Cars	Femur Force (kN)	Passenger Vans	Femur Force (kN)	Midsize Utility Vehicles	Femur Force (kN)
Chevrolet Lumina	5.8	Toyota Sienna	1.8	Mercedes M Class	3.5
Ford Taurus	2.1	Honda Odyssey	3.0	Toyota 4Runner	1.2
Buick LeSabre	2.8	Ford Windstar	2.1	Mitsubishi Montero	3.6
Chevrolet Impala	4.7	Mazda MPV	6.8	Nissan Xterra	1.2
Chrysler LHS	6.2	Chevrolet Astro	8.2	Ford Explorer	3.7
Pontiac Grand Prix	2.4	Nissan Quest	2.9	Jeep Grand Cherokee	6.7
Dodge Intrepid	4.7	Pontiac TransSport	6.7	Nissan Pathfinder	1.8

Source: Insurance Institute for Highway Safety

The researcher wants to test the claim that the mean femur forces for the three classes of vehicle are equal.

(a) State the null and alternative hypotheses.

(b) Verify that the requirements to use the one-way ANOVA procedure are satisfied. Normal probability plots indicate that the sample data come from a population that is normal.

(c) Test the claim that the mean femur forces for each vehicle type are the same at the $\alpha = 0.05$ level of significance.

(d) Draw boxplots of the three types of vehicle to support the analytic results obtained in part (c).

5. **Grading Timber**   With the desire to grade timber mechanically, engineers studied the modulus of rupture in lb/in^2, for a sample of untreated green, 7 inch by 9 inch, mixed oak timbers, sorted by grade: select structural, No. 1, No. 2, and below grade. A one-way analysis of variance was performed to test $H_0: \mu_{select} = \mu_{No.\,1} = \mu_{No.\,2} = \mu_{below\,grade}$. The null hypothesis was rejected with a *P*-value of 0.018. The engineers proceeded to conduct Tukey's test to determine which pairwise means differed using MINITAB. The results are presented below. Which pairwise means differ? What do you conclude from the test? **Note:** 1 = select structural, 2 = No. 1, 3 = No. 2, and 4 = below grade. *Source:* Mechanical Grading of Oak Timbers, *Journal of Materials in Civil Engineering*, Vol. 11, No. 2.

```
Tukey's Pairwise Comparison

Family error rate = 0.0500
Individual error rate = 0.0108

Critical value = 3.86

Intervals for (column level mean) - (row level mean)


                      1              2              3

        2         -6.914
                  13.064

        3          1.936         -1.139
                  21.914         18.839

        4         -2.989         -6.064        -14.914
                  16.989         13.914          5.064
```

6. **Seating Choice versus GPA** In a study of students at Lewis & Clark Community College, a professor compared the chosen seat location of students in mathematics courses to their overall grade point average (GPA). He randomly selected 10 students from each of three seat locations: (1) front, (2) middle, and (3) back. The GPAs follow:

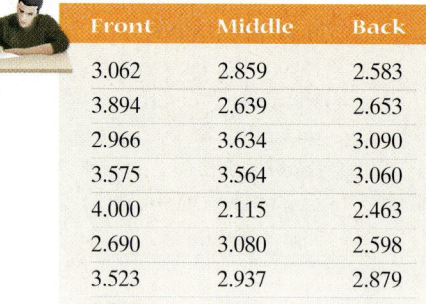

Front	Middle	Back
3.062	2.859	2.583
3.894	2.639	2.653
2.966	3.634	3.090
3.575	3.564	3.060
4.000	2.115	2.463
2.690	3.080	2.598
3.523	2.937	2.879
3.332	3.091	2.926
3.885	2.655	3.221
3.559	2.526	2.646

(a) Test the null hypothesis that the mean score for each location is the same at the $\alpha = 0.05$ level of significance. [**Note:** The requirements for a one-way ANOVA are satisfied.]

(b) If the null hypothesis is rejected in part (a), use Tukey's test to determine which pairwise means differ using a familywise error rate of $\alpha = 0.05$.

(c) Draw boxplots of the three treatment levels to support the analytic results obtained in parts (a) and (b).

7. **Catapults** Four different catapult designs are being tested for their ability to launch water balloons. Because the person firing the catapult could represent a source of variability, the experimenter decides to use a randomized complete block design and block by person firing. The distances achieved for each catapult are given next.

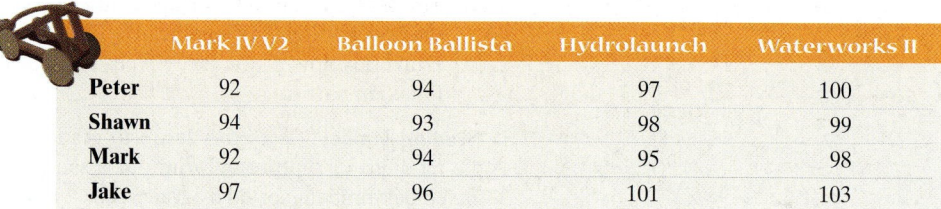

	Mark IV V2	Balloon Ballista	Hydrolaunch	Waterworks II
Peter	92	94	97	100
Shawn	94	93	98	99
Mark	92	94	95	98
Jake	97	96	101	103

(a) Test $H_0: \mu_1 = \mu_2 = \mu_3 = \mu_4$ against H_1: at least one of the means is different, where μ_1 is the mean for the Mark IV V2, μ_2 is the mean for the Balloon Ballista, and so on, at the $\alpha = 0.05$ level of significance.

(b) If the null hypothesis from part (a) was rejected, use Tukey's test to determine which pairwise means differ using a familywise error rate of $\alpha = 0.05$.

(c) Draw boxplots of the data for each catapult design using the same scale to support the analytical results obtained in parts (a) and (b).

8. **Rope Strength** A rock climber wishes to test the tensile strength of three different types of braided ropes. She knows that the rope thickness can make a difference, so she decides to block by thickness. The tensile strengths for each rope type are given next.

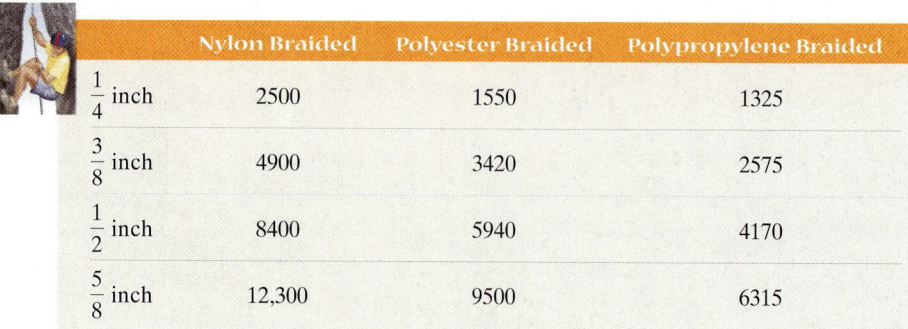

	Nylon Braided	Polyester Braided	Polypropylene Braided
$\frac{1}{4}$ inch	2500	1550	1325
$\frac{3}{8}$ inch	4900	3420	2575
$\frac{1}{2}$ inch	8400	5940	4170
$\frac{5}{8}$ inch	12,300	9500	6315

(a) Test $H_0: \mu_1 = \mu_2 = \mu_3$ against H_1: at least one of the means is different, where μ_1 is the mean for the nylon braided rope, μ_2 is the mean for the polyester braided rope, and μ_3 is the mean for the polypropylene braided rope, at the $\alpha = 0.05$ level of significance.

(b) If the null hypothesis from part (a) was rejected, use Tukey's test to determine which pairwise means differ using a familywise error rate of $\alpha = 0.05$.

(c) Draw boxplots of the data for each treatment using the same scale to support the analytical results obtained in parts (a) and (b).

9. **Defensive Driving**   An investigator for the state police wants to determine the effectiveness of three different defensive driving training programs to see if there are any gender differences. Five subjects of each gender who recently received speeding tickets are assigned to each program. At the end of the program, each is given a written test on his or her knowledge of defensive driving. The scores (out of 100) are given next.

		Program		
		One 8-hr session	**Two 4-hr sessions**	**Two 2-hr sessions**
Gender	**Male**	89	95	77
		96	87	78
		95	90	83
		90	91	78
		96	92	78
	Female	88	87	80
		92	92	82
		98	91	79
		99	94	86
		91	93	88

(a) The scores are approximately normally distributed. Verify the requirement of equal population variances.

(b) Determine whether there is significant interaction between program and gender.

(c) If there is no significant interaction, determine whether there is significant difference in the means for the three programs.

(d) Draw an interaction plot of the data to support the results of parts (a) and (b).

(e) If there is significant difference in the means for the three programs, use Tukey's test to determine which pairwise means differ using a familywise error rate of $\alpha = 0.05$.

10. **Learning time**   A manufacturing researcher wants to determine if age or gender significantly affects the time required to learn an assembly line task. He randomly selected 24 adults aged 20 to 65 years old, of whom 8 are 20 to 34 years old (4 male, 4 female), 8 are 35 to 49 years old (4 male, 4 female), and 8 are 50 to 64 years old (4 male, 4 female). He then measured the time (in minutes) required to complete a certain task. The data obtained are shown next.

		Age (years)		
		20–34	**35–49**	**50–64**
Gender	**Male**	5.2, 5.1	4.8, 5.8	5.2, 4.3
		5.7, 6.1	5.0, 4.8	5.5, 4.7
	Female	5.3, 5.5	5.0, 5.4	4.9, 5.5
		4.9, 5.6	5.6, 5.1	5.5, 5.0

(a) The learning times are approximately normally distributed. Verify the requirement of equal population variances.

(b) Determine whether there is significant interaction between age and gender.

(c) If there is not significant interaction, determine whether or not there is significant difference in the means for the three age groups.

(d) Draw an interaction plot of the data to support the results of parts (a) and (b).

(e) If there is significant difference in the means for the three age groups, use Tukey's test to determine which pairwise means differ using a familywise error rate of $\alpha = 0.05$.

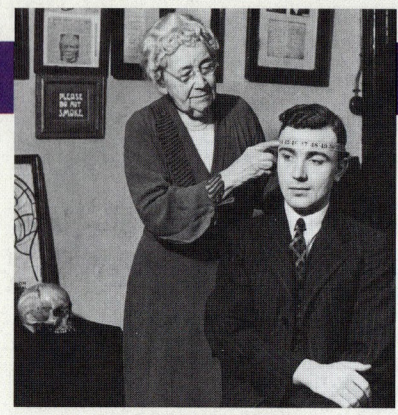

Hat Size and Intelligence

In the mid-19th century, Paul Broca, a professor of clinical surgery, effectively argued that the degree of a person's intelligence was directly related to the size of the brain. Broca concluded that

> In general, the brain is larger in mature adults than in the elderly, in men than in women, in eminent men than in men of mediocre talent.... Other things equal, there is a remarkable relationship between the development of intelligence and the volume of the brain. (Quoted in Stephen Jay Gould, *The Mismeasure of Man*. W. W. Norton & Company, New York, 1981, p. 83)

During the 20th century, arguments regarding brain size diminished as scientists turned to intelligence tests—a more direct method to measure and compare mental capacities, they claimed. Yet the older argument still manages to creep into modern discussions of intelligence. In a book published in 1978 that was designed to acquaint educators with modern brain research, H. T. Epstein declared,

> First we shall ask if there is any indication of a linkage of any kind between brain and intelligence. It is generally stated that there is no such linkage.... But the one set of data I have found seems to show clearly that there is a substantial connection. Hooton studied the head circumferences of white Bostonians as part of his massive study of criminals. The following table shows that the ordering of people according to head size yields an entirely plausible ordering according to vocational status. It is not at all clear how the impression has been spread that there is no such correlation. (Quoted in Stephen Jay Gould, *The Mismeasure of Man*. W. W. Norton & Company, New York, 1981, p. 109)

Epstein's table is reproduced here:

Mean and Standard Deviation (in mm) of Head Circumference for People of Vocational Status			
Vocational Status	n	Mean	S.D.
Professional	25	569.9	1.9
Semiprofessional	61	566.5	1.5
Clerical	107	566.2	1.1
Trades	194	565.7	0.8
Public service	25	564.1	2.5
Skilled trades	351	562.9	0.6
Personal services	262	562.7	0.7
Laborers	647	560.7	0.3

Source: Stephen Jay Gould, *The Mismeasure of Man*. (New York: W. W. Norton & Company, New York, 1981, p. 109)

At first glance, this table seems to support the assertion that those with more prestigious vocations have a larger head circumference. However, we can examine the alleged relationship more closely through a one-way analysis of variance. Specifically, we test the claim that all the head-circumference population means are equal. We use an $\alpha = 0.05$ significance level and obtain the following ANOVA table, derived from Epstein's statistics:

ANOVA Table for Epstein's Data on Head Circumference by Vocational Status				
Source of Variation	Sum of Squares	Degrees of Freedom	Mean Squares	F-Test Statistic
Treatment	7865.359	7	1123.6284	1998.6275
Error	935.450	1664	0.5622	
Total	8800.809	1671		

From these results, does it appear that at least one of the head-circumference means is different from the others? Explain.

Examine Epstein's table closely. Does it appear that the ANOVA procedure's requirement of equal variances is satisfied? Explain.

Subsequently, it was revealed that there was an error in Epstein's table. The column labeled S.D. (standard deviation) should have been labeled *standard error*. For each of the vocational categories, calculate the correct standard deviation by multiplying the tabled value by the square root of its sample size, n.

If the ANOVA table was recalculated with the correct standard deviations, predict the effect on the F-test statistic. Would you be more or less likely to reject the null hypothesis? Explain.

Using the formula for MSE, calculate a revised sum of squared errors, mean square due to error, and F-test statistic. Retest the claim that all of the head-circumference population means are equal. Use an $\alpha = 0.05$ significance level. From the corrected values for the standard deviation, does it appear that the ANOVA procedure's requirement of equal variances is satisfied? Did any of your conclusions change in light of the new information? Explain.

If the null hypothesis of equal population means was rejected, use Tukey's test to determine which of the pairwise means differ using a familywise error rate of $\alpha = 0.05$. Which of these categories appear to have significantly different means?

Further research revealed that results for three vocational status groups were not presented in Epstein's table. Inexplicably deleted were factory workers (rank 7 out of 11 in status), transportation employees (rank 8), and extractive trades (farming and mining, rank 11). The mean head-circumference sizes for these three groups were 564.7 mm, 564.9 mm, and 564.7 mm, respectively. Also, it was discovered that Epstein's table did not list each group in the order of Hooton's prestige ranking but, rather, listed the groups in order of mean head circumference; this manipulation suggested a perfect correlation where one does not exist. *(Source: Stephen Jay Gould, The Mismeasure of Man.* (New York: W. W. Norton & Company, New York, 1981, p. 110)

Using all this new information and your previous analyses, write a brief report summarizing your findings and conclusions.

Inference on the Least-Squares Regression Model and Multiple Regression

Outline

DECISIONS

You continue to find yourself in the market to purchase a car.
You have a pretty good idea as to how much your car will be worth over the years, but you would like to have some measure of confidence in your results. See the Decisions project on page 756.

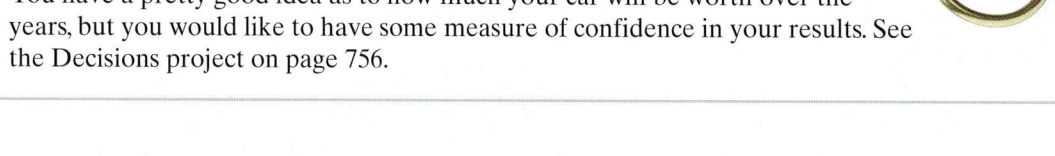

●●● Putting It All Together

We now extend inferential statistics to the least-squares regression line. Sections 14.1 and 14.2 introduce inferential methods that can be used on the least-squares regression line.

In Chapter 4, we presented methods for describing the relation between two variables-bivariate data. In addition, we performed diagnostic tests such as determining whether a linear model is appropriate, identifying outliers, and identifying influential observations.

In Section 14.1, we use the methods of hypothesis testing first presented in Chapter 10 to test the claim that a linear relation exists between two quantitative variables.

In Section 14.2, we construct confidence intervals about the predicted value of the least-squares regression line.

Section 14.3 introduces regression techniques used when there is more than one explanatory variable called *multiple regression*. We perform the same diagnostic tests for multiple regression that we did for *simple linear regression* (one explanatory variable). We also test whether the relation between the explanatory variables and response variable is significant, and we construct confidence intervals about the predicted value. The material in this section uses statistical software to do the "heavy lifting" so that we are not bogged down by calculation. This will allow us to concentrate on analyzing results, rather than getting results.

14.1 Testing the Significance of the Least-Squares Regression Model

Preparing for This Section   Before getting started, review the following:

- Scatter diagrams; correlation (Section 4.1, pp. 195–202)

- Least-squares regression (Section 4.2, pp. 212–221)

- Diagnostics on the least-squares regression line (Section 4.3, pp. 226–235)

- Sampling distribution of the sample mean $\bar{x}$ (Section 8.1, pp. 419–430)

- Testing a hypothesis about μ, σ unknown (Section 10.3, pp. 531–537)

- Confidence intervals about a mean (Section 9.2, pp. 466–473)

Objectives

1. **Understand the requirements of the least-squares regression model**
2. **Compute the standard error of the estimate**
3. **Verify that the residuals are normally distributed**
4. **Conduct inference on the slope and intercept**
5. **Construct a confidence interval about the slope of the least-squares regression model**

As a quick review of the topics discussed in Chapter 4, we present the following example:

EXAMPLE 1   **Least-Squares Regression**

Problem: A family doctor is interested in examining the relationship between a patient's age and total cholesterol. He randomly selects 14 of his female patients and obtains the data presented in Table 1. The data are based on results obtained from the National Center for Health Statistics.

Table 1

Age, x	Total Cholesterol, y	Age, x	Total Cholesterol y
25	180	42	183
25	195	48	204
28	186	51	221
32	180	51	243
32	210	58	208
32	197	62	228
38	239	65	269

Figure 1

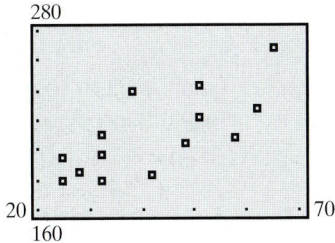

Figure 2

```
LinReg
 y=ax+b
 a=1.399064152
 b=151.3536582
 r²=.5152520915
 r=.7178106237
```

Draw a scatter diagram, compute the correlation coefficient, and find the least-squares regression equation and the coefficient of determination.

Approach: We will use a TI-84 Plus graphing calculator to obtain the information requested.

Solution: Figure 1 displays the scatter diagram. Figure 2 displays the output obtained from the calculator. The linear correlation coefficient is 0.7178. The least-squares regression equation for these data is $\hat{y} = 151.3537 + 1.3991x$, where $\hat{y}$ represents the predicted total cholesterol for a female whose age is x.

Figure 3

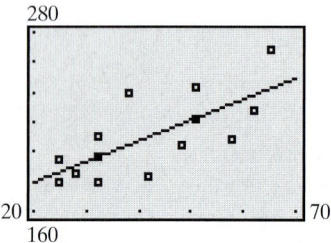

The coefficient of determination, R^2, is 0.515. So 51.5% of the variation in total cholesterol is explained by the regression line. Figure 3 shows a graph of the least-squares regression equation on the scatter diagram in order to get a feel for the fit.

The information obtained in Example 1 is descriptive in nature. Notice that the descriptions are both graphical (as in the scatter diagram) and numerical (as in the correlation coefficient and coefficient of determination).

In Other Words
Because b_0 and b_1 are statistics, they have sampling distributions.

① Understand the Requirements of the Least-Squares Regression Model

In the least-squares regression equation $\hat{y} = b_0 + b_1x$, the values for the slope, b_1, and intercept, b_0, are statistics, just as the sample mean, $\bar{x}$, and sample standard deviation, s, are statistics. The statistics b_0 and b_1 are estimates for the population intercept, β_0, and the population slope, β_1. The true linear relation between the explanatory variable, x, and the response variable, y, is given by $y = \beta_0 + \beta_1x$.

Because b_0 and b_1 are statistics, their values vary from sample to sample, so there is a sampling distribution associated with each of them. We use this sampling distribution to perform inference on b_0 and b_1. For example, we might want to test the claim that β_1 is different from 0. If we have evidence that supports this claim, we conclude that there is a linear relation between the explanatory variable, x, and response variable, y.

To find the sampling distributions of b_0 and b_1 we have some requirements about the population from which the bivariate data (x_i, y_i) were sampled. Just as in Section 8.1 when we discussed the sampling distribution of $\bar{x}$, we start by asking what would happen if we took many samples for a given value of the explanatory variable, x. For example, in looking back at Table 1, we notice that our sample included three women aged 32 years, so x has the same value, 32, for all three women in our sample, but the corresponding values of y for these three women are different: 180, 210, and 197. There is a distribution of total cholesterol levels for $x = 32$ years of age. Suppose we looked at *all* women aged 32 years. From these population data, we could find the population mean total cholesterol for *all* 32-year-old women, denoted $\mu_{y|32}$. The notation $\mu_{y|32}$ is read, "the mean value of the response variable y given that the explanatory variable is 32." We could repeat this process for any other age. In general, different ages will have different population mean total cholesterol. This brings us to our first requirement regarding inference on the least-squares regression model.

Requirement 1 for Inference on the Least-Squares Regression Model

For any particular value of the explanatory variable x (such as 32 in Example 1), the mean of the corresponding responses in the population depends linearly on x. That is,

$$\mu_{y|x} = \beta_0 + \beta_1x$$

for some numbers β_0 and β_1, where $\mu_{y|x}$ represents the population mean response when the explanatory variable is x.

We learned how to check this requirement in Section 4.3. If a plot of the residuals against the explanatory variable shows any discernible pattern (such as a U-shape), then the linear model is not appropriate and requirement 1 is violated.

We also have a requirement regarding the distribution of the response variable for any particular value of the explanatory variable.

In Other Words

When doing inference on the least-squares regression model, we require (1) that for any explanatory variable, x, the mean of the response variable, y, depends on the value of x through a linear equation, and (2) that the response variable, y, is normally distributed with a constant standard deviation, σ. The mean increases/decreases at a constant rate depending on the slope, while the variance remains constant.

Requirement 2 for Inference on the Least-Squares Regression Model

The response variables are normally distributed with mean $\mu_{y|x} = \beta_0 + \beta_1 x$ and standard deviation σ.

This requirement states that the mean of the response variable changes linearly, but the variance remains constant, and the distribution of the response variable is normal. For example, if we obtained a sample of many 32-year-old females and measured their total cholesterol, the distribution would be normal with mean $\mu_{y|32} = \beta_0 + \beta_1(32)$ and standard deviation σ. If we obtained a sample of many 43-year-old females and measured their total cholesterol, the distribution would be normal with mean $\mu_{y|43} = \beta_0 + \beta_1(43)$ and standard deviation σ. See Figure 4.

Figure 4

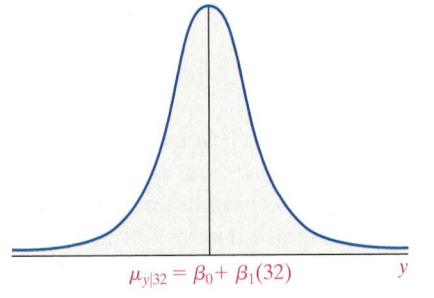
$\mu_{y|32} = \beta_0 + \beta_1(32)$ y

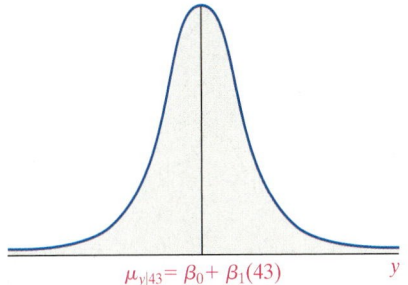
$\mu_{y|43} = \beta_0 + \beta_1(43)$ y

In Other Words

The larger σ is, the more spread out the data are around the regression line.

A large value of σ, the population standard deviation, indicates that the data are widely dispersed about the regression line, and a small σ indicates that the data lie fairly close to the regression line. Figure 5 illustrates the ideas just presented. The regression line represents the mean value of each normal distribution at a specified value of x. The standard deviation of each distribution is σ.

Figure 5

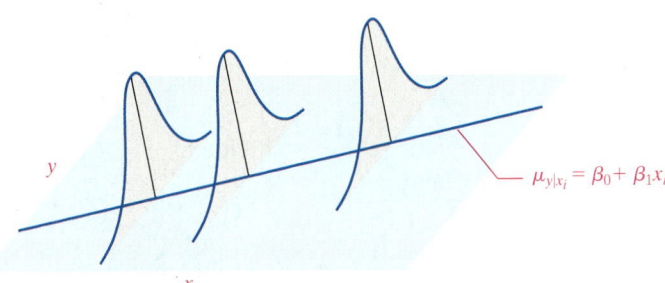
y
$\mu_{y|x_i} = \beta_0 + \beta_1 x_i$
x

Of course, not all the observed values of the response variable will lie on the true regression line, $\mu_{y|x} = \beta_0 + \beta_1 x$. The difference between the observed and predicted value of the response variable is an error term or residual, ε_i. We now present the least-squares regression model.

Definition

The **least-squares regression model** is given by

$$y_i = \beta_0 + \beta_1 x_i + \varepsilon_i \tag{1}$$

where

y_i is the value of the response variable for the ith individual,

β_0 and β_1 are the parameters to be estimated based on sample data,

x_i is the value of the explanatory variable for the ith individual,

ε_i is a random error term with mean 0 and variance $\sigma_{\varepsilon_i}^2 = \sigma^2$. The error terms are independent, and

$i = 1, \ldots, n$, where n is the sample size (number of ordered pairs in the data set).

Because the expected value or mean of y_i is $\beta_0 + \beta_1 x_i$ and the expression on the left side of Equation (1) equals the expression on the right side, the expected value or mean of the error term, ε_i, is 0.

Now Work Problem 13(a).

2 Compute the Standard Error of the Estimate

In Section 4.2, we learned how to obtain estimates for β_0 and β_1. We now present the method for obtaining the estimate of σ, the standard deviation of the response variable y for any given value of x. The unbiased estimator of σ is called the *standard error of the estimate*.

Remember the formula for the sample standard deviation presented in Section 3.2?

$$s = \sqrt{\frac{\Sigma(x_i - \bar{x})^2}{n - 1}}$$

We compute the deviations about the mean, square them, add up the squared deviations, and divide by $n - 1$. We divide by $n - 1$ because we lose 1 degree of freedom since one parameter, $\bar{x}$, is estimated. Exactly the same logic is used to compute the standard error of the estimate.

As we mentioned, the predicted values of y, denoted $\hat{y}_i$, represent the mean value of the response variable for any given value of the explanatory variable, x_i. So $y_i - \hat{y}_i$ = residual represents the difference between the observed value, y_i, and the mean value, $\hat{y}_i$. This calculation is used to get the standard error of the estimate.

Definition

The **standard error of the estimate**, s_e, is found using the formula

$$s_e = \sqrt{\frac{\Sigma(y_i - \hat{y}_i)^2}{n - 2}} = \sqrt{\frac{\Sigma \text{residuals}^2}{n - 2}} \qquad (2)$$

Notice that we divide by $n - 2$ because we have estimated two parameters, β_0 and β_1.

EXAMPLE 2 **Computing the Standard Error**

Problem: Compute the standard error for the data in Table 1.

Approach: We use the following steps to compute the standard error.

Step 1: Find the least-squares regression line.
Step 2: Obtain predicted values for each of the observations in the data set.
Step 3: Compute the residuals for each of the observations in the data set.
Step 4: Compute $\Sigma \text{residuals}^2$.
Step 5: Compute the standard error, using Formula (2).

Solution

Step 1: The least squares regression line was found in Example 1.
Step 2: Column 3 of Table 2 represents the predicted values for each of the $n = 14$ observations.
Step 3: Column 4 of Table 2 represents the residuals for each of the 14 observations.

Table 2				
Age, x	Total Cholesterol, y	$\hat{y} = 1.3991x + 151.3537$	Residuals, $y - \hat{y}$	Residuals^2, $(y - \hat{y})^2$
25	180	186.33	-6.33	40.0689
25	195	186.33	8.67	75.1689
28	186	190.53	-4.53	20.5209
32	180	196.12	-16.12	259.8544
32	210	196.12	13.88	192.6544
32	197	196.12	0.88	0.7744
38	239	204.52	34.48	1188.8704
42	183	210.12	-27.12	735.4944
48	204	218.51	-14.51	210.5401
51	221	222.71	-1.71	2.9241
51	243	222.71	20.29	411.6841
58	208	232.50	-24.50	600.25
62	228	238.10	-10.10	102.01
65	269	242.30	26.70	712.89

$$\Sigma \text{residuals}^2 = 4553.705$$

Step 4: Column 5 of Table 2 contains the squared residuals. We sum the entries in column 5 to obtain the sum of squared errors. So

$$\Sigma \text{residuals}^2 = 4553.705$$

Step 5: We use Formula (2) to compute the standard error.

$$s_e = \sqrt{\frac{\Sigma \text{residuals}^2}{n - 2}} = \sqrt{\frac{4553.705}{14 - 2}} = 19.48$$

CAUTION!
Be sure to divide by $n - 2$ when computing the standard error.

EXAMPLE 3   **Obtaining the Standard Error Using Technology**

Figure 6

Regression Statistics	
Multiple R	0.7178106
R Square	0.5152521
Adjusted R Square	0.4748564
Standard Error	19.480535
Observations	14

Problem: Obtain the standard error for the data in Table 1 using statistical software.

Approach: We will use Excel to obtain the standard error. The steps for obtaining standard errors using the TI-83/84 graphing calculators, MINITAB, and Excel are given in the Technology Step by Step on page 752.

Result: Figure 6 shows the partial output obtained from Excel. Notice that the results agree with the by-hand computation.

Now Work Problem 13(b).

---

③   **Verify That the Residuals Are Normally Distributed**

For the least-squares regression model $y_i = \beta_0 + \beta_1 x_i + \varepsilon_i$, we require that the response variable, y_i, be normally distributed. Because $\beta_0 + \beta_1 x_i$ is constant for any x_i, the requirement that y_i is normal means that the residuals, ε_i must also be normal. To perform statistical inference on the regression, we must verify that the residuals are normally distributed by examining a normal probability plot.

CAUTION!
The residuals must be normally distributed to perform inference on the least-squares regression line.

 **Verifying That the Residuals Are Normally Distributed**

Problem: Verify that the residuals obtained in Table 2 from Example 2 are normally distributed.

Approach: We construct a normal probability plot to assess normality. If the normal probability plot is roughly linear, the residuals are said to be normal.

Solution: Figure 7 contains the normal probability plot obtained from MINITAB.

Figure 7

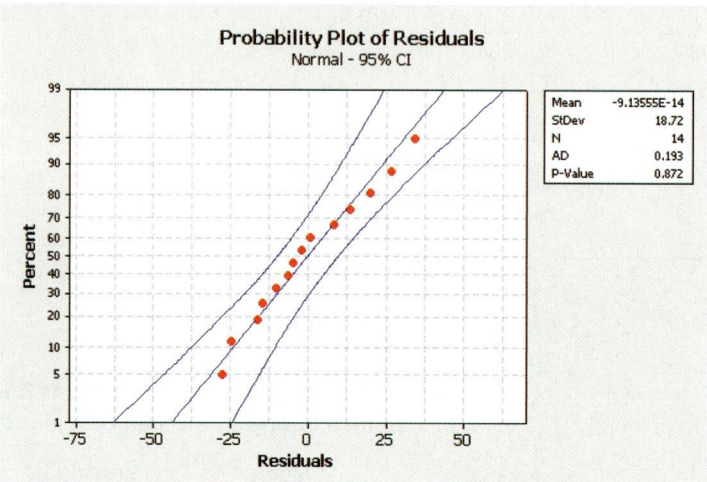

Because the points follow a linear pattern (all the points lie within the bands created by MINITAB), the residuals are normally distributed. We can perform inference on the least-squares regression equation. ▬▬

Now Work Problem 13(c).

 Conduct Inference on the Slope and Intercept

At this point, we know how to estimate the intercept and slope of the least-squares regression model. In addition, we can compute the standard error, s_e, which is an estimate of σ, the standard deviation of the response variable about the true least-squares regression model, and we know how to assess the normality of the residuals. We will now use this information to test the claim that a linear relation exists between the explanatory and the response variables.

This is the question that we want to answer: Do the sample data provide sufficient evidence to support the claim that a linear relation exists between the two variables? If there is no linear relation between the response and explanatory variables, the slope of the true regression line will be zero. Do you know why? A slope of zero means that information about the explanatory variable, x, does not change our "guess" as to the value of the response variable, y.

Using the notation of hypothesis testing, we can perform one of three tests

Two-Tailed	Left-Tailed	Right-Tailed
$H_0: \beta_1 = 0$	$H_0: \beta_1 = 0$	$H_0: \beta_1 = 0$
$H_1: \beta_1 \neq 0$	$H_1: \beta_1 < 0$	$H_1: \beta_1 > 0$

In the two-tailed test, we are testing the claim that a linear relation exists between two variables without regard to the sign of the slope. In the left-tailed

CAUTION

Before testing $H_0: \beta_1 = 0$, be sure to draw a residual plot to verify that a linear model is appropriate.

test, we are testing the claim that the slope of the true regression line is negative. In the right-tailed test, we are testing the claim that the slope of the true regression line is positive.

To test any one of these hypotheses, we need to know the sampling distribution of b_1. It turns out that

$$t = \frac{b_1 - \beta_1}{\dfrac{s_e}{\sqrt{\sum(x_i - \bar{x})^2}}} = \frac{b_1 - \beta_1}{s_{b_1}}$$

follows Student's t-distribution with $n - 2$ degrees of freedom, where n is the number of observations, b_1 is the estimate of the the slope of the regression line β_1, and s_{b_1} is the sample standard error of b_1.

Hypothesis Test Regarding the Slope Coefficient, β_1

To test the claim that two quantitative variables are linearly related, we use the following steps provided that

1. the sample is obtained using random sampling
2. the residuals are normally distributed with constant error variance.

Step 1: A claim is made regarding the linear relation between a response variable, y, and an explanatory variable, x. The claim is used to determine the null and alternative hypotheses. The hypotheses can be structured in one of three ways:

Two-Tailed	Left-Tailed	Right-Tailed
$H_0: \beta_1 = 0$	$H_0: \beta_1 = 0$	$H_0: \beta_1 = 0$
$H_1: \beta_1 \neq 0$	$H_1: \beta_1 < 0$	$H_1: \beta_1 > 0$

Step 2: Select a level of significance α, depending on the seriousness of making a Type I error.

Step 3: Compute the test statistic

$$t_0 = \frac{b_1 - \beta_1}{s_{b_1}} = \frac{b_1}{s_{b_1}}$$

which follows Student's t-distribution with $n - 2$ degrees of freedom. Remember, when computing the test statistic, we assume the null hypothesis to be true. So we assume $\beta_1 = 0$,

Classical Approach

Step 4: Use Table V to determine the critical value using $n - 2$ degrees of freedom.

	Two-Tailed	Left-Tailed	Right-Tailed
Critical value	$-t_{\alpha/2}$ and $t_{\alpha/2}$	$-t_\alpha$	t_α
Critical region			

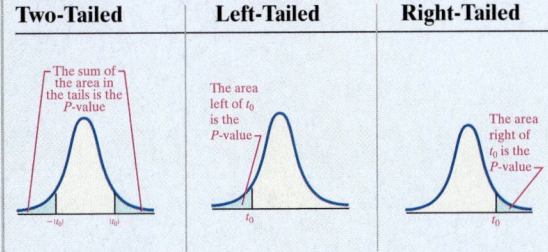

Step 5: Compare the critical value with the test statistic.

Two-Tailed	Left-Tailed	Right-Tailed
If $t_0 < -t_{\alpha/2}$ or $t_0 > t_{\alpha/2}$, reject the null hypothesis	If $t_0 < -t_\alpha$, reject the null hypothesis	If $t_0 > t_\alpha$, reject the null hypothesis

P-value Approach

Step 4: Use Table V to estimate the P-value using $n - 2$ degrees of freedom.

Two-Tailed	Left-Tailed	Right-Tailed

Step 5: If P-value $< \alpha$, reject the null hypothesis.

Step 6: State the conclusion.

The procedures just presented are **robust**, which means that minor departures from normality will not adversely affect the results of the test. In fact, for large samples ($n \geq 30$), inferential procedures regarding b_1 can be used even with significant departures from normality.

EXAMPLE 5   Testing for a Linear Relation

Problem: Test the claim that there is a linear relation between age and total cholesterol at the $\alpha = 0.05$ level of significance, using the data given in Table 1.

Approach: We verify that the requirements to perform the inference are satisfied. We then follow Steps 1–6.

Solution: In Example 1, we were told that the individuals were randomly selected. In Example 3, we confirmed that the residuals were normally distributed by constructing a normal probability plot. We can verify the requirement of constant error variance by plotting the residuals against the values of the explanatory variable as shown in Figure 8.

The errors are evenly spread around a horizontal line drawn at 0. The requirement of constant error variance is satisfied. We can now follow Steps 1–6 to test the claim.

Figure 8

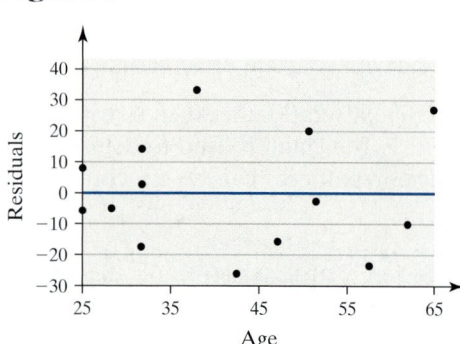

Age

Step 1: We are testing the claim that there is no linear relation between age and total cholesterol. Therefore, we are testing

$$H_0: \beta_1 = 0 \qquad \text{versus} \qquad H_1: \beta_1 \neq 0$$

Step 2: We are testing the claim at the $\alpha = 0.05$ level of significance.

Step 3: We obtained an estimate of β_1 in Example 1, and we computed the standard error, s_e, in Example 2. To determine the standard deviation of b_1, we need to compute $\sum(x_i - \bar{x})^2$, where the x_i are the values of the explanatory variable, age, and $\bar{x}$ is the sample mean. We compute this value in Table 3.

CAUTION
Use unrounded values of the sample mean in the computation of $\sum(x_i - \bar{x})^2$ to avoid round-off error.

Table 3			
Age, x	$\bar{x}$	$x_i - \bar{x}$	$(x_i - \bar{x})^2$
25	42.07143	−17.07143	291.4337
25	42.07143	−17.07143	291.4337
28	42.07143	−14.07143	198.0051
32	42.07143	−10.07143	101.4337
32	42.07143	−10.07143	101.4337
32	42.07143	−10.07143	101.4337
38	42.07143	−4.07143	16.5765
42	42.07143	−0.07143	0.0051
48	42.07143	5.92857	35.1479
51	42.07143	8.92857	79.7194
51	42.07143	8.92857	79.7194
58	42.07143	15.92857	253.7193
62	42.07143	19.92857	397.1479
65	42.07143	22.92857	525.7193
			$\sum(x_i - \bar{x})^2 = 2472.9284$

We have

$$s_{b_1} = \frac{s_e}{\sqrt{\sum(x_i - \bar{x})^2}} = \frac{19.48}{\sqrt{2472.9284}} = 0.3917$$

CAUTION
If we do not reject H_0, then we use the sample mean of y to predict the value of the response for any value of the explanatory variable.

The test statistic is

$$t_0 = \frac{b_1}{s_{b_1}} = \frac{1.3991}{0.3917} = 3.572$$

Classical Approach

Step 4: Because this is a two-tailed test, we determine the critical t-values at the $\alpha = 0.05$ level of significance with $n - 2 = 14 - 2 = 12$ degrees of freedom to be $-t_{0.05/2} = -t_{0.025} = -2.179$ and $t_{0.05/2} = t_{0.025} = 2.179$ The critical regions are displayed in Figure 9.

Figure 9

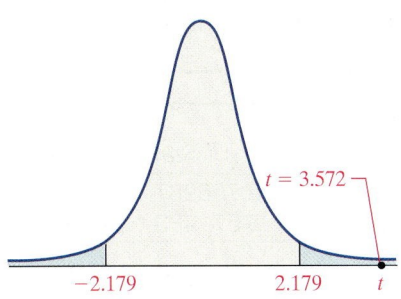

Step 5: The test statistic $t_0 = 3.572$. We label this point in Figure 9. Because the test statistic is greater than the critical value $t_{0.025} = 2.179$, we reject the null hypothesis.

P-value Approach

Step 4: Because this is a two-tailed test, the P-value is the area under the t-distribution with $14 - 2 = 12$ degrees of freedom to the left of the test statistic, $-t_0 = -3.572$, and to the right of $t_0 = 3.572$, as shown in Figure 10. That is, P-value $= P(t < -3.572) + P(t > 3.572) = 2P(t > 3.572)$, with 12 degrees of freedom.

Figure 10

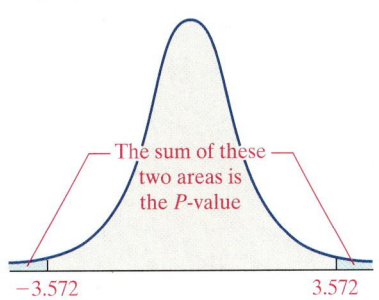

Using Table V, we find the row that corresponds to 12 degrees of freedom. The value 3.572 lies between 3.428 and 3.930. The value of 3.428 has an area under the t-distribution with 12 degrees of freedom of 0.0025 to the right. The area under the t-distribution with 12 degrees of freedom to the right of 3.930 is 0.001.

Because 3.572 is between 3.428 and 3.930, the P-value is between 2(0.001) and 2(0.0025). So $0.002 < P$-value < 0.005.

Step 5: Because the P-value is less than the level of significance $\alpha = 0.05$, we reject the null hypothesis.

Step 6: There is sufficient evidence at the $\alpha = 0.05$ level of significance to conclude that there is a linear relation between age and total cholesterol.

EXAMPLE 6   **Testing for a Linear Relation Using Technology**

Problem: Test the claim that a linear relation exists between age and total cholesterol at the $\alpha = 0.05$ level of significance using statistical software.

Approach: We will use MINITAB, Excel, and a TI-84 Plus graphing calculator to test the claim of a linear relation. The steps for testing the claim that a linear relation exists using the TI-83/84 graphing calculators, MINITAB, and Excel are given in the Technology Step by Step on page 752.

Result: Figure 11(a) shows the results obtained from MINITAB, Figure 11(b) shows the results obtained from Excel, and Figure 11(c) shows the results obtained from a TI-84 Plus graphing calculator.

Figure 11   **Regression Analysis**

```
The regression equation is
total cholesterol = 151 + 1.40 age
```

Predictor	Coef	StDev	T	P
Constant	151.35	17.28	8.76	0.000
Age	1.3991	0.3917	3.57	0.004

```
S = 19.48      R-Sq = 51.5%      R-Sq(adj) = 47.5%
```

Analysis of Variance

Source	Df	SS	MS	F	P
Regression	1	4840.5	4840.5	12.76	0.004
Residual Error	12	4553.9	379.5		
Total	13	9394.4			

(a) MINITAB output

Figure 11 (cont'd)

SUMMARY OUTPUT

Regression Statistics	
Multiple R	0.7178106
R Square	0.5152521
Adjusted R Square	0.4748564
Standard Error	19.480535
Observations	14

ANOVA

	df	SS	MS	F
Regression	1	4840.462	4840.462	12.75514
Residual	12	4553.895	379.4912	
Total	13	9394.357		

	Coefficients	Standard Error	t Stat	P-value
Intercept	151.35366	17.28376	8.756987	1.47E-06
Age	1.3990642	0.391737	3.571433	0.003842

(b) Excel output

P-value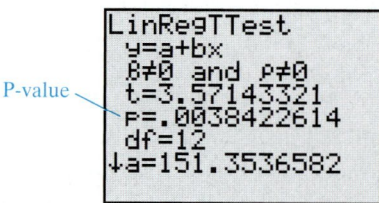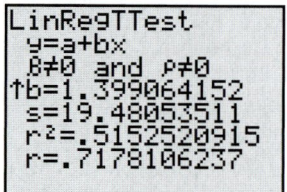

(c) TI-84 Plus output

In all three sets of output, the P-value for the slope is given as 0.004. This means that there is a 0.004 probability of obtaining a slope estimate as extreme as or more extreme than the one obtained if the null hypothesis of no linear relation was true. Because the P-value is less than the level of significance, $\alpha = 0.05$, we reject the null hypothesis of no linear relation.

Now Work Problems 13(d) and 13(e).

⑤ **Construct a Confidence Interval about the Slope of the Least-squares Regression Model**

We can also obtain confidence intervals for the slope of the least-squares regression line. The procedure is identical to that for obtaining confidence intervals about a mean. As was the case with confidence intervals about a population mean, the confidence interval for the slope of the least-squares regression line is of the form

$$\text{Point estimate} \pm \text{margin of error}$$

Definition

Confidence Intervals for the Slope of the Regression Line

A $(1 - \alpha) \cdot 100\%$ confidence interval for the slope of the true regression line, β_1, is given by the following formulas:

$$\text{Lower bound:} \quad b_1 - t_{\alpha/2} \cdot \frac{s_e}{\sqrt{\Sigma(x_i - \bar{x})^2}}$$

$$\text{Upper bound:} \quad b_1 + t_{\alpha/2} \cdot \frac{s_e}{\sqrt{\Sigma(x_i - \bar{x})^2}} \qquad (3)$$

Here, $t_{\alpha/2}$ is computed with $n - 2$ degrees of freedom.

Note: This interval can be computed only if the data are randomly obtained, the residuals are normally distributed, and there is constant error variance.

EXAMPLE 7   **Constructing a Confidence Interval about the Slope of the True Regression Line**

Problem: Compute a 95% confidence interval about the slope of the true regression line for the data presented in Table 1.

Approach

Step 1: Determine the least-squares regression line.

Step 2: Verify that the requirements for inference on the regression line are satisfied.

Step 3: Compute s_e.

Step 4: Determine the critical value $t_{\alpha/2}$ with $n - 2$ degrees of freedom.

Step 5: Compute the bounds on the $(1 - \alpha) \cdot 100\%$ confidence interval for β_1 using Formula (3).

Step 6: Interpret the result by stating, "We are 95% confident that β_1 is somewhere between *lower bound* and *upper bound*."

Solution

Step 1: The least-squares regression line was determined in Example 1 and is $\hat{y} = 151.3537 + 1.3991x$.

Step 2: The requirements were verified in Examples 1, 3, and 5.

Step 3: We computed s_e in Example 2 obtaining $s_e = 19.48$.

Step 4: Because we wish to determine a 95% confidence interval, we have $\alpha = 0.05$. Therefore, we need to find $t_{0.05/2} = t_{0.025}$ with $14 - 2 = 12$ degrees of freedom. Referring to Table V, we find that $t_{0.025} = 2.179$.

Step 5: We find the lower and upper bounds.

$$\text{Lower bound:} \quad b_1 - t_{\alpha/2} \cdot \frac{s_e}{\sqrt{\sum(x_i - \bar{x})^2}} = 1.3991 - 2.179 \cdot \frac{19.48}{\sqrt{2472.9284}}$$

$$= 1.3991 - 0.8536 = 0.5455$$

$$\text{Upper bound:} \quad b_1 + t_{\alpha/2} \cdot \frac{s_e}{\sqrt{\sum(x_i - \bar{x})^2}} = 1.3991 + 2.179 \cdot \frac{19.48}{\sqrt{2472.9284}}$$

$$= 1.3991 + 0.8536 = 2.2527$$

Step 6: We are 95% confident that the mean increase in cholesterol for each additional year of life is somewhere between 0.5455 and 2.2527.

CAUTION!

It is best that the explanatory variables be spread out when doing regression analysis.

In looking carefully at the formula for the standard deviation of b_1, we should notice that the larger the value of $\sum(x_i - \bar{x})^2$, the smaller the value of s_{b_1}. This result implies that whenever we are finding a least-squares regression line we should attempt to make the values of the explanatory variable, x, as evenly spread out as possible so that b_1, our estimate of β_1, is as precise as possible.

Now Work Problem 13(f).

Inference on the Linear Correlation Coefficient

Perhaps you are wondering why we have not presented hypothesis tests regarding the linear correlation coefficient. There are two basic reasons: (1) The hypothesis test on the slope and a hypothesis test on the linear correlation coefficient will yield the same conclusion, and (2) inferential methods on the linear correlation coefficient, ρ, require that the Y's at any given X be normally distributed and that the X's at any given Y be normally distributed. That is, testing a hypothesis such as $H_0: \rho = 0$ versus $H_1: \rho \neq 0$ requires that the two variables follow a **bivariate normal distribution** or be **jointly normally distributed**. Verifying this requirement is a difficult task. Although a normal probability plot of the x_i's and a separate normal probability plot of the y_i's generally means that the joint distribution is normal, it is not guaranteed. For these two reasons, we will be content in verifying the linearity of the data by performing inference on the slope coefficient only.

14.1 ASSESS YOUR UNDERSTANDING

Concepts and Vocabulary

1. State the requirements to perform inference on the least-squares regression model. How are these requirements verified?

2. Why is it important to perform graphical as well as analytical analysis when analyzing relations between two quantitative variables?

3. What do the y-coordinates on the least-squares regression line represent?

4. Why is it desirable to have the explanatory variables spread out to test a claim regarding β_1 or construct confidence intervals about β_1?

5. If $H_0: \beta_1 = 0$ is not rejected, what is the best guess for the value of the response variable for any value of the explanatory variable?

6. Why don't we conduct inference on the linear correlation coefficient?

Skill Building

In Problems 7–12, use the results of Problems 11–16, respectively, from Section 4.2 to answer the following questions:

(a) *What are the estimates of β_0 and β_1?*

(b) *Compute the standard error, the point estimate for σ.*

(c) *Assuming the residuals are normally distributed, determine s_{b_1}.*

(d) *Assuming the residuals are normally distributed, test $H_0: \beta_1 = 0$ versus $H_1: \beta_1 \neq 0$ at the $\alpha = 0.05$ level of significance.*

7.

x	3	4	5	7	8
y	4	6	7	12	14

8.

x	3	5	7	9	11
y	0	2	3	6	9

9.

x	-2	-1	0	1	2
y	-4		0	1	4	5

10.

x	-2	-1	0	1	2
y	7	6	3	2	0

11.

x	20	30	40	50	60
y	100	95	91	83	70

12.

x	5	10	15	20	25
y	2	4	7	11	18

Applying the Concepts

13. **Height versus Head Circumference** A pediatrician wants NW to determine the relation that may exist between a child's height and head circumference. She randomly selects 11 children from her practice, measures their height and head circumference, and obtains the following data:

Height (inches), x	Head Circumference (inches), y
27.75	17.5
24.5	17.1
25.5	17.1
26	17.3
25	16.9
27.75	17.6
26.5	17.3
27	17.5
26.75	17.3
26.75	17.5
27.5	17.5

Source: Denise Slucki, Student at Joliet Junior College

Use the results from Problem 17 in Section 4.2 to answer the following questions:

(a) Treating height as the explanatory variable, x, determine the estimates of β_0 and β_1.

(b) Compute the standard error of the estimate, s_e.

(c) Determine whether the residuals are normally distributed.

(d) If the residuals are normally distributed, determine s_{b_1}.

(e) If the residuals are normally distributed, test the claim that a linear relation exists between height and head circumference at the $\alpha = 0.01$ level of significance.

(f) If the residuals are normally distributed, construct a 95% confidence interval about the slope of the true least-squares regression line.

(g) A child comes in for a physical, and the nurse determines his height to be 26.5 inches. However, the child is being rather uncooperative, so the nurse is unable to measure the head circumference of the child. What would be a good guess as to this child's head circumference? Why is this a good guess?

14. Bone Length Research performed at NASA and led by Dr. Emily R. Morey-Holton measured the lengths of the right humerus and right tibia in 11 rats that were sent into space on Spacelab Life Sciences 2. The following data were collected:

Right Humerus (mm), x	Right Tibia (mm), y	Right Humerus (mm), x	Right Tibia (mm), y
24.80	36.05	25.90	37.38
24.59	35.57	26.11	37.96
24.59	35.57	26.63	37.46
24.29	34.58	26.31	37.75
23.81	34.20	26.84	38.50
24.87	34.73		

Source: NASA Life Sciences Data Archive

Use the results from Problem 20 in Section 4.2 to answer the following questions:

(a) Treating the length of the right humerus as the explanatory variable, x, determine the estimates of β_0 and β_1.

(b) Compute the standard error of the estimate.

(c) Determine whether the residuals are normally distributed.

(d) If the residuals are normally distributed, determine s_{b_1}.

(e) If the residuals are normally distributed, test the claim that a linear relation exists between the explanatory variable, x, and response variable, y, at the $\alpha = 0.01$ level of significance.

(f) If the residuals are normally distributed, construct a 99% confidence interval about the slope of the true least-squares regression line.

(g) What is the mean length of the right tibia on a rat whose right humerus is 25.93 mm?

15. Concrete As concrete cures, it gains strength. The following data represent the 7-day and 28-day strength (in pounds per square inch) of a certain type of concrete:

7-Day Strength, x	28-Day Strength, y	7-Day Strength, x	28-Day Strength, y
2300	4070	2480	4120
3390	5220	3380	5020
2430	4640	2660	4890
2890	4620	2620	4190
3330	4850	3340	4630

(a) Treating the 7-day strength as the explanatory variable, x, determine the estimates of β_0 and β_1.

(b) Compute the standard error of the estimate.

(c) Determine whether the residuals are normally distributed.

(d) If the residuals are normally distributed, determine s_{b_1}.

(e) If the residuals are normally distributed, test the claim that a linear relation exists between 7-day strength and 28-day strength at the $\alpha = 0.05$ level of significance.

(f) If the residuals are normally distributed, construct a 95% confidence interval about the slope of the true least-squares regression line.

(g) What is the estimated mean 28-day strength of this concrete if the 7-day strength is 3000 psi?

16. Tar and Nicotine Every year the Federal Trade Commission (FTC) must report tar and nicotine levels in cigarettes to Congress. The FTC obtains the tar and nicotine levels in over 1200 brands of cigarettes. A random sample from those reported to Congress is given in the following table:

Brand	Tar (mg), x	Nicotine (mg), y
Barclay 100	5	0.4
Benson and Hedges King	16	1.1
Camel Regular	24	1.7
Chesterfield King	24	1.4
Doral	8	0.5
Kent Golden Lights	9	0.8
Kool Menthol	9	0.8
Lucky Strike	24	1.5
Marlboro Gold	15	1.2
Newport Menthol	18	1.3
Salem Menthol	17	1.3
Virginia Slims Ultra Light	5	0.5
Winston Light	10	0.8

Source: Federal Trade Commission

(a) Treating the amount of tar as the explanatory variable, x, determine the estimates of β_0 and β_1.

(b) Compute the standard error of the estimate.

(c) Determine whether the residuals are normally distributed.

(d) If the residuals are normally distributed, determine s_{b_1}.

(e) If the residuals are normally distributed, test the claim that a linear relation exists between the explanatory variable, x, and response variable, y, at the $\alpha = 0.1$ level of significance.

(f) If the residuals are normally distributed, construct a 90% confidence interval about the slope of the true least-squares regression line.

(g) What is the mean amount of nicotine in a cigarette that has 12 mg of tar?

17. United Technologies versus the S&P 500 United Technologies is a conglomerate that includes companies such as Otis Elevators and Carrier Heating and Cooling. The ticker symbol of the company is UTX. The following data represent the rate of return of UTX stock for 11 months, compared with the rate of return of the Standard and Poor's Index of 500 stocks. Both are in percent.

Month	Rate of Return of S&P 500, x	Rate of Return in United Technologies, y
Aug-04	0.23	1.21
Sept-04	0.94	−0.57
Oct-04	1.40	−0.59
Nov-04	3.86	5.89
Dec-04	3.25	5.90
Jan-05	−2.53	−2.58
Feb-05	1.89	0.06
Mar-05	−1.91	1.78
Apr-05	−2.01	0.06
May-05	3.00	5.79
Jun-05	0.90	−1.57

Source: Yahoo! Finance

(a) Treating the rate of return of the S&P 500 as the explanatory variable, x, determine the estimates of β_0 and β_1.
(b) Compute the standard error of the estimate.
(c) Determine whether the residuals are normally distributed.
(d) If the residuals are normally distributed, determine s_{b_1}.
(e) If the residuals are normally distributed, test the claim that a linear relation exists between the explanatory variable, x, and response variable, y, at the $\alpha = 0.1$ level of significance.
(f) If the residuals are normally distributed, construct a 90% confidence interval about the slope of the true least-squares regression line.
(g) What is the mean rate of return for United Technologies stock if the rate of return of the S&P 500 is 4.2%?

18. **Fat-free Mass versus Energy Expenditure** In an effort to measure the dependence of energy expenditure on body build, researchers used underwater weighing techniques to determine the fat-free body mass in seven men. In addition, they measured the total 24-hour energy expenditure during inactivity. The results are as follows:

Fat-free Mass (kg), x	Energy Expenditure (Kcal), y
49.3	1894
59.3	2050
68.3	2353
48.1	1838
57.6	1948
78.1	2528
76.1	2568

Source: Webb, P. Energy expenditure and fat-free mass in men and women. *American Journal of Clinical Nutrition,* **34,** 1816–1826.

(a) What are the estimates of β_0 and β_1?
(b) Compute the standard error of the estimate.
(c) Determine whether the residuals are normally distributed.
(d) If the residuals are normally distributed, determine s_{b_1}.
(e) If the residuals are normally distributed, test the claim that a linear relation exists between the explanatory variable, x, and response variable, y, at the $\alpha = 0.01$ level of significance.
(f) If the residuals are normally distributed, construct a 99% confidence interval about the slope of the true least-squares regression line.
(g) What is the mean energy expenditure of a man if his fat-free mass is 57.3 kg?

19. **Calories versus Sugar** The following data represent the number of calories per serving and the number of grams of sugar per serving for a random sample of high-fiber cereals.

Calories, x	Sugar, y	Calories, x	Sugar, y
200	18	210	23
210	23	210	16
170	17	210	17
190	20	190	12
200	18	190	11
180	19	200	11

Source: Consumer Reports, October, 1999

(a) Draw a scatter diagram of the data, treating calories as the explanatory variable. What type of relation, if any, appears to exist between calories and sugar?
(b) Determine the least-squares regression equation from the sample data.
(c) Compute the standard error of the estimate.
(d) Determine whether the residuals are normally distributed.
(e) Determine s_{b_1}.
(f) If the residuals are normally distributed, test the claim that a linear relation exists between calories and sugar content at the $\alpha = 0.01$ level of significance.
(g) If the residuals are normally distributed, construct a 95% confidence interval about the slope of the true least-squares regression line.
(h) Suppose a high-fiber cereal is randomly selected. Would you recommend using the least-squares regression line obtained in part (b) to predict the sugar content of the cereal? Why? What would be a good guess as to the sugar content of the cereal?

20. **Age versus HDL Cholesterol** A doctor wanted to determine whether there is a relation between a male's age and his HDL (so-called good) cholesterol. He randomly selected 17 of his patients and determined their HDL cholesterol. He obtained the following data:

Age, x	HDL Cholesterol, y	Age, x	HDL Cholesterol, y
38	57	38	44
42	54	66	62
46	34	30	53
32	56	51	36
55	35	27	45
52	40	52	38
61	42	49	55
61	38	39	28
26	47		

Source: Data based on information obtained from the National Center for Health Statistics

(a) Draw a scatter diagram of the data, treating age as the explanatory variable. What type of relation, if any, appears to exist between age and HDL cholesterol?

(b) Determine the least-squares regression equation from the sample data.

(c) Plot the residuals against the explanatory variable, age. Does a linear model seem appropriate, on the basis of the residual plot? (*Hint:* See Section 4.3.)

(d) Are there any outliers or influential observations?

(e) Compute the standard error of the estimate.

(f) Determine whether the residuals are normally distributed.

(g) If the residuals are normally distributed, determine s_{b_1}.

(h) If the residuals are normally distributed, test the claim that a linear relation exists between age and HDL cholesterol levels at the $\alpha = 0.01$ level of significance.

(i) If the residuals are normally distributed, construct a 95% confidence interval about the slope of the true least-squares regression line.

(j) Suppose a 42-year-old male patient visits the doctor's office. Would you recommend using the least-squares regression line obtained in part (b) to predict the HDL cholesterol of this patient? Why? What would be a good guess as to the HDL cholesterol of this patient?

21. The U.S. Population The following data represent the population of the United States:

Year, x	Population, y	Year, x	Population, y
1900	76,212,168	1960	179,323,175
1910	92,228,496	1970	203,302,031
1920	106,021,537	1980	226,542,203
1930	123,202,624	1990	248,709,873
1940	132,164,569	2000	281,421,906
1950	151,325,798		

Source: U.S. Census Bureau

An ecologist is interested in finding an equation that describes the population of the United States over time.

(a) Determine the least-squares regression equation, treating year as the explanatory variable.

(b) A normal probability plot of the residuals indicates that the residuals are approximately normally distributed. Test the claim that there is a linear relation between year and population.

(c) Draw a scatter diagram, treating year as the explanatory variable.

(d) Plot the residuals against the explanatory variable, year.

(e) Does a linear model seem appropriate based on the scatter diagram and residual plot? (*Hint:* See Section 4.3.)

(f) What is the moral?

22. Kepler's Law of Planetary Motion The time it takes for a planet to complete its orbit around the sun is called the planet's *sidereal year*. Johann Kepler studied the relation between the sidereal year of a planet and its distance from the sun in 1618. The following data show the distances that the planets are from the sun and their sidereal years.

Planet	Distance from Sun, x (millions of miles)	Sidereal Year, y
Mercury	36	0.24
Venus	67	0.62
Earth	93	1.00
Mars	142	1.88
Jupiter	483	11.9
Saturn	887	29.5
Uranus	1785	84.0
Neptune	2797	165.0
Pluto	3675	248.0

(a) Determine the least-squares regression equation, treating distance from the sun as the explanatory variable.

(b) A normal probability plot of the residuals indicates that the residuals are approximately normally distributed. Test the claim that there is a linear relation between distance from the sun and sidereal year.

(c) Draw a scatter diagram, treating distance from the sun as the explanatory variable.

(d) Plot the residuals against the explanatory variable, distance from the sun.

(e) Does a linear model seem appropriate based on the scatter diagram and residual plot? (*Hint:* See Section 4.3.)

(f) What is the moral?

23. Influential Observations The following data represent the heights and weights of a random sample of professional baseball players.

Player	Height (inches)	Weight (pounds)
Alex Rodriguez	75	210
Derek Jeter	75	195
Greg Maddux	72	185
Randy Johnson	82	230
David Justice	75	200
Al Leiter	75	220
Barry Bonds	74	210
Ray Lankford	71	200
Jason Isringhausen	75	210

Source: Yahoo! Sports

(a) Draw a scatter diagram of the data, treating height as the explanatory variable and weight as the response variable.

(b) Determine the least-squares regression line. Test the claim that there is a linear relation between height and weight at the $\alpha = 0.05$ level of significance.

(c) Remove the values listed for Randy Johnson. Test the claim that there is a linear relation between height and weight. What effect does Randy Johnson have on the hypothesis test? Do you think that Randy Johnson is influential?

24. The output shown was obtained from MINITAB

```
The regression equation is
y = 12.4 + 1.40 x

Predictor      Coef    StDev       T       P
Constant     12.396    1.381    8.97   0.000
x            1.3962   0.1245   11.21   0.000

S = 2.167     R-Sq = 91.3%   R-Sq(adj) = 90.6%
```

(a) The least-squares regression equation is $\hat{y} = 12.396 + 1.3962x$. What is the predicted value of y at $x = 10$?

(b) What is the mean of y at $x = 10$?

(c) The standard error, s_e, is 2.167. What is the standard deviation of y at $x = 10$?

(d) If the requirements for inference on the least-squares regression model are satisfied, what is the distribution of y at $x = 10$?

Technology Step-by-Step | **Testing the Least-Squares Regression Model**

TI-83/84 Plus

Step 1: Enter the explanatory variable in L1 and the response variable in L2.

Step 2: Press STAT, highlight TESTS, and select E:LinRegTTest....

Step 3: Be sure that Xlist is L1 and Ylist is L2. Make sure that Freq: is set to 1. Select the direction of the alternative hypothesis. Place the cursor on Calculate and press ENTER.

MINITAB

Step 1: With the explanatory variable in C1 and the response variable in C2, select the **Stat** menu and highlight **Regression**. Highlight **Regression** . . . .

Step 2: Select the explanatory variable (MINITAB calls them predictors) and response variable and click OK.

Excel

Step 1: Make sure the Data Analysis Tool Pak is activated by selecting the **Tools** menu and highlighting **Add-Ins** . . . . Check the box for the Analysis ToolPak and click OK.

Step 2: Enter the explanatory variable in column A and the response variable in column B.

Step 3: Select the **Tools** menu and highlight **DataAnalysis** . . . .

Step 4: Select the **Regression** option.

Step 5: With the cursor in the Y-range cell, highlight the column that contains the response variable. With the cursor in the X-range cell, highlight the column that contains the explanatory variable. Click OK.

14.2 Confidence and Prediction Intervals

Preparing for This Section   Before getting started, review the following:

• Confidence intervals (Section 9.2, pp. 466–473)

Objectives

1 **Construct confidence intervals for a mean response**

2 **Construct prediction intervals for an individual response**

We know how to obtain the least-squares regression equation of best fit from data. We also know how to use the least-squares regression equation to obtain a predicted value. For example, the least-squares regression equation for the cholesterol data introduced in Example 1 from Section 14.1 is

$$\hat{y} = 151.3537 + 1.3991x$$

where $\hat{y}$ represents the predicted total cholesterol for a female whose age is x. The predicted value of total cholesterol for a given age x actually has two interpretations:

1. It represents the mean total cholesterol for all females whose age is x.

2. It represents the predicted total cholesterol for a randomly selected female whose age is x.

So, if we let $x = 42$ in the least-squares regression equation $\hat{y} = 151.3537 + 1.3991x$, we obtain $\hat{y} = 151.3537 + 1.3991(42) = 210.1$. We can interpret this result in one of two ways:

1. The mean total cholesterol for all females 42 years old is 210.1.

2. Our best guess as to the total cholesterol for a randomly selected 42-year-old female is 210.1.

Of course, there is a margin of error in making predictions, so we construct intervals about any predicted value to describe the accuracy of the prediction. The type of interval constructed will depend on whether we are predicting a mean total cholesterol for all 42-year-old females or the total cholesterol for an individual 42-year-old female. In other words, the margin of error is going to be different for predicting the mean total cholesterol for all females who are 42 years old versus the total cholesterol for one individual. Which prediction (the mean or the individual) do you think will be more accurate? It seems logical that the distribution of means should have less variability (and therefore a lower margin of error) than the distribution of individuals. After all, in the distribution of means, high total cholesterols can be offset by low total cholesterols.

Definitions

Confidence intervals for a mean response are intervals constructed about the predicted value of y, at a given level of x, that are used to measure the accuracy of the mean response of all the individuals in the population.

Prediction intervals for an individual response are intervals constructed about the predicted value of y that are used to measure the accuracy of a single individual's predicted value.

In Other Words
Confidence intervals are intervals for the mean of the population. Prediction intervals are intervals for an individual from the population.

If we use the least-squares regression equation to predict the mean total cholesterol for all 42-year-old females, we construct a confidence interval for a mean response. If we use the least-squares regression equation to predict the total cholesterol for a single 42-year-old female, we construct a prediction interval for an individual response.

Construct Confidence Intervals for a Mean Response

The structure of a confidence interval is the same as it was in Section 9.1. The interval is of the form

$$\text{Point estimate} \pm \text{margin of error}$$

The following formula can be used to construct a confidence interval about $\hat{y}$.

> **Confidence Interval for the Mean Response of y, $\hat{y}$**
>
> A $(1 - \alpha) \cdot 100\%$ confidence interval for $\hat{y}$, the mean response of y, is given by
>
> $$\text{Lower bound:} \quad \hat{y} - t_{\alpha/2} \cdot s_e \sqrt{\frac{1}{n} + \frac{(x^* - \overline{x})^2}{\sum (x_i - \overline{x})^2}}$$
>
> $$\text{Upper bound:} \quad \hat{y} + t_{\alpha/2} \cdot s_e \sqrt{\frac{1}{n} + \frac{(x^* - \overline{x})^2}{\sum (x_i - \overline{x})^2}} \qquad \textbf{(1)}$$
>
> where x^* is the given value of the explanatory variable, n is the number of observations, and $t_{\alpha/2}$ is the critical value with $n - 2$ degrees of freedom.

EXAMPLE 1 **Constructing a Confidence Interval for a Mean Response**

Problem: Construct a 95% confidence interval about the predicted mean total cholesterol of all 42-year-old females, using the data in Table 1.

Approach: We wish to determine the predicted mean total cholesterol at $x^* = 42$ using Formula (1) since our estimate is for the mean cholesterol of all 42-year-old females.

Solution: The least-squares regression equation is $\hat{y} = 151.3537 + 1.3991x$. To find the predicted mean total cholesterol of all 42-year-olds, let $x^* = 42$ in the regression equation and obtain $\hat{y} = 151.3537 + 1.3991(42) = 210.1$. From Example 2 in Section 14.1, we found that $s_e = 19.48$, and from Example 5 in Section 14.1, we found that $\sum (x_i - \overline{x})^2 = 2472.9284$ and $\overline{x} = 42.07143$. The critical t value, $t_{\alpha/2} = t_{0.025}$, with $n - 2 = 14 - 2 = 12$ degrees of freedom is 2.179. The 95% confidence interval about the predicted mean total cholesterol for all 42-year-old females is therefore

$$\text{Lower bound:} \quad \hat{y} - t_{\alpha/2} \cdot s_e \cdot \sqrt{\frac{1}{n} + \frac{(x^* - \overline{x})^2}{\sum (x_i - \overline{x})^2}} = 210.1 - 2.179 \cdot 19.48 \cdot \sqrt{\frac{1}{14} + \frac{(42 - 42.07143)^2}{2472.9284}}$$

$$= 198.8$$

$$\text{Upper bound:} \quad \hat{y} + t_{\alpha/2} \cdot s_e \cdot \sqrt{\frac{1}{n} + \frac{(x^* - \overline{x})^2}{\sum (x_i - \overline{x})^2}} = 210.1 + 2.179 \cdot 19.48 \cdot \sqrt{\frac{1}{14} + \frac{(42 - 42.07143)^2}{2472.9284}}$$

$$= 221.4$$

We are 95% confident that the mean total cholesterol of all 42-year-old females is between 198.8 and 221.4.

Now Work Problems 3(a) and 3(b).

In Other Words

Prediction intervals are wider than confidence intervals because it is tougher to guess the value of an individual than the mean of a population.

② Construct Prediction Intervals for an Individual Response

The procedure for obtaining a prediction interval for an individual response is identical to that for finding a confidence interval for a mean response. The only difference is the standard error. More variability is associated with individuals than with means. Therefore, the computation of the interval must account for this increased variability. Again, the form of the interval is

$$\text{Point estimate} \pm \text{margin of error}$$

The following formula can be used to construct a prediction interval about $\hat{y}$.

> **Prediction Interval for an Individual Response about $\hat{y}$**
>
> A $(1 - \alpha) \cdot 100\%$ prediction interval for $\hat{y}$, the individual response of y, is given by
>
> $$\text{Lower bound:} \quad \hat{y} - t_{\alpha/2} \cdot s_e \sqrt{1 + \frac{1}{n} + \frac{(x^* - \bar{x})^2}{\sum(x_i - \bar{x})^2}}$$
>
> $$\text{Upper bound:} \quad \hat{y} + t_{\alpha/2} \cdot s_e \sqrt{1 + \frac{1}{n} + \frac{(x^* - \bar{x})^2}{\sum(x_i - \bar{x})^2}} \qquad (2)$$
>
> where x^* is the given value of the explanatory variable, n is the number of observations, and $t_{\alpha/2}$ is the critical value with $n - 2$ degrees of freedom.

Notice the only differernce between Formula (1) and Formula (2) is the "1 + " under the radical in Formula (2).

EXAMPLE 2

Constructing a Prediction Interval for an Individual Response

Problem: Construct a 95% prediction interval about the predicted total cholesterol of a 42-year-old female.

Approach: We need to determine the predicted total cholesterol at $x^* = 42$ and use Formula (2) since our estimate is for a particular 42-year-old female.

Solution: The least-squares regression equation is $\hat{y} = 151.3537 + 1.3991x$. To find the predicted total cholesterol of a 42-year-old, let $x^* = 42$ in the regression equation and obtain $\hat{y} = 151.3537 + 1.3991(42) = 210.1$. From Example 2 in Section 14.1, we found that $s_e = 19.48$; from Example 5 in Section 14.1 we found that $\sum(x_i - \bar{x})^2 = 2472.9284$ and $\bar{x} = 42.07143$. We find $t_{\alpha/2} = t_{0.025}$ with $n - 2 = 14 - 2 = 12$ degrees of freedom to be 2.179.

The 95% confidence interval about the predicted total cholesterol for a 42-year-old female is

$$\text{Lower bound:} \quad \hat{y} - t_{\alpha/2} \cdot s_e \sqrt{1 + \frac{1}{n} + \frac{(x^* - \bar{x})^2}{\sum(x_i - \bar{x})^2}} = 210.1 - 2.179 \cdot 19.48 \cdot \sqrt{1 + \frac{1}{14} + \frac{(42 - 42.07143)^2}{2472.9284}} = 166.2$$

$$\text{Upper bound:} \quad \hat{y} + t_{\alpha/2} \cdot s_e \sqrt{1 + \frac{1}{n} + \frac{(x^* - \bar{x})^2}{\sum(x_i - \bar{x})^2}} = 210.1 + 2.179 \cdot 19.48 \cdot \sqrt{1 + \frac{1}{14} + \frac{(42 - 42.07143)^2}{2472.9284}} = 254.0$$

We are 95% confident that the total cholesterol of a randomly selected 42-year-old female is between 166.2 and 254.0.

Now Work Problems 3(c) and 3(d).

Notice that the interval about the individual (prediction interval for an individual response) is wider than the interval about the mean (confidence interval for a mean response). The reason for this should be clear: More variability is associated with individuals than with groups of individuals. That is, it is more difficult to predict a single 42-year-old female's total cholesterol than it is to predict the mean total cholesterol for all 42-year-old females.

EXAMPLE 3 Confidence and Prediction Intervals Using Technology

Problem: Construct a 95% confidence interval about the predicted total cholesterol of all 42-year-old females using statistical software. Construct a 95% prediction interval, about the predicted total cholesterol for a 42-year-old female using statistical software.

Approach: We will use MINITAB to obtain the intervals. The steps for obtaining confidence and prediction intervals using MINITAB and Excel are given in the Technology Step by Step on page 758.

USING TECHNOLOGY

The bounds for confidence and prediction intervals obtained using statistical software may differ from bounds computed by hand due to rounding error.

Result: Figure 12 shows the results obtained from MINITAB.

Figure 12

```
Predicted Values

       Fit    StDev Fit          95.0% CI              95.0% PI
    210.11         5.21    ( 198.77, 221.46)    ( 166.18, 254.05)
```

MAKING AN INFORMED DECISION

Buying a Car

You are still in the market to buy a car. As we all know, most cars lose value over time. Therefore, another item to consider when purchasing a car is its depreciation rate. The lower the depreciation rate, the less value the car loses each year. Using the same three cars that you used in the Chapter 3 and Chapter 4 Decisions, answer the following questions to determine the depreciation rate. Remember that, in Chapter 4 we found the exponential equation of best fit $y = ab^x$, where y is the price of the car and x is the age of the car. To do this, we linearized the data by computing the logarithm of the asking price, y, so that the least-squares regression equation was $\log y = \log a + (\log b)x$. We can rewrite this as $Y = A + Bx$, where $Y = \log y$, $A = \log a$ and $B = \log b$.

For each car that you are considering, answer the following:

1. What is the standard error of the estimate of the equation $Y = A + Bx$?
2. Are the residuals normally distributed?
3. If the residuals are normally distributed, test the claim that a linear relation exists between the explanatory variable, x, and the response variable, Y, at the $\alpha = 0.05$ level of significance.
4. What is the mean price of a 10-year-old car? Construct a 90% confidence interval about the mean price of a 10-year-old car.
5. Predict the price of a 7-year-old car.
6. Construct a 90% prediction interval about the price of a 7-year-old car.
7. Write a report that describes which car you would buy, given the information collected. Be sure to justify your conclusions.

14.2 ASSESS YOUR UNDERSTANDING

Concepts and Vocabulary

1. Explain the difference between a confidence interval and prediction interval.

2. Suppose a normal probability plot of residuals indicates that the requirement of normally distributed residuals is violated. Explain the circumstances under which confidence and prediction intervals could still be constructed.

Skill Building

In Problems 3–6, use the results of Problems 11–14 in Section 4.2 and Problems 7–10 in Section 14.1.

3. Using the sample data from Problem 7 in Section 14.1,
   NW (a) Predict the mean value of y if $x = 7$.
   (b) Construct a 95% confidence interval about the mean value of y if $x = 7$.
   (c) Predict the value of y if $x = 7$.
   (d) Construct a 95% prediction interval about the value of y if $x = 7$.
   (e) Explain the difference between the prediction in parts (a) and (c).

4. Using the sample data from Problem 8 in Section 14.1,
   (a) Predict the mean value of y if $x = 8$.
   (b) Construct a 95% confidence interval about the mean value of y if $x = 8$.
   (c) Predict the value of y if $x = 8$.

   (d) Construct a 95% prediction interval about the value of y if $x = 8$.
   (e) Explain the difference between the prediction in parts (a) and (c).

5. Using the sample data from Problem 9 in Section 14.1,
   (a) Predict the mean value of y if $x = 1.4$.
   (b) Construct a 95% confidence interval about the mean value of y if $x = 1.4$.
   (c) Predict the value of y if $x = 1.4$.
   (d) Construct a 95% prediction interval about the value of y if $x = 1.4$.

6. Using the sample data from Problem 10 in Section 14.1,
   (a) Predict the mean value of y if $x = 1.8$.
   (b) Construct a 90% confidence interval about the mean value of y if $x = 1.8$.
   (c) Predict the value of y if $x = 1.8$.
   (d) Construct a 90% prediction interval about the value of y if $x = 1.8$.

Applying the Concepts

7. **Height versus Head Circumference** Use the results of Problem 13 from Section 14.1 to answer the following questions:
   (a) Predict the mean head circumference of children who are 25.75 inches tall.
   (b) Construct a 95% confidence interval about the mean head circumference of children who are 25.75 inches tall.
   (c) Predict the head circumference of a randomly selected child who is 25.75 inches tall.
   (d) Construct a 95% prediction interval about the head circumference of a child who is 25.75 inches tall.
   (e) Explain the difference between the prediction in part (a) and the prediction in part (c).

8. **Bone Length** Use the results of Problem 14 in Section 14.1 to answer the following questions:
   (a) Predict the mean length of the right tibia of all rats whose right humerus is 25.83 mm.
   (b) Construct a 95% confidence interval about the mean length found in part (a).
   (c) Predict the length of the right tibia of a randomly selected rat whose right humerus is 25.83 mm.
   (d) Construct a 95% prediction interval about the length found in part (c).
   (e) Explain why the predicted lengths found in parts (a) and (c) are the same, yet the intervals constructed in parts (b) and (d) are different.

9. **Concrete** Use the results of Problem 15 from Section 14.1 to answer the following questions:
   (a) Predict the mean 28-day strength of concrete whose 7-day strength is 2550 psi.
   (b) Construct a 95% confidence interval about the mean 28-day strength of concrete whose 7-day strength is 2550 psi.
   (c) Predict the 28-day strength of a cylinder of concrete whose 7-day strength is 2550 psi.
   (d) Construct a 95% prediction interval about the 28-day strength of concrete whose 7-day strength is 2550 psi.
   (e) Explain the difference between the prediction in part (a) and the prediction in part (c).

10. **Tar and Nicotine** Use the results of Problem 16 in Section 14.1 to answer the following questions:
    (a) Predict the mean nicotine content of all cigarettes whose tar content is 12 mg.
    (b) Construct a 95% confidence interval about the tar content found in part (a).
    (c) Predict the nicotine content of a randomly selected cigarette whose tar content is 12 mg.
    (d) Construct a 95% prediction interval about the nicotine content found in part (c).
    (e) Explain why the predicted nicotine contents found in parts (a) and (c) are the same, yet the intervals constructed in parts (b) and (d) are different.

11. **United Technologies versus the S&P 500** Use the results of Problem 17 in Section 14.1 to answer the following questions:

   (a) What is the mean rate of return for United Technologies stock if the rate of return of the S&P 500 is 4.2%?

   (b) Construct a 90% confidence interval about the mean rate of return found in part (a).

   (c) Predict the rate of return on United Technologies stock if the rate of return on the S&P 500 for a randomly selected month is 4.2%.

   (d) Construct a 90% prediction interval about the rate of return found in part (c).

   (e) Explain why the predicted rates of return found in parts (a) and (c) are the same, yet the intervals constructed in parts (b) and (d) are different.

12. **Fat-free Mass versus Energy Expenditure** Use the results of Problem 18 in Section 14.1 to answer the following questions:

   (a) What is the mean energy expenditure for individuals whose fat-free mass is 57.3 kg?

   (b) Construct a 99% confidence interval about the mean energy expenditure found in part (a).

   (c) Predict the energy expenditure of a randomly selected individual whose fat-free mass is 57.3 kg.

   (d) Construct a 99% prediction interval about the energy expenditure found in part (c).

   (e) Explain why the predicted energy expenditures found in parts (a) and (c) are the same, yet the intervals constructed in parts (b) and (d) are different.

13. **Calories versus Sugar** Use the results of Problem 19 from Section 14.1 to answer the following:

   (a) Explain why it does not make sense to construct confidence or prediction intervals based on the least-squares regression equation.

   (b) Construct a 95% confidence interval for the mean sugar content of high-fiber cereals.

14. **Age versus HDL Cholesterol** Use the results of Problem 20 from Section 14.1 to answer the following:

   (a) Explain why it does not make sense to construct confidence or prediction intervals based on the least-squares regression equation.

   (b) Construct a 95% confidence interval for the mean HDL cholesterol of males.

Technology Step-by-Step | **Confidence and Prediction Intervals**

TI-83/84 Plus The TI-83/84 Plus does not compute confidence or prediction intervals.

MINITAB **Step 1:** With the predictor variable in C1 and the response variable in C2, select the **Stat** menu and highlight **Regression**. Highlight **Regression . . . .**

Step 2: Select the explanatory and response variables.

Step 3: Click the Options ... button.

Step 4: In the cell marked "Prediction intervals for new observations:", enter the value of x^*. Select a confidence level. Click OK twice.

Excel **Step 1:** Load the PhStat Add-in.

Step 2: Enter the values of the explanatory variable in column A and the corresponding values of the response variable in column B.

Step 3: Select the **PHStat** menu. Highlight **Regression**. Highlight **Simple Linear Regression**.

Step 4: With the cursor in the Y variable cell range, highlight the data in column B. With the cursor in the X variable cell range, highlight the data in column A. Select Confidence & Prediction Interval for X and enter the value of x^*. Choose a level of confidence. Click OK.

14.3 Multiple Regression

Preparing for This Section Before getting started, review the following:

- Correlation (Section 4.1, pp. 195–202)

- Least-Squares Regression (Section 4.2, pp. 212–221)

- Diagnostics on the Least-Squares Regression Line (Section 4.3, pp. 226–235)

Objectives

1. **Obtain the correlation matrix**

2. **Use technology to find a multiple regression equation**

3. **Interpret the coefficients of a multiple regression equation**

4 **Determine R^2 and adjusted R^2**
5 **Perform an *F*-test for lack of fit**
6 **Test individual regression coefficients for significance**
7 **Construct confidence and prediction intervals**
8 **Build a regression model**

1 Obtain the Correlation Matrix

The least-squares regression model $y_i = \beta_0 + \beta_1 x_i + \varepsilon_i$ introduced in Section 14.1 is a model for describing the linear relation between a single explanatory variable, x, and a response variable, y. However, it is often the case that more than one explanatory variable can be used to predict the value of a response variable. In these circumstances, we use a *multiple regression model*.

Definition

> A **multiple regression model** is given by
>
> $$y_i = \beta_0 + \beta_1 x_{1i} + \beta_2 x_{2i} + \cdots + \beta_k x_{ki} + \varepsilon_i \qquad (1)$$
>
> where
>
> > y_i is the value of the response variable for the ith individual,
> >
> > $\beta_0, \beta_1, \beta_2, \ldots, \beta_k$ are the parameters to be estimated based on sample data,
> >
> > x_{1i} is the ith observation for the first explanatory variable, x_{2i} is the ith observation for the second explanatory variable, and so on,
> >
> > ε_i is a random error term that is normally distributed with mean 0 and variance $\sigma_{\varepsilon_i}^2 = \sigma^2$.
> >
> > The error terms are independent, and $i = 1, 2, \ldots, n$, where n is the sample size.

If there are two explanatory variables, model (1) reduces to

$$y_i = \beta_0 + \beta_1 x_{1i} + \beta_2 x_{2i} + \varepsilon_i \qquad (2)$$

If there are three explanatory variables, model (1) reduces to

$$y_i = \beta_0 + \beta_1 x_{1i} + \beta_2 x_{2i} + \beta_3 x_{3i} + \varepsilon_i \qquad (3)$$

Because we typically do not have access to population data, the parameters, $\beta_0, \beta_1, \beta_2$, and so on, must be estimated using sample data. We will let b_0 represent the estimate of β_0, b_1 represent the estimate of β_1, and so on. Although formulas do exist for determining the values of $b_0, b_1, \ldots, b_k$, they are beyond the scope of this text. Therefore, we will be content to obtain estimates of the parameters using statistical software, such as MINITAB or Excel.

In obtaining a multiple regression model, we must concern ourselves with which variables to include in the model and which to exclude. One tool for determining which variables to include is a *correlation matrix*.

Definition

> A **correlation matrix** shows the linear correlation among all variables under consideration in a multiple regression model.

When choosing which explanatory variables to include in a multiple regression model, we choose the ones with the highest linear correlation with the response variable. However, some caution is in order. We must guard against including explanatory variables that are correlated among themselves.

Definition

When two explanatory variables have a high linear correlation between themselves, we say that **multicollinearity** exists between the explanatory variables.

!
CAUTION

If two explanatory variables in the regression model are highly correlated with each other, watch out for strange results in the regression output.

To help understand the idea of multicollinearity, consider the following scenario. Suppose someone is trying to predict sales of lemonade at their corner store. Certainly, one variable that might help to explain lemonade sales is the outside temperature. Another variable that might help explain lemonade sales is air-conditioning bills. If the researcher includes both explanatory variables in the model, he may get results that are a little strange because the two explanatory variables, temperature and air-conditioning bills, are highly correlated. As temperatures increase, so do air-conditioning bills. It would be silly to include both variables in the model because they are both doing the same job when it comes to explaining lemonade sales. Of course, identifying multicollinearity is not quite this obvious.

Multicollinearity does not necessarily cause problems for us, but it can. Some of the problems that result from multicollinearity include getting estimates of slope coefficients that are the opposite sign of what we would expect or obtaining estimates of slope coefficients that are not as large (or small) as we would expect.

There are a few factors, other than a high linear correlation between explanatory variables, that dictate whether multicollinearity is a problem, but the discussion is beyond the scope of this text. For our purposes, we will look at a correlation matrix to determine which, if any, explanatory variables have a high correlation between them. For the sake of having a general rule, we will say that a linear correlation between two explanatory variables less than -0.7 or greater than 0.7 may be cause for concern. If there are two explanatory variables in the regression model that are highly correlated, we will keep an eye out for unexpected results. We will have more to say about this at the end of the section.

EXAMPLE 1 | **Constructing a Correlation Matrix**

Problem: A family doctor wishes to further examine the variables that affect his female patients' total cholesterol. He randomly selects 14 of his female patients and asks them to determine their average daily consumption of saturated fat. He then measures their total cholesterol and obtains the data in Table 4. The data are based on results obtained from the National Center for Health Statistics.

	Table 4	
Age, x_1	**Saturated Fat (g), x_2**	**Total Cholesterol, y**
25	19	180
25	28	195
28	19	186
32	16	180
32	24	210
32	20	197
38	31	239
42	20	183
48	26	204
51	24	221
51	32	243
58	21	208
62	21	228
65	30	269

Approach: We enter the data into MINITAB and create the correlation matrix.

Solution: Figure 13 shows the correlation matrix from MINITAB.

Figure 13

Correlations: Age, Fat, Cholesterol

```
                   Age        Fat
Fat              0.324
Cholesterol      0.718      0.778

Cell Contents:  Pearson correlation
```

The linear correlation between total cholesterol and age is 0.718. The linear correlation between total cholesterol and average daily consumption of saturated fat is 0.778. Because the linear correlation between average daily consumption and age, the two explanatory variables, is only 0.324, we are not concerned with multicollinearity.

> **Now Work Problem 19(a).**

② Use Technology to Find a Multiple Regression Equation

When obtaining a multiple regression equation, we perform the same diagnostic tests that we performed for the least-squares regression model with one explanatory variable. We should always draw residual plots to verify that the model is appropriate. The residual plots that should be drawn are (1) a plot of residuals against fitted values, $\hat{y}$, to see if the linear model is appropriate and if the errors have constant variance, (2) a plot of residuals against each of the explanatory variables to make sure that the relation between the explanatory and response variables is linear, and (3) a boxplot of the residuals to check for outliers.

EXAMPLE 2   **Multiple Regression**

Problem: A family doctor wishes to further examine the variables that affect his female patients' total cholesterol. He randomly selects 14 of his female patients and asks them to disclose their average daily consumption of saturated fat. He then measures their total cholesterol and obtains the data in Table 4 presented in Example 1.

(a) Find the least-squares regression equation, $\hat{y} = b_0 + b_1 x_1 + b_2 x_2$, where x_1 represents the patient's age, x_2 represents the patient's daily consumption of saturated fat, and y represents the patient's total cholesterol.

(b) Draw residual plots and a boxplot of the residuals to assess the adequacy of the model.

Approach

(a) We will enter the data into MINITAB to obtain the least-squares regression equation.

(b) We use MINITAB to draw the residual plots and boxplot of the residuals.

Solution

(a) Figure 14 shows the output from MINITAB.

Figure 14

Regression Analysis: cholesterol versus age, fat

```
The regression equation is
cholesterol = 90.8 + 1.01 age + 3.24 fat

Predictor            Coef   SE Coef        T        P
Constant            90.84     15.99     5.68    0.000
Age                1.0142     0.2427     4.18    0.002
Fat                3.2443     0.6632     4.89    0.000

S = 11.42        R-Sq = 84.7%      R-Sq(adj) = 82.0%

Analysis of Variance

Source              DF        SS       MS        F        P
Regression           2    7960.3   3980.1    30.53    0.000
Residual Error      11    1434.1    130.4
Total               13    9394.4
```

The least-squares regression equation for these data is $\hat{y} = 90.84 + 1.0142x_1 + 3.2443x_2$.

(b) Figure 15(a) shows a plot of residuals against the predicted values, $\hat{y}$. To obtain this plot, we first find all the fitted values. For example, the predicted (or fitted) value for the first observation in Table 4 (age = 25, fat consumption = 19) is $\hat{y} = 90.84 + 1.0142(25) + 3.2443(19) = 177.8$. Therefore, residual = observed − predicted = $180 - 177.8 = 2.2$. We then plot the point (177.8, 2.2) on the graph. We repeat this for the remaining observations. Figure 15(b) shows a plot of residuals against the explanatory variable, age. Figure 15(c) shows a plot of residuals against the explanatory variable, saturated fat. Figure 15(d) shows a boxplot of the residuals. None of the residual plots show any discernible pattern, and the boxplot does not show any outliers. Therefore, the linear model is appropriate.

Figure 15

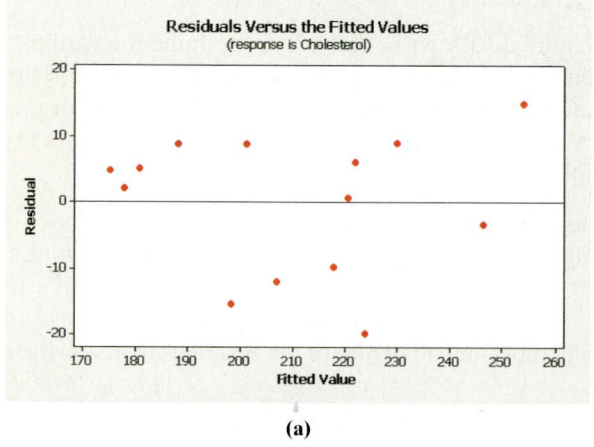

(a)

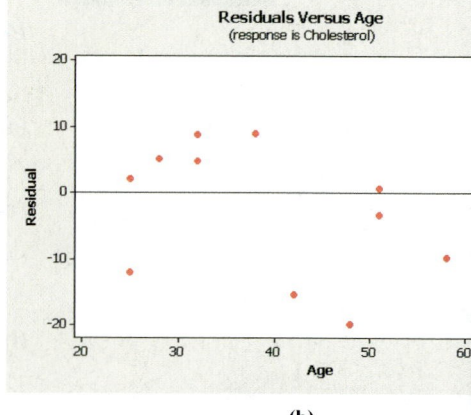

(b)

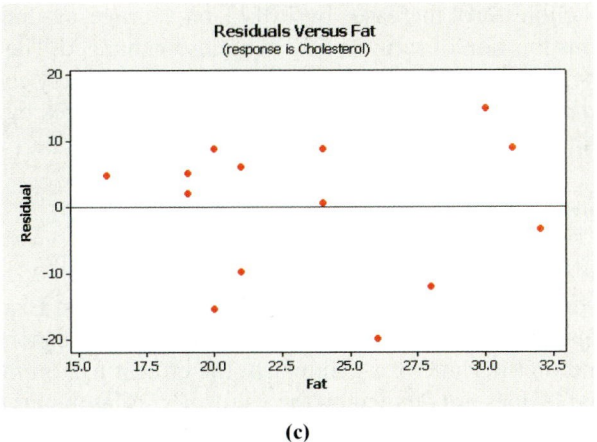

(c)

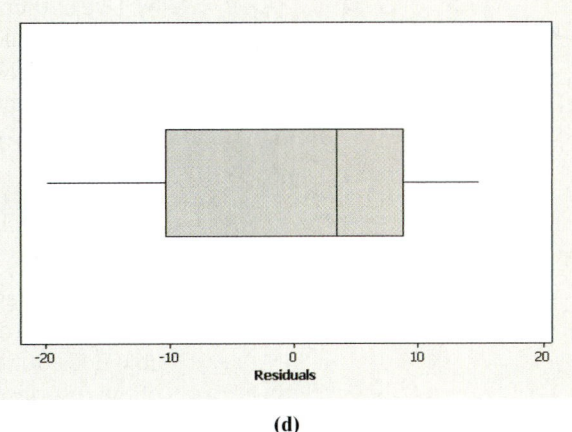

(d)

Now Work Problems 19(b) and 19(c).

3 Interpret the Coefficients of a Multiple Regression Equation

Recall, for the simple linear regression model, the slope, β_1, represented the *average* change in the value of the response variable for a 1-unit change in the value of the explanatory variable. For example, if income, y, and years of education, x, are related by the equation $y = 17{,}203 + 3423x$, then the slope, 3423, can be interpreted as follows: If the number of years of education increases by 1 year, income increases by an average of \$3423.

In the multiple regression model, we interpret the regression coefficients similarly. However, because there are other variables in the model, we assume that the remaining variables stay constant when interpreting the slope coefficient of a particular variable. If we have three explanatory variables, the multiple regression is $y_i = b_0 + b_1 x_{1i} + b_2 x_{2i} + b_3 x_{3i}$. So the interpretation of the regression coefficient b_1 is "If x_1 increases by 1-unit, then the response variable y would increase by b_1 units, on average, while x_2 and x_3 are held constant."

EXAMPLE 3 **Interpreting Regression Coefficients**

Problem: Interpret the regression coefficients for the least-squares regression equation found in Example 2.

Approach: We interpret the coefficients the same way as we did for a regression with one explanatory variable. However, we assume that the remaining explanatory variables are constant.

Solution: The least-squares regression equation found in Example 2 was $\hat{y} = 90.84 + 1.0142x_1 + 3.2443x_2$, where x_1 represents the patient's age and x_2 represents the patient's average daily consumption of saturated fat (in grams). The regression coefficient, $b_1 = 1.0142$, means that if the patient's age increases by 1 year then total cholesterol increases by 1.0142, on average, assuming that the average daily consumption of saturated fat remains unchanged. The regression coefficient, $b_2 = 3.2443$, means that if the patient's average daily consumption of saturated fat increases by 1 gram, total cholesterol increases by 3.2443, on average, assuming that age remains unchanged. _____

> **Now Work Problem 19(d).**

For the regression coefficients to be interpreted as described in Example 3, we have the requirement that the effect of x_1 on the value of the response variable does not depend on the value of x_2. Similarly, the effect of x_2 on the value of the response variable does not depend on the value of x_1. This means that the explanatory variables have an **additive effect** or **do not interact.** This is why it is important for the explanatory variables to have a low correlation. A positive correlation between two predictors, x_1 and x_2, implies that, if x_1 changes, then x_2 is likely to change as well. For example, if we are trying to describe the percentage of body fat in an individual we probably do not want our explanatory variables to be both skinfold thickness on the tricep and skinfold thickness on the thigh. Because they are highly correlated, if skinfold thickness on the tricep increases, so does skinfold thickness on the thigh.

④ Determine R^2 and Adjusted R^2

Recall from Section 4.3 that the coefficient of determination, R^2, measures the percentage of total variation in the response variable that is explained by the least-squares regression line. We also said that the computation of R^2 for the least-squares regression model with one explanatory variable was as easy as computing the square of the linear correlation coefficient, r, between the explanatory and response variable. That is, $R^2 = r^2$. While the interpretation of R^2 is the same for the least-squares regression model with two or more explanatory variables, its computation is by no means that straightforward. Once again, we will be content in using software to supply the value of R^2. Although we do not provide detail as to the computation of R^2, it is important that we recognize that R^2 is the ratio of variation in the response variable that is explained by the regression equation to the total variation in the response variable. That is,

$$R^2 = \frac{\text{explained variation}}{\text{total variation}} = 1 - \frac{\text{unexplained variation}}{\text{total variation}}$$

In Other Words

The value of R^2 always increases by adding one more explanatory variable.

Remember, unexplained variation is found by summing the squares of the residuals. The sum of the squared residuals will never increase if more explanatory variables are added. The total variation is constant regardless of the number of explanatory variables. Therefore, coefficient of determination, R^2, will never decrease with the addition of more explanatory variables. So, R^2 can often be made as large as we like by adding more explanatory variables. To compensate for the ability to artificially increase R^2 by adding more explanatory variable, it is recommended that the *adjusted* R^2 be used when working with least-squares regression models with two or more explanatory variables.

Definition

The **adjusted R^2,** denoted R_{adj}^2 or $\overline{R}^2$, is the adjusted coefficient of determination. It modifies the value of R^2 based on the sample size, n, and the number of explanatory variables, k. The formula for the adjusted R^2 is

$$R_{adj}^2 = 1 - \left(\frac{n-1}{n-k-1}\right)(1 - R^2)$$

The adjusted R^2 will decrease if an explanatory variable is added to the model that does little to explain the variation in the response variable. This is because the decrease in the unexplained variation is offset by the increase in the number of explanatory variables k.

EXAMPLE 4   Coefficient of Determination

CAUTION
Never use R^2 to compare regression models with a different number of explanatory variables. Rather, adjusted R^2 should be used.

Problem: For the model obtained in Example 2, determine the coefficient of determination and the adjusted R^2. Compare the R^2 with the two explanatory variables age and daily saturated fat to the R^2 with the single explanatory variable, age. Comment on the effect the additional explanatory variable has on the value of the model.

Approach: We will use MINITAB output to determine the values of the coefficient of determination and the adjusted coefficient of determination.

Solution: We can see from the output in Figure 14 that $R^2 = 0.847 = 84.7\%$. This means that 84.7% of the variation in the response variable is explained by the least-squares regression model. In addition, we find that $R^2_{\text{adj}} = 0.820 = 82.0\%$. If we look back to Figure 2 on page 737, we see that $R^2 = 51.5\%$ with age as the only explanatory variable. Based on the adjusted R^2 with two explanatory variables, age and daily saturated fat consumption, we conclude that the additional variable increases the proportion of explained variation.

Now Work Problem 19(e).

---

⑤ Perform an F-test for Lack of Fit

We now wish to determine whether there is a significant linear relation between the explanatory variables and response variables. That is, we want to test

$$H_0: \beta_1 = \beta_2 = \cdots = \beta_k = 0 \quad \text{versus} \quad H_1: \text{at least one } \beta_i \neq 0$$

In Other Words
The null hypothesis states that there is no linear relation between the explanatory variables and the response variable. The alternative hypothesis states that there is a linear relation between at least one explanatory variable and the response variable.

Just as in one-way analysis of variance, the computations in multiple regression can be cumbersome. Therefore, we are content to let statistical software perform the computations.

We use ANOVA to test if there is a significant linear relation between the explanatory variables and response variables. Recall from Section 13.1 that $F = \dfrac{\text{MST}}{\text{MSE}}$ follows Fisher's F-distribution, where MST is the mean square due to treatment and MSE is the mean square error. In the context of regression analysis, the treatment is the regression line itself and the residuals make up the error, which leads to the F-test statistic.

Test Statistic for Multiple Regression

$$F = \frac{\text{Mean Square Due to Regression}}{\text{Mean Square Error}} = \frac{\text{MSR}}{\text{MSE}}$$

with $k - 1$ degrees of freedom in the numerator and $n - k$ degrees of freedom in the denominator, where k is the number of explanatory variables and n is the sample size.

Because the coefficient of determination, R^2, represents the proportion of variation in the response variable and $1 - R^2$ represents the proportion of unexplained variation, we can express the F-test statistic using the coefficient of determination.

F-Test Statistic for Multiple Regression Using R^2

$$F = \frac{R^2}{1 - R^2} \cdot \frac{n - (k + 1)}{k}$$

where R^2 is the coefficient of determination

$\qquad$ k is the number of explanatory variables

$\qquad$ n is the sample size

A large F-test statistic provides evidence against the null hypothesis that all the slope coefficients are zero, because a large F-test statistic implies that the ratio of explained variation to unexplained variation is large. Care must be taken in using this as the sole judge in rejecting the null hypothesis, because a large number of explanatory variables (that is, a large value of k) will cause the ratio $\frac{n - (k + 1)}{k}$ to decrease and therefore cause the F-test statistic to decrease.

$\qquad$ Rather than comparing the F-test statistic to a critical value, we continue to use the P-value approach to testing hypotheses. We will use the P-value reported in the output of the statistical software package.

Decision Rule for Testing H_0: $\beta_1 = \beta_2 = \cdots = \beta_k = 0$

If the P-value is less than the level of signifance, α, reject the null hypothesis. Otherwise, do not reject the null hypothesis.

EXAMPLE 5 $\quad$ **Inference on the Regression Model**

Problem: Test H_0: $\beta_1 = \beta_2 = 0$ versus H_1: at least one $\beta_i \neq 0$ for the regression model found in Example 1.

Approach: To perform this inference, we need to determine whether it is reasonable to believe that the residuals are normally distributed with no outliers, because this is a requirement of the multiple regression model. To verify this requirement, we draw a normal probability plot. Once this requirement has been verified, we look at the P-value associated with the F-test statistic. If the P-value is less than the level of significance, α, we reject the null hypothesis.

Solution: Figure 16 shows the normal probability plot.

Figure 16

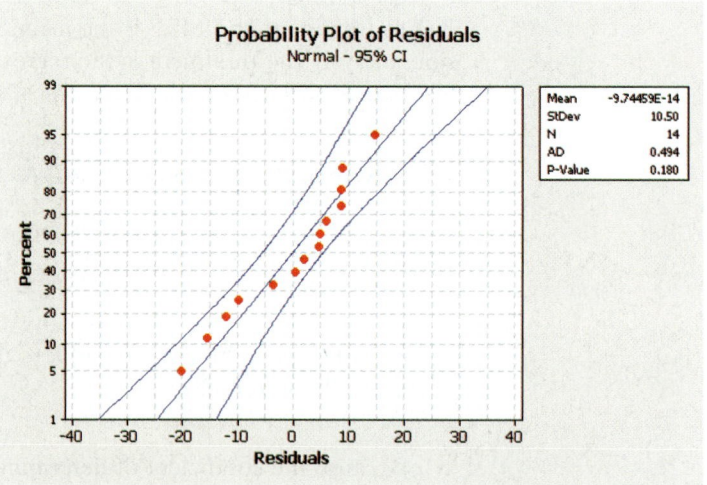

Because the points are roughly linear (or lie within the confidence bounds drawn by MINITAB), it is reasonable to conclude that the residuals are normally distributed. In addition, the boxplot of the residuals drawn in Figure 15(d) indicates that there are no outliers. Therefore, we can continue with the inference.

If we look back at Figure 14, we see that the F-test statistic is 30.53 with a P-value reported as 0.000. This small P-value is sufficient evidence against the null hypothesis. So we reject the null hypothesis and conclude that at least one of the regression coefficients is different from zero. There is a linear relation between at least one of the explanatory variables and the response variable. ▬

Now Work Problem 19(f).

> !
> **CAUTION**
> If we reject the null hypothesis that all the slope coefficients are zero, then we are saying that at least one of the slopes is different from zero, *not* that they *all* are different from zero.

By rejecting the null hypothesis in Example 5, we know *at least* one explanatory variable has a coefficient that is different from zero. Of course, this conclusion leads to this question: Which explanatory variables have a significant linear relation with the response variable?

⑥ Test Individual Regression Coefficients for Significance

We can use the t-distribution, along with P-values, to determine whether a specific explanatory variable has a significant linear relation with the response variable. Again, if the P-value is sufficiently small (say, less than $\alpha = 0.05$), we conclude that the explanatory variable has a significant linear relation with the response variable.

EXAMPLE 6 **Testing the Significance of Individual Predictor Variables**

Problem: For our model presented in Example 2, test the hypotheses

$$H_0: \beta_1 = 0 \qquad H_0: \beta_2 = 0$$

$$\text{vs.} \qquad \text{vs.}$$

$$H_1: \beta_1 \neq 0 \qquad H_1: \beta_2 \neq 0$$

where β_1 is the slope coefficient for the explanatory variable, age, and β_2 is the slope coefficient for the explanatory variable, saturated fat.

Approach: We will use MINITAB output to determine the P-values for each explanatory variable. If the P-value is less than the level of significance, we reject the null hypothesis and conclude the slope coefficient is different from zero.

Solution: If we look back at Figure 14, we see that the test statistic is $t = 4.18$ with a P-value $= 0.002$ for the explanatory variable, age. We also see that the test statistic is $t = 4.89$ with a P-value $= 0.000$ for the explanatory variable, saturated fat. Since both P-values are sufficiently small, we reject each null hypothesis. We conclude that both explanatory variables have a significant linear relation with the response variable, total cholesterol. ▬

Now Work Problem 19(g).

If the P-value for a slope coefficient is large, we should consider removing the explanatory variable from the model. If the P-value is small, we should consider keeping the explanatory variable in the model.

⑦ Construct Confidence and Prediction Intervals

Now that we know that both explanatory variables, age and saturated fat, have a significant linear relation with the response variable, total cholesterol, we can use the model to make predictions. Of course, because our model is based on sample data, we know that variability will be associated with any predictions. Therefore, we will construct intervals about our predicted value and describe the accuracy of our predictions.

Recall, from Section 14.2, that there are two interpretations to a predicted value:

1. It represents a mean value of the response variable.
2. It represents a predicted value for a specific individual.

Because of this, we require two types of intervals. One type, the confidence interval for a mean response, corresponds to an interval about the mean value; the other type, the prediction interval for an individual response, corresponds to an interval for a specific individual. Once again, we will leave the heavy work to MINITAB and let it construct the intervals. Our job will be to interpret the results.

EXAMPLE 7 Confidence and Prediction Intervals

Problem: Construct a 95% confidence interval for a mean response and a 95% prediction interval for an individual response for a 32-year-old female who consumes 23 grams of saturated fat daily using the model determined in Example 2. Be sure to interpret the results.

Approach: We will use MINITAB to construct the intervals.

Solution: Figure 17 shows the partial output from MINITAB.

Figure 17 **Predicted Values for New Observations**

```
New Obs       Fit   SE Fit         95.0% CI          95.0% PI
1           197.91   3.85   (189.45, 206.38)   (171.40, 224.43)

Values of Predictors for New Observations

New Obs      age    fat
1           32.0   23.0
```

Based on the MINITAB output, we are 95% confident that the mean total cholesterol of all 32-year-old females who consume 23 grams of saturated fat daily is between 189.5 and 206.4. We are 95% confident that the total cholesterol of a particular 32-year-old female who consumes 23 grams of saturated fat daily is between 171.4 and 224.4.

Now Work Problems 19(h), (i), and (j).

Multicollinearity Revisted

Suppose two explanatory variables that have a strong linear association with the response variable are highly correlated with each other. One consequece of this correlation can be that their t-test statistics are small and their P-values are large. The reason for this is that the t-test statistics are *marginal*. The word marginal means incremental. For example, if we had a regression model with three explanatory variables, X_1, X_2, and X_3, the t-statistic for X_1 is a measure for how much additional explanation the variable X_1 adds to the model when X_2 and X_3 are already present. Likewise, the t-statistic for X_2 is a measure for how much additional explanation the variable X_2 adds to the model when X_1 and X_3 are already present.

So a small t-test statistic for an explanatory variable means that a variable does not yield much additional explanation. Suppose that X_1 and X_2 are highly correlated to each other and to the response variable. If we include both explanatory variables in the model, the high correlation may lead to small t-test statistics for X_1 and X_2. This leads us to believe that neither variable is important.

In fact, the model may be "confused." This confusion results because the model believes that not much additional information is learned by adding X_1 to the model when X_2 is already in the model. In addition, X_2 does not provide much additional information when X_1 is already in the model. The model concludes that neither variable is important.

EXAMPLE 8 Effects of Multicollinearity

Table 5

X_1	X_2	y
11.4	−9.7	14.7
12.5	−11.5	38.8
16.4	−15.9	42.9
14.4	−13.9	45.7
15.3	−14.2	52.3
18	−18.5	55.9
19.5	−21.2	60.1
25.2	−27.2	72.6

Problem: Using the data presented in Table 5:

(a) Find the correlation matrix among all three variables.

(b) Find the least-squares regression model using both X_1 and X_2 as explanatory variables.

(c) Comment on the impact that including both X_1 and X_2 has on the t-test statistics.

Approach: We will use MINITAB to conduct the analysis.

Solution:

(a) Figure 18 shows the correlation matrix. There is an extremely high correlation between X_1 and X_2 so multicollinearity exists between the two variables.

(b) Figure 19 shows the regression output from MINITAB. First, we notice that the P-value for the F-test statistic is 0.018, indicating that at least one of the slope coefficients is different from zero. However, if we look at each individual t-test statistic, we see that each has a very high P-value, indicating that neither is different from zero.

Figure 18

Correlations (Pearson)

```
          x1         x2
x2    -0.996
y      0.891    -0.894
```

Figure 19

Regression Analysis

```
The regression equation is
y = 3.2 - 0.02 x1 - 2.72 x2

Predictor        Coef       StDev         T          P
Constant         3.20       35.47      0.09      0.932
x1             -0.015        8.829     -0.00      0.999
x2             -2.721        6.844     -0.40      0.707

S = 9.070        R-Sq = 80.0%      R-Sq(adj) = 72.0%

Analysis of Variance

Source           DF         SS         MS         F          P
Regression        2    1645.85     822.93     10.00      0.018
Residual Error    5     411.32      82.26
Total             7    2057.17
```

(c) The contradictory results of the regression output result because both X_1 and X_2 are related to the response variable, y, as indicated by the correlation matrix. However, X_1 and X_2 are also related to each other. So, with X_1 in the model, X_2 adds little explanation. Likewise, with X_2 in the model, X_1 adds little explanation. The solution is to use only one explanatory variable. Which explanatory variable we choose is up us. We can choose either the explanatory variable with the lower P-value or the explanatory variable that has the higher correlation with the response.

Now Work Problem 23.

8 Build a Regression Model

Now that we have all the skills needed to conduct inference on the multiple regression model, we can discuss an approach to determining the "best" model. We put the word best in quotes because there is really no best model. It is up to us, the researcher, to decide which model is best. Of course, we will have to justify the results presented. Nonetheless, we should follow some guidelines when developing a model.

Guidelines in Developing a Multiple Regression Model

1. Construct a correlation matrix to help identify the explanatory variables that have a high correlation with the response variable. In addition, look at the correlation matrix for any indication that the explanatory variables are correlated with each other. Remember, just because two explanatory variables have high correlation does not mean that multicollinearity is a problem, but it is a tip-off to watch out for multicollinearity.

2. If the multiple regression model using all the explanatory variables that have been identified by the researcher.

3. If the null hypothesis that all the slope coefficients are zero has been rejected, we proceed to look at the individual slope coefficients. Identify those slope coefficients that have small t-test statistics (and therefore large P-values). These are candidates for explanatory variables that may be removed from the model. We should only remove one explanatory variable at a time from the model before recomputing the regression model.

4. Repeat Step 3 until all slope coefficients are significantly different from zero.

5. Be sure that the model is appropriate by drawing residual plots.

This procdure is known as *backwards stepwise regression*.

EXAMPLE 9 **Building a Multiple Regression Model**

Problem: An engineer wants to develop a model to describe the gas mileage of an automobile. He collects the data presented in Table 6. Compression ratio is a measure of how much the air–fuel mixture is compressed before the spark plug ignites it (which turns the engine). A higher compression ratio means a more powerful engine. The coefficient of drag is a measure of how aerodynamic the car is. The higher coefficient of drag means that the car is less aerodynamic. Find the best model to describe miles per gallon.

Table 6

Car	Weight (lb)	Engine Size (cc)	Horsepower	Compression Ratio ($x:1$)	Coefficient of Drag	Miles per Gallon
Honda Accord	3470	2997	240	10.0	0.29	21
Lincoln LS	3674	2967	232	10.5	0.32	20
Nissan Altima	3460	3498	245	10.3	0.32	21
Toyota Camry	3520	2988	192	10.5	0.28	19
Volkswagen Passat	3610	2773	190	10.6	0.27	20
Lexus ES 300	3439	2995	210	10.5	0.28	21
Saab 93	3285	1988	210	9.5	0.28	23
Volvo S60	3276	2435	168	10.3	0.28	22
Ford Taurus	3343	2982	155	9.5	0.30	20
Infiniti G35	3336	3498	260	10.3	0.27	19
Saturn L Series	3033	2198	135	9.5	0.31	24
BMW 325	3560	2494	184	10.5	0.35	19
Chevrolet Impala	3389	3350	180	9.5	0.31	21
Cadillac CTS	3509	3175	220	10.0	0.31	18
Oldsmobile Alero	3085	3350	170	9.5	0.32	20

Source: vehix.com

Approach: We use MINITAB to obtain output and Steps 1–5 to guide us in choosing the best model.

Solution

Step 1: Figure 20 shows the correlation matrix.

Figure 20   **Correlations: weight, engine, horse, comp, drag, mpg**

	weight	engine	horse	comp	drag
engine	0.212				
horse	0.516	0.469			
comp	0.726	0.128	0.449		
drag	0.038	0.075	-0.174	-0.147	
mpg	-0.581	-0.600	-0.363	-0.442	-0.144

We notice the high correlation between weight and compression ratio (0.726). This may lead to strange results because of multicollinearity.

Step 2: Figure 21 shows the regression model with all explanatory variables included.

Figure 21   **Regression Analysis: MPG versus weight, engine, ...**

```
The regression equation is
MPG = 44.7 - 0.00409 weight - 0.00198 engine + 0.0083 horsepower
           - 0.46 compression - 4.9 drag
```

Predictor	Coef	SE Coef	T	P
Constant	44.664	9.441	4.73	0.001
weight	-0.004086	0.002945	-1.39	0.199
engine	-0.0019765	0.0008396	-2.35	0.043
horsepow	0.00832	0.01265	0.66	0.527
compress	-0.464	1.123	-0.41	0.689
drag	-4.91	15.45	-0.32	0.758

```
S = 1.248      R-Sq = 60.8%    R-Sq(adj) = 39.0%
```

Analysis of Variance

Source	DF	SS	MS	F	P
Regression	5	21.719	4.344	2.79	0.086
Residual Error	9	14.014	1.557		
Total	14	35.733			

Step 3: The *P*-value for the *F*-test statistic indicates that the model is reasonable (at least one slope coefficient is different from zero). However, looking at the individual *t*-test statistics, we see that some have *P*-values that are very large. We will remove the variable drag from the model because it has the largest *P*-value.

Figure 22 shows the new regression model.

Figure 22   **Regression Analysis: MPG versus weight, engine, horsepower, compression**

```
The regression equation is
MPG = 43.2 - 0.00433 weight - 0.00202 engine + 0.0093 horsepower
           - 0.39 compression
```

Predictor	Coef	SE Coef	T	P
Constant	43.173	7.815	5.52	0.000
weight	-0.004332	0.002711	-1.60	0.141
engine	-0.0020218	0.0007893	-2.56	0.028
horsepow	0.00935	0.01167	0.80	0.442
compress	-0.386	1.045	-0.37	0.719

```
S = 1.190      R-Sq = 60.3%    R-Sq(adj) = 44.5%
```

Analysis of Variance

Source	DF	SS	MS	F	P
Regression	4	21.562	5.391	3.80	0.039
Residual Error	10	14.171	1.417		
Total	14	35.733			

Step 4: The *P*-value for the *F*-test decreased to 0.039, but we still have some explanatory variables whose *P*-values are large. We will remove the explanatory variable compression from the model. We notice also that the adjusted R^2 increased from 39.0% to 44.5% when we removed drag from the model. This shows that the adjusted R^2 punishes a model that is overly complex.

Figure 23 shows the new regression model.

Figure 23   **Regression Analysis: MPG versus weight, engine, horsepower**

```
The regression equation is
MPG = 41.5 - 0.00498 weight - 0.00199 engine + 0.0087 horsepower

Predictor          Coef     SE Coef          T          P
Constant         41.530       6.170       6.73      0.000
weight         -0.004979    0.001988      -2.50      0.029
engine        -0.0019918   0.0007536      -2.64      0.023
horsepow         0.00867     0.01106       0.78      0.450

S = 1.143        R-Sq = 59.8%     R-Sq(adj) = 48.8%

Analysis of Variance

Source             DF          SS         MS          F          P
Regression          3      21.369      7.123       5.45      0.015
Residual Error     11      14.365      1.306
Total              14      35.733
```

Step 4: The *P*-value for the *F*-test statistic continues to decline as we eliminate variables! We notice that the slope coefficient for horsepower is positive. This implies that cars with more horsepower get better gas mileage. Of course, this contradicts common sense. Because the *P*-value for the explanatory variable horsepower is large, we remove it from the regression model. Again, adjusted R^2 increases. Also, we see that the *P*-value for weight dropped after we removed compress from the model. This is due to the fact that weight and compression ratio are highly correlated. The model was confused with both variables in the model.

Step 4 (again): Figure 24 shows the new regression model.

Figure 24   **Regression Analysis: MPG versus weight, engine**

```
The regression equation is
MPG = 40.0 - 0.00423 weight - 0.00174 engine

Predictor          Coef     SE Coef          T          P
Constant         39.964       5.743       6.96      0.000
weight         -0.004227    0.001713      -2.47      0.030
engine        -0.0017380   0.0006694      -2.60      0.023

S = 1.124        R-Sq = 57.6%     R-Sq(adj) = 50.5%

Analysis of Variance

Source             DF          SS         MS          F          P
Regression          2      20.567     10.283       8.14      0.006
Residual Error     12      15.166      1.264
Total              14      35.733
```

Now the *P*-value for the *F*-test is low (0.006), and the *P*-values for the individual slope coefficients are low. In addition, the estimates of the slope coefficients are reasonable. We expect both to be negative because (1) heavier cars should get worse gas mileage and (2) cars with bigger engines should get worse gas mileage. We believe that this is the "best" model.

Step 5: Figure 25 shows the residual plots as well as the normal probability plot of the residuals.

Figure 25

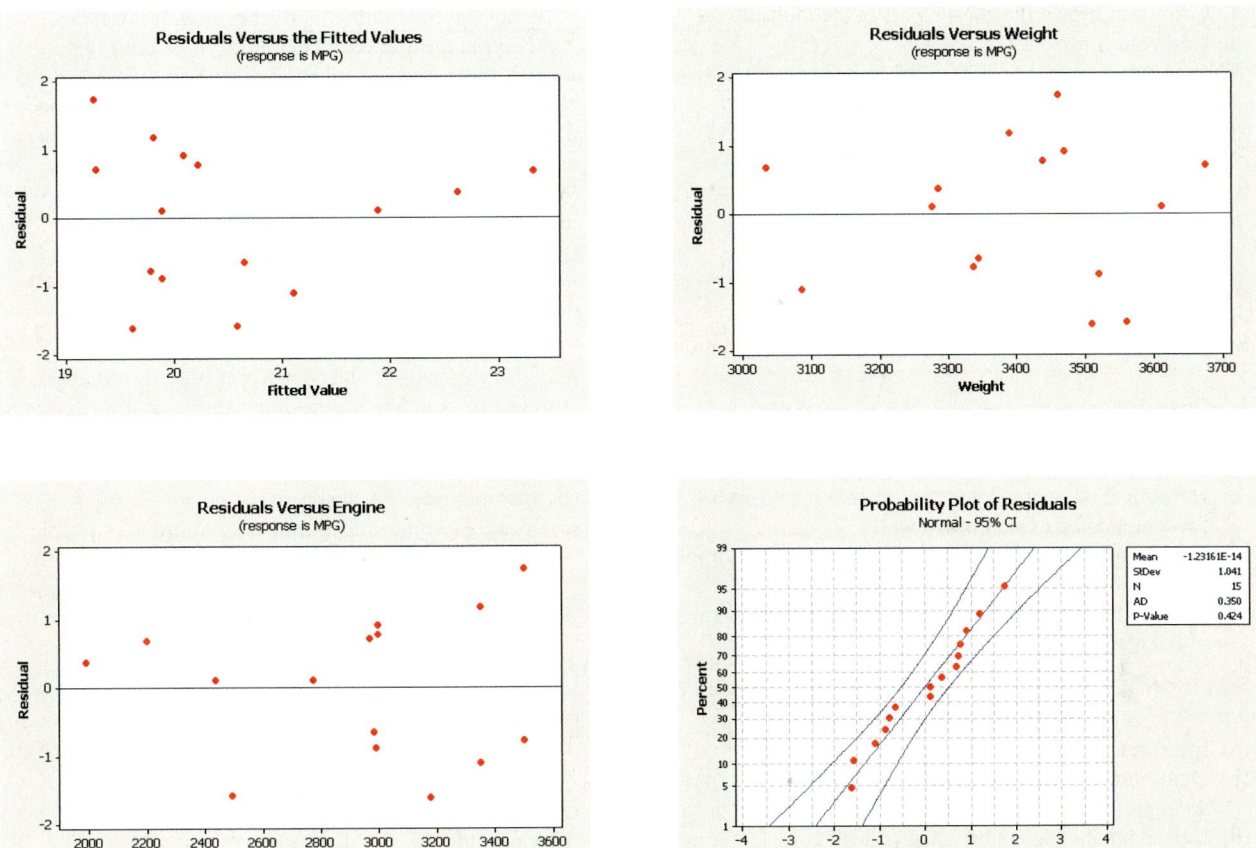

The residual plots do not indicate that there are any problems in the model, and the normal probability plot indicates that the residuals appear to be normally distributed, so we accept the model.

In-Class Activity: College Success

What variables affect student success in college, as determined by college GPA? It is widely accepted that no one single explanatory variable can be used to predict college success. However, several exlanatory variables taken together may yield better results.

(a) As a class, select five variables that might be related to college GPA and for which data can be easily obtained (such as ACT scores or high school GPA).

(b) Write your particular values for the selected variables on a piece of paper and submit the values to your instructor anonymously.

(c) Obtain the aggregate data for the class from your instructor.

(d) Using the data from Step 2, determine a model of best fit to predict college GPA.

(e) Discuss the selected model as a class. Were any of the selected variables removed from the model? Why might this be the case?

(f) How confident are you in the model's ability to predict college GPA? Why?

(g) Use the selected model to predict the college GPA for a student who is not in your class. How well did your model do? Is this surprising?

(h) Check the necessary requirements for your model. If any are not satisfied, how will that affect your previous answers?

14.3 ASSESS YOUR UNDERSTANDING

Concepts and Vocabulary

1. What is a correlation matrix?
2. How do we interpret the coefficients in the multiple regression equation?
3. When testing whether or not there is a linear relation between the response variable and the explanatory variables, we use an F-test. If the P-value indicates that we reject the null hypothesis, $H_0: \beta_1 = \beta_2 = \cdots = \beta_k = 0$, what conclusion should we come to? Is it possible that one of the β_i is zero if we reject the null hypothesis?

4. What does it mean when we say that the explanatory variables have an additive effect or do not interact?
5. Explain the difference between the coefficient of determination, R^2, and the adjusted coefficient of determination, R^2_{adj}. Which is better for determining whether an additional explanatory variable should be added to the regression model?
6. What is multicollinearity? How can we check for it? What are the consequences of multicollinearity?

Skill Building

7. Suppose that you obtain the multiple regression equation $\hat{y} = 5 + 3x_1 - 4x_2$ from a set of sample data.
   (a) Interpret the slope coefficients for x_1 and x_2.
   (b) Determine the regression equation with $x_1 = 10$. Graph the regression equation with $x_1 = 10$.
   (c) Determine the regression equation with $x_1 = 15$. Graph the regression equation with $x_1 = 15$.
   (d) Determine the regression equation with $x_1 = 20$. Graph the regression equation with $x_1 = 20$.
   (e) What is the effect of changing the value x_1 on the graph of the regression equation?

8. Suppose that you obtain the multiple regression equation $\hat{y} = -5 - 9x_1 + 4x_2$ from a set of sample data.
   (a) Interpret the slope coefficients for x_1 and x_2.
   (b) Determine the regression equation with $x_1 = 10$. Graph the regression equation with $x_1 = 10$.
   (c) Determine the regression equation with $x_1 = 15$. Graph the regression equation with $x_1 = 15$.
   (d) Determine the regression equation with $x_1 = 20$. Graph the regression equation with $x_1 = 20$.
   (e) What is the effect of changing the value x_1 on the graph of the regression equation?

9. Suppose that a multiple regression model has $k = 3$ explanatory variables. The coefficient of determination, R^2, is found to be 0.653 based on a sample of $n = 25$ observations.
   (a) Compute the adjusted R^2.
   (b) Compute the F-test statistic.
   (c) Suppose one additional explanatory variable is added to the model and R^2 increases to 0.665. Compute the adjusted R^2. Would you recommend adding the additional explanatory variable to the model? Why or why not?

10. Suppose that a multiple regression model has $k = 4$ explanatory variables. The coefficient of determination, R^2, is found to be 0.542 based on a sample of $n = 40$ observations.
    (a) Compute the adjusted R^2.
    (b) Compute the F-test statistic.
    (c) Suppose one additional explanatory variable is added to the model and R^2 increases to 0.579. Compute the adjusted R^2. Would you recommend adding the additional explanatory variable to the model? Why or why not?

11. For the data set

Observation	x_1	x_2	x_3	y
1	0.8	2.8	2.5	11.0
2	3.9	2.6	5.7	10.8
3	1.8	2.4	7.8	10.6
4	5.1	2.3	7.1	10.3
5	4.9	2.5	5.9	10.3
6	8.4	2.1	8.6	10.3
7	12.9	2.3	9.2	10.0
8	6.0	2.0	1.2	9.4
9	14.6	2.2	3.7	8.7
10	9.3	1.1	5.5	8.7

(a) Construct a correlation matrix between x_1, x_2, x_3, and y. Does the correlation matrix support the inclusion of the explanatory variables included in the model? Is there any evidence that multicollinearity exists? Why?

(b) Determine the multiple regression line with x_1, x_2, and x_3 as the explanatory variables.
(c) Assuming that the requirements of the model are satisfied, test H_0: $\beta_1 = \beta_2 = \beta_3 = 0$ versus H_1: at least one of the β_i is different from zero at the $\alpha = 0.05$ level of significance.
(d) Assuming that the requirements of the model are satisfied, test H_0: $\beta_i = 0$ for $i = 1, 2, 3$ versus H_1: $\beta_i \neq 0$ for $i = 1, 2, 3$ at the $\alpha = 0.05$ level of significance.

12. For the data set

Observation	x_1	x_2	x_3	y
1	24.9	13.5	3.7	59.8
2	26.7	15.7	11.4	66.3
3	30.6	13.8	15.7	76.5
4	39.6	8.8	8.8	77.1
5	33.1	10.6	18.3	81.9
6	41.1	9.7	21.8	84.6
7	25.4	9.8	16.4	87.3
8	33.8	6.8	25.9	88.5
9	23.5	7.5	15.5	90.7
10	39.8	6.8	30.8	93.4

(a) Construct a correlation matrix between x_1, x_2, x_3, and y. Is there any evidence that multicollinearity exists? Why?
(b) Determine the multiple regression line with x_1, x_2, and x_3 as the explanatory variables.
(c) Assuming that the requirements of the model are satisfied, test H_0: $\beta_1 = \beta_2 = \beta_3 = 0$ versus H_1: at least one of the β_i is different from zero at the $\alpha = 0.05$ level of significance.
(d) Assuming that the requirements of the model are satisfied, test H_0: $\beta_i = 0$ for $i = 1, 2, 3$ versus H_1: $\beta_i \neq 0$ for $i = 1, 2, 3$ at the $\alpha = 0.05$ level of significance. Should a variable be removed from the model? Why?
(e) Remove the variable identified in part (d) and recompute the regression model. Test the claim that at least one regression coefficient is different from zero. Then test whether each individual regression coefficient is significantly different from zero.

13. For the data set

x_1	x_2	x_3	x_4	y
43	19.6	7.1	32	200
44	13.1	58.5	37	204
40	24.7	2.1	32	215
35	30.4	41.4	39	229
38	28.2	7.7	30	231
39	24.9	25.0	26	243
39	45.7	28.5	25	266
40	38.4	27.7	24	278
47	36.9	26.2	17	287
35	66.3	4.2	23	298
36	112.8	26.2	21	339
44	108.4	22.3	24	359

(a) Construct a correlation matrix between x_1, x_2, x_3, x_4, and y. Is there any evidence that multicollinearity may be a problem?
(b) Determine the multiple regression line using all the explanatory variables listed. Does the F-test indicate that we should reject H_0: $\beta_1 = \beta_2 = \beta_3 = \beta_4 = 0$? Which explanatory variables have slope coefficients that are not significantly different from zero?

(c) Remove the explanatory variable with the highest *P*-value from the model and recompute the regression model. Does the *F*-test still indicate that the model is significant? Remove any additional explanatory variables on the basis of the *P*-value of the slope coefficient. Then compute the model with the variable re-moved.

(d) Draw residual plots and a boxplot of the residuals to assess the adequacy of the model.

(e) Use the model constructed in part (c) to predict the value of y if $x_1 = 34$, $x_2 = 35.6$, $x_3 = 12.4$, and $x_4 = 29$.

(f) Draw a normal probability plot of the residuals. Is it reasonable to construct confidence and prediction intervals?

(g) Construct 95% confidence and prediction intervals if $x_1 = 34$, $x_2 = 35.6$, $x_3 = 12.4$, and $x_4 = 29$.

14. For the data set

x_1	x_2	x_3	x_4	y
47.3	0.9	4	76	105.5
53.1	0.8	6	55	113.8
56.7	0.8	4	65	115.2
48.8	0.5	7	67	118.9
42.7	1.1	7	74	148.9
44.3	1.1	6	76	120.2
44.5	0.7	8	68	121.6
37.7	0.7	7	79	140.0
36.9	1.0	5	73	141.5
28.1	1.8	6	68	141.9
32.0	0.8	8	81	152.8
34.7	0.8	10	68	156.5

(a) Construct a correlation matrix between x_1, x_2, x_3, x_4, and y. Is there any evidence that multicollinearity may be a problem?

(b) Determine the multiple regression line using all the explanatory variables listed. Does the *F*-test indicate that we should reject H_0: $\beta_1 = \beta_2 = \beta_3 = \beta_4 = 0$? Which explanatory variables have slope coefficients that are not significantly different from zero?

(c) Remove the explanatory variable with the highest *P*-value from the model and recompute the regression model. Does the *F*-test still indicate that the model is significant? Remove any additional explanatory variables on the basis of the *P*-value of the slope coefficient. Then compute the model with the variable removed.

(d) Draw residual plots and a boxplot of the residuals to assess the adequacy of the model.

(e) Use the model constructed in part (c) to predict the value of y if $x_1 = 44.3$, $x_2 = 1.1$, $x_3 = 7$, and $x_4 = 69$.

(f) Draw a normal probability plot of the residuals. Is it reasonable to construct confidence and prediction intervals?

(g) Construct 95% confidence and prediction intervals if $x_1 = 44.3$, $x_2 = 1.1$, $x_3 = 7$, and $x_4 = 69$.

Applying the Concepts

15. Resisting the Computer Researchers Alfred P. Rovai and Marcus D. Childress asked the following question: "How can resistance to reduction of computer anxiety among teacher education students be explained and predicted?" (*Journal of Research on Technology in Education*, Vol. 35 No 2). To answer this research question, they identified 86 undergraduate teacher education students enrolled in a computer literacy course and administered a series of questionnaires that quantified various predictors of the response variable, computer anxiety, y. For example, computer anxiety was measured by administering each student the Computer Anxiety Scale. This score ranges from

20 to 100, with higher scores indicating higher levels of computer anxiety. The explanatory variables were these:

x_1: Computer confidence on a scale from 10 to 40, with higher scores indicating higher confidence

x_2: Computer knowledge on a scale from 0 to 33, with 0 indicating no computer knowledge and 33 indicating superior computer knowledge

x_3: Computer liking on a scale from 10 to 40, with higher scores indicating a higher like of computers

x_4: Trait anxiety on a scale from 20 to 80, with higher scores indicating a higher level of overall anxiety

The multiple regression model was

$$\hat{y} = 84.04 - 0.87x_1 - 0.51x_2 - 0.45x_3 + 0.33x_4.$$

(a) The reported P-value of the regression model was less than 0.0001. Would you reject the null hypothesis $H_0: \beta_1 = \beta_2 = \beta_3 = \beta_4 = 0$?

(b) Interpret the slope coefficients of the model in part (a). Are they all reasonable?

(c) Predict the computer anxiety score of an individual whose computer confidence score was 25, computer knowledge score was 19, computer liking score was 20, and trait anxiety score was 43.

(d) The coefficient of determination for this model is 0.69. Interpret this value.

(e) The article states that "regression assumptions were tested and found to be tenable." Explain what this means.

16. **Pistol Shooting** Researchers at Victoria University wanted to determine the factors that affect precision in shooting air pistols (Inter- and Intra-Individual Analysis in Elite Sport: Pistol Shooting, *Journal of Applied Biomechanics, 2003, 28–38*). The explanatory variables were

x_1: Percent of the time the shooter's aim was on target (a measure of accuracy)

x_2: Percent of the time the shooter's aim was within a certain region (a measure of consistency or steadiness)

x_3: Distance (in mm) the barrel of the pistol moves horizontally while aiming

x_4: Distance (in mm) the barrel of the pistol moves vertically while aiming

(a) One response variable in the study was the score that the individual received on the shot, with a higher score indicating a better shooter. The regression model presented was $\hat{y} = 10.6 + 0.02x_1 - 0.03x_3$. The reported P-value of the regression model was 0.05. Would you reject the null hypothesis $H_0: \beta_1 = \beta_3 = 0$?

(b) Interpret the slope coefficients of the model in part (a).

(c) Predict the score of an individual whose aim was on target $x_1 = 20\%$ of the time with a distance the pistol barrel moves horizontally of $x_3 = 12$ mm using the model from part (a).

(d) A second response variable in the study was the vertical distance that the bullet hole was from the target. The regression model for this response variable was $\hat{y} = -24.6 - 0.13x_1 + 0.21x_2 + 0.13x_3 + 0.22x_4$. The reported P-value of the regression model was 0.04. Would you reject the null hypothesis $H_0: \beta_1 = \beta_2 = \beta_3 = \beta_4 = 0$?

(e) Interpret the slope coefficients of the model in part (d).

(f) Based on your answer to part (e), do you think that the model is useful in predicting vertical distance from the target? Why?

17. **Wind Chill Temperature** A researcher wanted to determine if there was a linear relation among wind chill temperature, air temperature, and wind speed. The following data show wind chill temperature, air temperature (in degrees Fahrenheit) , and wind speed (in miles per hour) for various days.

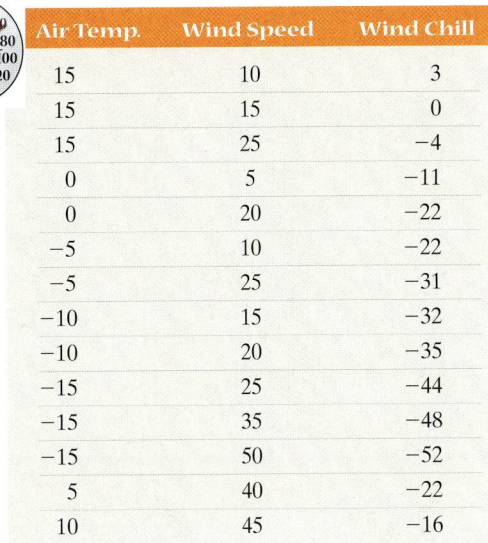

Air Temp.	Wind Speed	Wind Chill
15	10	3
15	15	0
15	25	−4
0	5	−11
0	20	−22
−5	10	−22
−5	25	−31
−10	15	−32
−10	20	−35
−15	25	−44
−15	35	−48
−15	50	−52
5	40	−22
10	45	−16

(a) Find the least-squares regression equation $\hat{y} = b_0 + b_1x_1 + b_2x_2$, where x_1 is air temperature and x_2 is wind speed, and y is the response variable, wind chill.

(b) Draw residual plots to assess the adequacy of the model. What might you conclude based on the plot of residuals against wind speed?

18. **Heat Index** A researcher wanted to determine whether there was a linear relation among heat index, air temperature, and dew point. The following data show the heat index, air temperature (in degrees Fahrenheit) , and dew point for various days.

Air Temp.	Dew Point	Heat Index
90	64	93
90	68	95
94	66	99
94	70	102
96	70	105
96	76	111
99	68	107
99	72	111
100	74	114
100	80	123
93	72	103
93	78	109
97	80	118
92	82	114
95	66	100
95	82	118

(a) Find the least-squares regression equation $\hat{y} = b_0 + b_1 x_1 + b_2 x_2$, where x_1 is air temperature and x_2 is dew point, and y is the response variable, heat index.
(b) Draw residual plots to assess the adequacy of the model. What might you conclude based on the residual plots?

19. Concrete A researcher wants to determine a model that can be used to predict the
NW 28-day strength of a concrete mixture. The following data represent the 28-day and 7-day strength (in pounds per square inch) of a certain type of concrete along with the concrete's slump. Slump is a measure of the uniformity of the concrete, with a higher slump indicating a less uniform mixture.

Slump (inches)	7-Day psi	28-Day psi
4.5	2330	4025
4.25	2640	4535
3	3360	4985
4	1770	3890
3.75	2590	3810
2.5	3080	4685
4	2050	3765
5	2220	3350
4.5	2240	3610
5	2510	3875
2.5	2250	4475

(a) Construct a correlation matrix between slump, 7-day psi, and 28-day psi. Is there any reason to be concerned with multicollinearity based on the correlation matrix?
(b) Find the least-squares regression equation $\hat{y} = b_0 + b_1 x_1 + b_2 x_2$, where x_1 is slump and x_2 is 7-day strength, and y is the response variable, 28-day strength.
(c) Draw residual plots and a boxplot of the residuals to assess the adequacy of the model.
(d) Interpret the regression coefficients for the least-squares regression equation.
(e) Determine and the interpret R^2 and the adjusted R^2.
(f) Test $H_0: \beta_1 = \beta_2 = 0$ versus H_1: at least one of the $\beta_i \neq 0$ at the $\alpha = 0.05$ level of significance.
(g) Test the hypotheses $H_0: \beta_1 = 0$ versus $H_1: \beta_1 \neq 0$ and $H_0: \beta_2 = 0$ versus $H_1: \beta_2 \neq 0$ at the $\alpha = 0.05$ level of significance.
(h) Predict the mean 28-day strength of all concrete for which slump is 3.5 inches and 7-day strength is 2450 psi.
(i) Predict the 28-day strength of a specific sample of concrete for which slump is 3.5 inches and 7-day strength is 2450 psi.
(j) Construct 95% confidence and prediction intervals for concrete for which slump is 3.5 inches and 7-day strength is 2450 psi. Interpret the results.

20. Income An economist was interested in modeling the relation among annual income, level of education, and work experience. The level of education is the number of years of education beyond eigth grade, so 1 represents completing 1 year of high school, 8 means completing 4 years of college, and so on. Work experience is the number of years employed in the current profession. From a random sample of 12 individuals, he obtained the following data:

Work Experience (years)	Level of Education	Annual Income (thousands of dollars)
21	6	30.2
14	3	15.6
4	8	19.7
16	8	54.9
12	4	28.7
20	4	36.0
25	1	18.0
8	3	12.7
24	12	84.6
28	9	62.7
4	11	42.7
15	4	45.3

(a) Construct a correlation matrix between work experience, level of education, and annual income. Is there any reason to be concerned with multicollinearity based on the correlation matrix?

(b) Find the least-squares regression equation $\hat{y} = b_0 + b_1x_1 + b_2x_2$, where x_1 is work experience and x_2 is level of education, and y is the response variable, annual income.

(c) Draw residual plots and a boxplot of the residuals to assess the adequacy of the model.

(d) Interpret the regression coefficients for the least-squares regression equation.

(e) Determine and interpret R^2 and the adjusted R^2.

(f) Test $H_0: \beta_1 = \beta_2 = 0$ versus H_1: at least one of the $\beta_i \neq 0$ at the $\alpha = 0.05$ level of significance.

(g) Test the hypotheses $H_0: \beta_1 = 0$ versus $H_1: \beta_1 \neq 0$ and $H_0: \beta_2 = 0$ versus $H_1: \beta_2 \neq 0$ at the $\alpha = 0.05$ level of significance.

(h) Predict the mean income of all individuals whose experience is 12 years and level of education is 4.

(i) Predict the income of a single individual whose experience is 12 years and level of education is 4.

(j) Construct 95% confidence and prediction intervals for income when experience is 12 years and level of education is 4.

21. Gas Mileage A researcher is interested in developing a model that describes the gas  mileage as measured in miles per gallon (mpg) of sport utility vehicles. Based on input from an engineer, she decides that the explanatory variables might be engine size (in cubic centimeters), horsepower (hp) rating of the engine, and weight (in pounds) . From a random sample of 11 SUVs, she obtains the following data:

Engine	HP	Weight	MPG
3471	260	4420	23
2979	225	4586	21
4195	275	4787	20
4701	235	4379	19
3471	240	4439	22
3960	195	3786	21
4701	235	3786	20
4701	265	3786	19
3311	230	3860	24
4664	235	5390	17
4605	302	4834	19

Source: msn.com

(a) Construct a correlation matrix. Is there any reason to be concerned about multicollinearity?

(b) Find the least-squares regression equation $\hat{y} = b_0 + b_1x_1 + b_2x_2 + b_3x_3$, where x_1 is engine size, x_2 is horsepower, x_3 is weight, and y is the response variable, miles per gallon.

(c) Test $H_0: \beta_1 = \beta_2 = \beta_3 = 0$ versus H_1: at least one of the $\beta_i \neq 0$ at the $\alpha = 0.05$ level of significance.

(d) Test the hypotheses $H_0: \beta_i = 0$ versus $H_1: \beta_i \neq 0$ for $i = 1, 2, 3$ at the $\alpha = 0.05$ level of significance. Should any of the explanatory variables be removed from the model? Which one?

(e) Determine the regression model with the explanatory variable identified in part (d) removed. Are each of the slope coefficients significantly different from zero?

(f) Draw residual plots, a boxplot of the residuals, and a normal probability plot of the residuals to assess the adequacy of the model found in part (e).

(g) Interpret the regression coefficients for the least-squares regression equation found in part (e).

(h) Determine and interpret R^2 and the adjusted R^2.

(i) Construct 95% confidence and prediction intervals for miles per gallon for a sport utility vehicle with a 4235-cc engine, 320 hp that weighs 3950 lb.

22. **Head Circumference** A pediatrician wants to determine the relation that may exist between a child's head circumference (in centimeters), height (in inches), and weight (in ounces). She randomly selects 14 three-year-old children from her practice and obtains the following data:

Height	Weight	Head Circumference
30	339	47
26.25	267	42
25	289	43
27	332	44.5
27.5	272	44
24.5	214	40.5
27.75	311	44
25	259	41.5
28	298	46
27.25	288	44
26	277	44
27.25	292	44.5
27	302	42.5
28.25	336	44.5

Source: Denise Slucki, student at Joliet Junior College

(a) Construct a correlation matrix. Is there any reason to be concerned with multicollinearity?

(b) Find the least-squares regression equation $\hat{y} = b_0 + b_1x_1 + b_2x_2$, where x_1 is height, x_2 is weight, and y is the response variable, head circumference.

(c) Test $H_0: \beta_1 = \beta_2 = 0$ versus H_1: at least one of the $\beta_i \neq 0$ at the $\alpha = 0.05$ level of significance.

(d) Test the hypotheses $H_0: \beta_1 = 0$ versus $H_1: \beta_1 \neq 0$ and $H_0: \beta_2 = 0$ versus $H_1: \beta_2 \neq 0$ at the $\alpha = 0.05$ level of significance.

(e) Compute the regression line after removing any explanatory variable that is not significant from the regression model.

(f) Draw residual plots, a boxplot of the residuals, and a normal probability plot of the residuals to assess the adequacy of the model found in part (e).

(g) Interpret the regression coefficients for the least-squares regression equation found in part (e).

(h) Determine and interpret R^2 and the adjusted R^2.

(i) Construct 95% confidence and prediction intervals for the head circumference of a child whose height is 27.5 inches and whose weight is 285 ounces. Interpret the results.

23. **Housing Prices** A realtor wanted to find a model that relates the asking price of a NW house to the square footage, number of bedrooms, and number of baths. The following data are from houses in Greenville, SC.

Square Footage	Bedrooms	Baths	Asking Price (thousands of dollars)
3632	4	2.5	419
4889	6	5	399
3000	5	3.5	395
3669	4	3.5	379
2800	4	3	359
3600	5	3.5	349
2800	5	2.5	320
2257	3	3	299
2000	3	3	295
2455	3	2.5	290
2250	3	2	285
2938	3	2	269
2399	3	2	260

Source: century21.com

(a) Construct the correlation matrix. Is there any reason to be concerned with multicollinearity?
(b) Find the least-squares regression equation $\hat{y} = b_0 + b_1 x_1 + b_2 x_2 + b_3 x_3$, where x_1 is square footage, x_2 is number of bedrooms, x_3 is number of baths, and y is the response variable, asking price.
(c) Test $H_0: \beta_1 = \beta_2 = \beta_3 = 0$ versus H_1: at least one of the $\beta_i \neq 0$ at the $\alpha = 0.05$ level of significance.
(d) Test the hypotheses $H_0: \beta_1 = 0$ versus $H_1: \beta_1 \neq 0$, $H_0: \beta_2 = 0$ versus $H_1: \beta_2 \neq 0$, and $H_0: \beta_3 = 0$ versus $H_1: \beta_3 \neq 0$ at the $\alpha = 0.05$ level of significance. What do you notice?
(e) Remove the explanatory variable with the highest P-value and compute the least-squares regression equation. Are all the slope coefficients significantly different from zero? If not, remove the explanatory variable with the higher P-value and compute the least-squares regression equation.
(f) Draw residual plots, a boxplot of the residuals, and a normal probability plot of the residuals to assess the adequacy of the model found in part (e).
(g) Interpret the regression coefficients for the least-squares regression equation found in part (e).
(h) Construct 95% confidence and prediction intervals for the asking price of a 2900-square-foot house in Greenville, SC, with 4 bedrooms and 3 baths. Interpret the results.

Problems 24 and 25 use the following definition:

A **dummy variable** or **indicator variable** is a method for including the influence of a qualitative variable in multiple regression analysis. A qualitative variable with k categories is represented by $0, 1, 2, \ldots, k - 1$, corresponding to the k classes of the variable. For example, the variable gender has two classes: male and female; so we might assign $X = 0$ for males and $X = 1$ for females.

24. **Using Dummy Variables** Researchers Willerman, Schultz, Rutledge, and Bigler wondered whether the size of a person's brain was related to the individual's mental capacity. They selected a sample of right-handed Anglo introductory psychology students who had Scholastic Aptitude Test scores higher than 1350. The subjects were administered the Wechsler (1981) Adult Intelligence Scale–Revised to obtain their IQ scores. The MRI scans, performed at the same facility, consisted of 18 horizontal MR images. The computer counted all pixels with nonzero gray scale in

each of the 18 images, and the total count served as an index for brain size. The resulting data are presented in the following table:

Gender	MRI Count	IQ	Gender	MRI Count	IQ
Female	816,932	133	Male	949,395	140
Female	951,545	137	Male	1,001,121	140
Female	991,305	138	Male	1,038,437	139
Female	833,868	132	Male	965,353	133
Female	856,472	140	Male	955,466	133
Female	852,244	132	Male	1,079,549	141
Female	790,619	135	Male	924,059	135
Female	866,662	130	Male	955,003	139
Female	857,782	133	Male	935,494	141
Female	948,066	133	Male	949,589	144

Source: Willerman, L., Schultz, R., Rutledge, J. N., and Bigler, E. (1991). In Vivo Brain Size and Intelligence, *Intelligence*, 15, 223–228.

(a) Find the least-squares regression equation $\hat{y} = b_0 + b_1x_1$, where x_1 is MRI count and y is the response variable, IQ.

(b) Test the hypotheses $H_0: \beta_1 = 0$ versus $H_1: \beta_1 \neq 0$. What do you conclude?

(c) Draw a scatter diagram, treating MRI count as the explanatory variable and IQ as the response variable, but use a different plotting symbol for males and females. For example, use a circle for males and a triangle for females.

(d) Find the least-squares regression equation $\hat{y} = b_0 + b_1x_1 + b_2x_2$, where x_1 is MRI count and $x_2 = 0$ for males and $x_2 = 1$ for females.

(e) Test the hypotheses $H_0: \beta_1 = 0$ versus $H_1: \beta_1 \neq 0$ and $H_0: \beta_2 = 0$ versus $H_1: \beta_2 \neq 0$.

(f) What do you conclude from this analysis?

25. Dummy Variables The following data represent the time (in minutes) it takes to drill an additional 5 feet from the depth indicated for both wet drilling and dry drilling conditions.

Depth	Conditions	Time
5	Wet	8.68
5	Wet	8.61
5	Dry	7.07
10	Wet	7.71
15	Dry	7.43
20	Wet	8.26
20	Dry	6.65
30	Wet	8.27
30	Dry	7.95
40	Wet	9.04
40	Dry	7.80
60	Wet	9.34
60	Dry	7.58
80	Wet	8.96
80	Dry	8.13
105	Wet	8.60
120	Dry	7.47
130	Wet	9.18
130	Dry	8.37
150	Wet	9.54
160	Dry	8.20

Source: American Statistician, Vol. 45, Issue 1.

(a) Draw a scatter diagram, treating depth as the explanatory variable and time as the response variable, but use a different plotting symbol for wet and dry. For example, use a circle for wet and a triangle for dry.

(b) Find the least-squares regression equation $\hat{y} = b_0 + b_1x_1 + b_2x_2$, where x_1 is depth and $x_2 = 0$ for wet and $x_2 = 1$ for dry.

(c) Test the hypotheses $H_0: \beta_1 = 0$ versus $H_1: \beta_1 \neq 0$ and $H_0: \beta_2 = 0$ versus $H_1: \beta_2 \neq 0$.

(d) Construct 95% confidence and prediction intervals for time to drill an additional 5 feet in dry conditions where drilling starts at 100 feet. Interpret the results.

Technology Step-by-Step

MINITAB   **Correlation Matrix**

Step 1: Enter the explanatory variables and response variable into the spreadsheet.

Step 2: Select the **Stat** menu and highlight **Basic Statistics**. Now select **Correlation** . . . .

Step 3: Highlight all the variables in the list of variables and click Select. Click OK.

Determining the Multiple Regression Line and Resdiual Plots

Step 1: Select the **Stat** menu and highlight **Regression.** Highlight **Regression** . . . .

Step 2: Select the explanatory variables (MINITAB calls them predictors) and response variable.

Step 3: Click GRAPHS. In the cell that says "Residuals versus the variables" enter the names of the explanatory variables. Select the box that says "Residuals versus fits". Click OK.

Step 4: Click STORAGE. Select the box that says "Residuals". The residuals are stored in the spreadsheet. Draw a normal probability plot of the residuals as indicated in Section 7.4.

Prediction and Confidence Intervals
See the steps given on page 758 in Section 14.2.

Chapter 14   Review

Summary

The first two sections of this chapter dealt with inferential techniques that can be used on the least-squares regression model $y_i = \beta_0 + \beta_1 x_i + \varepsilon_i$.

In Section 14.1, we used sample data to obtain estimates of an intercept and slope. The residuals are required to be normally distributed, with mean 0 and constant variance σ^2. The residuals for each observation should be independent. We verified these requirements through residual plots and a normal probability plot of the residuals. Provided that these requirements are satisfied, we can test claims regarding the slope to determine whether or not the relation between the explanatory and response variables is linear.

In Section 14.2, we learned how to construct confidence and prediction intervals about a predicted value. We construct confidence intervals about a mean response and prediction intervals about an individual response.

In Section 14.3, we learned how to build a multiple linear regression model $y_i = \beta_0 + \beta_1 x_{1i} + \beta_2 x_{2i} + \cdots + \beta_k x_{ki} + \varepsilon_i$. We used a correlation matrix to identify the explanatory variables that are linearly related to the response variable, y. We used technology to obtain estimates for the coefficients of each explanatory variable and the intercept.

The multiple linear regression model has the same requirements as the least-squares regression model with one explanatory variable. We used an F-test to test the claim that at least one coefficient is different from zero. If the null hypothesis that all coefficients are zero is rejected, we then use t-tests on each coefficient to determine which is different from zero.

When building a model, we remove explanatory variables one at a time. The explanatory variable with the highest P-value is removed, and the regression equation is obtained again. This process repeats until all coefficients in the model are significantly different from zero.

Watch out for multicollinearity among explanatory variables. It can easily distort results by giving us coefficients that are opposite in sign to what we would expect or have all coefficients appear to equal zero, even though the F-test leads us to reject the null hypothesis that all coefficients are zero.

Formulas

Standard Error of the Estimate

$$s_e = \sqrt{\frac{\sum(y_i - \hat{y}_i)^2}{n-2}} = \sqrt{\frac{\sum \text{residuals}^2}{n-2}}$$

Standard Error of b_1

$$s_{b_1} = \frac{s_e}{\sqrt{\sum(x_i - \bar{x})^2}}$$

Confidence Intervals for the Slope of the Regression Line
A $(1 - \alpha) \cdot 100\%$ confidence interval for the slope of the true regression line, β_1, is given by the following formulas:

Lower bound: $b_1 - t_{\alpha/2} \cdot \dfrac{s_e}{\sqrt{\sum(x_i - \bar{x})^2}} = b_1 - t_{\alpha/2} \cdot s_{b_1}$

Upper bound: $b_1 + t_{\alpha/2} \cdot \dfrac{s_e}{\sqrt{\sum(x_i - \bar{x})^2}} = b_1 + t_{\alpha/2} \cdot s_{b_1}$

Here, $t_{\alpha/2}$ is computed with $n - 2$ degrees of freedom.

Confidence Interval about the Mean Response of $\hat{y}$
A $(1 - \alpha) \cdot 100\%$ confidence interval for the mean response of y, $\hat{y}$, is given by the following formulas:

Lower bound: $\hat{y} - t_{\alpha/2} \cdot s_e \sqrt{\dfrac{1}{n} + \dfrac{(x^* - \bar{x})^2}{\sum(x_i - \bar{x})^2}}$

Upper bound: $\hat{y} + t_{\alpha/2} \cdot s_e \sqrt{\dfrac{1}{n} + \dfrac{(x^* - \bar{x})^2}{\sum(x_i - \bar{x})^2}}$

Here, x^* is the given value of the explanatory variable, and $t_{\alpha/2}$ is the critical value with $n - 2$ degrees of freedom.

Prediction Interval about an Individual Response, $\hat{y}$
A $(1 - \alpha) \cdot 100\%$ prediction interval for the individual response of y, $\hat{y}$, is given by

Lower bound: $\hat{y} - t_{\alpha/2} \cdot s_e \sqrt{1 + \dfrac{1}{n} + \dfrac{(x^* - \bar{x})^2}{\sum(x_i - \bar{x})^2}}$

Upper bound: $\hat{y} + t_{\alpha/2} \cdot s_e \sqrt{1 + \dfrac{1}{n} + \dfrac{(x^* - \bar{x})^2}{\sum(x_i - \bar{x})^2}}$

where x^* is the given value of the explanatory variable and $t_{\alpha/2}$ is the critical value with $n - 2$ degrees of freedom.

Vocabulary

Least-squares regression model (p. 739)
Standard error of the estimate (p. 740)
Robust (p. 744)
Bivariate normal distribution (p. 748)
Confidence interval about a mean response (p. 753)

Prediction interval about an individual response (p. 753)
Multiple regression model (p. 759)
Correlation matrix (p. 759)
Multicollinearity (p. 760)

Additive effect (p.764)
Adjusted R^2 (p. 764)

Objectives

Section	You should be able to . . .	Examples	Review Exercises
14.1	1 Understand the requirements of the least-squares regression model (p. 738)	pp. 738-740	9
	2 Compute the standard error of the estimate (p. 740)	2 and 3	1(b), 2(b), 3(b), 4(b), 5(b)
	3 Verify that the residuals are normally distributed (p. 741)	4	1(c), 2(c), 3(c), 4(c), 5(c)
	4 Conduct inference on the slope and intercept (p. 742)	5 and 6	1(e), 2(e), 3(e), 4(e), 5(e)
	5 Construct a confidence interval about the slope of the least-squares regression model (p. 746)	7	1(f), 2(f), 3(f), 4(f)
14.2	1 Construct confidence intervals for a mean response (p. 754)	1 and 3	1(g), 2(g), 3(g), 4(g)
	2 Construct prediction intervals for an individual response (p. 755)	2 and 3	1(i), 2(i), 3(i), 4(i)
14.3	1 Obtain the correlation matrix (p. 759)	1	7(a), 8(a)
	2 Use technology to find a multiple regression equation (p. 761)	2	7(b), 8(b)
	3 Interpret the coefficients of a multiple regression equation (p. 763)	3	7(d)
	4 Determine R^2 and adjusted R^2 (p. 764)	4	7(e), 8(h)
	5 Perform an F-test for lack of fit (p. 765)	5	7(f), 8(c)
	6 Test individual regression coefficients for significance (p. 767)	6	7(g), 8(d)
	7 Construct confidence and prediction intervals (p. 767)	7	7(i), 8(i)
	8 Build a regression model (p. 769)	9	8

Exercises

1. Engine Displacement versus Fuel Economy   The following data represent the size of a car's engine (in liters) versus its miles per gallon in the city for various 2005 domestic automobiles.

Car	Engine Displacement (liters), x	City MPG, y
Buick Century	3.1	20
Buick LeSabre	3.8	20
Cadillac DeVille	4.6	18
Chevrolet Cavalier	2.2	25
Chevrolet Impala	3.8	21
Chevrolet Malibu	2.2	24
Chrysler Sebring Sedan	2.7	22
Dodge Magnum	3.5	21
Ford Crown Victoria	4.6	18
Ford Focus	2.0	26
Ford Mustang	3.8	20
Mercury Sable	3.0	19
Pontiac Grand Am	3.4	20
Pontiac Sunfire	2.2	24
Saturn Ion	2.2	26

Source: www.roadandtrack.com

Using the results from Problems 1, 5, and 15 from the Chapter Review of Chapter 4, answer the following questions:

(a) What are the estimates of β_0 and β_1? What is the mean number of miles per gallon of all cars that have a 3.8-liter engine?

(b) Compute the standard error of the estimate, σ.

(c) Determine whether the residuals are normally distributed.

(d) If the residuals are normally distributed, determine s_{b_1}.

(e) If the residuals are normally distributed, test the claim that a linear relation exists between the explanatory variable, x, and the response variable, y, at the $\alpha = 0.05$ level of significance.

(f) If the residuals are normally distributed, construct a 95% confidence interval about the slope of the true least-squares regression line.

(g) Construct a 90% confidence interval about the mean miles per gallon of all cars that have a 3.8-liter engine.

(h) Predict the miles per gallon of a particular car with a 3.8-liter engine.

(i) Construct a 90% prediction interval about the miles per gallon of a particular car that has a 3.8-liter engine.

(j) Explain why the predicted miles per gallon found in parts (a) and (h) are the same, yet the intervals are different.

2. Temperature versus Cricket Chirps Crickets make a chirping noise by sliding their wings rapidly over each other. Perhaps you have noticed that the number of chirps seems to increase with the temperature. The following table lists the temperature (in degrees Fahrenheit) and the number of chirps per second for the striped ground cricket.

Temperature, x	Chirps per Second, y	Temperature, x	Chirps per Second, y
88.6	20.0	71.6	16.0
93.3	19.8	84.3	18.4
80.6	17.1	75.2	15.5
69.7	14.7	82.0	17.1
69.4	15.4	83.3	16.2
79.6	15.0	82.6	17.2
80.6	16.0	83.5	17.0
76.3	14.4		

Source: The Songs of Insects, Pierce, George W., *Cambridge*, MA: Harvard University Press, 1949. pp. 12–21.

Using the results from Problems 2, 6, and 16 from the Chapter Review of Chapter 4, answer the following questions:

(a) What are the estimates of β_0 and β_1? What is the mean number of chirps when the temperature is 80.2°F?

(b) Compute the standard error of the estimate, σ.

(c) Determine whether the residuals are normally distributed.

(d) If the residuals are normally distributed, determine s_{b_1}.

(e) If the residuals are normally distributed, test the claim that a linear relation exists between the explanatory variable, x, and response variable, y, at the $\alpha = 0.05$ level of significance.

(f) If the residuals are normally distributed, construct a 95% confidence interval about the slope of the true least-squares regression line.

(g) Construct a 90% confidence interval about the mean number of chirps found in part (a).

(h) Predict the number of chirps on a day when the temperature is 80.2°F.

(i) Construct a 90% prediction interval about the number of chirps found in part (h).

(j) Explain why the predicted number of chirps found in parts (a) and (h) are the same, yet the intervals are different.

3. Apartments The following data represent the square footage and rents for apartments in Queens, New York.

Square Footage, x	Rent per Month($), y
500	650
588	1215
1000	2000
688	1655
825	1250
1259	2700
650	1200
560	1250
1073	2350
1452	3300
1305	3100

Source: apartments.com

(a) What are the estimates of β_0 and β_1? What is the mean rent of a 900-square-foot apartment in Queens?

(b) Compute the standard error of the estimate, σ.

(c) Determine whether the residuals are normally distributed.

(d) If the residuals are normally distributed, determine s_{b_1}.

(e) If the residuals are normally distributed, test the claim that a linear relation exists between the explanatory variable, x, and response variable, y, at the $\alpha = 0.05$ level of significance.

(f) If the residuals are normally distributed, construct a 95% confidence interval about the slope of the true least-squares regression line.

(g) Construct a 90% confidence interval about the mean rent of all 900-square-foot apartments in Queens.

(h) Predict the rent of a particular 900-square-foot apartment in Queens.

(i) Construct a 90% prediction interval about the rent of a particular 900-square-foot apartment in Queens.

(j) Explain why the predicted rents found in parts (a) and (h) are the same, yet the intervals are different.

4. Boys' Heights The following data represent the height (in inches) of boys between the ages of 2 and 10 years.

Age, x	Boy Height, y	Age, x	Boy Height, y	Age, x	Boy Height, y
2	36.1	5	45.6	8	48.3
2	34.2	5	44.8	8	50.9
2	31.1	5	44.6	9	52.2
3	36.3	6	49.8	9	51.3
3	39.5	7	43.2	10	55.6
4	41.5	7	47.9	10	59.5
4	38.6	8	51.4		

Source: National Center for Health Statistics

(a) Treating age as the explanatory variable, determine the estimates of β_0 and β_1. What is the mean height of a 7-year-old boy?
(b) Compute the standard error of the estimate, σ.
(c) Determine whether the residuals are normally distributed.
(d) If the residuals are normally distributed, determine s_{b_1}.
(e) If the residuals are normally distributed, test the claim that a linear relation exists between the explanatory variable, age, and response variable, height, at the $\alpha = 0.05$ level of significance.
(f) If the residuals are normally distributed, construct a 95% confidence interval about the slope of the true least-squares regression line.
(g) Construct a 90% confidence interval about the mean height found in part (a).
(h) Predict the height of a 7-year-old boy.
(i) Construct a 90% prediction interval about the height found in part (h).
(j) Explain why the predicted heights found in parts (a) and (h) are the same, yet the intervals are different.

5. **Grip Strength** A researcher believes that as age increases the grip strength (in pounds per square inch) of an individual's dominant hand decreases. From a random sample of 17 females, he obtains the following data:

Age, x	Grip Strength, y	Age, x	Grip Strength, y
15	65	34	45
16	60	37	58
28	58	41	70
61	60	43	73
53	46	49	45
43	66	53	60
16	56	61	56
25	75	68	30
28	46		

Source: Kevin McCarthy, student at Joliet Junior College

(a) Treating age as the explanatory variable, determine the estimates of β_0 and β_1.
(b) Compute the standard error of the estimate, σ.
(c) Determine whether the residuals are normally distributed.
(d) If the residuals are normally distributed, determine s_{b_1}.
(e) If the residuals are normally distributed, test the claim that a linear relation exists between the explanatory variable, age, and response variable, grip strength, at the $\alpha = 0.05$ level of significance.
(f) Based on your answers to (d) and (e), what would be a good guess as to the grip strength of a randomly selected 42-year-old female?

6. **Depreciation** The following data represent the price of a random sample of used Chevy Camaros by age.

Age (years), x	Price (dollars) y	Age (years), x	Price (dollars) y
2	15,900	1	20,365
5	10,988	2	16,463
2	16,980	6	10,824
5	9,995	1	19,995
4	11,995	1	18,650
5	10,995	4	10,488

Source: www.onlineauto.com

(a) Determine the least-squares regression equation, treating age as the explanatory variable.
(b) A normal probability plot of the residuals indicates that the residuals are approximately normally distributed. Test the claim that there is a linear relation between age and price at the $\alpha = 0.05$ level of significance.
(c) Plot the residuals against the explanatory variable, age.
(d) Does a linear model seem appropriate based on the scatter diagram and residual plot? (*Hint:* See Section 4.3.) What is the moral?

7. **Course Grade** A statistics instructor wishes to investigate the relation between a student's final course grade and grades on a midterm exam and a major project. She selects a random sample of 10 statistics students and obtains the following information:

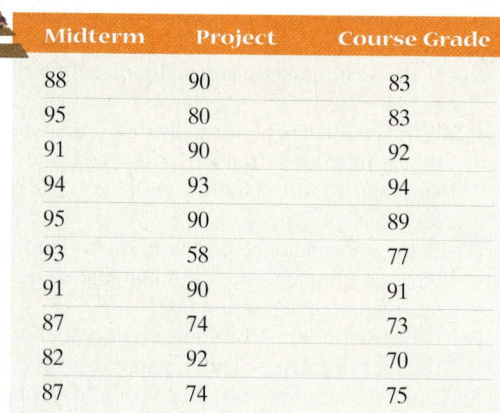

Midterm	Project	Course Grade
88	90	83
95	80	83
91	90	92
94	93	94
95	90	89
93	58	77
91	90	91
87	74	73
82	92	70
87	74	75

(a) Construct a correlation matrix between course grade, midterm grade, and project grade. Is there any reason to be concerned with multicollinearity based on the correlation matrix?
(b) Find the least-squares regression equation $\hat{y} = b_0 + b_1x_1 + b_2x_2$, where x_1 is the midterm exam score, x_2 is the project score, and y is the final course grade.

(c) Draw residual plots, a boxplot of residuals, and a normal probability plot of the residuals to assess the adequacy of the model.

(d) Interpret the regression coefficients for the least-squares regression equation.

(e) Determine and interpret R^2 and the adjusted R^2.

(f) Test $H_0: \beta_1 = \beta_2 = 0$ versus H_1: at least one $\beta_i \neq 0$ at the $\alpha = 0.05$ level of significance.

(g) Test the hypotheses $H_0: \beta_1 = 0$ versus $H_1: \beta_1 \neq 0$ and $H_0: \beta_2 = 0$ versus $H_1: \beta_2 \neq 0$ at the $\alpha = 0.05$ level of significance. Should any of the explanatory variables be removed from the model?

(h) Predict the mean final course grade of all statistics students who have an 85 on their midterm and a 75 on their project.

(i) Construct and interpret 95% confidence and prediction intervals for statistics students who score an 83 on their midterm and a 92 on their project. Interpret the results.

8. **Breakfast Cereal**   A nutritionist wants to model the relation between calories, protein, fat, and carbohydrates in breakfast cereal. Using a random sample of 12 ready-to-eat breakfast cereals, she obtained the following data per 100 g of cereal.

Calories	Protein (g)	Fat (g)	Carbohydrates (g)
373	5	1	89
380	7	3.8	83
389	4	3	88
370	9	3.7	81
355	10	3.2	81
381	3.4	3.2	87.1
357	9.1	5.5	73
387	6.4	1.8	89
347	7.9	1.8	78.1
385	3.6	3.9	87.9
386	6.4	3.6	84.9
377	6	1	86

(a) Construct a correlation matrix between calories, protein content, fat content, and carbohydrate content. Is there any reason to be concerned about multi-collinearity based on the correlation matrix?

(b) Find the least-squares regression equation $\hat{y} = b_0 + b_1 x_1 + b_2 x_2 + b_3 x_3$, where x_1 is protein content, x_2 is fat content, x_3 is carbohydrate content, and y is the number of calories.

(c) Test $H_0: \beta_1 = \beta_2 = \beta_3 = 0$ versus H_1: at least one $\beta_i \neq 0$ at the $\alpha = 0.05$ level of significance.

(d) Test the hypotheses $H_0: \beta_i = 0$ versus $H_1: \beta_i \neq 0$ for $i = 1, 2, 3$ at the $\alpha = 0.05$ level of significance. Should any of the explanatory variables be removed from the model? Which one?

(e) Determine the regression model with the explanatory variable identified in part (d) removed. Are each of the slope coefficients significantly different from zero?

(f) Draw residual plots, a boxplot of the residuals, and a normal probability plot of the residuals to assess the adequacy of the model found in part (e).

(g) Interpret the regression coefficients for the least-squares regression equation found in part (e).

(h) Determine and interpret R^2 and adjusted R^2.

(i) Construct and interpret 95% confidence and prediction intervals for calories in 100 g of ready-to-eat cereal with 4 grams of fat and 90 grams of carbohydrates.

9. What is the least-squares regression model? What are the requirements to perform inference on a least-squares regression line?

Housing Boom

During the early 2000s, the United States experienced a boom in the housing industry. This boom was in large part due to efforts by the government to boost consumer spending. The chairman of the Federal Reserve Board, Alan Greenspan testified before Congress on the economy (requoted from wallstreetwindow.com):

> The rapid rise in home prices over the past several years has provided households with considerable capital gains. Moreover, a significant increase in the rate of single-family home turnover has meant that many consumers have been able to realize gains from the sale of their homes. To be sure, such capital gains, largely realized through an increase in mortgage debt on the home, do not increase the pool of national savings available to finance new capital investment. But from the perspective of an individual household, cash realized from capital gains has the same spending power as cash from any other source.

For many, the lure of low interest rates put them in the market for a house. When house shopping, a natural question is "How much is the house worth?" This question is difficult to answer because it depends on what the market will bear. That is, the house is worth what someone else is willing to pay for it. So what are others willing to pay? A real estate agent wishes to help answer this question by examining several recent house sales in his territory and developing a formula that could be used to give a rough idea of a house's fair market value.

Articles on how to determine the value of a house often suggest comparing square footage, number of bedrooms, number of bathrooms, and size of the lot. The agent decided to examine these four variables, along with the age of the house and the number of rooms, in an effort to predict the house's selling price. The following data summarize his findings.

Price ($1,000s)	Acres	Bedrooms	Bathrooms	Sq. ft.	Age (yr)	Rooms
104.9	0.19	3	1.0	900	44	8
109.0	0.15	3	2.0	1431	34	6
94.9	0.20	3	1.5	1064	49	6
96.5	0.18	2	1.0	780	52	4
127.9	0.17	3	2.5	1140	47	6
129.9	0.18	3	1.5	1140	41	7
145.0	0.18	3	1.5	1845	45	6
199.9	0.17	4	3.5	1974	17	8
255.9	0.24	5	3.0	2460	22	8
310.0	0.23	4	3.5	2490	14	9
169.0	0.20	4	2.0	1896	37	8
344.5	0.24	4	4.5	2709	11	8
123.0	0.13	3	1.0	828	63	5
139.9	0.16	2	2.0	1131	75	5
169.9	0.15	2	2.0	1002	96	7
194.9	0.19	3	1.0	1024	55	6
210.0	0.23	3	1.0	1694	53	9
275.0	0.17	4	2.0	2380	10	8
299.5	0.17	4	2.0	1936	97	8
319.9	0.27	3	2.0	1648	77	8
397.5	0.30	4	2.5	2500	106	10

Price (\$1,000s)	Acres	Bedrooms	Bathrooms	Sq. ft.	Age (yr)	Rooms
189.9	0.18	2	1.0	1016	71	6
349.9	0.40	4	2.5	1816	70	8
454.9	0.96	3	3.0	2160	37	7
499.9	1.00	5	3.0	3104	48	10
615.0	0.66	4	3.5	3205	26	10
635.0	0.44	4	3.5	3084	27	10
929.0	0.90	5	4.5	4470	16	14

(a) Construct a correlation matrix with price, acres, bedrooms, bathrooms, square feet, age, and rooms. Is there any reason to be concerned with multicollinearity based on the correlation matrix?

(b) Find the least-squares regression equation $\hat{y} = b_0 + b_1x_1 + b_2x_2 + b_3x_3 + b_4x_4 + b_5x_5 + b_6x_6$, where x_1 is acres, x_2 is bedrooms, and so on.

(c) Test H_0: $\beta_i = 0$ versus H_1: at least one of the $\beta_i \neq 0$ at the $\alpha = 0.05$ level of significance.

(d) Test the hypotheses H_0: $\beta_i = 0$ versus H_1: $\beta_i \neq 0$ for all i; at the $\alpha = 0.05$ level of significance.

(e) Examine your regression results and remove any explanatory variable whose coefficient is not significantly different than 0 to obtain the model of best fit.

(f) Once you have obtained your model of best fit, draw residual plots and a boxplot of the residuals to assess the adequacy of the model.

(g) Determine and interpret R^2 and adjusted R^2. How well does your model appear to fit the data?

(h) Use your model to predict the selling price for another house from the agent's territory that has the following characteristics: 0.18 acre, 3 bedrooms, 1 bath, 1176 square feet, 47 years old, and 6 total rooms. Compare your prediction to the actual selling price: \$99,900.

Location, location, location! The agent knows that the location of the house can have a large impact on its selling price. He notes that the first 12 houses listed are from the same zip code, the next 10 houses are from a second zip code, and the last 6 houses listed are from a third zip code.

(i) Construct side-by-side boxplots of selling price for the three zip codes. Is there any reason to believe that selling prices vary from one zip code to the next within the agent's territory?

(j) Introduce a dummy explanatory variable to represent zip code and repeat parts (b)–(g) to find the model of best fit for price.

(k) Repeat part (h) assuming the house comes from the first zip code. Which model did a better job predicting the selling price?

(l) Use the model you obtained in part (e) to predict the selling price of a house in your area. How well did the model work?

(m) Explain the limitations of this model. Which, if any, can be dealt with, and how would you do so?

(n) What other variables might affect the cost of a house? Support your choices; then collect data from your area for the subset of variables you think best predicts house price. Find the model of best fit and test it with additional houses in your area, as well as houses in other areas. How well does your model work?

APPENDIX A TABLES

Table I

Random Numbers

Row Number	01–05	06–10	11–15	16–20	21–25	26–30	31–35	36–40	41–45	46–50
01	89392	23212	74483	36590	25956	36544	68518	40805	09980	00467
02	61458	17639	96252	95649	73727	33912	72896	66218	52341	97141
03	11452	74197	81962	48443	90360	26480	73231	37740	26628	44690
04	27575	04429	31308	02241	01698	19191	18948	78871	36030	23980
05	36829	59109	88976	46845	28329	47460	88944	08264	00843	84592
06	81902	93458	42161	26099	09419	89073	82849	09160	61845	40906
07	59761	55212	33360	68751	86737	79743	85262	31887	37879	17525
08	46827	25906	64708	20307	78423	15910	86548	08763	47050	18513
09	24040	66449	32353	83668	13874	86741	81312	54185	78824	00718
10	98144	96372	50277	15571	82261	66628	31457	00377	63423	55141
11	14228	17930	30118	00438	49666	65189	62869	31304	17117	71489
12	55366	51057	90065	14791	62426	02957	85518	28822	30588	32798
13	96101	30646	35526	90389	73634	79304	96635	06626	94683	16696
14	38152	55474	30153	26525	83647	31988	82182	98377	33802	80471
15	85007	18416	24661	95581	45868	15662	28906	36392	07617	50248
16	85544	15890	80011	18160	33468	84106	40603	01315	74664	20553
17	10446	20699	98370	17684	16932	80449	92654	02084	19985	59321
18	67237	45509	17638	65115	29757	80705	82686	48565	72612	61760
19	23026	89817	05403	82209	30573	47501	00135	33955	50250	72592
20	67411	58542	18678	46491	13219	84084	27783	34508	55158	78742

Table II

Binomial Probability Distribution

This table computes the probability of obtaining x successes in n trials of a binomial experiment with probability of success p.

n	x	0.01	0.05	0.10	0.15	0.20	0.25	0.30	0.35	0.40	0.45	0.50	0.55	0.60	0.65	0.70	0.75	0.80	0.85	0.90	0.95
2	0	0.9801	0.9025	0.8100	0.7225	0.6400	0.5625	0.4900	0.4225	0.3600	0.3025	0.2500	0.2025	0.1600	0.1225	0.0900	0.0625	0.0400	0.0225	0.0100	0.0025
	1	0.0198	0.0950	0.1800	0.2550	0.3200	0.3750	0.4200	0.4550	0.4800	0.4950	0.5000	0.4950	0.4800	0.4550	0.4200	0.3750	0.3200	0.2550	0.1800	0.0950
	2	0.0001	0.0025	0.0100	0.0225	0.0400	0.0625	0.0900	0.1225	0.1600	0.2025	0.2500	0.3025	0.3600	0.4225	0.4900	0.5625	0.6400	0.7225	0.8100	0.9025
3	0	0.9703	0.8574	0.7290	0.6141	0.5120	0.4219	0.3430	0.2746	0.2160	0.1664	0.1250	0.0911	0.0640	0.0429	0.0270	0.0156	0.0080	0.0034	0.0010	0.0001
	1	0.0294	0.1354	0.2430	0.3251	0.3840	0.4219	0.4410	0.4436	0.4320	0.4084	0.3750	0.3341	0.2880	0.2389	0.1890	0.1406	0.0960	0.0574	0.0270	0.0071
	2	0.0003	0.0071	0.0270	0.0574	0.0960	0.1406	0.1890	0.2389	0.2880	0.3341	0.3750	0.4084	0.4320	0.4436	0.4410	0.4219	0.3840	0.3251	0.2430	0.1354
	3	0.0000	0.0001	0.0010	0.0034	0.0080	0.0156	0.0270	0.0429	0.0640	0.0911	0.1250	0.1664	0.2160	0.2746	0.3430	0.4219	0.5120	0.6141	0.7290	0.8574
4	0	0.9606	0.8145	0.6561	0.5220	0.4096	0.3164	0.2401	0.1785	0.1296	0.0915	0.0625	0.0410	0.0256	0.0150	0.0081	0.0039	0.0016	0.0005	0.0001	0.0000
	1	0.0388	0.1715	0.2916	0.3685	0.4096	0.4219	0.4116	0.3845	0.3456	0.2995	0.2500	0.2005	0.1536	0.1115	0.0756	0.0469	0.0256	0.0116	0.0036	0.0005
	2	0.0006	0.0135	0.0486	0.0975	0.1536	0.2109	0.2646	0.3105	0.3456	0.3675	0.3760	0.3675	0.3456	0.3105	0.2646	0.2109	0.1536	0.0975	0.0486	0.0135
	3	0.0000	0.0005	0.0036	0.0115	0.0256	0.0469	0.0756	0.1115	0.1536	0.2005	0.2500	0.2995	0.3456	0.3845	0.4116	0.4219	0.4096	0.3685	0.2916	0.1715
	4	0.0000	0.0000	0.0001	0.0005	0.0016	0.0039	0.0081	0.0150	0.0256	0.0410	0.0625	0.0915	0.1296	0.1785	0.2401	0.3164	0.4096	0.5220	0.6561	0.8145
5	0	0.9510	0.7738	0.5905	0.4437	0.3277	0.2373	0.1681	0.1160	0.0778	0.0503	0.0313	0.0185	0.0102	0.0053	0.0024	0.0010	0.0003	0.0001	0.0000	0.0000
	1	0.0480	0.2036	0.3281	0.3915	0.4096	0.3955	0.3602	0.3124	0.2592	0.2059	0.1563	0.1128	0.0768	0.0488	0.0284	0.0146	0.0064	0.0022	0.0005	0.0000
	2	0.0010	0.0214	0.0729	0.1382	0.2048	0.2637	0.3087	0.3364	0.3456	0.3369	0.3125	0.2757	0.2304	0.1811	0.1323	0.0879	0.0512	0.0244	0.0081	0.0011
	3	0.0000	0.0011	0.0081	0.0244	0.0512	0.0879	0.1323	0.1811	0.2304	0.2757	0.3125	0.3369	0.3456	0.3364	0.3087	0.2637	0.2048	0.1382	0.0729	0.0214
	4	0.0000	0.0000	0.0005	0.0022	0.0064	0.0146	0.0284	0.0488	0.0768	0.1128	0.1563	0.2059	0.2592	0.3124	0.3602	0.3955	0.4096	0.3915	0.3281	0.2036
	5	0.0000	0.0000	0.0000	0.0001	0.0003	0.0010	0.0024	0.0053	0.0102	0.0185	0.0313	0.0503	0.0778	0.1160	0.1681	0.2373	0.3277	0.4437	0.5905	0.7738
6	0	0.9415	0.7351	0.5314	0.3771	0.2621	0.1780	0.1176	0.0754	0.0467	0.0277	0.0156	0.0083	0.0041	0.0018	0.0007	0.0002	0.0001	0.0000	0.0000	0.0000
	1	0.0571	0.2321	0.3543	0.3993	0.3932	0.3560	0.3025	0.2437	0.1866	0.1359	0.0938	0.0609	0.0369	0.0205	0.0102	0.0044	0.0015	0.0004	0.0001	0.0000
	2	0.0014	0.0305	0.0984	0.1762	0.2458	0.2966	0.3241	0.3280	0.3110	0.2780	0.2344	0.1861	0.1382	0.0951	0.0595	0.0330	0.0154	0.0055	0.0012	0.0001
	3	0.0000	0.0021	0.0146	0.0415	0.0819	0.1318	0.1852	0.2355	0.2765	0.3032	0.3125	0.3032	0.2765	0.2355	0.1852	0.1318	0.0819	0.0415	0.0146	0.0021
	4	0.0000	0.0001	0.0012	0.0055	0.0154	0.0330	0.0595	0.0951	0.1382	0.1861	0.2344	0.2780	0.3110	0.3280	0.3241	0.2966	0.2458	0.1762	0.0984	0.0305
	5	0.0000	0.0000	0.0001	0.0004	0.0015	0.0044	0.0102	0.0205	0.0369	0.0609	0.0938	0.1359	0.1866	0.2437	0.3025	0.3560	0.3932	0.3993	0.3543	0.2321
	6	0.0000	0.0000	0.0000	0.0000	0.0001	0.0002	0.0007	0.0018	0.0041	0.0083	0.0156	0.0277	0.0467	0.0754	0.1176	0.1780	0.2621	0.3771	0.5314	0.7351
7	0	0.9321	0.6983	0.4783	0.3206	0.2097	0.1335	0.0824	0.0490	0.0280	0.0152	0.0078	0.0037	0.0016	0.0006	0.0002	0.0001	0.0000	0.0000	0.0000	0.0000
	1	0.0659	0.2573	0.3720	0.3960	0.3670	0.3115	0.2471	0.1848	0.1306	0.0872	0.0547	0.0320	0.0172	0.0084	0.0036	0.0013	0.0004	0.0001	0.0000	0.0000
	2	0.0020	0.0406	0.1240	0.2097	0.2753	0.3115	0.3177	0.2985	0.2613	0.2140	0.1641	0.1172	0.0774	0.0466	0.0250	0.0115	0.0043	0.0012	0.0002	0.0000
	3	0.0000	0.0036	0.0230	0.0617	0.1147	0.1730	0.2269	0.2679	0.2903	0.2918	0.2734	0.2388	0.1935	0.1442	0.0972	0.0577	0.0287	0.0109	0.0026	0.0002
	4	0.0000	0.0002	0.0026	0.0109	0.0287	0.0577	0.0972	0.1442	0.1935	0.2388	0.2734	0.2918	0.2903	0.2679	0.2269	0.1730	0.1147	0.0617	0.0230	0.0036
	5	0.0000	0.0000	0.0002	0.0012	0.0043	0.0115	0.0250	0.0466	0.0774	0.1172	0.1641	0.2140	0.2613	0.2985	0.3177	0.3115	0.2753	0.2097	0.1240	0.0406
	6	0.0000	0.0000	0.0000	0.0001	0.0004	0.0013	0.0036	0.0084	0.0172	0.0320	0.0547	0.0872	0.1306	0.1848	0.2471	0.3115	0.3670	0.3960	0.3720	0.2573
	7	0.0000	0.0000	0.0000	0.0000	0.0000	0.0001	0.0002	0.0006	0.0016	0.0037	0.0078	0.0152	0.0280	0.0490	0.0824	0.1335	0.2097	0.3206	0.4783	0.6983

Table II (*continued*)

n	x	0.01	0.05	0.10	0.15	0.20	0.25	0.30	0.35	0.40	0.45	p 0.50	0.55	0.60	0.65	0.70	0.75	0.80	0.85	0.90	0.95
8	0	0.9227	0.6634	0.4305	0.2725	0.1678	0.1001	0.0576	0.0319	0.0168	0.0084	0.0039	0.0017	0.0007	0.0002	0.0001	0.0000	0.0000	0.0000	0.0000	0.0000
	1	0.0746	0.2793	0.3826	0.3847	0.3355	0.2670	0.1977	0.1373	0.0896	0.0548	0.0313	0.0164	0.0079	0.0033	0.0012	0.0004	0.0001	0.0000	0.0000	0.0000
	2	0.0026	0.0515	0.1488	0.2376	0.2936	0.3115	0.2965	0.2587	0.2090	0.1569	0.1094	0.0703	0.0413	0.0217	0.0100	0.0038	0.0011	0.0002	0.0000	0.0000
	3	0.0001	0.0054	0.0331	0.0839	0.1468	0.2076	0.2541	0.2786	0.2787	0.2568	0.2188	0.1719	0.1239	0.0808	0.0467	0.0231	0.0092	0.0026	0.0004	0.0000
	4	0.0000	0.0004	0.0046	0.0185	0.0459	0.0865	0.1361	0.1875	0.2322	0.2627	0.2734	0.2627	0.2322	0.1675	0.1361	0.0865	0.0459	0.0185	0.0046	0.0004
	5	0.0000	0.0000	0.0004	0.0026	0.0092	0.0231	0.0467	0.0808	0.1239	0.1719	0.2188	0.2568	0.2767	0.2786	0.2541	0.2076	0.1468	0.0839	0.0331	0.0054
	6	0.0000	0.0000	0.0000	0.0002	0.0011	0.0038	0.0100	0.0217	0.0413	0.0703	0.1094	0.1569	0.2090	0.2587	0.2965	0.3115	0.2936	0.2376	0.1488	0.0515
	7	0.0000	0.0000	0.0000	0.0000	0.0001	0.0004	0.0012	0.0033	0.0079	0.0164	0.0313	0.0548	0.0896	0.1373	0.1977	0.2670	0.3355	0.3647	0.3826	0.2793
	8	0.0000	0.0000	0.0000	0.0000	0.0000	0.0000	0.0001	0.0002	0.0007	0.0017	0.0039	0.0084	0.0168	0.0319	0.0576	0.1001	0.1678	0.2725	0.4305	0.6634
9	0	0.9135	0.6302	0.3874	0.2316	0.1342	0.0751	0.0404	0.0207	0.0101	0.0046	0.0020	0.0008	0.0003	0.0001	0.0000	0.0000	0.0000	0.0000	0.0000	0.0000
	1	0.0830	0.2985	0.3874	0.3679	0.3020	0.2253	0.1556	0.1004	0.0605	0.0339	0.0176	0.0083	0.0035	0.0013	0.0004	0.0001	0.0000	0.0000	0.0000	0.0000
	2	0.0034	0.0629	0.1722	0.2597	0.3020	0.3003	0.2668	0.2162	0.1612	0.1110	0.0703	0.0407	0.0212	0.0098	0.0039	0.0012	0.0003	0.0000	0.0000	0.0000
	3	0.0001	0.0077	0.0446	0.1069	0.1762	0.2336	0.2668	0.2716	0.2508	0.2119	0.1641	0.1160	0.0743	0.0424	0.0210	0.0087	0.0028	0.0006	0.0001	0.0000
	4	0.0000	0.0006	0.0074	0.0283	0.0661	0.1168	0.1715	0.2194	0.2508	0.2600	0.2461	0.2128	0.1672	0.1181	0.0735	0.0389	0.0165	0.0050	0.0008	0.0000
	5	0.0000	0.0000	0.0008	0.0050	0.0165	0.0389	0.0735	0.1181	0.1672	0.2128	0.2461	0.2600	0.2508	0.2194	0.1715	0.1168	0.0661	0.0283	0.0074	0.0006
	6	0.0000	0.0000	0.0001	0.0006	0.0028	0.0087	0.0210	0.0424	0.0743	0.1160	0.1641	0.2119	0.2508	0.2716	0.2668	0.2336	0.1762	0.1069	0.0446	0.0077
	7	0.0000	0.0000	0.0000	0.0000	0.0003	0.0012	0.0039	0.0098	0.0212	0.0407	0.0703	0.1110	0.1612	0.2162	0.2668	0.3003	0.3020	0.2597	0.1722	0.0629
	8	0.0000	0.0000	0.0000	0.0000	0.0000	0.0001	0.0004	0.0013	0.0035	0.0083	0.0176	0.0339	0.0605	0.1004	0.1556	0.2253	0.3020	0.3679	0.3874	0.2985
	9	0.0000	0.0000	0.0000	0.0000	0.0000	0.0000	0.0000	0.0001	0.0003	0.0008	0.0020	0.0046	0.0101	0.0207	0.0404	0.0751	0.1342	0.2316	0.3874	0.6302
10	0	0.9044	0.5987	0.3487	0.1969	0.1074	0.0563	0.0282	0.0135	0.0060	0.0025	0.0010	0.0003	0.0001	0.0000	0.0000	0.0000	0.0000	0.0000	0.0000	0.0000
	1	0.0914	0.3151	0.3874	0.3474	0.2684	0.1877	0.1211	0.0725	0.0403	0.0207	0.0098	0.0042	0.0016	0.0005	0.0001	0.0000	0.0000	0.0000	0.0000	0.0000
	2	0.0042	0.0748	0.1937	0.2759	0.3020	0.2816	0.2335	0.1757	0.1209	0.0763	0.0439	0.0229	0.0106	0.0043	0.0014	0.0004	0.0001	0.0000	0.0000	0.0000
	3	0.0001	0.0105	0.0574	0.1298	0.2013	0.2503	0.2668	0.2522	0.2150	0.1665	0.1172	0.0746	0.0425	0.0212	0.0090	0.0031	0.0008	0.0001	0.0000	0.0000
	4	0.0000	0.0010	0.0112	0.0401	0.0881	0.1460	0.2001	0.2377	0.2508	0.2384	0.2051	0.1596	0.1115	0.0689	0.0368	0.0162	0.0055	0.0012	0.0001	0.0000
	5	0.0000	0.0001	0.0015	0.0085	0.0264	0.0584	0.1029	0.1536	0.2007	0.2340	0.2461	0.2340	0.2007	0.1536	0.1029	0.0584	0.0264	0.0085	0.0015	0.0001
	6	0.0000	0.0000	0.0001	0.0012	0.0055	0.0162	0.0368	0.0689	0.1115	0.1596	0.2051	0.2384	0.2508	0.2377	0.2001	0.1460	0.0881	0.0401	0.0112	0.0010
	7	0.0000	0.0000	0.0000	0.0001	0.0008	0.0031	0.0090	0.0212	0.0425	0.0746	0.1172	0.1665	0.2150	0.2522	0.2668	0.2503	0.2013	0.1298	0.0574	0.0105
	8	0.0000	0.0000	0.0000	0.0000	0.0001	0.0004	0.0014	0.0043	0.0106	0.0229	0.0439	0.0763	0.1209	0.1757	0.2335	0.2816	0.3020	0.2759	0.1937	0.0746
	9	0.0000	0.0000	0.0000	0.0000	0.0000	0.0000	0.0001	0.0005	0.0016	0.0042	0.0098	0.0207	0.0403	0.0725	0.1211	0.1877	0.2684	0.3474	0.3874	0.3151
	10	0.0000	0.0000	0.0000	0.0000	0.0000	0.0000	0.0000	0.0000	0.0001	0.0003	0.0010	0.0025	0.0060	0.0135	0.0282	0.0563	0.1074	0.1969	0.3487	0.5987

Table II (continued)

n	x											p										
		0.01	0.05	0.10	0.15	0.20	0.25	0.30	0.35	0.40	0.45	0.50	0.55	0.60	0.65	0.70	0.75	0.80	0.85	0.90	0.95
11	0	0.8953	0.5688	0.3138	0.1673	0.0859	0.0422	0.0198	0.0088	0.0036	0.0014	0.0005	0.0002	0.0000	0.0000	0.0000	0.0000	0.0000	0.0000	0.0000	0.0000
	1	0.0995	0.3293	0.3835	0.3248	0.2362	0.1549	0.0932	0.0518	0.0266	0.0125	0.0054	0.0021	0.0007	0.0002	0.0000	0.0000	0.0000	0.0000	0.0000	0.0000
	2	0.0050	0.0867	0.2131	0.2866	0.2953	0.2581	0.1998	0.1395	0.0887	0.0513	0.0269	0.0126	0.0052	0.0018	0.0005	0.0001	0.0000	0.0000	0.0000	0.0000
	3	0.0002	0.0137	0.0710	0.1517	0.2215	0.2581	0.2568	0.2254	0.1774	0.1259	0.0806	0.0462	0.0234	0.0102	0.0037	0.0011	0.0002	0.0000	0.0000	0.0000
	4	0.0000	0.0014	0.0158	0.0536	0.1107	0.1721	0.2201	0.2428	0.2365	0.2060	0.1611	0.1128	0.0701	0.0379	0.0173	0.0064	0.0017	0.0003	0.0000	0.0000
	5	0.0000	0.0001	0.0025	0.0132	0.0388	0.0803	0.1321	0.1830	0.2207	0.2360	0.2256	0.1931	0.1471	0.0985	0.0568	0.0268	0.0097	0.0023	0.0003	0.0000
	6	0.0000	0.0000	0.0003	0.0023	0.0097	0.0268	0.0566	0.0985	0.1471	0.1931	0.2256	0.2360	0.2207	0.1830	0.1321	0.0803	0.0388	0.0132	0.0025	0.0001
	7	0.0000	0.0000	0.0000	0.0003	0.0017	0.0064	0.0173	0.0379	0.0701	0.1128	0.1611	0.2060	0.2365	0.2428	0.2201	0.1721	0.1107	0.0536	0.0158	0.0014
	8	0.0000	0.0000	0.0000	0.0000	0.0002	0.0011	0.0037	0.0102	0.0234	0.0462	0.0806	0.1259	0.1774	0.2254	0.2568	0.2581	0.2215	0.1517	0.0710	0.0137
	9	0.0000	0.0000	0.0000	0.0000	0.0000	0.0001	0.0005	0.0016	0.0052	0.0126	0.0269	0.0513	0.0887	0.1395	0.1998	0.2561	0.2953	0.2866	0.2131	0.0867
	10	0.0000	0.0000	0.0000	0.0000	0.0000	0.0000	0.0000	0.0002	0.0007	0.0021	0.0054	0.0125	0.0266	0.0516	0.0932	0.1549	0.2362	0.3248	0.3835	0.3293
	11	0.0000	0.0000	0.0000	0.0000	0.0000	0.0000	0.0000	0.0000	0.0000	0.0002	0.0005	0.0014	0.0036	0.0088	0.0196	0.0422	0.0659	0.1673	0.3138	0.5688
12	0	0.8864	0.5404	0.2824	0.1422	0.0687	0.0317	0.0138	0.0057	0.0022	0.0008	0.0002	0.0001	0.0000	0.0000	0.0000	0.0000	0.0000	0.0000	0.0000	0.0000
	1	0.1074	0.3413	0.3766	0.3012	0.2062	0.1267	0.0712	0.0368	0.0174	0.0075	0.0029	0.0010	0.0003	0.0001	0.0000	0.0000	0.0000	0.0000	0.0000	0.0000
	2	0.0060	0.0988	0.2301	0.2924	0.2835	0.2323	0.1678	0.1088	0.0639	0.0339	0.0161	0.0068	0.0025	0.0008	0.0002	0.0000	0.0000	0.0000	0.0000	0.0000
	3	0.0002	0.0173	0.0852	0.1720	0.2362	0.2581	0.2397	0.1954	0.1419	0.0923	0.0537	0.0277	0.0125	0.0048	0.0015	0.0004	0.0001	0.0000	0.0000	0.0000
	4	0.0000	0.0021	0.0213	0.0683	0.1329	0.1936	0.2311	0.2367	0.2128	0.1700	0.1208	0.0762	0.0420	0.0199	0.0078	0.0024	0.0005	0.0001	0.0000	0.0000
	5	0.0000	0.0002	0.0038	0.0193	0.0532	0.1032	0.1585	0.2039	0.2270	0.2225	0.1934	0.1489	0.1009	0.0591	0.0291	0.0401	0.0155	0.0006	0.0000	0.0000
	6	0.0000	0.0000	0.0005	0.0040	0.0155	0.0401	0.0792	0.1281	0.1766	0.2124	0.2256	0.2124	0.1766	0.1281	0.0792	0.1032	0.0532	0.0040	0.0005	0.0000
	7	0.0000	0.0000	0.0000	0.0006	0.0033	0.0115	0.0291	0.0591	0.1009	0.1489	0.1934	0.2225	0.2270	0.2039	0.1585	0.1936	0.1329	0.0193	0.0038	0.0002
	8	0.0000	0.0000	0.0000	0.0001	0.0005	0.0024	0.0078	0.0199	0.0420	0.0762	0.1208	0.1700	0.2128	0.2367	0.2311	0.2581	0.2362	0.0683	0.0213	0.0021
	9	0.0000	0.0000	0.0000	0.0000	0.0001	0.0004	0.0015	0.0048	0.0125	0.0277	0.0537	0.0923	0.1419	0.1954	0.2397	0.2323	0.2835	0.1720	0.0852	0.0173
	10	0.0000	0.0000	0.0000	0.0000	0.0000	0.0000	0.0002	0.0008	0.0025	0.0068	0.0161	0.0339	0.0639	0.1088	0.1678	0.1267	0.2062	0.2924	0.2301	0.0988
	11	0.0000	0.0000	0.0000	0.0000	0.0000	0.0000	0.0000	0.0001	0.0003	0.0010	0.0029	0.0075	0.0174	0.0368	0.0712	0.0317	0.2062	0.3012	0.3766	0.3413
	12	0.0000	0.0000	0.0000	0.0000	0.0000	0.0000	0.0000	0.0000	0.0000	0.0001	0.0002	0.0008	0.0022	0.0057	0.0138	0.0317	0.0687	0.1422	0.2824	0.5404

Table II (continued)

n	x											p									
		0.01	0.05	0.10	0.15	0.20	0.25	0.30	0.35	0.40	0.45	0.50	0.55	0.60	0.65	0.70	0.75	0.80	0.85	0.90	0.95
15	0	0.8601	0.4633	0.2059	0.0874	0.0352	0.0134	0.0047	0.0016	0.0005	0.0001	0.0000	0.0000	0.0000	0.0000	0.0000	0.0000	0.0000	0.0000	0.0000	0.0000
	1	0.1303	0.3658	0.3432	0.2312	0.1319	0.0668	0.0305	0.0126	0.0047	0.0016	0.0005	0.0001	0.0000	0.0000	0.0000	0.0000	0.0000	0.0000	0.0000	0.0000
	2	0.0092	0.1348	0.2669	0.2856	0.2309	0.1559	0.0916	0.0476	0.0219	0.0090	0.0032	0.0010	0.0003	0.0001	0.0000	0.0000	0.0000	0.0000	0.0000	0.0000
	3	0.0004	0.0307	0.1285	0.2184	0.2501	0.2262	0.1700	0.1110	0.0634	0.0318	0.0139	0.0052	0.0016	0.0004	0.0001	0.0000	0.0000	0.0000	0.0000	0.0000
	4	0.0000	0.0049	0.0428	0.1156	0.1876	0.2252	0.2186	0.1792	0.1268	0.0780	0.0417	0.0191	0.0074	0.0024	0.0006	0.0001	0.0000	0.0000	0.0000	0.0000
	5	0.0000	0.0006	0.0105	0.0449	0.1032	0.1651	0.2061	0.2123	0.1859	0.1404	0.0916	0.0515	0.0245	0.0096	0.0030	0.0007	0.0001	0.0000	0.0000	0.0000
	6	0.0000	0.0000	0.0019	0.0132	0.0430	0.0917	0.1472	0.1906	0.2066	0.1914	0.1527	0.1048	0.0612	0.0298	0.0116	0.0034	0.0007	0.0001	0.0000	0.0000
	7	0.0000	0.0000	0.0003	0.0030	0.0138	0.0393	0.0811	0.1319	0.1771	0.2013	0.1964	0.1647	0.1181	0.0710	0.0348	0.0131	0.0035	0.0005	0.0000	0.0000
	8	0.0000	0.0000	0.0000	0.0005	0.0035	0.0131	0.0348	0.0710	0.1181	0.1647	0.1964	0.2013	0.1771	0.1319	0.0811	0.0393	0.0138	0.0030	0.0003	0.0000
	9	0.0000	0.0000	0.0000	0.0001	0.0007	0.0034	0.0116	0.0298	0.0612	0.1048	0.1527	0.1914	0.2066	0.1906	0.1472	0.0917	0.0430	0.0132	0.0019	0.0000
	10	0.0000	0.0000	0.0000	0.0000	0.0001	0.0007	0.0030	0.0096	0.0245	0.0515	0.0916	0.1404	0.1859	0.2123	0.2061	0.1651	0.1032	0.0449	0.0105	0.0006
	11	0.0000	0.0000	0.0000	0.0000	0.0000	0.0001	0.0006	0.0024	0.0074	0.0191	0.0417	0.0780	0.1268	0.1792	0.2186	0.2252	0.1876	0.1156	0.0428	0.0049
	12	0.0000	0.0000	0.0000	0.0000	0.0000	0.0000	0.0001	0.0004	0.0016	0.0052	0.0139	0.0318	0.0634	0.1110	0.1700	0.2252	0.2501	0.2184	0.1285	0.0307
	13	0.0000	0.0000	0.0000	0.0000	0.0000	0.0000	0.0000	0.0001	0.0003	0.0010	0.0032	0.0090	0.0219	0.0476	0.0916	0.1559	0.2309	0.2856	0.2669	0.1348
	14	0.0000	0.0000	0.0000	0.0000	0.0000	0.0000	0.0000	0.0000	0.0000	0.0001	0.0005	0.0016	0.0047	0.0126	0.0305	0.0668	0.1319	0.2312	0.3432	0.3658
	15	0.0000	0.0000	0.0000	0.0000	0.0000	0.0000	0.0000	0.0000	0.0000	0.0000	0.0000	0.0001	0.0005	0.0016	0.0047	0.0134	0.0352	0.0874	0.2059	0.4633
20	0	0.8179	0.3585	0.1216	0.0388	0.0115	0.0032	0.0008	0.0002	0.0000	0.0000	0.0000	0.0000	0.0000	0.0000	0.0000	0.0000	0.0000	0.0000	0.0000	0.0000
	1	0.1652	0.3774	0.2702	0.1368	0.0576	0.0211	0.0068	0.0020	0.0005	0.0001	0.0000	0.0000	0.0000	0.0000	0.0000	0.0000	0.0000	0.0000	0.0000	0.0000
	2	0.0159	0.1887	0.2852	0.2293	0.1369	0.0669	0.0278	0.0100	0.0031	0.0008	0.0002	0.0000	0.0000	0.0000	0.0000	0.0000	0.0000	0.0000	0.0000	0.0000
	3	0.0010	0.0596	0.1901	0.2428	0.2054	0.1339	0.0716	0.0323	0.0123	0.0040	0.0011	0.0002	0.0000	0.0000	0.0000	0.0000	0.0000	0.0000	0.0000	0.0000
	4	0.0000	0.0133	0.0898	0.1821	0.2182	0.1897	0.1304	0.0738	0.0350	0.0139	0.0046	0.0013	0.0003	0.0000	0.0000	0.0000	0.0000	0.0000	0.0000	0.0000
	5	0.0000	0.0022	0.0319	0.1028	0.1746	0.2023	0.1789	0.1272	0.0746	0.0365	0.0148	0.0049	0.0013	0.0003	0.0000	0.0000	0.0000	0.0000	0.0000	0.0000
	6	0.0000	0.0003	0.0089	0.0454	0.1091	0.1686	0.1916	0.1712	0.1244	0.0746	0.0370	0.0150	0.0049	0.0012	0.0002	0.0000	0.0000	0.0000	0.0000	0.0000
	7	0.0000	0.0000	0.0020	0.0160	0.0545	0.1124	0.1643	0.1844	0.1659	0.1221	0.0739	0.0366	0.0146	0.0045	0.0010	0.0002	0.0000	0.0000	0.0000	0.0000
	8	0.0000	0.0000	0.0004	0.0046	0.0222	0.0609	0.1144	0.1614	0.1797	0.1623	0.1201	0.0727	0.0355	0.0136	0.0039	0.0008	0.0001	0.0000	0.0000	0.0000
	9	0.0000	0.0000	0.0001	0.0011	0.0074	0.0271	0.0654	0.1158	0.1597	0.1771	0.1602	0.1185	0.0710	0.0336	0.0120	0.0030	0.0005	0.0000	0.0000	0.0000
	10	0.0000	0.0000	0.0000	0.0002	0.0020	0.0099	0.0308	0.0686	0.1171	0.1593	0.1762	0.1593	0.1171	0.0686	0.0308	0.0099	0.0020	0.0002	0.0000	0.0000
	11	0.0000	0.0000	0.0000	0.0000	0.0005	0.0030	0.0120	0.0336	0.0710	0.1185	0.1602	0.1771	0.1597	0.1158	0.0654	0.0271	0.0074	0.0011	0.0001	0.0000
	12	0.0000	0.0000	0.0000	0.0000	0.0001	0.0008	0.0039	0.0136	0.0355	0.0727	0.1201	0.1623	0.1797	0.1614	0.1144	0.0609	0.0222	0.0046	0.0004	0.0000
	13	0.0000	0.0000	0.0000	0.0000	0.0000	0.0002	0.0010	0.0045	0.0146	0.0366	0.0739	0.1221	0.1659	0.1844	0.1643	0.1124	0.0545	0.0160	0.0020	0.0000
	14	0.0000	0.0000	0.0000	0.0000	0.0000	0.0000	0.0002	0.0012	0.0049	0.0150	0.0370	0.0746	0.1244	0.1712	0.1916	0.1686	0.1091	0.0454	0.0089	0.0003
	15	0.0000	0.0000	0.0000	0.0000	0.0000	0.0000	0.0000	0.0003	0.0013	0.0049	0.0148	0.0365	0.0746	0.1272	0.1789	0.2023	0.1746	0.1028	0.0319	0.0022
	16	0.0000	0.0000	0.0000	0.0000	0.0000	0.0000	0.0000	0.0000	0.0003	0.0013	0.0046	0.0139	0.0350	0.0738	0.1304	0.1897	0.2182	0.1821	0.0898	0.0133
	17	0.0000	0.0000	0.0000	0.0000	0.0000	0.0000	0.0000	0.0000	0.0000	0.0002	0.0011	0.0040	0.0123	0.0323	0.0716	0.1339	0.2054	0.2428	0.1901	0.0596
	18	0.0000	0.0000	0.0000	0.0000	0.0000	0.0000	0.0000	0.0000	0.0000	0.0000	0.0002	0.0008	0.0031	0.0100	0.0278	0.0669	0.1369	0.2293	0.2852	0.1887
	19	0.0000	0.0000	0.0000	0.0000	0.0000	0.0000	0.0000	0.0000	0.0000	0.0000	0.0000	0.0001	0.0005	0.0020	0.0068	0.0211	0.0576	0.1368	0.2702	0.3774
	20	0.0000	0.0000	0.0000	0.0000	0.0000	0.0000	0.0000	0.0000	0.0000	0.0000	0.0000	0.0000	0.0000	0.0002	0.0008	0.0032	0.0115	0.0388	0.1216	0.3585

Table III

Cumulative Binomial Probability Distribution

This table computes the cumulative probability of obtaining x successes in n trials of a binomial experiment with probability of success p.

											p										
n	x	0.01	0.05	0.10	0.15	0.20	0.25	0.30	0.35	0.40	0.45	0.50	0.55	0.60	0.65	0.70	0.75	0.80	0.85	0.90	0.95
2	0	0.9801	0.9025	0.8100	0.7225	0.6400	0.5625	0.4900	0.4225	0.3600	0.3025	0.2500	0.2025	0.1600	0.1225	0.0900	0.0625	0.0400	0.0225	0.0100	0.0025
	1	0.9999	0.9975	0.9900	0.9775	0.9600	0.9375	0.9100	0.8775	0.8400	0.7975	0.7500	0.6975	0.6400	0.5775	0.5100	0.4375	0.3600	0.2775	0.1900	0.0975
	2	1.0000	1.0000	1.0000	1.0000	1.0000	1.0000	1.0000	1.0000	1.0000	1.0000	1.0000	1.0000	1.0000	1.0000	1.0000	1.0000	1.0000	1.0000	1.0000	1.0000
3	0	0.9703	0.8574	0.7290	0.6141	0.5120	0.4219	0.3430	0.2746	0.2160	0.1664	0.1250	0.0911	0.0640	0.0429	0.0270	0.0156	0.0080	0.0034	0.0010	0.0001
	1	0.9997	0.9928	0.9720	0.9393	0.8960	0.8438	0.7840	0.7183	0.6480	0.5748	0.5000	0.4253	0.3520	0.2818	0.2160	0.1563	0.1040	0.0608	0.0280	0.0073
	2	1.0000	0.9999	0.9990	0.9966	0.9920	0.9844	0.9730	0.9571	0.9360	0.9089	0.8750	0.8336	0.7840	0.7254	0.6570	0.5781	0.4880	0.3859	0.2710	0.1426
	3	1.0000	1.0000	1.0000	1.0000	1.0000	1.0000	1.0000	1.0000	1.0000	1.0000	1.0000	1.0000	1.0000	1.0000	1.0000	1.0000	1.0000	1.0000	1.0000	1.0000
4	0	0.9606	0.8145	0.6561	0.5220	0.4096	0.3164	0.2401	0.1785	0.1296	0.0915	0.0625	0.0410	0.0256	0.0150	0.0081	0.0039	0.0016	0.0005	0.0001	0.0000
	1	0.9994	0.9860	0.9477	0.8905	0.8192	0.7383	0.6517	0.5630	0.4752	0.3910	0.3125	0.2415	0.1792	0.1265	0.0837	0.0508	0.0272	0.0120	0.0037	0.0005
	2	1.0000	0.9995	0.9963	0.9880	0.9728	0.9492	0.9163	0.8735	0.8208	0.7585	0.6875	0.6090	0.5248	0.4370	0.3483	0.2617	0.1808	0.1095	0.0523	0.0140
	3	1.0000	1.0000	0.9999	0.9995	0.9984	0.9961	0.9919	0.9850	0.9744	0.9590	0.9375	0.9085	0.8704	0.8215	0.7599	0.6836	0.5904	0.4780	0.3439	0.1855
	4	1.0000	1.0000	1.0000	1.0000	1.0000	1.0000	1.0000	1.0000	1.0000	1.0000	1.0000	1.0000	1.0000	1.0000	1.0000	1.0000	1.0000	1.0000	1.0000	1.0000
5	0	0.9510	0.7738	0.5905	0.4437	0.3277	0.2373	0.1681	0.1160	0.0778	0.0503	0.0313	0.0185	0.0102	0.0053	0.0024	0.0010	0.0003	0.0001	0.0000	0.0000
	1	0.9990	0.9774	0.9185	0.8352	0.7373	0.6328	0.5282	0.4284	0.3370	0.2562	0.1875	0.1312	0.0870	0.0540	0.0308	0.0156	0.0067	0.0022	0.0005	0.0000
	2	1.0000	0.9988	0.9914	0.9734	0.9421	0.8965	0.8369	0.7648	0.6826	0.5931	0.5000	0.4069	0.3174	0.2352	0.1631	0.1035	0.0579	0.0266	0.0086	0.0012
	3	1.0000	1.0000	0.9995	0.9978	0.9933	0.9844	0.9692	0.9460	0.9130	0.8688	0.8125	0.7438	0.6630	0.5716	0.4718	0.3672	0.2627	0.1648	0.0815	0.0226
	4	1.0000	1.0000	1.0000	0.9999	0.9997	0.9990	0.9976	0.9947	0.9898	0.9815	0.9688	0.9497	0.9222	0.8840	0.8319	0.7627	0.6723	0.5563	0.4095	0.2262
	5	1.0000	1.0000	1.0000	1.0000	1.0000	1.0000	1.0000	1.0000	1.0000	1.0000	1.0000	1.0000	1.0000	1.0000	1.0000	1.0000	1.0000	1.0000	1.0000	1.0000
6	0	0.9415	0.7351	0.5314	0.3771	0.2621	0.1780	0.1176	0.0754	0.0467	0.0277	0.0156	0.0083	0.0041	0.0018	0.0007	0.0002	0.0001	0.0000	0.0000	0.0000
	1	0.9985	0.9672	0.8857	0.7765	0.6554	0.5339	0.4202	0.3191	0.2333	0.1636	0.1094	0.0692	0.0410	0.0223	0.0109	0.0046	0.0016	0.0004	0.0001	0.0000
	2	1.0000	0.9978	0.9842	0.9527	0.9011	0.8306	0.7443	0.6471	0.5443	0.4415	0.3438	0.2553	0.1792	0.1174	0.0705	0.0376	0.0170	0.0059	0.0013	0.0001
	3	1.0000	0.9999	0.9987	0.9941	0.9830	0.9624	0.9295	0.8826	0.8208	0.7447	0.6563	0.5585	0.4557	0.3529	0.2557	0.1694	0.0989	0.0473	0.0159	0.0022
	4	1.0000	1.0000	0.9999	0.9996	0.9984	0.9954	0.9891	0.9777	0.9590	0.9308	0.8906	0.8364	0.7667	0.6809	0.5798	0.4661	0.3446	0.2235	0.1143	0.0328
	5	1.0000	1.0000	1.0000	1.0000	0.9999	0.9998	0.9993	0.9982	0.9959	0.9917	0.9844	0.9723	0.9533	0.9246	0.8824	0.8220	0.7379	0.6229	0.4686	0.2649
	6	1.0000	1.0000	1.0000	1.0000	1.0000	1.0000	1.0000	1.0000	1.0000	1.0000	1.0000	1.0000	1.0000	1.0000	1.0000	1.0000	1.0000	1.0000	1.0000	1.0000
7	0	0.9321	0.6983	0.4783	0.3206	0.2097	0.1335	0.0824	0.0490	0.0280	0.0152	0.0078	0.0037	0.0016	0.0006	0.0002	0.0001	0.0000	0.0000	0.0000	0.0000
	1	0.9980	0.9556	0.8503	0.7166	0.5767	0.4449	0.3294	0.2338	0.1586	0.1024	0.0625	0.0357	0.0188	0.0090	0.0038	0.0013	0.0004	0.0001	0.0000	0.0000
	2	1.0000	0.9962	0.9743	0.9262	0.8520	0.7564	0.6471	0.5323	0.4199	0.3164	0.2266	0.1529	0.0963	0.0556	0.0288	0.0129	0.0047	0.0012	0.0002	0.0000
	3	1.0000	0.9998	0.9973	0.9879	0.9667	0.9294	0.8740	0.8002	0.7102	0.6083	0.5000	0.3917	0.2898	0.1998	0.1260	0.0706	0.0333	0.0121	0.0027	0.0002
	4	1.0000	1.0000	0.9998	0.9988	0.9953	0.9871	0.9712	0.9444	0.9037	0.8471	0.7734	0.6836	0.5801	0.4677	0.3529	0.2436	0.1480	0.0738	0.0257	0.0038
	5	1.0000	1.0000	1.0000	0.9999	0.9996	0.9987	0.9962	0.9910	0.9812	0.9643	0.9375	0.8976	0.8414	0.7662	0.6706	0.5551	0.4233	0.2834	0.1497	0.0444
	6	1.0000	1.0000	1.0000	1.0000	1.0000	0.9999	0.9998	0.9994	0.9984	0.9963	0.9922	0.9848	0.9720	0.9510	0.9176	0.8665	0.7903	0.6794	0.5217	0.3017
	7	1.0000	1.0000	1.0000	1.0000	1.0000	1.0000	1.0000	1.0000	1.0000	1.0000	1.0000	1.0000	1.0000	1.0000	1.0000	1.0000	1.0000	1.0000	1.0000	1.0000

Table III (continued)

n	x	0.01	0.05	0.10	0.15	0.20	0.25	0.30	0.35	0.40	0.45	0.50	0.55	0.60	0.65	0.70	0.75	0.80	0.85	0.90	0.95
												p									
8	0	0.9227	0.6634	0.4305	0.2725	0.1678	0.1001	0.0576	0.0319	0.0168	0.0084	0.0039	0.0017	0.0007	0.0002	0.0001	0.0000	0.0000	0.0000	0.0000	0.0000
	1	0.9973	0.9428	0.8131	0.6572	0.5033	0.3671	0.2553	0.1691	0.1064	0.0632	0.0352	0.0181	0.0085	0.0036	0.0013	0.0004	0.0001	0.0000	0.0000	0.0000
	2	0.9999	0.9942	0.9619	0.8948	0.7969	0.6785	0.5518	0.4278	0.3154	0.2201	0.1445	0.0885	0.0498	0.0253	0.0113	0.0042	0.0012	0.0002	0.0000	0.0000
	3	1.0000	0.9996	0.9950	0.9786	0.9437	0.8862	0.8059	0.7064	0.5941	0.4770	0.3633	0.2604	0.1737	0.1061	0.0580	0.0273	0.0104	0.0029	0.0004	0.0000
	4	1.0000	1.0000	0.9996	0.9971	0.9896	0.9727	0.9420	0.8939	0.8263	0.7396	0.6367	0.5230	0.4059	0.2936	0.1941	0.1138	0.0563	0.0214	0.0050	0.0004
	5	1.0000	1.0000	1.0000	0.9998	0.9988	0.9958	0.9887	0.9747	0.9502	0.9115	0.8555	0.7799	0.6846	0.5722	0.4482	0.3215	0.2031	0.1052	0.0381	0.0058
	6	1.0000	1.0000	1.0000	1.0000	0.9999	0.9996	0.9987	0.9964	0.9915	0.9819	0.9648	0.9368	0.8936	0.8309	0.7447	0.6329	0.4967	0.3428	0.1869	0.0572
	7	1.0000	1.0000	1.0000	1.0000	1.0000	1.0000	0.9999	0.9998	0.9993	0.9983	0.9961	0.9916	0.9832	0.9681	0.9424	0.8999	0.8322	0.7275	0.5695	0.3366
	8	1.0000	1.0000	1.0000	1.0000	1.0000	1.0000	1.0000	1.0000	1.0000	1.0000	1.0000	1.0000	1.0000	1.0000	1.0000	1.0000	1.0000	1.0000	1.0000	1.0000
9	0	0.9135	0.6302	0.3874	0.2316	0.1342	0.0751	0.0404	0.0207	0.0101	0.0046	0.0020	0.0008	0.0003	0.0001	0.0000	0.0000	0.0000	0.0000	0.0000	0.0000
	1	0.9966	0.9288	0.7748	0.5995	0.4362	0.3003	0.1960	0.1211	0.0705	0.0385	0.0195	0.0091	0.0038	0.0014	0.0004	0.0001	0.0000	0.0000	0.0000	0.0000
	2	0.9999	0.9916	0.9470	0.8591	0.7382	0.6007	0.4628	0.3373	0.2318	0.1495	0.0898	0.0498	0.0250	0.0112	0.0043	0.0013	0.0003	0.0000	0.0000	0.0000
	3	1.0000	0.9994	0.9917	0.9661	0.9144	0.8343	0.7297	0.6089	0.4826	0.3614	0.2539	0.1658	0.0994	0.0536	0.0253	0.0100	0.0031	0.0006	0.0001	0.0000
	4	1.0000	1.0000	0.9991	0.9944	0.9804	0.9511	0.9012	0.8283	0.7334	0.6214	0.5000	0.3786	0.2666	0.1717	0.0988	0.0489	0.0196	0.0056	0.0009	0.0000
	5	1.0000	1.0000	0.9999	0.9994	0.9969	0.9900	0.9747	0.9464	0.9006	0.8342	0.7461	0.6386	0.5174	0.3911	0.2703	0.1657	0.0856	0.0339	0.0083	0.0006
	6	1.0000	1.0000	1.0000	1.0000	0.9997	0.9987	0.9957	0.9888	0.9750	0.9502	0.9102	0.8505	0.7682	0.6627	0.5372	0.3993	0.2618	0.1409	0.0530	0.0084
	7	1.0000	1.0000	1.0000	1.0000	1.0000	0.9999	0.9996	0.9986	0.9962	0.9909	0.9805	0.9615	0.9295	0.8789	0.8040	0.6997	0.5638	0.4005	0.2252	0.0712
	8	1.0000	1.0000	1.0000	1.0000	1.0000	1.0000	1.0000	0.9999	0.9997	0.9992	0.9980	0.9954	0.9899	0.9793	0.9596	0.9249	0.8658	0.7684	0.6126	0.3698
	9	1.0000	1.0000	1.0000	1.0000	1.0000	1.0000	1.0000	1.0000	1.0000	1.0000	1.0000	1.0000	1.0000	1.0000	1.0000	1.0000	1.0000	1.0000	1.0000	1.0000
10	0	0.9044	0.5987	0.3487	0.1969	0.1074	0.0563	0.0282	0.0135	0.0060	0.0025	0.0010	0.0003	0.0001	0.0000	0.0000	0.0000	0.0000	0.0000	0.0000	0.0000
	1	0.9957	0.9139	0.7361	0.5443	0.3758	0.2440	0.1493	0.0860	0.0464	0.0233	0.0107	0.0046	0.0017	0.0005	0.0001	0.0000	0.0000	0.0000	0.0000	0.0000
	2	0.9999	0.9885	0.9298	0.8202	0.6778	0.5256	0.3828	0.2616	0.1673	0.0996	0.0547	0.0274	0.0123	0.0048	0.0016	0.0004	0.0001	0.0000	0.0000	0.0000
	3	1.0000	0.9990	0.9872	0.9500	0.8791	0.7759	0.6496	0.5138	0.3823	0.2660	0.1719	0.1020	0.0548	0.0280	0.0106	0.0035	0.0009	0.0001	0.0000	0.0000
	4	1.0000	0.9999	0.9984	0.9901	0.9672	0.9219	0.8497	0.7515	0.6331	0.5044	0.3770	0.2616	0.1662	0.0949	0.0473	0.0197	0.0064	0.0014	0.0001	0.0000
	5	1.0000	1.0000	0.9999	0.9986	0.9936	0.9803	0.9527	0.9051	0.8338	0.7384	0.6230	0.4956	0.3669	0.2485	0.1503	0.0781	0.0328	0.0099	0.0016	0.0001
	6	1.0000	1.0000	1.0000	0.9999	0.9991	0.9965	0.9894	0.9740	0.9452	0.8980	0.8281	0.7340	0.6177	0.4862	0.3504	0.2241	0.1209	0.0500	0.0128	0.0010
	7	1.0000	1.0000	1.0000	1.0000	0.9999	0.9996	0.9984	0.9952	0.9877	0.9726	0.9453	0.9004	0.8327	0.7384	0.6172	0.4744	0.3222	0.1798	0.0702	0.0115
	8	1.0000	1.0000	1.0000	1.0000	1.0000	1.0000	0.9999	0.9995	0.9983	0.9955	0.9893	0.9767	0.9536	0.9140	0.8507	0.7560	0.6242	0.4557	0.2639	0.0861
	9	1.0000	1.0000	1.0000	1.0000	1.0000	1.0000	1.0000	1.0000	0.9999	0.9997	0.9990	0.9975	0.9940	0.9865	0.9718	0.9437	0.8926	0.8031	0.6513	0.4013
	10	1.0000	1.0000	1.0000	1.0000	1.0000	1.0000	1.0000	1.0000	1.0000	1.0000	1.0000	1.0000	1.0000	1.0000	1.0000	1.0000	1.0000	1.0000	1.0000	1.0000

Table III (continued)

n	x	0.01	0.05	0.10	0.15	0.20	0.25	0.30	0.35	0.40	0.45	0.50	0.55	0.60	0.65	0.70	0.75	0.80	0.85	0.90	0.95
11	0	0.8953	0.5688	0.3138	0.1673	0.0859	0.0422	0.0198	0.0088	0.0036	0.0014	0.0005	0.0002	0.0000	0.0000	0.0000	0.0000	0.0000	0.0000	0.0000	0.0000
	1	0.9948	0.8981	0.6974	0.4922	0.3221	0.1971	0.1130	0.0606	0.0302	0.0139	0.0059	0.0022	0.0007	0.0002	0.0000	0.0000	0.0000	0.0000	0.0000	0.0000
	2	0.9998	0.9848	0.9104	0.7788	0.6174	0.4552	0.3127	0.2001	0.1189	0.0652	0.0327	0.0148	0.0059	0.0020	0.0006	0.0001	0.0000	0.0000	0.0000	0.0000
	3	1.0000	0.9984	0.9815	0.9306	0.8389	0.7133	0.5696	0.4256	0.2963	0.1911	0.1133	0.0610	0.0293	0.0122	0.0043	0.0012	0.0002	0.0000	0.0000	0.0000
	4	1.0000	0.9999	0.9972	0.9841	0.9496	0.8854	0.7897	0.6683	0.5328	0.3971	0.2744	0.1738	0.0994	0.0501	0.0216	0.0076	0.0020	0.0003	0.0000	0.0000
	5	1.0000	1.0000	0.9997	0.9973	0.9883	0.9657	0.9218	0.8513	0.7535	0.6331	0.5000	0.3669	0.2465	0.1487	0.0782	0.0343	0.0117	0.0027	0.0003	0.0000
	6	1.0000	1.0000	1.0000	0.9997	0.9980	0.9924	0.9784	0.9499	0.9006	0.8262	0.7258	0.6029	0.4672	0.3317	0.2103	0.1146	0.0504	0.0159	0.0028	0.0001
	7	1.0000	1.0000	1.0000	1.0000	0.9998	0.9988	0.9957	0.9878	0.9707	0.9390	0.8867	0.8089	0.7037	0.5744	0.4304	0.2867	0.1611	0.0694	0.0185	0.0016
	8	1.0000	1.0000	1.0000	1.0000	1.0000	0.9999	0.9994	0.9980	0.9941	0.9852	0.9673	0.9348	0.8811	0.7999	0.6873	0.5448	0.3826	0.2212	0.0896	0.0152
	9	1.0000	1.0000	1.0000	1.0000	1.0000	1.0000	1.0000	0.9998	0.9993	0.9976	0.9941	0.9861	0.9698	0.9394	0.8870	0.8029	0.6779	0.6078	0.3028	0.1019
	10	1.0000	1.0000	1.0000	1.0000	1.0000	1.0000	1.0000	1.0000	1.0000	0.9996	0.9995	0.9986	0.9964	0.9912	0.9802	0.9578	0.9141	0.8327	0.6862	0.4312
	11	1.0000	1.0000	1.0000	1.0000	1.0000	1.0000	1.0000	1.0000	1.0000	1.0000	1.0000	1.0000	1.0000	1.0000	1.0000	1.0000	1.0000	1.0000	1.0000	1.0000
12	0	0.8864	0.5404	0.2824	0.1422	0.0687	0.0317	0.0138	0.0057	0.0022	0.0008	0.0002	0.0001	0.0000	0.0000	0.0000	0.0000	0.0000	0.0000	0.0000	0.0000
	1	0.9938	0.8816	0.6590	0.4435	0.2749	0.1584	0.0850	0.0424	0.0196	0.0083	0.0032	0.0011	0.0003	0.0001	0.0000	0.0000	0.0000	0.0000	0.0000	0.0000
	2	0.9998	0.9804	0.8891	0.7358	0.5583	0.3907	0.2528	0.1513	0.0834	0.0421	0.0193	0.0079	0.0028	0.0008	0.0002	0.0000	0.0000	0.0000	0.0000	0.0000
	3	1.0000	0.9978	0.9744	0.9078	0.7946	0.6488	0.4925	0.3467	0.2253	0.1345	0.0730	0.0356	0.0153	0.0056	0.0017	0.0004	0.0001	0.0000	0.0000	0.0000
	4	1.0000	0.9998	0.9957	0.9761	0.9274	0.8424	0.7237	0.5833	0.4382	0.3044	0.1938	0.1117	0.0573	0.0255	0.0095	0.0028	0.0006	0.0001	0.0000	0.0000
	5	1.0000	1.0000	0.9995	0.9954	0.9806	0.9456	0.8822	0.7873	0.6652	0.5269	0.3872	0.2607	0.1582	0.0846	0.0386	0.0143	0.0039	0.0007	0.0001	0.0000
	6	1.0000	1.0000	0.9999	0.9993	0.9961	0.9857	0.9614	0.9154	0.8418	0.7393	0.6128	0.4731	0.3348	0.2127	0.1178	0.0544	0.0194	0.0046	0.0005	0.0000
	7	1.0000	1.0000	1.0000	0.9999	0.9994	0.9972	0.9905	0.9745	0.9427	0.8883	0.8062	0.6956	0.5618	0.4167	0.2763	0.1576	0.0726	0.0239	0.0043	0.0002
	8	1.0000	1.0000	1.0000	1.0000	0.9999	0.9996	0.9983	0.9944	0.9847	0.9644	0.9270	0.8655	0.7747	0.6533	0.5075	0.3512	0.2054	0.0922	0.0256	0.0022
	9	1.0000	1.0000	1.0000	1.0000	1.0000	1.0000	0.9998	0.9992	0.9972	0.9921	0.9807	0.9579	0.9166	0.8487	0.7472	0.6093	0.4417	0.2642	0.1109	0.0196
	10	1.0000	1.0000	1.0000	1.0000	1.0000	1.0000	1.0000	0.9999	0.9997	0.9989	0.9968	0.9917	0.9804	0.9576	0.9150	0.8416	0.7251	0.5565	0.3410	0.1184
	11	1.0000	1.0000	1.0000	1.0000	1.0000	1.0000	1.0000	1.0000	1.0000	0.9999	0.9998	0.9992	0.9978	0.9943	0.9862	0.9683	0.9313	0.8578	0.7176	0.4596
	12	1.0000	1.0000	1.0000	1.0000	1.0000	1.0000	1.0000	1.0000	1.0000	1.0000	1.0000	1.0000	1.0000	1.0000	1.0000	1.0000	1.0000	1.0000	1.0000	1.0000

p

Table III (continued)

p

n	x	0.01	0.05	0.10	0.15	0.20	0.25	0.30	0.35	0.40	0.45	0.50	0.55	0.60	0.65	0.70	0.75	0.80	0.85	0.90	0.95
15	0	0.8601	0.4633	0.2059	0.0874	0.0352	0.0134	0.0047	0.0016	0.0005	0.0001	0.0000	0.0000	0.0000	0.0000	0.0000	0.0000	0.0000	0.0000	0.0000	0.0000
	1	0.9904	0.8290	0.5490	0.3186	0.1671	0.0802	0.0353	0.0142	0.0052	0.0017	0.0005	0.0001	0.0000	0.0000	0.0000	0.0000	0.0000	0.0000	0.0000	0.0000
	2	0.9996	0.9638	0.8159	0.6042	0.3980	0.2361	0.1268	0.0617	0.0271	0.0107	0.0037	0.0011	0.0003	0.0001	0.0000	0.0000	0.0000	0.0000	0.0000	0.0000
	3	1.0000	0.9945	0.9444	0.8227	0.6482	0.4613	0.2969	0.1727	0.0905	0.0424	0.0176	0.0063	0.0019	0.0005	0.0001	0.0001	0.0000	0.0000	0.0000	0.0000
	4	1.0000	0.9994	0.9873	0.9383	0.8358	0.6865	0.5155	0.3519	0.2173	0.1204	0.0592	0.0255	0.0093	0.0028	0.0007	0.0001	0.0000	0.0000	0.0000	0.0000
	5	1.0000	0.9999	0.9978	0.9832	0.9389	0.8516	0.7216	0.5643	0.4032	0.2608	0.1509	0.0769	0.0338	0.0124	0.0037	0.0008	0.0001	0.0000	0.0000	0.0000
	6	1.0000	1.0000	0.9997	0.9964	0.9819	0.9434	0.8689	0.7548	0.6098	0.4522	0.3036	0.1818	0.0950	0.0422	0.0152	0.0042	0.0008	0.0001	0.0000	0.0000
	7	1.0000	1.0000	1.0000	0.9994	0.9958	0.9827	0.9500	0.8868	0.7869	0.6535	0.5000	0.3465	0.2131	0.1132	0.0500	0.0173	0.0042	0.0006	0.0000	0.0000
	8	1.0000	1.0000	1.0000	0.9999	0.9992	0.9958	0.9848	0.9578	0.9050	0.8182	0.6964	0.5478	0.3902	0.2452	0.1311	0.0566	0.0181	0.0036	0.0003	0.0000
	9	1.0000	1.0000	1.0000	1.0000	0.9999	0.9992	0.9963	0.9876	0.9662	0.9231	0.8491	0.7392	0.5968	0.4357	0.2784	0.1484	0.0611	0.0168	0.0022	0.0001
	10	1.0000	1.0000	1.0000	1.0000	1.0000	0.9999	0.9993	0.9972	0.9907	0.9745	0.9408	0.8796	0.7827	0.6481	0.4845	0.3135	0.1642	0.0617	0.0127	0.0006
	11	1.0000	1.0000	1.0000	1.0000	1.0000	1.0000	0.9999	0.9995	0.9981	0.9937	0.9824	0.9576	0.9095	0.8273	0.7031	0.5387	0.3518	0.1773	0.0556	0.0055
	12	1.0000	1.0000	1.0000	1.0000	1.0000	1.0000	1.0000	0.9999	0.9997	0.9989	0.9963	0.9893	0.9729	0.9383	0.8732	0.7639	0.6020	0.3958	0.1841	0.0362
	13	1.0000	1.0000	1.0000	1.0000	1.0000	1.0000	1.0000	1.0000	1.0000	0.9999	0.9995	0.9983	0.9948	0.9858	0.9647	0.9198	0.8329	0.6814	0.4510	0.1710
	14	1.0000	1.0000	1.0000	1.0000	1.0000	1.0000	1.0000	1.0000	1.0000	1.0000	1.0000	0.9999	0.9995	0.9984	0.9953	0.9866	0.9648	0.9126	0.7941	0.5367
	15	1.0000	1.0000	1.0000	1.0000	1.0000	1.0000	1.0000	1.0000	1.0000	1.0000	1.0000	1.0000	1.0000	1.0000	1.0000	1.0000	1.0000	1.0000	1.0000	1.0000
20	0	0.8179	0.3585	0.1216	0.0388	0.0115	0.0032	0.0008	0.0002	0.0000	0.0000	0.0000	0.0000	0.0000	0.0000	0.0000	0.0000	0.0000	0.0000	0.0000	0.0000
	1	0.9831	0.7358	0.3917	0.1756	0.0692	0.0243	0.0076	0.0021	0.0005	0.0001	0.0000	0.0000	0.0000	0.0000	0.0000	0.0000	0.0000	0.0000	0.0000	0.0000
	2	0.9990	0.9245	0.6769	0.4049	0.2061	0.0913	0.0355	0.0121	0.0036	0.0009	0.0002	0.0000	0.0000	0.0000	0.0000	0.0000	0.0000	0.0000	0.0000	0.0000
	3	1.0000	0.9841	0.8670	0.6477	0.4114	0.2252	0.1071	0.0444	0.0160	0.0049	0.0013	0.0003	0.0000	0.0000	0.0000	0.0000	0.0000	0.0000	0.0000	0.0000
	4	1.0000	0.9974	0.9568	0.8298	0.6296	0.4148	0.2375	0.1182	0.0510	0.0189	0.0059	0.0015	0.0003	0.0000	0.0000	0.0000	0.0000	0.0000	0.0000	0.0000
	5	1.0000	0.9997	0.9887	0.9327	0.8042	0.6172	0.4164	0.2454	0.1256	0.0553	0.0207	0.0064	0.0016	0.0003	0.0000	0.0000	0.0000	0.0000	0.0000	0.0000
	6	1.0000	1.0000	0.9976	0.9781	0.9133	0.7858	0.6080	0.4166	0.2500	0.1299	0.0577	0.0214	0.0065	0.0015	0.0003	0.0000	0.0000	0.0000	0.0000	0.0000
	7	1.0000	1.0000	0.9996	0.9941	0.9679	0.8982	0.7723	0.6010	0.4159	0.2520	0.1316	0.0580	0.0210	0.0060	0.0013	0.0002	0.0000	0.0000	0.0000	0.0000
	8	1.0000	1.0000	0.9999	0.9987	0.9900	0.9591	0.8867	0.7624	0.5956	0.4143	0.2517	0.1308	0.0565	0.0196	0.0051	0.0009	0.0001	0.0000	0.0000	0.0000
	9	1.0000	1.0000	1.0000	0.9998	0.9974	0.9861	0.9520	0.8782	0.7553	0.5914	0.4119	0.2493	0.1275	0.0532	0.0171	0.0039	0.0006	0.0000	0.0000	0.0000
	10	1.0000	1.0000	1.0000	1.0000	0.9994	0.9961	0.9829	0.9468	0.8725	0.7507	0.5881	0.4086	0.2447	0.1218	0.0480	0.0139	0.0026	0.0002	0.0000	0.0000
	11	1.0000	1.0000	1.0000	1.0000	0.9999	0.9991	0.9949	0.9804	0.9435	0.8692	0.7483	0.5857	0.4044	0.2376	0.1133	0.0409	0.0100	0.0013	0.0001	0.0000
	12	1.0000	1.0000	1.0000	1.0000	1.0000	0.9998	0.9987	0.9940	0.9790	0.9420	0.8684	0.7480	0.5841	0.3990	0.2277	0.1018	0.0321	0.0059	0.0004	0.0000
	13	1.0000	1.0000	1.0000	1.0000	1.0000	1.0000	0.9997	0.9985	0.9935	0.9786	0.9423	0.8701	0.7500	0.5834	0.3920	0.2142	0.0867	0.0219	0.0024	0.0003
	14	1.0000	1.0000	1.0000	1.0000	1.0000	1.0000	1.0000	0.9997	0.9984	0.9936	0.9793	0.9447	0.8744	0.7546	0.5836	0.3828	0.1958	0.0673	0.0113	0.0026
	15	1.0000	1.0000	1.0000	1.0000	1.0000	1.0000	1.0000	1.0000	0.9997	0.9985	0.9941	0.9811	0.9490	0.8818	0.7625	0.5852	0.3704	0.1702	0.0432	0.0159
	16	1.0000	1.0000	1.0000	1.0000	1.0000	1.0000	1.0000	1.0000	1.0000	0.9997	0.9987	0.9951	0.9840	0.9556	0.8929	0.7748	0.5886	0.3523	0.1330	0.0755
	17	1.0000	1.0000	1.0000	1.0000	1.0000	1.0000	1.0000	1.0000	1.0000	1.0000	0.9998	0.9991	0.9964	0.9879	0.9645	0.9087	0.7939	0.5951	0.3231	0.2642
	18	1.0000	1.0000	1.0000	1.0000	1.0000	1.0000	1.0000	1.0000	1.0000	1.0000	1.0000	0.9999	0.9995	0.9979	0.9924	0.9757	0.9308	0.8244	0.6083	0.6415
	19	1.0000	1.0000	1.0000	1.0000	1.0000	1.0000	1.0000	1.0000	1.0000	1.0000	1.0000	1.0000	1.0000	0.9998	0.9992	0.9968	0.9885	0.9612	0.8784	1.0000
	20	1.0000	1.0000	1.0000	1.0000	1.0000	1.0000	1.0000	1.0000	1.0000	1.0000	1.0000	1.0000	1.0000	1.0000	1.0000	1.0000	1.0000	1.0000	1.0000	1.0000

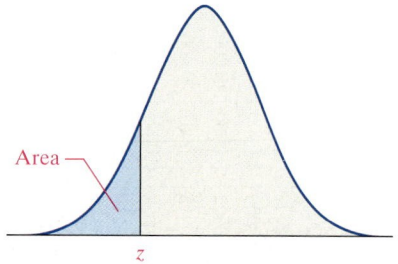

Area ─ z

Table IV

Standard Normal Distribution

z	.00	.01	.02	.03	.04	.05	.06	.07	.08	.09
−3.4	0.0003	0.0003	0.0003	0.0003	0.0003	0.0003	0.0003	0.0003	0.0003	0.0002
−3.3	0.0005	0.0005	0.0005	0.0004	0.0004	0.0004	0.0004	0.0004	0.0004	0.0003
−3.2	0.0007	0.0007	0.0006	0.0006	0.0006	0.0006	0.0006	0.0005	0.0005	0.0005
−3.1	0.0010	0.0009	0.0009	0.0009	0.0008	0.0008	0.0008	0.0008	0.0007	0.0007
−3.0	0.0013	0.0013	0.0013	0.0012	0.0012	0.0011	0.0011	0.0011	0.0010	0.0010
−2.9	0.0019	0.0018	0.0018	0.0017	0.0016	0.0016	0.0015	0.0015	0.0014	0.0014
−2.8	0.0026	0.0025	0.0024	0.0023	0.0023	0.0022	0.0021	0.0021	0.0020	0.0019
−2.7	0.0035	0.0034	0.0033	0.0032	0.0031	0.0030	0.0029	0.0028	0.0027	0.0026
−2.6	0.0047	0.0045	0.0044	0.0043	0.0041	0.0040	0.0039	0.0038	0.0037	0.0036
−2.5	0.0062	0.0060	0.0059	0.0057	0.0055	0.0054	0.0052	0.0051	0.0049	0.0048
−2.4	0.0082	0.0080	0.0078	0.0075	0.0073	0.0071	0.0069	0.0068	0.0066	0.0064
−2.3	0.0107	0.0104	0.0102	0.0099	0.0096	0.0094	0.0091	0.0089	0.0087	0.0084
−2.2	0.0139	0.0136	0.0132	0.0129	0.0125	0.0122	0.0119	0.0116	0.0113	0.0110
−2.1	0.0179	0.0174	0.0170	0.0166	0.0162	0.0158	0.0154	0.0150	0.0146	0.0143
−2.0	0.0228	0.0222	0.0217	0.0212	0.0207	0.0202	0.0197	0.0192	0.0188	0.0183
−1.9	0.0287	0.0281	0.0274	0.0268	0.0262	0.0256	0.0250	0.0244	0.0239	0.0233
−1.8	0.0359	0.0351	0.0344	0.0336	0.0329	0.0322	0.0314	0.0307	0.0301	0.0294
−1.7	0.0446	0.0436	0.0427	0.0418	0.0409	0.0401	0.0392	0.0384	0.0375	0.0367
−1.6	0.0548	0.0537	0.0526	0.0516	0.0505	0.0495	0.0485	0.0475	0.0465	0.0455
−1.5	0.0668	0.0655	0.0643	0.0630	0.0618	0.0606	0.0594	0.0582	0.0571	0.0559
−1.4	0.0808	0.0793	0.0778	0.0764	0.0749	0.0735	0.0721	0.0708	0.0694	0.0681
−1.3	0.0968	0.0951	0.0934	0.0918	0.0901	0.0885	0.0869	0.0853	0.0838	0.0823
−1.2	0.1151	0.1131	0.1112	0.1093	0.1075	0.1056	0.1038	0.1020	0.1003	0.0985
−1.1	0.1357	0.1335	0.1314	0.1292	0.1271	0.1251	0.1230	0.1210	0.1190	0.1170
−1.0	0.1587	0.1562	0.1539	0.1515	0.1492	0.1469	0.1446	0.1423	0.1401	0.1379
−0.9	0.1841	0.1814	0.1788	0.1762	0.1736	0.1711	0.1685	0.1660	0.1635	0.1611
−0.8	0.2119	0.2090	0.2061	0.2033	0.2005	0.1977	0.1949	0.1922	0.1894	0.1867
−0.7	0.2420	0.2389	0.2358	0.2327	0.2296	0.2266	0.2236	0.2206	0.2177	0.2148
−0.6	0.2743	0.2709	0.2676	0.2643	0.2611	0.2578	0.2546	0.2514	0.2483	0.2451
−0.5	0.3085	0.3050	0.3015	0.2981	0.2946	0.2912	0.2877	0.2843	0.2810	0.2776
−0.4	0.3446	0.3409	0.3372	0.3336	0.3300	0.3264	0.3228	0.3192	0.3156	0.3121
−0.3	0.3821	0.3783	0.3745	0.3707	0.3669	0.3632	0.3594	0.3557	0.3520	0.3483
−0.2	0.4207	0.4168	0.4129	0.4090	0.4052	0.4013	0.3974	0.3936	0.3897	0.3859
−0.1	0.4602	0.4562	0.4522	0.4483	0.4443	0.4404	0.4364	0.4325	0.4286	0.4247
−0.0	0.5000	0.4960	0.4920	0.4880	0.4840	0.4801	0.4761	0.4721	0.4681	0.4641

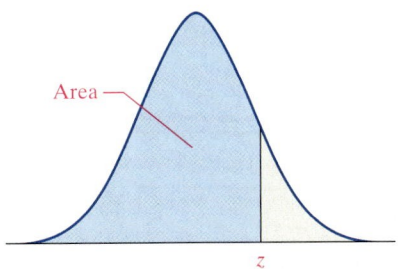
Area

Table IV (continued)

Standard Normal Distribution

z	.00	.01	.02	.03	.04	.05	.06	.07	.08	.09
0.0	0.5000	0.5040	0.5080	0.5120	0.5160	0.5199	0.5239	0.5279	0.5319	0.5359
0.1	0.5398	0.5438	0.5478	0.5517	0.5557	0.5596	0.5636	0.5675	0.5714	0.5753
0.2	0.5793	0.5832	0.5871	0.5910	0.5948	0.5987	0.6026	0.6064	0.6103	0.6141
0.3	0.6179	0.6217	0.6255	0.6293	0.6331	0.6368	0.6406	0.6443	0.6480	0.6517
0.4	0.6554	0.6591	0.6628	0.6664	0.6700	0.6736	0.6772	0.6808	0.6844	0.6879
0.5	0.6915	0.6950	0.6985	0.7019	0.7054	0.7088	0.7123	0.7157	0.7190	0.7224
0.6	0.7257	0.7291	0.7324	0.7357	0.7389	0.7422	0.7454	0.7486	0.7517	0.7549
0.7	0.7580	0.7611	0.7642	0.7673	0.7704	0.7734	0.7764	0.7794	0.7823	0.7852
0.8	0.7881	0.7910	0.7939	0.7967	0.7995	0.8023	0.8051	0.8078	0.8106	0.8133
0.9	0.8159	0.8186	0.8212	0.8238	0.8264	0.8289	0.8315	0.8340	0.8365	0.8389
1.0	0.8413	0.8438	0.8461	0.8485	0.8508	0.8531	0.8554	0.8577	0.8599	0.8621
1.1	0.8643	0.8665	0.8686	0.8708	0.8729	0.8749	0.8770	0.8790	0.8810	0.8830
1.2	0.8849	0.8869	0.8888	0.8907	0.8925	0.8944	0.8962	0.8980	0.8997	0.9015
1.3	0.9032	0.9049	0.9066	0.9082	0.9099	0.9115	0.9131	0.9147	0.9162	0.9177
1.4	0.9192	0.9207	0.9222	0.9236	0.9251	0.9265	0.9279	0.9292	0.9306	0.9319
1.5	0.9332	0.9345	0.9357	0.9370	0.9382	0.9394	0.9406	0.9418	0.9429	0.9441
1.6	0.9452	0.9463	0.9474	0.9484	0.9495	0.9505	0.9515	0.9525	0.9535	0.9545
1.7	0.9554	0.9564	0.9573	0.9582	0.9591	0.9599	0.9608	0.9616	0.9625	0.9633
1.8	0.9641	0.9649	0.9656	0.9664	0.9671	0.9678	0.9686	0.9693	0.9699	0.9706
1.9	0.9713	0.9719	0.9726	0.9732	0.9738	0.9744	0.9750	0.9756	0.9761	0.9767
2.0	0.9772	0.9778	0.9783	0.9788	0.9793	0.9798	0.9803	0.9808	0.9812	0.9817
2.1	0.9821	0.9826	0.9830	0.9834	0.9838	0.9842	0.9846	0.9850	0.9854	0.9857
2.2	0.9861	0.9864	0.9868	0.9871	0.9875	0.9878	0.9881	0.9884	0.9887	0.9890
2.3	0.9893	0.9896	0.9898	0.9901	0.9904	0.9906	0.9909	0.9911	0.9913	0.9916
2.4	0.9918	0.9920	0.9922	0.9925	0.9927	0.9929	0.9931	0.9932	0.9934	0.9936
2.5	0.9938	0.9940	0.9941	0.9943	0.9945	0.9946	0.9948	0.9949	0.9951	0.9952
2.6	0.9953	0.9955	0.9956	0.9957	0.9959	0.9960	0.9961	0.9962	0.9963	0.9964
2.7	0.9965	0.9966	0.9967	0.9968	0.9969	0.9970	0.9971	0.9972	0.9973	0.9974
2.8	0.9974	0.9975	0.9976	0.9977	0.9977	0.9978	0.9979	0.9979	0.9980	0.9981
2.9	0.9981	0.9982	0.9982	0.9983	0.9984	0.9984	0.9985	0.9985	0.9986	0.9986
3.0	0.9987	0.9987	0.9987	0.9988	0.9988	0.9989	0.9989	0.9989	0.9990	0.9990
3.1	0.9990	0.9991	0.9991	0.9991	0.9992	0.9992	0.9992	0.9992	0.9993	0.9993
3.2	0.9993	0.9993	0.9994	0.9994	0.9994	0.9994	0.9994	0.9995	0.9995	0.9995
3.3	0.9995	0.9995	0.9995	0.9996	0.9996	0.9996	0.9996	0.9996	0.9996	0.9997
3.4	0.9997	0.9997	0.9997	0.9997	0.9997	0.9997	0.9997	0.9997	0.9997	0.9998

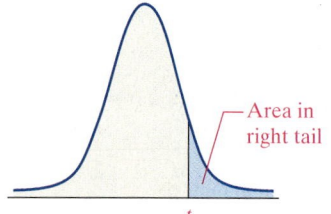
Area in right tail

t

Table V

t-Distribution
Area in Right Tail

df	0.25	0.20	0.15	0.10	0.05	0.025	0.02	0.01	0.005	0.0025	0.001	0.0005
1	1.000	1.376	1.963	3.078	6.314	12.706	15.894	31.821	63.657	127.321	318.289	636.558
2	0.816	1.061	1.386	1.886	2.920	4.303	4.849	6.965	9.925	14.089	22.328	31.600
3	0.765	0.978	1.250	1.638	2.353	3.182	3.482	4.541	5.841	7.453	10.214	12.924
4	0.741	0.941	1.190	1.533	2.132	2.776	2.999	3.747	4.604	5.598	7.173	8.610
5	0.727	0.920	1.156	1.476	2.015	2.571	2.757	3.365	4.032	4.773	5.893	6.869
6	0.718	0.906	1.134	1.440	1.943	2.447	2.612	3.143	3.707	4.317	5.208	5.959
7	0.711	0.896	1.119	1.415	1.895	2.365	2.517	2.998	3.499	4.029	4.785	5.408
8	0.706	0.889	1.108	1.397	1.860	2.306	2.449	2.896	3.355	3.833	4.501	5.041
9	0.703	0.883	1.100	1.383	1.833	2.262	2.398	2.821	3.250	3.690	4.297	4.781
10	0.700	0.879	1.093	1.372	1.812	2.228	2.359	2.764	3.169	3.581	4.144	4.587
11	0.697	0.876	1.088	1.363	1.796	2.201	2.328	2.718	3.106	3.497	4.025	4.437
12	0.695	0.873	1.083	1.356	1.782	2.179	2.303	2.681	3.055	3.428	3.930	4.318
13	0.694	0.870	1.079	1.350	1.771	2.160	2.282	2.650	3.012	3.372	3.852	4.221
14	0.692	0.868	1.076	1.345	1.761	2.145	2.264	2.624	2.977	3.326	3.787	4.140
15	0.691	0.866	1.074	1.341	1.753	2.131	2.249	2.602	2.947	3.286	3.733	4.073
16	0.690	0.865	1.071	1.337	1.746	2.120	2.235	2.583	2.921	3.252	3.686	4.015
17	0.689	0.863	1.069	1.333	1.740	2.110	2.224	2.567	2.898	3.222	3.646	3.965
18	0.688	0.862	1.067	1.330	1.734	2.101	2.214	2.552	2.878	3.197	3.611	3.922
19	0.688	0.861	1.066	1.328	1.729	2.093	2.205	2.539	2.861	3.174	3.579	3.883
20	0.687	0.860	1.064	1.325	1.725	2.086	2.197	2.528	2.845	3.153	3.552	3.850
21	0.686	0.859	1.063	1.323	1.721	2.080	2.189	2.518	2.831	3.135	3.527	3.819
22	0.686	0.858	1.061	1.321	1.717	2.074	2.183	2.508	2.819	3.119	3.505	3.792
23	0.685	0.858	1.060	1.319	1.714	2.069	2.177	2.500	2.807	3.104	3.485	3.768
24	0.685	0.857	1.059	1.318	1.711	2.064	2.172	2.492	2.797	3.091	3.467	3.745
25	0.684	0.856	1.058	1.316	1.708	2.060	2.167	2.485	2.787	3.078	3.450	3.725
26	0.684	0.856	1.058	1.315	1.706	2.056	2.162	2.479	2.779	3.067	3.435	3.707
27	0.684	0.855	1.057	1.314	1.703	2.052	2.158	2.473	2.771	3.057	3.421	3.690
28	0.683	0.855	1.056	1.313	1.701	2.048	2.154	2.467	2.763	3.047	3.408	3.674
29	0.683	0.854	1.055	1.311	1.699	2.045	2.150	2.462	2.756	3.038	3.396	3.659
30	0.683	0.854	1.055	1.310	1.697	2.042	2.147	2.457	2.750	3.030	3.385	3.646
31	0.682	0.853	1.054	1.309	1.696	2.040	2.144	2.453	2.744	3.022	3.375	3.633
32	0.682	0.853	1.054	1.309	1.694	2.037	2.141	2.449	2.738	3.015	3.365	3.622
33	0.682	0.853	1.053	1.308	1.692	2.035	2.138	2.445	2.733	3.008	3.356	3.611
34	0.682	0.852	1.052	1.307	1.691	2.032	2.136	2.441	2.728	3.002	3.348	3.601
35	0.682	0.852	1.052	1.306	1.690	2.030	2.133	2.438	2.724	2.996	3.340	3.591
36	0.681	0.852	1.052	1.306	1.688	2.028	2.131	2.435	2.719	2.990	3.333	3.582
37	0.681	0.851	1.051	1.305	1.687	2.026	2.129	2.431	2.715	2.985	3.326	3.574
38	0.681	0.851	1.051	1.304	1.686	2.024	2.127	2.429	2.712	2.980	3.319	3.566
39	0.681	0.851	1.050	1.304	1.685	2.023	2.125	2.426	2.708	2.976	3.313	3.558
40	0.681	0.851	1.050	1.303	1.684	2.021	2.123	2.423	2.704	2.971	3.307	3.551
50	0.679	0.849	1.047	1.299	1.676	2.009	2.109	2.403	2.678	2.937	3.261	3.496
60	0.679	0.848	1.045	1.296	1.671	2.000	2.099	2.390	2.660	2.915	3.232	3.460
70	0.678	0.847	1.044	1.294	1.667	1.994	2.093	2.381	2.648	2.899	3.211	3.435
80	0.678	0.846	1.043	1.292	1.664	1.990	2.088	2.374	2.639	2.887	3.195	3.416
90	0.677	0.846	1.042	1.291	1.662	1.987	2.084	2.368	2.632	2.878	3.183	3.402
100	0.677	0.845	1.042	1.290	1.660	1.984	2.081	2.364	2.626	2.871	3.174	3.390
1000	0.675	0.842	1.037	1.282	1.646	1.962	2.056	2.330	2.581	2.813	3.098	3.300
z	0.674	0.841	1.036	1.282	1.645	1.960	2.054	2.326	2.576	2.807	3.091	3.291

Table VI

Chi-Square (χ^2) Distribution
Area to the Right of Critical Value

Degrees of Freedom	0.995	0.99	0.975	0.95	0.90	0.10	0.05	0.025	0.01	0.005
1	—	—	0.001	0.004	0.016	2.706	3.841	5.024	6.635	7.879
2	0.010	0.020	0.051	0.103	0.211	4.605	5.991	7.378	9.210	10.597
3	0.072	0.115	0.216	0.352	0.584	6.251	7.815	9.348	11.345	12.838
4	0.207	0.297	0.484	0.711	1.064	7.779	9.488	11.143	13.277	14.860
5	0.412	0.554	0.831	1.145	1.610	9.236	11.071	12.833	15.086	16.750
6	0.676	0.872	1.237	1.635	2.204	10.645	12.592	14.449	16.812	18.548
7	0.989	1.239	1.690	2.167	2.833	12.017	14.067	16.013	18.475	20.278
8	1.344	1.646	2.180	2.733	3.490	13.362	15.507	17.535	20.090	21.955
9	1.735	2.088	2.700	3.325	4.168	14.684	16.919	19.023	21.666	23.589
10	2.156	2.558	3.247	3.940	4.865	15.987	18.307	20.483	23.209	25.188
11	2.603	3.053	3.816	4.575	5.578	17.275	19.675	21.920	24.725	26.757
12	3.074	3.571	4.404	5.226	6.304	18.549	21.026	23.337	26.217	28.299
13	3.565	4.107	5.009	5.892	7.042	19.812	22.362	24.736	27.688	29.819
14	4.075	4.660	5.629	6.571	7.790	21.064	23.685	26.119	29.141	31.319
15	4.601	5.229	6.262	7.261	8.547	22.307	24.996	27.488	30.578	32.801
16	5.142	5.812	6.908	7.962	9.312	23.542	26.296	28.845	32.000	34.267
17	5.697	6.408	7.564	8.672	10.085	24.769	27.587	30.191	33.409	35.718
18	6.265	7.015	8.231	9.390	10.865	25.989	28.869	31.526	34.805	37.156
19	6.844	7.633	8.907	10.117	11.651	27.204	30.144	32.852	36.191	38.582
20	7.434	8.260	9.591	10.851	12.443	28.412	31.410	34.170	37.566	39.997
21	8.034	8.897	10.283	11.591	13.240	29.615	32.671	35.479	38.932	41.401
22	8.643	9.542	10.982	12.338	14.042	30.813	33.924	36.781	40.289	42.796
23	9.260	10.196	11.689	13.091	14.848	32.007	35.172	38.076	41.638	44.181
24	9.886	10.856	12.401	13.848	15.659	33.196	36.415	39.364	42.980	45.559
25	10.520	11.524	13.120	14.611	16.473	34.382	37.652	40.646	44.314	46.928
26	11.160	12.198	13.844	15.379	17.292	35.563	38.885	41.923	45.642	48.290
27	11.808	12.879	14.573	16.151	18.114	36.741	40.113	43.194	46.963	49.645
28	12.461	13.565	15.308	16.928	18.939	37.916	41.337	44.461	48.278	50.993
29	13.121	14.257	16.047	17.708	19.768	39.087	42.557	45.722	49.588	52.336
30	13.787	14.954	16.791	18.493	20.599	40.256	43.773	46.979	50.892	53.672
40	20.707	22.164	24.433	26.509	29.051	51.805	55.758	59.342	63.691	66.766
50	27.991	29.707	32.357	34.764	37.689	63.167	67.505	71.420	76.154	79.490
60	35.534	37.485	40.482	43.188	46.459	74.397	79.082	83.298	88.379	91.952
70	43.275	45.442	48.758	51.739	55.329	85.527	90.531	95.023	100.425	104.215
80	51.172	53.540	57.153	60.391	64.278	96.578	101.879	106.629	112.329	116.321
90	59.196	61.754	65.647	69.126	73.291	107.565	113.145	118.136	124.116	128.299
100	67.328	70.065	74.222	77.929	82.358	118.498	124.342	129.561	135.807	140.169

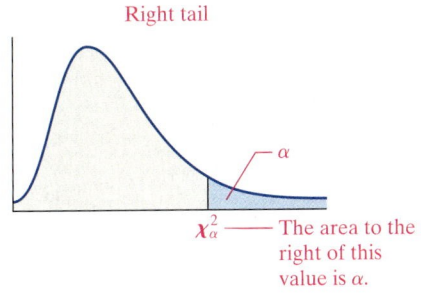

Right tail

χ^2_α —— The area to the right of this value is α.

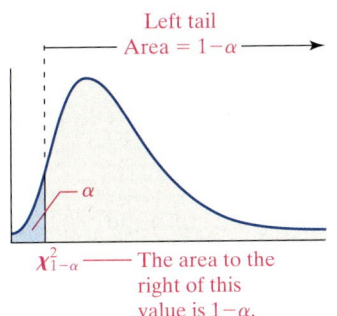

Left tail
Area = $1-\alpha$

$\chi^2_{1-\alpha}$ —— The area to the right of this value is $1-\alpha$.

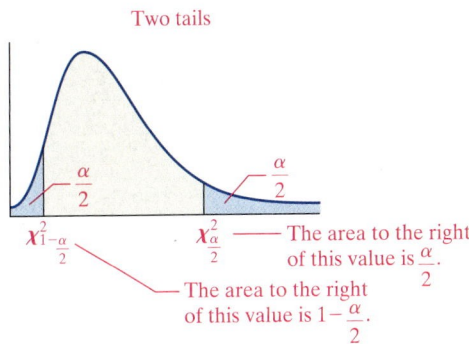

Two tails

$\chi^2_{1-\frac{\alpha}{2}}$

$\chi^2_{\frac{\alpha}{2}}$ —— The area to the right of this value is $\frac{\alpha}{2}$.

—— The area to the right of this value is $1-\frac{\alpha}{2}$.

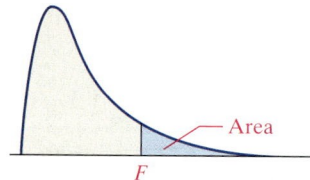
Area

F

Table VII

F-Distribution Critical Values
Degrees of Freedom in the Numerator

	Area in Right Tail	1	2	3	4	5	6	7	8
1	0.100	39.86	49.59	53.59	55.83	57.24	58.20	58.91	59.44
	0.050	161.45	199.50	215.71	224.58	230.16	233.99	236.77	238.88
	0.025	647.79	799.50	864.16	899.58	921.85	937.11	948.22	956.66
	0.010	4052.20	4999.50	5403.40	5624.60	5763.60	5859.00	5928.40	5981.10
	0.001	405284.00	500000.00	540379.00	562500.00	576405.00	585937.00	592873.00	598144.00
2	0.100	8.53	9.00	9.16	9.24	9.29	9.33	9.35	9.37
	0.050	18.51	19.00	19.16	19.25	19.30	19.33	19.35	19.37
	0.025	38.51	39.00	39.17	39.25	39.30	39.33	39.36	39.37
	0.010	98.50	99.00	99.17	99.25	99.30	99.33	99.36	99.37
	0.001	998.50	999.00	999.17	999.25	999.30	999.33	999.36	999.37
3	0.100	5.54	5.46	5.39	5.34	5.31	5.28	5.27	5.25
	0.050	10.13	9.55	9.28	9.12	9.01	8.94	8.89	8.85
	0.025	17.44	16.04	15.44	15.10	14.88	14.73	14.62	14.54
	0.010	34.12	30.82	29.46	28.71	28.24	27.91	27.67	27.49
	0.001	167.03	148.50	141.11	137.10	134.58	132.85	131.58	130.62
4	0.100	4.54	4.32	4.19	4.11	4.05	4.01	3.98	3.95
	0.050	7.71	6.94	6.59	6.39	6.26	6.16	6.09	6.04
	0.025	12.22	10.65	9.98	9.60	9.36	9.20	9.07	8.98
	0.010	21.20	18.00	16.69	15.98	15.52	15.21	14.98	14.80
	0.001	74.14	61.25	56.18	53.44	51.71	50.53	49.66	49.00
5	0.100	4.06	3.78	3.62	3.52	3.45	3.40	3.37	3.34
	0.050	6.61	5.79	5.41	5.19	5.05	4.95	4.88	4.82
	0.025	10.01	8.43	7.76	7.39	7.15	6.98	6.85	6.76
	0.010	16.26	13.27	12.06	11.39	10.97	10.67	10.46	10.29
	0.001	47.18	37.12	33.20	31.09	29.75	28.83	28.16	27.65
6	0.100	3.78	3.46	3.29	3.18	3.11	3.05	3.01	2.98
	0.050	5.99	5.14	4.76	4.53	4.39	4.28	4.21	4.15
	0.025	8.81	7.26	6.60	6.23	5.99	5.82	5.70	5.60
	0.010	13.75	10.92	9.78	9.15	8.75	8.47	8.26	8.10
	0.001	35.51	27.00	23.70	21.92	20.80	20.03	19.46	19.03
7	0.100	3.59	3.26	3.07	2.96	2.88	2.83	2.78	2.75
	0.050	5.59	4.74	4.35	4.12	3.97	3.87	3.79	3.73
	0.025	8.07	6.54	5.89	5.52	5.29	5.12	4.99	4.90
	0.010	12.25	9.55	8.45	7.85	7.46	7.19	6.99	6.84
	0.001	29.25	21.69	18.77	17.20	16.21	15.52	15.02	14.63
8	0.100	3.46	3.11	2.92	2.81	2.73	2.67	2.62	2.59
	0.050	5.32	4.46	4.07	3.84	3.69	3.58	3.50	3.44
	0.025	7.57	6.06	5.42	5.05	4.82	4.65	4.53	4.43
	0.010	11.26	8.65	7.59	7.01	6.63	6.37	6.18	6.03
	0.001	25.41	18.49	15.83	14.39	13.48	12.86	12.40	12.05

Degrees of Freedom in the Denominator

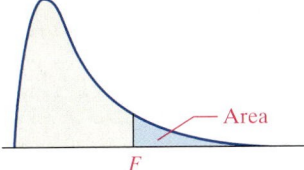
Area

F

Table VII *(continued)*

F-Distribution Critical Values
Degrees of Freedom in the Numerator

	Area in Right Tail	9	10	15	20	30	60	120	1000
1	0.100	59.86	60.19	61.22	61.74	62.26	62.79	63.06	63.30
	0.050	240.54	241.88	245.95	248.01	250.10	252.20	253.25	254.19
	0.025	963.28	968.63	984.87	993.10	1001.4	1009.8	1014.0	1017.7
	0.010	6022.5	6055.8	6157.3	6208.7	6260.6	6313.0	6339.4	6362.7
	0.001	602284.0	605621.0	615764.0	620908.0	626099.0	631337.0	633972.0	636301.0
2	0.100	9.38	9.39	9.42	9.44	9.16	9.47	9.48	9.49
	0.050	19.38	19.40	19.43	19.45	19.46	19.48	19.49	19.49
	0.025	39.39	39.40	39.43	39.45	39.46	39.48	39.49	39.50
	0.010	99.39	99.40	99.43	99.45	99.47	99.48	99.49	99.50
	0.001	999.39	999.40	999.43	999.45	999.47	999.48	999.49	999.50
3	0.100	5.24	5.23	5.20	5.18	5.17	5.15	5.14	5.13
	0.050	8.81	8.79	8.70	8.66	8.62	8.57	8.55	8.53
	0.025	14.47	14.42	14.25	14.17	14.08	13.99	13.95	13.91
	0.010	27.35	27.23	26.87	26.69	26.50	26.32	26.22	26.14
	0.001	129.86	129.25	127.37	126.42	125.45	124.47	123.97	123.53
4	0.100	3.94	3.92	3.87	3.84	3.82	3.79	3.78	3.76
	0.050	6.00	5.96	5.86	5.80	5.75	5.69	5.66	5.63
	0.025	8.90	8.84	8.66	8.56	8.46	8.36	8.31	8.26
	0.010	14.66	14.55	14.20	14.02	13.84	13.65	13.56	13.47
	0.001	48.47	48.05	46.76	46.10	45.43	44.75	44.40	44.09
5	0.100	3.32	3.30	3.24	3.21	3.17	3.14	3.12	3.11
	0.050	4.77	4.74	4.62	4.56	4.50	4.43	4.40	4.37
	0.025	6.68	6.62	6.43	6.33	6.23	6.12	6.07	6.02
	0.010	10.16	10.05	9.72	9.55	9.38	9.20	9.11	9.03
	0.001	27.24	26.92	25.91	25.39	24.87	24.33	24.06	23.82
6	0.100	2.96	2.94	2.87	2.84	2.80	2.76	2.74	2.72
	0.050	4.10	4.06	3.94	3.87	3.81	3.74	3.70	3.67
	0.025	5.52	5.46	5.27	5.17	5.07	4.96	4.90	4.86
	0.010	7.98	7.87	7.56	7.40	7.23	7.06	6.97	6.89
	0.001	18.69	18.41	17.56	17.12	16.67	16.21	15.98	15.77
7	0.100	2.72	2.70	2.63	2.59	2.56	2.51	2.49	2.47
	0.050	3.68	3.64	3.51	3.44	3.38	3.30	3.27	3.23
	0.025	4.82	4.76	4.57	4.47	4.36	4.25	4.20	4.15
	0.010	6.72	6.62	6.31	6.16	5.99	5.82	5.74	5.66
	0.001	14.33	14.08	13.32	12.93	12.53	12.12	11.91	11.72
8	0.100	2.56	2.54	2.46	2.42	2.38	2.34	2.32	2.30
	0.050	3.39	3.35	3.22	3.15	3.08	3.01	2.97	2.93
	0.025	4.36	4.30	4.10	4.00	3.89	3.78	3.73	3.68
	0.010	5.91	5.81	5.52	5.36	5.20	5.03	4.95	4.87
	0.001	11.77	11.54	10.84	10.48	10.11	9.73	9.53	9.36

Degrees of Freedom in the Denominator

Table VII (*continued*)

F-Distribution Critical Values

Degrees of Freedom in the Numerator

	Area in Right Tail	1	2	3	4	5	6	7	8	9	10
9	0.100	3.36	3.01	2.81	2.69	2.61	2.55	2.51	2.47	2.44	2.42
	0.050	5.12	4.26	3.86	3.63	3.48	3.37	3.29	3.23	3.18	3.14
	0.025	7.21	5.71	5.08	4.72	4.48	4.32	4.20	4.10	4.03	3.96
	0.010	10.56	8.02	6.99	6.42	6.06	5.80	5.61	5.47	5.35	5.26
	0.001	22.86	16.39	13.90	12.56	11.71	11.13	10.70	10.37	10.11	9.89
10	0.100	3.29	2.92	2.73	2.61	2.52	2.46	2.41	2.38	2.35	2.32
	0.050	4.96	4.10	3.71	3.48	3.33	3.22	3.14	3.07	3.02	2.98
	0.025	6.94	5.46	4.83	4.47	4.24	4.07	3.95	3.85	3.78	3.72
	0.010	10.04	7.56	6.55	5.99	5.64	5.39	5.20	5.06	4.94	4.85
	0.001	21.04	14.91	12.55	11.28	10.48	9.93	9.52	9.20	8.96	8.75
12	0.100	3.18	2.81	2.61	2.48	2.39	2.33	2.28	2.24	2.21	2.19
	0.050	4.75	3.89	3.49	3.26	3.11	3.00	2.91	2.85	2.80	2.75
	0.025	6.55	5.10	4.47	4.12	3.89	3.73	3.61	3.51	3.44	3.37
	0.010	9.33	6.93	5.95	5.41	5.06	4.82	4.64	4.50	4.39	4.30
	0.001	18.64	12.97	10.80	9.63	8.89	8.38	8.00	7.71	7.48	7.29
15	0.100	3.07	2.70	2.49	2.36	2.27	2.21	2.16	2.12	2.09	2.06
	0.050	4.54	3.68	3.29	3.06	2.90	2.79	2.71	2.64	2.59	2.54
	0.025	6.20	4.77	4.15	3.80	3.58	3.41	3.29	3.20	3.12	3.06
	0.010	8.68	6.36	5.42	4.89	4.56	4.32	4.14	4.00	3.89	3.80
	0.001	16.59	11.34	9.34	8.25	7.57	7.09	6.74	6.47	6.26	6.08
20	0.100	2.97	2.59	2.38	2.25	2.16	2.09	2.04	2.00	1.96	1.94
	0.050	4.35	3.49	3.10	2.87	2.71	2.60	2.51	2.45	2.39	2.35
	0.025	5.87	4.46	3.86	3.51	3.29	3.13	3.01	2.91	2.84	2.77
	0.010	8.10	5.85	4.94	4.43	4.10	3.87	3.70	3.56	3.46	3.37
	0.001	14.82	9.95	8.10	7.10	6.46	6.02	5.69	5.44	5.24	5.08
25	0.100	2.92	2.53	2.32	2.18	2.09	2.02	1.97	1.93	1.89	1.87
	0.050	4.24	3.39	2.99	2.76	2.60	2.49	2.40	2.34	2.28	2.24
	0.025	5.69	4.29	3.69	3.35	3.13	2.97	2.85	2.75	2.68	2.61
	0.010	7.77	5.57	4.68	4.18	3.85	3.63	3.46	3.32	3.22	3.13
	0.001	13.88	9.22	7.45	6.49	5.89	5.46	5.15	4.91	4.71	4.56
50	0.100	2.81	2.41	2.20	2.06	1.97	1.90	1.84	1.80	1.76	1.73
	0.050	4.03	3.18	2.79	2.56	2.40	2.29	2.20	2.13	2.07	2.03
	0.025	5.34	3.97	3.39	3.05	2.83	2.67	2.55	2.46	2.38	2.32
	0.010	7.17	5.06	4.20	3.72	3.41	3.19	3.02	2.89	2.78	2.70
	0.001	12.22	7.96	6.34	5.46	4.90	4.51	4.22	4.00	3.82	3.67
100	0.100	2.76	2.36	2.14	2.00	1.91	1.83	1.78	1.73	1.69	1.66
	0.050	3.94	3.09	2.70	2.46	2.31	2.19	2.10	2.03	1.97	1.93
	0.025	5.18	3.83	3.25	2.92	2.70	2.54	2.42	2.32	2.24	2.18
	0.010	6.90	4.82	3.98	3.51	3.21	2.99	2.82	2.69	2.59	2.50
	0.001	11.50	7.41	5.86	5.02	4.48	4.11	3.83	3.61	3.44	3.30
200	0.100	2.73	2.33	2.11	1.97	1.88	1.80	1.75	1.70	1.66	1.63
	0.050	3.89	3.04	2.65	2.42	2.26	2.14	2.06	1.98	1.93	1.88
	0.025	5.10	3.76	3.18	2.85	2.63	2.47	2.35	2.26	2.18	2.11
	0.010	6.76	4.71	3.88	3.41	3.11	2.89	2.73	2.60	2.50	2.41
	0.001	11.15	7.15	5.63	4.81	4.29	3.92	3.65	3.43	3.26	3.12
1000	0.100	2.71	2.31	2.09	1.95	1.85	1.78	1.72	1.68	1.64	1.61
	0.050	3.85	3.00	2.61	2.38	2.22	2.11	2.02	1.95	1.89	1.84
	0.025	5.04	3.70	3.13	2.80	2.58	2.42	2.30	2.20	2.13	2.06
	0.010	6.66	4.63	3.80	3.34	3.04	2.82	2.66	2.53	2.43	2.34
	0.001	10.89	6.96	5.46	4.65	4.14	3.78	3.51	3.30	3.13	2.99

Degrees of Freedom in the Denominator

Table VII (continued)

F-Distribution Critical Values
Degrees of Freedom in the Numerator

	Area in Right Tail	12	15	20	25	30	40	50	60	120	1000
9	0.100	2.38	2.34	2.30	2.27	2.25	2.23	2.22	2.21	2.18	2.16
	0.050	3.07	3.01	2.94	2.89	2.86	2.83	2.80	2.79	2.75	2.71
	0.025	3.87	3.77	3.67	3.60	3.56	3.51	3.47	3.45	3.39	3.34
	0.010	5.11	4.96	4.81	4.71	4.65	4.57	4.52	4.48	4.40	4.32
	0.001	9.57	9.24	8.90	8.69	8.55	8.37	8.26	8.19	8.00	7.84
10	0.100	2.28	2.24	2.20	2.17	2.16	2.13	2.12	2.11	2.08	2.06
	0.050	2.91	2.85	2.77	2.73	2.70	2.66	2.64	2.62	2.58	2.54
	0.025	3.62	3.52	3.42	3.35	3.31	3.26	3.22	3.20	3.14	3.09
	0.010	4.71	4.56	4.41	4.31	4.25	4.17	4.12	4.08	4.00	3.92
	0.001	8.45	8.13	7.80	7.60	7.47	7.30	7.19	7.12	6.94	6.78
12	0.100	2.15	2.10	2.06	2.03	2.01	1.99	1.97	1.96	1.93	1.91
	0.050	2.69	2.62	2.54	2.50	2.47	2.43	2.40	2.38	2.34	2.30
	0.025	3.28	3.18	3.07	3.01	2.96	2.91	2.87	2.85	2.79	2.73
	0.010	4.16	4.01	3.86	3.76	3.70	3.62	3.57	3.54	3.45	3.37
	0.001	7.00	6.71	6.40	6.22	6.09	5.93	5.83	5.76	5.59	5.44
15	0.100	2.02	1.97	1.92	1.89	1.87	1.85	1.83	1.82	1.79	1.76
	0.050	2.48	2.40	2.33	2.28	2.25	2.20	2.18	2.16	2.11	2.07
	0.025	2.96	2.86	2.76	2.69	2.64	2.59	2.55	2.52	2.46	2.40
	0.010	3.67	3.52	3.37	3.28	3.21	3.13	3.08	3.05	2.96	2.88
	0.001	5.81	5.54	5.25	5.07	4.95	4.80	4.70	4.64	4.47	4.33
20	0.100	1.89	1.84	1.79	1.76	1.74	1.71	1.69	1.68	1.64	1.61
	0.050	2.28	2.20	2.12	2.07	2.04	1.99	1.97	1.95	1.90	1.85
	0.025	2.68	2.57	2.46	2.40	2.35	2.29	2.25	2.22	2.16	2.09
	0.010	3.23	3.09	2.94	2.84	2.78	2.69	2.64	2.61	2.52	2.43
	0.001	4.82	4.56	4.29	4.12	4.00	3.86	3.77	3.70	3.54	3.40
25	0.100	1.82	1.77	1.72	1.68	1.66	1.63	1.61	1.59	1.56	1.52
	0.050	2.16	2.09	2.01	1.96	1.92	1.87	1.84	1.82	1.77	1.72
	0.025	2.51	2.41	2.30	2.23	2.18	2.12	2.08	2.05	1.98	1.91
	0.010	2.99	2.85	2.70	2.60	2.54	2.45	2.40	2.36	2.27	2.18
	0.001	4.31	4.06	3.79	3.63	3.52	3.37	3.28	3.22	3.06	2.91
50	0.100	1.68	1.63	1.57	1.53	1.50	1.46	1.44	1.42	1.38	1.33
	0.050	1.95	1.87	1.78	1.73	1.69	1.63	1.60	1.58	1.51	1.45
	0.025	2.22	2.11	1.99	1.92	1.87	1.80	1.75	1.72	1.64	1.56
	0.010	2.56	2.42	2.27	2.17	2.10	2.01	1.95	1.91	1.80	1.70
	0.001	3.44	3.20	2.95	2.79	2.68	2.53	2.44	2.38	2.21	2.05
100	0.100	1.61	1.56	1.49	1.45	1.42	1.38	1.35	1.34	1.28	1.22
	0.050	1.85	1.77	1.68	1.62	1.57	1.52	1.48	1.45	1.38	1.30
	0.025	2.08	1.97	1.85	1.77	1.71	1.64	1.59	1.56	1.46	1.36
	0.010	2.37	2.22	2.07	1.97	1.89	1.80	1.74	1.69	1.57	1.45
	0.001	3.07	2.84	2.59	2.43	2.32	2.17	2.08	2.01	1.83	1.64
200	0.100	1.58	1.52	1.46	1.41	1.38	1.34	1.31	1.29	1.23	1.16
	0.050	1.80	1.72	1.62	1.56	1.52	1.46	1.41	1.39	1.30	1.21
	0.025	2.01	1.90	1.78	1.70	1.64	1.56	1.51	1.47	1.37	1.25
	0.010	2.27	2.13	1.97	1.87	1.79	1.69	1.63	1.58	1.45	1.30
	0.001	2.90	2.67	2.42	2.26	2.15	2.00	1.90	1.83	1.64	1.43
1000	0.100	1.55	1.49	1.43	1.38	1.35	1.30	1.27	1.25	1.18	1.08
	0.050	1.76	1.68	1.58	1.52	1.47	1.41	1.36	1.33	1.24	1.11
	0.025	1.96	1.85	1.72	1.64	1.58	1.50	1.45	1.41	1.29	1.13
	0.010	2.20	2.06	1.90	1.79	1.72	1.61	1.54	1.50	1.35	1.16
	0.001	2.77	2.54	2.30	2.14	2.02	1.87	1.77	1.69	1.49	1.22

Degrees of Freedom in the Denominator

Table VIII

Critical Values for Tukey's Test

$\alpha = 0.05$

ν	k (or p): 2	3	4	5	6	7	8	9	10	11	12	13	14	15	16	17	18	19
1	17.97	26.98	32.82	37.08	40.41	43.12	45.40	47.36	49.07	50.59	51.96	53.20	54.33	55.36	56.32	57.22	58.04	58.83
2	6.085	8.331	9.798	10.88	11.74	12.44	13.03	13.54	13.99	14.39	14.75	15.08	15.38	15.65	15.91	16.14	16.37	16.57
3	4.501	5.910	6.825	7.502	8.037	8.478	8.853	9.177	9.462	9.717	9.946	10.15	10.35	10.53	10.69	10.84	10.98	11.11
4	3.927	5.040	5.757	6.287	6.707	7.053	7.347	7.602	7.826	8.027	8.208	8.373	8.525	8.664	8.794	8.914	9.028	9.134
5	3.635	4.602	5.218	5.673	6.033	6.330	6.582	6.802	6.995	7.168	7.324	7.466	7.596	7.717	7.828	7.932	8.030	8.122
6	3.461	4.339	4.896	5.305	5.628	5.895	6.122	6.319	6.493	6.649	6.789	6.917	7.034	7.143	7.244	7.338	7.426	7.508
7	3.344	4.165	4.681	5.060	5.359	5.606	5.815	5.998	6.158	6.302	6.431	6.550	6.658	6.759	6.852	6.939	7.020	7.097
8	3.261	4.041	4.529	4.886	5.167	5.399	5.597	5.767	5.918	6.054	6.175	6.287	6.389	6.483	6.571	6.653	6.729	6.802
9	3.199	3.949	4.415	4.756	5.024	5.244	5.432	5.595	5.739	5.867	5.983	6.089	6.186	6.276	6.359	6.437	6.510	6.579
10	3.151	3.877	4.327	4.654	4.912	5.124	5.305	5.461	5.599	5.722	5.833	5.935	6.028	6.114	6.194	6.269	6.339	6.405
11	3.113	3.820	4.256	4.574	4.823	5.028	5.202	5.353	5.487	5.605	5.713	5.811	5.901	5.984	6.062	6.134	6.202	6.265
12	3.082	3.773	4.199	4.508	4.751	4.950	5.119	5.265	5.395	5.511	5.615	5.710	5.798	5.878	5.953	6.023	6.089	6.151
13	3.055	3.735	4.151	4.453	4.690	4.885	5.049	5.192	5.318	5.431	5.533	5.625	5.711	5.789	5.862	5.931	5.995	6.055
14	3.033	3.702	4.111	4.407	4.639	4.829	4.990	5.131	5.254	5.364	5.463	5.554	5.637	5.714	5.786	5.852	5.915	5.974
15	3.014	3.674	4.076	4.367	4.595	4.782	4.940	5.077	5.198	5.306	5.404	5.493	5.574	5.649	5.720	5.785	5.846	5.904
16	2.998	3.649	4.046	4.333	4.557	4.741	4.897	5.031	5.150	5.256	5.352	5.439	5.520	5.593	5.662	5.727	5.786	5.843
17	2.984	3.628	4.020	4.303	4.524	4.705	4.858	4.991	5.108	5.212	5.307	5.392	5.471	5.544	5.612	5.675	5.734	5.790
18	2.971	3.609	3.997	4.277	4.495	4.673	4.824	4.956	5.071	5.174	5.267	5.352	5.429	5.501	5.568	5.630	5.688	5.743
19	2.960	3.593	3.977	4.253	4.469	4.645	4.794	4.924	5.038	5.140	5.231	5.315	5.391	5.462	5.528	5.589	5.647	5.701
20	2.950	3.578	3.958	4.232	4.445	4.620	4.768	4.896	5.008	5.108	5.199	5.282	5.357	5.427	5.493	5.553	5.610	5.663
24	2.919	3.532	3.901	4.166	4.373	4.541	4.684	4.807	4.915	5.012	5.099	5.179	5.251	5.319	5.381	5.439	5.494	5.545
30	2.888	3.486	3.845	4.102	4.302	4.464	4.602	4.720	4.824	4.917	5.001	5.077	5.147	5.211	5.271	5.327	5.379	5.429
40	2.858	3.442	3.791	4.039	4.232	4.389	4.521	4.635	4.735	4.824	4.904	4.977	5.044	5.106	5.163	5.216	5.266	5.313
60	2.829	3.399	3.737	3.977	4.163	4.314	4.441	4.550	4.646	4.732	4.808	4.878	4.942	5.001	5.056	5.107	5.154	5.199
120	2.800	3.356	3.685	3.917	4.096	4.241	4.363	4.468	4.560	4.641	4.714	4.781	4.842	4.898	4.950	4.998	5.044	5.086
∞	2.772	3.314	3.633	3.858	4.030	4.170	4.286	4.387	4.474	4.552	4.622	4.685	4.743	4.796	4.845	4.891	4.934	4.974

Table VIII (continued)

Critical Values for Tukey's Test

α = 0.05

v	k (or p): 20	22	24	26	28	30	32	34	36
1	59.56	60.91	62.12	63.22	64.23	65.15	66.01	66.81	67.56
2	16.77	17.13	17.45	17.75	18.02	18.27	18.50	18.72	18.92
3	11.24	11.47	11.68	11.87	12.05	12.21	12.36	12.50	12.63
4	9.233	9.418	9.584	9.736	9.875	10.00	10.12	10.23	10.34
5	8.208	8.368	8.512	8.643	8.764	8.875	8.979	9.075	9.165
6	7.587	7.730	7.861	7.979	8.088	8.189	8.283	8.370	8.452
7	7.170	7.303	7.423	7.533	7.634	7.728	7.814	7.895	7.972
8	6.870	6.995	7.109	7.212	7.307	7.395	7.477	7.554	7.625
9	6.644	6.763	6.871	6.970	7.061	7.145	7.222	7.295	7.363
10	6.467	6.582	6.686	6.781	6.868	6.948	7.023	7.093	7.159
11	6.326	6.436	6.536	6.628	6.712	6.790	6.863	6.930	6.994
12	6.209	6.317	6.414	6.503	6.585	6.660	6.731	6.796	6.858
13	6.112	6.217	6.312	6.398	6.478	6.551	6.620	6.684	6.744
14	6.029	6.132	6.224	6.309	6.387	6.459	6.526	6.588	6.647
15	5.958	6.059	6.149	6.233	6.309	6.379	6.445	6.506	6.564
16	5.897	5.995	6.084	6.166	6.241	6.310	6.374	6.434	6.491
17	5.842	5.940	6.027	6.107	6.181	6.249	6.313	6.372	6.427
18	5.794	5.890	5.977	6.055	6.128	6.195	6.258	6.316	6.371
19	5.752	5.846	5.932	6.009	6.081	6.147	6.209	6.267	6.321
20	5.714	5.807	5.891	5.968	6.039	6.104	6.165	6.222	6.275
24	5.594	5.683	5.764	5.838	5.906	5.968	6.027	6.081	6.132
30	5.475	5.561	5.638	5.709	5.774	5.833	5.889	5.941	5.990
40	5.358	5.439	5.513	5.581	5.642	5.700	5.753	5.803	5.849
60	5.241	5.319	5.389	5.453	5.512	5.566	5.617	5.664	5.708
120	5.126	5.200	5.266	5.327	5.382	5.434	5.481	5.526	5.568
∞	5.012	5.081	5.144	5.201	5.253	5.301	5.346	5.388	5.427

v	k (or p): 38	40	50	60	70	80	90	100
1	68.26	68.92	71.73	73.97	75.82	77.40	78.77	79.98
2	19.11	19.28	20.05	20.66	21.16	21.59	21.96	22.29
3	12.75	12.87	13.36	13.76	14.08	14.36	14.61	14.82
4	10.44	10.53	10.93	11.24	11.51	11.73	11.92	12.09
5	9.250	9.330	9.674	9.949	10.18	10.38	10.54	10.69
6	8.529	8.601	8.913	9.163	9.370	9.548	9.702	9.839
7	8.043	8.110	8.400	8.632	8.824	8.989	9.133	9.261
8	7.693	7.756	8.029	8.248	8.430	8.586	8.722	8.843
9	7.428	7.488	7.749	7.958	8.132	8.281	8.410	8.526
10	7.220	7.279	7.529	7.730	7.897	8.041	8.166	8.276
11	7.053	7.110	7.352	7.546	7.708	7.847	7.968	8.075
12	6.916	6.970	7.205	7.394	7.552	7.687	7.804	7.909
13	6.800	6.854	7.083	7.267	7.421	7.552	7.667	7.769
14	6.702	6.754	6.979	7.159	7.309	7.438	7.550	7.650
15	6.618	6.669	6.888	7.065	7.212	7.339	7.449	7.546
16	6.544	6.594	6.810	6.984	7.128	7.252	7.360	7.457
17	6.479	6.529	6.741	6.912	7.054	7.176	7.283	7.377
18	6.422	6.471	6.680	6.848	6.989	7.109	7.213	7.307
19	6.371	6.419	6.626	6.792	6.930	7.048	7.152	7.244
20	6.325	6.373	6.576	6.740	6.877	6.994	7.097	7.187
24	6.181	6.226	6.421	6.579	6.710	6.822	6.920	7.008
30	6.037	6.080	6.267	6.417	6.543	6.650	6.744	6.827
40	5.893	5.934	6.112	6.255	6.375	6.477	6.566	6.645
60	5.750	5.789	5.958	6.093	6.206	6.303	6.387	6.462
120	5.607	5.644	5.802	5.929	6.035	6.126	6.205	6.275
∞	5.463	5.498	5.646	5.764	5.863	5.947	6.020	6.085

Table VIII (continued)

Critical Values for Tukey's Test

α = 0.01

ν	k (or p): 2	3	4	5	6	7	8	9	10
1	90.03	135.0	164.3	185.6	202.2	215.8	227.2	237.0	245.6
2	14.04	19.02	22.29	24.72	26.63	28.20	29.53	30.68	31.69
3	8.261	10.62	12.17	13.33	14.24	15.00	15.64	16.20	16.69
4	6.512	8.120	9.173	9.958	10.58	11.10	11.55	11.93	12.27
5	5.702	6.976	7.804	8.421	8.913	9.321	9.669	9.972	10.24
6	5.243	6.331	7.033	7.556	7.973	8.318	8.613	8.869	9.097
7	4.949	5.919	6.543	7.005	7.373	7.679	7.939	8.166	8.368
8	4.746	5.635	6.204	6.625	6.960	7.237	7.474	7.681	7.863
9	4.596	5.428	5.957	6.348	6.658	6.915	7.134	7.325	7.495
10	4.482	5.270	5.769	6.136	6.428	6.669	6.875	7.055	7.213
11	4.392	5.146	5.621	5.970	6.247	6.476	6.672	6.842	6.992
12	4.320	5.046	5.502	5.836	6.101	6.321	6.507	6.670	6.814
13	4.260	4.964	5.404	5.727	5.981	6.192	6.372	6.528	6.667
14	4.210	4.895	5.322	5.634	5.881	6.085	6.258	6.409	6.543
15	4.168	4.836	5.252	5.556	5.796	5.994	6.162	6.309	6.439
16	4.131	4.786	5.192	5.489	5.722	5.915	6.079	6.222	6.349
17	4.099	4.742	5.140	5.430	5.659	5.847	6.007	6.147	6.270
18	4.071	4.703	5.094	5.379	5.603	5.788	5.944	6.081	6.201
19	4.046	4.670	5.054	5.334	5.554	5.735	5.889	6.022	6.141
20	4.024	4.639	5.018	5.294	5.510	5.688	5.839	5.970	6.087
24	3.956	4.546	4.907	5.168	5.374	5.542	5.685	5.809	5.919
30	3.889	4.455	4.799	5.048	5.242	5.401	5.536	5.653	5.756
40	3.825	4.367	4.696	4.931	5.114	5.265	5.392	5.502	5.559
60	3.762	4.282	4.595	4.818	4.991	5.133	5.253	5.356	5.447
120	3.702	4.200	4.497	4.709	4.872	5.005	5.118	5.214	5.299
∞	3.643	4.120	4.403	4.603	4.757	4.882	4.987	5.078	5.157

ν	k (or p): 11	12	13	14	15	16	17	18	19
1	253.2	260.0	266.2	271.8	277.0	281.8	286.3	290.4	294.3
2	32.59	33.40	34.13	34.81	35.43	36.00	36.53	37.03	37.50
3	17.13	17.53	17.89	18.22	18.52	18.81	19.07	19.32	19.55
4	12.57	12.84	13.09	13.32	13.53	13.73	13.91	14.08	14.24
5	10.48	10.70	10.89	11.08	11.24	11.40	11.55	11.68	11.81
6	9.301	9.485	9.653	9.808	9.951	10.08	10.21	10.32	10.43
7	8.548	8.711	8.860	8.997	9.124	9.242	9.353	9.456	9.554
8	8.027	8.176	8.312	8.436	8.552	8.659	8.760	8.854	8.943
9	7.647	7.784	7.910	8.025	8.132	8.232	8.325	8.412	8.495
10	7.356	7.485	7.603	7.712	7.812	7.906	7.993	8.076	8.153
11	7.128	7.250	7.362	7.465	7.560	7.649	7.732	7.809	7.883
12	6.943	7.060	7.167	7.265	7.356	7.441	7.520	7.594	7.665
13	6.791	6.903	7.006	7.101	7.188	7.269	7.345	7.417	7.485
14	6.664	6.772	6.871	6.962	7.047	7.126	7.199	7.268	7.333
15	6.555	6.660	6.757	6.845	6.927	7.003	7.074	7.142	7.204
16	6.462	6.564	6.658	6.744	6.823	6.898	6.967	7.032	7.093
17	6.381	6.480	6.572	6.656	6.734	6.806	6.873	6.937	6.997
18	6.310	6.407	6.497	6.579	6.655	6.725	6.792	6.854	6.912
19	6.247	6.342	6.430	6.510	6.585	6.654	6.719	6.780	6.837
20	6.191	6.285	6.371	6.450	6.523	6.591	6.654	6.714	6.771
24	6.017	6.106	6.186	6.261	6.330	6.394	6.453	6.510	6.563
30	5.849	5.932	6.008	6.078	6.143	6.203	6.259	6.311	6.361
40	5.686	5.764	5.835	5.900	5.961	6.017	6.069	6.119	6.165
60	5.528	5.601	5.667	5.728	5.785	5.837	5.886	5.931	5.974
120	5.375	5.443	5.505	5.562	5.614	5.662	5.708	5.750	5.790
∞	5.227	5.290	5.348	5.400	5.448	5.493	5.535	5.574	5.611

Table VIII (continued)

Critical Values for Tukey's Test

α = 0.01

v	k (or p): 20	22	24	26	28	30	32	34	36
1	298.0	304.7	310.8	316.3	321.3	326.0	330.3	334.3	338.0
2	37.95	38.76	39.49	40.15	40.76	41.32	41.84	42.33	42.78
3	19.77	20.17	20.53	20.86	21.16	21.44	21.70	21.95	22.17
4	14.40	14.68	14.93	15.16	15.37	15.57	15.75	15.92	16.08
5	11.93	12.16	12.36	12.54	12.71	12.87	13.02	13.15	13.28
6	10.54	10.73	10.91	11.06	11.21	11.34	11.47	11.58	11.69
7	9.646	9.815	9.970	10.11	10.24	10.36	10.47	10.58	10.67
8	9.027	9.182	9.322	9.450	9.569	9.678	9.779	9.874	9.964
9	8.573	8.717	8.847	8.966	9.075	9.177	9.271	9.360	9.443
10	8.226	8.361	8.483	8.595	8.698	8.794	8.883	8.966	9.044
11	7.952	8.080	8.196	8.303	8.400	8.491	8.575	8.654	8.728
12	7.731	7.853	7.964	8.066	8.159	8.246	8.327	8.402	8.473
13	7.548	7.665	7.772	7.870	7.960	8.043	8.121	8.193	8.262
14	7.395	7.508	7.611	7.705	7.792	7.873	7.948	8.018	8.084
15	7.264	7.374	7.474	7.566	7.650	7.728	7.800	7.869	7.932
16	7.152	7.258	7.356	7.445	7.527	7.602	7.673	7.739	7.802
17	7.053	7.158	7.253	7.340	7.420	7.493	7.563	7.627	7.687
18	6.968	7.070	7.163	7.247	7.325	7.398	7.465	7.528	7.587
19	6.891	6.992	7.082	7.166	7.242	7.313	7.379	7.440	7.498
20	6.823	6.922	7.011	7.092	7.168	7.237	7.302	7.362	7.419
24	6.612	6.705	6.789	6.865	6.936	7.001	7.062	7.119	7.173
30	6.407	6.494	6.572	6.644	6.710	6.772	6.828	6.881	6.932
40	6.209	6.289	6.362	6.429	6.490	6.547	6.600	6.650	6.697
60	6.015	6.090	6.158	6.220	6.277	6.330	6.378	6.424	6.467
120	5.827	5.897	5.959	6.016	6.069	6.117	6.162	6.204	6.244
∞	5.645	5.709	5.766	5.818	5.866	5.911	5.952	5.990	6.026

v	k (or p): 38	40	50	60	70	80	90	100
1	341.5	344.8	358.9	370.1	379.4	387.3	394.1	400.1
2	43.21	43.61	45.33	46.70	47.83	48.80	49.64	50.38
3	22.39	22.59	23.45	24.13	24.71	25.19	25.62	25.99
4	16.23	16.37	16.98	17.46	17.86	18.02	18.50	18.77
5	13.40	13.52	14.00	14.39	14.72	14.99	15.23	15.45
6	11.80	11.90	12.31	12.65	12.92	13.16	13.37	13.55
7	10.77	10.85	11.23	11.52	11.77	11.99	12.17	12.34
8	10.05	10.13	10.47	10.75	10.97	11.17	11.34	11.49
9	9.521	9.594	9.912	10.17	10.38	10.57	10.73	10.87
10	9.117	9.187	9.486	9.726	9.927	10.10	10.25	10.39
11	8.798	8.864	9.148	9.377	9.568	9.732	9.875	10.00
12	8.539	8.603	8.875	9.094	9.277	9.434	9.571	9.693
13	8.326	8.387	8.648	8.859	9.035	9.187	9.318	9.436
14	8.146	8.204	8.457	8.661	8.832	8.978	9.106	9.219
15	7.992	8.049	8.295	8.492	8.658	8.800	8.924	9.035
16	7.860	7.916	8.154	8.347	8.507	8.646	8.767	8.874
17	7.745	7.799	8.031	8.219	8.377	8.511	8.630	8.735
18	7.643	7.696	7.924	8.107	8.261	8.393	8.508	8.611
19	7.553	7.605	7.828	8.008	8.159	8.288	8.401	8.502
20	7.473	7.523	7.742	7.919	8.067	8.194	8.305	8.404
24	7.223	7.270	7.476	7.642	7.780	7.900	8.004	8.097
30	6.978	7.023	7.215	7.370	7.500	7.611	7.709	7.796
40	6.740	6.782	6.960	7.104	7.225	7.328	7.419	7.500
60	6.507	6.546	6.710	6.843	6.954	7.050	7.133	7.207
120	6.281	6.316	6.467	6.588	6.689	6.776	6.852	6.919
∞	6.060	6.092	6.228	6.338	6.429	6.507	6.575	6.636

Source: Jerrold Zar, *Biostatistical Analysis*, Prentice Hall, Inc., 1999.

ANSWERS

CHAPTER 1   Data Collection

1.1 Assess Your Understanding (page 9)

1. Statistics is the science of collecting, organizing, summarizing, and analyzing information to answer questions or draw conclusions.

3. information   **5.** Descriptive; inferential

7. An experimental group usually receives some sort of experimental drug or treatment, while the control group receives a placebo or a treatment whose effect on the response variable is known.

9. Variables

11. A discrete variable is a quantitative variable that has a finite or countable number of possible values. Continuous variables are also quantitative variables, but there are an infinite number of possible values that are not countable.

13. The process of statistics is to (1) identify the research objective, (2) collect the information needed to answer the questions posed in (1), (3) organize and summarize the information, and (4) draw conclusions from the information.

15. Qualitative   **17.** Qualitative   **19.** Quantitative

21. Quantitative   **23.** Quantitative   **25.** Qualitative

27. Discrete   **29.** Continuous   **31.** Discrete

33. Continuous   **35.** Discrete   **37.** Discrete

39. Population: teenagers 13 to 17 years of age who live in the United States; Sample: 1,028 teenagers 13 to 17 years of age who live in the United States

41. Population: entire soybean crop; Sample: 100 plants selected

43. Population: women 27 to 44 years of age with hypertension; Sample: 7,373 women 27 to 44 years of age with hypertension

45. (a) To determine if application of duct tape is effective as cryotherapy in the treatment of common warts
(b) 51 patients with warts
(c) 85% of patients in group 1 and 60% of the patients in group 2 had complete resolution of their warts.
(d) Duct tape is significantly more effective in treating warts than cryotherapy.

47. (a) To determine whether music cognition and cognitions pertaining to abstract operations such as mathematical or spatial reasoning and related
(b) Thirty-six college students
(c) The mean test score following the Mozart piece was 119, while the mean test score following silence was 110
(d) Subjects perform better on abstract and spatial reasoning tests after listening to Mozart.

49. (a) To determine the number of movies U.S. adults aged 18 years or older have attended in a movie theater in the past 12 months
(b) 1,003 U.S. adults aged 18 years or older
(c) About 65% of the 1,003 adults surveyed stated that they have attended at least one movie in a movie theater in the past 12 months.
(d) 65% of all adult Americans 18 years old or older have attended at least one movie in a movie theater in the past 12 months.

51. Individuals: Sanyo #PDP42H2W, Panasonic #PT-47WX54, Tatung #P50BSAT, RCA #HD50LPW42, RCA #D52W19, JVC #HD52Z575, Sony #KDF-60XS955. Variables: size, screen type price. Data for size: 42, 47, 50, 50, 52, 52, 60; data for screen type: plasma, projection, plasma, projection, projection, projection, projection; data for price: $2994, $1072, $4248, $2696, $1194, $2850, $4100. The variable size is continuous; the variable screen type is qualitative; the variable price is discrete.

53. Individuals: Colorado, Missouri, Montana, New York, Texas. Variables: minimum age for driver's license (unrestricted); blood-alcohol concentration limit, mandatory belt-use law seating positions, maximum allowable speed limit in 2003. Data for minimum age for driver's license: 17, 18, 15, 17, 16.5; data for blood-alcohol concentration limit: 0.10, 0.08, 0.08, 0.08, 0.08; data for mandatory belt-use law seating positions: front, front, all, all, front; data for maximum allowable speed limit in 2003: 75, 70, 75, 65, 75. The variable minimum age for driver's license is continuous; the variable blood-alcohol concentration limit is continuous; the variable mandatory belt-use law seating positions is qualitative; the variable maximum allowable speed limit in 2003 is continuous.

55. (a) Are levels of the substance BNP associated with heart trouble?
(b) All people (presumably adults)
(c) The 1,034 patients who were actually included in the study
(d) People with the highest protein levels were 2.5 times more likely to die than those with the lowest protein levels.
(e) Higher levels of BNP are associated with worse outcomes.

57. (a) Nominal   **(b)** Ordinal   **(c)** Nominal
(d) Ordinal   **(e)** Nominal

1.2 Assess Your Understanding (page 19)

1. An observational study does not attempt to manipulate or influence the individuals in the study. In an experiment, a treatment is applied to individuals in an attempt to isolate the effects of the treatment on the individuals.

3. Sampling is used because it is often difficult or impractical to attempt to study each individual in the population under consideration.

5. frame   **7.** True

9. Observational study   **11.** Experiment

13. Observational study   **15.** Experiment

17. Observational study   **19.** Answers will vary

21. (a) Answers will vary.   **(b)** Answers will vary.

23. (a) 83, 67, 84, 38, 22, 24, 36, 58, 34
(b) Answers may vary depending on the type of graphing calculator used. If using a TI-84, the sample will be 4, 20, 52, 5, 24, 87, 67, 86, 39.

25. (a) First, find a list of all currently enrolled students. This list will serve as the frame. Number the students on the list from 1 through 19,935. Using a random-number generator. Set a seed (or select a starting point if using Table 1). Generate 25 different numbers randomly. The students corresponding to these numbers will be the 25 students in the sample.
(b) Answers will vary.

1.3 Assess Your Understanding (page 30)

1. If the population can be divided into homogeneous, nonoverlapping groups

3. Convenience samples are not probability samples. Individuals that participate are often self-selected, so the results are not likely representative of the population.

5. stratified sample   **7.** False

9. Systematic   **11.** Cluster

13. Simple random   **15.** Cluster

17. Convenience   **19.** Systematic

21. Systematic   **23.** Answers will vary.

25. Answers will vary.

27. (a) 90
(b) Randomly select a number between 1 and 90. Suppose that we randomly select 15; then the individuals in the survey will be 15, 105, 195, 285, . . . , 4425.

29. Answers will vary. A good choice might be stratified sampling with the strata being commuters and noncommuters.

31. Answers will vary. One option would be cluster sampling. The clusters could be the city blocks. Randomly select clusters and then survey all the households in the selected city blocks.

33. Answers will vary. Simple random sampling will work fine here, especially because a list of 6,600 individuals who meet the needs of our study already exists (the frame).

35. The researcher should make a return visit or phone call. Allowing the spouse to answer is a form of convenience sampling and can lead to incorrect results.

1.4 Assess Your Understanding (page 37)

1. Frames are printed periodically, whereas populations are constantly changing.

3. A closed question has fixed choices for answers, whereas an open question is a free-response question. Closed questions are easier to analyze, but limit the responses. Open questions allow respondents to state exactly how they feel, but are harder to analyze due to the variety of answers and possible misinterpretation of answers.

5. Better survey results. Trained interviewers can elicit responses to difficult questions.

7. Pro: Higher chance of making contact with individuals. Con: People upset being interrupted during dinner.

9. Changing the order helps prevent bias due to previous question answers or situations where respondents are more likely to pick earlier choices.

11. (a) Flawed sampling method because only the first 50 students who enter the building have a chance of being surveyed
  (b) Could use systematic sampling, but a better choice would be cluster or simple random sampling because the vice president should have access to a complete frame.

13. (a) Flawed survey due to the wording of the question
  (b) The survey should begin by stating the current penalty for selling a gun illegally. The question might be rewritten as "Do you approve or disapprove of harsher penalties for individuals that sell guns illegally?" The words *approve* and *disapprove* should be rotated from individual to individual.

15. (a) Flawed sampling method because (assuming the survey is written in English), nonEnglish speaking homes will not read the survey. This is why the response rate may be low.
  (b) The survey can be improved through face-to-face or telephone interviews.

17. (a) Flawed sampling method because the sample of weekday shoppers may not represent the eating habits of all shoppers at the mall. There may be different tastes for those who shop in the evening or on the weekends.
  (b) A stratified sample would be better (e.g., weekday, weeknight, weekends).

19. (a) Flawed survey due to the wording of the question
  (b) The question should be reworded so that it doesn't imply the opinion of the editors. One possibility might be 'Do you believe that a marriage can be maintained after an extramarital relation?'

21. (a) Flawed survey because the students are not likely to respond truthfully when their teachers administer the survey
  (b) The survey should be administered by an impartial party so that the students will be more likely to respond truthfully.

23. The ordering of the questions is likely to affect the survey results. Perhaps question B should be asked first. Another possibility is to rotate the questions randomly.

25. The choices need to be rotated so that any response bias due to the ordering of the questions is minimized.

27. Poorly trained interviewers; interviewer bias, surveyed too many female voters

1.5 Assess Your Understanding (page 47)

1. (a) A person, object, or some other well-defined item upon which a treatment is applied
  (b) Any combination of the explanatory variables

  (c) A quantitative or qualitative variable that represents the variable of interest
  (d) The variable whose effect on the response variable is to be assessed by the experimenter
  (e) Neither the experimental unit nor the experimenter knows which treatment is being administered to the experimental unit.
  (f) An innocuous treatment such as a sugar tablet
  (g) The effect of two explanatory variables that cannot be distinguished.

3. Completely randomized; matched-pair

5. (a) Score on achievement test.
  (b) Method of teaching, grade level, intelligence, school district, teacher: Controlled: grade level, school district, teacher. Manipulated: method of teaching
  (c) New teaching method and traditional method; 2
  (d) Random assignment
  (e) Completely randomized design
  (f) 500 students
  (g)

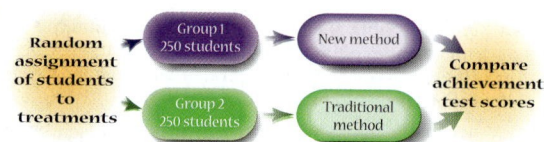

7. (a) Matched pair
  (b) Difference in test scores
  (c) Math course

9. (a) Completely randomized design
  (b) Adults with insomnia
  (c) Terminal wake time after sleep onset (WASO)
  (d) Type of intervention; CBT, RT, placebo
  (e) 75 adults with insomnia
  (f)

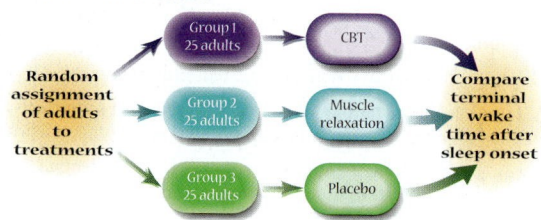

11. (a) Completely randomized design
  (b) Adults older than 60 years in good health
  (c) Score on a standardized test of learning and memory
  (d) Drug; 40 mg of ginkgo 3 times per day or a matching placebo
  (e) 98 men and 132 women older than 60 in good health
  (f)

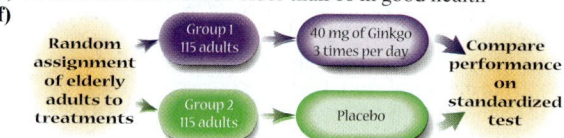

13. (a) Matched-pairs design
  (b) Distance yardstick falls
  (c) Hand dominance; dominant versus nondominant hand
  (d) 15 students
  (e) To eliminate bias due to starting on dominant or nondominant first for each trial
  (f)

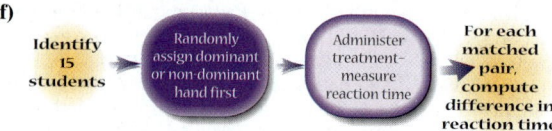

15. (a) Randomized block design
  (b) Score on recall exam
  (c) Type of advertising; 3
  (d) Region of country

(e)

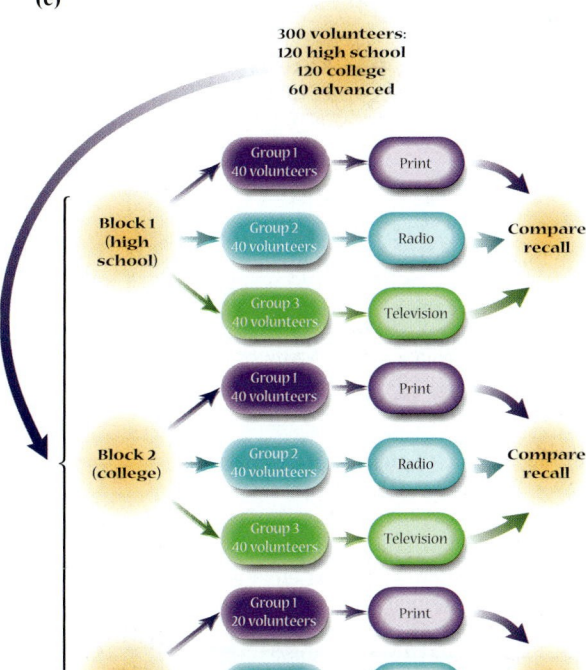

17. Answers will vary.

19. Answers will vary. Completely randomized design is likely best.

21. Answers will vary. Matched-pairs design is likely best (match by type of exterior finish).

23. Answers will vary. Randomized block design is likely best. Block by type of car.

25. (a) Blood pressure
   (b) Daily consumption of salt, daily consumption of fruits and veg-etables, body's ability to process salt
   (c) Salt: manipulate; Fruits/vegetables: control; body; cannot be con-trolled. To deal with variability in body types, randomly assign experimental units to each treatment group.
   (d) Answers will vary. Three might be a good choice; one level below RDA, one equal to RDA, one above RDA.

Chapter Review Exercises (page 53)

1. The science of collecting, organizing, summarizing, and analyzing data in order in draw conclusions or answer questions.

3. A subset of the population

5. Applies a treatment to the individuals in the study to isolate the ef-fect of the treatment on the response variable.

7. There are two main types of errors: nonsampling errors and sam-pling errors. Nonsampling errors are errors that result from the sur-vey process, such as an incomplete frame, poorly worded questions, inaccurate responses, and so on. Sampling errors are errors that re-sult from using a sample in estimate the characteristics of a popu-lation. They result because samples contain incomplete information regarding a population

9. Quantitative; discrete   **11.** Quantitative; continuous

13. Qualitative

15. Observational study, since no treatment has been imposed.

17. Experiment   **19.** Convenience sample

21. Cluster sample

23. (a) Undercoverage or nonrepresentative sample due to a poor sam-pling frame
   (b) Interviewer error
   (c) Data checks

25. Answers will vary.   **27.** Answers will vary.

29. Answers will vary. Label each goggle with pairs of digits from 00 to 99. Using row 12, column 1 of Table 1 and reading across, the se-lected labels would be 55, 36, 65, 10, 57, 90, 06, 51, 47, 91, 62, 42, 60. The goggles with these labels would be inspected for defects.

31. (a) Completely randomized design
   (b) Energy required to light the bulb
   (c) Integrated circuit; old circuit and new circuit
   (d) 200 fluorescent bulbs
   (e)

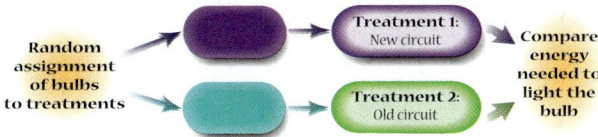

33. (a) Randomized block design
   (b) Exam grade
   (c) Notecard use: with notecard or without notecard
   (d) The instructor's statistics students
   (e)

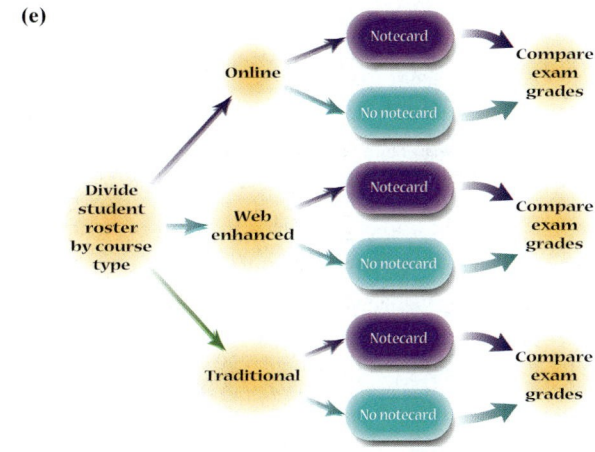

35. Answers will vary.

37. (a) Answers will vary.   **(b)** Answers will vary.

39. In a completely randomized design the experimental units are ran-domly assigned to one of the treatments. The value of the response variable is compared for each treatment. In a matched-pairs design, experimental units are matched up on the basis of some common characteristic (such as husband–wife, pre-post, or twins). The dif-ference in the paired experimental units is analyzed.

CHAPTER 2

2.1 Assess Your Understanding (page 67)

1. Raw data are data that are not organized.

3. It is a good idea to make sure that the frequencies add up to the correct number of observations.

5. A Pareto chart is a bar graph whose bars are drawn in decreasing order of frequency or relative frequency.

7. Answers vary.

9. (a) Large; 52%
   (b) X-Large; 9%
   (c) 23%

11. (a) United States has the most Internet users.
   (b) Approximately 20 million Internet users are in Canada.
   (c) Approximately 25 million more Internet users are in China than in Japan.
13. (a) Badnarik received 397,157 votes.
   (b) Bush received 50.7% of the vote; Kerry received 48.3%.
   (c) No. He only received 0.4% of the votes.
15. (a) 0.16; 0.30
   (b) Natural gas was the most popular source.
   (c) About 5.35 million households used LPG.
   (d) Answers will vary.
   (e) Fuel oil or kerosene; answers will vary.
   (f) Both natural gas and LPG have remained steady.
17. (a)

Source of Income	Percentage of Income
Individual income taxes	0.4453
Corporate income taxes	0.0739
Social insurance taxes	0.4000
Excise, estate and gift taxes, customs, and miscellaneous receipts	0.0807

   (b) 44.53% of the government's income is attributable to individual income taxes.
   (c)

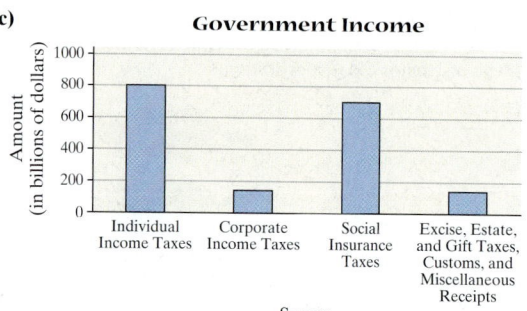

   (d)

   (e)

   (f) Answers will vary.
19. (a)

Response	Relative Frequency
Never	0.0262
Rarely	0.0678
Sometimes	0.1156
Most of the time	0.2632
Always	0.5272

   (b) 52.72%
   (c) 9.4%
   (d)

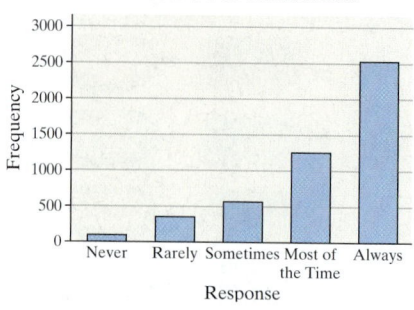

   (e)

   (f)

   (g) This is inferential statistics because it is taking a result of the sample and generalizing it to the population "all college students."
21. (a)

Region	Relative Frequency
Caribbean	0.1010
Central America	0.3690
South America	0.0630
Asia	0.2500
Europe	0.1370
Other Regions	0.0800

   (b) 25.00% of foreign-born residents were born in Asia.
   (c)

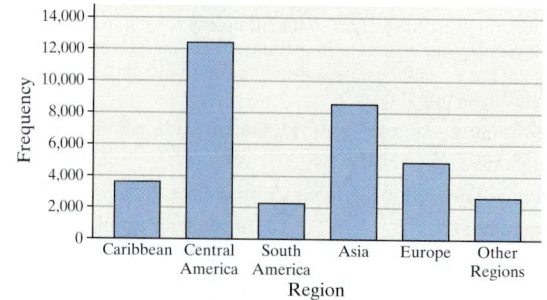

(d)

**Birthplace of
Foreign-Born Residents**

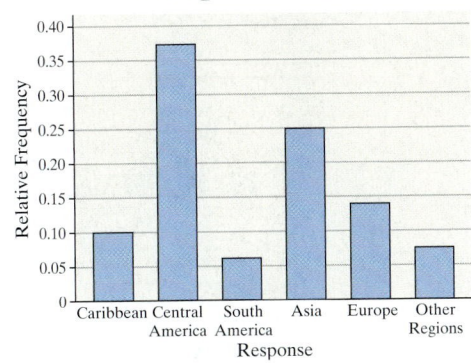

(b)

Age of Victim (female)	Relative Frequency
Less than 17	0.1314
17–24	0.2083
24–35	0.2083
35–54	0.3211
55 or older	0.1308

(c)

Murder Victims by Age - 2002

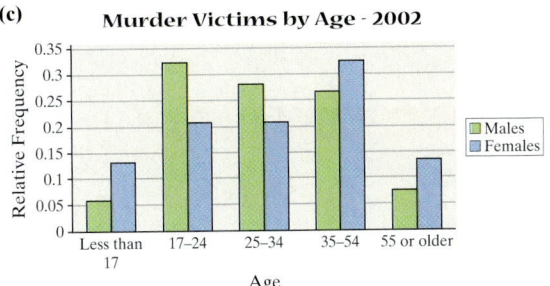

(e)

**Birthplace of
Foreign-Born Residents**

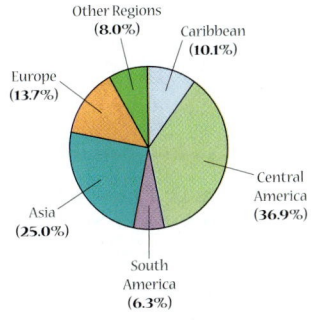

(d) Answers will vary.

27. (a)

Candidate	Frequency
Bush	21
Kerry	17
Nader	1
Badnarik	1

(b)

Candidate	Relative Frequency
Bush	0.525
Kerry	0.425
Nader	0.025
Badnarik	0.025

23. (a)

Educational Attainment	Relative Frequency (males)
Not a high school graduate	0.159
High school graduate	0.309
Some college, but no degree	0.171
Associate's degree	0.072
Bachelor's degree	0.185
Advanced degree	0.104

(b)

Educational Attainment	Relative Frequency (females)
Not a high school graduate	0.150
High school graduate	0.330
Some college, but no degree	0.172
Associate's degree	0.091
Bachelor's degree	0.175
Advanced degree	0.082

(c)

(c)

Votes in Los Alamos County, NM

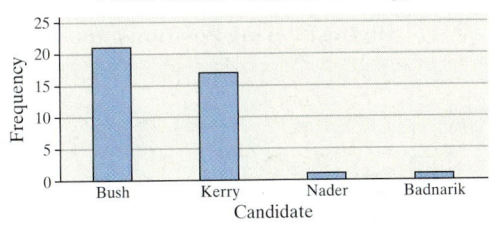

Educational Attainment

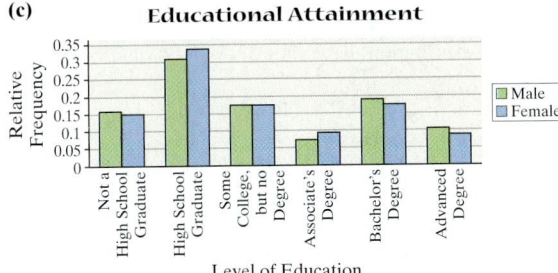

(d)

Votes in Los Alamos County, NM

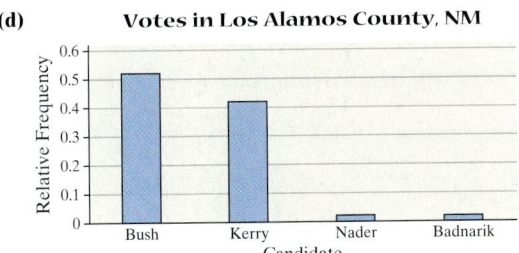

(d) Answers will vary.

25. (a)

Age of Victim (male)	Relative Frequency
Less than 17	0.0606
17–24	0.3204
24–35	0.2789
35–54	0.2667
55 or older	0.0734

(e)

Votes in Los Alamos County, NM

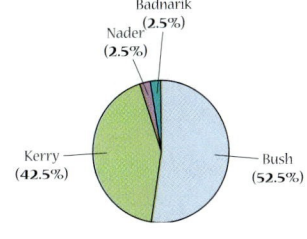

(f) Answers will vary. The conjecture is inferential.

29. (a)

Position	Frequency
First base	3
Second base	0
Third base	2
Shortstop	1
Pitcher	7
Catcher	1
Right field	6
Center field	2
Left field	2

(b)

Position	Relative Frequency
First base	0.1250
Second base	0.0000
Third base	0.0833
Short stop	0.0417
Pitcher	0.2917
Catcher	0.0417
Right field	0.2500
Center field	0.0833
Left field	0.0833

(c) Pitcher   **(d)** Second base

(e)

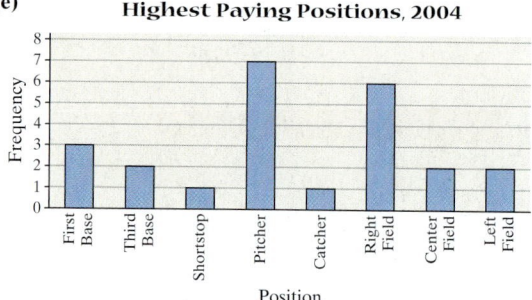

Highest Paying Positions, 2004

(f)

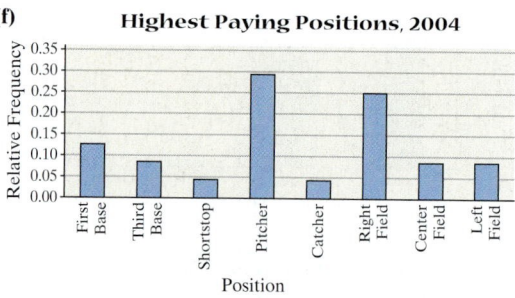

Highest Paying Positions, 2004

(g)

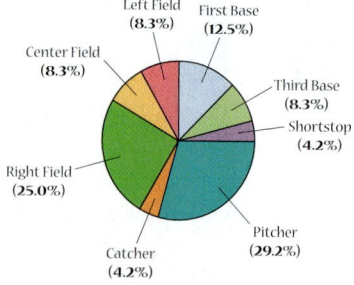

Highest Paying Positions, 2004

31. (a), (b)

Language Studied	Frequency	Relative Frequency
Chinese	3	0.100
French	3	0.100
German	3	0.100
Italian	2	0.067
Japanese	2	0.067
Latin	2	0.067
Russian	1	0.033
Spanish	14	0.467

(c)

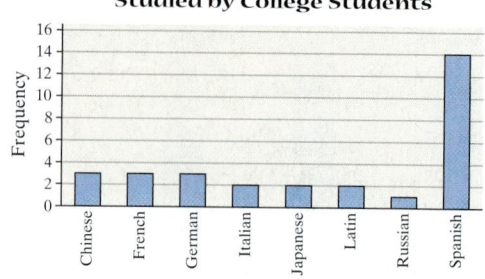

Foreign Languages Studied by College Students

(d)

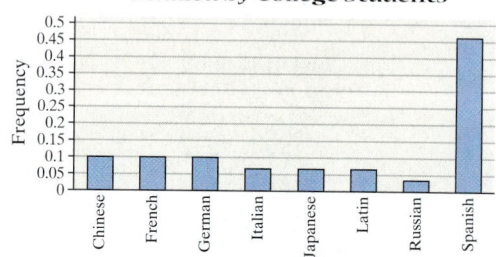

Foreign Languages Studied by College Students

(e)

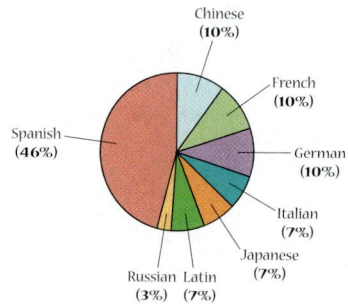

Foreign Languages Studied by College Students

2.2 Assess Your Understanding (page 87)

1. Answers will vary.

3. The class widths are not all the same, and the rectangles in a histogram should touch.

5. Answers will vary. Some possibilities follow: Histograms represent quantitative data, while bar graphs represent qualitative data. The bars in a histogram touch, but the bars in a bar graph do not touch. The width of the bars of a histogram has meaning, but the width of the bars in a bar graph is meaningless.

7. False. The distribution shape shown is skewed right.

9. (a) 8   **(b)** 2   **(c)** 15
(d) 15%   **(e)** Roughly symmetric

11. (a) 200   **(b)** 10
(c) 60–69, 2; 70–79, 3; 80–89, 13; 90–99, 42; 100–109, 58; 110–119, 40; 120–129, 31; 130–139, 8; 140–149, 2; 150–159, 1
(d) 100–109   **(e)** 150–159

13. (a)

Number of Children under 5	Relative Frequency
0	0.32
1	0.36
2	0.24
3	0.06
4	0.02

(b) 24%   **(c)** 60%

15. 10, 11, 14, 21, 24, 24, 27, 29, 33, 35, 35, 35, 37, 37, 38, 40, 40, 41, 42, 46, 46, 48, 49, 49, 53, 53, 55, 58, 61, 62

17. 1.2, 1.4, 1.6, 2.1, 2.4, 2.7, 2.7, 2.9, 3.3, 3.3, 3.3, 3.5, 3.7, 3.7, 3.8, 4.0, 4.1, 4.1, 4.3, 4.6, 4.6, 4.8, 4.8, 4.9, 5.3, 5.4, 5.5, 5.8, 6.2, 6.4

19. (a) Four classes
(b) Lower class limits: 25, 35, 45, 55; upper class limits: 34, 44, 54, 64
(c) Class width: 10 years

21. (a) Four classes
(b) Lower class limits: 100, 200, 300, 400; upper class limits: 199, 299, 399, 499
(c) Class width: 100 beds

23. (a)

Age	Relative Frequency
25–34	0.2323
35–44	0.2870
45–54	0.2822
55–64	0.1986

(b)

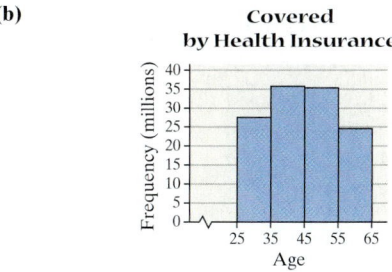

Covered by Health Insurance

(c)

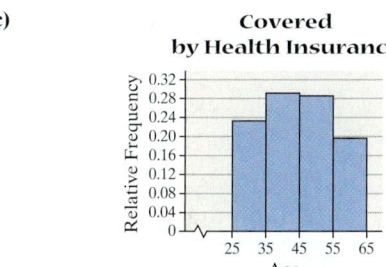

Covered by Health Insurance

23.23% is between the ages of 25 and 34; 51.93% is 44 years or younger.

25. (a)

Number of Beds	Relative Frequency
100–199	0.2990
200–299	0.2979
300–399	0.2456
400–499	0.1574

(b)

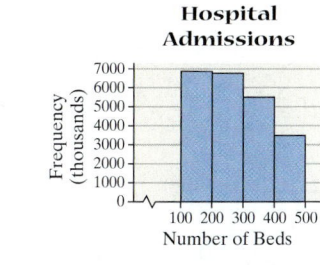

Hospital Admissions

(c)

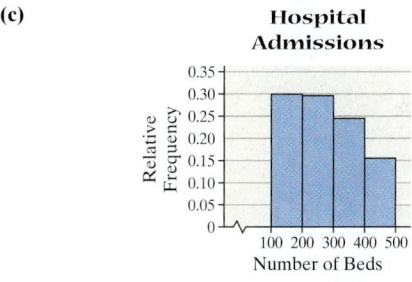

Hospital Admissions

24.56% of admissions was to hospitals with 300 to 399 beds; 40.30% of admissions was to hospitals with more than 300 beds.

27. (a),(b)

Number of Customers	Frequency	Relative Frequency	Number of Customers	Frequency	Relative Frequency
3	2	0.050	9	4	0.1
4	3	0.075	10	4	0.1
5	3	0.075	11	4	0.1
6	5	0.125	12	0	0
7	4	0.1	13	2	0.05
8	8	0.2	14	1	0.025

(c) 27.5%   **(d)** 20%

(e)

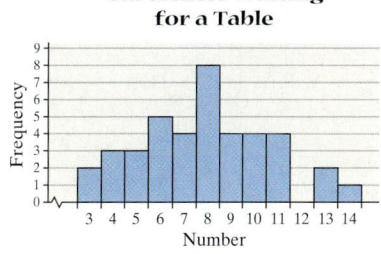

Customers Waiting for a Table

(f)

Customers Waiting for a Table

(g) Symmetric

29. (a), (b)

2003 per Capita Disposable Income		
Class Intervals	Frequency	Relative Frequency
20,000–22,499	3	0.0588
22,500–24,999	10	0.1961
25,000–27,499	14	0.2745
27,500–29,999	12	0.2353
30,000–32,499	7	0.1373
32,500–34,999	2	0.0392
35,000–37,499	2	0.0392
37,500–39,999	0	0
40,000–42,499	1	0.0196

(c)

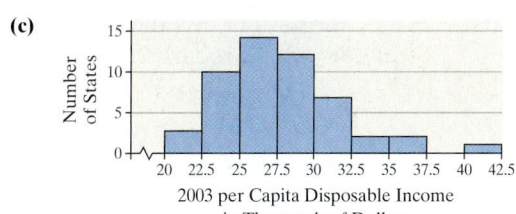

2003 per Capita Disposable Income in Thousands of Dollars

(d)

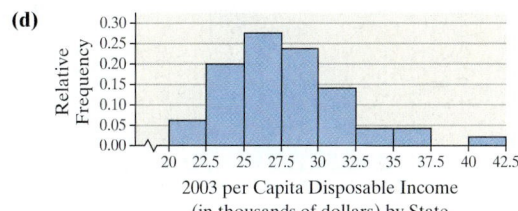

2003 per Capita Disposable Income (in thousands of dollars) by State

(e) The distribution is skewed right.

(f)

2003 per Capita Disposable Income		
Class Intervals	Frequency	Relative Frequency
20,000–23,999	10	0.1961
24,000–27,999	22	0.4314
28,000–31,999	14	0.2745
32,000–35,999	3	0.0588
36,000–39,999	1	0.0196
40,000–43,999	1	0.0196

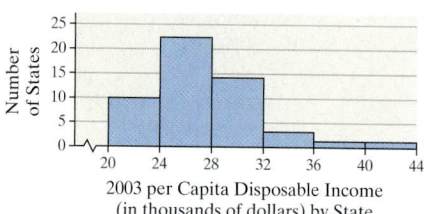

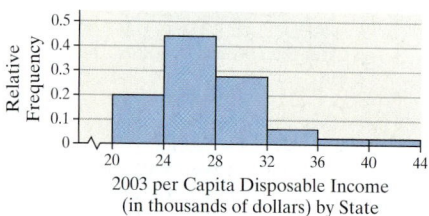

31. (a), (b)

Class	Frequency	Relative Frequency
20–29	1	$\frac{1}{40} = 0.025$
30–39	6	$\frac{6}{40} = 0.15$
40–49	10	$\frac{10}{40} = 0.25$
50–59	14	0.35
60–69	6	0.15
70–79	3	0.075

(c)

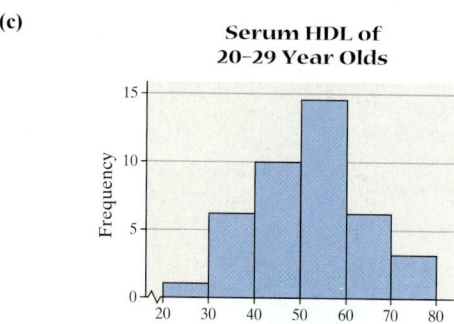

Serum HDL of 20–29 Year Olds

(d)

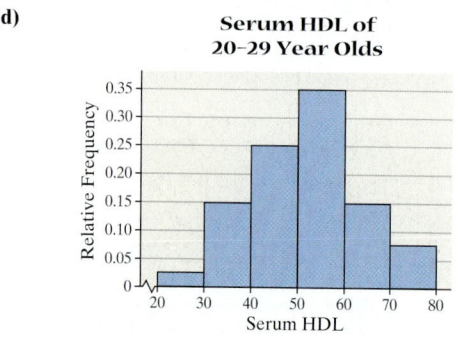

Serum HDL of 20–29 Year Olds

(e) Bell shaped

(f)

Class	Frequency	Relative Frequency
20–24	0	0
25–29	1	0.025
30–34	2	0.05
35–39	4	0.1
40–44	2	0.05
45–49	8	0.2
50–54	9	0.225
55–59	5	0.125
60–64	4	0.1
65–69	2	0.05
70–74	3	0.075
75–79	0	0

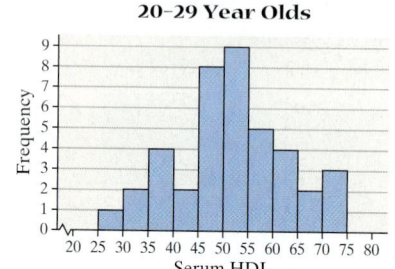

Serum HDL of 20–29 Year Olds

Serum HDL of 20–29 Year Olds

Bell shaped
(g) Answers will vary.

33. (a), (b)

Dividend Yield		
Class Interval	Number of Stocks	Relative Frequency
0–0.39	7	0.2500
0.40–0.79	4	0.1429
0.80–1.19	5	0.1786
1.20–1.59	2	0.0714
1.60–1.99	3	0.1071
2.0–2.39	4	0.1429
2.4–2.79	2	0.0714
2.8–3.19	1	0.0357

(c)

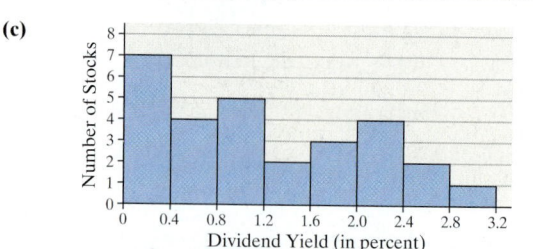

(d)

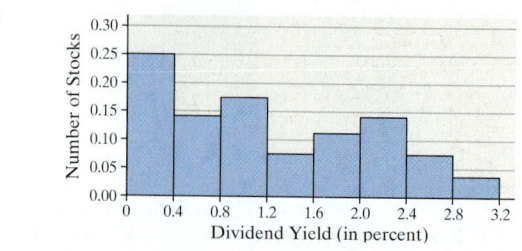

(e) The distribution is skewed right.

(f)

Dividend Yield (%)		
Class	Number of Stocks	Relative Frequency
0.0–0.79	11	0.3929
0.8–1.59	7	0.2500
1.6–2.39	7	0.2500
2.4–3.19	3	0.1071

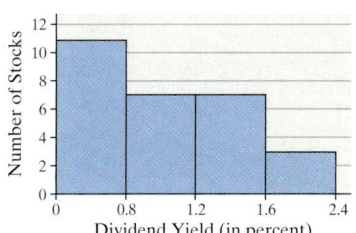

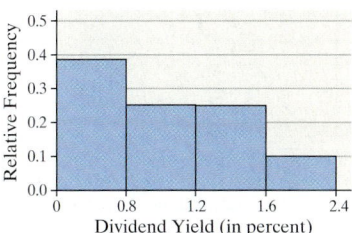

The distribution is skewed right.

35.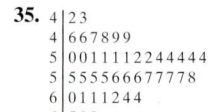
```
4 | 2 3
4 | 6 6 7 8 9 9
5 | 0 0 1 1 1 1 1 2 2 4 4 4 4 4
5 | 5 5 5 5 6 6 6 7 7 7 7 8
6 | 0 1 1 1 2 4 4
6 | 5 8 9
```

37.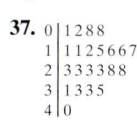
```
0 | 1 2 8 8
1 | 1 1 2 5 6 6 7
2 | 3 3 3 3 8 8
3 | 1 3 3 5
4 | 0
```

39. (a) The energy derived from coal rounded to the nearest percent:

54	0	75	64	16	8	94
9	32	90	35	88	56	0
41	62	23	83	17	0	51
49	13	3	60	60	38	8
1	1	60	63	95	42	98
78	46	27	51	90	62	68
10	94	57	23	61	37	96
58	83					

```
0 | 0 0 0 1 1 3 8 8 9
1 | 0 3 6 7
2 | 3 3 7
3 | 2 5 7 8
4 | 1 2 6 9
5 | 1 1 4 6 7 8
6 | 0 0 0 1 2 2 3 4 8
7 | 5 8
8 | 3 3 8
9 | 0 0 4 4 5 6 8
```

(b) Fairly uniform
(c) Answers will vary.

41. (a)

Make	Problems per 100 Vehicles	Make	Problems per 100 Vehicles	Make	Problems per 100 Vehicles
Lexus	160	Subaru	270	Audi	320
Infiniti	170	Nissan	270	Mercedes Benz	320
Buick	180	GMC	270	Jeep	320
Porsche	190	Chevrolet	270	Volvo	330
Acura	200	Saturn	270	Mitsubishi	340
Toyota	200	Oldsmobile	280	Hyundai	340
Cadillac	210	Mazda	290	Isuzu	370
Lincoln	210	Pontiac	290	Volkswagen	390
Honda	220	Chrysler	300	Suzuki	400
Mercury	240	Ford	300	Daewoo	420
Jaguar	250	Plymouth	300	Land Rover	440
Saab	260	Dodge	310	Kia	510
BMW	260				

(b)
```
1 | 6 7 8 9
2 | 0 0 1 1 2 4 5 6 6 7 7 7 7 7 8 9 9
3 | 0 0 0 1 2 2 2 3 4 4 7 9
4 | 0 2 4
5 | 1
```
Answers will vary.

(c)
```
1 | 6 7 8 9
2 | 0 0 1 1 2 4
2 | 5 6 6 7 7 7 7 7 8 9 9
3 | 0 0 0 1 2 2 2 3 4 4
3 | 7 9
4 | 0 2 4
4 |
5 | 1
```
Answers will vary.

43. (a)

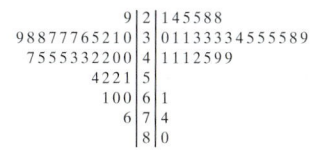

Best Actor Age   Best Actress Age
```
              9 | 2 | 1 4 5 5 8 8
    9 8 8 7 7 7 6 5 2 1 0 | 3 | 0 1 1 3 3 3 3 4 5 5 5 5 5 8 9
      7 5 5 5 3 3 2 2 0 0 | 4 | 1 1 1 2 5 9 9
                4 2 2 1 | 5 |
                  1 0 0 | 6 | 1
                      6 | 7 | 4
                        | 8 | 0
```

(b) Answers will vary.

45.

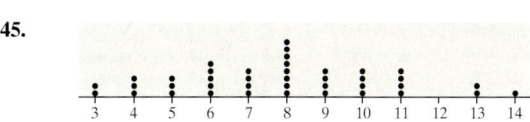

Number Waiting

2.3 Assess Your Understanding (page 101)

1. Answers will vary.

3. An ogive is a graph that represents the cumulative frequency or cumulative relative frequency of the class.

5. True

7. (a) The class width is 5.
(b) The lower class limit of the fourth class is 35.
(c) The upper class limit of the fourth class is 39.
(d) The highest frequency is in class 40–44 years.
(e) The lowest frequency is in class 75–79 years.

9. (a) The class width is 3.
(b) The lower class limit of the third class is 16.
(c) 90% of students had an ACT composite score of 27 or below.
(d) 1,054,314 students had a score less than or equal to 27.

11. (a), (b)

Age	Cumulative Frequency (millions)	Cumulative Relative Frequency
25–34	28.9	0.2323
35–44	64.6	0.5193
45–54	99.7	0.8104
55–64	124.4	1

(c)

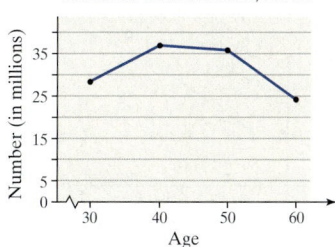

People Covered by Health Insurance, 2003

(d)

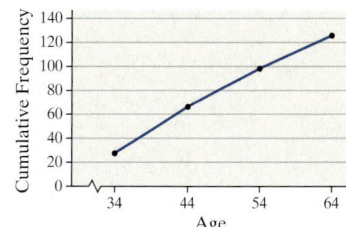

Number of People Covered by Health Insurance, 2003

(e)

Number of People Covered by Health Insurance, 2003

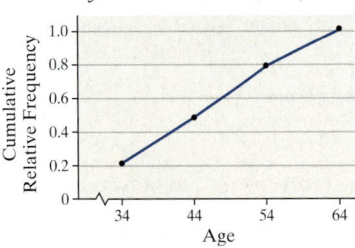

13. (a), (b)

Number of Beds	Cumulative Frequency (thousands)	Cumulative Relative Frequency (thousands)
100–199	6,826	0.2990
200–299	13,626	0.5970
300–399	19,233	0.8426
400–499	22,826	1

(c)

Community Hospital Admissions, 2002

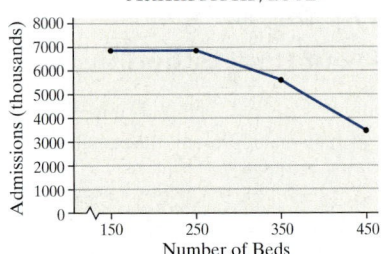

(d)

Community Hospital Admissions, 2002

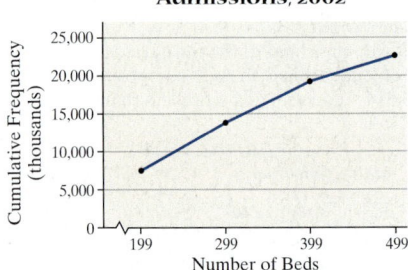

(e)

Community Hospital Admissions, 2002

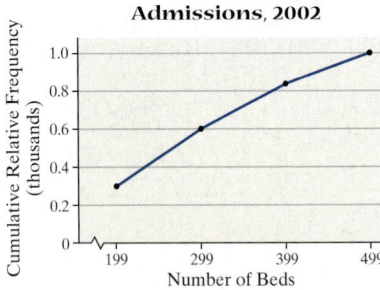

15. (a), (b)

Class	Cumulative Frequency	Cumulative Relative Frequency
20,000–22,499	3	0.0588
22,500–24,999	13	0.2549
25,000–27,499	27	0.5294
27,500–29,999	39	0.7647
30,000–32,499	46	0.9020
32,500–34,999	48	0.9412
35,000–37,499	50	0.9804
37,500–39,999	50	0.9804
40,000–42,499	51	1

(c)

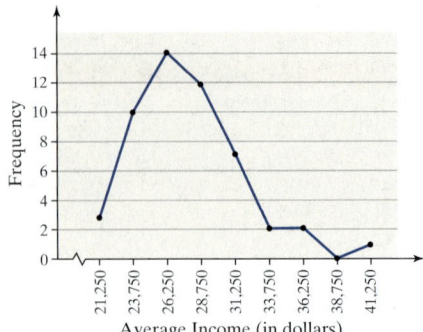

(d)

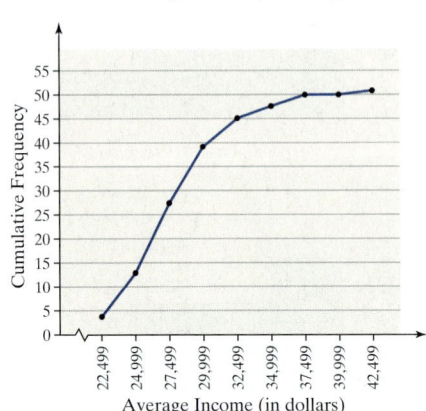

(e)

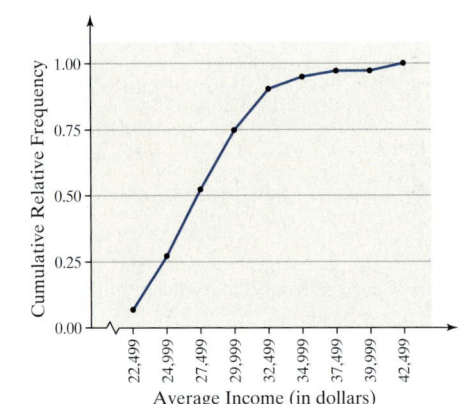

17. (a)

Closing Price of Pixar Stock

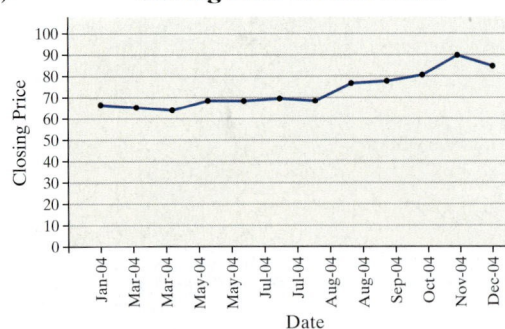

(b) The stock price increased the most during November. Explanations for why will vary.

19.

College Enrollment

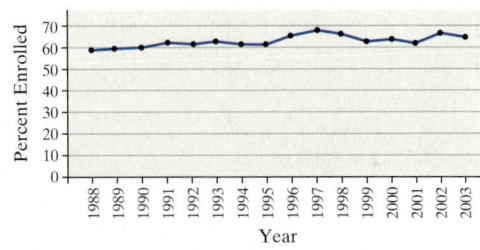

21. (a), (b)

Financial Stocks		
Class	Frequency	Relative Frequency
0–4.99	1	0.0313
5–9.99	7	0.2188
10–14.99	7	0.2188
15–19.99	8	0.2500
20–24.99	4	0.1250
25–29.99	4	0.1250
30–34.99	1	0.0313

Energy Stocks		
Class	Frequency	Relative Frequency
0–4.99	0	0
5–9.99	6	0.1875
10–14.99	2	0.0625
15–19.99	9	0.2813
20–24.99	4	0.1250
25–29.99	4	0.1250
30–34.99	1	0.0313
35–39.99	1	0.0313
40–44.99	3	0.0938
45–49.99	1	0.0313
50–54.99	1	0.0313

(c)

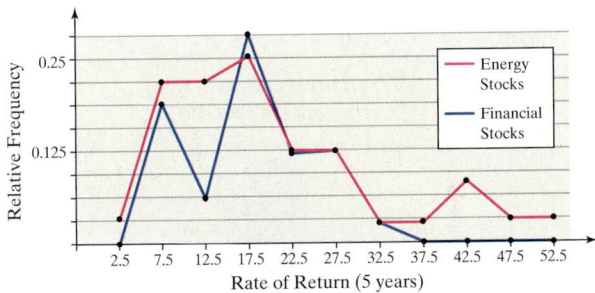

(d)

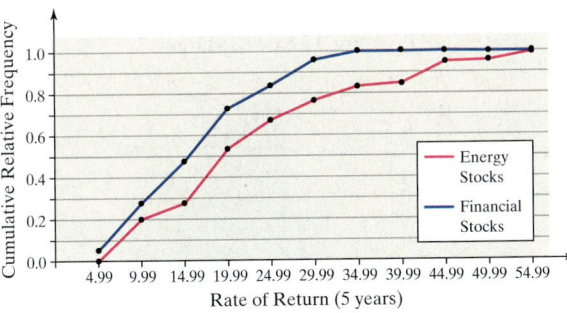

(e) Answers will vary.

2.4 Assess Your Understanding (page 107)

1. The lengths of the bars are not proportional. For example, the bar representing the cost of Clinton's inauguration should be slightly more than 9 times as long as the one for Carter's cost and twice as long as the bar representing Reagan's cost.

3. (a) The vertical axis starts at 28 instead of 0. This tends to indicate that the median earnings for females increased at a faster rate than it actually did.

(b) This graph indicates that the median earnings for females has increased slightly over the given 5-year period.

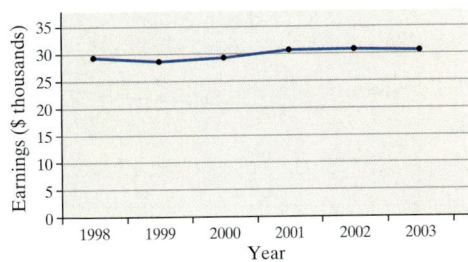

5. (a) The vertical axis starts at 0.1 instead of 0. This might cause the reader to conclude, for example, that the proportion of people aged 25 to 34 who are not covered by health insurance is more than 4 times the proportion for those aged 55 to 64.

(b)

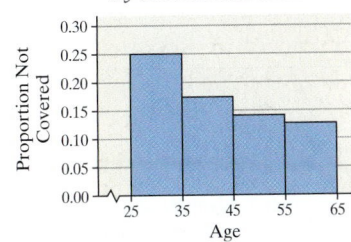

7. (a) The vertical axis starts at 35 without indicating a gap.
(b) It may convey that median household income is increasing more rapidly than it really is.

9. (a) The bar for housing should be a little more than twice the length of the bar for transportation, but it is not.
(b) Adjust the graphs so that the lengths of the bars are proportional.

11. (a)

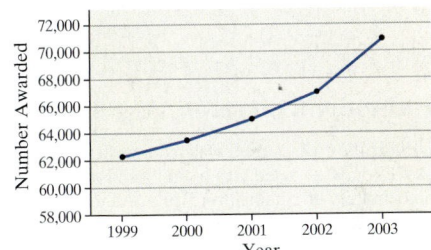

(b)

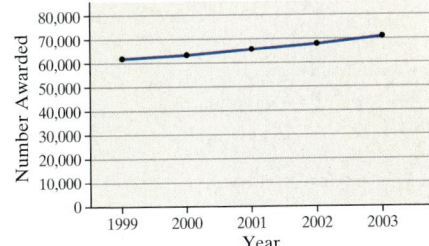

13. (a) The politician's view:

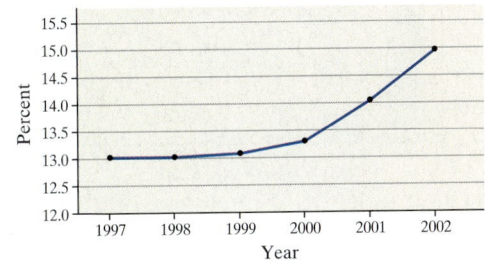

(b) The health care industry's view:

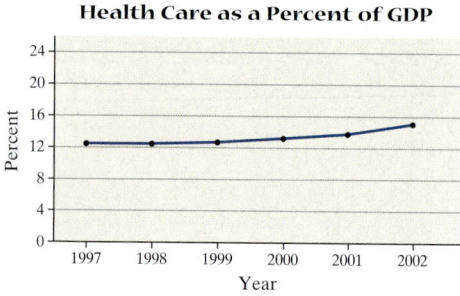

Health Care as a Percent of GDP

(c) An honest view:

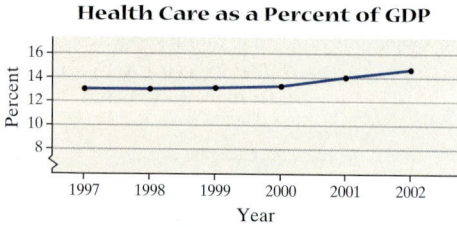

Health Care as a Percent of GDP

15. (a) Graphic that is not misleading:

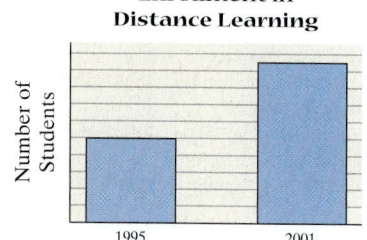

Enrollment in Distance Learning

(b) Graphics will vary.

Chapter Review Exercises (page 111)

1. (a) 22.5 quadrillion Btu's used was from natural gas.
  (b) 3 quadrillion Btu's used was from biomass.
  (c) Approximately 98 quadrillion Btu's of energy was consumed.
  (d) "Other" has the lowest frequency.
  (e) No, these are qualitative data.

3. (a)

Type of Weapon	Relative Frequency
Firearms	0.6617
Knives or cutting instruments	0.1248
Blunt objects (clubs, hammers, etc.)	0.0470
Personal weapons (hands, fists, etc.)	0.0659
Strangulation	0.0101
Fire	0.0073
Other weapon or not stated	0.0831

(b) 4.7% of homicides was committed with a blunt object.

(c)

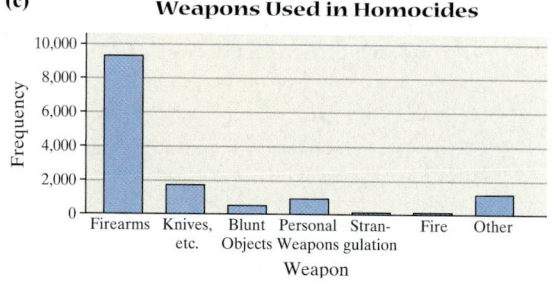

Weapons Used in Homicides

(d)

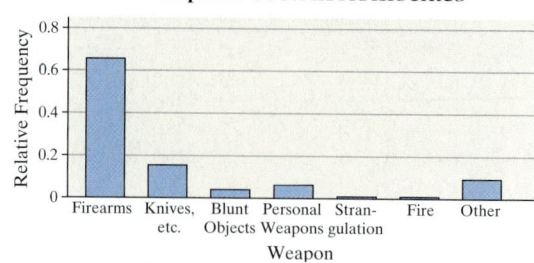

Weapons Used in Homocides

(e)

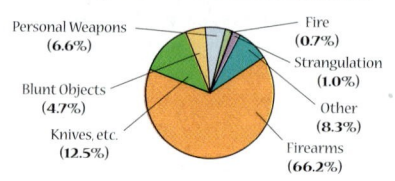

Weapons Used in Homicides

5. (a), (b), (c)

Age of Mother (years)	Relative Frequency	Cumulative Frequency	Cumulative Relative Frequency
10–14	0.0017	7	0.0017
15–19	0.1016	422	0.1033
20–24	0.2526	1454	0.3558
25–29	0.2660	2541	0.6219
30–34	0.2389	3517	0.8607
35–39	0.1145	3985	0.9753
40–44	0.0247	4086	1.0000

(d)

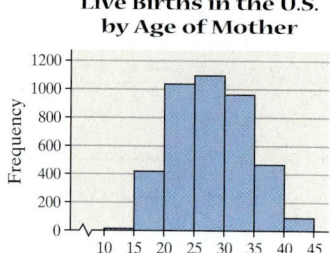

Live Births in the U.S. by Age of Mother

The distribution is symmetric.

(e)

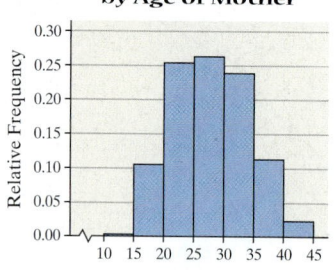

Live Births in the U.S. by Age of Mother

(f)

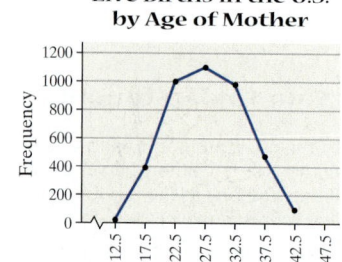

Live Births in the U.S. by Age of Mother

(g)

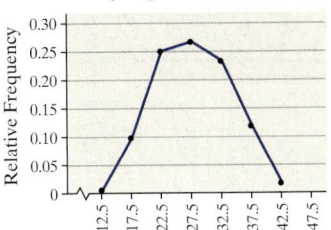

Live Births in the U.S. by Age of Mother

(h)

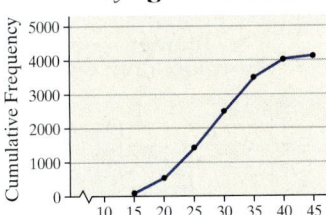

Live Births in the U.S. by Age of Mother

(i)

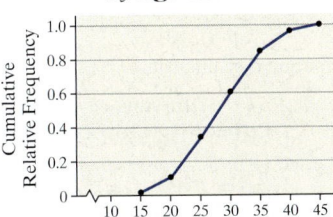

Live Births in the U.S. by Age of Mother

(j) 25.3% of the births was to mothers aged 20 to 24.
(k) 37.8% of the live births was to mothers age 30 or above.

7. (a), (b)

Affiliation	Frequency	Relative Frequency
Democrat	46	0.46
Independent	16	0.16
Republican	38	0.38

(c)

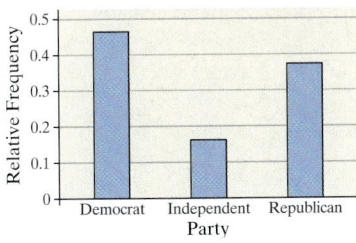

Political Affiliation

(d)

Political Affiliation

Republican (38%)   Democrat (46%)   Independent (16%)

(e) Democrat appears to be the most common affiliation.

9. (a), (b), (c), (d)

Number of Children	Frequency	Relative Frequency	Cumulative Frequency	Cumulative Relative Frequency
0	7	0.1167	7	0.1167
1	7	0.1167	14	0.2333
2	18	0.3000	32	0.5333
3	20	0.3333	52	0.8667
4	7	0.1167	59	0.9833
5	1	0.0167	60	1.0000

(e)

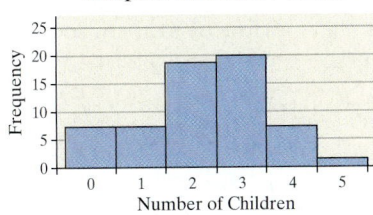

Number of Children for Couples Married 7 Years

The distribution is symmetric.

(f)

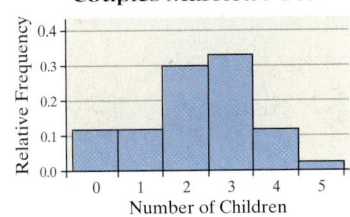

Number of Children for Couples Married 7 Years

(g) 30% of the couples has two children.
(h) 76.7% of the couples has at least two children.

(i)

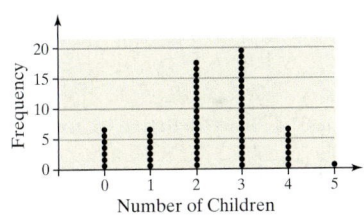

11. (a), (b), (c), (d)

		Crime Rate by State		
Class	Frequency	Relative Frequency	Cumulative Frequency	Cumulative Relative Frequency
2000–2399	2	0.0392	2	0.0392
2400–2799	4	0.0784	6	0.1176
2800–3199	8	0.1569	14	0.2745
3200–3599	6	0.1176	20	0.3922
3600–3999	4	0.0784	24	0.4706
4000–4399	8	0.1569	32	0.6275
4400–4799	8	0.1569	40	0.7843
4800–5199	6	0.1176	46	0.9020
5200–5599	2	0.0392	48	0.9412
5600–5999	0	0.0000	48	0.9412
6000–6399	2	0.0392	50	0.9804
6400–6799	0	0.0000	50	0.9804
6800–7199	0	0.0000	50	0.9804
7200–7599	0	0.0000	50	0.9804
7600–7999	0	0.0000	50	0.9804
8000–8399	1	0.0196	51	1.0000

(e)

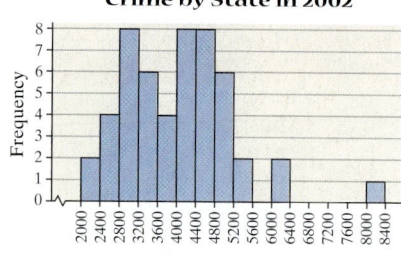

Crime by State in 2002

The distribution is skewed right.

(f)

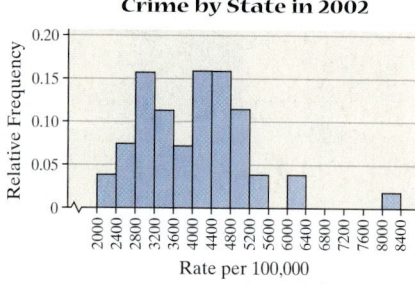

Crime by State in 2002

(g) For (a)–(d):

Crime Rate (per 100,000 population)

Class	Frequency	Relative Frequency	Cumulative Frequency	Cumulative Relative Frequency
2000–2999	10	0.1961	10	0.1961
3000–3999	14	0.2745	24	0.4706
4000–4999	17	0.3333	41	0.8039
5000–5999	7	0.1373	48	0.9412
6000–6999	2	0.0392	50	0.9804
7000–7999	0	0.0000	50	0.9804
8000–8999	1	0.0196	51	1.0000

For (e):

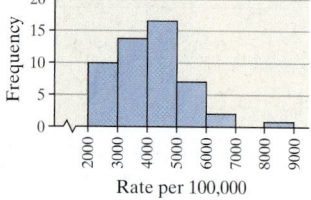

Crime by State in 2002

The distribution is skewed right.

For (f):

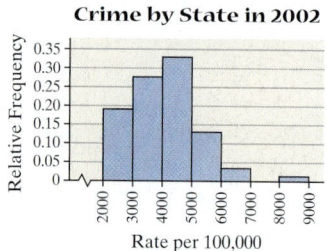

Crime by State in 2002

For (g): Answers will vary.

13. (a), (b), (c), (d)

Diameter of a Cookie

Class	Frequency	Relative Frequency	Cumulative Frequency	Cumulative Relative Frequency
2.2000–2.2199	2	0.0588	2	0.0588
2.2200–2.2399	3	0.0882	5	0.1470
2.2400–2.2599	5	0.1471	10	0.2941
2.2600–2.2799	6	0.1765	16	0.4706
2.2800–2.2999	4	0.1176	20	0.5882
2.3000–2.3199	7	0.2059	27	0.7941
2.3200–2.3399	5	0.1471	32	0.9412
2.3400–2.3599	1	0.0294	33	0.9706
2.3600–2.3799	1	0.0294	34	1.000

(e)

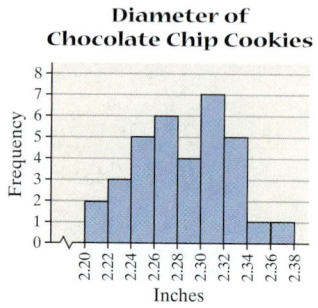

Diameter of Chocolate Chip Cookies

The distribution is symmetric.

(f)

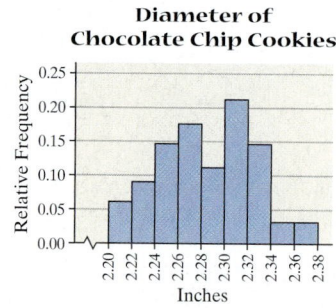

Diameter of Chocolate Chip Cookies

(g) For (a)–(d):

Diameter of a Chocolate Chip Cookie

Class	Frequency	Relative Frequency	Cumulative Frequency	Cumulative Relative Frequency
2.2000–2.2399	5	0.1471	5	0.1471
2.2400–2.2799	11	0.3235	16	0.4706
2.2800–2.3199	11	0.3235	27	0.7941
2.3200–2.3599	6	0.1765	33	0.9706
2.3600–2.3999	1	0.0294	34	1.0000

For (e):

Diameter of Chocolate Chip Cookies

The distribution is symmetric.

For (f):

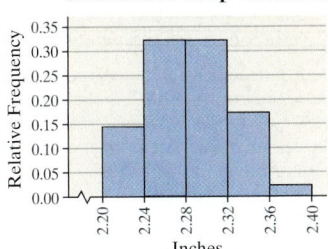

Diameter of Chocolate Chip Cookies

Answers will vary.

15.

```
 0 | 234456778
 1 | 47
 2 | 1334467
 3 | 2355589
 4 | 199
 5 | 13889
 6 | 017
 7 | 6
 8 | 3
 9 |
10 | 2
11 | 1
```

The distribution is skewed right.

17. (a)

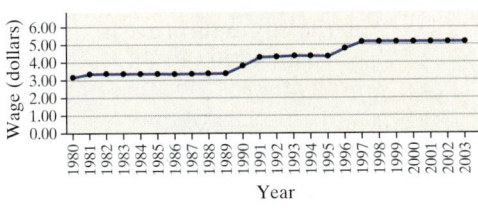

Federal Minimum Wage

(b) The trend indicates that the federal minimum wage remains constant.

19. No vertical scale.

21. (a) Graphs will vary.
(b) An example of a graph that does not mislead:

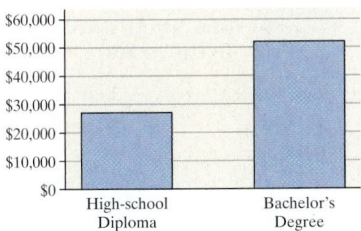

2002 Average Earnings

CHAPTER 3

3.1 Assess Your Understanding (page 130)

1. A statistic is resistant if it is not sensitive to extreme values. The median is resistant because it is a positional measure of central tendency, and increasing the largest value or decreasing the smallest value does not affect the position of the center. The mean is not resistant because it is a function of the sum of the values of data. Changing the magnitude of one value changes the sum of the values. The mode is a resistant measure of center.

3. $55,263 is the mean household income.

5. The mean will be larger. Explanations will vary.

7. The mode is used with qualitative data because they cannot be ordered to find the median or given numerical values needed to calculate the mean.

9. False   **11.** $\bar{x} = 11$ **13.** $\mu = 9$

15. Mean price per ad slot is $2.4 million.

17. Mean cost is $381.75; median cost is $414.50; there is no mode cost.

19. The mean, median, and mode strengths are 3670, 3830, and 4090 pounds per square inch, respectively.

21. (a) $\mu > M$ **(b)** $\mu = M$ **(c)** $\mu < M$
Justification will vary.

23. Los Angeles ATM fees: $\bar{x} = \$1.44$, $M = \$1.50$, mode $= \$2.00$. New York ATM fees: $\bar{x} = \$1.06$, $M = \$1.13$, mode $= \$1.00$. It appears that ATM fees in Los Angeles are higher than they are in New York. Explanations will vary.

25. (a) The mean pulse rate is 72.2 beats per minute.
(b) Samples and sample means will vary.
(c) Answers will vary.

27. (a) The mean number of goals she scored is 8.7 per year.
(b) Samples and sample means will vary.
(c) Answers will vary.

29. The mean is the better measure of central tendency because the distribution is symmetric.

31. (a) $\bar{x} = 51.1$; $M = 51$ **(b)** The distribution is symmetric.

33. $\bar{x} = 0.874$ gram; $M = 0.88$ gram. The distribution is symmetric; so the mean is the better measure of central tendency.

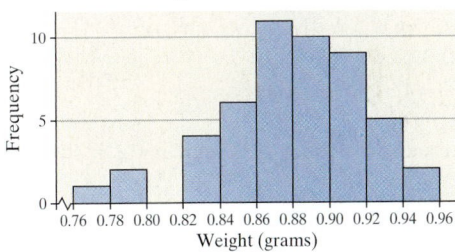

Weight of Plain M&Ms

35. $\bar{x} = 22$ hours; $M = 25$ hours. The distribution is skewed left, so the median is the better measure of central tendency.

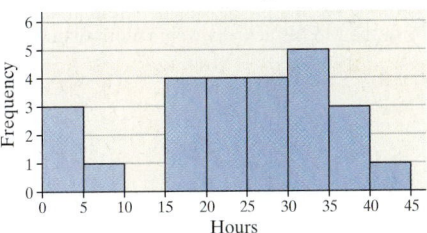

Hours Worked per Week

37. The mode region of birth is Central America.

39. The mode candidate is Bush.

41. Sample of size 5: All data recorded correctly: $\bar{x} = 99.8$; $M = 100$; 106 recorded as 160: $\bar{x} = 110.6$; $M = 100$

Sample of size 12: All data recorded correctly: $\bar{x} = 100.4$; $M = 101$; 106 recorded as 160: $\bar{x} = 104.9$; $M = 101$

Sample of size 30: All data recorded correctly: $\bar{x} = 100.6$; $M = 99$; 106 recorded as 160: $\bar{x} = 102.4$; $M = 99$

For each sample size, the mean becomes larger, but the median remains constant. As the sample size increases, the impact of the misrecorded data on the mean decreases.

43. Answers will vary.

45. The distribution is skewed right, so the median amount of money lost is less than the mean amount lost.

47. The sum is 204.

49. (a) Mean = $50,000; median = $50,000; mode = $50,000
(b) New data set: 32.5, 32.5, 47.5, 52.5, 52.5, 52.5, 57.5, 57.5, 62.5, 77.5; new mean = $52,500; new median = $52,500; new mode = $52,500. All three measures increased by $2500.
(c) New data set: 31.5, 31.5, 47.25, 52.5, 52.5, 52.5, 57.75, 57.75, 63, 78.75. new mean = $52,500; new median = $52,500; new mode = $52,500. All three measures increased by 5%.
(d) New data set: 30, 30, 45, 50, 50, 50, 55, 55, 60, 100; new mean = $52,500; new median = $50,000; new mode = $50,000; The mean increased by $2500, but the median and the mode remained at $50,000.

51. The trimmed mean is 0.875. Explanations will vary.

3.2 Assess Your Understanding (page 148)

1. It is not appropriate because the units are different.

3. No, none of the measures of dispersion is resistant.

5. A statistic is biased when it consistently overestimates or underestimates the population parameter.

7. The standard deviation is the square root of the variance.

9. True   **11.** $s^2 = 36$; $s = 6$ **13.** $\sigma^2 = 16$; $\sigma = 4$

15. $s^2 = 196$; $s = 14$ **17.** $R = \$226$; $s^2 = 9962.9$; $s = \$99.81$

19. $R = 1150$ psi; $s^2 = 211,275$ psi^2; $s = 459.6$ psi

21. (b) The data go from 30 to 75, in (a) they are clustered between 40 and 60.

23. Los Angeles: $R = \$2.00$, $s = \$0.68$; New York: $R = \$1.50$, $s = \$0.48$; Los Angeles has more dispersion based on both the range and standard deviation.

25. (a) $\sigma^2 = 58.8$, $\sigma = 7.7$ beats per minute
(b) (c) Answers will vary.

27. (a) $\sigma^2 = 41.0$, $\sigma = 6.4$ goals
(b) (c) Answers will vary.

29. (a) Ethan: $\mu = 10$ fish, $R = 19$ fish; Drew: $\mu = 10$ fish, $R = 19$ fish
(b) Ethan: $\sigma = 4.9$ fish, Drew: $\sigma = 7.9$ fish
(c) Answers will vary.

31. $R = 45$, $s^2 = 118.0$, $s = 10.9$

33. (a) $s = 0.04$ gram
(b) The histogram is approximately symmetric, so the Empirical Rule is applicable.
(c) 95% of the M&Ms should weigh between 0.79 and 0.95 gram.
(d) 98% of the M&Ms actually weigh between 0.79 and 0.95 gram.
(e) 16% of the M&Ms should weigh more than 0.91 gram.
(f) 14% of the M&Ms actually weigh more than 0.91 gram.

35.

	Car 1	Car 2	
Sample mean	$\bar{x} = 223.5$	$\bar{x} = 237.2$	mi./10 gal
Median	$M = 223$	$M = 230$	mi./10 gal
Mode	none	none	mi./10 gal
Range	$R = 93$	$R = 166$	mi./10 gal
Sample variance	$s^2 = 475.1$	$s^2 = 2406.9$	
Sample standard deviation	$s = 21.8$	$s = 49.1$	mi./10 gal

Answers will vary.

37. (a) Financial stocks: $\bar{x} = 15.716$, $M = 16.09$; energy stocks: $\bar{x} = 22.481$, $M = 19.585$. Energy stocks have the higher median rate of return.
(b) Financial stocks: $s = 7.378$; energy stocks: $s = 12.751$. Energy stocks are riskier.

39. (a) 95% of people has an IQ score between 70 and 130.
(b) 5% of people has an IQ score either less than 70 or greater than 130.
(c) 2.5% of people has an IQ score greater than 130.

41. (a) 95% of pairs of kidneys weighs between 265 and 385 grams.
(b) 99.7% of pairs of kidneys weighs between 235 and 415 grams.
(c) 0.3% of pairs of kidneys weighs either less than 235 or more than 415 grams.
(d) 81.5% of pairs of kidneys weighs between 295 and 385 grams.

43. (a) 88.9% of gas stations has prices within 3 standard deviations of the mean.

(b) 84% of gas stations has prices within 2.5 standard deviations of the mean. Gasoline priced from $1.245 to $1.495 is within 2.5 standard deviations of the mean.
(c) At least 75% of gas stations has prices between $1.27 and $1.47.

45. There is more variation among individuals than among means.

47. Sample of size 5: correct $s = 5.3$, incorrect $s = 27.9$
Sample of size 12: correct $s = 14.7$, incorrect $s = 22.7$
Sample of size 30: correct $s = 15.9$, incorrect $s = 19.2$
As the sample size increases, the impact of the misrecorded observation on standard deviation decreases.

49. (a) There is more variability in blood pressure after exercise.
(b) There is more variability in calcium concentration in the group with high blood pressure.

51. (a) Skewness = 3; distribution is skewed right.
(b) Skewness = 0; distribution is perfectly symmetric.
(c) Skewness = −2.5; distribution is skewed left.
(d) Skewness = −0.44; distribution is slightly skewed left.
(e) Skewness = 0.07; distribution is symmetric.

3.3 Assess Your Understanding (page 161)

1. The class midpoint is meant to represent the average value of all the data points in the class.

3. $\bar{x} = \$32.83$, $s = \$11.06$

5. $\mu = 17.3$ days, $\sigma = 12.6$ days

7. $\mu = 44.5$ years, $\sigma = 10.5$ years

9. (a) $\mu = 80.9°F$, $\sigma = 8.2°F$

(b)

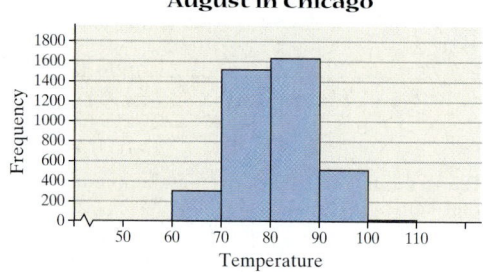

High Temperature in August in Chicago

(c) 95% of the time it is between 64.5° and 97.3°F.

11. (a) $\mu = 32.3$ years of age, $\sigma = 5.2$ years

(b)

Number of Multiple Births in 2002

(c) 95% of mothers of multiple births is between 21.9 and 42.7 years of age.

13. Grouped data: $\bar{x} = 51.8$, $s = 12.1$

15. GPA = 3.27

17. Cost per pound = $2.97

19. (a) Males: $\mu = 35.6$ years, $\sigma = 21.7$ years
(b) Females: $\mu = 38.1$ years, $\sigma = 23.0$ years
(c) Females have the higher mean age.
(d) Females have more dispersion in age.

21. $M = 14.9$ days   **23.** $M = 44.3$ years

25. Modal class: 10 to 19 days

27. Modal class: 35 to 44 years

29. Answers will vary.

3.4 Assess Your Understanding (page 172)

1. Answers will vary. The kth percentile separates the lower k percent of the data from the upper $(100 - k)$ percent of the data.

3. A four-star fund is in the 4th quintile of the funds. That is, it is above the bottom 60%, but below the top 20% of the ranked funds.

5. One qualifies for Mensa if one's intelligence is in the top 2% of people.

7. The 40-week gestation baby weighs less relative to the gestation period.

9. The woman is relatively taller.

11. Peavy had the better year because his ERA was 2.50 standard deviations below the National League's mean ERA, while Santana's ERA was 2.21 standard deviations below the American League's mean ERA.

13. **(a)** Approximately 40% of the states has violent crime rates less than 329.35 crimes per 100,000 population.
    (b) Approximately 95% of the states has violent crime rates less than 761,85 crimes per 100,000 population.
    (c) Approximately 10% of the states has violent crime rates less than 187.2 crimes per 100,000 population.
    (d) Florida is at the 94th percentile, meaning that approximately 94% of the states has violent crime rates less than 730.2 crimes per 100,000 population.
    (e) California is at the 78th percentile, meaning that approximately 78% of the states has violent crime rates less than 579.3 violent crimes per 100,000 population.

15. **(a)** $z = -1.70$. The rainfall in 1971 was 1.70 standard deviations below the mean April rainfall.
    (b) $Q_1 = P_{25} = 2.625$ in.; $Q_2 = P_{50} = M = 3.985$ in.; $Q_3 = P_{75} = 5.36$ in.
    (c) $IQR = 2.735$ in.
    (d) Lower fence $= -1.478$ in., upper fence $= 9.463$ in. There are no outliers according to this criterion.

17. **(a)** $z = 0.61$. The organic carbon concentration in this soil sample is 0.61 standard deviations above the mean concentration.
    (b) $Q_1 = P_{25} = 10.01$ mg/L, $Q_2 = P_{50} = M = 15.42$ mg/L, $Q_3 = P_{75} = 20.13$ mg/L
    (c) IQR $= 10.12$ mg/L
    (d) Lower fence $= -5.17$ mg/L, upper fence $= 35.31$ mg/L. There are no outliers according to this criterion.

19. The cutoff point is 574 minutes. If more minutes are used, the customer is contacted.

21. **(a)** \$12,777 is an outlier.

    (b)

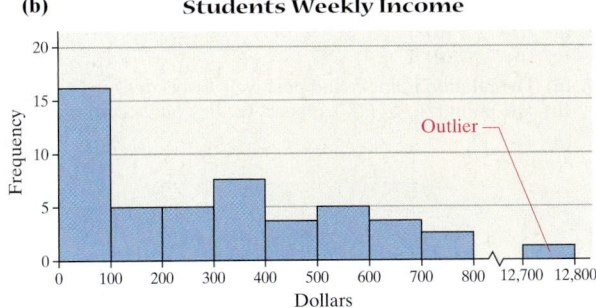

Students Weekly Income

    (c) Answers will vary.

23. Mean of the z-scores is 0.0.

    Standard deviation of the z-scores is 1.0.

Student	z-Score
Perpetual Bempah	0.49
Megan Brooks	−1.58
Jeff Honeycutt	−1.58
Clarice Jefferson	1.14
Crystal Kurtenbach	−0.03
Janette Lantka	1.07
Kevin McCarthy	1.01
Tammy Ohm	−0.55
Kathy Wojdyla	0.10

3.5 Assess Your Understanding (page 181)

1. The median and interquartile range are better measures of central tendency and dispersion if the data are skewed or if the data contain outliers.

3. **(a)** The distribution is skewed right.
   (b) Five-number summary: 0, 1, 3, 6, 16

5. Five-number summary: 42, 51, 55, 58, 69

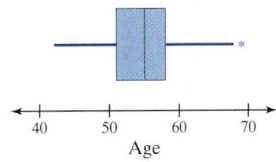

Age of Presidents at Inauguration

Graph is symmetric.

7. Five-number summary: 1, 7, 15, 22, 45

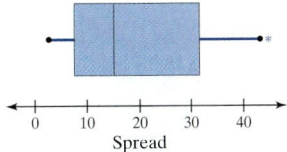

Super Bowl Point Spreads

The distribution is skewed right.

9. Five-number summary: 0.598, 0.604, 0.608, 0.610, 0.612

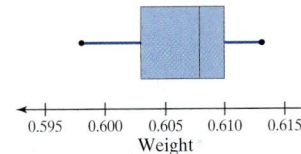

Weight (in grams) of Tylenol Tablets

The distribution is skewed left.

11. **(a)** Five-number summary: 28, 45, 51, 57.5, 73

    (b)

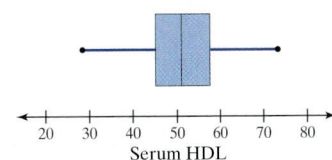

Serum HDL Cholesterol Level

    (c) The distribution is symmetric.
    (d) Mean and standard deviation

13. **(a)** Five-number summary: 0, 0.205, 1.055, 1.87, 2.83

    (b)

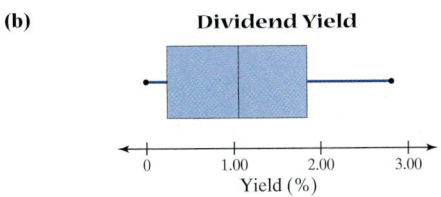

Dividend Yield

    (c) The distribution is skewed right.
    (d) Median and IQR.

15. Five-number summaries:

Keebler	20,	21.5,	25,	28,	33
Store brand	16,	21,	24,	27.5,	33

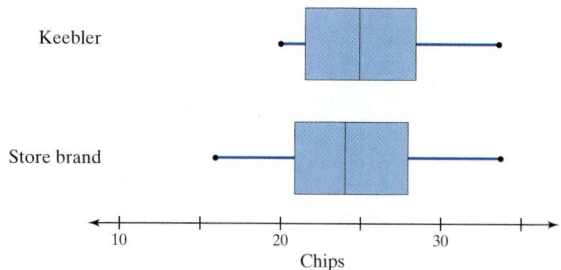

Answers will vary.

17. Five-number summaries:

McGwire:	340,	380,	420,	450,	550
Sosa:	340,	370.5	410,	430,	500
Bonds:	320,	380,	410,	420,	488

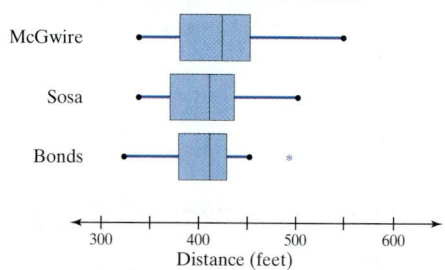

Chapter Review Exercises (page 186)

1. (a) Mean = 792.51 m/s, median = 792.40 m/s
   (b) Range = 4.8 m/s, s^2 = 2.03, s = 1.42 m/s

3. (a) $\bar{x}$ = $10,178.89, M = $9980
   (b) Range: R = $8550, s = $3074.92
   (c) $\bar{x}$ = $13,178.89, M = $9980; range: R = $35,550,
        s = $10,797.46
        The median is resistant.

5. (a) μ = 58.3 years, median = 58.5 years,
        bimodal: 56 and 62 years
   (b) Range = 25 years, σ = 6.9 years
   (c) Answers will vary.

7. (a) Mean = 2.2 children, median = 2.5 children
   (b) Range = 4 children, s = 1.3 children

9. (a) 441, 759
   (b) 95% of the light bulbs has a life between 494 and 706 hours.
   (c) 81.5% of the light bulbs has a life between 547 and 706 hours.
   (d) The firm can expect to replace 0.15% of the light bulbs.
   (e) At least 84% of the light bulbs has a life within 2.5 standard
        deviations of the mean.
   (f) At least 75% of the light bulbs has a life between 494 and
        706 hours.

11. (a) μ = 42.28 years of age
   (b) σ = 15.93 years

13. Michael's GPA is 3.33.

15. (a) Yankees: μ = $6,351,515.52; Mets: μ = $3,452,177.50
   (b) Yankees: M = $3,100,000; Mets: M = $900,000
   (c) The distributions are skewed right since their means are larger
        than their medians.
   (d) Yankees: σ = $6,242,767.24; Mets σ = $4,643,606.15
   (e) Yankees: $301,400, $837,500, $3,100,000, $11,623,571.50, $22,000,000
        Mets: $300,000, $318,375, $900,000, $4,666,666.50, $17,166,667

(f)

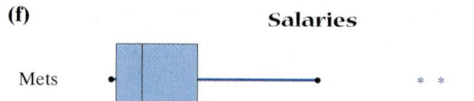

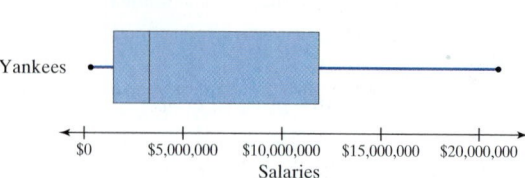

The Mets data contain two outliers: $17,166,667 (Mo Vaughn)
and $16,071,429 (Mike Piazza).

(g) Both distributions are skewed right.
(h) The median is the better measure of central tendency since the
      distributions are skewed.

17. (a) $368,817; approximately 40% of the drivers earned less than
        $368,817.
   (b) $5,957,165; approximately 95% of the drivers earned less than
        $5,957,165.
   (c) $67,862.50; approximately 10% of the drivers earned less than
        $67,862.50.
   (d) $4,117,750 is at the 83rd percentile.
   (e) $116,359 is at the 15th percentile.

19. The female is relatively heavier.

CHAPTER 4

4.1 Assess Your Understanding (page 203)

1. Univariate data measure the value of a single variable for each individual in the study. Bivariate data measure values of two variables
   for each individual.

3. Two variables that are linearly related are positively associated if
   increases in the explanatory variable tend to be associated with increases in the response variable.

5. If r = 0, then there is no *linear* relationship between the variables.

7. The linear correlation coefficient can only be calculated from bivariate *quantitative* data, and the gender of a driver is a qualitative variable.

9. Correlation is a unitless measure of association; it cannot be 0.93 *bushel.*

11. Nonlinear   **13.** Linear, positive

15. (a) III   **(b)** IV
   (c) II   **(d)** I

17. (a) The relation is linear and positively associated.
   (b) The point (24, 55,000) appears to stick out. Reasons will vary.

19. (a)

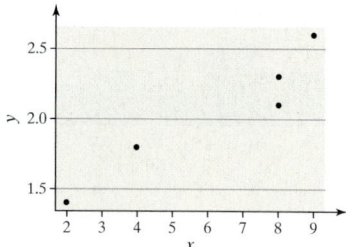

(b) r = 0.9572
(c) Positive association

21. (a)

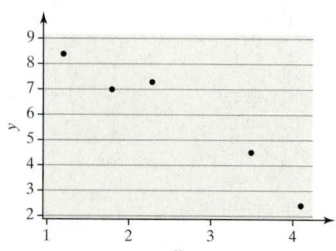

(b) $r = -0.9703$
(c) Negative association

23. (a) Explanatory height; Response; head circumference

(b)
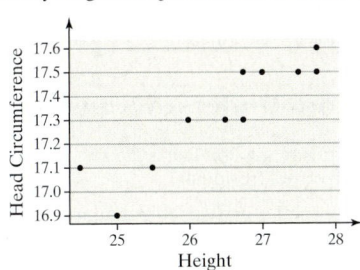

(c) $r = 0.9111$
(d) Positive association

25. (a) Explanatory variable: weight; Response variable: gas mileage

(b)
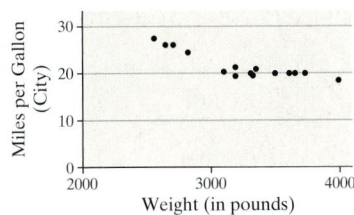

(c) $r = -0.892$
(d) There is a negative linear relation between the weight of a car and its gas mileage.

27. (a) Explanatory variable: Number of absences; Response variable: Average final grade

(b)
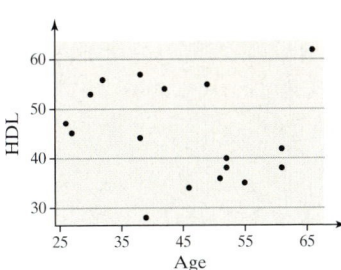

(c) $r = -0.947$
(d) There appears to be a strong negative linear relation between days absent and final grade.
(e) Going to class every day does not guarantee a passing grade. Factors may vary.

29. (a)
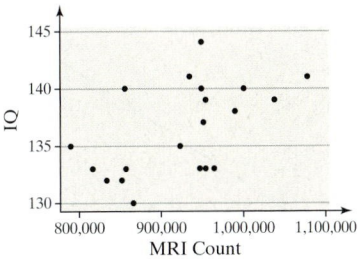

(b) -0.1637
(c) There appears to be little, if any, linear association between age and HDL cholesterol level.

31. (a)
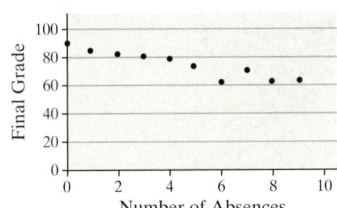

(b) $r = 0.548$. It would appear that there is a positive association between MRI count and IQ.

(c)
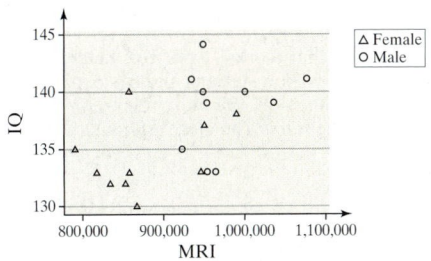

The females have lower MRI counts. Looking at each gender's plot, we see that the relation that appeared to exist between brain size and IQ disappears.
(d) Females: $r = 0.359$; Males: $r = 0.236$. There appears to be no linear relation between brain size and IQ.

33. (a)
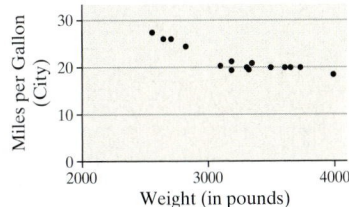

(b) Correlation coefficient (with Taurus included): $r = -0.879$
(c) The results are reasonable because the Taurus follows the overall pattern of the data.

(d)
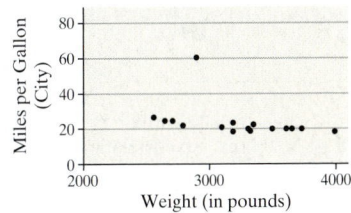

(e) Correlation coefficient (with Prius included): $r = -0.437$
(f) The Prius is a hybrid car; the other cars are not hybrids.

35. (a) $r = 0.82$ for all four data sets.
(b)

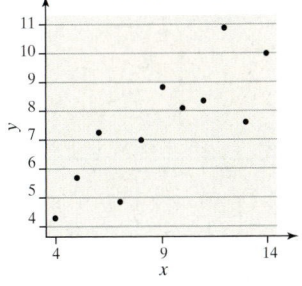
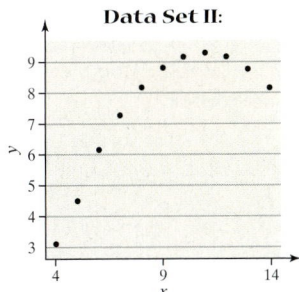

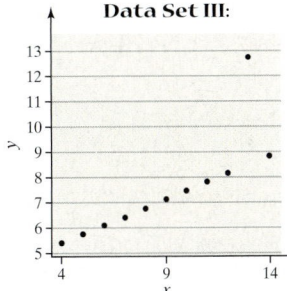
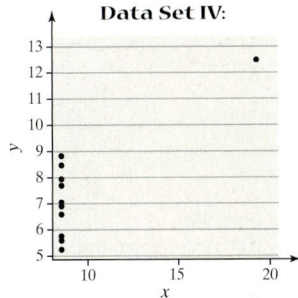

37. To have the lowest correlation between the two stocks, invest in General Electric and TECO Energy. $r = 0.0169$. If your goal is to have one stock go up when the other goes down, invest in Cisco Systems and TECO Energy: $r = -0.235$.

39. $r = 0.599$ implies that there is a positive linear relation between the number of television stations and life expectancy, but this is correlation, not causation. The more television stations a country has, the more affluent it is. The more affluent, the better the health care.

41. (a)

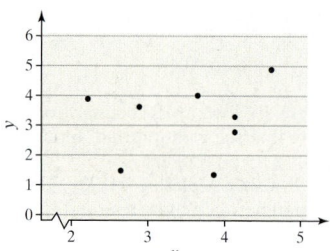

$r = 0.2280$

(b)

$r = 0.8598$

43. (a) Positive   **(b)** Negative   **(c)** Negative
(d) Negative   **(e)** No correlation

45. (a)

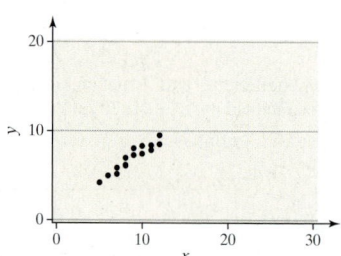

(b) 0.9518

x	10	12	14	14	16	16	16	16
y	8.4	10	10.4	11.8	12	12.4	12.2	13.8
x	18	18	20	20	22	22	24	24
y	14.4	16	16.6	14.8	16.8	15.6	17	19.0

(d)

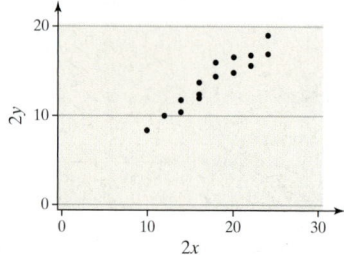

(e) 0.9518
(f) By doubling the values of x and y, we double the standard deviations and deviations about the mean. Therefore, the z scores for each point are unchanged.

47. (a) It appears that the SAT score combined with the high school GPA is the best predictor of college GPA. They have the highest correlation coefficient.
(b) It appears that SAT Verbal score is the worst predictor of college GPA. They have the lowest correlation coefficient.

4.2 Assess Your Understanding (page 221)

1. The least-squares regression line is the line that minimizes the sum of the squared errors (residuals).

3. Values of the explanatory variable that are much larger or much smaller than those observed are considered *outside the scope of the model*. It is dangerous to make such predictions because we do not know the behavior of the data for which we have no observations.

5. Answers will vary.

7. Rounding error can prevent the sum of the residuals from equaling 0.

9. (a)

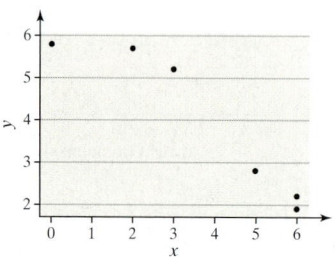

(b) $\hat{y} = -0.7136x + 6.5498$

(c)

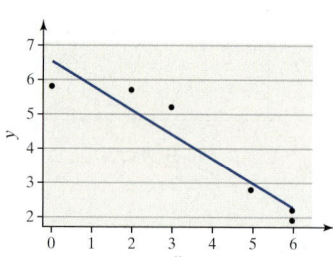

11. (a)

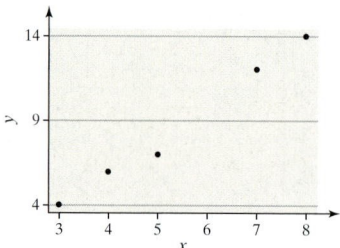

(b) Using points $(3, 4)$ and $(8, 14)$: $\hat{y} = 2x - 2$

(c)

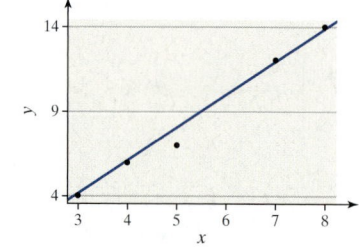

(d) $\hat{y} = 2.0233x - 2.3256$

(e)

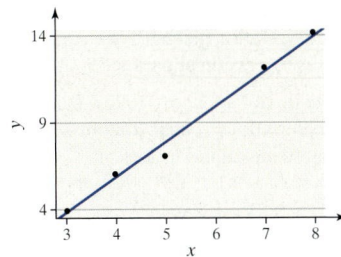

(f) Sum of squared residuals (computed line): 1
(g) Sum of squared residuals (least-squares line): 0.7907

13. (a)

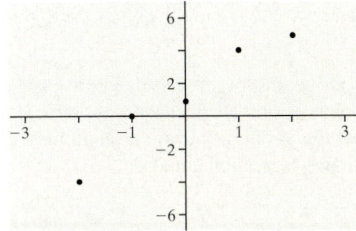

(b) Using points $(-2, -4)$ and $(2, 5)$: $\hat{y} = \dfrac{9}{4}x + \dfrac{1}{2} = 2.25x + 0.5$

(c)

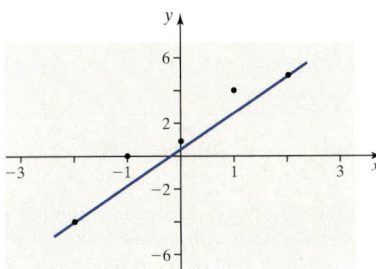

(d) $\hat{y} = 2.2x + 1.2$

(e)

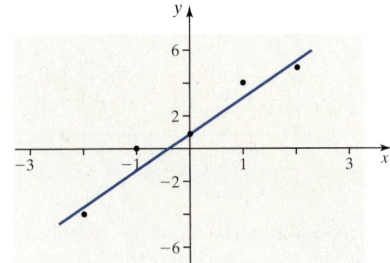

(f) Sum of squares residual from my line: 4.875
(g) Sum of squares residual from least-squares line: 2.4
(h) Answers will vary.

15. (a)

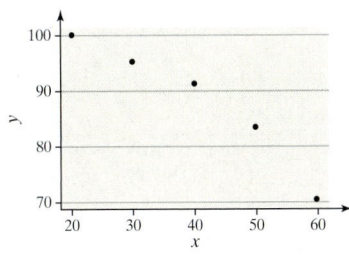

(b) Using points $(30, 95)$ and $(60, 70)$: $\hat{y} = -\dfrac{5}{6}x + 120$

(c)

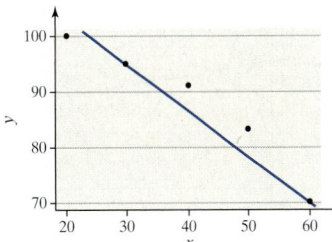

(d) $\hat{y} = -0.72x + 116.6$

(e)

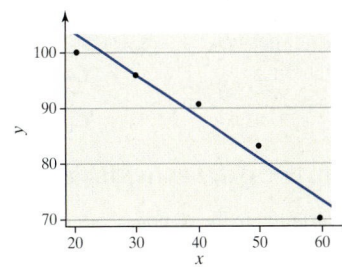

(f) Sum of squares residual from my line: 51.6667
(g) Sum of squares residual from least square line: 32.4

17. (a) $\hat{y} = 0.1827x + 12.4932$
(b) If height increases by 1 inch, head circumference increases by about 0.1827 inch, on average. It is not appropriate to interpret the y-intercept. It is outside the scope of the model.
(c) $\hat{y} = 17.06$ inches
(d) Residual $= -0.16$ inch; below

(e)

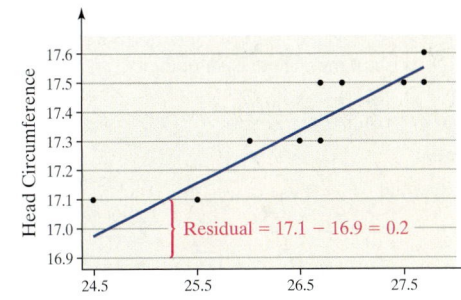

(f) For children who are 26.75 inches tall, head circumference varies.
(g) No, 32 inches is outside the scope of the model.

19. (a) $\hat{y} = -0.0062x + 41.4640$
(b) For every pound added to the weight of the car, gas mileage in the city will decrease by 0.0062 mile per gallon, on average. It is not appropriate to interpret the y-intercept.
(c) $\hat{y} = 21.0$ miles per gallon; residual: -1.0 miles per gallon; Mustang mileage is below average.

(d)

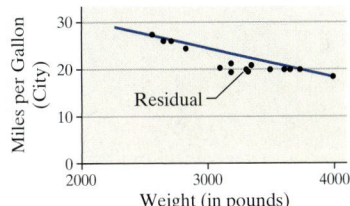

(e) It is not reasonable to use this least-squares regression to predict the miles per gallon of a Toyota Prius.

21. (a) $\hat{y} = -2.8273x + 88.7327$
(b) For every additional absence, the student's final grade drops 2.8273 points, on average. The final grade of students who miss no classes is 88.7327.

(c) $\hat{y} = 74.60$; residual is -0.70; the final grade is below average for 5 absences.

(d)

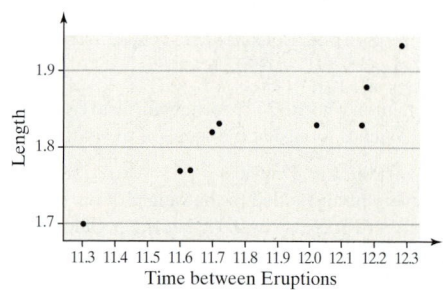

(e) No, 15 absences is out of the scope of the model.

23. (a) $\hat{y} = 0.00003x + 108.8940$

   (b) The slope is close to zero.

   (c) MRI count 1,000,000: $\hat{y} = \bar{y} = 136$; MRI count 830,000: $\hat{y} = \bar{y} = 136$

25. Answers will vary.

4.3 Assess Your Understanding (page 235)

1. 75% of total variation in the response variable is explained by the regression.

3. This indicates that the response and explanatory variables may not be linearly related.

5. An influential observation is one that significantly affects either the slope or the intercept of the least-squares regression line.

7. An influential observation can change either the slope or the intercept, or both.

9. If $R^2 = 0.81$, you can determine the magnitude of r, but not the sign of r. To find r, you need to know whether the relation between the variables is positive or negative.

11. Condition 2 is not met. The variance of the residuals is not constant.

13. Condition 3 is not met. There is an outlier.

15. (a) III   **(b)** II
   (c) IV   **(d)** I

17. Influential

19. Not influential

21. (a)

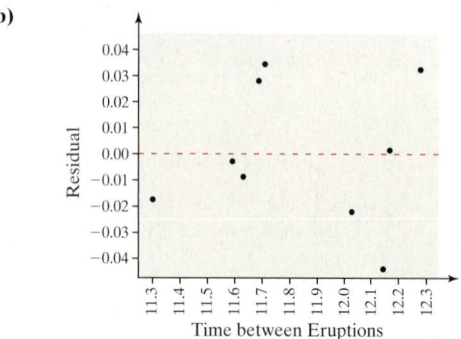

$\hat{y} = 0.1854x - 0.3779$

(b)

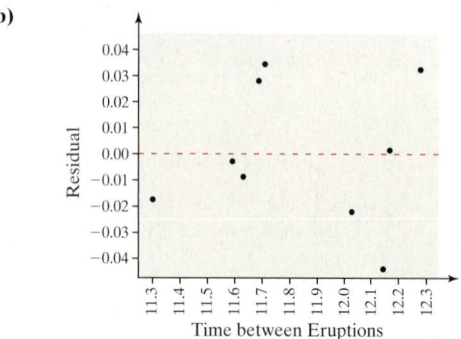

Length of eruption and time between eruptions appear to be linearly related, because the residual plot does not show any violations of the model.

   (c) 83.0% of the variation in length of eruption is explained by the least-squares regression equation.

23. (a) No, the data do not appear to follow a linear pattern.

   (b) 6.8% of the variation in sugar content is explained by the least-squares regression equation. Yes.

   (c)

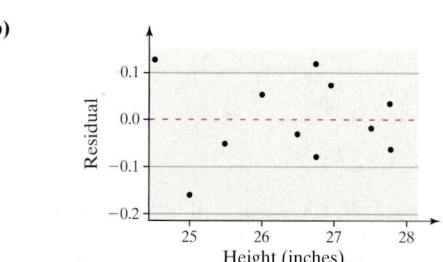

   Yes, it is influential, because the y-intercept of the least-squares regression equation changed substantially.

   (d) 42.1% of the variability in sugar content is explained by the least-squares regression equation.

25. (a) $R^2 = 83.0\%$

   (b)

   (c) 83.0% of the variation in head circumference is explained by the least-squares regression equation. The linear model appears to be appropriate, based on the residual plot.

27. (a) $R^2 = 0.796$

   (b)

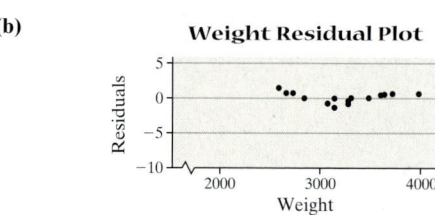

Weight Residual Plot

   (c) 79.6% of the variance in gas mileage is explained by the linear model. Residuals may be patterned.

29. (a)

   (b) $\hat{y} = 0.0657x - 12.4967$

(c)

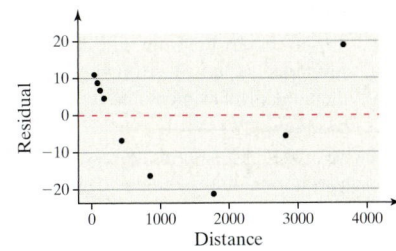

(d) No, the residuals follow a U-shaped pattern.

31. (a) The coefficient of determination with the Viper is $R^2 = 0.527$. Adding the Viper reduces the amount of variance explained by the model by approximately 33%.

(b) The Viper is not influential; neither the slope nor the y-intercept is significantly altered, but the Viper is an outlier.

33. (a)

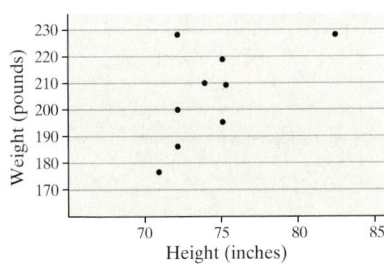

(b) $\hat{y} = 3.4631x - 51.091$; $r = 0.6076$

(c) $\hat{y} = 4.605x - 134.545$; $r = 0.4589$

(d) Randy Johnson's data are influential because removing them changes the y-intercept of the least-squares regression line significantly.

35. (a), (d)

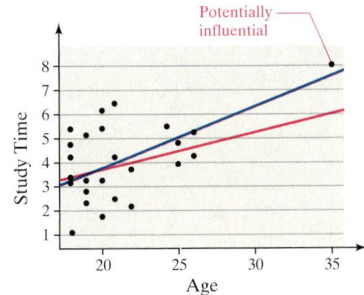

(b) $\hat{y} = 0.232x - 0.920$

(c) $\hat{y} = 0.155x + 0.635$

(e) The point $(35, 8.1)$ substantially affects both the slope and the y-intercept of the least-squares regression line.

Chapter Review Exercises (page 242)

1. (a)

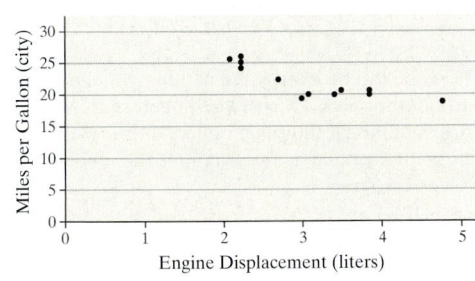

(b) $r = -0.8954$

(c) There appears to be a strong negative linear relation between engine displacement and fuel economy.

3. (a)

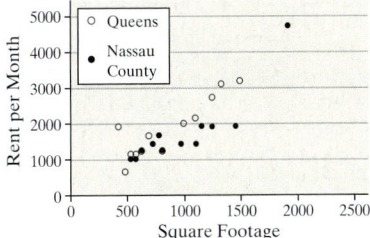

(b) Queens: $r = 0.9093$; Nassau County: $r = 0.8674$

(c) Both groups appear to have a positive association between square footage and monthly rent.

(d) For small apartments (those less than 1000 square feet in area), there seems to be no difference in rent between Queens and Nassau County. In larger apartments, Queens seems to have higher rents than Nassau County.

5. (a) $\hat{y} = -2.7977x + 30.3848$

(b)

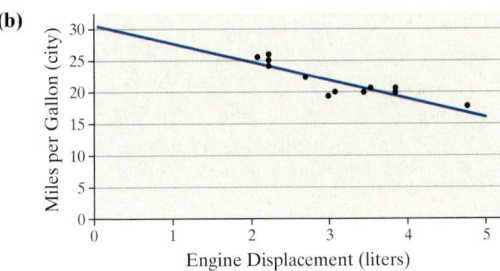

(c) The slope of the least-squares regression line indicates that, for each additional liter of engine displacement, fuel economy is lessened by 2.8 miles per gallon in city driving, on average. It is not appropriate to interpret the y-intercept because it is not possible to have zero displacement.

(d) A Ford Mustang with a 3.8-liter engine is predicted to get 19.8 miles per gallon when driven in the city.

(e) The residual is 0.2 mile per gallon.

(f) The miles per gallon is above average for a Ford Mustang.

7. (a) $\hat{y} = 2.2092x - 34.3148$

(b)

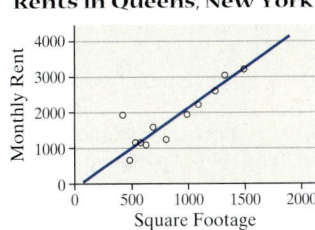

(c) The slope of the least-squares regression line indicates that, for each additional square foot of floor area, the rent increases by $2.21, on average. It is not appropriate to interpret the y-intercept since it is not possible to have an apartment with 0 square footage.

(d) For 825 square feet, $\hat{y} = \$1788.27$

(e) The residual $y - \hat{y} = -538.233$.

(f) This apartment's rent is below average.

9. (a)

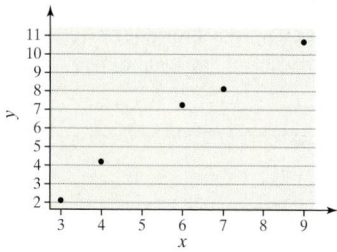

(b) Using points $(3, 2.1)$ and $(7, 8.1)$, $\hat{y} = 1.5x - 2.4$.

(c)

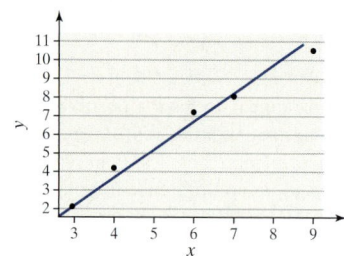

(d) $\hat{y} = 1.3877x - 1.6088$

(e)

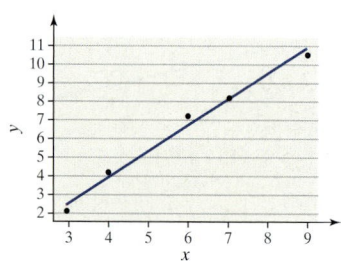

(f) Computed line: sum of squared residuals = 0.97
(g) Least-squares line: sum of squared residuals = 0.585
(h) Answers will vary.

11. Linear model is appropriate.

13. Linear model is not appropriate; nonconstant error variance.

15. (a) $R^2 = 0.8017$; 80.2% of the variation in gas mileage is explained by the least-squares regression model.

(b)

Engine Displacement Residual Plot

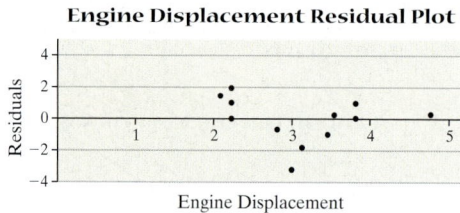

Engine Displacement

(c) The residuals are forming a discernable V-shaped pattern. The linear model may not be appropriate.
(d) The Ford Crown Victoria is an outlier.

(e) 1a., 5b

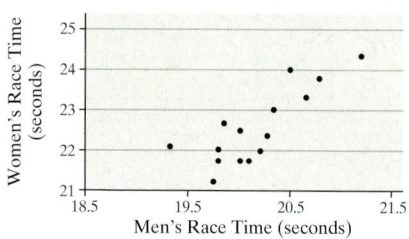

1b. $r = -0.9421$

1c. Based on the scatter diagram and the correlation coefficient, there appears to be a strong negative linear relation between engine displacement and fuel economy.

5a. $\hat{y} = -2.3834x + 29.1454$

5c. The slope of the least-squares regression line indicates that, for each additional liter of engine displacement, fuel economy is lessened by 2.4 miles per gallon in city driving, on average. It is not appropriate to interpret the y-intercept because it is not possible to have zero displacement.

5d. A Ford Mustang with a 3.8-liter engine is predicted to get 20.09 miles per gallon when driven in the city.

5e. The residual is -0.0884 mile per gallon.

5f. The miles per gallon are below average for a Ford Mustang.

15a-d. $R^2 = 88.87$. The viper is influential.

17. (a) $R^2 = 0.8268$; 82.7% of the variance in rent is explained by the least-squares regression model.

(b)

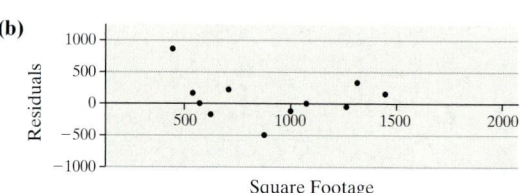

(c) The plot of residuals shows no discernible pattern, so a linear model is appropriate.
(d) The observation $(460, 1805)$ is both an outlier and an influential point. The least-squares line omitting the outlier is $\hat{y} = 2.5315x - 399.25$; the y-intercept is substantially different.

19. (a) $\hat{y} = 2.0995x - 347.364$
(b) The observation $(1906, 4625)$ is both an outlier and an influential point. The least-squares line omitting the outlier is $\hat{y} = 0.0779x + 675.295$; both the slope and the $\hat{y}$-intercept are substantially different.

21. (a)

(b) $r = 0.8352$
(c) The relation between men's and women's time seems to be positively associated. Men's times do not cause women's times.

23. No, correlation does not imply causation.

25. (a) Answers will vary.
(b) The slope can be interpreted as "the school day decreases by 0.01 hour for each 1% increase in percent low income, on average. additional thousand dollars of income." The intercept can be interpreted as the length of the school day for those with no income.
(c) $\hat{y} = 6.91$ hours
(d)–(i) Answers will vary.

CHAPTER 5

5.1 Asess Your Understanding (page 261)

1. Empirical probability is based on the outcome of a probability experiment and is approximately equal to the relative frequency of the event. Classical probability is based on counting techniques and is equal to the ratio of the number of ways an event can occur to the number of possible outcomes in the experiment.

3. Outcomes are equally likely when each outcome has the same probability of occurring.

5. True   **7.** Experiment

9. Rule 1: All probabilities in the model are greater than or equal to zero and less than or equal to one.
Rule 2: The sum of the probabilities in the model is 1.

The outcome blue is called an impossible event.

11. This is not a probability model because $P(\text{green}) < 0$.

13. The numbers $0, 0.01, 0.35$, and 1 could be probabilities.

15. The probability of 0.42 means that approximately 42 out of every 100 hands will contain two cards of the same value and three cards of different value.

17. The empirical probability is 0.95.

19. The $P(2) \neq \dfrac{1}{11}$ since the outcomes are not equally likely.

21. $S = \{1H, 1T, 2H, 2T, 3H, 3T, 4H, 4T, 5H, 5T, 6H, 6T\}$

23. 0.44   **25.** $P(E) = \dfrac{3}{10} = 0.3$ **27.** $P(E) = \dfrac{1}{2} = 0.5$

29. $P(\text{plays sports}) = \dfrac{288}{500} = 0.576$

31. (a) $P(\text{red}) = \dfrac{40}{100} = 0.4$ **(b)** $P(\text{purple}) = \dfrac{25}{100} = 0.25$

33. (a) $S = \{0, 00, 1, 2, 3, 4, \dots, 35, 36\}$

(b) $P(8) = \dfrac{1}{38} = 0.0263$; there is a 2.63% probability that the ball will land in slot 8.

(c) $P(\text{odd}) = \dfrac{18}{38} = 0.4737$

35. (a) $\{SS, Ss, sS, ss\}$

(b) $P(ss) = \dfrac{1}{4}$; there is a 25% probability that a randomly selected offspring will have sickle-cell anemia.

(c) $P(Ss \text{ or } sS) = \dfrac{1}{2} = 0.5$; there is 50% probability that a randomly selected offspring will be a carrier of sickle-cell anemia.

37. (a)

Response	Probability
Never	0.026
Rarely	0.068
Sometimes	0.116
Most of the time	0.263
Always	0.527

(b) It would be unusual to randomly find someone who never wears a seatbelt, because $P(\text{never}) < 0.05$.

39. (a)

Type of Larceny Theft	Probability
Pocket picking	0.008
Purse snatching	0.008
Shoplifting	0.198
From motor vehicles	0.331
Motor vehicle accessories	0.129
Bicycles	0.072
From buildings	0.176
From coin-operated machines	0.076

(b) Purse snatching larcenies are unusual.
(c) Larcenies from coin-operated machines are not unusual.

41. A, B, C, and F are consistent with the definition of a probability model.

43. Use model B if the coin is known to always come up tails.

45. (a) $S = \{JR, JC, JD, JM, RC, RD, RM, CD, CM, DM\}$

(b) $P(\text{Clarese and Dominique attend}) = \dfrac{1}{10} = 0.10$

(c) $P(\text{Clarice attends}) = \dfrac{4}{10} = 0.40$

(d) $P(\text{John stays home}) = \dfrac{6}{10} = 0.60$

47. (a) $P(\text{right field}) = \dfrac{24}{73} = 0.329$

(b) $P(\text{left field}) = \dfrac{2}{73} = 0.027$

(c) It is unusual for Barry Bonds to hit a home run to left field; the probability is less than 5%.

49. Answers will vary.

51. Answers will vary.

53. $P(\text{income is greater than } \$57,500) = \dfrac{1}{2}$

5.2 Assess Your Understanding (page 273)

1. Two events are mutually exclusive (disjoint) if they have no outcomes in common.

3. $P(E) + P(F) - P(E \text{ and } F)$

5. E and $F = \{5, 6, 7\}$; E and F are not mutually exclusive.

7. F or $G = \{5, 6, 7, 8, 9, 10, 11, 12\}$; $P(F \text{ or } G) = \dfrac{2}{3}$

9. There are no outcomes in event "E and G." Events E and G are mutually exclusive.

11. $E^c = \{1, 8, 9, 10, 11, 12\}$; $P(E^c) = \dfrac{1}{2}$

13. $P(E \text{ or } F) = 0.55$ **15.** $P(E \text{ or } F) = 0.70$

17. $P(E^c) = 0.75$ **19.** $P(F) = 0.30$

21. $P(\text{Titleist or Maxfli}) = \dfrac{17}{20} = 0.85$

23. $P(\text{not a Titleist}) = \dfrac{11}{20} = 0.55$

25. (a) All the probabilities are nonnegative; the sum of the probabilities equals 1.

(b) $P(\text{gun or knife}) = 0.793$. There is a 79.3% probability of randomly selecting a murder committed with a gun or a knife.

(c) $P(\text{knife, blunt object, or strangulation}) = 0.190$. There is 19.0% probability of randomly selecting a murder committed with a knife or blunt object or by strangulation.

(d) $P(\text{not a gun}) = 0.332$. There is 33.2% chance of randomly selecting a murder that was not committed with a gun.

(e) A murder by strangulation is unusual.

27. (a) $P(\text{30- to 39-year-old mother}) = 0.632$. There is a 63.2% probability that a mother involved in a multiple birth was between 30 and 39 years old.

(b) $P(\text{not 30 to 39 years old}) = 0.368$. There is a 36.8% probability that a mother involved in a multiple birth was either younger than 30 or older than 39.

(c) $P(\text{younger than 45 years}) = 0.984$. The probability that a randomly selected mother involved in a multiple birth was younger than 45 years old is 98.4%.

(d) $P(\text{at least 20 years old}) = 0.987$. There is a 98.7% probability that a randomly selected mother involved with a multiple birth was at least 20 years old.

29. (a) $P(\text{heart or club}) = \dfrac{1}{2} = 0.5$

(b) $P(\text{heart or club or diamond}) = \dfrac{3}{4} = 0.75$

(c) $P(\text{ace or heart}) = \dfrac{4}{13} = 0.3077$

31. (a) $P(\text{birthday is not November 8}) = \dfrac{364}{365} = 0.9973$

(b) $P(\text{birthday is not 1st of month}) = \dfrac{353}{365} = 0.9671$

(c) $P(\text{birthday is not 31st of month}) = \dfrac{358}{365} = 0.9808$

(d) $P(\text{birthday is not in December}) = \dfrac{334}{365} = 0.9151$

33. No, some people have both vision and hearing problems, but we don't know the proportion.

35. (a) $P(\text{only English or Spanish is spoken}) = 0.93$
(b) $P(\text{language other than only English or only Spanish}) = 0.07$
(c) $P(\text{not only English is spoken}) = 0.18$
(d) No, the sum of the probabilities would be greater than 1, and there would be no probability model.

37. (a) $P(\text{died from cancer}) = 0.007$
(b) $P(\text{current smoker}) = 0.057$
(c) $P(\text{died from cancer and smoked cigars until death}) = 0.001$
(d) $P(\text{died from cancer or was current cigar smoker}) = 0.064$

39. (a) $P(\text{student is satisfied}) = \dfrac{77}{125} = 0.616$

(b) $P(\text{student is a junior}) = \dfrac{94}{375} = 0.251$

(c) $P(\text{student is satisfied and is a junior}) = \dfrac{64}{375} = 0.171$

(d) $P(\text{student is satisfied or is a junior}) = \dfrac{87}{125} = 0.696$

41. (a) $P(\text{officer}) = 0.162$
(b) $P(\text{Navy}) = 0.261$
(c) $P(\text{Navy officer}) = 0.038$
(d) $P(\text{officer or in the Navy}) = 0.385$

5.3 Assess Your Understanding (page 273)

1. Independent
3. Addition
5. $P(E) \cdot P(F)$
7. (a) Dependent **(b)** Independent **(c)** Dependent
9. $P(E \text{ and } F) = 0.18$

11. $P(5 \text{ heads}) = \dfrac{1}{32} = 0.03125$; 3.125% of the time tossing a coin 5 times will result in 5 heads.
13. $P(\text{two left-handed people are chosen}) = 0.0169$

$P(\text{at least one person chosen is right-handed}) = 0.9831$

15. $P(\text{all five are negative}) = 0.9752$

$P(\text{at least one is positive}) = 0.0248$

17. (a) $P(\text{two will live to 41 years}) = 0.99437$
(b) $P(\text{five will live to 41 years}) = 0.98598$
(c) $P(\text{at least one of five dies}) = 0.01402$; it would be unusual that at least one of the five dies before age 41.

19. (a) $P(\text{two will have Rh}^+ \text{ blood}) = 0.9801$
(b) $P(\text{six will have Rh}^+ \text{ blood}) = 0.9415$
(c) $P(\text{at least out of six will have Rh}^- \text{ blood}) = 0.0585$; it would not be unusual to select a person with Rh$^-$ blood.

21. (a) $P(\text{batter makes 10 consecutive outs}) = 0.02825$
(b) A cold streak would be unusual.

23. $P(\text{at least 1}) = 0.9999$; answers will vary.

25. (a) $P(\text{two strikes in a row}) = 0.09$
(b) $P(\text{turkey}) = 0.027$
(c) $P(\text{strike followed by nonstrike}) = 0.21$

27. (a) $P(\text{all three have driven under the influence of alchohol}) = 0.0244$
(b) $P(\text{at least one has not driven under the influence of alcohol}) = 0.9756$
(c) $P(\text{no one has driven under the influence of alcohol}) = 0.3579$
(d) $P(\text{at least one has driven under the influence of alcohol}) = 0.6421$

5.4 Assess Your Understanding (page 289)

1. F occurring; E has occurred
3. $P(F|E) = 0.75$ **5.** $P(F|E) = 0.5676$
7. $P(E \text{ and } F) = 0.32$
9. The events are not independent.

11. $P(\text{club}) = \dfrac{1}{4}$; $P(\text{club}|\text{black card}) = \dfrac{1}{2}$

13. $P(\text{rains}|\text{cloudy}) = \dfrac{0.21}{0.37} = 0.568$

15. $P(\text{W}|\text{DO}) = \dfrac{58}{91} = 0.637$

17. (a) $P(\text{no health insurance}|\text{under 18 years of age}) = 0.110$
(b) $P(\text{under 18 years of age}|\text{no health insurance}) = 0.196$

19. (a) $P(\text{16 to 20 years old}|\text{male}) = 0.136$
(b) $P(\text{male}|\text{16 to 20 years old}) = 0.705$
(c) Probably the victim is male.

21. $P(\text{both TVs work}) = 0.40$; $P(\text{at least one TV doesn't work}) = 0.6$

23. $P(\text{Dave and then Neta is chosen}) = \dfrac{1}{20} = 0.05$

25. (a) $P(\text{like both songs}) = \dfrac{5}{39} = 0.128$; it is not unusual to like both songs.
(b) $P(\text{like neither song}) = \dfrac{14}{39} = 0.359$
(c) $P(\text{like exactly one song}) = \dfrac{20}{39} = 0.513$
(d) $P(\text{like both songs}) = \dfrac{25}{169} = 0.148$;

$P(\text{like neither song}) = \dfrac{64}{169} = 0.379$;

$P(\text{like exactly one song}) = \dfrac{80}{169} = 0.473$

27. (a) $P(\text{two red bulbs are chosen}) = 0.152$
(b) $P(\text{first a red and then a yellow bulb is chosen}) = 0.138$
(c) $P(\text{first a yellow and then a red bulb is chosen}) = 0.138$
(d) $P(\text{a red bulb and a yellow bulb are chosen}) = 0.276$

29. $P(\text{female and smokes}) = 0.051$; it would not be unusual to randomly select a female who smokes.

31. (a) $P(\text{10 people each have a different birthday}) = 0.883$
(b) $P(\text{at least two of the ten people have the same birthday}) = 0.117$

33. (a) $P(\text{being dealt five clubs}) = 0.000495$
(b) $P(\text{being dealt a flush}) = 0.0020$

35. (a) $P(\text{two defective chips}) = 0.0000245$
(b) Assuming independence: $P(\text{two defective chips}) = 0.000025$

37. $P(<18 \text{ years old}) = 0.246$; $P(<18 \text{ years old}|\text{no health insurance}) = 0.196$; the events are not independent.

39. $P(\text{female}) = 0.262$; $P(\text{female}|\text{16 to 20 years old}) = 0.295$; the events are not independent.

5.5 Assess Your Understanding (page 304)

1. Permutation   **3.** True   **5.** $5! = 120$
7. $10! = 3,628,800$ **9.** $0! = 1$ **11.** $_6P_2 = 30$
13. $_4P_4 = 24$ **15.** $_5P_0 = 1$ **17.** $_8P_3 = 336$
19. $_8C_3 = 56$ **21.** $_{10}C_2 = 45$ **23.** $_{52}C_1 = 52$
25. $_{48}C_3 = 17,296$

27. $ab, ac, ad, ae, ba, bc, bd, be, ca, cb, cd, ce, da, db, dc, de, ea, eb, ec, ed$; $_5P_2 = 20$

29. $ab, ac, ad, ae, bc, bd, be, cd, ce, de$; $_5C_2 = 10$

31. He can wear 24 different shirt-and-tie combinations.

33. Dan can arrange the songs $12! = 479,001,600$ ways.

35. The salesman can take $8! = 40,320$ different routes.

37. At most 18,278 companies can be listed on the NYSE.

39. (a) 10,000 different codes are possible.
(b) $P(\text{correct code is guessed}) = \dfrac{1}{10,000}$

41. 26^8 different user names are possible.

43. (a) There are 50^3 lock combinations.
(b) $P(\text{guessing the correct combination}) = \dfrac{1}{50^3}$

45. The top three cars can finish in $_{40}P_3 = 59{,}280$ ways.

47. The officers can be chosen in $_{20}P_4 = 116{,}280$ ways.

49. There are $_{25}P_4 = 303{,}600$ possible outcomes.

51. There are $_{50}C_5 = 2{,}118{,}760$ possible random samples of size 5.

53. $_6C_2 = 15$ different birth and gender orders are possible.

55. $\dfrac{10!}{3! \cdot 3! \cdot 2! \cdot 1! \cdot 1!} = 50{,}400$ different 10-letter words can be formed.

57. The trees can be planted $\dfrac{11!}{4! \cdot 5! \cdot 2!} = 6930$ ways.

59. $P(\text{winning}) = \dfrac{1}{_{30}C_5} = \dfrac{1}{142{,}506}$

61. (a) $P(\text{jury has only students}) = \dfrac{1}{153} = 0.0065$

(b) $P(\text{jury has only faculty}) = \dfrac{1}{34} = 0.0294$

(c) $P(\text{jury has two students and three faculty}) = \dfrac{20}{51} = 0.3922$

63. $P(\text{shipment is rejected}) = 0.1283$

65. (a) $P(\text{you like 2 of 4 songs}) = 0.3916$
(b) $P(\text{you like 3 of 4 songs}) = 0.1119$
(c) $P(\text{you like all 4 songs}) = 0.0070$

67. (a) Five cards can be selected from a deck $_{52}C_5 = 2{,}598{,}960$ ways.
(b) Three of the same card can be chosen $13 \cdot {_4C_3} = 52$ ways.
(c) The remaining two cards can be chosen $_{12}C_2 \cdot {_4C_1} \cdot {_4C_1} = 1056$ ways.
(d) $P(\text{three of a kind}) = \dfrac{52 \cdot 1056}{2{,}598{,}960} = 0.0211$

69. $P(\text{all 4 modems tested worked}) = 0.4912$

Chapter Review Exercises (page 308)

1. (a) Possible probabilities are $0, 0.75, 0.41$.

(b) Possible probabilities are $\dfrac{2}{5}, \dfrac{1}{3}, \dfrac{6}{7}$.

3. $P(F) = \dfrac{2}{5} = 0.4$ **5.** $P(E^c) = \dfrac{4}{5} = 0.8$

7. $P(E \text{ or } F) = 0.48$

9. Events E and F are not independent because $P(E \text{ and } F) \ne P(E) \cdot P(F)$.

11. $P(E|F) = 0.5$

13. (a) $P(\text{green}) = \dfrac{1}{19} = 0.0526$; there is a 5.26% probability that the next spin will end with the ball in a green slot.

(b) $P(\text{green or red}) = \dfrac{10}{19} = 0.5263$; there is a 52.63% probability that the next spin will end with the ball in either a green slot or a red slot.

(c) $P(00 \text{ or red}) = \dfrac{19}{38} = 0.5$; there is a 50% probability that the next spin will end with the ball either in 00 or in a red slot.

(d) $P(31 \text{ and black}) = 0$; this is called an impossible event.

15. (a) $P(\text{fatality was alcohol related}) = \dfrac{301}{575} = 0.5235$

(b) $P(\text{fatality was not alcohol related}) = \dfrac{274}{575} = 0.4765$

(c) $P(\text{both fatalities were alcohol related}) = 0.2736$
(d) $P(\text{neither fatality was alcohol related}) = 0.2266$
(e) $P(\text{at least one fatality was alcohol related}) = 0.7734$

17. (a)

Class of Worker	Probability
Private wage and salary worker	0.8428
Government worker	0.1236
Self-employed worker	0.0321
Unpaid family worker	0.0014

(b) It is unusual for a Louisville worker to be an unpaid family worker.
(c) It is unusual for a Louisville worker to be self-employed.

19. (a) $P(\text{baby was postterm}) = 0.0674$
(b) $P(\text{baby weighed 3000 to 3999 grams}) = 0.6591$
(c) $P(\text{baby weighed 3000 to 3999 grams and was postterm}) = 0.0484$
(d) $P(\text{baby weighed 3000 to 3999 grams or was postterm}) = 0.6781$
(e) $P(\text{baby weighed} <1000 \text{ grams and was postterm}) = 0.000008$; this event is not impossible.
(f) $P(\text{baby weighed 3000 to 3999 grams}| \text{baby was postterm}) = 0.7185$
(g) The events "postterm baby" and "weighs 3000 to 3999 grams" are not independent.

21. (a) $P(\text{complaint filed online}) = 0.63$
(b) $P(\text{complaint not filed online}) = 0.37$
(c) $P(\text{all five complaints filed online}) = 0.099$
(d) $P(\text{at least one of five complaints not filed online}) = 0.901$
(e) $P(\text{none of five complaints filed online}) = 0.007$
(f) $P(\text{at least one of five complaints filed online}) = 0.993$

23. $P(\text{matching the three winning numbers}) = 0.001$

25. $P(\text{shipment accepted}) = 0.8$

27. $26^2 \cdot 10^4 = 6{,}760{,}000$ different license plates can be formed.

29. There are $6! = 720$ different arrangements of the letters LINCEY.

31. There are $_{55}C_8 = 1{,}217{,}566{,}350$ possible random samples.

33. $P(\text{winning Fantasy 5}) = \dfrac{1}{_{35}C_5} = 0.000003$

35. (a) $P(\text{three Merlot}) = 0.0454$ **(b)** $P(\text{two Merlot}) = 0.3182$
(c) $P(\text{no Merlot}) = 0.1591$

37. Answers will vary.

39. (a) $P(\text{home run to left field}) = \dfrac{34}{70} = 0.4857$; there is a 48.6% probability that a randomly selected home run by McGwire went to left field.
(b) $P(\text{home run to right field}) = 0$
(c) No, it is not impossible for McGwire to hit a home run to right field.

CHAPTER 6

6.1 Assess Your Understanding (page 323)

1. A random variable is a numerical measure of the outcome of a probability experiment, so its value is determined by chance.

3. When $P(x)$ denotes the probability that the random variable X equals x, each probability must be between 0 and 1, inclusive, and the sum of the probabilities must equal 1.

5. A batting average of 0.300 means the probability of the player's getting a hit at any at-bat is 0.3, despite what the results of the previous at-bats.

7. (a) Discrete, $x = 0, 1, 2, \ldots, 20$ **(b)** Continuous, $t > 0$
(c) Discrete, $x = 0, 1, 2, \ldots$ **(d)** Continuous, $s \ge 0$

9. (a) Continuous, $r \ge 0$ **(b)** Discrete, $x = 0, 1, 2, 3, \ldots$
(c) Discrete, $x = 0, 1, 2, 3, \ldots$ **(d)** Continuous, $t > 0$

11. Yes, it is a probability distribution.

13. No, $P(50) < 0$.   **15.** No, $\Sigma P(x) \ne 1$ **17.** $P(4) = 0.3$

19. (a) Each probability is between 0 and 1, inclusive, and the sum of the probabilities equals 1.

(b)

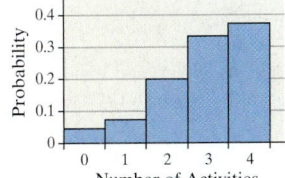

Parental Involvement in School (child grades K–5)

(c) $\mu_X = 2.9$; the mean of the random variable X is the mean outcome if the experiment were repeated many times.
(d) $\sigma_X^2 = 1.2$ **(e)** $\sigma_X = 1.1$
(f) $P(3) = 0.320$ **(g)** $P(3 \text{ or } 4) = 0.694$

21. (a) Each probability is between 0 and 1, inclusive, and the sum of the probabilities equals 1.

(b)

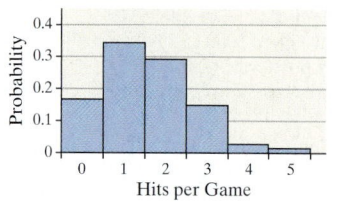

Ichiro's Hit Parade

(c) $\mu_X = 1.6$ **(d)** $\sigma_X = 1.2$
(e) $P(2) = 0.2857$ **(f)** $P(X > 1) = 0.4969$

23. (a)

x (games played)	P(x)
4	0.1852
5	0.1852
6	0.2222
7	0.4074

(b)

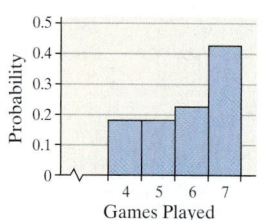

Games Played

(c) $\mu_X = 5.9$ **(d)** $\sigma_X = 1.1$

25. (a)

x (grade level)	P(x)
1	0.1246
2	0.1246
3	0.1259
4	0.1271
5	0.1269
6	0.1254
7	0.1243
8	0.1211

(b)

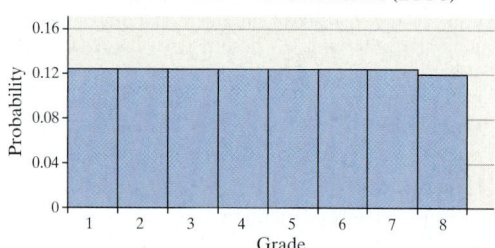

Grade School Enrollment (2000)

(c) $\mu_X = 4.5$; if the experiment were repeated many times, the mean grade would be 4.5.
(d) $\sigma_X = 2.3$

27. (a) $P(4) = 0.119$ **(b)** $P(4 \text{ or } 5) = 0.222$
(c) $P(X \geq 6) = 0.108$
(d) $\mu_X = 3.0$; you would expect that this is the mother's third live birth.

29. $E(X) = \$86.50$; the insurance company expects to make an average profit of $86.50 on every 20-year-old female it insures for 1 year.

31. The expected profit is $12,000.

33. $E(X) = -\$0.26$; if you played 1000 times, you would expect to lose about $260.

35. (a) The expected cash prize is $0.30. After paying $1.00 to play, your expected profit is $-\$0.70$.
(b) A grand prize of $118,000,000 has an expected profit greater than zero.
(c) The size of the grand prize does not affect the chance of winning provided the probabilities remain constant.

37. Answers will vary. The simulations illustrate the Law of Large Numbers.

6.2 Assess Your Understanding (page 340)

1. An experiment is binomial provided:
   1. The experiment consists of a fixed number, n, of trials.
   2. The trials are independent.
   3. Each trial has two possible mutually exclusive outcomes: success and failure.
   4. The probability of success, p, remains constant for each trial of the experiment.

3. When the binomial distribution is approximately bell shaped, about 95% of the outcomes are in the interval from $\mu - 2\sigma$ to $\mu + 2\sigma$. The empirical rule can be used when $np(1 - p) \geq 10$.

5. When n is small, the shape of the binomial distribution is determined by p. If p is close to zero, the distribution is skewed right, if p is close to 0.5, the distribution is approximately symmetric; and if p is close to 1, the distribution is skewed left.

7. Not binomial, because the random variable is continuous

9. Binomial

11. Not binomial, because the trials are not independent

13. Not binomial, because the number of trials is not fixed

15. Binomial

17. $P(3) = 0.2150$ **19.** $P(38) = 0.0532$ **21.** $P(3) = 0.2786$
23. $P(X \leq 3) = 0.9144$ **25.** $P(X > 3) = 0.5$
27. $P(X \leq 4) = 0.5833$

29. (a)

x	P(x)	x	P(x)
0	0.1176	4	0.0595
1	0.3025	5	0.0102
2	0.3241	6	0.0007
3	0.1852		

(b) $\mu_X = 1.8$; $\sigma_X = 1.1$ **(c)** $\mu_X = 1.8$; $\sigma_X = 1.1$

(d)

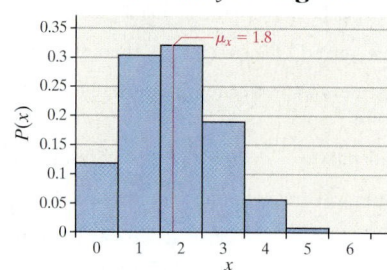

Probability Histogram

The distribution is skewed right.

31. (a)

x	P(x)	x	P(x)
0	0.0000038	5	0.1168
1	0.0001	6	0.2336
2	0.0012	7	0.3003
3	0.0087	8	0.2253
4	0.0389	9	0.0751

(b) $\mu_x = 6.75$; $\sigma_x = 1.3$ **(c)** $\mu_x = 6.75$; $\sigma_x = 1.3$

(d)

Probability Histogram

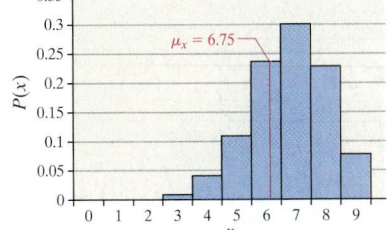

$\mu_x = 6.75$

The distribution is skewed left.

33. (a)

x	P(x)	x	P(x)
0	0.0010	6	0.2051
1	0.0098	7	0.1172
2	0.0439	8	0.0439
3	0.1172	9	0.0098
4	0.2051	10	0.0010
5	0.2461		

(b) $\mu_x = 5; \sigma_x = 1.6$ **(c)** $\mu_x = 5; \sigma_x = 1.6$

(d)

Probability Histogram

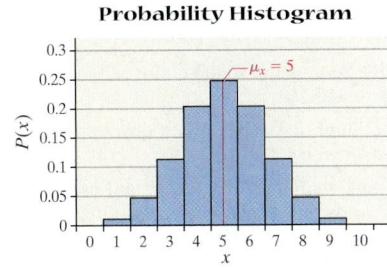

$\mu_x = 5$

The distribution is symmetric.

35. (a) This is a binomial experiment because:
   1. It is performed a fixed number, $n = 15$, of times.
   2. The trials are independent.
   3. For each trial, there are two possible mutually exclusive outcomes: on time and not on time.
   4. The probability of "on time" is fixed at $p = 0.90$.
   (b) $P(14) = 0.3432$ **(c)** $P(X \geq 14) = 0.5490$
   (d) $P(X < 14) = 0.4510$ **(e)** $P(12 \leq X \leq 14) = 0.7386$

37. (a) $P(5) = 0.1746$
   (b) $P(X \geq 10) = 0.0026$; it would be unusual for at least 10 of the 20 households selected to have high-speed Internet.
   (c) $P(X < 4) = 0.4114$ **(d)** $P(2 \leq X \leq 5) = 0.7350$

39. (a) $P(22) = 0.0120$ **(b)** $P(14 \leq X \leq 16) = 0.3648$
   (c) It would be unusual if 22 or more murders were committed using a firearm because $P(X \geq 22) < 0.05$.

41. (a) $P(6) = 0.1460$ **(b)** $P(X < 7) = 0.2241$
   (c) $P(X \geq 5) = 0.9803$ **(d)** $P(5 \leq X \leq 8) = 0.7363$

43. (a) $\mu_X = 90; \sigma_X = 3$
   (b) We expect that, in a random sample of 100 flights from Orlando to Los Angeles, 90 of them will be on time.
   (c) It would be unusual to have 80 on-time flights because $P(80) < 0.05$.

45. (a) $\mu_X = 20; \sigma_X = 4$
   (b) In a random sample of 100 U.S. households, we expect 20 of them to have a high-speed Internet connection.
   (c) It would not be unusual to observe 18 U.S. households with a high-speed Internet connection.

47. (a) $E(X) = 66.9$ murders.
   (b) It would not be unusual to observe 75 murders by firearm.

49. It would not be unusual to observe 86 patients who experience headaches because 86 is within two standard deviations of the mean.

51. (a), (b) Answers will vary
   (c) $P(10) = 0.1074$ **(d)** Answers will vary.
   (e) $P(X \geq 8) = 0.6778$
   (f) Simulated mean will vary. $E(X) = 8$ free throws

53. (a) You would select 38 residents 25 years and older.
   (b) You would select 65 residents 25 years and older.

55. (a) $P(15) = 0.2065$ **(b)** $P(15 \leq X \leq 17) = 0.5314$
   (c) $\dfrac{20}{500} = 0.04 = 4\%$; $P(15) = 0.2023$; $P(15 \leq X \leq 17) = 0.5259$

6.3 Assess Your Understanding (page 348)

1. To follow a Poisson process, a random variable X must meet the following:
   1. The probability of two or more successes in a sufficiently small subinterval is 0.
   2. The probability of success is the same for any two intervals of equal length.
   3. The number of success in any interval is independent of the number of success in any other disjoint interval.

3. $\lambda = 10$ per minute; $t = 5$ minutes

5. $\lambda = 0.07$ per linear foot; $t = 20$ linear feet

7. (a) $P(6) = 0.1462$ **(b)** $P(X < 6) = 0.6160$
   (c) $P(X \geq 6) = 0.3840$ **(d)** $P(2 \leq X \leq 4) = 0.4001$

9. (a) $P(4) = 0.0050$ **(b)** $P(X < 4) = 0.9942$
   (c) $P(X \geq 4) = 0.0058$ **(d)** $P(4 \leq X \leq 6) = 0.0057$
   (e) $\mu_X = 0.7; \sigma_X = 0.8$

11. (a) $P(7) = 0.1490$; about 15 of every 100 days, there will be exactly 7 hits to the Web site between 7:30 and 7:35 P.M.
   (b) $P(X < 7) = 0.4497$; about 45 of every 100 days, there will be fewer than 7 hits to the Web site between 7:30 and 7:35 P.M.
   (c) $P(X \geq 7) = 0.5503$; about 55 of every 100 days, there will be at least 7 hits to the Web site between 7:30 and 7:35 P.M.

13. (a) $P(2) = 0.2510$; in about 25 of every 100 five-gram samples of peanut butter, we expect to find 2 insect fragments.
   (b) $P(X < 2) = 0.5578$; in about 56 of every 100 five-gram samples of peanut butter, we expect to find fewer than 2 insect fragments.
   (c) $P(X \geq 2) = 0.4422$; in about 44 of every 100 five-gram samples of peanut butter, we expect to find at least 2 insect fragments.
   (d) $P(X \geq 1) = 0.7769$; in about 78 of every 100 five-gram samples of peanut butter, we expect to find at least 1 insect fragment.
   (e) It would not be unusual. About 7 of every 100 five-gram samples will contain at least 4 fragments.

15. (a) $P(0) = 0.9608$; there will be no deaths in approximately 96 of 100 randomly selected 100 million mile units of flight.
   (b) $P(X \geq 1) = 0.0392$; there will be at least one death in approximately 3.9 of 100 randomly selected 100 million miles of flight.
   (c) $P(X > 1) = 0.0008$; we expect more than one death in approximately 8 of 10,000 randomly selected 100 million miles of flight.

17. $P(X \geq 3) = 0.0017$; the 2004 hurricane season was unusual.

19. (a) (b)

Number of Cars, x	P(x)	Expected Number of Restaurants	Number of Cars, x	P(x)	Expected Number of Restaurants
0	0.0025	0.50	9	0.0688	13.77
1	0.0149	2.97	10	0.0413	8.26
2	0.0446	8.92	11	0.0225	4.51
3	0.0892	17.85	12	0.0113	2.25
4	0.1339	26.77	13	0.0052	1.04
5	0.1606	32.12	14	0.0022	0.45
6	0.1606	32.12	15	0.0009	0.18
7	0.1377	27.54	16	0.0003	0.07
8	0.1033	20.65			

   (c) The observed frequencies are lower for fewer cars (x small) and higher for more cars (x large), so the advertising campaign appears to be effective.

21. (a)

Number of Deaths, x	Proportion of Years
0	0.545
1	0.325
2	0.110
3	0.015
4	0.005

(b) $\mu_X = 0.61$

(c)

Number of Deaths, x	P(x)
0	0.5434
1	0.3314
2	0.1011
3	0.0206
4	0.0031

(d) Answers will vary.

23. (a) $E(X) = 119$ colds
(b)–(g) Answers will vary.

Chapter Review Exercises (page 353)

1. (a) Continuous, $s \geq 0$
(b) Discrete, $x = 0, 1, \ldots, 91$ (assuming 91 days of winter)
(c) Discrete, $x = 0, 1, 2, \ldots$

3. It is not a probability distribution because $\sum P(x) \neq 1$.

5. (a)

x	P(x)
4	0.3030
5	0.2424
6	0.2576
7	0.1970

(b)

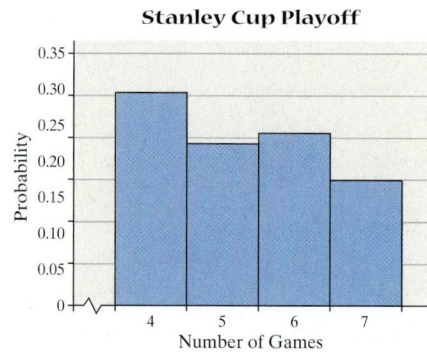

Stanley Cup Playoff

(c) $\mu_X = 5.3$ **(d)** $\sigma_X = 1.1$

7. $E(X) = 59.2$; the insurance company expects to earn an average profit of \$59.20 for each 35-year-old male that it insures for 1 year at \$100,000.

9. Binomial experiment

11. (a) $P(0) = 0.4344$ **(b)** $P(2) = 0.1478$
(c) $P(X \geq 2) = 0.1879$ **(d)** $P(1) = 0.3777$
(e) $\mu_X = 20; \sigma_X = 4.3$
(f) It would not be unusual to have 12 females in a sample of 250 with high cholesterol.

13. (a) $P(6) = 0.0454$ **(b)** $P(X < 4) = 0.6477$
(c) $P(X \geq 2) = 0.8244$ **(d)** $P(3) = 0.2428$
(e) $\mu_X = 75$
(f) It would be unusual to have 95 households in a sample of 500 tuned into the game.

15. (a)

x	P(x)	x	P(x)
0	0.3277	3	0.0512
1	0.4096	4	0.0064
2	0.2048	5	0.0003

(b) $\mu_X = 1; \sigma_X = 0.9$ **(c)** $\mu_X = 1; \sigma_X = 0.9$

(d)

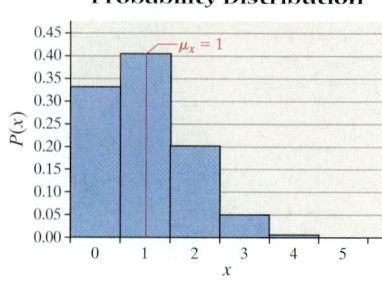

Binomial Probability Distribution

The distribution is skewed right.

17. (a) $P(3) = 0.0072$ **(b)** $P(X < 3) = 0.9921$
(c) $P(X \geq 3) = 0.0079$ **(d)** $P(3 \leq X \leq 5) = 0.0079$
(e) $\mu_X = 0.4; \sigma_X = \sqrt{0.4}$

19. (a) $P(4) = 0.1951$; on about 20 of every 100 randomly chosen days, exactly 4 cars will arrive at the bank's drive-through window between 4:00 P.M. and 4:10 P.M.
(b) $P(X < 4) = 0.4142$; on about 41 of every 100 randomly chosen days, fewer than 4 cars will arrive at the bank's drive-through window between 4:00 P.M. and 4:10 P.M.
(c) $P(X \geq 4) = 0.5858$; on about 59 of every 100 randomly chosen days, at least 4 cars will arrive at the bank's drive-through window between 4:00 P.M. and 4:10 P.M.

21. It would not be unusual to have more than 4 maintenance calls; $P(X > 4) = 0.0527$.

23. As a rule of thumb, if X is binomially distributed, the Empirical Rule can be used when $np(1 - p) \geq 10$.

25. We can sample without replacement and use the binomial probability distribution to approximate probabilities when the sample size is small in relation to the population size. As a rule of thumb, if the sample size is less than 5% of the population size, the trials can be considered nearly independent.

CHAPTER 7

7.1 Assess Your Understanding (page 367)

1. For a graph to be that of a probability density function, the area under the graph must equal 1, and the graph must be on or above the horizontal axis for all possible values of the random variable.

3. The area under the graph of a probability density function can be interpreted either as the proportion of the population with the characteristic described by the interval or as the probability that a randomly selected individual from the population has the characteristic described by the interval.

5. $\mu - \sigma; \mu + \sigma$

7. This graph is not symmetric; it cannot represent a normal density function.

9. This graph is not always greater than or equal to zero; it cannot represent a normal density function.

11. The graph can represent a normal density function.

13. $P(5 \leq X \leq 10) = \dfrac{1}{6}$ **15.** $P(X \geq 20) = \dfrac{1}{3}$

17. (a)

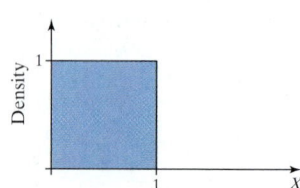

(b) $P(0 \leq X \leq 0.2) = 0.2$ **(c)** $P(0.25 \leq X \leq 0.6) = 0.35$
(d) $P(X \geq 0.95) = 0.05$ **(e)** Answers will vary.

19. Normal   **21.** Not normal

23. Graph A: $\mu = 10, \sigma = 2$; graph B: $\mu = 10, \sigma = 3$. A larger standard deviation makes the graph lower and more spread out.

25. $\mu = 2, \sigma = 3$ **27.** $\mu = 100, \sigma = 15$

29. (a)

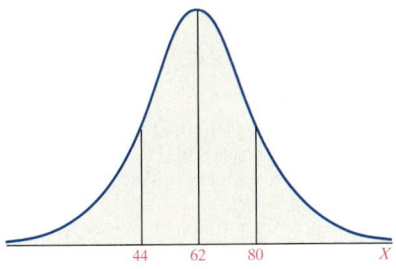

(b)

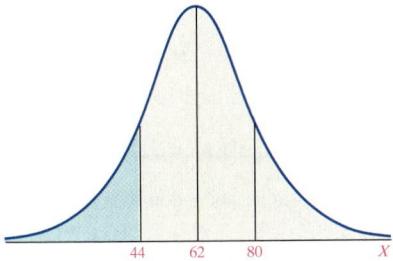

(c) (i) 15.87% of the cell phone plans in the United States is less than $44.00 per month.
(ii) The probability is 0.1587 that a randomly selected cell phone plan in the United States is less than $44.00 per month.

31. (a)

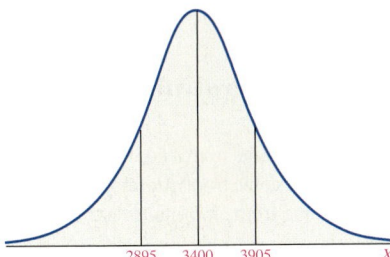

(b)

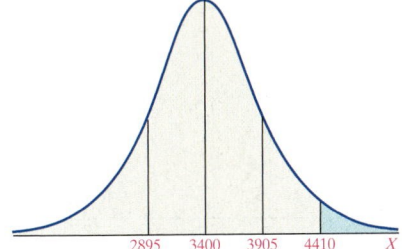

(c) (i) 2.28% of all full-term babies have a birth weight of at least 4410 grams.
(ii) The probability is 0.0228 that the birth weight of a randomly chosen full-term baby are at least 4410 grams.

33. (a) (i) The proportion of human pregnancies that last more than 280 days is 0.1908.
(ii) The probability that a randomly selected human pregnancy lasts more than 280 days is 0.1908.
(b) (i) The proportion of human pregnancies that last between 230 and 260 days is 0.3416.
(ii) The probability that a randomly selected human pregnancy lasts between 230 and 260 days is 0.3416.

35. (a) $Z_1 = -0.67$ **(b)** $Z_2 = 0.67$
(c) The area between Z_1 and Z_2 is 0.495

37. (c) Answers will vary.

7.2 Assess Your Understanding (page 381)

1. A standard normal curve has the following properties:
  1. It is symmetric about its mean, $\mu = 0$.
  2. Its highest point occurs at $\mu = 0$.
  3. It has inflection points at -1 and 1.
  4. The area under the curve is 1.
  5. The area under the curve to the right of $\mu = 0$ equals the area under the curve to the left of $\mu = 0$. Both equal $\frac{1}{2}$.
  6. As z increases, the graph approaches but never equals zero. When z decreases, the graph approaches but never equals zero.
  7. The Empirical Rule:
    Approximately 68% of the area under the standard normal curve is between -1 and 1. Approximately 95% of the area under the standard normal curve is between -2 and 2. Approximately 99.7% of the area under the standard normal curve is between -3 and 3.

3. False. Explanations will vary.

5. (a) Area $= 0.0071$ **(b)** Area $= 0.3336$
(c) Area $= 0.9115$ **(d)** Area $= 0.9998$

7. (a) Area $= 0.9987$ **(b)** Area $= 0.9441$
(c) Area $= 0.0375$ **(d)** Area $= 0.0009$

9. (a) Area $= 0.9586$ **(b)** Area $= 0.2088$
(c) Area $= 0.8479$

11. (a) Area $= 0.0456$ **(b)** Area $= 0.0646$
(c) Area $= 0.5203$

13. (a) Area $= 0.8877$ **(b)** Area $= 0.0500$
(c) Area $= 0.9802$

15. $z = -1.28$ **17.** $z = -2.05$ **19.** $z = 0.67$

21. $z = -1.23$ **23.** $z_0 = -1.28; z_1 = 1.28$

25. $z_0 = -2.575; z_1 = 2.575$ **27.** $z_{0.05} = 1.645$

29. $z_{0.01} = 2.33$ **31.** $z_{0.20} = 0.84$

33. $P(Z < 1.93) = 0.9732$ **35.** $P(Z < 1.93) = 0.9986$

37. $P(-1.20 \leq Z < 2.34) = 0.8753$

39. $P(Z \geq 1.84) = 0.0329$ **41.** $P(Z \leq 0.72) = 0.7642$

43. $P(Z < -2.56$ or $Z > 1.39) = 0.0875$

45. $P(z < 1) - P(z < -1) = 0.8413 - 0.1687 = 0.6826$. So about 68% of the data lies within 1 standard deviation of the mean.
$P(z < 2) - P(z < -2) = 0.9772 - 0.0228 = 0.9544$. So about 95% of the data lies within 2 standard deviations of the mean.
$P(z < 3) - P(z < -3) = 0.9987 - 0.0013 = 0.9974$. So about 99.7% of the data lies within 3 standard deviations of the mean.

47. Area $= 0.0054$ because the graph is symmetric around the mean $\mu = 0$.

49. Area $= 0.1906$ because the graph is symmetric around the mean $\mu = 0$.

7.3 Assess Your Understanding (page 390)

1. To find the area under any normal curve:
  1. Draw the curve, label the random variables, and shade the desired area.
  2. Convert the values of X to Z-scores using the formula
    $$Z = \frac{X - \mu}{\sigma}.$$
  3. Draw a standard normal curve, label the Z-scores, and shade the desired area.
  4. Find the area under the standard normal curve.

3.

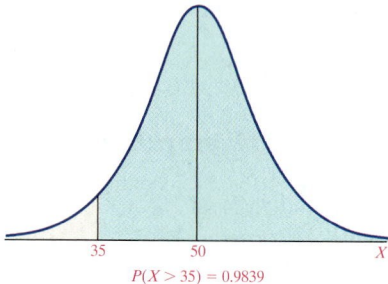

$P(X > 35) = 0.9839$

5.

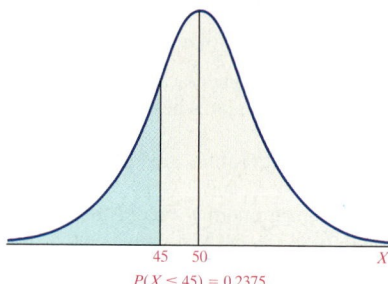

$P(X \le 45) = 0.2375$

7.

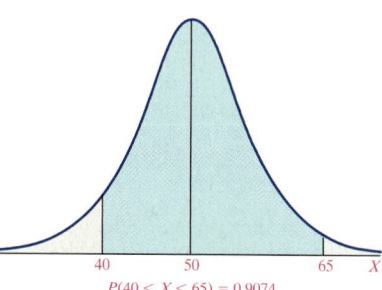

$P(40 < X < 65) = 0.9074$

9.

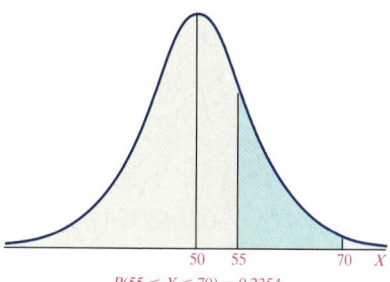

$P(55 \le X \le 70) = 0.2354$

11.

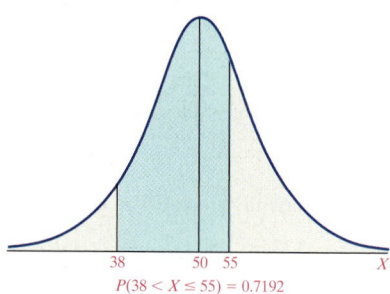

$P(38 < X \le 55) = 0.7192$

13. $X = 40.62$ is at the 9th percentile.

15. $X = 56.16$ is at the 81st percentile.

17. (a) $P(X < 20) = 0.1587$ **(b)** $P(X > 22) = 0.1587$
(c) $P(19 \le X \le 21) = 0.4772$ **(d)** Yes, $P(X < 18) = 0.0013$

19. (a) $P(1000 \le X \le 1400) = 0.8658$
(b) $P(X < 1000) = 0.0132$
(c) 0.7019 of the bags has more than 1200 chocolate chips.
(d) 0.1230 of the bags has fewer than 1125 chocolate chips.
(e) A bag that contains 1475 chocolate chips is at the 96th percentile.
(f) A bag that contains 1050 chocolate chips is at the 4th percentile.

21. (a) 0.1056 of the 2005 Honda Insights gets more than 60 miles per gallon.
(b) 0.0301 of the 2005 Honda Insights gets 50 miles per gallon or less.
(c) 0.2342 of the 2005 Honda Insights gets between 58 and 62 miles per gallon.
(d) $P(X < 45) = 0.0003$

23. (a) A 20- to 29-year-old female who is 60 inches tall is at the 7th percentile.
(b) A 20- to 29-year-old female who is 70 inches tall is at the 98th percentile.
(c) 0.9105 of 20- to 29-year-old females are between 60 and 70 inches tall.
(d) Yes, $P(X > 70) = 0.0176$

25. (a) 0.0764 of the rods has a length of less than 24.9 cm.
(b) 0.0324 of the rods is discarded.
(c) The plant manager expects to discard 162 of the 5000 rods manufactured.
(d) To meet the order, the plant manager should manufacture 11,804 rods.

27. (a) The 17th percentile for incubation time is 20 days.
(b) From 19 to 23 days make up the middle 95% of incubation times of the eggs.

29. (a) The 30th percentile for the number of chips in an 18-ounce bag is 1201 chips.
(b) The middle 99% of the bags contains between 958 and 1566 chocolate chips.

31. (a) The 97% gasoline mileage for the 2005 Honda Insight is 62 miles per gallon.
(b) From 51.3 to 60.7 miles per gallon is the middle 84% Honda Insight's gas mileage.

33. (a)

Age	Proportion
15–19.99	0.0110
20–24.99	0.0852
25–29.99	0.2701
30–34.99	0.3655
35–39.99	0.2117
40–44.99	0.0505

(b) Answers will vary.

7.4 Assess Your Understanding (page 399)

1. Explanations will vary.
3. The sample data do not come from a normally distributed population.
5. The sample data do not come from a normally distributed population.
7. The sample data come from a normally distributed population.
9. (a) The sample data come from a normally distributed population.
(b) $\bar{x} = 1247.4$ chips, $s = 101.0$ chips
(c)

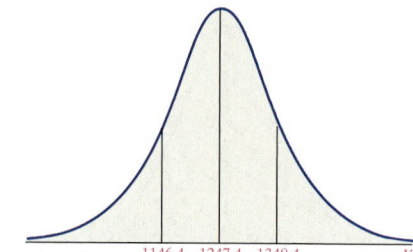

1146.4 1247.4 1348.4

(d) $P(X \ge 1000) = 0.9929$
(e) $P(1200 \le X \le 1400) = 0.6153$

11. The normal probability plot is approximately linear. So the sample data come from a normally distributed population.
13. The normal probability plot is not approximately linear. So the sample data do not come from a normally distributed population.

7.5 Assess Your Understanding (page 406)

1. A probability experiment is binomial if:
1. The experiment is performed n independent times.
2. For each trial there are two mutually exclusive outcomes—success and failure.
3. The probability of success p is the same for each trial of the experiment.

3. We use a correction for continuity because we are using a continuous density function to approximate a discrete probability.

5. Area under the normal curve to the right of $X = 39.5$

7. Area under the normal curve between $X = 7.5$ and $X = 8.5$

9. Area under the normal curve between $X = 17.5$ and $X = 24.5$

11. Area under the normal curve to the right of $X = 20.5$

13. Area under the normal curve to the right of $X = 500.5$

15. Using the binomial formula, $P(20) = 0.0616$; $np(1 - p) = 14.4$, the normal distribution can be used. Approximate probability is 0.0618.

17. Using the binomial formula, $P(30) < 0.0001$; $np(1 - p) = 7.5$, the normal distribution cannot be used.

19. Using the binomial formula, $P(60) = 0.0677$; $np(1 - p) = 14.1$, the normal distribution can be used. Approximate probability is 0.0630.

21. (a) $P(30) \approx 0.0444$
    (c) $P(X < 125) \approx 0.0021$
    (d) $P(125 \le X \le 135) \approx 0.5536$
    (b) $P(X \ge 130) \approx 0.9332$

23. (a) $P(20) \approx 0.0077$
    (c) $P(X \ge 22) \approx 0.0028$
    (d) $P(20 \le X \le 30) \approx 0.0143$
    (b) $P(X \le 20) \approx 0.9934$

25. (a) $P(20) \approx 0.0876$
    (c) $P(X \ge 30) \approx 0.0146$
    (d) $P(10 \le X \le 32) \approx 0.9899$
    (b) $P(X \le 15) \approx 0.1515$

27. (a) $P(X \ge 130) \approx 0.0028$
    (b) Result is unusual. Reasons will vary.

29. (a) $P(X \ge 80) \approx 0.0032$
    (b) Yes. Answers will vary.

Chapter Review Exercises (page 408)

1. (a) $\mu = 60$
    (b) $\sigma = 10$
    (c) (i) The proportion of random variables to the right of $X = 75$ is 0.0668.
    (ii) The probability that a randomly selected random variable is greater than $X = 75$ is 0.0668.
    (d) (i) The proportion of random variables between $X = 50$ and $X = 75$ is 0.7745.
    (ii) The probability that a randomly selected random variable is between $X = 50$ and $X = 75$ is 0.7745.

3. (a) $Z_1 = -0.5$
    (b) $Z_2 = 0.25$
    (c) The area between Z_1 and Z_2 is 0.2912.

5.

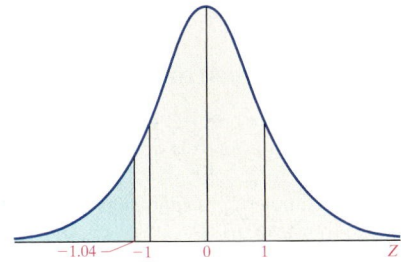

Area under the normal curve to the left of -1.04 is 0.1492.

7.

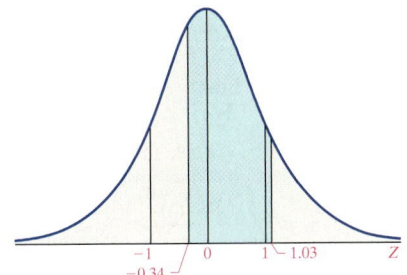

Area under the normal curve between -0.34 and 1.03 is 0.4816.

9. $P(Z < 1.19) = 0.8830$

11. $P(-1.21 < Z \le 2.28) = 0.8756$

13. $Z = 0.99$

15. $Z_1 = -1.75$ and $Z_2 = 1.75$

17. $z_{0.20} = 0.84$

19.

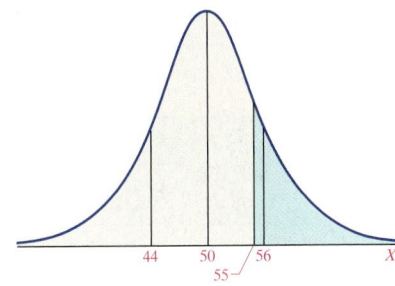

$P(X > 55) = 0.2023$

21.

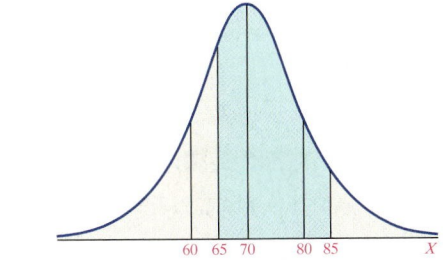

$P(65 < X < 85) = 0.6247$

23. (a) 12.71% of the tires lasts at least 75,000 miles.
    (b) 1.16% of the tires lasts at most 60,000 miles.
    (c) $P(65,000 \le X \le 80,000) = 0.8613$
    (d) The company should advertise 60,980 miles as its mileage warranty.

25. (a) 0.4090 of 16- to 19-year-old females has serum cholesterol above 180.
    (b) 0.4692 of 16- to 19-year-old females has serum cholesterol between 150 and 200.
    (c) $P(X < 140) = 0.2177$
    (d) A 16- to 19-year-old female with a cholesterol level of 120 is at the 10th percentile.
    (e) Yes, the 25th percentile on the normal curve is 144.

27. (a) 0.0119 of the baseballs produced is too heavy for use.
    (b) 0.0384 of the baseballs produced is too light to use.
    (c) 0.9497 of the baseballs produced is within acceptable weight limits.

29. (a) $np(1 - p) = 14.7 > 10$, so the normal distribution can be used to approximate the binomial probabilities.
    (b) $P(15) \approx 0.1000$
    (c) $P(X > 20) \approx 0.1210$
    (d) $P(X > 15) \approx 0.6517$
    (e) $P(X < 25) \approx 0.9864$
    (f) $P(15 \le X \le 25) \approx 0.6451$

31. The sample appears to be from a normally distributed population.

33. Not normal

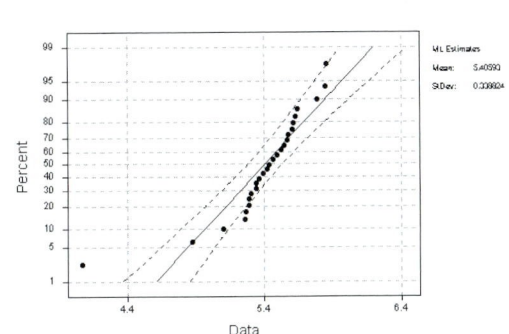

35. (a) 0.9772 of the Ford Escape HEVs gets more than 25 miles per gallon.
(b) 0.6915 of the Ford Escape HEVs gets less than 30 miles per gallon.
(c) $P(26 \leq X \leq 34) = 0.9270$ **(d)** $P(X > 35) = 0.0013$
(e) A Ford Escape HEV that gets 32 miles per gallon is at the 93rd percentile.
(f) A Ford Escape HEV that gets 25 miles per gallon is at the 2nd percentile.

37. (a)

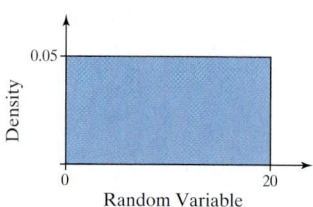

(b) $P(0 \leq X \leq 5) = 0.25$ **(c)** $P(10 \leq X \leq 18) = 0.4$

39. Answers will vary.

CHAPTER 8

8.1 Assess Your Understanding (page 431)

1. A sampling distribution is a probability distribution for all possible values of a statistic computed from a sample of size n.

3. Standard error; mean

5. $\mu_{\bar{x}} = \mu$, $\sigma_{\bar{x}} = \dfrac{\sigma}{\sqrt{n}}$ **7.** Four

9. The sampling distribution of $\bar{x}$ is normal with $\mu_{\bar{x}} = 30$ and $\sigma_{\bar{x}} = \dfrac{8}{\sqrt{10}}$.

11. $\mu_{\bar{x}} = 80$, $\sigma_{\bar{x}} = 2$ **13.** $\mu_{\bar{x}} = 52$, $\sigma_{\bar{x}} = \dfrac{10}{\sqrt{21}}$

15. (a) $\bar{x}$ is approximately normal with $\mu_{\bar{x}} = 80$, $\sigma_{\bar{x}} = 2$.
(b) $P(\bar{x} > 83) = 0.0668$ **(c)** $P(\bar{x} \leq 75.8) = 0.0179$
(d) $P(78.3 < \bar{x} < 85.1) = 0.7969$

17. (a) The population must be normally distributed to compute probabilities regarding the sample mean. If the population is normally distributed, then the sampling distribution of $\bar{x}$ is normally distributed with $\mu_{\bar{x}} = 64$ and $\sigma_{\bar{x}} = \dfrac{17}{\sqrt{12}}$.
(b) $P(\bar{x} < 67.3) = 0.7486$ **(c)** $P(\bar{x} \geq 65.2) = 0.4052$

19. (a) $P(x < 260) = 0.3520$ **(b)** $P(\bar{x} < 260) = 0.0465$
(c) $P(\bar{x} < 260) = 0.0040$ **(d)** Answers will vary.
(e) $P(256 \leq \bar{x} \leq 276) = 0.9844$

21. (a) $P(x > 95) = 0.3192$ **(b)** $P(\bar{x} > 95) = 0.0179$
(c) $P(\bar{x} > 95) = 0.0049$
(d) Increasing the sample size reduces the probability that $\bar{x} > 95$.
(e) Answers will vary.

23. (a) $P(x > 0) = 0.5675$ **(b)** $P(\bar{x} > 0) = 0.7291$
(c) $P(\bar{x} > 0) = 0.8051$ **(d)** $P(\bar{x} > 0) = 0.8531$
(e) Answers will vary.

25. (a) A sample size of at least 30 is needed to compute the probabilities.
(b) $P(\bar{x} < 10) = 0.0028$

27. (a) From the Central Limit Theorem, as the sample size increases, the sampling distribution of the mean becomes approximately normal.
(b) $\mu_{\bar{x}} = 3$, $\sigma_{\bar{x}} = \sqrt{\dfrac{3}{50}}$
(c) $P(\bar{x} \geq 3.6) = 0.0071$. This result is unusual. Conclusions will vary.

29. $P(\bar{x} \leq 45) = 0.0078$. It would be unusual for the random sample to have a mean number of blows less than or equal to 45.

31. (a) $\mu = 38.7$ years old
(b) Samples: (37,43);(37,29);(37,47);(37,36);(37,40);(43,29);(43,47); (43,36);(43,40);(29,47);(29,36);(29,40);(47,36);(47,40);(36,40)

(c)

$\bar{x}$	Probability	$\bar{x}$	Probability
32.5	$\dfrac{1}{15}$	39.5	$\dfrac{1}{15}$
33	$\dfrac{1}{15}$	40	$\dfrac{1}{15}$
34.5	$\dfrac{1}{15}$	41.5	$\dfrac{2}{15}$
36	$\dfrac{1}{15}$	42	$\dfrac{1}{15}$
36.5	$\dfrac{1}{15}$	43.5	$\dfrac{1}{15}$
38	$\dfrac{2}{15}$	45	$\dfrac{1}{15}$
38.5	$\dfrac{1}{15}$		

(d) $\mu_{\bar{x}} = 38.7$ years old
(e) $P(35.7 \leq \bar{x} \leq 41.7) = \dfrac{9}{15} = 0.6$

(f)

Samples		Sampling Distribution	
		$\bar{x}$	Probability
37, 43, 29	43, 29, 47	32	0.05
37, 43, 47	43, 29, 36	35	0.05
37, 43, 36	43, 29, 40	35.3	0.05
37, 43, 40	43, 47, 36	36	0.05
37, 29, 47	43, 47, 40	36.3	0.05
37, 29, 36	43, 36, 40	37.3	0.10
37, 29, 40	29, 47, 36	37.7	0.10
37, 47, 36	29, 47, 40	38.7	0.10
37, 47, 40	29, 36, 40	39.7	0.10
37, 36, 40	47, 36, 40	40	0.10
		41	0.05
		41.3	0.05
		42	0.05
		42.3	0.05
		43.3	0.05

$\mu_{\bar{x}} = 38.7$; $P(35.7 \leq \bar{x} \leq 41.7) = 0.7$

33. (a) Answers will vary.   **(b)** Answers will vary.
(c) Answers will vary.   **(d)** $\mu_{\bar{x}} = 100$, $\sigma_{\bar{x}} = \dfrac{8}{\sqrt{5}}$
(e) Answers will vary.   **(f)** $P(\bar{x} > 108) = 0.0125$
(g) Answers will vary.

35. (a) Answers will vary.   **(b)** Answers will vary.
(c) Answers will vary.   **(d)** Answers will vary.

8.2 Assess Your Understanding (page 439)

1. 0.44   **3.** False

5. The sampling distribution of $\hat{p}$ is approximately normal when $n \leq 0.05N$ and $np(1 - p) \geq 10$.

7. The sampling distribution of $\hat{p}$ is approximately normal with $\mu_{\hat{p}} = 0.4$ and $\sigma_{\hat{p}} = \sqrt{\dfrac{0.4(0.6)}{500}}$.

9. The sampling distribution of $\hat{p}$ is approximately normal with $\mu_{\hat{p}} = 0.103$ and $\sigma_{\hat{p}} = \sqrt{\dfrac{0.103(0.897)}{1000}}$.

11. (a) The sampling distribution of $\hat{p}$ is approximately normal with $\mu_{\hat{p}} = 0.8$ and $\sigma_{\hat{p}} = \sqrt{\dfrac{0.8(0.2)}{75}}$.
(b) $P(\hat{p} \geq 0.84) = 0.1922$ **(c)** $P(\hat{p} \leq 0.68) = 0.0047$

13. (a) The sampling distribution of $\hat{p}$ is approximately normal with $\mu_{\hat{p}} = 0.35$ and $\sigma_{\hat{p}} = \sqrt{\dfrac{0.35(0.65)}{1000}}$.
(b) $P(\hat{p} \geq 0.39) = 0.0040$ **(c)** $P(\hat{p} \leq 0.320) = 0.0233$

15. (a) The sampling distribution of $\hat{p}$ is approximately normal with

$$\mu_{\hat{p}} = 0.3 \text{ and } \sigma_{\hat{p}} = \sqrt{\frac{0.42(0.58)}{1460}}.$$

(b) $P(\hat{p} > 0.37) = 0.0630$; no   **(c)** $P(\hat{p} \leq 0.18) = 0.0043$; yes

17. (a) The sampling distribution of $\hat{p}$ is approximately normal with

$$\mu_{\hat{p}} = 0.26 \text{ and } \sigma_{\hat{p}} = \sqrt{\frac{0.26(0.74)}{500}}.$$

(b) $P(\hat{p} < 0.24) = 0.1539$

(c) It would be unusual for at least 150 adults to have no credit card. $P(\hat{p} \geq 0.3) = 0.0207$.

19. (a) $P(\hat{p} \leq 0.4) = 0.0436$

(b) It would not be unusual, $P(\hat{p} \geq 0.45) = 0.1271$.

21. (a) 62 more adult Americans must be polled to make $np(1-p) \geq 10$.

(b) 13 more adult Americans must be polled if $p = 0.2$.

23. (a) Answers will vary.   **(b)** Answers will vary.

(c) Answers will vary.   **(d)** Answers will vary.

(e) $\mu_{\hat{p}} = 0.3, \sigma_{\hat{p}} = \sqrt{\dfrac{0.3(0.7)}{765}}.$

25. (a) $\hat{p} = \dfrac{410}{500} = 0.82$ **(b)** $\sigma_{\hat{p}} = 0.017$

Chapter Review Exercises (page 443)

1. Answers will vary.

3. The sampling distribution of $\hat{p}$ is approximately normal when the sample size, n, is less than or equal to 5% of the population size, N, and $np(1-p) \geq 10$.

5. (a) $P(x > 2625) = 0.3085$. It is not unusual to randomly select a pregnant woman who needs more than 2625 kcal per day.

(b) $\bar{x}$ is normally distributed with $\mu_{\bar{x}} = 2600$ kcal and $\sigma_{\bar{x}} = \dfrac{25}{\sqrt{5}}$ kcal.

(c) $P(\bar{x} > 2625) = 0.0125$. It would be unusual to find a random sample of 20 pregnant women with a mean energy need of more than 2625 kcal per day.

7. (a) $\bar{x}$ is approximately normally distributed with $\mu_{\bar{x}} = 0.75$ inch and $\sigma_{\bar{x}} = \dfrac{0.004}{\sqrt{30}}$ inch.

(b) There is a 0.0061 probability that the inspector will conclude the machine needs adjustment.

9. (a) $\hat{p}$ is approximately normally distributed with $\mu_{\hat{p}} = 0.72$ and

$\sigma_{\hat{p}} = \sqrt{\dfrac{0.72(0.28)}{600}}.$ **(b)** $P(\hat{p} \leq 0.70) = 0.1379$

(c) It would a little be unusual for 450 of the 600 randomly selected individuals to prefer to start their own business. $P(\hat{p} \geq 0.75) = 0.0505$.

11. (a) $\hat{p}$ is approximately normally distributed with $\mu_{\hat{p}} = 0.09$ and

$\sigma_{\hat{p}} = \sqrt{\dfrac{0.09(0.91)}{200}}.$ **(b)** $P(\hat{p} \leq 0.06) = 0.0694$

(c) It would be unusual for 25 people in a random sample of 200 adults to have an advanced degree. $P(\hat{p} \geq 0.125) = 0.0418$.

13. $P(\bar{x} < -65,000) = 0.0071$

CHAPTER 9   Estimating the Value of a Parameter

9.1 Assess Your Understanding (page 458)

1. Level of confidence, sample size, and the standard deviation of the population.

3. The margin of error decreases as the sample size increases because the Law of Large Numbers states that as the sample size increases the sample mean approaches the value of the population mean.

5. The mean age of the population is a fixed value and so is not probabilistic.

7. No, the data are not normal and contain an outlier.

9. No, the data are not normal.

11. Yes, the data are normal, and there are no outliers.

13. $z_{\alpha/2} = z_{0.01} = 2.33$ **15.** $z_{\alpha/2} = z_{0.075} = 1.44$

17. (a) Lower bound: 32.44, upper bound: 35.96

(b) Lower bound: 32.73, upper bound: 35.67; increasing the sample size decreases the margin of error.

(c) Lower bound: 31.89, upper bound: 36.51; the interval in (c) is wider than that in (a). Increasing the confidence level increases the margin of error from 1.76 to 2.31.

(d) If the sample size is 15, the data must come from a normally distributed population.

19. (a) Lower bound: 102.67, upper bound: 113.33

(b) Lower bound: 99.57, upper bound: 116.43; decreasing the sample size increases the margin of error.

(c) Lower bound: 103.96, upper bound: 112.04; decreasing the level of confidence reduces the margin of error.

(d) We could not compute the confidence intervals because the samples are too small.

(e) The outliers would have increased the mean, shifting the confidence intervals to the right.

21. Lower bound: 8.10, upper bound: 8.24; we are 95% confident that the population mean time sleeping is between 8.10 and 8.24 hours.

23. Lower bound: 8.88, upper bound: 9.52; she is 90% confident that the population mean time spent with friends is between 8.88 hours and 9.52 hours.

25. (a) $\bar{x} = 542.25$ dollars

(b) The conditions are met. The data are approximately normal, and there are no outliers.

(c) Lower bound: 326.65, upper bound: 757.85; the institute is 95% confident that the population mean cost of repairs is between $326.65 and $757.85.

(d) Lower bound: 361.30, upper bound: 723.20; the institute is 90% confident that the population mean cost of repairs is between $361.30 and $723.20.

(e) When the level of confidence is decreased, the width of the confidence interval is decreased.

27. (a) $\bar{x} = 45.1$ years of age

(b) The conditions are met. The data are approximately normal, and there are no outliers.

(c) Lower bound: 41.1, upper bound: 49.1; the realtor is 95% confident that the mean age of a client is between 41.1 and 49.1 years of age.

(d) The realtor's clients do not appear to differ in age from the general population.

29. (a) $\bar{x} = 48.3$

(b) Lower bound: 42.0, upper bound: 54.6; Dr. Oswiecmiski is 95% confident that the mean HDL of his male patients is between 42.0 and 54.6.

(c) Dr. Oswiecmiski's patients do not appear to differ from the general population.

(d) To obtain a more precise (smaller) interval, he should either increase the sample size or decrease his confidence level.

31. (a) $\bar{x} = 1.64$ million shares

(b) Lower bound: 1.38, upper bound: 1.90; we are 90% confident that the mean number of shares traded is between 1.38 and 1.90 million.

(c) Lower bound: 1.48, upper bound: 2.00; We are 90% confident that the mean number of shares traded is between 1.48 and 2.00 million.

(d) The intervals are different because the sample mean in (c) is 100 thousand shares larger than the sample mean in (a).

33. (a) Since the length of dramas is not normally distributed, the sample must be large.

(b) Lower bound: 125.5; upper bound: 151.1; she is 99% confident that the mean length of a drama is between 125.5 minutes and 151.1 minutes.

35. (a) $\bar{x} = 49,477.7$ miles

(b) Lower bound: 40,607.2, upper bound: 58,308.2; she is 99% confident that the mean number of miles on 4-year-old Saturns in Chicago is between 40,607.2 and 58,308.2.

(c) Lower bound: 42,756.2, upper bound: 56,199.2; she is 95% confident that the mean number of miles on 4-year-old Saturns in Chicago is between 42,756.2 and 56,199.2.
(d) Decreasing the level of confidence decreases the width of the interval.
(e) No, because she only sampled cars in Chicago, the results cannot be generalized to the entire United States.

37. For a 99% confidence level, $n = 298$ patients; for a 95% confidence level, $n = 173$ patients. Decreasing the confidence level decreases the sample size needed.

39. For a 95% confidence level, $n = 1059$ subjects are needed.

41. (a) For a 90% confidence level, $n = 1051$ cars are needed in the sample.
(b) For a 90% confidence level, $n = 4201$ cars are needed in the sample.
(c) Doubling the required accuracy quadruples the required sample size.

43. (a)–(b) Answers will vary.
(c) Expect 19 intervals to include the population mean; actual number will vary.

45. Answers will vary.

47. The sample size must be increased by a factor of 4 to decrease the margin of error by one-half.

49. (a) Set I: $\bar{x} = 99.125$; set II: $\bar{x} = 99.1$; set III: $\bar{x} = 99.033$
(b) Set I: Lower bound: 88.73, upper bound: 109.52. Set II: Lower bound: 92.53, upper bound: 105.67. Set III: Lower bound: 93.67, upper bound: 104.40.
(c) As the size of the sample increases, the width of the confidence interval decreases.
(d) Set I: Lower bound: 77.48, upper bound: 98.27. Set II: Lower bound: 88.03, upper bound: 101.17. Set III: Lower bound: 90.67; upper bound: 101.40.
(e) Each interval except set I, which has an outlier, contains the population mean.

51. (a) Answers will vary. Expect 95% of the intervals to contain the population mean.
(b) Answers will vary.
(c) Answers will vary. Expect 95% of the intervals to contain the population mean.
(d) Confidence intervals with $n = 10$ should be wider than those with $n = 50$.

9.2 Assess Your Understanding (page 473)

1. A Z-interval should be constructed if the sample is random, the population from which the sample is drawn is normal or the sample size is large ($n \geq 30$), and if σ is known. A t-interval should be constructed if the sample is random, the population from which the sample is drawn is normal, but if σ is not known. Neither interval can be used if the sample is not random, the population is not normal and the sample size is small, or when there are outliers.

3. Robust means that the procedure is accurate when there are moderate departures from normality in the distribution of the population.

5. Similarities: Both are probability density distributions; both have $\mu = 0$; both are symmetric around their means.

Differences: t-distributions vary for different sample sizes, n; there is only one standard normal distribution; t-distributions have longer, thicker tails; the z-distribution has most of its area between -3 and 3.

7. (a) $t_{0.10} = 1.316$ **(b)** $t_{0.05} = 1.697$
(c) $t_{0.99} = -2.552$ **(d)** $t_{0.05} = 1.725$

9. (a) Lower bound: 103.7, upper bound: 112.3
(b) Lower bound: 100.4, upper bound: 115.6; decreasing the sample size increases the margin of error.
(c) Lower bound: 104.6, upper bound: 111.4; decreasing the level of confidence lessens the margin of error.
(d) No, the sample sizes were too small.

11. (a) Lower bound: 16.85, upper bound: 19.95
(b) Lower bound: 17.12, upper bound: 19.68; increasing the sample size decreases the margin of error.
(c) Lower bound: 16.32, upper bound: 20.48; increasing the level of confidence increases the margin of error.
(d) If $n = 18$, the population must be normal.

13. Lower bound: 12.05, upper bound: 14.75; we can be 99% confident that the mean number of books read by Americans during 2004 was between 12.05 and 14.75.

15. Lower bound: 1.08, upper bound: 8.12; we can be 95% confident that the mean incubation period of patients with SARS is between 1.08 and 8.12 days.

17. Lower bound: 12.86, upper bound: 13.14; the pollsters are 95% confident that the mean number of hours teenagers watch TV each week is between 12.86 and 13.14 hours.

19. Lower bound: 0.763, upper bound: 1.867; the stock analyst is 90% confident that the mean dividend yield of financial stocks is between 0.763% and 1.867%.

21. Lower bound: 12.40, upper bound: 19.44; we can be 99% confident that the mean concentration of dissolved organic carbon in organic soil is between 12.40 and 19.44 mg/L.

23. (a) Yes, the data are approximately normal, and there are no outliers.
(b) Lower bound: 31.9, upper bound: 44.7
(c) To increase the accuracy of the interval, increase the sample size.

25. (a) Yes, the data are approximately normal, and there are no outliers.
(b) Lower bound: 151.9, upper bound: 183.2
(c) To increase the accuracy of the interval, increase the sample size.

27. (a)

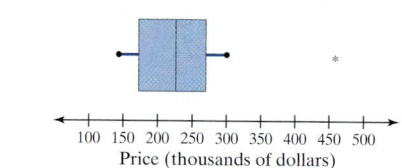

100   150   200   250   300   350   400   450   500
Price (thousands of dollars)

(b) Lower bound: \$158,567.7, upper bound: \$314,148.9
(c) Lower bound: \$165,219.6, upper bound: \$266,853.2
(d) Wider interval

29. Answers will vary. You would expect 19 of the 20 samples to include the population mean.

31. Answers will vary.

9.3 Assess Your Understanding (page 484)

1. $\hat{p}$ is the best point estimate of a population proportion.

3. Answers will vary.

5. Lower bound: 0.146, upper bound: 0.254

7. Lower bound: 0.191, upper bound: 0.289

9. Lower bound: 0.758, upper bound: 0.805

11. (a) $\hat{p} = 0.054$
(b) $n\hat{p}(1 - \hat{p}) = 44.44 \geq 10$, and the sample is less than 5% of the population.
(c) Lower bound: 0.041, upper bound: 0.067
(d) We are 90% confident that the proportion of Lipitor users who experience a headache as a side effect is between 0.041 and 0.067.

13. (a) $\hat{p} = 0.300$
(b) $n\hat{p}(1 - \hat{p}) = 211.68 \geq 10$, and the sample is less than 5% of the population.
(c) Lower bound: 0.266, upper bound: 0.334; the Gallup Organization is 98% confident that the proportion of adults aged 18 or older who believe the United States is spending too little on national defense and military purposes is between 0.266 and 0.334.

15. (a) $n\hat{p}(1 - \hat{p}) = 312.03 \geq 10$, and the sample is less than 5% of the population.
(b) Lower bound: 0.166, upper bound: 0.194; the Harris Organization is 90% confident that the proportion of adults who follow professional football whose favorite team is the Green Bay Packers is between 0.166 and 0.194.
(c) Lower bound: 0.158, upper bound: 0.202; the Harris Organization is 99% confident that the proportion of adults who follow professional football whose favorite team is the Green Bay Packers is between 0.158 and 0.202.
(d) Increasing the level of confidence widens the interval.

17. (a) Using $\hat{p} = 0.44, n = 1816$. **(b)** Using $\hat{p} = 0.5, n = 1842$.

19. (a) Using $\hat{p} = 0.55, n = 419$. **(b)** Using $\hat{p} = 0.5, n = 423$.

21. (a) Using $\hat{p} = 0.23, n = 2404$. **(b)** Using $\hat{p} = 0.5, n = 3394$.

23. $n = 984$ people were surveyed.

25. Answers will vary.

27. (a) Answers will vary.
 (b) Answers will vary.
 (c) Answers will vary. About 0.95 of the intervals should contain the population proportion.
 (d) As n increases, the proportion of intervals that capture p approaches the level of confidence.

9.4 Assess Your Understanding (page 491)

1. The chi-squared distribution is skewed to the right; its actual shape depends on the degrees of freedom of the distribution. As the degrees of freedom increase, the distribution becomes more symmetric, and the χ^2-values are nonnegative.

3. After the population is shown to be normal, a confidence interval for the standard deviation is obtained by completing the following steps.
 Step 1: Compute the sample variance.
 Step 2: Determine the critical values using the desired confidence level, the correct degrees of freedom, and the χ^2 distribution table.
 Step 3: Construct the confidence interval for the population variance by using the formulas

$$\text{Lower bound: } \frac{(n-1)s^2}{\chi^2_{\alpha/2}} \text{ and upper bound: } \frac{(n-1)s^2}{\chi^2_{1-\alpha/2}}$$

 Step 4: Compute the square root of the lower bound and the upper bound to find the confidence interval for the population standard deviation.

5. $\chi^2_{0.95} = 10.117, \chi^2_{0.05} = 30.144$

7. $\chi^2_{0.99} = 9.542, \chi^2_{0.01} = 40.289$

9. (a) Lower bound: 7.94, upper bound: 23.66
 (b) Lower bound: 8.59, upper bound: 20.63; the width of the interval decreases.
 (c) Lower bound: 6.61, upper bound: 31.36; the width of the interval increases.

11. Lower bound: 7.08, upper bound: 16.98; Essential Baby can be 95% confident that the population standard deviation of the time at which babies first crawl is between 7.08 and 16.98 weeks.

13. Lower bound: 13.51, upper bound: 49.83; the agricultural researcher can be 99% confident that the population standard deviation of the growing season in Chicago is between 13.51 and 49.83 days.

15. (a)

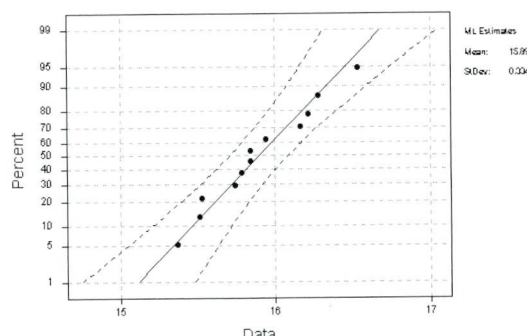

Normal Probability Plot for Ounces

 (b) $s = 0.349$ ounce
 (c) Lower bound: 0.261 oz, upper bound: 0.541 oz
 (d) No, 0.20 ounce is not in the confidence interval.

17. Fisher's approximation: $\chi^2_{0.975} \approx 73.772, \chi^2_{0.025} \approx 129.070$
 Actual values: $\chi^2_{0.975} = 74.222, \chi^2_{0.025} = 129.561$

9.5 Assess Your Understanding (page 494)

1. We construct a t-interval when we are estimating the population mean, we do not know the population standard deviation, and the underlying population is normally distributed. If the underlying population is not normally distributed, we can construct a t-interval to estimate the population mean provided the sample size is large ($n \geq 30$).

 We construct a Z-interval when we are estimating the population mean, we know the population standard deviation, and the underlying population is normally distributed. If the underlying population is not normally distributed, we can construct a Z-interval to estimate the population mean provided the sample size is large ($n \geq 30$). We also construct a Z-interval when we are estimating the population proportion, provided the sample size is smaller than 5% of the population and $np(1-p) \geq 10$.

3. Lower bound: 49.52, upper bound: 70.48

5. Lower bound: 0.069, upper bound: 0.165

7. Lower bound: 37.74, upper bound: 52.26

9. Lower bound: 114.98, upper bound: 126.02

11. Lower bound: 13.25, upper bound: 56.98

13. Lower bound: 51.4, upper bound: 56.6; we can be 95% confident that felons convicted of aggravated assault serve a mean sentence between 51.4 and 56.6 months.

15. Lower bound: 2693.8, upper bound: 3580.2; the Internal Revenue Service can be 90% confident that the mean additional tax owed is between \$2693.8 and \$3580.2.

17. Lower bound: 0.575, upper bound: 0.625; the Gallup Organization can be 90% confident that the proportion of adult Americans who are worried about having enough money for retirement is between 0.575 and 0.625.

19. Use a Z-interval. The distribution is normal, and we know the population standard deviation. Lower bound: 69.38, upper bound: 71.32; we are 95% confident that the mean height of 20- to 29-year-old males is between 69.38 and 71.32 inches.

21. Use neither interval. There is an outlier in the data set.

23. Use a t-interval. The distribution is normal, but we do not know the population standard deviation. Lower bound: 101.4, upper bound: 117.3; we are 95% confident that the mean pulse rate of women after 3 minutes of exercise is between 101.4 and 117.3 beats per minute.

Chapter Review Exercises (page 498)

1. $t_{0.005} = 2.898$

3. $\chi^2_{0.975} = 10.283, \chi^2_{0.025} = 35.479$

5. (a) Lower bound: 50.94, upper bound: 58.66
 (b) Lower bound: 51.65, upper bound: 57.95; increasing the sample size decreases the width of the interval.
 (c) Lower bound: 48.75, upper bound: 60.85; increasing the level of confidence increases the width of the interval.

7. (a) According to the Central Limit Theorem, the sampling distribution of $\bar{x}$ is approximately normal because the sample size is large.
 (b) Lower bound: 99,097, upper bound: 101,490; Michelin can be 90% confident that the mean mileage for its HydroEdge tire is between 99,098 and 101,490 miles.
 (c) Lower bound: 98,868, upper bound: 101,720; Michelin can be 95% confident that the mean mileage for its HydroEdge tire is between 98,868 and 101,720 miles.
 (d) Michelin would need to sample 63 tires to be within 1500 miles of the mean mileage with 99% confidence.

9. (a) Since the mean is larger than the median, the distribution is skewed right. A large sample is necessary to construct a confidence interval about the mean because the population is not normally distributed.
 (b) Lower bound: 12.70, upper bound: 13.30; the Gallup Organization is 90% confident that the mean number of hours that 13- to 17-year-olds watch television is between 12.7 and 13.3 hours per week.

11. (a) United States: $\bar{x} = 42.3$ hours; Canada: $\bar{x} = 40.8$ hours
 (b) Lower bound: 33.82, upper bound: 50.84
 (c) Lower bound: 33.6, upper bound: 48.0
 (d) It does not appear that Americans work more than Canadians.

13. (a) From the Central Limit Theorem, when the sample is large, $\bar{x}$ is approximately normally distribution with $\mu_{\bar{x}} = 2.27$ and
$$\sigma_{\bar{x}} = \frac{1.22}{\sqrt{60}}.$$
(b) Lower bound: 1.95, upper bound: 2.59; we can be 95% confident that couples who have been married for 7 years have a mean number of children between 1.95 and 2.59.
(c) Lower bound: 1.85, upper bound: 2.69; we can be 99% confident that couples who have been married for 7 years have a mean number of children between 1.85 and 2.69.

15. (a) $\bar{x} = 3.244$ liters; $s = 0.487$ liter
(b) Yes
(c) Lower bound: 2.935 liters, upper bound: 3.553 liters; the researchers are 95% confident that the mean blood plasma volume of women is between 2.935 and 3.553 liters.
(d) Lower bound: 2.807 liters, upper bound: 3.681 liters; the researchers are 99% confident that the mean blood plasma volume of women is between 2.807 and 3.681 liters.
(e) Lower bound: 0.312 liter, Upper bound: 1.001 liters. The researchers are 99% confident that the standard deviation blood plasma volume of women is between 0.312 liter and 1.001 liters.

17. (a) United States: Lower bound: 31.4 hours, upper bound: 53.2 hours
(b) Canada: Lower bound: 31.0 hours, upper bound: 50.6 hours
(c) It does not appear that Americans work more than Canadians.

19. (a) $\hat{p} = 0.086$
(b) Lower bound: 0.065, upper bound: 0.107; the Centers for Disease Control are 95% confident that the proportion of adult males 20–34-years old who have hypertension is between 0.065 and 0.107.
(c) 336 subjects would be needed if we use the point estimate of the proportion found in (a).
(d) 1068 subjects would be needed to obtain 95% confidence to be within 3% points if no prior estimate is available.

21. (a) United States: Lower bound: 10.39 hours, upper bound: 22.38 hours
(b) Canada: Lower bound: 9.29 hours, upper bound: 20.01 hours
(c) It doesn't appear that the hours worked per week by Canadians are less dispersed than those worked by Americans.

23. Area to the left of $t = -1.56$ is 0.0681 because the t-distribution is symmetric about zero.

25. The properties of the Student's t-distribution:
1. It is symmetric around $t = 0$.
2. It is different for different sample sizes.
3. The area under the curve is 1; half the area is to the right of 0 and half the area is to the left of 0.
4. As t gets extremely large, the graph approaches, but never equals, zero. Similarly, as t gets extremely small (negative), the graph approaches, but never equals, zero.
5. The area in the tails of the t-distribution is greater than the area in the tails of the standard normal distribution.
6. As the sample size n increases, the distribution (and the density curve) of the t-distribution becomes more like those of the standard normal distribution.

CHAPTER 10

10.1 Assess Your Understanding (page 511)

1. To reject the null hyothesis, when it is true. To not reject the null hypothesis when the alternative is true.

3. As the level of significance, α, decreases, the probability of making a Type II error, β, increases.

5. Answers will vary.

7. False

9. Right tailed, μ

11. Two tailed, σ

13. Left tailed, μ

15. (a) H_0: $p = 0.118$, H_1: $p < 0.118$
(b) Type I error: The sociologist rejects the hypothesis that the percentage of births to teenaged mothers is 11.8%, when in fact it is the true percentage of births to teenaged mothers.
(c) Type II error: The sociologist fails to reject the hypothesis that the percentage of births to teenaged mothers is 11.8%, when in fact the true percentage of births to teenaged mothers is less than 11.8%.

17. (a) H_0: $\mu = \$243{,}756$, H_1: $\mu > \$243{,}756$
(b) Type I error: The real estate broker rejects the hypothesis that the mean price of a single-family home is \$243,756, when in fact \$243,756 is the true mean cost of a single-family home.
(c) Type II error: The real estate broker fails to reject the hypothesis that the mean price of a single-family home is \$243,756, when in fact the mean cost of a single-family home is greater than \$243,756.

19. (a) H_0: $\sigma = 0.7$ psi, H_1: $\sigma < 0.7$ psi
(b) Type I error: We reject that the variability in the pressure required is 0.7 psi, when the true variability is 0.7 psi.
(c) Type II error: We fail to reject that the variability in the pressure required is 0.7 psi, when the variability is less than 0.7 psi.

21. (a) H_0: $\mu = \$49.91$, H_1: $\mu \neq \$49.91$
(b) Type I error: The researcher rejects the claim that the average monthly cell phone bill is \$49.91, when in fact the true mean cell phone bill is \$49.91.
(c) Type II error: The researcher fails to reject the claim that the average monthly cell phone bill is \$49.91, when the true mean cell phone bill is different from \$49.91.

23. There is sufficient evidence to support the claim that the percentage of births to teenage mothers is less than 11.8%.

25. There is not sufficient evidence to support the claim that the mean price of a single-family home has increased from \$243,756.

27. There is not sufficient evidence to support the claim that the variability in pressure has been reduced.

29. There is sufficient evidence to support the claim that the average monthly cell phone bill is different from \$49.91.

31. There is not sufficient evidence to support the claim that the percentage of births to teenage mothers is less than 11.8%.

33. There is sufficient evidence to support the claim that the mean price of a single-family home has increased from \$243,756.

35. (a) H_0: $\mu = 98.4$ pounds, H_1: $\mu > 98.4$ pounds
(b) There is sufficient evidence to support the claim that the mean per capita consumption of fruits and vegetables is more than 98.4 pounds.
(c) A Type I error was committed because the dietician rejected a true null hypothesis. When $\alpha = 0.05$, the probability of committing a Type I error is 0.05.

37. (a) H_0: $p = 0.102$, H_1: $p < 0.102$
(b) There is not sufficient evidence to support the claim that the percentage of high school students who tried marijuana for the first time before the age of 13 has decreased from 10.2%.
(c) A Type II error was committed because the SRO failed to reject a false null hypothesis.

39. H_0: $\mu = 0$, H_1: $\mu > 0$, where μ is the mean increase in gas mileage when using the device.

41. Answers will vary.

10.2 Assess Your Understanding (page 526)

1. We must have a simple random sample drawn from a normal population to test a claim regarding a population mean when we know σ. If the distribution of the population is not normal, or if it is unknown, the sample must be large.

3. When $\alpha = 0.05$ and σ is known, the critical values for a two-tailed test regarding a population mean are $Z = 1.96$ and $-Z = -1.96$.

5. P-value is the probability of observing a sample statistic as extreme or more extreme than the one observed under the assumption that the null hypothesis is true. If the P-value is less than the level of significance, α, the null hypothesis is rejected.

7. $P = 0.02$ is the probability of obtaining a sample mean more than $|Z_0|$ standard deviations from the hypothesized mean μ_0. Decisions and reasons will vary.

9. Answers will vary.

11. (a) $Z = -1.18$ **(b)** $-z_{0.05} = -1.645$

(c)

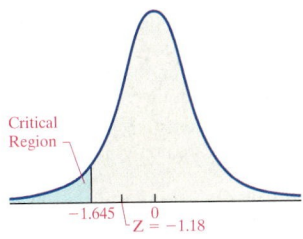

(d) No, the test statistic does not lie in the critical region.

13. (a) $Z = 3.29$

(b) $-z_{0.005} = -2.575$; $z_{0.005} = 2.575$

(c)

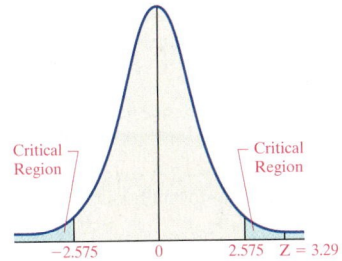

(d) Yes, the test statistic lies in the critical region.

15. (a) P-value $= 0.0082$, assuming $\mu = 20$; $P(\bar{x} \le 18.3) = 0.0082$

(b) Yes, the null hypothesis would be rejected at the $\alpha = 0.05$ level of significance, because the P-value is less than the level of significance ($0.0082 < 0.05$).

17. (a) The population does not need to be normally distributed to compute the P-value since the sample size is larger than 30.

(b) P-value $= 0.0614$, assuming $\mu = 105$; the probability of obtaining a sample mean that is more than 1.87 standard deviations from the hypothesized mean of 105 is 0.0614.

(c) No, the null hypothesis would not be rejected at the $\alpha = 0.02$ level of significance, because the P-value is greater than the level of significance ($0.0614 > 0.02$).

19. Hypotheses: H_0: $\mu = 63.7$, H_1: $\mu > 63.7$
Classical approach: $Z_0 = 0.38 < z_{0.05} = 1.645$; do not reject the null hypothesis.
P-value approach: P-value $= 0.3520 > \alpha = 0.05$; do not reject the null hypothesis.

21. (a) The scores must be normally distributed to test this claim because the sample size, $n = 20$, is small.

(b) Hypotheses: H_0: $\mu = 515$, H_1: $\mu < 515$
Classical approach: $Z_0 = -2.28 < -z_{0.10} = -1.28$; reject the null hypothesis.
P-value approach: P-value $= 0.0113 < \alpha = 0.10$; reject the null hypothesis.

23. (a) Yes, the conditions are satisfied. The data appear to be normally distributed, and there are no outliers.

(b) Hypotheses: H_0: $\mu = 5.03$, H_1: $\mu < 5.03$
Classical approach: $Z_0 = -4.74 < -z_{0.10} = -1.28$; reject the null hypothesis.
P-value approach: P-value $< 0.0001 < \alpha = 0.20$; reject the null hypothesis.

25. (a) Yes, the conditions are satisfied. The data appear to be normally distributed, and there are no outliers.

(b) Hypotheses: H_0: $\mu = 64.05$, H_1: $\mu \ne 64.05$
Classical approach: $Z_0 = -3.36 < -z_{0.005} = -1.645$; reject the null hypothesis.
P-value approach: P-value $= 0.0008 < \alpha = 0.01$; reject the null hypothesis.

(c) Yes, the machine should be recalibrated because the consequences for underfilling bottles are severe.

(d) Explanations will vary.

27. (a) Yes, the conditions are satisfied. The data appear to be normally distributed, and there are no outliers.

(b) Hypotheses: H_0: $\mu = 8.33$, H_1: $\mu < 8.33$
Classical approach: $Z_0 = -0.86 > -z_{0.10} = -1.28$; do not reject the null hypothesis.
P-value approach: P-value $= 0.1949 > \alpha = 0.10$; do not reject the null hypothesis.

29. Hypotheses: H_0: $\mu = 694$, H_1: $\mu > 694$
Classical approach: $Z_0 = 1.10 < z_{0.05} = 1.645$; do not reject the null hypothesis.
P-value approach: P-value $= 0.1357 > \alpha = 0.05$; do not reject the null hypothesis.

31. Hypotheses: H_0: $\mu = 31.8$, H_1: $\mu \ne 31.8$
Classical approach: $Z_0 = -3.32 < -z_{0.025} = -1.96$; reject the null hypothesis.
P-value approach: P-value $= 0.0010 < \alpha = 0.05$; reject the null hypothesis.

33. (63.974, 64.040); since the confidence interval does not include $\mu = 64.05$ ounces, reject the null hypothesis.

35. (18.60, 28.40); since the confidence interval does not include $\mu = 31.8$ million shares, reject the null hypothesis.

37. (a) H_0: $\mu = 514$, H_1: $\mu > 514$

(b) Classical approach: $Z_0 = 1.50 > z_{0.10} = 1.28$; reject the null hypothesis.
P-value approach: P-value $= 0.0668 < \alpha = 0.10$; reject the null hypothesis.

(c) Answers will vary.

(d) With $n = 400$ students:
Classical approach: $Z_0 = 0.71 < z_{0.10} = 1.28$; do not reject the null hypothesis.
P-value approach: P-value $= 0.2389 > \alpha = 0.10$; do not reject the null hypothesis.

39. (a), (c), (d) Answers will vary.

(b) Expect 5 samples to result in a Type I error.

41. (a) Hypotheses: H_0: $\mu = 150$, H_1: $\mu < 150$

(b) Answers will vary.

10.3 Assess Your Understanding (page 538)

1. A claim about a population mean with σ unknown can be tested provided that the sample is obtained using simple random sampling, and the population from which the sample is drawn is either normally distributed with no outliers or the sample size n is larger than 30.

3. The critical values for a two-tailed test at an $\alpha = 0.05$ level of significance when σ is unknown and there are 12 degrees of freedom are $t = 2.179$ and $-t = -2.179$.

5. (a) $t = -1.379$ **(b)** $-t_{0.05} = -1.714$

(c)

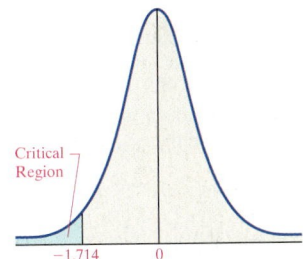

(d) There is not enough evidence for the researcher to reject the null hypothesis because it is a left-tailed test and the test statistic is greater than the critical value, $(-1.379 > -1.714)$.

7. (a) $t = 2.502$

(b) $-t_{0.005} = -2.819$; $t_{0.005} = 2.819$

(c)

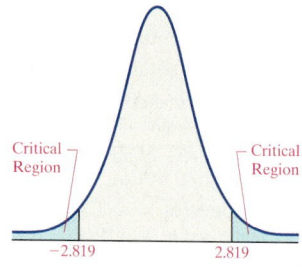

(d) There is not sufficient evidence for the researcher to reject the null hypothesis, since the test statistic is between the critical values $(-2.819 < 2.502 < 2.819)$.

9. (a) $t = -1.677$

(b)

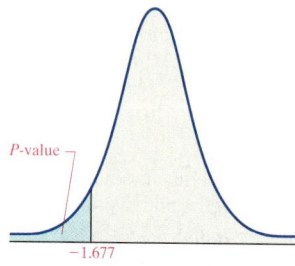

(c) $0.10 < P\text{-value} < 0.15$; the P-value is the area under the t-distribution with 17 degrees of freedom to the left of the test statistic, $t_0 = -1.677$.

(d) The researcher will not reject the null hypothesis at the $\alpha = 0.05$ level of significance because the P-value is greater than the level of significance $(0.10 > 0.05)$.

11. (a) No, $n \geq 30$ **(b)** $t = -3.108$

(c)

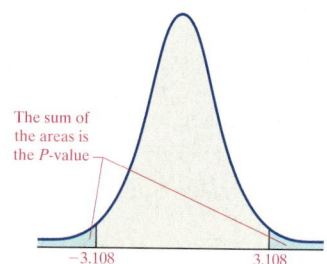

(d) $P\text{-value} = 0.0038$; the P-value is the sum of the areas under the t-distribution with 34 degrees of freedom to the left of the test statistic $-t_0 = -3.108$ and to the right of the test statistic $t_0 = 3.108$.

(e) The researcher will reject the null hypothesis at the $\alpha = 0.01$ level of significance because the P-value is less than the level of significance $(0.0038 < 0.01)$.

13. Hypotheses: $H_0: \mu = 9.02$, $H_1: \mu < 9.02$
Classical approach: $t_0 = -4.553 < -t_{0.01} = -2.718$; reject the null hypothesis.
P-value approach: $P\text{-value} = 0.0004 < \alpha = 0.01$; reject the null hypothesis.
There is sufficient evidence to support the researcher's claim that the mean hippocampal volume in alcoholic adolescents is less than the mean normal volume of 9.02 cm^3.

15. Hypotheses: $H_0: \mu = 1000$, $H_1: \mu > 1000$
Classical approach: $t_0 = 1.344 < t_{0.05} = 1.676$; do not reject the null hypothesis.
P-value approach: $P\text{-value} = 0.0825 > \alpha = 0.05$; do not reject the null hypothesis.
There is not sufficient evidence to support the IDFA's claim that male teenagers consume more than the recommended daily amount of calcium.

17. (a) $t_0 = -15.119 < -t_{0.01} = -1.282$; reject the null hypothesis.
(b) $P\text{-value} < 0.0001$; the P-value is the area under the t-distribution with 699 degrees of freedom to the left of the test statistic, $t_0 = -15.119$. It is less than 0.0001 square unit.

19. (a) Hypotheses: $H_0: \mu = 40.7$ years, $H_1: \mu \neq 40.7$ years
Classical approach: $t_0 = -1.061$ is between $-t_{0.05} = -1.696$ and $t_{0.05} = 1.696$ with 31 degrees of freedom; do not reject the null hypothesis.
P-value Approach: $P\text{-value} = 0.2970 > \alpha = 0.05$; do not reject the null hypothesis.
There is not sufficient evidence to support the sociologist's claim that mean age of a death row inmate is different from 40.7 years of age.
(b) Lower bound: 35.44, upper bound: 42.36; do not reject H_0 since 40.4 years is contained in the 95% confidence interval.

21. (a) Yes, the conditions are satisfied. The data appear to be normally distributed, and there are no outliers.
(b) $t_0 = 0.778$ is between $-t_{0.05} = -1.796$ and $t_{0.05} = 1.796$ with 11 degrees of freedom; do not reject the null hypothesis.
(c) $P\text{-value} = 0.4529$; the area under the t-distribution with 11 degrees of freedom to the right of the test statistic $t_0 = 1.796$ is 0.4529.

23. (a) Yes, the conditions are satisfied. The data appear to be normally distributed, and there are no outliers.
(b) Hypotheses: $H_0: \mu = 84.3$ seconds, $H_1: \mu < 84.3$ seconds
Classical approach: $t_0 = -1.310 > -t_{0.10} = -1.383$ with 9 degrees of freedom; do not reject the null hypothesis.
P-value approach: $P\text{-value} = 0.1113 > \alpha = 0.10$; do not reject the null hypothesis.
There is not sufficient evidence to support the manager's claim that the new system decreases the mean wait time to less than 84.3 seconds.

25. (a) The data appear to be normally distributed, and there are no outliers.

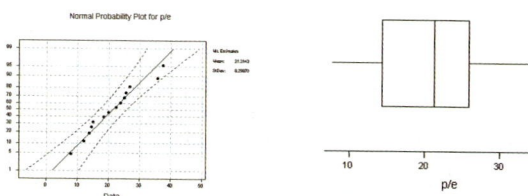

(b) Hypotheses: $H_0: \mu = 22.0$, $H_1: \mu < 22.0$
Classical approach: $t_0 = -0.2980 > -t_{0.05} = -1.771$ with 13 degrees of freedom; do not reject the null hypothesis.
P-value approach: $P\text{-value} = 0.3852 > \alpha = 0.05$; do not reject the null hypothesis.
There is not sufficient evidence to support the stock analyst's claim that the mean P/E ratio of the S&P 500 Index has decreased from its December 2000 level.
(c) The area under the t-distribution with 13 degrees of freedom to the left of the test statistic $t_0 = -0.2980$ is 0.3852 (the P-value).

27. (a) Yes, the conditions are satisfied. The data appear to be normally distributed, and there are no outliers.
(b) Hypotheses: $H_0: \mu = 130$, $H_1: \mu < 130$; the P-value $= 0.088$. The nursing student will not reject the null hypothesis at the $\alpha = 0.05$ level of significance. There is not sufficient evidence to support the student's claim that male surgical patients have a mean systolic blood pressure less than 130.

29. Answers will vary.

31. (a) Answers will vary. Experiment does not meet the requirements for hypothesis testing.
(b) Answers will vary. Expect 5 samples to result in a rejection of the null hypothesis at $\alpha = 0.05$.

10.4 Assess Your Understanding (page 550)

1. A hypothesis about a population proportion can be tested if the sample is obtained by simple random sampling and $np(1 - p) \geq 10$, with $n \leq 0.05N$ (the sample size is no more than 5% of the population size).

3. $np_0(1 - p_0) = 42 > 10$
(a) Classical approach: $Z_0 = 2.31 > z_{0.05} = 1.645$; reject the null hypothesis.
(b) P-value approach: $P\text{-value} = 0.0104 < \alpha = 0.05$; reject the null hypothesis. There is sufficient evidence at the $\alpha = 0.05$ level of significance to reject the null hypothesis.

5. $np_0(1 - p_0) = 37.1 > 10$
  (a) Classical approach: $Z_0 = -0.74 > -z_{0.10} = -1.28$; do not reject the null hypothesis.
  (b) P-value approach: P-value $= 0.2296 > \alpha = 0.10$; do not reject the null hypothesis. There is not sufficient evidence at the $\alpha = 0.10$ level of significance to reject the null hypothesis.

7. $np_0(1 - p_0) = 45 > 10$
  (a) Classical approach: $Z_0 = -1.49$ is between $-z_{0.025} = -1.96$ and $z_{0.025} = 1.96$; do not reject the null hypothesis.
  (b) P-value approach: P-value $= 0.1362 > \alpha = 0.05$; do not reject the null hypothesis. There is not sufficient evidence at the $\alpha = 0.05$ level of significance to reject the null hypothesis.

9. $np_0(1 - p_0) = 16.1 > 10$
  Hypotheses: $H_0: p = 0.019$, $H_1: p > 0.019$
  (a) Classical approach: $Z_0 = 0.65 < z_{0.01} = 2.33$; do not reject the null hypothesis.
  (b) P-value approach: P-value $= 0.2578 > \alpha = 0.01$; do not reject the null hypothesis. There is not sufficient evidence at the $\alpha = 0.01$ level of significance to support the claim that more than 1.9% of Lipitor users experience flulike symptoms as a side effect.

11. $np_0(1 - p_0) = 154.8 > 10$
  Hypotheses: $H_0: p = 0.81$, $H_1: p \neq 0.81$
  (a) Classical approach: $Z_0 = 1.62$ is between $-z_{0.025} = -1.96$ and $z_{0.025} = 1.96$; do not reject the null hypothesis.
  (b) P-value approach: P-value $= 0.1052 > \alpha = 0.05$; do not reject the null hypothesis. There is not sufficient evidence at the $\alpha = 0.05$ level of significance to support the claim that the proportion of adults who have read at least one book in the last year is different from 1990.

13. $np_0(1 - p_0) = 264.3 > 10$
  Hypotheses: $H_0: p = 0.38$, $H_1: p < 0.38$
  (a) Classical approach: $Z_0 = -5.52 < -z_{0.05} = -1.645$; reject the null hypothesis.
  (b) P-value approach: P-value $< 0.0001 < \alpha = 0.05$; reject the null hypothesis. There is sufficient evidence at the $\alpha = 0.05$ level of significance to support the claim that the proportion of families with children under the age of 18 who eat together 7 nights a week has decreased.

15. $np_0(1 - p_0) = 58.8 > 10$
  Hypotheses: $H_0: p = 0.30$, $H_1: p \neq 0.30$
  (a) Classical approach: $Z_0 = 3.65 > z_{0.05} = 1.645$; reject the null hypothesis.
  (b) P-value approach: P-value $= 0.0002 < \alpha = 0.10$; reject the null hypothesis. There is sufficient evidence at the $\alpha = 0.10$ level of significance to support the claim that the percentage of employed adults who are satisfied with their chances for promotion is significantly different from the percentage in 1998.

17. $np_0(1 - p_0) = 241 > 10$
  Hypotheses: $H_0: p = 0.40$, $H_1: p < 0.40$
  (a) Classical approach: $Z_0 = -1.91 < -z_{0.05} = -1.645$; reject the null hypothesis.
  (b) P-value approach: P-value $= 0.0281 < \alpha = 0.05$; reject the null hypothesis. There is sufficient evidence at the $\alpha = 0.05$ level of significance to support the claim that the proportion of 2005 adults aged 18 years or older having a great deal of confidence in the public schools is significantly lower than the 1995 proportion.

19. $np_0(1 - p_0) = 35 > 10$
  Hypotheses: $H_0: p = 0.37$, $H_1: p < 0.37$
  (a) Classical approach: $Z_0 = -0.25 > -z_{0.05} = -1.645$; do not reject the null hypothesis.
  (b) P-value approach: P-value $= 0.4013 > \alpha = 0.05$; do not reject the null hypothesis. There is not sufficient evidence at the $\alpha = 0.05$ level of significance to support the veterinarian's claim that less than 37% of pet owners speaks to their pets on the answering machine or telephone.

21. Hypotheses: $H_0: p = 0.04$, $H_1: p < 0.04$
  $np_0(1 - p_0) = 4.6 < 10$
  P-value $= 0.2887 > \alpha = 0.05$; do not reject the null hypothesis. There is not sufficient evidence at the $\alpha = 0.05$ level of significance to support the obstetrician's claim that less than 4% of mothers smokes 21 or more cigarettes during pregnancy.

23. Hypotheses: $H_0: p = 0.096$, $H_1: p > 0.096$
  $np_0(1 - p_0) = 6.9 < 10$
  P-value $= 0.0410 < \alpha = 0.10$; reject the null hypothesis. There is sufficient evidence at the $\alpha = 0.10$ level of significance to support the urban economist's claim that the percentage of Californians who spend more than 60 minutes traveling to work has increased since 2000.

25. (a) Answers will vary.
  (b) $H_0: p = 0.48$, $H_1: p < 0.48$; $np_0(1 - p_0) = 274.6 > 10$; $Z_0 = -1.66$; P-value $= 0.0918$. The results of the poll do not indicate that the proportion declined significantly.

10.5 Assess Your Understanding (page 556)

1. To test a claim regarding a population standard deviation, we must have a simple random sample taken from a normally distributed population.

3. (a) $\chi_0^2 = 20.496$ **(b)** $\chi_{0.95}^2 = 13.091$

(c)

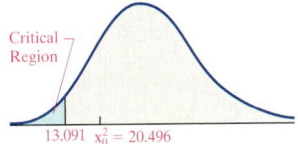

(d) Do not reject H_0, because $\chi_0^2 > \chi_{0.95}^2$.

5. (a) $\chi_0^2 = 30.222$ **(b)** $\chi_{0.1}^2 = 24.769$

(c)

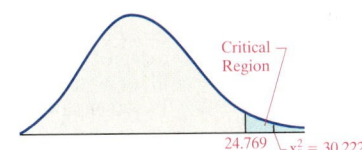

(d) Reject H_0, because $\chi_0^2 > \chi_{0.1}^2$.

7. (a) $\chi_0^2 = 13.707$
  (b) $\chi_{0.975}^2 = 3.816$; $\chi_{0.025}^2 = 21.920$

(c)

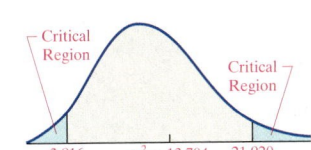

(d) Do not reject H_0, because $\chi_{0.975}^2 < \chi_0^2 < \chi_{0.025}^2$.

9. Hypotheses: $H_0: \sigma = 0.04$, $H_1: \sigma > 0.04$
  $\chi_0^2 = 13.590 < \chi_{0.05}^2 = 36.415$; do not reject the null hypothesis at the $\alpha = 0.05$ level of significance.

11. Hypotheses: $H_0: \sigma = 0.004$, $H_1: \sigma < 0.004$
  $\chi_0^2 = 9.375 < \chi_{0.99}^2 = 10.856$; reject the null hypothesis at the $\alpha = 0.05$ level of significance.
  There is sufficient evidence at the $\alpha = 0.01$ level of significance for the manager to conclude that the standard deviation has decreased.

13. Hypotheses: $H_0: \sigma = 0.2$, $H_1: \sigma \neq 0.2$
  $\chi_0^2 = 13.128$ is between $\chi_{0.975}^2 = 8.231$ and $\chi_{0.025}^2 = 31.526$; do not reject the null hypothesis at the $\alpha = 0.05$ level of significance.
  There is not sufficient evidence at the $\alpha = 0.05$ level of significance to support the claim that the standard deviation is not 0.2.

15. Hypotheses: $H_0: \sigma = 0.004$, $H_1: \sigma > 0.004$
  $\chi_0^2 = 13.614 < \chi_{0.01}^2 = 24.725$; do not reject the null hypothesis at the $\alpha = 0.01$ level of significance.
  There is not sufficient evidence at the $\alpha = 0.01$ level of significance to support the claim that Maxfli XS golf balls do not conform to USGA size requirements.

17. Hypotheses: $H_0: \sigma = 8.3$, $H_1: \sigma < 8.3$
$\chi_0^2 = 15.639 < \chi_{0.90}^2 = 15.659$; reject the null hypothesis at the $\alpha = 0.10$ level of significance.
There is sufficient evidence at the $\alpha = 0.10$ level of significance to support the claim that Iverson is a more consistent player than other shooting guards in the NBA.

19. (a)

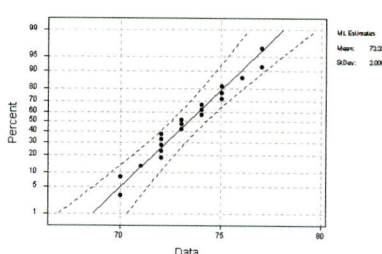

Normal Probability Plot for Height

(b) $s = 2.059$ inches
(c) Hypotheses: $H_0: \sigma = 2.9$, $H_1: \sigma < 2.9$
$\chi_0^2 = 9.5779 > \chi_{0.99}^2 = 7.633$; do not reject the null hypothesis at the $\alpha = 0.01$ level of significance.
There is not sufficient evidence at the $\alpha = 0.01$ level of significance to support the claim that the standard deviation of the heights baseball players is less than 2.9 inches.

21. P-value $= 0.9554$

10.6 Assess Your Understanding (page 559)

1. To test a claim about a population mean, we must use a simple random sample that is either drawn from a normally distributed population, or the sample must have at least 30 subjects. Assuming the prerequisites are met, use a normal model to test a claim if the population standard deviation (or variance) is known, and use a Student's t-distribution if the population standard deviation (or variance) is not known.

3. Hypotheses: $H_0: \mu = 70$, $H_1: \mu < 70$
Classical approach: $Z_0 = -1.87 < -z_{0.10} = -1.28$; reject the null hypothesis.
P-value approach: P-value $= 0.0307 < \alpha = 0.10$; reject the null hypothesis.
There is sufficient evidence at the $\alpha = 0.10$ level of significance to support the claim that the population mean is less than 70.

5. Hypotheses: $H_0: p = 0.5$, $H_1: p > 0.5$; $np_0(1 - p_0) = 50 > 10$
Classical approach: $Z_0 = 2.12 > z_{0.05} = 1.645$; reject the null hypothesis.
P-value approach: P-value $= 0.0170 < \alpha = 0.05$; reject the null hypothesis.
There is sufficient evidence at the $\alpha = 0.05$ level of significance to support the claim that more than half the individuals with valid drivers licenses drives an American-made automobile.

7. Hypotheses: $H_0: \mu = 25$, $H_1: \mu \neq 25$
Classical approach: $t_0 = -0.738$ is between $-t_{0.005} = -2.977$ and $t_{0.005} = 2.977$ with 14 degrees of freedom; do not reject the null hypothesis.
P-value approach: P-value $= 0.4729 > \alpha = 0.01$; do not reject the null hypothesis.
There is not sufficient evidence at the $\alpha = 0.01$ level of significance to support the claim that the population mean is different from 25.

9. Hypotheses: $H_0: \sigma^2 = 10$, $H_1: \sigma^2 > 10$
Classical approach: $\chi_0^2 = 20.55 < \chi_{0.05}^2 = 24.996$; do not reject the null hypothesis.
P-value approach: P-value $= 0.1518 > \alpha = 0.05$; do not reject the null hypothesis.
There is not sufficient evidence at the $\alpha = 0.05$ level of significance to support the claim that the population variance is greater than 10.

11. Hypotheses: $H_0: \mu = 100$, $H_1: \mu > 100$
Classical approach: $t_0 = 3.003 > t_{0.05} = 1.685$; reject the null hypothesis.
P-value approach: P-value $= 0.0023 < \alpha = 0.05$; reject the null hypothesis.
There is sufficient evidence at the $\alpha = 0.05$ level of significance to support the claim that the population mean is greater than 100.

13. (a) Hypotheses: $H_0: \mu = 2.5$, $H_1: \mu \neq 2.5$
Classical approach: $Z_0 = 2.64 > z_{0.05} = 1.645$; reject the null hypothesis.
P-value approach: P-value $= 0.0082 < \alpha = 0.10$; reject the null hypothesis.
There is sufficient evidence at the $\alpha = 0.10$ level of significance to support the claim that the mean ideal number of children has changed since 1985.
(b) Answers will vary.

15. Hypotheses: $H_0: \sigma = 7000$, $H_1: \sigma > 7000$
Classical approach: $\chi_0^2 = 21.81 < \chi_{0.01}^2 = 36.191$; do not reject the null hypothesis.
P-value approach: P-value $= 0.2938 > \alpha = 0.01$; do not reject the null hypothesis.
There is not sufficient evidence at the $\alpha = 0.01$ level of significance to support the claim that the standard deviation of yield strength is not acceptable.

17. Hypotheses: $H_0: \mu = 0.063$, $H_1: \mu < 0.063$
Classical approach: $t_0 = -0.9147 > -t_{0.05} = -1.684$; do not reject the null hypothesis.
P-value approach: P-value $= 0.1829 > \alpha = 0.05$; do not reject the null hypothesis.
There is not sufficient evidence at the $\alpha = 0.05$ level of significance to support the agent's claim that interest rates are lower than in 2001.

19. Hypotheses: $H_0: \mu = 774$, $H_1: \mu \neq 774$
Classical approach: $t_0 = -4.776 < -t_{0.005} = -2.728$; reject the null hypothesis.
P-value approach: P-value $< 0.0001 < \alpha = 0.01$; reject the null hypothesis.
There is sufficient evidence at the $\alpha = 0.01$ level of significance to support the salesman's claim that the mean expenditure for auto insurance is different from the 2002 amount.

21. Hypotheses: $H_0: \mu = 8$, $H_1: \mu < 8$
Classical approach: $t = -1.443$, $t_{0.01} = -2.403$; do not reject the null hypothesis.
P-value approach: P-value $= 0.0777$; do not reject the null hypothesis.
There is not, sufficient evidence at the $\alpha = 0.01$ level of significance to support the claim that the mean wait time is less than eight minutes.

10.7 Assess Your Understanding (page 565)

1. A Type II error is made when we fail to reject the null hypothesis and the alternative is true.

3. As the true mean gets closer to the hypothesized mean, the power of the test decreases.

5. (a) If the null hypothesis was not rejected and the true mean is less than 50, a Type II error has been made.
(b) $\beta = 0.7967$; power of the test: $1 - \beta = 0.2033$
(c) $\beta = 0.9066$; power of the test: $1 - \beta = 0.0934$

7. (a) If the null hypothesis was not rejected and the true mean is different from 100, a Type II error has been made.
(b) $\beta = 0.6985$; power of the test: $1 - \beta = 0.3015$
(c) $\beta = 0.9463$; power of the test: $1 - \beta = 0.0537$

9. (a) If the null hypothesis was not rejected and the true mean is less than 20, a Type II error has been made.
(b) $\beta = 0.0212$; power of the test: $1 - \beta = 0.9788$
(c) $\beta = 0.6950$; power of the test: $1 - \beta = 0.3050$

11. (a) If the null hypothesis was not rejected and the true mean is different from 105, a Type II error has been made.
(b) $\beta = 0.5273$; power of the test: $1 - \beta = 0.4727$
(c) $\beta = 0.7562$; power of the test: $1 - \beta = 0.2438$

13. (a) If the null hypothesis was not rejected and the true mean height of a woman is greater than 63.7 inches, a Type II error has been made.
(b) $\beta = 0.9082$; power of the test: $1 - \beta = 0.0198$
(c) $\beta = 0.2327$; power of the test: $1 - \beta = 0.7673$

15. (a) If the null hypothesis was not rejected and the true mean SAT score for students whose first language is not English is less than 515, a Type II error has been made.
(b) $\beta = 0.9582$; power of the test: $1 - \beta = 0.0418$
(c) $\beta = 0.8238$; power of the test: $1 - \beta = 0.1762$

17. $\beta = 0.9345$; power of the test: $1 - \beta = 0.0655$

19. (a) $\beta = 0.1762$
   (b) Power of the test: $1 - \beta = 0.8238$

Chapter Review Exercises   (page 568)

1. (a) $H_0: \mu = \$4277$, $H_1: \mu > \$4277$
   (b) If the null hypothesis is rejected, but H_0 is true, a Type I error is made.
   (c) A Type II error is made when the null hypothesis is not rejected but H_1 is true.
   (d) There is not sufficient evidence at the α level of significance to support the credit counselor's claim that the mean outstanding credit card debt per cardholder is more than the 2000 amount.
   (e) There is sufficient evidence at the α level of significance to support the credit counselor's claim that the mean outstanding credit card debt per cardholder is more than the 2000 amount.

3. The probability of a Type I error is 0.05.

5. (a) $Z = -1.08$ **(b)** $-z_{0.05} = -1.645$

(c)
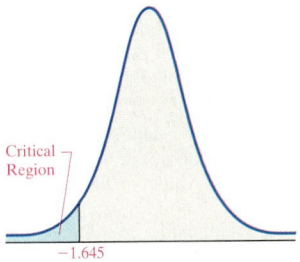

   (d) No, because $Z > -z_{0.05}$.   **(e)** P-value $= 0.1401$

7. (a) $t_0 = -1.506$
   (b) $-t_{0.01} = -2.624$; $t_{0.01} = 2.624$

(c)
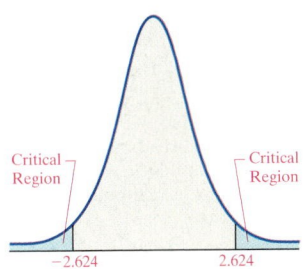

   (d) No, because $-t_{0.01} = -2.624 < t_0 = -1.506 < t_{0.01} = 2.624$.
   (e) P-value $= 0.1543$

9. Hypotheses: $H_0: p = 0.6$, $H_1: p > 0.6$; $np_0(1 - p_0) = 60 > 10$
   Classical approach: $Z_0 = 1.94 > z_{0.05} = 1.645$; reject the null hypothesis.
   P-value approach: P-value $= 0.0262 < \alpha = 0.05$; reject the null hypothesis.
   There is sufficient evidence at the $\alpha = 0.05$ level of significance to support the claim that $p > 0.6$.

11. (a) $\chi_0^2 = 15.095$
   (b) $\chi_{0.975}^2 = 7.564$; $\chi_{0.025}^2 = 30.191$

(c)
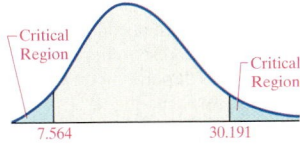

   (d) Do not reject H_0, because $\chi_{0.975}^2 < \chi_0^2 < \chi_{0.025}^2$.

13. Hypotheses: $H_0: \mu = 0.875$, $H_1: \mu > 0.875$
   Classical approach: $t_0 = 1.2 < t_{0.05} = 1.690$; do not reject the null hypothesis.
   P-value approach: P-value $= 0.1191 > \alpha = 0.05$; do not reject the null hypothesis.
   There is not sufficient evidence at the $\alpha = 0.05$ level of significance to support the claim that the mean distance between retaining rings is longer than 0.875 inch.

15. Hypotheses: $H_0: \mu = 474$, $H_1: \mu > 474$
   Classical approach: $Z_0 = 4.46 > z_{0.01} = 2.33$; reject the null hypothesis.
   P-value approach: P-value $< 0.0001 < \alpha = 0.01$; reject the null hypothesis.
   There is sufficient evidence at the $\alpha = 0.01$ level of significance to support the instructor's claim that students who use calculators frequently score better on the SAT math portion than those who use a calculator infrequently.

17. (a) Hypotheses: $H_0: \mu = 300$, $H_1: \mu > 300$
   Classical approach: $t_0 = 1.528 < t_{0.05} = 1.660$; do not reject the null hypothesis.
   P-value approach: P-value $= 0.0636 > \alpha = 0.05$; do not reject the null hypothesis.
   There is not sufficient evidence at the $\alpha = 0.05$ level of significance to support the claim that mean cholesterol consumption of 20- to 39-year-old males is greater than 300 mg.
   (b) Type I error: The nutritionist rejects the null hypothesis that the mean cholesterol consumption is 300 mg when, in fact, the mean consumption is 300 mg. Type II error: The nutritionist does not reject the null hypothesis that the mean cholesterol consumption is 300 mg when, in fact, the mean consumption is more than 300 mg.
   (c) The probability of making a Type I error is $\alpha = 0.05$.

19. (a) Yes, the conditions are satisfied. The data appear to be normally distributed, and there are no outliers.
   (b) Hypotheses: $H_0: \mu = 4.61$, $H_1: \mu < 4.61$
   Classical approach: $Z_0 = -2.48 < -z_{0.01} = -2.33$; reject the null hypothesis.
   P-value approach: P-value $= 0.0066 < \alpha = 0.01$; reject the null hypothesis.
   There is sufficient evidence at the $\alpha = 0.01$ level of significance to support the biologist's claim that the acidity of the rain has increased since 1990.

21. Hypotheses: $H_0: p = 0.56$, $H_1: p > 0.56$; $np_0(1 - p_0) = 73.9 > 10$
   Classical approach: $Z_0 = 0.24 < z_{0.01} = 2.33$; do not reject the null hypothesis.
   P-value approach: P-value $= 0.4052 > \alpha = 0.01$; do not reject the null hypothesis.
   There is not sufficient evidence at the $\alpha = 0.01$ level of significance to support the claim that the percentage of cases of tuberculosis of foreign-born residents has increased.

23. Hypotheses: $H_0: \sigma = 104$, $H_1: \sigma < 104$
   Classical approach: $\chi_0^2 = 25.5148 < \chi_{0.10}^2 = 30.813$; do not reject the null hypothesis.
   P-value approach: P-value $= 0.2733 > \alpha = 0.10$; do not reject the null hypothesis.
   There is not sufficient evidence at the $\alpha = 0.10$ level of significance to support the claim that the standard deviation of the students from low-income households is higher than that of higher-income households.

25. (a) $\beta = 0.4443$
   (b) Power of the test: $1 - \beta = 0.5557$

27. Hypotheses: $H_0: p = 0.4$, $H_1: p > 0.4$
   $np_0(1 - p_0) = 9.6 < 10$
   P-value $= 0.6885 > \alpha = 0.05$; do not reject the null hypothesis.
   There is not sufficient evidence at the $\alpha = 0.05$ level of significance to support the researcher's claim that the proportion of adolescents who prays daily has increased.

29. Hypotheses: $H_0: \mu = 73.2$, $H_1: \mu < 73.2$
   Classical approach: $t_0 = -2.018 < -t_{0.05} = -1.645$; reject the null hypothesis.
   P-value approach: P-value $= 0.0218 < \alpha = 0.05$; reject the null hypothesis.
   There is sufficient evidence at the $\alpha = 0.05$ level of significance to support the coordinator's claim that the mean test scores decreased.

CHAPTER 11

11.1 Assess Your Understanding (page 582)

1. Independent

3. $H_1: u_d < 0$ where $d_j = x_j - y_j$ **5.** Dependent

7. Independent   **9.** Independent

11. (a)

Observation	1	2	3	4	5	6	7
d_t	−0.5	1	−3.3	−3.7	0.5	−2.4	−2.9

(b) $\bar{d} = -1.614$; $s_d = 1.915$
(c) Classical approach: $t_0 = -2.230 < -t_{0.05} = -1.943$; reject the null hypothesis
P-value approach: P-value $= 0.0336 < \alpha = 0.05$; reject the null hypothesis
There is sufficient evidence at the $\alpha = 0.05$ level of significance to reject the null hypothesis that $\mu_d = 0$.
(d) We can be 95% confident that the mean difference is between −3.385 and 0.157.

13. (a) This is matched-pairs data because two measurements (A and B) are taken on the same round.
(b) Hypotheses: $H_0: \mu_d = 0$, $H_1: \mu_d \neq 0$
Classical approach: $t_0 = 0.8517$; do not reject the null hypothesis since $-t_{0.005} = -3.106 < 0.8517 < t_{0.005} = 3.106$
P-value approach: P-value $= 0.4125 > \alpha = 0.01$; do not reject the null hypothesis
There is not sufficient evidence at the $\alpha = 0.01$ level of significance to support the claim that there is a difference in the measurements of velocity between device A and device B.
(c) Lower bound: −0.309, upper bound: 0.543. We are 99% confident that the mean difference in measurement is between −0.309 and 0.543 feet per second.
(d)

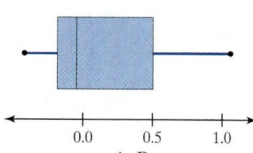

A–B

15. (a) Answers may vary.
(b) Hypotheses: $H_0: \mu_d = 0$, $H_1: \mu_d > 0$
Classical approach: $t_0 = 2.3837 > t_{0.05} = 1.895$; reject the null hypothesis
P-value approach: P-value $= 0.0243 < \alpha = 0.05$; reject the null hypothesis
There is sufficient evidence at the $\alpha = 0.05$ level of significance to support the claim that there has been an improvement in the clarity of the water in the 5-year period.
(c) We can be 95% confident that the mean difference in clarity between the initial measurements and those taken 5 years later is between 0.04 inch and 10.21 inches.
(d)

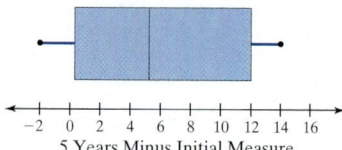

5 Years Minus Initial Measure

17. Hypotheses: $H_0: \mu_d = 0$, $H_1: \mu_d > 0$
Classical approach: $t_0 = 0.398 < t_{0.10} = 1.356$; do not reject the null hypothesis
P-value approach: P-value $= 0.3508 > \alpha = 0.10$; do not reject the null hypothesis
There is not sufficient evidence at the $\alpha = 0.10$ level of significance to support the researcher's claim that sons are taller than their fathers.

19. Hypotheses: $H_0: \mu_d = 0$, $H_1: \mu_d \neq 0$
Classical approach: $t_0 = 2.6 > t_{0.025} = 2.306$; reject the null hypothesis
P-value approach: P-value $= 0.0285 < \alpha = 0.05$; reject the null hypothesis

There is sufficient evidence at the $\alpha = 0.05$ level of significance to support the manufacturer's claim that the two different indenters result in different measurements.

21. Lower bound: 0.69, upper bound: 1.24. We can be 95% confident that the mean difference in reaction time when teenagers are driving impaired from when driving normally is between 0.69 second and 1.24 seconds.

23. (a), (b) Answers will vary.
(c) Neither variable is normally distributed.
(d) The difference in mileage appears to be approximately normally distributed.
(e) Classical approach: $t_0 = 1.14 < t_{0.05} = 1.812$; do not reject the null hypothesis
P-value approach: P-value $= 0.141 > \alpha = 0.05$; do not reject the null hypothesis
There is not sufficient evidence at the $\alpha = 0.05$ level of significance to support the claim that cars get better mileage when using 92-octane gasoline than when using 87-octane gasoline.

11.2 Assess Your Understanding (page 595)

1. To test a hypothesis regarding the difference of two means, we need two independent random samples that are drawn from normally distributed populations. If the populations are not normally distributed each sample must be large ($n_1 \geq 30$ and $n_2 \geq 30$).

3. (a) $H_0: \mu_1 = \mu_2$, $H_1: \mu_1 \neq \mu_2$
Classical approach: $t_0 = 0.8783$ is between $-t_{0.025} = -2.145$ and $t_{0.025} = 2.145$; do not reject H_0
P-value approach: P-value $= 0.3767 > \alpha = 0.05$; do not reject H_0
There is not sufficient evidence at the $\alpha = 0.05$ level of significance to support the claim that the population means are different.
(b) Lower bound: −1.526, upper bound: 3.726

5. (a) $H_0: \mu_1 = \mu_2$, $H_1: \mu_1 > \mu_2$
Classical approach: $t_0 = 3.081 > t_{0.10} = 1.333$; reject H_0
P-value approach: P-value $= 0.0024 < \alpha = 0.10$; reject H_0
There is sufficient evidence at the $\alpha = 0.10$ level of significance to support the claim that $\mu_1 > \mu_2$.
(b) Lower bound: 3.569, upper bound: 12.831

7. (a) $H_0: \mu_1 = \mu_2$; $H_1: \mu_1 < \mu_2$
Classical approach: $t_0 = -3.158 < -t_{0.02} = -2.172$; reject H_0
P-value approach: P-value $= 0.0013 < \alpha = 0.02$; reject H_0
There is sufficient evidence at the $\alpha = 0.02$ level of significance to support the claim that $\mu_1 < \mu_2$.
(b) Lower bound: −16.652, upper bound: −4.948

9. (a) $H_0: \mu_1 = \mu_2$, $H_1: \mu_1 > \mu_2$
Classical approach: $t_0 = 2.849 > t_{0.01} = 2.403$; reject H_0
P-value approach: P-value $= 0.0026 < \alpha = 0.01$; reject H_0
There is sufficient evidence at the $\alpha = 0.01$ level of significance to support the claim that mean improvement in the treatment group was greater than the mean improvement in the control group. The drug appears to be effective in improving the Young–Mania Rating Scale score.
(b) Lower bound: 1.98, upper bound: 11.42. The researchers are 95% confident that the mean Young–Mania Rating Scale score for the treatment group is between 1.98 and 11.42 points higher than that of the control group.

11. (a) This is an observational study.
(b) Answers will vary.
(c) $H_0: \mu_{\text{arrival}} = \mu_{\text{departure}}$, $H_1: \mu_{\text{arrival}} \neq \mu_{\text{departure}}$
Classical approach: $t_0 = 0.8456$ is between $-t_{0.025} = -2.032$ and $t_{0.025} = 2.032$; do not reject H_0
P-value approach: P-value $= 0.4013 > \alpha = 0.05$; do not reject H_0
There is not sufficient evidence at the $\alpha = 0.05$ level of significance to support the researcher's claim that travelers walk at different speeds depending on whether they are arriving or departing an airport.
(d) Lower bound: −12.63, upper bound: 30.63. We can be 95% confident that the mean difference in speed walked between passengers arriving and departing an airport is between −12.63 and 30.63 feet per minute.

13. (a) Yes, we can treat each sample as a simple random sample of all mixtures of each type. The samples were obtained independently. We are told that a normal probability plot indicates that the data could come from a population that is normal, with no outliers.

(b) H_0: $\mu_{67-0-301} = \mu_{67-0-400}$, H_1: $\mu_{67-0-301} < \mu_{67-0-400}$
Classical approach: $t_0 = -3.804 < -t_{0.05} = -1.860$; reject H_0
P-value approach: P-value $= 0.0007 < \alpha = 0.05$; reject H_0
There is sufficient evidence at the $\alpha = 0.05$ level of significance to support the engineer's claim that mixture 67-0-400 is stronger than mixture 67-0-301.

(c) Lower bound: 416, upper bound: 1212. We are 90% confident that the mean strength of mixture 67-0-400 is between 416 and 1212 psi stronger than the mean strength of mixture 67-0-301.

(d)

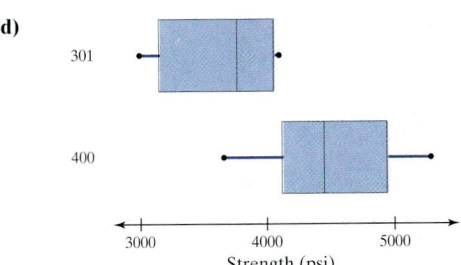

15. (a) H_0: $\mu_{carpet} = \mu_{nocarpet}$, H_1: $\mu_{carpet} > \mu_{nocarpet}$
Classical approach: $t_0 = 0.954 < t_{0.05} = 1.895$; do not reject H_0
P-value approach: P-value $= 0.1780 > \alpha = 0.05$; do not reject H_0
There is not sufficient evidence at the $\alpha = 0.05$ level of significance to support the claim that carpeted rooms have more bacteria than uncarpeted rooms.

(b) Lower bound: -2.09, upper bound: 4.91. We are 95% confident that the mean difference in the number of bacteria per cubic foot in a carpeted room versus that in an uncarpeted room is between -2.08 and 4.91.

17. H_0: $\mu_{AL} = \mu_{NL}$, H_1: $\mu_{AL} > \mu_{NL}$
Classical approach: $t_0 = 2.1356 > t_{0.05} = 1.699$; reject H_0
P-value approach: P-value $= 0.0187 < \alpha = 0.05$; reject H_0
There is sufficient evidence at the $\alpha = 0.05$ level of significance to support the claim that games played with a designated hitter result in more runs.

19. Lower bound: 0.71, upper bound: 2.33. We can be 90% confident that the mean difference in daily leisure time between adults without children and those with children is between 0.71 and 2.33 hours. Since the confidence interval does not include zero, we can conclude that there is a significant difference in the leisure time of adults without children and those with children.

21. (a) H_0: $\mu_{men} = \mu_{women}$ versus H_1: $\mu_{men} < \mu_{women}$
(b) P-value $= 0.0051$. Because P-value $< \alpha$, we reject the null hypothesis. There is sufficient evidence at the $\alpha = 0.01$ level of significance to support the claim that the mean step pulse of men is lower than the mean step pulse of women.
(c) Lower bound: -10.7, upper bound: -1.5. We are 95% confident that the mean step pulse of men is between 1.5 and 10.7 beats per minute lower than the mean step pulse of women.

11.3 Assess Your Understanding (page 609)

1. A pooled estimate of p is the best point estimate of the common population proportion p. However, when finding a confidence interval, the sample proportions are not pooled because no assumption about their equality is made.

3. (a) H_0: $p_1 = p_2$ versus H_1: $p_1 > p_2$ **(b)** $z_0 = 3.07$
(c) $z_{0.05} = 1.645$
(d) P-value $= 0.0011$. Because $z_0 > z_{0.05}$ (or P-value $< \alpha$), we reject the null hypothesis. There is sufficient evidence to support the claim that $p_1 > p_2$.

5. (a) H_0: $p_1 = p_2$ versus H_1: $p_1 \neq p_2$ **(b)** $z_0 = -0.37$
(c) $-z_{0.025} = -1.96$; $z_{0.025} = 1.96$
(d) P-value $= 0.7114$. Because $-z_{0.025} < z_0 < z_{0.05}$ (or P-value $> \alpha$), we do not reject the null hypothesis. There is not sufficient evidence to support the claim that $p_1 \neq p_2$.

7. Lower bound: -0.075, upper bound: 0.015

9. Lower bound: -0.063, upper bound: 0.043

11. (a) Each sample can be thought of as a simple random sample; $n_1\hat{p}_1(1 - \hat{p}_1) = 91 \geq 10$ and $n_2\hat{p}_2(1 - \hat{p}_2) = 60 \geq 10$; and each sample is less than 5% of the population size.
H_0: $p_1 = p_2$; H_1: $p_1 > p_2$
Classical approach: $z_0 = 2.20 > z_{0.05} = 1.645$; reject H_0
P-value approach: P-value $= 0.0139 < \alpha = 0.05$; reject H_0
There is sufficient evidence at the $\alpha = 0.05$ level of significance to support the claim that a higher proportion of subjects in the treatment group (taking Prevnar) experienced fever as a side effect than in the control (placebo) group.

(b) Lower bound: 0.01, upper bound: 0.07. We are 90% confident that the difference in the proportion of subjects who experience a fever as a side effect between the experimental and control groups is between 0.01 and 0.07.

13. (a) Each sample is a simple random sample; $n_1\hat{p}_1(1 - \hat{p}_1) = 73 \geq 10$ and $n_2\hat{p}_2(1 - \hat{p}_2) = 76 \geq 10$; and each sample is less than 5% of the population size.
H_0: $p_1 = p_2$, H_1: $p_1 > p_2$
Classical approach: $z_0 = 0.98 < z_{0.10} = 1.28$; do not reject H_0
P-value approach: P-value $= 0.1635 > \alpha = 0.10$; do not reject H_0
There is not sufficient evidence at the $\alpha = 0.1$ level of significance to support the claim that a higher proportion of individuals with at most an 8th-grade education consume too much cholesterol than of those with some college education.

(b) Lower bound: -0.04, upper bound: 0.11. We are 95% confident that the difference in the proportion of individuals with at most an 8th-grade education and individuals with some college education who consume too much cholesterol is between -0.04 and 0.11.

15. Provided we choose $\alpha \geq 0.01$, we can say the proportion of American adults who smoked at least one cigarette in the past week has decreased, because when testing the hypotheses H_0: $p_1 = p_2$ versus H_1: $p_1 < p_2$, where p_1 is the proportion of adults who smoked in 2005 and p_2 is the proportion of adults who smoked in 1990, we get a P-value $= 0.0041$ ($z_0 = -2.64$).

17. (a) Each sample is a simple random sample; $n_1\hat{p}_1(1 - \hat{p}_1) = 33 \geq 10$ and $n_2\hat{p}_2(1 - \hat{p}_2) = 115 \geq 10$; and each sample is less than 5% of the population size.
H_0: $p_1 = p_2$, H_1: $p_1 < p_2$
Classical approach: $z_0 = -6.74 < -z_{0.01} = -2.33$; reject H_0
P-value approach: P-value $< 0.0001 < \alpha = 0.01$; reject H_0
There is sufficient evidence at the $\alpha = 0.01$ level of significance to support the claim that the proportion of children in the experimental group who contracted polio is less than the proportion of children in the control group who contracted polio.

(b) Lower bound: -0.0005, upper bound: -0.0003. We are 90% confident that the difference in the proportion of children who contract polio with the vaccine versus without the vaccine is between 0.0003 and 0.0005 and that a smaller percentage of children contracted polio in the experimental group.

19. (a) Each sample is a simple random sample, $n_1\hat{p}_1(1 - \hat{p}_1) = 48 \geq 10$ and $n_2\hat{p}_2(1 - \hat{p}_2) = 30 \geq 10$, and each sample is less than 5% of the population size.
H_0: $p_1 = p_2$, H_1: $p_1 > p_2$
Classical approach: $z_0 = 2.13 > z_{0.05} = 1.645$; reject H_0
P-value approach: P-value $= 0.0166 < \alpha = 0.05$; reject H_0
There is sufficient evidence at the $\alpha = 0.05$ level of significance to support the claim that the proportion of individuals taking Clarinex and experiencing dry mouth is greater than that of those taking a placebo.

(b) No

21. Answers will vary. There is not a statistically significant difference in the percentage of gun owners from the 1999 poll to the 2004 poll. When testing the hypotheses H_0: $p_1 = p_2$ versus H_1: $p_1 > p_2$, we get a P-value of 0.1586 ($Z_0 = 1.00 < z_{0.10} = 1.28$).

23. (a) $n = n_1 = n_2 = 1406$ **(b)** $n = n_1 = n_2 = 2135$

25. Since the P-value is less than 0.05, there is sufficient evidence at the $\alpha = 0.05$ level of significance to support the claim that the drug is more effective in treating sleep apnea than the placebo.

11.4 Assess Your Understanding (page 620)

1. The test is not robust, so even small departures from normality adversely affect the test, making it unreliable. So, in the test for comparing two population standard deviations, the samples must come from populations that are normally distributed.

3. $F_{0.05,9,10} = 3.02$

5. $F_{0.975,6,8} = 0.18$, $F_{0.025,6,8} = 4.65$

7. $F_{0.90,25,20} = 0.58$ **9.** $F_{0.05,45,15} = 2.19$

11. Classical approach: $F_0 = 0.84$ is between $F_{0.975,15,15} = 0.35$ and $F_{0.025,15,15} = 2.86$; do not reject the null hypothesis
P-value approach: P-value $= 0.7330 > \alpha = 0.05$; do not reject the null hypothesis
There is not sufficient evidence at the $\alpha = 0.05$ level of significance to support the claim that $\sigma_1 \neq \sigma_2$.

13. Classical approach: $F_0 = 2.39 < F_{0.01,25,18} = 2.84$; do not reject the null hypothesis
P-value approach: P-value $= 0.0303 > \alpha = 0.01$; do not reject the null hypothesis
There is not sufficient evidence at the $\alpha = 0.01$ level of significance to support the claim that $\sigma_1 > \sigma_2$.

15. Classical approach: $F_0 = 0.40 < F_{0.90,50,25} = 0.65$; reject the null hypothesis
P-value approach: P-value $= 0.0026 < \alpha = 0.10$; reject the null hypothesis
There is sufficient evidence at the $\alpha = 0.10$ level of significance to support the claim that $\sigma_1 < \sigma_2$.

17. Hypotheses: $H_0: \sigma_1 = \sigma_2$, $H_1: \sigma_1 \neq \sigma_2$
Classical approach: $F_0 = 0.97$ is between $F_{0.975,54,59} = 0.58$ and $F_{0.025,54,59} = 1.75$; do not reject the null hypothesis
P-value approach: P-value $= 0.9087 > \alpha = 0.05$; do not reject the null hypothesis
There is not sufficient evidence at the $\alpha = 0.05$ level of significance to support the claim that the standard deviation in the treatment group is different from the standard deviation in the control group.

19. Hypotheses: $H_0: \sigma_1 = \sigma_2$, $H_1: \sigma_1 > \sigma_2$
Classical approach: $F_0 = 6.94 > F_{0.01,64,64} = 1.91$; reject the null hypothesis
P-value approach: P-value $< 0.0001 < \alpha = 0.01$; reject the null hypothesis
There is sufficient evidence at the $\alpha = 0.01$ level of significance to support the claim that the treatment group had a higher standard deviation for serum retinal concentration than did the control group.

21. (a) Hypotheses: $H_0: \sigma_1 = \sigma_2$, $H_1: \sigma_1 < \sigma_2$
Classical approach: $F_0 = 0.32 < F_{0.95,19,19} \approx 0.47$; reject the null hypothesis
P-value approach: P-value $= 0.0087 < \alpha = 0.05$; reject the null hypothesis
There is sufficient evidence at the $\alpha = 0.05$ level of significance to support the claim that the standard deviation for wait time in the single line is less than the standard deviation for wait time in the multiple lines.

(b)

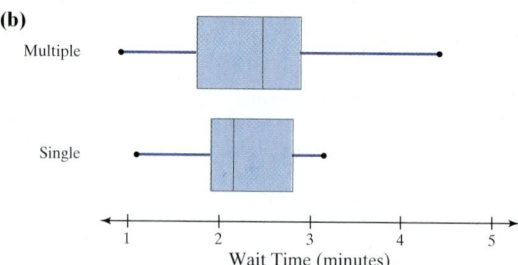

Wait Time (minutes)

23. (a) The normal probability plots are roughly linear, so we can conclude that the data are approximately normally distributed.
(b) Since P-value $= 0.169 > \alpha = 0.05$, which is the level of significance, the nursing student should conclude that there is not sufficient evidence to support the claim that men have more variability in their systolic blood pressure than women have.
(c) $F_0 = 0.50 > F_{0.95,19,16} = 0.45$, so do not reject the null hypothesis. The test statistic supports the previous decision.

Chapter Review Exercises (page 625)

1. Dependent **3.** Independent

5. (a) $F_{0.05,8,9} = 3.23$
(b) $F_{0.975,10,5} = 0.24$, $F_{0.025,10,5} = 6.62$

7. (a)

Observation	1	2	3	4	5	6
$d_i = x_i - y_i$	-0.7	0.6	0	-0.1	-0.4	-0.4

(b) $d = -0.167$; $s_d = 0.450$
(c) Hypotheses: $H_0: \mu_d = 0$, $H_1: \mu_d < 0$
Classical approach: $t_0 = -0.907 > -t_{0.05} = -2.015$; do not reject the null hypothesis
P-value approach: P-value $= 0.2030 > \alpha = 0.05$; do not reject the null hypothesis
There is not sufficient evidence to support the claim that the mean difference is less than zero.
(d) Lower bound: -0.79, upper bound: 0.45

9. (a) Hypotheses: $H_0: \mu_1 = \mu_2$, $H_1: \mu_1 \neq \mu_2$
Classical approach: $t_0 = 2.290 > t_{0.05} = 1.895$; reject the null hypothesis
P-value approach: P-value $= 0.0351 < \alpha = 0.10$; reject the null hypothesis
There is sufficient evidence at the $\alpha = 0.1$ level of significance to support the claim that $\mu_1 \neq \mu_2$.
(b) Lower bound: 0.73, upper bound: 7.67
(c) Hypotheses: $H_0: \sigma_1 = \sigma_2$, $H_1: \sigma_1 \neq \sigma_2$
Classical approach: $F_0 = 1.40$ is between $F_{0.975,12,7} = 0.28$ and $F_{0.025,12,7} \approx 4.76$; do not reject the null hypothesis
P-value approach: P-value $= 0.6734 > \alpha = 0.05$; do not reject the null hypothesis
There is not sufficient evidence at the $\alpha = 0.05$ level of significance to support the claim that the standard deviation in population 1 is different from the standard deviation in population 2.

11. (a) Hypotheses: $H_0: \mu_1 = \mu_2$, $H_1: \mu_1 > \mu_2$
Classical approach: $t_0 = 1.472 < t_{0.01} = 2.423$; do not reject the null hypothesis
P-value approach: P-value $= 0.0726 > \alpha = 0.01$; do not reject the null hypothesis
There is not sufficient evidence at the $\alpha = 0.01$ level of significance to support the claim that the mean of population 1 is larger than the mean of population 2.
(b) Lower bound: -0.43, upper bound: 6.43
(c) Hypotheses: $H_0: \sigma_1 = \sigma_2$, $H_1: \sigma_1 < \sigma_2$
Classical approach: $F_0 = 0.67 > F_{0.99,44,40} \approx 0.50$; do not reject the null hypothesis
P-value approach: P-value $= 0.0940 > \alpha = 0.01$; do not reject the null hypothesis
There is not sufficient evidence at the $\alpha = 0.01$ level of significance to support the claim that the standard deviation in population 1 is less than the standard deviation in population 2.

13. (a) $H_0: p_1 = p_2$ vs $H_1: p_1 \neq p_2$ **(b)** $z_0 = -1.68$
(c) $-z_{0.025} = -1.96$; $z_{0.025} = 1.96$
(d) P-value $= 0.0930$. Do not reject H_0. There is not sufficient evidence at the $\alpha = 0.05$ level of significance to support the claim that $p_1 \neq p_2$ (because $-z_{0.025} < z_0 < z_{0.025}$ or P-value $> \alpha$).

15. (a) The sampling method is dependent because the same individual is used for both measurements.
(b) Hypotheses: $H_0: \mu_d = 0$, $H_1: \mu_d \neq 0$
Classical approach: $t_0 = 0.506$ is between $-t_{0.025} = -2.262$ and $t_{0.025} = 2.262$; do not reject the null hypothesis
P-value approach: P-value $= 0.6209 > \alpha = 0.05$; do not reject the null hypothesis
There is not sufficient evidence at the $\alpha = 0.05$ level of significance to reject the student's claim that an individual's arm span equals the individual's height.

17. (a) The testing method is independent since the cars selected for the McDonald's sample had no bearing on the cars chosen for the Wendy's sample.
(b) Hypotheses: $H_0: \mu_{McD} = \mu_W$, $H_1: \mu_{McD} \neq \mu_W$
Classical approach: $t_0 = -4.059 < -t_{0.05} = -1.706$; reject the null hypothesis
P-value approach: P-value $= 0.0003 < \alpha = 0.10$; reject the null hypothesis

There is sufficient evidence at the $\alpha = 0.10$ level of significance to reject the claim that the mean wait time is the same at McDonald's and Wendy's drive-through windows.

(c)

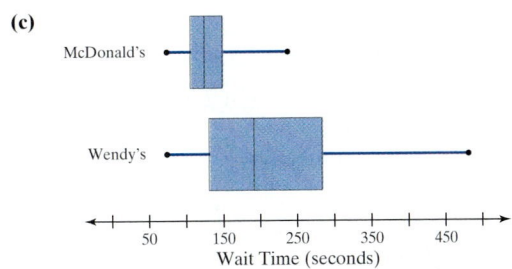

19. (a) Each sample is a simple random sample; $n_1\hat{p}_1(1 - \hat{p}_1) = 26 \geq 10$ and $n_2\hat{p}_2(1 - \hat{p}_2) = 45 \geq 10$; and each sample is less than 5% of the population size.

Hypotheses: H_0: $p_{\text{exp}} = p_{\text{control}}$, H_1: $p_{\text{exp}} < p_{\text{control}}$

Classical approach: $z_0 = -2.71 < -z_{0.01} = -2.33$; reject the null hypothesis

P-value approach: P-value $= 0.0037 < \alpha = 0.01$; reject the null hypothesis

There is sufficient evidence at the $\alpha = 0.01$ level of significance to support the claim that a lower proportion of women in the experimental group experienced a bone fracture than in the control group.

(b) Lower bound: -0.057, upper bound: -0.009. We are 95% confident that the difference in the proportion of women who experienced a bone fracture between the experimental and control group is between -0.06 and -0.01.

(c) This is a completely randomized design. The treatment is drug. It has two levels: 5 mg of Actonel versus a placebo.

21. (a) $n = n_1 = n_2 = 2136$ **(b)** $n = n_1 = n_2 = 3383$

23. Hypotheses: H_0: $\sigma_{\text{McD}} = \sigma_{\text{W}}$, H_1: $\sigma_{\text{McD}} < \sigma_{\text{W}}$

Classical approach: $F_0 = 0.15 < F_{0.95,29,26} \approx 0.45$; reject the null hypothesis

P-value approach: P-value $< 0.0001 < \alpha = 0.05$; reject the null hypothesis

There is sufficient evidence at the $\alpha = 0.05$ level of significance to support the claim that the standard deviation in wait time at Wendy's is more than the standard deviation in wait time at McDonald's.

25. Lower bound: -2.17, upper bound 12.37. Since the interval includes zero, we conclude that there is not sufficient evidence at the $\alpha = 0.05$ level of significance to reject the claim that arm span and height are equal.

27. Lower bound: -128.87, upper bound -42.02. Answers will vary.

CHAPTER 12

12.1 Assess Your Understanding (page 638)

1. Answers will vary.

3. To perform a goodness-of-fit test, all expected frequencies must be greater than or equal to 1, and at least 80% of the expected frequencies must be greater than or equal to 5.

5.

p_i	0.2	0.1	0.45	0.25
Expected counts	100	50	225	125

7. (a) $\chi_0^2 = 2.72$ **(b)** df $= 3$ **(c)** $\chi_{0.05}^2 = 7.815$
(d) Do not reject H_0, since $\chi^2 < \chi_{0.05}^2$.

9. (a) $\chi_0^2 = 12.56$ **(b)** df $= 4$ **(c)** $\chi_{0.05}^2 = 9.488$
(d) Reject H_0, since $\chi_0^2 > \chi_{0.05}^2$. There is sufficient evidence at the $\alpha = 0.05$ level of significance to reject the claim that the random variable X is binomial with $n = 4$, $p = 0.8$.

11. Classical approach: $\chi_0^2 = 18.738 > \chi_{0.05}^2 = 11.071$; reject the null hypothesis.
P-value approach: P-value $< 0.005 < \alpha = 0.05$; reject the null hypothesis.
There is sufficient evidence at the $\alpha = 0.05$ level of significance to reject the null hypothesis that there are 13% brown, 14% yellow, 13% red, 20% orange, 24% blue, and 16% green candies in a bag of M&Ms.

13. (a) Answers will vary depending on the desired probability of making a Type I error.
(b) Classical approach: $\chi_0^2 = 21.693$; compare χ_0^2 to your chosen χ_α^2.
P-value approach: P-value is between 0.10 and 0.005.
(c) Answers will vary.

15. Classical approach: $\chi_0^2 = 1.268 < \chi_{0.05}^2 = 19.675$; do not reject the null hypothesis.
P-value approach: P-value $> 0.95 > \alpha = 0.05$; do not reject the null hypothesis.
There is not sufficient evidence at the $\alpha = 0.05$ level of significance to reject the null hypothesis that the distribution of birth month is uniform.

17. Classical approach: $\chi_0^2 = 13.227 > \chi_{0.05}^2 = 12.592$; reject the null hypothesis.
P-value approach: P-value $< 0.05 < \alpha = 0.05$; reject the null hypothesis.
There is sufficient evidence at the $\alpha = 0.05$ level of significance to reject the null hypothesis that pedestrian deaths are uniformly distributed over the days of the week.

19. (a)

Grade	Relative Frequency	Observed Frequency	Expected Frequency
K-3	0.329	15	8.225
4-8	0.393	7	9.825
9-12	0.278	3	6.95

(b) Classical approach: $\chi_0^2 = 8.638 > \chi_{0.05}^2 = 5.991$; reject the null hypothesis.
P-value approach: P-value $< 0.025 < \alpha = 0.05$; reject the null hypothesis.
There is sufficient evidence at the $\alpha = 0.05$ level of significance to support the social worker's claim that the grade distribution of home-schooled children in her district is different from the national proportions.

21. (a) Answers will vary.
(b) Expect 20% of the numbers to be 1s, 20% to be 2s, 20% to be 3s, 20% to be 4s, and 20% to be 5s.
(c) Answers will vary.

23. (a) Expected number low birth weight: 8.52; expected number without low birth weight: 111.48.
(b) H_0: $p = 0.071$ vs H_1: $p > 0.071$
$\chi_0^2 = 1.225$; $\chi_{0.05}^2 = 3.841$; do not reject H_0, because $\chi_0^2 < \chi_{0.05}^2$.
(c) Do not reject, because $z_0 < z_{0.05}$ ($1.13 < 1.645$).

25. $\sum (O - E) = 0$

12.2 Assess Your Understanding (page 647)

1. A marginal distribution is a frequency or relative frequency distribution of either the row or column variable in a contingency table. A conditional distribution is the relative frequency of each category of one variable, given a specific value of the other variable in a contingency table.

3. Correlation is used with quantitative variables, association is used with categorical variables.

5. (a)

Poverty Level	Age				Marginal frequency (poverty level)
	<18	18-44	45-64	≥65	
Below poverty level	7571	3693	1718	1025	14,007
Above poverty	9694	4745	2282	2258	18,979
Marginal frequency (age)	17,265	8438	4000	3283	32,986

(b)

Poverty Level	Age				Marginal relative frequency (poverty level)
	<18	**18–44**	**45–64**	**≥65**	
Below poverty level	7571	3693	1718	1025	0.425
Above poverty	9694	4745	2282	2258	0.575
Marginal relative frequency (age)	0.523	0.256	0.121	0.100	1

(c) 42.5% of people receiving Medicaid are below the poverty level.

(d) 52.3% of people receiving Medicaid are under 18 years of age.

(e)

Poverty Level	Age				Total
	<18	**18–44**	**45–64**	**≥65**	
Below poverty level	0.541	0.264	0.123	0.073	1.000
Above poverty	0.511	0.250	0.120	0.119	1.000

(f)

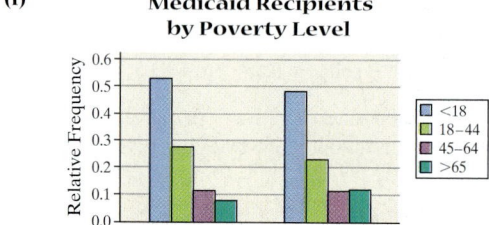

Medicaid Recipients by Poverty Level

(g) Answers will vary.

7. (a)

Age	Male	Female	Marginal distribution
16–24	4901	1601	6502
25–34	3693	1003	4696
35–44	3493	1143	4636
45–54	3032	979	4011
55–64	1866	679	2545
65–69	594	228	822
>69	2105	1100	3205
Unknown	17	4	21
Marginal Distribution	19,701	6737	26,438

(b)

Age	Male	Female	Relative frequency marginal distribution
16–24	4901	1601	0.246
25–34	3693	1003	0.178
35–44	3493	1143	0.175
45–54	3032	979	0.152
55–64	1866	679	0.096
65–69	594	228	0.031
>69	2105	1100	0.121
Unknown	17	4	0.001
Relative frequency marginal distribution	0.745	0.255	1

(c) 17.8% of drivers killed were between 25 and 34 years of age.

(d) 25.5% of drivers killed were female.

(e)

Age	Male	Female
16–24	0.249	0.238
25–34	0.187	0.149
35–44	0.177	0.170
45–54	0.154	0.145
55–64	0.095	0.101
65–69	0.030	0.034
>69	0.107	0.163
Unknown	0.001	0.001
	1	1

(f)

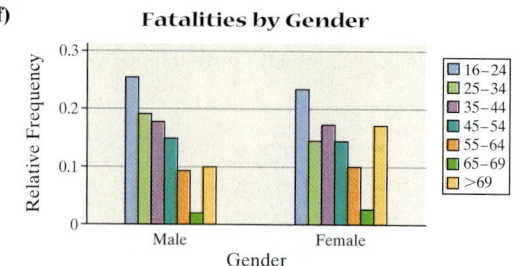

Fatalities by Gender

(g) Answers will vary.

9. (a)

Age	Cause of Death				Marginal frequency
	Accidents and Adverse Effects	Malignant Neoplasms	Heart Diseases	Cerebrovascular Diseases	
1–4	1,679	383	186	59	2,307
5–14	2,561	1,060	252	83	3,956
15–24	14,966	1,628	1,083	204	17,881
25–44	27,844	19,041	16,283	3,004	66,172
45–64	23,669	144,936	101,713	15,971	286,289
≥65	33,976	387,475	564,204	139,397	1,124,052
Marginal frequency	104,695	554,523	683,721	157,718	1,500,657

(b)

Age	Cause of Death				Marginal relative frequency
	Accidents and Adverse Effects	Malignant Neoplasms	Heart Diseases	Cerebrovascular Diseases	
1–4	1,679	383	186	59	0.002
5–14	2,561	1,060	252	83	0.003
15–24	14,966	1,628	1,083	204	0.012
25–44	27,844	19,041	16,283	3,004	0.044
45–64	23,669	144,936	101,713	15,971	0.191
≥65	33,976	387,475	564,204	139,397	0.749
Marginal relative frequency	0.070	0.370	0.456	0.105	1

(c) 7.0% of deaths were caused by accidents or adverse effects.

(d) 1.2% of deaths were to 15- to 24-year-olds.

(e)

Age	Cause of Death			
	Accidents and Adverse Effects	Malignant Neoplasms	Heart Diseases	Cerebrovascular Diseases
1–4	0.728	0.166	0.081	0.026
5–14	0.647	0.268	0.064	0.021
15–24	0.837	0.091	0.061	0.011
25–44	0.421	0.288	0.246	0.045
45–64	0.083	0.506	0.355	0.056
≥65	0.030	0.345	0.502	0.123

(f)

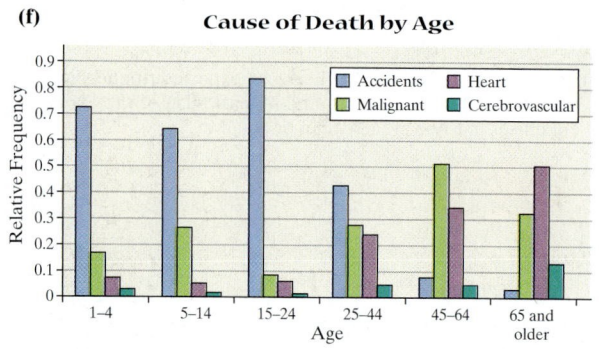

Cause of Death by Age

(g) Answers may vary.

11. (a)

	Abortions			
Age	1990	1995	2000	Marginal Distribution
≤19	369	274	244	887
20–24	532	441	430	1403
≥25	713	645	640	1998
Marginal distribution	1614	1360	1314	4288

(b)

	Abortions			
Age	1990	1995	2000	Relative Frequency Marginal Distribution
≤19	369	274	244	0.207
20–24	532	441	430	0.327
≥25	713	645	640	0.466
Relative frequency marginal distribution	0.376	0.317	0.306	1

(c) 37.6% of the abortions were performed in 1990.

(d) 20.7% of the women having abortions were 19 years of age or younger.

(e)

	Abortions		
Age	1990	1995	2000
≤19	0.229	0.201	0.186
20–24	0.330	0.324	0.327
≥25	0.442	0.474	0.487
Total	1	1	1

(f)

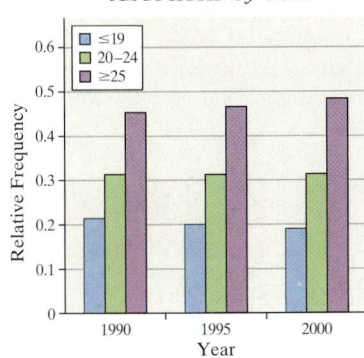

(g) Answers may vary.

13. (a)

	College Applicants		
	Accepted	Denied	Total
Male	0.158	0.842	1
Female	0.310	0.690	1

(b) 0.158 of the males who applied was accepted.
0.310 of the females who applied was accepted.

(c) Conclusion: a higher proportion of females is accepted.

(d) 0.15 of the males applying to the business school was accepted.
0.143 of the females applying to the business school was accepted.

(e) 0.40 of the males applying to the social work school was accepted.
0.364 of the females applying to the social work school was accepted.

(f) Answers will vary.

12.3 Assess Your Understanding (page 662)

1. Answers may vary, but differences should include: chi-square test for independence compares two characteristics from a single population, whereas the chi-square test for homogeneity compares a single characteristic from two (or more) populations. Similarities should include: the procedures of the two tests and the assumptions of the two tests are the same.

3. (a) $\chi_0^2 = 1.701$

(b) $\chi_{0.05}^2 = 5.991$. Since $\chi_0^2 < \chi_{0.05}^2$, do not reject H_0. There is evidence at the $\alpha = 0.05$ level of significance to support the belief that X and Y are independent. It appears that X and Y are not related.

(c) P-value $= 0.427$

5. $\chi_0^2 = 1.989$; $\chi_{0.01}^2 = 9.210$. Since $\chi_0^2 < \chi_{0.01}^2$, do not reject H_0. There is evidence at the $\alpha = 0.05$ level of significance to support the claim

that the proportions are equal. We don't have enough evidence to conclude that at least one of the proportions is different from the others. P-value $= 0.370$

7. (a)

	Both Biological/ Adoptive Parents	Single Parent	Parent and Stepparent	Nonparental Guardian
Had sexual intercourse	78.553	52.368	41.895	26.184
Did not have sexual intercourse	71.447	47.632	38.105	23.816

(b) (1) All expected frequencies are greater than or equal to 1, and (2) no more than 20% of the expected frequencies are less than 5.

(c) $\chi_0^2 = 10.357$

(d) $\chi_{0.05}^2 = 7.815$. Since $\chi_0^2 > \chi_{0.05}^2$, reject H_0. There is sufficient evidence at the $\alpha = 0.05$ level of significance to conclude that sexual activity and family structure are associated.

(e) The biggest difference between observed and expected occurs under the family structure in which both parents are present. Fewer females were sexually active than was expected when both parents were present. This means that having both parents present seems to have an impact on whether the child is sexually active.

(f)

	Both Biological/ Adoptive Parents	Single Parent	Parent and Stepparent	Nonparental Guardian
Had sexual intercourse	0.427	0.59	0.55	0.64
Did not have sexual intercourse	0.573	0.41	0.45	0.36

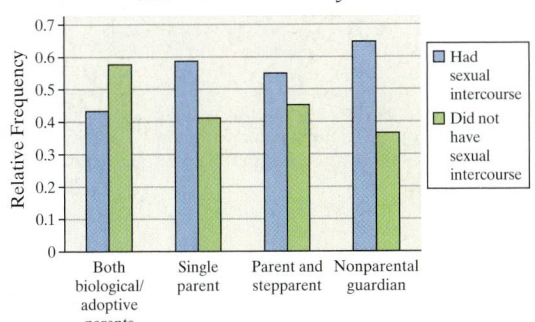

(g) P-value $= 0.016$

9. (a) $\chi_0^2 = 32.926$; $\chi_{0.05}^2 = 16.919$; P-value $= 0.0001$. Since $\chi_0^2 > \chi_{0.05}^2$, reject H_0. There is sufficient evidence at the $\alpha = 0.05$ level of significance to conclude that level of education and area of country are associated.

(b) The cell corresponding to Midwest and not a high-school graduate contributed most to the test statistic. The expected was less than the observed.

(c)

	Level of Education			
Area of Country	Not High School Graduate	High School Graduate	Some College	Bachelor's or Higher
Northeast	0.146	0.212	0.162	0.215
Midwest	0.346	0.252	0.237	0.219
South	0.334	0.352	0.343	0.329
West	0.174	0.183	0.258	0.237
Total	1	1	1	1

Education by Region of Country

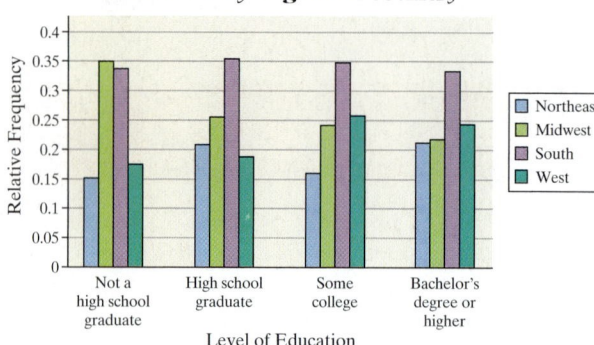

11. (a) $\chi_0^2 = 6.681$; $\chi_{0.05}^2 = 5.991$. Since $\chi_0^2 > \chi_{0.05}^2$, reject H_0. There is sufficient evidence at the $\alpha = 0.05$ level of significance to support the belief that age and opinion are associated. It appears to be the case that age plays a role in determining one's opinion regarding the legalization of marijuana. The 50 or older group is less likely to be in favor of the legalization of marijuana.
(b) P-value $= 0.035$
(c)

Opinion	18–29 Years Old	30–49 Years Old	50 Years or Older
For	0.768	0.752	0.684
Against	0.232	0.248	0.316
Total	1	1	1

Legalization of Marijuana

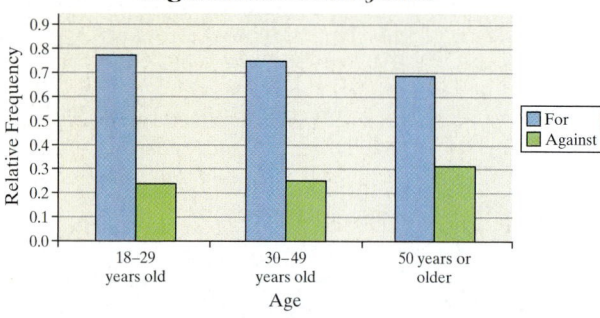

13. (a) $\chi_0^2 = 0.946$; $\chi_{0.05}^2 = 7.815$. Since $\chi_0^2 < \chi_{0.05}^2$, do not reject H_0. There is not sufficient evidence at the $\alpha = 0.05$ level of significance to support the belief that gender and delinquencies are associated.
(b) P-value $= 0.814$
(c)

Gender	Delinquency Person	Property	Drugs	Public Order
Female	0.198	0.188	0.152	0.214
Male	0.802	0.812	0.848	0.786
Total	1	1	1	1

Delinquencies by Type

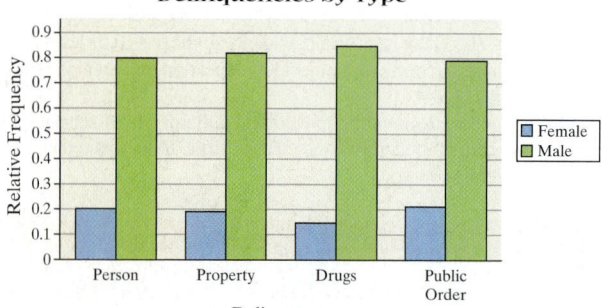

15. (a) H_0: $p_{18-29} = p_{30-49} = p_{50-64} = p_{65 \text{ and older}}$ versus
H_1: at least one of the proportions is not equal to the rest.
$\chi_0^2 = 6.000$; $\chi_{0.05}^2 = 7.815$. Since $\chi_0^2 < \chi_{0.05}^2$, do not reject H_0. There is not sufficient evidence at the $\alpha = 0.05$ level of significance to support the claim that at least one proportion is

different from the others. The evidence suggests that the proportion of individuals in each age group who smoked at least one cigarette in the past week is the same.
(b) P-value $= 0.112$
(c)

Smoking Status	Age 18–29	30–49	50–64	≥65
Smoked	0.3	0.2625	0.2875	0.15
Did not smoke	0.7	0.7375	0.7125	0.85
Total	1	1	1	1

Smoking Status by Age

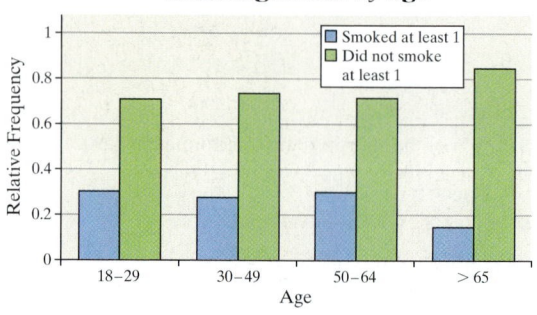

17. (a) H_0: $p_{\text{placebo}} = p_{50} = p_{100} = p_{200} = p_{\text{Naproxen}}$ versus
H_1: at least one of the proportions is not equal to the rest.
$\chi_0^2 = 49.703$; $\chi_{0.01}^2 = 13.277$. Since $\chi_0^2 > \chi_{0.01}^2$, reject H_0. There is sufficient evidence at the $\alpha = 0.01$ level of significance to support the belief that at least one proportion is different from the others. The evidence suggests that the subjects taking Naproxen experienced a higher incidence rate of ulcers than the other treatment groups.
(b) P-value < 0.001
(c)

Side Effect	Placebo	Treatment Celebrex (50 mg)	Celebrex (100 mg)	Celebrex (200 mg)	Naproxen (500 mg)
Ulcer	0.023	0.034	0.031	0.059	0.162
No ulcer	0.977	0.966	0.969	0.941	0.838
Total	1	1	1	1	1

Side Effects

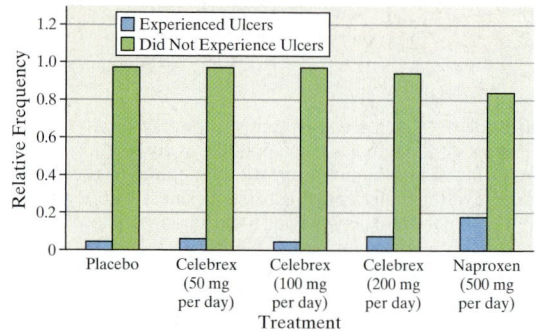

19. (a)

	Course	Personal	Work
Female	13	5	3
Male	10	6	13

(b) $\chi_0^2 = 5.595$; $\chi_{0.1}^2 = 4.605$. Since $\chi_0^2 > \chi_{0.1}^2$, reject H_0. The evidence suggests that there is some relation between gender and drop reason. Females are more likely to drop because of the course, while males are more likely to drop because of work.
(c) P-value $= 0.061$
(d)

	Reason Course	Personal	Work
Female	0.565	0.455	0.188
Male	0.435	0.545	0.813
Total	1	1	1

Why Did You Drop?

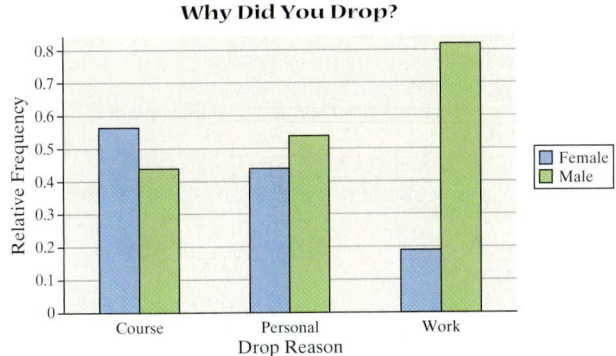

(f)

Law Enforcement Officers Killed

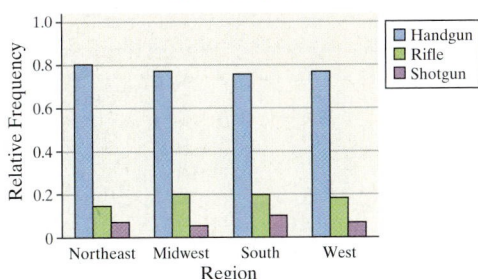

(g) Answers will vary.

21. (a) The expected number of adult Americans who have smoked is 252. The expected number of adult Americans who have not smoked is 755.
(b) $\chi_0^2 = 9.527$
(c) H_0: $p_{2004} = p_{1990}$ versus H_1: $p_{2004} \neq p_{1990}$
Since $\chi_0^2 = 9.527 > \chi_\alpha^2 = 3.841$, reject the null hypothesis. There is evidence to support the claim that in 2004 the proportion of adult Americans who smoke at least one cigarette a week is different from the proportion who smoked in 1990.
(d) $Z = 3.087$, $Z^2 = 9.5296$

Chapter Review Exercises (page 669)

1. $\chi_0^2 = 0.578$; $\chi_{0.05}^2 = 5.991$; since $\chi_0^2 < \chi_{0.05}^2$, do not reject H_0. There is not sufficient evidence at the $\alpha = 0.05$ level of significance to conclude that the wheel is out of balance.

3. $\chi_0^2 = 9.709$; $\chi_{0.1}^2 = 9.236$; since $\chi_0^2 > \chi_{0.1}^2$, reject H_0. There is sufficient evidence at the $\alpha = 0.1$ level of significance to conclude that the distribution of educational attainment is different today than it was in 1994.

5. (a)

Firearm	Region				Marginal frequency
	Northeast	**Midwest**	**South**	**West**	
Handgun	43	79	182	86	390
Rifle	7	21	45	20	93
Shotgun	3	5	20	6	34
Marginal frequency	53	105	247	112	517

(b)

Firearm	Region				Marginal relative frequency
	Northeast	**Midwest**	**South**	**West**	
Handgun	43	79	182	86	0.754
Rifle	7	21	45	20	0.180
Shotgun	3	5	20	6	0.066
Marginal relative frequency	0.103	0.203	0.478	0.217	1

(c) 75.4% of law enforcement officers were killed with a handgun.
(d) 10.3% of law enforcement officers were killed in the Northeast.

(e)

Firearm	Region			
	Northeast	**Midwest**	**South**	**West**
Handgun	0.811	0.752	0.737	0.768
Rifle	0.132	0.200	0.182	0.179
Shotgun	0.057	0.048	0.081	0.054

7. (a)

Type of Fatality	Gender		Marginal frequency
	Male	**Female**	
Passenger in vehicle	588	568	1156
Pedestrian	215	114	329
Bicycle	79	14	93
Marginal frequency	882	696	1578

(b)

Type of Fatality	Gender		Marginal relative frequency
	Male	**Female**	
Passenger in vehicle	588	568	0.733
Pedestrian	215	114	0.208
Bicycle	79	14	0.059
Marginal relative frequency	0.559	0.441	1

(c) 55.9% of children killed by a motor vehicle were male.
(d) 5.9% of children killed by a motor vehicle were riding bicycles.

(e)

Type of Fatality	Gender	
	Male	**Female**
Passenger in vehicle	0.667	0.816
Pedestrian	0.244	0.164
Bicycle	0.090	0.020
Total	1	1

(f)

Motor Vehicle Fatalities Children under 13 Years

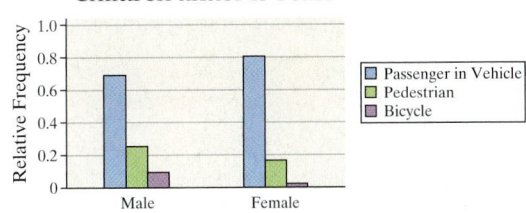

(g) Answers may vary.

9. (a)

	Belief			
	Developed with God	**Developed without God**	**Created in Present Form**	**No Opinion**
February 2001	365.82	116.94	467.27	65.97
June 1993	366.18	117.06	467.73	66.03

(b) All expected values are greater than 5.
(c) $\chi_0^2 = 2.203$
(d) $\chi_{0.05}^2 = 7.815$; do not reject H_0. The evidence indicates that it is reasonable to conclude that belief is independent of date.
(e) No opinion in 2001 contributed the most.

(f)

Date	Belief				
	Developed with God	Developed without God	Created in Present Form	No Opinion	Total
Feb 2001	0.370	0.120	0.450	0.060	1
June 1993	0.350	0.110	0.470	0.070	1

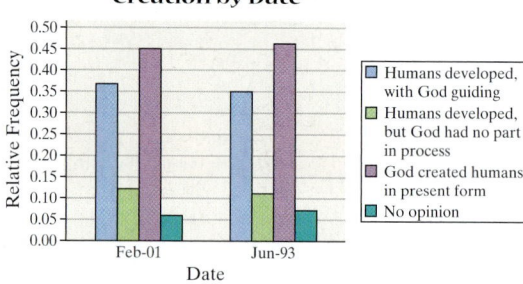

Evolution or Creation by Date

Legend:
- Humans developed, with God guiding
- Humans developed, but God had no part in process
- God created humans in present form
- No opinion

(g) P-value $= 0.531$

11. $\chi_0^2 = 242.829$; $\chi_{0.01}^2 = 30.578$; reject H_0. The evidence indicates that it is reasonable to conclude that race is not independent of region.

13. $\chi_0^2 = 3.877$; $\chi_{0.05}^2 = 7.815$; do not reject H_0. The evidence indicates that it is reasonable to conclude that the proportion of Americans who had a common cold is the same for each region of the country.

15. $\chi_0^2 = 126.168$; $\chi_{0.05}^2 = 7.815$; reject H_0. There is sufficient evidence to indicate that the proportion of Americans who are morning persons or night persons differs in at least one of the four age groups.

CHAPTER 13

13.1 Assess Your Understanding (page 685)

1. analysis of variance

3. The mean square due to treatment estimate of σ^2 is a weighted average of the squared deviations of each sample mean from the grand mean of all the samples. The mean square due to error estimate of σ^2 is the weighted average of the sample variances.

5.

Source of Variation	Sum of Squares	Degrees of Freedom	Mean Squares	F-Test Statistic
Treatment	387	2	193.5	0.650
Error	8042	27	297.852	
Total	8429	29		

7. $F = 1.154$

9. $F = \dfrac{MST}{MSE} = \dfrac{27.5833}{13.5278} = 2.039$

11. (a) H_0: $\mu_{\text{sludge plot}} = \mu_{\text{spring disk}} = \mu_{\text{no till}}$, H_1: At least one of the means is different.
(b) 1. Each sample is a simple random sample.
   2. The three samples are independent of each other.
   3. The samples are from normally distributed populations.
   4. The populations have equal variances.
(c) Since the P-value $= 0.007 < 0.05$, we reject the null hypothesis and conclude that at least one of the means is different.
(d) The boxplots support the result obtained in part (c). The boxplots indicate that significantly more plants are growing in the spring disk field.
(e) $F = \dfrac{MST}{MSE} = \dfrac{42.06}{5.92} = 7.10$

13. (a) H_0: $\mu_M = \mu_T = \mu_W = \mu_R = \mu_P$ versus H_1: At least one of the means is not equal.
(b) 1. Each sample is a simple random sample.
   2. The five samples are independent.
   3. The populations from which the samples are drawn must be normal.
   4. The populations have the same variance.
(c) P-value < 0.0001, which is less than $\alpha = 0.01$, so we reject the null hypothesis.
(d) Yes
(e) $F = \dfrac{2876908}{293451} = 9.80$
(f) Monday appears to have significantly fewer births than the other four days.

15. (a) H_0: $\mu_{\text{financial}} = \mu_{\text{energy}} = \mu_{\text{utilities}}$ versus H_1: At least one of the means is different.
(b) 1. Each sample is a simple random sample.
   2. The three samples are independent of each other.
   3. The samples are from normally distributed populations.
   4. The largest sample standard deviation is less than twice the smallest sample standard deviation ($5.12 < 2 \cdot 4.53$), so the requirement that the populations have equal variances is satisfied.
(c) $F = 2.077$, P-value $= 0.1502$. Since the P-value is greater than $\alpha = 0.05$, we do not reject the null hypothesis and conclude that there is not enough evidence to support the hypothesis that at least one of the means is different.
(d)

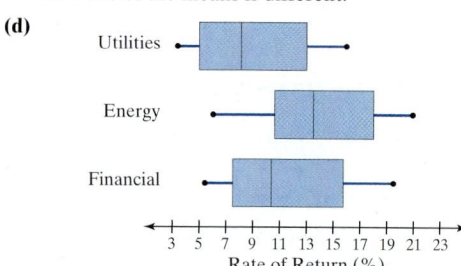

17. (a) H_0: $\mu_{\text{large}} = \mu_{\text{pass van.}} = \mu_{\text{mid. util. veh}}$ versus H_1: At least one of the means is different.
(b) 1. Each sample is a simple random sample.
   2. The three samples are independent of each other.
   3. The samples are from normally distributed populations.
   4. The largest sample standard deviation is less than twice the smallest sample standard deviation ($3.73 < 2 \cdot 3.26$), so the requirement that the populations have equal variances is satisfied.
(c) $F = 2.940 < F_{0.01,2,18} \approx 5.85$; P-value $= 0.0785 > \alpha = 0.01$. We do not reject the null hypothesis and conclude that there is not enough evidence to support the hypothesis that at least one of the means is different.
(d)

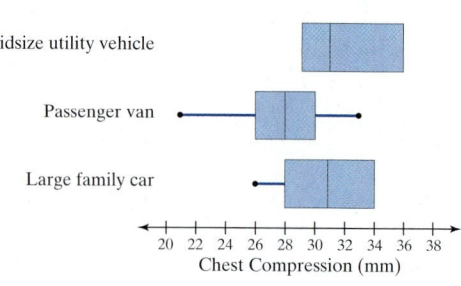

19. (a) H_0: $\mu_{\text{Alaska}} = \mu_{\text{Florida}} = \mu_{\text{Texas}}$ versus H_1: At least one of the means is different.
(b) 1. Each samples is a simple random sample.
   2. The three samples are independent of each other.
   3. The samples are from normally distributed populations.
   4. The largest sample standard deviation is less than twice the smallest sample standard deviation ($0.397 < 2 \cdot 0.252$), so the requirement that the populations have equal variances is satisfied.

(c) $F = 5.811 > F_{0.05,2,15} = 3.68$; P-value $= 0.0135 < \alpha = 0.05$. We reject the null hypothesis and conclude that there is sufficient evidence to support the hypothesis that at least one of the means is different.

(d)

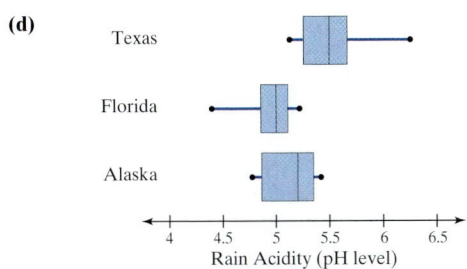

Rain Acidity (pH level)

21. (a) $H_0: \mu_{301} = \mu_{400} = \mu_{353}$ versus H_1: At least one of the means is not equal.

(b) The standard deviation for mixture 67-0-400 is more than two times larger than the standard deviation for mixture 67-0-301.

13.2 Assess Your Understanding (page 696)

1. Multiple comparison tests help to identify the means that differ after ANOVA results indicate that the null hypothesis should be rejected.

3. (a) $q_{0.05,10,3} = 3.877$ **(b)** $q_{0.05,24,5} = 4.166$
(c) $q_{0.05,32,8} \approx 4.602$ **(d)** $q_{0.05,60,5} = 3.977$

5. $q_{0.05,12,3} = 3.773$; $q_{\mu_3,\mu_2} = 3.932 > q_{0.05,12,3} = 3.773$, so reject H_0:
$\mu_2 = \mu_3$
$q_{\mu_3,\mu_1} = 3.757 < q_{0.05,12,3} = 3.773$, so do not reject H_0: $\mu_1 = \mu_3$
$q_{\mu_1,\mu_2} = 0.175 < q_{0.05,12,3} = 3.773$, so do not reject H_0: $\mu_1 = \mu_2$
Conclusion: Results are ambiguous, $\underline{\mu_2 \; \mu_1 \; \mu_3}$

7. (a) $H_0: \mu_1 = \mu_2 = \mu_3$; H_1: At least one of the means is different. $F = 9.72$; P-value $= 0.013 < \alpha = 0.05$. We reject the null hypothesis and conclude that there is sufficient evidence to support the hypothesis that at least one of the means is different.

(b) The result of Tukey's test indicates that $\mu_1 = \mu_2 \neq \mu_3$ ($\underline{\mu_1 \; \mu_2}$ μ_3).

(c)

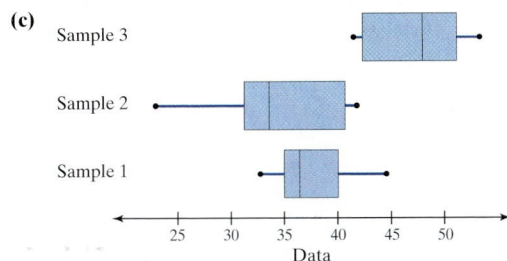

Data

9. The result of Tukey's test indicates that $\mu_{SP} = \mu_{NT} \neq \mu_{SD}$, ($\underline{\mu_{SP} \; \mu_{NT}} \; \mu_{SD}$). You should recommend using the spring disk method of planting.

11. The result of Tukey's test is ambiguous. It indicates that $\mu_{FL} = \mu_{AK}$ and $\mu_{TX} = \mu_{AK}$, but $\mu_{FL} \neq \mu_{TX}$ ($\underline{\mu_{FL} \; \mu_{AK} \; \mu_{TX}}$).

13. (a) $H_0: \mu_P = \mu_{R10} = \mu_{R17.5} = \mu_{A7.5} = \mu_{A12.5}$; H_1: At least one of the means is different. $F = 5.75$; P-value $= 0.001 < \alpha = 0.05$. We reject the null hypothesis and conclude that there is sufficient evidence to support the hypothesis that at least one of the mean abilities to follow rules is different.

(b) The result of Tukey's test indicates that $\mu_P \neq \mu_{R10} = \mu_{R17.5} = \mu_{A7.5} = \mu_{A12.5}$; ($\mu_P \; \underline{\mu_{R10} \; \mu_{R17.5} \; \mu_{A7.5} \; \mu_{A12.5}}$); that is, the mean ability to follow rules of a child taking the placebo is less than the mean ability to follow rules of a child taking any of the four drug treatments.

(c)

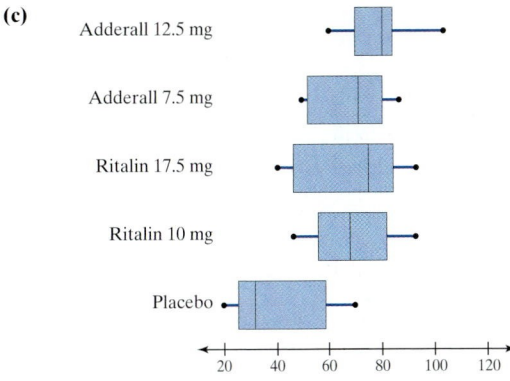

15. (a) $H_0: \mu_{nonsmoker} = \mu_{light} = \mu_{heavy}$ versus H_1: At least one of the means is different. $F = 17.09$; P-value $< 0.001 < \alpha = 0.05$. We reject the null hypothesis and conclude that there is sufficient evidence to support the hypothesis that at least one of the mean heart rates is different.

(b) The result of Tukey's test indicates that $\mu_{nonsmoker} \neq \mu_{light} = \mu_{heavy}$; $\mu_{nonsmoker} \; \underline{\mu_{light} \; \mu_{heavy}}$. We conclude that the mean heart rate of nonsmokers is lower than the mean heart rate of smokers, and that the mean heart rate of smokers is the same regardless if they are light smokers or heavy smokers.

(c)

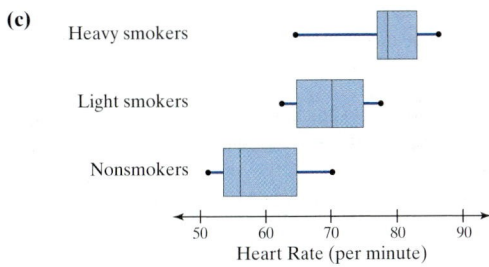

Heart Rate (per minute)

17. (a) $H_0: \mu_{skim milk} = \mu_{mixed milk} = \mu_{whole milk}$ versus H_1: At least one of the means is different. $F = 70.51$; P-value $< 0.001 < \alpha = 0.05$. We reject the null hypothesis and conclude that there is sufficient evidence to support the hypothesis that at least one of the mean calcium intake levels is different.

(b) The result of Tukey's test indicates that $\mu_{skim milk} = \mu_{whole milk} \neq \mu_{mixed milk}$; $\underline{\mu_{skim milk} \; \mu_{whole milk}} \; \mu_{mixed milk}$. We conclude that the mean calcium intake of children who drank mixed milk is different from the mean calcium intakes of children who drank skim or whole milk.

(c)

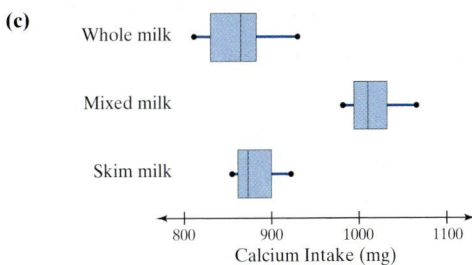

Calcium Intake (mg)

13.3 Assess Your Understanding (page 707)

1. In a completely randomized design, the researcher examines a single factor fixed at k levels and randomly distributes the experimental units among the levels. In a randomized complete block design, there is a factor whose level cannot be fixed. So the researcher partitions the experimental units according to this factor, forming blocks, which may form homogeneous groups. Within each block the experimental units are assigned to one of the treatments.

3. Randomized means that the order in which the treatment is applied within each block is random. Complete refers to the fact that each block gets every treatment.

5. A researcher can deal with explanatory variables by:
1. Controlling their levels so that they remain fixed throughout the experiment,
2. Manipulating or setting them at fixed levels, or
3. Randomizing so that any effects that cannot be identified or controlled are minimized.

7. (a) Since the *P*-value = 0.002, the researcher rejects the null hypothesis that the population means are equal.
(b) The mean square error MSE = 17.517.
(c) The Tukey test indicates that means 1 and 2 are the same, but mean 3 is different. ($\mu_1 = \mu_2 \neq \mu_3$ or $\underline{\mu_1\ \mu_2}\ \mu_3$)

9. (a) Since the treatment *P*-value = 0.108, the researcher does not reject the null hypothesis that the population means are equal.
(b) The mean square error MSE = 0.944.
(c) The Tukey test is only used if the null hypothesis is rejected.

11. (a) Since the treatment *P*-value = 0.002, we reject the null hypothesis that the population means are equal and conclude at least one of the means is different.
(b) The Tukey test indicates that means of treatment 2 and treatment 3 are equal, but the mean of treatment 1 is different.

(c)

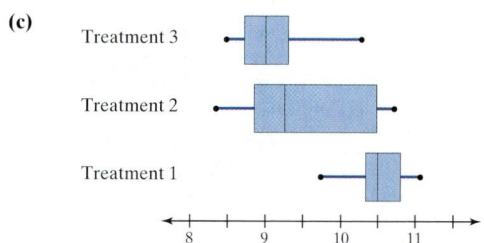

13. (a) The requirement for equal population variances is satisfied since the largest standard deviation is less than twice the smallest standard deviation, 1.439 < 2(1.225).
(b) Since the treatment *P*-value = 0.005, there is evidence that the mean miles per gallon is different among the three octane levels at the $\alpha = 0.05$ level of significance.
(c) The Tukey test indicates that $\mu_{87} = \mu_{89} < \mu_{92}$; $\underline{\mu_{87}\ \mu_{89}}\ \mu_{92}$.
(d) Conclude that the mean gas mileage is equal if 87- or 89-octane gasoline is used, but the mean miles per gallon increases when 92-octane gasoline is used.

15. (a) The requirement for equal population variances is satisfied since the largest standard deviation is less than twice the smallest standard deviation, 830 < 2(448).
(b) Since the treatment *P*-value = 0.105, there is not sufficient evidence to support the claim that the mean cost of repair is different among the car brands at the $\alpha = 0.05$ level of significance.
(c) Since the null hypothesis was not rejected, there is no need to do post hoc analysis.

17. (a) The requirement for equal population variances is satisfied since the largest standard deviation is less than twice the smallest standard deviation, 9.97 < 2(5.65).
(b) Since the treatment *P*-value = 0.001, there is sufficient evidence to support the claim that the mean water consumption of mice is different among the three experiment levels at the $\alpha = 0.05$ level of significance.
(c) The Tukey test indicates that $\mu_{LO-1} = \mu_{R+1} \neq \mu_{R+1\ Month}$; $\underline{\mu_{LO-1}\ \mu_{R+1}}\ \mu_{R+1\ month}$

19. (a) Since the treatment *P*-value is 0.438, there is not sufficient evidence to support the claim that there is a difference in mean wait time between the two loading procedures.
(b) Hypotheses: $H_0: \mu_d = 0$ versus $H_1: \mu_d \neq 0$
Classical approach: $t_0 = 0.83 < t_{0.025,6} = 2.447$, so we do not reject the null hypothesis.
P-value approach: *P*-value = 0.438 > $\alpha = 0.05$, so we do not reject the null hypothesis.
There is not sufficient evidence at the $\alpha = 0.05$ level of significance to support the quality-control manager's claim that the new loading procedure changes the wait time.

13.4 Assess Your Understanding (page 723)

1. In a completely randomized design, the researcher examines a single factor fixed at *k* levels and randomly distributes the experimental units among the levels. In a randomized complete block design, there

is a factor whose level cannot be fixed. So the researcher partitions the experimental units according to this factor, forming blocks. The experimental units are then randomly assigned to the blocks, being sure each block has each treatment. In a factorial design, the researcher has two variables A and B, called factor A and factor B. The factors are fixed, so factor A has *a* levels and factor B has *b* levels. The experimental units are then uniformly distributed among the cells formed by the factors, and the response variable is measured at each of the *ab* cells.

3. An interaction plot is a graph that illustrates the role of interaction in a factorial design.

5. The plots are relatively parallel, so there is no significant interaction.

7. The plots cross, so there is significant interaction.

9. (a) Main effect of factor A = 1.25; main effect of factor B = −9.25.
Interaction effects: If factor B is fixed at low and factor A changes from low to high, the response variable increases by 7 units. On the other hand, if factor B is fixed at high and factor A changes from low to high, the response variable increases by −4.5 (decreases by 4.5) units. So the change in the response variable for factor A depends on the level of factor B, meaning there is an interaction effect between factors A and B.
If factor A is fixed at low and factor B changes from low to high, the response variable increases by −3.5 (decreases by 3.5) units. On the other hand, if factor A is fixed at high and factor B changes from low to high, the response variable increases by −15 (decreases by 15) units. So the change in the response variable for factor B depends on the level of factor A, meaning there is an interaction effect between factors A and B.

(b)

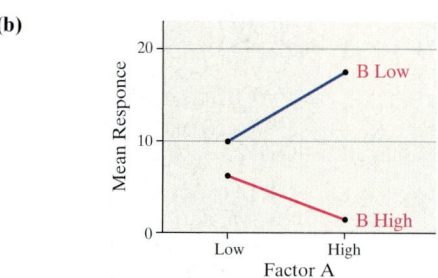

11. (a) There is no evidence of an interaction effect since interaction *P*-value is 0.836.
(b) Since factor A *P*-value = 0.003, there is evidence of difference in means from factor A. The factor B *P*-value < 0.001, giving evidence that there is also a difference in means from factor B.
(c) MSE = 45.3

13. (a) There is evidence of an interaction effect since interaction *P*-value < 0.001.
(b) Whenever there is evidence of an interaction effect, we do not consider main effects because they could be misleading.
(c) MSE = 31.9

15. (a) There is no evidence of an interaction effect since interaction *P*-value is 0.965.
(b) Since factor A *P*-value = 0.579, there is no evidence of difference in means from factor A. The factor B *P*-value = 0.021 is evidence that there is a difference in means from factor B.

(c)

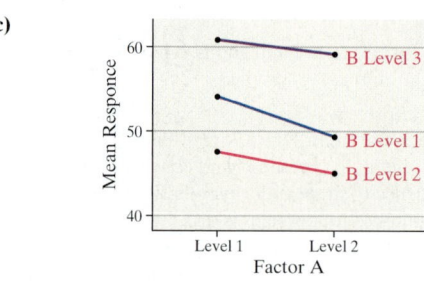

(d) The Tukey's test for factor A is unnecessary. The Tukey's test for factor B is ambiguous. It shows $B_1 = B_2 = B_3$, but $B_2 \neq B_3$; ($\underline{B_1\ B_2}\ B_3$).

17. (a) This is a 2×3 factorial design with two replications per cell.
   (b) There is no evidence of an interaction effect since interaction P-value is 0.159.
   (c) There is no significant difference in the means for the three age groups; age P-value = 0.016. There is no significance difference in the means for the two genders; gender P-value = 0.164.

   (d)

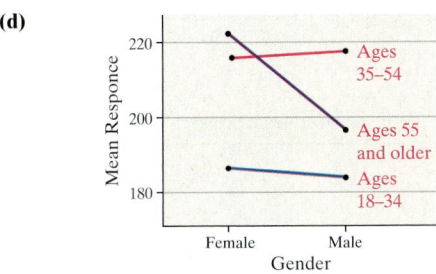

   (e) The mean cholesterol of the 18–34 year olds is significantly different from the other age groups.

19. (a) The requirement of equal population variances is satisfied, since the largest standard deviation is less than twice the smallest standard deviation [$320 < 2(168) = 336$].
   (b) There is no evidence of an interaction effect. Interaction P-value = 0.469.
   (c) There is evidence of a difference in the means for the three mixtures. There is evidence of a difference in the means for the three types of slump.

   (d)

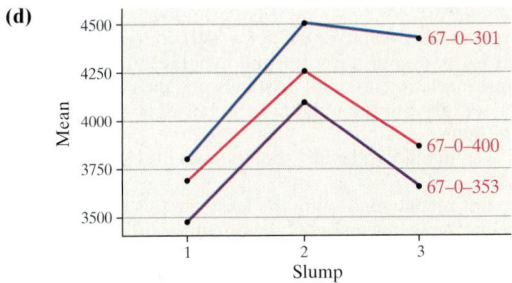

   (e) The Tukey's test indicates that the mean 28-day strength of mixture 67-0-400 is significantly different from the mean 28-day strength of the other two mixtures. The Tukey's test indicates that the mean 28-day strength for slump 3.75 is significantly different from the mean 28-day strength for slump 4 and slump 5. The mean 28-day strength for slump 4 is significantly different from the mean 28-day strength for slump 5.

21. (a) The requirement of equal population variances is satisfied, since the largest standard deviation is less than twice the smallest standard deviation [$4.00 < 2(2.5) = 5.0$].
   (b) There is no evidence of an interaction effect. Interaction P-value = 0.854.
   (c) There is no evidence of a difference in the means for three locations. Location P-value = 0.287. There is no evidence of a difference in the means for type of service station; type P-value = 0.804.

   (d)

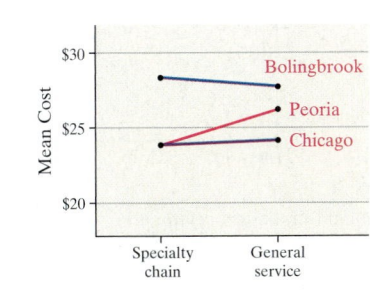

   (e) Since there were no significant main effects, there is no need to do Tukey's tests.

Chapter Review Exercises (page 729)

1. (a) $q_{0.05,16,7} = 4.741$
   (b) $q_{0.05,30,6} = 4.302$
   (c) $q_{0.01,42,4} \approx 4.696$
   (d) $q_{0.05,40,6} = 4.232$

3. (a) H_0: $\mu_{\text{water–organic soil}} = \mu_{\text{water–mineral soil}} = \mu_{\text{surface water}}$ versus H_1: At least one of the means is different.
   (b) To perform a one-way ANOVA, there must be k simple random samples drawn from k normally distributed populations all with the same variance. The k samples must be independent of each other.
   (c) Since P-value < 0.001, the researcher will reject the null hypothesis and conclude that the mean concentration of dissolved carbon is different in at least one of the collection areas.
   (d) Answers will vary.
   (e) $F = \dfrac{\text{MST}}{\text{MSE}} = \dfrac{235.7}{25.0} = 9.43$
   (f) The Tukey's test indicates that the mean dissolved carbon is equal in the surface water and in the ground water taken from organic soil, but it differs from the mean dissolved carbon concentration in ground water taken from mineral soil.
   $\underline{\mu_{\text{surface water}} \ \mu_{\text{water–organic soil}}} \ \mu_{\text{water–mineral soil}}$

5. The Tukey's test is ambiguous. It indicates that $\mu_{\text{select structural}} \neq \mu_{\text{no. 2}}$, but that the $\mu_{\text{below grade}}$ is equal to all other means.
   $\underline{\mu_1 \ \mu_2 \ \mu_4} \ \mu_3$

7. (a) The P-value < 0.001, so we reject the null hypothesis and conclude at least one of the means is different.
   (b) The Tukey's test indicates that $\mu_1 = \mu_2 \neq \mu_3 \neq \mu_4$; $\underline{\mu_1 \ \mu_2}$ $\underline{\mu_3 \ \mu_4}$.

   (c)

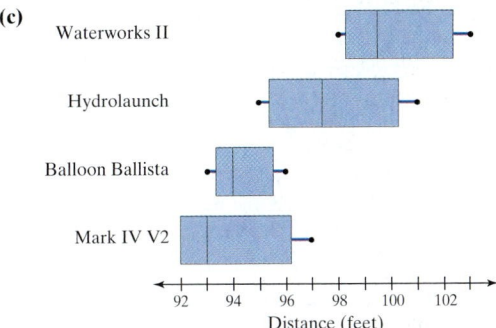

9. (a) Since the largest standard deviation is less than twice the smallest deviation, the requirement of equal population variances is satisfied [$4.72 < 2(2.39) = 4.78$].
   (b) There is no evidence of an interaction effect. Interaction P-value = 0.374.
   (c) There is no evidence of a difference in the means for gender; gender P-value = 0.196. There is evidence of a difference in the means for type of driving program; program P-value < 0.001. So the null hypothesis H_0: $\mu_{\text{8 hour}} = \mu_{\text{4 hour}} = \mu_{\text{2 hour}}$ is rejected.

   (d)

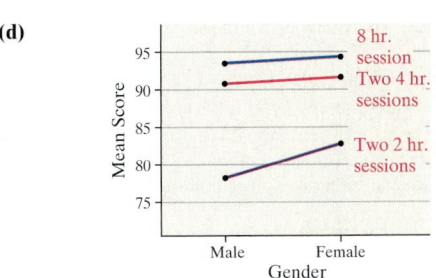

   (e) The Tukey's test indicates that the mean score after taking one 8-hour defensive driving class is equal to the mean score after taking two 4-hour classes, but that the mean score after taking two 2-hour classes is different. $\underline{\mu_{\text{8 hour}} \ \mu_{\text{4 hour}}} \ \mu_{\text{2 hour}}$

CHAPTER 14

14.1 Assess Your Understanding (page 750)

1. There are two requirements for conducting inference on the least squares regression model. First, for each particular value of the explanatory variable x, the corresponding responses in the population have a mean that depends linearly on x. Second, the response variables are normally distributed with mean $\mu_{y|x} = \beta_0 + \beta_1 x$ and standard deviation σ.

The first requirement is tested by examining a plot of the residuals against the explanatory variable. If the plot shows any discernible pattern, then the linear model is inappropriate. The normality requirement is checked by examining a normal probability plot of the residuals.

3. The y-coordinates on the least-squares regression line represent the mean value of the response variable for any given value of the explanatory variable.

5. If the null hypothesis, $H_0: \beta_1 = 0$, is not rejected, the best guess for the value of the response variable for any value of the explanatory variable is the sample mean of y.

7. (a) $\beta_0 \approx b_0 = -2.3256$, $\beta_1 \approx b_1 = 2.0233$
(b) $s_e = 0.5134$ is the point estimate for σ.
(c) $s_{b_1} = 0.1238$
(d) Since the P-value $< 0.001 < \alpha = 0.05$ (or $t_0 = 16.345 > t_{0.025} = 3.182$), we reject the null hypothesis and conclude that a linear relation exists between x and y.

9. (a) $\beta_0 \approx b_0 = 1.200$, $\beta_1 \approx b_1 = 2.200$
(b) $s_e = 0.8944$ is the point estimate for σ.
(c) $s_{b_1} = 0.2828$
(d) Because the P-value $= 0.004 < \alpha = 0.05$ (or $t_0 = 7.778 > t_{0.025} = 3.182$), we reject the null hypothesis and conclude that a linear relation exists between x and y.

11. (a) $\beta_0 \approx b_0 = 116.600$, $\beta_1 \approx b_1 = -0.7200$
(b) $s_e = 3.2863$ is the point estimate for σ.
(c) $s_{b_1} = 0.1039$
(d) Because the P-value $= 0.006 < \alpha = 0.05$ (or $t_0 = -6.928 < -t_{0.025} = -3.182$), we reject the null hypothesis and conclude that a linear relation exists between x and y.

13. (a) $\beta_0 \approx b_0 = 12.4932$, $\beta_1 \approx b_1 = 0.1827$
(b) $s_e = 0.0954$
(c) A normal probability plot of the residuals shows that they are approximately normally distributed.
(d) $s_{b_1} = 0.02756$
(e) Because the P-value $< 0.001 < \alpha = 0.01$ (or $t_0 = 6.63 > t_{0.005} = 3.250$), we reject the null hypothesis and conclude that a linear relation exists between a child's height and head circumference.
(f) 95% confidence interval: Lower bound: 0.1203; Upper bound: 0.2450
(g) A good estimate of the child's head circumference would be 17.34 inches.

15. (a) $\beta_0 \approx b_0 = 2675.6$, $\beta_1 \approx b_1 = 0.6764$
(b) $s_e = 271.04$
(c) A normal probability plot of the residuals shows that they are approximately normally distributed.
(d) $s_{b_1} = 0.2055$
(e) Because the P-value $= 0.011 < \alpha = 0.05$ (or $t_0 = 3.291 > t_{0.025} = 2.306$), we reject the null hypothesis and conclude that a linear relation exists between the 7-day strength and the 28-day strength of this type of concrete.
(f) 95% confidence interval: Lower bound: 0.2025; Upper bound: 1.1503
(g) The mean 28-day strength of this concrete if the 7-day strength is 3000 psi is 4704.8 psi.

17. (a) $\beta_0 \approx b_0 = 0.5975$, $\beta_1 \approx b_1 = 0.9764$
(b) $s_e = 2.3518$
(c) A normal probability plot of the residuals shows that they are approximately normally distributed.
(d) $s_{b_1} = 0.3379$

(e) Because the P-value $= 0.018 < \alpha = 0.10$ (or $t_0 = 2.890 > t_{0.05} = 1.833$), we reject the null hypothesis and conclude that a linear relation exists between the rate of return of the S&P 500 index and the rate of return of UTX.
(f) 90% confidence interval: Lower bound: 0.357; Upper bound: 1.598
(g) When the mean rate of return of the S&P 500 index is 4.2%, the mean rate of return for United Technologies is 4.698%.

19. (a)

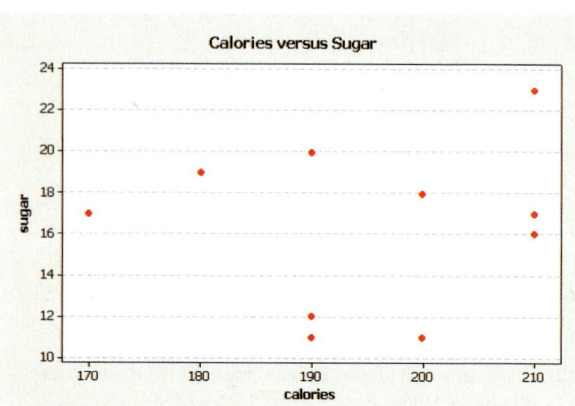

(b) $\hat{y} = 0.93 + 0.0821x$ **(c)** $s_e = 4.1512$
(d) A normal probability plot of the residuals shows that they are approximately normally distributed.
(e) $s_{b_1} = 0.0961$
(f) Because the P-value $= 0.41 > \alpha = 0.01$ (or $t_0 = 0.854 < t_{0.005} = 3.169$), we do not reject the null hypothesis and conclude that a linear relation does not exist between the number of calories per serving and the number of grams of sugar per serving in high-fiber cereal.
(g) 95% confidence interval: Lower bound: -0.1320; Upper bound: 0.2962
(h) Do not recommend using the least-squares regression line to predict the sugar content of the cereal because we did not reject the null hypothesis. A good guess to the sugar content would be $\bar{y} = 17.08$ grams.

21. (a) $\hat{y} = -3,771,542,830 + 2,018,994x$
(b) $t = 21.81$; P-value < 0.001; Reject $H_0: \beta_1 = 0$.

(c)

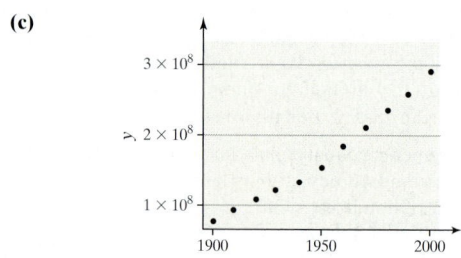

(d)

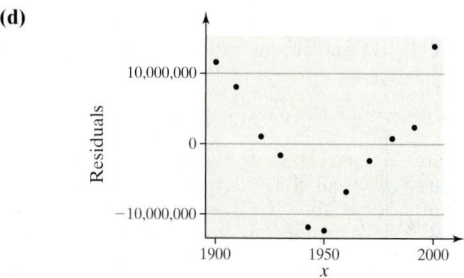

(e) Linear model is not appropriate.
(f) The moral is that the inferential procedures may lead us to believe that a linear relation between the two variables exists; even though diagnostic tools (such as residual plots) indicate that a linear model is inappropriate.

23. (a)

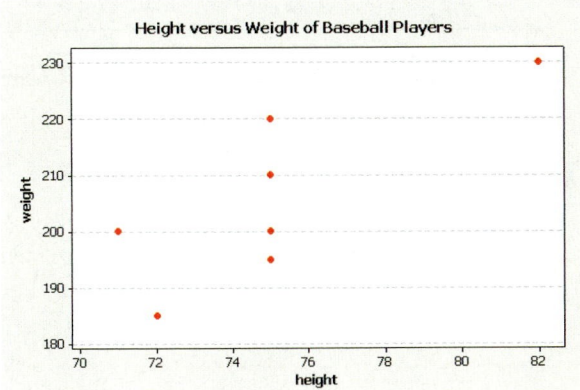

(b) $\hat{y} = -45 + 3.3605x$; Because the P-value $= 0.0166 < \alpha = 0.05$ (or $t_0 = 3.130 > t_{0.025} = 2.365$), we reject the null hypothesis and conclude that a linear relation exists between the weight and height of baseball players.

(c) $\hat{y} = -63.4722 + 3.611x$; Because the P-value $= 0.1772 > \alpha = 0.05$ (or $t_0 = 1.529 < t_{0.025} = 2.447$), we do not reject the null hypothesis and conclude that a linear relation does not exist between baseball players' weight and height.

14.2 Assess Your Understanding (page 759)

1. Confidence intervals are used to measure the accuracy of the mean response of all the individuals in the population, whereas, prediction intervals are used to measure the accuracy of a single individual's predicted value.

3. (a) $\hat{y} = 11.8$
  (b) Lower bound: 10.8; Upper bound: 12.8
  (c) $\hat{y} = 11.8$
  (d) Lower bound: 9.9; Upper bound: 13.7
  (e) The confidence interval is an interval estimate for the mean value of y at $x = 7$, whereas the prediction interval is an interval estimate for a single value of y at $x = 7$.

5. (a) $\hat{y} = 4.3$
  (b) Lower bound: 2.5; Upper bound: 6.1
  (c) $\hat{y} = 4.3$
  (d) Lower bound: 0.9; Upper bound: 7.7

7. (a) $\hat{y} = 17.20$ inches
  (b) 95% confidence interval: Lower bound: 17.12 inches; Upper bound: 17.28 inches
  (c) $\hat{y} = 17.20$ inches
  (d) 95% prediction interval: Lower bound: 16.97 inches; Upper bound: 17.43 inches
  (e) The confidence interval is an interval estimate for the mean head circumference of all children who are 25.75 inches tall. The prediction interval is an interval estimate for the head circumference of a single child who is 25.75 inches tall.

9. (a) $\hat{y} = 4400.4$ psi
  (b) 95% confidence interval: Lower bound: 4147.8 psi; Upper bound: 4653.1 psi
  (c) $\hat{y} = 4400.4$ psi
  (d) 95% prediction interval: Lower bound: 3726.3 psi; Upper bound: 5074.6 psi
  (e) The confidence interval is an interval estimate for the mean 28-day strength of all concrete cylinders that have a 7-day strength of 2550 psi. The prediction interval is an interval estimate for the 28-day strength of a single cylinder whose 7-day strength is 2550 psi.

11. (a) $\hat{y} = 4.698\%$
  (b) 90% confidence interval: Lower bound: 2.234%; Upper bound: 7.163%
  (c) $\hat{y} = 4.698\%$
  (d) 90% prediction interval: Lower bound: -0.267%; Upper bound: 9.6664%

(e) Although the predicted rate of return in parts (a) and (c) are the same, the intervals are different because the distribution of the mean rate of return, part (a), has less variability than the distribution of the individual rate of return of a particular month, part (c).

13. (a) It does not make sense to construct either a confidence interval or prediction interval based on the least squares regression equation because the evidence indicated that there was no linear relation between calories and sugar.
  (b) 95% confidence interval for the mean sugar content: Lower bound: 14.5 grams; Upper bound: 19.7 grams.

14.3 Assess Your Understanding (page 776)

1. A correlation matrix shows the linear correlation among all variables under consideration in a multiple linear regression model.

3. If the null hypothesis is rejected, we conclude that at least one of the explanatory variables is linearly related to the response variable. It is very possible that one (or more) of the $\beta_i = 0$ although the null hypothesis is rejected.

5. The difference between R^2 and adjusted R^2 is that R^2 becomes larger as more explanatory variables are added to the regression model regardless of their contribution to the explanation of the response variable. Adjusted R^2 compensates for this ability to artificially increase R^2 by simply adding more explanatory variables. It modifies R^2 based on the sample size, n, and the number of explanatory variables, k. Adjusted R^2 actually decreases if the additional variable does little to explain the variation in the response variable. So the adjusted R^2 is better for determining whether an additional explanatory variable should be added to the model.

7. (a) The slope coefficient of x_1 is 3. This indicates that $\hat{y}$ will increase 3 units, on average, for every one unit increase in x_1, provided x_2 remains constant. The slope coefficient of x_2 is -4. This indicates that $\hat{y}$ will decrease 4 units, on average, for every one unit increase in x_2, provided x_1 remains constant.

(b) $\hat{y} = 35 - 4x_2$

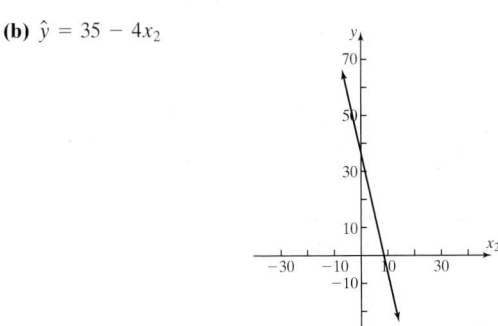

(c) $\hat{y} = 50 - 4x_2$ **(d)** $\hat{y} = 65 - 4x_2$

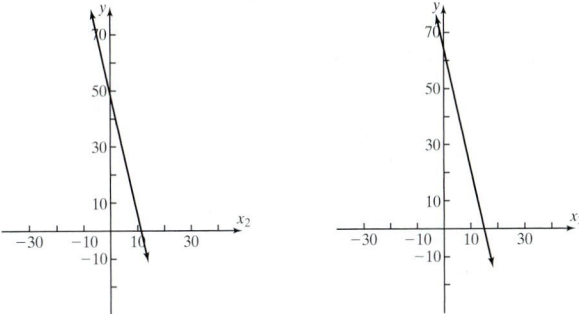

(e) Changing the value of x_1 has the effect of changing the y-intercept of graph of the regression line.

9. (a) $R^2_{\text{adj}} = 0.603$
  (b) $F_0 = 13.173$
  (c) $R^2_{\text{adj}} = 0.598$; Do not add the new variable. Its addition reduces the proportion of variance explained by the model.

11. (a)

Correlations: x1, x2, x3, y

	x1	x2	x3
x2	−0.460		
x3	0.183	−0.012	
y	−0.748	0.788	0.258

There is no evidence that multicollinearity will cause problems.

(b) $\hat{y} = 7.9647 - 0.1035x_1 + 0.9359x_2 + 0.1152x_3$

(c) $F_0 = 31.33$. Since the P-value < 0.001 we reject the null hypothesis and conclude that at least one of the explanatory variables is linearly related to the response variable.

(d) Reject H_0: $\beta_1 = 0$ since $t_{\beta1} = -4.98$ and the P-value $= 0.003 < \alpha = 0.05$.
Reject H_0: $\beta_2 = 0$ since $t_{\beta2} = 4.68$ and the P-value $= 0.003 < \alpha = 0.05$.
Reject H_0: $\beta_3 = 0$ since $t_{\beta3} = 3.61$ and the P-value $= 0.011 < \alpha = 0.05$.

13. (a)

Correlations: x1, x2, x3, x4, y

	x1	x2	x3	x4
x2	−0.184			
x3	0.182	−0.106		
x4	−0.166	−0.588	0.291	
y	−0.046	0.932	−0.063	−0.750

There is no evidence that multicollinearity will cause problems.

(b) $\hat{y} = 279.02 + 0.068x_1 + 1.146x_2 + 0.3897x_3 - 2.9378x_4$
$F_0 = 40.90$ and P-value < 0.001 so we reject the H_0 and conclude that at least one of the explanatory variables is linearly related to the response variable. The variables x_1 and x_3 have slope coefficients that are not significantly different from zero.

(c) $\hat{y} = 282.33 + 1.1426x_2 + 0.3941x_3 - 2.9570x_4$; The F-test still indicates that the model is significant (P-value < 0.001). Remove x_3; $\hat{y} = 281.62 + 1.1584x_2 - 2.6267x_4$

(d)

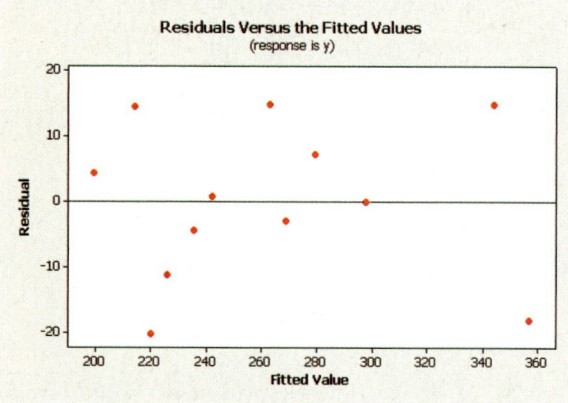

Residuals Versus the Fitted Values
(response is y)

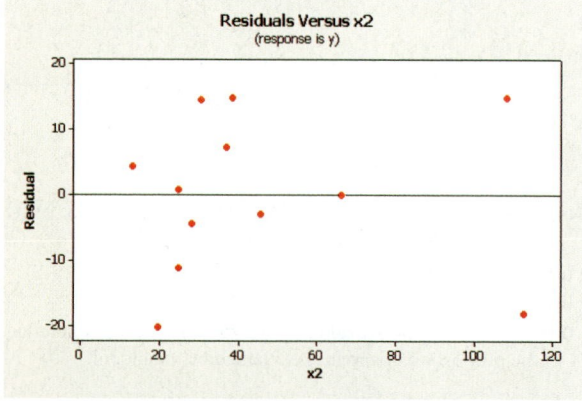

Residuals Versus x2
(response is y)

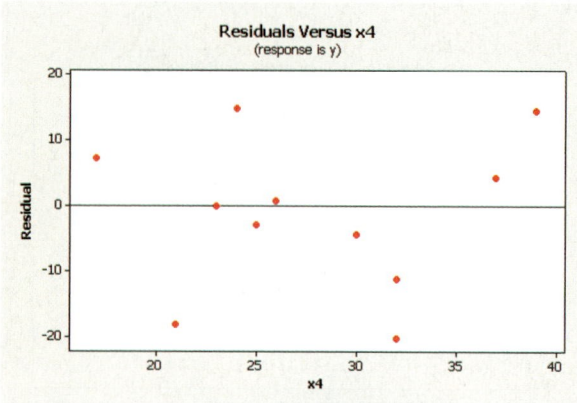

Residuals Versus x4
(response is y)

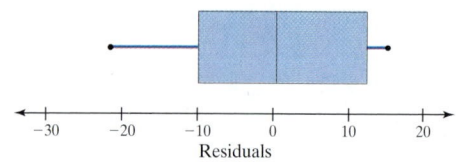

(e) $\hat{y} = 246.68$

(f)

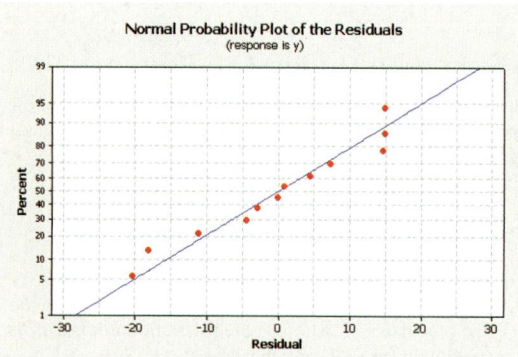

Normal Probability Plot of the Residuals
(response is y)

The residuals are normally distributed, so it is reasonable to construct confidence and prediction intervals.

(g) 95% confidence interval: Lower bound: 237.45; Upper bound: 255.91
95% prediction interval: Lower bound: 214.93; Upper bound: 278.43

15. (a) Yes, and conclude that at least one of the slope coefficients are linearly related to the response variable, computer anxiety.

(b) $\beta_1 = -0.87$: A one point increase in the computer confidence variable reduces the computer anxiety measurement by 0.87 point, on average, provided all other explanatory variables remain constant.
$\beta_2 = -0.51$: A one point increase in the computer knowledge variable reduces the computer anxiety measurement by 0.51 point, on average, provided all other explanatory variables remain constant.
$\beta_3 = -0.45$: A one point increase in the computer liking scale reduces the computer anxiety measurement by 0.45 point, on average, provided all other explanatory variables remain constant.
$\beta_4 = 0.33$: A one point increase in the trait anxiety scale increases the computer anxiety measurement by 0.33 point, on average, provided all other explanatory variables remain constant. They are all reasonable.

(c) $\hat{y} = 57.79$

(d) $R^2 = 0.69$ indicates that 69% of the variance in computer anxiety is explained by this model.

(e) The statement means that the requirements of a multiple linear regression model were checked and were met.

17. (a) $\hat{y} = -12.509 + 1.334x_1 - 0.414x_2$

(b)

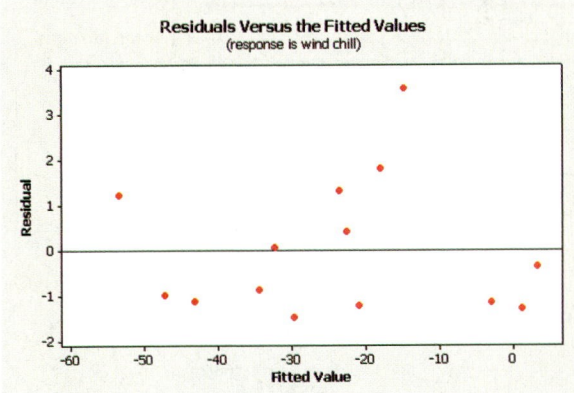

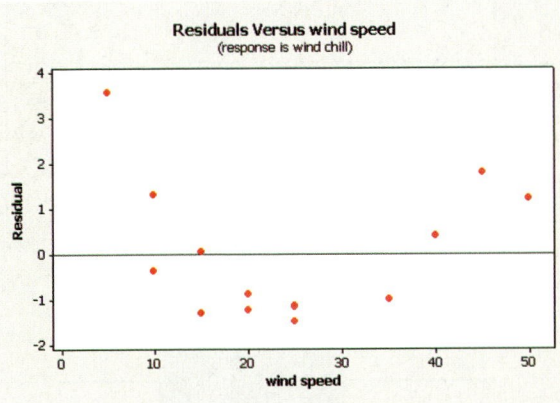

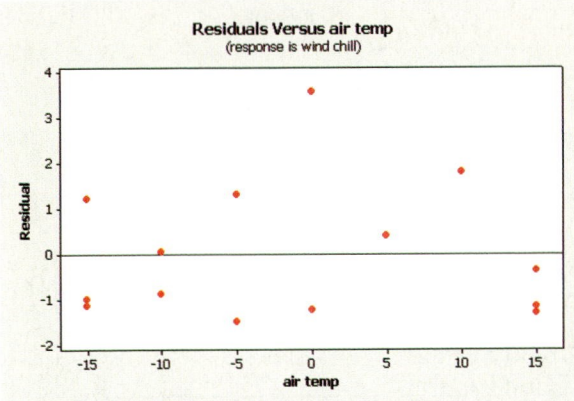

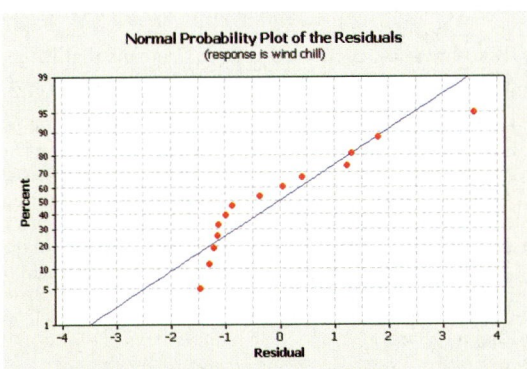

19. (a)   Correlations: SLUMP, 7 DAY, 28 DAY

```
        SLUMP       7 DAY
7 DAY    -0.460
28 DAY   -0.753      0.737
```

There is no reason to be concerned about multicollinearity based on the model.

(b) $\hat{y} = 3890.5 - 295.9x_1 + 0.552x_2$

(c)

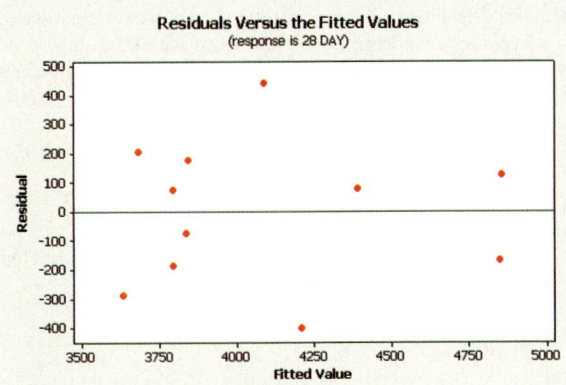

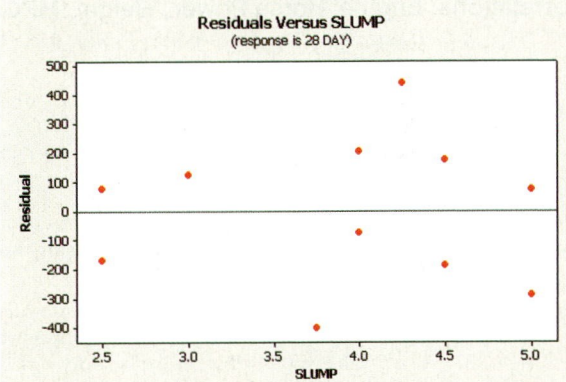

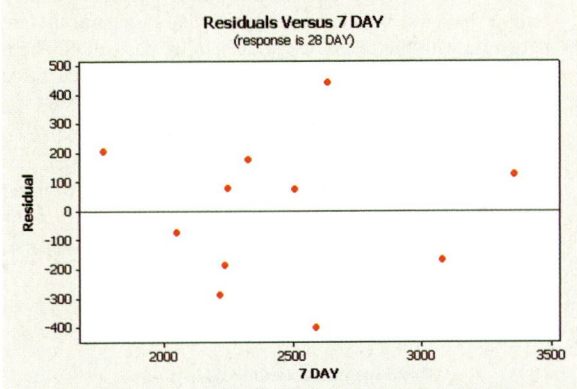

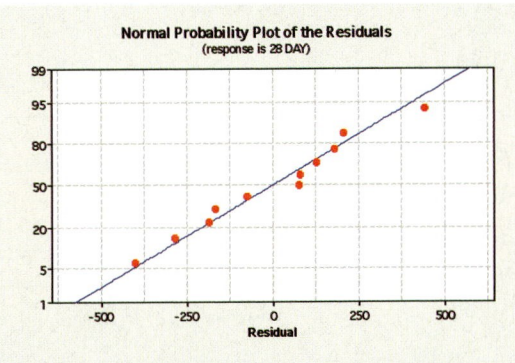

(d) $\beta_1 = -295.9$: A one unit increase in slump of the concrete decreases the 28-day strength of the concrete by 295.9 pounds per inch, on average, provided the other variable remains constant. $\beta_2 = 0.552$: A one pound per inch increase in 7-day strength increases the 28-day strength of the concrete by 0.552 pound per inch, on average, provided slump remains constant.

(e) $R^2 = 76.0\%$ represents the variance in 28-day strength explained by the model. $R^2_{\text{adj}} = 70.1\%$ modifies the value of R^2 based on the sample size, and the number of explanatory variables in the model.

(f) The F-test has a P-value $= 0.003 < 0.05$, so we reject the null hypothesis and conclude that at least one of the slope coefficients is linearly related to the 28-day strength of the concrete.

(g) Reject H_0: $\beta_1 = 0$ since $t_{\beta 1} = -2.69$ and the P-value $= 0.027 < \alpha = 0.05$.
Reject H_0: $\beta_2 = 0$ since $t_{\beta 2} = 2.54$ and the P-value $= 0.034 < \alpha = 0.05$.

(h) $\hat{y} = 4208.0$ pounds per inch.

(i) $\hat{y} = 4208.0$ pounds per inch

(j) 95% confidence interval: Lower bound: 3988.6 pounds; Upper bound: 4427.2 pounds per inch.
95% prediction interval: Lower bound: 3553.6 pounds; Upper bound: 4882.2 pounds per inch.

21. (a)

Correlations: Engine, Horse Power, Weight, MPG

	Engine	Horse Power	Weight
Horse Power	0.323		
Weight	0.092	0.370	
MPG	−0.812	−0.230	−0.477

There is no concern with multicollinearity.

(b) $\hat{y} = 35.181 - 0.00257x_1 + 0.0154x_2 - 0.00184x_3$

(c) Since the P-value $= 0.002$ ($F_0 = 14.34$), we reject the null hypothesis, and conclude that at least one of the explanatory variables is linearly associated with gasoline mileage.

(d) For the explanatory variable engine size, $t_0 = -5.61$ with a P-value $= 0.001$. For the explanatory variable horsepower, $t_0 = 1.37$ with a P-value $= 0.214$. For the explanatory variable weight, $t_0 = -3.16$ with a P-value $= 0.016$.
So we reject the first and the third null hypotheses, and conclude that both engine size and weight have a significant linear relation with the response variable, miles per gallon. The hypothesis $\beta_2 = 0$ is not rejected, so we conclude that the explanatory variable, horsepower, is not linearly related to mileage, and we remove x_2, horsepower, from the model.

(e) $\hat{y} = 36.904 - 0.00237x_1 - 0.00156x_3$; The slope coefficients of both explanatory variables are significantly different from zero. For the explanatory variable engine size, $t_0 = -5.19$ with a P-value $= 0.001$. For the explanatory variable weight, $t_0 = -2.71$ with a P-value $= 0.027$.

(f)

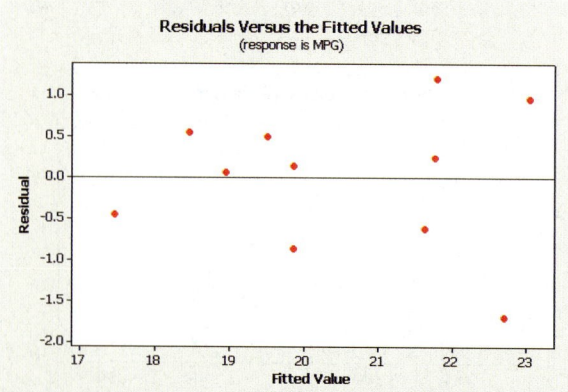

Residuals Versus the Fitted Values
(response is MPG)

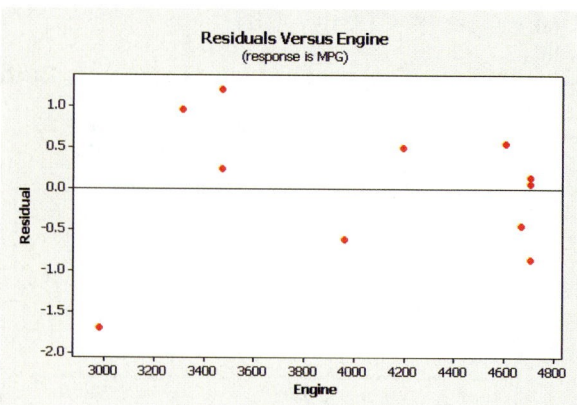

Residuals Versus Engine
(response is MPG)

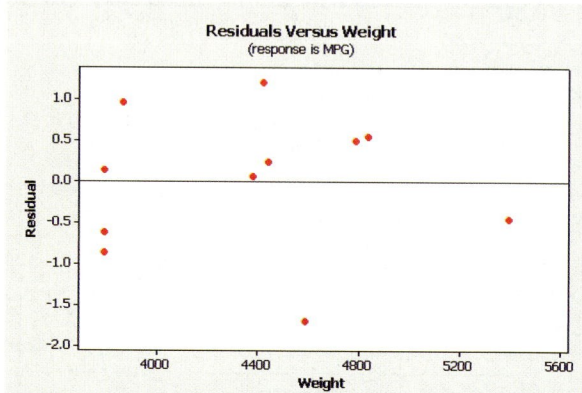

Residuals Versus Weight
(response is MPG)

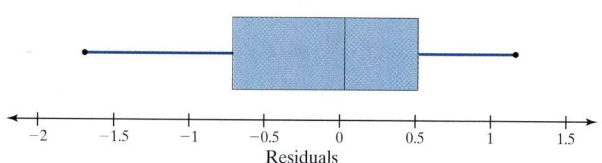

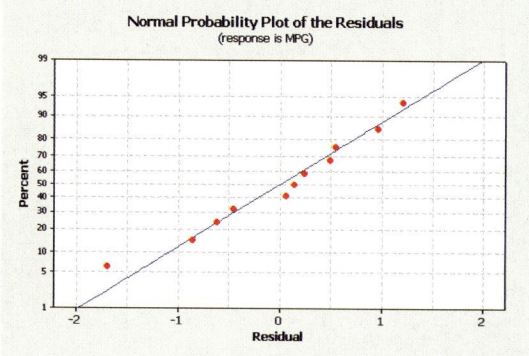

Normal Probability Plot of the Residuals
(response is MPG)

(g) The regression coefficient -0.00237 means that if the car's engine increases by 1 cubic centimeter then gas mileage decreases, on average, by 0.00237 mile per gallon, assuming that vehicle weight remains constant. The regression coefficient -0.00156 means that if the car's weight is increased by one pound, then gas mileage decreases, on average, by 0.00156 mile per gallon, assuming that vehicle's engine size remains constant.

(h) $R^2 = 0.823$; $R^2_{\text{adj.}} = 0.779$. R^2 and R^2_{adj} are measures of the proportion of the variance in gas mileage that is explained by the model.

(i) 95% confidence interval: (19.822, 21.601). 95% prediction interval: (18.348, 23.075)

23. (a)

Correlations: Square Footage, Bedrooms, Baths Asking Price

	Square Footage	Bedrooms	Baths
Bedrooms	0.791		
Baths	0.731	0.759	
Asking Price	0.749	0.738	0.669

There is a problem with multicollinearity among all three explanatory variables.

(b) $\hat{y} = 161.17 + 0.0268x_1 + 16.76x_2 + 8.70x_3$

(c) Since the P-value $= 0.026$ ($F_0 = 4.99$), we reject the null hypothesis, and conclude that at least one of the explanatory variables is linearly associated with asking price for a house.

(d) For the explanatory variable square footage, $t_0 = 1.12$ and the P-value $= 0.292$. For the explanatory variable number of bedrooms, $t_0 = 0.87$ and the P-value $= 0.407$. For the explanatory variable number of baths, $t_0 = 0.41$ and the P-value $= 0.694$. We do not reject any of the null hypotheses and conclude that none of the explanatory variables are linearly related to the asking price for a house. We notice that this contradicts the result found in part (b).

(e) Remove $x_3 =$ baths: $\hat{y} = 163.79 + 0.03x_1 + 20.16x_2$. Neither of the two slope coefficients are significantly different from zero. x_2, number of bedrooms, has the larger P-value, so we remove it from the model. $\hat{y} = 181.07 + 0.0508x_1$

(f)

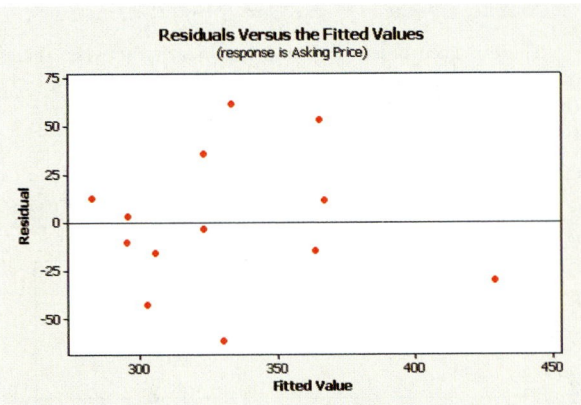

Residuals Versus the Fitted Values
(response is Asking Price)

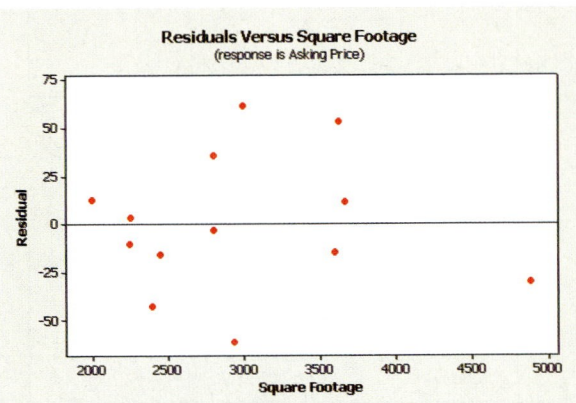

Residuals Versus Square Footage
(response is Asking Price)

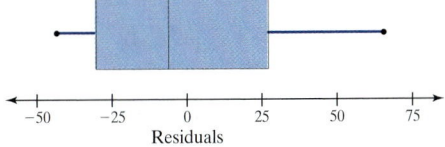

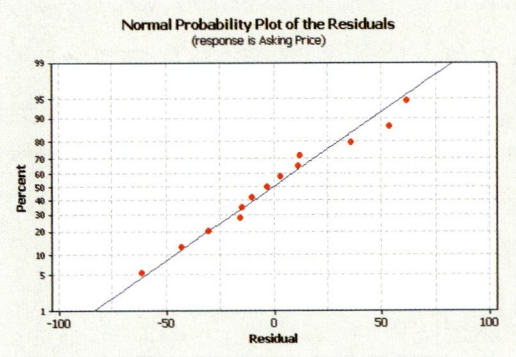

Normal Probability Plot of the Residuals
(response is Asking Price)

(g) The regression coefficient 0.0508 indicates that if the house increases by 1 square foot then the asking price will increase by \$50.80, on average.

(h) 95% confidence interval: (305.4, 351.2). We are 95% confident that mean asking price for all 2900-square foot houses is between \$305,400 and \$351,200. 95% prediction interval: (243.0, 413.6). We are 95% confident that the asking price of a randomly chosen 2900-square foot house is between \$243,000 and \$413,600.

25. (a)

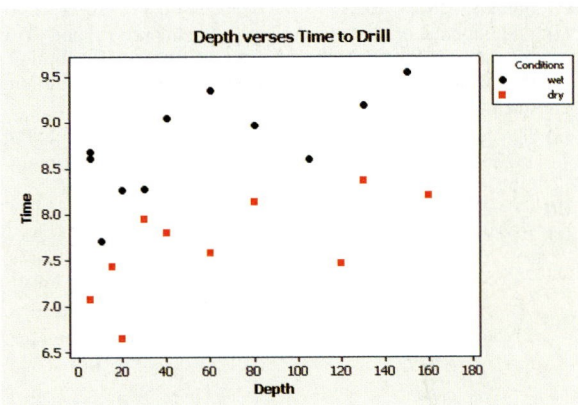

Depth verses Time to Drill

(b) $\hat{y} = 8.351 + 0.00682x_1 - 1.136x_2$

(c) For the explanatory variable depth, $t_0 = 3.86$ and the P-value $= 0.001$. So we reject the null hypothesis and conclude that there is a significant linear relation depth and time needed to drill another 5 feet.

For the explanatory variable condition, $t_0 = -6.36$ and the P-value < 0.001. So we reject the null hypothesis and conclude that there is a significant linear relation condition and time needed to drill another 5 feet.

(d) 95% confidence interval: (7.598, 8.196). We are 95% confident that mean time it takes to drill an additional 5 feet is between 7.598 minutes and 8.196 minutes. 95% prediction interval: (6.990, 8.804). We are 95% confident that for a randomly chosen drilling, it will take between 6.990 minutes and 8.804 minutes to drill an additional 5 feet.

Chapter Review Exercises (page 786)

1. (a) $\beta_0 \approx b_0 = 30.3848$; $\beta_1 \approx b_1 = -2.7977$;
$\hat{y} = 30.3848 - (2.7977)(3.8) = 19.75$, so the mean number of miles per gallon of cars with 3.8-liter engines is 19.8 miles per gallon.

(b) $s_e = 1.269$

(c)

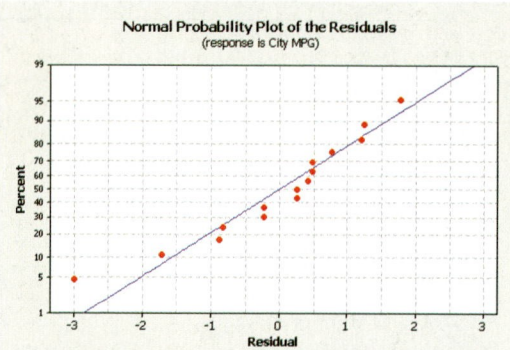

The normal probability plot indicates that the residuals are normally distributed.

(d) $s_b = 0.3859$

(e) There is evidence that a linear relation exists between engine displacement and city MPG (P-value < 0.001)

(f) 95% confidence interval about the slope of the true least-squares regression line: Lower bound: -3.6312, Upper bound: -1.9642.

(g) 90% confidence interval about the mean miles per gallon of all cars with a 3.8 liter engine: Lower bound: 19.065, Upper bound: 20.535.

(h) $\hat{y} = 19.8$

(i) 90% prediction interval about the miles per gallon of a particular car with a 3.8 liter engine: Lower bound: 17.552, Upper bound: 22.048.

(j) Although the predicted miles per gallon in parts (a) and (h) are the same, the intervals are different because the distribution of the means, part (a), has less variability than the distribution of the individuals, part (h).

3. (a) $\beta_0 \approx b_0 = -399.2$; $\beta_1 \approx b_1 = 2.5315$; $-399.2 + (2.5315) \cdot (900) = 1879.15$; so the mean rent of a 900-square-foot apartment in Queens is 1879.15.

(b) $s_e = 229.547$

(c)

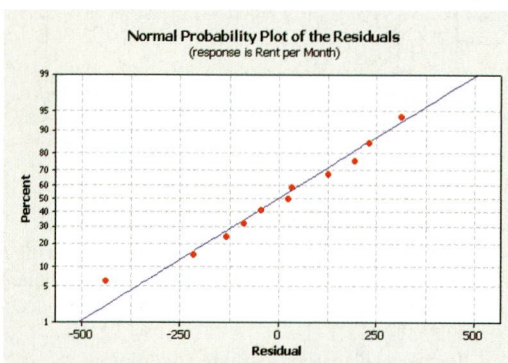

The residuals are normally distributed.

(d) $s_b = 0.2166$

(e) There is evidence that a linear relation exists between square footage of an apartment in Queens, NY and the monthly rent.

(f) 95% confidence interval about the slope of the true least-squares regression line: Lower bound: 2.0416, Upper bound: 3.0214.

(g) 90% confidence interval about the mean rent of 900 square foot apartments: Lower bound: $1752.20, Upper bound: $2006.00.

(h) When an apartment has 900 square feet, $\hat{y} = \$1879.15$.

(i) 90% prediction interval for the rent of a particular 900 square foot apartment: Lower bound: $1439.6, Upper bound: $2318.60.

(j) Although the predicted rents in parts (a) and (h) are the same, the intervals are different because the distribution of the means, part (a), has less variability than the distribution of the individuals, part (h).

5. (a) $\beta_0 \approx b_0 = 67.388$; $\beta_1 \approx b_1 = -0.263$

(b) $s_e = 11.1859$

(c)

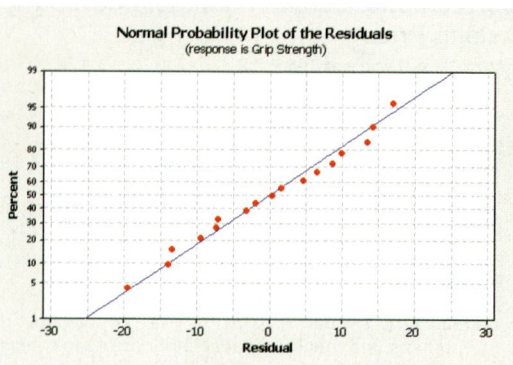

The residuals are normally distributed.

(d) $s_b = 0.1680$

(e) There is not sufficient evidence at the $\alpha = 0.05$ level of significance to support the claim that a linear relation exists between a woman's age and grip strength (P-value $= 0.138$).

(f) Based on the answers to (d) and (e) a good guess of the grip strength of a 42-year-old female would be the mean strength of the population.

7. (a) **Correlations: Midterm, Project, Course Grade**

	Midterm	Project
Project	-0.076	
Course Grade	0.704	0.526

There is no problem with multicollinearity between the explanatory variables.

(b) $\hat{y} = -91.29 + 1.5211x_1 + 0.4407x_2$

(c)

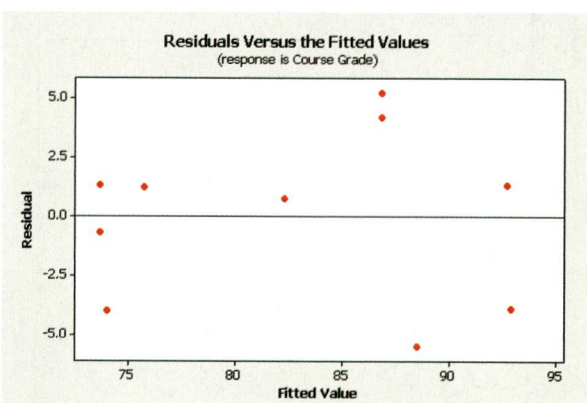

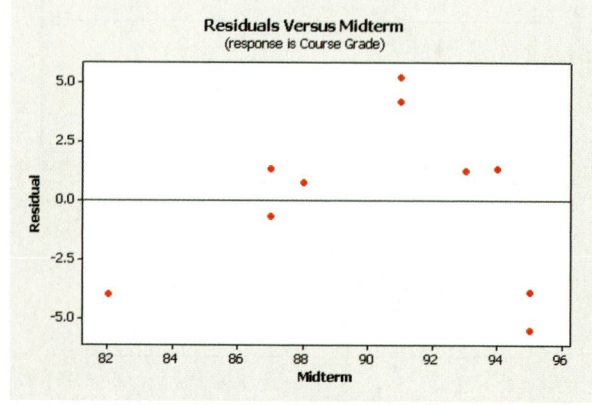

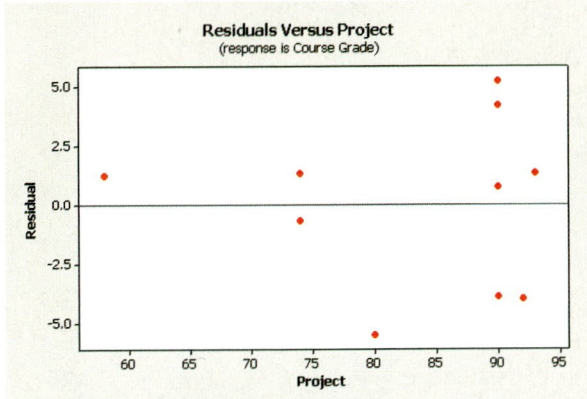

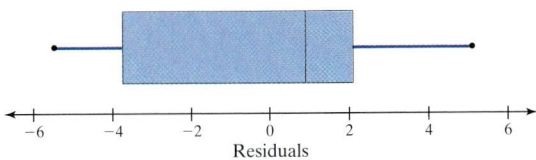

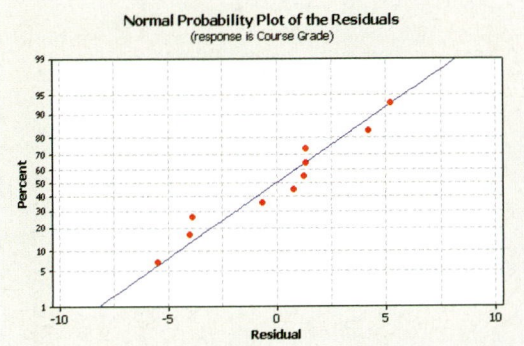

(d) The regression coefficient 1.5211 indicates when midterm grade increases by 1 point, the course grade increases by 1.5211 points, on average, assuming the other explanatory variables remain constant. The regression coefficient 0.4407 indicates when project grade increases by 1 point the course grade increases by 0.4407 points, on average, assuming the other explanatory variables remain constant.

(e) $R^2 = 0.833$; $R^2_{\text{adj.}} = 0.785$ are measures of the proportion of the variance in the course grade that is explained by the linear regression model.

(f) There is sufficient evidence at the $\alpha = 0.05$ level of significance to reject the null hypothesis and conclude that at least one $\beta_i \neq 0$. ($F_0 = 17.47$; P-value $= 0.002$)

(g) For the explanatory variable midterm, $t_0 = 4.83$ and the P-value $= 0.002$. For the explanatory variable project, $t_0 = 3.77$ and the P-value $= 0.007$. We reject both null hypotheses and conclude that both the explanatory variable midterm and the explanatory variable project are linearly related to course grade.

(h) $\hat{y} = 71$

(i) 95% confidence interval: Lower bound: 69, Upper bound: 82.04. We are 95% confident that the mean course grade for students who have an 85 on their midterm and a 75 on their project will be between 69 and 82.
95% prediction interval: Lower bound: 64.04, Upper bound: 87. We are 95% confident that randomly selected student who has an 85 on the midterm and a 75 on the project will have a course grade between 64 and 87.

INDEX

- Test Statistic for the Wilcoxon Matched-Pairs Signed-Ranks Test

The test statistic will depend on the size of the sample and the alternative hypothesis. Let n represent the sample size.

Small-Sample Case ($n \leq 30$)

Two-Tailed	Left-Tailed	Right-Tailed
$H_o: M_D = 0$	$H_o: M_D = 0$	$H_o: M_D = 0$
$H_1: M_D \neq 0$	$H_1: M_D < 0$	$H_o: M_D > 0$
Test Statistic: T is the smaller of T_+ or T_-	**Test Statistic:** $T = T_+$	**Test Statistic:** $T = \lvert T_- \rvert$

Large-Sample Case ($n > 30$)
The test statistic is given by

$$z_0 = \frac{T - \dfrac{n(n+1)}{4}}{\sqrt{\dfrac{n(n+1)(2n+1)}{24}}}$$

where T is the test statistic from the small-sample case.

- Test Statistic for the Mann–Whitney Test

The test statistic will depend on the size of the samples from each population. Let n_1 represent the sample size for population X and n_2 represent the sample size for population Y.

Small-Sample Case ($n_1 \leq 20$ and $n_2 \leq 20$)
If S is the sum of the ranks corresponding to the sample from population X, then the test statistic, T, is given by

$$T = S - \frac{n_1(n_1+1)}{2}$$

Note: The value of S is always obtained by summing the ranks of the sample data that correspond to M_x in the hypothesis.

Large-Sample Case ($n_1 > 20$) or ($n_2 > 20$)
Based on the Central Limit Theorem, the test statistic is given by

$$z_0 = \frac{T - \dfrac{n_1 n_2}{2}}{\sqrt{\dfrac{n_1 n_2(n_1 + n_2 + 1)}{12}}}$$

- Test Statistic for Spearman's Rank Correlation Test

The test statistic will depend on the size of the sample, n, and the sum of the squared differences.

$$r_s = 1 - \frac{6\sum d_i^2}{n(n^2 - 1)}$$

where d_i = the difference in the ranks of the two observations in the ith ordered pair.

- Test Statistic for the Kruskal–Wallis Test

The test statistic for the Kruskal–Wallis Test is

$$H = \frac{12}{N(N+1)} \sum \frac{1}{n_i}\left[R_i - \frac{n_i(N+1)}{2} \right]^2$$

A computational formula for the test statistic is

$$H = \frac{12}{N(N+1)}\left[\frac{R_1^2}{n_1} + \frac{R_2^2}{n_2} + \cdots + \frac{R_k^2}{n_k} \right] - 3(N+1)$$

where

R_1^2 is the sum of the ranks squared for the first sample, R_2^2 is the sum of the ranks squared for the second sample, and so on.

n_1 is the number of observations in the first sample, n_2 is the number of observations in the second sample, and so on.

N is the total number of observations $(N = n_1 + n_2 + \cdots + n_k)$.

k is the number of populations being compared.

CHAPTER 4   Describing the Relation between Two Variables

- Correlation Coefficient: $r = \dfrac{\sum \left(\dfrac{x_i - \bar{x}}{s_x} \right)\left(\dfrac{y_i - \bar{y}}{s_y} \right)}{n - 1}$

- The equation of the least-squares regression line is
$\hat{y} = b_1 x + b_0$, where $\hat{y}$ is the predicted value, $b_1 = r \cdot \dfrac{s_y}{s_x}$
is the slope, and $b_0 = \bar{y} - b_1 \bar{x}$ is the intercept.

- Residual = observed y − predicted $y = y - \hat{y}$

- Coefficient of Determination: $R^2 = $ the percent of total variation in the response variable that is explained by the least-squares regression line.

- $R^2 = r^2$ for the least-squares regression model
$\hat{y} = b_1 x + b_0$

- Exponential Equation of Best Fit
$$y = ab^x \qquad \text{Linear: } \log y = \log a + x \log b$$

- Power Equation of Best Fit
$$y = ax^b \qquad \text{Linear: } \log y = \log a + b \log x$$

CHAPTER 5   Probability

- Empirical Probability
$$P(E) \approx \frac{\text{frequency of } E}{\text{number of trials of experiment}}$$

- Classical Probability
$$P(E) = \frac{\text{number of ways that } E \text{ can occur}}{\text{number of possible outcomes}} = \frac{N(E)}{N(S)}$$

- Addition Rule for Disjoint Events
$$P(E \text{ or } F) = P(E) + P(F)$$

- Addition Rule for n Disjoint Events
$$P(E \text{ or } F \text{ or } G \text{ or } \cdots) = P(E) + P(F) + P(G) + \cdots$$

- General Addition Rule
$$P(E \text{ or } F) = P(E) + P(F) - P(E \text{ and } F)$$

- Complement Rule
$$P(E^c) = 1 - P(E)$$

- Multiplication Rule for Independent Events
$$P(E \text{ and } F) = P(E) \cdot P(F)$$

- Multiplication Rule for n Independent Events
$$P(E \text{ and } F \text{ and } G \cdots) = P(E) \cdot P(F) \cdot P(G) \cdot \ldots$$

- Conditional Probability Rule
$$P(F|E) = \frac{P(E \text{ and } F)}{P(E)} = \frac{N(E \text{ and } F)}{N(E)}$$

- General Multiplication Rule
$$P(E \text{ and } F) = P(E) \cdot P(F|E)$$

- Factorial
$$n! = n \cdot (n - 1) \cdot (n - 2) \cdot \cdots \cdot 3 \cdot 2 \cdot 1$$

- Permutation of n objects taken r at a time: $_nP_r = \dfrac{n!}{(n - r)!}$

- Combination of n objects taken r at a time:
$_nC_r = \dfrac{n!}{r!(n - r)!}$

- Permutations with Repetition: n_1 of one type, n_2 of a second type, $\ldots$, with $n_1 + n_2 + \cdots + n_k = n$

$$\frac{n!}{n_1! \cdot n_2! \cdot \cdots \cdot n_k!}$$

CHAPTER 12   Inference on Categorical Data

- Expected Counts (when testing for goodness of fit)

$$E_i = \mu_i = np_i \quad \text{for} \quad i = 1, 2, \ldots, k$$

- Expected Frequencies (when testing for independence or homogeneity of proportions)

$$\text{Expected frequency} = \frac{(\text{row total})(\text{column total})}{\text{table total}}$$

- Chi-Square Test Statistic

$$\chi^2 = \sum \frac{(\text{observed} - \text{expected})^2}{\text{expected}} = \sum \frac{(O_i - E_i)^2}{E_i}$$

$$i = 1, 2, \ldots, k$$

(1) All expected frequencies are greater than or equal to 1 and (2) no more than 20% of the expected frequencies are less than 5.

Use $k - 1$ degrees of freedom for goodness of fit.

Use $(r - 1)(c - 1)$ degrees of freedom when testing for independence or homogeneity of proportions (r is the number of rows, c is the number of columns).

CHAPTER 13   Comparing Three or More Means

- Test Statistic for One-Way ANOVA

$$F = \frac{\text{Mean square due to treatment}}{\text{Mean square due to error}} = \frac{\text{MST}}{\text{MSE}}$$

where

$$\text{MST} = \frac{n_1(\bar{x}_1 - \bar{x})^2 + n_2(\bar{x}_2 - \bar{x})^2 + \cdots + n_k(\bar{x}_k - \bar{x})^2}{k - 1}$$

$$\text{MSE} = \frac{(n_1 - 1)s_1^2 + (n_2 - 1)s_2^2 + \cdots + (n_k - 1)s_k^2}{n - k}$$

Test Statistic for Tukey's Test after One-Way ANOVA

$$q = \frac{(\bar{x}_2 - \bar{x}_1) - (\mu_2 - \mu_1)}{\sqrt{\frac{s^2}{2} \cdot \left(\frac{1}{n_1} + \frac{1}{n_2}\right)}} = \frac{\bar{x}_2 - \bar{x}_1}{\sqrt{\frac{s^2}{2} \cdot \left(\frac{1}{n_1} + \frac{1}{n_2}\right)}}$$

CHAPTER 14   Inference on the Least-squares Regression Model and Multiple Regression

- Standard Error of the Estimate

$$s_e = \sqrt{\frac{\sum(y_i - \hat{y}_i)^2}{n - 2}} = \sqrt{\frac{\sum \text{residuals}^2}{n - 2}}$$

- Standard error of b_1

$$s_{b_1} = \frac{s_e}{\sqrt{\sum(x_i - \bar{x})^2}}$$

- Test statistic for the Slope of the Least-Squares Regression Line

$$t_0 = \frac{b_1 - \beta_1}{s_e / \sqrt{\sum(x_i - \bar{x})^2}} = \frac{b_1 - \beta_1}{s_{b_1}}$$

- Confidence Interval for the Slope of the Regression Line

A $(1 - \alpha) \cdot 100\%$ confidence interval for the slope of the true regression line, β_1, is given by:

$$\text{Lower bound:} \quad b_1 - t_{\alpha/2} \cdot \frac{s_e}{\sqrt{\sum(x_i - \bar{x})^2}}$$

$$\text{Upper bound:} \quad b_1 + t_{\alpha/2} \cdot \frac{s_e}{\sqrt{\sum(x_i - \bar{x})^2}}$$

where $t_{\alpha/2}$ is computed with $n - 2$ degrees of freedom.

Tables and Formulas
for Sullivan, *Statistics: Informed Decisions Using Data*
© 2007 Pearson Education, Inc.

CHAPTER 10   Testing Claims Regarding a Parameter

Test Statistics

- $z_0 = \dfrac{\bar{x} - \mu_0}{\sigma / \sqrt{n}}$, provided that the population from which the sample was drawn is normal or the sample size is large ($n \geq 30$).

- $t_0 = \dfrac{\bar{x} - \mu_0}{s / \sqrt{n}}$ follows Student's t-distribution with $n - 1$ degrees of freedom, provided that the population from which the sample was drawn is normal or the sample size is large ($n \geq 30$).

- $z_0 = \dfrac{\hat{p} - p_0}{\sqrt{\dfrac{p_0(1 - p_0)}{n}}}$, provided that $np_0(1 - p_0) \geq 10$ and the sample size is less than 5% of the population size ($n < 0.05N$).

- $\chi_0^2 = \dfrac{(n - 1)s^2}{\sigma_0^2}$ follows the χ^2-distribution with $n - 1$ degrees of freedom, provided that the population from which the sample was drawn is normal.

CHAPTER 11   Inferences on Two Samples

- Test Statistic for Matched-Pairs data

$$t_0 = \dfrac{\bar{d}}{s_d / \sqrt{n}}$$

where $\bar{d}$ is the mean and s_d is the standard deviation of the differenced data.

- Confidence Interval for Matched-Pairs data:

Lower bound: $\bar{d} - t_{\alpha/2} \cdot \dfrac{s_d}{\sqrt{n}}$

Upper bound: $\bar{d} + t_{\alpha/2} \cdot \dfrac{s_d}{\sqrt{n}}$

Note: $t_{\alpha/2}$ is found using $n - 1$ degrees of freedom.

- Test Statistic Comparing Two Means (Independent Sampling):

$$t_0 = \dfrac{(\bar{x}_1 - \bar{x}_2) - (\mu_1 - \mu_2)}{\sqrt{\dfrac{s_1^2}{n_1} + \dfrac{s_2^2}{n_2}}}$$

- Confidence Interval for the Difference of Two Means (Independent Samples):

Lower bound: $(\bar{x}_1 - \bar{x}_2) - t_{\alpha/2}\sqrt{\dfrac{s_1^2}{n_1} + \dfrac{s_2^2}{n_2}}$

Upper bound: $(\bar{x}_1 - \bar{x}_2) + t_{\alpha/2}\sqrt{\dfrac{s_1^2}{n_1} + \dfrac{s_2^2}{n_2}}$

Note: $t_{\alpha/2}$ is found using the smaller of $n_1 - 1$ or $n_2 - 1$ degrees of freedom.

- Test Statistic Comparing Two Population Proportions

$$z_0 = \dfrac{\hat{p}_1 - \hat{p}_2}{\sqrt{\hat{p}(1 - \hat{p})}\sqrt{\dfrac{1}{n_1} + \dfrac{1}{n_2}}}$$

where $\hat{p} = \dfrac{x_1 + x_2}{n_1 + n_2}$.

- Confidence Interval for the Difference of Two Proportions

Lower bound: $(\hat{p}_1 - \hat{p}_2) - z_{\alpha/2}\sqrt{\dfrac{\hat{p}_1(1 - \hat{p}_1)}{n_1} + \dfrac{\hat{p}_2(1 - \hat{p}_2)}{n_2}}$

Upper bound: $(\hat{p}_1 - \hat{p}_2) + z_{\alpha/2}\sqrt{\dfrac{\hat{p}_1(1 - \hat{p}_1)}{n_1} + \dfrac{\hat{p}_2(1 - \hat{p}_2)}{n_2}}$

- Test Statistic for Comparing Two Population Standard Deviations

$$F_0 = \dfrac{s_1^2}{s_2^2}$$

- Finding a Critical F for the Left Tail

$$F_{1-\alpha, n_1-1, n_2-1} = \dfrac{1}{F_{\alpha, n_2-1, n_1-1}}$$

CHAPTER 6   Discrete Probability Distributions

- Mean (Expected Value) of a Discrete Random Variable

$$\mu_X = \sum x \cdot P(X = x)$$

- Variance of a Discrete Random Variable

$$\sigma_X^2 = \sum (x - \mu)^2 \cdot P(x) = \sum x^2 P(x) - \mu x^2$$

- Binomial Probability Distribution Function

$$P(x) = {}_nC_x p^x (1 - p)^{n-x}$$

- Mean of a Binomial Random Variable

$$\mu_X = np$$

- Standard Deviation of a Binomial Random Variable

$$\sigma_X = \sqrt{np(1 - p)}$$

- Poisson Probability Distribution Function

$$P(x) = \frac{(\lambda t)^x}{x!} e^{-\lambda t} \qquad x = 0, 1, 2, \ldots$$

- Mean and Standard Deviation of a Poisson Random Variable

$$\mu_X = \lambda t \qquad \sigma_X = \sqrt{\lambda t}$$

CHAPTER 7   The Normal Distribution

- Standardizing a Normal Random Variable

$$Z = \frac{X - \mu}{\sigma}$$

- Finding the Score:

$$X = \mu + Z\sigma$$

CHAPTER 8   Sampling Distributions

- Mean and Standard Deviation of the Sampling Distribution of $\bar{x}$

$$\mu_{\bar{x}} = \mu \quad \text{and} \quad \sigma_{\bar{x}} = \frac{\sigma}{\sqrt{n}}$$

- Sample Proportion:

$$\hat{p} = \frac{x}{n}$$

- Mean and Standard Deviation of the Sampling Distributon of $\hat{p}$

$$\mu_p = p \text{ and } \sigma_p = \sqrt{\frac{p(1 - p)}{n}}$$

- Standardizing a Normal Random Variable

$$Z = \frac{\bar{x} - \mu}{\frac{\sigma}{\sqrt{n}}} \qquad Z = \frac{\hat{p} - p}{\sqrt{\frac{p(1 - p)}{n}}}$$

CHAPTER 9   Estimating the Value of a Parameter Using Confidence Intervals

Confidence Intervals

- A $(1 - \alpha) \cdot 100\%$ confidence interval about μ with σ known is $\bar{x} \pm z_{\alpha/2} \cdot \frac{\sigma}{\sqrt{n}}$, provided the population from which the sample was drawn is normal or the sample size is large ($n \geq 30$).

- A $(1 - \alpha) \cdot 100\%$ confidence interval about μ with σ unknown is $\bar{x} \pm t_{\alpha/2} \cdot \frac{s}{\sqrt{n}}$, provided the population from which the sample was drawn is normal or the sample size is large ($n \geq 30$). *Note:* $t_{\alpha/2}$ is computed using $n - 1$ degrees of freedom.

- A $(1 - \alpha) \cdot 100\%$ confidence interval about p is $\hat{p} \pm z_{\alpha/2} \cdot \sqrt{\frac{\hat{p}(1 - \hat{p})}{n}}$, provided $n\hat{p}(1 - \hat{p}) \geq 10$.

- A $(1 - \alpha) \cdot 100\%$ confidence interval about σ^2 is $\frac{(n - 1)s^2}{\chi_{\alpha/2}^2} < \sigma^2 < \frac{(n - 1)s^2}{\chi_{1-\alpha/2}^2}$, provided the population from which the sample was drawn is normal.

Sample Size

- To estimate the population mean with a margin of error E at a $(1 - \alpha) \cdot 100\%$ level of confidence requires a sample of size $n = \left(\frac{z_{\alpha/2} \cdot \sigma}{E}\right)^2$ rounded up to the next integer.

- To estimate the population proportion with a margin of error E at a $(1 - \alpha) \cdot 100\%$ level of confidence requires a sample of size $n = \hat{p}(1 - \hat{p})\left(\frac{z_{\alpha/2}}{E}\right)^2$ rounded up to the next integer, where $\hat{p}$ is a prior estimate of the population proportion.

- To estimate the population proportion with a margin of error E at a $(1 - \alpha) \cdot 100\%$ level of confidence requires a sample of size $n = 0.25\left(\frac{z_{\alpha/2}}{E}\right)^2$ rounded up to the next integer when no prior estimate of p is available.

- Confidence Interval about the Mean Response of y, $\hat{y}$

  A $(1 - \alpha) \cdot 100\%$ confidence interval for the mean response of y, $\hat{y}$, is given by

  $$\text{Lower bound:} \quad \hat{y} - t_{\alpha/2} \cdot s_e \sqrt{\frac{1}{n} + \frac{(x^* - \bar{x})^2}{\sum(x_i - \bar{x})^2}}$$

  $$\text{Upper bound:} \quad \hat{y} + t_{\alpha/2} \cdot s_e \sqrt{\frac{1}{n} + \frac{(x^* - \bar{x})^2}{\sum(x_i - \bar{x})^2}}$$

  where x^* is the given value of the explanatory variable and $t_{\alpha/2}$ is the critical value with $n - 2$ degrees of freedom.

- Prediction Interval about an Individual Response, $\hat{y}$

  A $(1 - \alpha) \cdot 100\%$ prediction interval for the individual response of y, $\hat{y}$, is given by

  $$\text{Lower Bound:} \quad \hat{y} - t_{\alpha/2} \cdot s_e \sqrt{1 + \frac{1}{n} + \frac{(x^* - \bar{x})^2}{\sum(x_i - \bar{x})^2}}$$

  $$\text{Upper Bound:} \quad \hat{y} + t_{\alpha/2} \cdot s_e \sqrt{1 + \frac{1}{n} + \frac{(x^* - \bar{x})^2}{\sum(x_i - \bar{x})^2}}$$

  where x^* is the given value of the explanatory variable and $t_{\alpha/2}$ is the critical value with $n - 2$ degrees of freedom.

CHAPTER 15   Nonparametric Statistics

- Test Statistic for a Runs Test for Randomness

  Let n represent the sample size of which there are two mutually exclusive types.

  Let n_1 represent the number of observations of the first type.

  Let n_2 represent the number of observations of the second type.

  Let r represent the number of runs.

Small-Sample Case

If $n_1 \leq 20$ and $n_2 \leq 20$, the test statistic in the runs test for randomness is r, the number of runs.

Large-Sample Case

If $n_1 > 20$ or $n_2 > 20$, the test statistic in the runs test for randomness is

$$z_0 = \frac{r - \mu_r}{\sigma_r}$$

where

$$\mu_r = \frac{2n_1 n_2}{n} + 1 \quad \text{and} \quad \sigma_r = \sqrt{\frac{2n_1 n_2 (2n_1 n_2 - n)}{n^2 (n - 1)}}$$

- Test Statistic for a One-Sample Sign Test

The test statistic will depend on the structure of the hypothesis test and the sample size.

Small-Sample Case (n ≤ 25)

Two-Tailed	Left-Tailed	Right-Tailed
$H_0: M = M_0$	$H_0: M = M_0$	$H_0: M = M_0$
$H_1: M \neq M_0$	$H_1: M < M_0$	$H_1: M > M_0$
The test statistic, k, will be the smaller of the number of minus signs or plus signs.	The test statistic, k, will be the number of plus signs.	The test statistic, k, will be the number of minus signs.

Large-Sample Case (n > 25)

The test statistic, z, is

$$z_0 = \frac{(k + 0.5) - \dfrac{n}{2}}{\dfrac{\sqrt{n}}{2}}$$

where n is the number of minus and plus signs and k is obtained as described in the small sample case.

CHAPTER 2 Organizing and Summarizing Data

- Relative frequency $= \dfrac{\text{frequency}}{\text{sum of all frequencies}}$

- Class midpoint: The sum of consecutive lower class limits divided by 2.

CHAPTER 3 Numerically Summarizing Data

- Population Mean: $\mu = \dfrac{\sum x_i}{N}$

- Sample Mean: $\bar{x} = \dfrac{\sum x_i}{n}$

- Range = Largest Data Value − Smallest Data Value

- Population Variance: $\sigma^2 = \dfrac{\sum(x_i - \mu)^2}{N} = \dfrac{\sum x_i^2 - \dfrac{(\sum x_i)^2}{N}}{N}$

- Sample Variance: $s^2 = \dfrac{\sum(x_i - \bar{x})^2}{n-1} = \dfrac{\sum x_i^2 - \dfrac{(\sum x_i)^2}{n}}{n-1}$

- Population Standard Deviation $\sigma = \sqrt{\sigma^2}$

- Sample Standard Deviation: $s = \sqrt{s^2}$

- ***Empirical Rule:*** If the shape of the distribution is bell-shaped, then
  - Approximately 68% of the data lie within 1 standard deviation of the mean
  - Approximately 95% of the data lie within 2 standard deviations of the mean
  - Approximately 99.7% of the data lie within 3 standard deviations of the mean

- ***Chebyshev's Inequality:*** For any data set, regardless of the shape of the distribution, at least $\left(1 - \dfrac{1}{k^2}\right)100\%$ of the observations will lie within k standard deviations of the mean where k is any number greater than 1.

- Population Mean from Grouped Data: $\mu = \dfrac{\sum x_i f_i}{\sum f_i}$

- Sample Mean from Grouped Data: $\bar{x} = \dfrac{\sum x_i f_i}{\sum f_i}$

- Weighted Mean: $\bar{x}_w = \dfrac{\sum w_i x_i}{\sum w_i}$

- Population Variance from Grouped Data

$$\sigma^2 = \frac{\sum(x_i - \mu)^2 f_i}{\sum f_i} = \frac{\sum x_i^2 f_i - \dfrac{(\sum x_i f_i)^2}{\sum f_i}}{\sum f_i}$$

- Sample Variance from Grouped Data:

$$s^2 = \frac{\sum(x_i - \mu)^2 f_i}{(\sum f_i) - 1} = \frac{\sum x_i^2 f_i - \dfrac{(\sum x_i f_i)^2}{\sum f_i}}{\sum f_i - 1}$$

- Population Z-score: $z = \dfrac{x - \mu}{\sigma}$

- Sample Z-score: $z = \dfrac{x - \bar{x}}{s}$

- Percentile of $x = \dfrac{\text{Number of data values less than } x}{n} \cdot 100$

- Determining the kth percentile: $i = \left(\dfrac{k}{100}\right)(n + 1)$. If i is not an integer, find the mean of the observations on either side of i.

- Interquartile Range: $\text{IQR} = Q_3 - Q_1$

- Lower and Upper Fences: $\begin{aligned}\text{Lower fence} &= Q_1 - 1.5(\text{IQR})\\ \text{Upper fence} &= Q_3 + 1.5(\text{IQR})\end{aligned}$

- Five-Number Summary

$$\text{Minimum}, Q_1, M, Q_3, \text{Maximum}$$

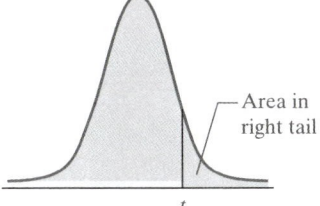
Area in right tail

t

Table V

t-Distribution
Area in Right Tail

df	0.25	0.20	0.15	0.10	0.05	0.025	0.02	0.01	0.005	0.0025	0.001	0.0005
1	1.000	1.376	1.963	3.078	6.314	12.706	15.894	31.821	63.657	127.321	318.289	636.558
2	0.816	1.061	1.386	1.886	2.920	4.303	4.849	6.965	9.925	14.089	22.328	31.600
3	0.765	0.978	1.250	1.638	2.353	3.182	3.482	4.541	5.841	7.453	10.214	12.924
4	0.741	0.941	1.190	1.533	2.132	2.776	2.999	3.747	4.604	5.598	7.173	8.610
5	0.727	0.920	1.156	1.476	2.015	2.571	2.757	3.365	4.032	4.773	5.893	6.869
6	0.718	0.906	1.134	1.440	1.943	2.447	2.612	3.143	3.707	4.317	5.208	5.959
7	0.711	0.896	1.119	1.415	1.895	2.365	2.517	2.998	3.499	4.029	4.785	5.408
8	0.706	0.889	1.108	1.397	1.860	2.306	2.449	2.896	3.355	3.833	4.501	5.041
9	0.703	0.883	1.100	1.383	1.833	2.262	2.398	2.821	3.250	3.690	4.297	4.781
10	0.700	0.879	1.093	1.372	1.812	2.228	2.359	2.764	3.169	3.581	4.144	4.587
11	0.697	0.876	1.088	1.363	1.796	2.201	2.328	2.718	3.106	3.497	4.025	4.437
12	0.695	0.873	1.083	1.356	1.782	2.179	2.303	2.681	3.055	3.428	3.930	4.318
13	0.694	0.870	1.079	1.350	1.771	2.160	2.282	2.650	3.012	3.372	3.852	4.221
14	0.692	0.868	1.076	1.345	1.761	2.145	2.264	2.624	2.977	3.326	3.787	4.140
15	0.691	0.866	1.074	1.341	1.753	2.131	2.249	2.602	2.947	3.286	3.733	4.073
16	0.690	0.865	1.071	1.337	1.746	2.120	2.235	2.583	2.921	3.252	3.686	4.015
17	0.689	0.863	1.069	1.333	1.740	2.110	2.224	2.567	2.898	3.222	3.646	3.965
18	0.688	0.862	1.067	1.330	1.734	2.101	2.214	2.552	2.878	3.197	3.611	3.922
19	0.688	0.861	1.066	1.328	1.729	2.093	2.205	2.539	2.861	3.174	3.579	3.883
20	0.687	0.860	1.064	1.325	1.725	2.086	2.197	2.528	2.845	3.153	3.552	3.850
21	0.686	0.859	1.063	1.323	1.721	2.080	2.189	2.518	2.831	3.135	3.527	3.819
22	0.686	0.858	1.061	1.321	1.717	2.074	2.183	2.508	2.819	3.119	3.505	3.792
23	0.685	0.858	1.060	1.319	1.714	2.069	2.177	2.500	2.807	3.104	3.485	3.768
24	0.685	0.857	1.059	1.318	1.711	2.064	2.172	2.492	2.797	3.091	3.467	3.745
25	0.684	0.856	1.058	1.316	1.708	2.060	2.167	2.485	2.787	3.078	3.450	3.725
26	0.684	0.856	1.058	1.315	1.706	2.056	2.162	2.479	2.779	3.067	3.435	3.707
27	0.684	0.855	1.057	1.314	1.703	2.052	2.158	2.473	2.771	3.057	3.421	3.690
28	0.683	0.855	1.056	1.313	1.701	2.048	2.154	2.467	2.763	3.047	3.408	3.674
29	0.683	0.854	1.055	1.311	1.699	2.045	2.150	2.462	2.756	3.038	3.396	3.659
30	0.683	0.854	1.055	1.310	1.697	2.042	2.147	2.457	2.750	3.030	3.385	3.646
31	0.682	0.853	1.054	1.309	1.696	2.040	2.144	2.453	2.744	3.022	3.375	3.633
32	0.682	0.853	1.054	1.309	1.694	2.037	2.141	2.449	2.738	3.015	3.365	3.622
33	0.682	0.853	1.053	1.308	1.692	2.035	2.138	2.445	2.733	3.008	3.356	3.611
34	0.682	0.852	1.052	1.307	1.691	2.032	2.136	2.441	2.728	3.002	3.348	3.601
35	0.682	0.852	1.052	1.306	1.690	2.030	2.133	2.438	2.724	2.996	3.340	3.591
36	0.681	0.852	1.052	1.306	1.688	2.028	2.131	2.435	2.719	2.990	3.333	3.582
37	0.681	0.851	1.051	1.305	1.687	2.026	2.129	2.431	2.715	2.985	3.326	3.574
38	0.681	0.851	1.051	1.304	1.686	2.024	2.127	2.429	2.712	2.980	3.319	3.566
39	0.681	0.851	1.050	1.304	1.685	2.023	2.125	2.426	2.708	2.976	3.313	3.558
40	0.681	0.851	1.050	1.303	1.684	2.021	2.123	2.423	2.704	2.971	3.307	3.551
50	0.679	0.849	1.047	1.299	1.676	2.009	2.109	2.403	2.678	2.937	3.261	3.496
60	0.679	0.848	1.045	1.296	1.671	2.000	2.099	2.390	2.660	2.915	3.232	3.460
70	0.678	0.847	1.044	1.294	1.667	1.994	2.093	2.381	2.648	2.899	3.211	3.435
80	0.678	0.846	1.043	1.292	1.664	1.990	2.088	2.374	2.639	2.887	3.195	3.416
90	0.677	0.846	1.042	1.291	1.662	1.987	2.084	2.368	2.632	2.878	3.183	3.402
100	0.677	0.845	1.042	1.290	1.660	1.984	2.081	2.364	2.626	2.871	3.174	3.390
1000	0.675	0.842	1.037	1.282	1.646	1.962	2.056	2.330	2.581	2.813	3.098	3.300
z	0.674	0.841	1.036	1.282	1.645	1.960	2.054	2.326	2.576	2.807	3.091	3.291

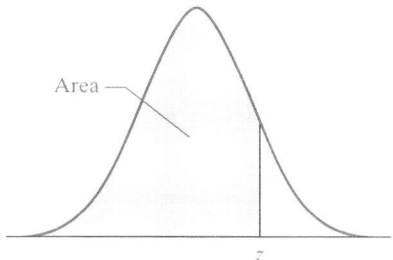

Area

z

Standard Normal Distribution

z	.00	.01	.02	.03	.04	.05	.06	.07	.08	.09
0.0	0.5000	0.5040	0.5080	0.5120	0.5160	0.5199	0.5239	0.5279	0.5319	0.5359
0.1	0.5398	0.5438	0.5478	0.5517	0.5557	0.5596	0.5636	0.5675	0.5714	0.5753
0.2	0.5793	0.5832	0.5871	0.5910	0.5948	0.5987	0.6026	0.6064	0.6103	0.6141
0.3	0.6179	0.6217	0.6255	0.6293	0.6331	0.6368	0.6406	0.6443	0.6480	0.6517
0.4	0.6554	0.6591	0.6628	0.6664	0.6700	0.6736	0.6772	0.6808	0.6844	0.6879
0.5	0.6915	0.6950	0.6985	0.7019	0.7054	0.7088	0.7123	0.7157	0.7190	0.7224
0.6	0.7257	0.7291	0.7324	0.7357	0.7389	0.7422	0.7454	0.7486	0.7517	0.7549
0.7	0.7580	0.7611	0.7642	0.7673	0.7704	0.7734	0.7764	0.7794	0.7823	0.7852
0.8	0.7881	0.7910	0.7939	0.7967	0.7995	0.8023	0.8051	0.8078	0.8106	0.8133
0.9	0.8159	0.8186	0.8212	0.8238	0.8264	0.8289	0.8315	0.8340	0.8365	0.8389
1.0	0.8413	0.8438	0.8461	0.8485	0.8508	0.8531	0.8554	0.8577	0.8599	0.8621
1.1	0.8643	0.8665	0.8686	0.8708	0.8729	0.8749	0.8770	0.8790	0.8810	0.8830
1.2	0.8849	0.8869	0.8888	0.8907	0.8925	0.8944	0.8962	0.8980	0.8997	0.9015
1.3	0.9032	0.9049	0.9066	0.9082	0.9099	0.9115	0.9131	0.9147	0.9162	0.9177
1.4	0.9192	0.9207	0.9222	0.9236	0.9251	0.9265	0.9279	0.9292	0.9306	0.9319
1.5	0.9332	0.9345	0.9357	0.9370	0.9382	0.9394	0.9406	0.9418	0.9429	0.9441
1.6	0.9452	0.9463	0.9474	0.9484	0.9495	0.9505	0.9515	0.9525	0.9535	0.9545
1.7	0.9554	0.9564	0.9573	0.9582	0.9591	0.9599	0.9608	0.9616	0.9625	0.9633
1.8	0.9641	0.9649	0.9656	0.9664	0.9671	0.9678	0.9686	0.9693	0.9699	0.9706
1.9	0.9713	0.9719	0.9726	0.9732	0.9738	0.9744	0.9750	0.9756	0.9761	0.9767
2.0	0.9772	0.9778	0.9783	0.9788	0.9793	0.9798	0.9803	0.9808	0.9812	0.9817
2.1	0.9821	0.9826	0.9830	0.9834	0.9838	0.9842	0.9846	0.9850	0.9854	0.9857
2.2	0.9861	0.9864	0.9868	0.9871	0.9875	0.9878	0.9881	0.9884	0.9887	0.9890
2.3	0.9893	0.9896	0.9898	0.9901	0.9904	0.9906	0.9909	0.9911	0.9913	0.9916
2.4	0.9918	0.9920	0.9922	0.9925	0.9927	0.9929	0.9931	0.9932	0.9934	0.9936
2.5	0.9938	0.9940	0.9941	0.9943	0.9945	0.9946	0.9948	0.9949	0.9951	0.9952
2.6	0.9953	0.9955	0.9956	0.9957	0.9959	0.9960	0.9961	0.9962	0.9963	0.9964
2.7	0.9965	0.9966	0.9967	0.9968	0.9969	0.9970	0.9971	0.9972	0.9973	0.9974
2.8	0.9974	0.9975	0.9976	0.9977	0.9977	0.9978	0.9979	0.9979	0.9980	0.9981
2.9	0.9981	0.9982	0.9982	0.9983	0.9984	0.9984	0.9985	0.9985	0.9986	0.9986
3.0	0.9987	0.9987	0.9987	0.9988	0.9988	0.9989	0.9989	0.9989	0.9990	0.9990
3.1	0.9990	0.9991	0.9991	0.9991	0.9992	0.9992	0.9992	0.9992	0.9993	0.9993
3.2	0.9993	0.9993	0.9994	0.9994	0.9994	0.9994	0.9994	0.9995	0.9995	0.9995
3.3	0.9995	0.9995	0.9995	0.9996	0.9996	0.9996	0.9996	0.9996	0.9996	0.9997
3.4	0.9997	0.9997	0.9997	0.9997	0.9997	0.9997	0.9997	0.9997	0.9997	0.9998

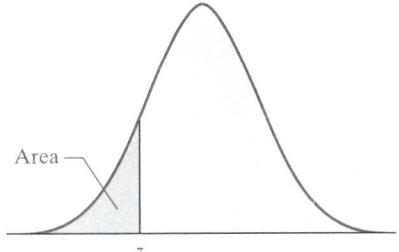

Area

z

Table IV

Standard Normal Distribution

z	.00	.01	.02	.03	.04	.05	.06	.07	.08	.09
−3.4	0.0003	0.0003	0.0003	0.0003	0.0003	0.0003	0.0003	0.0003	0.0003	0.0002
−3.3	0.0005	0.0005	0.0005	0.0004	0.0004	0.0004	0.0004	0.0004	0.0004	0.0003
−3.2	0.0007	0.0007	0.0006	0.0006	0.0006	0.0006	0.0006	0.0005	0.0005	0.0005
−3.1	0.0010	0.0009	0.0009	0.0009	0.0008	0.0008	0.0008	0.0008	0.0007	0.0007
−3.0	0.0013	0.0013	0.0013	0.0012	0.0012	0.0011	0.0011	0.0011	0.0010	0.0010
−2.9	0.0019	0.0018	0.0018	0.0017	0.0016	0.0016	0.0015	0.0015	0.0014	0.0014
−2.8	0.0026	0.0025	0.0024	0.0023	0.0023	0.0022	0.0021	0.0021	0.0020	0.0019
−2.7	0.0035	0.0034	0.0033	0.0032	0.0031	0.0030	0.0029	0.0028	0.0027	0.0026
−2.6	0.0047	0.0045	0.0044	0.0043	0.0041	0.0040	0.0039	0.0038	0.0037	0.0036
−2.5	0.0062	0.0060	0.0059	0.0057	0.0055	0.0054	0.0052	0.0051	0.0049	0.0048
−2.4	0.0082	0.0080	0.0078	0.0075	0.0073	0.0071	0.0069	0.0068	0.0066	0.0064
−2.3	0.0107	0.0104	0.0102	0.0099	0.0096	0.0094	0.0091	0.0089	0.0087	0.0084
−2.2	0.0139	0.0136	0.0132	0.0129	0.0125	0.0122	0.0119	0.0116	0.0113	0.0110
−2.1	0.0179	0.0174	0.0170	0.0166	0.0162	0.0158	0.0154	0.0150	0.0146	0.0143
−2.0	0.0228	0.0222	0.0217	0.0212	0.0207	0.0202	0.0197	0.0192	0.0188	0.0183
−1.9	0.0287	0.0281	0.0274	0.0268	0.0262	0.0256	0.0250	0.0244	0.0239	0.0233
−1.8	0.0359	0.0351	0.0344	0.0336	0.0329	0.0322	0.0314	0.0307	0.0301	0.0294
−1.7	0.0446	0.0436	0.0427	0.0418	0.0409	0.0401	0.0392	0.0384	0.0375	0.0367
−1.6	0.0548	0.0537	0.0526	0.0516	0.0505	0.0495	0.0485	0.0475	0.0465	0.0455
−1.5	0.0668	0.0655	0.0643	0.0630	0.0618	0.0606	0.0594	0.0582	0.0571	0.0559
−1.4	0.0808	0.0793	0.0778	0.0764	0.0749	0.0735	0.0721	0.0708	0.0694	0.0681
−1.3	0.0968	0.0951	0.0934	0.0918	0.0901	0.0885	0.0869	0.0853	0.0838	0.0823
−1.2	0.1151	0.1131	0.1112	0.1093	0.1075	0.1056	0.1038	0.1020	0.1003	0.0985
−1.1	0.1357	0.1335	0.1314	0.1292	0.1271	0.1251	0.1230	0.1210	0.1190	0.1170
−1.0	0.1587	0.1562	0.1539	0.1515	0.1492	0.1469	0.1446	0.1423	0.1401	0.1379
−0.9	0.1841	0.1814	0.1788	0.1762	0.1736	0.1711	0.1685	0.1660	0.1635	0.1611
−0.8	0.2119	0.2090	0.2061	0.2033	0.2005	0.1977	0.1949	0.1922	0.1894	0.1867
−0.7	0.2420	0.2389	0.2358	0.2327	0.2296	0.2266	0.2236	0.2206	0.2177	0.2148
−0.6	0.2743	0.2709	0.2676	0.2643	0.2611	0.2578	0.2546	0.2514	0.2483	0.2451
−0.5	0.3085	0.3050	0.3015	0.2981	0.2946	0.2912	0.2877	0.2843	0.2810	0.2776
−0.4	0.3446	0.3409	0.3372	0.3336	0.3300	0.3264	0.3228	0.3192	0.3156	0.3121
−0.3	0.3821	0.3783	0.3745	0.3707	0.3669	0.3632	0.3594	0.3557	0.3520	0.3483
−0.2	0.4207	0.4168	0.4129	0.4090	0.4052	0.4013	0.3974	0.3936	0.3897	0.3859
−0.1	0.4602	0.4562	0.4522	0.4483	0.4443	0.4404	0.4364	0.4325	0.4286	0.4247
−0.0	0.5000	0.4960	0.4920	0.4880	0.4840	0.4801	0.4761	0.4721	0.4681	0.4641

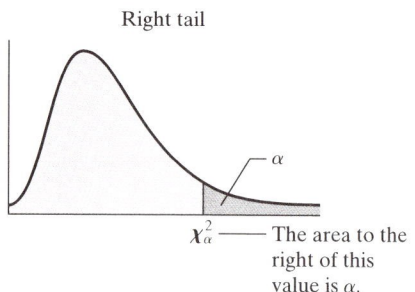

Right tail

χ_α^2 —— The area to the right of this value is α.

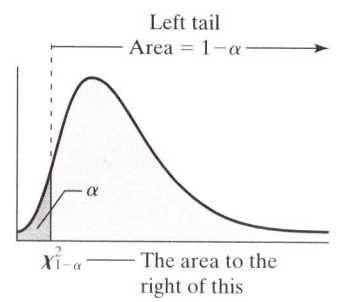

Left tail

Area $= 1-\alpha$

$\chi_{1-\alpha}^2$ —— The area to the right of this value is $1-\alpha$.

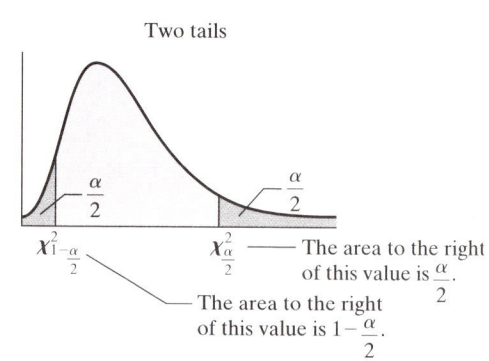

Two tails

$\chi_{1-\frac{\alpha}{2}}^2$ $\chi_{\frac{\alpha}{2}}^2$ —— The area to the right of this value is $\frac{\alpha}{2}$.

The area to the right of this value is $1-\frac{\alpha}{2}$.

Table VI

Chi-Square (χ^2) Distribution
Area to the Right of Critical Value

Degrees of Freedom	0.995	0.99	0.975	0.95	0.90	0.10	0.05	0.025	0.01	0.005
1	—	—	0.001	0.004	0.016	2.706	3.841	5.024	6.635	7.879
2	0.010	0.020	0.051	0.103	0.211	4.605	5.991	7.378	9.210	10.597
3	0.072	0.115	0.216	0.352	0.584	6.251	7.815	9.348	11.345	12.838
4	0.207	0.297	0.484	0.711	1.064	7.779	9.488	11.143	13.277	14.860
5	0.412	0.554	0.831	1.145	1.610	9.236	11.071	12.833	15.086	16.750
6	0.676	0.872	1.237	1.635	2.204	10.645	12.592	14.449	16.812	18.548
7	0.989	1.239	1.690	2.167	2.833	12.017	14.067	16.013	18.475	20.278
8	1.344	1.646	2.180	2.733	3.490	13.362	15.507	17.535	20.090	21.955
9	1.735	2.088	2.700	3.325	4.168	14.684	16.919	19.023	21.666	23.589
10	2.156	2.558	3.247	3.940	4.865	15.987	18.307	20.483	23.209	25.188
11	2.603	3.053	3.816	4.575	5.578	17.275	19.675	21.920	24.725	26.757
12	3.074	3.571	4.404	5.226	6.304	18.549	21.026	23.337	26.217	28.299
13	3.565	4.107	5.009	5.892	7.042	19.812	22.362	24.736	27.688	29.819
14	4.075	4.660	5.629	6.571	7.790	21.064	23.685	26.119	29.141	31.319
15	4.601	5.229	6.262	7.261	8.547	22.307	24.996	27.488	30.578	32.801
16	5.142	5.812	6.908	7.962	9.312	23.542	26.296	28.845	32.000	34.267
17	5.697	6.408	7.564	8.672	10.085	24.769	27.587	30.191	33.409	35.718
18	6.265	7.015	8.231	9.390	10.865	25.989	28.869	31.526	34.805	37.156
19	6.844	7.633	8.907	10.117	11.651	27.204	30.144	32.852	36.191	38.582
20	7.434	8.260	9.591	10.851	12.443	28.412	31.410	34.170	37.566	39.997
21	8.034	8.897	10.283	11.591	13.240	29.615	32.671	35.479	38.932	41.401
22	8.643	9.542	10.982	12.338	14.042	30.813	33.924	36.781	40.289	42.796
23	9.260	10.196	11.689	13.091	14.848	32.007	35.172	38.076	41.638	44.181
24	9.886	10.856	12.401	13.848	15.659	33.196	36.415	39.364	42.980	45.559
25	10.520	11.524	13.120	14.611	16.473	34.382	37.652	40.646	44.314	46.928
26	11.160	12.198	13.844	15.379	17.292	35.563	38.885	41.923	45.642	48.290
27	11.808	12.879	14.573	16.151	18.114	36.741	40.113	43.194	46.963	49.645
28	12.461	13.565	15.308	16.928	18.939	37.916	41.337	44.461	48.278	50.993
29	13.121	14.257	16.047	17.708	19.768	39.087	42.557	45.722	49.588	52.336
30	13.787	14.954	16.791	18.493	20.599	40.256	43.773	46.979	50.892	53.672
40	20.707	22.164	24.433	26.509	29.051	51.805	55.758	59.342	63.691	66.766
50	27.991	29.707	32.357	34.764	37.689	63.167	67.505	71.420	76.154	79.490
60	35.534	37.485	40.482	43.188	46.459	74.397	79.082	83.298	88.379	91.952
70	43.275	45.442	48.758	51.739	55.329	85.527	90.531	95.023	100.425	104.215
80	51.172	53.540	57.153	60.391	64.278	96.578	101.879	106.629	112.329	116.321
90	59.196	61.754	65.647	69.126	73.291	107.565	113.145	118.136	124.116	128.299
100	67.328	70.065	74.222	77.929	82.358	118.498	124.342	129.561	135.807	140.169

Table I

Random Numbers
Column Number

Row Number	01–05	06–10	11–15	16–20	21–25	26–30	31–35	36–40	41–45	46–50
01	89392	23212	74483	36590	25956	36544	68518	40805	09980	00467
02	61458	17639	96252	95649	73727	33912	72896	66218	52341	97141
03	11452	74197	81962	48443	90360	26480	73231	37740	26628	44690
04	27575	04429	31308	02241	01698	19191	18948	78871	36030	23980
05	36829	59109	88976	46845	28329	47460	88944	08264	00843	84592
06	81902	93458	42161	26099	09419	89073	82849	09160	61845	40906
07	59761	55212	33360	68751	86737	79743	85262	31887	37879	17525
08	46827	25906	64708	20307	78423	15910	86548	08763	47050	18513
09	24040	66449	32353	83668	13874	86741	81312	54185	78824	00718
10	98144	96372	50277	15571	82261	66628	31457	00377	63423	55141
11	14228	17930	30118	00438	49666	65189	62869	31304	17117	71489
12	55366	51057	90065	14791	62426	02957	85518	28822	30588	32798
13	96101	30646	35526	90389	73634	79304	96635	06626	94683	16696
14	38152	55474	30153	26525	83647	31988	82182	98377	33802	80471
15	85007	18416	24661	95581	45868	15662	28906	36392	07617	50248
16	85544	15890	80011	18160	33468	84106	40603	01315	74664	20553
17	10446	20699	98370	17684	16932	80449	92654	02084	19985	59321
18	67237	45509	17638	65115	29757	80705	82686	48565	72612	61760
19	23026	89817	05403	82209	30573	47501	00135	33955	50250	72592
20	67411	58542	18678	46491	13219	84084	27783	34508	55158	78742

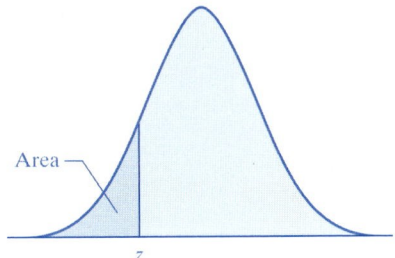

Area

z

Table IV

Standard Normal Distribution

z	.00	.01	.02	.03	.04	.05	.06	.07	.08	.09
−3.4	0.0003	0.0003	0.0003	0.0003	0.0003	0.0003	0.0003	0.0003	0.0003	0.0002
−3.3	0.0005	0.0005	0.0005	0.0004	0.0004	0.0004	0.0004	0.0004	0.0004	0.0003
−3.2	0.0007	0.0007	0.0006	0.0006	0.0006	0.0006	0.0006	0.0005	0.0005	0.0005
−3.1	0.0010	0.0009	0.0009	0.0009	0.0008	0.0008	0.0008	0.0008	0.0007	0.0007
−3.0	0.0013	0.0013	0.0013	0.0012	0.0012	0.0011	0.0011	0.0011	0.0010	0.0010
−2.9	0.0019	0.0018	0.0018	0.0017	0.0016	0.0016	0.0015	0.0015	0.0014	0.0014
−2.8	0.0026	0.0025	0.0024	0.0023	0.0023	0.0022	0.0021	0.0021	0.0020	0.0019
−2.7	0.0035	0.0034	0.0033	0.0032	0.0031	0.0030	0.0029	0.0028	0.0027	0.0026
−2.6	0.0047	0.0045	0.0044	0.0043	0.0041	0.0040	0.0039	0.0038	0.0037	0.0036
−2.5	0.0062	0.0060	0.0059	0.0057	0.0055	0.0054	0.0052	0.0051	0.0049	0.0048
−2.4	0.0082	0.0080	0.0078	0.0075	0.0073	0.0071	0.0069	0.0068	0.0066	0.0064
−2.3	0.0107	0.0104	0.0102	0.0099	0.0096	0.0094	0.0091	0.0089	0.0087	0.0084
−2.2	0.0139	0.0136	0.0132	0.0129	0.0125	0.0122	0.0119	0.0116	0.0113	0.0110
−2.1	0.0179	0.0174	0.0170	0.0166	0.0162	0.0158	0.0154	0.0150	0.0146	0.0143
−2.0	0.0228	0.0222	0.0217	0.0212	0.0207	0.0202	0.0197	0.0192	0.0188	0.0183
−1.9	0.0287	0.0281	0.0274	0.0268	0.0262	0.0256	0.0250	0.0244	0.0239	0.0233
−1.8	0.0359	0.0351	0.0344	0.0336	0.0329	0.0322	0.0314	0.0307	0.0301	0.0294
−1.7	0.0446	0.0436	0.0427	0.0418	0.0409	0.0401	0.0392	0.0384	0.0375	0.0367
−1.6	0.0548	0.0537	0.0526	0.0516	0.0505	0.0495	0.0485	0.0475	0.0465	0.0455
−1.5	0.0668	0.0655	0.0643	0.0630	0.0618	0.0606	0.0594	0.0582	0.0571	0.0559
−1.4	0.0808	0.0793	0.0778	0.0764	0.0749	0.0735	0.0721	0.0708	0.0694	0.0681
−1.3	0.0968	0.0951	0.0934	0.0918	0.0901	0.0885	0.0869	0.0853	0.0838	0.0823
−1.2	0.1151	0.1131	0.1112	0.1093	0.1075	0.1056	0.1038	0.1020	0.1003	0.0985
−1.1	0.1357	0.1335	0.1314	0.1292	0.1271	0.1251	0.1230	0.1210	0.1190	0.1170
−1.0	0.1587	0.1562	0.1539	0.1515	0.1492	0.1469	0.1446	0.1423	0.1401	0.1379
−0.9	0.1841	0.1814	0.1788	0.1762	0.1736	0.1711	0.1685	0.1660	0.1635	0.1611
−0.8	0.2119	0.2090	0.2061	0.2033	0.2005	0.1977	0.1949	0.1922	0.1894	0.1867
−0.7	0.2420	0.2389	0.2358	0.2327	0.2296	0.2266	0.2236	0.2206	0.2177	0.2148
−0.6	0.2743	0.2709	0.2676	0.2643	0.2611	0.2578	0.2546	0.2514	0.2483	0.2451
−0.5	0.3085	0.3050	0.3015	0.2981	0.2946	0.2912	0.2877	0.2843	0.2810	0.2776
−0.4	0.3446	0.3409	0.3372	0.3336	0.3300	0.3264	0.3228	0.3192	0.3156	0.3121
−0.3	0.3821	0.3783	0.3745	0.3707	0.3669	0.3632	0.3594	0.3557	0.3520	0.3483
−0.2	0.4207	0.4168	0.4129	0.4090	0.4052	0.4013	0.3974	0.3936	0.3897	0.3859
−0.1	0.4602	0.4562	0.4522	0.4483	0.4443	0.4404	0.4364	0.4325	0.4286	0.4247
−0.0	0.5000	0.4960	0.4920	0.4880	0.4840	0.4801	0.4761	0.4721	0.4681	0.4641

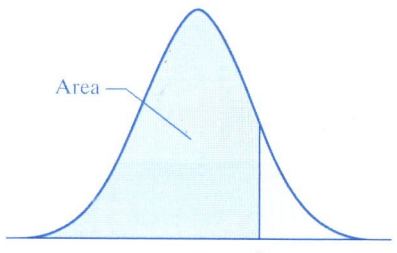

Area

z

Table IV (continued)

Standard Normal Distribution

z	.00	.01	.02	.03	.04	.05	.06	.07	.08	.09
0.0	0.5000	0.5040	0.5080	0.5120	0.5160	0.5199	0.5239	0.5279	0.5319	0.5359
0.1	0.5398	0.5438	0.5478	0.5517	0.5557	0.5596	0.5636	0.5675	0.5714	0.5753
0.2	0.5793	0.5832	0.5871	0.5910	0.5948	0.5987	0.6026	0.6064	0.6103	0.6141
0.3	0.6179	0.6217	0.6255	0.6293	0.6331	0.6368	0.6406	0.6443	0.6480	0.6517
0.4	0.6554	0.6591	0.6628	0.6664	0.6700	0.6736	0.6772	0.6808	0.6844	0.6879
0.5	0.6915	0.6950	0.6985	0.7019	0.7054	0.7088	0.7123	0.7157	0.7190	0.7224
0.6	0.7257	0.7291	0.7324	0.7357	0.7389	0.7422	0.7454	0.7486	0.7517	0.7549
0.7	0.7580	0.7611	0.7642	0.7673	0.7704	0.7734	0.7764	0.7794	0.7823	0.7852
0.8	0.7881	0.7910	0.7939	0.7967	0.7995	0.8023	0.8051	0.8078	0.8106	0.8133
0.9	0.8159	0.8186	0.8212	0.8238	0.8264	0.8289	0.8315	0.8340	0.8365	0.8389
1.0	0.8413	0.8438	0.8461	0.8485	0.8508	0.8531	0.8554	0.8577	0.8599	0.8621
1.1	0.8643	0.8665	0.8686	0.8708	0.8729	0.8749	0.8770	0.8790	0.8810	0.8830
1.2	0.8849	0.8869	0.8888	0.8907	0.8925	0.8944	0.8962	0.8980	0.8997	0.9015
1.3	0.9032	0.9049	0.9066	0.9082	0.9099	0.9115	0.9131	0.9147	0.9162	0.9177
1.4	0.9192	0.9207	0.9222	0.9236	0.9251	0.9265	0.9279	0.9292	0.9306	0.9319
1.5	0.9332	0.9345	0.9357	0.9370	0.9382	0.9394	0.9406	0.9418	0.9429	0.9441
1.6	0.9452	0.9463	0.9474	0.9484	0.9495	0.9505	0.9515	0.9525	0.9535	0.9545
1.7	0.9554	0.9564	0.9573	0.9582	0.9591	0.9599	0.9608	0.9616	0.9625	0.9633
1.8	0.9641	0.9649	0.9656	0.9664	0.9671	0.9678	0.9686	0.9693	0.9699	0.9706
1.9	0.9713	0.9719	0.9726	0.9732	0.9738	0.9744	0.9750	0.9756	0.9761	0.9767
2.0	0.9772	0.9778	0.9783	0.9788	0.9793	0.9798	0.9803	0.9808	0.9812	0.9817
2.1	0.9821	0.9826	0.9830	0.9834	0.9838	0.9842	0.9846	0.9850	0.9854	0.9857
2.2	0.9861	0.9864	0.9868	0.9871	0.9875	0.9878	0.9881	0.9884	0.9887	0.9890
2.3	0.9893	0.9896	0.9898	0.9901	0.9904	0.9906	0.9909	0.9911	0.9913	0.9916
2.4	0.9918	0.9920	0.9922	0.9925	0.9927	0.9929	0.9931	0.9932	0.9934	0.9936
2.5	0.9938	0.9940	0.9941	0.9943	0.9945	0.9946	0.9948	0.9949	0.9951	0.9952
2.6	0.9953	0.9955	0.9956	0.9957	0.9959	0.9960	0.9961	0.9962	0.9963	0.9964
2.7	0.9965	0.9966	0.9967	0.9968	0.9969	0.9970	0.9971	0.9972	0.9973	0.9974
2.8	0.9974	0.9975	0.9976	0.9977	0.9977	0.9978	0.9979	0.9979	0.9980	0.9981
2.9	0.9981	0.9982	0.9982	0.9983	0.9984	0.9984	0.9985	0.9985	0.9986	0.9986
3.0	0.9987	0.9987	0.9987	0.9988	0.9988	0.9989	0.9989	0.9989	0.9990	0.9990
3.1	0.9990	0.9991	0.9991	0.9991	0.9992	0.9992	0.9992	0.9992	0.9993	0.9993
3.2	0.9993	0.9993	0.9994	0.9994	0.9994	0.9994	0.9994	0.9995	0.9995	0.9995
3.3	0.9995	0.9995	0.9995	0.9996	0.9996	0.9996	0.9996	0.9996	0.9996	0.9997
3.4	0.9997	0.9997	0.9997	0.9997	0.9997	0.9997	0.9997	0.9997	0.9997	0.9998